T0338390

# Digital and Analog Fiber Optic COMMUNICATIONS for CATV and FTTx Applications

# Digital and Analog Fiber Optic COMMUNICATIONS for CATV and FTTx Applications

**Avigdor Brillant**

Bellingham, Washington USA

Library of Congress Cataloging-in-Publication Data

Brillant, Avigdor.
Digital and analog fiber optic communications for CATV and FTTx applications / Avigdor Brillant.
p. cm.
Includes bibliographical references and index.
"PM174."
ISBN 978-0-8194-6757-7 (alk. paper)
1. Optical fiber subscriber loops. 2. Cable television--Equipment and supplies. 3. Fiber optic cables. 4. Optoelectronic devices. I. Title.
TK5103.592.O68.B75 2008

621.382'75--dc22

2008019335

Published by

SPIE
P.O. Box 10
Bellingham, Washington 98227-0010 USA
Phone: +1 360 676 3290
Fax: +1 360 647 1445
Email: spie@spie.org
Web: http://spie.org

and

John Wiley & Sons, Inc.
111 River Street
Hoboken, New Jersey 07030
Phone +1 201 748 6000
FAX + 1 201 748 6088



Printed in the United States of America.

*To my wife Merav*
*and our children Guy, Rotem, and Eden*

*In the memory of my parents,*
*Rosita Brillant (Segal) and Edmond Wilhelm Brillant*

# Contents

# Nomenclature

| | |
|---|---|
| ABB | analog base-band |
| ABR | available bit rate |
| AC | alternating current or audio compression |
| ACI | adjacent channel interference |
| ADC | analog-to-digital converter |
| ADS | advanced design system (RF design software) |
| ADSL | asymmetric digital subscriber line |
| ADM | add–drop multiplexing |
| AGC | automatic gain control |
| AFC | automatic frequency control |
| ALC | automatic level control |
| AM | amplitude modulation |
| ANSI | american national standards institute |
| AOC | automatic offset compensation or automatic offset control |
| AON | active optical network |
| APC | automatic power control or angled physical contact |
| APD | avalanche photo detector |
| APE | annealed proton exchange |
| APPC | angled polished physical contact |
| AR | aspect ratio |
| ARPANET | advanced research projects agency network |
| ASC | automatic slope control |
| ASIC | application specific integrated circuit |
| ATC | automatic temperature control or automatic threshold control |
| ATIS | alliance for telecommunications industry solutions |
| ATM | asynchronous transfer mode |
| ATP | acceptance test procedure |
| ATR | acceptance test results |
| ATRC | advanced television research consortium |
| ATSC | advanced television systems committee |
| ATV | advanced television |
| AV | voltage gain |
| AWG | arrayed waveguide gratings |
| AWGN | additive white gaussian noise |

| | |
|---|---|
| BALUN | balanced to unbalanced |
| BBI | balanced bridge interferometer |
| BCD | binary coded decimal |
| BE | base emitter |
| BELCORE | Bell Labs communications research |
| BER | bit error rate |
| BERT | bit error rate tester |
| B-GT | birefringent analog |
| BGA | bragg grating arrays |
| BH | buried heterostructure |
| BJT | bipolar junction transistor |
| BiDi | bi directional |
| BLD | bottom level detector |
| BMR | burst mode receiver |
| BNC | Bayonet Neill-Concelman (inventors of the BNC connector) |
| BOM | bill of materials |
| BPF | band-pass filter |
| BPON | broadband passive optical network |
| BT | British Telecom |
| BTS | balanced twill structure |
| BW | bandwidth |
| BWER | bandwidth enhancement ratio |
| CAD | computer aided design |
| CATV | community access television |
| CB | common base |
| CBD | cumulative bit difference |
| CBR | constant bit rate |
| CC | common collector |
| CCD | coupled charged device |
| CCDP | cross-coupled differential pair |
| CCF | continuous carbon fiber |
| CCIR | committee consulatif international radiocommunications |
| CCITT | comite consultatif internationale de telegraphie et telephonie |
| CCN | carrier-to-composite noise |
| CDF | cumulative distribution function |
| CDR | clock data recovery |
| CE | common emitter |
| CFBG | chirp fiber bragg gratings |
| CID | consecutive identical data bits |
| CIN | carrier-to-intermodulation noise |
| CLDI | chrominance-to-luminance delay inequality |
| CLEC | competitive local exchange carrier |
| CLK | clock |
| CLR | cable loss ratio |

| | |
|---|---|
| CMA | constant modulus algorithm |
| CMOS | complementary metal oxide semiconductor |
| CMOT | common management interface protocol |
| CMRR | common mode rejection ratio |
| CMTS | cable modem termination system |
| CNR | carrier-to-noise ratio |
| CO | central office |
| CPLR | coupler |
| CPU | central processing unit |
| CRT | cathode ray tube |
| CS | component side |
| CSO | composite second order |
| CTB | composite triple beat |
| CTC | coefficient of thermal conductivity |
| CTN | carrier-to-thermal noise |
| CTU | coaxial termination unit |
| CW | continuous wave |
| CWDM | coarse WDM |
| DA | data aided |
| DAC | digital-to-analog converter |
| DAVIC | digital audio visual council |
| dB | decibels |
| dBc | decibels relative to carrier power |
| dBFS | decibels full scale |
| dBm | decibels relative to 1 mW |
| dBmv | decibels relative to 1 mV |
| DBR | distributed Bragg reflector |
| DBS | direct broadcast satellite |
| dBW | decibels relative to 1W |
| DCA | digital controlled attenuator |
| DCT | discrete cosine transform |
| DD | decision directed |
| DEMUX | demultiplexing |
| DFB | distributed feedback |
| DFF | delay flip flop |
| DFT | discrete Fourier transform |
| DG | differential gain |
| DH | double heterostructure |
| DIG | digital section |
| DIR | directional coupler |
| DLC | digital loop carrier |
| DLVA | detector log video amplifier |
| DML | direct modulated laser |
| DNS | domain name system |
| DOCSIS | data over cable service interface specifications |

| | |
|---|---|
| DOD | Department of Defense |
| DOE | diffractive optical element |
| DP | differential phase/dispersion penalty |
| DPCM | discrete pulse code modulation |
| DQPSK | differential QPSK |
| DSB | double side band |
| DSF | dispersion shift fiber |
| DSL | digital subscriber line |
| DSLAM | digital subscriber line access multiplexer |
| DSO | discrete second order |
| DSP | digital signal processing |
| DSSS | direct sequence spread spectrum |
| DTB | discrete triple beat |
| DTO | discrete third order |
| DTS | decode time stamp |
| DUR | desired to undesired ratio, or demodulated to unmodulated radio |
| DUT | device under test |
| DVB | digital video broadcasting |
| DVB–H | digital video broadcasting–handheld |
| DVB–T | digital video broadcasting–terrestrial |
| DWDM | dense WDM |
| ECL | external cavity laser or emitter coupled logic |
| EDFA | erbium doped fiber amplifier |
| EDTV | enhanced definition television (which is reduced resolution compared to HDTV) |
| EFMA | ethernet in the first mile alliance |
| EGSM | extended golbal system for mobile communication |
| EIA | electronic industries alliance |
| EiCN | equivalent input current noise |
| ELR | excess loss ratio |
| EMI | electromagnetic emission |
| EML | external modulated laser |
| ENR | excess noise ratio |
| EOL | end of life |
| EPON | ethernet passive optical network |
| EPM | external phase modulation |
| ER | extinction ratio |
| EVM | error vector magnitude |
| ETSI | European telecommunications standardization institute |
| FBG | fiber Bragg grating |
| FC | face contact |
| FC/APC | face contact/angled polished physical contact |
| FCC | federal communication commission |
| FDM | frequency division multiplexing |

| | |
|---|---|
| FDF | fiber distribution frame |
| FDH | fiber distribution hub |
| FEC | forward error correction |
| FET | field effect transistor |
| FF | feed forward or flip flop |
| FFT | fast Fourier transform |
| FHSS | frequency hopping spread spectrum |
| FIFO | first in first out |
| FIR | finite impulse response filter |
| FITL | fiber in the loop |
| FITs | functional integrity testing |
| FM | frequency modulation |
| FOC | fiber optical connector |
| FP | Fabry–Perot |
| FPD | frequency phase detector |
| FR | frame rate |
| FS | frame store |
| FSAN | full service access network |
| FSE | fractionally spaced equalizer |
| FSR | free spectra range/feedback shift register |
| FT | fourier transform |
| FTTB | fiber to the building |
| FTTC | fiber to the curb |
| FTTH | fiber to the home |
| FTTP | fiber to the premises |
| FTTx | fiber to the home, curb, building, business, premises |
| FWHM | full width at half maximum |
| Gb | gigabit |
| Gbps | gigabit per second |
| GBIC | gigabit interface converter |
| GBW | gain bandwidth |
| GBWP | gain bandwidth product |
| GCA | gain controlled amplifier |
| GCC | gain control circuit |
| GCSR | grating-assisted codirectional coupler with sampled reflector |
| GDR | group delay ripple |
| GF | Galois field |
| GHz | gigahertz (1,000,000,000 Hz) |
| GND | ground |
| GOP | group of pictures |
| GPON | gigabit passive optical network |
| GPRS | general packet radio service |
| GRIN | grated index or gradually variable reflection index |
| GS | gain state |
| GSM | global system for mobile communication |

| | |
|---|---|
| GT | Gires Tournois |
| GWS | grating waveguide structure |
| HBT | heterojunction bipolar transistor |
| HDTV | high definition television |
| HEMT | high electron mobility transistor |
| HFC | hybrid fiber coax |
| HFX | hybrid fiber X, where X can be either copper or coax |
| HP | home passed |
| HPF | high pass filter |
| HPNA | home phone networking alliance |
| HPPLA | home plug power line alliance |
| HRC | harmonically related carrier |
| IC | integrated circuit |
| $I^2C$-BUS | inter IC bus (philips invention) |
| ID | integration derivative |
| IE | ion exchange |
| IF | intermediate frequency |
| IFFT | inverse fast fourier transform |
| IIP2 | second order input intercept point |
| IIP3 | third order input intercept point |
| IIR | infinite impulse response filter |
| ILEC | incumbent local exchange carriers |
| IMD | inter-modulation |
| IO | image-orthicon |
| IP | intercept point or internet protocol |
| IP2 | second-order intercept point |
| IP3 | third-order intercept point |
| IRC | incrementally related carrier |
| IRE | Institute of Radio Engineers (former name of today's IEEE) |
| ISDB–T | terrestrial integrated services digital broadcasting |
| ISDN | integrated services digital networks |
| ISI | inter symbol interference |
| ISO | international standard organization |
| ITC | integrated triplexer to the curb |
| ITR | integrated triplexer |
| ITU | International Telecommunications Union |
| IEEE | Institute of Electrical and Electronic Engineers |
| IXC | interexchange carriers |
| JBIG | Joint Bilevel Image Experts Group |
| JPEG | Joint Photographic Experts Group |
| KCL | Kirchoff's current law |
| KHz | kilohertz (1000 Hz) |
| KVL | Kirchoff's voltage law |
| L1P | first layer protocol |
| LAN | local area network |

| | |
|---|---|
| LATA | local transport area |
| LC | inductance–capacitance |
| LCF | long carbon fiber |
| LCM | last common multiple |
| LCN | logical channels |
| LCP | liquid crystal polymer |
| LDS | laser driver stage |
| LEAF | large effective area fiber |
| LEC | local exchange carrier |
| LED | light emitting diode |
| LF | lattice filter |
| LFE | linear feedforward equalizer |
| LFSR | linear feedback shift register |
| LLC | logical link control |
| LMDS | local multipoint distribution system |
| LMS | least mean square |
| LNA | low-noise amplifier |
| LO | local oscillator |
| LPF | low-pass filter |
| LSB | low side band or least significant beat |
| LT | loop transmission (open loop gain) |
| LTE | linear transversal equalizer |
| LVCMOS | low-voltage CMOS |
| LVPECL | low-voltage pseudo-emitter coupled logic |
| MAC | media access control |
| Mb | megabit |
| MBE | molecular beam epitaxy |
| MCEF | modulation current efficiency |
| MCNS | multimedia cable network system |
| MDS | minimum detectable signal |
| MES | modified error signal |
| MEMS | micro electro mechanical system |
| MER | modulation error ratio |
| MESFET | metal-semiconductor field effect transistor |
| MFD | mode field diameter |
| M-FLO | media forward link only |
| MHN | mode hopping noise |
| MHz | megahertz |
| MIS | metal insulator semiconductor |
| ML | maximum likelihood |
| MMDS | multichannel multipoint distribution system |
| MMIC | monolithic microwave integrated circuit |
| MMSE | minimum mean square error |
| MOCVD | metal organic chemical vapor deposition |
| MODEM | modulator/demodulator |

| | |
|---|---|
| MOS | metal oxide semiconductor |
| MOSFET | metal-oxide-semiconductor field-effect transistor |
| M-PD | monitor photodiode or monitor photodetector |
| MPE | mode partition error |
| MPN | mode partition noise |
| MPEG | moving picture experts group |
| MQW | multiple quantum well |
| MSA | multisource agreement |
| MSR | mode suppression ratio |
| MTBF | mean time between failure |
| MUSE | multiple sub-Nyquist sampling encoding (japanese HDTV standard) |
| MUX | multiplexer, multiplexing |
| MZ | Mach-Zehnder |
| MZI | Mach-Zehnder interferometer |
| NA | network analyzer |
| NB | noise burst |
| NCO | numerically controlled oscillator |
| NCTA | national cable television association |
| NDA | none data aided |
| NEB | noise equivalent bandwidth |
| NF | noise figure 10 log($F$)/noise factor |
| NG-DLC | new/next generation digital loop carriers |
| NH | network header |
| NICAM | near instantaneous companding audio multiplex |
| NL | nonlinear |
| NLD | nonlinear distortions |
| NLG | nonlinear gain |
| NOG | noise over gain |
| NOL | number of levels |
| NOS | number of steps |
| NPDU | network protocol data units |
| NPR | noise power ratio |
| NRC | nyquist raised cosine |
| NRZ | nonreturn to zero |
| NTC | negative temperature coefficient |
| NTF | noise transfer function |
| NTT | Nippon telegraph and telephone |
| NTSC | National Television Systems Committee |
| OADM | optical add drop multiplexer |
| OC-1 | optical carrier level 1 (51.48 MB/Sec) |
| OC12 | optical carrier level 12 (12 × 51.48 MB/Sec) |
| OC-24 | optical carrier level 24 (24 × 51.48 MB/Sec) |
| OC-48 | optical carrier level 48 (48 × 51.48 MB/Sec) |
| OC-192 | optical carrier level 192 (192 × 51.48 MB/Sec) |

| | |
|---|---|
| OCDMA | optical code division multiple access |
| ODN | optical distribution network |
| ODP | optical double port |
| OFDM | orthogonal frequency division multiplexing |
| OFE | optical front end |
| OHI | optical hybrid integrated module |
| OIF | optical internetworking forum |
| OLT | optical line terminal |
| OMI | optical modulation index |
| ONT | optical network terminal |
| ONU | optical network unit |
| OOK | on–off keying |
| ORCAD | Oregon computer aided design |
| OSA | optical subassembly |
| OSI | open systems interconnection |
| OSR | over sampling ratio |
| OSNR | optical signal-to-noise ratio |
| OSP | outside plant |
| OTP | optical triple port |
| PAL | phase altered line |
| PANDA | polarization maintaining and absorption reducing |
| PAR | peak to average |
| PBS | polarized beamsplitter |
| PC | physical contact, power combiner, personal computer, peak comparator, polarization control |
| P/C | PECL to CMOS converter |
| PCB | printed circuit board |
| PCM | pulse coded modulation |
| PCR | program clock reference |
| PD | photodetector |
| PDD | polarization-dependent distortions |
| PDF | probability density function |
| PDL | polarization-dependent loss |
| PDU | protocol data unit |
| PD-λ | polarization-dependent wavelength |
| PECL | pseudo emitter coupled logic |
| PEL | picture element |
| PEP | peak envelope power |
| PES | packetized elementary stream |
| PHEMT | pseudo high electron mobility transistor |
| PHY | physical layer |
| PI | proportional integrator |
| PID | proportional integrative derivative |
| PLC | planar lightwave circuits |
| PLL | phase locked loop |

| | |
|---|---|
| PLOAM | operation and maintenance |
| PM | phase modulation |
| PMD | polarization mode dependent |
| PMMA | poly-methyl-meth-acrylate |
| PON | passive optical network |
| POTS | plain old telephone service |
| PPD | polarization dependent distortions |
| PPG | pulse pattern generator |
| PRBS | pseudo-random beat sequence |
| PRF | pulse repetition frequency |
| PRI | pulse repetition interval |
| PS | print side or power splitter or portal service |
| PSD | power spectral density |
| PSPICE | personal computer simulation program with integrated circuits emphasis (circuit simulation software) |
| PSRR | power supply rejection ratio |
| PSTN | public switched telephone network |
| PTFE | polytetrafluoroethylene |
| PTS | presentation time stamp |
| PW | pulse width |
| PWD | pulse width distortion |
| PWM | pulse width modulation |
| QAM | quadrature amplitude modulation |
| QE | quantum efficiency |
| QPSK | quadrature phase shift keying |
| QTP | qualification test procedure |
| QTR | qualification test results |
| QW | quantum well |
| R | resistor |
| RAP | radio access point |
| RAU | remote antenna unit |
| RC | resistance–capacitance or raised cosine or resistor capacitor network |
| RCF | receiver control field |
| RF | radio frequency |
| RFFE | RF front end |
| RFI | radio frequency interference |
| RFIC | radio frequency integrate circuit |
| RFC | radio frequency choke |
| RFL | reflection linearizer |
| RGB | red green blue |
| RHS | right hand side |
| RIN | relative intensity noise |
| R/L | return loss |
| RLC | resistor inductor capacitor network |

| | |
|---|---|
| RLL | retardation locked loop |
| rms | root mean square |
| ROF | radio over fiber |
| ROSA | receiver optical subassembly |
| RS | Reed–Solomon |
| RSSI | received signal strength indication |
| RZ | return to zero |
| ΣΔ | sigma delta |
| SA | spectrum analyzer |
| SAI | serving area interface |
| SAR | sample to average ratio |
| S&H | sample and hold |
| SAW | surface acoustic wave |
| SBS | stimulated Brillouin scattering |
| SC | square/subcarrier connector |
| SC/APC | angled physical contact |
| SCTE | Society of Cable Telecommunications Engineers |
| SCH | separate confinement heterostructure |
| SCM | subcarrier multiplexed |
| SDH | synchronous digital hierarchy |
| SDR | signal-to-distortion ratio |
| SDV | switched digital video |
| SE | shielding effectiveness |
| SECAM | sequential color with memory in french sequentiel couleur avec memoire |
| SER | symbol error |
| SFF | small form factor |
| SFP | small form pluggable |
| SGD | stochastic gradient descent |
| SG-DBR | sampled grating distributed Bragg reflector |
| SHB | spectral hole burning |
| SHIP | silicon hetero interface photodetector |
| SINAD | signal-to-noise and distortion |
| SL | strained layer |
| SLM | single longitudinal mode |
| SMA | sub miniature version A connector |
| SMB | sub miniature version B connector |
| SMF | single mode fiber |
| SMPTE | society of motion picture and television engineers |
| SMSR | side mode suppression ratio |
| SMT | surface mount technology |
| SMTP | simple mail transfer protocol |
| SNA | scalar network analyzer |
| SNMP | simple network management protocol |
| SNR | signal-to-noise ratio |

| | |
|---|---|
| SOA | semiconductor optical amplifier |
| SOI | second-order intermodulations |
| SONET | synchronous optical network |
| SOP | system on package |
| SoS | silica on silicon |
| SP | serial to parallel |
| SPC | super physical contact |
| SPICE | simulation program with integrated circuits emphasis (circuit simulation software) |
| SPM | self phase modulation |
| SQW | single quantum well |
| SRRC | square root raised cosine |
| SRS | stimulated raman scattering |
| SSB | single side band |
| SSC-LD | spot size converted laser |
| ST | straight tip connector |
| STF | signal transfer function |
| STB | set top box |
| STD | standard |
| SZ | step size |
| TCM | trellis coded modulation |
| TCP | transmission control protocol |
| TCP/IP | transmission control protocol/internet protocol |
| TDM | time division multiplexing |
| T-DMB | terrestrial digital multimedia broadcasting |
| TDMA | time division multiplexing access |
| TEC | thermo-electric cooler or thermally expanded core |
| TELNET | TELetype NETwork |
| TEM | thermo electric module/transverse electro-magnetic |
| THz | tera hertz (1,000,000,000,000 Hz) |
| TIA | trans impedance amplifier |
| TIG | trans impedance gain |
| TIR | total internal reflection |
| TLCG | tapered linearly chirped gratings |
| TO (TO-can) | transistor outline metal can package |
| TODC | turn-on delay comparator |
| TOG | take over gain |
| TOI | third-order intermodulations |
| TOSA | transmitter optical subassembly |
| T-PDU | transport protocol data units |
| TRL | transmission line linearizer |
| TR008 | telephone ring (008, 75 etc) |
| TS | transport stream |
| TSS | three-step search or tangential sensitivity |
| TV | television |
| TW | traveling wave |

| | |
|---|---|
| UDP | user data protocol |
| UHF | ultra high frequency |
| UMTS | universal mobile telecommunications system |
| UPC | ultra polish physical contact |
| USB | upper side band or universal serial bus |
| US-TX | up-stream transmit |
| UUT | unit under test |
| UV | ultraviolet |
| VB | video buffer |
| VBR | variable bit rate |
| VCO | voltage controlled oscillator |
| VCSEL | vertical cavity surface emitting laser |
| VDSL | very high speed digital subscriber line |
| VGA | variable gain amplifier |
| VHF | very high frequency |
| VITS | vertical interval test signal |
| VLC | variable length code |
| VLD | variable length coded words |
| VLSI | very large scale integration |
| VNA | vector network analyzer |
| VOD | video on demand |
| VSB + C | vestigial side band plus carrier |
| VSWR | voltage standing wave ratio |
| VT-WSPD | voltage tunable–wavelength selective photodetector |
| VVA | voltage variable attenuator |
| WAN | wide area network |
| WCDMA | wideband code division multiple access |
| WG | waveguide |
| WG-PD | waveguide photodiode |
| WDM | wavelength division multiplexing |
| WLAN | wireless LAN |
| WSP/WSPD | wavelength selective photodetector |
| XFMR | transformer |
| XFP | 10 gigabit small form pluggable |
| XOR | exclusive OR |
| XPM | cross phase modulation |
| XTAL | crystal |
| X-MOD | cross-modulation |
| X-Talk | cross talk |
| YAG | yttrium aluminum garnet |
| ZFE | zero forcing equalizer |

**Substances and Composites**

| | |
|---|---|
| Alumina | $Al_2O_3$ |
| Aluminum | Al |

| | |
|---|---|
| Arsenide | As |
| Gallium | Ga |
| Gallium–Arsenide | GaAs |
| Germanium | Ge |
| Gold | Au |
| Gold Tin (solder) | AuSn |
| Helium | He |
| Indium | In |
| Indium–Gallium–Arsenide | InGaAs |
| Indium–Gallium–Arsenide–Phosphate | InGaAsP |
| Indium–Phosphate | InP |
| Lead | Pb |
| Lithium | Li |
| Lithium–Niobate | $LiNbO_3$ |
| Neon | Ne |
| Niobate | Nb |
| Oxygen | O |
| Phosphate | P |
| Silicon | Si |
| Silicon–Germanium | SiGe |
| Tin | Sn |

# Constants and Symbols

| | |
|---|---|
| Å | Angstrom, 1 Å = 0.0001 μm = 0.1 nm |
| C | Capacitance |
| $c_0$ | Speed of light TEM velocity in free space $2.99792456 \times 10^8$ m/sec $\approx 3 \times 10^8$ m/sec |
| $e$ | Base of natural logarithms e = exp(1) = 2.71828183 |
| $f$ | Electrical frequency, Hz |
| $h$ | Planck's constant $6.626 \times 10^{-7}$ J/sec |
| $j$ | Imaginary unit, $j = \sqrt{-1}$ |
| $k$ | Boltzmann's constant $1.38 \times 10^{-23}$ J/K |
| L | Inductance |
| $N_0$ | Noise density at room temperature for BW of 1 Hz, $-174$ dBm/Hz $= 10 \log kT + 30$ |
| Q | Quality factor |
| $q$ | Electron charge $1.6021917 \times 10^{-19}$ Coulombs |
| R | Resistance |
| $T_0$ | Kelvin's absolute zero temperature $-273°\text{C} = 0$ K |
| $\varepsilon$ | Permittivity $\varepsilon = \varepsilon_r \times \varepsilon_0$ |
| $\varepsilon_0$ | Permittivity of vacuum $8.85 \times 10^{-12}$ F/m $\approx 1/(3.6\pi)$ pF/cm |
| $\varepsilon_r$ | Relative permittivity |
| $\eta_0$ | Characteristic impedance of free space $120\pi$, $\eta_0 = \mu_0/\varepsilon_0$ |
| $\lambda$ | Wavelength |
| $\mu$ | Permeability $\mu = \mu_r \times \mu_0$ |
| $\mu_0$ | Permeability of vacuum $12.56 \times 10^{-7}$ Hy/m $\approx 4\pi$ nHy/cm |
| $\mu_r$ | Relative permeability |
| $\upsilon$ | Optical frequency |
| $\Omega$ | Resistance units in Ohms |
| $\omega$ | Electrical radial frequency, $\omega = 2\pi f$ rad/sec |
| $\pi$ | $3.14159265 \approx 3.14$ |
| $\sigma$ | Conductance in Mho $(1/\Omega)$ |
| $\otimes$ | Convolution |

# Preface

As data rates increase, there is a higher requirement to combine microwave-engineering experience with digital design. The recent development of the Internet has created the need for wider knowledge and understanding of different aspects of system performance. For instance, the traditional digital and logic designer must be more familiar with the root cause for high-speed link performance tradeoffs such as sensitivity, BER (bit error rate), eye diagrams, jitter, etc. Some of these parameters require the background of an RF (radio frequency) engineer, and having, for instance, a fundamental understanding of passive and active network design. As an example consider the jitter problem, as all RF engineers are familiar with its spectral definition of phase noise. Phase noise, as we all know, is a stochastic process. However, jitter of an eye diagram is composed from both stochastic process and deterministic process. An experienced RF engineer or communications engineer would try to optimize the phase response of the data transmission line so that it would have a linear phase response vs. frequency. This way, the group delay is constant, and therefore all the harmonics of the digital signal propagate at the same velocity, and the deterministic jitter is minimized. There are many other parameters affecting the eye performance, such as matching, which creates reflections and double eye images or clock recovery phase locked loop phase noise. This example shows the essential wide background required for fast logic designers.

Any switch or router contains fast logic, and optics interface that operates at high speed. Moreover, as CATV (community aperture TV, cable TV) technology advanced, its video transport and return path were wired by fiber. Therefore, it is much more important to have a full understanding of all design aspects of fiber-optics transceivers in order to meet the system requirements. Modern CATV transmissions are shifting from traditional analog to higher modulation qualities such as high-order QAM (quadratrure amplitude modulation) modulation. In that case, the traditional RF engineer has to understand better the effects of designs on the signal quality and distortions. There is a need to understand the effects of AM-to-AM on the second-order distortions, and third-order distortions. In the CATV case, we are dealing with multitone transport; hence the designer has to understand the RF chain lineup tradeoffs such as CNR vs. compression and the effects on CSO (composite second order) and CTB (composite triple beat) in the receive channel. The RF engineer has to understand the effects of AM-to-AM and AM-to-PM on the

QAM signal constellation. Hence the RF engineer should have a wider background in digital communications and modulation techniques. Additionally, the RF engineer, as well as the digital design engineer, should have fundamental background in optical devices, at least their equivalent circuit and impedance matching, in order to reach high-spec system performance.

In some advanced designs, both disciplines, analog and digital, have to exist and operate in the same space and packaging enclosure. As the technology of semiconductors improves, the size of the components is getting smaller; PCB (printed circuit board) population density is increasing and becoming more crowded. Subassemblies such as optical transceivers have to be smaller one the one hand, and faster with higher data rates on the other. In the case of a fast digital transceiver packaged together with an analog CATV receiver, the challenge in creating an integrated optical triplexer module, ITR, is higher. ITR converts digital traffic from light into electronic signal and vice versa; when converting a sensitive analog signal from light into analog signal, it becomes a X-talk issue. Both designers should have full understanding of X-talk mechanisms, ground disciplines, radiation from transmission lines, the spectral content of digital signals at different series patterns, and shielding methods, as well as some background in other fields in order to solve the X-talk problem. The requirement for such a high level of integration is coexistence, meaning each system should operate without interference with the other. Consequently, the sensitive and susceptible channel is the analog channel. However, a proper design of such an integrated system yielding the required high performances is possible.

There are several excellent books covering many subjects related to fiber optics. However, the goal of this book is to guide young, as well as experienced, digital and RF engineers about fiber-optics transceiver electronics designs step by step, trying to focus on all design aspects and tradeoffs from theory to application as much as possible. This book tries to condense the all the needed information and design aspects into several structured subjects. It guides the engineer to have a proper, methodical design approach, by observing the component requirements given for a system design level. This way, the engineer will have a deep understanding of specifications parameters and the reasons behind it, as well as its effects and consequences on system performance, which are essential for proper component design. Further, a fundamental understanding of RF and digital circuit design aspects, linear and nonlinear phenomena is important in order to achieve the desired performances. Getting familiar with solid-state devices and passives used to build optical receivers and transmitters is important. This way, one can combat design limitations in an effective way.

The book is organized into six main sections covering the following subjects:

- Part 1: Overview
  This part contains three chapters that provide the reader a top-down structured approach to get familiar with hybrid fiber coax (HFC) systems. This part provides information about several architectures of data transport carried over fiber and interfaces, which includes MMDS, LMDS, CATV Return-Path, and Internet, with some glimpses of protocol stack and last

mile, last feet concepts. This section provides information about the ITU grid and optical bands and advantages of fiber as transmission lines and the WDM concept. This whole review leads to the FTTx architectures concept.

After the fundamental background about the system needs, there is an introduction to the structure of the last mile optical-to-coax interfacing. This review provides different topologies for digital and analog receivers, which lead to the FTTx integrated solution of access transponder, containing both CATV receiver and digital transceiver. Additionally, tunable-laser transponder architecture is explained as a variant of ordinary digital transceivers' solution for METRO WDM architectures.The last part is an introduction to TV and CATV standards and the concept of operation. The main idea behind the part, even though it looks unrelated, is to provide detailed information about this unique signal transport and the implication of system specifications on the FTTx platform and CATV receivers.

- Part 2: Semiconductor and Passives
  This section contains five chapters and provides detailed information about different optical building blocks of fiber-to-coax and coax-to-fiber converters, which were reviewed in the first part. In this section, the building blocks are categorized into lasers, photodetectors, and passives, such as couplers, WDM, filters, triplexers, duplexers, etc. Each type of device physics is explored and analyzed. Analogies to microwaves are provided at some points to guide those who are being introduced to fiber optics about the similarities.

- Part 3: RF and Control Concepts
  In this section, there are six chapters. This section deals in depth with RF topologies to design highly linear analog CATV receivers, and provides a wide background about the structure of devices for high-speed digital design. At first, basic RF definitions are provided and simple RF lineups are reviewed. CSO and CTB beat counts are explained. IMD effects on CATV picture are analyzed. An introduction to noise and limits is provided, and these are explored and investigated. Different kinds of RF amplifiers and front-end matching are investigated. Push–pull distortions and analysis techniques are explored. On the digital side, various TIAs are analyzed, such as distributed amplifiers for wideband data rates of 10 and 40 GBit (which can be a laser driver). The structure and limitations of operational amplifier TIA are investigated. Detailed AGC (automatic gain control) analysis is provided with analogies to APC (automatic power control) and TEC (thermoelectric cooler) loop designs.

- Part 4: Introduction to CATV MODEM and Transmitters
  This section contains four chapters that provide guidance on the CATV MODEM concept of operation. At first, the background about QAM modulators and impairments is reviewed. Then, the CATV MODEM

structure is explored, explaining the different building blocks, such as coding and synchronization, and limitations such as phase noise. Thereafter, the next part of linear transmission is investigated. Predistortion techniques such as optical and electrical are analyzed and reviewed. Link analysis and derived OMI specs are investigated and explained as a summary. Jitter and phase noise are reviewed. Fiber effects are introduced.

- Part 5: Digital Transceiver Performance
  This section contains two chapters structured top-down. It guides the reader from digital signal definitions to the concept of a digital transceiver and tunable laser transponder architecture. Performance analysis and synthesis are provided. At first, fundamental definitions of digital transport such as eye diagram, jitter, extinction ratio are reviewed using MathCAD. Data formats such as NRZ, RZ, and performances-over-fiber are investigated. CDR (clock data recovery) structure is analyzed. After providing a solid background, transceivers and tunable laser transponders are investigated. Burst-mode concepts and burst-mode AGC are explained in detail.

- Part 6: Integration and Testing
  This section contains two chapters and focuses on integration problems and methods to test performances. EMI RFI problems within the FTTx ITR platform are analyzed. X-talk between digital and analog parts in the FTTx transponder is investigated and methodologies to overcome interferences are provided. Analytical methods are given.The second chapter in this section provides original methods for testing and evaluating FTTx platform compliance to the NCTA specifications. At the end, a practical FTTx receiver specification is reviewed and analyzed.

At the end of each chapter, a summary of main points studied in that chapter is provided. This way one could condense key points in order to have the main idea and concepts behind each chapter.

# Acknowledgments

The seeds of this book were planted during my work on optical communications at MRV as head of analog group. The product design challenges required an extensive and in-depth acquisition of state-of-the-art knowledge and background, as well as pushing the technological edge of optical engineering. The outcome of this self learning process produced a wealth of know-how, which I applied to my work and then imparted to the R&D team at LuminentOIC—MRV. Mr. Rob Goldman, then director and creator of the FTTx FTTP business unit in LuminentOIC—MRV, noted that much of communicated know-how is unique in the way it is formulated and explained. It is Rob who inspired me to collect this mass of papers and notes into a book. It was a huge challenge.

I offered it to SPIE, and I wish to express my gratitude to Mr. Tim Lamkins who accepted the offer and patiently waited for the final delivery. I wish to express my appreciation to Merry Schnell, who devotedly orchestrated the project to its success.

The task of material collecting, searching references, learning and creating genuine models and simulations, editing, and desktop publishing took 3 years of intensive work in two continents and countries. Many friends and colleagues have assisted me. I would like to express my deep appreciation for the remarks and comments, discussions, material, and elaborations about important matters presented in this book, which made it a good reference design book.

I would like to thank Mr. Yonatan Biran for long discussions about noise mechanisms, front-ends matching topologies, and useful analysis approaches on that matter. I would like to thank Mr. David Cahana for taking the time and reading the analog section and analog analysis of optoelectronics semiconductors and sharing his remarks and suggestions. Thanks should go to my friend Mr. David Pezo for his review, notes, comments, and discussions about QAM and complex signals modulation transport.

Thanks go also to Dr. Amotz Shemi and Dr. David Brooks from Color–Chip for reviewing PLC technology and introducing to me ion-exchange implantation technology of PLC on glass.

Academically, I would like to express my thanks to Professor Dan-Sadot, the head of the optoelectronic department of the Ben-Gurion University in Beer-Sheba, Israel, and founder of X-Light, for his reviews, checking, remarks, and useful advice about passive and active optoelectronics devices and systems.

Thanks should go also to Professor Irving Kalet from Columbia University USA, and Tel Hai College in Israel for his assistance in reviewing the modem, coding, and decoding sections.

At this stage, I also want to express my gratefulness to Dr. James Farmer and Dr. John Kenny from Wave-7 Optics, for long technical discussions and guidance on the delicate nuances in analog optical design for CATV transport. Indeed, I had a lot to learn and educate myself.

I would like to express my gratitude and appreciation to Professor Shlomo Margalit and Dr. Zeev Rav Noy, founders of MRV Communications, USA, who recruited me in their company, trusted me, and gave me the exposure and the opportunity to create and innovate in a state-of-the-art optoelectronics company. I want to thank my friends and colleagues in MRV for wonderful times, technical discussions, and joy of creation. I wish to thank Dr. Mark Heimbuch, CTO, Luminent OIC, for long discussions about lasers and detectors, Dr. Sheng Z. Zhang for discussions about mode hopping, Dr. Mohammad Azadeh for discussions about mode hopping and high data rates problems as well as aiding in references search, Dr. Chriss LaBounty for sharing material about TEC, Dr. Mikael Tokhmakhian, Mr. Wenhai Yang, and Mr. Tom Ciplickas for analyzing digital and analog results throughout development, Mr. Moshe Amit for educating me about bulk optics, Dr. Near Margalit, CEO, LuminentOIC, for elevating the bar of challenge, and all of my friends and colleagues in LuminentOIC who aided me in my project by sharing references, papers, and books.

I also want to express my appreciation to my brother Ilan Brillant, director of business development in GILAT, previously director of the business development unit in X-Light, for his technical and business reviews about the fiber-optics market and tunable laser platforms, and for introducing me to Professor Dan Sadot and Dr. Amotz Shemi. Thanks also should go to my cousin Lea Stoller (Brillant), chief librarian of RAFAEL, Israel's Armament Development Authority, for helping me in IEEE references search.

I wish to dedicate this book to the memory of my parents Rosita Brillant (Segal) and Edmond Wilhelm Brillant, who took care of me and my education, which brought me to where I am.

Finally, I want to express my deepest appreciation to my wife Merav and our children Guy, Rotem, and Eden, who supported me, patiently were understanding the lack of family and father time in weekends and the long nights of working in order to complete this task.

*Avigdor Brillant*
*Zichron Ya'akov*
ISRAEL

# Part 1

# System Overview

## Chapter 1

# WDM, Fiber to the X, and HFC Systems: A Technical Review

## 1.1 Introduction

Fiber optics is mature technology. It was used at the beginning of the eighties for computers' local networking using light emitting diodes (LEDs). One of its major advantages over traditional "copper lines" and "coax lines" is the virtually infinite bandwidth of the fiber line, which translates into a higher data rate capacity and therefore more users per line. This advantage can be expanded when several wavelengths are transmitted through the same fiber, where each wavelength carries wide band data. This method is called wavelength division multiplexing or WDM. The second main advantage of a fiber optics line over a regular "copper line" is its low-loss nature, traditionally 0.15 dB/km. The traditional coax will lose half of the input power within a few hundred meters. In comparison, a good-quality fiber will lose half of its input power after 15 to 20 km. This means less retransmit and fewer nodes required to amplify the signal. It is known that the transmitted distance depends on the input power to the fiber losses and the receiver sensitivity. An additional advantage of fibers is their immunity to any kind of magnetic interference from the outside. Hence, there will be fewer problems related to surge protection to take care of during deployment. Furthermore, since the fiber does not emit any electromagnetic radiation, it is considered to be an ideal line that cannot be tapped. One more advantage over the coax is the fiber diameter of 10–50 microns. Thus, one fiber cable, which contains many fibers, results in a higher data rate per cable and higher data capacity.

As optics and dedicated integrated circuits (IC) technology progressed, more applications and content could be designed and implemented using fiber optics.[36,38] Traditional "community access TV" (CATV: cable TV) shifted to fiber optics; other applications, such as slow data transport "return path" for "video on demand" (VOD) charging, fast data transport such as the Internet, and digital data and multimedialike video-over internet protocol, created a need for different kinds of receivers and transmitters with MUX/deMUX blocks.[26] The MUX/deMUX methods involved using planar optical circuits (PLC) with low optical x-talk[3,12] and bulk optics, while size reduction effort was on both electronics and optics.[39,40] Mixed signal design strategies were emerging[21] in order to reduce costs, shrink size using system-on-package (SOP) technology and penetrate the market. Mature cost effective building blocks for optics and signal conversion from modulated light to electrical signal, and vice versa, resulted in optical deployments. These deployments are: fiber to the curb (FTTC), fiber to the home (FTTH), fiber to the business (FTTB) fiber to the premises (FTTP), or in general, fiber to the "x" (FTTx) implemented by a passive optical network (PON), ethernet passive optical network (EPON), gigabit passive optical network (GPON), broadband passive optical network (BPON), and asynchronous transfer mode (ATM) passive optical network (APON).

Another advantage of a fiber data transport system is its large guard band value. Laser modulation in optics is the dual of frequency up-conversion in RF. In optics, the information-bearing signal, digital data signal, CATV RF signal, or cellular channels[8,42] are up-converted into light frequencies. For instance, 950 MHz–2 GHz is up-converted to a 1550 nm wavelength or 193,548 GHz. In case of 1560 nm, the carrier frequency is 192,307 GHz. Assuming the CATV RF BW is from 50 to 870 MHz up-converted to 1560 nm, the relative BW percentage is narrow, approximately $0.4264 \times 10^{-4}$. Relative BW is a measure of $1/Q$, where $Q$ is the quality factor of the link known as fo/BW, where $f_o$ is the center frequency. Thus, absolute wideband data links such as 10GB/s are feasible. Hence, by using several colors and wavelengths, a larger number of channels with huge BW can be transmitted on the same fiber. The guard band in the fiber is on the order of thousands of GHz. The technology is called wavelength division multiplexing (WDM), which is the dual of frequency division multiple (FDM) access in RF.[1]

The above mentioned WDM advantages led to several receiver, transmitter, and transceivers concepts to be used in CATV and data networking over fiber. Figure 1.1 describes an FTTx optical receiver at the user end unit used for the last mile. The idea is to have a bidirectional integrated optical triplexer (B.D-ITR),[39] while integrating the electronics of video receivers and high-BW data transceivers in a compact, inexpensive package that consumes less than 2–3 W. This module receives 110 CATV channels and 192 DBS TV channels on the downstream path. The downstream video path is over a 1550-nm wavelength range. Additionally, it has a full duplex bidirectional data and voice path operating over the 1310-nm range. The 1310-nm channel supports up to OC3, 155.52 MB/s for a telephony return path and internet access. Today there are modules that supports OC12, 622.08 MB/s, and even half the rate of 1 GB ethernet (625 MB/s), the fastest modules support OC24 (1244.16 MB/s) and 1 GB ethernet (1250 MB/s). Today's integrated triplexer (ITR) supports the standard wavelengths (1310-nm receiving; 1310-nm transmitting; 1550-nm receiving[3]) and full service access network (FSAN) wavelengths (1310-nm transmitting; 1490-nm receiving; 1550-nm receiving). This approach made FTTx WDM a cost-effective solution in last-mile applications. In this approach, the CATV receiver is at 1550 nm, the up link is at 1310 nm, and the down link is at 1490 nm.[6]

Wavelength standards categorize ITU wavelength bands into six wavelength domains, as shown in Table 1.1. A standard ITR used for FTTH operates at the

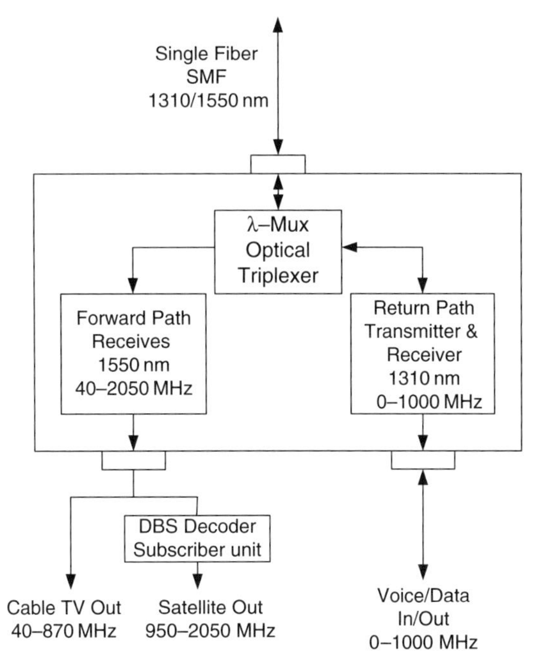


(a) Last mile topology of FTTx bidirectional optical unit using optical diplexer MUX.

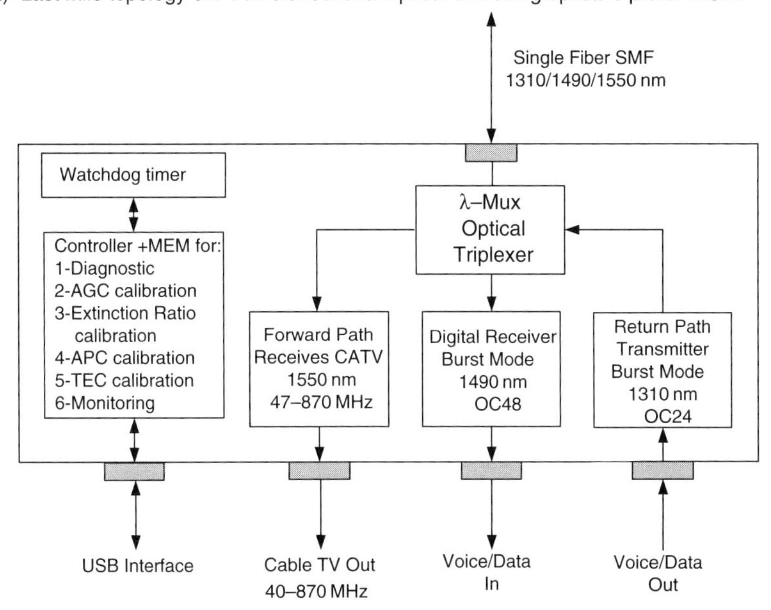


(b) Last mile topology of FSAN FTTx triple play optical unit using optical triplexer λ MUX for GPON applications. 1310 nm up-link is at OC24, 1490 nm is at OC48 and 1550 nm is an analog video channel that carries 79 AM—VSB NTSC signals and 31 QAM 64/256 HDTV. Watch-dog monitor for communications integrity and controller are shown

**Figure 1.1** Last-mile integrated triplexer topologies: (a) BiDi 1310 nm, return path Rx/Tx 1550 nm, CATV DBS; (b) Triple-play integrated FSAN triplexer known as ITR.

**Table 1.1** Proposed ITU wavelength bands.

| Name | Definition | Wavelength (nm) |
|---|---|---|
| O-Band | Original | 1260–1360 |
| E-Band | Extended | 1360–1460 |
| S-Band | Short | 1460–1530 |
| C-Band | Conventional | 1530–1565 |
| L-Band | Long | 1565–1625 |
| U-Band | Ultra-long | 1625–1675 |

O, S, and C bands for FSAN standard and at the O and C bands for the old wavelength plans of 1550 nm and 1310 nm. One of the technical problems in FTTx transducers is matching optics modules to the single-mode fibers (SMFs). For that purpose, large spot-size lasers were developed, improving the coupling efficiency in comparison to conventional lasers. A 1.2-dB coupling loss (75% coupling efficiency) was changed to a standard cleaved SMF and the 3-dB positional tolerance has been relaxed to a few microns.[2] Bulk optics sealing technologies such as laser welding and epoxies were improved and low-cost housings were introduced, as explained in Chapter 5.

The advantage of FTTx and its variants such as FTTC and FTTH over hybrid fiber coax (HFC), as a result from the above discussion, is due to fewer devices in the field, improved quality, and variety of services such as high definition television (HDTV) over IP or by using traditional RF modulation schemes, which all can be achieved due to greater BW. Homes become networked and data transport within it is distributed by the so-called "last 100 feet" method as explained in Table 1.2.[6]

**Table 1.2** Alternative home networking technologies.

| Standard | Technology | Max data rate | Advantage | Disadvantage |
|---|---|---|---|---|
| Ethernet | Wired (CAT 5) | 100 Mb/s | • Fast<br>• Inexpensive | • Requires new wiring<br>• Not portable |
| WLAN | Wireless | 54 Mb/s | • Portable<br>• Rapid cost reduction | • Converge issues<br>• Interface problems<br>• Range issues |
| HPNA | Wired (telephone lines) | 32 Mb/s | • Existing wiring in place | • Not portable<br>• Dependent on quality of phone wiring<br>• Limited number of locations |
| HomePlug | AC mains wiring | 14 Mb/s | • Somewhat portable<br>• Easy interface-to-line operated equipment | • Lower data rates<br>• Depends on quality of wiring |

**Table 1.3** Comparison of popular wireless protocols.

| Parameter | Bluetooth | IEEE 802.11(a) | IEEE 802.11(b) | IEEE 802.11(g) |
|---|---|---|---|---|
| Speed (Mb/s) | 1 | 54 | 11 | 22/54 |
| Launch | 1997 | 1999 | 1999 | 2002 |
| Range (ft) | 30 | 400 | 300 | 200 |
| Modulation | FHSS | OFDM | DSSS | DSSS/OFDM |
| Frequency (GHz) | 2.4 | 5 | 2.4 | 2.4 |

Ethernet-structured wiring appears in most new homes in the U.S. The infrastructure is for category-5 (Cat5). It is as follows, an FTTx box cable is brought to the basement or garage, and the cable inside it is connected to the router. This topology is often called "structured wiring." A user serial bus (USB) interface is an alternative way for connecting the cable modem or asymmetric digital subscriber line (ADSL) modem.

Wireless networking such as WLAN is an additional way to distribute home connections. The WLAN frequency is 2.4 GHz and its operation is defined by the IEEE standard 802.11(a) and (b). Version (b) was the first to see commercial implementation. Today chips are available to implement the 802.11(a) over 5 GHz. Table 1.3 provides a brief overview on wireless home networking protocols.

One more method to distribute service at home is by the Home Phone Networking Alliance (HPNA). In this approach, existing phone lines are used. HPNA had developed technology that can handle 4–10 MHz in a robust manner. Their latest standard, HPNA 2, uses 4 to 256 quadrature amplitude modulation (QAM) in this band, depending upon the quality of the phone lines. It is compatible with xDSL systems, which use the spectrum from above the voice band and up to 4 MHz.

The home plug is an additional distribution method and is relatively a new comer. This standard was created by the HPPLA (home plug power line alliance). One more standard of home networking is the IEEE 1349.

Cablelabs home networking initiative started with the publication of the CableHome 1.0 specification, which seeks to provide a unified framework for all cable industry–home interface devices. Those specifications did not address various physical layer interfaces as previously described. Rather it addresses the interface between the wide area network (WAN) and home network called portal services (PS). A number of the PS, but not all, are embodied in the current generation of WAN interfaces. To these services are added the ability to allow the cable operator to configure certain parameters of the PS, and remotely ping and perform loopback testing to the device. Quality of service parameters compatible with those provided in data-over-cable service interface specifications (DOCSIS) for modems may be configured in the portal service. FTTH systems do not use DOCSIS modems, but it is quite possible that some providers would like to use interfaces similar or identical to those prescribed in CableHome specifications.

The system design involves powering issues as well. One method is to have a battery at the home of the subscriber. The battery provides backup in case of power shutdown and disables all undesirable functionalities such as CATV, while keeping minimum power functions such as phone. Battery monitoring and maintenance is a complex issue. Not all batteries are the same nor do they have the same aging profiles. Most providers prefer to locate the battery at the subscriber's end; others install it at a central area. The disadvantage in the latter case is the voltage drop on the lines for a given current, which requires large batteries, larger voltage, and use of dc/dc at the subscriber's side. The advantage is simpler battery monitoring.

Other METRO applications use small increments of wavelength, for instance in C band (see Table 1.4) 1550.92 nm, 1550.12 nm, 1549.32 nm, etc. Tunable laser technologies[31] and fast switching between colors have led to new concepts and architectures, which suggest using one type of generic transmitter in a system and tuning it to a different wavelength. Hence the cost of a system with $N$ transmitters require only one more transmitter, rather than an additional $N$ transmitter—one per color. The spacing between each channel is 100 GHz per the ITU grid shown in Table 1.4. A denser spacing would be 50 GHz. Chapters 2 and 19 provide further elaboration. Those wavelength-multiplexing methods are called DWDM. Each wavelength at the user end is deMUXed by a phaser called arrayed waveguide gratings (AWG).[4] The AWG is a lens that has a single input and several outputs related to the wavelength. This way, each receiving side accepts its unique wavelength.[31] Further elaboration is provided in Chapter 4. This way, a central transmitter can operate in time division multiple access (TDMA), where at each burst, it transfers data on different wavelengths to the desired destination subscriber. Digital data transport is done by on–off keying (OOK) modulation. Thereby, OOK is AM and its modulation index is defined as the amount of modulation current applied on the laser known as extinction ratio (ER); further information is provided in Sec. 6.9.1.

The above introduction describes how WDM optical linking with modulated light is converted to data and video, and applied within the home's local networking. WDM is used to accomplish high-capacity digital and analog video-trunking applications.[10] However, there are several architectures of how to deploy an optical network whose content can be CATV DBS and data for FTTx, video over IP, internet, and radio over fiber. These topologies are PON, EPON, GPON, BPON, and APON. Additionally, CATV and other services are distributed in a so-called HFC architecture. These definitions describe the physical layer. When describing hardware and layers model it is refereed to first two layers. The first layer is the hardware and its management coding, sometimes called L1P, and firmware. The connection between the hardware bits, frames symbols etc., to higher layers is done by layer 2. This layer contains the media access control (MAC) and logical link control (LLC). When using burst mode, layer 1 (Data Link) contains all required algorithms for synchronized forward error corrections (FEC), delay compensation, and chromatic dispersion compensation using DSP equalizers. These algorithms refer to the so called physical layer DSP and algorithm codes. Hence hardware (physical layer 1) requires burst-mode operation firmware to control burst-mode transceivers within the digital section of the FTTx-integrated triplexer located at the subscribers' home and at the central physical layer (PHY) which includes the modem too. In HFC topology, the return path monitoring up-link channel and its impairments, such as the noise funnel, will be introduced.[18]

**Table 1.4** ITU frequencies and wavelengths for L and C band, 100 GHz spacing, 100 channels.

| Frequency (THz) | Wavelength (nm) | Frequency (THz) | Wavelength (nm) |
|---|---|---|---|
| 186.00 | 1611.79 | 190.70 | 1572.06 |
| 186.10 | 1610.92 | 190.80 | 1571.24 |
| 186.20 | 1610.06 | 190.90 | 1570.42 |
| 186.30 | 1609.19 | 191.00 | 1569.59 |
| 186.40 | 1608.33 | 191.10 | 1568.77 |
| 186.50 | 1607.47 | 191.20 | 1567.95 |
| 186.60 | 1606.60 | 191.30 | 1567.13 |
| 186.70 | 1605.74 | 191.40 | 1566.31 |
| 186.80 | 1604.88 | 191.50 | 1565.50 |
| 186.90 | 1604.03 | 191.60 | 1564.68 |
| 187.00 | 1603.17 | 191.70 | 1563.86 |
| 187.10 | 1602.31 | 191.80 | 1563.05 |
| 187.20 | 1601.46 | 191.90 | 1562.23 |
| 187.30 | 1600.60 | 192.00 | 1561.42 |
| 187.40 | 1599.75 | 192.10 | 1560.61 |
| 187.50 | 1598.89 | 192.20 | 1559.79 |
| 187.60 | 1598.04 | 192.30 | 1558.98 |
| 187.70 | 1597.19 | 192.40 | 1558.17 |
| 187.80 | 1596.34 | 192.50 | 1557.36 |
| 187.90 | 1595.49 | 192.60 | 1556.55 |
| 188.00 | 1594.64 | 192.70 | 1555.75 |
| 188.10 | 1593.79 | 192.80 | 1554.94 |
| 188.20 | 1592.95 | 192.90 | 1554.13 |
| 188.30 | 1592.10 | 193.00 | 1553.33 |
| 188.40 | 1591.26 | 193.10 | 1552.52 |
| 188.50 | 1590.41 | 193.20 | 1551.72 |
| 188.60 | 1589.57 | 193.30 | 1550.92 |
| 188.70 | 1588.73 | 193.40 | 1550.12 |
| 188.80 | 1587.88 | 193.50 | 1549.32 |
| 188.90 | 1587.04 | 193.60 | 1548.51 |
| 189.00 | 1586.20 | 193.70 | 1547.72 |
| 189.10 | 1585.36 | 193.80 | 1546.92 |
| 189.20 | 1584.53 | 193.90 | 1546.12 |
| 189.30 | 1583.69 | 194.00 | 1545.32 |
| 189.40 | 1582.85 | 194.10 | 1544.53 |
| 189.50 | 1582.02 | 194.20 | 1543.73 |
| 189.60 | 1581.18 | 194.30 | 1542.94 |
| 189.70 | 1580.35 | 194.40 | 1542.14 |
| 189.80 | 1579.52 | 194.50 | 1541.35 |
| 189.90 | 1578.69 | 194.60 | 1540.56 |
| 190.00 | 1577.86 | 194.70 | 1539.77 |
| 190.10 | 1577.03 | 194.80 | 1538.98 |
| 190.20 | 1576.20 | 194.90 | 1538.19 |
| 190.30 | 1575.37 | 195.00 | 1537.40 |
| 190.40 | 1574.54 | 195.10 | 1536.61 |
| 190.50 | 1573.71 | 195.20 | 1535.82 |
| 190.60 | 1572.89 | 195.30 | 1535.04 |
| 195.40 | 1534.25 | 195.70 | 1531.90 |
| 195.50 | 1533.47 | 195.80 | 1531.12 |
| 195.60 | 1532.68 | 195.90 | 1530.33 |

## 1.2 Cable TV and Networks System Overview

Cable TV DBS and other services can be distributed in two ways: (1) HFC networks as downstream and return path for signaling plain old telephone service (POTS) in upstream,[18,19,22] and (2) FTTx architecture utilizing WDM PON[1,7,14] and its variants, which have become popular in Korea,[17] U.S., and Japan.[35] There are several methods or landmarks in twisted-pair network-development technologies:[30]

- ADSL: The first step in interactive broadband services for the residential subscriber. The bit rate is affected by the distance, up to 6 Mb/s in the downstream and 640 kb/s in the upstream.
- Long range very high speed digital subscriber line (VDSL): The next step in the evolution of a twisted-pair network. Here, a fiberlink is used in order to reduce the copper drop to 1 – 1.5 km. This is called FTTCab or fiber to the cabinet. As a consequence, the transmission capacity can be increased to 25 MB/s downstream and 3 Mb/s upstream.
- Short range VDSL: In FTTC and FTTB, the last 300 m are copper drop, resulting in broadband service onto the customer's premises.
- HFC: Adopted during the 1990s by phone and cable companies. The rationale was merging services and reducing costs.[17,19,23,24] Cable companies deployed fiber to a remote node and from there, a coaxial cable was used for video for POTS. Telephone lines could be used for Internet.
- HFX: The interstage between HFC and FTTx.[15] Suggested by SBC for upgrading infrastructure of HFC with minimum investment.
- FTTH: The last copper drop replaced by an optical fiber. This step is currently accomplished by cost reduction of optical transceivers, such as using integrated triplexers and planar lightwave circuits (PLC) technologies[3,9] as well as system cost-optimization models.[17]

### 1.2.1 Hybrid Fiber Coax (HFC) networks

An HFC network is structured like a tree, where the head-end central office (CO) or primary hub transmits on the downstream, where analog and digital signals from various sources are multiplexed and transmitted over single mode fiber (SMF).[29,33,37] The data are received at the node and converted into an electrical signal. The TV signal sources may be satellite transponders, terrestrial broadcasting, and video servers. Additionally there are data services from Internet sources. These signals are transferred to the home by a coax and amplified by a line amplifier. Signal distribution to the subscriber's home is done by taps. The coax line from the tap is connected to a coaxial termination unit from where it is distributed the home terminals such as a set-top box for the CATV services, videophone, and PC for Internet.

The frequency spectrum in a two-way HFC/CATV system is divided between forward and return paths. The forward path is called "down link" and represents the traffic from the node to the tap and then to the home. The return path is called "up link" and is opposite to the down link.

The frequency spectrum of HFC is divided asymmetrically. It is asymmetric in the sense that the available down link spectrum is larger than the return path. This means different bandwidths.

The HFC return path spectrum has three standards:

- 5–42 MHz: U.S. standard,
- 5–55 MHz: Japanese standard, and
- 5–65 MHz: European standard.

There is a guard band between the return path (up link) and the CATV (down link). The U.S. standard supports 50–550 MHz for NTSC AM—VSB analog video and 550–750 MHz for HDTV QAM. This band was expanded to 870 MHz and even to 1 GHz. The allocated BW for the up stream (return path) is 5–42 MHz. This range is divided into channels having a BW of 1 to 3 MHz. Depending on the modulation scheme, the BW will vary. Chapter 3, Fig. 3.26 depicts the CATV frequency plan. The broadband HFC/CATV network supports both the analog AM-VSB and M-QAM. The QAM transmitter and the QAM receiver are located at the central office and the subscriber respectively and known as QAM modems. The standard for digital video and audio compression are set by Motion Picture Expert Group (MPEG). The high-speed internet access digital data are transmitted using cable modem based on data over cable interface specifications (DOCSIS). Other standards for broadband digital data services include digital audio visual council (DAVIC) or digital video broadcasting (DVB).

The distribution system is based on a main hub or primary hub ring[27,33] (Fig. 1.2). Each primary hub serves about 100,000 homes. The primary hub or head-end is connected to up to four or five secondary hubs, each of which serves up to 25,000 homes. These hubs are used for additional distributions of the data and video signals. The head-end or primary hub is connected in a backbone network through which it shares video sources with all other hubs. That way, multiple broadcast video sources are saved. The backbone network has a ring architecture that uses a synchronous optical network (SONET). The SONET primary hubs have add–drop multiplexers (ADMs) or some other methods to distribute the data off the ring. SONET specifications were defined by Bellcore. Figure 1.2 describes a typical ring architecture.

The hierarchy standardization of digital data was made by American National Standards Institutes (ANSI). This defines the digital data rates as follows: 51.48 Mb/s are defined as optical carrier level-1 (OC-1). Integer multiplications of

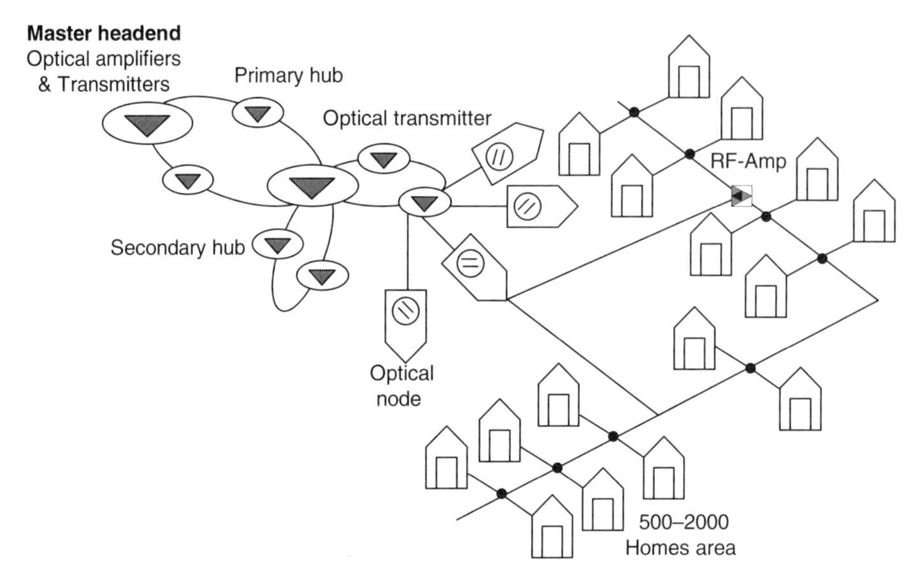


**Figure 1.2** HFC ring architecture.[44] (Reprinted with permission from *Journal of Lightwave Technology* © IEEE, 1998.)

this rate define other speeds. For instance, 51.84 Mb/s $\times$ 12 = 622.08 Mb/s. In the same way, OC48 is 2488.32 Mb/s. To increase spectrum efficiency or BW efficiency, statistical time division multiplexing access (TDMA) methods can be used instead of synchronous TDM. Thus in the statistical TDM, the time slot is provided per demand statistics.[11] In order to reduce deployment costs, TV operators made a choice to select and use SONET-compatible equipment. The distribution from the secondary hub is done at a node previously defined.

At that point, the optical signals are converted into electrical signals. These signals are transmitted to a subscriber; in the CATV analog signals case, these signals are amplified by various kinds of RF amplifiers. The node size is measured by the number of homes it supports, or homes passed (HP). A small node supports 100 HP and a large one supports above 2000 HP. Since the node has to support many subscribers, and it transmits many analog CATV channels and QAM channels, the RF amplifiers of the optical receiver should be linear with enough back-off in order to maintain low CSO of $-65$ dB and CTB 65 dBC and cross modulation (X-mod) of 65 dBc distortion levels, which is a hard specification. Additionally, in order to have a proper carrier-to-noise (CNR) of 53 dB and sensitivity at the subscriber end, the receiver should have a very low noise density of decibels relative to 1 mv at the optical receiver output.[24] The system specifications are degraded somewhat when compared to the optical receiver, which is required to meet CSOs of 60 dB and 53 dB CTB and X-mod. Typical values are between 4.5 and $-6$ pA/$\sqrt{\text{Hz}}$ at dark current state, with some variations if the receiver has an automatic gain control (AGC). To overcome generation of composite second order (CSO) products, a push–pull method is used to construct optical receivers and line RF amplifiers; additional options for RF line amplifiers are feedforward or parallel hybrid using two push–pulls in parallel, which are driven by a booster push–pull.[23] Further elaboration is provided in Chapters 2 and 12, and Sec. 16.2. The number of splitters and losses between the node and the subscriber defines the maximum distance of the subscriber from the node, where proper minimum CNR performance exists to satisfy the condition of minimum detectable signal (MDS). That will be demonstrated in link budget calculations in Chapter 19.

### 1.2.2 Return path

As was previously mentioned and is discussed elaborately in Sec. 2.3, a return path transmitter delivers POTS signals and data in the upstream. The upstream BW is negotiated between the head-end and the terminals. An active terminal sends to the head-end its BW requirement in an upstream frame. A BW allocation algorithm is executed in the head-end. This algorithm allocates the BW based upon several criteria such as available BW, user type, type of service, and number of users. Once the decision is made, the user who requested the BW is informed of the status. The limitation of this method is that it is prone to collisions. Such a case occurs when two users request BW while using the same upstream frame. For this reason, the dynamic reservation protocol for integrating constant bit rate (CBR)/variable bit rate (VBR)/available bit rate (ABR) traffic over an 802.14 HFC network is used. In this protocol, an upstream BW is slotted and framed. Each slot has a 48-byte payload and a 6-byte header. A 48-byte payload is used as it is compatible with ATM, so that the packet can be sent through the ATM network just by changing the header. Additionally whenever a slot is not being used for carrying data, it can be used as minislots by dividing into slots of 8 bytes each. These minislots can be used to send control messages, thereby preventing waste of the entire slot. The protocol operates in different manners for different services while they are simultaneously present.

When the node asks for CBR, it receives the highest priority. When the traffic is VBR, statistical multiplexing is done. VBR sources generate data in bursts. The data is buffered and sent when BW is allocated to the particular source. For ABR, the requests are received by the head-end and stored in a queue. The requests are processed in FIFO. When CBR and ABR sources need to be serviced at the same time, the protocol operates as follows. The CBR has highest priority. Arbitrating between VBR and ABR requires some kind of a flag bit. The head-end maintains this flag on which decisions arc based. The flag is set when there are requests from a VBR source; when there are no requests from the VBR it is reset. Decisions are made by the set of conditions given below:

- If the flag is set, a slot upstream is assigned to VBR.
- If the flag is reset and the data request, queue, and ABR are not empty, the slots are allocated to the ABR sources.
- If the flag is not set and the data-request queue is empty, the slot is converted to minislots.

In this way, the protocol provides a dynamic BW and it is more effective than the 802.14 media access control (MAC) layer protocols.

The return path suffers from a phenomenon called noise funnel. This results from several users' upstream transmission. Additionally it is susceptible to the following noise sources:[18]

- Narrowband shortwave signals propagated through the atmosphere and coupled to the return path at the subscriber location of the distribution plant.
- Impulse noise due to electrical appliances and motors.
- Common mode distortions originating from nonlinearities in the plane; i.e., oxidized connectors that behave as diodes.
- Location-specific interface generated by a device at the subscriber's location.

One of the ways to prevent noise from leaking out at the user's end to use a low-pass filter (LPF) that allows the 5–15 MHz out but prevents the ingress coming from inside the subscriber's residence from leaking out in the range of 15–40 MHz. A modem that provides telecommunication services such as POTS, video-telephony, and high-speed data VOD utilizes 15–40 MHz. Thus it is located after the blocking filter as shown in Fig. 1.3.

### 1.2.3 CATV and data in HFX

The HFX architecture is depicted in Fig. 1.4. The nodes C and R/L are the network elements. Node C is the CO-based fiber node/multiplexer connected to the xDSL modem ports of a digital subscriber line access multiplexer (DSLAM), the cable modem termination system (CMTS), and an analog video system, which is not depicted in Fig. 1.4.

The R/L node is considered as two different types of equipment with similar functionality: the R node is the fiber node deployed in a remote terminal/cabinet collocated with digital loop carrier (DLC)/next generation (NG)-DLC. The R/L node is connected to the C node via fiber. The DLC/NG-DLC supports telephony services to the customer's premises via copper coax, and to the remote antenna unit (RAU) for wireless access. Both nodes C and R/L transparently support xDSL connections, analog and digital video, DOCSIS, and wireless across a fiber network, taking modulated signals and using the subcarrier multiplexed (SCM) as opposed to terminating the physical connection of each technology separately and multiplexing at an alternative layer such as the inception point (IP), ATM, or TDM. The HFX network architecture and technology provides the local exchange carriers (LEC) an opportunity to deploy video services in selective

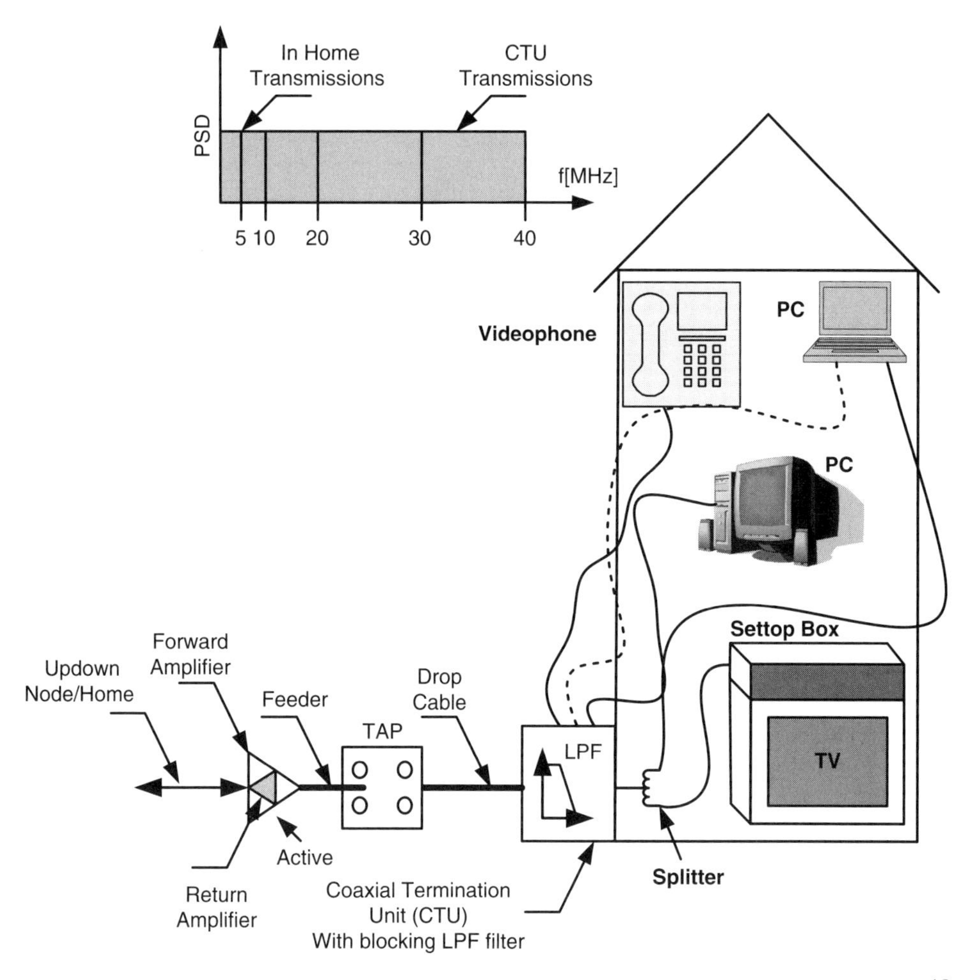


**Figure 1.3** HFC structure where the return path LPF is shown at the user end.[18] (Reprinted with permission from *Communications Magazine*. © IEEE, 1995).

markets through their IP-based transport connection and controls through xDSL, or using the same video platform as a cable company.

HFX technology has the following advantages:

- It pushes the equipment complexity back toward the CO and supports existing legacy customer equipment.
- All DSLAM functionality can be currently located in COs, supporting very high pools of XDSL modems.
- It leverages the installed base of copper plant, the DLC, and local switch-based infrastructure already installed by the local exchange carriers (LECs).

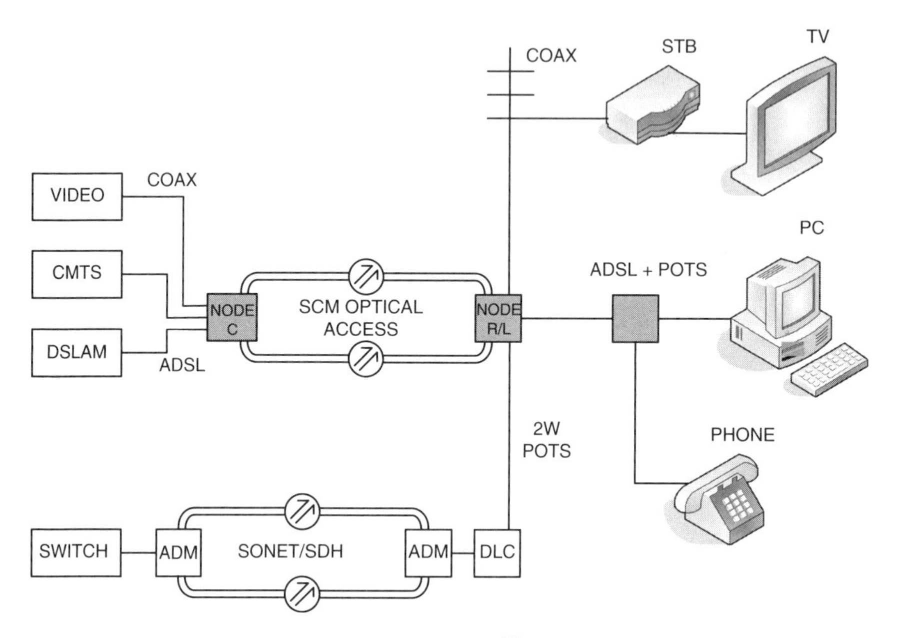


**Figure 1.4** HFX access network architecture.[15] (Reprinted with permission from *Communications Society* © IEEE, 2004.)

- It enables coexistence of xDSL, DOCSIS, analog and digital video, and wireless through the same common access network platform with minimal disruption to the existing LEC telephony infrastructure.
- It enables overlay of video via subcarrier multiplexing (SCM) in cable networks as an alternative to supporting VOD via the IP layer through ADSL and VDSL.
- It enables deployment of RAU in a wireless access network as opposed to the completely equipped base stations at the cell sites that reduce the complexity of the equipment deployed at the cell sites. This also facilitates dynamic carrier allocation and reduces the cost of equipment deployed at the cell site.

The tradeoff for reducing the complexity in the access network is that there is reduced statistical gain of customer traffic and concentration of broadband services in the plant, especially above the transport layer, either the IP, ATM, or MAC layer.

Because each customer or small group of customers have dedicated RF channels through the HFX network from the CO, there is an obvious concern with respect to the impact on the fiber capacity in the outside plant. Thus, WDM can be used to combine SCM signals between nodes C and R/L.

### 1.2.4 CATV and data transport in FTTx

Service providers activities in the U.S. date back to as early as the 1990s when several incumbent LECs (ILEC) started the experiments of FTTH. In those days, the user-end equipment was expensive, it did not result in widespread deployment. Fiber-to-the-curb-type deployments were developed by BellSouth. Cable companies deployed HFC networks as a cost-effective default per the aforementioned sections. This solution had released the so-called "last mile copper bottleneck"[36] since it limited BW.

With the development of cost-effective technology such as optical triplexers, made by ORTEL and MRV, as well as PLC technology, it became feasible to reconsider FTTx deployments. A major deployment trial of FTTH/P conducted by TelLabs AFC and Harmonic Inc. for Verizon[35] started at the end of 2003 for more than 500,000 homes. The advantage of FTTx over HFC is that the broader BW enables the transfer of HDTV over IP, as done in Japan, using passive architectures and requiring less equipment to deliver the data. In the FTTx structure, the TV signal transmitted by the central office (CO) is from satellites or microwave facilities that are received by the CO. As an example for video–data rates, HDTV requires compressed bit rate of 19 Mb/s per channel compared to 4 Mb/s of MP-2. TV services can be delivered as TV over IP or as AM–VSB and QAM over 1550 nm.[15] The main advantages of TV over IP are as follows: first, saving the wavelength of 1550 nm and using only two wavelengths; second, having a robust digital data, transport compared to analog RF over fiber.

The FTTx deployment utilizes WDM PON to distribute the media to the residential area, homes, office, and premises. The evolution of FTTx naturally started from HFC, where the head-end transmitted and received from a node that converted the signals bidirectionally to electrical/light and then distributed the service to homes by coax and copper, covering 200–500 homes for all services mentioned previously. The next stage was FTTC, where the CO/optical line terminal (OLT) replaced the HE; at the other side of the fiber is an optical network unit (ONU), covering 10–100 homes. The next stage was FTTH, where the COs distributed fiber optical services point-to-point; lastly, the PON architecture was used where the CO/OLT distributed the services by a main fiber split to homes using PON. The services delivered by FTTx to the subscriber became known as the triple-play set of data, voice, and video.[1,32,34,36]

FTTx uses WDM FSAN wavelengths of 1550 nm for video downstream, 1490 nm for data downstream, and 1310 nm for data upstream.[35] The video frequency plan remains as was previously mentioned: 79 analog channels and 53 digital channels covering the frequency range between 50 and 550 MHz for the analog channels with a frequency spacing of 6 MHz, and the digital 256QAM occupies the range of 550–870 MHz[34] over all 132 channels. As an example, the HDTV data rate over RF, 64 QAM traffic of 34 QAM channels carries 6 bits per symbol, and the transmit rate is 5 Msymbol/s or 30 Mbps per line. Hence the total 36 DTV carries an average of 1 Gbps, which can support 50 HDTV MPEG compressed broadcast channels. These channels are produced into a video by a cable modem transmission system (CMTS).[5]

**Table 1.5** Summary of last mile technologies and standards.[36]

| Service | Medium | Intrinsic Bandwidth | Per User Offered Peak Bit Rate (down/up) | Standard | Issued by |
|---|---|---|---|---|---|
| DSL | 24 gauge | 10 KHz | ADSL: 1//0.1; Mb/s $<$ 6 km[a] | G.992 | ITU |
| DSL | Twisted pair | 10 KHz | VDSL: 1//10; Mb/s $<$ 1 km[a] | Emerging | ITU/ETSI |
| Cable modem | Coax (HFC) | 1 GHz | 2//0.4 Mb/s[a] | DOCSIS 1.1 | CableLabs |
| BPON | Fiber | 75,000 GHz | 622 or 155 Mb/s//155 Mb/s | G.983 FSAN | ITU |
| GPON | Fiber | 75,000 GHz | 2.4 or 1.2 Gb/s// 622 or 155 | G.984 Draft | ITU |
| EPON + | Fiber | 75,000 GHz | 10–1000 Mb/s// 10–1000 Mb/s | 802.3ah & ae | IEEE |

[a]User rates delivered for significantly loaded systems in New York area.

Table 1.5 summarizes the last-mile technologies that compare performances between digital subscriber line (DSL) HFC cable-modem and FTTx PON diversities.[36]

FTTP services are called the triple-play communication services of voice, video and data. These services are required to meet the needs of residential and small business market places. An example of FTTP triple-play deployment is provided in Fig. 1.5.[35] In this case, the access portion, CO to the end subscriber, as well as the core network, is in support of the FTTP. This is based on ATM PON as specified by the ITU G.982.x. It basically consists of 622 Mb/s downstream at 1490 nm, and

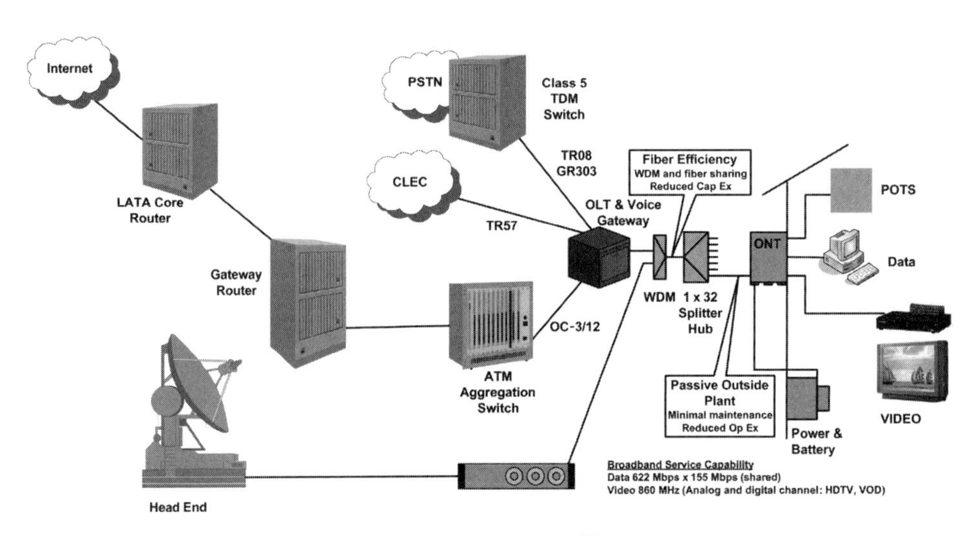


**Figure 1.5** Verizon FTTP PON network architecture.[35] (Reprinted with permission from *Journal of Lightwave Technology* © IEEE, 2005.)

155 Mb/s upstream at 1310 nm. The video is transported at 1550 nm according to the FSAN wavelength plan, which is coupled to the feeder fiber and the data transmission from the optical line terminal (OLT) via the WDM device.

The optical network terminal (ONT) is supported by a battery backup. This together with a sleep-mode function at the ONT allows the usage of POTS lines even with an extended outage of more than 8 h. The POTS signals are carried to the OLT and then split off via a GR-303 or TR008 interface to the PSTN. The OLT can support a TR-57 analog interface. The data is aggregated via an ATM switch before being transferred through a gateway router and local transport area (LATA) core router before going on to the Internet. The video has an upstream channel, connected to the ONT via an internal or external STB ONT that transfers the data channel and can be sent to the video head-end to support advance video services and two-way video interactions.

There are several methodologies to deploy a FTTP access network: the first is an overlay and the second is a full build, as shown in Figs. 1.6 and 1.7.[35]

In the full build, where a special arrangement is made by the residential area developer, each home is equipped with ONT together with a backup battery. This is in concert with some novel sleep-mode features of the ONT, which facilitates the support of POTS services during extended outages of more than 8 h.

In the case of an overlay deployment, a fiber passes all homes and businesses in the distribution area. However, only a small percentage of homes and businesses are connected to it and served with FTTP. Adding a new customer requires connecting the ONT to the service pole, whose drop terminal is connected to a distribution fiber coming from the splitter hub. A full build case is implemented in new residential areas where overlay is cheaper and easier to deploy in existing neighborhoods.

An additional architecture is fiber in the loop (FITL).[20] This architecture will give the network operator the opportunity to deploy fiber technology in an access

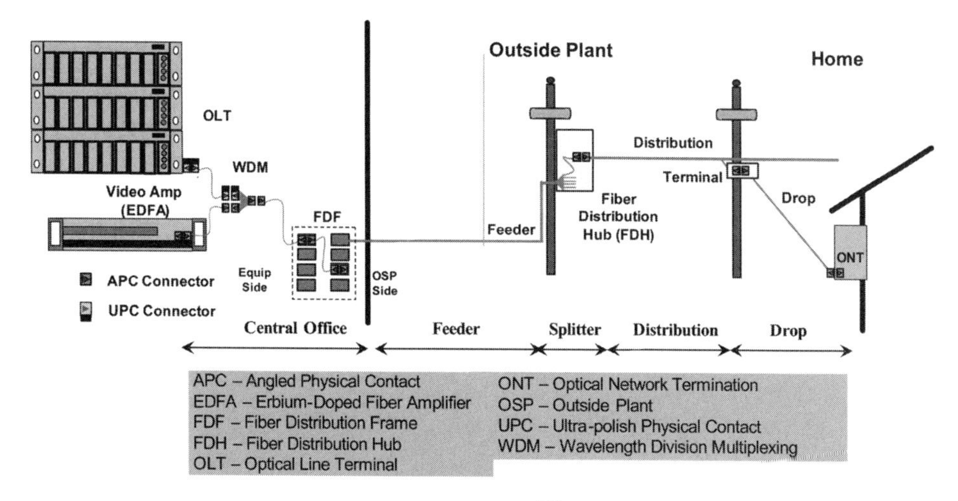


**Figure 1.6** FTTP access network deployment.[35] (Reprinted with permission from *Journal of Lightwave Technology* © IEEE, 2005.)

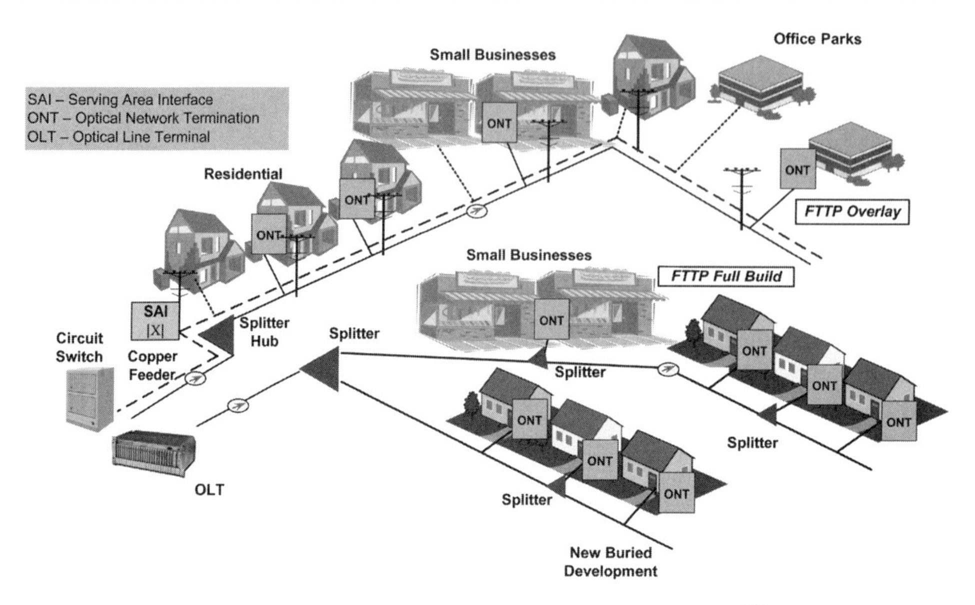


**Figure 1.7** FTTP deployment example of overlay and full build.[35] (Reprinted with permission from *Journal of Lightwave Technology* © IEEE, 2005.)

network. It is an old concept; it is in fact in a family of concepts such as FTTN, FTTC, and FTTH. FITL systems were intended to be compatible with typical LEC service, transmission, and operation.

### 1.2.5 ONT structure in FTTx

This section will describe, in brief, the concept of an ONT that is installed at the house premises. An illustration is provided in Fig. 1.8, which shows ALCATEL's ONT.[45] As it can be observed, the ONT consists of four main sections:

1. On the top left is the SMF redundancy, ended with an angled SC/APC connector to prevent optical reflections.

2. On the bottom left, there is the shielded FTTx ITR, whose pigtail is connected to the SC-APC input fiber. The ITR outputs are connected to an electronic PCB where the MAC ASIC is integrated. The video output is connected to the home coax distribution through the ONT, while the ITR output is through an SMB connector.

3. The top right of the ONT contains two more PCB sections, the upper side of which can be the power management section regulation, battery charger, etc.

4. On the bottom right is an interface PCB with RJ connectors.

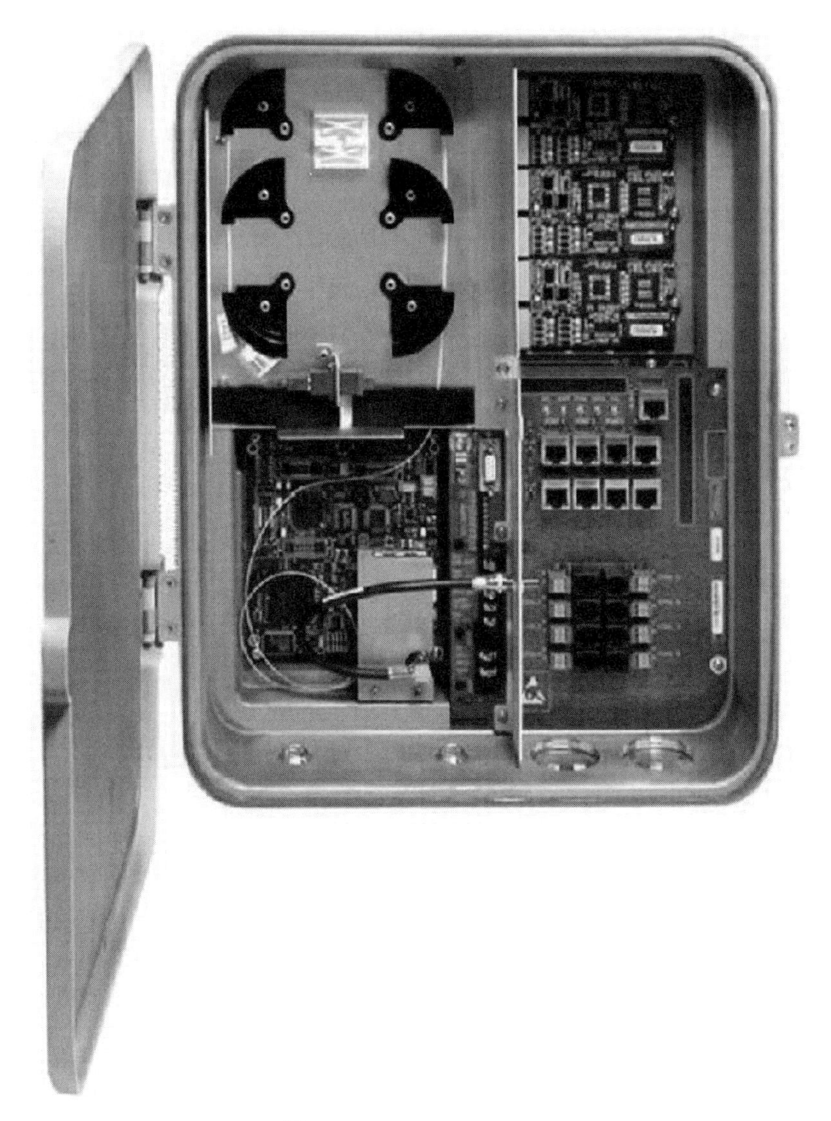

**Figure 1.8** FTTX ONT unit.[45] (Courtesy of ALCATEL with permission © 2005.)

The ONT is generally installed inside the home garage so that it is less exposed to extreme temperature effects. Housing is made of polycarbonate with antihumidity sealing.

## 1.3 PON and Its Variants

### 1.3.1 Standardization

So far, FTTx was described in a conceptual way; however, the distribution network architecture utilized for FTTx as well as HFC is based on PON architecture and its variants as well as WDM.[13,14] The standardization started in 1995 when several of the world's largest carriers, NTT, BT, France Telecom, etc., and their equipment vendors started discussing a complete video services solution. At that time, the two logical choices for protocol and physical planting were ATM and PON: ATM, because it was thought to be suited for multiple protocols; PON, because it is the most economical broadband optical solution. The ATM PON format proposed by the FSAN committee has been accepted as an ITU standard known as ITU–T Rec G.983.[28,43]

Although FSAN is often touted as a standard, it is, in fact, a guideline since it does not define equipment interoperability. It is not like a SONET or Ethernet, in which two cards from different vendors can communicate.

There are three standardized versions of PON described in brief in Table 1.6:[13,35]

- ***Ethernet PON (EPON)***: Ethernet equipment vendors formed Ethernet in the First Mile alliance (EFMA) worked on architecture for FTTH as Ethernet is a dominant protocol in local area network (LAN). EPON-based FTTH was adopted by IEEE standard IEEE 802.3ah in September 2004. Adopting Ethernet technology in access networks would make a uniform protocol at the customer end and simplify the network management. A single protocol in LAN, access network, and backbone network enables easy rollout of FTTH.

- ***Broadband PON (BPON)***: Generalized by the G.983.1, G.983.2, and G983.2, BPON has two key advantages: first, it provides a third wavelength for video services, and secondly, it is a stable standard that reuses ATM infrastructure. The ITU–T recommendation G.981.1 defines clauses of performance, namely class A, class B, and class C.

- ***Gigabit PON (GPON)***: The progress in technology and the need for larger BWs as well as the complexity of ATM forced the FSAN group to look for better technology. A standardization was initiated by FSAN in 2001 for designing networks over 1 GBps. GPON offers services up to 2.5 GPS. GPON enables the transport of multiple services in their native format, mainly TDM and data. For soft transition from BPON to GPON, many functions of BPON are reused in GPON. In January 2003, the GPON standards were ratified by ITU–T and are known as ITU–T recommendations G.984.1, G.984.2, and G.984.3.

### 1.3.2 Protocol model background: the seven-layer protocol stack

The Internet network of today, which is delivered over optical fiber, started as an experiment of the US department of defense (DOD) at the late 1960s. The idea was to rout data with high quality service and with high security between various sites.

**Table 1.6** APON/BPON, GPON, EPON, WDM-PON comparison.[13,35]

| Parameter | APON/BPON | EPON | GPON | WDM-PON |
|---|---|---|---|---|
| Standard | ITU G.983 | IEEE 803.ah | ITU-T G.984 | None |
| Data packet cell size | 53 bytes (48 payload. 5 overhead) | 1518 bytes | Variable size (from 53 bytes up to 1518) | |
| Bandwidth | 1.2 Gps downstream; 622 M upstream; 622 Mbps/622 Mbps; 622 Mbps/155 Mbps | Up to 1.25 Mbps | Downstream configurable from 1.24 Gps to 2.48 Gps; upstream configurable in 155 Mbps, 622 Mbps, 1.24 Gbps, or 2.48 Gps | 1-10 Gbps |
| Downstream wavelength | 1480-1500 nm | 1500 nm | 1480-1500 nm | |
| Upstream wavelength | 1260-1360 nm | 1310 nm | 1480-1500 nm | |
| Traffic modes | ATM | Ethernet | ATM, Ethernet, TDM | Protocol independent |
| Voice | TDM | VoIP or TDM | Native TDM | |
| Video | 1550 nm overlay | 1550 nm overlay | 1550 nm | |
| Video transport | RF | RF/IP | RF/IP | RF/IP |
| ODN classes support OLT to ONT losses | A, B, and C | A and B | A, B, and C | |
| Maximum PON splits | 32 | 16 | 64 | |
| Efficiency | 72% | 49% | 94% | |

In 1969 the computers of four universities were connected forming the ARPANET network after the research group of "advanced research projects agency." In 1973 a team was created to develop a method to connect all computers, while using various transmissions and rates. In 1978 the program was mostly completed and it resulted in the TCP/IP protocols. In 1983 it was decided that TCP/IP would be the internet protocol. Since then the transmission control protocol (TCP)/internet protocol (IP) known as TCP/IP appears as the main internet communication protocol. Originally, the protocol had 3 to 4 layers, IP, and TCP, which is the connection orientation. These protocols are improving all time.

When one wants to establish a connection, for instance make a call, all one needs to do is dial a number. Hence the phone network looks transparent. However, this is done through a communication protocol that routs the call to its destination. The same thing applies to IP services and cellular as well as optical networks, but with some modifications[25] but 1977, the International Standard Organization (ISO) had established a working group to define functions and protocols for open systems without being dependent on the equipment vendors. The model was called open system interconnection (OSI) and it was released in

1983 under ISO 7498, and later under CCITT X.200. The idea for the "protocol stack" was first suggested by IBM in 1977 and again later on by DEC.

The OSI model consists of a stack of seven layers. Each layer is defined by functions that enable it to receive information from an upper layer, do additional processing, and add a tag to it by a title. In the second layer, a title and ending are added. At the end, the first layer, the physical layer, transmits the data as a bit stream to the network. On the receive side, an inverse process is done. The title added at the transmit side is removed after the processing, and this way, all shells are removed until the data is recovered. The protocol stack layers are provided in Table 1.7.

Each layer receives from the above layer protocol information called protocol data units (PDU). For instance, the third layer receives PDUs from the fourth layer transport PDU (T-PDU). After adding a title to it, a network header (NH), it is transmitted as a network PDU (N-PDU). The process of adding the title or header is called "protocol." The layers according to OSI have the following functions:

- Layer 1: *The physical layer*, also marked as L1P (first layer protocol). It defines the connectors and interfaces, voltages levels, and the way of operating the physical interface. For instance EIA-RS232, USB, EIA–RS 449. and other LAN interfaces. The following chapters review the physical layer implementations and interfaces.

- Layer 2: *The data link layer*, which is responsible for frame transfer between two elements connected through data line. At this layer, data packets are encoded and decoded into *bits*. It furnishes transmission protocol knowledge and management, and handles errors in the physical layer, flow control, and frame synchronization. The data link layer is divided into two sublayers: the media access control (MAC) layer and the logical link control (LLC) layer. The MAC sublayer controls how a computer on the network gains access to the data and permission to transmit it. The LLC layer controls frame synchronization, flow, and error checking.

**Table 1.7** Protocol stack seven-layers model and its equality to TCP/IP.

<table>
<tr><td>Tx/Tx</td><td>Information</td><td colspan="6">TCP/IP</td></tr>
<tr><td>DATA</td><td>DATA</td><td rowspan="3">FTP</td><td rowspan="3">TELNET</td><td rowspan="3">SMTP</td><td rowspan="3">CMOT</td><td rowspan="3">DNS</td><td rowspan="3">SNMP</td></tr>
<tr><td>7—Application</td><td>Data+ annex</td></tr>
<tr><td>6—Presentation</td><td>A—PDU</td></tr>
<tr><td>5—Session</td><td>P—PDU</td><td colspan="4">TCP</td><td colspan="2">UDP</td></tr>
<tr><td>4—Transport</td><td>S—PDU</td><td colspan="6">IP</td></tr>
<tr><td>3—Network</td><td>T—PDU</td><td colspan="3" rowspan="3">LAN</td><td colspan="3" rowspan="3">WAN</td></tr>
<tr><td>2—Data Link</td><td>N—PDU</td></tr>
<tr><td>1—Physical</td><td>Bits</td></tr>
</table>

- Layer 3: *The network layer*, which routes the data and provides addresses within the network. For instance, the IP address in transmission control protocol (TCP)/IP protocol or logical channels (LCN) in X.25 protocol.
- Layer 4: *The transport layer* is responsible for reliable transport, error correction, and flow control. Protocols at this layer are TCP and UDP from Internet protocols family.
- Layer 5: *The session layer* is responsible for initialization and ending of a session, and defines its nature as full or half duplex.
- Layer 6: *The display layer* performs coding and compressing of text and video.
- Layer 7: *The application layer* provides applications such as e-mail, etc.

### 1.3.3 Implementation

A PON is an optical-access architecture that facilitates broadband voice, data, and video communications between an OLT and multiple remote ONUs over a purely passive optical distribution network and uses two wavelengths for data, 1310 nm and 1490 nm, and an additional 1550 nm for video.[28,32,41] A PON has no active elements in the network path that require optical-to-electrical converters. PON systems use passive fiber optics couplers or splitters to optically route data traffic. In contrast, an active optical network (AON) such as a SONET/synchronous digital hierarchy (SDH) infrastructure requires optical-to-electrical-to-optical conversion at each node. The PON can aggregate traffic from up to 32 ONUs back to the CO using a tree, bus, or fault-tolerant ring architecture. Like SONET/SDH, PON is a layer 1 transport technology. Until recently, most telecommunications fiber rings used SONET/SDH transport equipment. These rings—using regeneration at each node—are optimized for long-haul and metropolitan infrastructures, but are not the best choice for local access networks. A PON offers a cost-effective solution as an "optical collector loop" for metropolitan and long-distance SONET/SDH infrastructures. A PON has a lower initial cost because deployment of the fiber is required upfront. ONUs can then be added to the PON incrementally as demand for services increases, whereas active networks require installation of all nodes upfront because each node is a regenerator. To further reduce costs, a WDM layer can be added to a PON because the nodes sit "off" the backbone. When employing WDM on a SONET/SDH ring, demultiplexing and remultiplexing is required to bypass each node. A PON is also a cost-effective way to broadcast video because it is a downstream point-to-multipoint architecture. The broadcast video, either analog or digital, is added to the TDM or the WDM layer of the PON. Unlike SONET/SDH, a PON can also be asymmetrical. For example, a PON can broadcast OC-12 (622 Mb/s) downstream and access OC-3 (155 Mb/s) upstream, as in

FSAN. The FSAN defines two levels of loss budget, class B and class C. The laser should operate with an extinction ratio $>10$ dB and with a mean launch of power between $-4$ dBm and $+2$ dBm for class B, and between $-2$ dBm and $+4$ dBm for class C at 155 Mbps. For 622 Mbps, the launch power should be between $-2$ dBm and $+4$ dBm for both classes. The BER should be $<10^{-10}$ for a minimum receiver sensitivity of $-30$ dBm for class B and $-33$ dBm for class C operation at 155 Mbps. In the case of 622 Mbps, the receiver sensitivity for class B is relaxed to $-28$ dBm, with no changes for class C.[43] Hence, various classes require different photodetection such as PIN or APD. Since the 155 Mbps baseband BW is shared by up to 32 ONUs, there is a limitation on the number of video channels that can be simultaneously delivered as switched digital video (SDV) to users who want to use HDTV. This limitation is partially resolved by increasing the downstream rate to 622 Mbps; however, this will require an upgrade of the CO equipment.[41] Table 1.8 provides the standard requirements of a PON transceiver in order to comply to the ITU-T G.983 specifications.[41]

An asymmetrical local loop allows for lower-cost ONUs, which use less expensive transceivers. SONET/SDH, however, is symmetrical. Thus, in an OC-12 SONET/SDH ring, all line cards must have an OC-12 interface. For local-loop applications, a PON can be more fault-tolerant than a SONET/SDH. The PON node resides off the network, so that loss of power to a node does not affect any other node. This is not true for SONET/SDH, where each node performs regeneration. The ability of a node to lose power without network disruption is important in the local loop, because telephone companies cannot guarantee power backup to each remote terminal. For these reasons, PON technology is the most likely transport candidate for the local telephone network, as shown in Fig. 1.9. In a sense, PON technology does not compete with SONET/SDH, because it can use SONET/SDH-compatible interfaces and serve as the short-haul "optical collector loop" for both metropolitan and long-haul SONET/SDH

**Table 1.8** ITU-T G983 specifications for transceivers for an ATM PON.

| Item | Unit | ITU-T G983 Specification | | |
|---|---|---|---|---|
| Direction | | Downstream | | Upstream |
| Wavelength | nm | 1480–1580 | | 1260–1360 |
| Extinction ratio | dB | $>10$ | | $>10$ |
| BER | | $<10^{-10}$ | | $<10^{-10}$ |
| Bit rate | Mbps | 155.22 | 622 | 155.52 |
| Min mean launched optical power class B/C | dBm | $-4/-2$ | $-2/-2$ | $-4/-2$ |
| Max mean launched optical power class B/C | dBm | $+2/+4$ | $+4/+4$ | $+2/+4$ |
| Min receiver sensitivity class B/C | dBm | $-30/-33$ | $-28/-33$ | $-30/-33$ |
| Min receiver overload class B/C | dBm | $-8/-11$ | $-8/-11$ | $-8/-11$ |

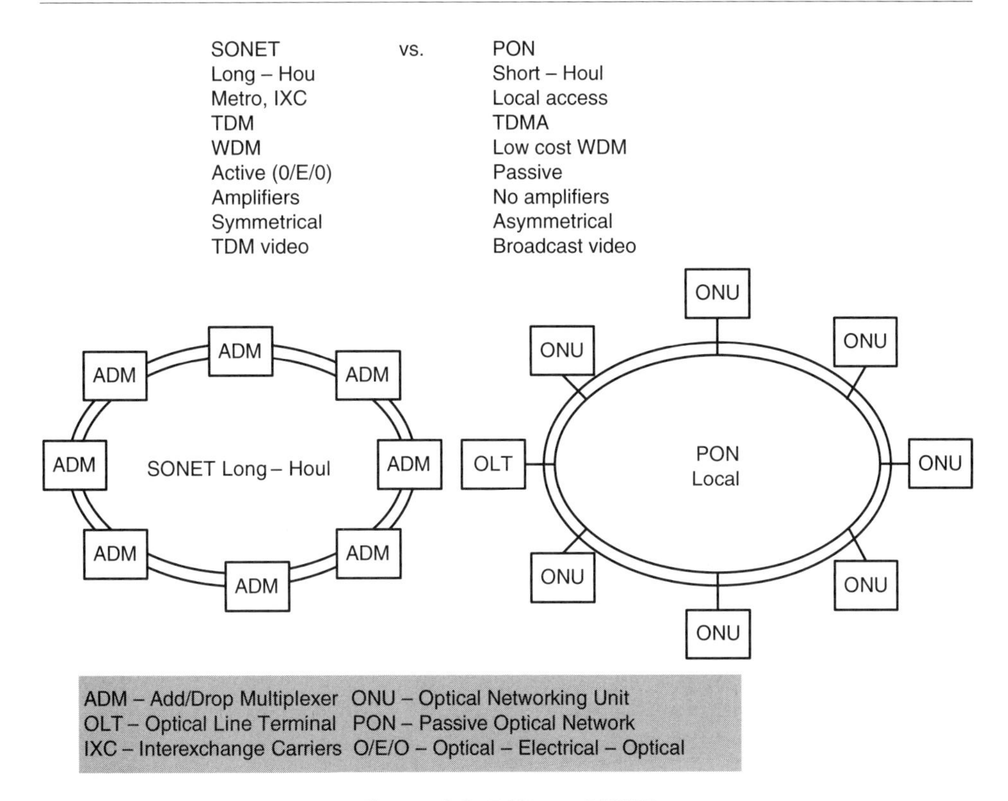


**Figure 1.9** PON vs. SONET.

rings. PON can serve as a feeder to extend fiber from the local exchange to the neighborhood or curb, where copper, coaxial, or wireless systems provide the last-mile connections to subscribers. However, in a feeder network, reliability is crucial since if the feeder network fails, all customers are potentially left without service. Restoration in the feeder should ideally be implemented in both optical and electronic layers with coordination between the two domains. For instance, cable cut protection and complete node failure is accomplished by $2 \times 2$ optical switches that are deployed per fiber such that the service traffic can be looped back onto protection capacity in order to avoid the point of failure. Additionally, redundancy can be provided against failures in the electronic domain, i.e., laser failure.[16] One of the most promising applications for a PON is as a feeder for DLC systems. DLC systems multiplex voice and data channels over copper or fiber to remote terminals that distribute services to subscribers. Typically, DLC systems have been used to reduce copper-cable deployments and extend POTS to remote subscribers. Recently, DLC vendors introduced integrated access platforms that carry voice, data, and video traffic. These systems respond to the growing demand for broadband capabilities from telephone companies and enable a variety of services, including POTS, integrated services digital network (ISDN), DSL, and digital carrier services such as T1/E1 (1.544 Mb/s, 2.048 Mb/s). In rural areas, this new generation of DLC products will help

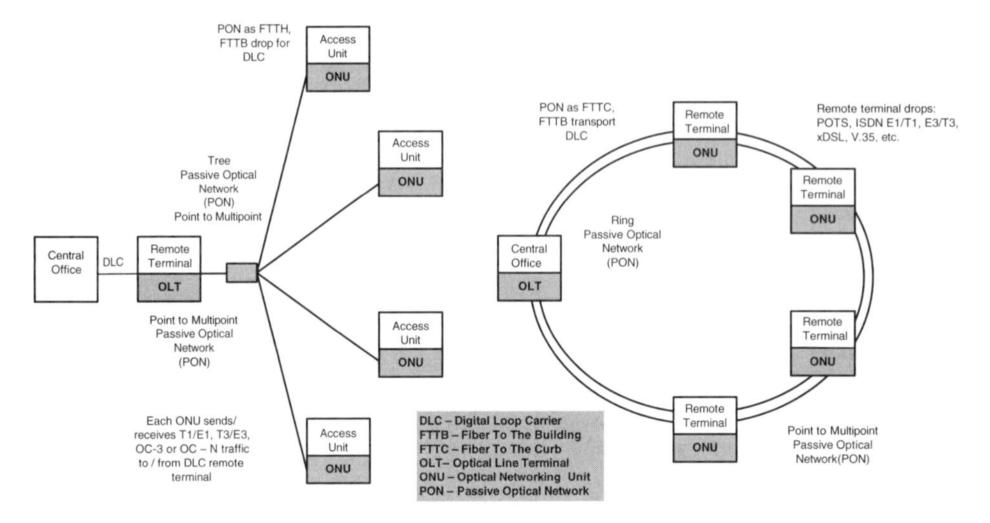


**Figure 1.10** Using PON as a DLC system feeder.

extend broadband services to remote subscribers. In urban settings, these DLC systems will improve network efficiency by expanding the BW of existing copper and fiber facilities. This means that integrated access DLC systems could become the preferred solution for telephone companies that want to simplify their CO and provide broadband services over shorter drops of twisted-pair copper.

As DLC systems evolve toward a broadband architecture, new deployments will require FTTC or FTTB solutions. One alternative is to multiplex traffic to and from remote terminals or integrated access units using a low-cost, high-BW PON. Using a PON as a feeder for DLC systems can significantly lower the cost and improve the performance of DLC deployment. As shown in Fig. 1.10, a PON can be used between the CO and the remote terminal to provide FTTC or FTTB. It can also be used between the remote terminal and the customer terminal equipment to provide FTTB or FTTH.

PONs offers a number of benefits over traditional feeder solutions for DLC systems:

- Using a point-to-multipoint ring architecture dramatically reduces the cost of the outside plant by minimizing the amount of fiber cable required. Because the cost of deploying fiber can be more than \$5 per foot, using a point-to-multipoint topology can save more than \$1 million compared to a network with point-to-point links.

- The PON's point-to-multipoint architecture broadcasts signals from the OLT at the CO downstream to every ONU. This makes it an appealing solution for telephone companies that plan to introduce cable TV.

- A PON can deliver a POTS channel for less than \$10. While the DLC remote terminal still requires a POTS line card, the additional cost of a point-to-multipoint network is small. If the DLC multiplexer and OLT are merged into one box, then a PON becomes the lowest-cost narrowband DLC transport solution. When broadband is considered, PON technology easily wins the cost war.

- Because a PON features a point-to-multipoint architecture with typically 32 or fewer nodes, low-cost WDM solutions can be used to deliver dedicated high-speed data to each ONU. Using a WDM overlay, a PON is capable of any OC-N speed. This allows the service provider to carry high-speed leased lines and gives competitive local-exchange carriers (CLECs) the ability to lease a high-speed access path.
- As a point-to-point link, a traditional DLC system is subject to cable failure. When using a fault-tolerant PON ring, a DLC system is fully fault tolerant.

In response to competition and a growing demand for broadband services, telephone companies are attempting to leverage their existing copper infrastructure through deployment of xDSL technologies. Very-high-speed digital subscriber line (DSL VDSL) technology will allow for up to 50-Mb/s transmission. It can only provide this BW, however, over twisted copper pairs of less than 1000 feet. Thus, only a very small percentage of subscribers can benefit from this technology. To overcome this distance/BW problem, xDSL systems will eventually be deployed with a fiber optic feeder network. With this in mind, major telephone companies are backing an initiative to develop standards for a broadband local loop called full-services access network (FSAN).[36] FSAN advocates a network architecture using a shared PON with dedicated VDSL drops. By extending a fiber feeder to within 3000 feet of subscribers, telephone companies can provide 25-Mb/s broadband service, including video over the "drop" of twisted-pair copper into the subscriber's premises, as depicted in Fig. 1.11.

Broadband wireless systems facilitate rapid local-loop deployment using technologies that will deliver megabit-per-second service. Several broadband wireless technologies are now available, including LMDS, a multichannel multipoint distribution system (MMDS), and digital microwave. These systems typically consist of a single hub or CO linked to multiple base stations that deliver services to buildings or residences.

Broadband wireless systems require line-of-sight access, operate over limited distances, and are subject to attenuation from rain. Given these limitations, a critical issue facing wireless operators is how to aggregate traffic from multiple base stations back to the CO. One alternative is to interconnect base stations and the CO using a low-cost, high-BW PON, as shown in Fig. 1.12.

Cable operators also face the need to upgrade their networks to introduce interactive broadband services. To date, the cable network uses an analog coaxial bus that can broadcast large amounts of BW. However, it is a unidirectional medium with high maintenance and coaxial amplifier costs. To address these limitations,

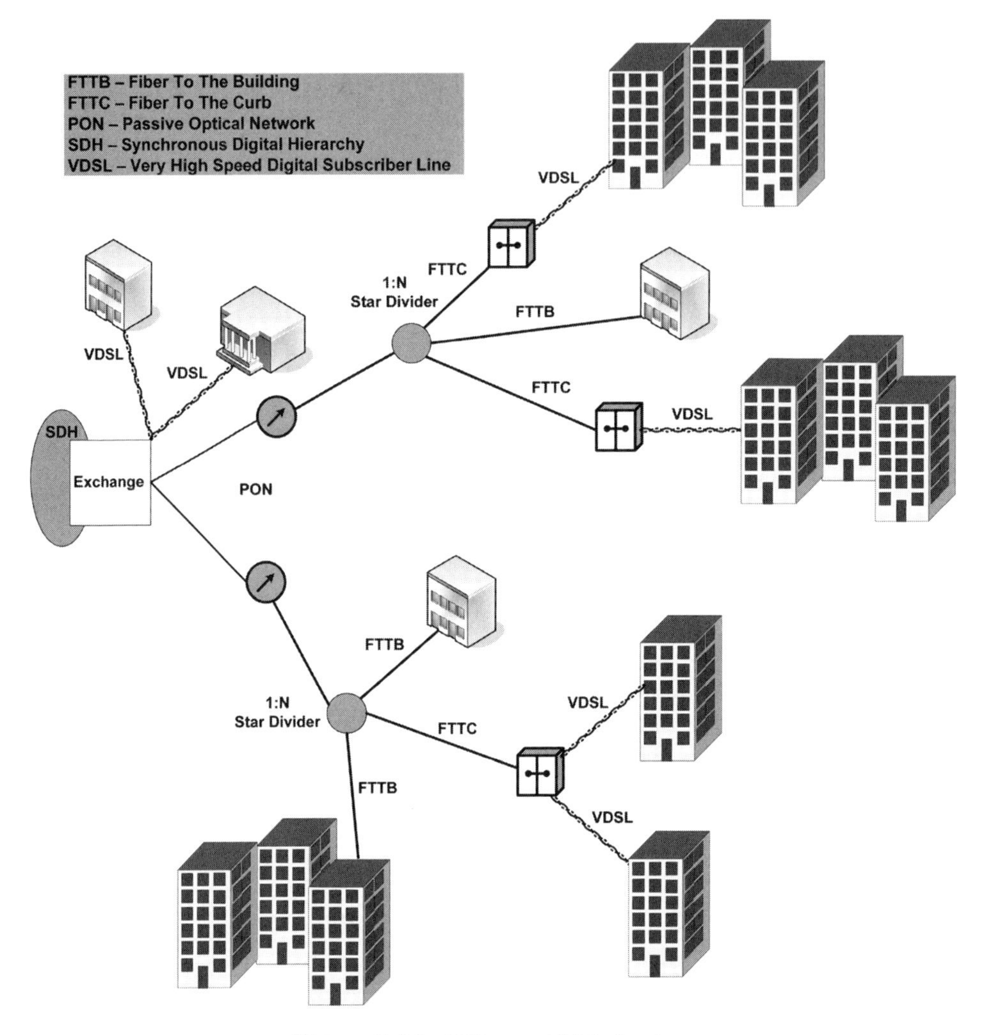


**Figure 1.11** PON as a VDSL feeder.

many cable TV operators have adopted HFC architecture with optical fibers penetrating deep into the network. HFC implementations replace part of the coaxial access network by running a fiber optic cable from the cable-TV head-end to the distribution nodes. The broadcast signal is translated into an electrical analog signal that is transmitted to the subscriber via a coaxial connection. The primary shortcoming of this approach is its inability to provide fault-tolerant, symmetric, bidirectional service, which is required for truly interactive broadband applications.

To address this problem, cable TV operators can deploy a low-cost PON between the head-end station and fiber nodes. Like the cable-TV network, a PON is a downstream broadcast architecture, so it distributes video downstream from a single hub to multiple fiber nodes, along with bidirectional voice and data traffic, as shown in Fig. 1.13.

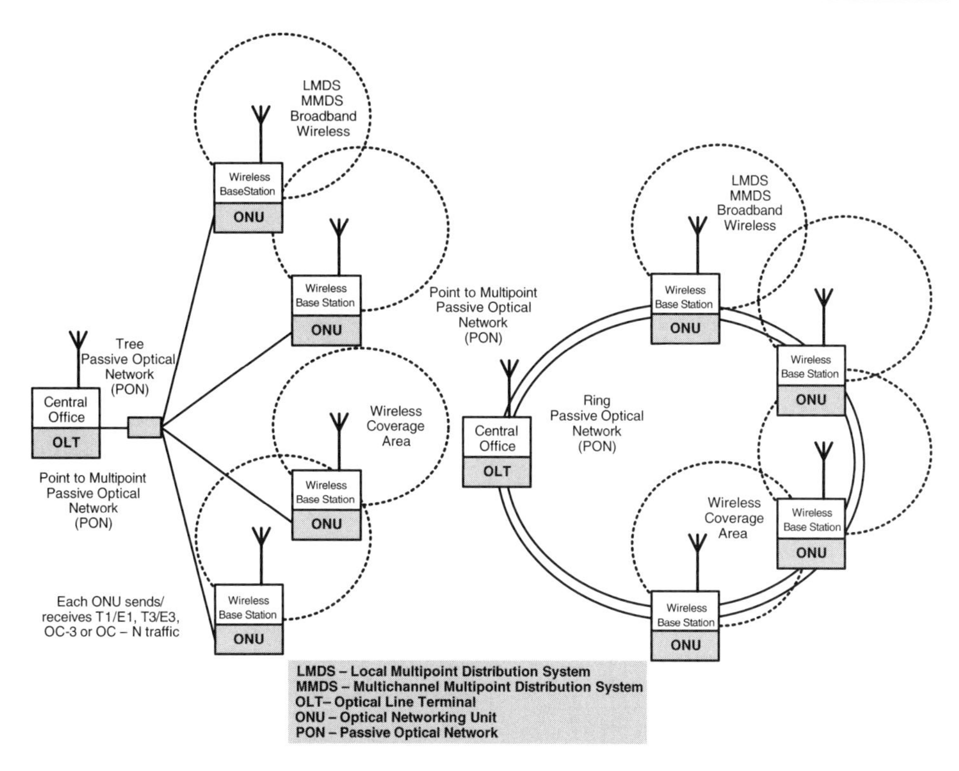


**Figure 1.12** PON as wireless feeder.

A PON can serve as the feeder for broadband copper, wireless, and coaxial drops, but it can also be used for the "home run." A PON can support an FTTx architecture; the "x" is where the ONU resides: at the curb, neighborhood, building, premises, or home. Prior to FTTH deployments, PON technology initially was used in FTTC and FTTB architectures such as SBC. These architectures are usually hybrid networks with PON technology serving as the low-cost, high-BW feeder; traditional copper and coaxial cables are used as the customer device interface. When a PON is deployed as an FTTB system, an ONU located in a building's telecommunications closet can provide full broadband services to all occupants. In this type of an optical collector loop, the PON can use standard SONET/SDH rates and WDM interfaces. A PON can provide any type of service, including voice, data, and video. Like SONET/SDH, it is a transport network, but one that is designed for the cost-sensitive local loop. What makes PON so attractive is that while it can enable DLC systems, DSL, HFC, and wireless to reach broadband capacity, it is not a new access solution. In that sense, the PON does not replace copper, coaxial, and wireless embedded networks, but allows these traditional infrastructures to be used in shorter, higher BW drops.

PON was used also to distribute HFC services as was elaborated in previous sections and depicted in Fig. 1.13.

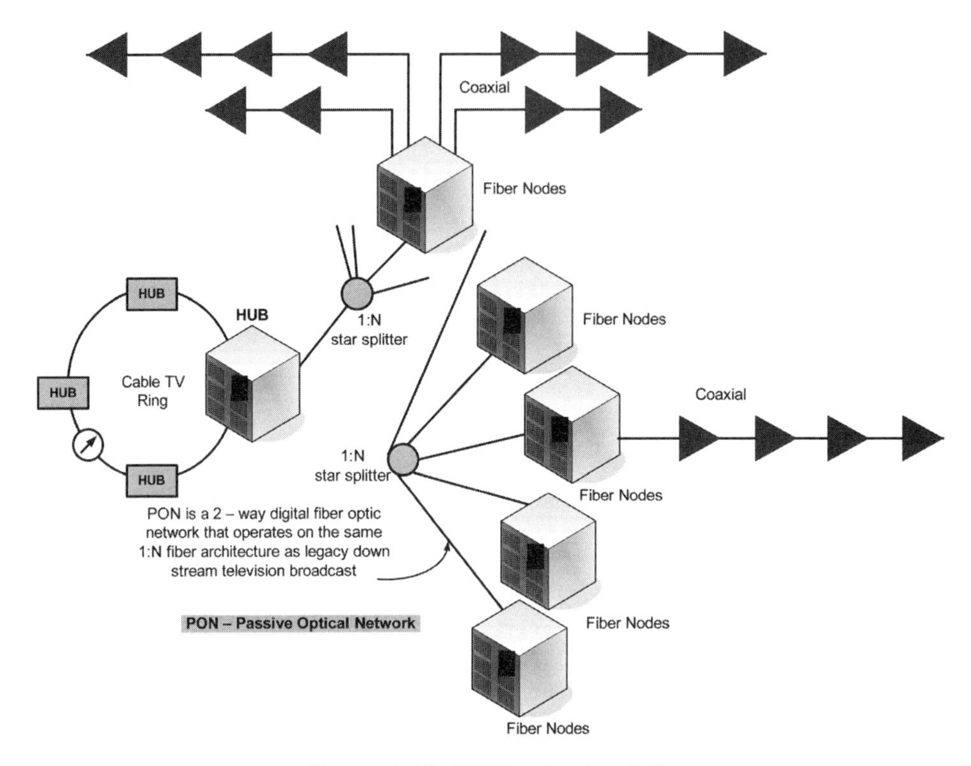


**Figure 1.13** PON feeder for HFC.

## 1.4 Main Points of this Chapter

1. The advantage of the optical fiber is its low loss of about 0.15 dB/km, its infinite BW, and its immunity to magnetic interference.

2. WDM is wave division multiple access, where a single fiber carries several wavelengths with spacing of 50 to 100 GHz between each wavelength.

3. A fiber bundle can carry many wavelengths and thus an optical network supports a large data rate.

4. Services on fiber are video, voice, and data, called triple play.

5. Architectures used today are FTTx, which stands for fiber to the home, business, premises, building, or curb.

6. FTTx carries CATV data, transferred over 1550 nm and carries 79 AM–VSB channels and QAM channels, uplink data of 155 Mbps over 1310 nm, and down link at 625 Mbps over 1490 nm.

7. The three-wavelength service of 1310 nm transmission, 1490 nm receiving, and 1550-nm receiving is defined by FSAN.

8. The center that delivers triple-play services is called CO and the receiver side box is called ONT.

9. At the ONT is located the optical transceiver called ITR, which converts the data from optical to electrical and vice versa.

10. Home is called the last 100 feet defined by HPPLA (home plug power line alliance) and HPNA as well as wireless LAN (WLAN) and other standards.

11. Prior to the emergence of FTTx internet data and TV, signals were delivered over fiber hybrid coax (HFC) using DOCSIS protocol initiated by CableLabs.

12. The main issue of ONT is power backup and battery lifetime. Efficient design of ONT with sleep mode for power failure can support POTS services up to 8 hours and more.

13. An HFC network consists of providing downstream CATV and data from a primary hub ring. This hub is the master head-end; the split from the primary hub is the secondary hub ring from which fiber nodes are distributed to homes.

14. The distribution to home from the node fiber is done by a tap that converts optical signals to electrical signals carried by coax. From each tap, there is a drop cable to the coaxial termination at the subscriber's home.

15. The line from the tap to home is bidirectional due to the return path upstream.

16. The return-path line suffers from a noise issue called "noise funneling" caused by several users are sharing the same uplink line.

17. Additional noise sources injected into the return path are in-house noises such as radiative noise due to appliances, narrowband short waves propagating through the atmosphere, and common mode noise due to connector oxidation.

18. The return path noise can be minimized by placing a low-pass filter at the return-path outlet, thereby blocking the noise.

19. HFC data communication is done by DOCSIS protocol initiated by CableLabs.

20. An HFX solution is an interstage solution between HFC and PON FTTx architecture, and consists of SCM optical access and SONET/SDH for POTS.

21. There are several levels of carriers: incumbent ILEC, CLEC, etc.

22. The current architecture to deliver FTTx variants is PON.

23. A PON variant is defined by the data rate and method of data rate and transfer such as gigabit PON (GPON), ethernet PON (EPON), etc.

24. PON can be used as wireless feeder, LMDS feeder, etc.

25. There are two main PON network topologies: ring and tree.

26. The method of communication and sequence is defined by the protocol stack.

27. The hardware implementations and limitations in the Layer 1 (physical layer per the OSI model), and algorithms related to layer 2, generally made by algorithm and DSP group called Physical Layer Team, are to be treated in the next chapters.

## References

1. An, F.T.K., S. Kim, Y.L. Hsueh, M. Rogge, W.T. Shaw, and L. Kazovsky, "Evolution, challenges and enabling technologies for future WDM-based optical access networks," *Proceedings of JCIS*, pp. 1449–1453 (2003).
2. McDonna, A.P., and B.M. MacDonald, "Optical component technologies for FTTH applications," *Proceedings of IEEE Electronic Components and Technology Conference*, pp. 1087–1091 (1995).
3. Okano, H., A. Hiruta, H. Komano, T. Nishio, R. Takahashi, T. Kanaya, and S. Tsuji, "Hybrid integrated optical WDM transceiver module for FTTH systems." Hitachi U.D.C. 621.372.88.029.72.049.776 (2005).
4. Mukherjee, B., "WDM optical communication networks: progress and challenges." *Journal of Selected Areas in Communications*, Vol. 18 No. 10, pp. 1810–1824 (2000).
5. Wood, T.H., G.W. Wilson, R.D. Feldman, and J.A. Stiles. "Fiber vista: a cost effective fiber to the home (FTTH) system providing broadband data over cable modems along with analog and digital video." *Photonics Technology Letters*, Vol. 11 No. 4, pp. 475–477 (1999).
6. Farmer, J.O., "Delivering video, voice and data to consumers via an all-fiber network." *Transactions on Consumer Electronics*, Vol. 48 No. 3, pp. 548–555 (2000).
7. Park, S. J., S. Kim, K.H. Song, and J.P. Lee. "DWDM based FTTC access network." *Journal of Lightwave Technology*, Vol. 19 No. 12, pp. 1851–1855 (2001).
8. Lin, Y.M. and W.I. Way, "The feasibility study of transporting IS-95 CDMA signals over HFC networks." *IEEE Photonics Technology Letters*, Vol. 11 No. 6, pp. 736–738 (1999).
9. Ikushima, I., S. Himi, T. Hamaguchi, M. Suzuki, N. Medea, H. Kodera, and K. Yamashita, "High performance compact optical WDM transceiver

module for passive double star subscriber systems." *Journal of Lightwave Technology*, Vol. 13 No. 3, pp. 517–524 (1995).

10. Ho, K.P., H. Aai, C. Lin, S.K. Liaw, H. Gysel, and M. Ramachandran, "Hybrid wavelength division multiplexing systems for high capacity digital and analog video applications." *IEEE Photonics Technology Letters*, Vol. 10 No. 2, pp. 297–299 (1998).
11. Shankar, M.M.M., R. Parthasarathy, and S. Ramasubbu, "Quality of service for the last mile." Virginia Tech (2001).
12. Kim, K.Y., S.Y. Kim, M.W. Kim, and S. Jung, "Development of compact and low-crosstalk PLC WDM filters for hybrid integrated bidirectional optical transceivers." *Journal of Lightwave Technology*, Vol. 23 No. 5, pp. 1913–1917 (2005).
13. Gutierrez, D., K.S. Kim, S. Rotolo, F.T. An, and L.G. Kazovsky, "FTTH standards, deployments and research issues." *Proceedings of JCIS*, pp. 1358–1361 (2005).
14. Park, S.J., C.H. Lee, K.T. Jeong, H.J. Park, J.G. Ahn, and K.H. Song, "Fiber to the home services based on wavelength division multiplexing passive optical network." *Journal of Lightwave Technology*, Vol. 22 No. 11, pp. 2582–2591 (2004).
15. Bye, S.J., "The HFX architecture for local access: supporting voice, video and data services." *IEEE Communications Society*, pp. 105–108 (2004).
16. Saleh, A.M. and J.M. Simmons, "Architectural principles of optical regional and metropolitan access networks." *Journal of Lightwave Technology*, Vol. 17 No. 12, pp. 2431–2448 (1999).
17. Kim, J., M.S. Lee, S. Choi, S.S. Lee, H.H. Song, and H.S. Hong, "Modeling and optimization of subscriber loop deployment strategy for Hanaro Telecom." *Proceedings of IEEE Systems, Man, and Cybernetics Conference*, Vol. 4, pp. 3971–3976 (1998).
18. CATV Return Path Characterization for Reliable Communications. C. A. Eldering, N. Himayat, and F. M. Gardner. *IEEE Communications Magazine*, pp. 62–69, August (1995).
19. Lee, J.J., and V. Shukla, "Cost analysis for HFC networks: installed first cost (IFC) for video and incremental cost (IC) for telephony." *Proceedings of IEEE Global Telecommuncations Conference*, Vol. 1, pp. 257–261 (1996).
20. Schmid, W., "Fiber in the loop cable TV system for standard TV signals and DCV signals using analog optical transmission with optical amplifiers." *IEEE Proceedings on Optical/Hybrid Access Networks*, pp. 8.03/01–8.03/07 (1993).
21. Iyer, M.K., P.V. Ramana, K. Sudharsanam, C.J. Leo, M. Sivakumar, B.L.S. Pong, and X. Ling, "Design and development of optoelectronic mixed signal system on package (SOP)." *IEEE Transactions on Advanced Packaging*, Vol 27 No. 2, pp. 278–285 (2004).
22. Gobl, G., C. Lundquist, B. Hillerich, and M. Perry, "Fiber to the residential customer." *IEEE Proceedings on Global Telecommunications*, pp. 165–169 (1992).

23. Chiddix, J.A., J.A. Vaughan, and R.W. Wolfe, "The use of fiber optics in cable communications networks." *Journal of Lightwave Technology*, Vol. 11 No. 1, pp. 154–166 (1993).
24. Chiddix, J.A., and W.S. Ciciora, "Introduction of optical fiber transmission technology into existing cable television networks and its impact on the consumer electronics interface." *IEEE Transactions on Consumer Electronics*, Vol. 35 No. 2, pp. 51–62 (1982).
25. Chan, V.W., K.L. Hall, E. Modiano, and K. Rauschenbach, "Architectures and technologies for high speed optical data networks." *Journal of Lightwave Technology*, Vol. 16 No. 12, pp. 2146–2168 (1998).
26. Kawata, H., T. Ogawa, N. Yoshimoto, and T. Sugie, "Multichannel video and IP signal multiplexing system using CWDM technology." *Journal of Lightwave Technology*, Vol. 22 No. 6, pp. 1454–1462 (2004).
27. Papannareddy, R., and G. Bodeep, "HFC/CATV Transmission Systems." *Proceedings of 7th Annual Conference on Optoelectronics, Fiber Optics, and Photonics* (2004).
28. Kramer, G., and G. Pesavento, "Ethernet passive optical network (EPON): building a next generation optical access network." *IEEE Communications Magazine*, pp. 66–73 (2002).
29. Bisdikian, C., K. Maruyama, D. Seidman, and D.N. Serpanos, "Cable access beyond the hype: on residential broadband data services over HFC networks." *IEEE Communications Magazine*, pp. 128–135 (1996).
30. Vetter, P., and G. Van der Plas, "Competing access technologies." *IEEE Proceedings of Integrated Optics and Optical Fiber Communications Conference*, Vol. 3 No. 22–25, pp. 279–282 (1997).
31. Borella, M.S., J.P. Jue, D. Banerjee, B. Ramamurthy, and B. Mukherjee, "Optical components for WDM lightwave networks." *Proceedings of IEEE*, Vol. 85 No. 8, pp. 1274–1307 (1997).
32. Shaik, J.S., and N.R. Patil, "FTTH deployment options for telecom operators." White Paper, Sterlite Optical Technologies (2007).
33. Paff, A., "Hybrid fiber/coax in the public telecommunications infrastructure." *IEEE Communications Magazine*, pp. 40–45 (1995).
34. Perkins, B., "The art of overlaying video services on a BPON." *Bechtel Telecommunications Technical Journal*, Vol. 2 No. 2, pp. 61–69 (2004).
35. Abrams, M., P.C. Becker, Y. Fujimoto, V. O'Byrne, and D. Piehler, "FTTP deployments in the United States and Japan—equipment choices and service provider imperatives." *Journal of Lightwave Technology*, Vol. 23 No. 1, pp. 236–246 (2005).
36. Green, P.E., "Fiber to the home: the next big broadband thing." *IEEE Communications Magazine*, pp. 100–106 (2004).
37. Chiddix, J.A., H. Laor, D.M. Pangrac, L.D. Williamson, and R.W. Wolfe, "AM video on fiber in CATV systems: need and implementation." *Journal of Selected Areas in Communications*, Vol. 8 No. 7, pp. 1229–1239 (1990).
38. Tan, A.H.H., "Super PON: a fiber to the home cable network for CATV and POTS/ISDN/VOD as economical as a coaxial cable network." *Journal of Lightwave Technology*, Vol. 15 No. 2, pp. 213–218 (1997).

39. Zheng, R., and D. Habibi, "Emerging architectures for optical broadband access networks." *Proceedings of the Australian Telecommunications, Networks and Applications Conference* (2003).
40. Bell, J., D. Snyder, M. Yanushefski, R. Yang, J. Osenbach, and M. Asom, "OE transmitter packaging for broadband access systems (FTTx)." *IEEE Proceeding of Lasers and Electro-Optics Society Annual Meeting*, Vol. 2, pp. 299–300 (1997).
41. Chand, N., P.D. Magill, S.V. Swaminathan, and T.H. Daugherty, "Delivery of digital video and other multimedia services (>1 Gb/s bandwidth) in passband above the 155Mb/s baseband services on a FTTx full service access network." *Journal of Lightwave Technology*, Vol. 17 No. 12, pp. 2449–2460 (1999).
42. Sabella, R., "Performance analysis of wireless broadband systems employing optical fiber links." *IEEE Transactions on Communications*, Vol. 47 No. 5, pp. 715–721 (1999).
43. Chand, N., P. Magill, V. Swaminathan, and T. H. Daugherty, "Delivery of >1 Gb/s (digital video, data and audio), downstream in passband above 155Mb/s baseband services on FTTx full service access network." *IEEE Photonics Technology Letters*, Vol. 11 No. 9, pp. 1192–1194 (1999).
44. Ovadia, S., et al., "Performance characteristics and applications of hybrid multichannel AM–VSB/M–QAM video lightwave transmission system." *Journal of Lightwave Technology*, Vol. 16 pp. 1171–1185 (1998).
45. Granger, A., E. Ringoot, D. Wang, and T. Pfeiffer, "Optical fibers pave the way to faster broadband access." White Paper, ALCATEL publication (2005).

# Chapter 2

# Basic Structure of Optical Transceivers

The role of an optical receiver is to convert an optical signal into an electrical signal. The goal of an optical transmitter is to convert an electrical signal into a modulated optical signal. These requirements define digital transceivers as well as analog receivers and transmitters. However, they differ from each other in respect of design requirements and design considerations, irrespective of digital or analog. The digital transceivers deal with large signals, while the analog receivers handle smaller signals per channel and overall large loading due to multi tone transport. On the other hand, both topologies have common requirements that can be analyzed similarly, such as sensitivity and jitter, with minor differences.

Digital transceivers, however, differ from analog with respect to interfaces, controls, status, reports, and indications due to the differences in mode of operation. As for transmission, an analog transmitter handles fewer signals per channel since the optical modulation index (OMI) is low, about 4% per channel at maximum. But it has to transmit many channels; hence, the loading is similar to that of a large signal. The modulation depth for a digital transmitter is that of a large signal, driving the laser between threshold to high conduction and high optical power. Further discussion on OMI is provided in Secs. 6.9, 8.2.4, and 21.2. Additional material about community access TV (CATV) signals, standards, and broadcast methods are provided in Chapter 3.

## 2.1 Analog CATV Receiver and Coax Cables

An analog receiver consists of a photodetector (PD), input matching network, and RF chain [front end low-noise amplifier (LNA), automatic gain control (AGC, if needed), and an output stage]. The statuses from the analog receiver are; PD monitor, AGC voltage, and optional received signal strength indication (RSSI). The PD monitor indicates responsivity. Generally voltage is read over a photocurrent sampling resistor of 1 KΩ at 1 mW input power. Then the responsivity in mA/mW is calculated. The RSSI monitor provides the RF reading at a given optical level. It is commonly used to sample a portion of the low band CATV frequency carriers. This way the RSSI refers to a known number of channels, and therefore provides the power level estimation of each channel. A third indicator is the AGC control voltage. The AGC, in case of a feedback topology, senses a limited bandwidth (BW) of the low frequency channels, hence the AGC voltage is proportional to RF level and can indicate the RF power per channel at a given optical level. These indicators can provide the management information system for OMI estimation at any given optical level for any place where the receiver is installed. This is achieved by a lookup table that maps the RF level indicator readings, RSSI, and AGC, with respect to the PD monitor. Additionally, the PD indicator voltage provides the management system with information about the optical power impinging the receiver at any place it is installed, assuming the receiver was calibrated for PD responsivity during production. Photodiode dc voltage monitoring may be used for power leveling when using a feedforward AGC, as well as for reporting the optical level if the photodiode responsivity is known. Sections 8.2, 12.1, and 12.2.7 provide more detailed explanations and numerical examples about photodiodes and feedforward AGC. Some more control functions are used to operate the receiver, such as an on/off function that turns the receiver on or off for low-power mode for battery-save operation as well as CATV service inhibition for late billing. One of the ways to realize an optical CATV receiver is the single ended approach. The detected RF is sampled from the photo diode as depicted by Fig. 2.1. RF power level vs. optical level is measured by the RF-RSSI.

The RF-RSSI sampling coupler is located before the AGC attenuator; hence, it is out of the AGC-controlled power loop and senses the RF as a function of optical level. In some applications, the management system requires AGC-lock-detect indication. In this case, the AGC status is provided by the AGC RF root mean square (rms) detector or after the AGC rms detector dc amplifier. The lock-status

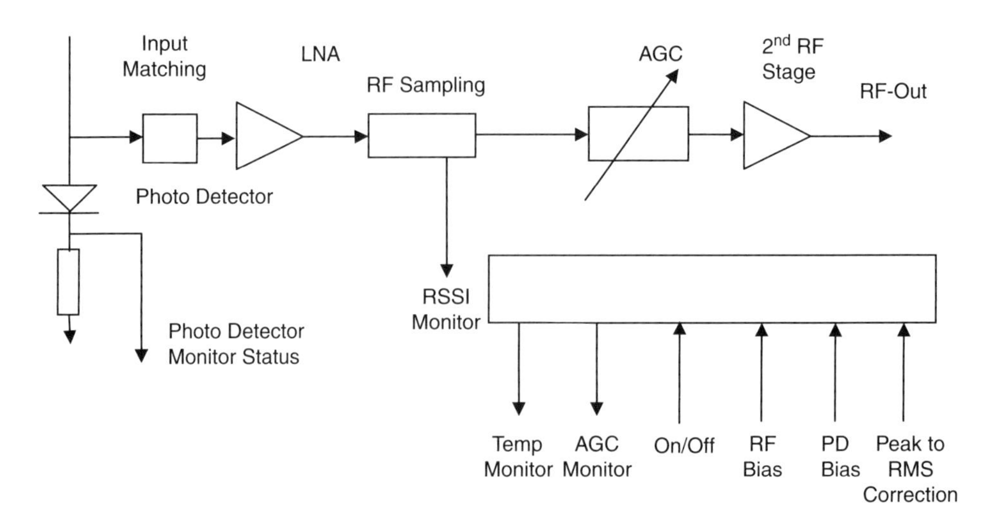


**Figure 2.1** Typical block diagram of a single-ended CATV analog receiver.

flag is achieved in the following way: when the feedback AGC system is locked onto the desired RF level, the voltage at the output of the RF rms detector is constant throughout the entire AGC dynamic range. In case the RF level is below the required level or above it for some reason, the voltage at the AGC RF detector output is not at the nominal locking value. This information is an indication for locking status. In analog feedback AGC, there is an additional port that compensates the peak-to-rms error. This error is due to the fact that, in production, the AGC is calibrated at continuous wave (CW), whereas an actual deployment under live video signal is characterized by a varying peak-to-rms ratio. Thus the peak-to-rms compensation reduces the average level of the modulated video; so its peak level equals the CW calibration level. Further explanation is provided in Sec. 12.2.8. AGC topologies can be feedback or feedforward, as explained in Chapter 12. The bias voltages for the electronics and the PD are separated: one supply is for the active RF circuit, and the second supply is for the PD. The PD bias is generally higher, in the region of 15 V, in order to improve the PD linearity and BW performance, as explained in Chapter 8.

The block diagram in Fig. 2.2 describes a higher linear performance approach for a CATV receiver by using push–pull configuration.[6,8] An output combiner BALUN is used in that case to sum both arms of the push–pull receiver. In this case, amplifiers are class A, as in any CATV receiver.

The noise density can be optimized to 3 pA/$\sqrt{\text{Hz}}$ up to 800 MHz using a pickup inductor to compensate for the PD stray capacitance.

Generally, fiber to the home (FTTH) CATV optical receivers are tuned to have 2–4 dB up-tilt response versus frequency. The reason for this specification is to compensate for the coax cable frequency response.[1] Coax cables increase their losses versus frequency; hence an up-tilt versus frequency provides a flat response. Fiber to the curb (FTTC) modules are tuned with an up-tilt of up to 8 dB for the same reason of cable loss compensation. Up-tilt gain is generally achieved by

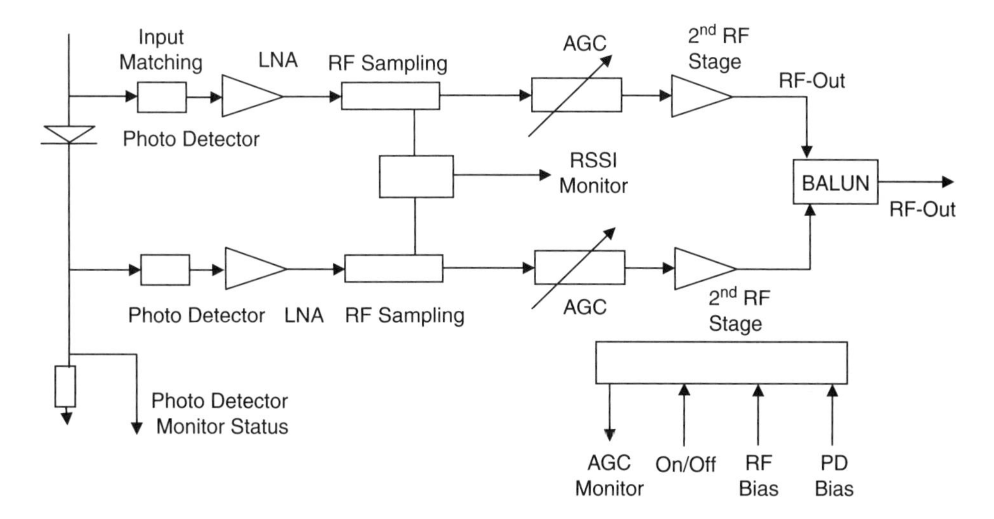


**Figure 2.2** Typical block diagram of a push–pull CATV analog receiver.

using equalizers in the optical receiver. More about equalization techniques will be discussed in Sec. 11.9. Since in FTTC the cable is longer, losses are higher; hence, it is required to have a higher tilt spec. The coax cable is the second stage of data transport to the subscriber in hybrid fiber coax (HFC) networks, after conversion from optical to electrical signal. In addition to their losses versus frequency characteristics, coax cables suffer from propagation distortions such as group delay. Group delay arises from the low-pass filter (LPF) nature of the coax cable and its cutoff frequency. The result is that the cable phase response at high frequencies is not as linear as at low frequencies. Therefore, the phase derivative versus frequency is not a constant number, which results in group delay distortions. For this reason, the group delay variation needs to be specified for a 6-MHz bandwidth. The impact of this parameter, along with the phase jitter, is especially critical to high-level modulations such as QAM, and is manifested in the eye-pattern and BER measurements of such signals, and is further elaborated in Chapters 14, 15, and 18.

Coax cables used in network distribution typically consist of a copper-clad aluminum wire, which is the inner central conductor, an insulating dielectric layer such as foam, polyethylene, a solid aluminum shield, which is the outer conductor generally referred as ground (GND) potential, and a covering made of PVC. The ratio of the inner conductor diameter to the inner diameter of the solid aluminum shield and the type of dielectric define the characteristic impedance of the coax cable.

Subscriber-drop coax cables are manufactured with a copper-clad steel center conductor and a combination of aluminum braid and aluminum polypropylene–aluminum tap shield. In some installation applications such as plenum installations, the coaxial cable jackets are generally made with polytretrafluoroethylene (PTFE) material, which is used in high-frequency printed circuit board (PCB) applications as well.

As was explained earlier, the nature of the cable depends on its characteristic impedance dimensions and is derived from the diameter ratio and dielectric properties of the insulation material. The cable losses are due to dielectric losses and ohmic resistive losses. The signal energy flows through the inner central conductor. A skin-effect phenomenon is directly related to the conductor characteristic resistance and the frequency of operation. At low frequencies or dc, the entire cross section of the conductor transfers the current and it is uniformly distributed. As frequency is increased, the current travels on the surface of the conductor, and thus the resistive losses increase versus frequency.[2]

There are three types of coax cables used in a distribution system: trunk, feeder, and drop cables. Cable types differ by diameter: the largest are the trunk cables, with a typical diameter range of 0.5 in. up to 1 in. and losses of 0.89 dB at 50 MHz up to 3.97 dB at 750 MHz, measured per 100 ft at a diameter of 1 in. The second largest is the feeder, followed by the drop at the subscriber's home. Cable losses increase slowly versus temperature. One of the reasons is that as the conducting metal expands, it becomes longer, and thus the resistance increases. Table 2.1 provides typical losses for drop cables in dB per 100 ft, with four different cable diameters versus frequency.[1]

**Table 2.1** Cable losses vs. frequency with four different diameters at 68°F in dB/100 ft.[1]

| Frequency (MHz) | 59 Series Foam (dB/100 ft) | 6 Series Foam (dB/100 ft) | 7 Series Foam (dB/100 ft) | 11 Series Foam (dB/100 ft) |
|---|---|---|---|---|
| 5 | 0.86 | 0.58 | 0.47 | 0.38 |
| 30 | 1.51 | 1.18 | 0.92 | 0.71 |
| 40 | 1.74 | 1.37 | 1.06 | 0.82 |
| 50 | 1.95 | 1.53 | 1.19 | 0.92 |
| 110 | 2.82 | 2.24 | 1.73 | 1.36 |
| 174 | 3.47 | 2.75 | 2.14 | 1.72 |
| 220 | 3.88 | 3.11 | 2.41 | 1.96 |
| 300 | 4.45 | 3.55 | 2.82 | 2.25 |
| 350 | 4.80 | 3.85 | 3.05 | 2.42 |
| 400 | 5.10 | 4.15 | 3.27 | 2.60 |
| 450 | 5.40 | 4.40 | 3.46 | 2.75 |
| 550 | 5.95 | 4.90 | 3.85 | 3.04 |
| 600 | 6.20 | 5.10 | 4.05 | 3.18 |
| 750 | 6.97 | 5.65 | 4.57 | 3.65 |
| 865 | 7.52 | 6.10 | 4.93 | 3.98 |
| 1000 | 8.12 | 6.55 | 5.32 | 4.35 |

A useful relation to estimate cable losses is called cable loss ratio (CLR):

$$\mathrm{CLR} = \sqrt{\frac{f_1}{f_2}}. \tag{2.1}$$

By knowing the CLR and losses of a cable at a given frequency, the cable loss at a desired frequency can be calculated as

$$\mathrm{CL[dB]} = \mathrm{loss[dB]}\sqrt{\frac{f_1}{f_2}} = \mathrm{loss[dB]} \times \mathrm{CLR}, \tag{2.2}$$

where $f_2$ represents the high desired frequency, and $f_1$ is the known frequency with its corresponding loss[dB].

## 2.2 Analog CATV Return-Path Receiver and Transmitter

The return-path receiver described in Fig. 2.3 is located at the central office (infrastructure central where upstream is for video on demand (VOD), requests, phone, etc.), as shown in the system concept provided in Fig. 2.5. The return-path receiver receives plain old telephone service (POTS) signals occupying the BW of 10 KHz to 3 MHz and 23 RF signals of 1.5 MHz quadrature phase shift keying (QPSK) each from 5 to 42 MHz in the U.S., and 5 to 65 MHz in Europe. The traditional standard calls for 1310 nm for the return path and 1550 nm for the CATV transport. The signals are detected by the photodiode and then passed through a duplexer, which consists of low-pass and high-pass filters, and are separated

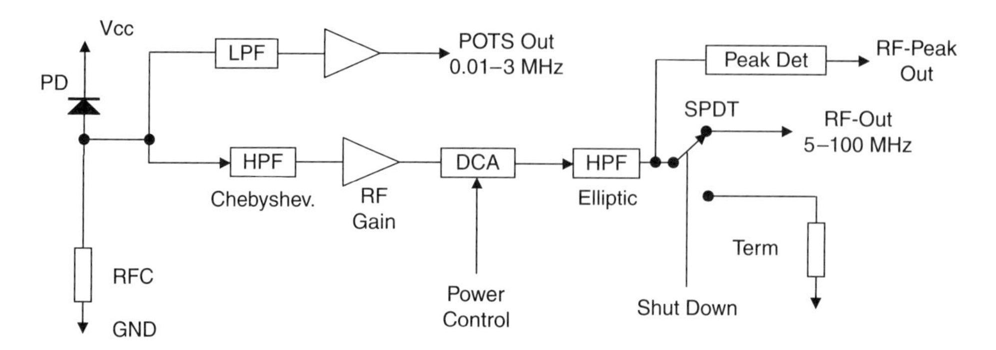


**Figure 2.3** Typical structure of analog return-path receiver with POTS shut-down gain control and RF peak level telemetry outputs.

into RF and POTS. The POTS signals are amplified and transmitted out via copper. The RF signals pass through a Chebyshev filter, are amplified, and then pass through a digital control attenuator (DCA). The power level of these signals is controlled in order to provide a high dynamic range. The link dynamic range is characterized by the noise power ratio known as the noise power ratio (NPR) test.[7] An NPR test simulates multichannel loading by injecting noise energy in a specific BW; then, by a notch filter, the noise is rejected at a specific frequency, creating a white noise—limited BW with a notch. The test measures the notch depth degradation via the link. Link degradation results from nonlinearities at high gain and noise figure effects at low gain. The difference between the two extremes provides the NPR dynamic range. There is a specific link budget where the NPR notch depth reaches its maximum. This state is called peak NPR. For a U.S. band of a given design, the NPR dynamic range would be higher compared to a European bandwidth. This is because of the higher noise loading of the transmitter and the receiver for the European standard, for the same transmitting power level. The second filter in the return-path receiver is LPF with an elliptic response that further isolates the POTS channel from the RF channel. Typical specs require 50 dB of POTS rejection with respect to RF level. The POTS impedance level standard calls for 1000 Ω and RF impedance is 75 Ω. A shut-down switch is used to turn off service, and peak detector output indicates the signal level.

The return-path transmitter, shown in Fig. 2.4, is generally located at the curb or main node, as depicted in Fig. 2.5, where the electrical signals are converted into optical. The input signals to the return-path transmitter are POTS signals occupying the BW of 10 KHz to 3 MHz and 23 RF signals of 1.5 MHz QPSK each from 5 to 42 MHz. Both these modulating signals are ac coupled to the laser bias and AM–modulate the laser. A duplexer consisting of an LPF for the POTS and high pass fillter (HPF) for the RF signals separates and isolates the signals. In the RF section of the return-path transmitter, sometimes there are two stages of filtering and isolation from the POTS. Prior to modulating

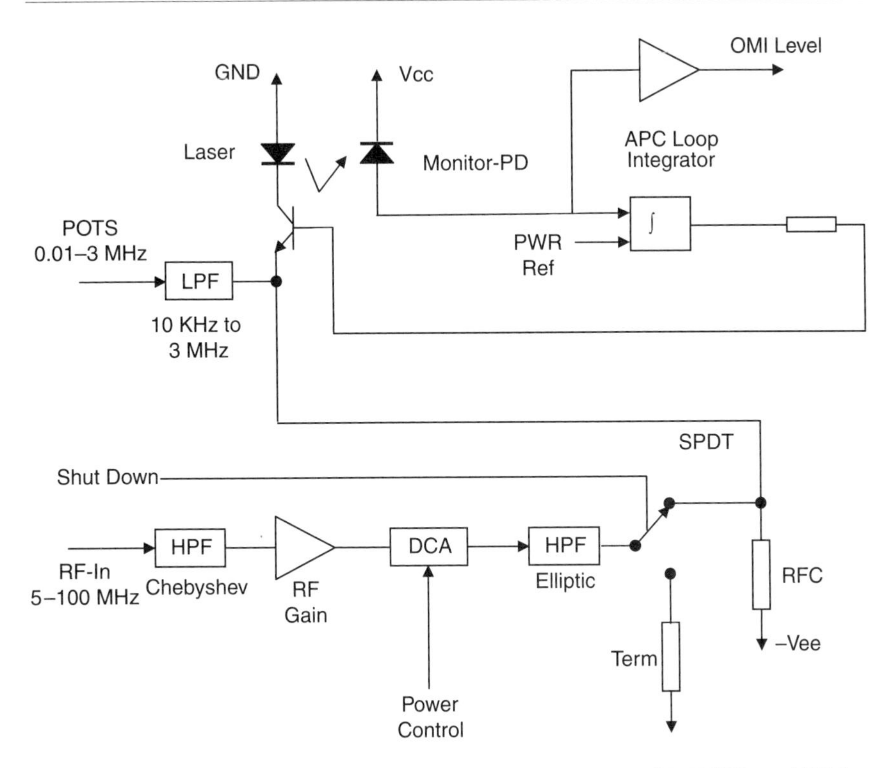


**Figure 2.4** Structure of analog return-path transmitter with POTS and RF inputs, OMI control, OMI level indication, and shutdown.

the laser, the RF signal passes via a digital controlled attenuator (DCA) in order to control the optical modulation index (OMI) level of the laser and prevent distortions. Link linearity and dynamic range is characterized by the NPR test. Typically, the link degradation in an NPR dynamic range results from the return-path laser compression.

The return-path transmitter has an automatic power control (APC) leveling loop that sets the laser optical power by locking its bias current. A back-facet monitor PD diode senses the optical level, and the photocurrent, which is linearly proportional to the responsivity of the photodiode, is compared with the reference point. The loop amplifier corrects the bias point of the laser by controlling the transistor's current. These types of control topologies should have a very high open loop gain in order to increase the accuracy and should have very narrow 3 dB loop BW in order to filter out fast transitions that may affect optical power control such as modulation. If the loop BW is wider than 1 KHz, it would start to suppress the AM of the POTS. The return-path transmitter in this case uses an isolated distributed feedback (DFB) laser to achieve a highly linear performance and low reflections. Chapter 13 provides more details about APC and further investigation is provided later in Chapter 12 concerning feedback power control.

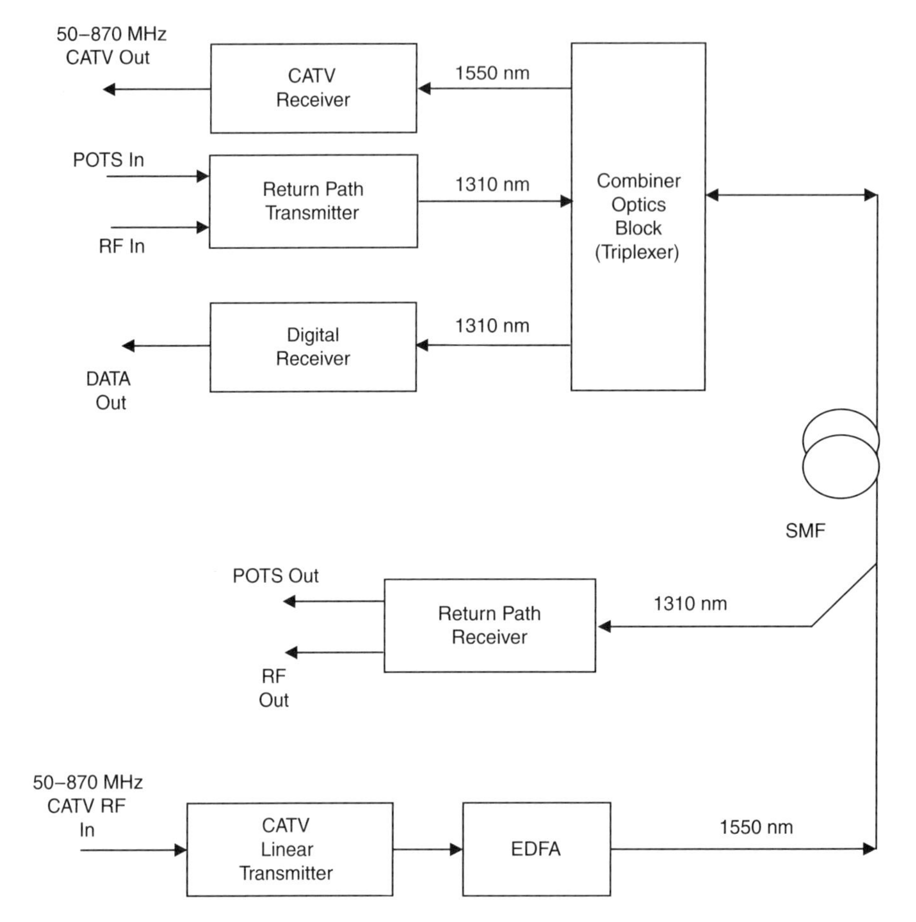


**Figure 2.5** HFC CATV link structure with analog return-path receiver at the central point and analog return-path transmitter CATV receiver and digital receiver at the curb. CATV down-stream is on the 1550 nm and return-path up-stream is at 1310 nm.

## 2.3 Digital Transceiver

Digital transceivers consist of three main blocks, which are elaborately analyzed in Chapters 18 and 19:

1. the optics section, which can be wavelength division multiple access/ multiplexing (WDM) duplexer module, triplexer module,[5] or laser PD block transmitter optical subassembly (TOSA) and PD module receiver optical subassembly (ROSA),
2. the receive section, which is an electronic circuit, and
3. the transmit section, which is an electronic circuit.

Additional functionalities are added to the transceiver module in order to provide the system management indications, control, and diagnostics such as

1. transmitter burst control,
2. laser temperature sensing,
3. laser thermoelectric cooler (TEC) control loop, and
4. receiver signal detection.

In new transceivers, the diagnostic is managed by a microcontroller with inter ic bus ($I^2C$-BUS) The $I^2C$-BUS standard is used for reading the temperature of the laser, for programming the laser bias point and power levels, and for "1" and "0" logic levels that determine the extinction ratio.

The $I^2C$-BUS is a bidirectional serial bus providing a link between ICs. Philips introduced that standard 20 years ago for mass production of products such as TVs. There are several data rates under this standard:

1. standard, 100 KB/s,
2. fast, 400 KB/s, and
3. high speed, 3.4 MB/s.

Digital transceivers are well defined by standards. The TUV standard defines the means for eye safety and laser shut down for eye protection. This eye safety circuit turns off the laser and prevents high current to bias the laser; high optical power emission that can damage the eye is prevented. Data rates are defined as OC12, 24, 48, etc.

Packaging standards were defined by the multisource agreement (MSA) as listed below. However, packaging standards for CATV receivers or FTTx integrated triplexers (ITR) or integrated triplexers for the curb (ITC) form factors are not defined by standards.

1. GBIC gigabit interface converter.
2. SFP small form, pluggable.
3. SFF small form, factor.
4. 2 × 9, two rows of 9 pins each.
5. XFP 10 gigabit.
6. Butterfly.

Digital transceivers cover broadband data rates up to 10 GHz XFP.

In high-data-rate transmitters, an external modulator (EM) is used. This device is used to solve the laser chirp problem, which results in chromatic dispersion. Chirp effects in a laser is due to a change in the refractive index in the active region because of the modulation current. Additional problems solved by EM is data rate limitation due to the relaxation–oscillation effect in direct modulated lasers. Relaxation–oscillation also occurs in dispersion and eye overshoots. Digital transceivers in optical networks operate in TDM (time division multiplexing) in order to serve more users when utilizing the same wavelength. A digital transceiver block diagram is provided by Fig. 2.6. Today's advanced technology enables minimizing the size of the packaging. A high level of integration using application specific integrated circuit (ASIC) technology includes all the necessary

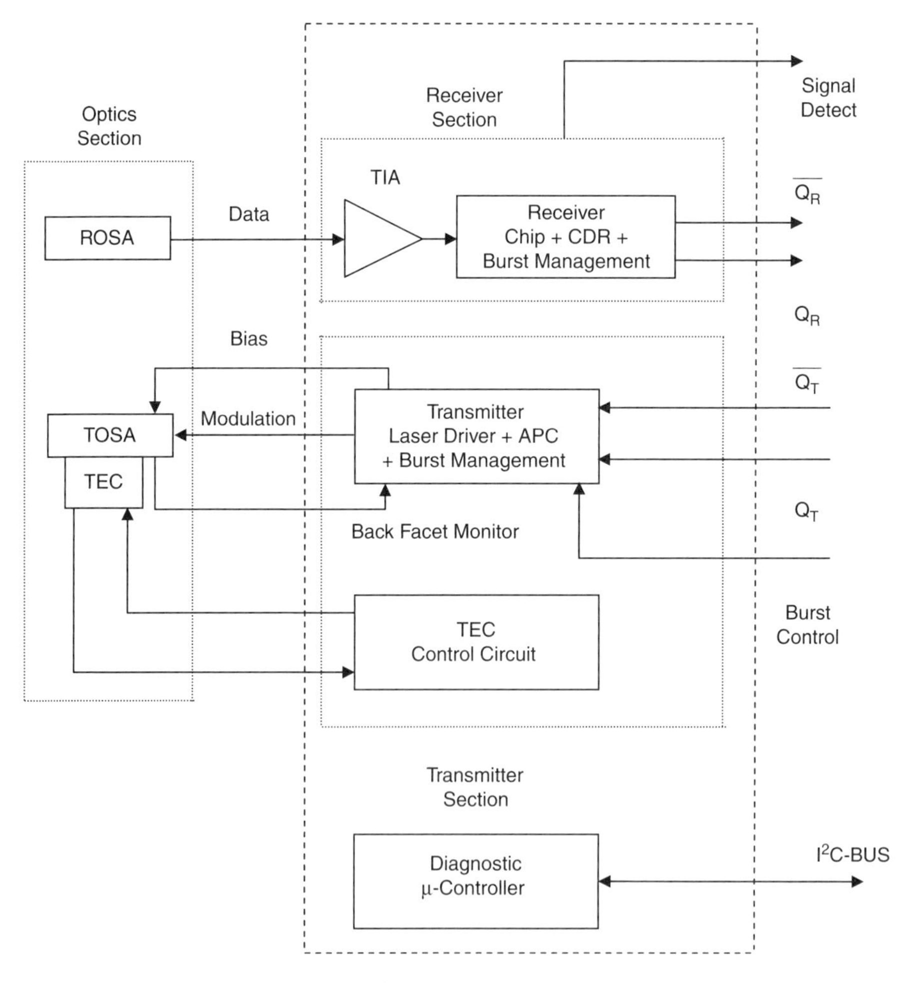


**Figure 2.6** Typical block diagram of a digital transceiver with burst mode, TEC, APC, diagnostic via $I^2C$-BUS, and differential data in/out using TOSA housing for the laser and ROSA for the receiver photodiode.

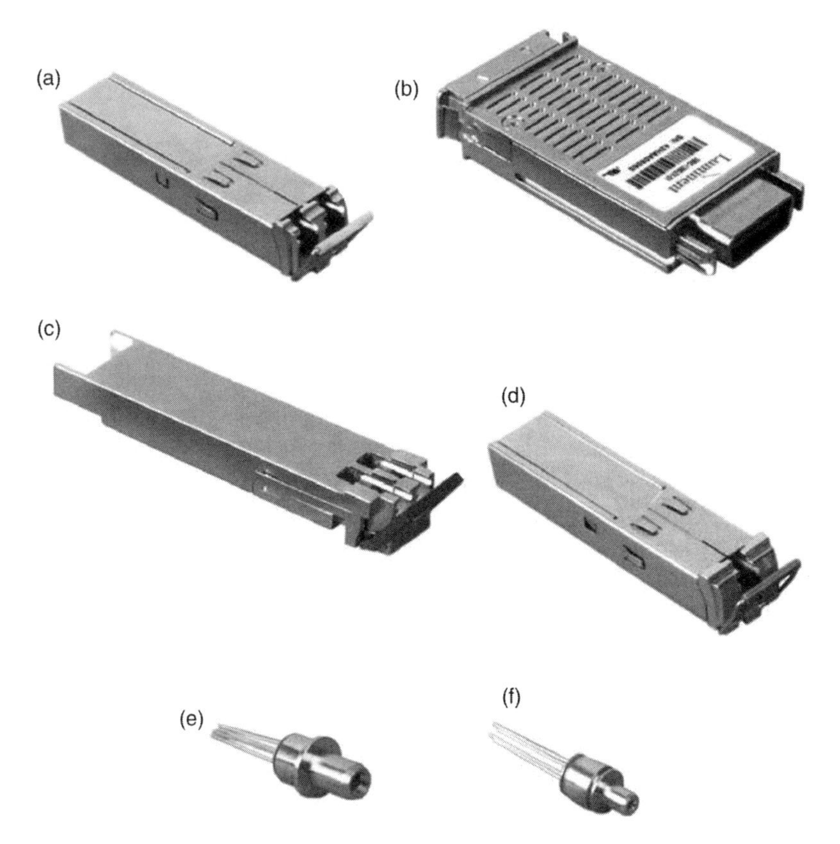


**Figure 2.7** Several digital transceiver and optics packaging standards: (a) GBIC; (b) SFP; (c) SFF; (d) XFP; (e) TOSA; (f) ROSA. (Courtesy of Luminent OIC, Inc./Source Photonics.)

functionalities for control and management of each section, receiving or transmitting, in one chip. Using the "chip-on-board" technique allows minimizing the size of the module's electronic card, reducing the number of components and increasing the reliability of the whole transceiver as well as reducing its cost. Figure 2.7 describes several digital transceivers and optics packaging standards. Further integration is introduced by having the ROSA modules with integrated PIN TIA as part of the receive optics. This is done by using chip and bonding technology. This way, sensitivity is increased and also deterministic distortions such as reflections or X-talk are minimized, since parasitics are reduced. Further discussion on this is provided in Chapters 19 and 20.

## 2.4 ITR Digital Transceiver and Analog Receiver

An integrated triplexer (ITR) is a platform containing an analog receiver (described in Figs. 2.1 and 2.2), digital transceiver electronics (described in Fig. 2.6), and a single fiber triplexer optics module [shown in Figs. 2.9(b)

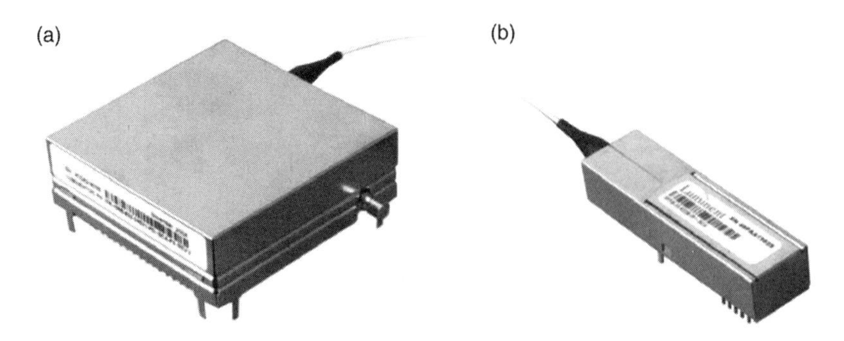


**Figure 2.8** (a) Integrated FTTx triplexers ITR and ITC. (b) Digital BiDi 1310 nm/ 1490 nm FTTP transceiver. (Courtesy of Luminent OIC, Inc./Source Photonics.)

and 2.10]. This module is the fiber to the x (FTTx), FTTP-ITR platform solution [shown in Fig. 2.8(a)], the block diagram, which is shown in Fig. 1.1. The main idea behind ITR, as described in the introduction, is to reduce the system complexity and costs for FTTx. The major challenge in such a small form factor design is to overcome the X-talk effects due to the leakage from the digital section into the analog section.

The strict system specifications for linearity as defined by the Society of Cable Telecommunications Engineers (SCTE) standard requires that the X-talk spectral lines be lower than the maximum allowable CSO or CTB levels. Further discussion on the X-talk design approach is provided in Chapter 20. A second challenge is to have high isolation in the optics between the digital receiving and transmitting, and the analog detector as well as between the digital reception and transmission itself. The filtering is done optically and by light traps that catch undesired transmitter light with the same wavelength as the receiver. The transmitter optical leakage and desensitization is due to reflections as elaborated in Sec. 5.2. The PDs are wide band and can detect either the analog channel wavelength or the digital channels. Hence, the optical X-talk specification for an optical module is an important factor that affects the sensitivity performance of an ITR. There are two kinds of integrated platforms: ITR that is used for FTTH, providing RF

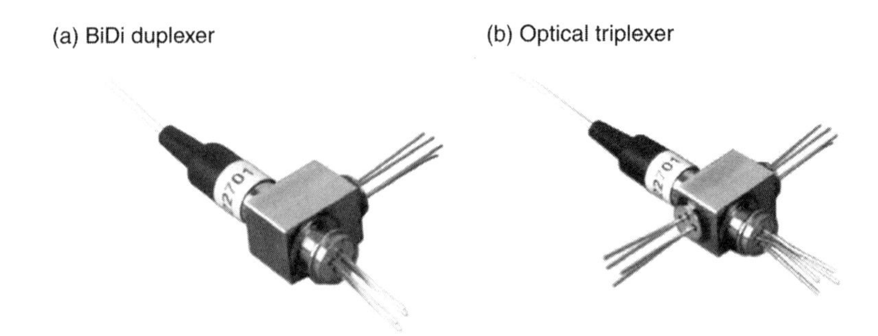


**Figure 2.9** (a) BiDi duplexer. (b) Optical triplexer. (Courtesy of Luminent OIC, Inc./ Source Photonics.)

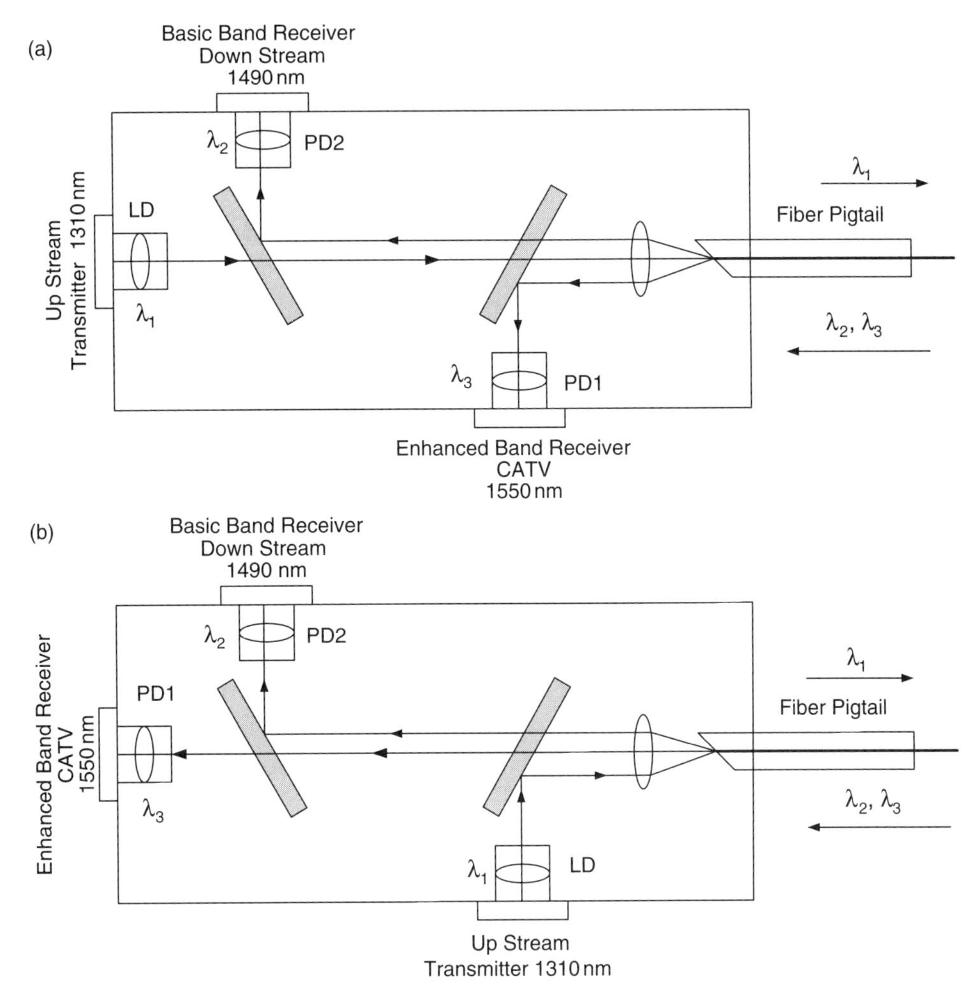


**Figure 2.10** Concept of full service access network (FSAN) triplexer: (a) laser at main axis; (b) laser on side.[5] (Reprinted with permission from Edith Cowan University, Australia.)

levels of 14–18 dBmV and ITC (integrated triplexer to the curb) that is used for FTTC/P (fiber to the curb or premises), providing RF levels of 30–34 dBmV. Eventually, such a platform would have an AGC and diagnostics. The advantage in such a design is the use of common resources such as CPU, digital to analog, and analog to digital, to report analog diagnostics and digital diagnostics (see Sec. 2.1). The other side of the FTTx link that receives the up-link digital data, such as voice over IP and digital return path, is a digital BiDi transceiver shown in Fig. 2.8(b) with the BiDi optics module shown in Fig. 2.9(a). In modern FTTx systems, the return path is fully digital, thus the analog receiver transmitter described in Sec. 2.3 is avoided. Furthermore, advanced FTTx systems have full digital video transport, thus the analog design issues of CSO, CTB, and CNR as well as the need for 1550-nm PD are avoided.

## 2.5 Architecture of Tunable Wavelength Transmitters

A tunable laser transponder is a different approach in WDM networking that allocates a specific wavelength to an optical network unit (ONU). It is similar to a digital transceiver except for the laser- and wavelength-locking ability. Additionally, since such a transmitter requires more supporting electronics and optics for the wavelength-locking loop, it is larger; hence it is called a transponder.[3,4] As a transponder, it contains MUX/deMUX converters for serializing/deserializing the nonreturn to zero (NRZ) data. A tunable laser transponder is comprised of a tunable laser as the lasing source, and an EM for high-data-rate modulation, thus avoiding chirp in the laser wavelength as directional modulation. A wavelength locker indicates the wavelength deviation from the desired wavelength (an analog-to-phase detector is in phase-locked loop (PLL) but is not a phase detector), and temperature control units set the constant base plate temperature profiles for having higher accuracy in wavelength emission from the laser with higher wavelength discrimination. It also has an automatic power control (APC) loop. For size conservation, the reference PD in the wavelength locker can be used as a back-facet monitor for optical power control-feedback looping. The APC operation is analyzed in Chapter 13.

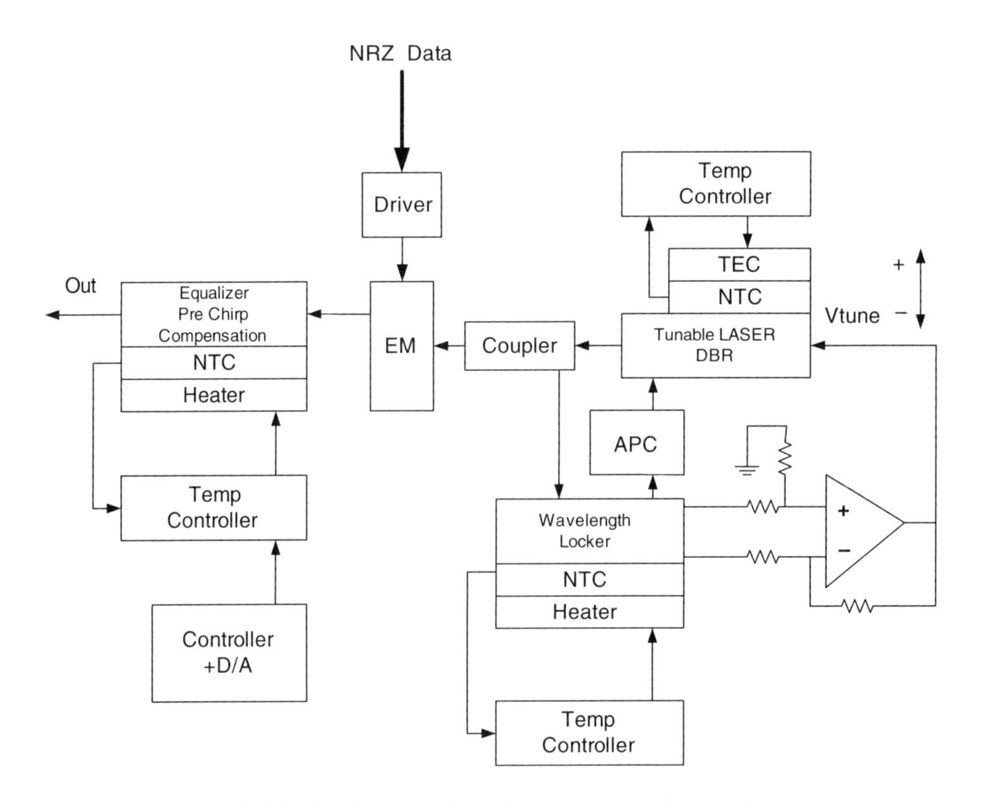


**Figure 2.11** Tunable wavelength transponder block diagram.

Modern architectures contain equalizers to remove chirp effects resulting from the Mach Zehnder EM due to sensitivity of the refractive index to modulating voltage; this is further discussed in Chapter 7. This compensation concept is based on the idea that the phase response derivative in filters compensates for the resultant dispersion from the EM. The center frequency of the chirp compensator filter is tuned thermally by a heater. Sections 7.3 and 7.5 provide more information about the physics of these devices. In order to improve optical return loss, an isolator can be used at the output of the chirp compensator.

A conceptual block diagram of a tunable transponder is depicted in Fig. 2.11. Operation details of the wavelength locker are elaborated in Chapter 19.

The transponder's receiver side is not shown here; however, 10 GB/s receivers are similar to those in Fig. 2.6. A 10 Gbit receiver of a transponder contains a few additional blocks to combat jitter. Such blocks are integrated circuit equalizers, which are realized in DSP techniques to remove jitter. This jitter may result from chromatic dispersion of the fiber. CDR (clock data recovery) is then used for coherent detection.

## 2.6 Main Points of this Chapter

1. An analog CATV receiver converts modulated light into an RF signal.
2. There are two topologies for CATV realization: single-ended and push–pull.
3. In the CATV receiver, there is an AGC that is either feedforward or feedback.
4. The PD in a CATV receiver operates at a high voltage of 12–15 volts to minimize distortions and extend its bandwidth.
5. A digital transceiver converts modulated light into a digital signal and modulates the laser to convert a digital signal into modulated light.
6. A laser's wavelength is stabilized by a TEC.
7. Digital transceivers and integrated triplexers provide diagnostics via $I^2C$-BUS. The diagnostics report is for temperature, modulation depth, responsivity, and extinction ratio. Additionally, $I^2C$-BUS communication is used for programming the transmitter's extinction ratio and AGC for the CATV. The data is sent via the $I^2C$-BUS to a controller used in the transceiver's circuit. Communication integrity is monitored by a watch-dog timer. The watch-dog timer is a counter that operates as a one shot. Once its count has elapsed and it is not reset within the grace period, it hard resets the controller and sends an interruption via the $I^2C$-BUS to the MAC.

8. Small optical modules used are called TOSA and ROSA. These modules are used in SFP XFP transceivers.

9. The integrated platform combining a digital transceiver and a CATV receiver is called ITR or ITC (to the home or to the curb).

10. RF power level to the home is 14–18 dBmV and to the curb is 30–34 dBmV. Both have up-tilt to compensate for the coax-cable losses versus frequency.

## References

1. Electronics Industries Association (EIA) "Cable television channel identification plan." EIA IS-132 (1994).
2. Taylor, A.S., "Characterization of cable TV networks as transmission media for data." *Journal of on Selected Areas in Communications*, SAC-3, pp. 255–265 (1985).
3. Brillant, I., "OPEN widely tunable laser technology." White Paper, GWS Photonics (2004).
4. Simsarian, J.E., and L. Zhang, "Wavelength locking a fast switching tunable laser." *IEEE Photonics Technology Letters*, Vol. 16 No. 7, pp. 1745–1747 (2004).
5. Zheng, R., and D. Habibi, "Emerging architectures for optical broadband access networks." Edith Cowan University (2003).
6. Pophillat, L., and R. David, "Less than $3pA/\sqrt{Hz}$ over 1–800 MHz noise current spectral density optical receiver for SCM–CATV Systems." *IEEE Electronics Letters*, Vol. 29 No. 19, pp. 1721–1722 (1993).
7. Germanov, V., "The notch filter test method simulation for the intermodulation noise of return path CATV amplifiers." *IEEE Transactions on Broadcasting*, Vol. 46 No. 1, pp. 88–92 March (2000).
8. Childs, R.B., T.A. Tatlock, and V.A. O'Byrne, "AM: video distribution system with 64-way passive optical splitting." *IEEE Photonics Technology Letters*, Vol. 4 No. 1, pp. 86–88 January (1992).

# Chapter 3

# Introduction to CATV Standards and Concepts of Operation

A general introduction about hybrid fiber coax (HFC) systems has been provided in previous chapters. HFC system transport consists of digital data and video data.[9] The video data can be in either analog or digital modulation scheme. In this chapter, the concept of TV and its signal structures will be reviewed. Community access television (CATV) tests will be reviewed in relation to system-level design approaches affecting the structure and architecture of analog optical receivers. In addition, the meaning and effects of system level tests over TV image performance will be explained.

## 3.1 Television Systems Fundamentals

Analog broadcast TV is phasing out in 2009. New digital standards of digital transmission are phasing in for cellular applications and terrestrial TV such as digital video broadcasting–terrestrial (DVB–T), digital video broadcasting–handheld (DVB–H), terrestrial digital multimedia broadcasting (T–DMB), terrestrial integrated services digital broadcasting (ISDB), and forward link only (FLO). However, analog CRT TV is not going to be phased out at that time; conversion set top boxes (STBs) are going to be used. This section provides an introduction to analog TV broadcasting and receiving.

Television's main advantage is transmission of visual images through electrical signals. The picture consists of several small squares known as picture elements (PELs). In digital pictures they are defined as pixels. A large number of PELs in a given image means higher resolution and cleaner reproduction at the TV receiver. There are mainly three standards in use throughout the world:

1. National Television Systems Committee (NTSC),
2. phase altered line (PAL), and
3. sequential color with memory; in French, Sequentiel Couleur Avec Memoire (SECAM).

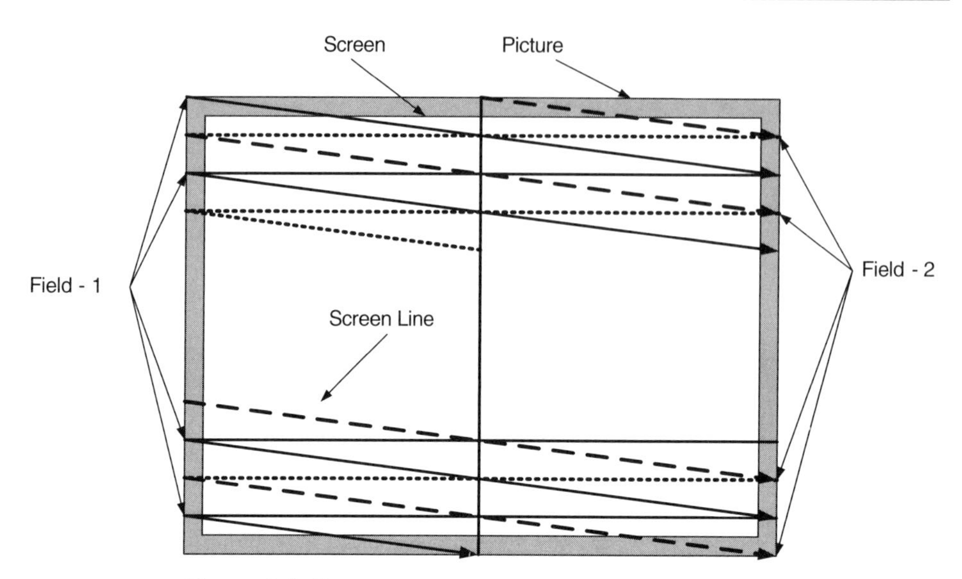


**Figure 3.1** Frame structure from two interlaced fields.

### 3.1.1 Analog NTSC standard

Television signals in the U.S. are broadcast using the NTSC-M standards, who initiated the basic monochrome TV standards in 1941 and were designated as system M by the Committee Consultatif International Radiocommunications (CCIR). In 1953, the NTSC of Electronic Industries Alliance (EIA) established the NTSC for color TV standards, which is used today for terrestrial broadcasting and cable TV transmission systems in North America, Japan, and many other countries. This system was designed to be compatible with the monochrome black and white TV system previously used, and calls for 525 horizontal scans (interlaced lines) per frame. The number of frames (2 fields) per second is 59.94–60 frames per second. Two interlacing fields form one frame (Fig. 3.1).

Usually the image is scanned along the lines as shown in Fig. 3.1 until the entire image frame is completed. This type of scanning is called progressive scanning. In order to reduce the effect of flickering, the frame in Fig. 3.1 is divided into two fields, and each field is used to show consecutive images. The odd field represents the first image and the even field represents the image that follows. This type of scanning is called interlaced scanning, which reduces and minimizes the flicker effect to a considerable extent. Therefore, the following arithmetic is applicable to estimate the picture frame scan frequency and time. There are 525 horizontal scan paths per frame; hence, there are 262.5 horizontal scan paths per field.

Scan rate is 30 (29.97) frames (complete pictures) per second, or 60 (59.94) fields (half picture) per second, so the horizontal scan frequency yields $f = 525 \times 29.97 = 15734.25$ Hz. As a result, the time for each scan is $1/f = 63.55$ μs. The scanning pattern of a frame is illustrated in Fig. 3.2.

The reason for using a 60-Hz vertical scan is to synchronize the monochrome TV sets with the main power supply and prevent power-related distortions. In color

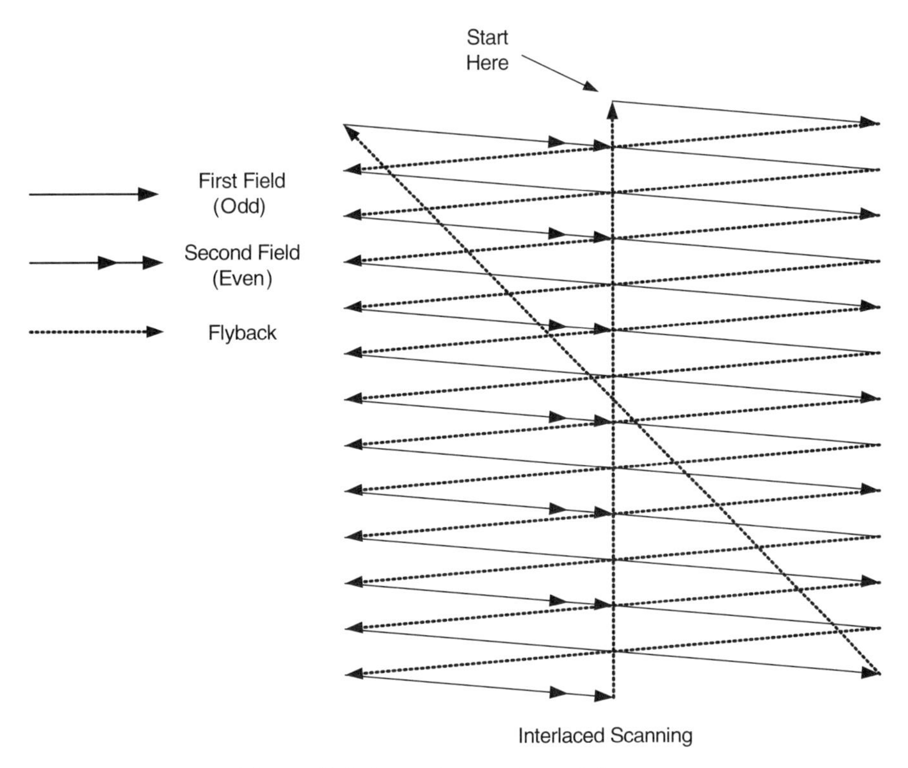


**Figure 3.2** Scanning pattern of frame.

TV, the vertical and horizontal frequencies were slightly reduced to allow the interference beat between the chrominance carrier and the aural carrier to be synchronized with the video signal. The signal terminology is as follows:

1. The brightness of the video signal (containing the picture information details) is called luminance, visual carrier, or "luma."
2. The color portion of the video signal (containing the information of the picture hue or tint and color saturation) is called the chrominance or "chroma" carrier.
3. The sound audio carrier is called the "aural" carrier.

### 3.1.2 Video camera tubes

The image is furnished by the television camera tube [5]. The first tube was the iconoscope. In the image-orthicon (IO) tube, the optical system generates a focused image on the photocathode, which eventually generates a charged image on another surface, known as the target mosaic. Every point on the target mosaic surface acquires a positive electric charge proportional to the brightness of the corresponding spot in the image. In other words, instead of a light image

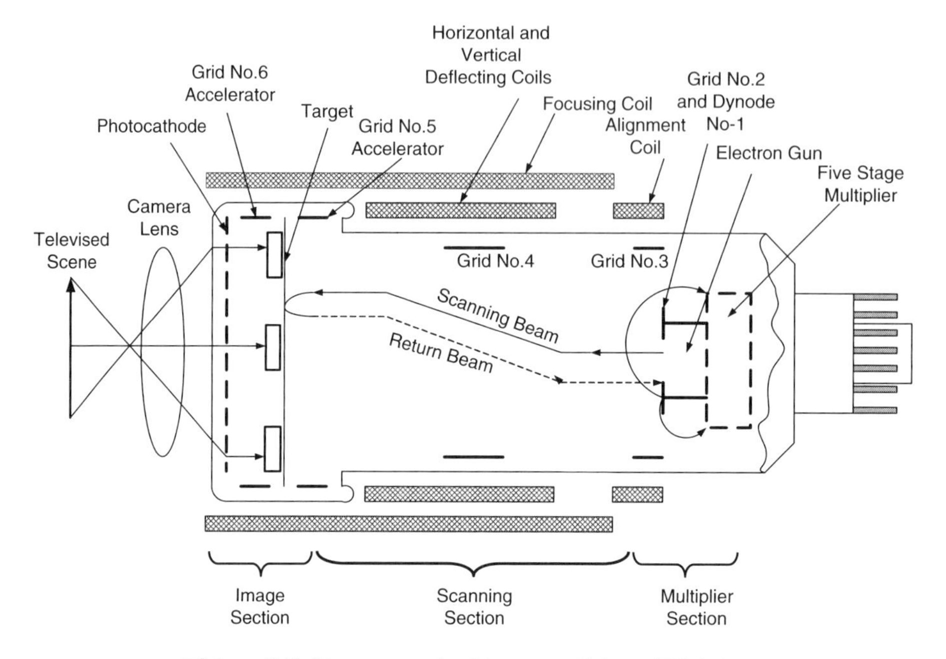


**Figure 3.3** The concept of Image-orthicon (IO) tube.

there is a charge image. An electron gun scans the mosaic charge back surface with an electron beam in the manner as shown in Fig. 3.1. The areas with no charges bounce the electrons back to the tube. These areas are the darker areas of the image. Areas that have a positive charge due to the picture brightness, light absorbs the electrons until none of the electrons are be left to move back to the tube. Before the electrons are collected, they pass through a series of positively charged dynode or electron multiplier plates. Impinging electrons liberate two or more electrons before they move to the next positive plate. In this way the video image is amplified by dynatron action. The concept of IO tubes was very large and heavy by modern terms; the filament warm-up time was long. Several other tubes were developed such as vidicon, plumbicon, vistacon, saticon, newvicon, and solid state coupled charged device (CCD). In Fig. 3.3, the concept of the IO tube is provided in a diagram.

Image section consists of photo-cathode, image accelerator, target with wire mesh, held-in-target cap and backed by field mesh, decelerator, and the entire unit is held in position by the shoulder base with locating bushes. The central scanning section of the tube comprises the persuader, and beam-focus electrode, with external scanning coils to deflect beam (arrowed). The return beam, after striking target, enters the multiplier section of tube, which consists of the first dynode, multiplier, and anode.

### 3.1.3 Scanning method

The camera tube scans the mosaic image in the following manner. The electron beam is controlled by a set of voltages across the horizontal and vertical

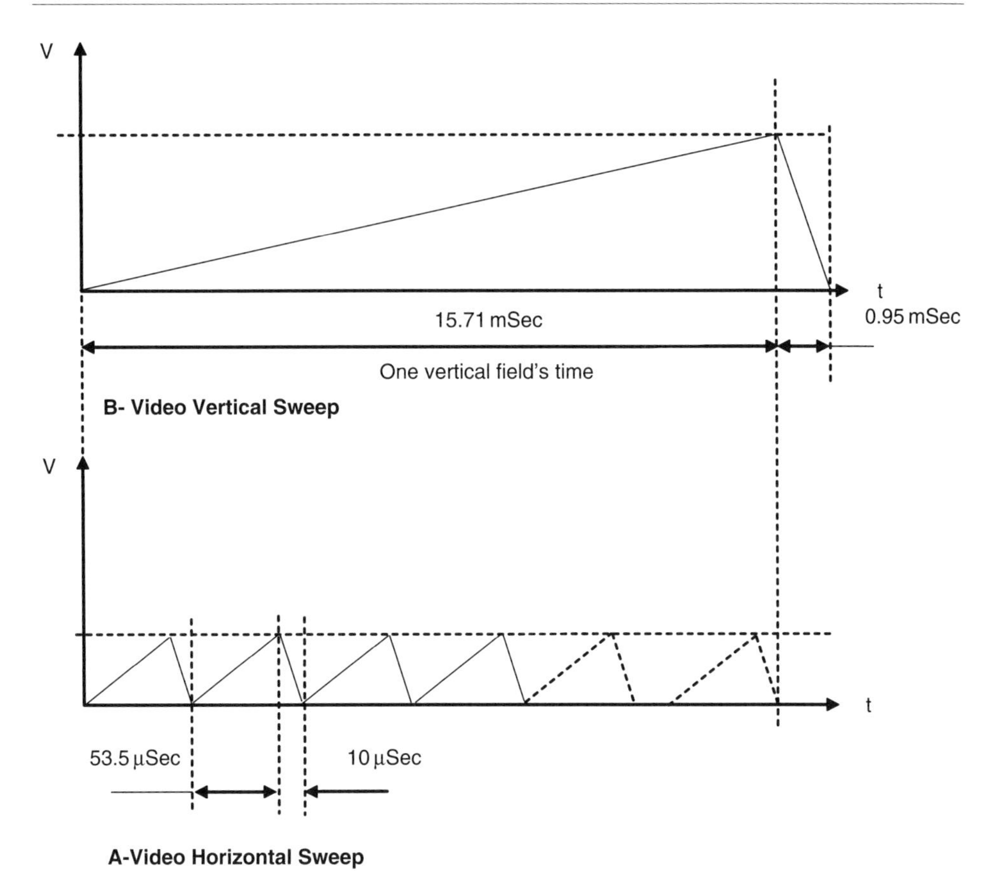


**Figure 3.4** Horizontal scanning saw-tooth pulses vs. vertical scan saw-tooth signal. The period of 0.95 ms is the vertical blanking period. The fly-back horizontal blanking period is 10 μm.

deflection plates. Periodic saw-tooth signals are applied on these plates as shown in Fig. 3.4. The beam scans the first and second horizontal lines of the first field within 53.5 μs, then, within 10 μs, returns back to the next row to scan third and fourth lines, and so on. Meanwhile, the beam is deflected down by the vertical scan signal until it finishes the first field scan. Therefore, the scanned lines are not perfectly horizontal but have a down tilt. The period of the vertical scan is 15.71 ms and the return time for the next field scan is 0.95 ms; during the return time the vertical scan is blanked. Within a single vertical scan the beam would scan 247.4 field lines due to the following relation: The number of scanned lines is equal to 15.71 ms/(53.5 μs + 10 μs). After scanning the first field, the beam starts to scan the next field, creating a complete frame scan of 495 lines. Out of the 525 lines, 495 lines are the active scan lines and 30 lines are inactive (Figs. 3.1 and 3.2). Scanning is continuous at a rate of 60 fields per second. The resultant electrical signal is the video signal corresponding to the visual image. This signal with some AM modifications called vestigial side band (VSB) (to be

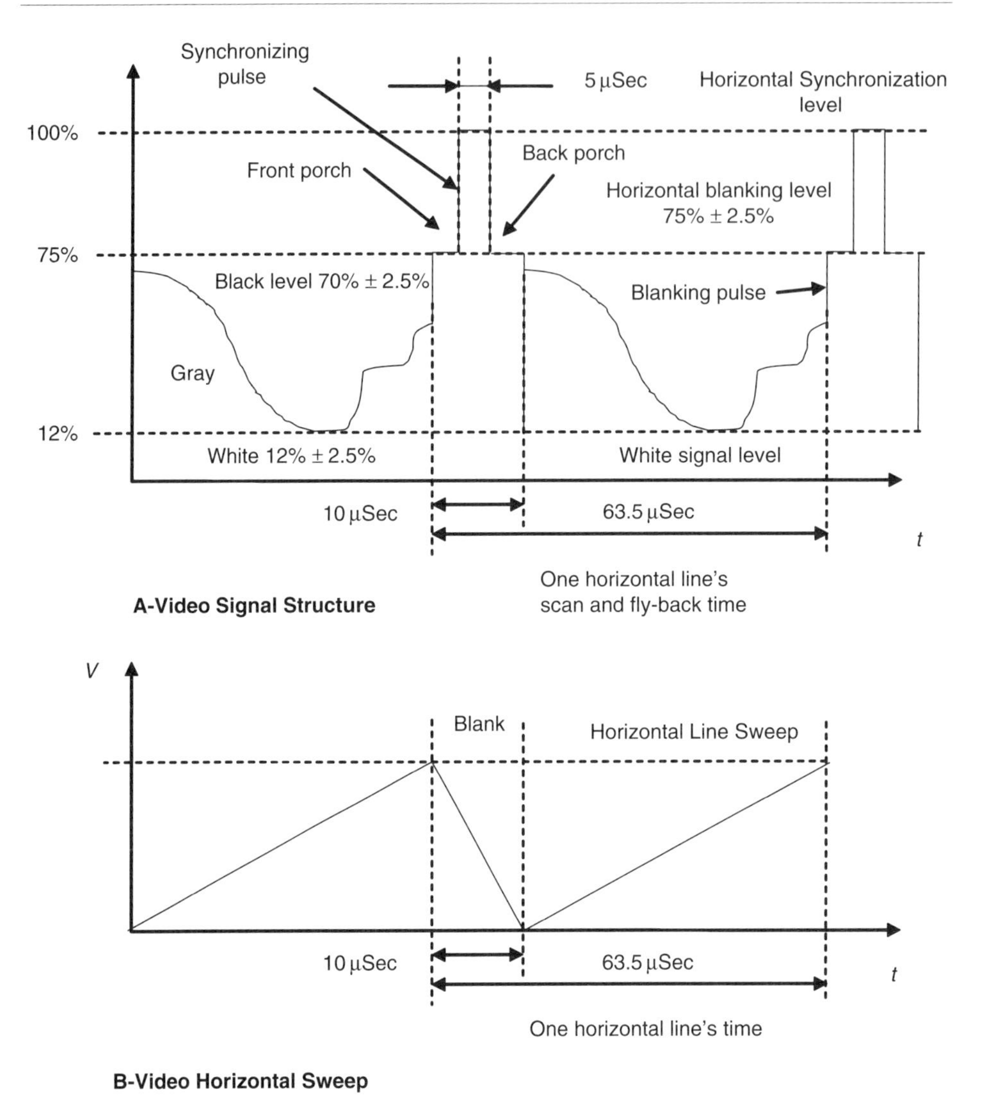


**Figure 3.5** Video signal and modulation indices vs horizontal saw tooth sweep signal.

explained in Sec. 3.2.3) modulates the video carrier. This carrier is transmitted along with the FM-modulated audio (to be explained later in the section about the TV spectrum). The TV receiver is similar to an oscilloscope. An electron gun with horizontal and vertical deflection plates generates an electron beam that scans the screen exactly in the same pattern and synchronization of the scanning at the transmitter (Fig. 3.5).

To prevent a fly-back trace by the scanning beam at the receiver after each line, a blanking pulse is added during the fly-back interval. This pulse is known as the horizontal blanking pulse (Figs. 3.4 and 3.5), which is activated at the end of each horizontal sweep. Similarly, a vertical blanking pulse is added at the end of each vertical scan to eliminate the unwanted vertical retrace (Fig. 3.4). These horizontal blanking, vertical blanking, and synchronization pulses are added to the video signal by the transmitter. The resultant signal is called composite video.

## 3.2 Video Bandwidth and Spectrum Considerations of Color TV

### 3.2.1 Image bandwidth

The screen resolution is determined by the number of picture elements (PELs). The ratio between the horizontal and the vertical dimensions of the image is 4/3 and is called the aspect ratio. Therefore, the number of PELs increases by a factor of 4/3. Because the scanning pattern is not perfectly aligned to the resolution grid shown in Fig. 3.1, the resolution is decreased by a factor of 0.7; this correction factor is known as the "Kerr-Factor."

The picture frame provided by the NTSC method consists of two fields of 262.5 rows each. Each row is a single horizontal scan containing 262.5 PELs; therefore, a single field contains 262.5 × 262.5 PELs. The field scan rate is 30 fields per second, and single frame contains two interlaced fields. The uncorrected PEL scanning rate will be 262.5 × 262.5 × 30 × 2 = 4,134,375 Hz. This result provides a rough video bandwidth (BW) estimate for the NTSC signal, which is about 4.13 MHz. This example demonstrates the advantage of the two-field scanning method, which has a better spectral efficiency than the progressive scan with a single field containing 525 horizontal lines. In this case, the required bandwidth will be 525 × 525 × 30 = 8,268,750 Hz, which is about 8.26 MHz.

As a conclusion, the trade-offs of choosing the scanning standards for a video system are given by[1]

$$\mathrm{BW} = \frac{1}{2}\frac{\mathrm{AR} \times \mathrm{FR} \times N_{\mathrm{L}} \times R_{\mathrm{H}}}{C_{\mathrm{H}}}, \tag{3.1}$$

where AR is the aspect ratio, FR the frame rate, $N_{\mathrm{L}}$ the number of scanning lines per frame, $R_{\mathrm{H}}$ the horizontal resolution, and $C_{\mathrm{H}}$ the net horizontal scanning time without the fly-back blanking time. In Table 3.1, details of different TV scanning standards are provided.[1]

Now the video spectrum can be explored. For the sake of simplicity, assume a still image. The case then becomes scanning an array with a repeatable pattern in both dimensions as shown in Fig. 3.6.

The brightness level $b$ for Fig. 3.6 is a function of both directions $x$ (horizontal) and $y$ (vertical) and can be expressed as $b(x, y)$. Since the picture repeats itself in both $x$ and $y$ dimensions, $b(x, y)$ is a periodic function with periods $\alpha$ and $\beta$, respectively. For this reason, $b(x, y)$ can be represented by a 2D Fourier series with fundamental frequencies $2\pi/\alpha$ and $2\pi/\beta$, respectively. The exponential transformation is given by[3,11]

$$b(x, y) = \sum_{m=-\infty}^{\infty} \sum_{n=-\infty}^{\infty} B_{m,n} \exp\left[j2\pi\left(\frac{mx}{\alpha} + \frac{ny}{\beta}\right)\right]. \tag{3.2}$$

**Table 3.1** Different scanning standards.

| System | Aspect ratio | Interlace | Frames/s | Total/active lines | Lines/s | Bandwidth (MHz) |
|---|---|---|---|---|---|---|
| USA | | | | | | |
| Mono | 4:3 | 2:1 | 30 | 525/480 | 15750 | 4.2 |
| Color NTSC | 4:3 | 2:1 | 29.97 | 525/480 | 15734 | 4.2 |
| Color HDTV | 16:9 | No | 60 | 750/720 | 45000 | 6.0[1] |
| Color HDTV | 16:9 | 2:1 | 30 | 1125/1080 | 33750 | 6.0[1] |
| UK | | | | | | |
| Color PAL | 4:3 | 2:1 | 25 | 625/580 | 15625 | 5.5 |
| Japan | | | | | | |
| Color NTSC | 4:3 | 2:1 | 29.97 | 525/480 | 15734 | 4.2 |
| France | | | | | | |
| Color SECAM | 4:3 | 2:1 | 25 | 625/580 | 15625 | 6.0 |
| Germany | | | | | | |
| Color PAL | 4:3 | 2:1 | 25 | 625/580 | 15625 | 5.0 |
| Russia | | | | | | |
| Color SECAM | 4:3 | 2:1 | 25 | 625/580 | 15625 | 6.0 |
| China | | | | | | |
| Color PAL | 4:3 | 2:1 | 25 | 625/580 | 15625 | 6.0 |

[1]Digital transmission; scanning standards shown are typical; other variations are possible.

Assuming that the scanning beam moves with a velocity $v_x$ and $v_y$ in the $x$ and $y$ directions, respectively, $x = v_x t$ and $y = v_y t$ and video signal voltage $e(t)$ is given by

$$e(t) = \sum_{m=-\infty}^{\infty} \sum_{n=-\infty}^{\infty} B_{m,n} \exp\left[ j2\pi\left( m\frac{v_x}{\alpha}t + n\frac{v_y}{\beta}t \right)\right]. \tag{3.3}$$

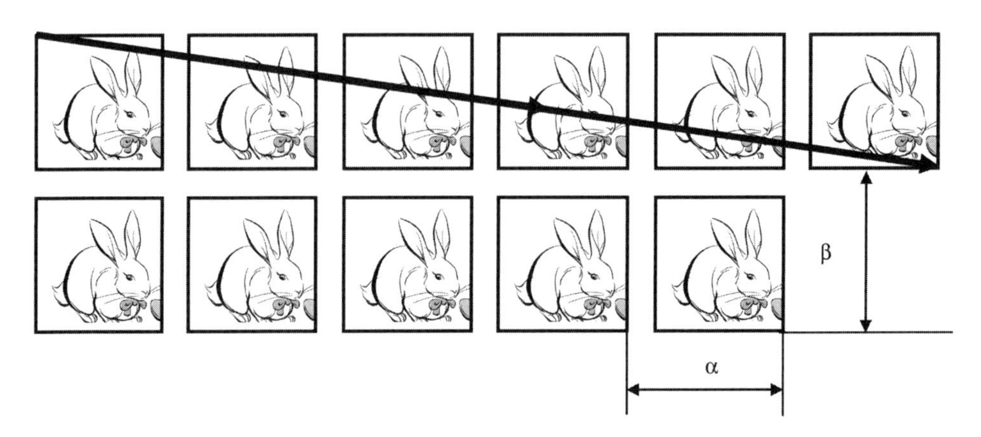


**Figure 3.6** Scanning model process using a doubly periodic image field.

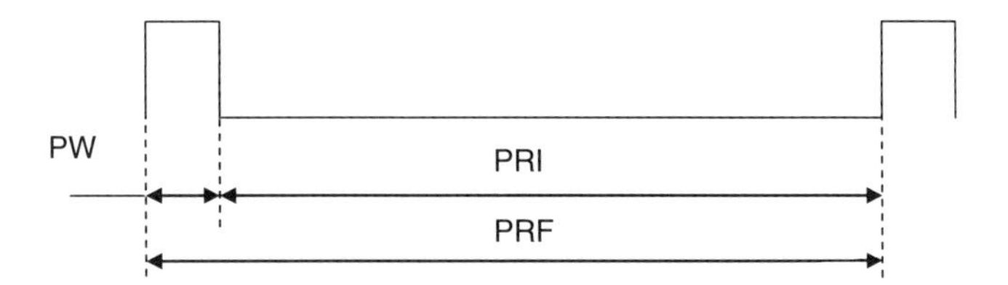


**Figure 3.7** Vertical pulse PRI, PRF, and PW definitions; PW = PRF − PRI, $T = 1/\text{PRF}$, $f = \text{PRF}$.

However, the time required to scan a single horizontal and vertical line is given by $T_x = \alpha/v_x$ and $T_y = \beta/v_y$, respectively. It is known that single horizontal scan time is $1/(60 \times 262.5)$ s per field and the complete image scan time is $1/30$ s (two fields), but for single field, scan rate is $1/60$ s. Then Eq. (3.3) becomes

$$e(t) = \sum_{m=-\infty}^{\infty} \sum_{n=-\infty}^{\infty} B_{m,n} \exp[j2\pi(15750\,m + 60n)t]. \tag{3.4}$$

In conclusion, it can be stated that the video signal is periodic with a fundamental frequency of 15.75 kHz for the horizontal sweep and around each harmonic clustered satellite spaced 60 Hz apart; which is related to the vertical sweep. The timing of the TV transmitted lines is controlled by the synchronization-pulse (sync-pulse) generator. Each vertical sync pulse is characterized by its pulse repetition interval (PRI) and pulse repetition frequency (PRF) (Fig. 3.7). The PRF of the vertical blank has the lowest frequency, which is 60 Hz. The blanking pulse width (PW) is 950 μs. The PRF parameter is an important factor for determining the CATV optical receiver automatic gain control (AGC) bandwidth, and it should be lower than 0.5 Hz. This is for preventing the feedback AGC circuit frequency response from operating as an envelope detector of the vertical blanking signals. In addition, the horizontal blanking and sync together with varying color leveling requires AGC compensation of peak to rms as will be explained later in Chapters 14 and 21.

Luminance spectral energy lines are clustered around harmonics of horizontal scans as shown in the upper part of Fig. 3.8. Chrominance spectral energy lines are clustered around odd multiples of half horizontal scan frequency. Both spectral lines are multiplexed in the frequency domain, creating a video signal as shown in the lower part of Fig. 3.8. Figure 3.9 provides the spectrum of the 6.35 μs horizontal sync that appears in Fig. 3.5. The spectrum shows 15.75 KHz. Figure 3.10 depicts the vertical sync spectrum of 15.71 μs, which appears in Fig. 3.4. The spectrum in Fig. 3.10 shows ~60 Hz. Those video signals appear as video, aural, and sync signals as shown by Fig. 3.15. More information is provided in Ref. [33].

### 3.2.2 Color transmission

The TV camera consists of a lens that focuses the picture to be televised onto and through special dichroic glass semimirrors. The dichroic semimirror operates as

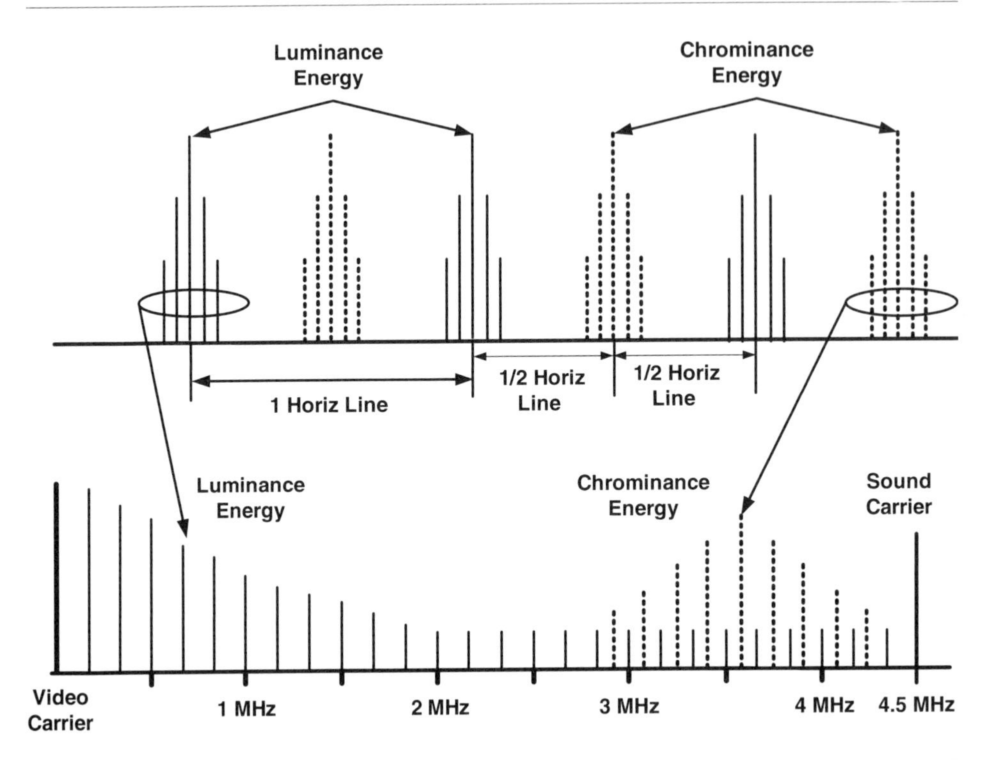


**Figure 3.8** Frequency domain of interleaving of the chrominance and luminance signal spectra.

both a beamsplitter and an optical filter. The TV camera contains three camera tubes for the three basic colors red, blue, and green (RGB). The reason to have green color rather than yellow is the availability of phosphors that glow with these colors. The color triangle[2] shown in Fig. 3.11 provides a rough approximation of the color distribution. None of the colors can actually be reproduced in a 100% saturated form. But the possible percent is equal to or better than that obtained with printing ink.

With proper levels of RGB, perception of all colors in the eye is activated and white is seen. It means that to transmit a picture of full color, it will be only necessary to scan the picture simultaneously for its RGB content. The problem is to transmit three signals and synthesize a color. It would require three times more BW compared with the calculation in Sec. 2.4.4, which shows the assessment for 4.13-MHz monochromatic transmission. The problem is solved by using signal matrixing. The camera RGB $m_{\mathrm{r}}(t), m_{\mathrm{g}}(t), m_{\mathrm{b}}(t)$ information is transmitted as three linear combinations of all three signals that are linearly independent:[3]

$$\begin{aligned} m_{Y}(t) &= 0.30\, m_{\mathrm{r}}(t) + 0.59\, m_{\mathrm{g}}(t) + 0.11\, m_{\mathrm{b}}(t), \\ m_{I}(t) &= 0.60\, m_{\mathrm{r}}(t) + 0.28\, m_{\mathrm{g}}(t) - 0.32\, m_{\mathrm{b}}(t), \\ m_{Q}(t) &= 0.21\, m_{\mathrm{r}}(t) - 0.52\, m_{\mathrm{g}}(t) + 0.31\, m_{\mathrm{b}}(t). \end{aligned} \tag{3.5}$$

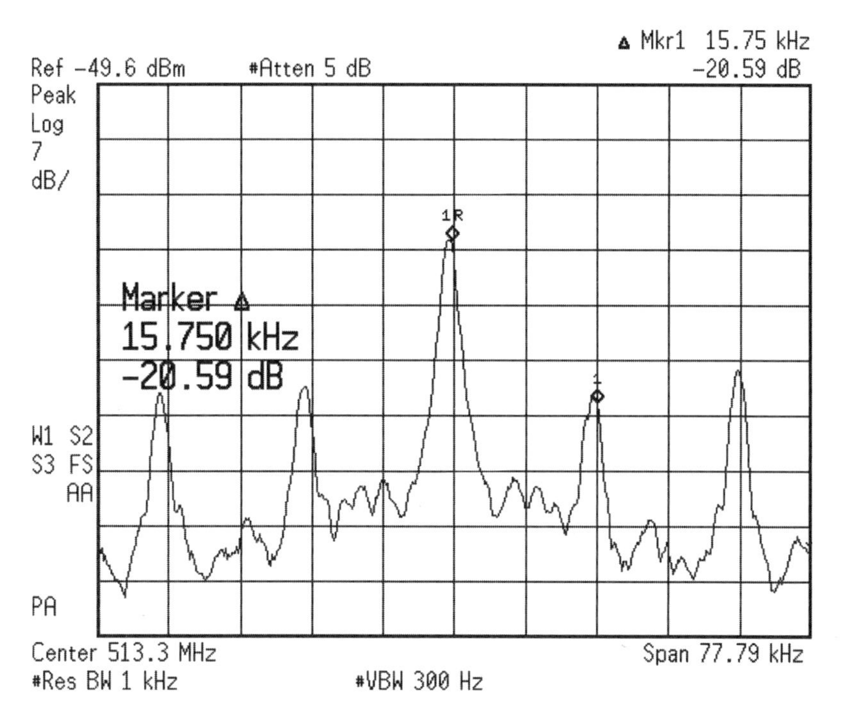


**Figure 3.9** Spectrum presentation of the 15.75-kHz horizontal sync.

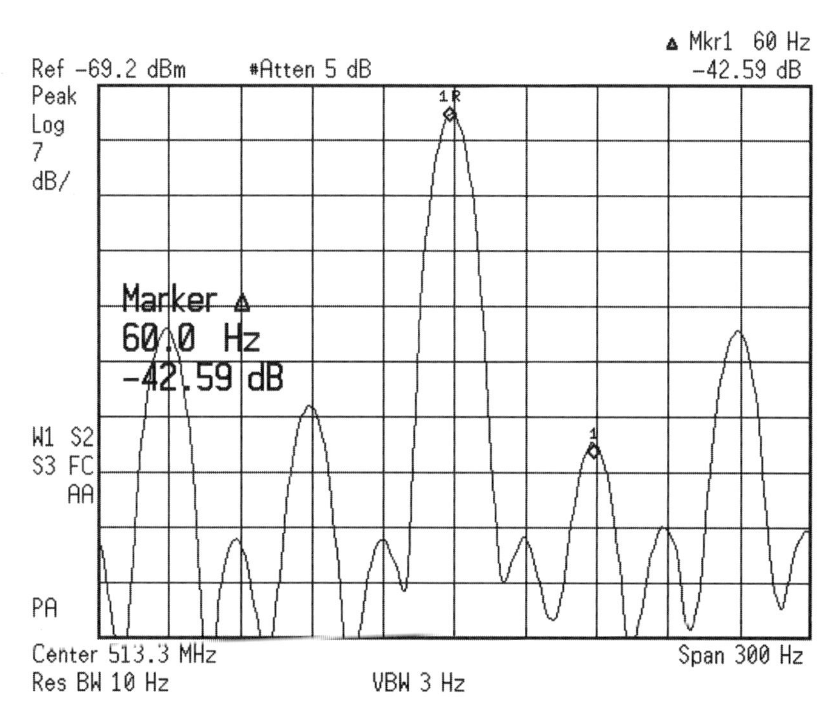


**Figure 3.10** Spectrum presentation of the 60-Hz vertical sync.

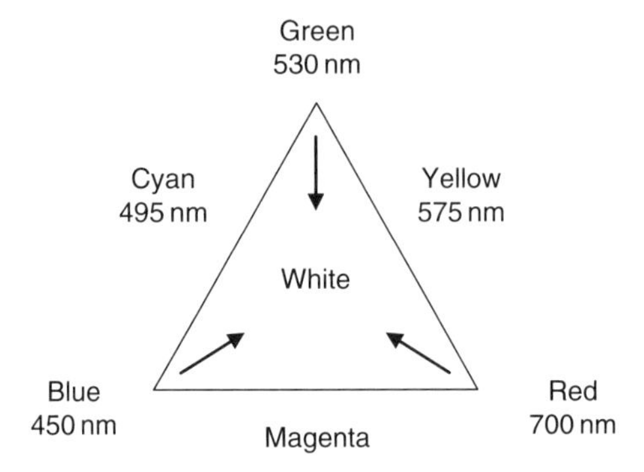


**Figure 3.11** Color triangle with wavelength of hues.

The signals $m_r(t)$, $m_g(t)$, and $m_b(t)$ are normalized to maximum value of 1, so that the amplitude of the color signal is within the range of 0 to 1. That leads to the conclusion that $m_Y(t)$ is always positive, while $m_I(t)$ and $m_Q(t)$ are bipolar. The signal $m_Y(t)$ is known as luminance signal or contrast since it closely matches the conventional monochrome luminance signal. The signals $m_I(t)$ and $m_Q(t)$ are known as chrominance. These signals have an interesting interpretation in terms of hue and saturation colors. Hue refers to attributes of colors that make them red, yellow, green, blue, or any other color. Saturation or color intensity refers to the color's purity. For instance, deep red has 100% saturation, but pink, which is a combination of red and white, has lower amount of red saturation. The saturation and the hue angles are given by Eqs. (3.6) and (3.7), respectively:

$$\text{sat} = \sqrt{m_I^2(t) + m_Q^2(t)} \tag{3.6}$$

and

$$\text{hue} = tg^{-1}\left[\frac{m_Q(t)}{m_I(t)}\right]. \tag{3.7}$$

Note that white, gray, and black are not hues.

The summation of chrominance signals, $m_I(t)$ and $m_Q(t)$, is occupying the BW of 4.2 MHz for both the $I$ and the $Q$. The luminance $Y$ or $m_Y(t)$ also uses 4.2-MHz BW and is transmitted as monochrome video signal. Now, the BW needs to be optimized to comply the same BW of monochrome transmission. Subjective tests show that the human eye is not perceptive to changes in chrominance (hue and saturation) over smaller areas. That means that the BW can be cut and optimized by filtering out high-frequency components without affecting the quality of the picture since the eye would not be sensitive to them anyway. Therefore, BW of $I$ and $Q$ were limited to 1.6 and 0.6 MHz, respectively. Both filtered $Q$ and the 0–0.6-MHz portion of $I$ are modulated by a QPSK modulator, generating a quadrature amplitude modulation (QAM) signal at its output while the upper portion of the $I$ signal $m_{I-H}(t)$, occupying the BW between 0.6 and 1.6 MHz, is sent by the low-side band (LSB). The subcarrier frequency is $f_{cc} = 3.583125$ MHz (see Figs. 3.8 and 3.15). The result of this process provides the relation for the multiplexed signals $Q$ and $I$, respectively:

$$X_Q(t) = m_Q(t)\sin(\omega_{cc}t), \tag{3.8}$$

and

$$X_I(t) = [m_i(t) - m_{I-H}(t)]\cos(\omega_{cc}t) + m_{I-H}(t)\cos(\omega_{cc}t) + m_{I-Hh}(t)\sin(\omega_{cc}t), \tag{3.9}$$

where $\omega_{cc} = 2\pi f_{cc}$. The first argument in Eq. (3.9) is a double side band (DSB) QAM component, while the last two are the LSB.

The composite multiplexed video is given by

$$m_v(t) = m_\gamma(t) + m_Q(t)\sin(\omega_{cc}t) + m_I(t)\cos(\omega_{cc}t) + m_{I-Hh}(t)\sin(\omega_{cc}t). \tag{3.10}$$

In addition, a color burst is added on the "back-porch" of the above multiplexed signal (Figs. 3.5 and 3.12) of the horizontal blanking pulse for frequency and phase synchronization of the locally generated subcarrier at the receiver side. The composite video is transmitted by a VSB + C modulation.

The Institute of Radio Engineers (IRE) standard (Fig. 3.12) for NTSC TV defines a modulation swing of 1 Vptp from the horizontal sync tip to white level as 140 IRE division units. The horizontal blanking is a 0 IRE reference level. The white level is 100 IRE units and the horizontal synchronization pulse tip is $-40$ IRE.

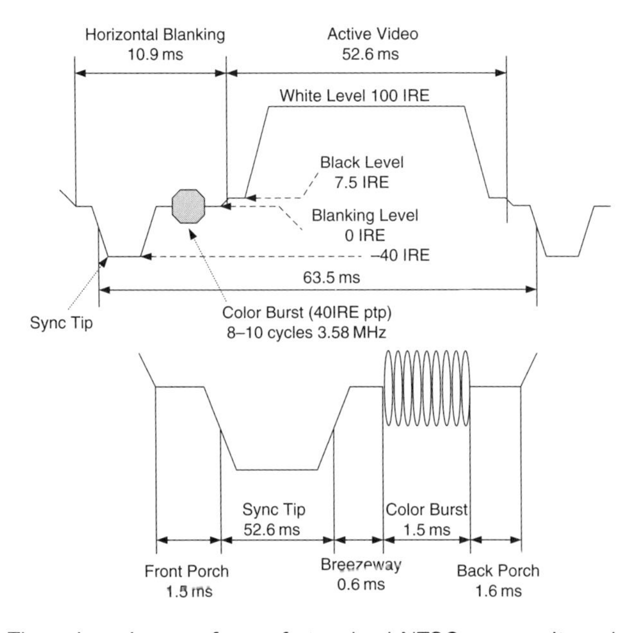


**Figure 3.12** Time domain waveform of standard NTSC composite color video signal.

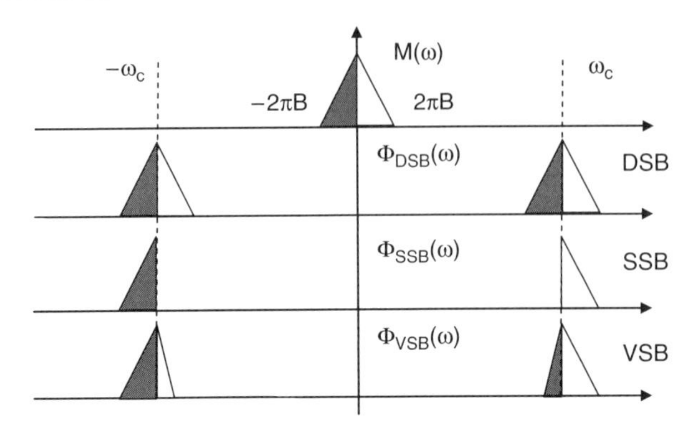


**Figure 3.13** Spectra of a modulating signal and corresponding DSB, SSB, and VSB signals.

There are several types of PAL standards, which is mostly used in European countries. The PAL systems assigned a particular letter for each country. For instance, "PAL-I" is for England, PAL-B, -G, and -H are for the continent of Europe, and PAL-M is for Brazil. As in the NTSC method, the PAL method uses QAM of the chroma carrier to transmit the color-difference information as a single composite chroma signal. The $R - Y$ signal at the $Q$ vector input of the QPSK modulator is phase inverted on alternate lines. It means that each parallel line differs by 180 deg from the next or previous horizontal line. This way, the picture quality is improved for different conditions. This phase alternation gives PAL its name. In PAL, there is no color burst for horizontal synchronization. This is achieved by the phase alternation. The simple PAL system relies on the human eye to average the color switching process line by line. Thus, the picture degrades by the 50-Hz line beats caused by the system nonlinearities, introducing visible luminance changes at the line rate. This problem was solved by an accurate delay element that stores the chroma signal for one complete line period. This method is called PAL-deluxe (PAL-D). Similar to the NTSC system method, the PAL has an equal BW 1.3 MHz at 3 dB for the two color differences $U$ and $V$ [where $V = R - Y$, $U = B - Y$, see Fig. 3.14 and Eq. (3.5)]. There are minor variations among PAL systems, mainly where 7 and 8-MHz BW are used. In PAL-I, where 8-MHz BW per channel is used, a new digital sound carrier called near instantaneous companding audio multiplex (NICAM) was added.

The NICAM carrier is located 6.552 MHz above the visual carrier, nine times higher than the bit rate, it uses differential quadrature phase shift keying (DQPSK) modulation at a bit rate of 728 kB/s, and its level is 20 dB below the visual carrier peak sync power. Since the new sound carrier is located closer to the analog FM aural carrier as well as the adjacent channel luminance carrier, the digital sound signal is scrambled before it is modulated.

The SECAM system is similar to NTSC by having the same luminance carrier $m_Y(t)$ equation as in Eq. (3.5) and same color difference $U$ and $V$ components. However, this method differs from PAL and NTSC in two main ideas. The aural carrier in SECAM is AM, while in PAL and NTSC the sound is FM. The color

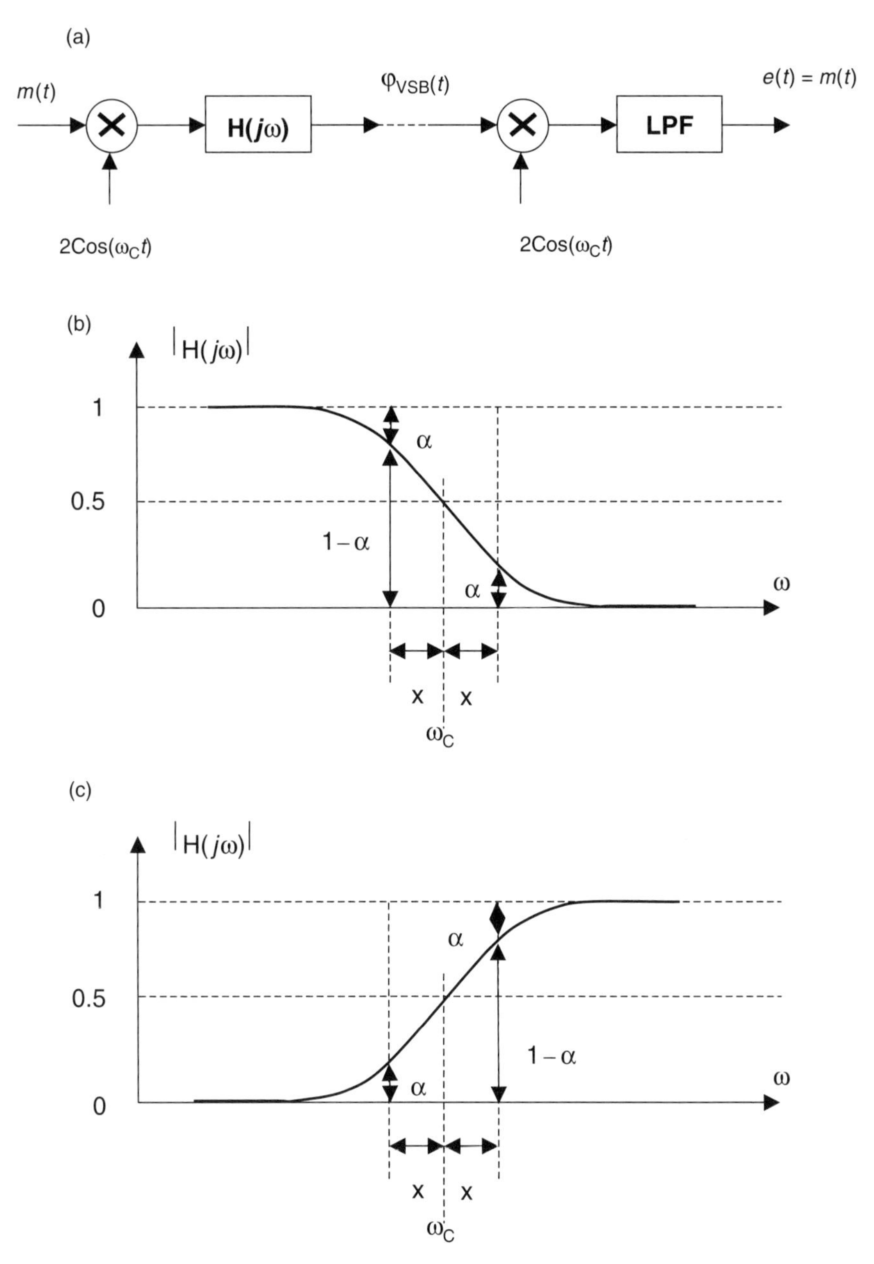


**Figure 3.14** (a) VSB modulator and demodulator, (b) VSB filter characteristic for LSB, (c) VSB filter characteristic for upper side band (USB).

differences are transmitted alternately in time sequences from one successive line to the next with the same visual carrier for every line. Since there are an odd number of lines in a frame, any given line carries the $R - Y$ component on one field and $B - Y$ for the next. In addition, $R - Y$ and $B - Y$ use FM. Like in PAL-D, an accurate delay component is used for the synchronization with the

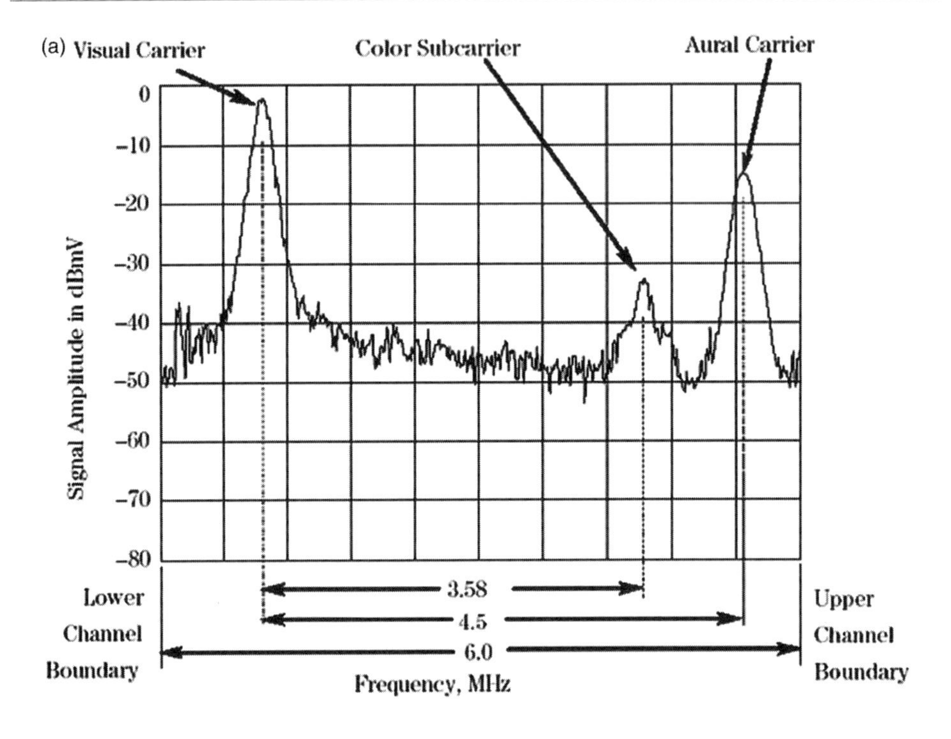


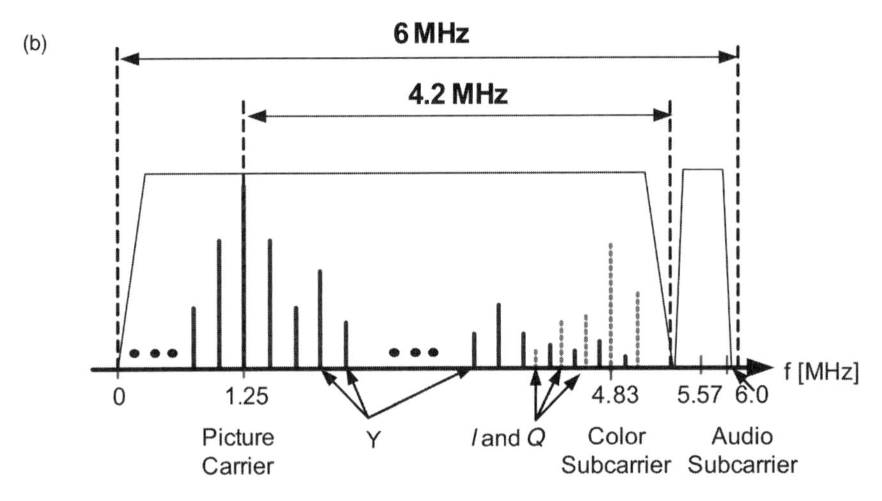


**Figure 3.15** (a) NTSC composite video RF spectrum structure showing the luma, chroma, and aural carriers. The audio (aural) carrier is placed 4.50 MHz above the visual carrier (or 250 kHz from the upper edge of the channel). The audio is an FM signal with a 50-kHz bandwidth (i.e., peak deviation of $\pm 25$ kHz).[17] (Reprinted with permission from *HP Cable Television System Measurement Handbook* © HP, 1994.) (b) Interleaving of luminance (*Y*), chrominance *I* and *Q* components, and audio (aural) FM carrier.

switching process to have simultaneous existence of $R - Y$ and $B - Y$ in the linear matrix to form the $G - Y$ color difference component. The BW of the chroma signals $U$ and $V$ is reduced to 1 MHz. More detailed information can be found in Ref. [27].

### 3.2.3 Vestigial side band

In TV NTSC systems, the video is modulated by using a vestigial side band (VSB) modulation scheme. It is a compromise between double side band (DSB) and single side band (SSB). It inherits the advantages of both DSB and SSB, but avoids their disadvantages (Fig. 3.11). VSB signals are relatively easy to create and their BW is slightly more than that of SSB, approximately by 25%. In VSB, the rejection is not as sharp as in SSB but has a moderate roll-off and cutoff response. As was explained before, the TV video signal exhibits a large BW and significant low frequency content, which suggests the use of VSB. In addition, the circuit for demodulation in the receiver should be simple and cheap. VSB demodulation is done by simple envelope detector signal recovery.

Now there is a need to determine the shape of a vestigial filter $H(\mathrm{W})$ for producing VSB from DSB as in Fig. 3.13. Therefore, according to Fourier,

$$\Phi_{\mathrm{VSB}} = [M(\omega + \omega_c) + M(\omega - \omega_c)]H(\omega). \tag{3.11}$$

The requirement is that $m(t)$ be recoverable from the VSB signal $\varphi_{\mathrm{VSB}}(t)$ using synchronous demodulation of the latter. This is done by multiplying the incoming VSB signal $\varphi_{\mathrm{VSB}}(t)$ by $2\cos\omega_c t$ to produce the information signal $e_d(\mathrm{t})$:

$$e_d(t) = 2\varphi_{\mathrm{VSB}}(t)\cos\omega_c t \leftrightarrow [\Phi_{\mathrm{VSB}}(\omega + \omega_c) + \Phi_{\mathrm{VSB}}(\omega - \omega_c)], \tag{3.12}$$

where $\Phi_{\mathrm{VSB}}$ is the Fourier images of recovered $\varphi_{\mathrm{VSB}}(t)$.

From Eqs. (3.11) and (3.12), after filtering the higher harmonics of $\pm 2\omega_c$, the output voltage $e_0(t)$ at the output of the low pass filter (LPF), see Fig. 3.14, is given by

$$e_0(t) = M(\omega)[H(\omega + \omega_c) + H(\omega - \omega_c)]. \tag{3.13}$$

For distortion free response, the following constraint should be applied:

$$e_0(t) \leftrightarrow CM(\omega), \tag{3.14}$$

where $M$ is a constant chosen, so that $C = 1$; thus, Eq. (3.13) becomes

$$[H(\omega + \omega_c) + H(\omega - \omega_c)] = 1 \quad \text{and} \quad |\omega| \leq 2\pi B, \tag{3.15}$$

where $B$ is the bandwidth.

For any real filter, it is known that $H(-\omega) = H^*(\omega)$. Hence, Eq. (3.15) becomes

$$H(\omega_c + \omega) + H^*(\omega_c - \omega) = 1 \quad \text{and} \quad |\omega| \leq 2\pi B, \tag{3.16}$$

where the symbol * indicates the complex conjugate.

In a more general way, Eq. (3.16) turns into

$$H(\omega_c + x) + H^*(\omega_c - x) = 1 \quad \text{and} \quad |x| \leq 2\pi B. \tag{3.17}$$

This is precisely a vestigial filter. Assuming a filter function form of $|H(\omega)|\exp(-j\omega t_d)$, the phase term $\exp(-j\omega t_d)$ represents pure delay; thus, only the magnitude $|H(\omega)|$ needs to satisfy Eq. (3.17). Since $|H(\omega)|$ is real,

$$|H(\omega_c + x)| + |H^*(\omega_c - x)| = 1 \quad \text{and} \quad |x| \leq 2\pi B. \tag{3.18}$$

It can be seen from Fig. 3.13 that the filters in Figs. 3.14(b) and (c) are used to retain the LSB and USB, respectively.

It is not required to realize the desired VSB shape in a single $|H(\omega)|$ transfer function filter. It can be done in two stages: one at the transmitter and one at the receiver. This is exactly how it is done in broadcast. If the two filters do not match the VSB shape, the remaining equalization is done by other LPF in the receiver.

### 3.2.4 Color television transmitter and receiver structure

Color TV transmitter camera tubes (or CCDs) provide RGB outputs. These outputs are amplified by video amplifiers and are passed through gamma correction blocks. This correction is to compensate for the camera brightness region, which does not correspond to human eye brightness recognition. The gamma-corrected signals are fed to a transmitter luminance matrix, creating $m_Y(t)$ as per Eq. (3.5). This process is done to produce a signal that discriminates better against noise and also produces a better rendition of whites, grays, and blacks when watched in monochrome. This output is known as $Y$ or luminance as marked in Fig. 3.16 and contains frequencies up to 4.2 MHz and would produce black and white picture on any monochrome receiver. Hence, the color transmission is compatible to monochrome. This $Y$ signal is fed in two directions: one to the adder, where luminance, color, sync, and blanking pulses and a sample of the color subcarrier frequency called a color burst of eight cycles are added to form a modulating signal for the color TV transmitter.[2] The $Y$ signal is fed into a phase shifter of 180 deg that creates a $-Y$ signal. The $-Y$ is fed into an adder, which generates $B - Y$ and $R - Y$ signals. These signals are fed into a QPSK modulator matrix, generating two quadrature signals $I$ and $Q$, which are marked as $m_I(t)$, $m_Q(t)$, respectively. The QPSK modulator matrix consists of a 90-deg power splitter and a linear combination matrix for amplitude and phase adjustments. These $I$ and $Q$ outputs are fed into the $I$ and $Q$ arms of the QPSK modulator, where the quadrature vector $Q$ is modulated by the local oscillator (LO) frequency +90 deg and the in-phase vector $I$ is modulated by the LO directly. The $I$ and $Q$ signals carry the color information, and the amplitudes of $I$ and $Q$ determine the color saturation (purity) as indicated

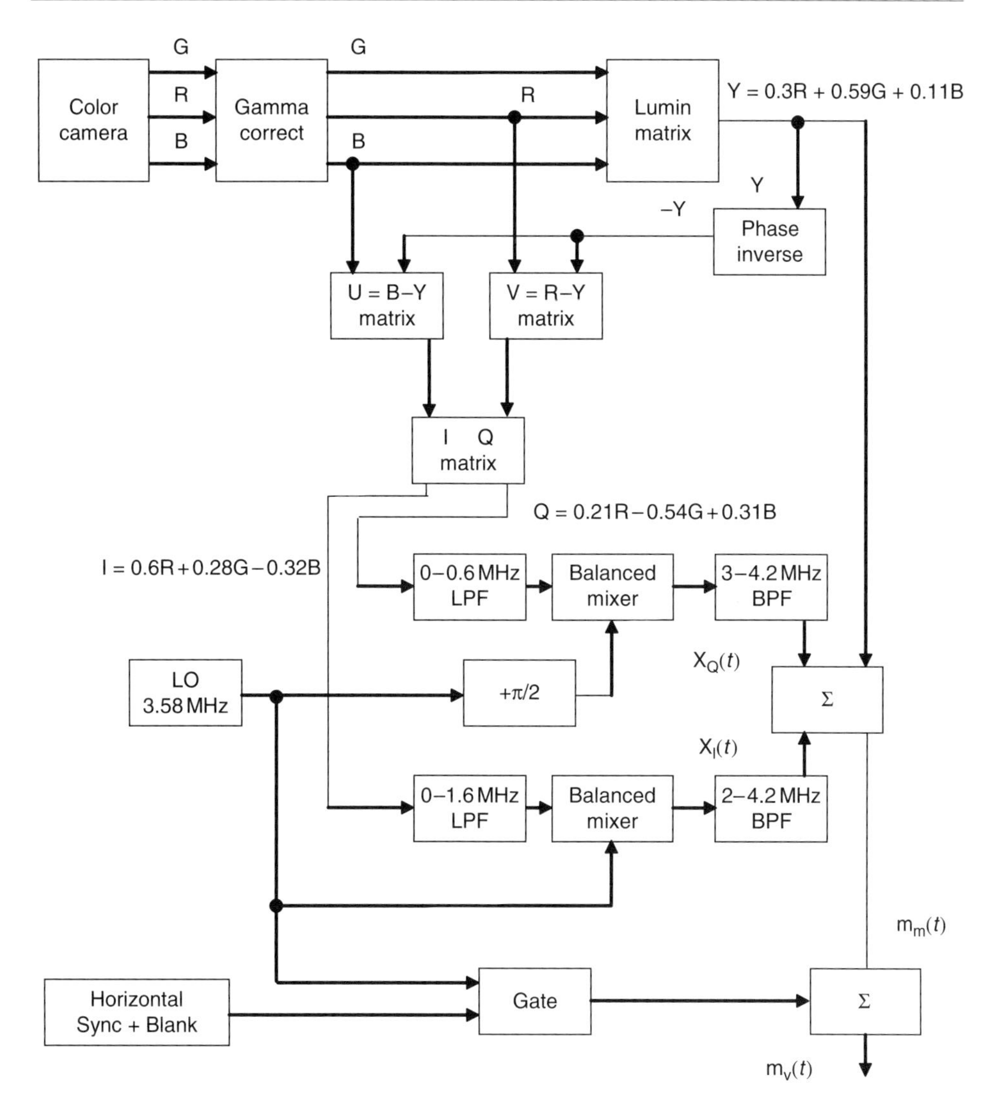


**Figure 3.16** Block diagram of luminance and chrominance multiplexing on subcarrier of 3.58 MHz, including blanking and sync generation for single horizontal row synchronization and fly-back blanking.

in Eq. (3.6). The phase between $I$ and $Q$ determines the hue (the actual color, which is a proper mixture of the RGB) as indicated by Eq. (3.7). The $Y$ signal controls the brightness.

The TV receiver in Fig. 3.17 is a super heterodyne receiver. The RF converter converts and shifts the entire spectrum of the desired channel, video, and sound, and AM and FM into IF frequency. The LO is synthesized by a frequency synthesizer. The image signal is detected by an envelope detector and the sound (aural) is detected by FM phase locked loop (PLL) detector. The receiver's IF frequency range is 41 to 46 MHz and provides the vestigial shape.

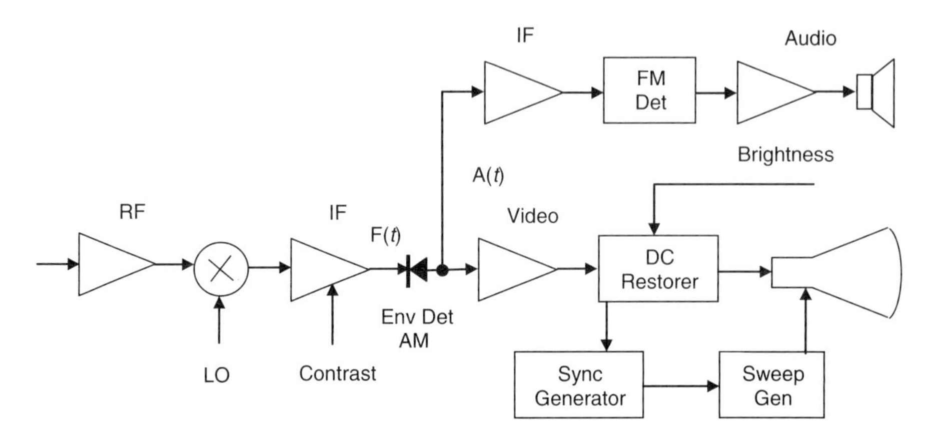


**Figure 3.17** Block diagram concept of a monochrome TV from the RF to the CRT and sound.

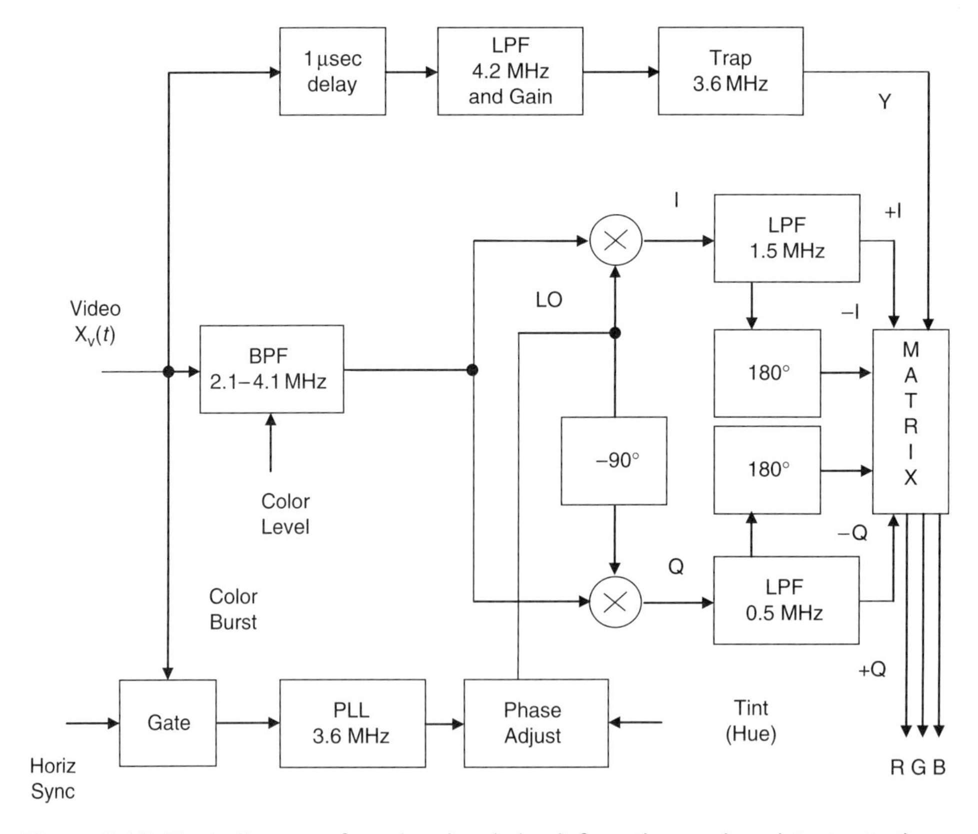


**Figure 3.18** Block diagram of a color signal circuit from the receiver detector to three signals feeding the picture tube (CRT) color guns.

The total signal at the input to the envelope detector is given by[11]

$$F(t) = A_{CV}[1 + \mu X(t)] \cos \omega_{CV} t - A_{CV} \mu X_Q(t) \sin \omega_{CV} t + A_{CA} \cos[(\omega_{CV} + \omega_A)t + \varphi(t)], \quad (3.19)$$

where $A_{CV}$ is the video carrier amplitude, $A_{CA}$ is the audio carrier amplitude, $\mu$ the AM index (approximately 0.85), $\omega_{CV}$ is the video carrier frequency, $\omega_A$ the aural frequency, $\omega_{CV} + \omega_A$ is the aural carrier that is 4.5 MHz above the video, $X(t)$ is the video signal amplitude (AM), and $\varphi(t)$ is the FM audio.

Since $\mu X_Q(t) \ll 1$ and $A_{CA}$; $A_{CV}$, the resulting output from the envelope detector, is given by:[11]

$$A(t) = A_{CV}[1 + \mu X(t)] + A_{CA} \cos[\omega_A t + \varphi(t)]. \quad (3.20)$$

The video amplifier has a low pass filter to remove the audio signal from $A(t)$ as well as a dc restorer that electronically clamps the blanking pulses and hence restores the correct dc level to the video signal. The amplified and restored video signal is applied to the picture tube and to the sync pulse separator that provides synchronization for the sweep generators. The brightness control is achieved by manual adjustment of the dc level and the contrast is done by controlling the IF gain. The aural frequency is 4.5 MHz, FM, and the frequency of the image is 4.5 MHz, AM.

In Eq. (3.20), it is shown that the envelope detector contains the modulated audio by the value $\varphi(t)$. This component is picked up and amplified by another IF amplifier at 4.5 MHz. Even though the transmitted composite video-audio in Eq. (3.19) is from the type of frequency division multiplexing (FDM), there is no need for separate conversion for the audio because the video operates like an LO for the audio at the envelope detection process. This method is called an inter-carrier-sound system and has the advantage that audio and video are always tuned together. Successful operation is made by the condition of larger video component with respect to audio, as well as keeping the video larger than the audio on the transmitter side.

The function of a color TV receiver is based on the same concept as shown in Fig. 3.17 with slight differences in the video processing. As was explained previously, any transmitted color TV signal contains three separate systems of side bands, all within a band of 4.2 MHz. The $Y$ signal side bands occupy the whole 4.2 MHz. The $I$ and $Q$ quadrature sets of chrominance side bands occupy the space between the $Y$ side-band clusters in the upper 2.5 MHz of the side band spectrum. The purpose of the receiver is to detect signals that are equal to the three original RGB signals picked up by the camera and reassemble them in their own component colors and identities in their proper places on the screen.

The video signal $X_v(t)$ detected by the envelope detector is described by[11]

$$X_v(t) = X_Y(t) + X_Q(t) \sin(\omega_{cc} t) + X_I(t) \cos(\omega_{cc} t) + \hat{X}_{IH}(t) \sin(\omega_{cc} t), \quad (3.21)$$

where $\hat{X}_{IH}(t)$ is the Hilbert transform of the high-frequency portion of $X_I(t)$ and accounts for the asymmetric side bands. This baseband replaces the monochrome baseband signal after the AM detector shown in Fig. 3.17. Additionally, an eight-cycle piece of the color subcarrier is placed on the "back porch" of the blanking pulses for synchronization as shown in Fig. 3.10. Demultiplexing is accomplished in a color TV receiver after the envelope detector as in Figs. 3.17 and 3.18. The whole composite video component is passed through a 1-μs delay and is then fed to the $Y$ (luminance) amplifier. This delay is necessary since the chrominance signals are passed through a narrow pass-band filter that creates a delay of 1 μs. Then the $Y$ signal is passed through a 3.6-MHz trap (notch filter) to remove a major flicker component, which is the frequency of the chroma subcarrier. Then the resultant $Y$ signal is fed into the signal matrix adderblock. There are three adder-block networks to regenerate the RGB components for the color tube:

$$R(t) = Y(t) - 0.96Q(t) + 0.62I(t), \tag{3.22}$$

$$G(t) = Y(t) - 0.28Q(t) - 0.64I(t), \tag{3.23}$$

and

$$B(t) = Y(t) - 1.10Q(t) + 1.70I(t). \tag{3.24}$$

Both the $I$ and $Q$ signals are produced by a QPSK demodulator, and the $-I$ and $-Q$ signals are produced by a 180-deg phase shifter as shown in Fig. 3.16. The LO of the QPSK demodulator is synchronized and locked on the color-burst signal using a PLL, which locks the 3.6-MHz voltage control oscillator (VCO) that provides the LO frequency. The color burst is gated by the horizontal sync pulse. Additional manual controls are provided to control the color level, i.e., saturation or the color purity level, and phase control between $I$ and $Q$ to define "tint" or "hue" to control the chrominance angle as was explained by Eqs. (3.6) and (3.7).

## 3.3 Digital TV and MPEG Standards

### 3.3.1 The motivation for HDTV

The first impetus for high-definition television (HDTV) came from wide screen movies. Soon after wide screen was introduced, movie producers discovered that spectators seated in the first rows enjoyed a higher level of participation in the action, which was not possible with conventional movies. Evidently, having a wide field of view of the screen increased significantly the feeling of "being there." In the early 1980s, movie producers were offered a high-definition television system developed by SONY and NHK.

In the late 1970s, this system was called NHK Hi-Vision and it was equal to 35-mm film.[16] Following the introduction of HDTV to film industry, interest began to build and develop HDTV system for commercial broadcasting. Such a system would have roughly twice the number of vertical and horizontal lines than the conventional system. The most significant problem facing the HDTV standard is similar to the problem faced by color TV in 1954. There are approximately

600 million TVs in the world and the critical question is whether the new HDTV standard should be compatible with the current color TV standards or supplant the existing standard, or whether it should be simultaneously broadcast with existing standards, knowing that existing standards will phase out over time. In 1957, the U.S. set a precedent by choosing compatibility when developing the color TV standard. The additional chrominance signal created some minor carrier interface problems; however, both monochrome and color TVs could process the image and sound of the same signal.

The basic concept behind HDTV is not to increase the definition per unit of area, but increase the percentage of visual field contained by the image. The majority of proposed analog and digital HDTV are working toward an approximate 100% increase in the number of horizontal and vertical pixels (Sec. 3.2.1). This translates to 1 MB per frame with roughly 1000 lines containing 1000 horizontal points per line or one million PELs. This results in an improvement by a factor of two to three in the angle of vertical and horizontal field; furthermore, HDTV proposes to change the aspect ratio from $4/3$ to $16/9$, which makes the screen image look more like an image on a movie screen. Note that the aspect ratio of a picture is defined as the ratio of the picture width $W$ to its height $H$.

### 3.3.2 Technical limitations and concept development of HDTV

In previous sections it was explained that a conventional NTSC image of 525 horizontal interlaced lines and a resolution of 427 pixels or image elements, and a scan rate of 29.97 Hz would require a BW of 3.35 MHz. In the case of HDTV with 1050 lines containing 600 pixels per line, keeping the same frame rate with no interlace would require a BW of 18.88 MHz, which is a problem. The current terrestrial (regular air transmission rather than satellite) channel allocations are limited to 6 MHz only. The question is what options are available in the case of 20 MHz BW for HDTV:[15]

1. Change channel allocation from 6 to 20 MHz.
2. Compress the signal to fit inside the 6-MHz BW.
3. Allocate multiple channels, two with compression or three without, for HDTV signal.

The first two options are virtually incompatible with current NTSC service. Hence, the only remaining option is to have separate channels for NTSC and HDTV. This way compatibility is maintained with current NTSC since the first 6 MHz of a signal could be dedicated to the standard NTSC and remaining would be the additional argumentation signal for HDTV. There are several opinions about HDTV transport: the first view is that these systems will be ultimately successful outside the conventional channels of terrestrial broadcasting, and another view is that HDTV must use existing terrestrial broadcast channels.

In 1987, the Federal Communication Commission (FCC) issued a ruling indicating that HDTV standards to be issued would be compatible with existing NTSC service and would be confined to the existing very high frequency (VHF) and ultra high frequency (UHF) frequency plans. In 1988 the FCC received about 23 proposals for HDTV and enhanced definition television (EDTV), which reduced resolution. Those proposals were all for analog and mixed analog/digital systems like multiple sub-Nyquist sampling encoding (MUSE) and offered a variety of options for resolution, interlace, and BW. In 1990, the FCC announced that HDTV would be simultaneously broadcast and that its preference would be for a full HDTV standard. The two decisions contradicted each other. The 1987 decision leaned toward an augmentation type format where NTSC and new channels provide HDTV augmentation to those already existing. The 1990 decision was a radical verdict to phase out NTSC. On the other hand, the FCC did not have any jurisdiction over channel allocation in cable networks. Therefore, there was freedom for the CATV companies to have their plans, of which there were several options. They could continue to broadcast conventional NTSC, they could install 20-MHz MUSE-type HDTV systems (Japanese HDTV standard), or they could go with the digital Grand-Alliance systems [Grand Alliance are AT&T, General Instrument, MIT, Philips, Sarnoff, Thomson, and Zenith, the partners who developed the digital high-definition television system underlying the advanced television (ATV) standard recommended to the FCC by its Advisory Committee], which resulted in two HDTV standards: one for terrestrial and one for CATV. In May 31, 1990, General Instruments (GI) Corp. submitted the first standard of specifications proposal for all digital HDTV systems. Later, on December 1990, Advanced Television Research Consortium (ATRC), an organization of several large consumer electronics companies, research facilities, and broadcast entities that developed U.S. HDTV standards, announced its digital entry, followed by Zenith, AT&T, and then MIT. As a result, there were four serious candidates for digital HDTV as well as modified narrow MUSE and EDTV. During 1991, these systems were tested. In 1993, the FCC made a key decision for an all-digital technology and accepted the recommendations from the Grand Alliance. During 1994, the HDTV system was constructed, and a detailed system review and modification followed. The Grand Alliance and the technical subgroup recommended the system parameters:

- The system would support two, and only two, scanning rates: (1) 1080 active lines with 1920 square pixels-per-line interlace scanned at 59.94 and 60 fields/s, and (2) 720 active lines with 1280 pixels-per-line progressively scanned at 59.94 and 60 frames/s. Both formats would also operate in the progressive scanning mode at 30 and 24 frames/s.
- The system would employ MPEG-2 compatible video compression and transport systems.
- The system would use the Dolby AC-3, 384 Kb/s audio system.[26]

**Table 3.2** Various MUSE standards.[16]

| Standard and year | Lines per frame | Field rate (Hz) | Y bandwidth (MHz) | C bandwidth, wide (MHz) | C bandwidth, narrow (MHz) | Aspect ratio |
|---|---|---|---|---|---|---|
| NHK, 1980 | 1125 | 60 | 20 | 7 | 5.5 | 5/3 |
| MUSE, 1986 | 1125 | 60 | 20 | 6.5 | 5.5 | 5/3 |
| SMPTE, 1987, Studio | 1125 | 60 | 20 | 30 | 30 | 16/9 |

Following the subsystem transmission tests of the VSB[14] system and the QAM system, the VSB system was approved on February 24, 1994[12,13] (more details are in Chapters 16 and 17).

Currently, Japan is the pioneer country that has full HDTV transport to the home over fiber via digital channels, which means there is no need for an analog optical-to-coax converter, rather a regular digital transceiver is needed. The early seeds of Japanese HDTV started at 1968 when Japan's NHK started a huge intensive project to develop a new TV standard. This is an 1125-line analog system utilizing digital compression techniques. It is a satellite broadcast, which is not compatible to the current Japanese NTSC terrestrial broadcast. The reason for this is the fact that Japan is a group of islands covered by one or two satellites. The MUSE system, originally developed by NHK has a 1125-line interlaced scan 60-Hz system with an aspect ratio of 5/3 and with an optimal viewing distance of about 3.3*H*. The BW of the *Y* signal prior to compression is 20 MHz, and the chrominance (*C*) BW before compression was 7 MHz. This standard was upgraded. The various MUSE standards are provided in Table 3.2.

The Japanese rejected the conventional VSB, which was similar to NTSC, and chose satellite broadcast. They also explored the idea of a conventionally constructed composite FM signal, which would be similar in structure to *Y*/*C* NTSC, while *Y* is at the lower frequencies and *C* is at the higher frequencies. The satellite power in this case would be approximately 3 KW for getting a 40-dB signal-to-noise ratio (SNR) for composite FM signal in a 22-GHz satellite.[16] This was incompatible with satellite broadcast. Hence, the other idea was to use separate transmission for *Y* and *C*. This method drops the effective frequency range and reduced the transponder power to about 570 W, where 360 W for *Y* and 210 W for *C* would be required in order to have the same SNR of 40 dB for a separate *Y*/*C* FM signal using 22-GHz satellite band, which is a much more realistically feasible system.

Additional power savings is due to the nature of the human eye. Its lack of response to low frequency noise allows significant reduction in the transponder power in case the higher video frequencies are emphasized prior to modulation at the transmitter and deemphasized at the receiver. This method was adopted with crossover frequencies for emphasize/deemphasize at 5.2 MHz for *Y* and 1.6 MHz for C. This reduces the transponder power to 190 W for *Y* and 69 W for C.[16] The BW fitting of the *Y*/*C* signal combination into 8.15-MHz satellite

BW was solved by digital compression. The NHK HDTV signal is initially sampled at 48.6 Megasamples/sec. This signal controls two filters: one is responsive to stationary parts of the image and the other to the moving parts. The outputs of the two filters are combined and then sampled at the sub-Nyquist frequency of 16.2 MHz, which is one-third of the initial sampling rate. The resultant pulse train is converted by a digital to analog with a base frequency of 8.1 MHz.[16]

### 3.3.3 Digital video audio signals, 8VSB 16VSB 64QAM 256QAM and MPEG

The Grand-Alliance proposed 8VSB, and it was accepted by the FCC. It is the RF modulation format utilized by the DTV (ATSC, Advanced Television Systems Committee) digital television standard to transmit digital bits via RF. Since the terrestrial TV system must overcome numerous channel impairments such as ghosts, echoes in fiber and coax, noise bursts, fading, and interferences, the selection of the correct RF modulation scheme is critical. Moving Picture Experts Group (MPEG) is a formatting standard that was developed to overcome BW limitations for digital video. Data compression of audio and video was developed by the MPEG committee under the International Standard Organization (ISO). The MPEG technology includes many patents from many companies, it deals with the technical standards, and it is not involved and does not address intellectual property issues. MPEG-II is the video compression/packetization format used for DTV. The video processing steps comprise MPEG-II encoding and 8VSB modulation for terrestrial broadcast, and 64 QAM is set for cable operators. Hence, the two main blocks are MPEG-II encoder and 8VSB or QAM exciter. The broadcast standard also includes 16VSB, which is proposed for cables and can carry two HDTV programs. On the other hand, 256 QAM is demonstrated as well on cable, showing it is capable of carrying two HDTV programs. The 8VSB method is specifically designed for terrestrial broadcast with a reference pilot. This is done due to the fact that the broadcast environment must overcome cochannel interference between multiple DTV signals and between DTV and NTSC. The reference pilot carrier helps the DTV tuner to acquire the signal when the channel is changed. In Secs. 3.3.6 and 3.3.7, the structure of HDTV VSB/QAM exciter is reviewed for general background. Other related standards for still-image compression are called Joint Photographic Experts Group (JPEG) and Joint Bilevel Image Experts Group (JBIG), which uses binary image compression for faxes. For audio compression, the AC3 algorithm is used and will be reviewed in Sec. 3.3.6.

### 3.3.4 MPEG-1 standard

The main goal of MPEG-1 compression algorithm is to improve spectrum efficiency. Generally speaking, video sequences contain a significant amount of *statistical* and *subjective* redundancy[29,33] within and between frames. The ultimate goal of video source coding is bit-rate reduction for storage and transmission by exploring both statistical and subjective redundancies and to encode a "minimum set" of information using entropy coding techniques. For instance, one aspect is to have clever statistical prediction of the motion vector of the image. The performance of video compression techniques depends on the amount of redundancy within the image as well as on the concrete compression techniques used for coding. There is a trade-off between coding performance, i.e., high compression with decent quality, and implementation complexity. An MPEG compression algorithm also involves state-of-the-art very large scale integration (VLSI) technology for realization, which is crucial for that process. Further detailed information is provided in Refs. [4 and 18].

In the previous sections, it was explained that each PEL or pixel composed from three signals $Y$, $U$, and $V$, where $V = R - Y$, $U = B - Y$, and $Y$ is given by Eq. (3.5) or can be noted as $Y = 0.3R + 0.59G + 0.11B$. The MPEG-1 algorithm operates on video in the *YUV* domain; hence, if the image is stored in the RGB domain, it is transformed into the *YUV*($Y$, Cr, Cb) domain first. Consequently, the image can be referred to as a grid of PELs where each element has three coordinates. As a result, there are three samples for each pixel. Originally, MPEG-1 started as a low resolution video sequence compression algorithm of approximately 352 by 240 pixels, at a rate of 29.97 (approximately 30) frames per second. Remember that two frames compose a picture for interlaced scan mode. The MPEG quality, however, is equal to that of a high-quality original audio CD. The main idea behind video compression is that natural human eye response is similar to a low-pass filter. The human eye cannot resolve high-frequency color changes in the picture, meaning that image redundancies can be minimized for compression purposes. In other words, since humans see color with much less spatial resolution than they see black and white, it makes sense to "decimate" the chrominance signal, which is the color with respect to illumination, i.e., luminance. The outcome, then, is that color images are converted to $Y$, $U$, $V$ vector space, while the two chroma components are decimated to a lower resolution of 176 by 120 pixels (half of original). The same image can be displayed with the same eye observation quality but with lower number of PELs. The resolution $352 \times 240$ and the sampling rate of $(352 \times 240) \times 29.97$ was derived from the CCIR-601 DTV standard, which is used in professional digital video equipment. In the U.S. it calls for $720 \times 243 \times 60$ fields (not frames) per second. The fields are interlaced when displayed. As was explained in previous sections, an image is composed of two fields, each field is acquired and displayed within $\sim$1/60 s apart, it can be said though that within one second, approximately 60 fields are displayed or approximately 30 frames are acquired and viewed, because of the interlacing method (Sec. 3.1.1). The chrominance channels are $360 \times 243$ by 60 fields interlaced again. This ratio of 2:1 decimation in horizontal direction is defined as 4:2:2 sampling.

The idea behind the *a*:*b*:*c* notation is that given numbers are stating how many pixel values, per four original pixels, are actually sent. The *a*:*b*:*c* sampling notations are relative to luminance signal $Y$ and can be found in the CCIR-610.

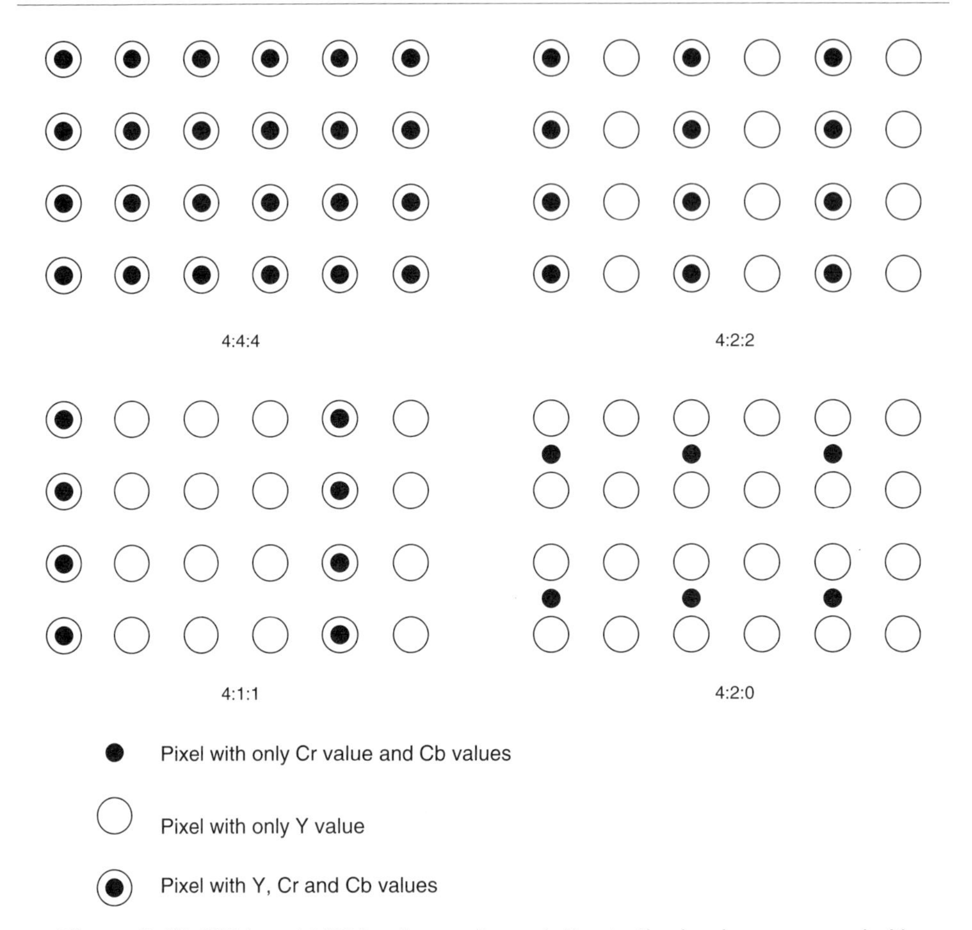


**Figure 3.19** *U*(Cr) and *V*(Cb) subsampling relative to the luminance sample *Y*.

They have the following meaning:

- The scheme 4:2:2 means 2:1 horizontal down sampling, and no vertical down sampling. In other words, "4 *Y* samples for every 2 Cb and 2 Cr samples are in a scanline." That is, of four pixels horizontally labeled as 0 to 3, all four *Y*'s are sent, and two Cb's and two Cr's are sent, as ($Cb_0$, $Y_0$), ($Cr_0$, $Y_1$), ($Cb_2$, $Y_2$), ($Cr_2$, $Y_3$), ($Cb_4$, $Y_4$), and so on (or averaging is used).

- The scheme 4:1:1 means 4:1 horizontal down sampling, no vertical. In other words, "4 *Y* samples for every 1 Cb and 1 Cr samples in a scanline."

- The scheme 4:2:0 means 2:1 horizontal and 2:1 vertical down sampling. Theoretically, an average chroma pixel is positioned between the rows and columns as shown in Fig. 3.19. The scheme 4:2:0, along with other schemes, is commonly used in JPEG and MPEG-1.

- The scheme 4:4:4 means that no chroma subsampling is made. Each pixel's *Y*, Cb, and Cr values are transmitted, 4 for each of *Y*, Cb, and Cr.

A video stream is a sequence of consecutive video frames showing motion; each frame is a still image. A video player displays the frames at a rate of 30 frames per second. The frames are digitized in the RGB domain as 24 bits, 8 bits for each color red, green, and blue. Assume MPEG-1 is designed for a bit rate of 1.5 MB/s and can be used for images at sizes of 352 × 288 at a rate of 24 to 30 frames per second, hence the resultant data rate is 55.7 MB/s up to 69.9 MB/s. (In fact MPEG-1 is not limited to a specific frame size and rate as noted by the CCIR-601. It supports higher resolutions and rates such as 704 × 480 up to as high as 4095 × 4095 × 60 frames per second or 1 GB/s). This RGB format is converted to the *YUV* domain and its content is preserved. The MPEG-1 algorithm operates on the *YUV* domain. In the *YUV* domain, format images are represented by 24 bits per pixel in the following way: 8 bits for luminance information *Y* and 8 bits each for chrominance *U*(Cr) and *V* (Cb). This *YUV* format is subsampled at the ratio of 2:1 for both horizontal and vertical directions. Therefore, there are two bits per each pixel of *U* and *V* information. This subsampling does not degrade the image quality since the human eye is more sensitive to luminance rather than chrominance. The input source format for MPEG-1 is called SIF, which is a CCIR-601 decimated by 2:1 in the horizontal direction, 2:1 in the time direction, and an additional 2:1 in the chrominance vertical direction. In order to make the scale as divisions of 8 or 16, some lines were cut off if required. In Europe, where PAL and SECAM are used, the display standard is 50 Hz, which means 50 fields or 25 frames per second. The number of lines per field is changed from 243 or 240 to 288. The NTSC low resolution decimated chrominance is changed from 120 lines to 144 lines. The reason for this is the fact that the source data rate is same for PAL, SECAM, and NTSC: $288 \times 50 = 240 \times 60$.

The fundamental building block of an MPEG picture is called a macroblock. Frames are divided into 16 × 16 pixel macroblocks. Each macroblock comprises four 8 × 8 luminance (*Y*) and two 8 × 8 chrominance blocks, which are the *U* and *V* vectors. In other words, there are two pairs of 16 × 16 luminance (*Y*) samples and two corresponding 8 × 8 blocks of chrominance samples, which are the *U* and *V* vectors.

After frames were divided into macroblocks, the next step is coding. The fundamental idea in MPEG coding is to predict statistically the motion from frame to frame in a sequential, temporal direction and then to use discrete cosine transforms (DCTs) to organize the redundancy in the spatial directions. The DCT process is done on 8 × 8 blocks, and the motion prediction is done in the luminance (*Y*) signal, which is 16 × 16 blocks. This means that for a given 16 × 16 block in the current frame, the algorithm compares it with the most similar previous or future frame; there are backward prediction modes where later frames are sent first to allow interpolation between frames. Hence, the first step of the coding algorithm is "temporal redundancy reduction," which explores the maximum redundancy reduction using temporal predictions. There are three types of coded frames. Intra, marked as "I," is coded as still frame without using past history, and it provides access points for random access and has moderate compression. The second frame is called the predicted and marked as "P." It has a past and, therefore, was predicted either from Intra frame or from a previous P frame. The last frame is the bidirectional and is marked as "B." This frame is interpolated to have the lowest bit rate; furthermore, the B frame is interpolated either from the previous or from the future reference frames. Hence, B frames are never used as a reference frame and are an average of two reference frames. The MPEG standard does not limit the number of B frames between any two references I or P. Hence, a sequence of "IBBPBBPBBPBBIBBI" is a valid one. Basically, the number of B frames defines the quantization of a motion between two reference frames. The motion prediction explained above is achieved by comparing PELs between two consecutive frames. This assumption results in a high temporal correlation between frames. This rule breaks down under unique circumstances such as a scene change in a movie, which is a video sequence change. The temporal correlation is minimized and vanishes, hence intraframes are used to rebuild new video sequences and achieve data compression.

Three main conclusions can be derived from the above discussion.

- Temporal correlation between PELs is maximal between consecutive frames and this factor is used for generating B frames by marinating maximum temporal correlation.
- The quantization level between I and P images is defined by the number of B frames.
- Temporal correlation goes to minimum or zero in case of a video sequence brake such as a scene change, which results in an extreme sharp change of the image content (new background and figures which is color "chrominance" illumination "luminance" and motion).

One of the tools of compression in the MPEG algorithm is motion compensation prediction. As was explained, the frames are predicted from previous and future frames from macroblocks. Hence, a matching technique is used between blocks by measuring the mismatch between the reference block and current block. The most commonly used function is absolute difference (AE):

$$AE(d_x, d_y) = \sum_{i=0}^{i=7} \sum_{j=0}^{j=7} \left| f(i,j) - g(i - d_x, j - d_y) \right|, \tag{3.25}$$

where $f(i, j)$ represents a block of $16 \times 16$ pixels (macroblock) generated from the current picture, and the $g(i, j)$ represents the same macroblock from a reference frame. The reference macroblock is marked by a vector $(d_x, d_y)$, which represents search location. Hence, the best matching macroblock search will have the lowest mismatch error. The AE function is calculated at several locations within the search range. In order to reduce computing time and extent, an algorithm called the "three step search" (TSS) is used. The algorithm first evaluates the AE at the center and eight surrounding locations of the $32 \times 32$ search area. The location that provides the lowest AE value becomes the next center of the next stage and search range is reduced by half. This sequence is repeated three times.

The next mathematical tool is the spatial redundancy reduction DCT function. This process is implemented in each I frame, or used for error prediction in P and B frames. This technique can be 1D on the columns or 2D on the rows.

For an $8 \times 8$ matrix, a 2D DCT is noted, while $f(i, j)$ are pixel values and the frequency domain transformation is $F(u, v)$. In fact, this is 2D trigonometric Fourier transform:[29]

$$F(u, v) = \frac{1}{4} C(u)C(v) \sum_{i=0}^{i=7} \sum_{j=0}^{i=7} f(i, j) \cos\left[\frac{(2i+1)u\pi}{16}\right] \cos\left[\frac{(2j+1)v\pi}{16}\right], \quad (3.26)$$

where

$$C(x) = \begin{cases} 1/\sqrt{2} \ldots x = 0 \\ 0 \ldots\ldots \text{otherwise} \end{cases}. \quad (3.27)$$

The transformed DCT coefficients, $F(u, v)$, are quantized to reduce the number of bits representing them as well as to increase the number of zero-value coefficients.

In other words, high-frequency energy components are removed, i.e., pixels are taken out. This is the extra redundancy in the picture the human eyes do not see due to the inherent LPF nature of the human eyes. The above mathematical processes are ended by tables of parameters. The quantized DCT coefficients are rearranged into a 2D array by scanning them in zigzag order. This process is called entropy coding, in which the dc coefficient is placed at the first location of the array and the remaining ac coefficients are arranged from low to high frequency in both horizontal and vertical directions. It is assumed that the high-frequency coefficient value is most likely to be zero, which is then separated from other coefficients. This rearranged array is coded into a sequence of run level pair. The run array is defined as the distance between zero and nonzero values. The run level pair information and the information about the macroblock such as the motion vectors and prediction types are further compressed using entropy coding.

This process is done by the video encoder, which packs the compressed video, and the data then has a layered structure.

A structured layer packet of an MPEG block contains the following information:

- picture data of the compressed sequence of images I, B, and P, which is defined as group of pictures (GOP),

- packet of additional information, which includes program clock reference (PCR), optional encryption, packet priority levels, all of which are collectively called packetized elementary stream (PES),
- the encoder or the MPEG multiplexer adds to the PES a label, which is called a presentation time stamp (PTS), and
- the encoder adds an additional label called the decode time stamp (DTS).

Prior to PES transmission, a transport stream (TS) is formed. In the decoding process, the DTS tag informs the decoder when to decode each frame, while the PTS informs the decoder when to display each frame. MPEG decoding and image displaying and processing should be done in a pipeline, which is parallel processing. Assume the sequence "IBBPBBPBBPBBIBBI" is to be decoded. The packet has 12 images between I to I frames. Since each frame time is $1/30$ s, frames each starting point will be most likely $2/5$ s $= 0.4$ s. Hence, during transmission and decoding, when the pipeline is full, the processing sequence will be as follows:

1. The decoder receives the (intra) I image, then it has to process the (predicted) P image and keep them in the memory. The reason for this process is that both I and P images are reference images the B bidirectional images.
2. Then the decoder will display the first I image, then the two B images, and last will be the P image.
3. When I is displayed, P is stored after analysis, then B is analyzed, and so on.

All is done in parallel while the next packet is analyzed in the pipeline. Now it is easier to understand the correlation loss since this will create a process of recharging the decoder pipeline.

MPEG-1 originally was developed to process progressive scan such as in computer monitors. However, in TV, fields are interlaced and two fields create a frame or image. The idea is to convert an interlaced source into a progressive intermediate scan format. In fact, this is the way a synthetic MPEG compatible field is encoded. Thus, there is a preprocessing prior to encoding and postprocessing after decoding. Then the field is displayed as an even field while the odd field is interpolated.

### 3.3.5 MPEG-2 standard

MPEG-2 standard is used for compression of both video and audio. It is similar to MPEG-1 and is an extension of it.[25] MPEG-2, however, covers broadcast

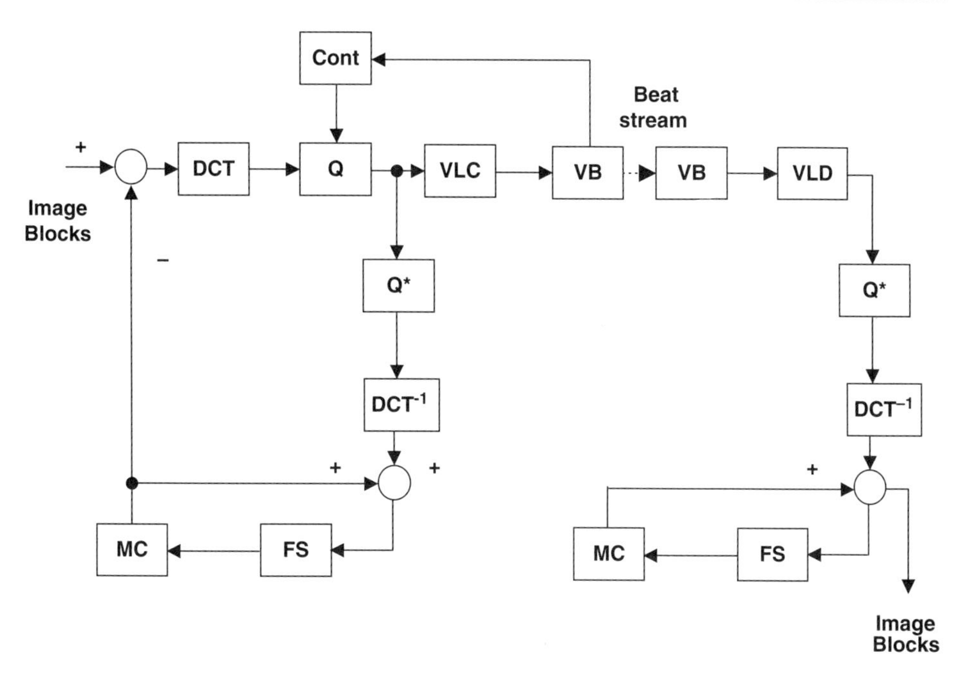


**Figure 3.20** Block diagram of basic hybrid DCT/DPCM encoder and decoder structure concept used for both MPEG-1 and MPEG-2.

transmissions of HDTV and digital storage, and it answers the need for interlaced video compression as in video cameras. MPEG-2 coding algorithm is based on general hybrid DCT/DPCM (differential pulse code modulation) coding scheme shown in Fig. 3.20. It incorporates a macroblock structure, motion compensation, and coding for conditional replenishment of macroblocks. It uses the same technique of I, B, and P frames. MPEG-2 has additional $Y{:}U{:}V$ luminance and chrominance sampling ratios to assist video applications with the highest video quality. Hence, to the 4:2:0 MPEG-1 sampling format were added the 4:2:2 format, suitable for studio video coding applications, and the 4:4:4 format, which results in a higher video quality, i.e., images with better SNR performance and higher resolution.

Basically, the MPEG-2 compression process is similar to MPEG-1. The first frame in the video sequence is I, which is encoded as an intra mode with no reference. DCT is applied at the encoder on each $8 \times 8$ luminance and chrominance block. After the DCT process, the 64 coefficients of the DCT are uniformly quantized at the Q block. The quantizer step size (SZ) is used to quantize the DCT coefficients within a macroblock that is transmitted to the receiver. After quantization, the lowest DCT frequency coefficient (dc) has special processing, differing it from other ac coefficients. The dc coefficient refers to average intensity of the component block and is encoded by using differential dc prediction method. The nonzero ac quantizer values of the remaining DCT coefficients and their locations are then zigzag-scanned and passed through length-entropy coding using variable length code (VLC) tables.

The decoder has a feedback system. First it extracts and decodes the variable length coded words (VLD) from the beat-stream to find locations and quantizer values of the nonzero DCT coefficients for each block. Then it uses the reconstruction block marked as Q* for all nonzero DCT coefficients belonging to the same clock. These subsequently pass through an inverse DCT ($DCT^{-1}$) and the quantizer block pixels are obtained. By processing the entire bit-stream, all image blocks are recovered. When coding P image, the previous ($N - 1$) I or P image is stored in the frame-store (FS) memory in both the decoder and the encoder. The motion compensation (MC) is performed on a macroblock basis. Only one motion vector between the ($N$th $- 1$) frame and the current $N$th frame is estimated and encoded for a particular macroblock. The motion vectors are coded and transmitted to the receiver. Then the motion compensated prediction error is calculated by subtracting each PEL to PEL in the macroblock comparing $N$ to $N - 1$. An $8 \times 8$ DCT is applied to each of the $8 \times 8$ blocks contained in the macroblock, and then a quantization (Q) of the DCT coefficients is done, followed by a VLC entropy coding. A video buffer (VB) is needed to ensure constant bit rate at the encoder output. The quantization SZ is adjusted for each macroblock in a frame to achieve and maintain the targeted bit-stream rate and to prevent the VB overflow or underflow.

The decoder uses a reverse process to produce a macroblock of the $N$th frame in the receiver side.

From above encoding–decoding discussion, it is obvious that the bottleneck of the process will be the memory-buffer capacity of the VB block. MPEG-2 committee understood that a universal compression system capable of fulfilling and meeting the requirements of every application was an unrealistic goal. Hence, the solution was to define several operation profiles and levels that create some degree of commonality and compatibility among them, as provided by Table 3.3.

Even though the MPEG-2 profiles were defined, there are still 12 compliances. In Table 3.3, the CCIR sampling notation of chrominance with respect to luminance, frame type used, resolution, and data rate and abbreviation for commonly used levels and profiles are provided. Table 3.4 is a subtable summarizing the characteristics of the MPEG-2 profiles.[4]

The MPEG2 bounds presented in Table 3.3 are the upper bounds of performance but except VBV size which is given in Table 3.4. The use of B frames increases the resolution of motion between I and P frames. For given GOP containing 12 frames per 0.4 sec, and due to the flexibility of the rearrangement of I, P, and B frames there is very low correlation between compression ratio and picture quality; hence, quality is preserved.

MP@ML is a very good way to distribute video; however, it had some weakness in postproduction. The $720 \times 480$ and $720 \times 526$ sampling structures defined for ML (main level) disregarded the fact that there are 486 active lines for 525 lines in NTSC and 625 lines in PAL.

By the possible exception of transition cuts and overlay limits, lossy compressed video could not be postprocessed, i.e., zoomed, rotated, or resized while in its compressed state. Tilt should be decoded first to some baseband per ITU-R601. Without specific decoders and encoders that have the ability by design to exchange information regarding previous compression operations, the MP@ML

**Table 3.3** MPEG-2 main profiles and levels.

| Level \ Profile | Simple | Main | 4:2:2 | SNR (scalable) | Spatial (scalable) | High |
|---|---|---|---|---|---|---|
| High | – | 4:2:0; 1920 × 1152; 80 MB/s; I, P, B; MP@HL | – | – | – | 4:2:0, 4:2:2; 1920 × 1152; 100 MB/s; I, P, B; HP@HL |
| High (1440) | – | 4:2:0; 1440 × 1152; 60 MB/s; I, P, B; MP@H14 | – | – | 4:2:0; 1440 × 1152; 60 MB/s; I, P, B; SSP@H14 | 4:2:0, 4:2:2; 1440 × 1152; 80 MB/s; I, P, B; HP@H14 |
| Main | 4:2:0; 720 × 576; 15 MB/s; I, P; SP@ML | 4:2:0; 720 × 576; 15 MB/s; I, P, B; MP@ML | 4:2:2; 720 × 608; 15 MB/s; I, P, B; 422P@ML | 4:2:0; 720 × 576; 15 MB/s; I, P, B; SNRSP @ML | – | 4:2:0, 4:2:2; 720 × 576; 20 MB/s; I, P, B; HP@ML |
| Low | – | 4:2:0; 352 × 288; 4 MB/s; I, P, B; MP@LL | – | 4:2:0; 352 × 288; 4 MB/s; I, P, B; SNRSP @LL | – | – |

quality deteriorates rapidly when 4:2:0 color sampling is repeatedly used while decoding and re-encoding during postproduction. Long GOP, with each frame heavily dependent on others in the group, makes editing a complex task. Hence, 15-MB/s MP@ML at its upper data rate limit makes it impossible to achieve good quality pictures in the case of short GOP having one or two frames. The 4:2:2 profile 422 P@ML (Table 3.3) utilizes 4:2:2 color sampling, which provides more robust reencoding. The maximum video lines are raised to 608 and the

**Table 3.4** MPEG-2 levels bounds for main profiles.

| MPEG parameter | Luma rate (samples/s) | VBV buffer size (bits) |
|---|---|---|
| MP@HL | 62668800 | 9781248 |
| MP@H14 | 47001600 | 7340032 |
| MP@ML | 10368000 | 1835008 |
| MP@LL | 3041280 | 475136 |

maximum data rate reaches 50 MB/s. Moving picture experts group (MPEG) 422 P@ML was used as a foundation for SMPTE-308M (SMPTE: Society of Motion Picture and Television), which is a compression standard used for HDTV. The outcome of this discussion shows that the use of B frames increases computational complexity, bandwidth, delay, and picture buffer size of encoded video. That is because some macroblocks require averaging between two macroblocks. At the worst case memory, BW is increased by an extra 16 MB/s for additional prediction. Hence, extra delay is introduced because backward prediction needs to be transmitted to the decoder prior to the decoding and display of the B frame. The extra picture buffer pushes the decoder DRAM memory requirements beyond the 1-MB threshold. Several companies such as AT&T and GI debated if there is any need for B frames at all. In 1991, GI introduced DigiCipher-II that supports full MPEG-2 main profile syntax as well as Dolby AC-3 audio compression algorithm.[19]

The MPEG-2 is then modulated by an 8VSB RF scheme and then delivered through the HFC access networks to the subscribers' home or curb (FTTx). Then it is decoded at the set top box (STB) and displayed on screen.

### 3.3.6 MPEG AC3 and 8VSB baseband processing versus QAM

The modulation scheme used for terrestrial[30] HDTV in the U.S. is 8VSB. The conceptual MPEG 8VSB sequence is given in Fig. 3.21.

MPEG-2 packet contains some additional tags and bits. For instance, MPEG-2 length is 188 bytes. The first byte in each packet is always the sync byte. When the MPEG-2 byte stream reaches the 8VSB exciter, these packets are synchronized to the internal circuits of the 8VSB exciter. Prior to any signal processing, the 8VSB exciter identifies starting points and ending points of each MPEG-2 data packet. After all the processes of receiving synchronization and identification of the MPEG-2 packet are completed, the sync byte is discarded.

The next step is data randomization, which is done at the data randomizer block. The purpose of this process is to have a spread spectrum response,

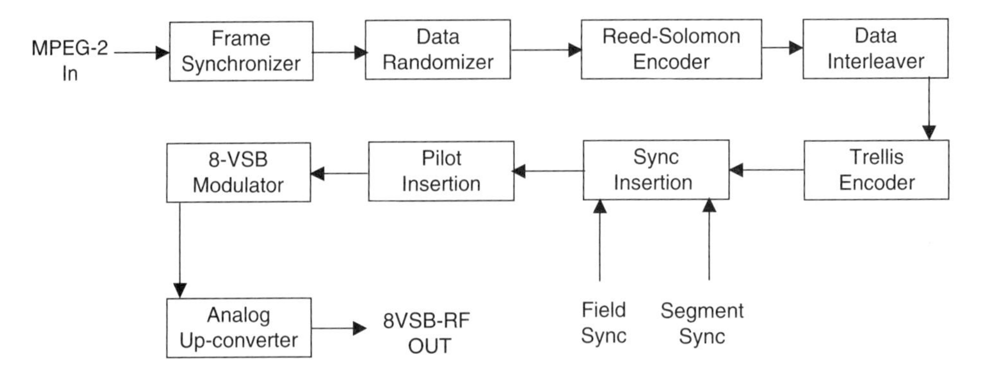


**Figure 3.21** 8-VSB exciter block diagram. In CATV case the pilot insertion block is not needed and the 8-VSB modulator is replaced by a 64QAM or 256QAM modulator.

making the data stream look like a random noise. This is done in order to have the maximal spectrum efficiency of the allotted RF channel. The randomization process is done by a known pseudo-random pattern; in the receiver side, there is also the same known pseudo-random pattern that recovers the spread spectrum data. Prior to data randomization, the modulated RF pattern would have energy concentration at certain points and holes on others. This is due to the reoccurring rhythms of the MPEG-2 patterns.

After randomization of the baseband data, it passes through Reed–Solomon encoding process, which is a forward error correction code (FEC). The FEC process is used to prevent and improve bit error rate (BER) due to the link interferences and nonidealities. These nonidealities may come from reflections, echoes, multipath, signal fading, and transmitter nonlinearities. The Reed–Solomon process captures the entire incoming MPEG-2 data packet after sync byte was removed, and mathematically manipulates them as a block, creating a kind of a digital "ID tag" of the block content. This ID tag occupies additional 20 bytes, which are linked to the end of the original 187-byte packet that came out of the MPEG-2 encoder. These 20 bytes are known as the Reed–Solomon parity bytes. At the DTV receiver side, the receivers compare the 187-byte block to the 20 parity bytes and determine the data validity. In case of an error at the receiver, the algorithm defines no relation between the Reed–Solomon parity to the data and searches for a similar packet most closely related to the ID tag. This is an error correction process. However, the Reed–Solomon error correction process is limited; the greater the discrepancy between the ID tag and the received packet, the greater the chance of inability to correct the error. In fact, the Reed–Solomon error correction process can correct up to 10 bytes of error per packet, which is about 5.3% of its length. In case of too many bytes of errors, the algorithm can no longer match the ID tag. The data and the entire MPEG-2 packet must be discarded; therefore, an additional redundancy called trellis coding is added.

Following the block diagram in Fig. 3.21, the next processing step is the "data-interleaving" done by the "data-interleaver" block. The data-interleaver scrambles the sequential order of the data stream and scatters the MPEG-2 packet data throughout time over a range of approximately 4.5 ms by using memory buffers. The main idea for this process is to increase the data immunity against burst-type interferences. The "data-interleaver" creates new smaller data packets, which contain small fragments from many MPEG-2 preinterleaved packets. These newly reconstructed data packets have the same length as the original MPEG-2 packets, i.e., 187 bytes plus the 20 bytes of the Reed–Solomon ID-tag. This process generates a time diversity since not the entire coded MPEG-2 is transmitted but only parts of it at a time; hence, in case of signal corruption by bursts of interfering noise, only some parts of the information are lost and the error correction done by Reed–Solomon can be processed on the receiver side.

After the "data-interleaving" process, an additional redundancy is added to the data recovery process by the trellis FEC. Trellis coding differs from Reed–Solomon in that the Reed–Solomon coding treats the entire MPEG-2 packet simultaneously as a block, while trellis coding is an evolving code that tracks the progression of the bit stream as it develops through time. Reed–Solomon is a block code, while trellis coding is a convolution code. The trellis coding algorithm splits each byte (8 bits) into a stream of 2 bits. The trellis coder compares each 2 bits arriving into their past history of the previous 2 bits. Then a 3-bit mathematical code is generated, describing the transition from the previous 2 bits to the current one. These 3 bits substitute the original 2 bits and are transmitted as eight level symbols of 8VSB since $2^3 = 8$. In other words, for each 2 bits coming into the trellis encoder, 3 bits come out. Hence, it is defined as $2/3$ rate encoder. On the receiver side, the trellis decoder uses the transition codes of the 3 received bits to reconstruct the evolution from 2 bits to the next 2 bits. In this manner, the trellis decoder tracks the trail, while the signal progresses from 2 bits to the next 2 bits. In this way, the trellis coding tracks the signal history throughout time and remove potential faulty information and errors, based on the signal's past and future behavior. When sometimes, the 3 bits are corrupted, it is impossible to connect between the past and the future of the signal. The trellis error correction coding algorithm will search for the most likely combination that connects between the past and the future of the 3-bit code.

After processing and encoding the MPEG-2 baseband, the next step is to insert signals to aid the HDTV modem to actually locate and demodulate the RF transmitted signal. These signals are Advanced Television Systems Committee (ATSC) pilot (for 8VSB) and segment sync. These guiding signals are inserted purposely after the Reed–Solomon, interleaving, and trellis coding in order to prevent any signal corruption of time and amplitude relationships that these signals should possess to be effective for detection.

In coherent digital demodulation, recovering a clock signal from the RF transmitted signal is essential. The data must be sampled by the receiver with accurate timing to have a proper recovery. The clock is generated accurately from the recovered data. Then, if the noise level rises to where a significant amount of errors occur, the whole clocking recovery crashes and data is lost. Therefore, the following concepts were implemented to overcome the data susceptibility to channel interferences, enabling the receiver to properly lock on the 8VSB signal and begin decoding. The trellis encoder in the HDTV system performs 12 symbols leapfrogging ahead to determine the next symbol transition by learning its future. Hence, symbol 0 is linked with 12, 24, 36, ..., symbol 1 is linked with 13, 25, 37, ..., symbol 2 is linked with 14, 26, 38, ..., and so on. This creates an additional form of interleaving, which adds an additional burst of noise protection. This method was designed to work well in conjugation with an NTSC interference rejection filter in the receiver, which uses a 12-symbol-tapped delay line. In addition, in NTSC there was a need for the amplitude of the sync pulse to be higher than the RF signal envelope (Fig. 3.5). This way, the receiver synchronization circuits could lock on the sync pulses and maintain the correct image framing even though the content of the picture was a bit snowy due to a poor carrier-to-noise ratio (CNR). In addition, the NTSC signal had the advantage of a large residual visual carrier, resulting from the dc component of the modulating video. This carrier is mixed with an LO with the same frequency and is used by the TV receiver tuners to zero in on the transmitted carrier. Similar concepts are adopted by the 8VSB for sync pulses and residual carriers, which

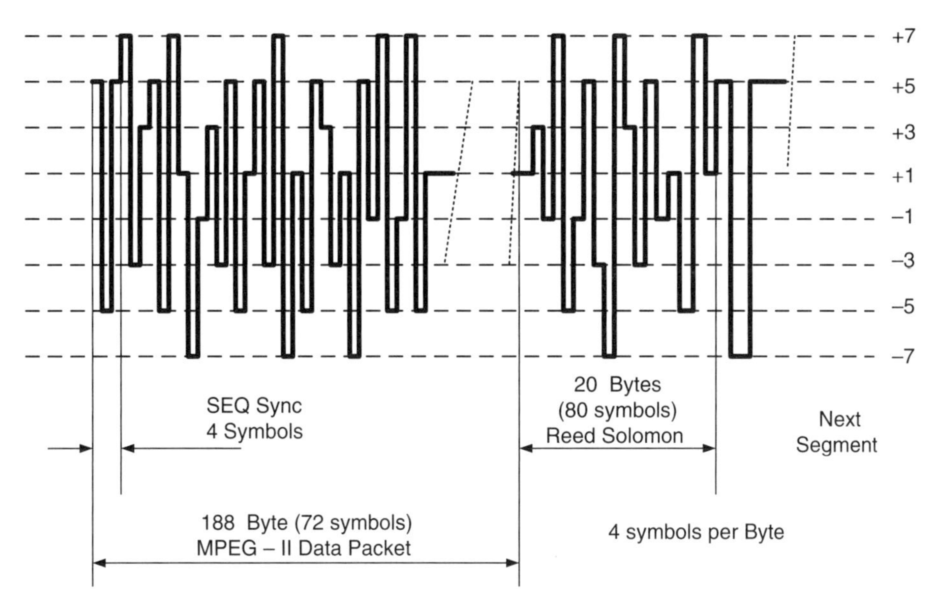


**Figure 3.22** ATSC 8-VSB baseband data segment structure.

enable the HDTV receiver's modem to lock on the incoming signal and start decoding even with the presence of heavy ghosting and poor CNR conditions. The ATSC pilot is inserted, then, by adding dc offset on both $I$ and $Q$ signals of the coded baseband prior to modulation (more details on Sec. 16.4.1). A carrier leakage at zero frequency of the baseband MASK results. This signal is the ATSC pilot. This carrier signal is used by the HDTV receiver RF PLL as reference signal to lock on independently without any relation to the signal itself. The ATSC pilot consumes only 0.3 dB of the transmitted power, which is about 7%.

The next processing stage is inserting the ATSC segment and field sync pulses. As was explained before, an ATSC data segment comprises 207 bytes or 1656 bits, which is equal to 828 symbols of 2 bits each that after trellis coding becomes 3 bits each with eight levels per symbol (hence the total segment size is 2484 bits). The ATSC segment sync is a four-symbol pulse that is attached to the front of each data segment and replaces the missing first byte, which is the packet sync of the original MPEG-2 data packet. As a consequence, the sync label will appear once every 832 symbols and will always have a transition from positive pulse swinging between the signal levels of +5 and −5 of the 8VSB levels (Fig. 3.22).

In the receiver side, 8VSB correlation circuits detect the repetitive characteristics of the segment sync, which is distinguished adjacent to the background of the pseudo-random data. The recovered segment sync is used to restore the system clock (CLK) and sample the received signal. Due to the sync high repetition frequency (similar to PRF, see Fig. 3.7), large signal swing from +5 to −5, which is a 5-level swing depth with an extended duration (similar to PW, see Fig. 3.7), making the segments easy to find. As a consequence, an accurate CLK

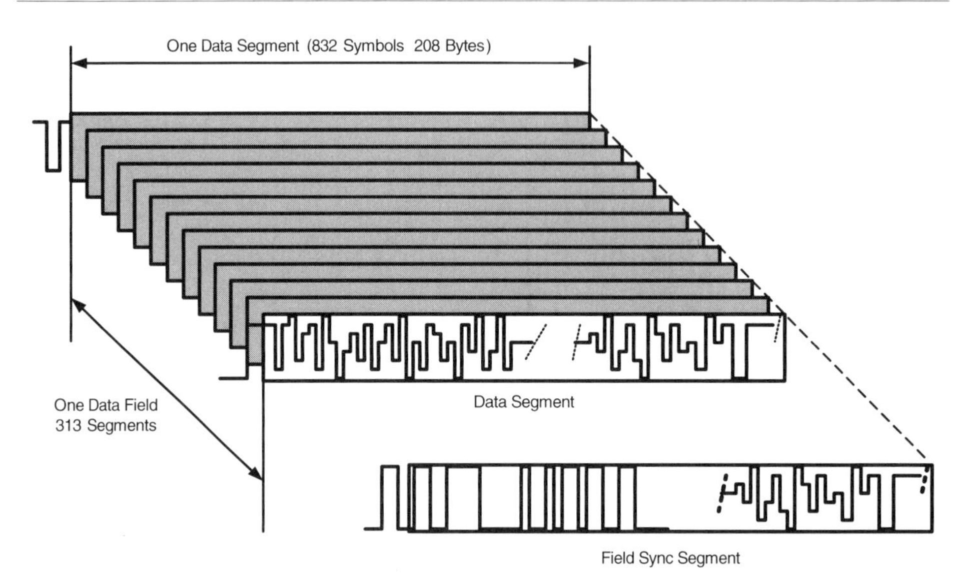


**Figure 3.23** HDTV field structure showing sync segment at the end of 313 data segments.

recovery is possible at high levels of noise interferences. Hence, data recovery is possible too. This robust approach enables detection at 0-dB SNR, while data recovery requires 15-dB SNR in case none of the above-mentioned coding was made. As a comparison to analog color NTSC TV, each ATSC segment sync PW equals to 0.37 μs, while NTSC sync duration is 4.7 μs. In addition, an ATSC data segment lasts 77.3 μs compared with NTSC line duration of 63.6 μs. Therefore, HDTV sync PW is narrower compared with that of the analog NTSC, while the segment duration is longer compared with the line period in the NTSC system. This way the active data payload is maximized, while the sync overhead time is minimized.

One field of HDTV contains 313 consecutive data segments as shown in Fig. 3.23. The ATSC field sync as a whole segment appears once every 314 segments, which is once every 24.2 ms. The field sync has a positive-to-negative pulse transition, which is a known pattern. This is used by the receiver's modem to eliminate shadow signals generated by poor reception. This is done by comparing the received field sync with errors of the known field sync that occurs prior to transmission. The resulting error vectors are used to adjust the receiver's shadow (ghost) canceling the equalizer. As the segment sync, the large signal swing and repetitive nature allow them to be successfully recovered at very high noise and interference levels (up to 0 dB SNR). The robustness of the segment and field sync permits accurate clock recovery and ghost cancelling is completely in the 8VSB receiver even when the active data payload is corrupted due to bad reception. However, ATSC syncs do not have any role in framing the displayed image on the CRT picture tube or plasma screen. The information is digitally coded as part of the MPEG algorithm. In addition, at the end of each field sync segment, the last 12 symbols of the data segment number 313 are repeated in

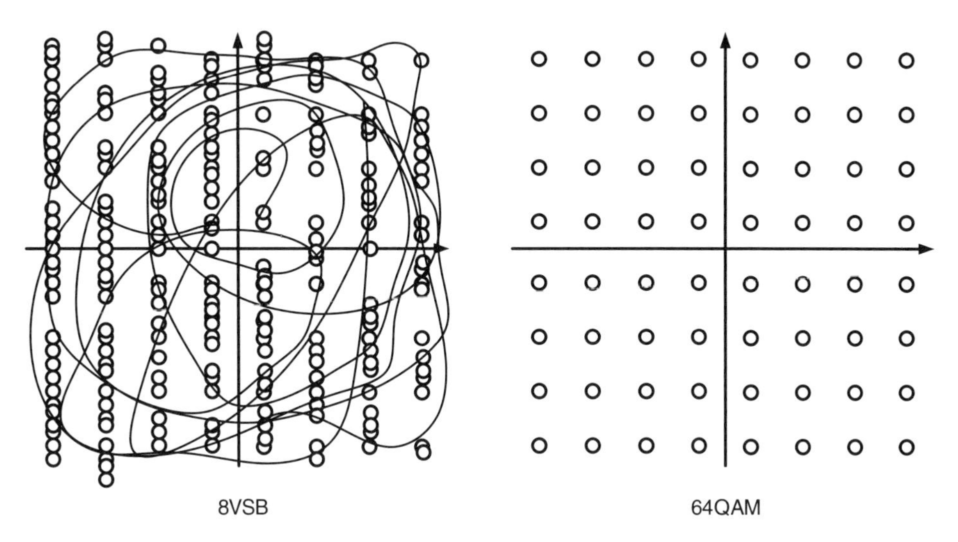


**Figure 3.24** Comparison between 8VSB and 64QAM constellation presentations.

order to restart the trellis decoder of the receiver for the first frame of the next field to come.

Since the next chapters deal with the QAM scheme only, which is adopted by the CATV, the differences between the QAM scheme and 8VSB are provided in this chapter. The FCC has designated 8VSB modulation as the terrestrial or air modulation scheme in the U.S.

The 8VSB is very similar to 64QAM and 16VSB is very similar to 256QAM, but in the VSB scheme only one phase of the carrier is used. There is no quadrature component and there is no use of QPSK modulator. As it is explained in Chapters 16 and 17, the QAM scheme has both $I$ and $Q$ vectors. Only the 8VSB uses the in-phase $I$ axis; however, both modulation schemes have eight levels, hence both modulation schemes' constellation presentation shows eight vertical lines. Since 8VSB has no phase information as in 64QAM, due to the lack of use of the $Q$ vector in 8VSB, the symbols are not arranged in a matrix as in the 64QAM. It is known that the QAM scheme has 64 constellation points, and the difference between QAM and VSB signals can be understood by the fact that eight levels are described by 3 bits. Therefore, in 8VSB there are 3 bits and in 64QAM there are 6 bits. There are $2^3$ states for the in-phase $I$ and $2^3$ states for quadrature $Q$ to indicate the phase information. In other words, 8VSB constellation has only one coordinate in the $I$, $Q$ field, thus at each vertical line the phase of the symbol is arbitrary, while 64QAM has two coordinates, which describe 64 ($2^3 \times 2^3 = 2^6 = 64$) combinations or locations in the constellation plane. In Fig. 3.24, the differences between 8VSB and 64QAM constellations are illustrated.

As was explained previously, in order to detect and synchronize the signal, in a 8VSB scheme an ATSC pilot is inserted for carrier recovery. In QAM, however, the carrier recovery method takes advantage of the symmetrical structure of its spectrum. Hence, by using a frequency doubler, the second harmonic of the carrier can be used. Further discussion on that technique is provided in Chapters

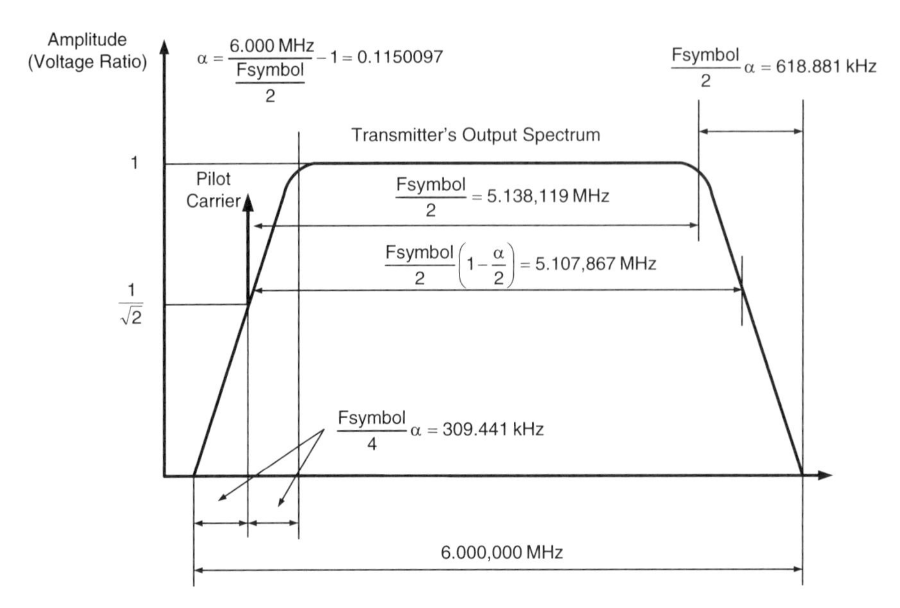


**Figure 3.25** 8VSB spectrum mask structure.[24] (Reprinted with permission from *IEEE Transactions on Broadcasting* © IEEE, 2003.)

16 and 17. Both modulation schemes have the same spectral efficiency. This results in a mask shape and BW. After the up-conversion process and filtering the adjacent undesired noise, it results in a minimum spectrum mask emission for a 8VSB scheme, which is illustrated in Fig. 3.25.

## 3.4 NTSC Frequency Plan and Minimum System Requirements

The CATV frequency plan in the U.S., provided in Table 3.7, is specified by the FCC. In Part 76, the FCC had assigned a channel plan in accordance with the channel set plan of the EIA. The nominal channel spacing is 6 MHz, except for 4 MHz frequency gap between channels 4 and 5. As explained previously, the appearance of color transmission on a spectrum analyzer shows that the video carrier is 1.25 MHz above the lower edge of the channel boundary (Fig. 3.15). This carrier is the main RF carrier. Therefore, the video carrier frequency is given by $f = (1.25 + 6N)$ MHz, where $N$ is the channel number. This frequency plan is called the standard (STD) plan. The visual video carrier for channels 4 and 5 are located 0.75 MHz below the 6 MHz multiples, i.e., 0.5 MHz above the lower edge of the channel boundary. In addition, the visual carrier in the following channel groups, 14–15, 25–41, and 43–53, has a 12.5 kHz frequency offset relative to the rest of the channels. The incrementally related carrier (IRC) frequency plan calls for $f = (1.625 + 6N)$ MHz, since the carriers there are located at 1.625 MHz above the lower edge of the channel boundary. The exceptions are channels 42, 60, and 61. The IRC plan has a

12.5-kHz frequency offset compared to the STD plan. This selection was done to minimize the interference in the 25-kHz radio links, which are used in airport control towers and aircraft navigation based on FCC regulations (FCC 76.612 Cable Television Frequency Separation Standards). The harmonically related carrier (HRC) plan, except for channels 60 and 61, calls for 6.0003-MHz multiplies. The reason for that is derived from the same motivation as in the IRC plan. The development and deployment of optical trunks opened competition for various vintages of equipment from a variety of cable equipment providers such as trunk amplifiers, optical-to-RF converting receivers, and return path equipment as well as splitters, filters, etc. Equipment is provided for various frequency ranges depending on the deployment and architecture of the cable services. Optical receivers cover the bands of 50–370, 50–470, 50–550, 50–870, and 50–1000 MHz. Other equipment covers the DBS range as well and provides the coverage of 50–870 and 950–2050 MHz in a single integration. The optical modulation index (OMI) plan is 3–4% for the channels in the frequency range between 50 and 550 MHz and half of the modulation index mentioned for channels in the frequency range between 550 and 870 MHz and DBS band (Fig. 3.26).

The above discussion provides test methodology and minimum pass or fail criteria for any analog CATV HFC receiver that converts optical information into multichannels of RF video signals. When multichannels are transmitted via a

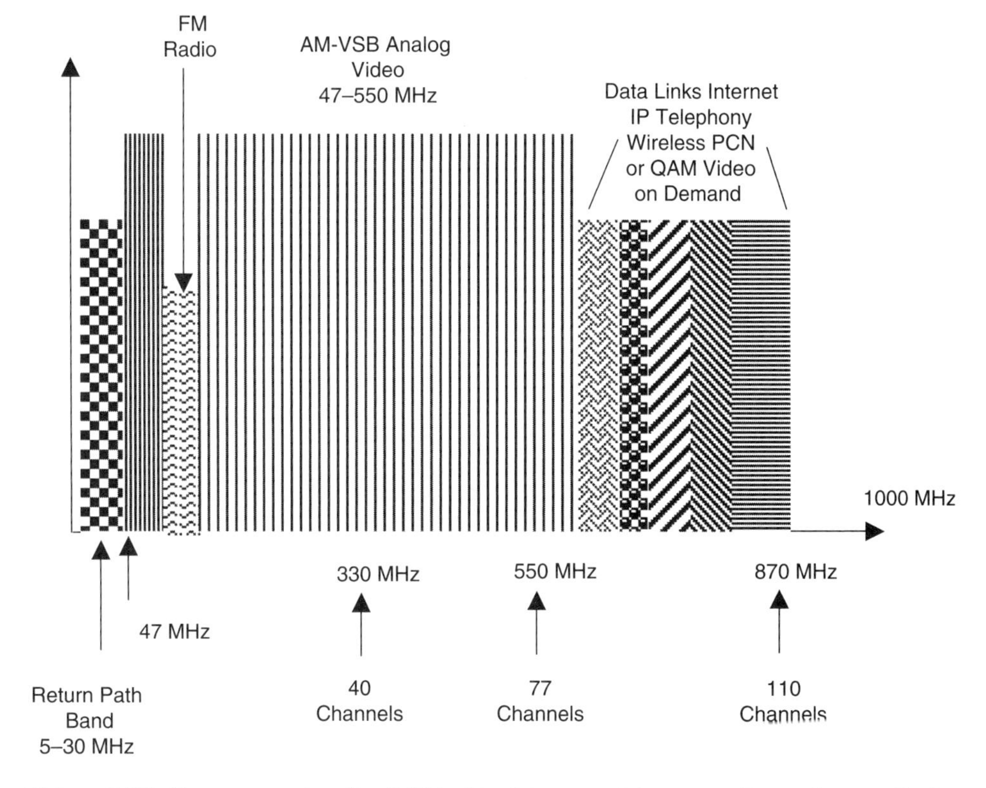


**Figure 3.26** Frequency plan for CATV signal transport incorporating return path for video-on-demand CATV channels and IP telephony.

nonlinear device such as a trunk amplifier and optical analog FFTx receiver, laser transmitter generates various nonlinear distortions (NLDs). Further discussion is provided in Chapters 9, 10, and 17. These discrete distortion products may degrade the television image on screen. The multichannel distortions are categorized into composite second order (CSO) and composite triple beat (CTB). The CTB products are similar to the familiar third-order two tones in a sense that the intermodulation (IMD) product is at the same frequency of the desired channel. In CATV channels, it means the CTB products will have the same frequency of the visual carrier as appears in Figs. 3.15(a) and (b). The visual carrier is the main carrier in the NTSC standard. Hence, a CTB distortion would act as a legal channel at the detection circuits and may take the image out of synchronization. This will appear as horizontal streaks, covering one or more lines of video. For CSO products, the worst distortion products are located at $\pm 0.75$ and $\pm 1.25$ MHz from the visual carrier. The low CSO distortions marked by any multitone tester as CSO_L are at the low side of the visual carrier. Therefore, they are out of band, since the 6-MHz filter starts to roll off, as shown in (Fig. 3.15). However, CSO marked as CSO_H are in band and are within the luminance energy band as shown in Fig. 3.15. CSO distortions generally appear as swimming diagonal strips in the TV picture. In addition, CNR/SNR affects image quality by creating "snow." The minimum standard is defined by the FCC[8] regulations governing the specifications for and testing of CTB CSO CNR/SNR as follows: Regulation Text for Technical Standard and Regulation Test for Measurements.

FCC 76.605 (a) (8) and 76.609 (f) define minimum acceptable performances for CSO/CTB:

- CSO/CTB distortions must be more than 51 dBc below the visual carrier level in systems where the carrier levels are not tied to a single synchronizing generator.
- CSO/CTB distortion products for IRC/ICC (ICC: incremental coherent carriers) systems need to be 47 dBc down.

It has been measured statistically that CTB can be perceived with little distortions as 57 dBc under ideal program and receiver conditions. These system specs drive the FTTx optical receiver to have a CSO/CTB requirement of 60 dBc as minimum standard over temperature, which means higher production margin of about 2 to 5 dB on top, bringing the specification level between 62 and 65 dBc for pass fail criteria for CSO/CTB and cross-modulation [X-mod FCC 76.605 (c) (8) and 76.609 (f)]. In Table 3.4, the NLD frequency combinations for CSO and CTB are provided.

FCC 76.605 (a) (7) and 76.609 (e) define minimum acceptable performances for CNR and weighted SNR. SNR is defined as the ratio between the peak luma carrier level at normal picture level of 714 mv or 100 IRE (Figs. 3.12 and 3.15) and the rms amplitude level of weighted noise contained in the signal. In the U.S., this measurement is performed by using a special luma weighted filter called Network Transmission Committee-7 (NTC-7), according to the CCIR recommendation 567.

- The requirements of CNR define the threshold of acceptable image and the compliance level is 43 dB. This is the weighted SNR. National cable television association (NCTA) requires that SNR minimum standard should be 53 dB.
- The CNR test is carried out at the output of the set top box converter and it is a system level test.
- Noise levels are tested at 4-MHz BW.

System wise, it has been measured statistically that a CNR required for acceptable image, the noise level should be 50 dB below the carrier at the output of an analog optical receiver at an optical level of −6 dBm and an OMI of 3.5–4%.

The above requirements ultimately define the approach to architecture when designing CATV optical receivers as will be discussed in Chapters 12–14. The testing recommendations that are defined by the American National Standards Institute (ANSI)/Society of cable telecommunications engineers (SCTE) (formerly SCTE IPS TP 206) are for composite distortion measurements.[21] The signal-to-noise tests are per ANSI/SCTE 17 2001 (formerly SCTE IPS TP 216) and the Test Procedure for Carrier to Noise (CNR, CCN, CIN, CTN).[22] These tests call for spectrum analyzer parameters as shown in Table 3.5. Additional correction factors are added for CNR tests as described in Ref. [22]. X-mod tests are defined in Ref. [23]. These tests are performed by using a multitone tester, where the measurement is done at the output of the FTTx optical receiver. These tests are the so-called RF tests and are described in Chapter 21. Table 3.6 provides the frequency combination of each distortion type and number of components for each term. An elaborate explanation for CSO CTB is provided in Section 9.6. Figure 3.27 provides the spectrum for the CSO and CTB products appearing

**Table 3.5** SCTE recommendations for spectrum analyzer settings for NLD and C/N tests.

| Parameter | Test conditions |
|---|---|
| Center frequency | Carrier frequency under test |
| Span | 3 MHz (300 kHz/div) |
| Detector | Peak |
| Resolution bandwidth | 30 kHz |
| Video bandwidth | 30 Hz |
| Input attenuation | ≥10 dB |
| Vertical scale | 10 dB/div |
| Log detect Rayleigh correction factor | Subtract 2.5 dB |
| Video filter shape factor correction | Subtract 0.52 dB |

**Table 3.6** Frequency combinations for CSO and CTB products.

| Distortion order | Distortion type term | Number of components per term | Term level above harmonic in dB |
|---|---|---|---|
| Second order | 2A | 1 | 0 |
| | A + B | 2 | 6 |
| | A − B | 2 | 6 |
| | 3A | 1 | 0 |
| | 2A + B | 3 | 9.54 |
| | 2A − B | 3 | 9.54 |
| Third order | A − 2B | 3 | 9.54 |
| | A + B − C | 6 | 15.56 |
| | A − B − C | 6 | 15.56 |
| | A + B + C | 6 | 15.56 |

Pay attention that IP3 terms of the type 2A ± B are lower by 6 dB compared with the triple beat terms of A ± B ± C. Further discussion is provided in Chapter 9.

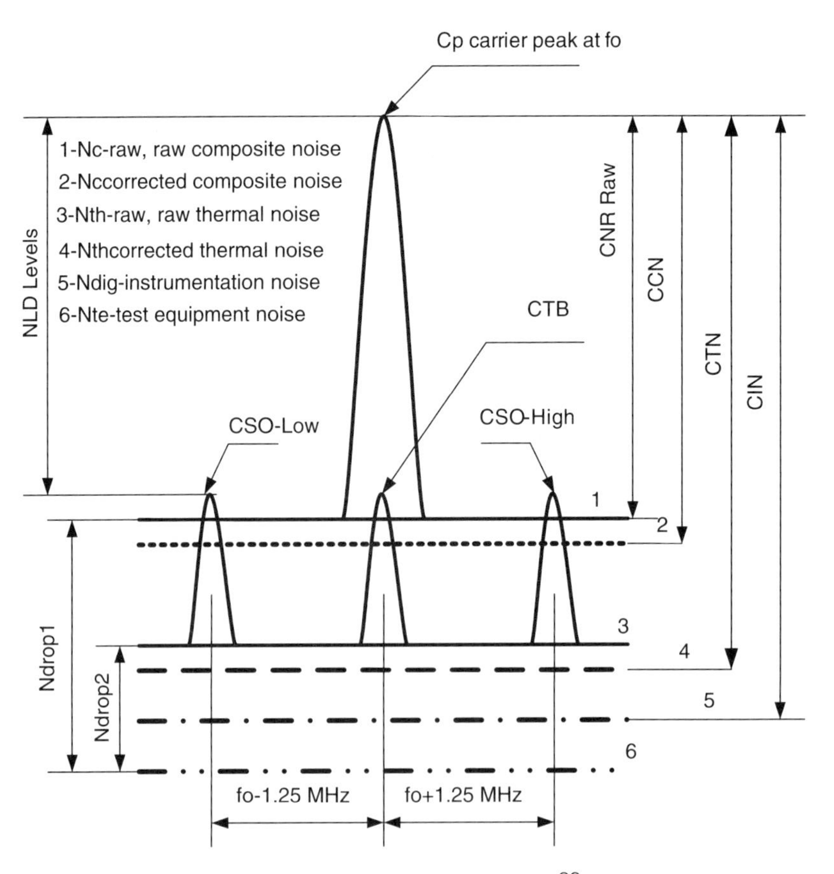


**Figure 3.27** CSO, CTB, CCN test on a spectrum analyzer.[22] (Reprinted with permission from *SCTE* © SCTE, 2001.)

in Table 3.6. The spectrum analyzer settings for CNR and NLD tests are provided by Table 3.5 and are described by the standard.[28] Additional tests are related to the video and signal performance as will be reviewed in Sec. 3.5.

## 3.5 Basic NTSC TV Signal Testing

In previous sections an introductory was provided about TV schemes standards and system RF level tests such as CSO, CTB, CNR, and X-mod. These tests provide information about system performance that may affect video quality. Additional tests refcrring the video baseband performance and video signals are needed as well to provide the entire information about the link. These tests are:

- differential gain marked as DG;
- differential phase marked as DP;
- chrominance to luminance delay inequality marked as CLDI; and
- hum noise.

### 3.5.1 Differential gain

Differential gain (DG) is defined as the difference in amplitude response at the color (chroma) subcarrier measured at 3.58 MHz (Fig. 3.15) as a result of a change of the luminance carrier level on which it rides. The source test signal used is a "staircase" signal generator with chroma added to risers. It consists of five steps in luminance *Y* covering the entire range from blanking to peak white. The *Y* steps are 20 IRE units each from zero to 100 IRE. To each step is added a color-burst phase, −(B − Y), which swings 40 IRE peak to peak (ptp), identical to the NTSC burst. In case of no DG distortion, the signal will last with all subcarrier packets at 20 IRE (ptp).

The signal can be seen on the monitor when the chroma filter is switched on. The filter removes the luminance structure on which the chroma signal is carried. Chroma packets should be within 0 to 100 IRE units. The standard calls for the measurement of the ratio between the difference of the largest chroma packet to the smallest chroma packet and the largest chroma packet. The ratio result is expressed in percent or decibels. According to FCC section 76.605, the maximum DG distortion value cannot exceed 20%. In the case of point-to-point terrestrial links, the NCTA[7] requirements for maximum DG is 15%. This value indicates the extent of drop in color saturation due to an increase in the luminance subcarrier power. The staircase signal range is up to 100 IRE units, which correspond to 12.5% modulation index as shown in Fig. 3.5. The reason for that minimum is to prevent a phenomena called "intercarrier buzz" in TV sets. This appears when one wears a white shirt in front of a camera in the studio. That is why staircase

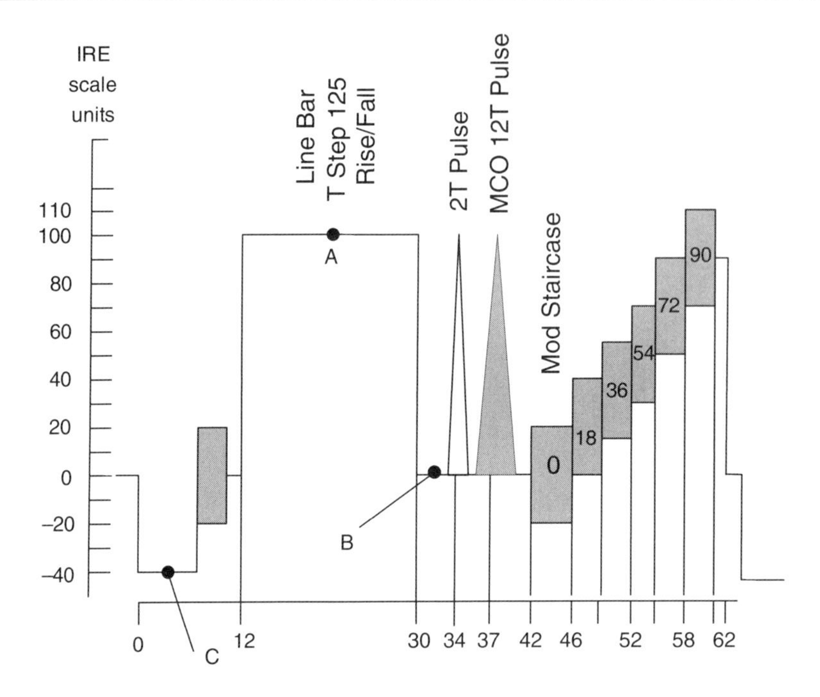


**Figure 3.28** FCC composite test signal during single horizontal scan. Vertical interval test signal (VITS) for ANSI T1.502, 1988 (NTS-7) is shown here with modulated staircase that reaches 110 IRE on chroma packets. The horizontal axis shows nominal timing in microseconds.[7] (Reprinted with permission from NCTA © NCTA, 1993.)

modulated signal is used with an amplitude modulator, i.e., TV and CATV, and the signal is held within 100 IRE. In Fig. 3.28, one vertical interval test signal (VITS), recognized as the FCC composite radiated signal, which is passed through an amplitude modulator, for VITS filed 1 line 18 is shown. The signal consists of half a line of the modulated staircase and half a line of pulse bar, i.e., two test signals at one line. However, careful observation shows that the top luminance step is 80 IRE, and 20 chroma peaks reach just up to 100 IRE. Hence, a standard amplitude-modulated video staircase will look almost white, with no saturation on a TV set at the fifth bar of the staircase signal.

Another VITS test signal was developed in 1988 by the NTC called standard EIA/TIA-250C and ANSI T 1.502. It is similar to FCC composite radiated signal, with the exception that the staircase and the pulse-bar signals are reversed in time sequence. In this case the *Y* top step is at 90 IRE and the peak subcarrier is at 110 IRE. This signal is used to check network transmission such as the transmission through satellites that have more dynamic range.

## 3.5.2 Differential phase

Differential phase (DP) is defined as a distortion of the color subcarrier phase (hue) depending on the luminance (*Y*) level on which the subcarrier is carried on. It shows up as the hue shift. The method involves using the same staircase test signal as was described in DG test and measuring the phase in a vector scope. In fact, a vector scope provides information about both DG and DP: DG is a change in the radius (magnitude), and the phase change refers to DP distortion. In case of no DP and DG, the vector scope will show only one single dot. According to FCC section 76.605, the DP for color carrier is measured at the largest phase difference between the chroma component and a reference component at 0 IRE, which is the blanking level. Eventually the largest phase shift will be at the fifth stair level. The maximum phase shift should not exceed $\pm 10$ deg.

### 3.5.3 Chrominance to luminance delay inequality

The chrominance to luminance delay inequality (CLDI) test measures the change in time delay of the chroma subcarrier signal with respect to the luminance $Y$ signal. The measurement is taken at the modulator output or processing unit at the cable head end. The minimum standard defined by FCC section 76.605 states that CLDI should be within 170 ns. The test is performed by inserting VITS signal prior to its reception at the cable TV head end.

### 3.5.4 Hum test

The definition of hum noise is that it results from the amplitude modulation of the carriers over the cable. As was explained before, the visual carrier in NTSC and PAL is amplitude modulated. The test is done by using a spectrum analyzer. It is caused by low frequency amplitude modulation and its measurement description is provided in Ref. [24]. The source of hum may be the poor filtering of power supply that results in a ripple over the dc bias at the frequency of 50 to 60 Hz and its harmonics. Another source may be a loose RF connector that forms a diode due to corrosion. Hum noise will generate rolling bars on the TV screen. The FCC standard section 76.605 (a) (10) states that the peak-to-peak variations caused by undesired low frequency disturbances, which are generated within the system or due to poor filtering and frequency response, should not exceed 3% of the visual signal level. Measurement is performed on a single channel by using nonmodulated carrier and checking the AM index as a function of the residual modulation on the dc input bias.

As a summary for the above discussion, it is important to note that any fiber-to-coax translator or receiver would have full compliance to the minimum standard. NLD distortions will result in some effects as explained in Sec. 3.4; however, the second-order effects may result in poor DG, DP, and CLDI. In addition to the traditional low-frequency sources, hum noise may also result from the instability of AGC circuits in the optical analog receiver. Since AGCs are narrow band feedback systems or feed forward system with hysteresis, improper design would result in a low frequency residual AM with the same effects as hum noise. Further discussion on AGC is given in Chapters 12 and 21 (Sec. 21.5).

**Table 3.7** CATV frequency plan CEA-542-B.[10] (Reprinted with permission of EIA/CEA © all rights reserved EIA/CEA 2002).

| | Channel edge | | | Analog carrier | | | Digital carrier | | | | | |
|---|---|---|---|---|---|---|---|---|---|---|---|---|
| CH | Low (MHz) (nominal, not HRC) | High (MHZ) | Note | STD (MHz) | IRC (MHz) | HRC (MHz) | STD 8/16VSB (MHz) | STD QAM (MHz) | IRC 8/16VSB (MHz) | IRC QAM (MHz) | HRC 8/16VSB (MHz) | HRC QAM (MHz) |
| 1 | 72.0000 | 76.0000 | | | 73.2625 | 72.0036 | | | 72.3100 | 75.0000 | 71.0600 | 73.7500 |
| 2 | 54.0000 | 60.0000 | | 55.2500 | 55.2625 | 54.0027 | 54.3100 | 57.0000 | 54.3100 | 57.0000 | 53.0600 | 55.7500 |
| 3 | 60.0000 | 66.0000 | | 61.2500 | 61.2625 | 60.0030 | 60.3100 | 63.0000 | 60.3100 | 63.0000 | 59.0600 | 61.7500 |
| 4 | 66.0000 | 72.0000 | | 67.2500 | 67.2625 | 66.0033 | 66.3100 | 69.0000 | 66.3100 | 69.0000 | 65.0600 | 67.7500 |
| 5 | 76.0000 | 82.0000 | | 77.2500 | 79.2625 | 78.0039 | 76.3100 | 79.0000 | 78.3100 | 81.000 | 77.0600 | 79.7500 |
| 6 | 82.0000 | 88.0000 | | 83.2500 | 85.2625 | 84.0042 | 82.3100 | 85.0000 | 84.3100 | 87.000 | 83.0600 | 85.7500 |
| 7 | 174.0000 | 180.0000 | | 175.2500 | 175.2625 | 174.0087 | 174.3100 | 177.0000 | 174.3100 | 177.0000 | 173.0600 | 175.7500 |
| 8 | 180.0000 | 186.0000 | | 181.2500 | 181.2625 | 180.0090 | 180.3100 | 183.0000 | 180.3100 | 183.0000 | 179.0600 | 181.7500 |
| 9 | 186.0000 | 192.0000 | | 187.2500 | 187.2625 | 186.0093 | 186.3100 | 189.0000 | 186.3100 | 189.0000 | 185.0600 | 187.7500 |
| 10 | 192.0000 | 198.0000 | | 193.2500 | 193.2625 | 192.0096 | 192.3100 | 195.0000 | 192.3100 | 195.0000 | 191.0600 | 193.7500 |
| 11 | 198.0000 | 204.0000 | | 199.2500 | 199.2625 | 198.0099 | 198.3100 | 201.0000 | 198.3100 | 201.0000 | 197.0600 | 199.7500 |
| 12 | 204.0000 | 210.0000 | | 205.2500 | 205.2625 | 204.0102 | 204.3100 | 207.0000 | 204.3100 | 207.0000 | 203.0600 | 205.7500 |
| 13 | 210.0000 | 216.0000 | | 211.2500 | 211.2625 | 210.0105 | 210.3100 | 213.0000 | 210.3100 | 213.0000 | 209.0600 | 211.7500 |
| 14 | 120.0000 | 126.0000 | | 121.2625 | 121.2625 | 120.0060 | 120.3100 | 123.0000 | 120.3100 | 123.0000 | 119.0600 | 121.7500 |
| 15 | 126.0000 | 132.0000 | | 127.2625 | 127.2625 | 126.0063 | 126.3100 | 129.0000 | 126.3100 | 129.0000 | 125.0600 | 127.7500 |
| 16 | 132.0000 | 138.0000 | | 133.2625 | 133.2625 | 132.0066 | 132.3100 | 135.0000 | 132.3100 | 135.0000 | 131.0600 | 133.7500 |
| 17 | 138.0000 | 144.0000 | | 139.2500 | 139.2625 | 138.0069 | 138.3100 | 141.0000 | 138.3100 | 141.0000 | 137.0600 | 139.7500 |
| 18 | 144.0000 | 150.0000 | | 145.2500 | 145.2625 | 144.0072 | 144.3100 | 147.0000 | 144.3100 | 147.0000 | 143.0600 | 145.7500 |
| 19 | 150.0000 | 156.0000 | | 151.2500 | 151.2625 | 150.0075 | 150.3100 | 153.0000 | 150.3100 | 153.0000 | 149.0600 | 151.7500 |
| 20 | 156.0000 | 162.0000 | | 157.2500 | 157.2625 | 156.0078 | 156.3100 | 159.0000 | 156.3100 | 159.0000 | 155.0600 | 157.7500 |
| 21 | 162.0000 | 168.0000 | | 163.2500 | 163.2625 | 162.0081 | 162.3100 | 165.0000 | 162.3100 | 165.0000 | 161.0600 | 163.7500 |
| 22 | 168.0000 | 174.0000 | | 169.2500 | 169.2625 | 168.0084 | 168.3100 | 171.0000 | 168.3100 | 171.0000 | 167.0600 | 169.7500 |
| 23 | 216.0000 | 222.0000 | | 217.2500 | 217.2625 | 216.0108 | 216.3100 | 219.0000 | 216.3100 | 219.0000 | 215.0600 | 217.7500 |
| 24 | 222.0000 | 228.0000 | | 223.2500 | 223.2625 | 222.0111 | 222.3100 | 225.0000 | 222.3100 | 225.0000 | 221.0600 | 223.7500 |
| 25 | 228.0000 | 234.0000 | | 229.2625 | 229.2625 | 228.0114 | 228.3100 | 231.0000 | 228.3100 | 231.0000 | 227.0600 | 229.7500 |
| 26 | 234.0000 | 240.0000 | | 235.2625 | 235.2625 | 234.0117 | 234.3100 | 237.0000 | 234.3100 | 237.0000 | 233.0600 | 235.7500 |

| | | | | | | | | | | | | |
|---|---|---|---|---|---|---|---|---|---|---|---|---|
| 27 | 240.0000 | 246.0000 | | 241.2625 | 241.2625 | 240.0120 | 240.3100 | 243.0000 | 240.3100 | 243.0000 | 239.0600 | 241.7500 |
| 28 | 246.0000 | 252.0000 | | 247.2625 | 247.2625 | 246.0123 | 246.3100 | 249.0000 | 246.3100 | 249.0000 | 245.0600 | 247.7500 |
| 29 | 252.0000 | 258.0000 | | 253.2625 | 253.2625 | 252.0126 | 252.3100 | 255.0000 | 252.3100 | 255.0000 | 251.0600 | 253.7500 |
| 30 | 258.0000 | 264.0000 | | 259.2625 | 259.2625 | 258.0129 | 258.3100 | 261.0000 | 258.3100 | 261.0000 | 257.0600 | 259.7500 |
| 31 | 264.0000 | 270.0000 | | 265.2625 | 265.2625 | 264.0132 | 264.3100 | 267.0000 | 264.3100 | 267.0000 | 263.0600 | 265.7500 |
| 32 | 270.0000 | 276.0000 | | 271.2625 | 271.2625 | 270.0135 | 270.3100 | 273.0000 | 270.3100 | 273.0000 | 269.0600 | 271.7500 |
| 33 | 276.0000 | 282.0000 | | 277.2625 | 277.2625 | 276.0138 | 276.3100 | 279.0000 | 276.3100 | 279.0000 | 275.0600 | 277.7500 |
| 34 | 282.0000 | 288.0000 | | 283.2625 | 283.2625 | 282.0141 | 282.3100 | 285.0000 | 282.3100 | 285.0000 | 281.0600 | 283.7500 |
| 35 | 288.0000 | 294.0000 | | 289.2625 | 289.2625 | 288.0144 | 288.3100 | 291.0000 | 288.3100 | 291.0000 | 287.0600 | 289.7500 |
| 36 | 294.0000 | 300.0000 | | 295.2625 | 295.2625 | 294.0147 | 294.3100 | 297.0000 | 294.3100 | 297.0000 | 293.0600 | 295.7500 |
| 37 | 300.0000 | 306.0000 | | 301.2625 | 301.2625 | 300.0150 | 300.3100 | 303.0000 | 300.3100 | 303.0000 | 299.0600 | 301.7500 |
| 38 | 306.0000 | 312.0000 | | 307.2625 | 307.2625 | 306.0153 | 306.3100 | 309.0000 | 306.3100 | 309.0000 | 305.0600 | 307.7500 |
| 39 | 312.0000 | 318.0000 | | 313.2625 | 313.2625 | 312.0156 | 312.3100 | 315.0000 | 312.3100 | 315.0000 | 311.0600 | 313.7500 |
| 40 | 318.0000 | 324.0000 | | 319.2625 | 319.2625 | 318.0159 | 318.3100 | 321.0000 | 318.3100 | 321.0000 | 317.0600 | 319.7500 |
| 41 | 324.0000 | 330.0000 | | 325.2625 | 325.2625 | 324.0162 | 324.3100 | 327.0000 | 324.3100 | 327.0000 | 323.0600 | 325.7500 |
| 42 | 330.0000 | 336.0000 | * | 331.2750 | 331.2750 | 330.0165 | 330.3100 | 333.0000 | 330.3100 | 333.0000 | 329.0600 | 331.7500 |
| 43 | 336.0000 | 342.0000 | | 337.2625 | 337.2625 | 336.0168 | 336.3100 | 339.0000 | 336.3100 | 339.0000 | 335.0600 | 337.7500 |
| 44 | 342.0000 | 348.0000 | | 343.2625 | 343.2625 | 342.0171 | 342.3100 | 345.0000 | 342.3100 | 345.0000 | 341.0600 | 343.7500 |
| 45 | 348.0000 | 354.0000 | | 349.2625 | 349.2625 | 348.0174 | 348.3100 | 351.0000 | 348.3100 | 351.0000 | 347.0600 | 349.7500 |
| 46 | 354.0000 | 360.0000 | | 355.2625 | 355.2625 | 354.0177 | 354.3100 | 357.0000 | 354.3100 | 357.0000 | 353.0600 | 355.7500 |
| 47 | 360.0000 | 366.0000 | | 361.2625 | 361.2625 | 360.0180 | 360.3100 | 363.0000 | 360.3100 | 363.0000 | 359.0600 | 361.7500 |
| 48 | 366.0000 | 372.0000 | | 367.2625 | 367.2625 | 366.0183 | 366.3100 | 369.0000 | 366.3100 | 369.0000 | 365.0600 | 367.7500 |
| 49 | 372.0000 | 378.0000 | | 373.2625 | 373.2625 | 372.0186 | 372.3100 | 375.0000 | 372.3100 | 375.0000 | 371.0600 | 373.7500 |
| 50 | 378.0000 | 384.0000 | | 379.2625 | 379.2625 | 378.0189 | 378.3100 | 381.0000 | 378.3100 | 381.0000 | 377.0600 | 379.7500 |
| 51 | 384.0000 | 390.0000 | | 385.2625 | 385.2625 | 384.0192 | 384.3100 | 387.0000 | 384.3100 | 387.0000 | 383.0600 | 385.7500 |
| 52 | 390.0000 | 396.0000 | | 391.2625 | 391.2625 | 390.0195 | 390.3100 | 393.0000 | 390.3100 | 393.0000 | 389.0600 | 391.7500 |
| 53 | 396.0000 | 402.0000 | | 397.2625 | 397.2625 | 396.0198 | 396.3100 | 399.0000 | 396.3100 | 399.0000 | 395.0600 | 397.7500 |
| 54 | 402.0000 | 408.0000 | | 403.2500 | 403.2625 | 402.0201 | 402.3100 | 405.0000 | 402.3100 | 405.0000 | 401.0600 | 403.7500 |
| 55 | 408.0000 | 414.0000 | | 409.2500 | 409.2625 | 408.0204 | 408.3100 | 411.0000 | 408.3100 | 411.0000 | 407.0600 | 409.7500 |
| 56 | 414.0000 | 420.0000 | | 415.2500 | 415.2625 | 414.0207 | 414.3100 | 417.0000 | 414.3100 | 417.0000 | 413.0600 | 415.7500 |
| 57 | 420.0000 | 426.0000 | | 421.2500 | 421.2625 | 420.0210 | 420.3100 | 423.0000 | 420.3100 | 423.0000 | 419.0600 | 421.7500 |
| 58 | 426.0000 | 432.0000 | | 427.2500 | 427.2625 | 426.0213 | 426.3100 | 429.0000 | 426.3100 | 429.0000 | 425.0600 | 427.7500 |
| 59 | 432.0000 | 438.0000 | | 433.2500 | 433.2625 | 432.0216 | 432.3100 | 435.0000 | 432.3100 | 435.0000 | 431.0600 | 433.7500 |

*(Continued)*

**Table 3.7** *Continued.*

| CH | Channel edge Low (MHz) | High (MHZ) (nominal, not HRC) | Note | Analog carrier STD (MHz) | Analog carrier IRC (MHz) | Analog carrier HRC (MHz) | Digital carrier STD 8/16VSB (MHz) | Digital carrier STD QAM (MHz) | Digital carrier IRC 8/16VSB (MHz) | Digital carrier IRC QAM (MHz) | Digital carrier HRC 8/16VSB (MHz) | Digital carrier HRC QAM (MHz) |
|---|---|---|---|---|---|---|---|---|---|---|---|---|
| 60 | 438.0000 | 444.0000 | | 439.2500 | 439.2625 | 438.0219 | 438.3100 | 441.0000 | 438.3100 | 441.0000 | 437.0600 | 439.7500 |
| 61 | 444.0000 | 450.0000 | | 445.2500 | 445.2625 | 444.0222 | 444.3100 | 447.0000 | 444.3100 | 447.0000 | 443.0600 | 445.7500 |
| 62 | 450.0000 | 456.0000 | | 451.2500 | 451.2625 | 450.0225 | 450.3100 | 453.0000 | 450.3100 | 453.0000 | 449.0600 | 451.7500 |
| 63 | 456.0000 | 462.0000 | | 457.2500 | 457.2625 | 456.0228 | 456.3100 | 459.0000 | 456.3100 | 459.0000 | 455.0600 | 457.7500 |
| 64 | 462.0000 | 468.0000 | | 463.2500 | 463.2625 | 462.0231 | 462.3100 | 465.0000 | 462.3100 | 465.0000 | 461.0600 | 463.7500 |
| 65 | 468.0000 | 474.0000 | | 469.2500 | 469.2625 | 468.0234 | 468.3100 | 471.0000 | 468.3100 | 471.0000 | 467.0600 | 469.7500 |
| 66 | 474.0000 | 480.0000 | | 475.2500 | 475.2625 | 474.0237 | 474.3100 | 477.0000 | 474.3100 | 477.0000 | 473.0600 | 475.7500 |
| 67 | 480.0000 | 486.0000 | | 481.2500 | 481.2625 | 480.0240 | 480.3100 | 483.0000 | 480.3100 | 483.0000 | 479.0600 | 481.7500 |
| 68 | 486.0000 | 492.0000 | | 487.2500 | 487.2625 | 486.0243 | 486.3100 | 489.0000 | 486.3100 | 489.0000 | 485.0600 | 487.7500 |
| 69 | 492.0000 | 498.0000 | | 493.2500 | 493.2625 | 492.0246 | 492.3100 | 495.0000 | 492.3100 | 495.0000 | 491.0600 | 493.7500 |
| 70 | 498.0000 | 504.0000 | | 499.2500 | 499.2625 | 498.0249 | 498.3100 | 501.0000 | 498.3100 | 501.0000 | 497.0600 | 499.7500 |
| 71 | 504.0000 | 510.0000 | | 505.2500 | 505.2625 | 504.0252 | 504.3100 | 507.0000 | 504.3100 | 507.0000 | 503.0600 | 505.7500 |
| 72 | 510.0000 | 516.0000 | | 511.2500 | 511.2625 | 510.0255 | 510.3100 | 513.0000 | 510.3100 | 513.0000 | 509.0600 | 511.7500 |
| 73 | 516.0000 | 522.0000 | | 517.2500 | 517.2625 | 516.0258 | 516.3100 | 519.0000 | 516.3100 | 519.0000 | 515.0600 | 517.7500 |
| 74 | 522.0000 | 528.0000 | | 523.2500 | 523.2625 | 522.0261 | 522.3100 | 525.0000 | 522.3100 | 525.0000 | 521.0600 | 523.7500 |
| 75 | 528.0000 | 534.0000 | | 529.2500 | 529.2625 | 528.0264 | 528.3100 | 531.0000 | 528.3100 | 531.0000 | 527.0600 | 529.7500 |
| 76 | 534.0000 | 540.0000 | | 535.2500 | 535.2625 | 534.0267 | 534.3100 | 537.0000 | 534.3100 | 537.0000 | 533.0600 | 535.7500 |
| 77 | 540.0000 | 546.0000 | | 541.2500 | 541.2625 | 540.0270 | 540.3100 | 543.0000 | 540.3100 | 543.0000 | 539.0600 | 541.7500 |
| 78 | 546.0000 | 552.0000 | | 547.2500 | 547.2625 | 546.0273 | 546.3100 | 549.0000 | 546.3100 | 549.0000 | 545.0600 | 547.7500 |
| 79 | 552.0000 | 558.0000 | | 553.2500 | 553.2625 | 552.0276 | 552.3100 | 555.0000 | 552.3100 | 555.0000 | 551.0600 | 553.7500 |
| 80 | 558.0000 | 564.0000 | | 559.2500 | 559.2625 | 558.0279 | 558.3100 | 561.0000 | 558.3100 | 561.0000 | 557.0600 | 559.7500 |
| 81 | 564.0000 | 570.0000 | | 565.2500 | 565.2625 | 564.0282 | 564.3100 | 567.0000 | 564.3100 | 567.0000 | 563.0600 | 565.7500 |
| 82 | 570.0000 | 576.0000 | | 571.2500 | 571.2625 | 570.0285 | 570.3100 | 573.0000 | 570.3100 | 573.0000 | 569.0600 | 571.7500 |
| 83 | 576.0000 | 582.0000 | | 577.2500 | 577.2625 | 576.0288 | 576.3100 | 579.0000 | 576.3100 | 579.0000 | 575.0600 | 577.7500 |
| 84 | 582.0000 | 588.0000 | | 583.2500 | 583.2625 | 582.0291 | 582.3100 | 585.0000 | 582.3100 | 585.0000 | 581.0600 | 583.7500 |
| 85 | 588.0000 | 594.0000 | | 589.2500 | 589.2625 | 588.0294 | 588.3100 | 591.0000 | 588.3100 | 591.0000 | 587.0600 | 589.7500 |

| | | | | | | | | | | | | |
|---|---|---|---|---|---|---|---|---|---|---|---|---|
| 86 | 594.0000 | 600.0000 | | 595.2500 | 595.2625 | 594.0297 | 594.3100 | 597.0000 | 594.3100 | 597.0000 | 593.0600 | 595.7500 |
| 87 | 600.0000 | 606.0000 | | 601.2500 | 601.2625 | 600.0300 | 600.3100 | 603.0000 | 600.3100 | 603.0000 | 599.0600 | 601.7500 |
| 88 | 606.0000 | 612.0000 | | 607.2500 | 607.2625 | 606.0303 | 606.3100 | 609.0000 | 606.3100 | 609.0000 | 605.0600 | 607.7500 |
| 89 | 612.0000 | 618.0000 | | 613.2500 | 613.2625 | 612.0306 | 612.3100 | 615.0000 | 612.3100 | 615.0000 | 611.0600 | 613.7500 |
| 90 | 618.0000 | 624.0000 | | 619.2500 | 619.2625 | 618.0309 | 618.3100 | 621.0000 | 618.3100 | 621.0000 | 617.0600 | 619.7500 |
| 91 | 624.0000 | 630.0000 | | 625.2500 | 625.2625 | 624.0312 | 624.3100 | 627.0000 | 624.3100 | 627.0000 | 623.0600 | 625.7500 |
| 92 | 630.0000 | 636.0000 | | 631.2500 | 631.2625 | 630.0315 | 630.3100 | 633.0000 | 630.3100 | 633.0000 | 629.0600 | 631.7500 |
| 93 | 636.0000 | 642.0000 | | 637.2500 | 637.2625 | 636.0318 | 636.3100 | 639.0000 | 636.3100 | 639.0000 | 635.0600 | 637.7500 |
| 94 | 642.0000 | 648.0000 | | 643.2500 | 643.2625 | 642.0321 | 642.3100 | 645.0000 | 642.3100 | 645.0000 | 641.0600 | 643.7500 |
| 95 | 90.0000 | 96.0000 | | 91.2500 | 91.2625 | 90.0045 | 90.3100 | 93.0000 | 90.3100 | 93.0000 | 89.0600 | 91.7500 |
| 96 | 96.0000 | 102.0000 | | 97.2500 | 97.2625 | 96.0048 | 96.3100 | 99.0000 | 96.3100 | 99.0000 | 95.0600 | 97.7500 |
| 97 | 102.0000 | 108.0000 | | 103.2500 | 103.2625 | 102.0051 | 102.3100 | 105.0000 | 102.3100 | 105.0000 | 101.0600 | 103.7500 |
| 98 | 108.0000 | 114.0000 | * | 109.2750 | 109.2750 | 108.0250 | 108.3100 | 111.0000 | 108.3100 | 111.0000 | 107.0600 | 109.7500 |
| 99 | 114.0000 | 120.0000 | * | 115.2750 | 115.2750 | 114.0250 | 114.3100 | 117.0000 | 114.3100 | 117.0000 | 113.0600 | 115.7500 |
| 100 | 648.0000 | 654.0000 | | 649.2500 | 649.2625 | 648.0324 | 648.3100 | 651.0000 | 648.3100 | 651.0000 | 647.0600 | 649.7500 |
| 101 | 654.0000 | 660.0000 | | 655.2500 | 655.2625 | 654.0327 | 654.3100 | 657.0000 | 654.3100 | 657.0000 | 653.0600 | 655.7500 |
| 102 | 660.0000 | 666.0000 | | 661.2500 | 661.2625 | 660.0330 | 660.3100 | 663.0000 | 660.3100 | 663.0000 | 659.0600 | 661.7500 |
| 103 | 666.0000 | 672.0000 | | 667.2500 | 667.2625 | 666.0333 | 666.3100 | 669.0000 | 666.3100 | 669.0000 | 665.0600 | 667.7500 |
| 104 | 672.0000 | 678.0000 | | 673.2500 | 673.2625 | 672.0336 | 672.3100 | 675.0000 | 672.3100 | 675.0000 | 671.0600 | 673.7500 |
| 105 | 678.0000 | 684.0000 | | 679.2500 | 679.2625 | 678.0339 | 678.3100 | 681.0000 | 678.3100 | 681.0000 | 677.0600 | 679.7500 |
| 106 | 684.0000 | 690.0000 | | 685.2500 | 685.2625 | 684.0342 | 684.3100 | 687.0000 | 684.3100 | 687.0000 | 683.0600 | 685.7500 |
| 107 | 690.0000 | 696.0000 | | 691.2500 | 691.2625 | 690.0345 | 690.3100 | 693.0000 | 690.3100 | 693.0000 | 689.0600 | 691.7500 |
| 108 | 696.0000 | 702.0000 | | 697.2500 | 697.2625 | 696.0348 | 696.3100 | 699.0000 | 696.3100 | 699.0000 | 695.0600 | 697.7500 |
| 109 | 702.0000 | 708.0000 | | 703.2500 | 703.2625 | 702.0351 | 702.3100 | 705.0000 | 702.3100 | 705.0000 | 701.0600 | 703.7500 |
| 110 | 708.0000 | 714.0000 | | 709.2500 | 709.2625 | 708.0354 | 708.3100 | 711.0000 | 708.3100 | 711.0000 | 707.0600 | 709.7500 |
| 111 | 714.0000 | 720.0000 | | 715.2500 | 715.2625 | 714.0357 | 714.3100 | 717.0000 | 714.3100 | 717.0000 | 713.0600 | 715.7500 |
| 112 | 720.0000 | 726.0000 | | 721.2500 | 721.2625 | 720.0360 | 720.3100 | 723.0000 | 720.3100 | 723.0000 | 719.0600 | 721.7500 |
| 113 | 726.0000 | 732.0000 | | 727.2500 | 727.2625 | 726.0363 | 726.3100 | 729.0000 | 726.3100 | 729.0000 | 725.0600 | 727.7500 |
| 114 | 732.0000 | 738.0000 | | 733.2500 | 733.2625 | 732.0366 | 732.3100 | 735.0000 | 732.3100 | 735.0000 | 731.0600 | 733.7500 |
| 115 | 738.0000 | 744.0000 | | 739.2500 | 739.2625 | 738.0369 | 738.3100 | 741.0000 | 738.3100 | 741.0000 | 737.0600 | 739.7500 |
| 116 | 744.0000 | 750.0000 | | 745.2500 | 745.2625 | 744.0372 | 744.3100 | 747.0000 | 744.3100 | 747.0000 | 743.0600 | 745.7500 |
| 117 | 750.0000 | 756.0000 | | 751.2500 | 751.2625 | 750.0375 | 750.3100 | 753.0000 | 750.3100 | 753.0000 | 749.0600 | 751.7500 |

(Continued)

**Table 3.7** *Continued.*

| CH | Channel edge Low (MHz) (nominal, not HRC) | Channel edge High (MHZ) (nominal, not HRC) | Note | Analog carrier STD (MHz) | Analog carrier IRC (MHz) | Analog carrier HRC (MHz) | Digital carrier STD 8/16VSB (MHz) | Digital carrier STD QAM (MHz) | Digital carrier IRC 8/16VSB (MHz) | Digital carrier IRC QAM (MHz) | Digital carrier HRC 8/16VSB (MHz) | Digital carrier HRC QAM (MHz) |
|---|---|---|---|---|---|---|---|---|---|---|---|---|
| 118 | 756.0000 | 762.0000 | | 757.2500 | 757.2625 | 756.0378 | 756.3100 | 759.0000 | 756.3100 | 759.0000 | 755.0600 | 757.7500 |
| 119 | 762.0000 | 768.0000 | | 763.2500 | 763.2625 | 762.0381 | 762.3100 | 765.0000 | 762.3100 | 765.0000 | 761.0600 | 763.7500 |
| 120 | 768.0000 | 774.0000 | | 769.2500 | 769.2625 | 768.0384 | 768.3100 | 771.0000 | 768.3100 | 771.0000 | 767.0600 | 769.7500 |
| 121 | 774.0000 | 780.0000 | | 775.2500 | 775.2625 | 774.0387 | 774.3100 | 777.0000 | 774.3100 | 777.0000 | 773.0600 | 775.7500 |
| 122 | 780.0000 | 786.0000 | | 781.2500 | 781.2625 | 780.0390 | 780.3100 | 783.0000 | 780.3100 | 783.0000 | 779.0600 | 781.7500 |
| 123 | 786.0000 | 792.0000 | | 787.2500 | 787.2625 | 786.0393 | 786.3100 | 789.0000 | 786.3100 | 789.0000 | 785.0600 | 787.7500 |
| 124 | 792.0000 | 798.0000 | | 793.2500 | 793.2625 | 792.0396 | 792.3100 | 795.0000 | 792.3100 | 795.0000 | 791.0600 | 793.7500 |
| 125 | 798.0000 | 804.0000 | | 799.2500 | 799.2625 | 798.0399 | 798.3100 | 801.0000 | 798.3100 | 801.0000 | 797.0600 | 799.7500 |
| 126 | 804.0000 | 810.0000 | | 805.2500 | 805.2625 | 804.0402 | 804.3100 | 807.0000 | 804.3100 | 807.0000 | 803.0600 | 805.7500 |
| 127 | 810.0000 | 816.0000 | | 811.2500 | 811.2625 | 810.0405 | 810.3100 | 813.0000 | 810.3100 | 813.0000 | 809.0600 | 811.7500 |
| 128 | 816.0000 | 822.0000 | | 817.2500 | 817.2625 | 816.0408 | 816.3100 | 819.0000 | 816.3100 | 819.0000 | 815.0600 | 817.7500 |
| 129 | 822.0000 | 828.0000 | | 823.2500 | 823.2625 | 822.0411 | 822.3100 | 825.0000 | 822.3100 | 825.0000 | 821.0600 | 823.7500 |
| 130 | 828.0000 | 834.0000 | | 829.2500 | 829.2625 | 828.0414 | 828.3100 | 831.0000 | 828.3100 | 831.0000 | 827.0600 | 829.7500 |
| 131 | 834.0000 | 840.0000 | | 835.2500 | 835.2625 | 834.0417 | 834.3100 | 837.0000 | 834.3100 | 837.0000 | 833.0600 | 835.7500 |
| 132 | 840.0000 | 846.0000 | | 841.2500 | 841.2625 | 840.0420 | 840.3100 | 843.0000 | 840.3100 | 843.0000 | 839.0600 | 841.7500 |
| 133 | 846.0000 | 852.0000 | | 847.2500 | 847.2625 | 846.0423 | 846.3100 | 849.0000 | 846.3100 | 849.0000 | 845.0600 | 847.7500 |
| 134 | 852.0000 | 858.0000 | | 853.2500 | 853.2625 | 852.0426 | 852.3100 | 855.0000 | 852.3100 | 855.0000 | 851.0600 | 853.7500 |
| 135 | 858.0000 | 864.0000 | | 859.2500 | 859.2625 | 858.0429 | 858.3100 | 861.0000 | 858.3100 | 861.0000 | 857.0600 | 859.7500 |
| 136 | 864.0000 | 870.0000 | | 865.2500 | 865.2625 | 864.0432 | 864.3100 | 867.0000 | 864.3100 | 867.0000 | 863.0600 | 865.7500 |
| 137 | 870.0000 | 876.0000 | | 871.2500 | 871.2625 | 870.0435 | 870.3100 | 873.0000 | 870.3100 | 873.0000 | 869.0600 | 871.7500 |
| 138 | 876.0000 | 882.0000 | | 877.2500 | 877.2625 | 876.0438 | 876.3100 | 879.0000 | 876.3100 | 879.0000 | 875.0600 | 877.7500 |
| 139 | 882.0000 | 888.0000 | | 883.2500 | 883.2625 | 882.0441 | 882.3100 | 885.0000 | 882.3100 | 885.0000 | 881.0600 | 883.7500 |
| 140 | 888.0000 | 894.0000 | | 889.2500 | 889.2625 | 888.0444 | 888.3100 | 891.0000 | 888.3100 | 891.0000 | 887.0600 | 889.7500 |
| 141 | 894.0000 | 900.0000 | | 895.2500 | 895.2625 | 894.0447 | 894.3100 | 897.0000 | 894.3100 | 897.0000 | 893.0600 | 895.7500 |
| 142 | 900.0000 | 906.0000 | | 901.2500 | 901.2625 | 900.0450 | 900.3100 | 903.0000 | 900.3100 | 903.0000 | 899.0600 | 901.7500 |
| 143 | 906.0000 | 912.0000 | | 907.2500 | 907.2625 | 906.0453 | 906.3100 | 909.0000 | 906.3100 | 909.0000 | 905.0600 | 907.7500 |

| | | | | | | | | | | | | |
|---|---|---|---|---|---|---|---|---|---|---|---|---|
| 144 | 912.0000 | 918.0000 | | 913.2500 | 913.2625 | 912.0456 | 912.3100 | 915.0000 | 912.3100 | 915.0000 | 911.0600 | 913.7500 |
| 145 | 918.0000 | 924.0000 | + | 919.2500 | 919.2625 | 918.0459 | 918.3100 | 921.0000 | 918.3100 | 921.0000 | 917.0600 | 919.7500 |
| 146 | 924.0000 | 930.0000 | | 925.2500 | 925.2625 | 924.0462 | 924.3100 | 927.0000 | 924.3100 | 927.0000 | 923.0600 | 925.7500 |
| 147 | 930.0000 | 936.0000 | | 931.2500 | 931.2625 | 930.0465 | 930.3100 | 933.0000 | 930.3100 | 933.0000 | 929.0600 | 931.7500 |
| 148 | 936.0000 | 942.0000 | | 937.2500 | 937.2625 | 936.0468 | 936.3100 | 939.0000 | 936.3100 | 939.0000 | 935.0600 | 937.7500 |
| 149 | 942.0000 | 948.0000 | | 943.2500 | 943.2625 | 942.0471 | 942.3100 | 945.0000 | 942.3100 | 945.0000 | 941.0600 | 943.7500 |
| 150 | 948.0000 | 954.0000 | | 949.2500 | 949.2625 | 948.0474 | 948.3100 | 951.0000 | 948.3100 | 951.0000 | 947.0600 | 949.7500 |
| 151 | 954.0000 | 960.0000 | ++ | 955.2500 | 955.2625 | 954.0477 | 954.3100 | 957.0000 | 954.3100 | 957.0000 | 953.0600 | 955.7500 |
| 152 | 960.0000 | 966.0000 | ++ | 961.2500 | 961.2625 | 960.0480 | 960.3100 | 963.0000 | 960.3100 | 963.0000 | 959.0600 | 961.7500 |
| 153 | 966.0000 | 972.0000 | ++ | 967.2500 | 967.2625 | 966.0483 | 966.3100 | 969.0000 | 966.3100 | 969.0000 | 965.0600 | 967.7500 |
| 154 | 972.0000 | 978.0000 | | 973.2500 | 973.2625 | 972.0486 | 972.3100 | 975.0000 | 972.3100 | 975.0000 | 971.0600 | 973.7500 |
| 155 | 978.0000 | 984.0000 | | 979.2500 | 979.2625 | 978.0489 | 978.3100 | 981.0000 | 978.3100 | 981.0000 | 977.0600 | 979.7500 |
| 156 | 984.0000 | 990.0000 | | 985.2500 | 985.2625 | 984.0492 | 984.3100 | 987.0000 | 984.3100 | 987.0000 | 983.0600 | 985.7500 |
| 157 | 990.0000 | 996.0000 | | 991.2500 | 991.2625 | 990.0495 | 990.3100 | 993.0000 | 990.3100 | 993.0000 | 989.0600 | 991.7500 |
| 158 | 996.0000 | 1002.0000 | | 997.2500 | 997.2625 | 996.0498 | 996.3100 | 999.0000 | 996.3100 | 999.0000 | 995.0600 | 997.7500 |

*Excluded from comb due to FCC offset.

+Use of this channel for priority programming is not recommended. It is used as the second local oscillator frequency for some television sets. The possibility exists that local oscillator leakage from the set may cause interference to another TV viewing this channel. The interference may be independent of the channel to which the subject TV (i.e., the one containing the double conversion tuner) is tuned.

++Use of these channels for any programming is not encouraged. They are used as the first IF in some television sets. When such a set is tuned to any channel in this part of the spectrum, it may experience interference from carriers on these channels. If this occurs, the only solution may be to provide a bandstop filter tuned to these channels. Such a filter will, of necessity, remove several additional channels on either side of channels 151–153.

## 3.6 Main Points of this Chapter

1. There are three main standards in TV transmission: NTSC, PAL, and SECAM.

2. NTSC and PAL systems use AM for the video and FM for the sound. SECAM uses FM for the video color information and AM for the sound and luminance.

3. The reasons for using interlaced scanning are to save video bandwidth and to prevent picture flickering.

4. NTSC has 525 horizontal lines, and PAL and SECAM have 625 horizontal lines each. Active horizontal lines are 480 for NTSC and 580 for both PAL and SECAM.

5. The significant signal that defines a CATV feedback AGC bandwidth in an optical receiver is the vertical scan frequency sync signal, which is 50–60 Hz per field. Thus, the AGC bandwidth should be below 0.5 Hz. This rate of the vertical sync is called PRF.

6. VSB+C is the modulation type for the video signal in PAL and NTSC. It combines the benefits of DSB and SSB modulations. Its advantage is its spectrum efficiency. VSB demodulation is cheap and is done by a simple envelope detector.

7. The CATV frequency plan calls for an OMI level between 3% and 4% for the channels between 50 and 550 MHz and between 1.75% and 2% to the channels from 550 to 870 MHz.

8. CATV channel spacing is typically 6 MHz and the channel plans are typically with 110 channels in the frequency range of 50 to 870 MHz.

9. The three main colors in TV are RGB (red, green, and blue).

10. Horizontal scan color sync in NTSC is done by color burst on the back porch. Phase inversion of the $R - Y$ between alternating horizontal scans are used in PAL. In SECAM, $R - Y$ and $B - Y$ are transmitted alternately and are FM signals.

11. CATV minimum specifications call for 51-dBc for CSO and CTB, and 43-dB for CNR. Practically, minimum specifications call for a standard of 60 to 65-dBc for CSO and CTB, and 48- to 50-dB for CNR.

12. CSO_H are the most critical second-order distortions since they are located both within the channel BW at +0.75 MHz and at +1.25 MHz off the luminance carrier.

13. CSO_L are the least critical second-order distortions since they are located both at the low end of the channel BW at $-0.75$ MHz and at $-1.25$ MHz off the luminance carrier. Hence, they do not affect the previous channel signal, which is frequency-modulated sound signal.

14. CTB distortion signal drops at the same frequency of the luminance carrier and may cause loss of sync in case it has high enough in power.

15. IRE units are defined as 1 V = 140 IRE units.

16. NTSC BW is 6 MHz, visual carrier (luma $Y$) is at 1.25 MHz from the low BW edge, color (chroma) is at 4.38 MHz from the low BW edge, and the sound (aural) is at 5.75 MHz from the low BW edge.

17. Color information is modulated by a QPSK modulator having $I$ and $Q$ vectors.

18. Scanning process of an image is 2D Fourier transform.

19. HDTV uses MPEG-2 and AC3 for image and sound compression.

20. MPEG compression is done by a Fourier transform called DCT, where the low-energy, high-frequency pixels are dropped.

21. Down sampling is the first step prior to MPEG compression process. There are several down sampling schemes, 4:2:2, 4:1:1, and 4:2:0.

22. There are three types of frames in MPEG: I, intra, not predicted fundamental image; P, predicted; and B, bidirectional, can be predicted from I and P and used between I and P.

23. A sequence of I, B, and P images is called GOP, which stands for group of pictures.

24. Prior to HDTV transmission, MPEG2 passes through frame synchronizer, data randomizer, Reed–Solomon encoding, data interleaver, Trellis encoding, field and segment sync insertion, pilot insertion in case of 8VSB, and modulation (8VSB for terrestrial broadcast or 64QAM for CATV).

25. Trellis encoding is convolution coding and Reed–Solomon encoding is a block coding.

26. 8VSB modulation scheme corresponds to 3 bits or eight modulation levels, and 64QAM modulation scheme corresponds to 8 bits or 64 modulation levels.

27. Both 8VSB and 64QAM have similar spectrum efficiency.

28. The pilot in 8VSB assists the HDTV tuner to lock in quickly to a new channel.

29. HDTV contains 313 segments in a field plus one additional segment called the field sync segment.

30. Differential gain and differential phase are video tests checking the distortions affecting the color saturation and hue, respectively.

31. Hum noise can be a result of ac ripple leakage into the dc bias circuit, which results in a low frequency residual AM on the video signal, but can be also a result of the optical receiver's AGC oscillation, which is a low frequency oscillation.

32. Distortions in digital TV may reduce BER performance.

33. HDTV transmission is much more robust and can handle lower CNR for BER of $10^{-10}$. In 64QAM, CNR = 28 dB and for 256QAM, CNR = 35 dB.[6]

## References

1. Luther, A., and A. Inglis, *Video Engineering*. McGraw-Hill, New York, NY (1999).
2. Shrader, R.L., *Electronic Communication*. McGraw-Hill, New York, NY (1991).
3. Lathi, B.P., *Modern Digital and Analog Communication Systems*. Oxford University Press, USA (1998).
4. Ovadia, S., *Broadband Cable TV Access Networks*. Prentice Hall (1998).
5. Ciciora, W., J. Farmer, and D. Large, *Modern Cable Television Technology*. Morgan Kaufman Publishers, Inc. (1991).
6. Way, W.I., *Broadband Hybrid Fiber/Coax Access System Technologies*. Academic Press (1998).
7. *NCTA Recommended Practices for Measurement on Cable Television Systems, 2nd Edition*, National Cable Television Association, Washington D.C. (1993).
8. *Code of Federal Regulations, Title 47, Telecommunications*, Part-76, Cable Television Service, Federal Communications Commissions Rules and Regulations (1990).
9. Olshansky, R., "Options for video delivery," *OFC Tutorial* (1994).
10. Cable Television Channel Identification Plan, EIA/CEA-542-B (2003).
11. Carlson, A.B., *Communication Systems*. McGraw-Hill (1986).
12. Flaherty, J.A., "A perspective on digital TV and HDTV." HDTV Magazine, http://ilovehdtv.com/hdtv/flahertyhist.html.
13. Donnely, D.F., "HDTV standard setting: politics technology and industry." http://www.tfi.com/pubs/ntq/articles/view/95Q3_A4.pdf.

14. Sparano, D., "What exactly is 8-VSB, anyway?" http://www.broadcastpapers.com/tvtran/Harris8VSB-print.htm.
15. Kuhn, K.J., "An introduction to EE498." HDTV Television, http://www.ee.washington.edu/conselec/CE/kuhn/hdtv/95x5.htm.
16. Benson, K., and D. Fink, *HDTV: Advanced Television for the 1990s*, McGraw Hill, New York (1991).
17. *HP Cable Television System Measurement Handbook, NTSC Systems*, Hewlett Packard (1994).
18. MPEG.org, http://www.mpeg.org (2007).
19. Mitchell, J.L., W.B. Pennebaker, C.E. Fogg, and D.J. LeGall, "*MPEG Video Compression Standard*," Chapman and Hall, USA (1996).
20. Sgrignoli, G., "DTV repeater emission mask analysis." *IEEE Transactions on Broadcasting*, Vol. 49 No. 1, pp. 32–80 (2003).
21. Engineering Committee Interface Practices Subcommittee, Society of Cable Telecommunications Engineers, *Composite Distortion Measurements (CSO & CTB).* American National Standard ANSI/SCTE 06 (Formerly IPS TP 216) (1999).
22. Engineering Committee Interface Practices Subcommittee, Society of Cable Telecommunications Engineers. *Test Procedure for Carrier to Noise (C/N, CCN, CIN, CTN)*, American National Standard ANSI/SCTE 17 (Formerly IPS TP 216) (2001).
23. Engineering Committee Interface Practices Subcommittee, Society of Cable Telecommunications Engineers, *AM Cross Modulation Measurements.* American National Standard ANSI/SCTE 58 (Formerly IPS TP 208) (2003).
24. Society of Cable Telecommunications Engineers, Engineering Committee Interface Practices Subcommittee, *Test Procedure for Hum Modulation.* American National Standard ANSI/SCTE 16 (Formerly IPS TP 204) (2001).
25. Haskell, B.G., P.G. Howard, Y.A. LeCun, A. Puri, J. Ostermann, M.R. Civanlar, L. Rabiner, L. Bottou, and P. Haffner, "Image and video coding-emerging standards and beyond." *IEEE Transactions on Circuits and Systems for Video Technology*, Vol. 8 No. 7 (1998).
26. Advanced Television Systems Committee, *ATSC Standard, Digital Audio Compression (AC-3), Revision A, Doc.A/52A* (2001).
27. Patchett, G.N., *Colour Television: With Particular Reference to the PAL System, 3rd edition.* Norman Price, London (1974)
28. Thomas, J.L., *Cable Television: Proof of Performance.* Prentice Hall PTR (1995).
29. Ciciora, W., J. Farmer, and D. Large, *Modern Cable Television Technology: Voice, Video, and Data Communications.* Morgan Kaufmann Publishers, Inc. (2004).
30. Wu, Y., E. Pliszka, B. Caron, P. Bouchard, and G. Chouin, "Comparison of terrestrial DTV transmission system: the ATSC 8-VSB, the DVB-T COFDM and the ISDB-T BST-OFDM." *IEEE Transactions on Broadcasting*, Vol. 46 No. 2, pp. 101–113 (2000).

# Part 2

# Semiconductors and Passives

# Chapter 4

# Introduction to Optical Fibers and Passive Optical Fiber Components

Optical fibers are one of the latest additions to signal-link submillimeter-wavelength transmission line technology. Optical fibers are cylindrical waveguides with significantly low loss. Since frequencies of light are very high compared to that of microwaves, the bandwidth (BW) of an optical fiber is high, and high data rate transports can be delivered through it. The fundamental difference between a fiber and a coax is the material fiber versus metal and the nature of the signal: photons versus electrons. In coax, the information signal travels in the central conductor, whereas in the fiber, light travels in the core. In coax, the central conductor is surrounded by an insulation dielectric material and then by a conductive sleeve. In the fiber, the core is covered by a cladding, and the core refractive index and diameter define the type of the fiber. However, in both cases, the fiber and the coax can be used to transfer electromagnetic waves.[1,2] Similar to circular or rectangular waveguides used in microwaves, the fiber is a circular waveguide. Hence, propagation modes depend on the dimensions of the fiber. These modes are marked as $EH_{np}$ or $HE_{np}$, where $n$ and $p$ indices are the mode order of the Bessel equation solution.[1] There are three main types of optical fibers.

1. *Single mode fiber* (SMF) *step index*. Single mode fiber is a low-loss optical fiber commonly used for long haul. These fibers are characterized by a small core radius ranging from 1 to 16 μm. The differential refractive index between the core and the cladding is about 0.6%. These fibers are used with lasers only because their acceptance numerical aperture (NA) is narrow.

2. *Multimode step index fiber*. Multimode step index fibers have a large core diameter and a larger NA (acceptance cone), so light emitting diodes (LEDs) can be coupled efficiently into them. The core radius range is 25 to 60 μm. The differential in its refractive indices is from 1% to 10%. It

is a low-bit-rate fiber and low-BW short haul and is used for low-cost applications such as short links between computers.

3. *Multimode-graded index fiber*. Graded index fiber has a gradual monotonic decrease in the refractive index between the core center until it converges into a plateau at the cladding region. For the purposes of wide band performance, multimode fibers are designed and manufactured with graded index profiles in order to reduce the intermode dispersion. This type of fiber has a core radius range of 10 to 35 μm and a cladding radius range of 50 to 80 μm (Fig. 4.1).

In addition, passive components such as couplers, connectors, and filters and optical isolators, fiber Bragg grating (FBG), and wave division multiplexing (WDM), multiplexers (MUXs) will be reviewed. All these components have a major role in passive optical networks (PONs) deployment.

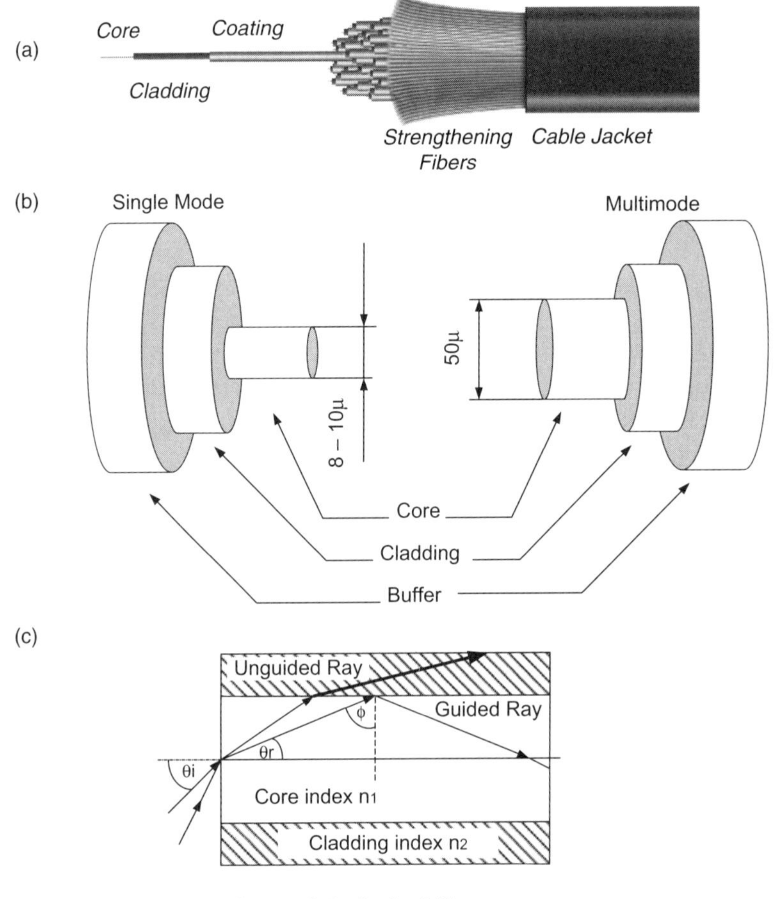


**Figure 4.1** Optical fiber types.

## 4.1 Single Mode Fiber

Single mode fiber is commonly used in community access television (CATV), quadrature amplitude modulation (QAM), amplitude modulation (AM), and vestigial side band (VSB) transport, as well as in high-speed data rate channels due to its lower loss and dispersion compared to multimode fibers. Group velocity dispersion (chromatic dispersion) is a primary cause of concern in high bit rate (>2.5 Gbps), single-mode WDM systems. Dispersion in an optical pulse creates pulse broadening such that the pulse spreads into the slots (in the time domain) and interferes with adjacent pulses. In addition to intersymbol interference (ISI), it also introduces a power penalty, which can cause degradation of the overall system's signal-to-noise ratio (SNR). In addition, even though it is a passive device, these characteristics introduce nonlinear effects when driven by multiwavelength. This can be analyzed by Volterra series.[3] The nature of SMF response versus optical wavelength is well demonstrated by Fig. 4.2.

Optical loss in commercial high-quality SMF is caused mainly by the Rayleigh scattering of the silica glass. The losses are inversely proportional to the fourth power of the wavelength and decrease as wavelength increases:[2]

$$\begin{aligned} P(L) &= P(0) \times 10^{-\alpha L/10} \\ P(L)[\text{dBm}] &= P(0)[\text{dBm}] - \alpha L, \end{aligned} \tag{4.1}$$

where $\alpha$ is given by

$$\alpha = \frac{C}{\lambda^4}, \tag{4.2}$$

where $C$ is a constant in the range of 0.7 to 0.9 (dB/Km)/$\mu$m$^4$.

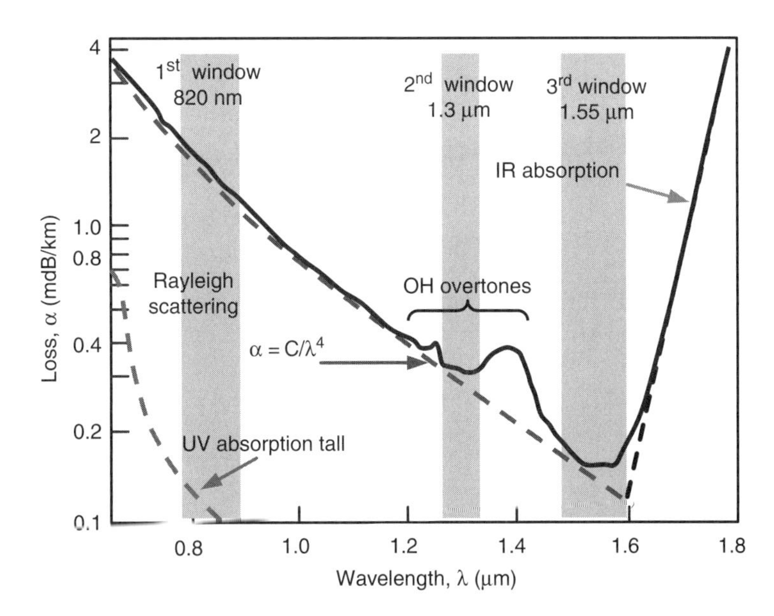


**Figure 4.2** Optical attenuation rate in dB/Km as a function of wavelength showing the three wavelength windows of operation.

In the logarithmic plot scale in Fig. 4.2, the Rayleigh asymptotic rate reaches the 1600-nm zone minimum where the intrinsic absorption dominates. This common minimum is used for the 1310 and 1550 nm wavelengths. The reason for Rayleigh scattering is because the inner core of the fiber is composed of glass with a slightly higher refraction index than the outer cladding. This phenomenon is because of local microscopic fluctuations in silica density. In silica, the molecules vibrate randomly during the fiber fabrication between solid and freeze states, which lead to refractive index fluctuations. Hence, while propagating through the fiber, some of the light is lost due to optical absorption that results from the glass impurities, scattering due to nonuniformities in the material, and bend losses. The intrinsic material absorption is divided into two categories. Losses are caused by pure silica, where extrinsic absorptions are related to losses generated by impurities. Silica molecule electronic resonance is in the ultraviolet region of $\lambda < 400$ nm and vibrational resonance occurs in the infrared region where $\lambda > 700$ nm. Due to the amorphous nature of fused silica, these absorptions are at the band edges, where its tail continues into the visible wavelength. Hence, short wavelengths can be used at the first window (Fig. 4.2). Extrinsic absorption results due to the presence of impurities such as Fe, Cu, Co, Ni, Mn, and Cr, where Cr absorbs highly in the 600- to 1600-nm range. Modern fabrication methods have the ability to produce pure silica. However in state-of-the-art pure silica, the main absorption is in the presence of residual water vapors. From Fig. 4.2, it is observed that 1550 nm has a loss advantage over the 1310-nm wavelength; however, system performance also depends on the dispersion factor, which is lower for 1310 nm. Dispersion also depends on the wavelength of the signal in multimode fibers, which have higher core diameters. That raises many modes to propagate through the fiber, since their wavelength satisfies the boundary conditions of this circular waveguide. However, dispersion occurs to some extent in SMF, since the glass refraction index is dependent on wavelength too.

The definition of chromatic dispersion is the negative change of the group travel time per fiber length unit per wavelength change. Hence, chromatic dispersion is always provided in units of ps/(km × nm). The SMF propagation mode is determined by an effective refractive index $n_{\text{eff}}$, since the indices of both the core and cladding influence the propagation:

$$n_{\text{eff}} = n_2 + b(\lambda)(n_1 - n_2), \tag{4.3}$$

where $n_1$ is the core refractive index and $n_2$ is the cladding refractive index. Assume the following well-known relations:

$$V_{\text{gr}} = \frac{d\omega}{d\beta}, \tag{4.4}$$

$$\omega = \frac{2\pi C}{\lambda}, \tag{4.5}$$

$$\beta = \frac{2\pi n_{\text{eff}}}{\lambda}, \tag{4.6}$$

where $V_{gr}$ is the group velocity, $\omega$ is the radian optical frequency, $\beta$ is the propagation constant along the fiber, $C$ is the speed of light, and $\lambda$ is the wavelength in vacuum. The propagation time in picoseconds can be defined as

$$t_{gr} = \frac{L}{V_{gr}} = L\frac{d\beta}{d\omega}\ [\text{ps}]. \tag{4.7}$$

Substituting Eqs. (4.3) to (4.6) into Eq. (4.7) results in

$$t_{gr} = \frac{L}{C}\left[n_{2gr} + \left(n_{1gr} - n_{2gr}\right)(b - \lambda)\frac{db}{d\lambda}\right], \tag{4.8}$$

where

$$n_{1gr} = n_1 - \lambda\frac{dn_1}{d\lambda} \tag{4.9}$$

and

$$n_{2gr} = n_2 - \lambda\frac{dn_2}{d\lambda}. \tag{4.10}$$

Subtracting Eqs. (4.9) and (4.10), assuming small change in the derivative parts, results in the following justified approximation:

$$n_1 - n_2 = n_{1gr} - n_{2gr}. \tag{4.11}$$

This approximation in Eq. (4.8) and the definition of $b(\lambda)$, together with the IEEE definition provided in Ref. [4], results in the chromatic dispersion expression given by

$$D_{CHR} = -\frac{dt_{gr}}{Ld\lambda}\left[\frac{\text{ps}}{\text{km} \times \text{nm}}\right], \tag{4.12}$$

$$D_{CHR} \approx \frac{\lambda}{C}\frac{d^2n_2}{d\lambda^2} + \frac{\lambda}{C}(n_1 - n_2)\frac{d^2b}{d\lambda^2}. \tag{4.13}$$

The first argument in Eq. (4.13) describes the material dispersion, while the second part describes the waveguide dispersion. The waveguide dispersion, however, makes a very small contribution to the overall dispersion performance of the fiber. One of the methods to compensate for dispersion is to use dispersion-shifted fibers. This is done by segmenting the core of the fiber. Hence, larger waveguide dispersion is achieved and cancellation of the dispersion at one wavelength or more can be accomplished. Assuming a negligible multimode dispersion, the 3-dB optimal fiber BW $B_{CHR}$ due to chromatic dispersion effect is given by:

$$B_{CHR} = \frac{0.44}{L\Delta\lambda|D_{CHR}|}, \tag{4.14}$$

where $L$ is the total length of the fiber and $\Delta\lambda$ is spectral BW at full width at half maximum (FWHM) of the source. The above expression is based on fiber response to an optical impulse, narrow zero width pulse, having a Gaussian spectrum, where the spectral FWHM width is equal to $\Delta\lambda$. This is a Gaussian pulse, which has an FWHM pulse width equal to the denominator of Eq. (4.14) at the fiber output.

It is observed in Fig. 4.3 that zero chromatic dispersion occurs at the 1310-nm wavelength domain for silica SMF.

Using a dispersion-shifted fiber may solve this problem; however, the 1310 would then result in dispersion of about $-30$ ps/(km $\times$ nm). In FTTx, an integrated triplexer (ITR) modules, where standard 1310-nm Rx/Tx are used, and a 1550 nm is used, the use of a shifted dispersion factor would penalize the other wavelength. Hence, the solution is somewhere between the two wavelengths, which fits for the full service access network (FSAN) standard and the regular 1310/1550.

In addition to the inherent natural distortions associated with SMF, there are distortions and losses generated by mechanical deformations of the fiber such as bending. Bending a fiber in a radius $R$ is called macrobending.[6,7] Cable bending is essential during deployment. The bending radius defines the amount of associated losses and additional dispersion.[5,6]

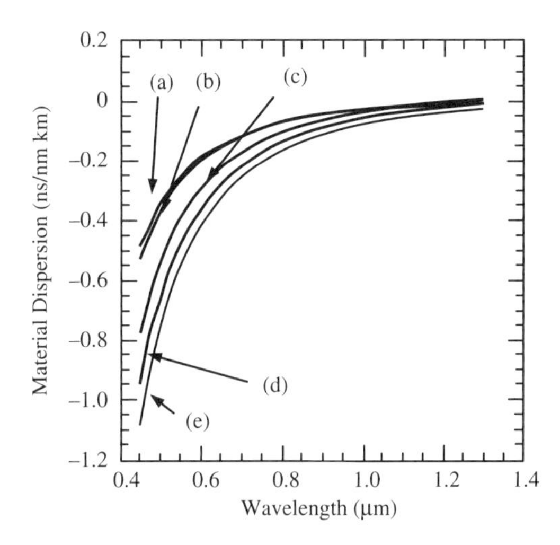


**Figure 4.3** Dispersion as a function of wavelength, while the fiber material is a parameter; material dispersion of polymethylmethacrylate (PMMA), PF polymer, and silica. (a) PF homopolymer; (b) 15 wt.% PF dopant-added PF polymer; (c) pure SiO; (d): 13.5 mol% GeO-doped SiO. (e) PMMA homopolymer.[5] (Reprinted with permission from the *Journal of Lightwave Technology* © IEEE, 2000.)

As was explained above, group velocity dispersion (chromatic dispersion) is the primary cause for concern in high bit rate (>2.5 Gbps) single-mode WDM systems. Dispersion in an optical pulse creates pulse broadening in such a way that the pulse spreads into neighboring slots (in the time domain) of the other pulses. In addition to the ISI it creates, it also introduces a power penalty, which can cause a degradation of the system's SNR. Optical SNR is a true figure of merit for optical communications. The dispersion power penalty dual port (DP) is given by:[8]

$$\mathrm{DP} = \frac{10\log(\sigma/\sigma_0)}{\sqrt{1+[D_{\mathrm{CHR}}L(\sigma_\lambda/\sigma_0)]}}, \tag{4.15}$$

where $\sigma_\lambda$ is the spectral width, $\sigma_0$ is the pulse width, and $\sigma$ is the spectral width variance. Furthermore, for SMF fibers, the chromatic dispersion parameter $D = 17$ ps/(km × nm). The limit on transmission distance is

$$B^2L < 16\frac{\lambda^2 D}{2\pi C} \quad \text{or} \quad L < \frac{16\lambda^2 D}{2\pi C B^2}. \tag{4.16}$$

Therefore, the following equation is true:

$$L < \frac{K}{B^2}. \tag{4.17}$$

Fiber BW is provided by Eq. (4.18), but when calculating BW, it is important to define what mode it refers to. In SMF, fundamental mode $HE_{1,1}$, i.e., $n = 1$, $p = 1$ is supported:

$$X_{n,p} = \frac{2\pi a}{\lambda_0}\sqrt{\left(n_1^2 - n_2^2\right)}. \tag{4.18a}$$

The condition for a single mode is $X_{n,p} = X_{1,1} < 2.405$, where $\lambda_0$ is the free space cut-off wavelength.

## 4.2 Optical Fiber Connectors

The fiber optical connector (FOC) is used to connect and disconnect between digital transceivers to the fiber, between a WDM analog receiver and an ITR/integrated triplexer to the curb (ITC) pigtail to the fiber. The FOC should have high mechanical reliability for multiconnects and disconnects without degrading its loss performance, reflections, and mechanical joint locking between connectors. The main design goal of SMF connectors is to have very low insertion loss (IL) of an average of 0.2 dB and an optical max of lower than 0.6 dB. The return loss, which indicates optical reflections, is lower than $-43$ dB, while the maximum value is targeted to be $-40$ dB in order to minimize distortions created by reflections in the analog section, and minimize deterministic jitter and dispersion in the digital section. Hence, it is necessary to minimize the air gap between the two fibers. Connection impairments can be divided into two categories, intrinsic factors such as a core diameter mismatch, cladding diameter mismatch, core eccentricity, and core ellipticity (ovality). These defects are related to the fiber fabrication process. The second impairment is extrinsic and depends on the connection and splicing quality. Surfaces of contact should be clean of dirt so that both contact surfaces are aligned on their axes and are parallel to each other. In addition, it should be verified that both fibers are not scratched, nor have any mechanical defect such as hackles, burrs, and fractures. This way, the geometrical propagation patterns of the light rays are not disrupted and deflected; so they will not enter the next fiber.

The optical connector structure is generally made of alignment sleeve and interconnecting plugs, as shown in Fig. 4.4. The plug is essentially a ferrule made of ceramic or metal that has an accurately drilled hole at its center, where the SMF fiber itself is inserted and then polished at its ending surface. There are several ways of polishing, as is explained later.

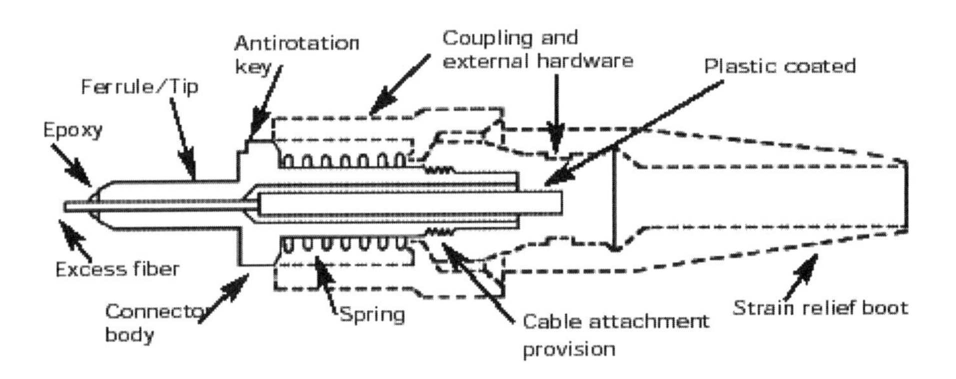


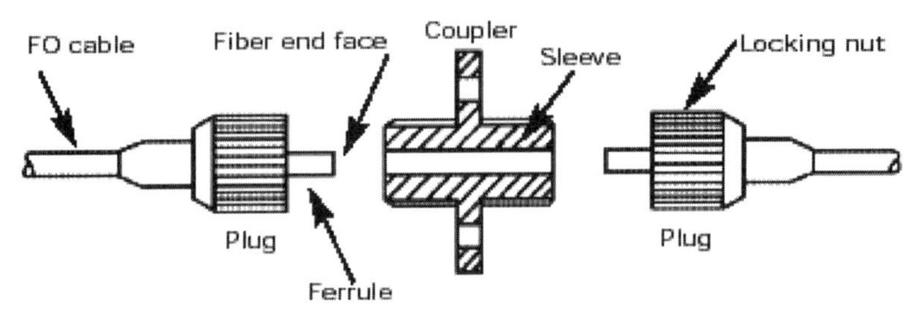


**Figure 4.4** A generic fiber optic connector.[9] (Reprinted with permission from *Optics and Photonics News* © OSA, 1995.)

The quality of the polished surface, which is the contact surface, is an important parameter in defining the connection quality with respect to loss and reflection. The optical return loss is similar to the RF return loss definition, as given by Eq. (4.18b):[10]

$$R/L[\mathrm{dB}] = 20\log\left(\frac{n_2 - n_1}{n_2 + n_1}\right). \tag{4.18b}$$

Hence, if there is an air gap between the two surfaces, then referring to air as the "characteristic impedance" analogy, $n_1 = 1$ and $n_2$ is the fiber refractive index. The Bell Labs Communication Research (BELCORE)[11] standard defines the criteria of return loss and IL at the connector contact surface.

There are about 10 popular types of FOC in the fields of communications, video, and datacom, as shown in Fig. 4.5.

The most popular in the field of hybrid fiber coax are the subscriber connector (SC), face contact (FC)/PC, Straight tip (ST), and SC/APC. The optical connectors were introduced to the market by the main connector factories and R&D labs

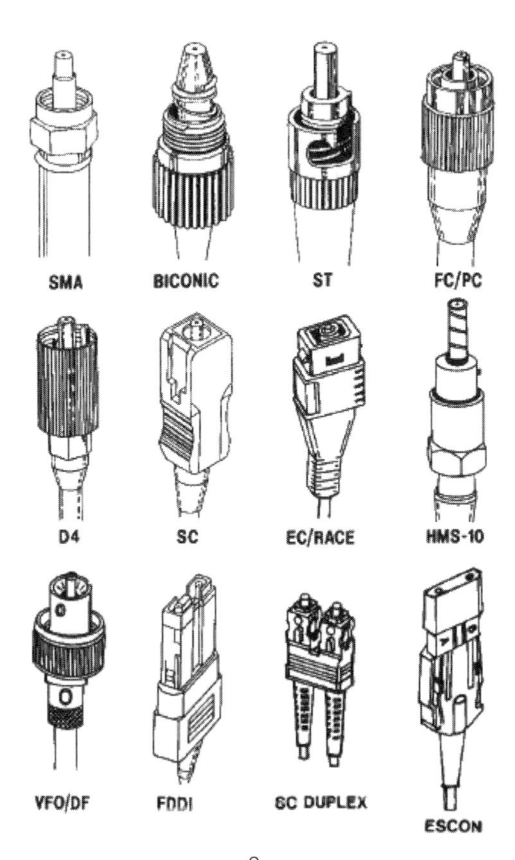


**Figure 4.5** Popular commonly used FOCs.[9] (Reprinted with permission from *Optics and Photonics News* © OSA, 1995.)

of communications companies; so these connectors became a standard. The connector names also indicate the polishing finish of each. For instance, the abbreviation PC refers to "physical contact" and APC refers to "angled polished physical contact." Moreover, among connector types, FOC/PC connectors have a better return loss than those that have no PC between surfaces. FOC/APC is even better in reflection performances than the FOC/PC. As data rates increase, the specifications for optical return loss are more crucial and FOC/APC connectors are preferred; for 10 GBit rate, the spec calls for a 50- to 70-dB return loss. In this case, the ferrule combines 8 deg to 10 deg angled geometric design with PC convex polishing at the end face. These kinds of APC optical connectors are ideal for high-speed data rates (GBits). Since connectors are angled and polished, with some tolerance on the angle, IL might be a bit higher, but back reflections are low due to the return loss. That is a key parameter in WDM systems when an optical triplexer or duplexer is used. Due to finite isolation value between the Tx laser and the Rx detector in the optical block, additional reflected energy may reduce the sensitivity performance of the receive side, for instance, a standard 1310-nm transmitter, 1310-nm receiver, and 1550-nm receiver for CATV. In this case, the 1310-nm transmitter leaks into the 1310-nm receiver due to the optical structure of the triplexer; however, bad return loss would cause more energy to leak into the receive side of the 1310-nm detector, which is harder to isolate than at the 1550-nm detector port.

The FC/PC connector was developed by Nippon Telegraph and Telephone Corp (NTT). It has a flat end face on the ferrule that provides face surface contact between joining connectors. The PC between surfaces reduces the IL even further and improves the return loss. To increase performance and improve return loss, FC/APC connectors were introduced. The goal of the HFC use of such connectors is to reduce reflections and hence improve digital eye performance, thus getting a better bit error rate (BER) result, which essentially refers to better sensitivity. In addition, in CATV signal aspect, reduction in reflections improves AM VSB and QAM, since, for instance, optical distortions are reduced.

FC/super physical contact (SPC): These connectors are identical to the FC/APC except that the polish on the end of the ferrule is not angled. Instead, the ferrule is polished to form a convex surface so that when two connectors are mated, the centers of their tips physically contact each other to provide a smooth optical path. Depending on the quality of the polish, which is indicated by the PC or SPC name, the optical back reflections are better than −30 or −40 dB, respectively.

SC/PC (square/subscriber connector) was developed by NTT around 1986. It employs a rectangular cross section of molded plastic. The connector is a push-to-insert and pull-to-remove with a locking mechanism that prevents rotational misalignment. This kind of connector and ones similar to the SC/APC are commonly used by FTTx ITR (integrated triplexers) and FTTx modules. It has a locking mechanism; therefore, it achieves low IL as well as return loss. The locking mechanism is advantageous, since it is a fast mating feature, preventing loss fluctuations in case of mechanical interferences. Therefore the SC types were adopted by the standards of ANSI/TIA/EIA-568 of commercial building wiring FTTP.

**Table 4.1** Connector-type performances.

| Connector type | Back reflections | Insertion loss |
|---|---|---|
| FC/PC | <−30 dB | <0.5 dB, 0.25 dB typ |
| FC/SPC | <−40 dB | <0.5 dB, 0.25 dB typ |
| FC/APC | <−60 dB | <0.5 dB, 0.25 dB typ |

The ST connector was introduced in early 1985 by AT&T Bell Laboratories. Its design uses a spring-loaded twist and a lock bayonet coupling that is similar to Bayonet Neil Connectors (BNCs) for coax cables. It has the same advantages as the SC types. The ST has a cylindrical 2.5 mm diameter ferrule. The ferrule is made of ceramic, stainless steel, glass, or plastic. It is a popular connector owing to its compatibility between manufacturers.

When selecting an optical fiber connector, there are several key parameters that should be looked for:

- minimum IL, for CATV, it is desirable to have less than 0.5 dB;
- repeatability over multiple connections/disconnections;
- better than 40 dB for CATV and better than 50 dB for high GBit rate; and
- compatibility with other connectors from the same type.

Two popular standards that are often referred by FOC manufacturers are as follows:

- Bellcore (Bell Communications Research) Technical reference TR-NWT-000326 Issue 3 titled *Generic Requirements for Optical Fiber Connectors and Connectorized Jumper Cables*.
- Japanese Industrial Standard JIS C 5970 1987, titled FO1 *Type Connectors for Optical Fiber Cords*. It has an English translated version.

Table 4.1 demonstrates a specification for an FTTx pigtail finish, where parameters such as IL, return loss, connector type, are referred to.

## 4.3 Optical Couplers

The optical fiber coupler is composed of two optical fibers or waveguides and has a fixed coupling length as shown in Fig. 4.8. Optical fibers are analyzed by the even and odd mode equations where at one part of the analysis, both inputs are excited with the same signal amplitudes and phases.[69,70] Figure 4.6 depicts a waveguide coupler where the distance between the two waveguides defends the coupling.[69]

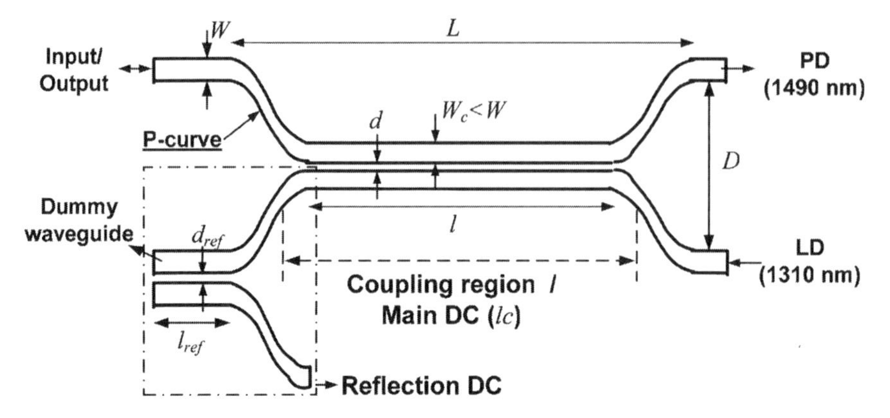


**Figure 4.6** WDM wave guide filter based on directional couplers (DC). Concept used in PLC (planar lightwave circuit). Coupling region is $d$. Coupling is determined by $d$.[69] (Reprinted with permission from *Lightware Technology* © IEEE, 2005.)

In the second part of the analysis, the ports are excited with the same signal amplitudes but the signals are out of phase, i.e., 0 deg and 180 deg. The common method of analysis is to present the coupler by $4 \times 4$ $S$ parameter matrix and solve $S$ parameters constraints. In both cases of the analysis, the signals have the same wavelength $\lambda$. The $2 \times 2$ port coupler is identified by four ports. Input, which is the S11 port, output (throughput), which is the S22 port, coupled, which is the S44 port, and isolated or cross-talk, which is the S33

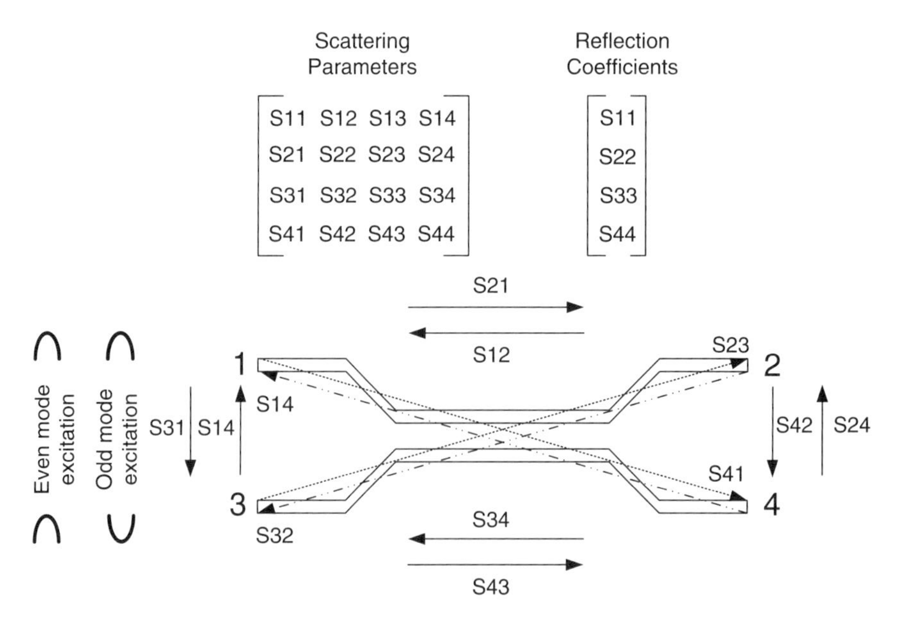


**Figure 4.7** 4 ports network S parameters for analyzing optical coupler. An Even and odd mode of excitations for solving the coupler transmission function is presented.

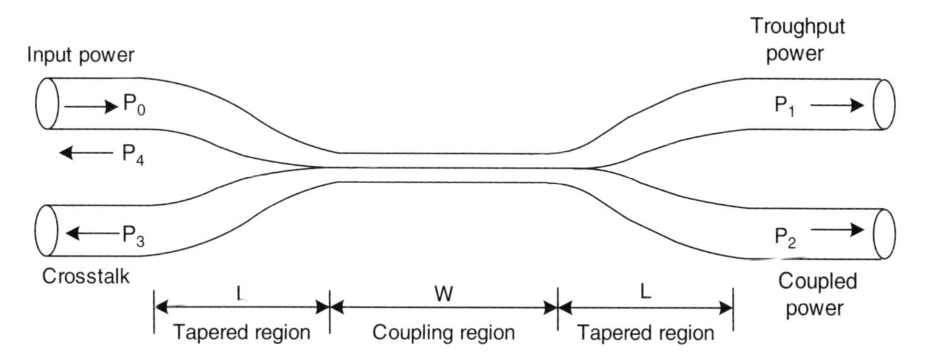


**Figure 4.8** Cross-sectional view of a fused-fiber coupler with a coupling region *W* and two tapered regions of length *L*. The total span $L_t = 2L + W$ is the coupler draw length.[13] (Reprinted with permission from G. Keiser, *Optical Fiber Communications* © McGraw-Hill, 2000.)

isolated port.[7] Figure 4.7 illustrates a four-port coupler with *S* parameters. This method is very similar to the analysis of transmission lines couplers and filters in microwaves.[12] In microwaves, the analysis goes for *V* and *I*, while in optics, the equivalents are *E* and *H*, respectively, where the power is proportional to $E^2$.

The manner in which a 2 × 2 coupler is fabricated by fusing together two SMFs over a uniform section of length *W*, as shown in Fig. 4.8.

Each input and output fiber has a long tapered section of length *L*. The total representative length is $L_t = L + W$. This device is known as a fused biconical tapered coupler. By making the tapers (changing the radius of the fiber to a narrower one) very gradual, only a negligible fraction of the incoming optical power is reflected back into either of the input ports. Thus, these devices are also known as directional couplers, similar to those used in microwaves.

The coupled optical power is varied through three parameters: the axial length of the coupling region; the size of the reduced radius *r* in the coupling region; and the difference $\Delta r$ in the radii of the two fibers in the coupling region. By making a fused-fiber coupler, only *L* and *r* change, as the coupler is elongated. The power $P_2$ coupled from one fiber to a coupled port S22 over an axial distance *z* is given by[13]

$$P_2 = P_0 \sin^2(k \cdot z). \tag{4.19}$$

Conservation of power, for identical-core fibers, the direct arm power is given by

$$P_2 = P_0 - P_1 = P_0\left[1 - \sin^2(k \cdot z)\right] = P_0 \cos^2(k \cdot z), \tag{4.20}$$

where *k* is the coupling coefficient, *z* is the electromagnetic distance coordinate from S11 port and is given by $z = \pi/(2k)$ for the spot, and $P_2 = 0$; $P_0$ is the incident input power, $P_2$ is the coupled power, and $P_1$ is the direct arm power. It can be seen that the optical coupler behaves as a lag–lead network since the coupled arm has a 90-deg lag with respect to the direct arm. The conclusion is that the resonance length of the coupler is given by $z = L_R = \pi/(2k)$. Hence, by changing the $k$ value, the resonance period and the coupling ratios are changed. The coupling ratio is given by

$$\mathrm{CPLR} = \frac{P_2}{P_2 + P_1} \cong \frac{P_2}{P_0}. \tag{4.21}$$

However, since there are power losses, the above equality sign is theoretical only but is a good approximation. The excess loss ratio (ELR) is defined as the ratio between the incident power $P_0$ to the sum of the output powers at the two arms:

$$\mathrm{ELR} = \frac{P_0}{P_2 + P_1} > 1. \tag{4.22}$$

The isolation or cross-talk is given by

$$\mathrm{ISO} = \frac{P_0}{P_3}. \tag{4.23}$$

Hence, the directivity would be the ratio of Eqs. (4.21) to (4.23) or, in dB, the difference:

$$\mathrm{DR} = 10\log(\mathrm{CPLR}) - 10\log(\mathrm{ISO}). \tag{4.24}$$

When deploying a network or HFC network, couplers and splitters are used to form the distribution network. For this purpose, a star coupler is used. The prime role of star couplers is to combine the powers from $N$ inputs and divide them equally among $M$ output ports. The fiber-fusion technique has been a popular construction method for $N \times N$ star couplers. Figure 4.9 shows a generic $4 \times 4$ fused-fiber star coupler.

However to build a $1 \times N$ splitter, which divides 1 input into $N$ outputs, the overall couplers needed are $N - 1$. That is because the structure splits from one

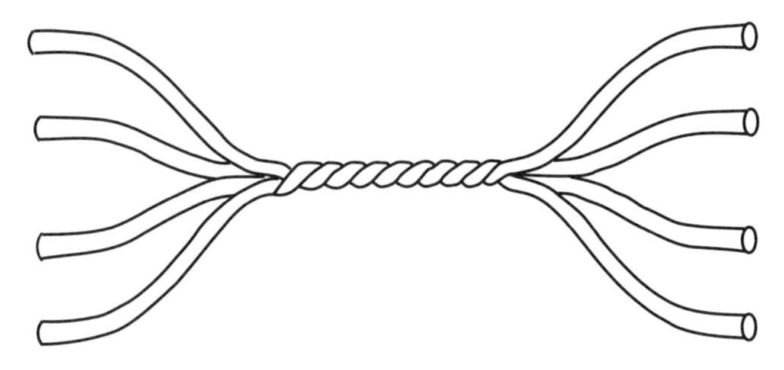

**Figure 4.9** Generic $4 \times 4$ fused-fiber star coupler fabricated by twisting, heating, and pulling on four fibers to fuse them together.[13] (Reprinted with permission from G. Keiser, *Optical Fiber Communication* © McGraw-Hill, 2000.)

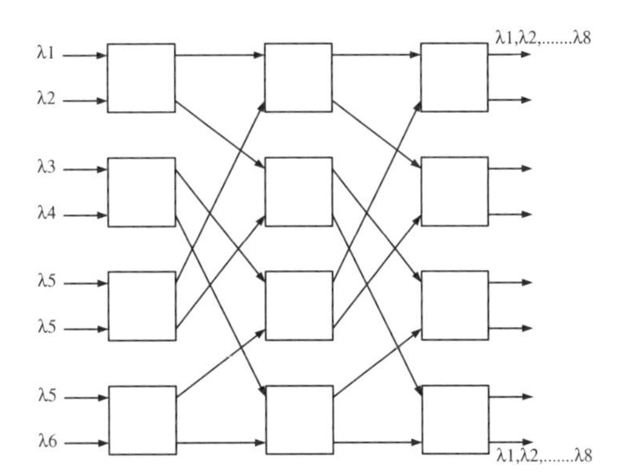


**Figure 4.10** An 8 × 8 star coupler made of twelve 2 × 2 couplers. The first two columns from the left compose two 4 × 4 star couplers. In fact, swapping places between couplers of $\lambda_3$, $\lambda_4$, $\lambda_5$, $\lambda_6$, would create a full separation between the two 4 × 4 stars. This way, the first two columns create the two 4 × 4 sections and the last column creates the four 2 × 2 sections in this star matrix.[13] (Reprinted with permission from G. Keiser, *Optical Fiber Communication* © McGraw-Hill, 2000.)

arm into two arms; these are cascaded with additional two couplers creating four arms, etc., so the split becomes a structure of $1+N/2+ N/4+\cdots+N/2^{n-1}$. In such a case, the split loss ratio is $10\log N$. One way to create an $N \times N$ distribution is to connect back-to-back two $1 \times N$ couplers. That essentially would double the split loss by 2 or would raise it by $3\text{ dB} \times 2\log_2 N$. To overcome this problem, a different structure is suggested.[13] Assume $N = n \times m$ and $mn \times n$ couplers are used in a cascade with $nm \times m$ couplers; then the signal is routed from any input port to any output port and goes through $\log_2 N$ stages. Thus, the losses are reduced to $3\text{ dB}\log_2 N$. The total number of 2 × 2 couplers would be $(N/2)\log_2 N$. An example is provided in Fig. 4.10.

The star matrix structure results in $N/2$ elements in the vertical direction and $\log_2 N$ elements horizontally. Realizing such a distribution coupler for large $N$ values becomes a space problem due to the large amount of couplers. Moreover, since each coupler in the matrix has different access, loss, and slightly different reflections and variations, the matrix performance is not uniform. The advanced technology of today enables the use of silica glass waveguide couplers, which are more compact than the ordinary fused-fiber couplers, that are used in PON applications, etc.

## 4.4 WDM Multiplexers

In Sec. 1.1, the concept of WDM was introduced. The enormous BW of the optical fiber has a great potential for large amount of optical wavelength high-data rate transport with a relatively narrow BW percentage but very a wide absolute BW. Note that $\text{BW}\% = \text{BW}/f_0$, where BW is the signal absolute BW and $f_0$ is the wavelength frequency. The wavelength response and characteristic of an SMF was provided in Sec. 8.1. From Fig. 8.2, it can be seen that an optical fiber has two long wavelength minima located at 1.3 and 1.5 μm. The frequency $f$ at a given wavelength is calculated by the relation $f = c/\lambda$. Hence the incremental BW $\Delta f$ can be calculated as the first-order differential approximation term:

$$\frac{\mathrm{d}f}{\mathrm{d}\lambda} = -c\frac{1}{\lambda^2} \Rightarrow \Delta f \approx -\left(c\frac{1}{\lambda^2}\right)\Delta\lambda. \tag{4.25}$$

This expression is a good linear approximation to evaluate the absolute RF or electrical BW by knowing the relative optical BW or relative optical BW as a percentage as

$$\Delta f = -\frac{c}{100 \times \lambda} \times \mathrm{BW}[\%]. \tag{4.26}$$

For example, assume a C-band center frequency of 1547.5 nm and optical BW of $\pm 17.5$ nm. The relative optical BW is 2.26%, and the absolute RF BW or electrical BW is ~4384.5 GHz, which is about ~4.4 THz. One of the applications of METRO WDM networks is to use tunable-wavelength transponders and reduce the cost of redundancy. The current technology of tunable-wavelength transponders for METRO applications enables the transfer of 10 GB/s over a single wavelength, using a lithium niobate ($LiNbO_3$) external modulator (EM) to avoid chirp and hence chromatic dispersion. A single-chip solution using Mach–Zehnder EM is presented in Ref. [14]. Wavelength MUXs and demultiplexers (DeMUXs) are essentially needed. A selective-wavelength DeMUX can be used, where the amount of selectivity, or wavelength filter BW, and shape-factor sharpness is determined by the wavelength's transport density. From Eq. (4.25), it can be seen that the increments in $\Delta\lambda$ define the absolute electrical $\Delta f$ and delineate the data rate, which is available at that particular wavelength. In other words, $\Delta\lambda$ defines the relative BW% for a given wavelength and the wavelength spacing scale. The trend of lowering network costs by using PON makes WDM, coarse WDM (CWDM), and DWDM passive components key players in current passive network architectures.

### 4.4.1 Mach–Zehnder interferometer concept for CWDM MUX–DeMUX

CWDM wavelength splitters are based on coupled optical transmission lines. The coupled optical transmission lines can be either fused-fiber couplers or planar lightwave circuits. In both cases, the physical concept is the same and is based on a selective wavelength coupling. The wavelength selectivity increments of $\Delta\lambda$ in a Mach–Zehnder interferometer (MZI) are at the range of 5 to 100 nm. The main idea in this concept is to create different electromagnetic paths for each wavelength, so that each wavelength resonates at a selective length. This way, wavelength separation is achieved. In another approach, the coupling

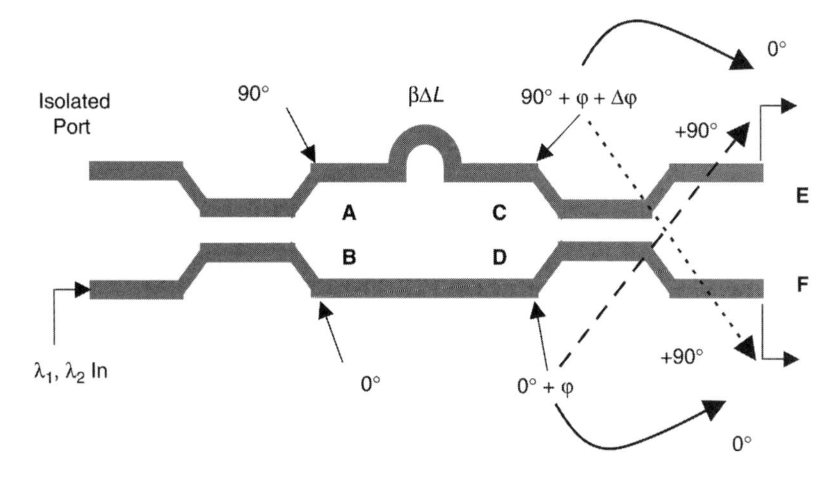


**Figure 4.11** MZI concept.

factor is wavelength dependent. The idea here is to create a coupled line filter by using two identical cascading directional couplers, while creating a phase difference between one arm to the other as shown in Fig. 4.11.

Both wavelengths, $\lambda_1$ and $\lambda_2$, are applied at the input port. Since the coupler is a 3-dB coupler, both wavelength powers appear at ports A and B, while port A has a phase lead over port B by 90 deg ($\pi/2$). Both arms' output powers have to travel from the A and B output ports to the C and D inputs of the next coupler. However, the optical length of the A to C path is longer than the B to D path by the optical angle $\beta\Delta L$. Analyzing the power phasors summation and phases results in the following relations. The phasor of wavelength $\lambda_1$ at port D adds to the other phasor with the same wavelength $\lambda_1$ from port C, resulting in the following phases and amplitudes at port F:

$$P_{\text{out F}} = P_{\text{D}} < \varphi + P_{\text{C}} < \left(\varphi + \frac{\pi}{2} + \Delta\varphi + \frac{\pi}{2}\right). \tag{4.27}$$

As for the power phasors' summation to port E, the result is

$$P_{\text{out E}} = P_{\text{D}} < \left(\varphi + \frac{\pi}{2}\right) + P_{\text{C}} < \left(\varphi + \frac{\pi}{2} + \Delta\varphi\right). \tag{4.28}$$

From the power phasor diagram for output $F$, it can be seen that the relative phase between phasors C and D is $\pi + \Delta\varphi$. The relative phase between phasors C and D at output E arm is $\Delta\varphi$.

Placing the phasors in Cartesian coordinates, as shown in Fig. 4.12, and calculating the equivalent power value results in the following for outputs $F$ and

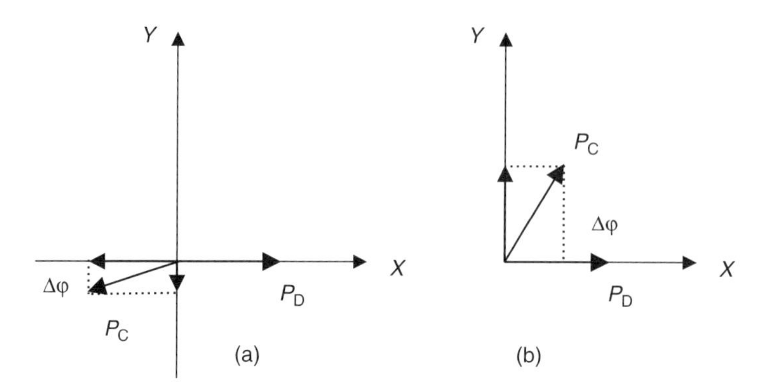


**Figure 4.12** Phasor diagram of power summation at the output ports of MZI, showing relative angles between power vectors. (a) Output port F; (b) output port E.

$E$, respectively

$$\begin{aligned} P_{\mathrm{EQ\_F}} &= \sqrt{P_{\mathrm{D}}^2 - 2P_{\mathrm{C}}P_{\mathrm{D}}\cos(\Delta\varphi) + P_{\mathrm{C}}^2\cos^2(\Delta\varphi) + P_{\mathrm{C}}^2\sin^2(\Delta\varphi)} \\ P_{\mathrm{EQ\_E}} &= \sqrt{P_{\mathrm{D}}^2 + 2P_{\mathrm{C}}P_{\mathrm{D}}\cos(\Delta\varphi) + P_{\mathrm{C}}^2\cos^2(\Delta\varphi) + P_{\mathrm{C}}^2\sin^2(\Delta\varphi)} \end{aligned}. \tag{4.29}$$

Since the magnitude of $P_{\mathrm{D}}$ is equal to $P_{\mathrm{C}}$, the power at port F is

$$\begin{aligned} P_{\mathrm{EQ\_F}} &= \sqrt{P^2 - 2P^2\cos(\Delta\varphi) + P^2\cos^2(\Delta\varphi) + P^2\sin^2(\Delta\varphi)} \\ &= P\sqrt{2 - 2\cos(\Delta\varphi)} = 2P\sin^2\left(\frac{\Delta\varphi}{2}\right). \end{aligned} \tag{4.29a}$$

As for the E output, the following result is applicable:

$$\begin{aligned} P_{\mathrm{EQ\_E}} &= \sqrt{P^2 + P^2\cos^2(\Delta\varphi) + 2P^2\sin^2(\Delta\varphi) + 2P^2\cos(\Delta\varphi)} \\ &= P\sqrt{2 + 2\cos(\Delta\varphi)} = 2P\cos^2\left(\frac{\Delta\varphi}{2}\right). \end{aligned} \tag{4.30}$$

Knowing that the electrical angle $\Delta\varphi = \beta\Delta L = n(2\pi/\lambda)\Delta L$, the following condition applies for resonance:

$$n\frac{2\pi}{\lambda}\Delta L = k\pi. \tag{4.30a}$$

Hence, for even values of $k$, the output at F is null, and at port E the output is equal to the sum of the two powers $2P$. For odd values of $k$, the port E becomes null, and port F is the summation of the two powers equal to $2P$. Now the design approach would be that both wavelengths have orthogonal transfer functions to each other in order to create the wavelength separation. That means null for $\lambda_1$ is a peak for $\lambda_2$ at the same port. Consequently, this leads to the following conditions for cavity resonance modes under different wavelengths:

$$n\frac{2\pi}{\lambda_1}\Delta L = 2k\pi + \pi \tag{4.31}$$

and under the same port,

$$n\frac{2\pi}{\lambda_2}\Delta L = 2k\pi. \tag{4.32}$$

This provision results in a separate condition for the wavelengths, since for any integer value of $k$, both arms are null for each wavelength and the opposite arms are peaks.

Subtracting Eq. (4.32) from Eq. (4.31) results in

$$\Delta\lambda = \lambda_2 - \lambda_1 = \frac{\lambda_2\lambda_1}{2n\Delta L}. \tag{4.33}$$

Hence, the frequency increment is given by

$$\Delta f = \frac{c}{2n\Delta L}. \tag{4.34}$$

The CWDM concept is commonly used as an optical front-end building block for FTTx modules. Cascading MZI blocks are used to build a planar optical triplexer for the FSAN wavelength standard of 1550-nm receiving, 1310-nm transmitting, and 1490-nm receiving (see Chapter 5).

### 4.4.2 Concept of DWDM filters

The interference phenomenon of light passing through slits was demonstrated in the 1800s by the English scientist Thomas Young. This experiment was crucial in establishing the wave nature of light.[15] The DWDM channel spacing resolution is kept at the range of 0.5 to 5 nm, based on the facts uncovered by Young. The operation in the case presented here is based on the diffraction grating phenomena (Fig. 4.13). The idea is to create a reflection plane with parallel angular-shaped reflector lines etched on a wafer or a reflection surface. The light hitting this surface is reflected at a specific angle, $\theta_d$, according to its wavelength $\lambda$ and incident angle $\theta_i$. The idea here is to create multiple sources that will add constructive interference. The light beam that hits the triangular slits on the grating plane is refracted. Hence, the grating plane has multiple refracted light sources. The period of the grating slit is marked as $\Lambda$. This length of interval defines the wavelengths' separation step size or increments. In other words, the slit period $\Lambda$ defines the optical filter selectivity.

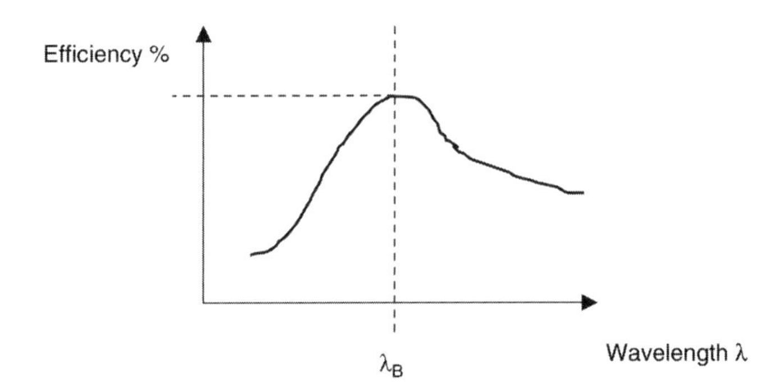


**Figure 4.13** Characteristic efficiency curve for given diffraction order intensity. The maximum efficiency indicates blazing condition.

The condition for constructive interference is that the optical path $\Lambda$ between the incident wavelength and the adjacent wavelength, reflected from the grating plane, would be integer multiplications of its wavelength. This is the fundamental condition for creative diffraction when light strikes a grating diffraction plane:

$$\Lambda[\sin(\theta_i) - \sin(\theta_d)] = n\lambda \Rightarrow [\sin(\theta_i) - \sin(\theta_d)] = \frac{n\lambda}{\Lambda}. \tag{4.35}$$

The waves hitting the Bragg diffraction plane are traveling waves; the impinging wavelength is a forward-traveling wave, while the scattered one is a backward-traveling wave. Hence, Eq. (4.35) becomes a phase condition for the difference between the propagation constants $\beta$ of the forward- and backward-traveling waves. Knowing that $\sin(-\theta) = -\sin(\theta)$, and for small values of $\theta$, $\sin(\theta) \cong \theta$, and assuming $n = 1$, Eq. (4.35) can be rewritten as:

$$\frac{2\pi}{\lambda}[\sin(\theta_i) + \sin(-\theta_d)] = \frac{2\pi}{\Lambda} \Rightarrow \frac{2\pi}{\lambda}[\theta_i - \theta_d] = \frac{2\pi}{\Lambda}. \tag{4.36}$$

This notation is called the momentum conservation law.

The diffraction grating can, in certain configurations, concentrate a large percentage of the incident energy impinging on the grating plane into a specific order. This phenomenon is called blazing.

The measure of the light intensity diffracted from a grating is called efficiency. The blaze wavelength is defined as the one that yields at a given diffraction order $n$, the maximum efficiency, as shown in Fig. 4.13.

Back again to Fig. 4.14: on examining the incident and reflection angles $\theta_i$ and $\theta_d$ on the grating plane, we find that there is a specific condition that occurs under a specific geometrical arrangement where $\theta_i = \theta_d$. This situation is called the "Littrow condition" or "Littrow configuration." At that point, the intensity and the diffraction efficiency is maximized and the lens astigmatism is minimized.

System applications of DWDM filters can be realized where multiple colors are transmitted, for instance, in a METRO application where add–drop filtering

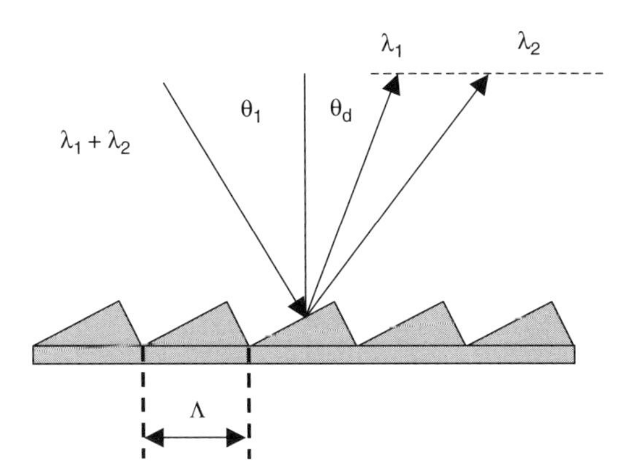


**Figure 4.14** Diffraction-grating plane showing the slit period $\Lambda$, angle of incident waves $\lambda_1 + \lambda_2$, which is $\theta_1$ and reflected wave $\lambda_1$ at $\theta_d$.

is used. Additional applications include lasers [distributed feedback (DFB), distributed Bragg reflector (DBR)], gain equalization dispersion compensation fiber coupling (mode size transformers), and noise filtering, by using grating filters.[16]

High-finesse resonance can occur in a grating waveguide structure (GWS). Changing the refractive index and taking advantage of the high-finesse property can lead to a tuning range larger than the resonance frequency.[17] These filter types are used to create an external cavity laser (ECL) by cleaving the filter at one side of the laser and selecting the resonance mode by voltage control of the refractive index. The tenability range achieved with this technique in TE polarization mode, and with a reverse tuning voltage range of 0 to 20 V, was 0.5 nm at 1520 nm, and 0.3 nm in the case of 1586 nm. For TM polarization, a wider range was observed: 0.8 nm for 1515 nm and 0.5 nm for 1589 nm, with a tuning voltage range of 0 to 30 V (Fig. 4.15).

A different approach to building a DWDM filter is by utilizing the MZI method. As was explained previously, the frequency increments and wavelength separation is given by Eqs. (4.33) and (4.34). Therefore, the longer the delay line $\Delta L$, the finer the frequency increment step. For instance $\Delta L = 20$ nm can provide an optical frequency spacing of about 5 GHz. This is equal to 0.04-nm spacing at 1550 nm. The MZI WDM can practically provide channel spacing from 0.01 up to 250 nm for various $\Delta L$ lengths.[18] Pay attention to the fact that a thin film heater is integrated on one of the MZI arms, the $\Delta L$ arm, to compensate for phase variations. When the light path is heated, the refraction index varies in about 10 PPM/°C[19] (Fig. 4.18). The other MZI arm is loaded sometimes with mechanical stress applied by a film to balance the wavelength birefringence between the long and the short arms.[20] The MZI approach is used for instance for interleavers and deinterleavers. An interleaver is a periodic optical filter that combines or separates a comblike dense wavelength transport of DWDM.[21] There are several types of interleavers, as indicated by Fig. 4.16.

The requirements for performances in an optical filter within the pass band are low IL, sharp roll-off or shape factor (SF), strong adjacent channel rejection, and

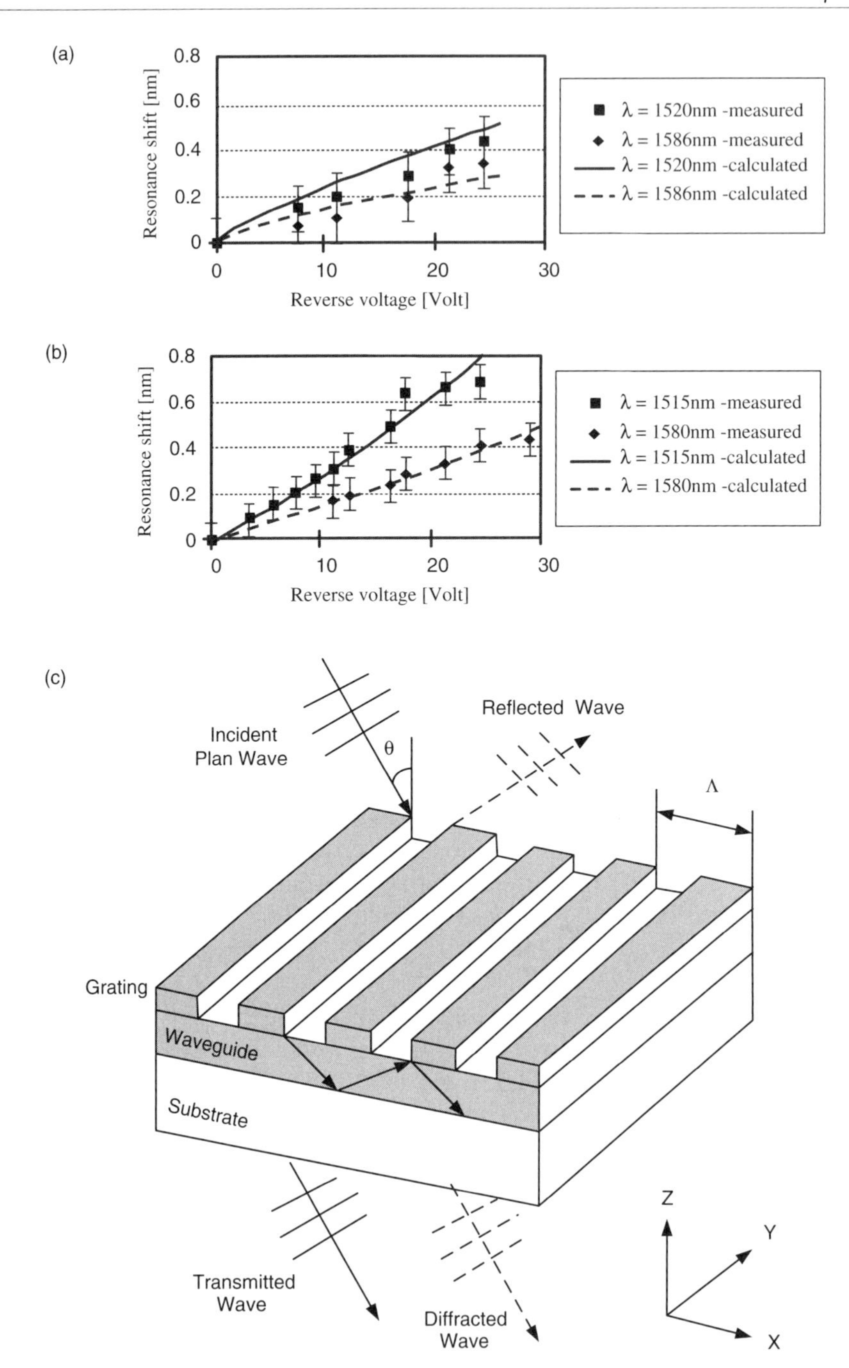


**Figure 4.15** Calculated and measured resonance wavelength shift as a function of the applied reverse voltage for two different resonance wavelengths: (a) TE polarization. (b) TM polarization. (c) Basic geometry of a grating waveguide structure and relevant incident, transmitted, and diffracted waves. At resonance, there is destructive interference between the transmitted wave and the diffracted wave.[17] (Reprinted with permission from the *Journal of Lightwave Technology* © IEEE, 2004.)

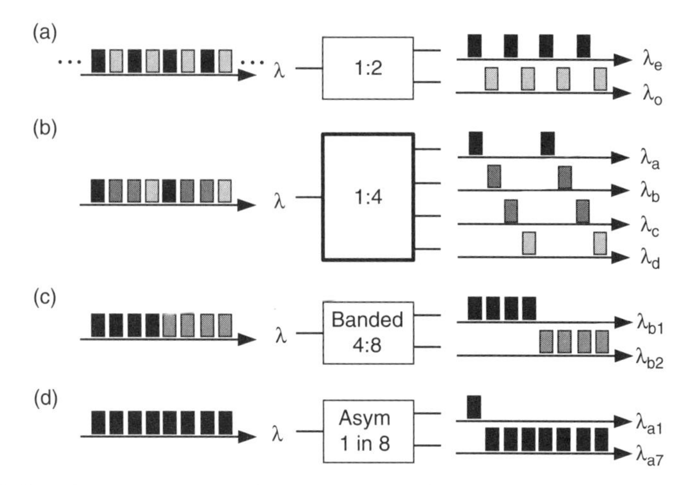


**Figure 4.16** Interleaver types: (a) 2:1 even–odd channels' separation onto two ports. (b) 4:1 separation deinterleaver or higher. (c) Banded interleaver; separates even and odd channels out to 4:1 or higher. (d) Asymmetric interleaver separates one channel in $N$.[21] (Reprinted with permission from the *Journal of Lightwave Technology* © IEEE, 2004.)

minimum chromatic dispersion within the pass band. These are contradicting requirements and a design compromise is made on the tradeoffs between the rejection depth and rejection BW of the adjacent channels. Interleavers have been demonstrated that resolve comblike DWDM frequencies on 100, 50, 25, and 12.5 GHz center frequencies. The period is governed by the free spectral range (FSR) of the core element $\Delta L$. The FSR parameter describes the spacing between the center frequency of one filter and the adjacent filter, or from the center of the lower adjacent rejection band to the higher one, as shown in Fig. 4.17.

The interleaver filters suffer from several types of optical losses:

- polarization-dependent loss;
- polarization-dependent wavelength defines the frequency shift between the $s$ and the $p$ states (Fig. 4.18); and
- polarization-dependent mode dispersion, which is differential group delay between two paths.

The design goal is to essentially minimize these parameters. The ITU standard defines a comblike frequency plan with an anchor at 193.1 THz and frequency spacing of 100 GHz[22] (Table 1.2, Chapter 1). However, many transport systems use 50-GHz channel spacing, which is a much denser comb and leaves less room for FSR errors. Furthermore, due to the denser comb plan, the filter SF becomes tougher, since the relative BW is smaller and the rejection is sharper.

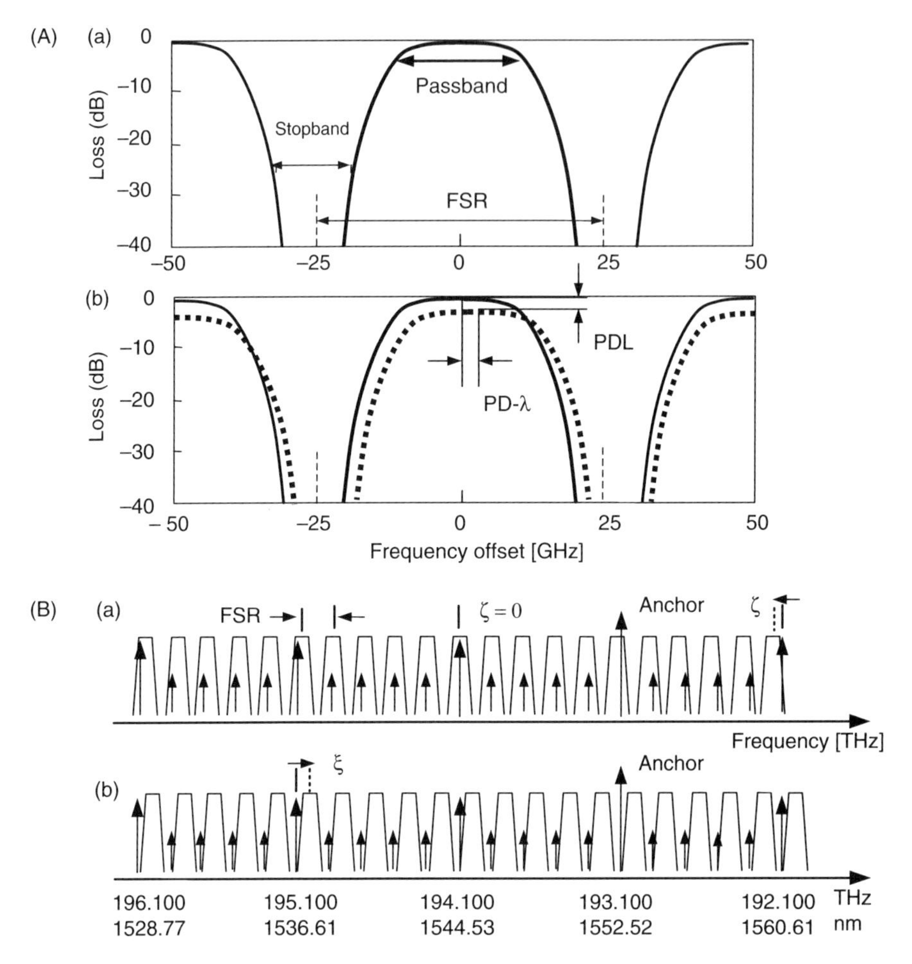


**Figure 4.17** (A) Narrow-band design parameters of interleaver showing FSR, PDL, and PD-λ errors and misalignments. (B) These misalignments at a given wavelength grid are the reason for leakage to the adjacent channel resulting in cross-talk.

There are three main design approaches for realization of an interleaver:

(1) Lattice filter with the following design approaches:

- The birefringent filter known as Loyt or Solc filter.
- Mach-Zehnder filter as in Fig. 4.19. Lattice filters are typically used for 2:1 interleavers and can be cascaded to realize $1{:}2^N$ filters. On the other hand, $1{:}2^N - 1$ is required to build the whole $1{:}2^N$ and terminate one output; hence this approach is less economic in such a case.

(2) Gires–Tournois (GT) with the following design approaches:

- The interference filter.
- The birefringent analog (B-GT), which is used for 1:2 inteleavers and asymmetrical interleavers too.

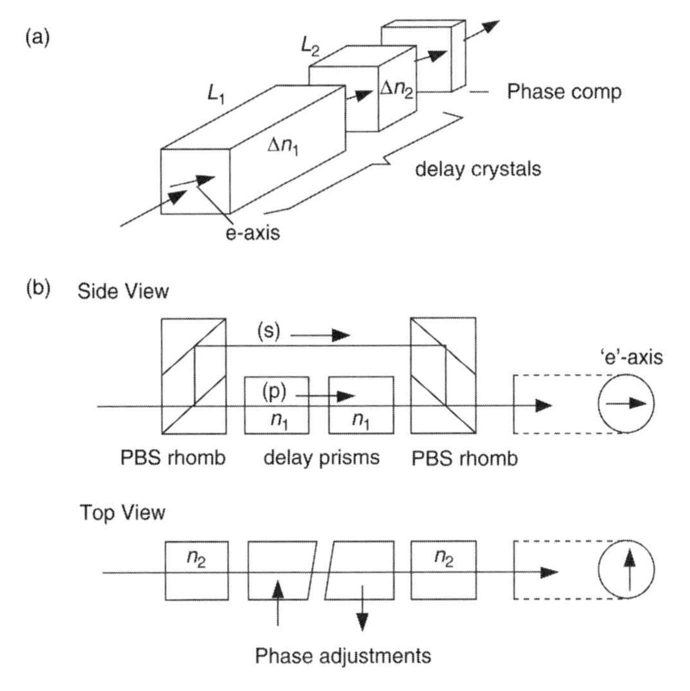


**Figure 4.18** LF filter cell unit technology. (a) Complementary birefringent crystals impart differential delay on orthogonal polarization states, marked as $n_1$ and $n_2$ at path *p*. A crystal pair is used to passively temperature-compensate the birefringent phase. The phase compensation plate makes a fine adjustment onto the International Telecommunications Union (ITU) grid. The e-axes are aligned within a unit cell and cut at an angle determined by the filter synthesis. (b) Differential delay is imparted on orthogonal polarization states by an all-glass unit cell. The leading and following PBS rhombs separate the polarizations, and the delay prisms determine the delay and phase.[21] (Reprinted with permission from the *Journal of Lightwave Technology* © IEEE, 2004.)

(3) The arrayed waveguide router, which is well suited for $1{:}2^N$ and deinterleaving in a single stage.

The FSR of the filter cell in Eq. [4.18(b)] is given by:

$$\mathrm{FSR} = \frac{c}{\Delta n_1 L_1 \pm \Delta n_2 L_2}, \tag{4.37}$$

where $c$ is the speed of light, $\Delta n_{1,2}$ is the group birefringence of the associated crystal, and $L_{1,2}$ is the respective crystal length. The $\Delta n$ is a signed quantity, positive for uniaxial crystals and is negative otherwise. The $\pm$ sign indicates alignment (+) or crossing ( ) of the extraordinary axes.

Hitachi-Cable[23,24] had first demonstrated the Mach–Zehnder dispersion-compensated interleaver, where, in some cases of planar waveguides, phase adjustment is done by thermal-optic heating pads located above the two interferometer pads.[25]

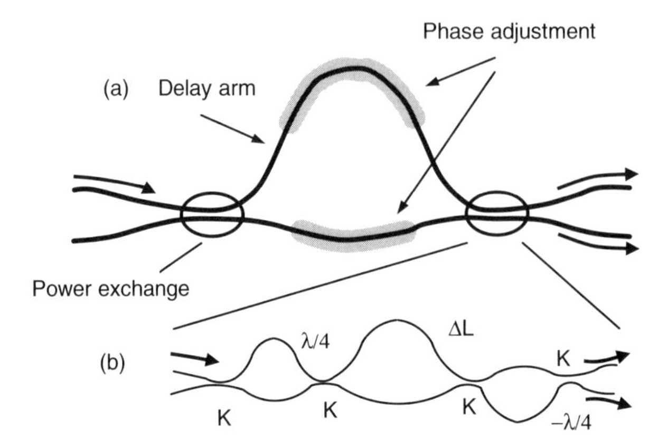


**Figure 4.19** The Mach–Zehnder unit cell consists of planar waveguides. (a) The path-length difference of the arms between the power couplers imparts differential delay, which results in wavelength separation, as was explained in Sec. 8.4.1. The heating pads adjust the phase permanently. (b) Power coupler structure for relaxing the fabrication tolerance. Cascade of four couplers with equal coupling ratios and three appropriately designed delay transfers coupling dependence from fabrication-sensitive to fabrication-insensitive L.[21] (Reprinted with permission from the *Journal of Lightwave Technology* © IEEE, 2004.)

An arrayed waveguide grating (AWG) is one more way to realize interleavers and deinterleavers. R&D has also focused on AWGs for DWDM and MUX and DeMUX.[26–30] The main idea behind this topology is to create a phased array whose beam direction or azimuth varies as a function of wavelength, as shown in Figs. 4.20 and 4.21. This way, for several wavelengths, there are several beams, where each one points to a different waveguide in the receive-side aperture. Realization is made by using a power splitter where all the input light power from the input waveguide splits in a free propagation region (FPR) to several waveguides. The power splitter is a kind of lens whose aperture is the plane where the waveguide array is connected. The beam of light is no longer laterally confined and becomes divergent in the lens. The waveguide array length is chosen to be normalized to the central wavelength of the band of interest. Each waveguide in the array is incrementally longer than its adjacent waveguide by $\Delta L$, which equals integer multiplications of wavelengths[27] at the center frequency:

$$\Delta L = m\frac{\lambda_c}{N_g} = m\frac{c}{N_g f_c}, \tag{4.38}$$

where $\lambda_c$ and $f_c$ are the central wavelength and frequency respectively, $m$ is the order of the phased array waveguide, and $N_g$ is the effective index of the waveguide mode. The result is a dispersive response of the phased array. In the center frequency, all individual fields will arrive at the lens aperture of the receiver, equal in phase and separate from each other by integer multiples of $2\pi$.

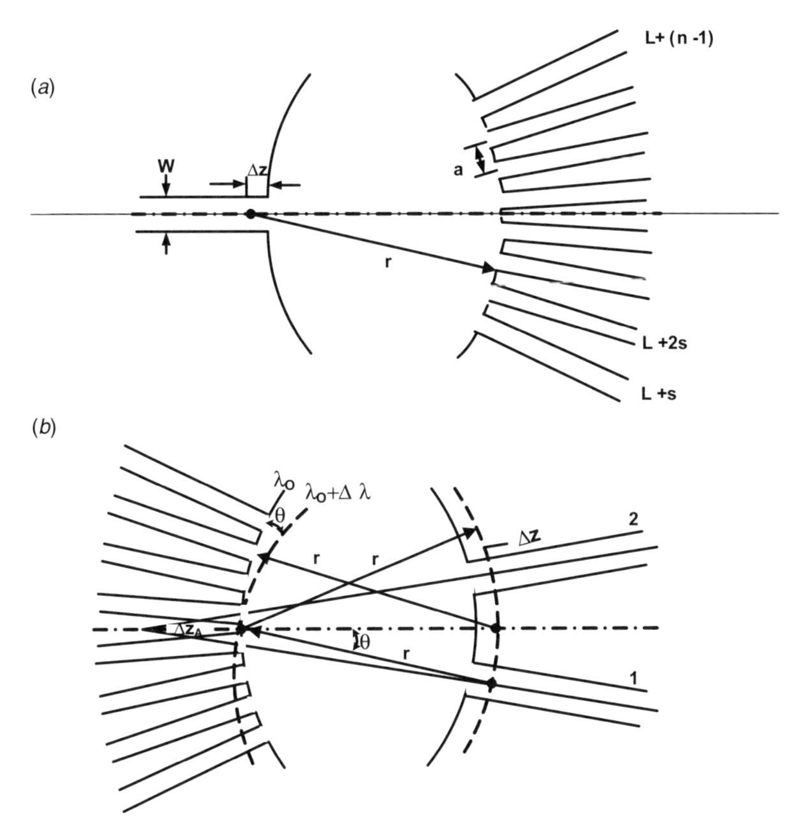


**Figure 4.20** AWG diagram of an array coupler showing how the wavefront plane tilts at the angle $\theta$ as a result of wavelength change of $\Delta\lambda$ off $\lambda_0$.[32] (Reprinted with permission from the *Journal of Lightwave Technology* © IEEE, 1993.)

However, when the wavelength is off from the center frequency wavelength, each waveguide in the phased array would provide an electrical field at the receiver's lens with a different phase. Hence, the beam would be tilted along the image plane of the receiver's lens, and the beam azimuth angle will be changed from its main focal point. The dispersion of the phaser is accomplished by linearly increasing the length of the array waveguides, essentially causing a phase change due to change in the wavelength. In microwaves, this technique is very similar to a phased array antenna, where the frequency is fixed and the feeders of the array elements have a phase control; here, the feeders have a fixed length and the frequency varies. Hence, in both cases, there is the same effect of azimuth divergence of the beam.

Simulation tools for complex integrated optical circuits can be based on microwave computer-aided design (CAD) software. As was explained in Sec. 4.3, there is duality and similarity between microwave analysis methods and optics. Today's professional microwave CAD platforms, such as HP ADS, are suitable for simulation analysis and synthesis of AWGs, DeMUXs, and MZI.[31]

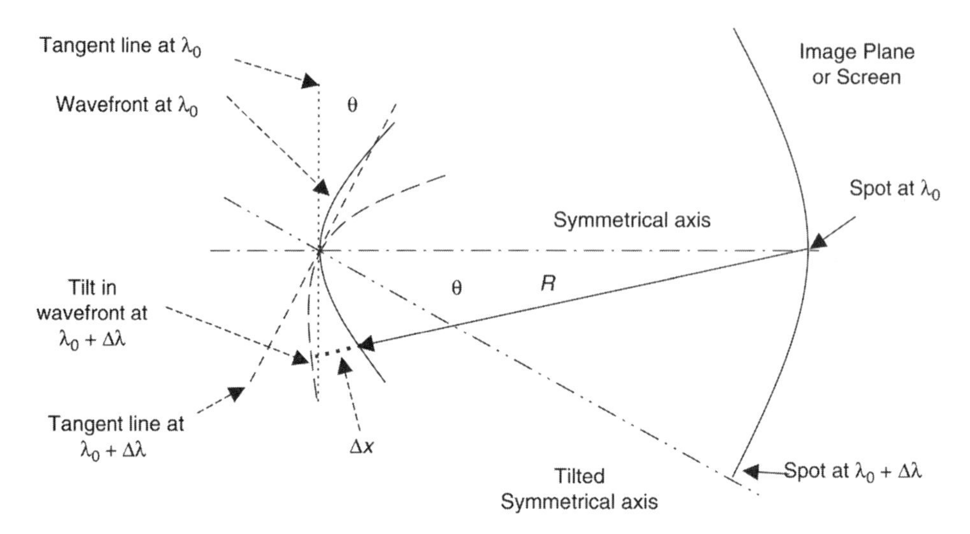


**Figure 4.21** Beam geometry between input and output apertures (planes or screens) of the array lens.

From the geometry of wave propagation shown in Fig. 4.20(b), it is easy to see that the dispersion angle $\theta$ resulting from the phase difference $\Delta\Phi$ between adjacent waveguides is composed of two wave indices, the waveguide $\beta$ index and the FPR lens index $\beta_{FPR}$. Since between two adjacent waveguides, the length increment is $\Delta L$, the phase difference is the difference between the indices of the two adjacent waveguides $\beta(R)$ and $\beta(R+\Delta L)$ (Figs. 4.20 and 4.21), where $R$ is the path length of the lens from the waveguide path to the image plane.

However, the phase occurring in the propagation of a common phase front can differ by integer multiplications of $2\pi$, i.e., $2m\pi$. This is the condition for constructive interference, which states that all wavefront powers are summed in phase.[32] Hence, the following conditions are applicable at the center wavelength $\lambda_0$. At the center wavelength $\lambda_0$, the spot of all-powers summation will be at the symmetrical axis line of the lens coupler. As was explained above, deviation from $\lambda_0$ will cause an azimuth tilt of the spot along the focal line at the image plane (Fig. 4.20):

$$\Delta L = m\lambda_0. \tag{4.39}$$

This condition states that the length increment $\Delta L$ must equal integer multiplications of center wavelength $\lambda_0$ in order to have constructive interference at the center wavelength (frequency). The wave propagation index $\beta$ is defined as follows:

$$\beta = \frac{2\pi}{\lambda_{eff}} = \frac{2\pi}{(\lambda/N)} = \frac{2\pi N}{\lambda}, \tag{4.40}$$

where $N$ is the refractive index of the media, $\lambda$ is the free-space wavelength, and $\lambda_{eff}$ is the wavelength at the media. Using this notation, the wave index of lens, sometimes called array or coupler, is marked as $\beta_{FPR}$, the refractive index is marked as $N_{FPR}$, and the wavelength is marked as $\lambda_{FPR}$. The wave parameters at the waveguide are marked with the subscript "g."

Assume a frequency shift of $\Delta f$ from the center frequency, resulting in a wavelength change of $\Delta\lambda$. This will result in a wavefront tilt of $\theta$ with respect to the input phaser plane, as shown in Fig. 4.20. As a consequence, the summation spot at the focal line travels from the symmetrical axis by the same degree of angle, $\theta$. However, the condition of constructive interference is preserved; therefore, $\Phi$ satisfies the integer multiplications of $2m\pi$. By observing the geometry of the calculation provided in Fig. 4.21, and calculating the path differences from two adjacent waveguides until the focal line or image plane of the array, $\Phi$, is quantified. Special attention should be paid to the media of propagation, owing to the transitioning from the waveguide to the lens, which is reflected by a change in the refraction indices.

As can be observed from Fig. 4.21, at the center frequency, the wavefront merges with the array-input plane. However, when there is a change in the wavelength, the wavefront tilts off the array-input plane around the symmetrical axis. This results in a $\Delta x$ reduction from $R$ or addition to $R$, as shown in Fig. 4.21. Assuming that the output aperture period is marked as $d_a$, which is the lateral spacing, on center lines, of the waveguides in input array aperture, as shown in Fig. 4.22, then $\Delta x$ is given by the following geometrical relation:

$$\Delta x = d_a \sin\theta \tag{4.41}$$

and for small $\theta$ values, it is approximated to

$$\Delta x = d_a\theta.$$

The optical path of the beam is then $R - \Delta x$, and the adjacent beam travels the distance $R$. This optical path occurs in the array media with the FPR refraction index. On the other hand, the initial phase condition for $\Phi$ at the center frequency is given by using wave propagation index of the waveguide $\beta_g$:

$$\Delta\phi = \beta_g \Delta L = 2m\pi. \tag{4.42}$$

This condition should be preserved to have the spot on the image aperture at all wavelengths. That means, the phase difference at the array satisfies the same condition. Hence, the phase condition defines the $\Delta x$ equation:

$$\Delta\phi = \beta_{FPR}\Delta x + \beta_g \Delta L = \beta_{FPR} d_a \sin\theta + \beta_g \Delta L. \tag{4.43}$$

Using the conditions in Eqs. (4.42) and (4.43), the tilt angle $\theta$ is given by[27]

$$\theta = a\sin\left[\frac{(\Delta\phi - m2\phi\pi)}{d_a\beta_{FPR}}\right] \approx \frac{\Delta\phi - m2\pi}{d_a\beta_{FPR}}. \tag{4.44}$$

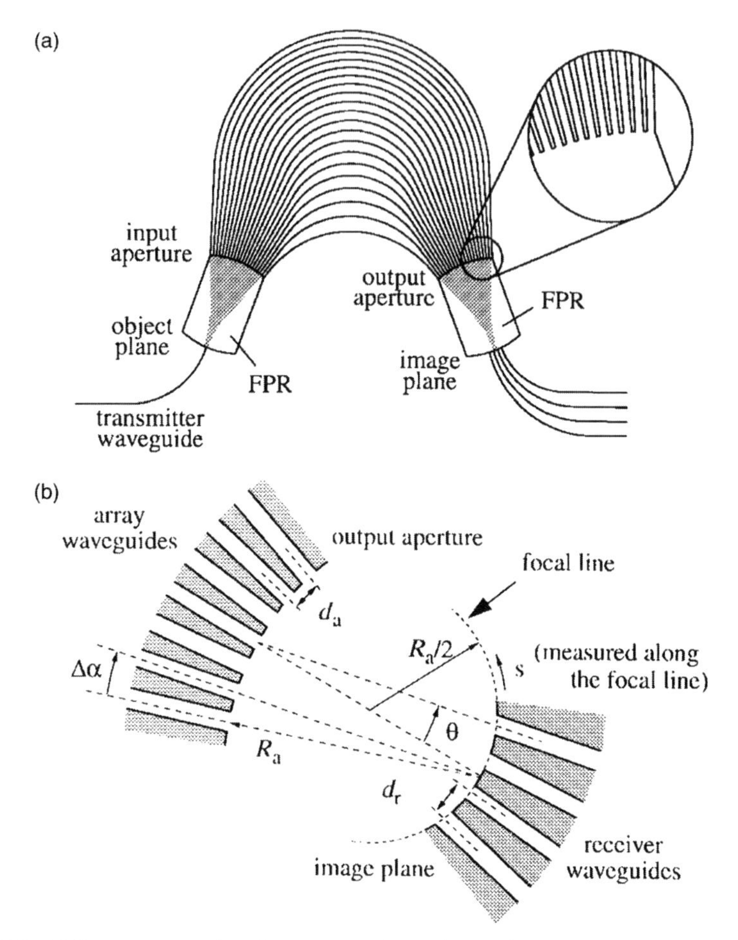


**Figure 4.22** AWG concept showing transmit and receive sides. (a) Layout of the PHASAR demultiplexer. (b) Geometry of the receiver side.[27] (Reprinted with permission from the *Journal of Selected Topics in Quantum Electronics* © IEEE, 1996.)

Equation (4.44) describes the tilt in the focal axis along the image aperture. Now there is a need to find the displacement of the focal spot as a function of frequency deviation from the central wavelength frequency $f_0$. Observing Fig. 4.22, the motivation is to find an expression that describes the change of $S$ versus the frequency $f$. This parameter is called the dispersion factor $D$. From Fig. 4.22, the divergence angle $\Delta\alpha$ between two adjacent waveguides at the input aperture is given by:

$$\Delta\alpha = \frac{d_a}{R}. \tag{4.45}$$

Using the detention of $D$ and the geometrical relations in Fig. 4.22, we get

$$D = \frac{dS}{df} = R\frac{d\theta}{df}. \tag{4.46}$$

Using the relations in Eqs. (4.38), (4.40), (4.42), and (4.44) and some manipulations, Eq. (4.46) becomes[27]

$$D = \frac{1}{f_0} \cdot \frac{\tilde{N}_g}{N_{FPR}} \cdot \frac{\Delta L}{d_a} R = \frac{1}{f_0} \cdot \frac{\tilde{N}_g}{N_{FPR}} \cdot \frac{\Delta L}{\Delta \alpha}, \tag{4.47}$$

where

$$\tilde{N}_g = N_g + f \frac{dN_g}{df}. \tag{4.48}$$

Equation (4.48) describes the waveguide refraction index dispersion for the same reason as described in Sec. 4.1 for an optical fiber. One more interesting observation from Eq. (4.47) is that $R$ is absorbed in the ratio describing $\Delta\alpha$. Hence the dispersive properties of the MUX are not affected by the filling-in of the space between the array waveguides near the apertures due to a finite resolution of the lithography mask.

FSR can be obtained from Eqs. (4.38) and (4.42), which describe the phase condition of $\Delta L$. It was explained that the phase delta, $\Delta\phi$, between two adjacent waveguides equals $2\pi$. That means that there is a frequency period that satisfies this condition. The minimum difference between two frequencies satisfying this condition is called FSR and is given by

$$\text{FSR} = f_2 - f_1 = \Delta f = \frac{c}{N_g \Delta L}. \tag{4.49}$$

AWGs are used as building blocks for $N \times N$ MUXs based on $SiO_2$/Si waveguides. The IL for $N = 7$ is typically lower than 2.5 dB and the cross-talk is less than $-25$ dB.[29] The MUX response is approximately periodic. In each period, the MUX accepts from each waveguide $N$ channels. Similarly each output port receives $N$ channels, one from each input port. Hence, the total channel number transmitted simultaneously equals $N^2$.

Another application is an add–drop MUX on an InP substrate consisting of a $5 \times 5$ phaser DeMUX integrated with MZI, and an electro-optical switch at a total size of $3 \times 6$ mm$^2$ is reported in Ref. [30]. Reported cross-talk values are better than $-20$ dB. The on-chip losses for dropped or added signals and for signals coupled from the input to the output port are lower than 7 and 11 dB, respectively. Optimization and design approaches for achieving low loss, nearly ideal response, low cross-talk, and small size in order to build efficient cost-effective networking WDM building blocks such as star couplers and optical cross connect are described in Refs. [33–37]. Further switching arrangements and optimization of couplers' connections is described in Ref. [38]. Cross-talk mechanisms are important to understand and optimize for having the desired sensitivities for each wavelength channel. Cross-talk may result from the following reasons:[27]

1. Receiver cross-talk due to coupling between receivers through the exponential tails of the field distributions. This may result from the overlap between fields of different wavelengths.

2. Truncation of the field, which results from the finite width of the aperture. The relative spacing of the receiver is given by the ratio $dr/w$, where $dr$ is the spacing between two adjacent waveguides, and $w$ is the waveguide width.

3. Mode conversion resulting from the cross section of the array waveguides. In case it does not satisfy the single-mode condition, the first-order mode is excited at the junctions between the straight and the curved waveguides. This mode can propagate coherently through the array and create "ghost" images. Owing to the difference in the propagation constant between the fundamental and the first-order mode, these images will occur at different spot locations, which may couple the "ghost" image into an undesired receiver.

4. Coupling in the array may result in phase distortion. This phenomenon results from filling the gaps in the array aperture between adjacent waveguides. Hence, this may affect the spot-focusing point and, as a consequence, results in leakage to an undesired receiver.

5. Phase-transfer incoherence, resulting from imperfections in the fabrication process, reduces cross-talk performance as well. It is also one of the sources for inconsistency in the propagation constant. This may be due to small deviations in effective index, which results from local variations in composition, film thickness, or waveguide width, or by inhomogeneous filling in of the gap near the apertures of the phased array.

6. Background radiation caused by scattered light out of the waveguides at the junction or due to rough edges. Beside the waveguides, the light is also guided in shallowly etched ridge guides, or in waveguide structures on heavily doped substrates. In this case, the undoped buffer layer may also act as a waveguide.

In conclusion cross-talk is not a design limitation, but it is due to imperfections in the fabrication process. Typical cross-talk values are in the range of −25 dB for InP and better than −30 dB for silica-based devices. Improving the fabrication process and technology may reduce power leakage between channels due to cross-talk.

### 4.4.3 Optical fiber Bragg gratings

The fiber grating concept is similar to the Bragg grating, described in Sec. 8.4.2. The fabrication process uses a photosensitive fiber.[7,39] For instance, a conventional silica fiber doped with germanium becomes extremely photosensitive. Exposing the fiber to extensive ultraviolet (UV) radiation would change the refractive index of the fiber core. Hence, when the core is exposed to higher UV intensity, the refractive index is increased. This phenomenon was demonstrated in 1978 at the Canadian Communication Research Center (Ottawa, Ontario, Canada).

There are two methods to make a grating in a fiber. The first method, called the "two-beam interferometer method" uses two UV sources, which causes a periodic change in the refractive index.[40–42] The second method of grating writing uses a "phase-mask."[43] A phase mask is an optical grid element that creates diffraction. When the grid is illuminated by a UV beam, it splits the beam into the different diffractive orders used for grating writing. Similar to a Bragg gating, the fiber Bragg period $\Lambda$ is defined as the distance between two refractive indices along the fiber. The fiber Bragg can be divided into two categories, short-period and long-period gratings. In the short-period grating, the period value of $\Lambda$ is of the order of wavelength, while in the long-period, $\Lambda$ is much higher than the wavelength and its range is from a few hundred microns to a few millimeters. The short-grating method is used for filters, add–drop filters, and tunable filters for tunable lasers, while the long-period fibers are used for erbium-doped fiber amplifiers as an equalizer to compensate for their nonflat gain spectrum. This section focuses mainly on short-period fiber gratings. Figure 4.23 provides an illustrative concept of refraction index behavior of fiber grating versus fiber length and lightwave, incident and reflected light.

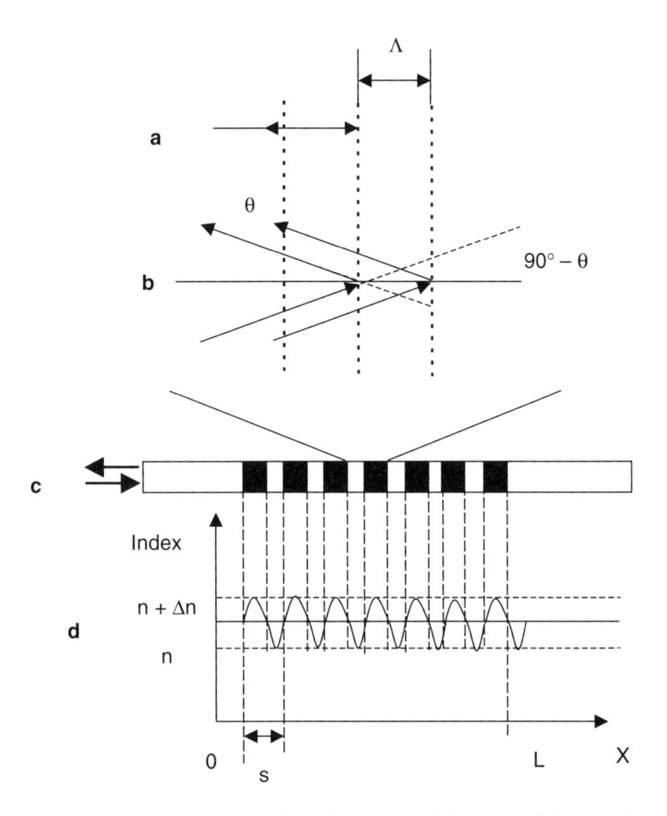


**Figure 4.23** Fiber Bragg grating. (a) Wavelength incident at right angle, $\theta = 90$ deg. (b) Wavelength incident at arbitrary angle $\theta$. (c) Fiber grating length with the periodic-changed refractive index vs. fiber grating length.

From Fig. 4.23(b), it is clear that in order for both waves to recombine constructively, the phase condition should satisfy that the extra path created by the other beam of light be equal to the effective wavelength at the propagation media. Observing the geometry, the condition for constructive diffraction is given by the Bragg condition:

$$2\Lambda \sin\theta = \frac{\lambda}{n}, \tag{4.50}$$

where $n$ is the refractive index of the media and $\lambda$ is the free space wavelength. A special definition case is when $\theta = 90$ deg or $\pi/2$. This condition determines the Bragg wavelength or frequency:

$$\lambda_B = \frac{\lambda}{n} = 2\Lambda. \tag{4.51}$$

Using the result in Eq. (4.51), the Bragg propagation constant or Bragg wave propagation index is given by:

$$\beta_B = \frac{\pi}{\Lambda}. \tag{4.52}$$

The motivation is to use the above results to synthesis WDM filters such as add–drop elements in optical networking applications. Each propagating wave can be described as a summation, superposition of forward-propagating and backward-propagating wave. These waves are represented by the following angular phasors: $\exp(-j\beta z)$ and $\exp(j\beta z)$, respectively. There are several ways to solve the wave equation in fiber grating. One of the methods calls for square profile index approximation of the refraction index and solving the boundary conditions numerically, using the transfer matrix method.[44] The second method is the coupled mode theory.[45]

In the transfer matrix method, consider a plane wave polarized in the $y$ direction and with a refraction index $n$ incident perpendicularly on a film with thickness of $2a$ and refraction index $n'$, and emerging to a medium with refractive index $n$. This case can be presented as a traveling wave within three domains, presented for the $E$ field by:

$$E = \begin{cases} \hat{y}E_1^+ \exp(+j\beta x) + \hat{y}E_1^- \exp(-j\beta x), & x < -a, \\ \hat{y}E_2^+ \exp(+j\beta_1 x) + \hat{y}E_2^- \exp(-j\beta_1 x), & -a < x < a, \\ \hat{y}E_3^+ \exp(+j\beta x) + \hat{y}E_3^- \exp(-j\beta x), & a < x, \end{cases} \tag{4.53}$$

where $\hat{y}$ is a unit vector in the direction of $y$.

The corresponding $H$ field can be found by the relation between $E$ and $H$ in a planar wave:

$$H = kx\frac{E}{\omega\mu}. \tag{4.54}$$

Hence, the $H$ field is given by:

$$H = \begin{cases} \hat{z}\dfrac{n}{c\mu_0}E_1^+\exp(+j\beta x) - \hat{z}\dfrac{n}{c\mu_0}\hat{y}E_1^-\exp(-j\beta x), & x < -a, \\ \hat{z}\dfrac{n'}{c\mu_0}E_2^+\exp(+j\beta_1 x) - \hat{z}\dfrac{n'}{c\mu_0}E_2^-\exp(-j\beta_1 x), & -a < x < a, \\ \hat{z}\dfrac{n}{c\mu_0}E_3^+\exp(+j\beta x) - \hat{z}\dfrac{n}{c\mu_0}E_3^-\exp(-j\beta x), & a < x, \end{cases} \quad (4.55)$$

where $\mu_0$ is the magnetic permeability of free space, and $c$ is the speed of light in vacuum.

Observing Fig. 4.23, it can be seen that Eqs. (4.53) and (4.55) describe one piece of cascading incident and reflected fields within the fiber-grating array. Therefore, the motivation is to find the transfer matrix of one-array piece and find the power of $N$ of that matrix in order to solve the array transfer function. Finding the matrix of the zone of $-a < x < a$, using boundary conditions of tangential field components, and continuity of fields, we get

$$\begin{aligned} m &= \begin{bmatrix} (\cos 2\beta_1 a - j\varepsilon_+ \sin 2\beta_1 a)\exp(2j\beta a) & j\varepsilon_- \sin 2\beta_1 a \\ -j\varepsilon_- \sin 2\beta_1 a & (\cos 2\beta_1 a + j\varepsilon_+ \sin 2\beta_1 a)\exp(-2j\beta a) \end{bmatrix} \\ &\equiv \begin{bmatrix} w & z \\ z^* & w^* \end{bmatrix}, \end{aligned} \quad (4.56)$$

where

$$\varepsilon_\pm = \frac{1}{2}\left(\eta \pm \frac{1}{\eta}\right), \quad \eta = \frac{n}{n'}. \quad (4.57)$$

Thus, the following notation and relations are valid:

$$\begin{bmatrix} E_1^+ \\ E_1^- \end{bmatrix} = m \begin{bmatrix} E_3^+ \\ E_3^- \end{bmatrix}. \quad (4.58)$$

The matrix $m$, as was explained above, refers to a transfer of a single part of the grating array, as shown in Fig. 4.23. The goal is to find the matrix of the whole interval. The electric field between the boundaries of the $m$ element of the refraction index $n$ can be written as follows:

$$E_m(x) = A_m \exp[j\beta(x - ms)] + B_m \exp[-j\beta(x - ms)], \quad (4.59)$$

where $(m-1)s + a < x < ms - a$; and $0 < m < N$.

This means duplicating the basic boundary case $N$ times at a regular interval $s$, and knowing that $s > 2a$. It also should be noted that within $-a < x < a$, the refractive index is described by $n' = n + \Delta n$. A whole cycle period of the refraction factor modulation is "$s$," and $\beta$ is the wave index at the $n$-zone index of refraction, where $\beta_1$ refers to the $n'$ zone (see Fig. 4.23). In this case, the transfer function of two consecutive grating cells is given by:

$$\begin{bmatrix} A_m \\ B_m \end{bmatrix} = \begin{bmatrix} (\cos 2\beta_1 a - j\varepsilon_+ \sin 2\beta_1 a)\exp(2j\beta a) & j\varepsilon_- \sin 2\beta_1 a \\ -j\varepsilon_- \sin 2\beta_1 a & (\cos 2\beta_1 a + j\varepsilon_+ \sin 2\beta_1 a)\exp(-2j\beta a) \end{bmatrix} \times \begin{bmatrix} \exp(-j\beta s) & 0 \\ 0 & \exp(+j\beta s) \end{bmatrix} \times \begin{bmatrix} A_{m+1} \\ B_{m+1} \end{bmatrix}. \tag{4.60}$$

By defining the $m$ transfer function at the cascaded grating as

$$P = \begin{bmatrix} (\cos 2\beta_1 a - j\varepsilon_+ \sin 2\beta_1 a)\exp(2j\beta a) & j\varepsilon_- \sin 2\beta_1 a \\ -j\varepsilon_- \sin 2\beta_1 a & (\cos 2\beta_1 a + j\varepsilon_+ \sin 2\beta_1 a)\exp(-2j\beta a) \end{bmatrix} \times \begin{bmatrix} \exp(-j\beta s) & 0 \\ 0 & \exp(+j\beta s) \end{bmatrix} = m \times \begin{bmatrix} \exp(-j\beta s) & 0 \\ 0 & \exp(+j\beta s) \end{bmatrix}, \tag{4.61}$$

the transfer function from the fiber grating input to the $N$ cell is given by:

$$\begin{bmatrix} A_0 \\ B_0 \end{bmatrix} = P^N \times \begin{bmatrix} A_N \\ B_N \end{bmatrix}. \tag{4.62}$$

The approximation result of the $P^N$ matrix is given by Cayley–Hamilton theorem[46]:

$$P^N = \begin{bmatrix} \begin{matrix} (\cos\beta_1 r - j\varepsilon_+ \sin\beta_1 r)\exp(j\beta r) \\ \times \exp(-j\beta s)U_{N-1}(\xi) - U_{N-2}(\xi) \end{matrix} & j\varepsilon_- \sin(\beta_1 r)U_{N-1}(\xi)\exp(+j\beta s) \\ -j\varepsilon_- \sin(\beta_1 r)U_{N-1}(\xi)\exp(-j\beta s) & \begin{matrix} (\cos\beta_1 r + j\varepsilon_+ \sin\beta_1 r)\exp(-j\beta r) \\ \times \exp(+j\beta s)U_{N-1}(\xi) - U_{N-2}(\xi) \end{matrix} \end{bmatrix} \tag{4.63}$$

where $U_N$ is the $N$th Chebychev polynomial of the second kind,[47] and the $r$ notations symbolize the width of the $n'$ refraction value and $l$ symbolizes the $n$ refraction value. Thus, $s = l + r$, and $l = r = a$, then $s = 2a$. The value of $\xi$ is given by:

$$\xi = \frac{1}{2}\mathrm{tr}(P) = \frac{1}{2}[(\cos\beta_1 r - j\varepsilon_+ \sin\beta_1 r)\exp(j\beta r)\exp(-j\beta s) + (\cos\beta_1 r + j\varepsilon_+ \sin\beta_1 r)\exp(-j\beta r)\exp(j\beta s)], \tag{4.64}$$

where tr($P$) is the transmission of $P$. This expression can be simplified using Euler's identities.[48] From Eqs. (4.62) and (4.63), the transfer transmission $T_N$

and the reflectivity function $R_N$ can be obtained while assuming an ideal reflection without absorption:

$$T_N = \frac{1}{1+\left[\left|j\varepsilon_- \sin(\beta_1 r)\right| U_{N-1}(\xi)\right]^2}, \tag{4.65}$$

$$R_N = 1 - T_N, \tag{4.66}$$

where

$$\xi = \cos(\beta_1 r)\cos(\beta l) - \varepsilon_+ \sin(\beta_1 r)\sin(\beta l), \tag{4.67}$$

$$U_N(\xi) = \frac{\sin(N+1)\gamma}{\sin\gamma}, \tag{4.68}$$

$$\gamma \equiv \cos^{-1}\xi. \tag{4.69}$$

Since the refraction factor $\Delta n$ is minute, $\varepsilon_+ \cong 1$, and Eq. (4.67) can be simplified as the cosine of the sum of angles. The main idea in creating a fiber-grating filter is to maximize the reflection at the desired wavelength. That means minimizing $T_N$ to zero and $R_N$ goes to a value of 1. Hence $\gamma$ should be equal to $k\pi$, where $k$ is an integer number ranging from 0, $\pm 1$, $\pm 2$ and so on, to maximize $U_N(\xi)$. $\xi$ values range between $-1$ and 1; then by the L'Hospital theory, the values of $U_N(\xi)$ are $U_N(1) = N+1$, and $U_N(-1) = (-1)^N(N+1)$. In addition, the function inside the absolute value is at a maximum value when $\beta_1 r$ equals odd multiples of $\pi/2$ or is close to that value. These yield several conditions for tuning. At the region of $r$, $n'$ is fixed and constant value; hence the degree of freedom is to optimize $r$ for getting $\beta_1 r = (\pi/2) + k\pi$. In this case when $\gamma = k\pi$, then after all limits are investigated, it is obvious by observing Eq. (4.65) that $T_N$ converges to zero approximately as $T_N = C/N^2$, where $C$ is a constant. Under these conditions, $R_N = 1 - C/N^2$. Following the above analysis, the maximum reflectivity condition at each domain of refractive index requires the following constraints:

$$\beta_1 r = \frac{n' r 2\pi}{\lambda_0} = \frac{\pi}{2} \Rightarrow r = \frac{\lambda_0}{4n'}, \tag{4.70}$$

$$\beta l = \frac{n l 2\pi}{\lambda_0} = \frac{\pi}{2} \Rightarrow l = \frac{\lambda_0}{4n}. \tag{4.71}$$

The refractive index modulation period is $s = l + r$; by using the conditions given in Eqs. (4.70) and (4.71), the period is given by:

$$s = r + l = \frac{\lambda_0}{4}\left(\frac{1}{n'} + \frac{1}{n}\right) = \frac{\lambda_0}{2n_{\text{eff}}}. \tag{4.72}$$

This defines the effective refraction factor of the grating at the tuned wavelength $\lambda_0$:

$$n_{\mathrm{eff}} = \frac{2n'n}{n+n'}. \tag{4.73}$$

In case the wavelength is out of the tuned range, this filter is supposedly swept by a tunable laser sweeper; the question is what is the filter frequency response. The sweeping process is from a wavelength lower than the tuned value to one higher than the tuned value. The constraints in Eqs. (4.70) and (4.71) are fixed values for $r$ and $l$ due to the geometry of the grating. Thus the wave index varies as the wavelength changes due to the change in its denominator. The results of these conditions are

$$\widehat{\beta_1 r} = \frac{\pi}{2}\left(\frac{\lambda_0}{\lambda}\right) = \frac{\pi}{2}\rho, \tag{4.74}$$

$$\widehat{\beta_1 l} = \frac{\pi}{2}\left(\frac{\lambda_0}{\lambda}\right) = \frac{\pi}{2}\rho. \tag{4.75}$$

During the sweep of $\lambda$, the value of $\rho$ varies from a value greater than 1 to smaller than 1; however, it is higher than 0 but converges to 0. The value of $\sin(\beta_1 r)$ varies from 1 to 0 monotonically, as the $\lambda$ sweep travels from $\lambda_0$ to a higher or lower value within the range $0 \leq \rho \leq 2$. $\xi$ varies from $-1$ to 1. For the range of $-1$ to $+1$, the Chebychev polynomial $U_{N-1}(\xi)$ has $N-1$ zeros[49] as given below:

$$\xi_m^{N-1} = \cos\left(\frac{m}{N}\pi\right), \tag{4.76}$$

where $m$ ranges from 1 to $N-1$. By using Eqs. (4.57), and (4.67) changes to:

$$\xi = \cos(\beta_1 r + \beta l) - \frac{\Delta n^2}{2n(n+\Delta n)}\sin(\beta_1 r)\sin(\beta l). \tag{4.77}$$

The minima wavelengths are given by combining Eqs. (4.76) and (4.77):

$$\lambda_m = \frac{\pi}{2}\frac{\lambda_0}{\cos^{-1}\sqrt{\dfrac{B+\cos(m\pi/N)}{1+B}}}, \tag{4.78}$$

where

$$B \equiv 1 + \frac{\Delta n^2}{2n(n+\Delta n)}. \tag{4.79}$$

Using these results with those of Eqs. (4.65) and (4.66) may provide the wavelength response of the grating filter. Realization of an optical band-pass filter (BPF) using an FBG is based on the concept of routing the desired reflected wavelength to a specific port. All other wavelengths will travel through the fiber except the reflected wavelength. Therefore, the reflected wavelength port acts as a BPF. The through port, which routes all other wavelengths, serves as a band-stop filter for the specific wavelength reflected by the gratings. Figure 4.24 illustrates three typical methods of optical BPF realization.

The advantagc of the circulator method over the other two configurations is its low-optical IL of 0.7 to 0.8 dB. This method of BPF realization can be used for add–drop applications; in this case, the reflected wave is the dropped wavelength. The add wavelength can be realized by using an additional circulator at the output of the Bragg gratings, as shown in Figure 4.24.

An optical BPF, as described above, is a building block for Bragg grating arrays (BGAs) and optical-code division multiple access (OCDMA).[41] The main idea of BGAs is based on writing long Bragg gratings fibers, where it may be considered as several-wavelength gratings cascaded in a series. Each one of these is tuned to a different wavelength spaced by 100 GHz per the ITU grid, enabling the transfer of 10 GB/s per channel. This concept may be used for OCDMA. The concept of OCDMA is provided in Figs. 4.25 and 4.26.

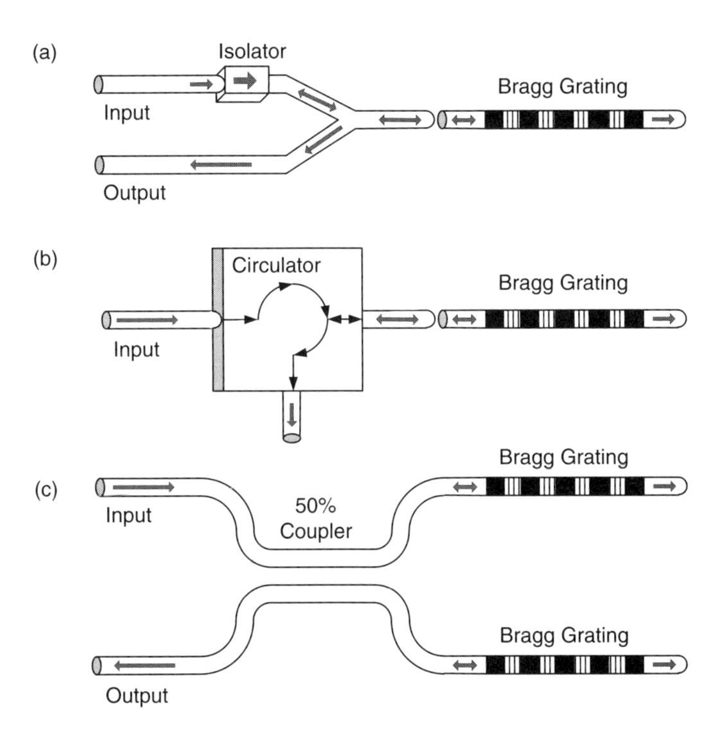


**Figure 4.24** Three main methods to realize an optical BPF. (a) Using an optical combiner. (b) Using optical circulator. (c) Using 50% directional coupler; in this case both gratings are tuned to the same wavelength $\lambda_0$.

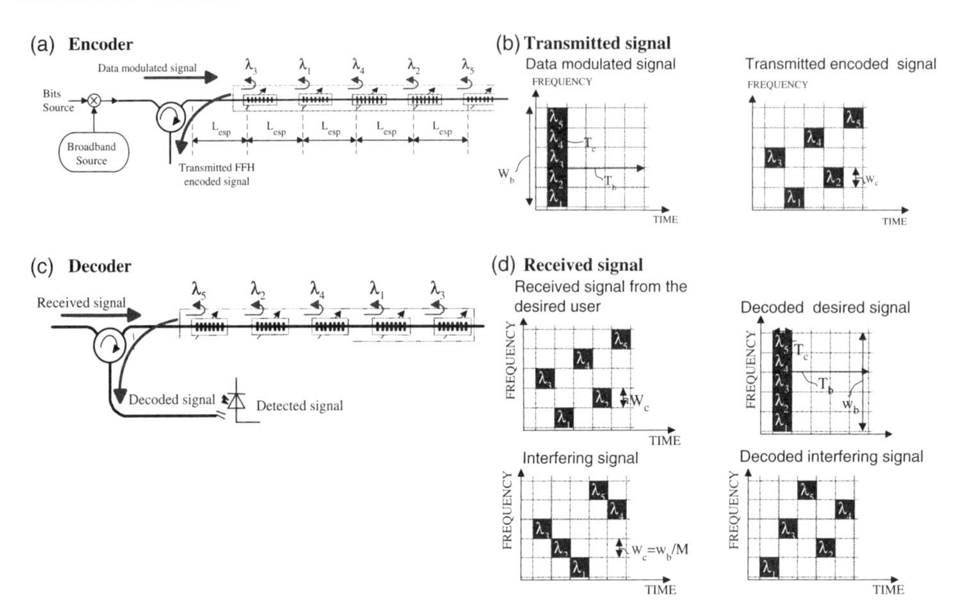


**Figure 4.25** Encoding and decoding of OCDMA using BGA, showing building blocks and signal. (a) Encoder. (b) Transmitted signal. (c) Decoder. (d) Received signal and orthogonal-coded interfering signal.[41]

The beat sequences modulate a broadband CW source that is a tunable laser transponder. The BGA encoder selectively reflects predetermined bands and therefore introduces delays between the reflected pulses at each wavelength. At the receiver section, the decoder consists also of a BGA identical to the transmitter's BGA. The pulses from the desired-encoder BGA matching the decoder BGA consequently are synchronized, and the BGA is now connected in opposite order. Pulses transmitted from an interfering source, which does not match the encoder BGA, are dispersed. Sharing frequency space is achieved by assigning orthogonal or nearly orthogonal wavelength coding for each user. This method can be expanded, as shown in Fig. 4.26.

BGA-designed systems have improved bit rates owing to efficient spectrum utilization. The bit rate of the system is directly related to the encoder length, because the data pulse has to exit the decoder prior to the entrance of the next pulse. As a consequence, the higher the data rate, the shorter the encoder, and fewer the users. These two facts are opposing limitations, since for higher modulation rates, the BGA would have fewer wavelengths and the number of orthogonal codes is reduced. In addition, high matching between the encoder and the decoder is required in order to accomplish higher SNR by improving the decoding correlation. An increase of users will decrease BER performance.[50,51]

As was explained above, BGAs consist of several gratings tuned for different wavelengths. These fiber gratings are written over the same fiber, eventually creating a series of connected wavelength reflectors, which results in long grating writing. In addition, the requirement is to have accuracy so the desired

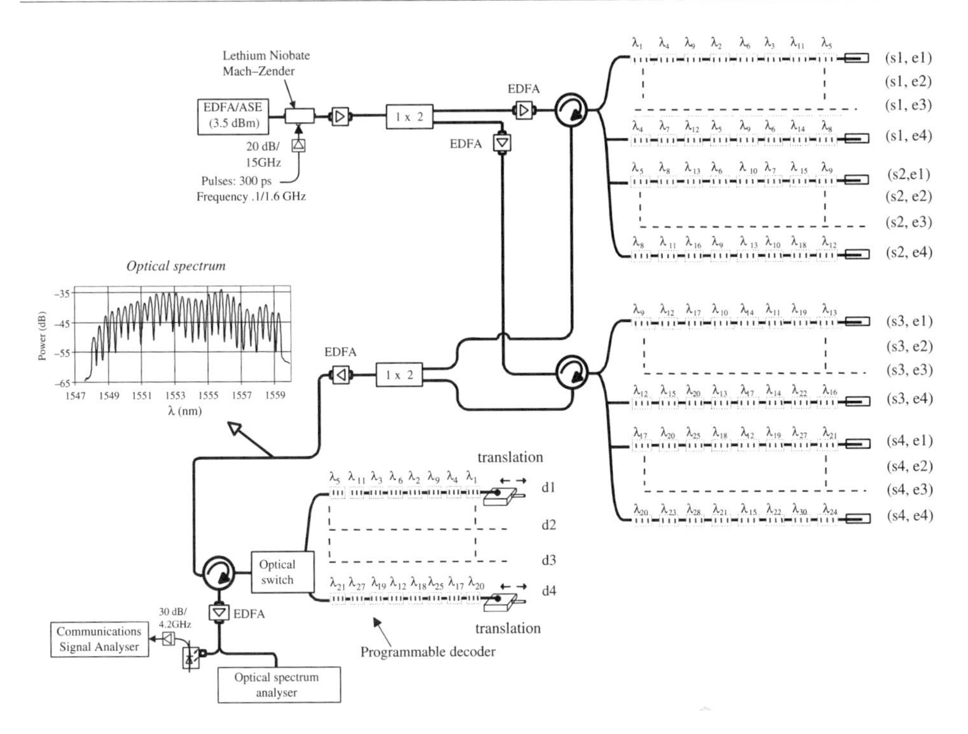


**Figure 4.26** OCDMA system application with 16 BGA encoders and 4 BGA decoders.[41] (© SPIE.)

wavelengths' reflections do not any cross-talk with the adjacent wavelength channel. There are several methods, as was previously mentioned, to tune the Bragg wavelength of the fiber grating during UV exposure. The interferometric method, using the two-mirrors interferometer, is generally used for grating writing with lengths generally shorter than 1 cm. On the other hand, when using a phase mask, tuning of the Bragg wavelength can be achieved over a limited range, typically less than 10 nm.

Several methods have been explored and suggested to control the Bragg wavelength, which include tilting the fiber, rotating the scanning mirrors, starching the fiber, varying the curvature of the writing wavefront, and moving the fiber relative to the phase mask. For writing BGAs, a Sagnac-type interferometer, originally proposed by Ouellette and Krug, is used. This interferometer writer combines the flexibility of the interferometer technique by allowing the Bragg wavelength to be tuned over a wide range, together with the phase mask scanning advantage, good-quality longer gratings can be produced. Figures 4.27 to 4.29 illustrate these three methods of UV writing.

The Sagnac interferometer used to write BGAs operates as follows: the phase mask diffracts the UV light coming from the laser in the $\pm 1$ order, thereby producing two counter-propagating beams in the interferometer. The optical fiber is placed slightly above the phase mask and the two mirrors are adjusted with a small out-of-plane tilt to recombine the UV beams at the optical fiber position.

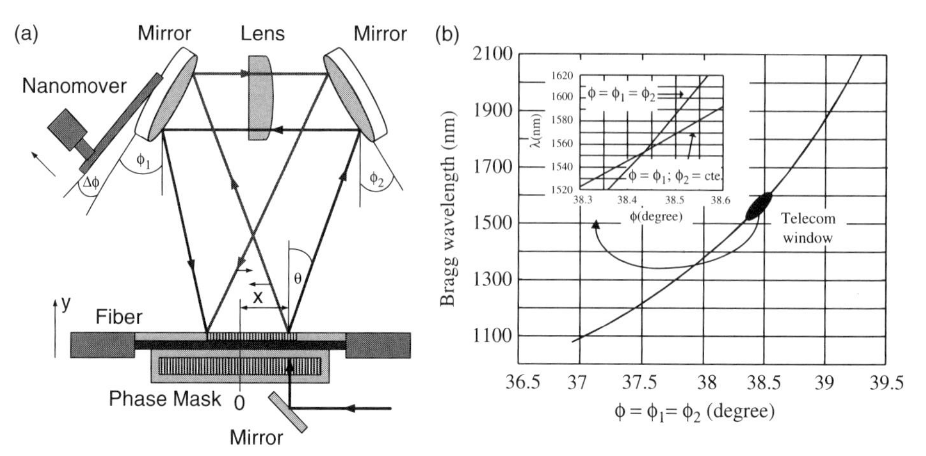


**Figure 4.27** (a) Sagnac interferometer. (b) Tuning of Bragg wavelength as a function of mirror angle.[41]

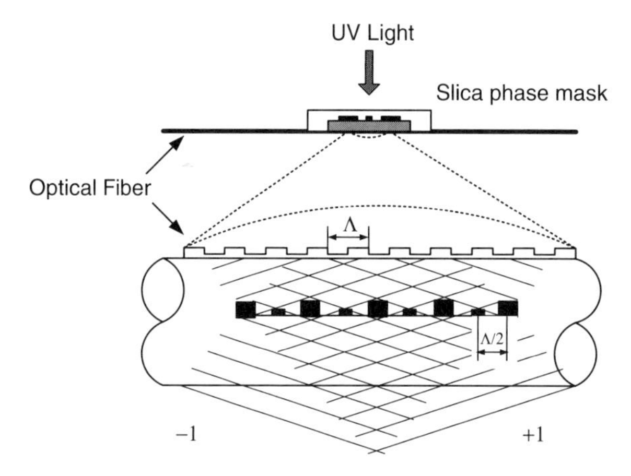


**Figure 4.28** Phase mask UV writing process.[41]

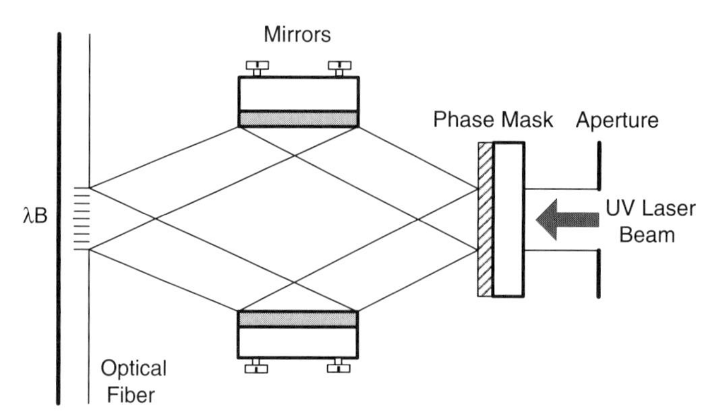


**Figure 4.29** Two-beam interferometer UV writing process. The Bragg wavelength $\lambda_B$ is determined by the adjustment of $\theta_B$.[40,42] (Reprinted with permission from *NASA Langley Research Center.*)

Due to the fact that both beams interact with two mirrors, this interferometer is immune against some vibration modes of the mirrors, and a stable writing interferometer is realized. The cylindrical lens is introduced between the two mirrors to focus the light along the optical fiber axis. When scanning of the phase mask is performed, the exposed region of the fiber moves in the direction opposite to the translation of the UV beam (see Fig. 4.25). From the geometry in Fig. 4.25, the Bragg wavelength changes according to

$$\lambda_B = \frac{n_{\text{eff}}\lambda_{\text{UV}}}{\sin(\pi - 2\varphi_1 - 2\varphi_2 - \theta)}, \tag{4.80}$$

where $n_{\text{eff}}$ is the effective index of the guided mode, $\lambda_{\text{UV}}$ is the writing wavelength, $\varphi_1$ and $\varphi_2$ are the mirrors' angles. Taking the differential of $\Delta\lambda_B$ with respect to the incremental change of $\Delta\varphi$, while simultaneously moving the two mirrors under the condition of $\varphi_1 = \varphi_2 = \varphi$, tuning the relative Bragg wavelength is given by expression (4.81)

$$\frac{\Delta\lambda_B}{\lambda_B} = 4\cot(\pi - 4\varphi - \theta)\Delta\varphi, \tag{4.81}$$

and the angular sensitivity in nm/rad is given by

$$\frac{\Delta\lambda_B}{\Delta\varphi} = n_{\text{eff}}\lambda_{\text{UV}}\frac{4\cot(\pi - 4\varphi - \theta)}{\sin(\pi - 4\varphi - \theta)}. \tag{4.82}$$

A special case occurs when $\varphi_1 = \varphi_2 = \varphi = (\pi/4) - (\theta/2)$, where the interference takes place directly above the phase mask and the Bragg wavelength is given by the common expression of $\lambda_B = n_{\text{eff}}\Lambda_{\text{PM}}$, where $\Lambda_{\text{PM}}$ is the phase mask period.

The increasing application of DWDM transmission schemes with bit rates of 10 Gb/s and 50 GHz channel spacing, which is denser than the standard ITU grid, places a strict requirement on dispersion management. The use of chirped FBGs to compensate for the dispersion of fiber links has become increasingly well known over the recent years. In subcarrier multiplexed (SCM) lightwave systems such as OCDMA, it is important to maintain a linear group delay within the channel BW, which is opposite to the one caused by the fiber. On the analog side, tapered linearly chirped gratings (TLCGs) are used to reduce harmonics and intermodulations in long-distance haul broadband SCM systems. Composite second orders (CSOs) and composite triple beats (CTBs) are improved. For this purpose, FBGs should be linearly chirped, as presented in Fig. 4.30.[52]

The analysis of a chirped fiber grating is done by solving the coupled wave equations using the perturbation theory method.[52] The propagation of light in a fiber grating is defined by the refractive index variation, which is governed by the coupled equation[45,52]

$$\left.\begin{aligned} \frac{du(z)}{dz} &= +j[\beta(z)u(z) + k(z)v(z)] \\ \frac{dv(z)}{dz} &= -j[\beta(z)v(z) + k(z)u(z)] \end{aligned}\right\}, \tag{4.83}$$

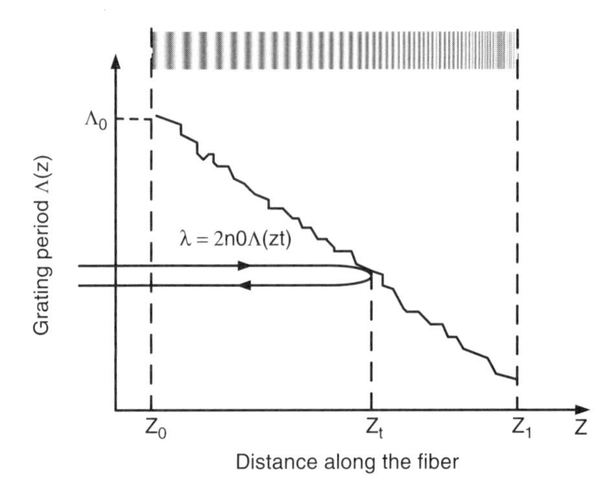


**Figure 4.30** Reflection of light from chirped Bragg fiber grating. The lower frequency, which refers to a longer wavelength, is near the input of the fiber while the shortest wavelength with the highest frequency is at the end of the fiber. The grating period becomes shorter at the higher frequencies.[52] (Reprinted with permission from *Optics Express* © OSA, 2002.)

while the two traveling waves are $u(z)$ backward and $v(z)$ forward

$$\left.\begin{aligned} u(z) &= U(z)\exp\left(j\frac{2\pi}{\Lambda(z)}\right)z \\ v(z) &= V(z)\exp\left(-j\frac{2\pi}{\Lambda(z)}\right)z \end{aligned}\right\}. \tag{4.84}$$

The coupling coefficient is $k(z)$ and is given by[52]

$$k(z) = \frac{\pi\Delta n(z)}{2n_0\Lambda_0}, \tag{4.85}$$

where $\Delta n(z)$ is the amplitude of the refraction index modulation. Substitution of the variables given by Eq. (4.84) into Eq. (4.83) results in the conditions of Eq. (4.86). This assumes that $\Lambda(z) = \Lambda$ and that no dependency of $z$ exists; hence, there is no chirp:

$$\left.\begin{aligned} \frac{dU(z)}{dz} &= +j\left[\left(\beta(z) - \frac{\pi}{\Lambda}\right)U(z) + k(z)V(z)\right] \\ \frac{dV(z)}{dz} &= -j\left[\left(\beta(z) - \frac{\pi}{\Lambda}\right)V(z) + k(z)U(z)\right] \end{aligned}\right\}. \tag{4.86}$$

The next assumption is that $\beta(z)$ is the Bragg wave index. This parameter can be expanded using a Taylor first-order term around a certain BW of the Bragg frequency. From Eq. (4.84), the assumption is $\Lambda(z) = \Lambda$ for no chirp; thus the following relation is applicable:

$$\beta = \beta(\omega_B) + \left.\frac{d\beta}{d\omega}\right|_{\omega=\omega_B} \times (\omega - \omega_B) = \beta(\omega_B) + \frac{\omega - \omega_B}{v_g}, \tag{4.87}$$

where $\omega_B$ is the Bragg frequency, and $v_g$ is the group velocity of the unperturbed medium. Defining the parameter $\psi$ as

$$\psi(\omega) = \frac{\omega - \omega_B}{v_g}. \tag{4.88}$$

From Eq. (4.83), the value $[\beta(z) - \pi/\Lambda]$ measures the deviation $\beta(z)$ from the wave index at the Bragg period; this may be the detuning value $\psi$. Hence placing $\psi$ in Eq. (4.86) instead results in a second-order differential equation:[45]

$$\frac{d^2V}{dz^2} = \left(|k|^2 - \psi^2\right)V = \gamma^2 V. \tag{4.89}$$

The solution of this equation is from the type of $V = A \times \exp[-(\sqrt{|k|^2 - \psi^2})z]$. This solution is somewhat similar to a microwave waveguide. It has a cut-off frequency at $|k| = \psi$, and when $|k| > \psi$, it has attenuation and all the energy is reflected. When $|k| < \psi$, it is a pass band with harmonic solution of $-\sin(\sqrt{|k|^2 - \psi^2})z$. The reflection factor $\Gamma$ is the ratio of the reflected wave to the propagating wave, given by $\Gamma = V/U$. The goal is to manipulate these results for chirped FBGs (CFBGs). One intuitive conclusion is that in order to create a chirp, $\gamma$ should have several cut-off frequencies. That means $k$ and $\psi$ should be equal at different wavelengths at different $z$ locations over the fiber. Hence, the grating period $\Lambda$ should vary as a function of $z$, described by Ref. [52], describes Fig. 4.30:

$$\Lambda(z) = \Lambda_0 + C(z - z_0) + \Delta\Lambda(z), \tag{4.90}$$

where $\Delta\Lambda(z)$ is the grating period noise and $C$ is the grating period chirp.

In addition to reducing group-delay ripple in CFBG filters used for dispersion compensation, the amplitude function must be dc apodized (tapered). This method eliminates side-lobes by maintaining a constant average index in the gratings, and thus the Bragg wavelength is unchanged through the length of the device, which are called TLCGs. This requires a change in Eq. (4.86) by multiplying it by additional phase function:[53]

$$\left.\begin{aligned} \frac{dU(z)}{dz} &= +j[\psi(\omega)U(z) + k(z)V(z)\exp(j\varphi z)] \\ \frac{dV(z)}{dz} &= -j[\psi(\omega)V(z) + k(z)U(z)\exp(-j\varphi z)] \end{aligned}\right\}, \tag{4.91}$$

where $\varphi(z)$ is given by Ref. [53]:

$$\varphi(z) = C\left(\frac{z}{Z_1}\right)^2, \tag{4.92}$$

in which $C$ is a chirp coefficient and $Z_1$ is FBG's length as it appears in Fig. 4.31, assuming $Z_0 = 0$. There are several tapers or adoption functions. To compare between taper functions, the following function is defined as

$$k_0 L_{\text{eq}} = \int_{-z_1/2}^{z_1/2} k(z)\text{d}z. \tag{4.93}$$

The result of this expression provides a different equivalent length for a given $k_0$. As for the coupling coefficients, there are several options[53] as provided below:

- Gauss function, where $G$ is a taper-adjustable parameter and $k_0$ is at $z = 0$, i.e., the middle of the fiber-grating length since the integral is from $-Z_1/Z_2$

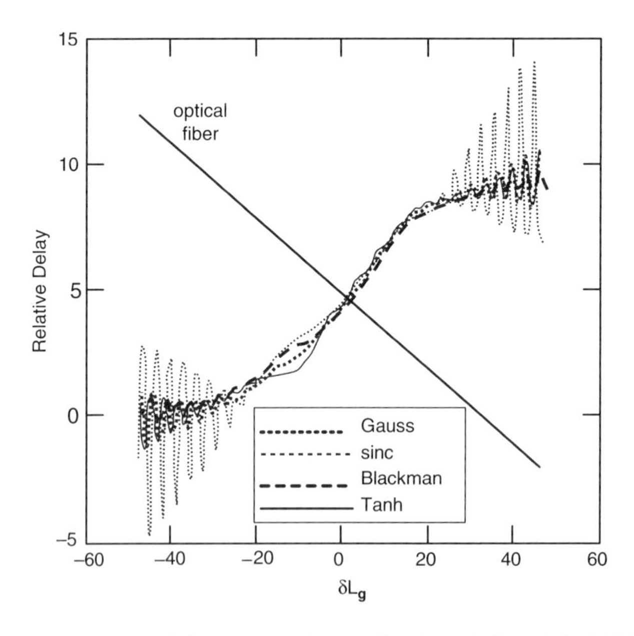


**Figure 4.31** Relative group delay versus the grating length $L_g$, while $k(z)$ is a parameter. Fiber spool length is 100 km; the dispersion $D = 17$ ps/(km × nm); $n_0 = 1.45$. It can be observed for 1550 nm, the sinc and Blackman can provide almost linear compensation, while the other two functions may introduce some ripple at the BW edges. Note that the sum of fiber delay and the TLCG delay results in a flat line with fringes at both of the BW edges.[53] (Reprinted with permission from the *Journal of Lightwave Technology* © IEEE, 1997.)

to $Z_1/Z_2$, so the origin is shifted:

$$k(z) = k_0 \exp\left(-G\frac{z}{Z_1}\right). \quad (4.94)$$

- Hyperbolic tangent:

$$k(z) = \frac{k_0}{2}\left(1 + \frac{\tan h\{4 - 8\exp[A\log(2z/Z_1)]\}}{\tan h(4)}\right), \quad (4.95)$$

where $A$ is a taper-adjustable parameter:

$$k(z) = k_0 \mathrm{sinc}^A\left[\frac{1}{2}\left(\frac{2z}{Z_1}\right)\right]. \quad (4.96)$$

- Blackman:

$$k(z) = k_0 \frac{1 + (1 + B)\cos(2\pi z/Z_1) + B\cos(4\pi z/Z_1)}{2 + 2B}, \quad (4.97)$$

where $B$ is a taper-adjustable parameter.

The minimum length $L_{g\text{-min}}$ of fiber grating that can provide sufficient delay compensation for a given 3 dB BW ($B$) and a given optical fiber link distance ($L$) is calculated as:

$$L_{g\text{-min}} = LB\left(\frac{\lambda^2 D}{2n_0}\right). \quad (4.98)$$

Once the optimal length $L_{g\text{-min}}$ is defined, then the optimal chirp coefficient can be adjusted by:

$$F_{opt} = \frac{8Z_1^2 n_0^2 \pi}{\lambda^2 LDc}, \quad (4.99)$$

where $n_0$ is the refractive index, $\lambda$ is the center optical frequency wavelength, $c$ is the speed of light in vacuum, and the dispersion is given by $D$ [ps/(km × nm)]. In case the length is shorter than $F_{opt}$, there is not full fiber delay compensation. On the other hand, a tapered linearly chirped grating (TLCG) longer than $F_{opt}$ results in a delay introduced by the fiber, which also results in nonlinear distortions. Figure 4.32 illustrates TLCG application.

## 4.5 Optical Isolators and Circulators

Optical isolators and circulators are essential components in optical modules and optical networks. As was demonstrated previously in Sec. 4.4.3, circulators may be used for add–drop in WDM networks;[54–56] this is a three-port passive device and it has a direction. It is not a reciprocal passive device. Similarly the isolator has a

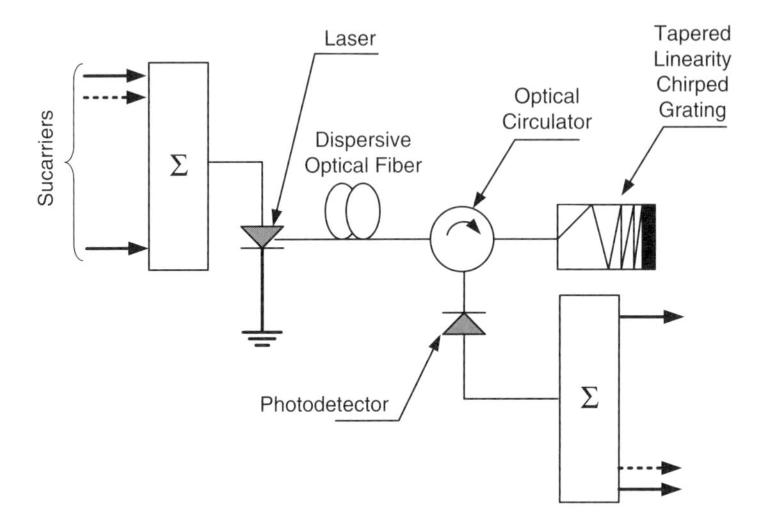


**Figure 4.32** SCM utilizing a tapered linearly chirped grating.[52] (Reprinted with permission from the *Journal of Lightwave Technology* © IEEE, 1997.)

direction and it is not a reciprocal passive device. Isolators are commonly used in optical blocks such as optical triple ports (OTPs), modules such as triplexers, or optical dual port (ODP) modules such as duplexers that improve optical return loss of the transmitting laser or laser modules such as thermo-electric–cooled DFB lasers. The structure of optical blocks are explained in Chapter 9. Both devices, isolators and circulators, are based on the same Faraday effect, discovered by Michael Faraday in 1842. The phenomenon is that a plane of polarized light is rotated when it is transmitted through glass that is contained in a magnetic field. The direction of rotation is independent of the direction of the light propagation, but depends only on the direction of the magnetic field. A polarity rotation device is a key building block of isolators and circulators, and is called a Faraday rotator.

### 4.5.1 Optical isolators

Optical isolators are categorized into two types, polarization-sensitive and polarization-insensitive,[57] as shown in Fig. 4.33.

The concept of a polarized sensitive isolator is given by Fig. 4.33(a). It is composed of three building blocks. At the input of the isolator, there is a front polarizer, then a Faraday rotator, and then a back polarizer. When the input beam enters the isolator, its vertical or horizontal polarized component passes via the first linear front polarizer, depending on the polarizer type. The Faraday rotator shifts the polarity by rotating it 45 deg. The rotator applies a magnetic field on the light and its length and field strength determines the amount of rotation added to the input beam. The back polarizer is placed at the output of the rotator and is aligned to it with 45 deg of polarization as well. Hence, the rotated beam passes

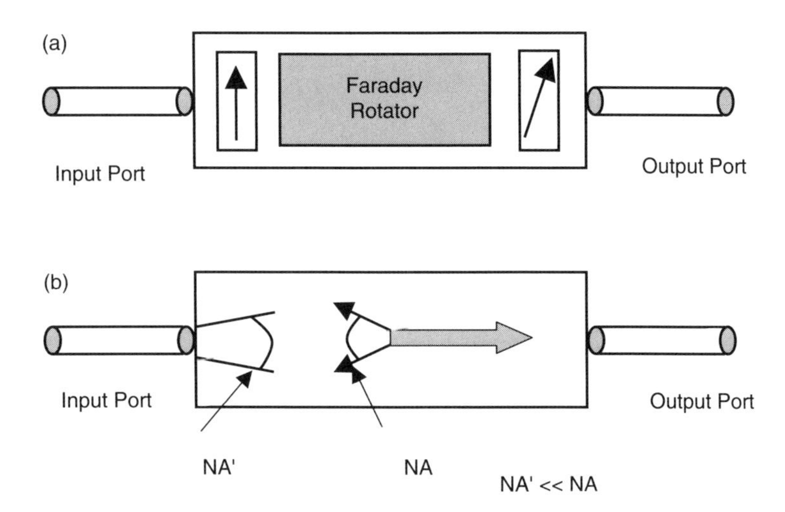


**Figure 4.33** Conceptual block diagram of isolator types. (a) Polarized-sensitive using Faraday rotator. (b) Polarized-insensitive. NA symbolizes numerical aperture.

is it too. In fact, the polarizer is a kind of filter that allows specific polarities to pass. For undesired beams that travel back into the isolation output, the 45-deg polarized beam passes the back polarizer and the rotator adds another 45 deg. Hence, the beam will have a 90-deg polarity offset with respect to the input and it will be orthogonal to the front linear polarizer. As a consequence, it will not pass to the isolator input port and an optical isolation is accomplished. The disadvantage of these types of isolators is due to random polarization of the light. Observing Fig. 4.33(a), it can be seen that since the first polarizer is vertical or horizontal, this may occur in a 3-dB loss in the case of circular polarization or light input.

The second type of isolator is a polarization-insensitive optical isolator as shown in Fig. 4.33(b). The main idea of this isolation structure involves the arrangement of lenses in the isolator. It must ensure that the effective NA of the reflected light is much higher than that of the input light. In other words, the input light is concentrated due to collimating and focusing it into the output fiber while the reflected light is dispersed. This way, the transmission from the input fiber to the output fiber has a low loss with IL of about 0.5 dB. On the other hand, the reverse transmission from the output to input has a high loss, which is defined as isolation. As miniaturization requirements have become more essential for optical modules, size reduction, cost reduction, simplicity, and mass production demands have to be considered; isolator chips are used in combination with thermally expanded core (TEC) fibers.[58,59] Direct embedding of the isolator chip into SMF causes a diffraction loss. For standard SMF, the spot diameter in optical communications marked as $2w$ is about 10 μm. Therefore, the gap between the two fibers should be at a maximum 100 μm. It was observed that diffraction losses are reduced significantly when the spot diameter is increased. TEC fibers with $2w$ equal to 40 μm can embed a 2-mm thick chip. The fabrication of TEC is demonstrated in Fig. 4.34, where the fibers are heated by an electrical furnace or microburner.

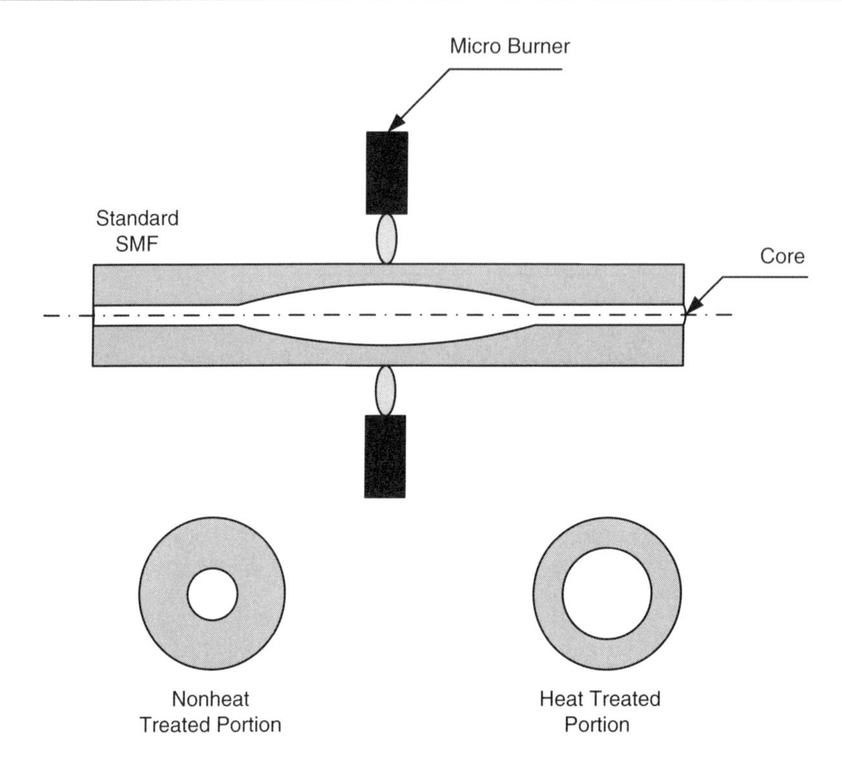


**Figure 4.34** Schematic diagram of TEC fabrication using microburners.[58] (Reprinted with permission from *Transactions on Magnetics* © IEEE, 1996.)

The heat rate is higher than 1300°C for about thirty minutes. The dopant $GeO_2$ is diffused in the silica SMF and the spot diameter at the heated portion is expanded in proportion to the dopant diffusion.

The structure and concept of TEC fiber-based isolator operation is as follows. The isolator is built from four building blocks, as shown in Fig. 4.35:

1. input birefringent plate, which is the offset compensator;
2. birefringent wedge;
3. Faraday rotator;
4. backward birefringent wedge.

The directions of the birefringent wedge are set at 45 deg around the fiber axis. In order to avoid reflected light from the isolator itself, the chip is inserted between the TEC fibers with a small tilt angle. Consider the forward-traveling wave of a light beam. Prior to the input birefringent plate, both polarizations may exist in the input TEC fiber. After passing through the input birefringent plate, the ordinary wave component passes the plate with refraction at the interface, marked as "o." The extraordinary wave component has both refraction and displacement "walk-off" in

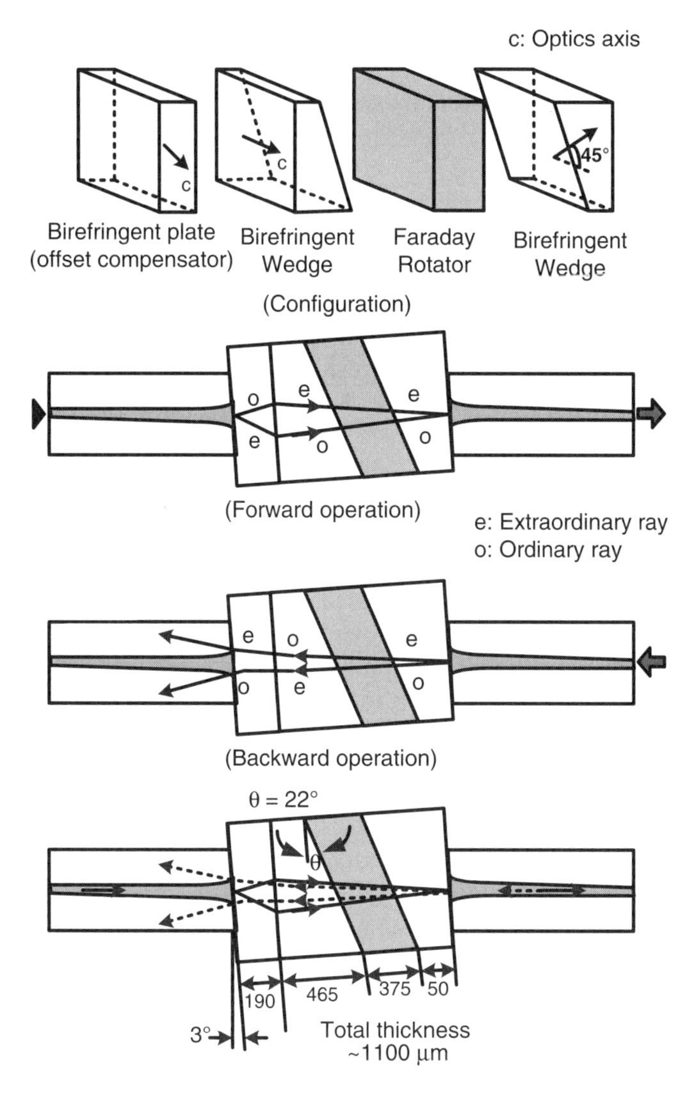


**Figure 4.35** Configuration of isolator chip and its forward and backward operation.[58] (Reprinted with permission from *Transactions on Magnetics* © IEEE, 1996.)

the lower direction, marked as "e" (Fig. 4.35). At the first birefringent wedge, the extraordinary beam marked as "e" turns into the ordinary wave marked as "o," and the ordinary wave beam marked as "o" turns into the extraordinary wave marked as "e." The polarization directions of the rays rotate by 45 deg after passing the Faraday rotator; however, rays maintain the same ordinary and extraordinary states through the rotator and the output wedge, respectively. Therefore, the angles of the rays against the fiber axis in the first wedge are the same as those in the second wedge. The light rays are converged into the endface of the output TEC without a tilt. There is no offset between the two beams at the output endface of the chip. The input birefringent plate operates as an offset compensator.

A backward-traveling beam of light observed from the output endface of the TEC as in Fig. 4.35 reaches the input TEC endface with a tilt angle of $\phi$ with an offset due to the nonreciprocity of the rotator and refraction effects of the paired birefringent wedges. The tilt causes higher coupling losses between output and input.

The optical behavior of uniaxial birefringent crystals provides unique opportunities to manipulate light that are not available or are cumbersome with isotropic materials such as optical glass and plastics. Mainly, beam doubling into ordinary and extraordinary waves can result in two distinct phenomena, spatial and polarization separation. Spatially, the light can be merely displaced or offset, or angular separation can be achieved, which results in ray separation that varies with distance. From the polarization point of view, uniaxial birefringent crystals split the beam into two orthogonal linear polarization states. There are several textbooks describing light propagation through uniaxial crystals.[60–63] CAD tools utilizing ray tracing are used in designing and analyzing light propagation through such complex optical systems and components. Generally speaking, rays are referred to as sampling the wavefront propagation vector, which is perpendicular to the wavefront. In isotropic materials, the propagation vector is the energy direction given by the flux Poynting vector $\mathbf{S} = \mathbf{E} \times \mathbf{H}$. That is not the case in anisotropic media such as uniaxial birefringent crystals, where the ray doubling into ordinary and extraordinary waves occurs at uniaxial interfaces.[64] An ordinary wave is similar in isotropic media as was previously described. Moreover, it follows Snell's law, using an ordinary index of refraction. However, in extraordinary waves, the energy flow is not perpendicular to the wavefront. It is along the Poynting vector $\mathbf{S}$. Hence the ray vector $\boldsymbol{\rho}$ represents the energy flow direction, while the wave vector $\boldsymbol{\kappa}$ is normal to the wavefront. The refraction index of an extraordinary wave is not a constant. It is a function of the wave direction, the crystal orientation, which is specified by the crystal axis $c$, and the ordinary and extraordinary indices of the crystal, $n_o$ and $n_e$, respectively, given by:

$$\frac{1}{n^2} = \frac{\sin^2\theta}{n_e^2} + \frac{\cos^2\theta}{n_o^2}, \tag{4.100}$$

where $\theta$ is the angle subtended by the crystal axis $c$ and the extraordinary wave vector $\boldsymbol{\kappa}_e$.

The plane that contains $c$ and $\boldsymbol{\kappa}_e$ is called the principal section.[65] Using Maxwell's equations and the relationships between the electric field $\mathbf{E}$, the electric displacement $\mathbf{D}$, vectors in uniaxial media and the definition of the Poynting vector $\mathbf{S} = \mathbf{E} \times \mathbf{H}$, the vectors $\mathbf{E}$, $\mathbf{D}$, $\boldsymbol{\kappa}_e$, $\mathbf{S}$, and c are all coplanar and lie in the principal section. From the polarization perspective, the extraordinary wave is linearly polarized in the plane of the principal section, while the ordinary wave is linearly polarized, orthogonal to the principal section. Birefringent materials that are often used in telecom at the 1500 nm region are $LiNbO_3$, rutile ($TiO_2$), and yttrium vanadate ($YVO_4$). More details about design methods of isolators and circulators can be found in Refs. [66] and [67]. As a general remark, polarization-sensitive isolators are used in optics blocks such as OTP and ODP because they have higher isolation compared to polarization-insensitive isolators. Moreover, polarization-insensitive isolators are used as a part of the fiber externally to the optical modules.

### 4.5.2 Optical circulators

An optical isolator is a three-port component; it has a direction, and therefore, it is a nonreciprocal device. Its operation is similar to a microwave circulator in the aspect of signal flow. It has an input port, output port, and reflector port. When a beam of light enters the input port, it travels to the output port. In the case of reflection at the output due to an open fiber or specific wavelength such as FBG, the light is reflected to the reflected port.

An example of an optical circulator design using birefringent crystals is provided by Fig. 4.36.

In this case, a four-port quasi-circulator is analyzed. It is composed of two polarizing beamsplitters (PBSs), two pentagonal prisms, two nonreciprocal rotators (NRRs), and two walk-off PBSs. The structure is not dependent on the number of ports. The purpose of each NRR is to rotate the light polarization plane by $-45$ deg for wave propagation in the $+z$ direction, and by 45 deg for wave propagation in the $-z$ direction. The walk-off PBSs are birefringent crystal blocks that have the same length; however, their optical axes are not parallel. The first birefringent crystal PBS1 is placed such that the $+45$ deg (resp. $-45$ deg) azimuth linearly polarized light propagating in the +z (resp. −z) direction is an extraordinary wave. Hence this extraordinary wave propagating in the $-z$ direction walks-off simultaneously along the $x$ and $y$ axes by $-d$ and $+d$, respectively. In contrast to this arrangement, the second birefringent crystal block PBS2 is set in such manner that +45 deg (resp. −45 deg) azimuth linearly polarized light propagating in the +z (resp. −z) direction is an extraordinary wave. Hence, this extraordinary wave propagating in the $-z$ direction walks-off simultaneously along the $x$ and $y$ axes by $-d$ and $-d$, respectively. The unpolarized beam of light from Fiber 1 is divided first into two linearly polarized light rays after PBS1. The path of these rays is given by the dashed lines in Fig. 4.36(A). The first beam, noted as beam-1, passes through PBS1, while the second beam, marked as beam-1′, is reflected by PBS1 and the pentagonal prism 1. Both beams then enter NRR1 and their azimuths are rotated by −45 deg. Eventually, beam-1 enters walk-off PBS1 with −45 deg azimuth, while beam-1′ enters walk-off PBS2 with +45 deg. Beam-1 and beam-1′ pass through walk-off PBSs 1 and 2 without walk-off, respectively. Because NRR2 rotates the light azimuth propagating in $+z$ direction by −45 deg, beam-1 and beam-1′ come out from NRR2 with −90 deg and 0 deg polarizations, respectively. After that, beam-1 is reflected by the pentagonal prism 2 and PBS2, and beam-1′ passes through PBS2. In this way, these beams are combined and coupled into the output port of fiber 2.

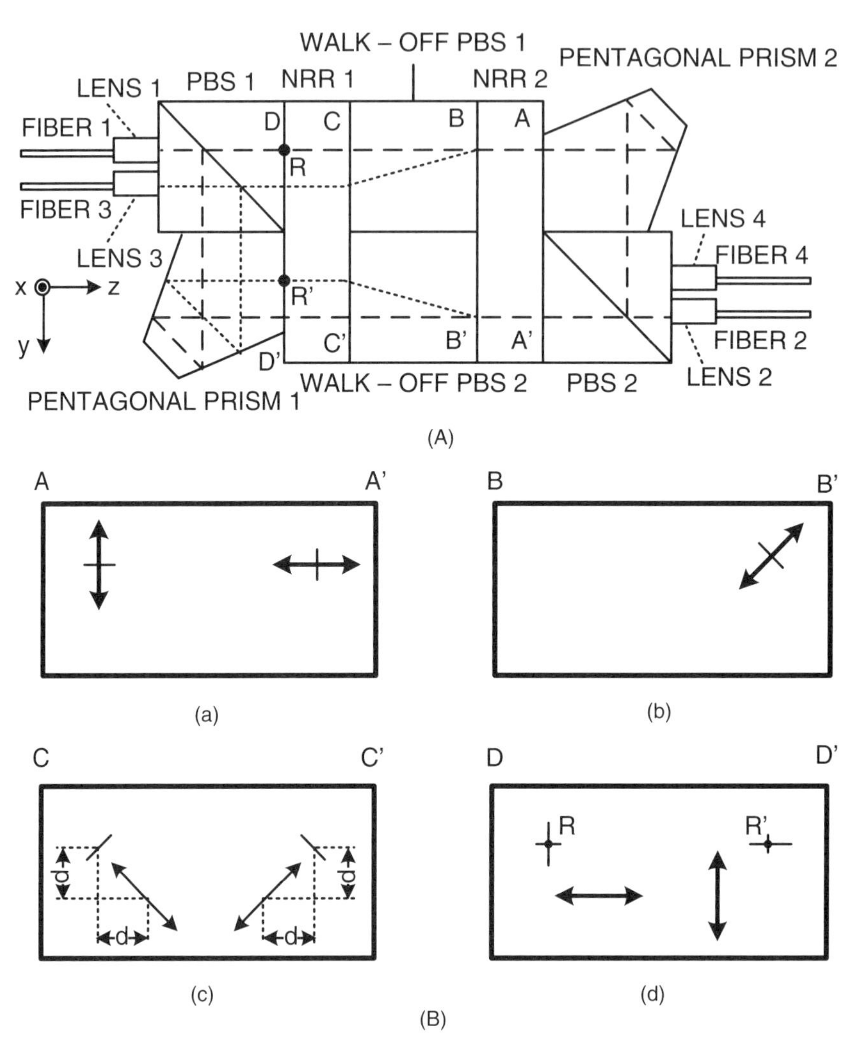


**Figure 4.36** (A) Structure of high-isolation polarization-independent quasi-circulator. (B) Positions and polarization states of beam-2 and beam-2′ on the same element surfaces. (a) A – A′. (b) B – B′. (c) C – C′. (d) D – D′.[68] (Reprinted with permission from the *Journal of Lightwave Technology* © IEEE, 1992.)

From the polarization vectors in Fig. 4.36(B), the polarization and wave manipulations coupling from fiber 2 to fiber 3 can be explained. It is a vector presentation of the beam power. Hence, the longer the vector arrow, the higher the power is. Thus the main beam is presented by the pair of the longest vectors in each subframe in Fig. 4.36(B). The beam path from fiber 2 is shown by the dotted line in Fig. 4.36(B). Note that it may merge with the dashed line in some instances. The reflected unpolarized ray from fiber 2 is divided into linearly polarized light beams by PBS2. The first beam (beam-2) passes through PBS2 and the other beam (beam-2′) is reflected by the pentagonal prism 2 [Fig. 4.36(B)(a)].

These two beams pass via NRR2, and the polarizations of beam-2 and beam-2′ are changed to +45 deg and −45 deg, respectively [Fig. 4.36B(b)]. Both beam-2 and beam-2′ enter the walk-off PBSs 1 and 2, respectively. Because beam-2 has +45 deg polarization, it is an extraordinary wave of walk-off PBS2. Beam-2 comes out from it with walk-off along the $x$ and the $y$-axes by $-d$ and $-d$, respectively. For the same reason, beam-2′ appears from walk-off PBS1, with walk-off along the $x$ and $y$ axes by $-d$ and $+d$, as shown in Fig. 4.36(B)(c). NRR1 rotates the polarity azimuth of rays propagating in the $-z$ direction by +45 deg. As a result, beam-2 and beam-2′ come out from NRR1 with 90 deg and 0 deg polarizations, respectively, and are orthogonal to each other. As shown by Fig. 4.36(B)(d), as beam-2 is reflected by the pentagonal prism 1, the axis displacement of beam-2 is changed as follows: $-d$ along the $x$ axis and $+d$ along the $z$ axis from the path of beam-1′. Thus beam-2 and beam-2′ fall on beamsplitting layers of PBS1 at the same point. This way, both beams are combined into port 3. Similarly, reflected light from fiber 3 is coupled into fiber 4.

When designing such a component, it is optimized around a specific wavelength to have the maximum isolation between the output port and the input port. For instance, an isolator with a center wavelength of 1300 nm and BW of $\pm 0.05$ nm has an isolation of about 50 dB or better. Over temperature changes of $\pm 20°$C, isolation ranges better than 40 dB. The IL is better than 1.7 dB. Advanced isolators have even better IL, which is lower than 0.7 dB.

## 4.6 Main Points of this Chapter

1. An optical fiber is a circular transmission line and hence its solution is a Bessel function.

2. There are three main kinds of optical fibers:

   *Single-mode step index*. The fiber-core diameter is from 1 to 6 μm. Only the fundamental mode is excited. These fibers are used for long distances since they have low loss and low dispersion. The lowest loss occurs at 1500 nm and is about 0.2 dB/km; lowest dispersion is at 1300 nm.

   *Multimode step index-fiber*. The fiber-core diameter is from 25 to 60 μm. It is used for low-cost, short-distance communications with LEDs.

   *Multimode graded index fiber*. It has a monotonic decrease in refractive index between the core center and the cladding The fiber-core diameter is from 10 to 35 μm, and cladding range is 50 to 80 μm. It is used for wideband performance in order to overcome dispersion.

3. There are several popular optical fiber connectors:
   - FC/PC: face contact/physical contact.
   - FC/APC: face contact/angled polished physical contact.
   - SC/PC: square/subscriber connector physical contact.

- SC/APC: square/subscriber connector angled polished physical contact.
- ST: an AT&T Bell Labs connector.

4. An optical coupler is a four-port passive optical component. The four arms are direct coupled and isolated arms. The directivity of the coupler is the difference in decibels between the coupling and the isolation. The additional coupler is $4 \times 4$. Both of these are the main building blocks of star couplers used in PON networks.

5. Wave division multiplexing (WDM) is a technique to multiplex several wavelengths on a fiber. CWDM stands for coarse wave division multiplexing; and DWDM stands for dense wave division multiplexing of 50 to 100 GHz spacing between each wavelength.

6. MZI is used in CWDM. It is a cavity that resonates at a specific wavelength. The cavity is a delay line equal to $2k\pi$ at wavelength $A$ and $(2k+1)\pi$ at wavelength $B$. It is used to separate and decode wavelength out of a CWDM transmission. Hence MZI is a wavelength DeMUX. It is composed of two 3-dB couplers and a delay line cavity that is resonated at the desired wavelength. $n(2\pi/\lambda_2)\Delta L = 2\,k\pi$ and $n(2\pi/\lambda_1)\Delta L = 2\,k\pi + \pi$ are null and peak conditions for wavelength $A$ and $B$ at both ports of the MZI.

7. DWDM filters are based on diffraction gratings. The idea is to have creative summation so that the optical path between each slit will be equal to an integer multiplication of wavelength divided by the slit length, which is called the slit period marked denoted as $\Lambda$. $\Lambda[\sin(\theta_i) - \sin(\theta_d)] = n\lambda \Rightarrow [\sin(\theta_i) - \sin(\theta_d)] = n\lambda/\Lambda$.

8. The FSR between two adjacent wavelengths is measured between center-to-center wavelengths of the BPFs, or between stop-band center wavelengths on both sides of the optical BPF.

9. An MZI can be realized by using an optical fused-fiber coupler and fiber delay line, or by birefringent crystals.

10. High-finesse resonance can occur in a GWS. By changing the refractive index, and taking advantage of the high-finesse property may result in a tuning range larger than the resonance frequency. These filter types are used to create ECL by cleaving the filter at one side of the laser and selecting the resonance mode by voltage control of the refractive index.

11. AWG is a wavelength decoder used in DWDM. The main concept is to create a phased array that is composed from a power splitter with a single waveguide at the input feeding port and multiple output ports. The splitter is the FPR type. The output ports of the FSR are connected to waveguides. The waveguide length is increased incrementally from

each port to the next one by $\Delta L$. The length $\Delta L$ between the first port and port $n$ satisfies $\Delta L = m\lambda_0$. At the output of the waveguides, another FSR is connected. This FSR has multiple inputs and outputs. At the center wavelength, the wavefront beam will point to the output port on the focal point of the multiport FSR image plane. In case of a different wavelength, or deviation from the center frequency, a phase shift will occur because of the waveguide array connecting the input FSR splitter to the multiport FSR. An azimuth tilt will result, pointing the wavefront beam to the appropriate output port on the image plane in the multiport FSR.

12. A fiber Bragg grating is the filter used in DWDM. Its concept is similar to that of the diffraction grating. The condition for creative summation is given by the Bragg wavelength, $\lambda_B = \lambda/n = 2\Lambda$ is the gratings period and the propagation constant is $\beta_B = \pi/\Lambda$.

13. There are two methods to analyze FBGs. The first one calls for square profile index approximation of the refraction index and numerically solving the boundary conditions using the transfer matrix method. The second method is the coupled mode theory.

14. Fiber Bragg gratings are used for add–drop networks, optical CDMA architectures, and optical filters.

15. CFBGs are written as FBGs but have a varying $\Lambda$ period. They are used for compensating group delay to overcome dispersion.

16. Optical isolator is a nonreciprocal component. It has a direction. Forward or S21 is transfer, low IL. Reverse S12 is isolation, high IL. The optical isolator is used to improve optical return loss between the transmitting laser and the fiber, and consequently, reduce distortions associated with reflections.

17. There are two types of isolators, polarized-sensitive and polarized-insensitive.

18. Circulator is a three-port nonreciprocal network component, input, output, and coupled or reflected. The transfer function is S21 forward, S32 reflected, and S12 isolation. These components are used in add–drop networks and optical CDMA together with FBGs, and group delay compensation networks together with chirped FBGs.

19. The optical behavior of uniaxial birefringent crystals provides unique opportunities to manipulate light, which are not available or are cumbersome with isotropic materials such as optical glass and plastics. Mainly, the beam doubles into ordinary and extraordinary waves.

20. An ordinary wave is similar to a planar wave, where the wave propagation vector $\boldsymbol{\kappa}$ is with the flux Poynting vector $\mathbf{S}$.

21. In extraordinary wave, the energy flow is not perpendicular to the wavefront. It is along the Poynting vector **S**. Hence, the ray vector $\boldsymbol{\rho}$, represents the energy flow direction, while the wave vector $\boldsymbol{\kappa}$ is normal to the wavefront.

22. Principal plane is the plane containing the birefringent crystal axis *c* and wave propagation vector $\boldsymbol{\kappa}_e$.

23. From the polarization perspective, the extraordinary wave is linearly polarized in the plane of the principal section, while the ordinary wave is linearly polarized orthogonal to the principal section.

## References


1. Liao, S.Y., *Microwaves Devices and Circuits 2nd Ed.* Prentice-Hall, Inc. (1985).
2. Agrawal, G.P., *Fiber Optics Systems*, *2nd Ed.*, John Wiley & Sons (1992).
3. Peddanarappagari, K.V., and M. Brandt-Pearce, "Volterra series transfer function of single-mode fibers." *Journal of Lightwave Technology*, Vol. 15 No. 12, pp. 2232–2241 (1997).
4. IEEE, *Standard 812: Definition of Terms Relating to Fiber-Optics*. IEEE, New York (1984).
5. Ishigure, T., Y. Koike, and J.W. Fleming, "Optimum index profile of the perfluorinated polymer-based GI polymer optical fiber and its dispersion properties." *Journal of Lightwave Technology*, Vol. 18 No. 2, pp. 178–184 (2000).
6. Renner, H., "Bending losses of coated single-mode fibers, a simple approach." *Journal of Lightwave Technology*, Vol. 10 No. 5, pp. 544–551 (1992).
7. Ramaswami, R., and K.N. Sivarajan, *Optical Networks*. Morgan Kaufmann (1998).
8. Gumaste, A., and T. Antony, *DWDM Network Design and Engineering solution*. Cisco Press (2002).
9. Ajemian, G.R., "A selection guide for fiber optic connectors." *Optics and Photon News*, Vol. 6 No. 6, pp. 32–36 (1995).
10. Hobbs, P.C.D., *Building Electro-Optical Systems, Making It All Work*. John Wiley & Sons, Inc. (2000).
11. Bellcore Technical Reference, "TR-NWT-000326: Generic requirements for optical fiber connectors and connectorized jumper cables." No. 1 (1991).
12. Collin, R.E., *Foundations for Microwave Engineering*. Wiley-IEEE Press (2000).
13. Keiser, G., *Optical Fiber Communications*. McGraw-Hill (2000).
14. Coldren, L.A., G.A. Fish, Y.A. Akulova, J.S. Barton, L. Johansson, and C.W. Coldren, "Tunable semiconductor lasers a tutorial." *Journal of Lightwave Technology*, Vol. 22 No.1, pp. 193–202 (2004).

15. Sears, F.W., and M.W. Zemansky, *College Physics, Part 2, 3rd Ed.* Addison-Wesley (1966).
16. Murphy, T.E., "Design, fabrication, and measurement of integrated Bragg grating filters." Ph.D. Thesis, MIT (2001).
17. Dudovich, N., G. Levy-Yurista, A. Sharon, A.A. Friesem, and H.G. Weber, "Active semiconductor-based grating waveguide structures." *Journal of Quantum Electronics*, Vol. 37 No. 8, pp. 1030–1039 (2001).
18. Takato, N., T. Kominato, A. Sugita, K. Jinguji, H. Toba, and M. Kawachi, "Silica-based integrated optic Mach–Zehnder multi/demultiplexer family with channel spacing of 0.01–250 nm." *Journal of Selected Areas in Communication*; Vol. 8, pp. 1120–1127 (1990).
19. Kawachi, M., "Recent progress in silica-based planar lightwave circuits on silicon." *Proceedings of IEEE Conference on Optoelectronics*, Vol. 143, pp. 257–262 (1996).
20. Kawachi, M., "Silica waveguides on silicon and their application to integrated optic components." *Optical and Quantum Electronics*, Vol. 22, pp. 391–416 (1990).
21. Cao, S., J. Chen, J.N. Damask, C.R. Doerr, L. Guiziou, G. Harvey, Y. Hibino, H. Li, S. Suzuki, K.Y. Wu, and P. Xie, "Interleaver technology: comparisons and applications requirements." *Journal of Lightwave Technology*, Vol. 22 No.1, pp. 281–289 (2004).
22. International Telecommunication Union, "Spectral grids for WDM applications: DWDM frequency grid Series G: Transmission Systems and Media, Digital Systems and Networks Transmission media characteristics: characteristics of optical components and subsystems" STD ITU-T G.694.1 (2002).
23. Chiba, T., H. Arai, K. Ohira, H. Nonen, H. Okano, and H. Uetsuka, "Novel architecture of wavelength interleaving filter with Fourier transform-based MZIs," *Proceedings of the Optical Fiber Communications Conference*, Vol. 3, pp. WB5-1–WB5-3 (2001).
24. Chiba, T., H. Arai, K. Ohira, S. Kashimura, H. Okano, and H. Uetsuka, "Chromatic dispersion free Fourier transform-based wavelength splitter for DWDM," *Optoelectronics and Communications Conference Technical Digest*, 13B2-2, pp. 374–375 (2000).
25. Kohtoku, M., T. Mizuno, T. Kitoh, M. Oguma, T. Shibata, Y. Inoue, and Y. Hibino, "Low loss and low crosstalk PLC-based interleave filter fabricated with automatic phase trimming method." *European Conference on Communication Technical Digest* (2002).
26. Capmany, J., C. Doerr, K. Okamoto, and M.K. Smit, "Introduction to the special issue on arrayed grating routers/WDM MUX/DEMUXs and related applications/uses." *Journal of Select Topics Quantum Electronics*, Vol. 8 No. 6, pp. 1087–1089 (2002).
27. Smit, M.K., and C. van Dam, "PHASAR-based WDM-devices: principles, design, and applications." *Journal of Select Topics Quantum Electronics*, Vol. 2 No. 2, pp. 236–250 (1996).

28. Dargone, C., "An N × N optical multiplexer using planar arrangement of two-star couplers." *IEEE Photonics Technology Letters*, Vol. 3 No. 9, pp. 812–815 (1991).
29. Dargone, C., C.A. Edwards, and R.C. Kistler, "Integrated optics N × N multiplexer on silicon." *IEEE Photonics Technology Letters*, Vol. 3 No. 10, pp. 896–899 (1991).
30. Vreeburg, C.G.M., T. Uitterdijk, Y.S. Oei, M.K. Smit, F.H. Groen, E.G. Metaal, P. Demeester, and H.J. Frankena, "First InP-based reconfigurable integrated add-drop multiplexer." *IEEE Photonics Technology Letters*, Vol. 9 No. 2, pp. 188–190 (1997).
31. Leijtens, X.J.M., P. Le Lourec, and M.K. Smit, "S-matrix oriented CAD-tool for simulating complex integrated optical circuits." *Journal of Select Topics in Quantum Electro*, Vol. 2 No. 2, pp. 257–262 (1996).
32. Adar, R., C.H. Henry, C. Dragone, R.C. Kistler, and M.A. Milbrodt, "Broad-band array multiplexers made with silica waveguides on silicon." *Journal of Lightwave Technology*, Vol. 11 No. 2, pp. 212–219 (1993).
33. Dragone, C., "Efficient N × N star couplers using fourier optics." *Journal of Lightwave Technology*, Vol. 7 No. 3, pp. 479–489 (1989).
34. Doerr, C.R., and C. Dragone, "Proposed optical cross connect using a planar arrangement of beam steerers." *IEEE Photonics Technology Letters*, Vol. 11 No. 2, pp. 197–199 (1999).
35. Chen, J.C., and C. Dragone, "A proposed design for ultralow-loss waveguide grating routers." *IEEE Photonics Technology Letters*, Vol. 10 No. 3, pp. 379–381 (1998).
36. Dragone, C., "Planar 1 × N optical multiplexer with nearly ideal response." *IEEE Photonics Technology Letters*, Vol. 14 No. 11, pp. 1545–1547 (2002).
37. Bernasconi, P., C. Doerr, C. Dragone, M. Cappuzzo, E. Laskowski, and A. Paunescu, "Large N × N waveguide grating routers." *Journal of Lightwave Technology*, Vol. 18 No. 7, pp. 985–991 (2000).
38. Dragone, C., "Optimum nonblocking networks for photonics switching." *Journal of Select Topics Quantum Electron*, Vol. 6 No. 6, pp. 1029–1039 (2000).
39. Sadot, D., and E. Boimovich, "Tunable optical filters for dense WDM networks." *IEEE Communications Magazine*, Vol. 6 No. 12, pp. 50–55 (1998).
40. Wu, M.C., and W.H. Prosser, "Simultaneous temperature and strain sensing for cryogenic applications using dual wavelength fiber Bragg gratings." *Proceedings of SPIE*, Vol. 5191, pp. 208–213 (2003).
41. LaRochelle, S., P.Y. Cortes, H. Fathallah, L.A. Rusch, and B. Jaafar, "Writing and applications for fiber Bragg grating arrays." *Proceedings of SPIE*, Vol. 4087, pp. 140–149 (2000).
42. Wu M.C., R.S. Rogowski, and K.K. Tedjojuwono, "Fabrication of extremely short length fiber Bragg gratings for sensor applications." *Proceedings of IEEE Sensors*, Vol. 1, pp. 49–55 (2002).
43. André, P.S., J. L. Pinto, I. Abe, H. J. Kalionowski, O. Frazaõ, and F. M. Araújo, "Fiber Bragg grating for telecommunications applications: tunable

thermally stress enhanced OADM." *Journal of Microwave Optoelectronics*, Vol. 2 No. 3, pp. 32–45 (2001).

44. Tai, H., "Theory of fiber optical Bragg grating—revisited." *Proceedings of SPIE*, Vol. 5178, pp. 131–138 (2004).
45. Haus, H.A., *Waves and Fields in Optoelectronics*. Prentice Hall (1984).
46. Finkbeiner, D.T., *Introduction to Matrices and Linear Transformation*. W.H. Freeman and Company (1960).
47. Arfken, G.B., and H.J. Weber, *Mathematical Methods for Physicists, 4th Edition*. Academic Press (1995).
48. Spiegel, M.R., *Mathematical Handbook of Formulas and Tables, Second Edition*. McGraw Hill (1998).
49. Abramowitz, M., and I.A. Segun, *Handbook of Mathematical Functions with Formulas Graphs and Mathematical Tables*. Dover Publications (1965).
50. Marulanda, J.L., and P. Torres, "Performance analysis of an optical CDMA communications system based on fiber Bragg gratings." *Revisita Colombiana Fisica*, Vol. 36 No. 2, pp. 358–361 (2004).
51. Wei, Z., H. Ghafouri-Shiraz, and H.M.H. Shalaby, "New code families for fiber-Bragg grating based spectral amplitude coding optical CDMA systems." *IEEE Photonics Technology Letters*, Vol. 13 No. 8, pp. 890–892 (2001).
52. Sumetsky, M., and B.J. Eggleton, "Theory of group delay ripple generated by chirped fiber gratings." *OSA Optic Express*, Vol. 10 No. 7, pp. 332–340 (2002).
53. Marti, J., D. Pastor, M. Tortola, J. Campany, and A. Montero, "On the use of tapered linearly chirped gratings as dispersion-induced distortion equalizers in SCM systems." *Journal of Lightwave Technology*, Vol. 15 No. 2, pp. 79–187 (1997).
54. Kim, J., and B. Lee, "Bidirectional wavelength add-drop multiplexer using multiport optical circulators and fiber Bragg gratings." *IEEE Photonics Technology Letters*, Vol. 12 No. 5, pp. 561–563 (2000).
55. Wu, X., Y. Shen, C. Lu, T.H. Cheng, and M.K. Rao, "Fiber Bragg grating based rearrangeable nonblocking optical cross connects using multiport optical circulators." *IEEE Photonics Technology Letters*, Vol. 12 No. 6, pp. 696–678 (2000).
56. Tran, A.V., W. De Zhong, R.S. Tucker, and K. Song, "Reconfigurable multichannel optical add-drop multiplexers incorporating eight-port optical circulators and fiber Bragg gratings." *IEEE Photonics Technology Letters*, Vol. 13 No. 10, pp. 1100–1102 (2001).
57. Kikushima, K., K. Sato, H. Yoshinaga, and E. Yoneda, "Polarization-dependent distortion in AM-SCM video transmission systems." *Journal of Lightwave Technology*, Vol. 12 No. 4, pp. 650–657 (1994).
58. Shirashi, K., T. Irie, T. Sato, R. Kasahara, O. Hanaizumi, and S. Kawakami, "Integration of in-line optical isolators." *IEEE Transactions on Magnetics*, Vol. 13 No. 5, pp. 4108–4112 (1996).
59. Sato, T., R. Kasahara, J. Sun, and S. Kawakami, "In-line optical isolators integrated into a fiber array without alignment." *IEEE Photonics Technology Letters*, Vol. 9 No. 7 pp. 943–945 (1997).

60. Born, M., and E. Wolf, *Principles of Optics, 6th Ed.*, Pergamon Press (1980).
61. Yeh, P., *Optical Waves in Layered Media*. John Wiley & Sons (1988).
62. Collett, E., *Polarized Light*. Marcel Dekker, Inc. (1993).
63. O'Neill, E.L., *Introduction to Statistical Optics*, Dover Publications, Inc. (1991).
64. McClain, S.C., L.W. Hillman, and R.A. Chipman, "Polarization ray tracing in anisotropic optically active media." *Proceedings of SPIE*, Vol. 1746, pp. 107–118 (1992).
65. Trolinger Jr, J.D., R.A. Chipman, and D.K. Wilson, "Polarization ray tracing in birefringent media." *Optical Engineer*, Vol. 30 No. 4, pp. 461–466 (1991).
66. Shirasaki, M., and K. Asmam, "Compact optical isolators for fibers using birefringent wedges." *Applied Optics*, Vol. 21 No. 23, pp. 4296–4299 (1982).
67. Lee, R., S. Lin, C.H. Chiu, and J.Y. Lin, "Design of optical circulators." *Proceedings of SPIE*, Vol. 3420, pp. 311–317 (1998).
68. Fujii, Y., "High isolation polarization independent quasi-optical isolator." *Journal of Lightwave Technology*, Vol. 10 No. 9, pp. 1226–1229 (1992).
69. Kim, K.Y., S.Y. Kim, M.W. Kim, and S. Jung, "Development of compact and low-crosstalk PLC-WDM filters on hybrid-integrated bidirectional optical transceivers." *IEEE Journal of Lightwave Technology*, Vol. 23 No. 5, pp. 1913–1917 (2005).
70. Riziotis, C. and N. Zervas, "Design considerations in optical add/drop multiplexers based on grating-assisted null couplers." *IEEE Journal of Lightwave Technology*, Vol. 19 No. 1, pp. 1913–1917 (2001).

# Chapter 5

# Optics, Modules, and Lenses

Optical modules are the interface between electronics components and optical fiber. These optical modules building blocks convert the electrical signal into modulated light in the transmit section and the modulated light into an electrical signal, analog or digital, in the receiving side. An introduction to these modules was provided in Chapter 2, explains block diagrams of different receivers and transceivers. This chapter provides review and design approaches to these building blocks. An introduction to the optical design of complex modules such as duplexer's optical dual port (ODP) and triplexer's optical triple port (OTP) are made. Lens types and optical retracing analysis is explained and investigated by showing error tracking and isolation optimization. Planar lightwave circuits (PLC) and free-space modules are also reviewed.

## 5.1 Planar Lightwave Circuits

Optical modules such as triplexers and duplexers are the key components in optical transceiver applications used in various applications such as FTTx. These platforms include both reception of wide band analog community access television (CATV) signals, and high data digital transports for transmission and reception. The front end of such platforms is an optical block used as a DeMUX in the receive side and a MUX in the transmit side. These modules usually operate at 1550-nm for CATV analog reception and at 1310 nm for digital receiving–transmitting. As per the full service access network (FSAN) standard, optical modules are used to receive 1550-nm analog CATV and 1490-nm digital downlink, and transmit 1310-nm digital uplink. Observing the structure of such an integrated platform (ITR), the main effort at cost reduction stresses on optics. Traditionally, such optical front-end, free-space, bulk-optics blocks are realized from encapsulated lasers and detectors in a transistor outline metal (TO) can, beamsplitters, optical filters, lenses, and metal or plastic housing. The process of alignment and integration adds costs and complexities as well as more optical parts to the bill of materials (BOM). Additionally, the packaging and sealing process, the temperature stabilization to prevent power degradation due to tracking error, sensitivity reduction because of reflections from the digital transmitter into the receiver for 1550/1310-nm, as well as repeatability in production, led to the integrated optics PLC design. The motive was to target the market needs for optimum commodity price level of FTTx platforms; to reduce deployment cost at homes, offices, etc., as well as to have mass production repeatability and small size and space occupied by the optical block. Several technologies are used to develop or to grow these modules. Throughout the last decade, numerous technologies have been explored. The common design challenge to all PLC technologies is the accurate design of an accurate mask and the use of the photolithographic process in a high-class clean room. Once the mask is ready, production of PLC is completed and the wafer is produced, diced, and then integrated to a complete module. The leading technologies for such wafers are:

- III-V(InP, GaAs),
- silica on silicon (SoS),
- SiON on silicon,
- silicon on insulator,
- polymer waveguide,
- lithium niobate, and
- ion exchange (IE).

The most common PLC technology is SoS, which evolved from the semiconductor industry, and the most competitive one is based on IE.[18]

IE in glass-based waveguides is a simple low-cost method for realizing surface mount flip-chip PLC FSAN triplexer.[1] Ion-exchanged glass exhibits many desirable features such as low propagation loss and compatibility with optical fibers.[11] Therefore, the glass material should be compatible with IE chemistry and with a refractive index close to that of optical fibers. IE process involves the exchange of monovalent cations such as $Li^+$, $K^+$, $Rb^+$, $Cs^+$, $Tl^+$, or $Ag^+$, with $Na^+$ present in the glass.[7,8] Annealed channel waveguides fabricated on a BK7 glass showed propagation loss of 0.4 dB/cm. The benefits of easy fabrication and low cost[7] made this technology a future option for low-cost FTTx platforms.

In PLC devices, the wavelength separation is achieved by using the Mach-Zehnder interferometer (MZI) reviewed in Sec. 4.4.1. The fundamental concept presented in Ref. [1] utilizes the surface normal light beams that are 90-deg folded and coupled into a waveguide using total internal reflection (TIR). These perpendicular optoelectronic semiconductors are bump bonded and coupled to the optical substrate, with the optical aperture facing the optical interconnect substrate. In this application of a PLC triplexer, transparent optical materials are used to couple the bump-bonded devices such as the laser and the two detectors. These materials replace the lenses in free space, which characterizes bulk-optics topology. PLC on glass is a cost-effective process. It is composed of four stages. At first, a waveguide glass wafer is fabricated using IE technology. Using photolithography, deposition IE areas are exposed. The width of openings in the mask must be controlled precisely since it determines the core diameter of the IE waveguide. There are three criteria for the choice of the mask: (1) the mask must have good adhesion to the glass; (2) the mask must be chosen for resistance to the salts of the IE baths, which means, it has to be chemically inert with respect to the salt, and; (3) it must act as a localized barrier to ion diffusion. Titanium, $SiO_2$, and combinations of these have been reported as candidates for this.[13] After completion of the IE process, electrical lines and contacts are printed on the PLC glass layer and a 45-deg beam folder slot is diced, which sometimes can be filled with metal. The next stage is to fabricate the support glass wafer (bulk). This layer involves drilling or etching via holes; the via holes are coated with a conductive metal, and conductive lines and pads are printed on both sides. After completion of the support bulk, the bulk is attached to the PLC optical glass, semiconductors are aligned using an active alignment beam, and then bump bonded to the optical wafer. Finally, the components are encapsulated by a thermal conductive polymer. The final step is packaging, including dicing the wafer at the optical-fiber facet side, and creating double bars. The polishing process of the optical-fiber side at 8 deg to minimize reflections, attaching the pigtail fibers to the waveguide, and dicing the double bars to create two separate electro-optical modules correspond to a PLC triplexer. Pigtailing loss at the fiber-to-substrate waveguide is controlled to 0.1–0.2 dB. Figure 5.1 presents a PLC triplexer on a glass using beam folding by IE process.

The advantage of the IE process is its simplicity and writing (deposition) time of not more then 60 min. Parameters that influence the IE process are time and temperature of the IE, intensity and duration of the electric field, and composition and concentration of the salt baths.[13] The step after the deposition stage is burying the ions deep inside the glass by electric field technology process at high temperature. This stage is a diffusion process that lasts about 15–20 min at a temperature

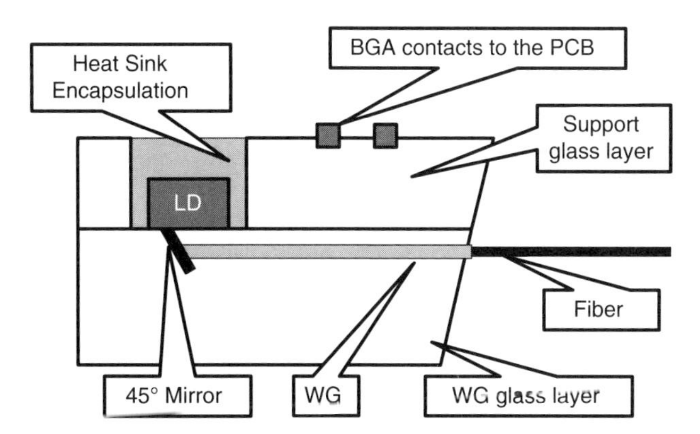


**Figure 5.1** PLC Opto-electronic chip: cross section (left) and isometric view of a single bar (right).[1] (© all rights reserved Colorchip Inc.,)

of 350°C with an electric field of 30 V/mm produced by a dc power supply.[8] The IE technology is superior to Si in terms of strength of materials and immunity against cracks as well as low-cost fabrication in lower class of clean rooms versus higher class. Commercial simulation tools and aids such as BeamPROP or MATLAB finite element method analysis enable accurate prediction of performances and obtain design parameters required for fabricating PLC couplers and Mach-Zehnder devices for wavelength separation for fabricating BiDi components.[1,9]

Figure 5.2 presents simulated and measured results for BiDi PLC coupler responses for separating and combining two different wavelengths of 1310/1490-nm with the typical length between 7 and 12 mm. Measured results of a BiDi PLC coupler of 1310/1520-nm exhibit shifted peaks, with response close to the theoretical prediction.

The results in Fig. 5.2 were measured by using a white light source and spectrum analyzer in order to scan the transmission function and identify the wavelength peaks.[1] In order to accurately measure the peaks isolation, 1310-nm laser source was used and resulted in better than 25-dB isolation, with good agreement with simulation-predicted results.

Coupling of the laser diode beam into the PLC optical circuit is done by using a tapered waveguide. This is one of the best methods of using an intermediate device for coupling light into a single mode fiber (SMF). Tapered waveguides are manufactured by a unique process of burying the silver waveguides that were developed for this purpose. The tapered waveguide is realized with a small mode field diameter (MFD) and a low depth at one side (low electric field side) and a high MFD and deep waveguide at the other side (high electric field side) as shown in Fig. 5.3 (right). IE enables conical taper to 2-$\mu$m diameter by a unique control of diffusion stage, with no additional production steps. In Fig. 5.3 (left), recent experimental images of the results of the tapering process are given. Measured coupling losses were 3.5–4 dB from a laser die with $1.2 \times 3$ $\mu$ MFD to the output fiber, using this tapered waveguide with elliptical MFD of $4 \times 5$ $\mu$. In comparison, direct coupling of laser to fiber yields a 7.5 dB loss. Improving the tapering process would achieve better coupling efficiency of the laser diode to the SMF, with losses lower than 3 dB.

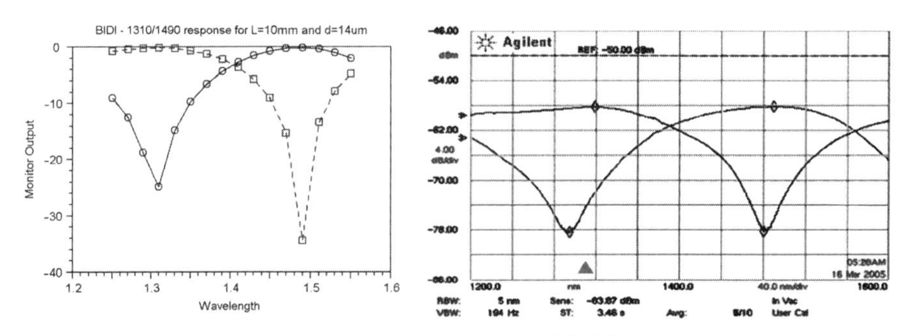


**Figure 5.2** A response of the PLC couplers: 1310/1490 nm theoretical (left) and 1316/1530 nm white light experimental (right).[1] (Courtesy of Colorchip, Inc., all rights reserved.)

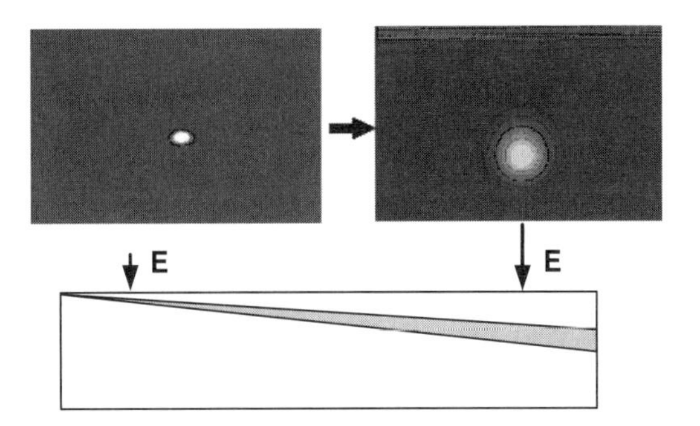


**Figure 5.3** Tapered waveguide from 4 μ (elliptical) to 10 μ (circular) MFD: experimental beam cross sections (left) and waveguide depth in the glass scheme as function of MFD (right).[1] (Courtesy of Colorchip, Inc., all rights reserved.)

One of the most critical problems in such a PLC process is coupling the light in and out of the optical plane. There are various techniques to fabricate 45-deg TIR micromirrors. Laser-ablated micromirrors are already reported and can be fabricated using exciter laser. The facets are based on TIR at the interface between the glass and the air gap.

In passive optical network (PON) systems, PLC modules based on silicon were reported.[2,3,5] In this case, a triplexer is realized by cascading a directional coupler to a Mach-Zehnder PLC wavelength interferometer. This would result in a 3-dB loss in both the receive section and the transmit section. Observing the light path from the pigtail to the receiver, it first goes through the directional coupler, which is a 3-dB coupler, resulting in a loss of 3 dB. Then, the received light reaches the Mach-Zehnder and is separated into two wavelengths, 1490 and 1550 nm. Observing the transmit path, the 1310-nm light is coupled to the pigtail fiber. The coupler port, which was the output in the receive path is now the isolated port for the transmit path. For the same reason, there is a 3-dB loss in the transmitting laser. Hence, in the design of such FSAN PLC triplexer, the isolation requirement should be high enough to prevent desensitization of the 1490- and 1550-nm receiver section. Further review is provided in Fig. 5.4.

In the same manner, as shown in Fig. 5.4(a), PLC for old 1310-nm receiver-transmitter/1550-nm receiver can be realized. In this case, the first directional coupler connected will be a WDM MZI separating the 1550 nm from the 1310 nm. The 1550 nm detector is connected at port A. Then the 1310-nm

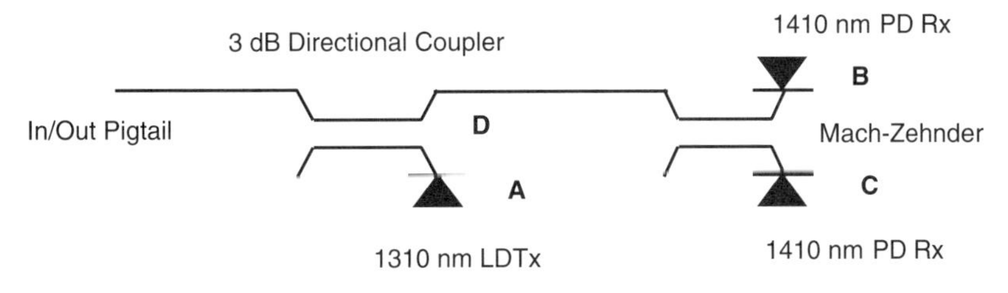


**Figure 5.4** Concept option for FSAN PLC triplexer realization.

receiver-transmitter will be located at the back ports marked as B and C. This is a 3-dB directional coupler, where the 1310-nm at port B with respect to port C is at the isolated port. The constraint here is to have enough isolation between ports B and C in order to prevent desensitization of the receiver's path by the transmitter's path. This is an equal requirement as in FSAN for isolation from ports A to D. Some other applications use Y branch rather than a directional coupler.[15] Y branch is a power splitter based on tapered waveguide branching from the input waveguide.[17] The two branches' core cross section or width is narrow at the input port and gradually increases toward the output port waveguides. Power ratio is defined by the core width ratio between the two output ports. An additional design parameter of the Y-branch splitter is the tapered velocity $dW/dL$, which defines the rate of the core width change versus length. Recent published research reports about low-cost PLC for fiber to the home (FTTH) suggest the use of VCSEL (vertical cavity surface emitting lasers) because these are a low cost, easy to fabricate light source.[4] Novel BiDi triplexers were fabricated of 200 MB/s data rate for small form pluggable (SFP) configuration. In this application, WDM filters were 780-nm LPF, 815-nm BPF, and 850-nm HPF. These filters were thin film filters coated on a glass substrate. Filters were fabricated by sputtering $Ta_2O_5/SiO_2$. This new optical subassembly (OSA) structure is realized with satisfactory coupling efficiency and total passive alignment process without any lens system. Therefore, the VCSEL BiDi triplexers would be expected to be a cost-effective solution for optical transceivers used in a set top box (STB) at homes for active optical networks (AON) FTTH systems. Figure 5.5 reviews several integrations of PLC in FTtransmitter transceivers. In Fig. 5.5(a), 1.3/1.5-μm BiDi Triplexer is introduced. A spot-size converted laser diode (SSC-LD), a waveguide photodiode (WG-PD), and a monitor (WG-PD) are flip-chip mounted onto a PLC platform. This platform also contains a 1.3/1.5-μm WDM dielectric filter and an asymmetric Y-branch silica waveguide. This PLC platform is an embedded silica waveguide with $8 \times 8$-μm core, having refractive index difference of 3%. A Si bench is used as the alignment plane and heat sink for the optical chips. Index marks for the passive alignment are formed on the surface of the Si bench. The WDM filter separating 1.3/1.5-μm is a multilayered dielectric filter. The advantage of such a filter structure is its ability to arrange 1.3/1.5-μm and 1.5-μm output waveguides to the same side of the PLC platform; however, it has a disadvantage, that is, the trench of inserting this WDM filter. The trench has to be set at the right position without any angle deflection to the waveguide. Any deviation and deflection in positioning the WDM filter result in impossible insertion loss from the 1.3/1.5-μm port to the 1.5-μm port. Reported isolation between 1.3/1.5- and 1.5-μm ports for this PLC topology is 52-dB. Figure 5.5(b) provides information about a full duplex 1.3/1.5-μm WDM optical transceiver used in ATM/PON. This is a 156 MB/s full duplex optical transceiver that meets the FSAN class B and C specifications simultaneously. It is an optical hybrid integrated module (OHI). In this architecture, the optical devices are mounted on the PLC as was presented previously.

This module contains 3.3 V CMOS ICs such as burst-mode laser diode and automatic gain control (AGC) amplifier all mounted on a glass epoxy electrical

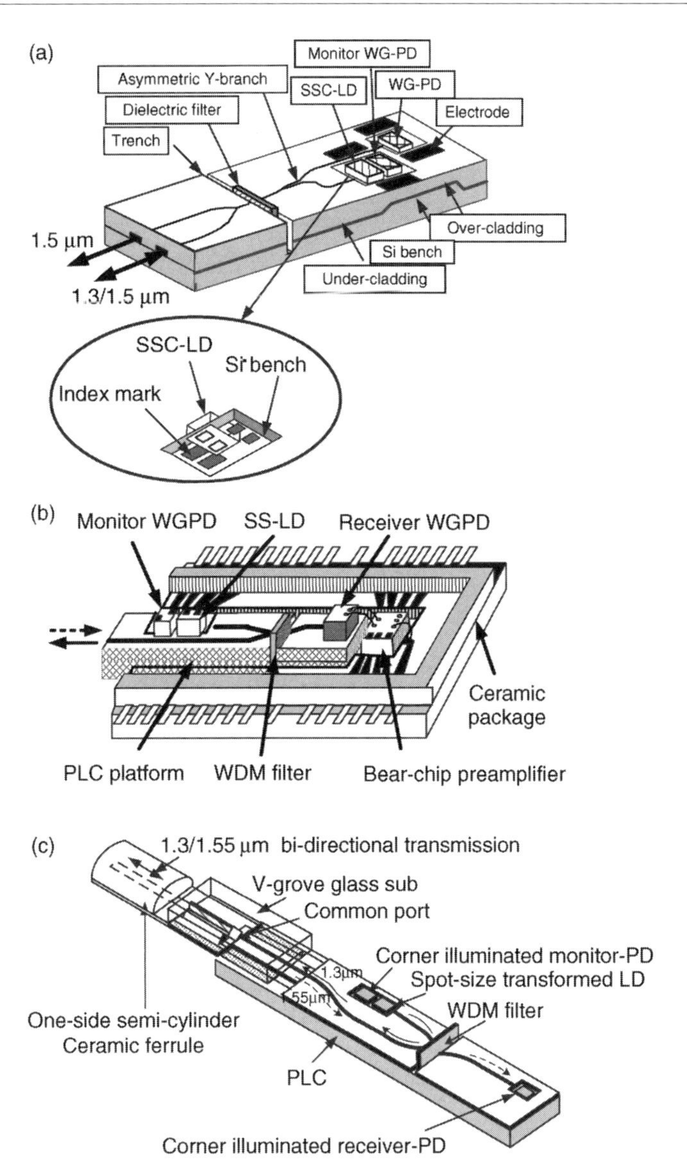


**Figure 5.5** Several applications of FTTx BiDi modules using PLC technology. [(a). Reprinted with permission from IEEE LEOS © 1996]. [(b)[2] Reprinted with permission from IEEE, ECOC ©1998]. [(c)[5] Reprinted with permission from IEEE © 1999.]

circuit board. Such a design also contains a multistage feed-forward automatic threshold control for the AGC IC in order to procure an instantaneous response for the burst-mode data. Electrical cross talk between the transmitter and receiver is better than 100 dB. Figure 5.5(c) shows a PLC module where the WDM filter separates between 1.3- and 1.5-µm wavelengths. This PLC was mounted on a glass substrate and the semicylindrical-type ferrule made this structure compact.

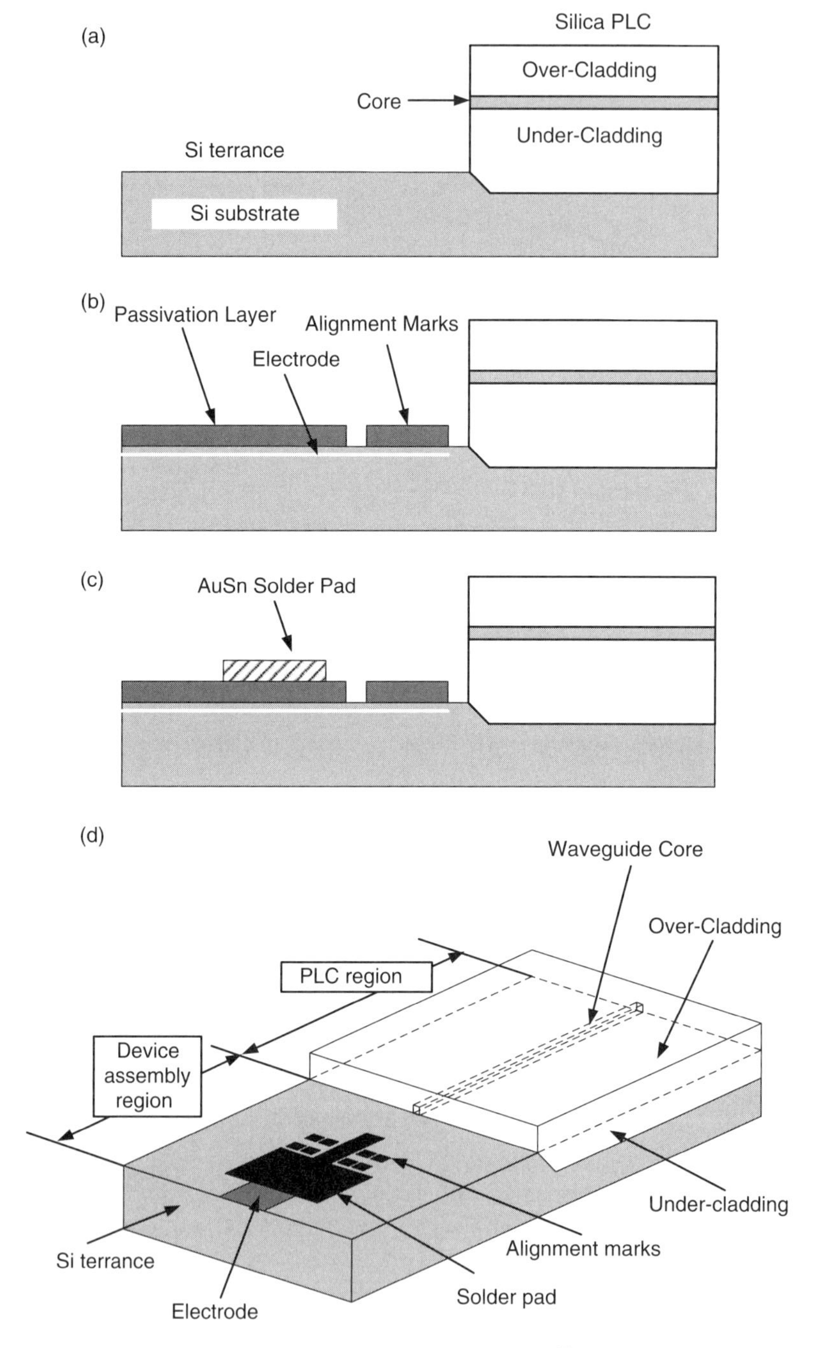


**Figure 5.6** Fabrication process stages of PLC platform.[16] (Reprinted with permission from *Proceedings of IEEE Electronic Components and Technology Conference* © IEEE, 1996.)

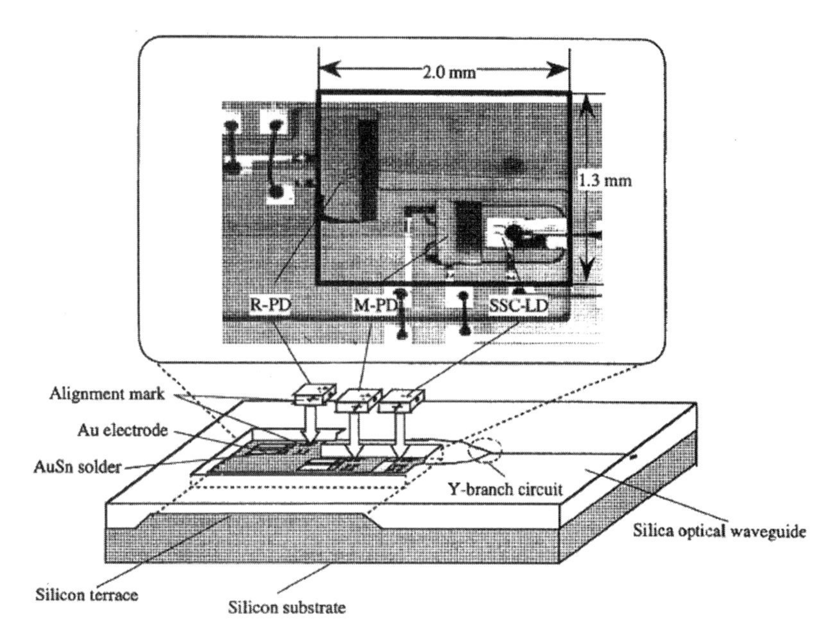


**Figure 5.7** Integration of optics semiconductors R-PD (receive photodetector), M-PD (monitor photodetector), and SSC-LD on a PLC platform. The area of optical components is recessed in the silica of the optical circuit.[16] (Reprinted with permission from *Proceedings of IEEE Electronic Components and Technology Conference* © IEEE, 1996.)

This module transmitter had a power of 2 mW with $\pm 0.7$ dB tracking error max over $-40^{\circ}$C to $+95^{\circ}$C. Optical cross talk is better than 60-dB. Electrical cross talk below 156 MHz was better than 100 dB.

Silicon PLC BiDi triplexer structure can be divided into two sections. The first one is the PLC region itself, where the optical circuit is located, and the second one is where the active components are assembled using alignment markers.[16] This kind of PLC arrangement is called multichip integration on PLC. Si PLC is fabricated on a Si terrace. The first half of the Si bulk is occupied by the PLC silica as shown in Fig. 5.6(a). The residual Si terrace has $SiO_2$ passivation layer of 0.5-μm. Electrodes and alignment markers are formed simultaneously by lifting off the evaporated Au layer as shown in Fig. 5.6(b). Finally, AuSn solder film about 2–3-μm thick is evaporated and patterned as shown in Fig. 5.6(c). Figure 5.6(d) shows the basic PLC platform structure and Fig. 5.7 provides more details about Si multichip hybrid PLC.

## 5.2 Free-Space Bulk-Optic WDM Modules

### 5.2.1 General concept

The front end of any FTtransmitter triple-play transceiver is a bulk optic component.[36] This component is a WDM optical module multiplexing and demultiplexing several wavelengths going in and out of an SMF. Unlike the future technology of PLC, this optical component is a free-space optical system containing discrete building blocks such as lenses, beamsplitters (dichroic mirrors that are multilayer wavelength filters or just beamsplitters), photodetectors for the two wavelength channels, 1310-nm digital channel and 1550-nm analog CATV in old systems, and 1490-nm digital channel and 1550-nm analog CATV in FSAN, as well as a laser diode typically at 1310 nm for both FSAN and old wavelength plan. This kind of triplexer design requires wavelength filters for the FSAN wavelength and for the old wavelength plan. Special care for designing the trapping light is required to prevent the 1310-nm transmission leakage into the 1310-nm receiving channel in the old wavelength plan. Figure 5.8 describes FSAN and old optics wavelength plan's internal structure of such bulk triplexers. Early designs suffered from low yield and a long alignment process of the optics to and pigtailing of the SMF fiber, as well as complicated assembly problems. Special design emphasis was placed on tight mechanical tolerances, making the production of optical housings difficult. The problem of tight mechanical design tolerances for machined and die-cast housing is critical and cannot be compromised even today with advanced bulk optics. Early optics had large housing and the optical block was not in full compliance with its parameters over temperature. Creep of adhesives such as epoxies and metal temperature coefficients resulted in optical power degradation from the laser into the SMF because of tracking error. A closing process of older module TO headers was created by Allen screws; thus, it was sensitive to mechanical stress over temperature. This reduced in tracking-error performance more. Tracking error occurs because of the optical light beam-spot motion from SMF input. Numerous modules were not immune to humidity. One of the reasons for the problem of humidity limitation was the attempt to reduce the analog detector capacitance in order to gain BW. Some of the approaches used Kapton as the TO header base for the analog PD. Since the Kapton's relative dielectric constant was nearly at unity, it was not building up parasitic capacitance from the photodiode to the housing of the optical triplexer. This method left the analog photodiode uncapped and exposed within the optical housing of the triplexer. Additionally, costs of lenses were too high. These technical problems and the low-yield, high-BOM cost provided the motive for the development of PLC technology. PLC technology has the advantage of mass production and automatic pick-and-place assembly using SMT footprints. However, bulk optics products reached a design maturity and the above mentioned problems were mostly solved.

Optical modules such as triplexers became small sized and hermetically sealed using laser welding and advanced epoxy glues. Active optical components were hermetically sealed. Currently, TO was made with glass or advanced low-cost plastics integral lenses. Yield increased to 90% and the manufacturing process became repeatable and robust. Some of the optical detectors contain a PIN transimpedance amplifier (TIA) inside the TO package to minimize parasitics and improve sensitivity of the digital section. Advanced research efforts are invested to integrate highly linear analog PIN TIAs within the photodetector housings to improve performance of the analog channel. New technologies of manufacturing lenses made of advanced plastic materials as well as casting the optical triplexers housing reduced the BOM costs even further.

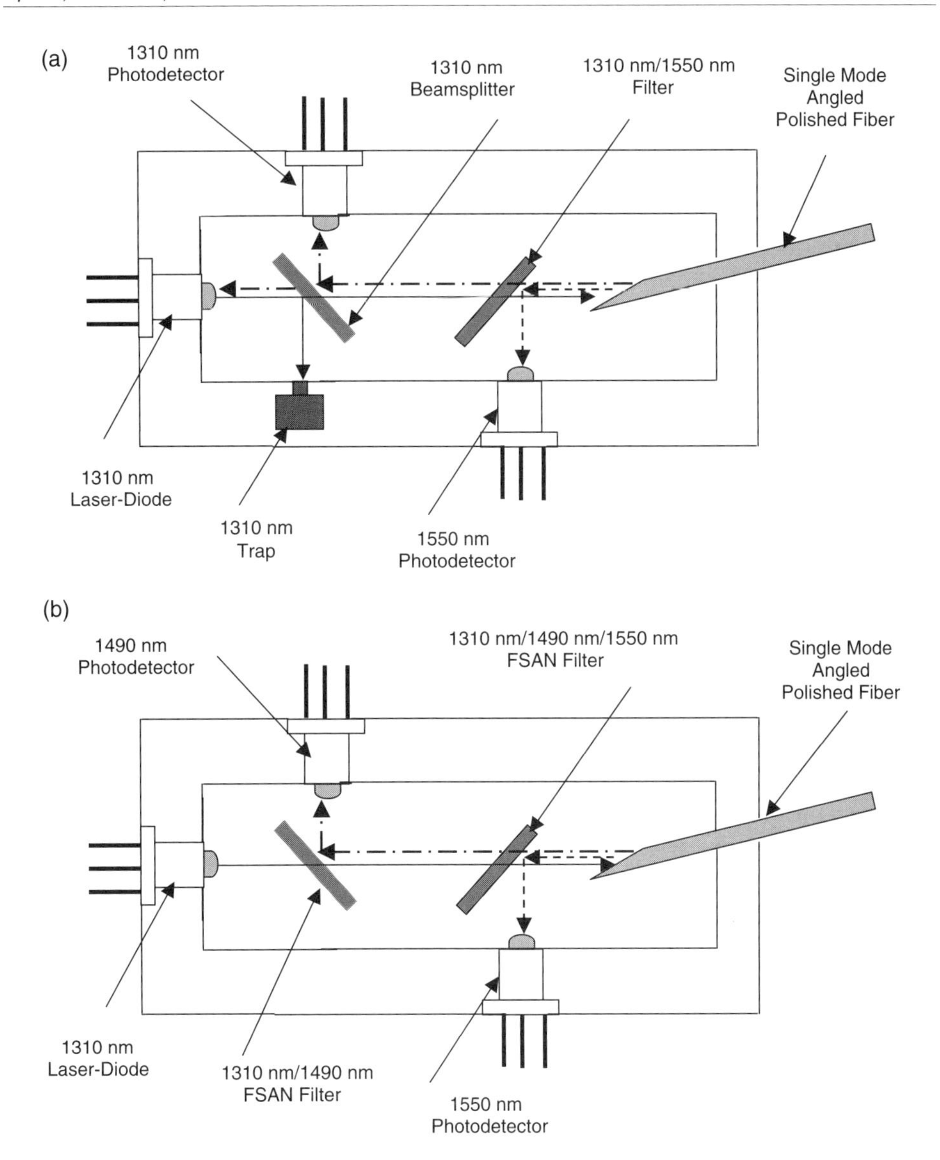


**Figure 5.8** Optical triplexer structure. (a) An old wavelength plan for a 1310-nm digital receiver-transmitter and a 1550-nm analog receiver, showing the 1310-nm light trap. (b) Three FSAN wavelengths standard 1310-nm digital transmitter, 1490-nm digital receiver, and 1550-nm analog receiver.

Transmitter optical subassembly (TOSA) and receiver optical subassembly (ROSA) modules are made from advanced plastic materials. Moreover, this technology has solved the problems of parasitic capacitance between the photodiode and the optical housing. Advanced CAD tools such as the ZEMAX[35] design program, which provides detailed accurate libraries of optical components used as building blocks for bulk optics, such as lenses and beamsplitters, enables the designer to optimize design parameters and tolerances, thus reaching tight specification performances. Optical parameter design control is important in order to reach highly linear performances in CATV channels. For instance, spot size or defocusing of the beam over the analog PD may improve linear performance by reducing distortions. This is done by observing and optimizing beam tracings and aberration parameters that result from the lenses. The subject is well covered in Sec. 8.6.3. As a conclusion from these facts, bulk optics components play a major role and are the key components in modern FTTx designs such as BiDi and integrated triplexer platforms. This progress also provided a major stride in the development of TOSA and ROSA optics for SFP and XSFP technology at high data rates of 10 GBit/s.

The concept of the optical triplexer is well understood by observing Figs. 5.8(a) and (b). Looking at the received beam of light incoming from the pigtail fiber into the optical system, it first passes through a selective beamsplitter. This beamsplitter operates as a wavelength filter. It directs the 1550-nm into the analog photodetector because of its refractive index sensitivity to that wavelength. The other wavelength ray (1310-nm in old plan or 1490-nm in case of FSAN) travels through this beamsplitter until it hits the second. There, the received ray (1310 nm as in the old wavelength plan or 1490 nm in case of FSAN) is directed into the digital PD. So far this is the ray tracing on the receiving side. Both photodetectors are cupped in a TO can, which has a ball lens. Since the diameters of photodetectors range from 20 to 40 μm, depending on the BW required, there is no problem of losing power of light due to lens aberration. This is because the optical spot created on the image plane, which is the detector active surface, is within the active photonic area. Therefore, there is hardly any degradation in CNR. Observing the transmitted ray, the following design consideration should be paid attention to. The 1310-nm transmitted ray impinges at first on the last beamsplitter. This beamsplitter directs the received ray into the digital PD. The transmitted ray passes through this element in the case of the FSAN wavelength plan. In this design case, this optical element is an optical filter based on thin film layers of glass. It has a wavelength sensitivity of refractive index; thus, it directs the received 1490 nm into the digital PD and allows the 1310-nm transmitter to path through. However, the design becomes more complicated when both digital receiver and transmitter wavelengths are 1310 nm, per the old wavelength plan. In this case, the optical element is a beamsplitter with a defined splitting ratio. Therefore, the splitting ratio creates a power penalty in both receive and transmit paths. On the receiver side, it is desired to have more reflection to the photodiode and less path through the splitter into the laser diode direction. Nevertheless, on the transmitter side, it is desirable to transmit more and reflect less in order to meet transmission power levels at the fiber. Observing Fig. 5.8(a), it is clear that for higher reflection and less transmission power, the reflected transmission beam hits the opposing side of the digital receiving photodetector. This situation can create optical cross talk between the 1310-nm receiver and transmitter. The reason this happens is that the optical reflections from the counter wall veers off into the digital receiving detector. The reflected beam from the wall leaks into the 1310-nm PD per the beamsplitters power ratio. The solution is to create an optical trap on the counter wall that locks the transmitter's reflected beam inside and does not allow it to exit into the receiving photodetector.

Moreover, special passivation or absorptive paint is used to plate the inside housing cavity of the optical block.

This design approach is difficult to manufacture, and controlling the optical cross talk by creating a trap cavity is not possible in large-scale production. A different solution for cross talk is provided by creating an absorptive conical washer around the 1550-nm lens. Bear in mind that in reality the optical distance between the 1310-nm beamsplitter in Fig. 5.8(a) and the 1550-nm photodetector is very short, so reflected light from the 1310-nm beamsplitter is at the proximity of the 1550-nm detector. As a conclusion, if the transmitter's 1310-nm reflected beam hits the conical washer surrounding the lens of the 1550-nm photodiode, it may be deflected to some desired absorption point. Since this washer is conical, the reflected light ray trace is toward the laser diode. Given that the absorptive washer attenuates the energy of light, the remaining power should be absorbed at the laser diode zone. One of the solutions is to insert between the laser and the 1310-nm beam, a splitting polarization-dependent isolator and then an absorptive rod. The rod attenuates the light reflected from the conical washer at the 1550-nm area. The isolator completes the attenuation of the remaining 1310-nm reflected energy and absorb it even more. Further, cross-talk optimization is done by commercial ray tracing CAD software.

In conclusion, from the aforementioned discussion of cross talk, a design compromise is made as a trade-off between the receiver's sensitivity, optical cross talk, and laser's transmitted power. These design parameters define the beamsplitter's power ratio for the 1310-nm transmitter and receiver.

Generally, design sets are for 50%/50% power ratio or 70%/30% in favor of receiver sensitivity. The last splitting ratio makes isolation requirements harder since a larger amount of 1310-nm reflected energy has to be attenuated.

After passing the last beamsplitter, the transmission ray passes through the first beamsplitter toward the SMF core surface, which is a wavelength sensitive refractive index filter, the old 1310-nm/1550-nm wavelength plan reflects the 1550-receiver ray into the analog PD and lets the 1310-nm path through in both directions without any reflections. Figure 5.8 describes rays tracing in both directions that receive and transmit. In the old wavelength plan, detail "a," 1310-nm ray, is given by a dashed dotted line. Portion of the received energy passes through the beamsplitter and reaches the laser diode. The 1550-nm wavelength is indicated by the dashed line. The laser diode wavelength of the 1310-nm is shown by continuous line. It can be seen that the reflected light from the 1310-nm beamsplitter is trapped within the 1310-nm light trap. In the FSAN design, this filter is sensitive to 3 wavelengths; it reflects the 1550-nm receiver into the analog PD and allows it to pass through the both 1310-nm transmitter and the 1490-nm receiver. In conclusion, it is clear that the FSAN triplexer has no need for optical trapping since wavelengths are isolated by optical filters. The other design goal is to focus the transmitting beam into a narrow spot on the SMF core surface. Therefore, aspherical lenses are used to correct and compensate or minimize aberration created by the laser ball lens, meaning aspherical lenses are used as a part of the laser TO, rather than a ball lens, which is a spherical lens. Aberration is critical in the transmitting beam's path as the SMF fiber active area diameter is 8-μm. Consequently, a large optical spot results in loss of transmission power since only a portion of the optical field flux is passing through the fiber. At some instances where transmission power can be compromised against cost, ball lenses are used and not aspherical lenses. Figure 5.8 describes rays tracing in both receiving and transmitting directions. For the FSAN wavelength plan, detail "b," the 1490-nm ray is shown by the dash-dotted line; it is directed to the 1490-nm digital PD. The 1550-nm wavelength is shown by the dashed line. The laser diode wavelength of the 1310-nm is indicated by the continuous line. Generally, in order to prevent reflections, the pigtail SMF is polished at 8 deg. Thus, the pigtailing is done at a tilt of 4 deg and there is design compensation of 4 deg per Snell's law.

When optical return loss performance is required, and in case of an old analog return path, when high linearity is required too for meeting decent link NPR (noise power ratio) specifications, the isolator is inserted as was described above as a part of the optical trap. For the FSAN frequency plan, the isolator is then inserted between the fiber and the first filter (beamsplitter), where the 1310- and 1490-nm pass through it and 1550 nm is reflected into the analog photodiode.

As a general conclusion, bulk optics is a strong technology competition to PLC. It has a small size low profile and covers frequencies up to 10 GB/s. It is a low-cost, robust solution. PLC modules are as long in size; thus, there is no edge in FTtransmitter for PLC.

Pigtailing is done with a tilt to compensate the polishing angle of the SMF and is specially made at approximately 8 deg to improve optical return loss.

More details can be found in Ref. [36].

### 5.2.2 Lenses

The theory of lenses' wave optics and geometrical optics is well explored with sufficient elaboration in basic physics literature.[19,23,24,34] Traditional basic analysis of optical systems involving lenses refers to thin lenses.[19] In this section, the discussion focuses on thick and thin lenses, taking into account their physical properties such as refractive index, spherical shape, and ray tracing, showing their approximation to thin lenses. Commonly used lenses in bulk optics are ball lenses (spherical lenses) and aspheric lenses. Those lenses' characteristics were explored in the literature.[20,23,24] In general, lenses are described in a Cartesian axis system, where the $z$ axis represents the common optical axis of the refracting and reflecting surfaces; it is positive when it is directed to the right. The $x$ axis is positive when going into the diagram and the $y$ axis is positive when it is directed upward and is in the plane of the diagram[23] as shown in Fig. 5.9.

Optical design is based on several fundamental laws that enable the prediction of ray traces. Snell's law formulized the law of refraction in 1621. It states the following. If the angle created by a ray between the point of incidence and the normal to the surface of incidence is $\theta_1$, and if the angle of refraction created between the refracted ray and the same normal is $\theta_2$, then the angles are related by the following equation:

$$n_1 \sin\theta_1 = n_2 \sin\theta_2, \tag{5.1}$$

where $n_1$ and $n_2$ are the refractive indices of the two materials.

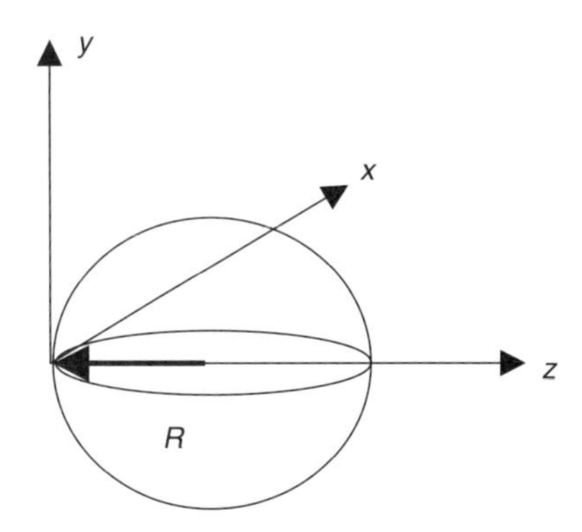


**Figure 5.9** A sphere with a radius of $R$ tangent to the 3D Cartesian system origin.

Furthermore, Snell's law states that the incident ray, the refracted ray, and the normal-to-the-surface ray are coplanar. Moreover, the refractive index describes the ray velocity with respect to the speed of light in vacuum. Therefore, the following relation emerges:

$$n = \frac{c}{v}, \tag{5.2}$$

where $v$ is the speed of light in the material and $c$ is the speed of light in vacuum. Thus, it is understood from the theory of waves that the refractive index refers to the root square of the relative dielectric constant of the material.

A ball lens, or spherical lens, is commonly used for coupling optics between SMF and photodiode TO or laser TO can, where optical power is a compromise or as an aid for optical add–drop multiplexers (OADM) modules.[21,22] Ray tracing in such lenses is shown in Fig. 5.11. It is constructed from two surfaces that are nominally spherical in form and have some thickness between the surfaces. Each sphere has its own radius $R$, where each sphere is a portion of a ball. By controlling the radii of the lens spheres, ray tracings can be optimized up to a certain level. There are ball lenses in which the radii are not identical and are made of two spheres. These lenses are called spherical lenses.

Since many ball lenses are constructed from spherical surfaces, it is convenient to describe these surfaces by an equation describing a sphere that is tangent to the origin as shown in Fig. 5.9.

The equation describes such a sphere as in Fig. 5.9 given in its canonic form by the following equation:[23]

$$R^2 = x^2 + y^2 + (z - R)^2. \tag{5.3}$$

Replacing $x^2 + y^2$ by $r^2$ and solving this 3D square equation for $z$ as a function of $r$ and $R$ would provide

$$z = R\left[1 \pm \sqrt{1 - \left(\frac{r}{R}\right)^2}\right]. \tag{5.4}$$

The value $R$ is presented sometimes by $1/c$, where $c$ is called the curvature of the surface or vertex curvature. Then by multiplying by the conjugate sign, Eq. (5.4) is written as

$$z = \frac{1 \pm \sqrt{1-(cr)^2}}{c} = \frac{cr^2}{1 \pm \sqrt{1-(cr)^2}}. \tag{5.5}$$

Equation (5.5) describes the sphere coordinates with a curvature $c$. The reason for $\pm$ symbol is that for a given $x$ and $y$ coordinate or fixed distance of a radius $r$ from the $z$ axis, a plane that intersects the sphere parallel to the $z$ axis would have two intersection points. Those intersection points are at two different $z$ locations; however, they are having the same $x$ and $y$ coordinates. Meaning, the solution is the same distance from the $z$ axis. In 3D geometry, this can be a flat plane intersecting the sphere parallel to the $z$ axis or a cylindrical shape such that its symmetry axis is the $z$ axis. Figure 5.10 provides detailed elaboration. The goal in this solution is to describe a moving point along the sphere as a function of $r$ or $x$ and $y$ coordinates. Thus, Eq. (5.4) refers to the first intersection point by observing only the $-$ or $+$ sign in the denominator of Eq. (5.5). In other words, it provides the lower value solution of $z$ for a given $x$ and $y$ coordinate or radius $r$. Equation (5.5) can be rearranged and expanded as a power series in term of $r^2$:

$$\begin{aligned} zc &= 1 - \sqrt{1 - c^2r^2} \\ &= 1 - \left(1 - \frac{1}{2}c^2r^2 - \frac{1}{8}c^4r^4 - \frac{1}{16}c^6r^6 - \cdots - \frac{1}{2^n}c^{2(n-1)}r^{2(n-1)}\right). \end{aligned} \tag{5.6}$$

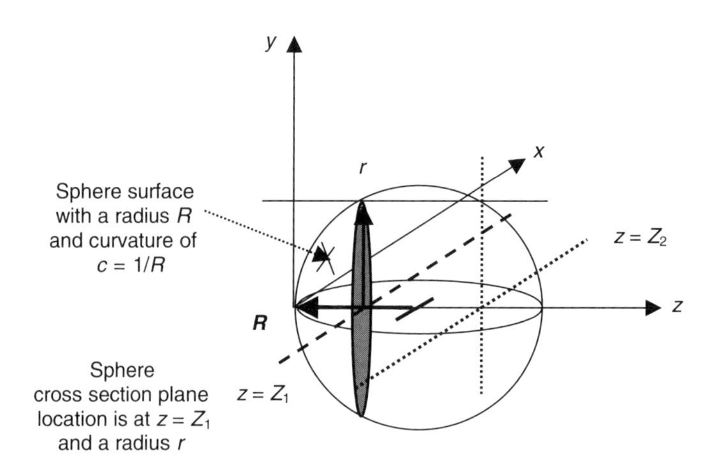


**Figure 5.10** A sphere surface with cross section of radius of $r$ at $z = Z_1$ and curvature of $c = 1/R$.

Note that the indices are now modified; the first variable in the parentheses is $n = 1$ thus it equal to 1, the next variable is $n = 2$, etc. Thus, Eq. (5.6) results in

$$z = \frac{1}{2}cr^2 + \frac{1}{8}c^3r^4 + \frac{1}{16}c^5r^6 + \cdots + \frac{1}{2^n}c^{2n-1}r^{2n}. \tag{5.7}$$

Note that the indices of Eq. (5.7) are modified; the first variable is $n = 1$ was previously $n = 2$ in Eq. (5.6), etc.

Equation (5.7) is useful when defining a paraxial region of a spherical lens:

$$z = \frac{1}{2}cr^2. \tag{5.8}$$

This approximation states that $R$ is relatively large and that the portion of the sphere observed from the left side of the $y$ axis is approximately a flat plane, with hardly any curvature, since the vertex distance $Z_1$ is small. Figure 5.10 explains Eq. (5.8) in a 3D $x$, $y$, $z$-Cartesian system. In addition, the object and the image surfaces are proportional to the square of the object and the image dimensions. In other words, it should be no larger than the spheres surface cross section. From Fig. 5.10 it is concluded that $z = Z_1$ is the surface height. It is also assumed that any angle of incidence of the surface is small enough, so the approximations of $\sin\theta \cong \theta$ and $\cos\theta \cong 1$ are valid. Thus, every object plan is perfectly imaged in the image space. This method of analyzing lens surfaces is used for other shapes such as paraboloids, ellipsoids, hyperboloids, and oblate and prolate ellipsoids.[23]

Aspheric surfaces are also called diffractive optical elements (DOEs). They are especially powerful for correction of chromatic aberration. This surface is described by

$$z_{\mathrm{ASPHE}}(r) = \frac{cr^2}{1 + \sqrt{1 - (K+1)c^2r^2}} + A_4r^4 + A_6r^6 + \cdots + A_{2j}r^{2j}, \tag{5.9}$$

where $z_{\mathrm{ASPHE}}$ is the sagittal surface height, $c$ is the vertex curvature $1/R$ (also called reciprocal vertex radius), $r$ is a radial coordinate as was shown in Fig. 5.10, and sometimes marked as $\rho$, $A_4$, $A_6$, etc., are the surface deformation coefficients, and $K$ is the conic constant that determines the shape of the basic conic section. Various values for $K$ determine the surface shape as given by Table 5.1.[23,25]

The slope, at any point of the sphere, can be calculated by the derivative of Eq. (5.9) with respect to $r$.

**Table 5.1** Shapes representation vs. $K$ values.

| Constant | Shape |
|---|---|
| $K = 0$ | Sphere |
| $K < -1$ | Hyperbola |
| $K = -1$ | Parabola |
| $-1 < K < 0$ | Ellipse (Prolate) |
| $K > 0$ | Ellipse (Oblate) |

Thus tan(α) is given by

$$\tan(\alpha) = \frac{1}{\frac{\mathrm{d}z_{\mathrm{ASPHE}}(r)}{\mathrm{d}r}}. \tag{5.10}$$

Therefore, the normal-to-the-lens surface at a cylindrical coordinate value of $r$ is given by

$$\cot(\alpha) = \frac{\mathrm{d}z_{\mathrm{ASPHE}}(r)}{\mathrm{d}r}. \tag{5.11}$$

Knowing the normal angle to the lens surface and the angle created by the ray with respect to the positive direction of $z$ provides the incidence angle between the normal and the incident ray hitting the surface.

After exploring, in brief, lens surfaces and surface approximations, the next step is to review simple ray tracing in a thin lens and then fix this calculation for a thick lens. It is well known that plane waves in homogeneous media travel in straight lines. This fact allows the use of ray optics in thin lenses. The rules of thin lenses are

1. Rays passing through the center of the lens are undeviated.
2. Rays entering parallel to the axis pass through the focus.
3. The locus of ray bending is the plane of the center of the lens.

From Fig. 5.11, the lens magnification $M$ is given by

$$M = \frac{d_i}{d_o}. \tag{5.12}$$

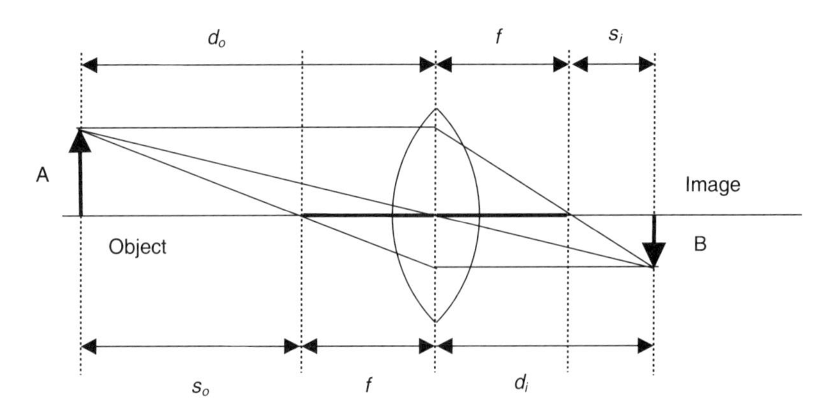


**Figure 5.11** Ray tracing in a thin lens.

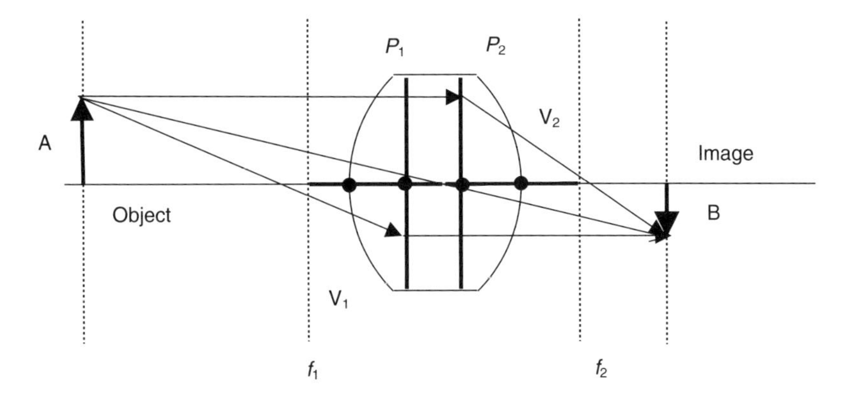


**Figure 5.12** Ray tracing in a thick lens.

Again, from Fig. 5.11 it can be proven by planar geometry that the following identities are valid:

$$\frac{1}{f} = \frac{1}{d_o} + \frac{1}{d_i}, \tag{5.13}$$

$$f^2 = s_o s_i. \tag{5.14}$$

Figure 5.12 describes a thick lens. $P_1$ and $P_2$ are the principal planes of the lens and intersect at the lens axis as shown. $f_1$ and $f_2$ are the front and the back foci. The distance from the vertex of the left side of the lens to the focal point, $f_1$, is called the front focal distance or working distance. The same applies for the back focal point $f_2$. The question then is how to define and determine a thin lens? A thin lens has to have a thickness that is small compared with the depth of the focus of the beam used. In a thin lens limit, the two principal planes are merged into one at the center of the lens as shown in Fig. 5.11.

The general lens maker's equation for a thick lens is given by

$$P = \frac{1}{f} = (n-1)\left(\frac{1}{R_1} - \frac{1}{R_2} + \frac{t}{n}\frac{n-1}{R_1 R_2}\right), \tag{5.15}$$

where $t$ is the lens thickness from vertex to vertex, and $n$ is the refractive index. The front and back focal points are given by

$$l_1 = f_1\left[1 - \frac{t(n-1)}{nR_1}\right], \tag{5.16}$$

$$l_2 = f_2\left[1 - \frac{t(n-1)}{nR_2}\right]. \tag{5.17}$$

The separation between the two principal planes is given by

$$2\delta = t + l_2 - l_1 = \frac{t(n-1)}{n}. \tag{5.18}$$

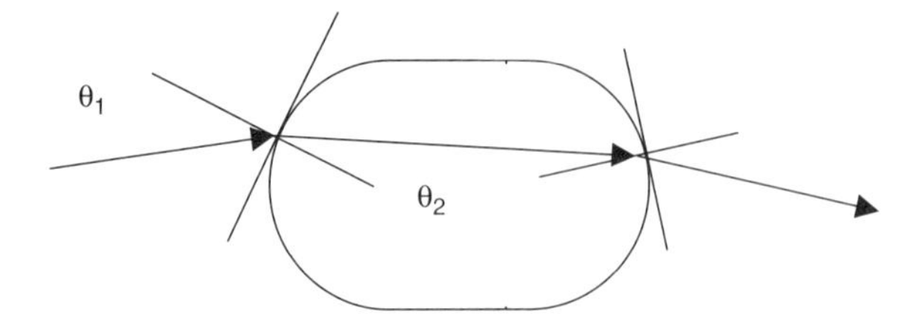


**Figure 5.13** Ray tracing in a ball lens per Snell's law. Note that the normal-to-the-lens sphere is calculated per Eq. (5.11) for a sphere surface given by Eq. (5.6).

For the thin lens approximation, the following terms are valid for $t \ll R$. In addition, it is assumed that the refractive indices of the media on both sides of the lens are the same. This condition results in equality between the focal points. Using these conditions, Eq. (5.15) becomes

$$P = \frac{1}{f} = (n-1)\left(\frac{1}{R_1} - \frac{1}{R_2}\right). \tag{5.19}$$

Moreover, both focal points at Eqs. (5.16) and (5.17) may have the same value. Thus, it is clear that the separation between the two principal planes at Eq. (5.18) goes to zero. In practice, the design of an optical module such as triplexers is done by using the thin lens approximation.

Optimization of free-space optical blocks for optical coupling is done by optimizing spot size and minimum aberration for laser, and in the case of photodetectors, large spots are preferred for improving linearity by minimizing distortions as explained in Sec. 8.6.3. Geometrical optics and spherical or aspherical lens equations are useful methods for such optimization done by commercial CAD Fig. 5.13 depics the rays traces in a ball lens.

### 5.2.3 Optical calculations

Assume the following design problem of a free-space triplexer: a TO56 laser with a ball lens, two beamsplitters, and an SMF fiber are needed to be optimized for a precise spot of the Tx beam on the fiber. The fiber is angled, polished at 8 deg, and inserted into the optics block with up tilt as shown in Figs. 5.8(a) and (b). In addition, the refractive indices of the two beamsplitters are $n_1$ and $n_2$, respectively, as a general case, and the lens has a refractive index $n$ and a radius $R$. This is the design problem facing optimization of each optical-block ray.

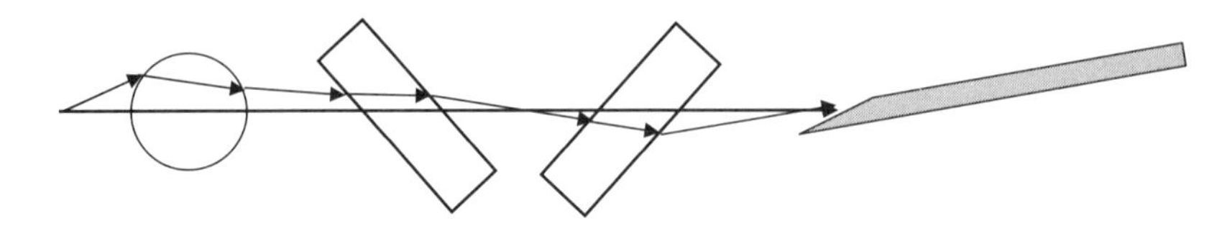

**Figure 5.14** Tx laser ray tracing inside a triplexer.

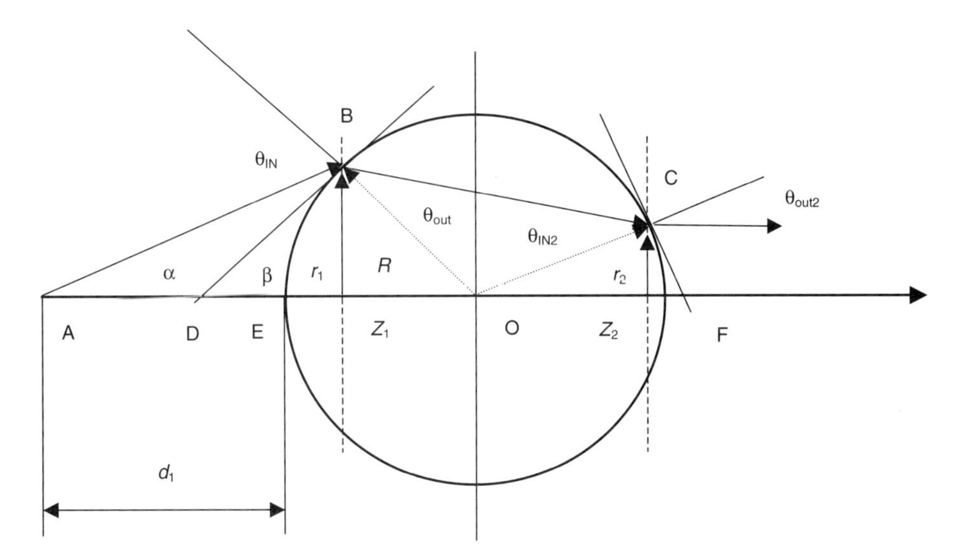


**Figure 5.15** Ray tracing inside a ball lens with a radius $R$ showing ABC ray path.

This kind of problem can be solved with geometric optics and by drawing the ray tracings along the optical axis of the triplexer system. Figure 5.14 provides an illustrative ray scheme of this design challenge. The solution is based on the concepts of Sec. 5.2.2.

Ray tracing CAD programs are based on solving geometrical and trigonometric identities as well as using Snell's law and other optical concepts. Figure 5.15 illustrates the ray-tracing problem within a ball lens. It is assumed in this case that the light source at point A at a distance $d_1$ from the lens vertex emits a ray of light with an angle of $\alpha$ with respect to the optical positive axis direction. The ray emitted out of the source impinges on the lens surface at point B, travels within the lens, and is refracted again at point C. It is clear that $\angle ABD = \beta - \alpha$. Thus, $\theta_{IN} = 90\text{ deg} - (\beta - \alpha)$.

In the same way, $\angle BOE = 90\text{ deg} - \beta$ and $\angle BOC = 180\text{ deg} - 2\theta_{out}$. This yields $\theta_{out} = \theta_{IN2}$. Knowing this, it becomes clear that $\angle COF = 2\theta_{out} + \beta - 90$ deg. After all relations between angles are arranged, values of angles are calculated. For the given values of $R$, $d_1$, and $\alpha$, the values of $\beta$, $Z_1$, and $r_1$ are solved by the following set of equations:

$$R\cos\beta = (Z_1 + d_1)\tan\alpha, \tag{5.20}$$

$$Z_1 = R(1 - \sin\beta). \tag{5.21}$$

The values of refractive angles are solved by Snell's law. Having all this information, the exact location of points C, and the values of $Z_2$ and $r_2$ are solved too by the relations

$$Z_2 = 2R[1 - \cos(2\theta_{out} + \beta - 90)], \tag{5.22}$$

$$r_2 = R\sin(2\theta_{out} + \beta - 90). \tag{5.23}$$

It is clear from the discussion that in a ball lens, angles of incidence at the in-ray surface and the out-ray surface with respect to the normal are preserved. Using this technique in a CAD program, optimization of the beamsplitter's location is done to bring the spot of light to the fiber surface. In general, beamsplitters are tilted at $\pm 45$ deg and they have the same refractive index. Thus, the optimization process is straightforward. This analysis becomes more complicated when dealing with aspheric lenses.

### 5.2.4 Aberration and tracking error

In the previous sections, lenses were analyzed and the thin lens approximation was driven. Object and image distances were connected to focal points, radii of curvature, refractive index, etc. This analysis is based on the approximation that all rays make small angles with respect to their optical axis. However, the image created by rays is based not only on points which are on the axis but also on points that lie off the axis, Fig. 5.16 represents such a core. Nonparaxial rays proceeding from a given object source do not intersect at the same point after refraction by a lens. Thus, the image formed by such a lens is not a sharp one. Moreover, the focal length of a lens depends upon its refractive index, which varies with wavelength. Therefore, if a light emitted from an object is not monochromatic, the lens forms a

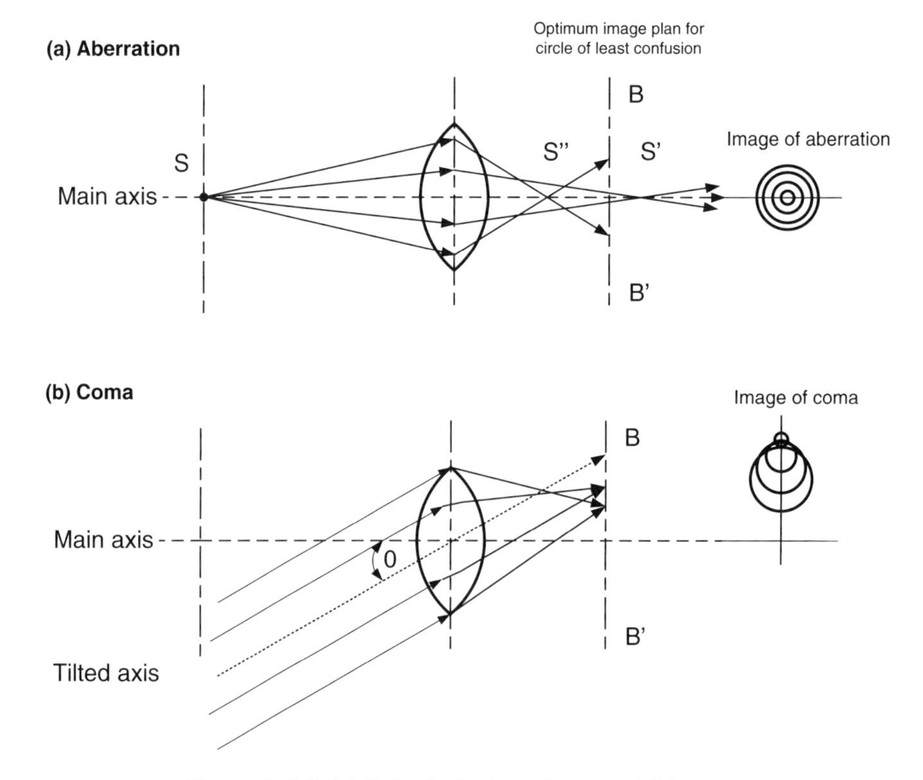


**Figure 5.16** (a) Spherical aberration, and (b) coma.

number of colored images that lie at different places and have different sizes, even if formed from paraxial rays. The departure of an actual image from its predicted calculated position is called aberration. In addition, there are aberrations that arise even from a monochromatic light source such as a laser. This kind of aberration is called spherical aberration. This type of aberration is the main design concern in bulk optics structure.

Figure 5.16 demonstrates spherical aberration. Paraxial rays from point S are imaged at S′. Rays incident to the lens near its rim are imaged at S″ closer to the lens. Rays incident to the lens surface between the rim and its optical axis are imaged between S′ and S″. Thus, there is no plane where a sharp clear spot of S is formed. The spot S is formed on perpendicular plane B B′ which is located between the two spots S″ and S′ called the image plane. This plane travels between the two spots S″ and S′ since each image of S is created at a different distance. The cross section of the beam coming out from the lens provides a circular image shape. The design goal is to find the smallest spot, known as the circle of least confusion. This is where the maximum light flux is delivered; thus in bulk optics design the SMF should be placed for maximum transmit power at the zone of least confusion where the spot is optimized.

Another lens imperfection is coma. Coma affects rays that are not on the lens axis. It is similar to spherical aberration, where the aberration is on the lens axis, and both coma and spherical aberration arise from the failure of the lens to image rays at the same point. Coma differs from spherical aberration in the imaging result of the object's image: the object is not imaged as a circle but as a comet shape; hence, the term coma.

The term "tracking error" describes power delivered to the SMF as a function of spot size and location on the active area of the fiber. Spot size is defined by the optical ray path design shown in Fig. 5.14 and lens aberration. It is desirable to have the spot center aligned with the SMF active area center. Moreover, it is a design goal to have a spot diameter smaller than the SMF active area diameter. Tracking error can result from spot misalignment on the fiber; thus, the power delivered to the fiber will be proportional to $(S - \Delta S)/S$, where $S$ is the spot area and $\Delta S$ is the spot area out of the fiber active surface. This situation results from mechanical errors as a function of temperature of the optical block. Such a case describes the spot drift from its target point, resulting in transmit power degradation. For example, laser cup position is varying because the glue is sensitive to temperature thus the spot drifts. In old triplexer technologies, laser TO can closing was done with screws; hence, mechanically it was much more susceptible to temperature changes thus the laser spot drifts. Moreover, a larger spot than the SMF active area delivered less power to the fiber. All these phenomena resulted in power sensitivity versus mechanical parameters. This sensitivity function of power alignment to fiber is called tracking error.

### 5.2.5 Packaging and integration technology

Packaging bulk optics can be divided into two main efforts. The first is the development of compact housing and packaging technology of the optical system presented in Fig. 5.8, and the second is the development of cost-effective, high-performance transistor outline metal can package (TO-can) integration. Both are challenging. Optical integration involves all design aspects presented in previous sections in regard to optical performance, such as optical cross talk, tracking error, optical transmit power, optical coupling, aberration, sealing, and complying with environmental specifications.

The other part involves integration of optical semiconductors into a TO-can and complying with required performances for high data rates and wide bandwidths for analog receivers.

Laser coupling to SMF by using molded aspheric lenses was reported about a decade ago.[31] Use of aspheric lenses is common in bulk optics to overcome spherical aberration. This technology has a coupling loss of 2.2 dB and an assembly loss of 0.24 dB. Further research analyzed the coupling of lasers to SMF by using spherical lenses.[22] MacSIGMA CAD software is available for ray tracing and optimization of the optical design.

Recent research[30] reports show miniaturization of laser packaging, going from traditional TO56 used for DFB lasers to TO38. The main idea is to build a small OSA using a prealigned Si microlens and a laser diode integrated on a Si V-groove substrate with a footprint of less than 1 mm by 1 mm. The lens is based on a Si DOE. The DOE has a binary structure to simulate the smooth profile of a Fresnel lens. The Fresnel lens profile can be obtained by quantizing the conventional refractive lens in units of wavelength as shown in Fig. 5.17. The advantage of DOE is strict control of focal length determined by the pitch of fringe structure. Silicon is transparent for wavelengths longer than 1100 nm. Tight control of focal length of the lens is required because of alignment of the lens to the small dimension OSA. This technology enables 4-f coupling by placing a combination of lenses on the same V-groove substrate. Between these two lenses, the propagating beam is collimated up to 2 mm without coupling efficiency degradation. Consequently, any functional elements such as an isolator or a WDM filter can be inserted, making this OSA a versatile optical platform. Reliability of optical components as a function of solders was investigated at the first strides of optical modules.[32] Creep of soldered joints or other material may affect coupling performance of optical components to SMFs in the long term. The creep mechanism results because of the build in mechanical stress during the construction of optical modules. Mechanical stress may result from temperature, mechanical mounting of optical modules to the whole optical electrical platform, etc.

Results showed median life of approximately $10^7$ hours in case of laser diode and SMF system. The failure rate at 25 years of service is approximately $10^{-3}$ of functional integrity testing. The epoxy cure process in bulk optics is an important parameter too.[33] Epoxies have residual stress as a result of their solidification. The epoxy layer between the TO–CAN, in which the laser or PD are capped, and the housing is a linear viscoelastic (time dependent) material. The epoxy layer create a cylindrical shape and axial symmetry is used to solve the strain problem resulting during solidification. Cure process after solidification is one of the methods to overcome strain build in epoxy.

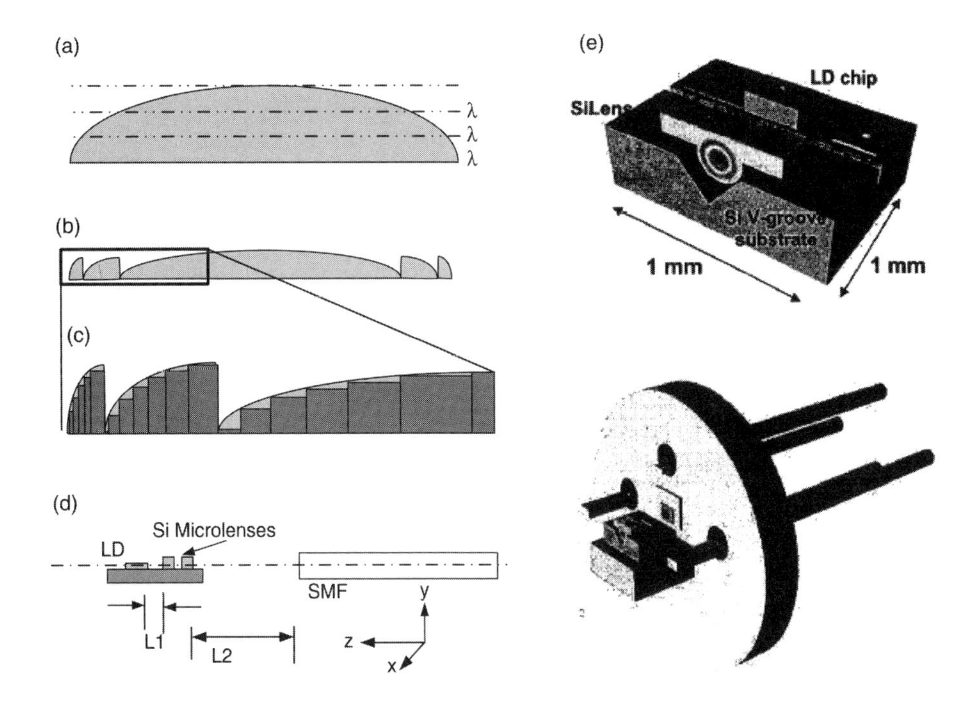


**Figure 5.17** Integrated optics on a TO can principle of DOE: (a) refractive lens; (b) Fresnel lens; (c) DOE represents the lens surface by multilevel etched profile; (d) optical system for OSA to SMF coupling; (e) OSA bench with V groove and integration to TO can.[30] (Reprinted with permission from *Proceedings of IEEE Electronic Components and Technology Conference* © IEEE, 2004.)

## 5.3 Main Points of this Chapter

1. The two preferred technologies of PLC are IE and SoS
2. Wavelength separation in PLC is accomplished by MZI.
3. Directivity of couplers in PLC is the key to accomplish isolation of transmitted power leakage into the receive section. This is one of the sources for optical cross talk in PLC.
4. The PLC technology edge, by its repeatability of performance, has less human involvement in assembly. There is accurate control of parameters such as power isolation, return loss, and wavelength separation as well as isolation.
5. PLC packaging is SMT rather than long leads of TO can. Thus, it has potential to operate in high data rates for digital applications and wide BW for analog applications.

6. IE process is a fast process in which a core is created by changing the refractive index of a glass. The process has no need for a high-class clean room. The IE process has 11 steps:

   - mask preparation;
   - electrons deposition process of about 60 minutes;
   - burying the electrons in glass using diffusion process, for 15 min, at 350°C and under electrical field;
   - printing conductors using lithography;
   - connecting the bulk support glass to the optical circuit wafer;
   - etching via holes;
   - assembly of components using alignment and marking techniques;
   - dicing;
   - polishing;
   - packaging; and
   - pigtailing.

7. One of the advantages of PLC is low coupling losses to fiber since it has a direct contact to the fiber and similar refractive indices. In IE technology, it has the optimal matching to fiber. Thus, mismatch loss is minimal.

8. PLC technology is less susceptible to environmental thermal changes since it has no mechanical deformation and changes in dimensions that affect the optical transfer functions from the laser to the fiber, from the fiber to the photodiodes, and between the laser and the photo detectors; for this reason, there is hardly any reduction in cross talk over temperature.

9. Bulk optics technology is an integration of discrete components into a mechanical optical system, which is composed of the following:

   - laser diode in a TO can with integral ball or aspherical lens;
   - photodetectors in a TO can with integral ball lens;
   - beamsplitters that are dichroic mirrors; and
   - housing.

10. Bulk optics triplexers for 1310 nm, 1310 nm, 1550 nm that were used for old wavelength plans have an internal optical trap to prevent leakage of the 1310-nm transmitter into the 1310-nm receiver photodetector because of internal reflections.

11. FSAN plan is 1310-nm digital transmitting, 1490-nm digital receiving, 1550-nm analog CATV receiving.

12. In FSAN bulk optics isolation between 1310-nm and 1490-nm receivers is achieved by the optical filters.

13. The transmitted power in bulk optics is defined by the optimization of the laser beam spot on the SMF active area.

14. Tracking error in bulk optics results because of the movement of the transmitted light spot from the core of the SMF. Thus there is a reduction in optical power delivered to the SMF core surface.

15. Tracking error results due to changes in mechanical dimensions because of a change in temperature or because of an external mechanical stress such as mounting screw torque.

16. Aberration results from a lack of an ideal lens in which paraxial beams have different focal points than the rays incident to the lens surface at a larger distance from the optical axis or near the lens rim.

17. Spot optimization of the transmitted light is done by selecting the image circle of minimum point of confusion of a lens.

18. One of the methods to reduce aberration is by using aspherical lens and optimizing the aspheric plane of the lens.

19. For receiving, it is desirable to defocus the beam and create a large spot on the photodetector area since it minimizes distortions.

20. Paraxial approximation $z = \frac{1}{2}cr^2$ defines the distance from a sphere in which it looks as a surface or a plane. In case of a small lens with a small radii, the distance from the sphere surface, is very small. An analogy for that is observing earth from earth or from space; the earth looks like a flat plane for us on earth, and from space it looks like a sphere.

21. In a high magnification lens, a small error on the object side results in a large error on the image side.

22. In bulk optics design, thin lens approximation is accepted and further optimization is done by ray tracing with CAD tools.

23. Bulk optics closing technology is by advanced epoxies and laser welding.

## References

1. Malinovich, Y., E. Arad, D. Brooks, and A. Shemi, "Novel opto-electronic hybrid concept for high performance and low cost modules based on ion exchange in glass PLC." ColorChip LTD (2004).
2. Kurosaki, T., T. Hashimoto, N. Ishihara, Y. Suzuki, M. Yanagisawa, H. Kimura, M. Nakamura, Y. Tohmori, K. Kato, Y. Kawaguchi, Y. Akahori, Y. Yamada, K. Kato, H. Toba, and J. Yoshida, "Full duplex 1300/1550 nm WDM optical transceiver modules for ATM-PON systems using PLC-hybrid integration and CMOS IC technologies." pp. 631–632, *Proceedings from the 24th European Conference on Optical Communication* (1998).
3. Okano, H., M. Okawa, H. Uetsuka, T. Teraoka, S. Aoki, and S. Tsuji, "Passive aligned hybrid integrated optical module using planar lightwave circuit platform." pp. 73–74, *Proceedings of IEEE, LEOS Annual Meeting*, Vol 1, pp. 74–75 (1996).
4. Ahn, J., Y. Lee, K. Kwak, D. Shin, S. Kim, and T. Kim, "Development of novel cost-effective bidirectional optical triplexer based on polymer PLC platform." pp. 1396–1400, *Proceedings of IEEE, Electronic Components and Technology Conference* (2004).
5. Yamamoto, T., G. Nakagawa, K. Terada, K. Shibata, T. Naruse, H. Nobuhara, and K. Tanaka, "Highly temperature stable, low-cost PLC optical transceiver module for global (FSAN) ATM-PON systems." *Proceedings of IEEE APCC-OECC 99 Fifth Asia-Pacific Conference*, pp. 1665–1666 (1999).
6. Anthamatten, O., R.K. Battig, B. Valk, P. Vogel, C. Marxer, M. Gretillar, and N.F. de Rooji, "Packaging of a Reflective Optical Duplexer Based on Silicon Micromechanics." *Proceedings of IEEE LEOS Summer Meeting*, pp. 61–62 (1996).
7. Liu, K., and E.Y.B. Pun, "Single mode Rb þ -Kþ ion-exchanged BK7 glass waveguide with low losses at 1550 nm." *IEEE Photonics Technology Letters*, Vol. 16 No. 1, pp. 120–122 (2004).
8. Valles-Villarreal, N., A. Villalobos, and H. Marquez, "Stress in copper ion-exchanged glass waveguides." *Journal of Lightwave Technology*, Vol. 17 No. 4, pp. 606–612 (1999).
9. Guntau, M., A. Brauer, W. Karthe, and T. Pobner, "Numerical simulation of ion-exchange in glass for Q39 integrated optical components." *Journal of Lightwave Technology*, Vol. 10 No. 3, pp. 312–315 (1992).
10. Miliou, A.N., R. Srivastava, and R.V. Ramaswamy, "A 1.3 mm directional coupler polarization splitter by ion-exchange." *Journal of Lightwave Technology*, Vol. 11 No. 2, pp. 220–225 (1993).
11. Morshed, A.H., and D.A. Khalil, "Fabrication and characterization of surface and buried optical waveguides by ion-exchange on glass." *Proceedings of the Electrotechnical Conference MELECON*, pp. 463–466 (2002).

12. Albert, J., and G.L. Yip, "Wide single-mode channels and directional coupler by two-step ion-exchange in glass." *Journal of Lightwave Technology*, Vol. 6 No. 4, pp. 552–563 (1988).
13. Beguin, A., T. Dumas, M.J. Hackert, R. Jansen, and C. Nissim, "Fabrication and performance of low-loss optical components made by ion-exchange in glass." *Journal of Lightwave Technology*, Vol. 6 No. 10, pp. 1483–1487 (1988).
14. Cichalewski, W., A. Napieralski, H. Camon, and B. Estibals, "Analytical modeling and simulations of a MEMS micro-mirror MATLAB implementation." *Proceedings of the 7th International Conference CAD Systems in Microelectronics*, pp. 360–365 (2003).
15. Yoshida, J., "Low cost optical modules for fiber to the home." *Proceeding of IEEE LEOS Annual Meeting*, Vol. 1, pp. 69–70 (1996).
16. Nakasuga, Y., T. Hashimoto, Y. Yamada, H. Terui, M. Yanagisawa, K. Moriwaki, Y. Akahori, Y. Tohmori, Kato, S. Sekine, and M. Hriguchi, "Multi-chip hybrid integration on PLC platform using passive alignment technique." pp. 20–24, *Proceedings of Electronic Components and Technology Conference*, Vol. 46 No. 28–31, pp. 20–25 (1996).
17. Uetsuka, H., H. Okano, and T. Shiota, "Novel asymmetric Y-branches for planar lightwave circuits." *OFC Technical Digest*, pp. 176–177 (1996).
18. ColorChip, Ltd., "Selecting the proper waveguide." White Paper (2004).
19. Sears, F.W., and M.W. Zemansky, *College Physics Part-2, Third Edition.* Addison Wesley Publishing Company, Inc. (1966).
20. Kato, K., and I. Nishi, "Low loss laser diode module using molded aspheric glass lens." *IEEE Photonics Technology Letters*, Vol. 2 No. 7, pp. 473–474 (1990).
21. Jiang, W., Y. Sun, R.T. Chen, B. Guo, J. Horwitz, and W. Moerey, "Ball lens based optical add-drop multiplexers: design and implementation." pp. 825–827, *IEEE Photonics Technology Letters*, Vol. 14 No. 6 (2002).
22. Wilson, R. G., "Numerical aperture limits on efficient ball lens coupling of laser diodes to single-mode fibers with defocus to balance spherical aberration." *NASA Technical Memorandum 4578*, Langley Research Center (1994).
23. Kidger, M.J., *Fundamental Optical Design.* SPIE Press, Bellingham, WA (2001).
24. Hobbs, P.C.D., *Building Electro-Optical Systems: Making It All Work.* John Wiley & Son, Inc. (2000).
25. Riedl, M.J., "Diamond turned diffractive optical elements for the infrared: suggestion for specification standardization and manufacturing remarks." *Proceedings of SPIE*, Vol. 2540, pp. 257–269 (1995).
26. Riedl, M.J., "Predesign of diamond turned refractive/diffractive elements for IR objectives." *SPIE Critical Review*, Vol. CR 41, pp. 140–156 (1992).
27. Blacke, P.N., and R.O. Scattergood, "Ductile regime machining of germanium and silicon." *Journal American Ceramic Society*, Vol. 73 No. 4, pp. 949–957 (1990).

28. Swanson, G.J., "The theory and design of multilevel diffractive elements." *Binary Optics Technology*, Lincoln Laboratory Technical Report 854, p. 6 (1989).
29. Hudyma, R., et al. "Design and fabrication of an infrared wide field of view objective utilizing a hybrid refractive/diffractive optical element." *Proceedings of SPIE*, Vol. 1970, pp. 68–78 (1993).
30. Shimura, D., M. Uekawa, R. Sekikawa, K. Kotani, Y. Maeno, H. Sasaki, and T. Takamori, "Ultra compact optical subassembly using integrated laser diode and silicon microlens for low-cost optical component," *Proceedings of IEEE Electronic Components and Technology Conference*, pp. 219–224 (2004).
31. Kato, K., and I. Nishi, "Low-loss laser diode module using a molded aspheric glass lens," *IEEE Photonics Technology Letters*, Vol. 2 No. 7, pp. 473–474 (1990).
32. Mitomi, O., T. Nozawa, and K. Kawano, "Effects of solder creep on optical component reliability," *IEEE Transactions on Components, Hybrids, and Manufacturing Technology*, Vol. 9 No. 3, pp. 265–271 (1986).
33. Broadwater, K., and P.F. Mead, "Experimental and numerical studies in the evaluation of epoxy-cured fiber optics connectors." *Proceedings of IEEE Electronic Components and Technology Conference*, pp. 981–988 (2000).
34. Hecht, E., *Optics, 2nd Edition*. Addison Wesley Publishing Company (1984).
35. ZEMAX Optical Design Program User's Guide Version 8.0, Focus Software (1999).
36. Zheng, R., and D. Habibi, "Emerging architectures for optical broadband access networks." *Proceedings of Australian Telecommunications, Networks and Applications Conference* (2003).

# Chapter 6

# Semiconductor Laser Diode Fundamentals

This chapter reviews laser physics and its concept of operation. A short introduction to the quantum phenomenon of stimulated emission is presented, explaining the process of lasing and emission types for general background. This is followed by modes of excitation and structure of gain-guided lasers versus index-guided lasers. Common lasers used by the industry are reviewed and background is provided about longitudinal modes, distributed feedback (DFB) lasers, Fabry–Perot (FP) lasers, quantum well lasers, tunable lasers, and vertical cavity surface-emitting lasers (VCSELs). After a basic solid foundation of laser physics is laid, a practical RF modeling of the laser is presented, followed by modulation terminologies for RF and digital schemes. An introduction to the optical isolator is provided at the end of this chapter to explain how to overcome and improve the reflection coefficient of the laser to the fiber. Reflections in both cases of digital and analog transport introduce distortions. In the case of digital transport, the observation might be double-eye transitions, implying deterministic jitter. For analog, reflections may cause a reduction in carrier-to-noise ratio (CNR) and intermodulation (IMD) performance is explained in Chapter 17.

## 6.1 Basic Laser Physics—Concepts of Operation

Semiconductor lasers are similar to other lasers such as Maiman's ruby laser or the He-Ne gas laser since in both cases the emitted radiation has spatial and temporal coherency. Laser radiation is highly monochromatic, which means it has a narrow wavelength bandwidth (BW) and a highly directional beam of light. In other words, the energy of laser light or flux has a high density in a small area. In field theory, it means that the pointing vector $\mathbf{S} = \mathbf{E} \times \mathbf{H}$ of the electromagnetic wave has high value. However, the difference between a semiconductor laser and the other laser is as follows:

1. In conventional lasers, the electron quantum transitions occur between discrete energy levels. In a semiconductor laser, the transitions are related to the band-level properties of the material.

2. A semiconductor laser chip is very compact in size compared to a conventional laser, and it is within the size order of 0.1 mm. Moreover, since the active region of a semiconductor laser is very narrow, the divergence of *d* has a very short photon lifetime, high-frequency modulations are achievable to a certain point, where second-order effects such as wavelength chirp and relaxation oscillation start to have an effect (Chapter 7).

Because of the above advantages, semiconductor laser is one of the most important light sources for optical-fiber communications. There are three wavelengths that are commonly used: VCSELs of 900 nm are used, and a Si avalanche photodiode photodetector (APD) is used on the receiving side; where the fiber has a low loss of 0.6 dB/Km and low dispersions, 1310 nm is used; and where the fiber losses are at the minimum values of about 0.2 dB/Km, 1550 nm is used. In both cases, the InGaAs photodetectors are used on the receiver side.

LASER is an acronym for light amplification by stimulated emission of radiation. When the PN junction is formed, several physical phenomena are created. There is a PN barrier due to the potential difference between the P type and the N type.[1,2] This potential barrier creates an electrical field *E*, which creates a depletion region at the PN boundary. The width of the entire depletion region depends on the P- and the N-type doping concentrations.[1,2] Therefore, to operate the laser, there is a need to apply a forward-drive current to overcome this potential barrier. This current is the threshold where the diode starts to conduct; an additional threshold current, which this discussion refers to, is where the diode starts to emit.

The current mechanism at a PN junction is diffusion of electrons and holes across the junction.[1,2] In this case, both electrons and holes are present simultaneously in the depletion region and can recombine by creating spontaneous and stimulated emission, generating a photon in an appropriate semiconductor material. Stimulated emission is a result of optical gain and optical feedback. Consequently, the lasing process is a result of positive optical feedback that enables oscillations. Thus, two conditions should be satisfied:

1. The optical gain should be equal to the optical loss at the active region.
2. Positive optical feedback should exist. In an FP laser, this feedback is provided by cleaved facets that create multiple reflections or by distributed gratings in a DFB laser.

Assume two energy Fermi levels $E_1$ and $E_2$ are in an atom of a semiconductor. $E_1$ is the basic low-level energy and $E_2$ is the excited level. Any transitioning between the two energy levels involves emission or absorption of a photon with an optical frequency $\nu_{12}$ according to $h\nu_{12} = E_2 - E_1$, where $h$ is Plank's constant. Generally, at room temperature, all the atoms are in the low-level energy state. This equilibrium is disturbed when a photon of energy that is exactly equal to the quantum energy difference $E_2 - E_1$ impinges on the atom and transfers an electron from $E_1$ to $E_2$. This is an exciting process and the photon energy is absorbed by the atom. The electron at the $E_2$ level is in an unstable energy state and after a short dwell time, it returns to its original energy level and emits a photon with the optical frequency equal to $h\nu_{12}$. This process is called spontaneous emission. The lifetime of such a spontaneous emission varies typically from nanosecond to millisecond, depending on the semiconductor material type, i.e., the energy band gap and density of recombination centers. The spontaneous emission is an incoherent process since photons are emitted randomly and in random directions; thus, there is no phase relationship or spatial coherency. An interesting and important thing happens when a photon hits an atom when its electron is excited and is in a higher energy level. At the instant an excited atom is impinged on by a photon, the atom is immediately stimulated; the excited electron returns to its basic energy level $E_1$ and emits a photon. Consequently, two photons are emitted. One is the incident photon and one is due to the quantum transition. The excited photon is coherent in the phase with the incident radiation. That process is called "stimulated emission," and is described in Fig. 6.1.

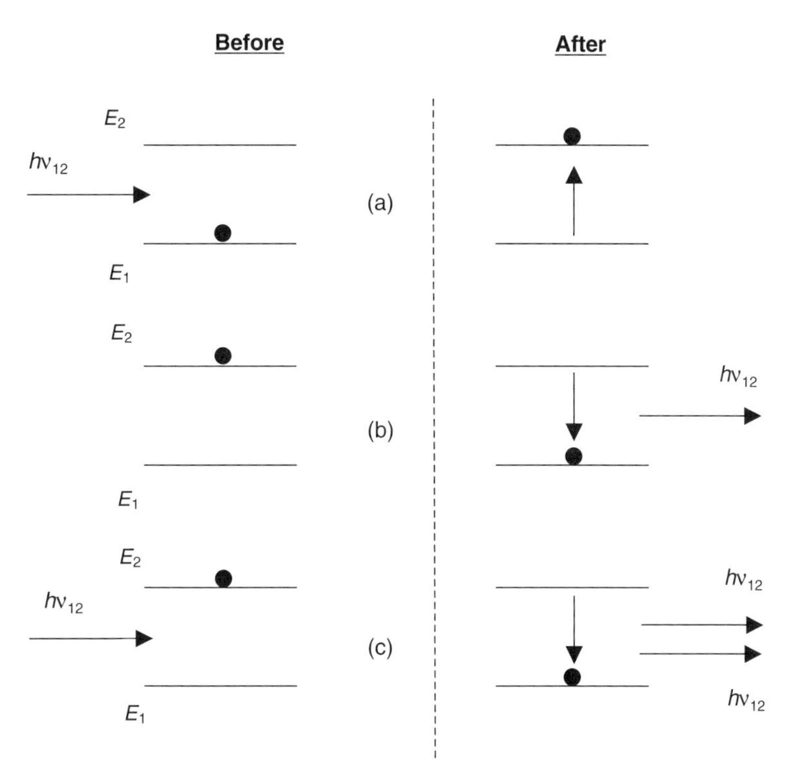


**Figure 6.1** The three basic states of transition of Fermi energy levels $E_2$ and $E_1$. (a) Absorption of the photon quantum energy $h\nu_{12} = E_2 - E_1$. (b) Spontaneous emission. (c) Stimulated emission in the phase due to the quantum energy $h\nu_{12} = E_2 - E_1$.

One of the best ways to explain the basics of laser operation is by looking at the first small Ruby laser[6,10] shown in Fig. 6.2. This laser type can be referred to as the foundation of the FP solid-state laser. The ruby rod is the gain medium of the laser. The resonance structure or cavity as defined in lasers as accomplished by having mirrors on both ends of the ruby rod. The rod cleaved sides is covered with mirrors which are using one full mirror and one side has a half-silvered mirror. The laser is excited by optically pumping the ruby rod using a flash tube. This process brings electrons to a high-energy state. As some electrons in the rod spontaneously drop from this high-energy state to a lower fundamental level, they emit photons that trigger further stimulated emission. The photons bounce between the mirrors, generating more stimulated emissions. The process of gain and positive feedback results in oscillation between the two mirrors. Only certain wavelengths have a resonance and lasing occurs through the half-silvered mirror. In solid-state lasers such as FP, the replacement for the flash tube is the forward current across the PN junction, which creates the population inverse. As in the ruby laser, FP laser has two cleaved facets with mirrors (see Fig. 6.3).

Stimulated emission is the basis of laser operation. However, there is a need to overcome the material absorption in order to have a significant amount of light emission. In the solid-state semiconductor laser, this is achieved only when there is a greater population in the excited energy level $E_2$ than in $E_1$. This situation is called "population inversion" and is a prerequisite for laser operation. The phenomenon is achieved by current injection across the PN junction. At the instance the population inversion state is accomplished, the quasi-Fermi-level, the energy level with an occupation probability of 50%, separation between the conduction bands, and the valence level exceed the band gap under forward biasing of the PN junction. That means

$$E_{\mathrm{Fc}} - E_{\mathrm{Fv}} > h\nu > E_{\mathrm{g}}, \tag{6.1}$$

where $E_{\mathrm{Fc}}$ is the quasi-Fermi-level energy, $E_{\mathrm{Fv}}$ is the valence level, and $E_{\mathrm{g}}$ is the band-gap energy.

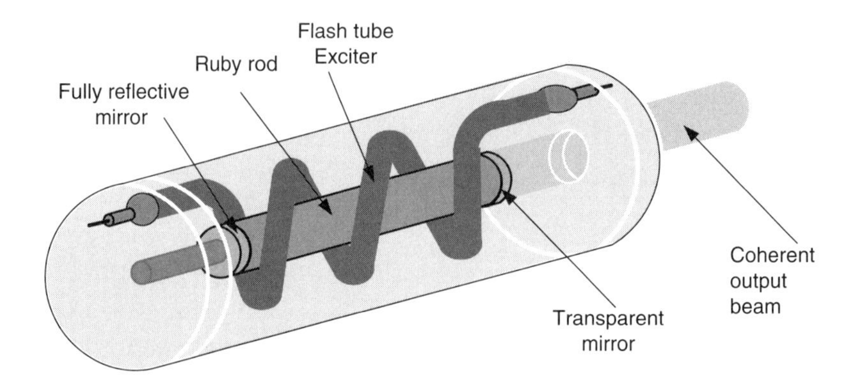


**Figure 6.2** The Ruby laser of Theodore Maiman.

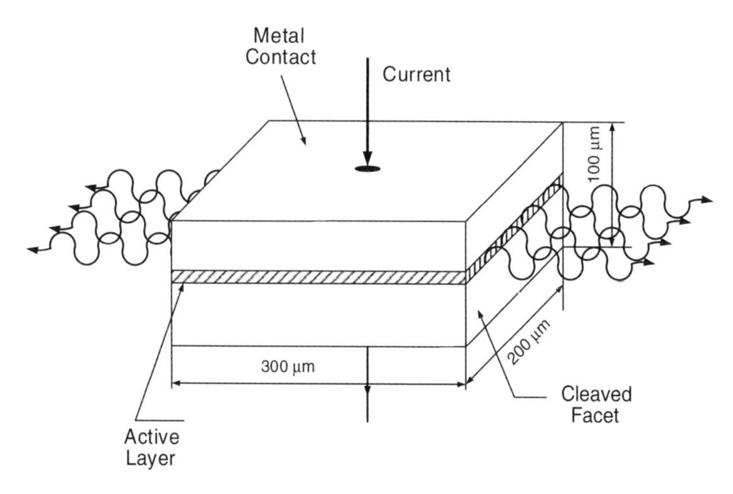


**Figure 6.3** Concept of FP laser in a double heterostructure (DH).[3] (Reprinted with permission of Van Nostrand/Springer © 1986.)

Equation (6.1) states that in order to have emission, the photon energy has to be higher than the band-gap energy but lower than the difference between quasi-Fermi-level energies in the conductance state (population inverse state) and the valance level. Optical gain is achievable in the active layer when the injected carrier density is high enough to overcome the absorption in the semiconductor. When the gain is sufficient, the peak optical gain dependency upon the carrier density $N$ can be expressed by using the linear approximation given by[3]

$$g_P = \overline{g_0}[N - N_0], \tag{6.2}$$

where $N_0$ is the carrier density at the transparency, typically $1-1.5 \times 10^{18}$ cm$^{-3}$ for an InGaAs laser, $N$ is the injected carrier density, and $\overline{g_0}$ is the spatial gain constant in centimeters squared. It can be observed that when the injected density exceeds $N_0$, a positive gain is obtained.

Not all semiconductors generate emission during the recombination process and populate inverse conditions at the forward bias state. For instance, Si and Ge do not emit under the above mentioned conditions. The reason for that is that it is a nonradiative recombination process in these materials. This process includes recombination at defects, surface recombination, and Auger recombination. This last one is important in the case of long wavelengths such as 1300–1600-nm semiconductor laser sources.[3,4] In the Auger recombination process, the energy released during electron–hole recombination is transferred to another electron or hole as kinetic energy rather than producing light.[3] Because of nonradiative recombination existence, there is a reduction in the quantum efficiency (QE) of producing and emitting light. This effect or phenomenon is defined by internal QE factor $\eta_{\text{int}}$, which indicates that only a fraction of the injected carriers during forward biasing and population inversion is converted into photons and is described by

$$\eta_{\text{int}} = \frac{R_{\text{rr}}}{R_{\text{rr}} + R_{\text{nr}}} = \frac{\tau_{\text{rr}}}{\tau_{\text{rr}} + \tau_{\text{nr}}}, \tag{6.3}$$

where $R_{\text{rr}}$ is the radiative recombination, which was explained previously, composed from both mechanisms of spontaneous and stimulated recombination, and $R_{\text{nr}}$ is the nonradiative recombination process. In case of Si, Ge, which means $\eta_{\text{int}} = 10^{-5}$, mainly due to the fact that they have an indirect band gap. A direct band gap is when the conduction band's minimum level and valence is maximum level occur at the same level of the electron wave vector, **k**. That means the pointing vector of the radiation is created by recombination with no defect loss. In direct band-gap, materials such as GaAs or InP $\eta_{\text{int}}$ approaches 100%, which means stimulated emission dominates. That means high probability of radiative recombination since energy and momentum of electron–hole recombination are easily preserved. In case of indirect band-gap semiconductors, the released energy from electron–hole recombination is transferred to lattice phonons by lattice vibration in the crystal; hence, the vibration energies are potential and kinetic energies result in very low energy for radiation.

## 6.2 Semiconductor Laser Structure—Gain Guided Versus Index Guided

Laser diodes are divided into two main categories: gain guided and index guided. The early semiconductor laser chips, which were homostructure injection lasers, were gain guided. Laser structure was improved to a double heterostructure (DH) as shown in Fig. 6.3.

These lasers exhibit high-threshold current densities; for instance, *J*th values for GaAs homostructure injection lasers are about 50–100 $\text{kA/cm}^2$ at room temperature. The laser structure was a regular PN junction, which was cleaved by two facets perpendicular to the junction plane. These facets provided the optical-positive feedback to the spontaneous emission. The feedback generated the stimulated emission by using the spontaneous emission as a trigger for optical oscillation and amplification. The emission was coplanar with the PN junction active plane. In other words, emission was from the active PN junction plane. However, this structure concept suffered from several disadvantages. It had no carrier and hole confinement for recombination of the lateral injected carriers. This, of course, reduced the efficiency of the emitting recombination, which resulted in high current consumption. In other words, it had no optical confinement, which is a poor optical cavity. This added optical loss during the optical feedback and amplification process. To overcome losses, a DH was used. However, there was no lateral confinement of photons and the entire PN area was used as a gain section.

One of the methods to overcome the large area anode was by creating a narrow long stripe as the anode, shown in Fig. 6.4. An insolating layer of $SiO_2$ was deposited on top of the P-type material. The $SiO_2$ was then etched to create a recess and

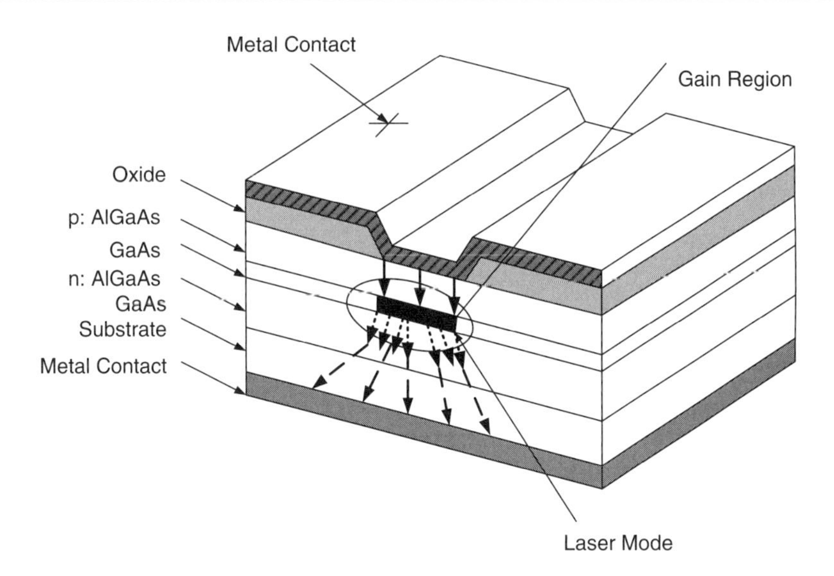


**Figure 6.4** Gain-guided laser in strip geometry showing P contact, N contact, active layer lasing region, and cleaved mirrors in a DH stripe laser.

a P-contact was deposited on top of the insulator, creating a long narrow stripe that has contact to the P type. This contact is the anode, which created a confined *E* field under the stripe. This created a waveguide or recombination channel under the strip. Thus, the *E*-field distribution in that topology was laterally getting weaker at the stripe-width edges and peaked to its max at the middle of the stripe width. Hence, the generated beam was narrow with very low spatial gain variations. The injected carriers were already confined under the narrow stripe since the anode stripe, with respect to the cathode, is a narrow capacitor plate. This plate is connected to a positive voltage potential that generates an *E* field only under that striped area with fringes at the edges. Hence, the current density *J* is given by the following equation:

$$J = \frac{I}{W \cdot L}, \tag{6.4}$$

where *I* is the bias current, *L* is the stripe length, and *W* is the stripe width. As a consequence, the active area at the PN junction is narrow and confined mostly under the stripe. That method is called "gain guided" since the confinement of the optical modes is governed by the lateral optical gain under the stripe. The stripe dimensions in such a design approach are 1–5 μm for *W* and 200–500 μm for *L*. The current density distribution changes laterally at the stripe edges, while flowing through the cladding layer since it has some fringes out of the W boundary. This is because of the carriers' lateral diffusion and drift at the active layer. Hence, the *E* field would have a similar distribution. The current mechanism in a PN junction is diffusion-mechanism solved by the differential diffusion equation pairs.[1] As a result, there is a vertical diffusion but also lateral diffusion

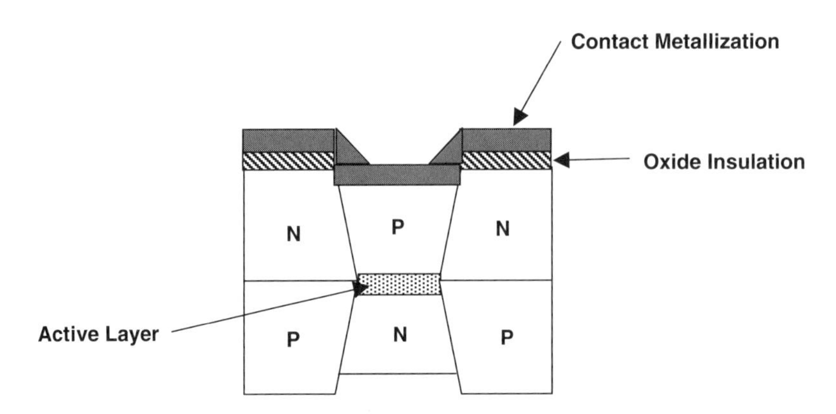


**Figure 6.5** Index-guided buried heterostructure laser.

creating a "bell-shaped" density distribution. The accomplishment of this method was a gain increase on one hand and a current reduction on the other. This improved the laser efficiency as well as providing a narrower lasing radiation.

The concept of index-guided lasers is based on creating a waveguide, which is confined due to variation in refraction index. Figure 6.5 illustrates such a buried heterostructure.

The active layer (gain section) is confined because of the semiconductor structure. The active section is a forward biased PN-junction diode sandwiched between two reversed biased NP-junction diodes. These reversed diodes are insulated from the metal contact by oxide insulation. The forward-biased diode has a photon confinement mechanism similar to the gain-guided one. However, the lateral confinement observed in the index-guided structure is minimized due to the reverse bias on the NP cladding. This, of course, generates a step change in the index of refraction, consequently creating a buried waveguide. In this structure, the confinement area under the anode metal contact is narrower and its lateral boundaries are controlled better compared to the gain-guided laser. Moreover, the isolated NP junction, which is in reverse bias, is depleted, thus creating a clearer lateral boundary at the active layer. As a result, the beam width and spectral width of the index-guided laser is narrower compared to a gain-guided laser. Further, better confinement means a higher quality factor (Q) of the cavity and lower losses. This reflects directly on current consumption and QE. Section 6.3 reviews these aspects in detail.

## 6.3 Longitudinal Modes and a Fabry Perot (FP) Lasers

From a review of Sec. 6.1, it is clear that the fundamental laser structure is a cavity that resonates at the lasing wavelength. From wave theory and boundary conditions of partial differential equations describing this case, there are several cavity modes. The cavity mode wavelengths are longitudinal modes in which the waves are constructively interfering. This condition states that a wave injected into the cavity has a maximum transmission under this condition. Figure 6.6 illustrates the longitudinal-mode excitation mechanism.

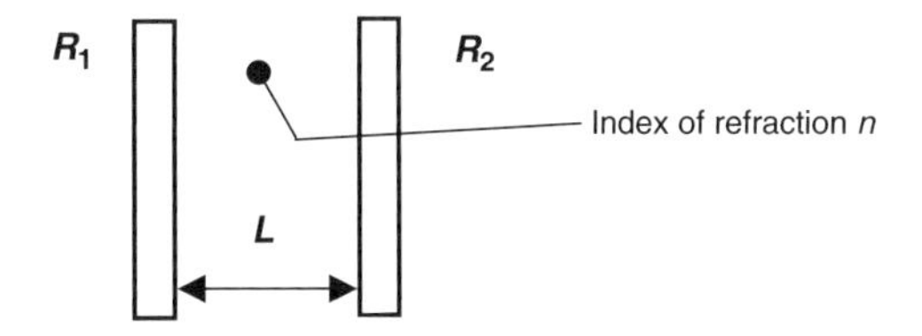


**Figure 6.6** FP laser cavity with length of $L$ and reflectivities $R_1$ and $R_2$.

The threshold condition required to create oscillation is to have a minimum optical gain that is equal to the cavity losses and can operate the laser. This condition, in analogy, is similar to a feedback oscillator assuming a cavity with length of $L$, loss factor per length unit of $\alpha$, and two cleaved mirrors with reflectivities of $R_1$ and $R_2$. To check the threshold condition, the electric field is examined within one round trip. The initial field and the reflected field after a single round trip should be equal in order to satisfy creative interference. Now, assume an initial field with amplitude $E_0$, angular frequency $\omega$, propagation constant $\beta$, and the refraction index of the cavity $n$. During one cavity round trip, the field amplitude changes by

$$\sqrt{R_1 R_2}\exp(-\alpha_i L), \tag{6.5}$$

where $\alpha_i$ is the internal cavity loss due to free absorption and scattering of carriers.

However, at the same time, the cavity adds gain and thus the amplitude of the electric field is increased. The cavity gain is referred to as one round trip and a round trip equals $2L$. Thus, one round-trip gain is given by

$$\exp\left(\frac{g}{2}\cdot 2L\right) = \exp(g_{\text{th}}L), \tag{6.6}$$

where $g_{\text{th}}$ is the threshold gain. In addition, the electric field phase after one round trip is denoted by $2\beta L$. These three conditions result in a round-trip-threshold condition of creative interference:

$$E_0 = E_0 \exp(g_{\text{th}}L)\sqrt{R_1 R_2}\exp(-\alpha_i L)\exp(2j\beta L). \tag{6.7}$$

This complex equation has two conditions in its real and imaginary parts. Solving the real part of Eq. (6.7) and multiplying both sides Eq. (6.7) and multiplying both sides by the complex conjugate results in the feedback condition for lasing threshold gain. As was expected, the feedback should be equal to the gain:

$$g_{\text{th}} = \alpha_i - \frac{1}{2L}\ln(R_1 R_2) = \alpha_{\text{cavity}}. \tag{6.8}$$

The phase condition of Eq. (6.7), which is the imaginary part, provides the essential state for creative interference.

$$2\beta_k L = 2k\pi, \tag{6.9}$$

where $k$ is an integer. Since the cavity refractive index is $n$, the wave propagation index is given by $\beta = n\omega/c$ or $2\pi\nu n/c$ or $2\pi n/\lambda$, where $c$ is the speed of light in free space, $\nu$ is the wavelength frequency in free space, and $\lambda$ is the wavelength in free space. Consequently, the wavelength must satisfy the following condition for creative interference:

$$\lambda_k = \frac{2nL}{k}. \tag{6.10}$$

The wavelength separation of the FP laser is given by checking the modes' sensitivity to wavelength:

$$\frac{dk}{d\lambda_k} = \frac{d}{d\lambda_k}\left(\frac{2nL}{\lambda_k}\right) = \frac{2L}{\lambda_k^2}\left(\lambda_k \frac{dn}{d\lambda_k} - n\right) = -\frac{2n_g L}{\lambda_k^2}, \tag{6.11}$$

where $n_g$ is the group index. Equation (6.11) can be written in a different way:

$$\Delta\lambda = \frac{\lambda_k^2}{2n_g L}\Delta k. \tag{6.12}$$

However, $\Delta k$ is the index increment and equals unity. Thus, the FP mode separation is given by

$$\Delta\lambda = \frac{\lambda^2}{2n_g L} \approx \frac{\lambda^2}{2nL}. \tag{6.13}$$

The quality factor, $Q$, of FP is a measure of the device's resolution. This is calculated from Eq. (6.7) by finding the wavelength in which the electrical field amplitude drops by $1/\sqrt{2}$.

$$Q = \frac{2nL}{\lambda_0}\frac{\pi\sqrt[4]{R_1 R_2}}{1 - \sqrt{R_1 R_2}}. \tag{6.14}$$

Another measure of FP is the finesse $F$. The finesse is the ratio between the wavelength separation and its BW at half-power level:

$$F = \frac{\pi\sqrt[4]{R_1 R_2}}{1 - \sqrt{R_1 R_2}}. \tag{6.15}$$

Thus, the ratio between the quality factor and the finesse provides the dominant longitudinal mode $k$ of a given FP laser:

$$k = \frac{Q}{F}. \tag{6.16}$$

## 6.4 Distributed Feedback (DFB) and Distributed Bragg Reflector (DBR) Lasers

As was explained in Sec. 6.4, FP lasers' disadvantage is in their multiple modes of oscillation and lasing. This phenomenon affects eye purity in digital transport because of chromatic dispersion in fibers, thus making more jitter in the eye mask. To overcome this problem, which becomes crucial in long distance and high data rate communications, other laser cavity structures were developed. The idea was to create a single longitudinal mode (SLM) cavity laser. This way only one lasing dominant mode exists. Therefore, information reaches the receiving section in a uniform group velocity, unlike in FP where each mode reaches the receiver at a different group velocity. Nonuniform group velocity is the root cause for dispersion over a single-mode fiber with an FP laser. Those lasers are the DFB and distributed Bragg reflector (DBR) lasers. In such structures, the feedback is accomplished with an internal grating that causes spatially periodic variations in the effective refractive index of the oscillating mode. The phenomena of Bragg diffraction causes coupling between forward and backward propagating modes. These modes should satisfy the Bragg condition (further details and theory are provided in Sec. 4.4.3):

$$\Lambda = m\left(\frac{\lambda_B}{2n_{\mathrm{eff}}}\right), \tag{6.17}$$

where $\Lambda$ is the grating period, $n_{\mathrm{eff}}$ is the effective mode index, $\lambda_B$ is the Bragg wavelength at free space, and $m$ is oscillation mode that typically equals unity, but in some occasions $m = 2$ is used as well. As shown in Sec. 4.4.3, Eq. (4.86), the stop band is formed and centered at the Bragg wavelength. The BW depends upon the coupling coefficient $k$. That mechanism makes the DBR and DFB lasers operate in a single dominant mode, while suppressing other modes of oscillation as exists in a FP laser. In a DBR laser, the internal grating is placed at one or both ends of the gain medium as shown by Fig. 6.7.

One structure of DBR laser contains a single grating at one end, while the other end contains coated or uncoated cleaved facet edge. Another structure of DBR has two gratings at both sides of its gain section; the feedback is distributed, but it occurs only within the Bragg wavelength, where reflection is at its maximum. This is like having FP structure with thick mirrors, reflecting at a selective wavelength only. The DFB laser has a grating feedback element within its gain section.

SLM lasers such as DFB and DBR are initially designed with loss profiles that are mode dependent. This way the lowest cavity loss will reach threshold first while neighboring modes are filtered out because of higher cavity losses. Another way of looking at this is as an equivalent circuit of gain $A$ and feedback $\beta$. The Berkhousen oscillation condition occurs where $\beta A + 1 = 0$. Thus, in case of high loss $\beta$ value, there are no oscillations. Consequently, the grating mechanism is equivalent to the "tank" circuit of the laser and a laser is an optically coherent oscillator as was explained earlier. Having a narrow BW grating feedback

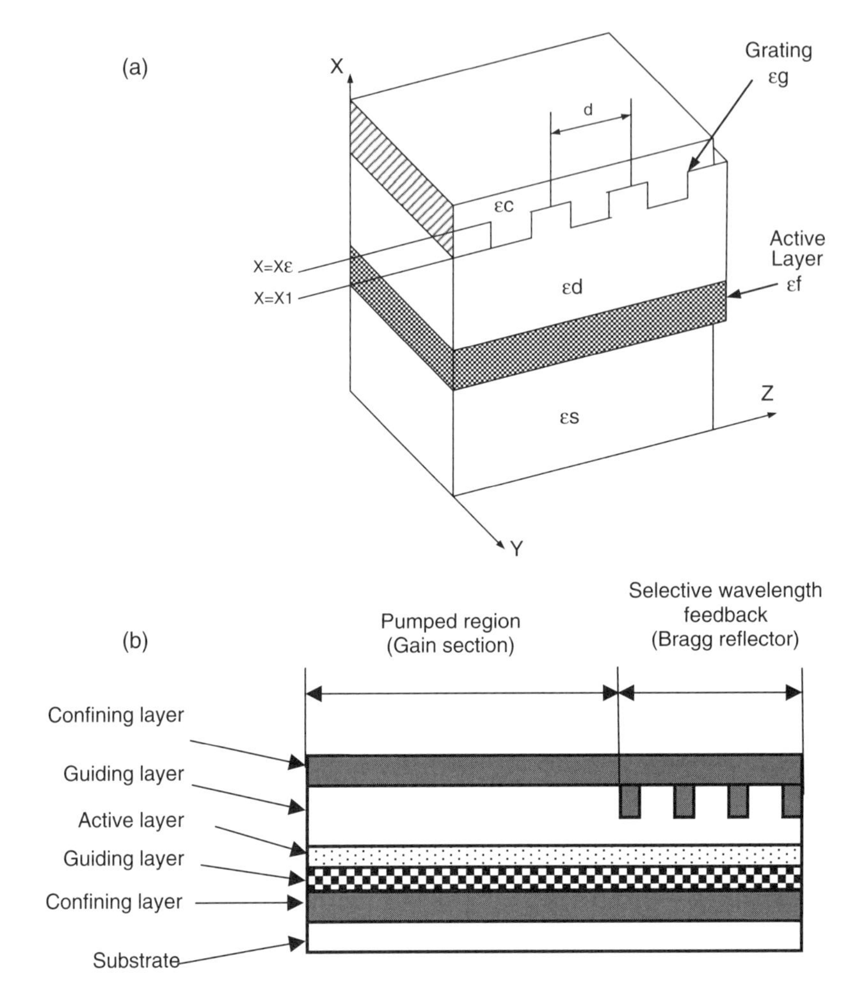


**Figure 6.7** (a) Typical DFB laser structure showing the grating feedback along the active area and the periodic fluctuating dielectric constant. (b) DBR lasers with a grating at one facet and a gain section.

equals to narrow band high Q resonator "tank" circuit in an oscillator. Consequently, from this discussion, SLM lasers' performance or mode purity is defined by a mode suppression ratio (MSR):[4]

$$\mathrm{MSR} = \frac{P_{\mathrm{mm}}}{P_{\mathrm{sm}}}, \tag{6.18}$$

where the main mode is $P_{\mathrm{mm}}$ power and $P_{\mathrm{sm}}$ is the side mode power. Generally, in order to have pure single-mode operation, MSR should exceed 30 dB. Again, looking at this definition and comparing it to an oscillator, the requirement is to have a feedback filter with sharp-shape factor. The above description is analog to an oscillator with SAW or crystal tank compared to a low-phase noise oscillator with low Q parallel inductance-capacitance (LC) tank. Since grating feedback is a selective wavelength reflective feedback, it operates as a band-stop filter where its attenuation is maximum at $\lambda_B$. Hence, it has low loss at both sides that may satisfy the threshold condition for lasing. Eventually, this may result in two modes emitted from a DBR or a DFB laser with power difference of $\sim$10 dB between the two. To overcome this problem, the grating stop band is tuned to the undesired mode. This method is called detuning of the grating stop band with respect to the gain peak of the laser. Another design method is to have at the gain peak a grating pass band with minimum loss and high loss out of the pass band. That means creating a band-pass filter feedback at the desired wavelength. The method of creating a pass band with minimum loss at the gain peak is called quarter wavelength detune or grating shift. Observing the forward propagating wave and backward propagating wave in Eq. (4.81), adding $j\pi/2$ and $-j\pi/2$, the arguments, respectively, would result in a pass-band and stop-band condition, thus changing the solution of Eq. (4.86). DFB and DBR lasers are analyzed using the theory of coupled mode equations.[11–13] Numerical analysis is used to map the $E$ field along the grating.[14] It shows there that the feedback attenuation factor in the direction of propagation increases as the grating thickness increases until it reaches a saturation and remains constant starting from a relative thickness of 10% of the grating thickness with respect to the center wavelength $\lambda_0$, where $\lambda_0/2$ is the grating period.

As was explained in Chapters 4 and 5, the refractive index refers to the relative dielectric constant $\varepsilon_r$ root square, while its dispersion versus wavelength provides wavelength separation by refraction. For this reason, the dielectric constant $\varepsilon = \varepsilon_o\varepsilon_r$ is replaced in such analysis by the effective refractive index. Analysis is in the same manner as in Chapter 4.

The analysis starts from the wave equation of the electric field $E$. Analysis is along the laser cavity length $L$ of both forward and backward propagating waves, with a propagation constant at vacuum value $k_O$:

$$\frac{d^2E}{dz^2} + \varepsilon k_O^2 E = 0. \tag{6.19}$$

The dielectric constant in a grating varies periodically by $\varepsilon$ and can be expanded by the Fourier series:

$$\varepsilon = \bar{\varepsilon} + \sum_i \Delta\varepsilon_i \exp\left(j\frac{2\pi}{\Lambda} iz + j\varphi\right), \tag{6.20}$$

where $\bar{\varepsilon}$ is the average dielectric constant, $\Delta\varepsilon_i$ is the dielectric constant perturbation, and $\varphi$ is the grating phase at $z = 0$. The general solution of Eq. (6.19) is given by the forward and backward propagating wave:

$$E = A\exp(-j\beta z) + B\exp(-j\beta z), \tag{6.21}$$

where the propagating constant β is given by

$$\beta = n_{\text{eff}} k_{\text{O}} + \frac{jg_{\text{th}}}{2}, \tag{6.22}$$

where $n_{\text{eff}}$ is the effective refractive index, and $g_{\text{th}}$ is the power threshold. Substituting Eqs. (6.20) and (6.21) in (6.19) and using the slowly varying envelope approximation (SVEA), neglecting second derivatives, and using

$$\left(\bar{\varepsilon} k_{\text{O}}^2 - \beta^2\right) E = 0 \tag{6.23}$$

results in

$$\begin{aligned} \frac{dA}{dz}\exp(-j\beta z) - \frac{dB}{dz}\exp(j\beta z) = -j[&kA\exp(-2j\Delta\beta z + j\varphi + j\beta z) \\ &+ k^* B\exp(2j\Delta\beta z - j\varphi - j\beta z)], \end{aligned} \tag{6.24}$$

where ∗ marks the complex conjugate, and $k$ is the coupling factor given by

$$k = \frac{\Delta\varepsilon_{\text{m}} k_{\text{O}}^2}{2\beta}. \tag{6.25}$$

$\Delta\beta = \beta - \beta_{\text{O}}$ shows the fluctuations of the propagation, where $\beta_{\text{O}} = \pi \text{m}/\lambda$ is the Bragg propagation constant. Note that $k$ increases with increasing corrugation depth. Using partial differential equation boundary conditions at both sides of the cavity provide the following relations and solution for a DFB transmittance with a uniform grating:

$$\gamma = \sqrt{(\Delta\beta)^2 - kk^*} = \sqrt{(\Delta\beta)^2 - |k|^2}, \tag{6.26}$$

$$r_{\text{G1}} = \frac{-\gamma + \Delta\beta}{-k\exp(-j\varphi)} = \frac{k\exp(j\varphi)}{-\gamma - \Delta\beta}, \tag{6.27}$$

$$r_{\text{G2}} = \frac{\gamma - \Delta\beta}{k\exp(j\varphi)} = \frac{-k^*\exp(-j\varphi)}{\gamma + \Delta\beta}, \tag{6.28}$$

$$E_{\text{b1}} = r_{\text{G1}} E_{\text{f1}}, \tag{6.29}$$

$$E_{\text{f2}} = r_{\text{G2}} E_{\text{b2}}, \tag{6.30}$$

where $E$ marks the electric fields at both sides of the cavity, $b$ index marks the backward propagating wave, and $f$ index marks the forward propagating wave.

The cavity's boundary conditions are

$$E_f(0) = r_1 E_b(0), \tag{6.31}$$

$$E_b(L) = r_2 E_f(L), \tag{6.32}$$

where $r_1$ and $r_2$ are facet reflectivities of the electric field at $z = 0$ and $z = L$, respectively, and are given by

$$r_i = |r_i| \exp(-j\phi_i), \tag{6.33}$$

where the index $i = 1, 2$, and $\phi$ is the phase shift introduced by the cavity facet reflection.

These relations and further manipulations lead to set off differential equations, which provide the DFB oscillation condition

$$\frac{(r_{G2} - r_1)(r_{G1} - r_2)}{(1 - r_1 r_{G1})(1 - r_2 r_{G2})} \exp(-2j\gamma L) = 1. \tag{6.34}$$

Assuming $E_b(L) = 0$ since facets reflectivities are zero, the transmittance for uniform grating is given by

$$T = \left|\frac{E_f(L)}{E_f(0)}\right|^2 = \left|\frac{1 - r_{G1} r_{G2}}{\exp(j\gamma L) - r_{G1} r_{G2} \exp(-j\gamma L)}\right|^2. \tag{6.35}$$

The last two equations provide the spectral behavior of a DFB laser. Transmittance $T$ is measured versus detuning $\delta L$, normalized coupling $kL$, and various normalized gains $gL$ values.

DBR laser is analyzed in the same manner and is similar to the analysis of the FP laser with a grating as facets to create selective wavelength feedback. The reflection condition in a DBR at $z = L$ is given by

$$r_1 r_{eq} = \exp(2j\beta L) = 1, \tag{6.36}$$

where $r_1$ is the facet reflectivity and $r_{eq}$ is the equivalent reflectivity from the grating at the DBR region. The propagation constant for DBR is given by

$$\beta = n_{eff} k_0 + \frac{j\alpha_G}{2}, \tag{6.37}$$

where $\alpha_G$ (replaces the power threshold gain $g_{th}$) is the power absorption and $r_{eq}$ is given by

$$r_{eq} = C_P r_{DBR}, \tag{6.38}$$

where $C_P$ is the coupling efficiency between the active DBR grating regions and the DBR reflectivity $r_{DBR}$ is given by

$$r_{DBR} = \frac{E_b(0)}{E_f(0)}. \quad (6.39)$$

Using the same solution technique as for DFB, while $E_b(L_{eff}) = 0$ results in

$$r_{DBR} = C_P \frac{r_{G1}[1 - \exp(-2j\gamma L_{eff})]}{1 - r_{G1} r_{G2} \exp(-2j\gamma L_{eff})}. \quad (6.40)$$

Substituting Eq. (6.40) in (6.38) provides the oscillation condition of a DBR laser:

$$r_1 C_P \frac{r_{G1}[1 - \exp(-2j\gamma L_{eff})]}{1 - r_{G1} r_{G2} \exp(-2j\gamma L_{eff})} \exp(2jn_{eff}k_0 L - \alpha_G L) = 1. \quad (6.41)$$

## 6.5 Multiple Quantum Well Lasers

It was previously explained that both FP DFB and DBR lasing is from the active layer. The active layer is characterized by its lateral width and its thickness. The cross section of this active layer defines the beam intensity. The thickness of the active layer of DH laser ranges between 1000 and 3000 Å. A single quantum well (SQW) corresponds to 7 to 15 atomic layers, or 50–100 Å, per well. These thin layers are grown exclusively using the technique of molecular beam epitaxy (MBE) and metal organic chemical vapor deposition (MOCVD).[18] Thus, if more active layers are grown, more QWs are created.[21,22] From solid-state-device physics,[1,16,17,37] it is known that each electron has several spherical quantum states. Thus, limiting the thickness of the active layer will limit the number of states and certain thickness corresponding to about one atomic layer. Eventually 3D quantum states would transform into 2D. De Broglie's statement is that an electron is a particle (photon), and a wave and its momentum relation $P$ is

$$P = h\nu, \quad (6.42)$$

where $\nu$ is the optical frequency. Using the above mentioned equation, momentum identities and kinetic energy, and substituting them into the wave equations results in Schrödinger equation $\psi$. Schrödinger solution of $|\psi|^2$ describes the energy statistics of an electron in a box or potential well:

$$E = E_0 + \frac{\hbar^2}{2m^*}\left(k_x^2 + k_y^2 + k_z^2\right), \quad (6.43)$$

where $m^*$ is the effective electron mass, $h$ is the Plank constant $\hbar = h/2\pi$, and $k$'s projections in $x$, $y$, and $z$ directions are the wave vectors or numbers. Observing the confined particle energy levels in the $z$ direction, and noting the $z$-direction wave vector as a function of quantum level results in

$$E_n = \frac{\hbar^2}{2m^*}\left(\frac{n\pi}{L_z}\right)^2, n = 1, 2, 3. \tag{6.44}$$

This notation describes the quantum energy levels in a finite potential well with a thickness of $L_z$; thus, it describes a particle motion normal to the well. Since, in this discussion, the $x$ and $y$ directions are infinite, there is a continuity of energy states in those directions. However, the $z$ direction is quantized. This results in a step response of energy states. Hence, $E_2 = 4E_1$, $E_3 = 9E_1$, and $E_n = n^2E_1$. The conclusion is that for given $x$ and $y$ energy states, there is a split or substate in the $z$ direction. Since $\psi$ is a statistical function, $\psi\psi^*$ is the state's probability density function, and the integration of $\psi\psi^*$ over the well in the $z$ direction should equal 1; the normalization factor is calculated. The density of state in the energy interval d$E$ is

$$p(E)\mathrm{d}E = \frac{m^*}{\pi \cdot \hbar^2 L_z}\mathrm{d}E \text{ when } E > E_1. \tag{6.45}$$

As a consequence, the density of states is a constant independent of energy provided $E$ is larger than the first allowed state $E_1$. Thus, energy steps will be on the parabolic line in Fig. 6.8 and as described by Eq. (6.46). In a conclusion, by choosing the dimension $L_z$, one can design the energy state and eventually manipulate and engineer the band gap:

$$p(E)\mathrm{d}E = \frac{1}{2\pi^2}\left(\frac{2m^*}{\hbar^2}\right)^{3/2}\sqrt{E}\cdot \mathrm{d}E. \tag{6.46}$$

In summary, QWs and multiquantum wells (MQWs) were extensively investigated in the last 15 years.[18–22] Properties such as modulation BW, relaxation oscillation, threshold gain, current high data rate modulation, spectral purity, and distortions were investigated. The advantages of QWs and MQWs were explored and reported. In the case of fiber in the loop (FITL) applications, ordinary laser transmitters require thermoelectric cooling (TEC) since they had shown poor characteristics to meet the Belcore's Generic Requirement. The $Al_xGa_yIn_{1-x-y}As/InP$ had shown 3-dB modulation BW of 19.6 GHz for compressive strained lasers and 17 GHz for tensile strained lasers by an optical modulation technique.[21] The strong carrier confinement also resulted in a small $k$ factor, which indicates the potential for high-speed modulation of up to 35 GHz. Low-threshold current was accomplished because of low-cavity losses and efficient mirror coatings.[19] From the above results, it is clear that the relaxation-oscillation frequency is higher compared to a regular laser structure. Thus, the distortions of such a laser

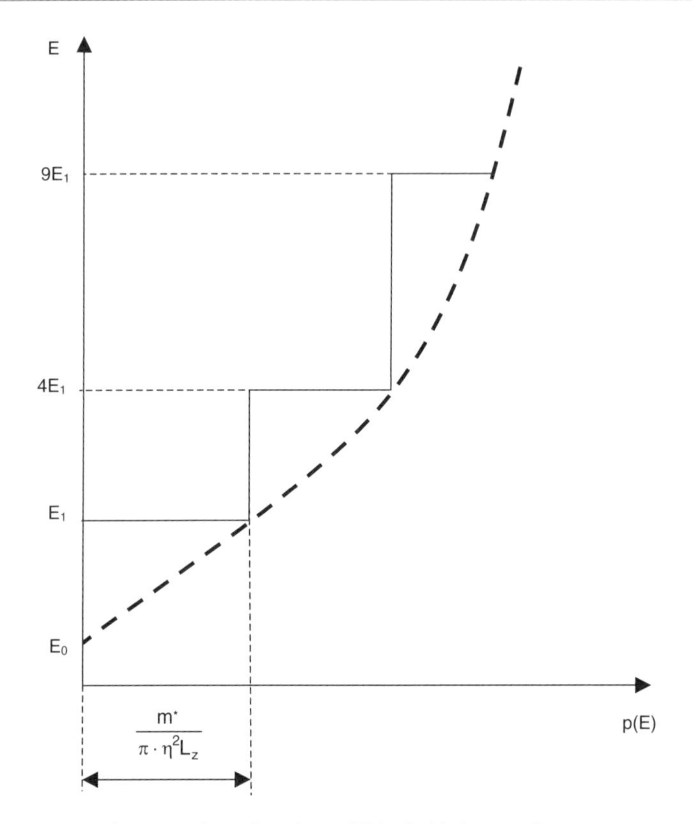


**Figure 6.8** Density-of-states function in a QW of thickness $L_z$, vs energy of QW laser. The dashed line is described by Eq. (6.46).

are lower, making it an ideal laser for hybrid fiber coax (HFC) community access television (CATV) transmitters.

A high temperature of 170°C, high CW 300-mW power lasers were reported with no TEC in a grated index (GRIN) separate confinement heterostructure (SCH) 1.3 μm GaInAs/InP.[20] A small signal direct modulation model was provided with respect to photon lifetime, spectral-hole burning (SHB), carrier heating, and energy transfer to photons and phonons were explored, as well as carrier transport processes.[22] Evidently, because of excellent uniformity of epitaxial QW material, and since the active layer well has low loss and thus high Q for the resonance and the confinement efficiency, the spectral line width of MQW and QW lasers is narrower even for FP cavity. Figs. 6.9 and 6.10 show the LI curves of such lasers and their structure and the quantum process of small signal modulation.

## 6.6 Vertical Cavity Surface-Emitting Laser

In previous sections, edge-emitting lasers, where in semiconductor topology the photons' emission is along the active area were reviewed. However, there is an additional semiconductor laser structure in which the emission is orthogonal to

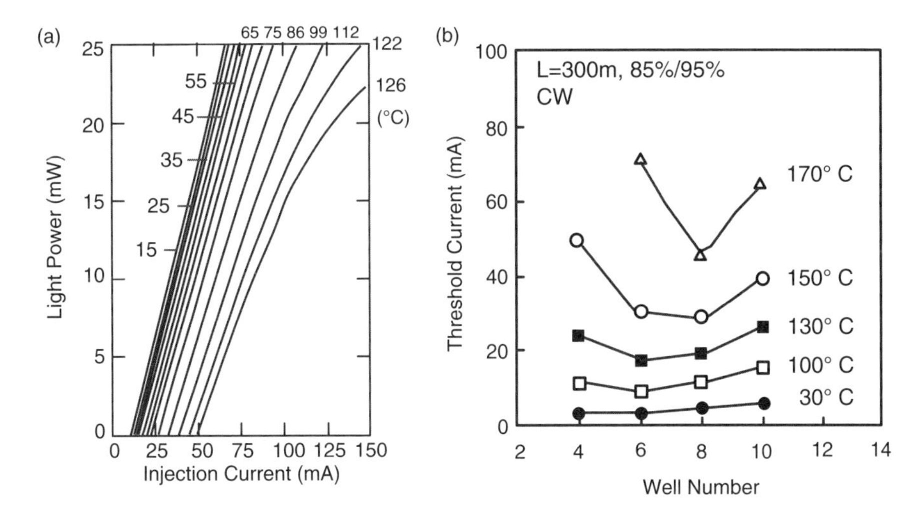


**Figure 6.9** LI curves of MQW lasers showing threshold points (a) AlGaInAs/InP Light current characteristics of a 508-μm long tensile-strained QW laser with 98% high reflection coating on the rear facet at various heat-sink temperatures. (b) Temperature dependence of threshold current and characteristic temperature.[20,21] (Reprinted with permission from the *Journal of Quantum Electronics* © IEEE, 1994.)

the NP active area.[35] This laser structure is called a verticle cavity surface-emitting laser, or VCSEL. As in edge-emitting lasers, the cavity is made by using mirrors or gratings. Because the gratings are used as selective wavelength reflectors instead of a mirror, this reflector is considered to be a DBR. Thus, in a VCSEL, there are

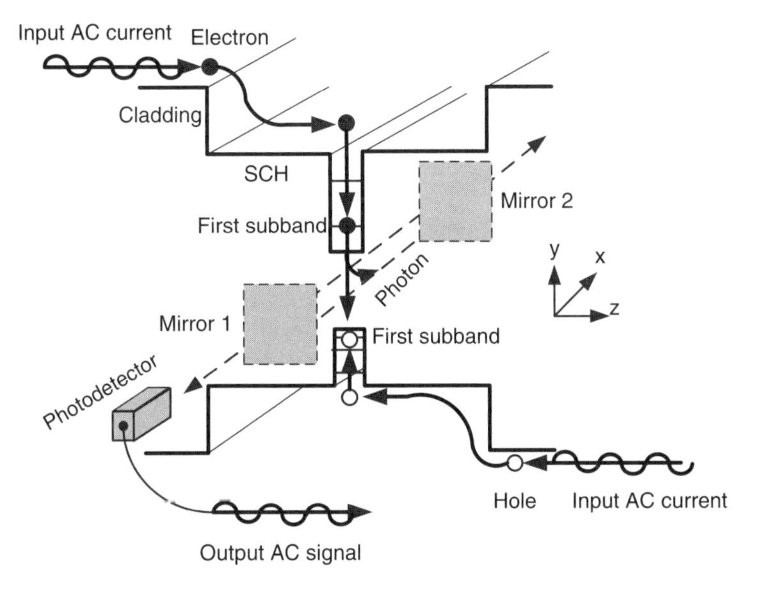


**Figure 6.10** Quantum process in case of a small signal modulation of a SQW laser.[22] (Reprinted with permission from the *Journal of Quantum Electronics* © IEEE, 1997.)

cavity reflector similarities to FP and DBR edge-emitting lasers.[28,30] Mode stabilization is optimized by improving reflectors.[33] As in edge-emitting laser technology, the VCSEL active layer can be grown as a QW or MQW.[25] VCSELs are popular in the use of short wavelengths CWDM such as Agilent 820, 835, 850, 865 nm. Efforts on InP-based lasers' longer wavelengths of 1310, 1490, and 1550 nm favored for long-haul full service access network (FSAN) fibers to the home-curb-building business-premises (FTTx) applications have met with slower progress because of inherent difficulties in constructing highly reflecting mirrors as well as high-gain active region layers.[26,27,35] In cellular applications for wireless on fiber, VCSELS are used for distributed antenna transponder applications and for IEEE802 AH bidirectional link applications of 1310 nm upstream and 1310 nm downstream.

VCSELs became popular and attractive due to several technical progresses in their fabrication.[24] This enabled the operation of VCSELs at higher temperatures such as 88°C with InGaAlAs/InP, GaAsSb/GaAs, GaAsSb/GaSb, and InGaAsP/InP for 1550 nm, and InGaAsN/GaAs, InGaAsNP/GaAs, InGaAsNSb/GaAs, as well as highly strained InGaAs/GaAs for 1310 nm with operation of up to 125°C.[24,34]

The VCSEL structure is described in Fig. 6.11. From a manufacturing point of view, it has a simpler fabrication process, Similar to that of light emitting diodes (LEDs). Thus, it is a low-cost, high-yield mass production process. Fabrication

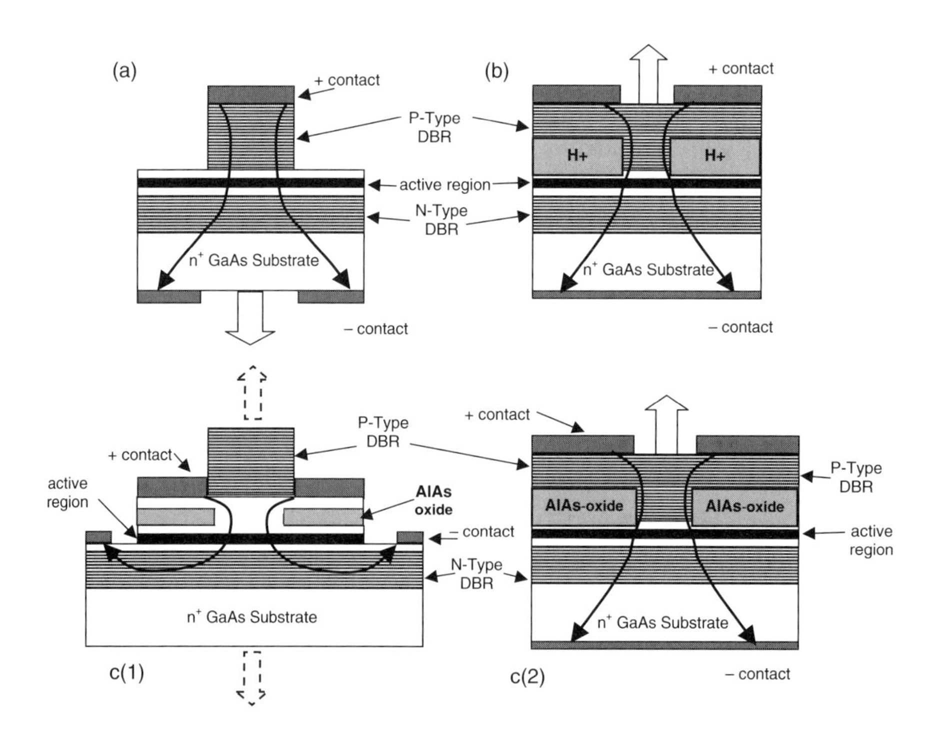


**Figure 6.11** Three types of VCSEL structures commonly used: (a) Mesa etched, (b) Proton implanted, (c) Dielectric aperture [mesa c(1) and lateral c(2).]

of long wavelength VCSELs is based on GaAs or InP substrates. There are several structures for realizing VCSELs depending on the applications.

- *Etched mesa.* This is analogous to the edge-emitting ridge structure. However, in this case, a round or square postlike mesa is formed. The etching is generally stopped just above the active layer to prevent surface recombination of carriers and reliability problems. Consequently, the current is confined to the lateral dimensions of the mesa. However, carriers are free to diffuse laterally in the active region. Therefore, a lateral leakage current becomes important for dimensions less than 10 μm in diameter. Generally, such a structure emits from the bottom surface and the top mesa is used as a contact electrode. One of the problems with such a structure is excessive resistance or conduction through the mirrors. The series voltage develops across the mirrors and results in a potential barrier at numerous heterointerfaces. This is problematic in P-type mirrors since small barriers inhibit thermionic emission for heavy holes. This is solved by optimizing the band gap. The mesa structure is considered to be gain-guided laser topology.
- *Proton implanted.* This is a gain-guided structure. This configuration is simple to manufacture and thus has been considered to be one of the most important VCSELs. It has a planar structure, allowing for relatively large-area electrodes with low-contact resistance. The proton implant apertures the current to provide a desired current confinement. It emits from both the top and bottom surface. However, it generally has a ring-contacted, top-emitting configuration to allow 850-nm emission with GaAs substrates. The additional advantage of this structure is lower thermal impedance compared to etched mesa. On the other hand, it suffers from the same problems as the mesa of conduction through mirrors. In addition, there are two problems resulting from the implantation. The first is that the implant stops above the active region and the other is that it penetrates the region. To prevent the implant from penetrating the active area, the implant is stopped above the active region. As a consequence, a lateral current occurs above the active region. This adds a nonproductive shunt current and power dissipates. In the case of the implant penetrating the active region, the current is effectively apertured. However, since its life goes to zero at the implant edge, the result is effective infinite recombination velocity leading higher carrier losses compared to the mesa. Modulation tests on these configurations show decent results. Small signal BW showed relaxation oscillation of 14 GHz at 8 mA and 12 GHz at 2.5 mA.[26–28] These results are very attractive and economic in power consumption compared to edge-emitting laser.
- *Dielectric apertured.* In this topology, previous technical problems were addressed.[48] First, it provides a solution to the extra losses between the positive contacts and the active layer. In the case of the mesa structure,

these losses are because the current has to flow through the dielectric mirrors. In addition, its purpose is to block the shunt current that may flow between p and n regions. The advantage of the dielectric current aperturing over the proton implant aperturing is that it does not reduce the carrier lifetime in the active region. Thus, this design approach is a preferred one since it answers the technical problems presented previously without losing the design flexibility.

- Visible in this design approach, short visible wavelengths, $<700$ nm, are excited. This is done by adding aluminum to the GaAs QW to increase the band gap. These lasers are not used in HFC and long- and short-haul links. Thus, this chapter will not provide further elaboration on these types of VCSEL lasers.
- *InP based.* These lasers are used for 1.3–1.55-μm applications. The main difficulty in the evolution of these wavelength lasers was to develop highly reflective mirrors and high-gain active regions. The familiar InGaAsP/InP and InGaAlAs/InP endure a small range of achievable optical indices. Consequently, very long epitaxial mirror stacks are necessary to build up the required reflectivity. Because of optical losses, the remaining reflectivity is only 98%. Thus, epitaxial mirrors are not successfully used in CW devices. More effort is being made in that regard. The low gain problem results from Auger process of nonradiative recombination of carriers. MQW have been found to provide an increased gain for lower current densities.

Figure 6.11 describes several VCSEL structures. In long wavelength lasers, similar structure as in c(1) and c(2) is used. Figure 6.11c(1) shows two options of emitting; it can be top or bottom. To solve the lateral current, a current-blocking layer is grown, confining the active layer between the two DBRs. Thus, the current flows through the active layer to the cathode. This current-blocking layer can be the oxide insulation.[24] The bulk substrate for long wavelength lasers is InGaAs ternary substrate. Carbon doping is used in top P mirrors, reducing their loss and heating.[26] In CATV VCSEL applications, modulation BW and electron photon resonance frequency $f_r$ grow linear with the square root of the injected current above threshold:[27]

$$f_r^2 = b(I - I_{th}), \tag{6.47}$$

$$b = \frac{1}{4\pi^2} \cdot \frac{dg}{dN} \cdot \frac{\Gamma V_g \eta_i}{qV}, \tag{6.48}$$

where $N$ is the carrier's density, $I_{th}$ is the threshold current, $dg/dN$ is the differential gain, $V$ is the active volume, $\eta_i$ is the injection efficiency, $\Gamma$ is the optical confinement factor, $q$ is the electron charge, and $V_g$ is the group velocity. Hence, $b$ [GHz$^2$/mA] describes the modulation current efficiency (MCEF). Experimentally,[27] a factor called MCEF [GHz/$\sqrt{\text{mA}}$] (MCEF factor) is used and relates to the modulation BW:

$$\text{MCEF} = 1.55\sqrt{b}. \tag{6.49}$$

In conclusion, the modulation efficiency $b$ improves substantially at higher differential gain. Furthermore, a current increase in a single-mode laser would improve its BW.

## 6.7 Tunable Lasers

In previous sections, several laser topologies were reviewed and presented. Lasing mechanisms and resonance cavity structures were analyzed. These concepts of laser realizations are used as foundations for tunable laser devices. Creating a tunable cavity using different methods and means would result in a tunable wavelength laser. Tunable lasers have been in development for over a decade but only during the late 90s did technologies start to become mature enough to serve the demanding applications of telecomm DWDM equipment. DWDM technology greatly improves the transmission capacity in optical fibers and employs multiple wavelength grids separated by the 100-GHz International Telecommunications Union (ITU) standard or 50 GHz. The inventory cost and redundancy volume can be saved by the use of a tunable laser transponder. Various laser designs are available in the market, each with its own "pros and cons." With the slow, yet growing rate of tunable lasers currently available, system vendors and service providers have started to ramp up the adoption of tunable lasers into system design requirements. The following section will cover the various tunable laser technologies that are currently available.[42] Chapter 4 provides a glance at optical code division multiple access (OCDMA) that uses tunable wavelength transponders and fiber Bragg gratings as encoders and decoders. Chapter 19 will discuss tunable wavelength transmitters and transponders.

### 6.7.1 Tunable laser types

There are four families of tunable lasers in the market today. These are:

1. DFB—distributed feedback
2. DBR—distributed Bragg reflector
3. VCSEL—vertical cavity surface-emitting laser
4. ECL—external cavity laser

These four basic laser types have evolved to create variations of the general structure in order to achieve a wide range of tunability covering the full C or L bands. There are three tuning mechanisms, which are being used in tunable lasers:

- Thermal tuning (very slow, 5–15 s)
- Electrical tuning (fast, 1–10 ms)
- Mechanical tuning (slow, 50 ms–1 s)

#### 6.7.1.1 Tunable distributed feedback (DFB) laser

The DFB laser has been serving the WDM industry for over 20 years and was the laser of choice for system vendors due to narrow line width and higher temperature stability. The tuning mechanism of the DFB laser is temperature tuning and the key drawbacks of the DFB laser is its narrow tuning range in the area of 5 nm.

The main problem is that in order to tune a DFB laser over 5 nm, the laser needs to be heated to high temperatures of over 40 and even 50°C. These temperatures are high and are beyond the long-term safe operation point of the laser. In order to overcome the narrow tuning range of the a DFB laser, solutions evolved in the form of a DFB laser array. In these solutions, an array of 10 or more DFB lasers were packed together on the same substrate, where each of the DFB elements can tune over a range of 30 nm around a continuous central wavelength.[38] This way, by selecting a different laser element each time, it is possible to achieve continuous tuning over the C or L bands. Coupling the light out of the DFB laser element to the tunable laser package output is done by using either a microelectromechanical system (MEMS), micromirror, or a waveguide as shown in Fig. 6.12. Under normal operation, there are three control loops that are maintained by electronics. For coupling the MEMS, voltages are set to their calibrated values. However, they are continually optimized to maximize the fiber-coupled power as measured by the wavelength locker. During a wavelength switching event, the MEMS mirror moves to an extreme position to blank the output by about 50 dB. After new current and temperature values are stabilized, the MEMS is then unblanked and the locker can provide fine wavelength control.

#### 6.7.1.2 Distributed Bragg reflector (DBR) lasers and multisection DBR

DBR lasers use a grating structure in a similar way to the DFB laser. However, in DBR the grating is located on a passive element (cavity section) rather than in the main (gain) laser element as shown in Fig. 6.13. Tuning is made by applying

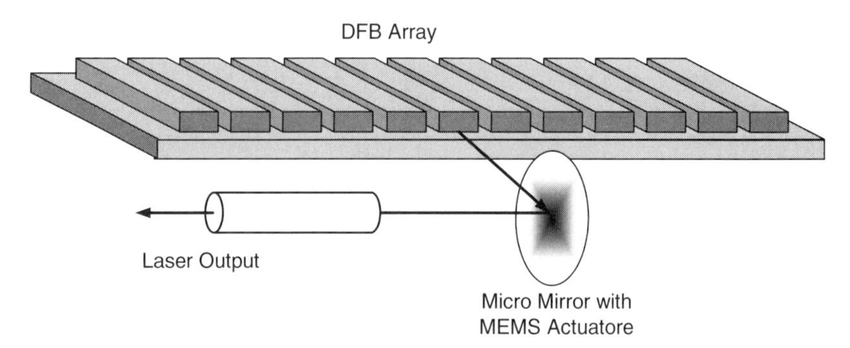


**Figure 6.12** A widely tunable laser based on DFB array. The mirror is a MEMS tilt mirror. A typical structure like this has 12 DFB elements with wavelength spacing of 3 nm. A collimating lens is inserted between the array, and the mirror and focusing lens is placed between the mirror and the SMF fiber.[7] (Image Courtesy of GWS Photonics, Israel.)

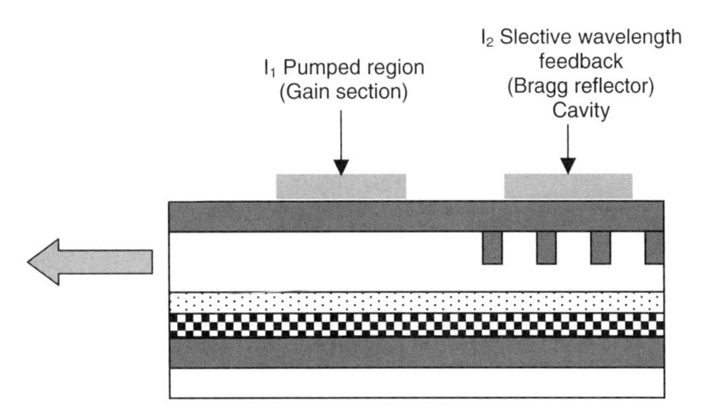


**Figure 6.13** Tunable DBR laser showing the two current phases to control the gain and the refraction index.

current into the cavity section, which in turn changes the refractive index and causes tuning.

Again, similar to the DFB laser, a single grating section in the DBR is limited in its tuning range, and in order to achieve a wider tuning range covering the C or L bands, several (three) grating sections are needed as shown in Fig. 6.14. By applying current to the three grating sections, they act as a hierarchal filter structure where one grating section selects the "super mode" of operation, the second section aligns to it, and the phase section selects the FP mode. When all the three are aligned, the selected mode is lased out. A significant advantage of the multisection DBR laser is its being made of a monolithic semiconductor material, which allows for mass production at low costs. In addition, it allows integration on the same chip additional components such as a semiconductor optical amplifier (SOA) and external modulated laser (EML) modulator. Electrical tuning allows fast tuning between wavelengths in the area of a few milliseconds. Among the weaknesses of this laser are its optical properties, which are somewhat lower than those of a DFB.

The tuning control of a multisection DBR is quite complex. It involves three independent variables, which are the currents injected to the grating sections. Different "current vectors" can produce the same wavelength and a complex

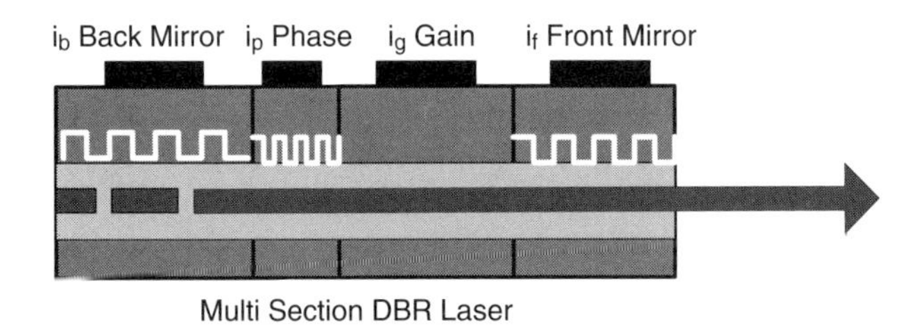


**Figure 6.14** Widely tunable laser based on multisection-sampled-grating DBR (SG-DBR).

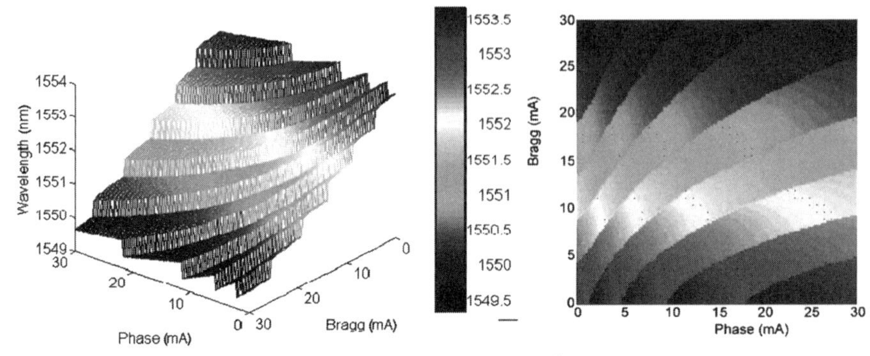


**Figure 6.15** Mode map of a DBR laser. It shows a 3D control matrix of wavelength versus phase current in milliampheres and Bragg current in milliampheres.[7] (Image Courtesy of GWS Photonics, Israel.)

characterization process is needed to map the wavelengths and current vectors and to select the final set of operating vectors (Fig. 6.15).

A key area of concern for system vendors and service providers is the laser's long-term stability and aging. The concern is that the laser's will experience mode hops as its operation point ages and drifts over time.

### 6.7.1.3 VCSELs, MEMS, and ECLs

VCSELs tunability is accomplished by changing the vertical cavity using MEMS technology. Figure 6.16 illustrates the concept using electrostatic tuning. The VCSEL is made of two PN junctions. The first PN junction is the forward-biased VCSEL. The second junction is a reversed bias PN junction utilized as the top reflector plate of the VCSEL. This way a free-space capacitor is created between the P side of the laser and the N side of the reversed biased PN junction. By changing the voltage across the reversed biased junction, the electrostatic field

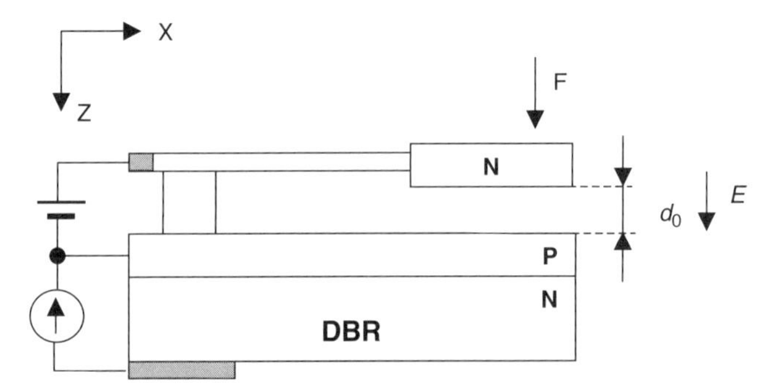


**Figure 6.16** Concept of tunable VCSEL showing attraction force F and electrostatic electrical field $E$. The movable cantilever plate is N type and this part is a reversed biased diode.

between the plates varies, thus affecting the electrostatic attraction force between the plates. Consequently, the distance between the movable cantilever plate, which is the N side of the junction, and the P-type plate changes.

Using electrostatic methods and strength of materials theory,[51,52] it is easy to analyze the electrical and mechanical forces applied on the movable cantilever. The electrostatic force, $F_E$, between the capacitor plates is given by the derivative of the capacitor-stored energy. On the other hand, this force applies a counter torque on the N-plate rod. Thus, this process reaches equilibrium at the point where the electrostatic attraction force equals the mechanical force $F_M$. The mechanical force is given by Hook's law:

$$F_E = \frac{\partial}{\partial z}\left(\frac{CV^2}{2}\right) = \frac{\varepsilon A V^2}{2(d_0 - z)^2}, \tag{6.50}$$

$$F_M = kz, \tag{6.51}$$

where, $\varepsilon$ is the dielectric permittivity, $A$ is the capacitor area, $z$ is the incremental change between the plates in $z$ direction, and $k$ is an elastic constant of the beam or rod, sometimes called the spring constant. Solving Eqs. (6.50) and (6.51) provides two states for $z$. However, only one is a stable solution, showing that the maximum change in cantilever equals $Z_{max} = d_0/3$ independent of $k$. Using the strength of materials theory, the beam is analyzed according to Young's law:

$$\frac{\partial^2 Z(x)}{\partial x^2} = \frac{M(x)}{E \cdot I(x)}, \tag{6.52}$$

where $E$ is Young's modulus, $I(x)$ is the beam cross section, and $Z(x)$ $z$ is the incremental change between the plates in $z$ direction as a function of the beam length. This equation describes the distributed load over the beam and the stress resulting in it. An approximation can be made by postulating a concentrated force at the beam applied on the capacitor plate. This assumption is justified since the capacitor area is larger compared to the beam's active capacitive area:

$$Z(L) = \frac{F}{k_{\text{eff}}} = -\frac{\left(\frac{\varepsilon V^2}{2d_0^2}\right)\left(A + \frac{3}{8}wL\right)}{\left(\frac{Ewt^3}{4L^3}\right)}, \tag{6.53}$$

where $w$ is the beam width, $L$ is the length, $A$ is the capacitor area, $t$ is the beam thickness, and $E$ is Young's module. This relation provides two design rules for cavity sensitivity of plate spacing versus tune voltage.

The effective elastic constant $k_{\text{eff}}$ is proportional to $w\left(\frac{t}{L}\right)^3$ and the tuning voltage $V$ is proportional to $\sqrt{w\left(\frac{t}{L}\right)^3}$.

From that model, it is clear that several design compromises are made. For instance, a thicker or wider beam would require more force to bend it and thus

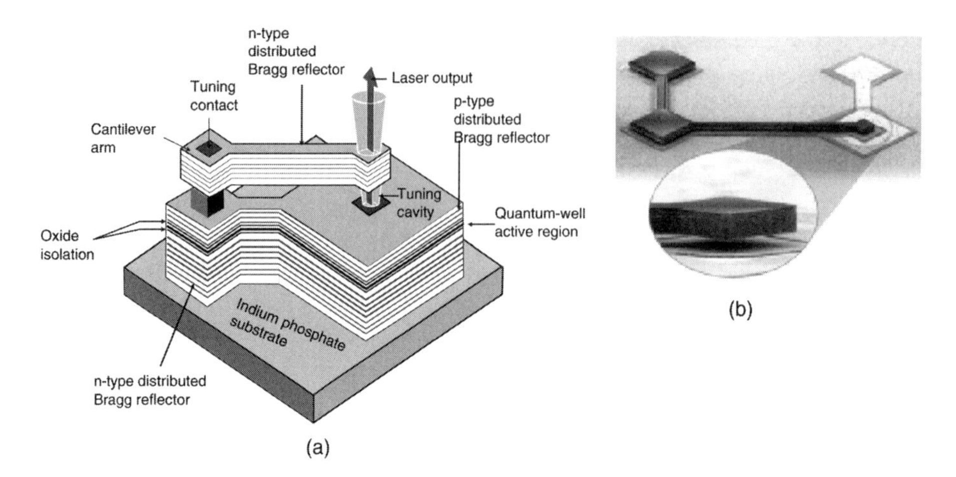


**Figure 6.17** VCSEL on the left side, vertical cavity surface-emitting laser is tuned by moving the cantilever arm to change the dimensions of optical cavity. Right side actual magnification of tunable VCSEL chip.[6,8] [(a) Reprinted with permission from IEEE Spectrum © IEEE 2002. (b) Reprinted with permission from *Optics and Photonics News* © Optics and Photonics News 2001.]

a higher voltage would be necessary. Therefore, a longer beam would result in a larger torque and this would require lower voltage to tune it to the same wavelength as a shorter beam with a smaller cross section. However, a very long cantilever is harder to manufacture. Thus, a compromise in design between the beam's cross section, length, tuning voltage range, and tuning range should be considered. Figure 6.17 illustrates MEMS tunable VCSEL.

#### 6.7.1.4 MEMS and ECLs

In ECLs, the tunability effect is achieved through a cavity situated outside the laser element (gain section). For that reason, it is possible to use a standard FP laser chip with an antireflection coating on the output facet of the laser chip. Currently existing ECLs are based on the mechanical tuning apparatus and an architecture known as the Litman configuration. However, a new generation of ECLs soon to be offered to the market will provide an improved design of ECLs without any moving parts using an electrically tuned grating-waveguide structure (GWS) as a reflective optical filter.

In a Litman-type tunable laser, the cavity is created between the external moving mirror and the reflective facet of the laser chip. A diffractive grating is used as a depressive element. A lens is used to collimate the light coming out of the laser chip. When the light hits the diffractive grating, it spreads in different directions. The laser's operative wavelength is the one that is incident perpendicular to the mirror out of the diffractive grating. Tuning is achieved by moving the mirror up and down around the pivot point. This way, each time a different wavelength is in a normal path toward the mirror. The mirror motion is

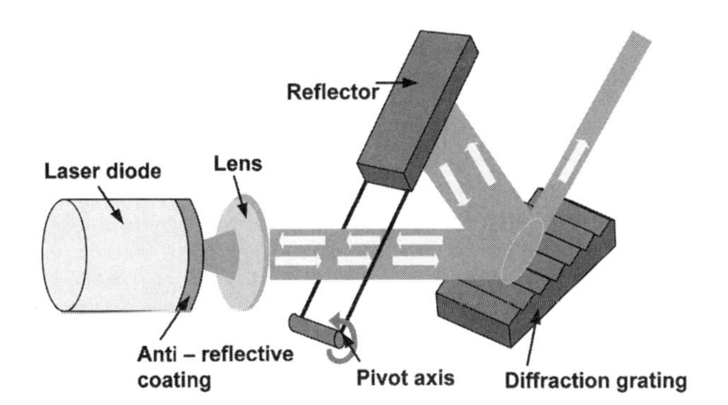


**Figure 6.18** ECL with Litman – Metcalf configuration (mechanical tuning).

accomplished by using a MEMS device like that shown in Fig. 6.18, which can be based upon the same concept of electrostatic force provided in Fig. 6.16. The drawback of this design is the mechanical inertia of the moving parts, which reduce its long-term reliability somewhat and provide slow tuning. Assembly of the free-space optical elements is also complex and costly. On the other hand, ECLs are known to have very high optical fidelity, similar or better than standard DFB lasers, with an ability to produce high optical power.

Recent reports[39,40] provide detailed information on tunable FP lasers with MEMS-controlled mirrors. A laser is made by the integration of a surface MEMS 3D mirror as external cavity reflector, FP laser, and a butt-coupling optical fiber. Figure 6.19 provides details on the assembly of this configuration. In Fig. 6.19(a), the optical fiber is aligned very close to the exit of the laser diode with the intention of direct coupling to the laser output beam. There is no need for an additional coupling lens between the laser and the fiber. To maximize the transmittance and to prevent unnecessary feedback into the laser diode, the surface end of the fiber is antireflection coated. The 3D mirror is aligned near the other exit window of the FP laser chip to provide an external variable cavity. The 3D mirror is placed on a translating stage that can be driven to translate by the comb drive. The suspension beams are used to keep the movable stage above the substrate to avoid stiction and to eliminate wobbling during lateral movement. The mirror is assembled in the vertical position and kept in this position by the microfabricated position holder at the back of the micromirror. By applying different voltages to the comb drive, the 3D mirror translates and shifts from one state to other, varying the cavity length of the FP laser. The assembly of such a structure is done manually using a probe station. The initial separation between the laser and the mirror is 10 μm. The distance between the laser and the fiber is 15 μm. In Figs. 6.19(b) and (c), guide rails are shown. These guide rails are used to align the fiber to the diode exit facet. Figure 6.19(d) shows the comb drive actuator. The overall size of this tunable laser, including the 3D mirror and the laser diode, is $2 \times 1.5 \times 1$ mm.

The comb drive is a micromotion motor that varies the 3D mirror position. It has two rows of comb fingers as shown in Fig. 6.19(d), with 156 moving fingers.

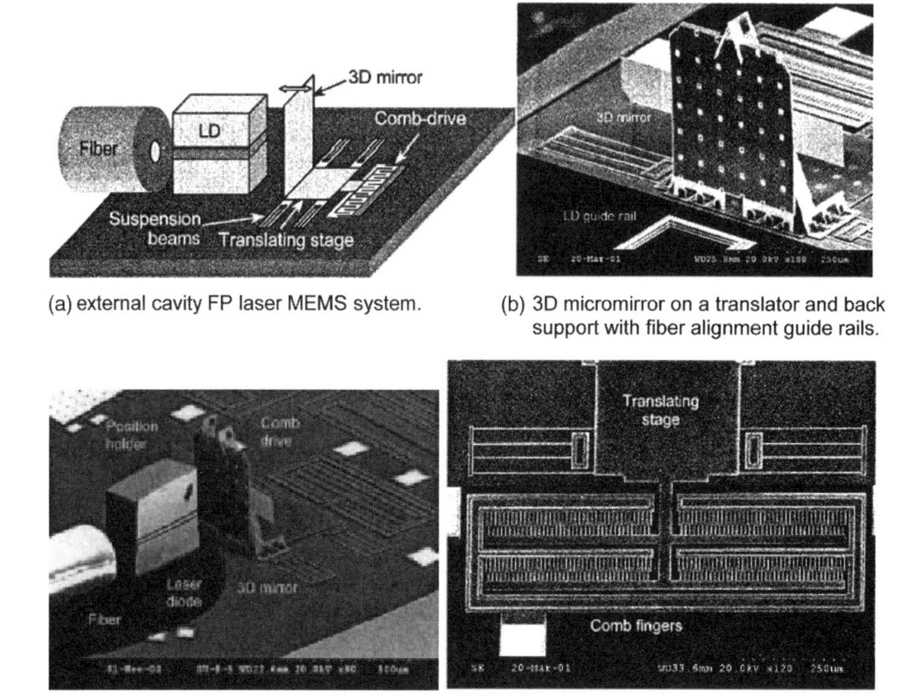


(a) external cavity FP laser MEMS system.

(b) 3D micromirror on a translator and back support with fiber alignment guide rails.

(c) external cavity FP laser MEMS system SEM micro graph photo.

(d) SEM micro graph photo of a comb actuator.

**Figure 6.19** MEMS ECL FP laser. (a) External cavity FP laser MEMS system. (b) 3D micromirror on a translator and back support with fiber alignment guide rails. (c) External cavity FP laser MEMS system. (d) SEM micrograph photo of a comb SEM micrograph photo actuator.[39] (Reprinted with permission from the *Journal of Selected Topics of Quantum Electronics* © IEEE, 2002.)

Each finger is 30-μm long, 2-μm wide, and 2-μm thick. The gap between the fingers is 2 μm. The moving and the fixed fingers initially overlap by 10 μm.

This way, the comb driveway is within its linear range. The comb motion is accomplished by electrostatic force. The displacement, $\delta$, of the comb drive is given by[41]

$$\delta = \frac{\varepsilon_0 N h}{g k} V^2, \qquad (6.54)$$

where $N$ is the number of moving fingers of the comb, $h$ and $g$ are the finger thickness and gap, respectively, and $k$ represents the stiffness of the suspension beams. Applying tuning voltage shifts the wavelength continually until it jumps to the adjacent mode. Thus, the wavelength tuning is a combination of both continuous wavelength change and mode hopping. The tuning range of such a structure showed a 1526–1548-nm wavelength change for a 3D mirror displacement from 0.3 to a 0.9 μm and a tuning control range of 10 V. Larger displacements of 3.5 μm required a tuning range of 30 V.

#### 6.7.1.5 A new open ECL technology based on reflective GWS tunable filter

A new design approach of ECL[7] is based on a tunable GWS reflective filter.[9] This allows solving the Litman ECL drawbacks while keeping the high level of optical fidelity. In a GWS ECL laser, a standard FP laser chip is used and attached to its back is the GWS reflective tunable filter. Tuning is achieved by applying electrical voltage on the reflective filter. As a result, one wavelength is selected since the finesse selection of resonance frequency of the filter and that wavelength is reflected back to the laser, thus creating the tuned cavity. An etalon, located in the cavity between the GWS tunable reflective filter and the back facet of the laser chip, guarantees precise ITU grid wavelength operation and eliminates the need for a separate wavelength locker. Key benefits of such ECL designs are its simple structure, very small size, and ability to further reduce the laser size, enhanced optical properties and fidelity, low production costs, wide and continuous tuning range (can tune beyond the C or L bands), high reliability (no moving parts), fast and simple tuning (electrical tuning), and high output power with a low thermal load if used in a transponder or transceiver. The ECL based on GWS reflective tunable filters[9] is an open architecture and the reflective filters are available in a die format. This makes it easy for leading optical component vendors to design their specific widely tunable lasers based on the GWS ECL design kit (Fig. 6.20).

#### 6.7.1.6 An open tunable laser technology—ECL based on GWS

Major benefit of the ECL-GWS laser is that it is not a vendor-specific singular solution laser with all the risk involved in a unique and singular solution. Rather the

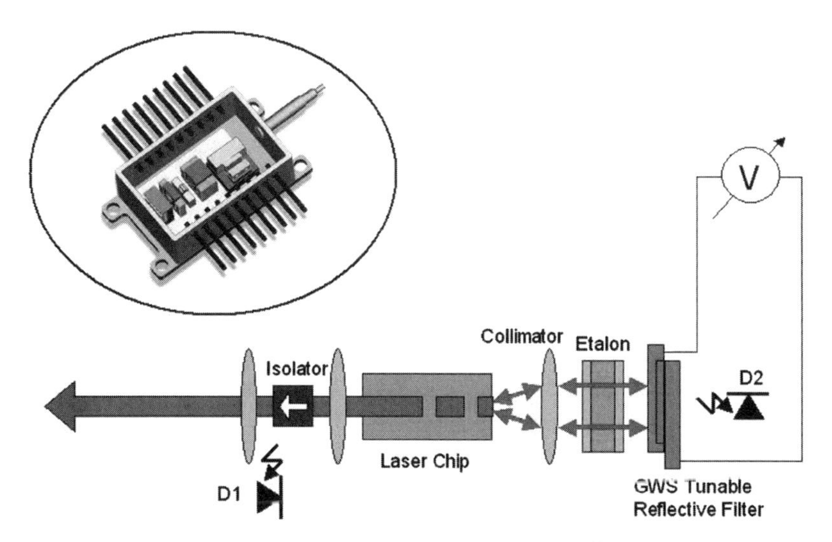


**Figure 6.20** ECL based on GWS refractive tunable filter.[7] (Image Courtesy of GWS Photonics, Israel.)

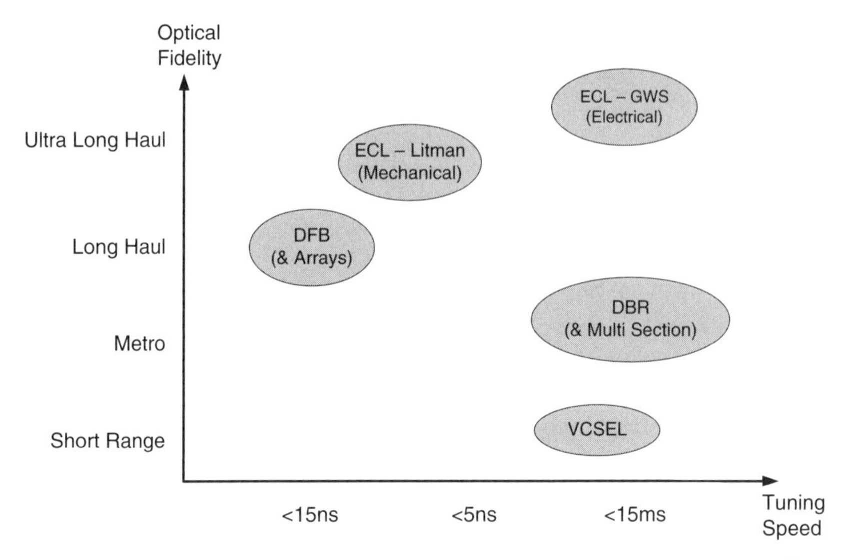


**Figure 6.21** Lasers positioning according to tuning speed and optical fidelity.[7] (Image Courtesy of GWS Photonics, Israel.)

new ECL-GWS technology is an open technology, which is freely available for implementation by the industry and by leading optical components vendors. Using an open technology can act as a catalyst to achieve what the market is looking for; i.e., a "standard" tunable laser design with optical fidelity and performance that can address all the applications including metro, long haul, and ultralong haul, and with availability from multiple sources. Having such a "standard" tunable laser can also help reduce its costs through economy of scale and thus drive the desired mass adoption of the widely tunable laser, making it as simple to use as a standard DFB laser is today.

The design of an ECL based on a GWS reflective filter is flexible enough to the allow the leading tier 1 optical components vendors to differentiate and evolve their tunable laser designs to meet the evolving roadmap of transponders and transceivers. The following table summarizes the key features of each laser type. Figure 6.21 presents the general positioning of each tunable laser technology according to the vectors used by the optical internetworking forum (OIF) to differentiate between the laser types (i.e., optical fidelity and the tuning speed). It can be seen that the new open ECL technology based on GWS reflective filter technology can address and provide an answer to the various applications both in optical fidelity and in tuning speed Table 6.1.

### 6.7.2 Switching speed limitations in a GCSR laser

In previous sections, varieties of tunable laser concepts were introduced. These lasers are divided into two main categories: fast tuning and slow tuning. Slow tuning lasers are MEMS and thermal tuned. These lasers are set once for a specific

**Table 6.1** Comparison between different tunable laser technologies.

| Type Parameters | DFB | DFB array | DBR | Multisection DBR | VCSEL | FP ECL Litman | ECL GWS |
|---|---|---|---|---|---|---|---|
| Tuning mechanism | Thermal | Thermal and mechanical (MEMS) | Electrical | Electrical | Mechanical (MEMS) | Mechanical (MEMS) | Electrical |
| Tuning control | Simple | Fairly simple | Complex | Very complex | Simple | Fairly simple | Simple |
| Availability | Multiple sources | Vendor specific | Multiple sources | Vendor specific | Vendor specific | Vendor specific | OPEN technology (enable multiple sources) |
| Tuning range | ~5 nm | >40 nm | ~10 nm | >40 nm | >40 nm | >40 nm | >40 nm |
| Tuning time | ~10 ms | ~5 s | ~1 ms | ~10 ms | ~10 ms | ~50 ms | ~10 ms |
| Side mode suppression ratio | | ~50 | | ~45 | | ~50 | >60 |
| Output power | >10 mW | >20 mW | | >20 mW (with SOA) | ~1 mW | >20 mW | >25 mW |
| Packaging | Butterfly | Larger than butterfly | Butterfly | Butterfly | | | Butterfly or smaller |

wavelength and rarely retuned to another wavelength. The fast-tuned lasers are the electrically controlled, such as multisection DBR named SG-DBR, grating-assisted codirectional coupler with sampled reflector (GCSR) lasers, and digital supermode distributed Bragg reflectors (DS-DBR). These lasers can be used in optical packet switching, OCDMA, which are considered a key technology in the next generation of optical communication networks. Fast-tunable lasers can be used to color bursts of data according to their destination. Fast-tuning time on the order of nanoseconds is essential for implementing efficient burst switching.[45,46] One of the popular lasers for fast tuning is GCSR, whose structure is given in Fig. 6.22.

The laser has four sections: MQW gain, codirectional coupler, phase matching, and a sampled Bragg reflector. The tuning range of such a laser is 1529–1561 nm with 3-dBm optical power coupled to the pigtail. The combination of sampled reflector and codirectional coupler section ensures only one cavity mode lasing by providing wavelength-selective loss. The spectral response of these sections is tuned by current injection. The current injection affects the carrier concentration in the waveguide gain section, and the refractive index is affected in the reflectors. The reflection peaks of the sampled reflector are 4-nm apart. The coupler operates as a filter with 8-nm BW. The current applied to the coupler section is controlled to select one reflector peak. Fine tuning is accomplished by tuning the reflector peak to a desired wavelength and adjusting the phase current to align the cavity mode comb to the selected wavelength as described in Sec. 6.8.1.2. The inherent physical limitation of such lasers' tuning time is due to the carrier lifetime in the tuning sections on the order of 1 ns. To achieve fast tuning time, a fast electronic driving section combined with accurate selection of tuning current operation points is needed.[47,48] Hence, during the process, the carrier concentration in each section changes continuously until it reaches a steady state. During wavelength tuning, the frequency does not change continuously and mode hops occur between the different peaks of the sampled reflector and the longational modes of the cavity. Consequently, large optical power variations occur. Thus, the laser is considered ready for data transmission

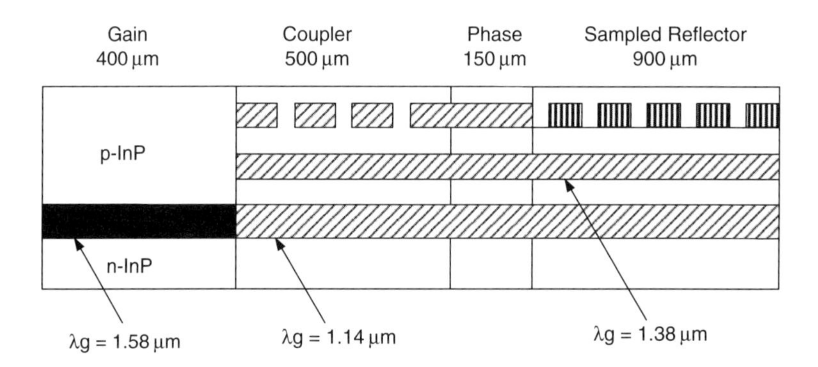


**Figure 6.22** GCSR laser structure with three tuning sections.[43] (Reprinted with permission from IEEE, *Journal of Lightwave Technology* © IEEE, 2004.)

only after the last wavelength mode is terminated. Therefore, in communication protocols, the overhead time is allocated for the laser tuning that includes some guard time. This process time is a fixed value and not a dynamic variable. The reason behind it is that tuning time between any two arbitrary wavelengths is not equal. Eventually, the longest tuning time between any two wavelengths defines the tuning time for the entire system.

Modeling and analysis of GCSR tuning time is based on an equivalent circuit of a forward-biased diode driven by a current source in Fig. 6.23(a), where a series resistor $R$ is located between the current source and the anode of the diode.[43] A shunt capacitor $C_p$ represents the laser parasitic capacitance. The tuning voltage $v(t)$ is sampled on the laser anode. The forward bias current $i(t)$ drives the laser. The rate Eqs. (6.55) and (6.46) provide the relation between the carrier concentration in the junction of the tuned section and the divining in

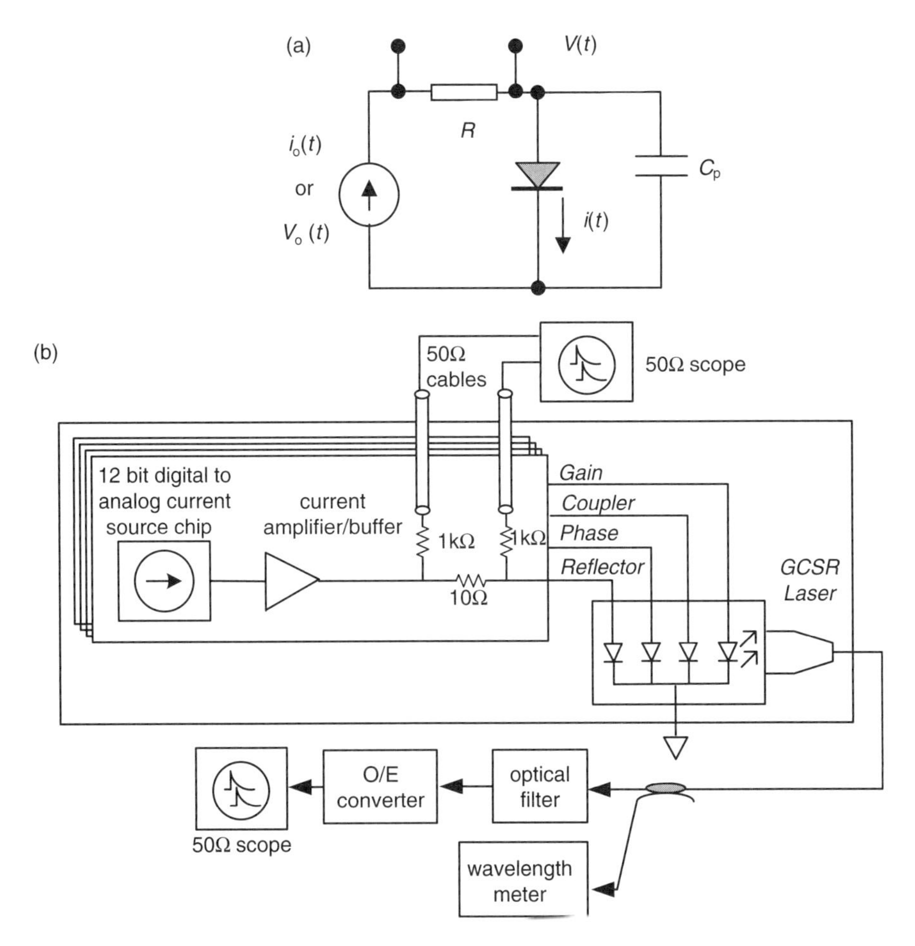


**Figure 6.23** (a) GCSR laser model of tuning section driving circuit. (b) Ultrafast GCSR transmitter and switching time measurement setup.[43] (Reprinted with permission from the *Journal of Lightwave Technology* © IEEE, 2004.)

current $i(t)$ or voltage $v(\mathrm{t})$:[47]

$$\frac{\mathrm{d}N}{\mathrm{d}t} = \frac{\eta i(t)}{qV_{\mathrm{j}}} - \left(\mathrm{AN} + \mathrm{BN}^2 + \mathrm{CN}^3\right), \tag{6.55}$$

$$V = \frac{kT}{q}\left[\ln\left(\frac{N^2}{N_{\mathrm{C}}N_{\mathrm{V}}}\right) + \frac{N}{\sqrt{8}}\left(\frac{1}{N_{\mathrm{C}}} + \frac{1}{N_{\mathrm{V}}}\right)\right] + \frac{E_{\mathrm{g}}}{q}, \tag{6.56}$$

where $N$ is the carrier concentration, $\eta$ is the current confinement factor, $i(t)$ is the tuning current, $V_{\mathrm{j}}$ is the junction volume, $q$ is the electron charge, $A$, $B$, and $C$ are nonradiative bimolecular and Auger recombination coefficients, respectively, $k$ is the Boltzman's constant, $T$ is temperature in Kelvin, $N_{\mathrm{c}}$ and $N_{\mathrm{v}}$ are the donor and acceptors concentrations of the N and the P type of the junction, and $E_{\mathrm{g}}$ is the band gap energy of the junction.

Replacing $i(t)$ in Eq. (6.55) with $i(t) = [V_{\mathrm{o}}(t) - V(t)]/R$ will provide $N(t)$. Then substituting $V(t)$ in Eq. (6.56) will result is a solution for $N(t)$. Taking into account $C_{\mathrm{p}}$ and replacing $i(t)$ by $i(t) = i_{\mathrm{o}}(t) - i_{\mathrm{c}}(t)$ and knowing that $i_{\mathrm{c}}(t) = C_{\mathrm{p}}\ (\mathrm{d}V_{\mathrm{c}}/\mathrm{d}t)$ and taking the derivative of Eq. (6.56), the value of $\mathrm{d}V_{\mathrm{c}}/\mathrm{d}t$ can be placed and solved as in Eq. (6.57).

$$\frac{\mathrm{d}N}{\mathrm{d}t} = \left\{1 + \left(\frac{C_{\mathrm{p}}kT}{q^2V_{\mathrm{j}}}\right)\left[\frac{2}{N} + \frac{1}{\sqrt{8}}\left(\frac{1}{N_{\mathrm{C}}} + \frac{1}{N_{\mathrm{V}}}\right)\right]\right\}^{-1} \times \left[\frac{\eta i(t)}{qV_{\mathrm{j}}} - \left(AN + BN^2 + CN^3\right)\right]. \tag{6.57}$$

The steady state of Eq. (6.55) is given by Eq. (6.58), which used to plot the output power of the GCSR laser as a function of the coupler and the reflector currents as appears in Fig. 6.24. The axis can be transformed from current to carrier concentration using

$$I = \left(\frac{qV_{\mathrm{j}}}{\eta}\right)\left(AN + BN^2 + CN^3\right). \tag{6.58}$$

The different shapes or modes classify different longitudinal modes, which are ordered vertically in stacks, each stack associated with a different sampled reflector peak. At the center of each mode, the coupler transmission peak, one of the reflector peaks, and one of the FP cavity modes are all aligned. The operating points are selected to be at the center of the mode map for optimal stability. A small deviation in carrier concentration in each of the sections from the operating point value will cause a small index change through free plasma-tuning coefficients.[47] This change results in a shift of spectral response of each section of the laser. A large change in carriers' concentration results in a mode hop. The laser is mapped by three tuning parameters: coupler, phase, and reflector reflectivity called CPR. These parameters affect the wavelength in the following way: $\alpha_i = \{\partial f/\partial N_i\}$, where the mode $I$ refers to $i = \mathrm{C}$, P, and R.

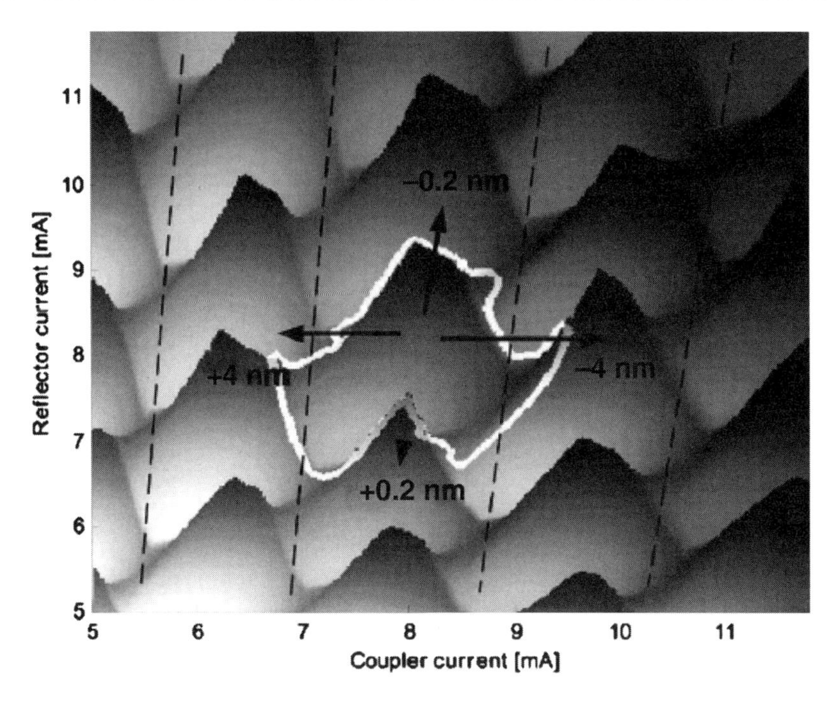


**Figure 6.24** GCSR laser map of output power versus coupler and reflector section currents showing lasing modes.[43] (Reprinted with permission from the *Journal of Lightwave Technology* © IEEE, 2004.)

Tuning speed limitations of GCSR lasers depend upon the current tuning range and parasitic capacitance. Tuning current range is defined by $I_{max}$ and $I_{min}$, which are the high and the low limits of each section of the laser. It was found that the longest tuning time occurs when switching between these values. Moreover, switching downward from $I_{max}$ to $I_{min}$ results in a longer tuning time than switching the current upward. The investigation process is done by changing the current of one tuning element, while freezing the others. The tuning time is defined as the time when one set of tuning signals begins until the laser frequency settles to within 6GHz of its final frequency. The calculation is done by using the relation that $\Delta f(t) = \sum_{i=C,P,R} \alpha_i \cdot \Delta N_i(t)$, where $\alpha_i$ was defined previously, $\Delta N_i(t)$ and $\Delta f(t)$ are the carrier concentration deviations relative to their respective values, and the center of destination mode and frequency change, respectively. Note that the equation for $\Delta f(t)$ holds only when $\Delta N_i(t)$ are within the neighborhood defined by the modes boundaries. Observation shows that the current tuning range is defined between $I_{max}$ and $I_{min}$. $I_{min}$ is 2 mA and $I_{max}$ varies between 25 and 50 mA. A second test was freezing $I_{max}$ at 25 mA and varying $I_{min}$ between 1 and 15 mA. It was observed that $I_{min}$ had a strong affect on the tuning time, whereas $I_{min}$ had very little. Parasitic capacitance, which is typically 10 pF for the laser and the contacts, has the same effect on both rise and fall time. Tuning time to 6 GHz hop is about 100 ns.

Figure 6.24 provides a map of the output power of a GCSR laser as a function of the coupler and the reflector currents. The axis can be transformed from current to carrier concentration by using Eq. (6.58), which is the steady state solution of Eq. (6.55). The distinct shapes or "modes" define different longitudinal modes. The modes are ordered vertically in stacks. Each stack is associated with a different sampled reflector peak. At the center of each mode, the coupler transmission peak, one of the reflector peaks, and one of the FP cavity modes are all aligned. The operating points are selected at the center of the mode for optimal stability. The map in Fig. 6.24 is taken for a fixed phase current. At higher phase currents the map appears to be shifted upwards.

## 6.8 LASER Characteristics in RF

Laser diode is the front end in any optical transmitter, whether it is an analog CATV, upstream HFC transmitter with or without a predistortion circuit, or if it is a digital data or return path transmitter or a digital transport transceiver. In both cases, there is a need to match the output of the transmitter electronics to the laser load in order to achieve high-power efficiency as well as a high-quality signal.

The laser is a low impedance load when biased and has to operate in a wide band frequency. Therefore, the matching circuit should be wide-band matching in order to provide a decent return loss between the laser and the transmitter electronics. Furthermore, in a digital transport modulation case, the laser represents a time varying impedance $Z(t)$. This happens when dealing with a large digital signal or a high optical modulation index (OMI) analog signal. The time varying impedance $Z(t)$ may affect other parameters such as jitter, extinction ratio (ER), and matching due to the dynamic change in the loading conditions. Poor return loss in the digital mode of operation would result in deterministic pattern-dependent jitter. This is because of the nature of the digital signal that consists of the harmonics sum according to Fourier transform (see Chapters 12 and 16). In the analog case, poor return loss affects ripple and power due to mismatch. Laser RF characteristic, being forward biased, is given in a Smith chart in Fig. 6.25.

As the frequency increases, the laser impedance gets more inductive until its first resonance point $Z = \infty$. A better understanding of the laser S11 is given by the laser's RF equivalent model described in Fig. 6.26. The intrinsic laser in a transistor outline metal can (TO) header is coupled to the package parasitic capacitances and inductances. These excessive inductances and capacitances affect the laser performance by reducing its BW. For this reason, the laser is considered an RF short. Simple matching is achieved by a series 45–47 $\Omega$ resistor. Reactive characteristics are minimized by wide-band reactive matching techniques.

### 6.8.1 Slope efficiency and extinction ratio

Laser slope efficiency $\eta$[mA/mW] or [A/W] is defined as the ratio between the output power increment and the current increment (see Fig. 6.27).

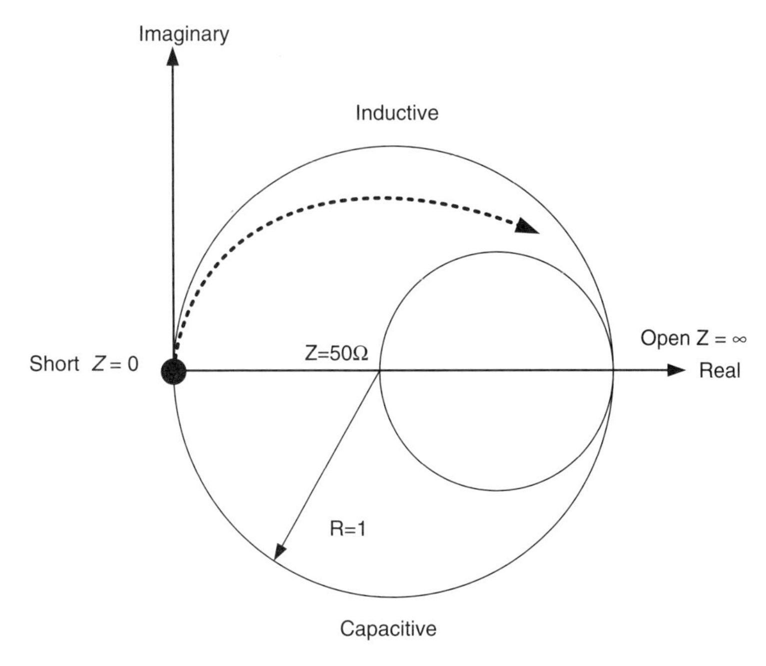


**Figure 6.25** Smith chart presentation of a Laser S11.

That value measures the laser characteristics and its performance for different applications:

$$\eta = \Delta P/\Delta I, \tag{6.59}$$

$$\Delta P/\Delta I = (P_2 - P_1)/(I_2 - I_1), \tag{6.60}$$

$$\eta = t_g\theta. \tag{6.61}$$

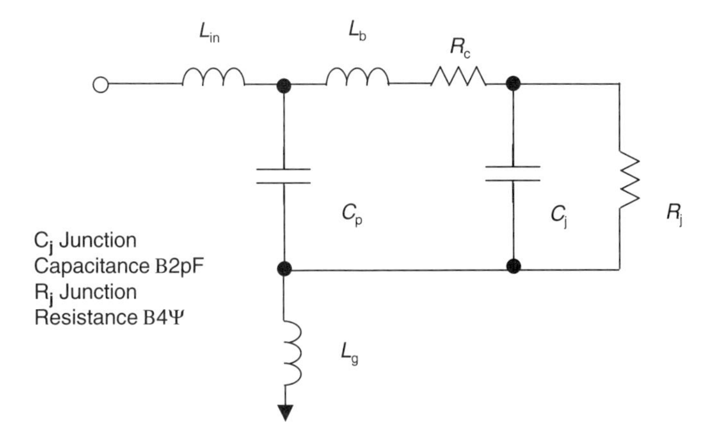


**Figure 6.26** Laser input impedance model showing lead inductance $L_{IN}$, bond inductance $L_b$, contact resistance $R_c$, junction capacitance and resistance $R_j$ and $C_j$ and TO return ground (GND) lead $L_g$.

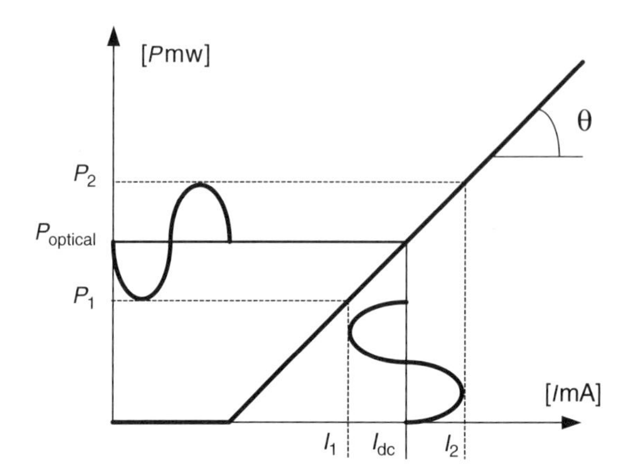


**Figure 6.27** Laser modulation definition for analog in LI curve.

A high-slope efficiency value, $\eta$, means a sensitive laser, which requires a lower modulation current in order to achieve a high-modulation index. Additionally, it is much more sensitive to temperature changes since a slight change in its dc bias would result in a higher change of its output optical power. Hence, in advanced transmitters, there are TEC loops (TECs) that lock the temperature of the laser to be at a constant level, and, as a consequence, stabilizes the laser wavelength and power. Some other transmitters use an automatic power control (APC) loop to control the power or have both TEC and APC.

In the case of digital modulation, when a high-modulation index and speed are needed, a high $\eta$ characteristic laser is beneficial. That kind of laser would have good eye diagram opening which represents high signal-to-noise ratio (SNR) and low jitter on one hand, and a fast rise and fall time on the other. Hence, in the digital case, the term ER value is used to measure the laser modulation index and eye opening for a given rise and fall time requirements.

"Extinction ratio" (ER) is defined as the ratio of optical level when "1" to "0" logic level data is at the laser input (threshold level) (see Fig. 6.28).

For ER, the notation would be

$$\mathrm{ER} = 10\log(P_1/P_0), \tag{6.62}$$

or

$$\mathrm{ER} = 10\log(1 + \Delta P/P_0), \tag{6.63}$$

where,

$$\Delta P[\mathrm{dB}] = P_1[\mathrm{dB}] - P_0[\mathrm{dB}]. \tag{6.64}$$

The value $\Delta P$ measures the eye opening and power margin between high-level logic and low-level logic state. However, high laser slope efficiency is not the only requirement for speed. Further optimization of the laser driver and its charge

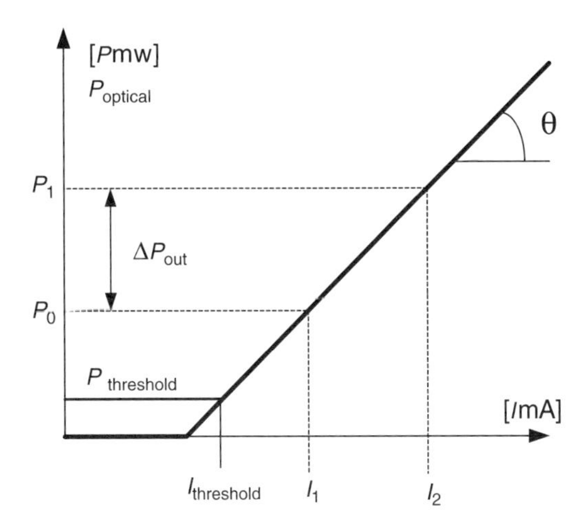


**Figure 6.28** Laser digital modulation definition in LI curve.

pumps as well as the matching circuit to the laser is needed. Moreover, ER, rise specifications, and fall time contradict each other. Hence, a trade-off value should be noted. Ultrafast rise and fall time refers to a low ER value and vice versa.

### 6.8.2 Bias and modulation

From the above definitions, the conclusion is that laser bias in analog cases should be in its linear operation point. The dc-bias point must be high above the threshold level to prevent any distortions (see Fig. 6.27). In a digital case, the laser is biased close to the threshold level in order to meet the ER requirement for digital eye performance (see Fig. 6.28).

In analog and CATV systems, OMI is defined as follows:

$$\mathrm{OMI}[\%] = (I_{\mathrm{peak}}/I_{\mathrm{dc}}) \times 100. \tag{6.65}$$

This modulation index measures the peak deviation in a percentage of an RF signal from a dc-bias quiescent point. It is a similar definition to the AM index.

The dc-bias point of a laser should be far away from the threshold "knee" or the nonlinear region. It should be such that the peak of the RF signal is far off from the nonlinear region.

By selecting the correct dc point, distortions are avoided since there is no clipping of the RF signal. This assumption is true at first approximation at a single CW tone. However, the dc-bias point is not a sufficient condition to have a linear CATV transmitter that meets the CATV standards for distortions. In professional CATV transmitters, special predistortion techniques are used in addition to the laser-bias point. That is a multichannel loading of the laser.

Multichannel loading of a laser increases the overall OMI, which means a larger modulation depth and hence more distortions. Sometimes there are requirements to measure and test receiver performance with two tones for second- and third-order distortions. The test tone in such a case equals half the number of channels occupying the receiver's BW. Hence, the two-tone test simulates full channel loading conditions up to a certain accuracy as will be discussed in Chapter 10. By using a single laser, the measured result would be the laser distortion and not the receiver. Thus, the two-tone test requires two similar lasers, each laser having the same OMI representing half the channel number. Generally, in such a case, the wavelength separation between the lasers is 10 nm.

### 6.8.3 Some modulation analysis

Typical laser matching circuit uses series matching resistor described in Fig. 6.29. Observing Fig. 6.29 and OMI definitions, the following relations are obtained:

$$P_{\mathrm{RF}} = I_{\mathrm{MOD}}^2 Z_{\mathrm{O}}, \tag{6.66}$$

$$\frac{P_{\mathrm{RF}}}{Z_{\mathrm{O}}} = I_{\mathrm{MOD}}^2 = \frac{I_{\mathrm{PEAK}}^2}{2}, \tag{6.67}$$

$$I_{\mathrm{MOD}} \equiv I_{\mathrm{rms}} \equiv \frac{I_{\mathrm{PEAK}}}{\sqrt{2}}, \tag{6.68}$$

where $Z_{\mathrm{O}}$ is the characteristic impedance, in our case 50 Ω. $I_{\mathrm{MOD}}$ is the rms modulating current marked as $I_{\mathrm{rms}}$, $I_{\mathrm{PEAK}}$ is the peak current or the modulation amplitude, and η is the slope efficiency of the laser (see Fig. 6.27).

Observing Fig. 6.27 in first approximation, the laser has a linear LI curve equation approximately as

$$Y = aX + B, \tag{6.69}$$

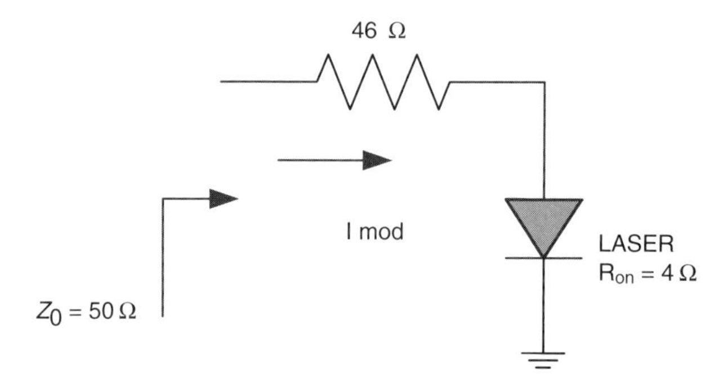


**Figure 6.29** Modulation circuit model.

where $B \approx 0$. $B$ is the optical power at zero bias. Hence, $a = \eta$ and therefore $\eta \times I_{dc} = P_{OPTICAL}$.

The goal of the calculation is to define the RF power needed to satisfy the OMI spec for a given optical power and slope efficiency $\eta$. A different way of presenting this question is to define what is the required RF level needed to meet the OMI requirements under specific bias conditions. After some simple manipulation of Eqs. (6.66)–(6.68), $I_{PEAK}$ and $I_{dc}$ are derived as a function of RF level OMI and slope efficiency $\eta$:

$$\sqrt{2 \cdot \frac{P_{RF}}{Z_0}} = I_{PEAK}, \tag{6.70}$$

$$\mathrm{OMI} = \frac{I_{PEAK}}{I_{dc}} \Rightarrow \frac{I_{PEAK}}{\mathrm{OMI}} = I_{dc}, \tag{6.71}$$

$$\frac{\sqrt{2 \cdot \frac{P_{RF}}{Z_0}}}{\mathrm{OMI}} = I_{dc}, \tag{6.72}$$

$$\eta \cdot \frac{\sqrt{2 \cdot \frac{P_{RF}}{Z_0}}}{\mathrm{OMI}} = P_{OPTICAL}. \tag{6.73}$$

Changing Eq. (6.73), while OMI is in decimal units optical power is in watts results in

$$\frac{P_{OPTICAL}^2 \cdot \mathrm{OMI}^2}{\eta^2} = \frac{2P_{RF}}{Z_0}, \tag{6.74}$$

$$P_{RF}[W] = \left(\frac{P_{OPTICAL} \cdot \mathrm{OMI}}{\eta \cdot \sqrt{2}}\right)^2 Z_0, \tag{6.75}$$

$$P_{RF}[mW] = \left(\frac{P_{OPTICAL} \cdot \mathrm{OMI}}{\eta\sqrt{2}}\right)^2 Z_0 \cdot 1000. \tag{6.76}$$

A different notation yields

$$P_{RF} = \frac{\left(\frac{P_{OPTICAL} \cdot \mathrm{OMI}}{\eta\sqrt{2}}\right)^2 Z_0}{0.001}. \tag{6.77}$$

Hence, $P_{RF}$[dBm] is given by

$$P_{RF}[\mathrm{dBm}] = 10 \log \frac{\left(\frac{P_{OPTICAL} \cdot \mathrm{OMI}}{\eta\sqrt{2}}\right)^2 Z_0}{0.001}. \tag{6.78}$$

The last two equations are most important and are used to calibrate RF on fiber setups. This means that the RF power needs to satisfy OMI for a given LASER with known slope efficiency. In case many channels drive the laser, the overall RF power is given by

$$P_{\text{TOT RF}} = N \times P_{\text{RF}}, \tag{6.79}$$

where $P_{\text{RF}}$ is a single channel power and $N$ is the number of channels. Hence, the following relation between OMI and multichannel loading $\text{OMI}_{\text{TOT}}$ is applicable. $I_{\text{PEAK\_TOT}}$ represents the equivalent RF current of $N$ channels:

$$\sqrt{2 \cdot \frac{P_{\text{RF}} \cdot N}{Z_0}} = I_{\text{PEAK\_TOT}}, \tag{6.80}$$

$$I_{\text{PEAK\_TOT}} = \sqrt{N} \cdot I_{\text{PEAK}}, \tag{6.81}$$

$$\text{OMI}_{\text{TOT}} = \frac{I_{\text{PEAK\_TOT}}}{I_{\text{DC}}} \Rightarrow \frac{I_{\text{PEAK\_TOT}}}{\text{OMI}_{\text{TOT}}} = I_{\text{dc}}, \tag{6.82}$$

$$\frac{\sqrt{2 \cdot \dfrac{N \cdot P_{\text{RF}}}{Z_0}}}{\text{OMI}_{\text{TOT}}} = I_{\text{dc}}, \tag{6.83}$$

$$\text{OMI}_{\text{TOT}} = \text{OMI} \cdot \sqrt{N}. \tag{6.84}$$

Equation (6.84) provides the equivalent modulation depth in the case of $N$ channels with equal OMI. Therefore, if a two-tone test is needed or more tones are required, the OMI per laser is given by

$$\text{OMI}_{\text{TOT}} = \frac{\text{OMI} \cdot \sqrt{N}}{\sqrt{\text{num}}}, \tag{6.85}$$

where num is the number of lasers in the test.

In conclusion, in a multilaser test, the RF power for each laser given by

$$P_{\text{RF}}[\text{dBm}] = 10 \log \frac{\left( \dfrac{P_{\text{OPTICAL}} \cdot \dfrac{\text{OMI} \cdot \sqrt{N}}{\sqrt{\text{num}}}}{\eta\sqrt{2}} \right)^2 Z_0}{0.001}. \tag{6.86}$$

The above OMI calibration method suffers from an accuracy disadvantage. It does not take into account the laser frequency response, only its dc characteristic. As the frequency of the RF modulation current increases, a portion of the modulating current is shunted by the capacitive components of the laser package and the matching network. In some other cases, it is impossible to measure the laser slope efficiency. A different OMI measurement method and calibration approach is needed and will be presented by using a photodetector.

## 6.9 Quantum Efficiency

In the previous section, a brief introduction to laser-modulation definition using LI curve was provided. Usually, the laser's optical-output power versus its injection current is presented on the LI curve. However, the laser-slope efficiency, also named quantum efficiency (QE), depends on temperature. As the temperature increases, the slope efficiency decreases and the threshold increases exponentially. The definition of QE is given by

$$\frac{dP_O}{dI} = \eta_F \frac{h\nu}{q} \cdot \frac{\eta_{\mathrm{int}}\alpha_{\mathrm{m}}}{\alpha_m + \alpha_{\mathrm{int}}}, \tag{6.87}$$

where $\eta_F$ is the portion of optical power emitted from the laser's front facet. $\eta_{\mathrm{int}}$ is given in Eq. (6.3). This value decreases due to the Auger recombination rate and heterobarrier leakage increases exponentially with temperature in InGaAsP. As a result, the slope efficiency decreases if the laser in the LI curve decreases accordingly. The factors $\alpha_{\mathrm{m}}$ and $\alpha_{\mathrm{int}}$ are mirror loss and the internal loss of the laser cavity, respectively.

In general literature, the slope efficiency is defined as a differential QE:

$$\eta_{\mathrm{d}} = \frac{\eta_{\mathrm{int}}\alpha_{\mathrm{m}}}{\alpha_{\mathrm{m}} + \alpha_{\mathrm{int}}}. \tag{6.88}$$

The laser's threshold-current temperature characteristic is given by

$$I_{\mathrm{th}} = I_0 \exp\left(\frac{T}{T_0}\right). \tag{6.89}$$

$T_0$ is a constant value that defines the temperature sensitivity at the threshold current. This value is in the range of 50–100 K for InGaAsP lasers. For GaAs lasers it is 100–150 K. The exponential rise in the threshold current may result due to the decrease in the optical gain $g$. The internal loss increases. The Auger recombination increases and the carrier leakage also increases in an exponential nature.[48] In long-wave lasers, the temperature dependence of optical gain is considered a dominant factor.[49,50]

## 6.10 Main Points of this Chapter

1. Lasing is a quantum phenomenon where stimulated emission is generated when a photon collides with an excited electron. The outcome of this is the return of the electron to its original energy level, emission of a photon, and emission of a second photon due to the collision of the electron on the excited atom having a high-energy electron. This process describes optical gain.

2. A laser is an optical oscillator with optical gain and optical feedback. The optical feedback is accomplished by the facet's reflectors, which are semiconductor mirrors.

3. In a semiconductor laser, the prerequisite conditions for lasing are population inversion, optical feedback, optical gain, radiative recombination, and low optical cavity losses.

4. The cavity modes of a FP laser are called longitudinal modes. The cavity causes several modes to occur where one mode governs, and other modes laser in lower power. Thus, this laser is called multimode laser known as Fabry–Perot or FP. Examples of lasers are the early ruby laser of Theodore Maiman and solid-state lasers with two mirrors. Even the high-power industry lasers are the FP type.

5. A solid-state laser is a PN junction where the active layer is at the junction.

6. There are two types of lasers: gain-guided and index-guided lasers. The difference between the two is the confinement method of the carriers in the active layer.

7. A gain-guided laser is a laser where the mechanism of carrier confinement is done electrically by creating a narrow metal contact stripe on the anode. The stripe results in a limited area of electrical field. This is a kind of capacitor above the PN active area. Thus, the carriers are confined under the stripe electrically. The reason for the "gain-guided" name is that the pumping current and field create the carriers' confinement.

8. Index-guided laser is a laser whose layer structure creates an optical waveguide at the active area. The waveguiding is achieved by creating refraction index differences on the waveguide boundaries. Thus, the carriers are confined to a specific area because of the index difference and not because of the electrical field.

9. The motivation for a narrow confinement cross-section area of the carriers is to have a narrow spatial beam, narrower spectral width, and higher power per unit of area as well as lower loss. A lower loss laser would require a lower current. Nonconfined lasers, such as early lasers, require a higher current in order to meet the same optical power as a carrier-confined laser like the gain- and the index-guided lasers.

10. The advantage of an index-guided laser over the gain-guided laser is the sharper boundaries of the confinement cross section and waveguiding. In the gain-guided laser, there is a lateral leakage out of the confinement area because of the current lateral diffusion and the electrical field fringes. However, the index-guided laser has well-defined, buried-waveguide boundaries. Therefore, the beam emitted from an index-guided

laser may have a narrower spectral width and higher power per unit of area, meaning less spatial distribution. This may result even in a lower operating current.

11. The disadvantage of a multimode laser is the chromatic dispersion. This results from the propagation of several modes in a fiber with different group velocities. This dispersion may result in chromatic jitter and distortion on the receiving side. Hence, these lasers are used for short-haul applications.

12. There are two popular single-mode lasers known as distributed feedback (DFB) and distributed Bragg reflector (DBR) lasers. The concept of these lasers is to create a wavelength-selective optical feedback. The feedback is based on Bragg gratings. The Bragg filter creates reflections at the desired wavelength and filters out undesired modes.

13. In a wavelength-selective optical-feedback filter, the shape factor of the feedback defines the conditions for oscillations. This shape factor is called mode suppression ratio (MSR).

14. The lasing oscillation conditions in FP, DFB, or DBR laser can be simplified to $AB - 1 = 0$, where $A$ is the optical gain and $B$ is the optical feedback. It refers to the reflections of the mirror or the selective wavelength filter or mirror.

15. In a DFB laser, the optical feedback is within the active area acting as a selective wavelength filter on the longitudinal modes allowing only one mode to excite.

16. In a DBR laser, the optical feedback is due to a wavelength-selective mirror, which reflects only the desired mode of excitation lasing. Thus, DBR can be referred to as a "single-mode FP laser." The mirrors are based on Bragg reflections.

17. Since the DFB feedback creates a large loss at the desired wavelength and operates as a band-stop filter, BPF, the DFB may emit two wavelengths or two modes. Hence, the feedback is detuned so only one mode is excited. One of the detuning methods that the Bragg feedback into a BPF is called $\lambda_0/4$ shift.

18. A QW laser is a laser in which the active area or waveguide is grown in a controlled manner to a specific thickness. This way the confinement of quantum states are defined. The motivation behind that method is to have better temperature stability, higher operating temperature, better linearity for CATV lasers, and higher efficiency by reducing the power consumption.

19. Laser can have MQWs for higher power purposes.

20. VCSEL operates on the thickness or depth modes of the active layers. As in longitudinal laser or edge-emitting lasers, as they are called, there are FP VCSELs and DBR VCSELs. There is no DFB VCSEL since it is not possible to manufacture.

21. There are several structures of VCSEL growth, named mesa etched, proton implanted, and dielectric aperture.

22. The motive behind these three methods is to create the injection current confinement through the active layer. The second goal is to block nonproductive lateral current to the active layer.

23. VCSELs suffer from low optical power and resistive losses through the mirror causing a voltage drop. Thus, to overcome this problem, the dielectric aperture was developed.

24. Recent development introduced long-wavelength VCSELs.

25. Tunable lasers are wavelength-controlled lasers. The wavelength control is accomplished by changing the cavity dimensions affecting the mode of lasing or by controlling the Bragg-feedback filter by changing its refractive index.

26. Tunable laser's control can be divided into two main concepts, electrical control and mechanical control using MEMS technology. The electronic controlled lasers are tuned to different wavelengths by changing the filters and the cavity modes due to gain and refractive index change, whereas the MEMS methods are controlled mechanically by changing the cavity length. Another method used to control the laser wavelength is thermal control utilized on DFB lasers.

27. Wavelength control of a laser, depending on the laser's cavity type (internal or external) is categorized into two methods. The ECL uses MEMS technology and electronic methods to achieve wavelength control, whereas for the internal cavity lasers, wavelength control is done electronically.

28. MEMS and thermal tuning methods are slower compared to electronically tuned lasers.

29. An electrostatic actuator acting as a tiny motor accomplishes MEMS nanomotion control. This motor displaces the micromirrors at one end of the laser facet. There mirror reflects the beam of light to the output fiber.

30. One of the ECL electronic methods to tune the wavelength is called "open" architecture, where tunable filters are attached to the facet of the laser creating an example tunable FP laser.

31. Laser characteristics are measured on LI curve light power versus current curve. The slope of that curve is called slope efficiency or QE.

32. A laser in forward bias represents a load of about 4 Ω; hence, RF-wise it is almost a complete short.

33. Laser OMI is the ratio of the ac-current swing to the quiescent dc-bias point current. This ratio is preserved in the optical power as well. Thus, the ac optical power is given by $P = \text{OMI} \times P_{dc}$.

34. The laser modulation index in digital transport is measured by the term extinction ratio (ER). This factor describes the ratio between the maximum optical-power state and the minimum optical-power state.

35. Laser slope efficiency defines the input swing required to have a desired ER and affects the optical swing. The steeper the laser, the more sensitive it is to input to output changes.

36. QE of a laser degrades as temperature increases and the laser threshold current increases. This is the root cause of its increase in nonradiative recombination. The reduction in QE is exponential. The increase in threshold current is also exponential.

## References


1. Bar-Lev, A., *Semiconductors and Electronic Devices, Third Edition.* Prentice Hall International (1993).
2. Sze, S.M., *Physics of Semiconductor Devices, 2nd Edition.* John Wiley & Sons (1981).
3. Agrawal, G.P., and N.K. Dutta, *Long Wavelength Semiconductor Lasers, 2nd Edition.* Van Nostrand Reinhold Company, New York (1986).
4. Agrawal, G.P., *Fiber-Optic Communications Systems, 2nd Edition.* John Wiley & Sons (1997).
5. Pezeshki, B., G. Pahar, and N. Bahar, "Tunable lasers, an array based approach for short-term inventory needs and new dynamic networks." *IEEE Circuits and Devices Magazine*, pp. 36–40 (2003).
6. Bruce, E., "Tunable Lasers." *IEEE Spectrum*, pp. 35–39 (2002).
7. Brillant, I., "GWS photonics, OPEN widely tunable laser technology." White Paper (2004).
8. Pezeshki, P., "New approaches to laser tuning." *Optics and Photonics News*, pp. 34–38 (2001).
9. Dudovich, N., G. Levy-Yurista, A. Sharon, A.A. Friesem, and H.G. Weber, "Active semiconductor based grating waveguide structure." *IEEE Journal of Quantum Electronics*, Vol. 37 No. 8, pp. 1030–1039 (2001).
10. Maiman, T., *The Laser Odyssey.* Laser Press (2000).

11. Chinone, N., and M. Okai, *Semiconductor Lasers Past Present and Future*, G. P. Agrawal, Ed. AIP Press, Woodbury, NY (1995).
12. Wang, J.Y., M. Cada, and J. Sun, "Theory for optimum design and analysis of distributed-feedback lasers." *IEEE Photonics Technology Letters*, Vol. 11 (1999).
13. Huang, Y.C. and Y.Y. Lin, "Coupled wave theory for Distributed feedback optical parametric amplifiers and oscillators." *Journal of the Optical Society of America B*, Vol. 21 No. 4, pp. 777–780 (2004).
14. Akbari, M., M. Shahabadi, and K. Schüemann, "A rigorous two-dimensional field analysis of DFB structures." *Progress in Electromagnetics Research*, PIER 22, pp. 197–212, EMW Publishing (1999).
15. Chen, N., Y. Nakano, K. Okamoto, K. Tada, G.I. Morthier, and R.G. Baets, "Analysis fabrication and characterization of tunable DFB lasers with chirped gratings." *IEEE Journal of Selected Topics in Quantum Electronics*, Vol. 3 No. 2, pp. 541–546 (1997).
16. Solymar, L., and D. Walsh, *Lectures on the Electrical Properties of Material, 2nd Edition*, Oxford Science Publications (1988).
17. Verdeyen, J.T., *Laser Electronics, 2nd Edition.* Prentice Hall (1989).
18. Yariv, A., "Quantum well semiconductor lasers are taking over." *IEEE Circuits and Devices Magazine*, pp. 25–28 (1989).
19. Nagarajan, R., M. Ishikawa, T. Fukushima, R.S. Geels, and J.E. Bowers, "High speed quantum-well lasers and carrier transport effect." *IEEE Journal of Quantum Electronics*, Vol. 28 No. 10, pp. 1990–2003 (1992).
20. Namegya, T., N. Matsumoto, N. Yamanaka, N. Iwai, H. Nakayama, and A. Kasukawa, "Effects of well number in 1.3 mm GaInAsP/InP GRINSCH strained-layer quantum well lasers." *IEEE Journal of Quantum Electronics*, Vol. 30 No. 2, pp. 578–584 (1994).
21. Zah, C.E., R. Bhat, B.N. Pathak, F. Favire, W. Lin, M. C. Wang, N.C. Andreadakis, D.M. Hwang, M.A. Koza, T.P. Lee, Z. Wang, D. Darby, D. Flanders, and J. Hsieh, "High-performance uncooled 1.3-mm AlxGay-In1-x-yAs/InP strained-layer quantum-well lasers for subscriber loop applications." *IEEE Journal of Quantum Electronics*, Vol. 30 No. 2, pp. 511–523 (1994).
22. Tsai, C.Y., F.P. Shih, T.L. Sung, T.Y. Wu, C.H. Chen, and C.Y. Tsai, "A small-signal analysis of the modulation response of high-speed quantum-well lasers: effects of spectral hole burning, carrier heating, and carrier diffusion-captureescape." *IEEE Journal of Quantum Electronics*, Vol. 33 No. 11, pp. 2084–2096 (1997).
23. Lee, T.P., "Recent advances in long-wavelength semiconductor lasers for optical fiber communications." *Proceedings of the IEEE*, Vol. 79, No. 3, pp. 253–276, March (1991).
24. Otsubo, K., Y. Nishijima, H. Ishikawa. "Long-wavelength semiconductor lasers on InGaAs ternary substrates with excellent temperature characteristics." *Fujitsu Science and Technology Journal*, Vol. 34 No. 2, pp. 212–222 (1998).

25. Piprek, J., A. Black, P. Abraham, N.M. Margalit, E.L. Hu, and J.E. Bowers, "1.55 mm laterally oxidized vertical cavity lasers." MICRO Project Report 97-092 (1998).
26. Black, K.A., P. Abraham, N.M. Margalit, E.R. Hegblom, Y.J. Chiu, J. Piprek, J.E. Bowers, and E.L. Hu, "Double fused 1.5 mm vertical cavity lasers with record high of 132K at room temperature." *Electronics Letters*, Vol. 34 No. 20 (1998).
27. Piprek, J., K. Takiguchi, J.R. Wesselmann, and J.E. Bowers. "Analog transmission using long-wavelength vertical cavity lasers." MICRO Project Report 97-014 (1998).
28. Sceats, R., N. Balkan, M.J. Adams, J. Masum, A.J. Dann, S.D. Perrin, I. Reid, J. Reed, P. Cannard, M.A. Fisher, D.J. Elton, and M.J. Halow. "A GaInAsP/InP vertical cavity surface emitting laser for 1.5 mm operation." *Turkish Journal of Physics*, Vol. 23, pp. 781–787 (1999).
29. Klein B., L.F. Register, M. Grupen, and K. Hess, "Numerical simulation of vertical cavity surface emitting lasers." *Optics Express*, Vol. 2, No. 4, pp. 163–167 (1998).
30. Choon, W., Y. Liu, and K. Hess, "Resonant wavelength control and optical confinement analysis for graded SCH VCSELs using a self-consistent effective index method." *Journal of Lightwave Technology*, Vol. 31, No. 2, pp. 555–560 (2003).
31. Gustavson, J.S., A. Haglund, J. Bengtsson, and A. Lasson, "High speed digital modulation characteristics of oxide-confined vertical cavity surface emitting lasers numerical simulations consistent with experimental results." *IEEE Journal of Quantum Electronics*, Vol. 8, No. 8, pp. 1089–1096 (2002).
32. Hanamaki, Y., H. Akiyama, and Y. Shiraki, "Spontaneous emission alteration in InGaAs/GaAs vertical cavity surface emitting laser (VCSEL) structures." *Semiconductor Science and Technology*, Vol. 14, pp. 797–803 (1999).
33. Park, S.H., Y. Park, and H. Jeon, "Theory of mode stabilization mechanism in concave-micromirror capped vertical cavity surface emitting lasers." *Journal of Applied Physics*, Vol. 94, No. 3, pp. 1312–1317 (2003).
34. Johnston, A.H., T.F. Miyahira, and B.G. Rax, "Proton damage in advanced laser diodes." *IEEE Transaction on Nuclear Science*, Vol. 48, No. 6, pp. 1764–1772 (2001).
35. Kaminow, I.P., and T.L. Koch, *Optical Fiber Telecommunications IIIB*. Academic Press (1997).
36. Gustasson, J.S., J.A. Vukusic, J. Bengtsson, and A. Larsson, "A comprehensive model for the modal dynamics of vertical-cavity surface emitting lasers." *IEEE Journal of Quantum Electronics*, Vol. 38, No. 2, pp. 203–212 (2002).
37. Wang, S., *Fundamentals of Semiconductors Theory and Device Physics*. Prentice-Hall, College Division (1989).
38. Pezeshki, B., et al., "20 mW Widely Tunable Laser Module using DFB Array and MEMS Selection." *Photonics Technology Letters*, Vol. 14, No. 10, pp. 1457–1459 (2002).

39. Liu, A.Q., X.M. Zhang, V.M. Murukeshan, C. Lu, and T.H. Cheng, "Micromachined wavelength tunable laser with an extended feedback model." *IEEE Journal of an Selected Topics in Quantum Electronics*, Vol. 8, No. 1, pp. 73–79 (2002).
40. Liu, A.Q., X.M. Zhang, V.M. Murukeshan, and Y. Lam, "A novel integrated micromachined tunable laser using polysilicon 3D mirror." *IEEE Photonics Technology Letters*, Vol. 13, No. 5, pp. 427–429 (2001).
41. Hirano, T., T. Furuhata, K.J. Gabriel, and H. Fujita, "Design fabrication and operation of submicron gap comb-drive microactuators." *Journal of Microelectromechanical System*, Vol. 1, No. 1, pp. 52–59 (1992).
42. Sadot, D., and E. Voinovich, "Tunable optical filters for dense WDM networks." *IEEE Communication Magazine*, pp. 50–55 (1998).
43. Buimovich, E., and D. Sadot, "Physical limitation of tuning time and system considerations in implementing fast tuning of GCSR lasers." *IEEE Journal of Lightwave Technology*, Vol. 22, No. 2, pp. 582–588 (2004).
44. Joseph, H., and D. Sadot, "A novel self-heterodyne method for combined temporal and spectral high-resolution measurement of wavelength transients in tunable lasers." *IEEE Photonics Technology Letters*, Vol. 16, No. 8, pp. 1921–1923 (2004).
45. Sadot, D., and I. Elhanany, "Optical switching speed requirements for terabit/sec packet over WDM networks." *IEEE Photon Technology Letters*, Vol. 2, pp. 440–442 (2000).
46. Rubin, S., E. Buimovich, G. Ingbar, and D. Sadot, "Implementation of an ultra-fast widely tunable burst-mode 10 Gbps transceiver." *Electron Lett*, Vol. 38, No. 23, pp. 1462–1463, November (2002).
47. Amann, M.C., J. Buus, and Suematsu, Y., *Tunable Laser Diodes*. Artech House, Norwood, MA (1998).
48. Coldren, L., and S.W. Corzine, "Diode lasers and photonic integrated circuits." John Wiley and Sons (1995).
49. Ackerman, D.A., P.A. Morton, G.E. Shtengel, M.S. Hybertsen, R.F. Kazarinov, T. Tanbum-Ek, and R.A Logan, "Analysis of $T_o$ in 1.3 mm multiple quantum well and bulk active lasers." *Applied Physics Letter*, Vol. 66, pp. 2613–2615 (1995).
50. Zou, Y., J.S. Osinski, P. Grodzinski, P.D. Dapkus, W.C. Rideout, W.F. Sharfin, J. Schlafer, and F.D. Crawford, "Experimental study of auger recombination gain and temperature sensitivity of 1.5 mm compressively strained semiconductor laser." *IEEE Journal of Quantum Electronics*, Vol. 29, pp. 1565–1575 (1993).
51. Li, M.Y., W. Yuen, and C.J. Chang-Haanian, "Top emitting micromechanical VCSEL with a 31.6 nm tuning range." *IEEE Photonics Technology Letters*, Vol. 10, No. 1, pp. 18–20 (1998).
52. Li, M.Y., W. Yuen, and C.J. Chang-Haanian, "High performance continuously tunable top emitting vertical cavity laser with 20 nm wavelength range." *Electronics Letters*, Vol. 33, No. 12, pp. 1501–1502 (1997).

## Chapter 7

# Laser Dynamics: External Modulation for CATV and Fast Data Rates

Direct modulation of a semiconductor laser is the most convenient way of using these devices in communication circuits. However, there are certain fundamental limitations for a direct-current modulation approach. In digital data transport, this limitation starts to become important as the speed of modulation increases beyond 10 GBPS. The main limitations result from the wavelength chirping of the laser output as well as relaxation–oscillation. The first one causes jitter whereas the other parameters result in overshooting the eye pattern. Jitter phenomena of fiber dispersion due to wavelength chirp affects the digital transport eye quality because of group delay. When dealing with laser analysis, there are two parameters that are examined, intensity modulation (IM) in Secs. 7.1 and 7.2 and frequency modulation (FM) known as wavelength chirp in Sec. 7.3.

In analog community access television (CATV) transmitters, the bandwidth (BW) or modulation frequencies are much lower; however, the key problem there is nonlinear distortions (NLD) due to the interaction between the chirp phenomena of the distributed feedback (DFB) 1550-nm laser and the dispersion of a standard single-mode fiber (SMF). Additionally, analog channels at microwave frequencies and millimeter waves may be affected by relaxation–oscillation BW limitation due to reduction in distortion performance. In CATV, transmission at 1550 nm is desirable due to the advantage of low fiber losses and the use of erbium-doped fiber amplifiers (EDFAs). The power required from a 1310-nm DFB laser is higher due to the inherent higher losses of the SMF. These constraints impose more limitations. Moreover, the performance of a direct modulated laser (DML) is affected by multiple optical reflections and chirp. These problems limit the transmission distance over standard SMF.

The solution to these problems both in digital and in multicarrier analog CATV modulation schemes is using external modulation (EM) techniques. When EM is used, the laser itself is biased on continuous mode with an external device modulating the optical output.

First, this chapter provides a review of the physical principles behind the dynamic behavior of laser diodes described by the rate equations. Mode hopping and modal noise are reviewed. Next, the most commonly used EM schemes are explained. Finally, some commercial devices available in the market are discussed. These user phenomena will be used later in Chapter 19 when analyzing digital and analog links.

## 7.1 Dynamic Response of Semiconductor Laser

As was reviewed in Chapter 6, a semiconductor laser is a cavity where the P-type and the N-type carriers are generated by injecting a current into a junction of known volume called the active region. These carriers are then recombined producing coherent stimulated emission. Assume a single-longitudinal-mode (SLM) DFB with uniform distribution of electrons, with density $N$ over a volume $V = d \times L \times W$ of the active layer, and photon density $S$ for the given lasing mode.

The dynamic behavior of such a laser is normally explained through the following phenomenological differential equations, which have several notations,[1,2,16,21] and is given by Eqs. (7.1–7.3). For the active region of a double heterostructure device (DH), the injected current provides a generation term and various radiative and nonradiative recombination processes as well as nonproductive carrier-leakage recombination. Hence, the carrier's change over time is the difference between generation and recombination. This fact results in[21]

$$\frac{\mathrm{d}N}{\mathrm{d}t} = G_{\mathrm{gen}} - R_{\mathrm{rec}}, \tag{7.1}$$

where $G_{\mathrm{gen}}$ is the rate of injected electrons and $R_{\mathrm{rec}}$ is the recombination rate per unit volume in the active region. It is well known that $I = \mathrm{d}Q/\mathrm{d}t$ or $I \times t = Q = Nq$. Thus, the net number of electrons per second injection from a current source is given by $I/q = N/t$. Since there is some leakage, the net current used for generation is multiplied by the quantum efficiency $\eta$ resulting in $\eta \times (I/q)$. Hence, the overall generation per unit of volume is $\eta \times (I/qV)$. $R_{\mathrm{rec}}$ can be broken into several recombination mechanisms[21] that finally lead to the rate equations:[2]

$$\frac{\mathrm{d}N}{\mathrm{d}t} = \frac{I}{qV} - \frac{N}{\tau_{\mathrm{n}}} - \Gamma v_{\mathrm{g}} g_0 (N - N_0)(1 - \varepsilon S)S, \tag{7.2}$$

$$\frac{\mathrm{d}S}{\mathrm{d}t} = \Gamma v_{\mathrm{g}} g_0 (N - N_0)(1 - \varepsilon S)S - \frac{S}{\tau_{\mathrm{p}}} + \Gamma \gamma_{\mathrm{sp}} \frac{N}{\tau_{\mathrm{n}}}, \tag{7.3}$$

where $\tau_{\mathrm{n}}$ is the electron lifetime (ns), $\tau_{\mathrm{p}}$ is the photon lifetime (ps), $\Gamma$ is the carrier confinement factor in the active layer; i.e., it shows the fractional radiation coupled in the lasing mode, $v_{\mathrm{g}}$ is the group velocity, $\gamma_{\mathrm{sp}}$ is the fraction of spontaneous emission coupled to the cavity mode, $V$ is the volume of the active region ($m^3$), $q$ is the electron charge, $N$ is the electron density ($m^{-3}$), $S$ is the photon density inside the cavity ($m^{-3}$), $\varepsilon$ is the gain compression coefficient or the nonlinear gain parameter ($m^3$), $N_0$ is the carrier density at which the semiconductor becomes transparent, $g_0$ is the differential gain coefficient at the central mode $\lambda_0$ ($m^3/s$), and $I$ is the injection current (amp).

A nice way to describe the dynamics of this process below and above threshold is by water flow to a reservoir. The net injection ($I/qV$) is the pipe entrance flow. The leakage current is $(1-\eta)\times(I/qV)$. The water flow into the reservoir is $\eta\times(I/qV)$. Now, if the laser is above threshold, the reservoir is full of liquid. Thus, an increase of input results in increased output, i.e., overflow, but no increase in carrier density, which is the water level. The reservoir may have some drain holes, so besides overflow, there is some other leakage. Those leakages from the reservoir are the recombination mechanisms. Now, assume a case of low threshold. Then the overflow is likely low. In this case, an increase of input would not create an output increase but would affect the carrier density in the active region by filling the reservoir volume. Moreover, the flow through the drain holes in the reservoir will not change below and above threshold; these drain holes represent spontaneous and nonradiative recombination. This reservoir is the population inversion. The same analogy applies for the carriers reservoir. The carriers create spontaneous and stimulated emission as described by Eq. (7.3). The leakage is spontaneous emission and recombination; the overflow is the stimulated emission filling the reservoir. Coming back again to the rate equations, Eq. (7.2) shows that the electron-density rate of change is proportional to the injected rate of electrons, which is the first argument in the equation. The electron density rate of reduction is due to spontaneous emission, which is the second argument in the equation. The rate reduction is due to stimulated emission of the lasing mode, which is the third argument in the equation. In the same way, Eq. (7.3) shows that the photons density rate of change is proportional to fraction $\Gamma$ of the photons stimulated emission of the lasing mode, which is the first argument in the equation. The rate of photons loss due to the photons lifetime $\tau_p$ can be referred to as the average time it spends in the cavity before it is lost by internal absorption or transmission through the facets. This is the second argument in the equation. The rate of photon density increases due to spontaneous emission in the active layer. This is the third argument in the equation. The peak mode gain due to stimulated emission is given by

$$G = v_g g_0 (N - N_0). \tag{7.4}$$

However, due to an important phenomenon of spectral-hole burning (SHB), the optical gain is compressed. Therefore, the net gain in the equation is reduced. SHB is a small, localized optical gain curve reduction at the lasing mode wavelength.[3] This occurs when the stimulated emission rate is very high and a large number of available electrons in the conduction band cannot immediately fill the localized reduction in the number of conduction band states since the intraband scattering process such as electron–electron and electron–phonon have relaxation time on the order of 1 ps.[4] The reduction in gain rate due to SHB is proportional to the stimulated emission rate and is given by Taylor expansion of $G(N, S)$ around the points $N = N_0$ and $S = S_0$.

$$\Delta G = v_g g_0 (N - N_0) S \varepsilon. \tag{7.5}$$

Hence, the subtraction of Eq. (7.5) from (7.4) provides the total net gain, including the compression reduction:

$$G(N, S) = v_g g_0 (N - N_0)(1 - \varepsilon S). \tag{7.6}$$

The rate equations are a large signal model and can be solved numerically. An analytical steady-state solution can be obtained using small-signal linearization approximation as provided by

$$I = I_0 + i_1 \exp(j\omega t), \tag{7.7}$$

$$N = N_0 + n_1 \exp(j\omega t), \tag{7.8}$$

$$S = S_0 + s_1 \exp(j\omega t), \tag{7.9}$$

where $\omega$ is the angular frequency of the modulating signal, $I_0$, $N_0$, and $S_0$ are the steady-state values of the modulating current, electron density, and photon density, respectively, and $i_1$, $n_1$, and $s_1$ are the small signal components. Substituting these phasors in the set of differential equations describing the density rate changes versus time results in the normalized-at-zero-frequency small-signal response of the laser. Note that second-order values were neglected:[5]

$$H(j\omega) = \frac{s(j\omega)/S_0}{i(j\omega)/I_0} = \frac{(1 - \varepsilon S_0)\omega_0^2}{-\omega^2 + j\omega\left[\dfrac{\Gamma \gamma_{sp} I_{TH}}{qVS_0} + \dfrac{1}{\tau_n} + S_0\left(v_g g_0 + \dfrac{\varepsilon}{\tau_p}\right)\right] + \dfrac{\Gamma \gamma_{sp} I_{TH}}{qVS_0 \tau_n} + \dfrac{\gamma_{sp} + \varepsilon S_0}{\tau_n \tau_p} + (1 - \varepsilon S_0)\omega_0^2}. \tag{7.10}$$

This equation shows that the small signal response of the laser resembles that of a second-order low-pass filter (LPF) transfer function with resonance frequency $\omega_0$ called the "relaxation–oscillation of the laser" at which the response peaks, after which it begins to roll off at approximately 40 dB/dec. It has a damping factor $\xi$ and all other parameters of the second-order LPF denominator:

$$-\omega^2 + 2j\xi\omega_0\omega + \omega_0^2. \tag{7.11}$$

Note that $I_{TH}$ in Eq. (7.10) is the threshold current. Solving the denominator of Eq. (7.10) for $\omega$ would result in the "relaxation–oscillation" frequency:

$$\omega_0 = \left[\frac{v_g g_0 S_0}{\tau_p}\right]^{1/2} = \left[\frac{\Gamma v_g g_0}{qV}(I - I_{TH})\right]^{1/2}. \tag{7.12}$$

Assuming the laser operation point is well above the threshold, $\gamma_{sp} = 0$ and the damping coefficients $1/\tau_n$ and $g_0 S_0$ are negligible, then Eq. (7.10) becomes

$$H(j\omega) = \frac{(1 - \varepsilon S_0)\omega_0^2}{(1 - \varepsilon S_0)\omega_0^2 - \omega^2 + j\omega(S_0\varepsilon/\tau_p)}. \tag{7.13}$$

The transfer-function magnitude or spectral power is given by multiplying Eq. (7.13) by the complex conjugate, resulting in

$$|H(j\omega)|^2 = \frac{(1 - \varepsilon S_0)^2\omega_0^4}{[(1 - \varepsilon S_0)\omega_0^2 - \omega^2]^2 + \omega^2[S_0\varepsilon/\tau_p]^2}. \tag{7.14}$$

Solving Eq. (7.14) for half power, assuming that the damping coefficients are much smaller than the resonance frequency (i.e., $1 - \varepsilon S_0 \cong 1$ and $\varepsilon S_0 \cong 0$),[6] results in

$$\omega_{3dB} = \omega_0\sqrt{1 + \sqrt{2}} = 1.55\omega_0. \tag{7.15}$$

Several experiments had indicated that the 3-dB-modulation frequency is proportional to $(I - I_{TH})^{1/2}$.[7] Sometimes it is more common to characterize the laser with respect to the normalized current $(I/I_{TH} - 1)^{1/2}$. In Fig. 7.1,[1,6] several resonance frequencies are measured under four different biasing-level conditions, which essentially define four power levels (1–1 mW, 2–2 mW, 3–2.7 mW, 4–5 mW). The test was taken on a BH (buried heterostructure) semiconductor laser with a

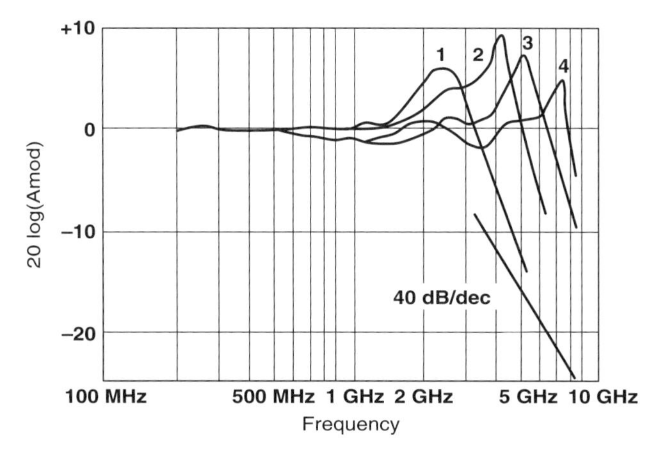


**Figure 7.1** Measured BW characteristics of 120-μm cavity at various bias points.[1,7,64] Emitted power: 1- 1 mW, 2- 2 mW, 3- 2.7 mW, 4- 5 mW (Reprinted with permission from the *Journal of Quantum Electronics* © IEEE, 1985.)

120-$\mu$m cavity. Note that the lowest power peaks earlier and the highest peaks the latest. However, all responses are flat up to 1 GHz. But, as the biasing gets higher, the roll off becomes steeper.[1]

Observing Fig. 7.1, there are several parameters characterizing the small signal frequency response. Overshoot is directly related to the damping factor having the overshoot radian frequency $\omega p$. In time domain, this overshoot is translated into ringing with converging time or decay depending on the damping factor. Observing Eq. (7.12), there are three possible ways to increase the laser BW. The first is to increase the average photon density $S_o$. This can be done by decreasing the width of the optical field distribution in the transverse direction parallel to the junction plane.[24] Creating a laser with a narrow active region and tight optical field confinement would have a wide BW and strong damping factor. The second method for increasing $\omega_o$ is by increasing the gain slope, $g_o$. This can be accomplished by decreasing the temperature[8] and increasing the doping level in the active layer. Quantum-well structures show great promise for having a high $g_o$. The third method for increasing $\omega_o$ is by reducing the photon lifetime, $\tau_p$. This is done by shortening the device cavity length. In addition, a low-threshold current and high-bias current would be also a way to increase the laser BW and essentially increase the laser power.

Small signal approximation is useful to derive an equivalent lumped-element parallel-resister inductor capacitor (RLC) circuit, which models the laser.[22–27] The modeling method is by the small signal approximation of the rate equations, which are two linearly coupled differential equations similar to the voltage and the current equations of the parallel-RLC circuit. In the laser cavity, the electrons and photons exchange energy through absorption and emission, with various loss mechanisms dissipating energy in the cavity. These losses are leakage-Auger-recombination heat and absorption. Similarly, in the RLC circuit, the capacitor and the inductor exchange energy and the resistor dissipates energy out of the circuit. Furthermore, the continuity conditions of the electrical circuit, voltage across the capacitor, and current through the inductor are equivalent to those of the laser cavity; i.e., the changes in the electron and photon numbers inside the cavity are continuous as shown by the rate equations. Hence, the excited electrons $N$ in the laser cavity are represented by a charge across the capacitor, and the photon density $S$ at a given laser mode are represented by magnetic flux linkage of an inductor. When the analysis case refers to a multimode FP laser, the equivalent circuit would have several parallel inductors. Each inductor represents a specific lasing mode. In fact, each inductor of each mode is made of two series inductors. The first inductor represents the mode whereas the second one represents the reduction in coupling coefficient of that mode with the electrons. The coupling of the mode to the electrons is simulated by the series resistor to the mode inductors. Thus, the mode with the strongest coupling to an electron has the lowest values for $L_j$, $L_x$, $R_s$, and $R_{pj}$; the mode power is proportional to $L_j$. Figure 7.2 illustrates equivalent circuits for DFB and FP lasers. This circuit can be cascaded to the laser package.[24] Small-signal analysis for a

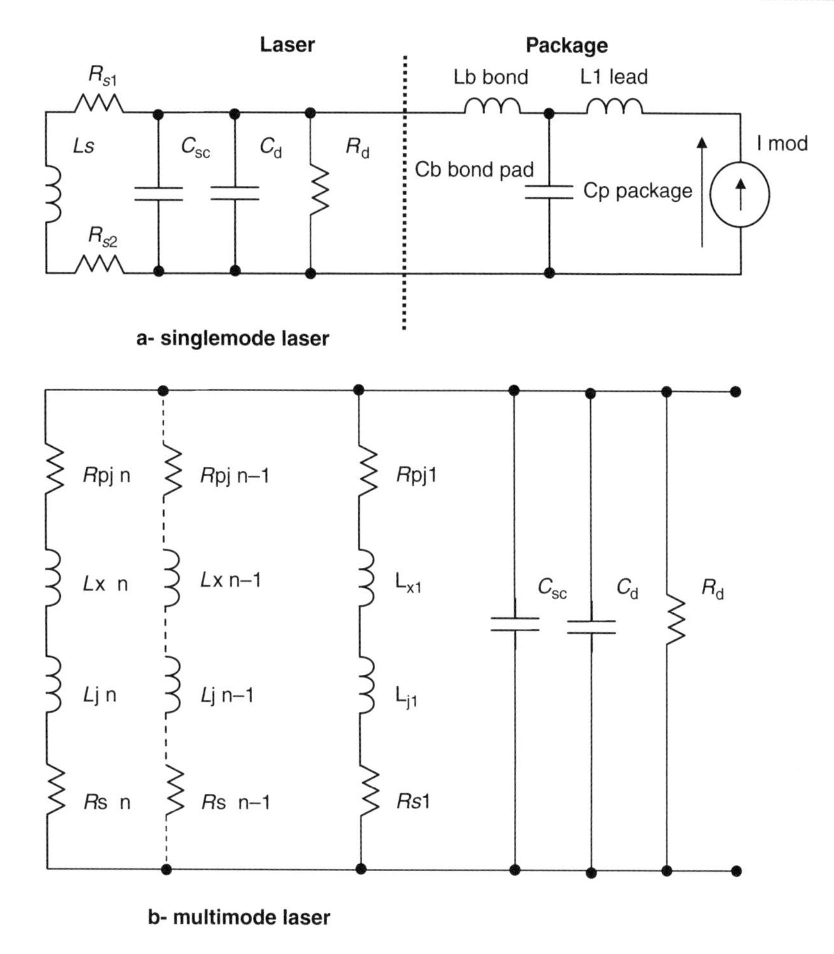


**Figure 7.2** Equivalent circuit modeling for relaxation-oscillation. (a) An SLM laser including cascaded package parasitics. (b) An FP multimode laser showing all *n*-mode-coupled inductors.

ridge laser, shown in Fig. 7.3, with a strong carrier confinement suggests the following elements relations above threshold current:[25,26]

$$R_1 \cong \frac{R_d}{1 + g_0 \tau_n S_0}, \tag{7.16}$$

$$C_d = \frac{\tau_n}{R_d}, \tag{7.17}$$

$$R_d \cong \frac{2kT}{qI_{tA}}, \tag{7.18}$$

$$L_S \cong \frac{R_d \tau_p}{g_0 \tau_n S_0}, \tag{7.19}$$

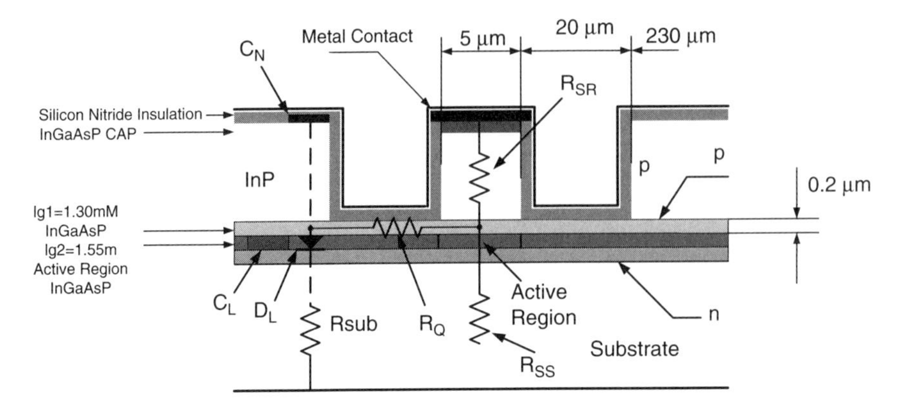


**Figure 7.3** Cross section of a ridge SLM laser with strong carriers confinement.[24] (Reprinted with permission from the *Journal of Lightwave Technology* © IEEE, 1984.)

$$R_{S1} \cong \frac{\varepsilon R_d}{g_0 \tau_n}, \tag{7.20}$$

$$R_{S2} \cong \frac{\gamma_{sp} \Gamma R_d \tau_p I_{tA}}{q V g_0 \tau_n S_0^2}, \tag{7.21}$$

where the threshold current of the active layer is given by

$$I_{tA} \cong \frac{qV\left(\dfrac{1}{\tau_p \Gamma g_0} + N_0\right)}{\tau_n}, \tag{7.22}$$

and the steady-state photon density above threshold is given by

$$S_0 \cong \frac{\Gamma \tau_p}{qV}(I_0 - I_{th}), \tag{7.23}$$

where $I_o$ is the dc component of the total drive current, that is, $I_A + I_L$, $I_A$ is the active current, and $I_L$ is the leakage current; thus, $I_{th} \cong I_{tA} + I_L$. The LC model relaxation–oscillation frequency is given by

$$f_r = \frac{1}{2\pi\sqrt{L_S(C_d + C_{SC})}} \cong \frac{1}{2\pi\sqrt{L_S C_d}}, \tag{7.24}$$

where $C_{SC}$ is the space-charge capacitance that is smaller compared to $C_d$, nanofarads compared to picofarads. Substituting Eqs. (7.17) and (7.19) in Eq. (7.24) and then substituting $S_0$ from Eq. (7.23) results in

$$f_r = \frac{1}{2\pi}\sqrt{\frac{g_0 S_0}{\tau_P}} = \frac{1}{2\pi}\sqrt{\frac{\Gamma g_0}{qV}(I_0 - I_{th})}. \tag{7.25}$$

The damping time constant $\tau_1$ associated with the relaxation–oscillation is approximately given by

$$\tau_1 = \frac{L_S}{R_{S1} + R_{S2}}, \tag{7.26}$$

where

$$\tau_1 = \frac{\tau_p}{\varepsilon S_0}. \tag{7.27}$$

Based on the above relations, the small-signal impedance of the RLC equivalent circuit can be derived:[26]

$$Z(j\omega) = \frac{(R_{S1} + R_{S2}) + j\omega L_S}{-\omega^2(C_d + C_{SC})L_S + j\omega[(C_d + C_{SC})(R_{S1} + R_{S2}) + L_S/R_1] + (R_{S1} + R_{S2} + R_1)/R_1}. \tag{7.28}$$

This equation can be noted with respect to the quality factor of the inductor and the capacitor as well as the damping factor of the circuit,[28,29] by using the resonance relation in Eq. (7.24) as radian frequency $\omega_0$.

Now for the sake of notation as second-order transfer function, assume the two following cases. At first, assume the circuit in Fig. 7.2 has a zero-ohm-series-loss resistor for the inductor $L_s$, e.g., $R_{s1}$ and $R_{s2}$ are zero. This case is equal to a parallel RLC circuit with a loss resistor of $R_1$. The second case, $R_1$, goes to infinity; i.e., open and $R_{s1}$ and $R_{s2}$ are not zero. This will be a case of series RLC. Moreover, both circuits have the same resonance radian frequency $\omega_0$. Knowing that, the Q-factor damping rate $\alpha$ and damping factor $\xi$ can be noted for parallel RLCs:

$$\alpha_P = \frac{1}{2(C_d + C_{SC})R_1}, \tag{7.29}$$

$$\xi = \frac{\alpha}{\omega_0}, \tag{7.30}$$

$$Q_p = \frac{R_1}{\omega_0 L_S} = \omega_0 R_1 (C_d + C_{SC}), \tag{7.31}$$

and for a series of RLCs:

$$\alpha_S = \frac{R_{S1} + R_{S2}}{2L_S}, \tag{7.32}$$

$$Q_S = \frac{\omega_0 L_S}{R_{S1} + R_{S2}} = \frac{1}{\omega_0 (C_d + C_{SC})(R_{S1} + R_{S2})}. \tag{7.33}$$

Knowing that $1/Q = 2\xi$, and using the above relations, Eq. (7.29) can be modified to

$$Z(j\omega) = \frac{R_1}{Q_P Q_S} \frac{1 + j\frac{\omega}{\omega_0} Q_S}{-\left(\frac{\omega}{\omega_0}\right)^2 + j\frac{\omega}{\omega_0} 2(\xi_P + \xi_S) + 4\xi_P \xi_S + 1}. \tag{7.34}$$

From this presentation and Fig. 7.2, it can be understood that the damping factor of the relaxation–oscillation depends on damping due to spontaneous emission coupled to the lasing mode $\xi_S$ and damping due to the spontaneous and stimulated recombination terms in the rate equation $\xi_P$. This means that the quality factor of the modulation impedance depends on the charge storage efficiency and the coupling mode efficiency. Moreover, the equivalent quality factor is made $\xi_S$ and $\xi_P$ to mechanisms, which also defines the equivalent damping factor. The relaxation–oscillation BW describes the ability of the carriers reservoir to fluctuate and release charges for emission synchronously with the pumping current. Hence, as a conclusion from this analogy, the larger the cavity, the slower it is. In addition, the coupling of this reservoir to the active mode is important too. Thus, the structure of a laser should minimize losses and transport time.

That is why a tight confinement structure and QW laser would have a better response. Observing FP lasers in Fig. 7.2, each mode has a different coupling. Thus, on an average, the BW of such a laser would be determined by the dominant mode, which would be lower compared to a single mode.

The structure of the ridge waveguide laser chip is depicted in Figs. 7.3 and 7.6. The dominant electrical parasitics are the resistance in series with the active region and the shunt capacitance between the metal contacts. The total series resistance $R_s$ is the sum of the ridge and resistance $R_{SR}$, including the contact resistance at the interface between the metal and the AA layer and the resistance $R_{SS}$ of the substrate under the active region. $R_{SR}$ is the larger of these two resistances and dominates $R_s$. For a ridge dimension of $2 \times 5 \times 250\ \mu m^3$, a hole mobility of $70\ cm^2\ V^{-1} s^{-1}$, doping density of $3 \times 10^{17} cm^{-3}$, and a dc value of $R_{SR}$ is calculated to be $5\ \Omega$, assuming zero-contact resistance.[24]

The shunt parasitic capacitance $C_s$ arises from capacitance $C_N$, associated with the silicon nitride insulator layer in series with the space-charge capacitance $C_L$ at the heterojunction diode $D_L$ at the quaternary P–N interface. This interface is distributed across the entire chip, but most of it is electrically isolated from the active region by thin, $\sim$0.2 μm, P-type quaternary cladding layer in the channels that extend 20 μm on each side of the ridge. When the laser is forward biased, a small amount of current is injected into $D_L$ via the leakage resistance $R_Q$ through the cladding layer. Since $R_Q$ is large, about 1–10 kΩ, $D_L$ remains essentially unbiased, even when the laser is driven above threshold. The substrate resistance $R_{SUB}$ is in series with $C_L$.[24]

It is important to note that the metal-insulator-semiconductor (MIS) capacitor $C_N$ and the space-charge capacitance $C_L$ are distributed across the entire 500-μm width of the laser chip. Therefore, they can have quite a large value. Similarly,

**Table 7.1** Typical values for a CATV DFB transmit laser

| Parameter | Typical value |
|---|---|
| Active layer volume $V$ | $W \approx 1-2$ μm; $H \sim 0.2$ μm; $L \sim 300$ μm |
| Electron life time $\tau_n$ | 1–3 ns |
| Photon life time $\tau_p$ | 1–2 ps |
| Optical mode confinement factor $\Gamma$ | $\Gamma \approx 0.3-0.5$ |
| Differential gain coefficient $g_0$ | $g_0 \approx 1-3 \times 10^{-12}$ m$^3$/s |
| Gain compression coefficient $\varepsilon$ | $\varepsilon \approx 1-5 \times 10^{-23}$ m$^3$ |
| Electron density at transparency $N_0$ | $N_0 \approx 1-1.5 \times 10^{24}$ m$^{-3}$ |
| Fraction of spontaneous emission coupled to the cavity mode $\gamma_{sp}$ | $\gamma_{sp} \approx 10^{-4}$ |

$R_{SUB}$ is small since the substrate has a doping density in excess of $10^{18}$ cm$^{-3}$. Typical values are $C_N \approx 8-15$ pF, depending on the silicon nitride stoichiometry and thickness of ~0.2 μm, $C_N \approx 40-400$ pF, depending on the doping densities of the quaternary layers and $R_{SUB} \sim 1\ \Omega$.[24]

The laser chip is normally mounted P-side down on a metal stud and N-side is connected to a 1.3 × 0.8-mm contact pad on a 0.5-mm thick alumina standoff via a gold bond wire. The bond wire is approximately 1-mm long and has a diameter of 50 μm. The dominant electrical parasitics associated with the stud are the bond wire inductance $L_P$ and resistance $R_P$, and the standoff shunt capacitance to ground $C_P$.[24] Typical DFB laser parameters are given in Table 7.1.

Laser behavior is extremely nonlinear near its resonance frequency. For subcarrier multiplexing (SCM) lightwave transmission, the RF carrier frequencies should be sufficiently far away from the relaxation–oscillation resonance frequency. Following the small signal model, it is clear that this system has a memory, meaning carrier charges. Now, if the signal is faster than the time constant of the laser, i.e., it cannot track the pump current changes, the next state will discharge the previous state, thus distorting the signal. An intuitive explanation for the jitter distortions at the relaxation–oscillation point is due to abrupt phase change versus frequency, which is not constant. Thus, its derivative—that is, the group delay—changes versus frequency-creating dispersion. That effect is observed well in digital transmission in higher values of eye jitter. Several results have shown that strained-layer multiple-quantum well (SL-MQW) DFB lasers have 30% and 90% higher resonance frequency compared with MQW DFB lasers and conventional DFB lasers, respectively. Moreover, at that point, the laser has higher intensity noise as reviewed by Chapter 10.

## 7.2 Large Signal Deviation from a Basic Model

Small-signal analysis of a laser as was reviewed in Sec. 7.1 is not applicable when dealing with full a CATV channel plan. The standard optical modulation index (OMI) per channel calls for a modulation index range of 3–4% per channel.

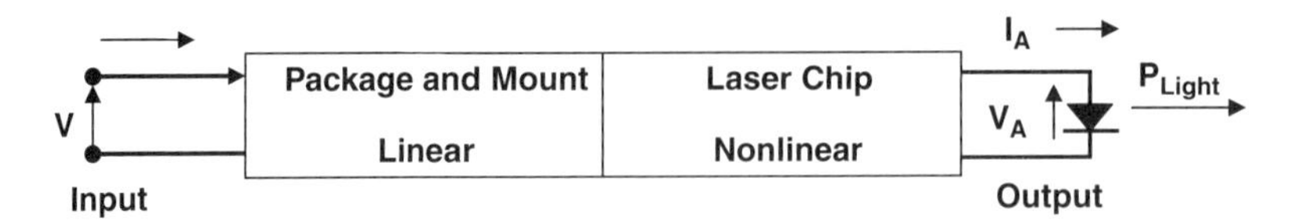


**Figure 7.4** A two-port model of a semiconductor laser taking into account package and device parasitics. The package is linear and the device includes linear and nonlinear components to model leakage current that creates distortions. Used for harmonic-balance analysis.

Therefore, a load of 79 channels, as was explained in Sec. 6.9, would result in total OMI load in the range of 27–35.5%. Full 110-channel plan would increase this result even further to 28–37%. In case of digital modulation, the modulation index is 100%. This, of course, requires a large-signal model.[10–13]

The traditional way to describe device modeling, taking into account all the parameters affecting its performance, is to explain it as a two-port system built of three sections as described by Fig. 7.4:[10,13]

1. Package and mount accesses parasitics that represents a linear network. These are the package feed-through capacitance and bond-wire inductance and resistance. These components can be omitted if the laser-drive circuitry is integrated with the laser.

2. The device nonlinear equivalent network and linear access parasitics of the chip. The linear parasitics are the chip-contact resistance and capacitance and nonlinear values associated with the semiconductor surrounding the active region.

3. The intrinsic laser is the active region and the cavity. This is the large-signal equivalent model.

In case of large signal, the response Eqs. (7.2) and (7.3) are solved by numerical methods. However, an analytic approach is available by creating an equivalent large-signal circuit for the laser.[12–14] The advantage of this method is the usage of conventional-circuit-analysis methods for solving it. The nonlinear components in the circuit are represented by nonlinear relationships. Figure 7.5 shows an equivalent model of the chip package access parasitics and access parasitics of the device itself. $C_P$, $L_P$, and $R_P$ are the feed-through capacitance and the bond-wire impedance and resistance. $C_S$ and $R_S$ are the intrinsic laser parasitic capacitance and resistance, respectively. The current source $I_L$ is the dc-leakage current around the active region, generating severe distortions on the second order.[15] The output impedance of the exciting source $I_S$ is $R_{IN}$. The intrinsic device is the nonlinear model. Figure 7.6 shows the nonlinear intrinsic-laser model. Therefore, the overall response is the cascade of the two models. The intrinsic-laser model is described by the following equations for each component in the schematics. The source signal current driving the intrinsic laser is marked as $I_A$.

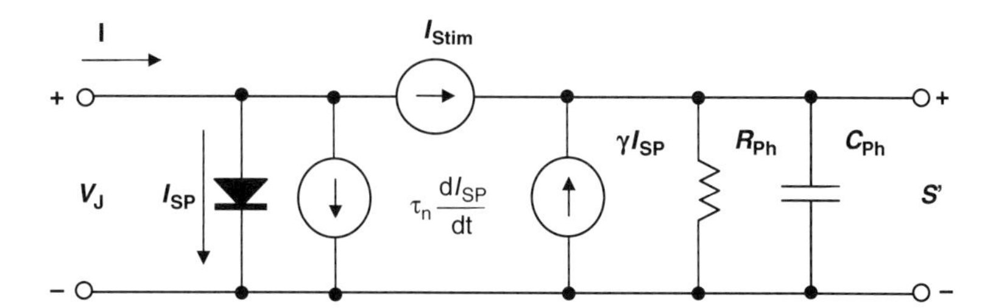


**Figure 7.5** Large-signal modeling of intrinsic laser showing the photons reservoir $C_{PH}$, Auger recombination loss $R_{PH}$, spontaneous current $I_{sp}$, simulation current, and coupling of spontaneous current to the lasing mode.[24] (Reprinted with permission from the *Journal of Lightwave Technology* © IEEE, 1984.)

This is the net current for the lasing process in the intrinsic laser. The transformation of rate Eqs. (7.2) and (7.3) for the modeling is defined as below. The model equivalent spontaneous recombination current is given by

$$I_{SPN} = \frac{qVN}{\tau_n} = I_s \exp(qV_A/\theta kT), \tag{7.35}$$

where $\theta \approx 2$ and $I_s$ is the saturation current. The equivalent stimulated emission current is defined by

$$I_{stim} = qVv_g g_0(N - N_0)(1 - \varepsilon S). \tag{7.36}$$

The photon loss and photon storage are modeled by equivalent resistor $R_{PH}$ and capacitor $C_{PH}$, which are defined by the following:

$$R_{PH} = \frac{\tau_p}{qV}, \tag{7.37}$$

$$C_{PH} = qV. \tag{7.38}$$

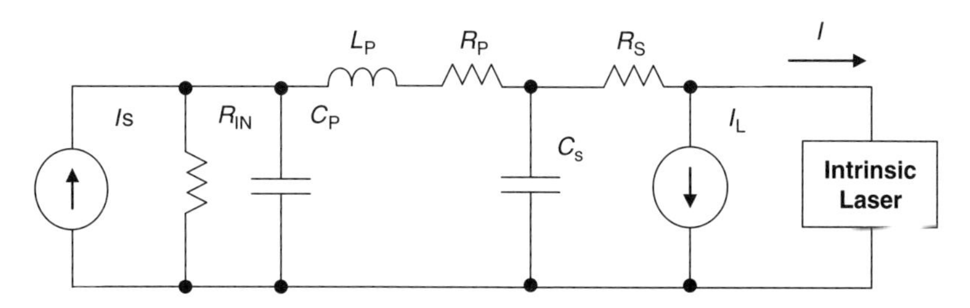


**Figure 7.6** Large signal modeling of intrinsic laser, including package effects.[10] (Reprinted with permission from *Springer Verlag* © Springer Verlag, 1990.)

Using the above transformation relations, the rate Eqs. (7.2) and (7.3) become

$$\tau_n \frac{dI_{SPN}}{dt} = I - I_{SPN} - I_{stim}, \tag{7.39}$$

$$C_{PH}\frac{dS}{dt} = I_{stim} - \frac{S}{R_{PH}} + \gamma_{SP} I_{SPN}. \tag{7.40}$$

## 7.3 Amplitude-Phase Coupling (Chirp)

Amplitude-phase coupling is a result of the direct modulation of a laser. The reason for that phenomenon is the electron density $N$ in the active region by the modulation process. As a consequence, the refractive index of the medium is changed. The refractive index change affects the lasing wavelength and its complex number. The imaginary part of the refraction index $n$ is related to the spatial exponential gain (or loss) constant of the laser medium. The real part of the index $n$ is related to the time dependency of the medium gain. Letting the refractive index $n$ to be presented as a complex number[16]

$$n = n_1 + jn_2. \tag{7.41}$$

Then, it can be noted that a change in the medium gain is given by[16]

$$\Delta g = -4\pi \Delta n_2/\lambda. \tag{7.42}$$

According to Eq. (7.29), the change in the optical frequency is[16]

$$\frac{\Delta\nu}{\nu} = \frac{\Delta n_1}{n_1}. \tag{7.43}$$

Now it is possible to define the $\alpha$ parameter of the laser, which is commonly called the line-width-enhancement factor, by Eq. (7.44),[1,16–18] and it is defined as the incremental ratio between the real part and the imaginary part of the refractive index[1,19] or the ratio between the derivative of the real part of $n$ as a function of the change in the electrons density $N$ and the gain change as a function of a change in the electrons density $N$:

$$\alpha = \frac{\Delta n_1}{\Delta n_2} = \frac{\Delta n_1}{\frac{\Delta g \lambda}{-4\pi}} = -\frac{4\pi}{\lambda}\left[\frac{\frac{dn_1}{dN}}{\frac{dg}{dN}}\right]. \tag{7.44}$$

Using the photon rate Eq. (7.3) and using the Taylor expansion around a dc operating point, the frequency chirp $\Delta\nu$ can be expressed as a function of the variations in the output power under a direct-current-modulation laser known as DML:[20]

$$\Delta\nu(t) = -\left(\frac{\alpha}{4\pi}\right)\left\{\frac{d}{dt}\ln[P(t)] + kP(t)\right\}, \tag{7.45}$$

where $P(t)$ is the time variation of the optical power and $k$ is a parameter given by

$$k = \frac{2\Gamma\varepsilon}{\eta_d h\nu V}, \qquad (7.46)$$

where $V$ is the volume of the laser's active region.

For the harmonic RF modulation, the RF component of Eq. (7.45) becomes[16]

$$\frac{\Delta\nu}{\Delta P} = -\frac{j\alpha}{2}\frac{f}{P} + k. \qquad (7.47)$$

It can be observed from Eq. (7.47) that at high-modulation frequencies, the chirp increases with the $\alpha$ parameter and becomes worse linearly with frequency. In addition, it gets larger at large optical variations or during relaxation–oscillation. On the other hand, the strong damping factor of the relaxation–oscillation would reduce the extent of frequency chirp. The second term in Eqs. (7.45) and (7.47) dominates at low frequencies and is called "adiabatic chirp" and arises from the damping factor of the relaxation–oscillation. It can be minimized by increasing the volume of the active region in the laser structure. The result of it is a wavelength shift between the high and the low power points in the optical waveform. On the other hand, excessive damping would reduce the modulation BW. Essentially, the outcome of it is a compromise design between the damping factor of the relaxation–oscillation and the wavelength transient chirp. A typical chirp-width value for commercial DFB lasers is about 0.4–0.6 nm at 1–2 Gbit/s modulation index. This chirp allows transmission of data at 2.5 Gbit/s (OC48) up to distances of 100 km with SMF at the 1550-nm wavelength. But, the transmission distance is limited to only a few kilometers at 10 Gbit/s because of the chromatic dispersion in single-mode fibers. The chromatic dispersion results from the difference in the group velocities at different wavelengths of the traveling optical light. These wavelength differences within the light pulse are the result of the frequency chirp of DML modulation method for a laser. The result of this in digital data transport is excessive eye jitter. In the case of CATV data transport, the chirp and the chromatic dispersion in a single-mode fiber result in unacceptable distortion levels.

The chirp phenomenon described above creates a residual FM noise due to amplitude modulation (AM). This term is defined as AM to FM and results in line-width broadening as a function of OMI. The electric field of such a modulated optical carrier is described by

$$E = E_0\sqrt{1 + \mathrm{OMI}\cos(\omega_m t)}\exp\{j[\omega_c t + \beta\sin(\omega_m t)]\}, \qquad (7.48)$$

where the index $m$ is the modulating carrier, and $c$ is the optical carrier. $\beta$ stands for the FM-modulation index given by[30]

$$\beta = \left(2\pi\frac{A_m}{\omega_m}\right)\Delta\nu. \qquad (7.49)$$

While $A_m$ represents the peak modulation current and can be extracted from the OMI definition, $\Delta v$ is the peak frequency deviation:

$$I_{m-max} = \frac{P_{opt}OMI}{\eta} = OMI\frac{E_0^2}{2\eta Z_0}, \tag{7.50}$$

where $\eta$ is the laser-slope efficiency, and $Z_0$ is the medium characteristic impedance. Now, for weak-varying amplitude or small OMI, the amplitude part of Eq. (7.48) can be omitted and the exponent part of it can be expanded in the Bessel series.[31] For low OMI, $\beta < 1$, and according to Carson under this condition, the FM BW is equal to $2(\beta + 1)\omega_m/2\pi \cong \omega_m/\pi$. This means that there are two undesirable FM sidebands. Moreover, the higher the modulation index the more the Bessel sidebands. Therefore, Eq. (7.48) becomes[32]

$$\varphi_{FM} = A\sum_{n=-\infty}^{\infty} J_n(\beta)\exp[j(\omega_c + n\omega_m)t], \tag{7.51}$$

or

$$\varphi_{FM} = A\Bigg(\sum_{n=1}^{\infty} J_n(\beta)[\cos(\omega_c - n\omega_m)t + (-1)^n\cos(\omega_c + n\omega_m)t](-1)^n. \\ + J_0(\beta)\cos(\omega_c t)\Bigg), \tag{7.52}$$

where the Bessel function $J_n$ is given by

$$J_n(\beta) = \sum_{k=1}^{\infty}\frac{(-1)^k(\beta/2)^{(n+2k)}}{k!(n+k)!}, \tag{7.53}$$

and $A$ is approximated for the small OMI to be per Taylor series

$$A = E_0\sqrt{1 + OMI\cos(\omega_m t)} \approx E_0\left(1 + \frac{1}{2}OMI\cos(\omega_m t)\right) \\ \approx E_0[1 + OMI\cos(\omega_m t)]. \tag{7.54}$$

Substituting Eq. (7.54) into Eq. (7.52) and using trigonometric identities, the AM-modulated laser chirp results in a residual narrow-band FM with the following amplitudes:

carrier:

$$E_0 J_0(\beta), \tag{7.55}$$

upper sideband:

$$E_0\left\{J_1(\beta) + \frac{OMI}{2}[J_2(\beta) + J_0(\beta)]\right\}, \tag{7.56}$$

and lower sideband:

$$E_0\left\{-J_1(\beta)+\frac{\text{OMI}}{2}[J_2(\beta)+J_0(\beta)]\right\}. \tag{7.57}$$

Using an optical-spectrum analyzer with fine-resolution BW, the difference between the two sideband magnitudes normalized to the carrier would be $\Delta dB = 20\log[2J_1(\beta)/J_0(\beta)]$. From Carson's rule for FM, the formula for FM BW and the two sidebands initial value for β can be found. Placing this initial value into the above expression, β can be figured in an iterative way. Then, from the linear sum of the two sidebands and normalizing it to the carrier $\Delta\text{dB} = 20\log\{\text{OMI}[J_2(\beta)+J_0(\beta)]/J_0(\beta)\} = 20\log\{\text{OMI}[J_2(\beta)/J_0(\beta)]+1\}$, the modulation index is obtained. It is clear too from Eqs. (7.56) and (7.57) that the sideband amplitudes are not equal. Moreover, in digital schemes, where the modulation index is higher, or in the case of full CATV load, where the OMI is about 35%, there are many FM sidebands as well as lasers that have wider line broadening. These sidebands propagate at different velocities through the SMF and result in eye jitter or intermodulation (IMD) for analog channels. Further discussion will be provided in Sec. 7.4.3.

## 7.4 Laser Distortions

Semiconductor diode lasers have several distortion mechanisms[33] as was reviewed above:

- NLD due to relaxation–oscillation phenomena is referred to as dynamic nonlinearity. This mechanism governs modulation frequencies higher than 1–2 GHz, and is related to local multipoint-distribution system (LMDS) and multichannel multipoint distribution system (MMDS) applications.
- Distortions are related to the bias point on the L–I curve called static distortions.
- A combination of static and dynamic distortions are due to a high-modulation index.

The last two mechanisms are related to low frequencies that are related to CATV and direct broadcast satellite (DBS) bands.

In HFC FTTx analog transport, the channel loading reaches a high OMI of almost 35% to 40%, which is equal to 110 channels of loading plan. These limitations led advanced designs of predistortion and for analog transport and EM for both analog and digital transport fidelity. This section focuses on all three mechanisms.

### 7.4.1 Dynamic nonlinearity modes

Assume a laser is biased at a dc point, as shown in Fig. 6.27, and fed by two small RF current signals around the quiescent point. Assume that-both tones have the same amplitude $i$. Hence, the pumping current is given by

$$I(t) = I_0 + I_{\text{ac1}} + \Delta I_{\text{ac2}} = I_0 + i[\exp(\omega_1 t) + \exp(\omega_2 t)], \tag{7.58}$$

assuming a power of series expansions at the quiescent point for the photons density $S$ and carrier density $N$ are justified since these values are governed by the modulation current as denoted by the rate Eqs. (7.2) and (7.3). Then $S(t)$ and $N(t)$ it is given by

$$\begin{cases} S(t) = S_0 + \Delta S^1 + \Delta S^2 + \Delta S^3 + \cdots + \Delta S^k \\ N(t) = N_0 + \Delta n^1 + \Delta n^2 + \Delta n^3 + \cdots + \Delta n^k, \end{cases} \tag{7.59}$$

where $\Delta n^k$ and $\Delta S^k$ are the two-tone swings resulting from the modulation current.

Equations (7.58) and (7.59) are substituted in the rate Eqs. (7.2) and (7.3) and solved for $k < 3$; i.e., $k = 2$ is second order only.[34] The photons carrier densities for the second and third order are given by

$$\frac{S_{2\omega_1}}{S_{\omega_1}} = \text{OMI} \times H(j2\omega_1)\frac{\omega_1^2}{\omega_0^2}, \tag{7.60}$$

$$\frac{S_{2\omega_1-\omega_1}}{S_{\omega_1}} = \frac{1}{2}\text{OMI}^2 \times H(j2\omega_1)H(j2\omega_1) \times \sqrt{\left[\left(\frac{\omega_1}{\omega_0}\right)^4 - \frac{1}{2}\left(\frac{\omega_1}{\omega_0}\right)^2\right]^2 + \left\{\left(\frac{\omega_1}{\omega_0}\right)^2\left[\frac{1}{2\omega_0\tau_n} - \left(\frac{\omega_1}{\omega_0}\right)^2 \left(\omega_0\tau_p + \frac{3}{2\omega_0\tau_n} + \frac{3\varepsilon S_{\omega_1}}{2\omega_0\tau_p}\right)\right]^2\right\}}, \tag{7.61}$$

where $\omega_0$ is the relaxation–oscillation frequency and $H(j\omega)$ is the relaxation–oscillation transfer function developed in Eqs. (7.10)–(7.13). Thus, from the above equations, the RF signal is related to the their square power. This is because the optical flux generates a photocurrent, which generates RF power referred to as its square power. Hence, the second order is proportional to $\text{OMI}^2$ and the third order to OMI.[4] It is clear too that second and third orders become low when the operating frequency is far below the relaxation–oscillation frequency $\omega_0$. Several interesting parameters can be observed from these equations.

- At low-gain compression $\varepsilon$ values, third-order distortions increase at the region of $\omega_0$ and above.

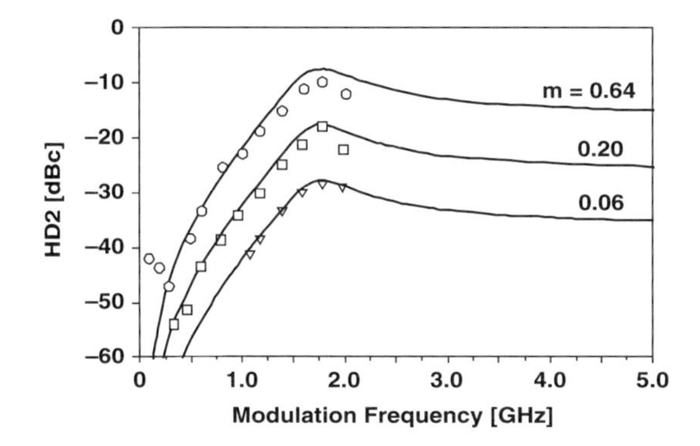


**Figure 7.7** Second-order harmonic distortion vs. modulation frequency, while OMI is a parameter. Note that it peaks at 1.7 GHz, which is equal to half of the relaxation–oscillation frequency per Eq. (7.60) prediction. This calculation shows good agreement to test results.[34] (Reprinted with permission from the *Journal of Lightwave Technology* © IEEE, 1991.)

- Second-order and third-order IMD rolls off at a rate of 40 dB/dec or $40\log(f_1/f_0)$ when operating below $\omega_0$ and will increase at the same rate in the opposite case.

The difference between the theory and the measured results at low-modulation frequencies, $f < 250$ MHz and OMI = 64% in Fig. 7.7 is due to static nonlinearities of the laser, e.g., current leakage, nonlinear gain resulting from spatial-hole burning (SBH), which are not included in this analysis. The deviation of theory results in Fig. 7.8 at high OMI is due to clipping,[35] which was not included in

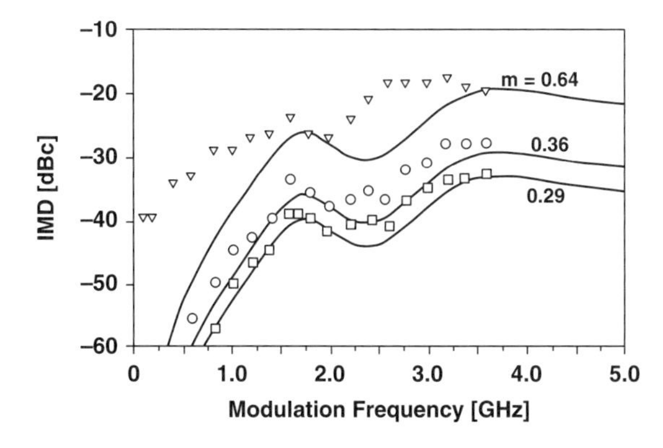


**Figure 7.8** Third-order IMD distortion vs. modulation frequency, while OMI is a parameter. Note that it peaks at 1.7 GHz, which is equal to half of the relaxation–oscillation frequency per Eq. (7.61) prediction. This calculation shows good agreement to test results.[34] (Reprinted with permission from the *Journal of Lightwave Technology* © IEEE, 1991.)

this model. Similar results were reported using the Volterra series.[36] Analyzing these results for digital transport can be as follows. The digital transport is a high-modulation index of the square wave. This square-wave Fourier transform is sincX with the spectral line spaced per the pseudo-random-beat sequence (PRBS) or data pattern. Thus, the resulting distortions add undesirable spectral lines, which would add to eye jitter.

### 7.4.2 Static nonlinearity modes and spatial hole-burning effects

In order to satisfy the stringent CATV distortion requirements, the light power versus current input, L–I curve must be highly linear for frequencies of interest. Further, if the laser L–I curve are linear above the threshold current, signal distortions occur because the laser intensity is occasionally cut off or clipped by large negative excursions of the broadband modulation current. Gorfinkel and Luryi[37] reported about the new L–I model. The main point in this model results from carrier-heating effects, namely the free-carrier absorption of the coherent radiation in the cavity, and the power flux into the active layer associated with the input current. This report claims that the composite second order (CSO) could not be better than −62 dBc in the case of an 80-channel load between 60 and 540 MHz with 3% OMI/Ch. This limit is a concern for SCM. Using rate Eqs. (7.2) and (7.3), the slope efficiency is given by Refs. [37] and [38]. Using the stationary case, the changes over time in Eqs. (7.2) and (7.3) equal zero, e.g., $dN/dt = 0$. This way the bar-marked variables at no modulation are given by[37]

$$\overline{J} = \overline{g}\overline{S} + \frac{\overline{N}}{\tau_n}, \tag{7.62}$$

where $J = I/qV$, $\overline{g} = v_s g_0(N - N_0)$, and $\Delta\overline{g} = v_s g_0(N - N_0)S\varepsilon$. The gain notation $\overline{g} - \Delta\overline{g}$ represents gain compression due to absorption of free carriers[37] and is modified to

$$\overline{g} = g - \alpha N. \tag{7.63}$$

Thus, the second-rate equation at the stationary mode becomes

$$0 = \Gamma S(\overline{g} - \alpha n) - \frac{\overline{S}}{\tau_p}, \tag{7.64}$$

which yields the gain expression of

$$\overline{g} = \alpha\overline{N} + \frac{1}{\Gamma\tau_p}. \tag{7.65}$$

One more, energy balance equation is given by the carrier quantum-temperature equation:

$$\frac{3}{2}\frac{\mathrm{d}T_e}{\mathrm{d}t} = \frac{\alpha S\hbar\nu_s}{2} + \frac{JE_k}{2N} - \frac{3}{2}\frac{T_e - T}{\tau_e}, \tag{7.66}$$

Thus, at the stationary mode, $\overline{T_e}$ is given by

$$\overline{T_e} = \frac{\alpha\overline{S}\hbar\nu_s\tau_e}{3} + \frac{JE_k\tau_e}{3\overline{N}} + T, \tag{7.67}$$

where $\hbar = h/2\pi$, $\tau_e$ is the energy relaxation time, $\alpha$ is the free-carrier-absorption factor, and $\alpha > 0$. $T_e$ is the carrier temperature in energy units and $E_k$ is the kinetic energy per carrier injected into the active region. This energy depends on the laser structure, i.e., the fraction of carriers getting in the active region. Generally, $E_k \leq E_g$, where $E_g$ is the band gap energy between cladding and active layers. The gain function $g$ is in the form of

$$g(T_e, N, \nu_s) = g_{\max}(f_e + f_h - 1), \tag{7.68}$$

where $f_e, f_h$ are Fermi functions of electrons and holes. This gain function can be approximated and at the threshold state where $\overline{S} = 0\overline{T}_{e-th}$, $N_{th}$ and $g_{th}$ are found from Eqs. (7.62)–(7.67). The gain function in Eq. 7.68 can be presented as a complete differential approximation according to bidimensional Taylor series.[37] Using these relations brings the slope-efficiency expression to zero modulation by

$$E_k\overline{J}(1 - \tau_J/\tau_n) = \overline{S}g_{th}(1 - \tau_S/\tau_n) + \overline{S}^2\alpha g_{th}\tau_S + \overline{S}E_k\overline{J}\alpha\tau_J, \tag{7.69}$$

$$\frac{\mathrm{d}\overline{S}}{\mathrm{d}\overline{J}} = \frac{1}{g_{th}} \times \frac{1 - \tau_J/\tau_n - \tau_J\alpha\overline{S}}{1 + \tau_S/\tau_n + 2\tau_S\alpha\overline{S} + \tau_J(\alpha/g_{th})E_k\overline{J}}. \tag{7.70}$$

It can be seen that in the absence of free-carrier absorption $\alpha = 0$ or $\tau_s = 0$, the light current characteristic is linear and the slope efficiency is constant too. However, absorption affects the slope efficiency, making it a function of the operating point rather than a slope of a linear function. Moreover, as an increase of $\overline{J}$ heats the carriers, their concentration rises to compensate for the effect of $\overline{T}_e$ on the steady state. It can be observed from Eq. (7.65) that $g$ is constant for $\alpha = 0$.

In addition to the above L–I nonlinearities, it was shown that FP lasers have better CSO performance compared to DFB lasers. The reason for that is spatial hole burning (SHB). This effect causes nonuniform distribution of light intensity along the laser's active-region layer axis. This effect results from the following:

- the DFB normalized coupling coefficient kL in the laser cavity,[40,41,65]
- the facet reflectivity,[39,42]
- carriers heating.[37,38]

Assume a constant current injection along the laser axis and assume it is an index-guided laser. Thus, the laser exhibits good carrier confinement. Now because of nonoptimized kL, the light power at a certain point becomes higher than in other point. This of course results in higher stimulated recombination, hence the reduction in the net carrier density near that location. Thus, there is a carrier-density change along the axis. This supports the arguments of Eqs. (7.2), (7.4), (7.62), and (7.65), while the compression factor $\varepsilon = 0$ and stationary mode where $dN/dt = 0$ are valid. Thus, from Eq. (7.2), the carrier density along the active layer becomes

$$N(z) = N_c - N_{st}(z) = \frac{\tau_n I}{qV} - \tau_n g(z)S(z). \tag{7.71}$$

The first variable on the right-hand side (RHS) is the fixed number of equally distributed injected carriers. This current is a uniform injection along the laser anode electrode (see Figs. 6.4 and 6.5). The other RHS term shows spatially dependent stimulated emission. Thus, the gain varies along the $z$-axis as noted by Eq. (7.64). Alternatively, by a different notation $g(z) = g_0(N_Z(z) - N_0)$, as shown in Fig. 7.9,[43] the gain variation is related to the carrier's density distribution. The reduction in $N_Z(z)$ is the cause for gain compression. This is the same phenomena presentation of Eqs. (7.4) and (7.5). Thus, the higher the gain-compression factor $\varepsilon$ of a laser, the higher the second-order distortions.

Several works were published about the kL factor's influence on CSO in a DFB laser.[39–41,43] The conclusions are that for a kL factor equal to 0.5, the light power is higher near the front facet; for a kL factor of 1, it is equally distributed with still some higher intensity near the front facet; and for a kL value of 2, it reversed to higher intensity near the rear facet. These optimum values of kL vary from laser to laser depending on their topology and structure. Flat response for kL provides the best CSO performance as shown in Fig. 7.10. In addition, the bias point, with respect to threshold level, may affect CSO 1 distortion levels as well.[44] The chart in Fig. 7.11 describes a two-tone DSO test result of $f_1 + f_2$ and $f_2 - f_1$ as a parameter. The test tones are equal in OMI to full channel loading. OMI/channel is 25% of $\mathrm{DSO}_{\mathrm{low}}$ and is lower than $\mathrm{DSO}_{\mathrm{high}}$. Thus, the higher the bias point above threshold, the lower the CSOs are (see Fig. 6.27). There is an optimum point where CSOs are the lowest. At this point, the laser bias is twice the threshold. Looking at the analysis of FP laser Eq. (6.8), the threshold gain becomes a function of the carrier's lifetime. Under modulation, the carrier's lifetime and threshold gain become functions of carrier density. When spatial hole burning happens, this may affect CSO performance. Equation (7.59) can be used for the second harmonic-distortion ratio using Eq. (7.8) and both analyses at small-signal modulation.

### 7.4.3 Mode partition

In Sec. 7.1, the dynamics of FP lasers were presented by the rate equations. In the equivalent circuit describing relaxation–oscillation, it shows as multiple inductors

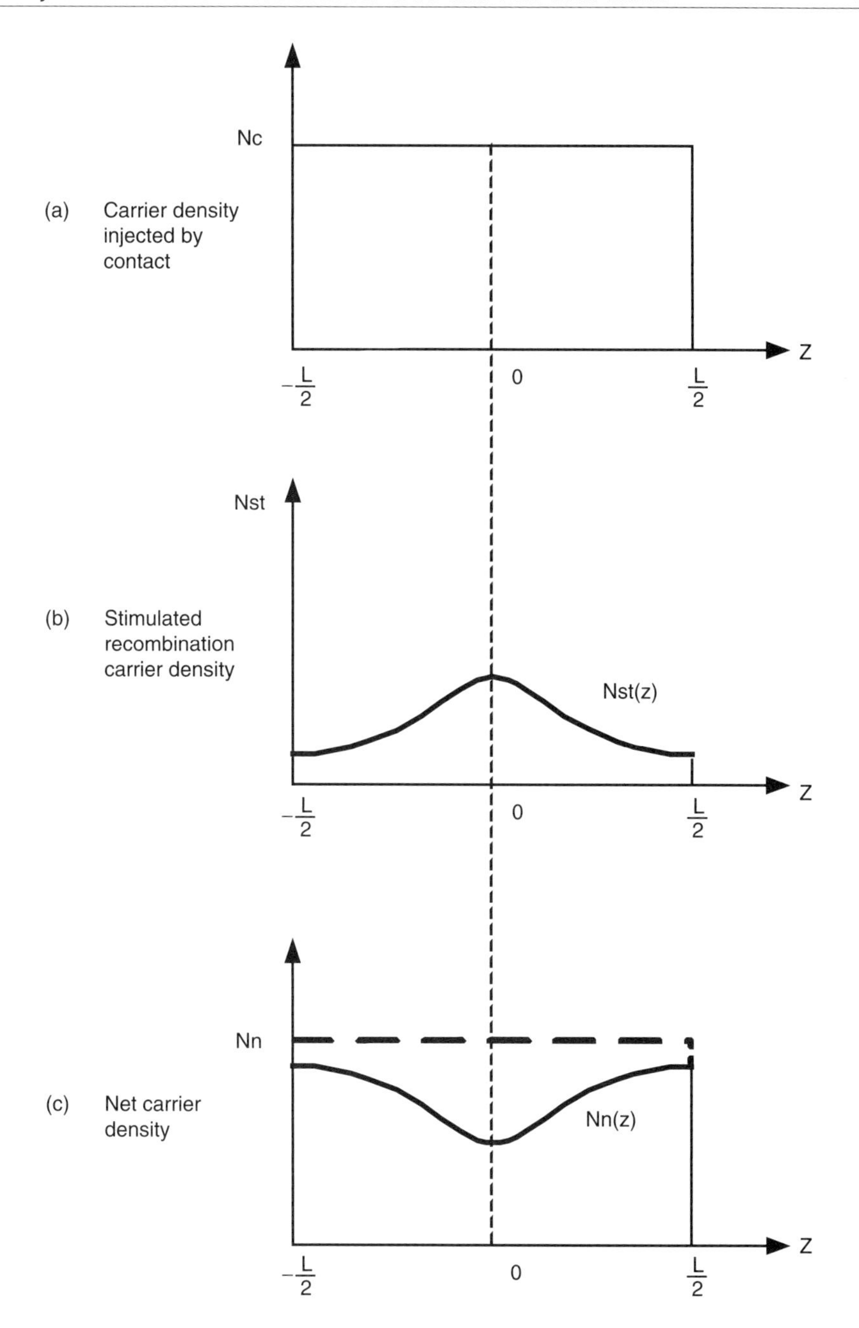


**Figure 7.9** Carrier-density distribution along the active region of a DFB laser.[43] (Reprinted with permission from *IEEE Quantum Electron* © IEEE, 1987.)

with different coupling resistors, which describe the coupling mode. Spectrum-wise, the FP lasing appears as multiple wavelengths spaced $\Delta\lambda$ apart. The dominant mode is the highest in its power and is at the center of the spectrum whereas the other modes are around it. Generally, the lasing of FP lasers

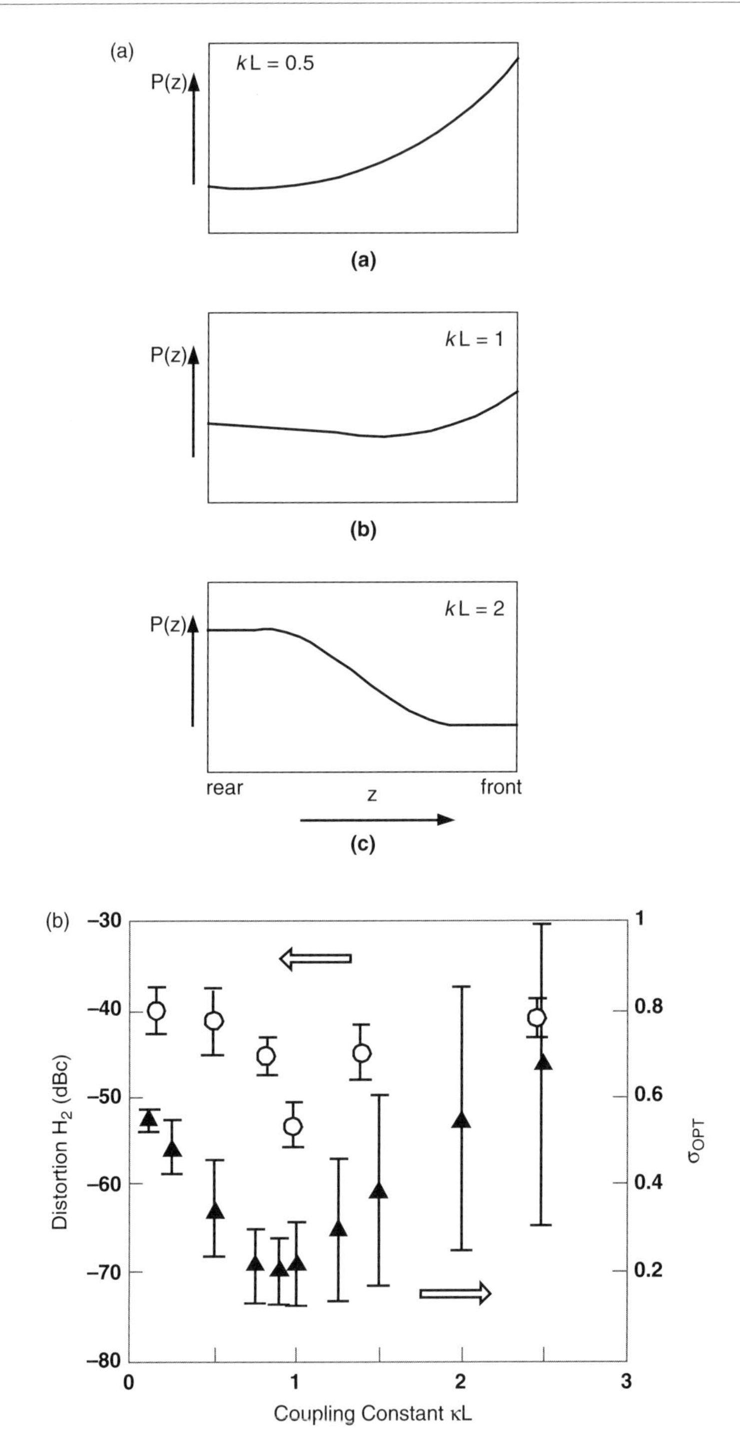


**Figure 7.10** (a) Optical intensity along a DFB laser cavity where kL = 0.5 (*top*) kL = 1 and kL = 2 (*bottom*) (b) CSO performance as function of kL.[40] (Reprinted with permission from *IEEE Selected Areas Communication* © IEEE, 1990.)

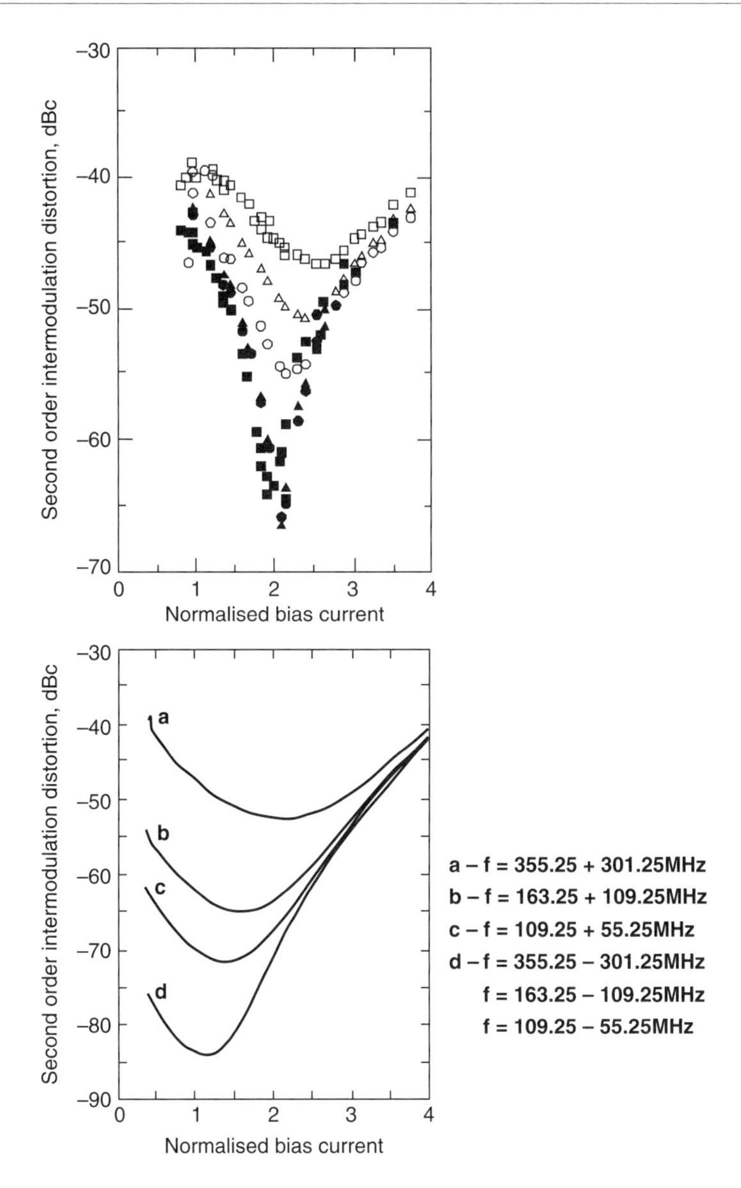


**Figure 7.11** CSO performance vs laser normalized bias point simulated (a) measured (b).[44] (Reprinted with permission from *Electron Letters* © IEEE, 1990.)

appears to be a Gaussian shape of a discrete wavelength. The lasing power measured is the average of all lasing modes. The FP laser's instantaneous emission, however, is one wavelength at a time. Thus, there is a mode hopping mechanism in the FP laser. In other words, there is mode competition between the lasing modes on the mode gain buildup.[16] The highest likelihood mode is the dominant lasing mode. Thus, this mode has the highest power. The process itself is statistical. Now, consider a digital transceiver with a direct modulation FP laser (DML). Hence, within modulation-bit duration, the FP laser would hop between several

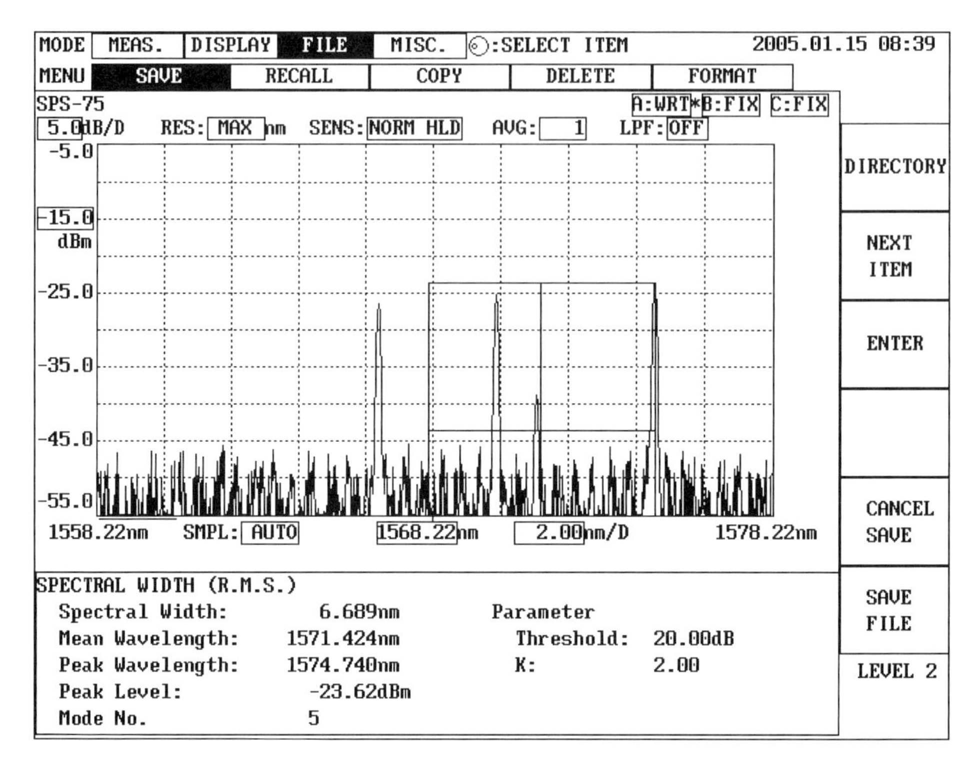


**Figure 7.12** 1550-nm FP-modulated laser-spectrum dispersion after 10 km of fiber.

modes. This means that the information is carried by several wavelengths of the multimode laser. Each wavelength carries a portion of it. Thus, on the receiving side, it is essential to have data from all modes arrive at the same time without dispersion in order to recover the data signal properly. This becomes a problem because of SMF fiber dispersion, which results in different delays for each mode. Therefore, on the receiving side, after 10 km of fiber, the highest frequency will arrive first while the lowest will be last (see Fig. 7.12).

Looking at the time-domain fast-sampling scope as in Fig. 7.13, the pulse structure can be observed. At first, on the right appears to be a high-energy pulse, which corresponds to the highest spectral line with the highest optical frequency. After the large pulse, a narrow pulse appears, which refers to the second spectral line, and last is a wide low pulse related to the lowest wavelength. Note that the pulse amplitude and width refer to its spectral energy and the duration of each mode. This phenomenon creates eye jitter and bit-error rate (BER) degradation. Moreover, it can be observed on nearly single-mode lasers[45,47] such as high mode selectivity DFB-DBR lasers, even though these lasers were developed for this purpose with a mode rejection of 30 dB or more. The mode partitioning is attributed to turn on transient stage and biasing of the laser, e.g., above or below the threshold. This may limit system performance at high data rates due to the turn-on jitter. The laser output always consists of fluctuations because of the quantum nature of the absorption and emission processes. These fluctuations are described by statistical distributions, which are called photon statistics.

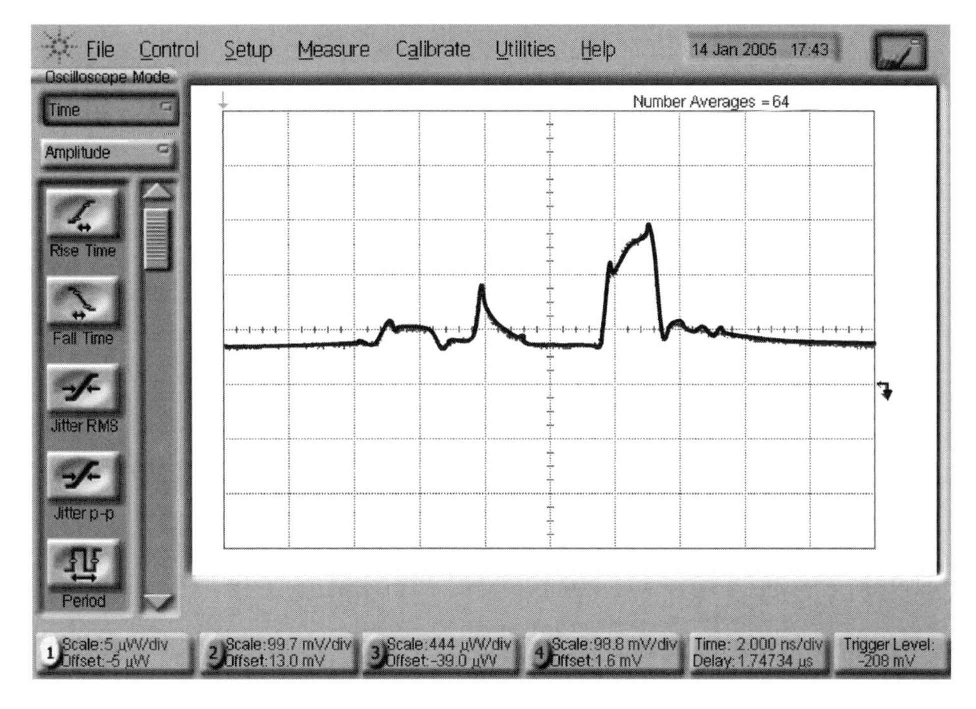


**Figure 7.13** 1550-nm FP modulated laser pulse dispersion after 10 km of fiber.

When the laser is biased below threshold or nonlasing, the output has a broad noise distribution. But, when the laser is biased well above threshold, the distribution approaches a narrow Poisson distribution. The probability density decays exponentially for higher output power according to.[47]

$$p(s) = \frac{1}{N}\exp(-S/\langle S\rangle), \tag{7.72}$$

where $S$ is the photon density in the laser or the instantaneous output power, $\langle S\rangle$ is its average value, and $N$ is a normalization factor so the density integral equals unity. In this case, the normalization was selected to $N = 1$. Otherwise, it means that $S = 0$ has the highest likelihood.

Mode partition phenomena can be explained in terms of carrier-to-noise ratio (CNR) and $E_b/N_0$ in the following manner.[48] The concept of analysis is based on two basic assumptions:

- Total laser output power is constant.
- The partition probability function is based on a time-average spectrum.

Assume an ideal laser lasing at a fixed wavelength and modulated by analog or digital signal. Since the laser is an ideal one, the CW power is constant. The CW term refers to the average power. But since the laser is single-mode, CNR and $E_b/N_0$ do not vary in time.

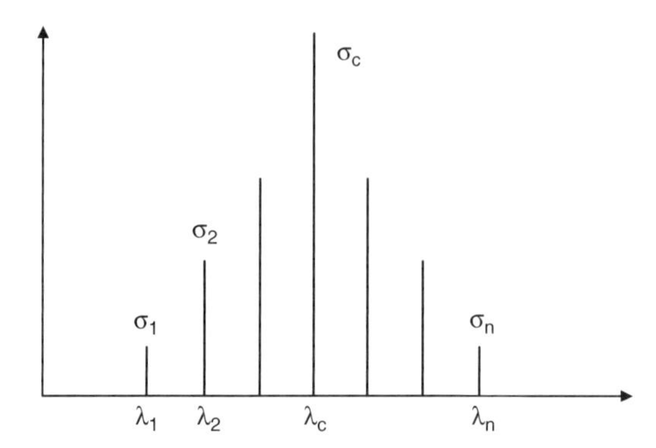


**Figure 7.14** Spectrum distribution of multimode laser of *N* longitudinal modes and amplitudes.[48] (Reprinted with permission from the *Journal of Quantum Electronics* © IEEE, 1982.)

Now, consider a laser with mode hopping. The time-averaged spectrum distribution shows N longitudinal modes when directly modulated. Those wavelengths are $\lambda_i$ $(i = 1 \text{ to } N)$ and the normalized power carried by each longitudinal mode is $a_i$ $(i = 1 \text{ to } N)$. Hence, the following notations are valid for constant power- and time-averaged Gaussian spectrum Fig. 7.14.[48]

$$\sum_{i=1}^{N} a_i = 1, \tag{7.73}$$

$$\begin{aligned}\langle a_i \rangle &= \int a_i p(a_1 \ldots\ldots a_i \ldots\ldots a_N) \mathrm{d}a_1 \ldots \mathrm{d}a_N \\ &= \frac{\exp\left[-(\lambda_i - \lambda_c)^2/2\sigma^2\right]}{\sum\limits_{i=1}^{N} \exp\left[-(\lambda_i - \lambda_c)^2/2\sigma^2\right]} = p(\lambda_i),\end{aligned} \tag{7.74}$$

where $p(\lambda_i)$ is the spectrum distribution, $\sigma$ is the half RMS spectrum width, $\lambda_c$ is the center wavelength, and $p(a_1 \ldots\ldots a_i \ldots\ldots a_N)$ is the joint probability distribution of the amplitudes of the various longitudinal modes, where each longitudinal mode $(i)$ has the amplitude $a_i$, which is the partition-probability function. At the receiving section, the eye diagram is a result of a decision circuit, which determines the logic level of 0 or 1. The "mark" and "space" sequence is described by $\cos(\pi Bt)$, where $B$ is the data rate and the decision circuit samples of the received signal at times of $(1/B + n/B)$, where $n = 0 \ldots \infty$. Assume a misalignment in sampling because of the delayed $i$th mode and the central mode $\Delta\tau_i$. This may result in amplitude degradation:

$$\Delta_i = [1 - \cos(\pi B \Delta\tau_i)]a_i \approx \frac{1}{2}(\pi B \Delta\tau_i)^2 a_i. \tag{7.75}$$

If the misalignment in sampling is $\pi B\Delta\tau_i \ll 1$, then the total amplitude fluctuation denoted as $\Delta$ during time at the decision circuit input would be

$$\Delta = \sum \Delta_i = \frac{1}{2}(\pi B)^2 \sum (\Delta\tau_i)^2 a_i. \tag{7.76}$$

The average noise power due to this fluctuation is marked as $\sigma_{\text{pc}}^2$, which is the average partition noise. This value is defined as the difference between variance and average values of $\Delta$:

$$\sigma_{\text{pc}}^2 = \langle \Delta^2 \rangle - \langle \Delta \rangle^2. \tag{7.77}$$

The variance and the average values of $\Delta$ are obtained from Eqs. (7.73) and (7.74) by replacing $a_i$ with $\Delta$ given by Eq. (7.76) and the square value of $\Delta$ for the variance. Hence,

$$\langle \Delta \rangle = \frac{1}{2}(\pi B)^2 \sum_{i=1}^{N} (\Delta\tau_i)^2 p(\lambda_i), \tag{7.78}$$

which states that the average value of $\Delta$ can be calculated by the time averaged spectrum distribution even though the partition probability function is unknown. However, this partition-probability function is required for solving the variance of $\Delta$ and it can vary from one laser to another:

$$\begin{aligned} \langle \Delta^2 \rangle &= \int \Delta^2 p(a_1 \ldots\ldots a_i \ldots\ldots a_N) \mathrm{d}a_1 \ldots\ldots \mathrm{d}a_N \\ &= \frac{1}{4}(\pi B)^4 \sum_{i=1}^{N} (\Delta\tau_i)^4 \int a_i^2 p(a_1 \ldots\ldots a_N) \mathrm{d}a_1 \ldots\ldots \mathrm{d}a_N \\ &\quad + \frac{1}{4}(\pi B)^4 \sum_{i} \sum_{j \neq i}^{1} (\Delta\tau_i)^2 (\Delta\tau_j) \int a_i a_j^2 p(a_1 \ldots\ldots a_N) \mathrm{d}a_1 \ldots\ldots \mathrm{d}a_N. \end{aligned} \tag{7.79}$$

This is a hard equation to solve and an upper bound limit is obtained by replacing the term $a_i(1 - \sum_{j \neq i} a_j)$ and then obtaining the maximal second moment of $\Delta$ expressed by $a_j$ and $a_i a_j$.

The condition of the maximum moment is given by

$$\int a_i a_j p(a_1 \ldots\ldots a_N) \mathrm{d}a_1 \ldots\ldots \mathrm{d}a_N = 0, \quad i \neq j. \tag{7.80}$$

This condition of maximum second moment indicates that modes within one pulse are mutually exclusive. Each pulse contains a single dominant longitudinal mode and its wavelength varies from pulse to pulse. Hence, the variance of $\Delta$ can be defined by the spectrum distribution function $p(\lambda_i)$ without knowing the partition-probability function. That way the variance upper bound is given by

$$\langle \Delta^2 \rangle \leq \frac{1}{4}(\pi B)^4 \sum_{i=1}^{N} (\Delta\tau_i)^4 p(\lambda_i). \tag{7.81}$$

The functions $p(\lambda_i)$ and $\Delta\tau_i$ can be measured from the optical spectrum analyzer, etc. Thus, the SNR caused by the mode-partition noise (MPN) at the laser output can be expressed as

$$\frac{S}{N} = \frac{1}{\sigma_{\mathrm{pc}}^2} = \frac{4}{(\pi B)^4 \sum_{i=1}^{N} \left[(\Delta\tau_i)^4 p(\lambda_i) - \left[(\Delta\tau_i)^2 p(\lambda_i)\right]^2\right]}, \tag{7.82}$$

where for simplicity, the spectrum distribution is approximated as a continuous Gaussian function rather than a discrete Gaussian distribution:

$$p(\lambda_i) = \frac{1}{\sqrt{2\pi\sigma^2}} \exp\left[\frac{-(\lambda_{\mathrm{i}} - \lambda_{\mathrm{c}})^2}{2\sigma^2}\right]. \tag{7.83}$$

Information may provide a statistical modeling to estimate the BER versus CNR or $E_{\mathrm{b}}/N_0$. This can be expanded when taking into account other statistics such as dispersion fiber losses, etc.[48]

As explained previously, the fiber is characterized by its length $z$ and propagation constant $\beta(\Omega)$ where $\Omega = 2\pi c/\lambda$ is the radian light frequency and $c$ is the speed of light. This way the differential delay between the center wavelength and the side mode can be expressed by the Taylor-series expansion of $\Omega_{\mathrm{i}}$ around $\Omega_{\mathrm{C}}$. This describes the group velocity difference with respect to the central frequency. Since $T = (\mathrm{d}\beta/\mathrm{d}\omega)L = L/\mathrm{Vg}$ and $\Delta T = [L/(\mathrm{Vg} + \Delta\mathrm{Vg}) - L/\mathrm{Vg}]$, the delay is obtained by

$$\Delta\tau_{\mathrm{i}} = \left| \ddot{\beta}(\Omega_{\mathrm{i}} - \Omega_{\mathrm{C}}) + \frac{1}{2}\dddot{\beta}(\Omega_{\mathrm{i}} - \Omega_{\mathrm{C}})^2 \right| \cdot Z, \tag{7.84}$$

where the symbol $\ddot{\beta} = \mathrm{d}^2\beta/\mathrm{d}\omega^2 = (1/V_{\mathrm{g}})'$ and $\dddot{\beta} = \mathrm{d}^3\beta/\mathrm{d}\omega^3 = (1/V_{\mathrm{g}})''$. More details about fiber propagation are given by Eqs. (4.4)–(4.6), and (4.25) in Chapter 4.

There is now a need to calculate $\Omega_{\mathrm{i}} - \Omega_{\mathrm{C}}$:

$$\Delta\Omega = \Omega_{\mathrm{i}} - \Omega_{\mathrm{C}} = 2\pi c\left(\frac{1}{\lambda + \Delta\lambda} - \frac{1}{\lambda}\right) = -2\pi c\Delta\lambda \frac{1}{\lambda^2 + \lambda \cdot \Delta\lambda}$$
$$\approx -\frac{2\pi c\Delta\lambda}{\lambda^2}. \tag{7.85}$$

Using Eqs. (7.73), (7.74), (7.78), (7.81), (7.84), and (7.85), the MPN provided by Eq. (7.77) can be written as

$$\sigma_{pc}^2 = \frac{N}{S} = \frac{1}{2}(\pi B)^4\left[A_1^4\sigma^4 + 48A_2^4\sigma^8 + 42A_1^2A_2^2\sigma^6\right], \tag{7.86}$$

where

$$\begin{cases} A_1 = \left(\dfrac{2\pi C}{\lambda_C^2}\right)\ddot{\beta}z \\ A_2 = \left(\dfrac{2\pi C}{\lambda_C^2}\right)^2\dddot{\beta}z. \end{cases} \tag{7.87}$$

Equation (7.86) claims that the signal-to-partition-noise ratio depends on the half width of the laser diode spectrum and the chromatic dispersion of the fiber and it is independent of $n$, the signal power. Thus, even with an unlimited large-signal power, the overall system SNR cannot be improved beyond the limit imposed by the partition noise. These results will be used in Chapter 10 to determine BER limits.

Theoretical modeling of the dynamics and fluctuations of nearly single-mode lasers are done by the rate equations with Langevin noise[51] terms.[45,47] For simplicity, the analysis refers to two modes, main mode "$m$" and side mode "$s$."[45] Hence, the rate Eqs. (7.2) and (7.3) are denoted as

$$\begin{cases} \dfrac{dN}{dt} = \dfrac{I}{qV} - \dfrac{N}{\tau_{SP}} - \dfrac{C}{\Gamma n_g}\sum_{i=s,m} gI_i \\ \dfrac{dI_i}{dt} = \dfrac{C}{n_g}(g-\alpha_i)\cdot I_i + \dfrac{\gamma_{sp}}{\tau_{SP}}D_n + \sqrt{\dfrac{2\gamma_{sp}}{\tau_{SP}}D\cdot I_iN\cdot F_i(t)}\Big|_{i=s,m} \end{cases} \tag{7.88}$$

$$g = \frac{\Gamma n_g A(Dn - n_0)}{c[1 + s(I_s + i_m)]}, \tag{7.89}$$

where

| | |
|---|---|
| $\Gamma$ | is the mode-confinement factor, |
| $D$ | is the line-shape factor, |
| $\tau_{SP}$ | is the spontaneous lifetime, |
| $n_g$ | is the group index, |
| $n_0$ | is the carrier density at transparency, |
| $\gamma_{sp}$ | is the fraction of spontaneous emission coupled into the mode, |
| $c$ | is the speed of light in vacuum, |
| $V$ | is the volume of the active region, |
| $A$ | is the differential gain, |
| $C_{th}$ | is the threshold current, |
| $s$ | is the saturation parameter, |
| $\alpha_m$ | is the loss of main mode, and |
| $\lambda$ | is the optical wavelength. |

It can be seen from Eqs. (7.88) and (7.89) that the total laser output power is proportional to the sum of the side and main mode currents. The laser threshold current is given by

$$C_{\mathrm{th}} = \frac{2}{\tau_{\mathrm{sp}}}\left(\frac{\alpha_{\mathrm{m}}}{A\Gamma n_{\mathrm{g}}} + n_0\right). \tag{7.90}$$

The next step is to minimize the likelihood of excitation on the undesired side mode in a nearly single-mode laser. This can be investigated by setting the laser bias under various data sequences using a statistical model and the above rate equations.

A typical digital transmitter is designed to bias the laser above threshold or below threshold at no lasing state. This case is referred as "0" logic. At "1" logic state, the laser is forward bias and the power and bias point are defined by the extension ratio. Bias point and reflections may result in a mode jump.[46] Moreover, "0" logic level can be biased above threshold or below. This may affect the mode partition error (MPE) as well.[45] Rise time and fall time as well as pulse duration may affect MPE too.

So far, the MPN discussion was limited to CNR and digital modulation. On CATV applications, mainly at low-cost return path FP laser transmitters without isolators such as shown by Figs. 2.4 and 2.5, MPN and mode hopping noise (MHN) are observed. From Refs. [45], [49] and [50], it is observed that the relative noise power is flat at the low-frequency range up to 50–100 MHz. Then it starts to roll off at 6 dB/octave; bias current may affect it too (see Fig.7.15).

It was denoted by Eqs. (7.72) and (7.77) that the sum of intensities is zero. Thus, the intensity noise of each mode at any frequency before traveling

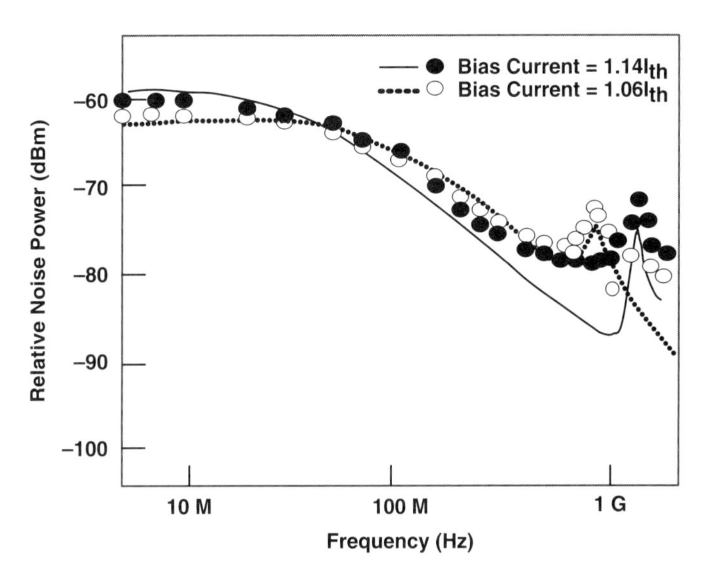


**Figure 7.15** Measured MPN of Al GaAs laser.[50] (Reprinted with permission from the *Journal of Lightwave Technology* © IEEE, 1994.)

through SMF is given by

$$\begin{cases} N_{P1} = A_n \cos(\omega_n t + \phi) \\ N_{P2} = -A_n \cos(\omega_n t + \phi), \end{cases} \quad (7.91)$$

where $\omega_n$ is the noise frequency. After traveling through SMF and having dispersion, the two modes are delayed with respect to each other. Thus, Eq. (7.76) becomes

$$\begin{cases} N_{P1} = A_n \cos\left[\left(\omega_n t + \frac{\tau}{2}\right) + \phi\right] \\ N_{P2} = -A_n \cos\left[\left(\omega_n t - \frac{\tau}{2}\right) + \phi\right]. \end{cases} \quad (7.92)$$

When the two modes are combined, the dispersion-enhanced mode-partition noise (EMPN) becomes

$$N_{P1} + N_{P2} = 2A_n \sin(\omega_n \tau/2) \sin(\omega_n t + \phi) \approx A_n \omega_n \tau \sin(\omega_n t + \phi). \quad (7.93)$$

This approximation is valid when $\omega_n \tau \ll 1$. As a conclusion, the EMPN is proportional to the noise frequency $\omega_n$, the fiber dispersion $\tau$, and the mode spectral noise density $A_n$. The MHN origin is from the random hopping of the laser oscillations between various modes and thus the output difference of those modes due to coupling effects among the modes. This kind of noise results in higher relative-intensity noise (RIN) $> -130$ dB/Hz at the frequency range below a few tens of megahertz. However, the $f_1 \pm f_2$ second-order distortion may stimulate "skirt" noise similar to phase noise. Hence, FP lasers are not used for HFC systems downstream where RIN (AM noise) below $-155$ dB/Hz is needed.

### 7.4.4 Laser Relative Intensity Noise (RIN)

The power emitted from a laser diode fluctuates around its steady-state value.[45,51] These power fluctuations result in quantum fluctuations associated with the lasing process and are described by the modified rate equations, which include the Langevin noise source $F(t)$:

$$\frac{dP}{dt} = (G - \gamma)P + R_{SP} + F_P(t), \quad (7.94)$$

$$\frac{dN}{dt} = \frac{I}{q} + \gamma_e - GP + F_N(t), \quad (7.95)$$

$$\frac{d\phi}{dt} = -(\omega - \omega_{th}) + \frac{1}{2}\beta_c(G - \gamma) + F_\varphi(t). \quad (7.96)$$

$F_P$ and $F_\varphi$ arise from spontaneous emission, whereas $F_N$ has origin of discrete carrier generation and recombination shot noise; $R_{SP} = \gamma_{SP} n_{sp} N/\tau_N$ is the spontaneous emission rate; $\beta_C = \Delta n'/\Delta n''$ and $\Delta n = \Delta n' + j\Delta n''$; $N$ and $P$ represent the steady-state average values of carrier and photon population, respectively; and $G$ is the stimulated emission rate and is a function of $N$. Thus, it becomes a stochastic differential equation. The problem is simplified if the Markovian process is assumed, where the correlation time of noise sources is much shorter than the relaxation time $\gamma^{-1} = \tau_p$ and $\gamma_e^{-1} = \tau_e$, which means the system has no memory. The solution is simplified by assuming Langevin forces are Gaussian random processed with a zero mean. Hence, under this assumption, the Langevin forces satisfy

$$\langle F_i(t) \rangle = 0, \tag{7.97}$$
$$\langle F_i(t) F_j(t') \rangle = 2D_{i,j}\delta(t - t'), \tag{7.98}$$

where $D_{i,j}$ is the diffusion coefficient associated with the corresponding noise source.

These quantum fluctuations are related to spontaneous emission and fluctuations of the electron density. The changes affect the amplitude and phase of steady-state oscillation conditions as will be elaborated in Sec. 7.4.5. The phase change outcome is phase noise, or frequency noise, known as FM noise, while the amplitude changes are related to AM noise. This AM noise is defined as RIN, which is defined by

$$\text{RIN} = \frac{S_P(\omega)}{P^2}, \tag{7.99}$$

where the spectral density $S_P(\omega)$ is defined by

$$S_P(\omega) = \int_{-\infty}^{\infty} \langle \delta P(t+\tau) \delta P(t) \rangle \exp(-j\omega\tau) d\tau, \tag{7.100}$$

where

$$S_P(\omega) = \operatorname*{Lim}_{T\to\infty} \frac{1}{T} \left| \delta\tilde{P}(\omega) \right|^2. \tag{7.101}$$

The ensemble average in Eq. (7.100) was replaced by a time-domain average over the interval $T$ assuming an ergodic-stochastic process:

$$\lim_{T\to\infty} \frac{1}{T} \left[ \tilde{F}_i^*(\omega) \tilde{F}_j(\omega) \right] = 2D_{i,j}. \tag{7.102}$$

Thus, at a steady-state, RIN is defined as

$$\text{RIN} = \frac{\langle \delta P(\omega) \rangle}{P^2}, \tag{7.103}$$

where $\delta P(\omega)$ is the output-power fluctuation from the average power $P$.

The laser RIN versus frequency can be written as.[51]

$$\mathrm{RIN}(\omega) = \frac{2R_{\mathrm{SP}}\left[\left(\Gamma_{\mathrm{N}}^2 + \omega^2\right) + G_{\mathrm{N}}^2 P^2\left(1 + \gamma_e N/R_{\mathrm{SP}}P\right) - 2\Gamma_{\mathrm{N}} G_{\mathrm{N}} P\right]}{P\left[(\Omega_{\mathrm{R}} - \omega)^2 + \Gamma_{\mathrm{R}}^2\right] \cdot \left[(\Omega_{\mathrm{R}} + \omega)^2 + \Gamma_{\mathrm{R}}^2\right]}, \tag{7.104}$$

where $\Omega_R$ is the relaxation–oscillation radian frequency, $\Gamma_{\mathrm{R}} \equiv (\Gamma_{\mathrm{N}} + \Gamma_{\mathrm{S}})/2$ is the relaxation–oscillation decay rate, and $\Gamma_{\mathrm{N}} = \gamma_{\mathrm{N}} + N\left(\partial\gamma_{\mathrm{N}}/\partial N\right) + G_{\mathrm{N}}P$ is the small signal-carrier decay rate.

Equation (7.104) shows that the RIN is flat below relaxation–oscillation $\omega \ll \Omega_{\mathrm{R}}$ and peaks at $\Omega_{\mathrm{R}}$. To have a better understanding of RIN($\omega$), assume $\omega \ll \Omega_R$ limit. Since $\Gamma_R \ll \Omega_R$, the denominator of Eq. (7.103) is replaced by $P\Omega_{\mathrm{R}}^4$, and in the numerator are terms proportional to $P$ and $P^2$.

Since

$$\Omega_{\mathrm{R}} \approx \sqrt{G\frac{\partial G}{\partial N}P} = \sqrt{GG_{\mathrm{N}}P}, \tag{7.105}$$

$$P = (I - I_{\mathrm{th}})/(qG), \tag{7.106}$$

$$\Omega_{\mathrm{R}} = \sqrt{\frac{G_{\mathrm{N}}(I - I_{\mathrm{th}})}{q}}, \tag{7.107}$$

it follows that at a given frequency, the RIN decreases with bias as $(I - I_{\mathrm{th}})^{-3}$. As the bias current increases, the RIN decreases slower as $(I - I_{\mathrm{th}})^{-1}$. As will be reviewed in Chapter 10, the RIN imposes a quantum limit on achievable CNR. Eventually, DFB laser transmitters used for video transport have a limit of $-155$ dB/Hz as proved in Chapter 10.

### 7.4.5 Laser-phase noise

Phase noise of a semiconductor laser is a result of quantum fluctuations associated with the lasing process.[51] A phase change leads to a frequency shift $\delta\omega_{\mathrm{L}} = \delta d\phi/dt$. The phase noise is one of the parameters for evaluating the performance of coherent optical communication systems.

There are two mechanisms that contribute to phase fluctuations:

$$\delta\tilde{\phi} = \frac{1}{j\omega}\left(\tilde{F}_{\phi} + \frac{1}{2}\beta_{\mathrm{C}} G_{\mathrm{N}} \delta\tilde{N}\right). \tag{7.108}$$

- Spontaneous emission $\tilde{F}_{\varphi}$. Each spontaneously emitted photon changes the optical phase by a random amount.
- Fluctuations in the carrier populations $\beta_{\mathrm{C}} G_{\mathrm{N}} \delta\tilde{N}/2$. This is because a change in $N$ affects not only the optical gain but also the refractive index, which eventually affects the phase. The parameter $\beta_{\mathrm{C}}$ provides

proportionality between the gain and the index changes. This parameter sometimes is referred to as the line-width enhancement factor found to increase the line width by a factor $1 + \beta_C^2$.[51,54]

These terms are defined by Eqs. (7.2) and (7.3), and (7.79)–(7.81) as an effect on the differential gain and the confinement factor as well as the spontaneous emission-coupling factor to the lasing mode $\gamma_{sp}$. The analysis of laser oscillation conditions in Secs. 6.3 and 6.4 provide a complex equation. The real part of it relates to the amplitude oscillation conditions whereas the imaginary part of it refers to the phase. Thus, the amplitude fluctuation term relates to the AM noise mentioned previously as RIN, and the phase fluctuation refers to the FM noise known as phase noise. The phase-noise skirt shape defines the spectral line width, which depends on the active area.

The spectral density of frequency noise (FM noise, phase noise) is defined by

$$S_{\dot{\phi}} = \left\langle \left| \omega \delta \tilde{\phi}(\omega) \right|^2 \right\rangle, \tag{7.109}$$

and is approximated by the following equation:

$$S_{\dot{\phi}}(\omega) \cong \frac{R_{SP}}{2P} \left( 1 + \frac{\beta_C^2 \Omega_R^4}{\left[ \left( \Omega_R^2 - \omega^2 \right)^2 + (2\omega\Gamma_R)^2 \right]} \right), \tag{7.110}$$

where $\Omega_R$ is the relaxation–oscillation radian frequency, $R_{SP}$ is the spontaneously emitted photons rate, $P$ is the photon population (proportional to power), $\Gamma_R$ is the relaxation–oscillation frequency decay rate, and $\beta_C = \Delta n'/\Delta n''$ and $\Delta n = \Delta n' + j\Delta n''$, so the factor $\beta_C$ is showing the $tg\theta$ of the loss due to the imaginary part of the refractive index. It is also important to note that $\Gamma_R$ increases with the bias current as the relaxation–oscillation frequency increases. However, the relaxation–oscillation is damped faster at high currents. Moreover, this equation shows that the phase noise is flat at $\omega \ll \Omega_R$.

The ratio $R_{SP}/P$ also defines how many decibels below carrier the noise density is. Thus, the spontaneous emission minimization would reduce the phase noise or, in other words, improve it.

The line width of the laser is given by

$$\Delta f = \frac{1}{2\pi} S_{\dot{\phi}}(0) = \frac{R_{SP}\left(1 + \beta_C^2\right)}{4\pi P} = \Delta f_0\left(1 + \beta_C^2\right). \tag{7.111}$$

This equation shows that the line width of a laser is enhanced by the factor $1 + \beta_C^2$. Again, the ratio $R_{SP}/P$ defines the $\Delta f_0$ line width. The contribution of $\beta_C^2 \Delta f_0$ to the phase noise is mainly due to the $\delta N$, because each spontaneously emitted photon changes the laser power. This changes the gain and thus the carrier population. As a result, it affects the refraction index since the dielectric changes. Thus, it affects the optical phase. The

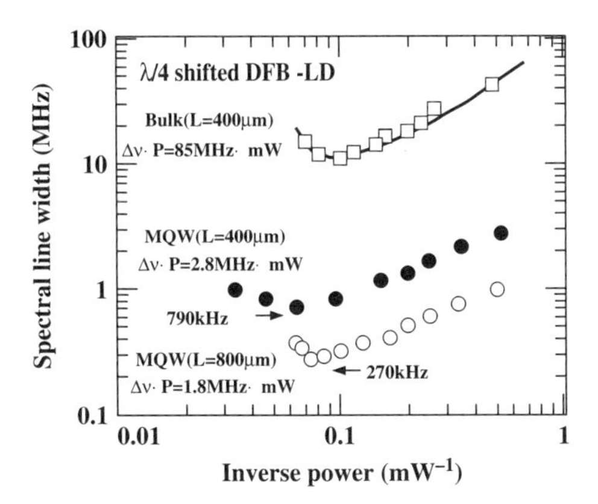


**Figure 7.16** Spectral line width as a function of power for two MQW DFB lasers of 400 and 800-μm cavity lengths in comparison to a thick active layer laser marked as bulk. Note that MQW line width is narrower since $\beta_C$ is lower, meaning lower losses.[55] (Reprinted with permission from the *Journal of Quantum Electronics* © IEEE, 1991.)

resulting delayed phase fluctuation is affected by relaxation–oscillation and leads to satellite peaks at multiples of $\Omega_R$, as well as broadening the central mode peak by $1 + \beta_C^2$.

Optical phase noise and AM noise are measured by special test equipment with low phase noise in order to improve the measurement accuracy.[53] The optical signal is converted into an RF signal and then measured on an RF spectrum analyzer.

Currently used MQW lasers for HFC applications and direct modulation exhibit a spectral line width of 1–10 MHz at a power level of 5–10 mW. Moreover, as the laser power increases, the line width decreases. Figure 7.16 shows a line width saturation for several 1550-nm DFB lasers.[55] For $\beta_C = 5$, optical power of 2 mW, $R_{SP} = 1.28 \times 10^{12}\ \mathrm{S}^{-1}$, and $P = 7.76 \times 10^4$, then $\Delta f \approx 70$ MHz.

## 7.5 External Modulation

In order to achieve high-speed modulation for high-data-rate transport, the DML method is limited due to several effects. The first is frequency chirping that is shown by Eq. (7.45). There, it can be observed that the faster the transition, the wider the frequency chirp is. In addition, there is the relaxation–oscillation limit of the laser. In this case, high-speed data transport involves fast transitions that may reach the relaxation–oscillation frequency. Under these operation conditions, the chirp RIN and other parameters mentioned previously, laser performance is reduced. Further, on analog transport, this may result in additional NLD. Thus, EM schemes must be used. In this scheme, the laser is biased at a CW level and a different device modulates the optical output of the laser. It should be clarified that the term "external" in this context means external to the laser cavity and not necessarily a separate device. This external modulator (EM) can be integrated with the laser monolithically, resulting in a single device as far as the outside world is concerned.

There are several schemes available to modulate a light beam. In general, these schemes fall into one of the two categories: index modulation and absorption coefficient modulation.

In the first category, usually the electro-optic effect is utilized where the refractive index of a medium or crystal is changed as the function of an applied electric field. If a light beam is passing through the medium, this causes a phase shift or phase modulation of the beam. To convert this phase modulation to an IM, usually an interference effect should be utilized. Such an effect can be implemented by a Mach–Zehnder (MZ) interferometer. Its concept is mentioned in Sec. 4.4.1. In Fig. 4.11 the phase shift $\beta\Delta L$ is done electrically.

In the second category, the applied electric field modifies the absorption coefficient of the medium. As a result, a light beam passing through the medium experiences various degrees of attenuation as a function of the applied field causing a modulated output power. The electro-absorption (EA) modulator is based on the bulk Franz-Keldysh effect.

### 7.5.1 Mach-Zehnder (MZ) Lithium-Niobate modulator

The MZ modulators rely on the electro-optic properties of certain types of crystals. In several crystals, and depending on the internal symmetries within the lattice, the index of refraction can be a function of the applied electric field. If the crystal lacks inversion symmetry, the index of refraction varies linearly with the electric field. This is called the "linear electro-optic effect." In most practical situations, MZ modulators are realized in lithium–niobate ($LiNbO_3$) because the electro-optic effect is very strong in it. This is a key parameter to producing large changes of refractive index for $f_o$, a fixed applied electrical field. The waveguide is created by locally doping the crystal, which result in waveguides that have low propagation losses. The boundary of the waveguide is where the refractive index is changed.

Optical waveguides are delineated on $LiNbO_3$ by standard photolithography. This process was reviewed in Chapter 5 for PLC technology. There are two leading methods for fabricating optical waveguides on $LiNbO_3$: the annealed proton exchange (APE), and titanium indiffused. In APE, only the extraordinary index is increased. This term was explained in Chapter 4 when dealing with optical isolators. This means that only single polarization light is supported and a better than 50-dB polarization extinction ratio is achieved. This is an advantage since it may avoid the problem of having different modulation efficiencies for different polarization modes in the waveguide. The disadvantage is that the light polarization state must be aligned with the waveguide. The second advantage is that APE can produce larger refractive index change compared to titanium indiffused. This is attractive since it enables waveguides with small bend radiuses. Figure 7.17 describes the Mach–Zehnder interferometer (MZI) modulator.

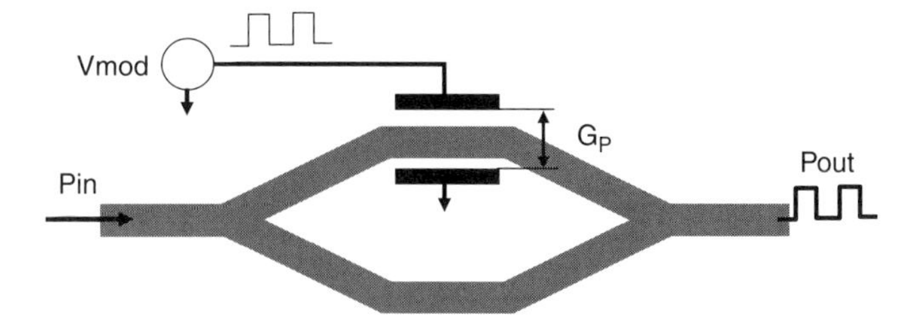


**Figure 7.17** MZI modulator concept.

The input optical field is divided into two branches using a Y-branch splitter. The applied electric field changes the index of refraction for the upper branch waveguide. This change in the refractive index affects the optical length, which results in a creative or destructive interference. In practical situations, the BW of such a modulator configuration is often limited by parasitics, including the electrode capacitance.

The geometry in Fig. 7.17 is sufficient to explain the physics of operation of the MZI modulator. A voltage, $V$, marked as $V_{\text{mod}}$ is applied between the two electrodes. The distance between the electrodes is $G_{\text{p}}$; the electric field applied on the waveguide is $|E| = V/G_{\text{p}}$. Since the MZI wafer is birefringent crystal, the largest optic coefficient is accomplished when the field is applied along the $Z$ direction ($E_Z$). Thus, the maximum induced index change is given by

$$\Delta n = -n^3 r_{33} \frac{E_z}{2}, \tag{7.112}$$

where $r_{33}$ is the largest electro-optic tensor and $n$ is either the ordinary or the extraordinary refractive-index value. The electrode orientation with respect to the waveguide is such that it generates an electrical field in the $Z$ direction. This depends on the used-crystal orientation. The orientation is generally specified as the "cut" direction, which is perpendicular to the flat surface on which the waveguide is fabricated. In other words, the "cut" is the surface of the waveguide cross section. Hence, the normal to the waveguide cross section surface defines the light-propagation direction and the electric-field propagation. This is demonstrated by Fig. 7.18.

Equation (7.112) is modified by a correction factor $\Gamma$ for the fact that the optical and the electrical fields are not aligned and overlap completely. Thus, the induced refractive index becomes

$$\Delta n = -n^3 r_{33} \frac{V}{2G_{\text{P}}}. \tag{7.113}$$

Using the relation from Sec. 4.4.1, the delta change in the phase due to the change in the refraction index can be written as

$$\Delta\varphi = \Delta\beta L = \frac{2\pi \Delta n L}{\lambda} = -\pi n^3 r_{33} \Gamma \frac{VL}{\lambda G_{\text{P}}}. \tag{7.114}$$

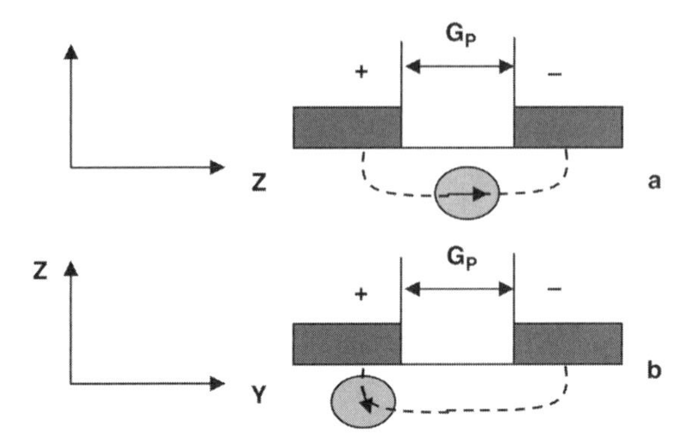


**Figure 7.18** Cross section of an embedded optical waveguide and the electrodes with respect to the "cut" direction. (a) *X*-cut; the waveguide cross-section normal is in the *Y* direction, which is the RF-field propagation; *X* is normal to the surface where the waveguide is implanted. (b) *Z* cut; the waveguide cross-section normal is in the *X* direction, which is the RF-field propagation; *Z* is normal to the surface where the waveguide is implanted.

From this equation, it can be seen that a phase shift can be optimized by controlling the ratio $V \cdot L$ to $\Gamma/G_p$. Large phase shifting can be accomplished by a long electrode or small gap. In the case of a small gap, the coupling between the two electrodes would be higher; i.e., it would be a microwave-coupled line that may have high capacitance. Thus, by the theory of microwave-coupled lines, the modulation BW would decrease. As a conclusion, a large phase-shift modulation with low voltage would require long electrodes, whereas a wide BW would be made by short electrodes with a higher modulation voltage. Long electrodes define the RF quarter-wavelength resonance and center frequency of microwave-coupled lines. Strong coupling means a higher insertion loss and a narrower BW. An additional parameter affecting the transmission-line-electrodes characteristic impedance as well as the coupled line is the odd-mode impedance, zoo, and even mode impedance, zoe, as well as the relative dielectric constant and substrate thickness.[60] In both cases, the coupling gap $G_P$ is fixed. The voltage that creates a phase shift of 180 deg is marked as $V\pi$. The appropriate figure of merit that allows meaningful comparison between EMs is V/*f*_3 dB. In CATV, this BW is below 1 GHz; thus, $3 \leq V\pi \leq 4$, for higher data rates the modulation voltage and power would be higher. The next step is to define the modulation depth as a function of the modulation voltage. Assuming a linear phase dependency of the control voltage with respect to the 180 deg $V\pi$, the phase versus control voltage is given by the ratio $\varphi = V/V\pi$. In addition, the output average optical power coming out from a lossy optical waveguide is given by

$$\langle P \rangle = \frac{P_{\text{in}}}{2} L_{\text{ex}}, \tag{7.115}$$

where $L_{\text{ex}}$ is the excess loss of the modulator. Using these notations and the results from Sec. 4.4.1 given by Eqs. (4.29) and (4.30) with some trigonometric identities,

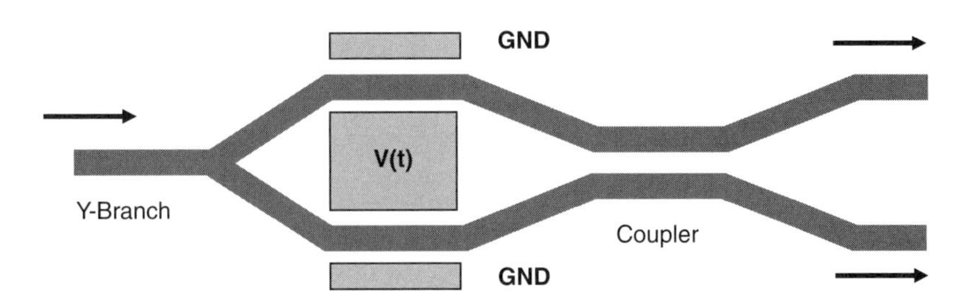


**Figure 7.19** Balanced bridge travelling wave interferometer external modulator.

the optical output power at the MZI is given by

$$\begin{cases} P(V) = \langle P \rangle \left[1 + \sin\left(\pi \frac{V}{V_\pi} + \Delta\phi\right)\right] \\ \text{or} \\ P(V) = \langle P \rangle \left[1 + \cos\left(\pi \frac{V}{V_\pi} - \Delta\phi\right)\right], \end{cases} \tag{7.116}$$

Coming back again to Eq. (7.114) and realizations of MZI modulators by using Y branch splitter, balanced bridge interferometer (BBI) per Fig. 7.19, or dual Y branch per Fig. 7.17, the value of the phase delta resulting from the difference between the two wave propagation indices is $\Delta\varphi = \Delta\beta L = (\beta_1 - \beta_2)\mathrm{L} = V/V\pi$ and $\phi$ is the inherent phase difference between the two traces.

External modulators can be divided into two categories based on the electrodes configuration as shown in Fig. 7.18: lumped circuit element modulator, and traveling wave (TW) modulator.

In the lumped-element case, the input feeder is connected to a pair of electrodes and terminated by a parallel matched load. The electrode pair is a capacitive-load parallel to the termination impedance (see Fig. 7.20) Thus, the BW limitation arises from the RC time constant. To overcome this problem, a TW modulator is used. In this topology, the modulating RF signal is fed into a microwave transmission line parallel coupled to the optical waveguide. The end of the transmission line is terminated by a matching load with the same value of the RF/data characteristic impedance. The interaction between the optical propagating light wave and the RF/data along the modulator length results in efficient modulation, which is not limited any more by the RC time constant. The BW of TW EM is limited by the velocity difference between the lightwave and the RF/data signal. Note that the dielectric constant satisfies the following relation with respect to the refraction index $n$, $\sqrt{\varepsilon_r \varepsilon_0} = n$.

An additional method to extend the MZI EM BW is by the differential input modulation signal and using the TW scheme. This modulator would look like the one in Fig. 7.17, where the upper branch is the transmission and the lower arm is the transmission input. To obtain chirping characteristic of such MZI EM, the electrical field at the output of the modulator should be examined:[56]

$$E = \frac{E_0}{2}\exp(-j\beta_1 L) + \frac{E_0}{2}\exp(-j\beta_2 L). \tag{7.117}$$

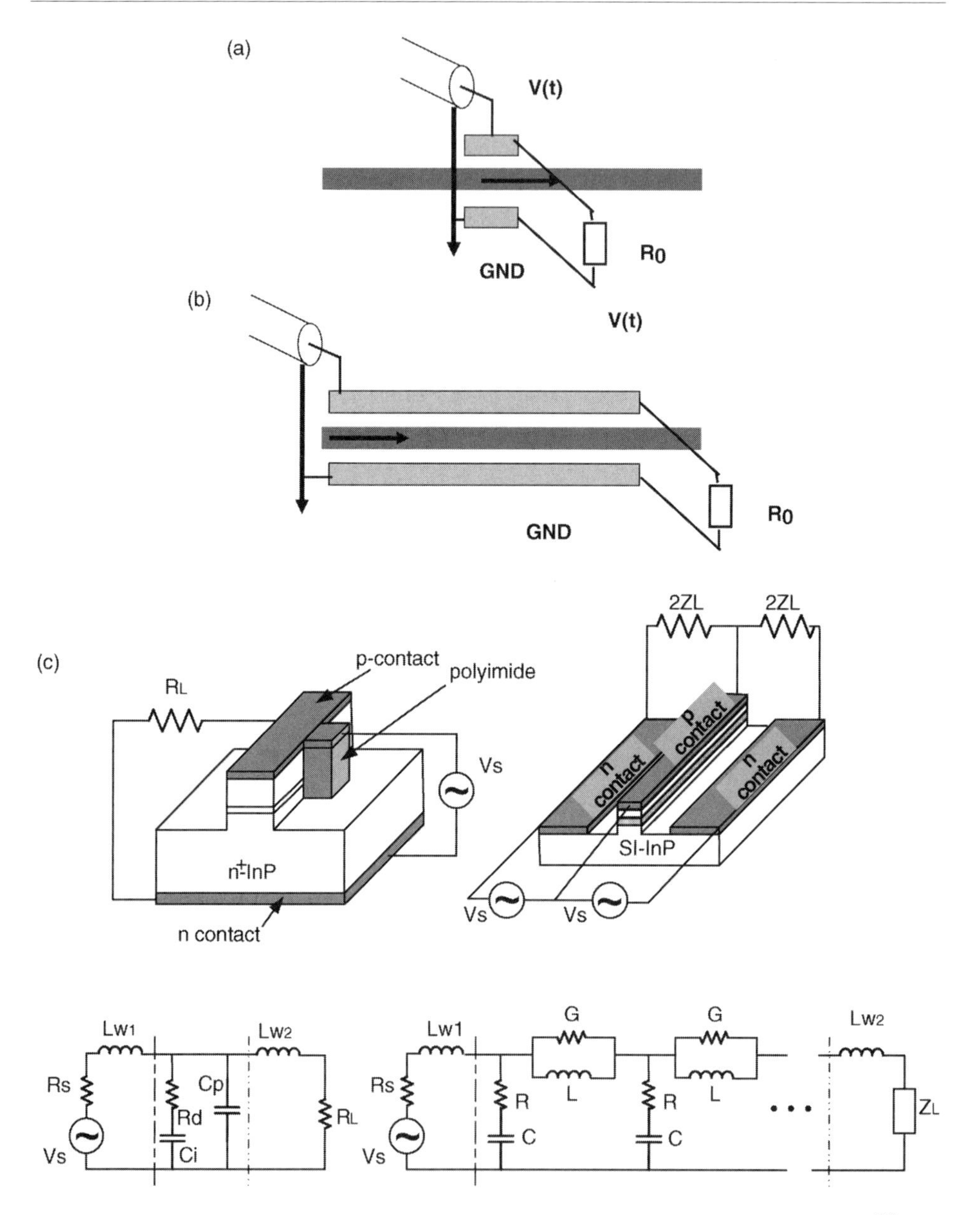


**Figure 7.20** (a) Lumped element vs. (b) TW electrode (c) equivalent circuit.[62]

This equation can be noted as

$$E = E_0 \cos\left(\frac{\beta_1 - \beta_2}{2} L\right) \exp\left(j \frac{\beta_1 + \beta_2}{2} L\right). \tag{7.118}$$

The exponential term determines the instantaneous phase or frequency of the output. To obtain zero-phase variations, the output should have $\Delta\beta_1 = -\Delta\beta_2$. This happens when the voltages applied on the two branches are equal and of opposite signs. In other words, it is theoretically possible to eliminate the chirp in MZ modulators. That is an important advantage of MZI EM over electroabsorption modulators.

### 7.5.2 Electro Absorption (EA) modulators

Unlike the MZ modulators that rely on the change in the refractive index, the electro-absorption EA modulators rely on the change in the absorption coefficient with the applied electric field. These devices can further be divided to different subcategories depending on the physical principle behind them. These categories include the bulk Franz–Keldysh, the Stark–Shift MQW, and the Wannier–Stark modulators. However, the most practical modulators are the bulk Franz–Keldysh type that is reviewed here.

The principle behind the Franz–Keldysh bulk modulators can be described as below. In a bulk semiconductor, the band gap energy $E_g$ can be changed slightly with an applied electric field. On the other hand, the absorption coefficient of a semiconductor for an incident light is strongly dependent on the photons energy $E_p = h\nu$.[56] Thus, the absorption coefficient is expressed as

$$\alpha(h\nu) \approx \alpha_0 \exp\left[\frac{E_g - h\nu}{\Delta E_g(E)}\right], \tag{7.119}$$

where, $\alpha_0$ is the absorption coefficient with no external electric field, and $E$ is the electric field. If $E_p > E_g$, the electromagnetic radiation and the semiconductor lattice strongly interact; the photons are absorbed and electron-hole pairs are generated. On the other hand, if $E_p < E_g$, the photons cannot generate electron-hole pairs and the photons do not interact with the semiconductor. Hence, the semiconductor becomes transparent to those photons. Because $E_g$ can be varied with the applied electric field, it is possible to make the semiconductor absorptive or transparent to optical wavelengths close to the band-gap energy by applying an external electric field.

In practical devices, usually a reverse-biased PIN-diode structure is used. In such a device, the intrinsic region can act as waveguide because of the difference between the indices of refraction of the core (I region) and the P and N layers. The reverse bias across the junction creates the desired electric field, which is used to modulate the optical wave. This can be an integrated part of the DFB laser as shown in Fig. 7.21.

In these structures, it is important to keep the laser and the modulator as electrically isolated as possible to prevent wavelength shifts in the laser due to the modulator operation. These shifts result from reflections that create mode hopping. One of the main advantages of the EA modulators is that structurally they are very similar to the laser used in communications and therefore they can

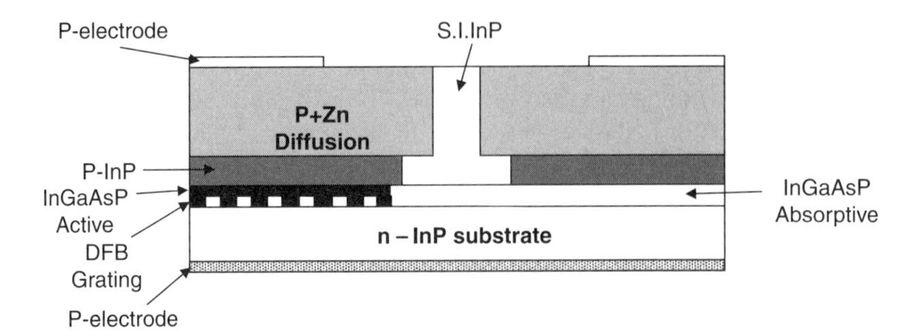


**Figure 7.21** An integrated DFB laser with EA EM.

be monolithically integrated with the laser. This can reduce costs and eliminate insertion losses and the need for additional alignment. The output of a typical EA modulator as a function of the applied field is given by:[56]

$$P_{\text{out}} = P_{\text{in}} \exp[-\alpha V(t)]. \tag{7.120}$$

The chirping characteristics of an EA modulator result from the changes in the index of refraction in the absorption section. This is because absorption and the phase change in a medium are related to the imaginary and real parts of the refractive index as were introduced in Sec. 7.4.5. Related to each other are $n''$ and $n'$, according to the Kramers–Kronig relation and the $\beta_C$ factor mentioned in Sec. 7.4.5. The chirp is given by Eq. (7.45). It is the same chirp that a DML would experience. For Franz–Keldysh modulators, the value of the $\beta$-enhancement factor is about 0.5, whereas in DML it is about 3–5. EA modulators are reported with high dynamic range for analog purposes.[57] These EM were based on the TW method with strain-compensated InGaAsP MQW, $V\pi = 0.37$ V, and high extinction ratio of $>30$ dB/V. By optimizing the bias voltage and the optical input power, the SFDR (spurious free dynamic range) was improved by 1 to 30 dB. SFDR at 10 GHz was 128 dB–Hz$^{4/5}$.

### 7.5.3 Comparison between EA and MZI modulators

EA and MZI modulators can be compared based on several parameters.[21,58] In terms of monolithic integration,[63] EA modulators are preferred because the MZI modulators are usually fabricated separately. Monolithic integration reduces insertion loss between the laser and the modulator significantly as well as simplifies the alignment process. Another advantage of EA is a lower driving voltage, which make them easier to drive. The major disadvantage of EA modulators lies in chirping effects, which cannot be eliminated. In MZI, the chirping can be eliminated using a differential drive. The BW of both EA and MZI is similar although the fastest modulators available tend to be MZI, probably due to the option to eliminate chirp. Moreover, in EA modulators, pumping the optically induced carriers out of the junction may be a potential band-limiting problem. With careful design, it is

**Table 7.2** General comparison of EA and MZI modulators.

| Modulator parameter | MZI | EA |
|---|---|---|
| Monolithic integration with laser | No | Yes |
| Driving voltage | Higher | Lower |
| Chirping | Can be eliminated | Cannot be eliminated |
| Extinction ratio | Higher | Lower |
| BW | Can be higher | Slightly lower |
| Heating | Less problematic | More problematic |

possible to get good extinction ratios in both cases, even though MZI modulators generally tend to have higher extinction ratios. The other potential problem with EA modulators is heating[61] since the input signal has to be absorbed in the modulator. These points are summarized in Table 7.2.

### 7.5.4 Commercial modulators

There are a number of commercial modulators available in the market. Generally, lasers designed for speeds of 10 GBPS and higher are integrated with an EA modulator, although MZI are also used. Table 7.3 summarizes some of the specifications from various manufacturers.

As can be seen, the specifications are generally comparable except for the last two rows that are 40 GBPS MZI modulators.

**Table 7.3** Specifications of various modulators.

| Manufacturer | Modulator/speed | Extinction ratio | Rise/fall time | Mod voltage |
|---|---|---|---|---|
| Nortel LCM155EW-64 | MZ-10 G | 13 dB | 50 ps max | 4V typ ptp |
| NEC NX8560LJ-CC | EA-10 G | 11 dB | 40 ps max | 2.5V typ ptp |
| Samsung ML48G2A | EA-10 G | 10 dB | 40 ps max | 2V typ ptp |
| Hitachi LE7602-L | EA-10 G | 10 dB | 50 ps max | 2.6V max |
| Toshiba TOLD387S | EA-10 G | 10 dB | 50 ps max | 2.5V max |
| Lucent E2560 | EA-10 G | 10 dB | 40 ps max | 1V max ptp |
| Lucent 2623 Series | MZ-10 G Non integrated | 13 dB | 10 GHz | 5V max |
| Lucent | MZ-40 G | | 30 GHz | 6V typ |
| Sumitomo[59] | MZ-40 G | 40 dB | 40 GHz | 5V |

## 7.6 Main Points of this Chapter

1. Laser dynamics are described by the rate equations for both photons and carriers.
2. Laser differential gain results from difference in carriers in the active area.
3. Laser-gain compression results from nonproductive recombination, absorption Auger recombination, and spontaneous emission.
4. Rate equations are differential equation sets that are solved numerically for a large signal. However, a small harmonic signal approximation is used to analyze the laser dynamics in order to find the laser BW limits.
5. One of the laser parameters analyzed using the small signal approximation of the rate equations is the "relaxation–oscillation" frequency of the laser.
6. The relaxation–oscillation of a laser defines its modulation BW for DML applications. This is a second-order low-pass response.
7. The relaxation–oscillation of a laser directly depends on the bias point and output power. The higher the power and higher the bias above threshold, the higher the relaxation–oscillation frequency is. The higher the differential optical gain of a laser, the higher the relaxation–oscillation frequency is.
8. The small-signal-equivalent RLC circuit is used to describe the laser small-signal response. The capacitor describes the carriers reservoir; the resistor describes the coupling mode loss and the inductor, the photon reservoir for emission in the active area.
9. The multimode-laser-equivalent circuit may be described by a capacitor and multiple-coupled inductors in parallel for each mode.
10. The large-signal modeling of a laser is composed of a passive-linear network, generally describing the package, and the laser chip, which is the nonlinear part. This model can be analyzed using harmonic balance method.
11. Laser chirp is a result of the amplitude coupling to the optical resonator. It affects the phase condition's gain and population, which affects the refractive index. This eventually results in an optical frequency shift. Thus, the AM of a laser results in residual FM modulation.
12. Laser chirp in an analog application is directly related to the OMI value. In a digital case, this chirp is related to the extension ratio. The result of

the FM modulation is Bessel spectral lines, which results in IMD and chromatic dispersion due to optical wavelength shift. This affects CNR and IMD dynamic range in analog link and BER as well as eye jitter and noise in digital link.

13. Laser chirp creates spectral-line broadening.

14. Laser distortion sources may result from the bias point defined as the static mechanism, relaxation–oscillation defined as dynamic mechanism, and a combination of both.

15. MPN results from the nonideality of the laser that emits power while hopping between several lasing modes (different lasing wavelengths).

16. Mode partition hopping creates delay time between the modes. This generates sampling misalignment of digital data.

17. Mode partition may affect distortions, create chromatic dispersion, sampling misalignment of digital data sampling that may reduce BER, close the eye diagram, and increase jitter because of the different group delay for each mode.

18. The statistics of BER versus $E_b/N_0$ for MPN is directly related to the mode hopping statistics.

19. Laser RIN results from optical power fluctuations around the average optical power and is referred to as AM noise. These fluctuations are because of quantum fluctuations associated with the lasing process.

20. Laser-phase noise results from optical frequency fluctuations around the center frequency and is referred to as FM noise. These fluctuations are because of quantum fluctuations associated with the lasing process affecting the oscillations phase conditions. Laser-phase noise is similar to chirp but occurs even without modulation.

21. Laser-phase noise, RIN, and distortions are worst at the relaxation–oscillation frequency.

22. To overcome the direct modulation impairments such as frequency chirp, relaxation–oscillation, residual FM, and spectral line broadening, the method is to use special device to modulate the optical signal externally to the laser cavity.

23. There are two main types of optical modulators: index and absorption coefficient.

24. The concept for index modulators is based on the change of refractive index as a function of the electrical field.

25. Absorption coefficient modulation is based on changing the absorption coefficient of the medium as a function of electrical field.

26. The Mach Zehnder modulator is based on the Mach Zehnder interferometer (MZI) used for CWDM while controlling the phase delay by changing the refractive index as a function of the modulating voltage, which changes the electric field in the medium.

27. The MZI modulator should create constrictive and distractive interferences by the modulating signal due to the change of the refractive index, which affects the preparation factor of the optical waveguide of the modulator.

28. There are two methods to designing MZI modulators: lumped-element and distributed-element circuit.

29. The advantage of the distributed-element circuit MZI modulator is its wider BW and data-rate abilities.

30. The advantage of the EA modulator is its integration to the laser.

31. The disadvantage of electroabsorption (EA) modulator with respect to index MZI modulators is that it has residual chirp. This chirp can be minimized in MZI modulators.

32. An additional disadvantage of EA modulator is heat since it absorbs light.

33. An additional advantage of the EA modulator with respect to index-MZI modulators is its lower-modulation voltage requirements.

## References


1. Yariv, A., *Optical Electronics in Modern Communication.* Oxford University Press (1996).
2. Agrawal, G.P., and N.K. Dutta, *Long-Wavelength Semiconductor Lasers.* Van Nostrand Reinhold Company, New York (1986).
3. Agrawal, G.P., "Effects of gain and index nonlinearities on single-mode dynamics in semiconductor lasers." *IEEE Journal of Quantum Electronics*, Vol. 26, pp. 1901–1909 (1990).
4. Asada, M., and Y. Suematsu, "Density matrix theory of semiconductor lasers with relaxation broadening modal-gain and gain suppression in semiconductor lasers." *IEEE Journal of Quantum Electronics*, Vol. 21 No. 5, pp. 434–442 (1985).
5. Tucker, R.S., "High speed modulation of semiconductor lasers." *IEEE Journal of Lightwave Technology*, Vol. 3 No. 6, pp. 1180–1192 (1985).
6. Lau, K.Y., and A. Yariv, "Ultra-high speed semiconductor lasers." *IEEE Journal of Quantum Electronics*, Vol 21 No. 2, pp. 121–138 (1985).

7. Derry, P.L., T.R. Chen, Y.H. Zhuang, J. Paslaski, M. Mittelstine, K. Vahala, and A. Yariv, "Spectral and dynamic characteristics of buried-heterostructure single quantum well (Al,Ga) as lasers." *Applied Physics Letters*, Vol. 53 No. 4, pp. 271–273 (1988).
8. Bowers, J.E., B.R. Hemenway, A.H. Gnauck, T.J. Bridges, and E.G. Burkhardt, "High-frequency constricted MESA lasers." *Applied Physics Letters*, Vol. 47, pp. 78–80 (1985).
9. Nagarian, R., M. Ishikawa, T. Fukushima, R.S. Geels, and J. Bowers, "High speed quantum-well lasers and carriers transport effects." *IEEE Journal of Quantum Electronics*, Vol. 28, pp. 1990–2007 (1992).
10. Kaminow, I.P., and R.S. Tucker, *Mode-Controlled Semiconductors Lasers in Guided-Wave Optoelectronics, 2nd Edition*. T. Tamir, Ed., Springer-Verlag, Berlin (1990).
11. Way, W.I., "Large signal nonlinear distortions predictions for a single-mode laser diode under microwave intensity modulation." *IEEE Journal of Lightwave Technology*, Vol. 9, pp. 1331–1333 (1997).
12. Tucker, R.S., and I.P. Kaminow, "High frequency characteristics of directly modulated InGaAsP ridge waveguide and buried heterostructure lasers." *IEEE Journal of Lightwave Technology*, Vol. 2 No. 4, pp. 385–393 (1984).
13. Tucker, R.S., "High-speed modulation of semiconductor lasers." *IEEE Journal of Lightwave Technology*, Vol. 3 No. 6, pp. 1180–1192 (1985).
14. Habermayer, I., "Nonlinear circuit model for semiconductor lasers." Optical and Quantum Electronics, Vol. 13 No. 6, pp. 461–468 (1984).
15. Lin, M.S., S.J. Wang, and N.K. Dutta, "Measurement and modeling of the harmonic distortion in InGaAs distributed feedback lasers." *IEEE Journal of Quantum Electronics*, Vol. 26 No. 6, pp. 998–1004 (1990).
16. Sze, S.M., *Modern Semiconductor Device Physics, 2nd Edition*. John Wiley & Sons (2001).
17. Henry, C.H., "Theory of the linewidth of semiconductor lasers." *IEEE Journal of Quantum Electronics*, Vol. 18 No. 2, pp. 259–264 (1982).
18. Osinski, M., and J. Buus, "Linewidth broadening factor in semiconductor lasers: an overview." *IEEE Journal of Quantum Electronics*, Vol. 23 No. 1, pp. 9–29 (1987).
19. Harder, C., Vahala, K.J., and A. Yariv, "Measurement of the linewidth enhancement factor of a semiconductor lasers." *Applied Physics Letters*, Vol. 42 No. 4, pp. 328–330 (1983).
20. Koch, T.L., and R.A. Linke, "Effects of nonlinear gain reduction on semiconductor laser wavelength chirping." *Applied Physics Letters*, Vol. 48 No. 10, pp. 613–615 (1986).
21. Coldren, L.A., and S.W. Corzine, *Diode Laser and Photonic Integrated Circuits*. John Wiley and Sons (1995).
22. Ozyazici, M.S., "The complete electrical equivalent circuit of a double heterojunction laser diode using scattering parameters." *Journal of Optoelectronics and Advanced Materials*, Vol. 6 No. 4, pp. 1243–1253 (2004).

23. Kibar, O., D.V. Blerkom, C. Fan, P.J. Marchand, and S.C. Esener, "Small signal equivalent circuits for a semiconductor laser." *Applied Optics*, Vol. 37 No. 26, pp. 6136–6139 (1998).
24. Tucker, R.S., and I.P. Kaminow, "High frequency characteristics of directly modulated InGaAsP ridge waveguide and buried heterostructure lasers." *Journal of Ligthwave Technology*, Vol. 2 No. 4, pp. 385–393 (1984).
25. Tucker, R.S., and D.J. Pope, "Circuit modeling of the effect of diffusion on damping in a narrow stripe semiconductor laser." *IEEE Journal of Quantum Electronics*, Vol. 19, No. 7, pp. 1179–1183 (1983).
26. Katz, J., S. Margalit, C. Harder, D. Wilt, and A. Yariv, "The intrinsic electrical equivalent circuit of a laser diode." *IEEE Journal of Quantum Electronics*, Vol. 17, No. 1, pp. 4–7 (1981).
27. Tucker, R.S., and D.J. Pope, "Microwave circuit models of semiconductor injection lasers." *IEEE Transactions on Microwave Theory and Techniques*, Vol. 31, No. 3, pp. 289–294 (1983).
28. Desoer, C.A., and E.S. Kuh, *Basic Circuit Theory*. McGraw-Hill (1969).
29. Edminister, J.A., *Electric Circuits*. Schaum Publishing Co., NY (1965).
30. Carlson, B., *Communications Systems, Third Edition*. McGraw-Hill (1987).
31. Taub, H., and D.L. Schilling, *Principles of Communications Systems, 2nd Edition*. McGraw-Hill (1986).
32. Lathi, B.P., *Modern Digital and Analog Communication Systems*. Oxford University Press (1998).
33. Sadhwani, R., J. Basak, and B. Jalali, "Adaptive electronic linearization of fiber optics links." *Proceedings of the Optical Fiber Conference*, Vol. 2, pp. 477–479 (2003).
34. Helms, J., "Intermodulation and Harmonic Distortions of Laser Diodes with Optical Feedback." *IEEE Journal of Lightwave Technology*, Vol. 9 No. 11, pp. 1567–1575 (1991).
35. Saleh, A.A.M., "Fundamental limit on number of channels in subcarrier-multiplexed lightwave CATV system." *Electronics Letters*, Vol. 25 No. 12, pp. 776–777 (1989).
36. Biswas, T.K., and W.F. McGee, "Volterra series analysis of semiconductor laser diode." *IEEE Photonics Technology Letters*, Vol. 3 No. 8, pp. 706–708 (1991).
37. Gorfinkel, V.B., and S. Luryi, "Fundamental limits of linearity of CATV lasers." *Journal of Lightwave Technology*, Vol. 13 No. 2, pp. 252–260 (1995).
38. Rainal, A.J., "Limiting distortions of CATV lasers." *IEEE Journal of Lightwave Technology*, Vol. 14 No. 3, pp. 474–479 (1996).
39. Okuda, T., H. Yamada, T. Toriaki, and T. Uji, "DFB laser intermodulation distortion analysis taking longitudinal electrical field distribution into account." *IEEE Photon Technology Letters*, Vol. 6, pp. 27–30, January (1994).
40. Takemoto, A., H. Wantanabe, Y. Nakajima, Y. Sakakibara, S. Kakimoto, J. Yamashita, T. Hatta, and Y. Miyake, "Distributed feedback laser diode and module for CATV systems." *IEEE Selected Areas in Communications*, Vol. 8 No. 7, pp. 1359–1364 (1990).

41. Yonetani, H., I. Ushijima, T. Takada, and K. Shima, "Transmission characteristics of DFB lasers modules for analog applications." *IEEE Journal of Lightwave Technology*, Vol. 11 No. 1, pp. 147–153 (1993).
42. Okuda, T., H. Yamada, T. Torikai, and T. Uji, "Novel partially corrugated waveguide laser diode with low modulation distortions characteristics for subcarrier multiplexing." *Electronics Letters*, Vol. 30 No. 11, pp. 862–863 (1994).
43. Soda, H., Y. Kotaki, H. Sudo, H. Ishikawa, S. Yamakoshi, and H. Imai, "Stability in single longitudinal mode operation GaInAsP/InP phase-adjusted DFB lasers." *IEEE Journal of Quantum Electronics*, Vol. 23 No. 6, pp. 804–814, June (1987).
44. Kawamura, H., K. Kamite, and H. Yonetani, "Effect of varying threshold gain on second-order intermodulation distortion in distributed feedback lasers." *Electronic Letters*, Vol. 26 No. 20, pp. 1720–1721 (1990).
45. Valle, A., C.R. Mirasso, and L. Pesquera, "Mode partition noise of nearly single-mode semiconductor lasers modulated at ghz rates." *IEEE Journal of Quantum Electronics*, Vol. 31 No. 5, pp. 876–885 (1995).
46. Alalusi, M.R., and R.B. Darling, "Effects of nonlinear gain on mode-hopping in semiconductor laser diodes." *IEEE Journal of Quantum Electronics*, Vol. 31 No. 7, pp. 1181–1192 (1995).
47. Liu, P.L., and M.M. Choy, "Modeling rare turn-on events of injection lasers." *IEEE Journal of Quantum Electronics*, Vol. 25 No. 8, pp. 1767–1770 (1989).
48. Ogawa, K., "Analysis of mode partition noise in laser transmission systems." *IEEE Journal of Quantum Electronics*, Vol. 18 No. 5, pp. 849–855 (1982).
49. Ito, T., S. Machida, K. Nowata, and T. Ikegami, "Intensity fluctuations in each longitudinal mode of a multimode AlGaAs laser." *Journal of Quantum Electronics*, Vol. 13 No. 8, pp. 574–579 (1977).
50. Meslener, G.J., "Mode partition noise in microwave subcarrier transmission systems." *IEEE Journal of Lightwave Technology*, Vol. 12 No. 1, pp. 118–126 (1994).
51. Agarwal, G.P., and N.K. Dutta, *Semiconductor Lasers*, 2nd Edition. Kluwer Academic (1993).
52. Nakamura, M., N. Suzuki, Y. Uematsu, T. Ozeki, and S. Takahashi. "Laser linewidth requirement for eliminating modal noise in pulse frequency modulation video transmission." *IEEE Journal of Lightwave Technology*, Vol. 2 No. 5, pp. 735–740 (1984).
53. Scott, R.P., C. Langrock, and B.H. Kolner, "High dynamic range laser amplitude and phase noise measurement techniques." *IEEE Journal on Selected Topics in Quantum Electronics*, Vol. 7 No. 4, pp. 641–655 (2001).
54. Henry, C.H., "Theory of linewidth of semiconductor lasers." *IEEE Journal of Quantum Electronics*, Vol. 18 No. 2, pp. 259–264 (1982).
55. Aoki, M., K. Uomi, T. Tsuchiya, S. Sasaki, M. Okai, and N. Chinone, "Quantum size effect on longitudinal spatial hole burning in mqw m4-shifted dfb lasers." *Journal of Quantum Electronics*, Vol. 27 No. 6, pp. 1782–1789 (1991).

56. Sabella, R., and P. Lugli, *High Speed Optical Communication*. Kluwer Academic Publishers (1999).
57. B. Liu, J. Shim, Y.J. Chiu, A. Keating, J. Piprek, and J.E. Bowers, "Analog characterization of low-voltage MQW traveling wave electroabsorption modulators." *IEEE Journal of Lightwave Technology*, Vol. 21 No. 12, pp. 3011–3019 (2003).
58. Dagenais, M., R.F. Leheny, and J. Crow, *Integrated Optoelectronics*. Academic Press (1995).
59. Chen, J.C., "Comparison and linearization of $LiNbO_3$ and semiconductor modulators." *IEEE Journal of Optical Communication*, Vol. 22 No. 2, pp. 2–8 (2001).
60. Gupta, K.C., R. Garg, and I.J. Bahl, *Microstrip Lines and Slot Lines*. Artech House (1996).
61. Bian, Z., J. Christofferson, A. Shakouria, and P. Kozodoy, "High-power operation of electroabsorption modulators." *Applied Physics Letters*, Vol. 83 No. 17, pp. 3605–3607 (2003).
62. Chiu, Y., S.Z.Zhang, V. Kaman, J. Piprek, and J.E. Bowers, "High-speed traveling-wave electro-absorption modulators." *Proceedings of SPIE*, Vol. 4490, pp. 1–10 (2001).
63. Aimeza, V., J. Beauvais, J. Beerens, S.L. Ng, and B.S. Ooi, "Monolithic intracavity laser-modulator device fabrication using postgrowth processing of 1.55 mm heterostructures." *Applied Physics Letters*, Vol. 79 No. 22, pp. 3582–3584 (2001).
64. Lau, K.Y., and A. Yariv, "Ultra high speed semiconductor lasers." *Journal of Quantum Electronics*, Vol. 21 No. 2 (1985).
65. Streifer, W., R.D. Burnham, and D.R. Scifres, "Effect of external reflectors on longitudinal modes of distributed feedback lasers." *IEEE Journal of Quantum Electronics*, Vol. QE-11 No. 4, pp.154–161 (1975).

# Chapter 8
# Photodetectors

## 8.1 Photodetectors and Detection of General Background

Photodetectors are semiconductor devices that are capable of detecting light through a quantum electronic process. There are a variety of photodetectors, such as infrared, ultraviolet, and those used for optical communications, which operate near the infrared wavelength region of 0.8–1.6 μm. In communications applications, optical detectors are also categorized by their semiconductor materials and the bandgap energies used to realize them (Table 8.1).

A different way to categorize photodetectors is by their junction structure, PN, PIN, or avalanche photodetector (APD). In the PN photodiode (PD), electron-hole pairs are created in the depletion region of the NP junction proportional to the optical power. The electron and hole pairs are swept out by the electric field, creating a photocurrent. In the PIN photodetector, the electric field is concentrated in a thin intrinsic layer *I* sandwiched between N and P types. APD is similar to a PIN but has an additional layer (for instance, $P^+LPN^+$ where L is a lightly doped layer). Thus, this kind of diode experiences a photocurrent gain. APDs are used for high-sensitivity, long-haul systems, such as small form pluggable (SFP) transceivers or wireless free-space optical links (optical data telescopes).[17]

The performance of a photodetector is measured by three parameters: quantum efficiency (QE) $\eta$, the response time to generate carriers, and the sensitivity that defines the carrier to noise ratio (CNR) limits of a link. The last parameter will be reviewed in Chapter 11 for all types of detectors discussed in this chapter.

### 8.1.1 Quantum efficiency and responsivity

The optical signal arriving at the detector is a modulated light signal. Consequently, the rate of photons impinging on the photodetector is proportional to the optical modulation index (OMI). Hence, the optical signal beating is counted by the photodetector while generating the photocurrent. In order to detect each individual photon and to convert it into a photocurrent, the detector should have infinite bandwidth (BW). It is also known that the photon generation

**Table 8.1** Photodetector families per semiconductor materials.

| Material | Bandgap Energy (eV) | Wavelength Range (nm) | Peak Response Wavelength (nm) | Responsivity Max Value ($A/W$) (mA/mW) |
|---|---|---|---|---|
| Si | 1.17 | 300–1100 | 800 | 0.5 |
| Ge | 0.775 | 500–1800 | 1550 | 0.7 |
| InGaAs | 0.75–1.24 | 1000–1700 | 1700 | 1.1 |

by an optical transmitter is a random process. Thus, the arrival of photons to the photodetector is a random process as well. Therefore, photodetection can be analyzed as a receiver front end with random signal and noise.[14] A photodetector detects and responds to the electrical field impinging on its active area. Consider a beam of light's instantaneous optical electric field $E(t)$ with optical power $P_s$ and impedance of the medium $Z_0$. It is also known that the optical power measured is an average power. Hence, the following relation is applicable:

$$E(t) = \sqrt{2P_s Z_0}\cos(\omega_c t + \varphi). \tag{8.1}$$

Taking the above expression for the optical power $P_s$, it is given in units of $V^2/(\Omega \cdot \mathrm{m}^2)$ or $\mathrm{W/m^2}$, which are units of flux $S$ or Poynting vector $\mathbf{E} \times \mathbf{H}$ of a propagating electromagnetic wave. The absorption of the beam power by the photodetector generates the photocurrent.[15] It is known from solid-state devices theory[1,15] that the conductivity of an intrinsic region is given by

$$\sigma = q\left(\mu_n n + \mu_p p\right), \tag{8.2}$$

where $q$ is the electron charge, $\sigma$ is the conductivity in units of $1/(\Omega \times \mathrm{m})$, $\mu_n$ and $\mu_p$ are the electron and hole motilities, respectively in units of $\mathrm{m}^2/(\mathrm{V} \times \mathrm{s})$, and $n$ and $p$ are the electron and hole concentrations in units of $1/\mathrm{m}^3$. Thus, the photodetector under illumination increases its conductivity because the carriers' number increases. Knowing these facts, it is valid to state that the increase in conductivity arising from photon flux $\Phi$ (photons per second) illuminating a semiconductor volume $W \times A$ with a length of $W$ and cross section $A$ can be calculated.[18] However, when this photodetector is illuminated by a long wavelength, its cutoff wavelength is given by the photoelectric relation of

$$\lambda_C = \frac{hc}{E_g} \tag{8.3}$$

where $h$ is Planck's constant, $c$ is the speed of light, and $E_g$ is the bandgap energy of the semiconductor. In case of a wavelength shorter than $\lambda_C$, the incident radiation is absorbed by the semiconductor and electron-hole pair is created. As for extrinsic case, photoexcitation may result between the band edge and the energy level in the energy gap. Photoconductivity can take place by absorption of photons with energy equal to or greater than the energy separation of the bandgap levels and the conduction or valence band. Thus, in long wavelength, cutoff is defined by the depth of the forbidden gap energy level as shown in Fig. 8.1.

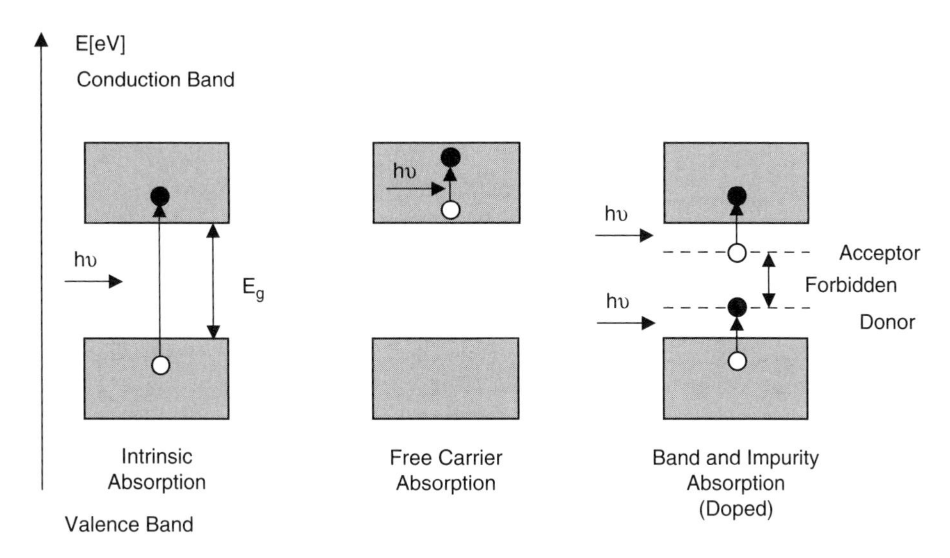


**Figure 8.1** Photon absorption process is a semiconductor with and without doping.

The QE $\eta$, $(0 \leq \eta \leq 1)$, of a photodetector is defined as the probability that a single photon, incident on the device, would generate a photocarrier pair of hole and electron that contributes to the detector current. When a flux of many photons impinges the surface of a semiconductor, $\eta$ is the ratio between the effective flux, which generates an electron-hole pair to the total flux incident the device active surface. To create electron-hole pairs, the flux of photons must be absorbed by the device. For this reason, not all photons generate electron-hole pairs, since the entire flux is not absorbed totally by the device. The reasons for that are the statistics of absorption process and reflection from the device surface. Some other pairs recombine fast and, additionally, the beam spot sometimes is not aligned well on the device active area. Hence, the QE is given by[18]

$$\eta = (1 - \Gamma)\zeta[1 - \exp(-\alpha d)], \tag{8.4}$$

where $\Gamma$ is the surface optical reflectance coefficient, $\zeta$ is the fraction of electron-hole pairs that contributes to the current, and $\alpha$ is the absorption coefficient of the material in $\text{cm}^{-1}$. Parameters in Eq. (8.4) can be optimized for better performance. $(1 - \Gamma)$ can be maximized by having antireflection coating on the active surface of the PD. $\zeta$ can be improved by careful material growth reducing nonproductive surface recombination. Finally, the last factor can be maximized by having a sufficiently large value of depth for $d$ or $L$, as noted in other references. Additionally, $\eta$ is a function of wavelength. This limitation is noted by Eq. (8.3). If $\lambda_0 \geq \lambda_C$, absorption cannot occur. The photon energy will not overcome the bandgap energy $E_g$ since $E_g \geq hc/\lambda_0$. On the other hand, a wavelength that is too short would cause nonproductive pairs that would recombine at the surface due to the short lifetime of the carrier pairs. Most of the light is absorbed within a distance of $1/\alpha$.

Now the responsivity $r$ can be defined. It is known that each photon generates an electron-hole pair. A flux of photons $\Phi$ [1/s] produces electric current:

$$i_p = q\Phi. \tag{8.5}$$

This flux has an optical power given by

$$P = h\upsilon\Phi, \tag{8.6}$$

where $\upsilon$ is the optical frequency. Hence, using Eqs. (8.4)–(8.6), the photocurrent can be noted as a function of the responsivity $r$:

$$i_p = \eta q\Phi = \frac{\eta qP}{h\nu} = rP. \tag{8.7}$$

Responsivity units are in A/W, mA/mW, or any other factor of current to power. The responsivity expression given by Eq. (8.7) can be written in terms of wavelength:

$$r = \frac{\eta q}{h\nu} = \eta\frac{q\lambda_0}{hc} = \eta\lambda_0\frac{1}{1.24}. \tag{8.8}$$

It can be observed that $r$ increases with respect to wavelength. However, $\lambda_0$ should satisfy the bandgap and absorption conditions. Figure (8.2) provides the absorption coefficient values.

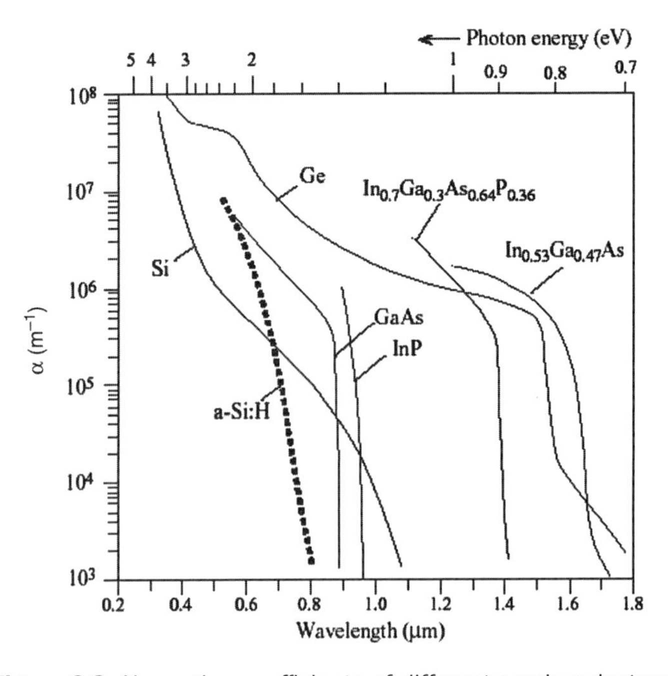


**Figure 8.2** Absorption coefficients of different semiconductors.

### 8.1.2 Time response

In the previous section, it was explained that the electron-hole pair is created by a photon incident to the semiconductor. The charge of this hole-electron pair with a value of $q$ is delivered to a circuit. This charge travels through the semiconductor at different times, since a hole and an electron have different mobilities. Hence, the hole and electron velocities are given by $v_h = E \cdot \mu_h$ and $v_e = E \cdot \mu_e$, respectively. This process is called transit time spread. This is one of the factors limiting the speed and BW of photodetectors. When dealing with a photocurrent carrier, the charge $Q$ is composed of either a hole $q$ or an electron $-q$ charges. This carrier motion is at a velocity $v(t)$. Assuming $e$ length of $W$, the instantaneous current is given by Ramo's theorem:[19]

$$i(t) = -\frac{Q}{w} v(t). \tag{8.9}$$

Assuming a charge of hole and electron created at a spot $x$ in the semiconductor, the electron travels the distance $w - x$, while the hole travels the distance of $x$. Their velocities under the electric field are as mentioned above. Thus, the total charge coming out of the semiconductor is given by

$$\begin{aligned} q = i_h \cdot t_1 + i_e \cdot t_2 &= \left(q\frac{v_h}{w}\right) \cdot \left(\frac{x}{v_h}\right) + \left(q\frac{v_e}{w}\right) \cdot \left(\frac{w - x}{v_h}\right) \\ &= q\left(\frac{x}{w} + \frac{w - x}{w}\right). \end{aligned} \tag{8.10}$$

The conclusion from the above discussion and Eq. (8.10) is that the hole and electron do not generate twice the charge of 29 but the charge of 9 is preserved.

Figure 8.3 illustrates the carrier transport inside the semiconductor explained above. Going on with this discussion, in the presence of an electric field $E$, a charge carrier will drift in mean velocity given by mean mobility $\mu$; hence, $v = \mu E$. Because of this, $J = \sigma E$ and $\sigma = \mu Q$, which justifies the claim of Eq. (8.2). The conductivity increment can be explained as follows: A hole-electron pair is created by a given photons flux $\Phi$ incident to the semiconductor volume $W \times A$. The rate of carrier creation $R$ is proportional to the QE $\eta$. This extra charge of holes and electrons is given by $\Delta n$ and under steady state $R = \Delta n/\tau$, where $\tau$ is the excess carrier recombination lifetime. In other words, the creation of new carriers is equal to the recombination rate under a steady state, so $\Delta n = \eta\tau\Phi/(W \times A)$. This results in an expression for the increase in conductivity:

$$\Delta\sigma = q\Delta n(\mu_e + \mu_h) = \frac{q\eta\tau(\mu_e + \mu_h)}{WA}\Phi. \tag{8.11}$$

As a conclusion, the photodetection process can be described as using a resistor sensitive to light. A flux of photons impinging on the resistor reduces its resistance,

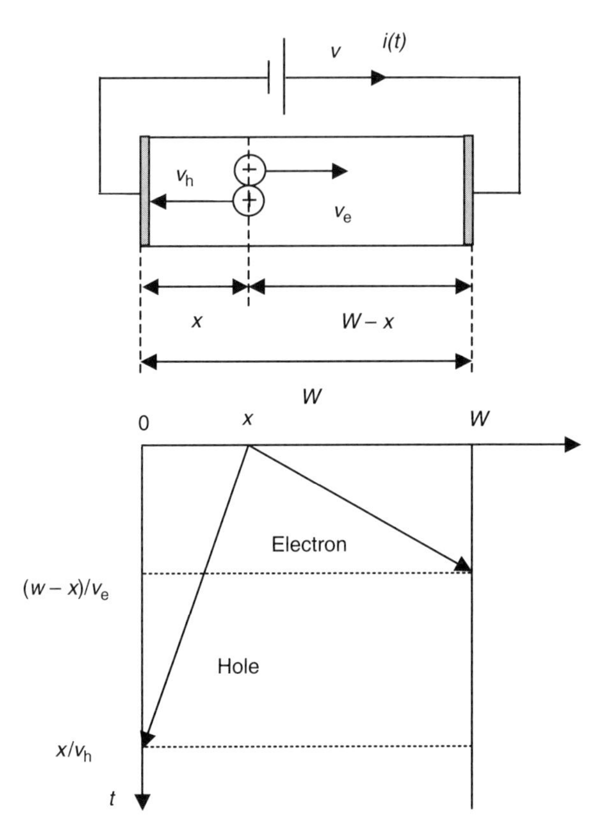


**Figure 8.3** Creation of an electron-hole pair in a semiconductor at a random position $x$. The hole drifts across the distance $x$ slowly, whereas the electron travels faster down the path $w - x$.

therefore, more current passes through it. This extra current results due to the reduction of resistance in the photocurrent. Now it is easier to understand BW limitations of photodetectors in a different perspective. When a modulated flux of light impinges on the photodetector, it modulates the detector resistance and varies it around an average value. The ability of the semiconductor to follow these fluctuations defines its BW. This limitation refers to the transport time. The other parameter limiting the speed of photodetectors is its internal series resistance $R_s$ and depletion capacitance $C_d$. This is a different mechanism that refers to the detector structure and bias method defining $\tau = C_d R_s$.

### 8.1.3 Sensitivity and noise in photodetectors

From a review of previous sections, it is understood that a photodetector is a device that measures photons' flux. In fact, a photodetector responds to the photon flux rather than the energy or field. As was reviewed, the flux energy to move an electron from its valence level relates to its wavelength, and the detector response to wavelength is directly related to the bandgap energy. Thus, the photocurrent is a result of the electron-hole statistical count by the photodetector with a Poisson distribution process.[14,18] However, through the process of detection and photocurrent generation, there is additive noise. This noise is an additional random fluctuation around the current value. Thus, $\sigma^2 = \langle (i - \hat{i})^2 \rangle$. Assuming this process has a zero mean $\hat{i} = 0$ and standard deviation $\sigma$, the standard deviation is the rms value of the noise current $\sigma = i_{\mathrm{rms}} = \sqrt{\langle i^2 \rangle}$. There are several noise sources that are an inherent part of photon detection:

- Photon noise: This results from the random process of photon detection described by the Poisson distribution process of random photon arrival and count.
- Photoelectron noise: This describes the probability of detection. In the case of QE $\eta < 1$, a photon will generate an electron-hole pair with probability $\eta$. However, it fails to generate an electron-hole pair with the complementary probability. The photon failing this transfers its energy to phonons and vibration, resulting in noise.
- Gain noise: This mechanism exists in gain photodetectors such as APD. In this case, each detected photon generates a random number of electron-hole pairs with an average value. The deviation from the mean value is random, resulting in noise.
- Receiver's electronics noise: This is the circuit noise of the components.
- Dark current noise: This is the noise created by the leakage current of the photodetector without the input of incidenting photon flux.

These noise sources define the system performance by setting the limits for signal-to-noise ratio (SNR). Further detailed investigation of noise statistics and system performance is given in Chapter 11 of noise modeling.

### 8.1.4 Gain process

It was explained that in the steady state, the rate of generation equals the recombination.

$R = \Delta n/\tau$, where $\tau$ is the carrier lifetime. The number of carriers in a unit of volume at $t = 0$ equals $n_0$. The decay is given by $n = n_0 \exp(-t/\tau)$. Thus, $dn/dt$ is the recombination rate. Assume a photoconductor with length $W$, cross section $A$, width $H$, and thickness $D$, where $A = H \times D$. Assume now that $D >> 1/\alpha$ thickness and is larger than the light penetration depth or absorption depth. Then, the total steady-state generation rate of carriers per unit of volume is given by

$$R = \frac{n}{\tau} = \frac{\eta(P_S/h\nu)}{WHD}. \tag{8.12}$$

The photocurrent between the electrodes is given by

$$I_{\mathrm{p}} = (\sigma E) \cdot A = (q\mu n E) \cdot A = (qnV_{\mathrm{d}}) \cdot A, \tag{8.13}$$

where $V_{\mathrm{d}}$ is the drift velocity under the influence of the field $E$. Taking the value of $n$ in Eq. (8.12) and substituting it in Eq. (8.13) results in

$$I_{\mathrm{p}} = q\left(\eta \frac{P_{\mathrm{S}}}{h\nu}\right)\left(\frac{\mu \tau E}{W}\right). \tag{8.14}$$

That is because of differences in mobilities. When a voltage is applied across the semiconductor, there is an electric field $E$, and the charge velocity is $v = \mu E$. The time to travel the distance $W$ is $T = W/\mu E$, which is defined as transit time.

Now, if the photocurrent is defined by

$$I_{\mathrm{ph}} = q\left(\eta \frac{P_{\mathrm{S}}}{h\nu}\right). \tag{8.15}$$

The photocurrent gain is defined as the ratio between $I_{\mathrm{p}}$ and $I_{\mathrm{ph}}$ and is given by

$$G = \frac{I_{\mathrm{p}}}{I_{\mathrm{ph}}} = \frac{\mu \tau E}{W} = \frac{\tau V_{\mathrm{d}}}{W} = \frac{\tau}{t_{\mathrm{r}}}, \tag{8.16}$$

where $t_{\mathrm{r}}$ is the carrier transit time. Thus, the gain depends upon the ratio between lifetime and transit time. That means that for a high enough $\tau$ value, there are several trips of electrons within the semiconductor, until the hole finishes its electrode and recombines with the electron. Hence, the number of trips until recombination is the gain $G$. Regular photodetectors such as PN and PIN have a gain $G = 1$. APD gain range is 100–10,000 approximately.

Consider now AM optical signal with, given OMI value of

$$P_{\mathrm{Opt}}(\omega) = P_{\mathrm{s}}[1 + \mathrm{OMI} \cdot \exp(j\omega t)]. \tag{8.17}$$

Then the average detected signal, including the detector frequency response, is given by

$$i_{\mathrm{p}} \approx \frac{q\eta \cdot \mathrm{OMI} \cdot P_{\mathrm{Opt}}}{\sqrt{2}\, h\nu}\left(\frac{\tau}{t_r}\right)\frac{1}{\sqrt{1 + (\omega\tau)^2}}. \tag{8.18}$$

## 8.2 Junction Photodetector

The ability to respond to light is a fundamental requirement of all optical receivers. Essentially a photodetector is a front-end device of any fiberoptic communications receiver. There are several types of detectors, such as PN, PIN, APD, and metal semiconductor metal (MSM), described in the literature.[18] Advanced semiconductor technology enables the growth of heterojunctions made of several layers of semiconductors. This way, parameters of photodetectors (such as bandgap energy) can be controlled and manipulated through the absorption path of a photon. Moreover, the absorption depth is controlled too.[16] Photodetectors can be classified into two categories, photoconductive and photovoltaic. Photoconductive detectors operate as photosensitive resistors. Under this family are all communications photodetectors, which are going to be reviewed here, and their quantum physics concept of operation will be briefly introduced. Photovoltaic detectors are solar detectors that produce voltage in the presence of light. Further information about photovoltaic detectors can be found in Refs. [1], [15], [16], and [18]. Simply, photoconductive photodetectors are homogeneous semiconductor slabs with ohmic contacts as shown in Fig. 8.3. Reverse biased PN junction and its diverse fall in this category. It has a region known as the depletion region, where a large build in electric field is created due to the bandgap difference between the PN materials. Such a zone is the place where high velocity carriers create the drift current when there is illumination.

### 8.2.1 PN photodetector

Before embarking on a study of deep concepts of PN photodetectors, a brief review about abrupt PN junction structure is provided.[1] In order to grow a junction, it is essential to have at one side of the junction an excessive concentration of donors, and on the other side an excessive concentration of acceptors. This is created by the doping process. Under thermodynamic equilibrium, without any external voltage being applied on the junction, there is electrical charge neutrality. Thus, at those neutral zones of the semiconductor, the total charge equals zero:

$$N_D + \bar{p} = N_A + \bar{n}. \tag{8.19}$$

At the junction boundary, there is a large gradient of carriers from both sides due to the difference in concentrations between P and N sides. This will cause diffusion from both sides of the junction. N electrons migrate to the P side, P holes migrate to the N side, and the charge is built up. As a result, a potential difference is created in the junction and an electric field is created. This field prevents more migration of carriers. Consequently, a depletion region is created. There are two types of current for each type of doping diffusion, due to the gradient in concentrations and drift current in the opposite direction due to the electric field. At thermodynamic equilibrium, the hole and the electron current is zero. This is called "principle of detailed balance." Following this process description, the junction equations are noted:

$$\begin{cases} J_h = q_e \mu_h \bar{p} E - q_e D_h \dfrac{d\bar{p}}{dx} = 0 \\ J_e = q_e \mu_e \bar{n} E + q_e D_e \dfrac{d\bar{n}}{dx} = 0, \end{cases} \tag{8.20}$$

where $D_e$ is the electrons diffusion coefficient, $D_h$ is the holes diffusion coefficient, $N_D$ is the donors doping concentration, and $N_A$ is the acceptors doping concentration. Integrating the Eq. (8.20) the pair of equations will provide the internal junction potential:

$$\begin{cases} V_B = \dfrac{D_h}{\mu_h}\ln\left(\dfrac{\bar{p}_p}{\bar{p}_n}\right) = \dfrac{D_h}{\mu_h}\ln\left(\dfrac{N_A N_D}{n_i^2}\right) \\ V_B = \dfrac{D_e}{\mu_e}\ln\left(\dfrac{N_A N_D}{n_i^2}\right). \end{cases} \tag{8.21}$$

Since potential $V_B$ is the same, Einstein's law applies:

$$\frac{D_e}{\mu_e} = \frac{D_h}{\mu_h} = \frac{kT}{q}. \tag{8.22}$$

Hence, Eq. (8.21) state that this voltage is the thermodynamic voltage with a correction factor related to the doping concentrations. In thermodynamic equilibrium, it is also known that $np = n_i^2$. The method to solve the junction parameters (such as electric field, potential, and bandgap) is by using the Poisson relation between field gradient and charge throughout the diode depth. Figure 8.4 illustrates the PN junction under reverse bias:

$$-\frac{d^2V}{dx^2} = \frac{dE}{dx} = \frac{q(N_D - N_A + p - n)}{\varepsilon_r\varepsilon_0}. \tag{8.23}$$

Integrating Eq. (8.23) twice and using the boundary conditions of charge, field, and voltage continuity will provide the depletion depth, field, and potential at the PN junction.

The depletion depth in the N region is given by

$$d_n = \sqrt{\frac{2\varepsilon_0\varepsilon_r V_t}{q} \cdot \frac{N_A}{N_D(N_A + N_D)}}. \tag{8.24}$$

The depletion depth in the P region is given by

$$d_p = \sqrt{\frac{2\varepsilon_0\varepsilon_r V_t}{q} \cdot \frac{N_D}{N_A(N_A + N_D)}}. \tag{8.25}$$

Thus, the total depletion region length is the sum of both $d_p$ and $d_n$. The electric field $E$ at both N and P sides is given by

$$\begin{cases} E = -\dfrac{dV}{dx} = \dfrac{qN_D}{\varepsilon_0\varepsilon_r}(d_n + x) & -d_n \le x \le 0 \\ E = -\dfrac{dV}{dx} = \dfrac{qN_A}{\varepsilon_0\varepsilon_r}(d_p - x) & 0 \le x \le d_p. \end{cases} \tag{8.26}$$

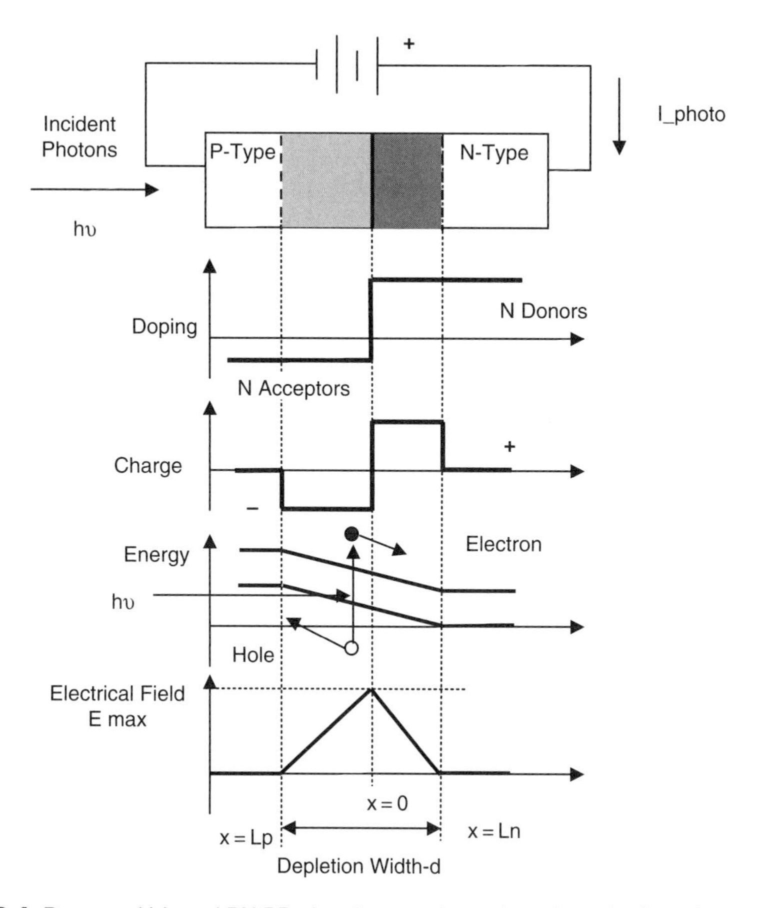


**Figure 8.4** Reversed biased PN PD showing an abrupt junction, doping, charge, energy, and electric field. Note that depletion depth goes deeper to areas of lower doping densities.

The maximum value of the $E$ field occurs at $x = 0$ and is given by

$$E_{\mathrm{max}} = \sqrt{\frac{2qV_{\mathrm{t}}}{\varepsilon_0\varepsilon_{\mathrm{r}}} \cdot \frac{N_{\mathrm{A}}N_{\mathrm{D}}}{N_{\mathrm{A}} + N_{\mathrm{D}}}}. \tag{8.27}$$

The potentials with respect to $x = 0$ are given by

$$\begin{cases} V = \dfrac{qN_{\mathrm{D}}}{\varepsilon_0\varepsilon_{\mathrm{r}}}\left(\dfrac{x^2}{2} + d_{\mathrm{n}}x\right) & x \leq 0 \\ V = \dfrac{qN_{\mathrm{A}}}{\varepsilon_0\varepsilon_{\mathrm{r}}}\left(\dfrac{x^2}{2} - d_{\mathrm{p}}x\right) & 0 \leq x. \end{cases} \tag{8.28}$$

Because of the build-up charge and depletion region created in the PN junction, a junction depletion capacitance results:

$$C = \frac{\varepsilon_0 \varepsilon_r A}{d} = \frac{\varepsilon_0 \varepsilon_r A}{\sqrt{\frac{2\varepsilon_0 \varepsilon_r V_t}{q(N_A + N_D)} \left( \sqrt{\frac{N_A}{N_D}} + \sqrt{\frac{N_D}{N_A}} \right)}}, \tag{8.29}$$

where $A$ is the cross section of the junction and $d$ is the total depletion region width.

So far, the introduction above described the junction under thermodynamic equilibrium. However, while illuminating, this equilibrium is disturbed and photocurrent is created as explained previously. Excessive carriers diffuse and then drift through the depletion region. This process defines the dimensions of the junction. There are two types of PN junction diodes: thick and thin. The definition is by comparing the N and P width to the diffusion depth. Thus, in a photodetector, it is desired that the absorption depth be sufficient to enable carriers to diffuse toward the depletion region and be attracted by the depletion field creating the drift current. In fact, optimum performance of QE requires the absorption depth to overlap the depletion region. This way, carriers create a larger amount of drift current, which means larger QE. In other words, the photons' flux generates current that is more productive. Observing Eq. (8.4) for an ideal diode where surface reflections are zero and assuming all photons produce productive current, the optimum depletion region for desired QE can be calculated:

$$L_d = \frac{-1}{\alpha} \ln(1 - \eta), \tag{8.30}$$

where $L_d$ is the depletion region depth, $\eta$ is the QE, and $\alpha$ is the absorption factor.

The other aspect in designing a PD is to minimize its junction capacitance, known as $C_j$ or depletion capacitor. When observing Eqs. (8.24), (8.25), and (8.29), it is clear that the nonbiased depletion region depends on the doping concentrations. Moreover, this defines the PD inherent junction capacitance. One of the methods to increase the depletion region, thereby reducing the junction capacitance, is reverse biasing. The goal is to increase BW and speed by minimizing the RC constant. On the other hand, by increasing the depletion region and minimizing $C_j$, the transport time increases, since the carrier has to drift through a larger depletion region. Because of this, BW will be reduced. The outcome of this discussion is that it is desired to have a larger depletion region, lower $C_j$, optimal transport time, and high QE in order to accommodate all design goals of a PD. Thus, design trade-offs are made, and a different PD topology (such as PIN) with larger depletion region is used.

### 8.2.2 PIN photodetector

One of the popular photodetectors is the PIN detector which is a PN junction with an intrinsic layer of semiconductor sandwiched between P and N layers

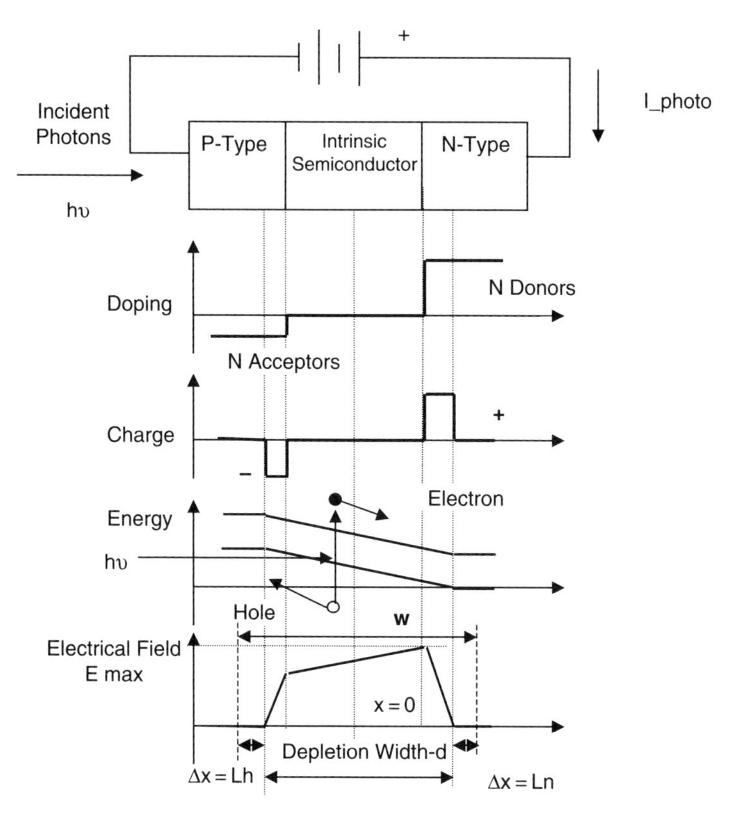


**Figure 8.5** Reversed bias PIN photodetector. *W* is the active region including diffusion regions.

(see Fig. 8.5). The intrinsic layer is a lightly doped layer with respect to both P and N types. The PIN diode junction is governed by the same laws of thermodynamics as a PN junction. However, because of the intrinsic region (I–region), the electric field in the depletion region is more uniform, as shown in Fig. 8.5. The PIN diode is reversed biased and thus operates at the third quadrant of the IV curve, as shown in Fig. 8.6. When dark or illuminated, the diode is in its reversed bias condition. The diode operates below its breakdown voltage.

Illuminating the diode excites the generation of the minority carriers and increases their concentration and the photonic leakage current through the diode. The device should have a thin doping area, so the radiation that hits the N side is absorbed and excites the minority carriers near and within the depletion region. Pairs of holes and electrons generated near the depletion are separated and attracted by the electrical field and generate a field-drift current. Minority carriers that are generated within diffusion range to the depletion zone have a high probability to generate photocurrent before their recombination. Pairs that are far from the depletion zone would have recombination and would not generate any current. Hence, in the first approximation, the photocurrent equals the

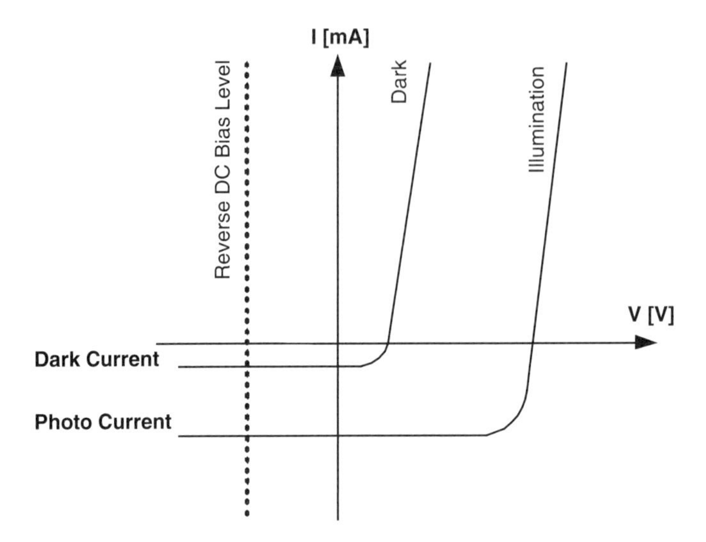


**Figure 8.6** Photodetector in reversed bias.

excitement rate $R[1/\text{s} \times \text{cm}^3]$ multiplied by the active volume with area $A$ and length $W$. The photocurrent is given by[1,15]

$$I_{\text{ph}} = q_{\text{e}}AWR. \tag{8.31}$$

Since the active length $W$ (Fig. 4.1) equals the sum of depletion depth $L_{\text{d}}$ in both N and P and the intrinsic region and diffusion length $L_{\text{e}}$ and $L_{\text{h}}$ of both electrons and holes, the photocurrent, $I_{\text{ph}}$, notation is given by

$$I_{\text{ph}} = qAR(L_{\text{e}} + L_{\text{h}} + L_{\text{d}}). \tag{8.32}$$

where the diffusion distances for electrons and holes are given by[1,15]

$$\begin{cases} L_{\text{e}} = \sqrt{D_{\text{e}}\tau_{\text{e}}} \\ L_{\text{h}} = \sqrt{D_{\text{h}}\tau_{\text{h}}} \end{cases}, \tag{8.33}$$

where $D_{\text{h}}$ and $D_{\text{e}}$ are the hole and electron diffusion factors, and $\tau_{\text{h}}$ and $\tau_{\text{e}}$ are the hole and electron lifetime, respectively. In order to improve $I_{\text{ph}}$ response, it can be seen from the expression of $I_{\text{ph}}$ that $W$ should be enlarged. Making the depletion depth over the entire volume where there is an excitement of minority carriers by light would result by the increase in the photocurrent. This is the advantage of a PIN junction over a PN. Applying reversed bias, even in a low value, increases the depletion depth throughout the entire intrinsic layer, creating a larger active region. However, it is impossible to have an undoped layer. Assuming a weak concentration ND, the field is going to be as shown in Fig. 8.5. Observing the Poisson relation in Eq. (8.23) between field gradient and charge throughout the diode depth and solving it results in the solution for the field, junction potential, and depletion regions for the PIN diode case. As was shown before, it can be easily seen that the higher the reverse voltage, the wider the depletion depth. $V_J$ is the junction bandgap potential, $V_{dc}$ is the reverse bias, and $\psi = Vt = V_J - V_{dc}$, where $V_{dc} < 0$ because it is reverse biased. The calculation concept is to solve the PIN boundary conditions of its three sections and calculating the depletion region in the P side and the N side. Thus, $L_d = x_p + d + x_n$ where $d$ is the depletion length. The solution of the $I$ region for a PIN as a function of bias is given by:[19]

$$\begin{cases} x_n = -\xi + \sqrt{\xi^2 - \xi d \dfrac{N_I}{N_D} + \dfrac{2\varepsilon N_A \psi}{q N_D \left(1 + \dfrac{N_A}{N_D}\right)}} \\ x_p = \dfrac{N_D}{N_A} x_n + \dfrac{N_I}{N_A} d \\ \xi = \dfrac{d(N_I \varepsilon_I + N_A \varepsilon)}{\varepsilon_I (N_A + N_D)}, \end{cases} \tag{8.34}$$

where $\varepsilon_I$ is the dielectric constant of the $I$ region, and $\varepsilon$ is the dielectric constant of the P and the N sides. The depletion region always penetrates more and will be thicker at the lower doping concentrations, as was explained in the previous section and can be seen from Eq. (8.34). Therefore, the entire $I$ region would be depleted in a PIN diode, whereas the depleted regions in N and P sides are thinner.

For $W \approx 1/\alpha$, where $\alpha(\lambda)[\text{cm}^{-1}]$ is the light absorption factor of the semiconductor, generation of minority carriers would be throughout the length $W$, and $I_{ph}$ would be higher. Charges generated within the diffusion distances $L_e$ and $L_h$ would require more time to reach the depletion layer, and then they will drift by the $E$ field attraction. That is why, if the illumination suddenly stops, those areas would still provide minority carriers, and the current $I_{ph}$ would decay slowly.

The PD frequency response is composed of a fast one that results from the depletion area's electrical field drift velocity in the hole electron pair, and a slow response due to diffusion charges. One more conclusion is that the photodetector sensitivity or light responsivity would increase for wider active area $W$. The larger depletion area reduces the PD junction capacitance $C_j$. The junction capacitance $C_j$ is given by the Eq. (8.29), where $d = L_d$, the depletion depth including the $I$-region width. The junction series resistance is marked as $R_s$; therefore, the cutoff frequency of the diode due to the junction RC time constant is given by

$$\omega_{RCJ} = \frac{1}{R_S C_J}, \tag{8.35}$$

$$f_{RC} = \frac{\omega_{RC}}{2\pi}. \tag{8.36}$$

The value of the radial 3-dB transit frequency for a case of $1/\alpha \leq w \leq 2/\alpha$[14,15] is given by

$$\omega_t = \frac{2.78}{\tau_t}; \tag{8.37}$$

hence,

$$f_t = \frac{\omega_t}{2\pi} = \frac{1}{w/0.44V_s} \approx 0.4\alpha V_s. \tag{8.38}$$

$V_s$ is the saturation speed, and $W$ is the depleted $I$-region thickness. Minimizing the diode depletion capacitance by increasing $W$ would improve junction BW, but on the other hand, it would increase the transit time of minority carriers, thus reducing its BW for high frequency, such as millimeter-wave frequencies. The carriers' velocity at the depletion area due to the applied $E$ field is at saturated speed $V_s$ because of the lattice scattering. As soon as the carriers reach the P and N concentrations, they contribute to the photocurrent. The two conditions of transport delay and junction capacitance, and series resistance derive the combined BW of the photodetector as

$$\mathrm{BW}_{\mathrm{PD}} = \frac{1}{\sqrt{\left(\frac{1}{f_{\mathrm{RC}}}\right)^2 + \left(\frac{1}{f_t}\right)^2}}, \tag{8.39}$$

where $f_{\mathrm{RC}}$ and $f_t$ are the 3-dB cutoff frequencies for the junction RC and transit time, respectively.

After some substitutions in Eq. (8.39), the PIN photodetector diode BW is expressed by its physical dimensions:

$$\mathrm{BW}_{\mathrm{PD}} = \frac{1}{\sqrt{\left(2\pi R_S \varepsilon_0 \varepsilon_r \frac{\pi d^2}{4L_d}\right)^2 + \left(\frac{w}{0.44V_s}\right)^2}}, \tag{8.40}$$

where $d$ is the PIN diode photodetector diameter, and $L_d$ is the depletion depth.

If the diffusion depth is thin enough (i.e., $L_e + L_h \ll L_d$) then the above notation is given by[14,31]

$$\mathrm{BW}_{\mathrm{PD}} = \frac{1}{\sqrt{\left(2\pi R_S \varepsilon_0 \varepsilon_r \frac{\pi d^2}{4L_d}\right)^2 + \left(\frac{L_d}{0.44V_s}\right)^2}}, \tag{8.41}$$

Writing Eq. (8.41) for $d$ would give a measure of detector diameter selection per required BW:

$$d = \sqrt{\frac{2L_d}{\pi^2 R_S \varepsilon_0 \varepsilon_r} \sqrt{\left(\frac{1}{\mathrm{BW}_{\mathrm{PD}}}\right)^2 - \left(\frac{L_d}{0.44V_s}\right)^2}}. \tag{8.42}$$

Additional important parameters to consider while selecting a photodetector are its responsivity and QE. Those define the amount of detection ability of the photodetector. Higher responsivity $r$ and QE $\eta$ would save electronics for amplifying the signal and would result in a better CNR performance.

From the above relations, there are several important conclusions referring to the photodetector dimensions and frequency performance (Figs. 8.7 and 8.8).

There is an optimum point for the cutoff frequency depending on the junction RC and transit time. Large depletion depth improves RC BW, but the penalty is in transit BW. Large depletion depth increases responsivity and QE but reduces transit time BW. A small-diameter photodetector can provide higher BW, and data rates. As a consequence, photodetectors with a very high data rate would

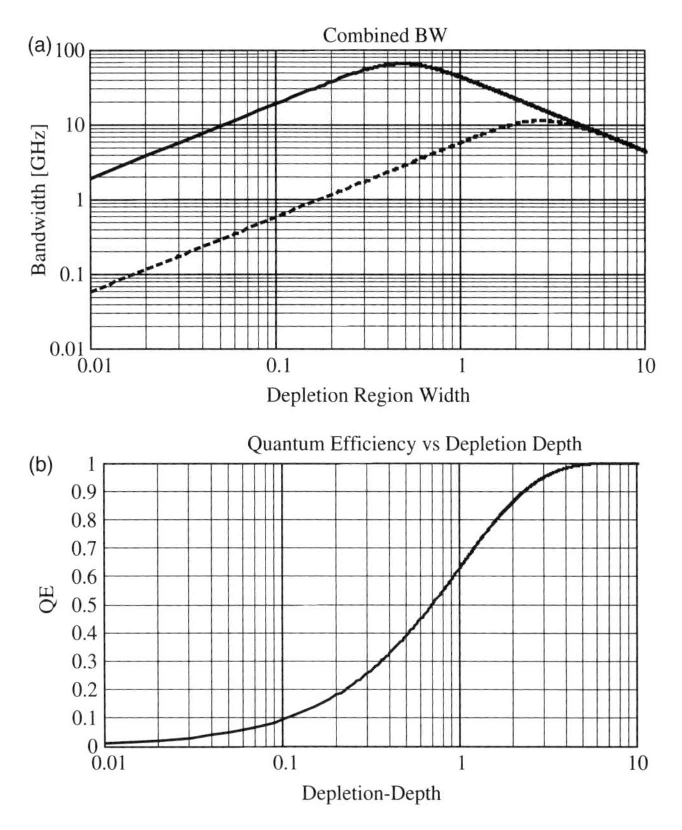


**Figure 8.7** Photodetector combined BW [GHz] (a) and QE (b) as a function of the depletion region length. Simulation parameters: Series resistance $R_s = 5\ \Omega$; the diameter of the first PD is 20 μm with 20 Ω load transimpedance amplifier (TIA) (*solid line*) and the second photodetector is 40 μm with 200 Ω matched load (*dashed line*). The relative dielectric constant is $\varepsilon_r = 12$, $V_s = 10^7$ cm/s, and absorption factor $\alpha = 10^6\ \mathrm{m}^{-1}$, which describes InGaAs at 1600 nm per Fig. 8.2. Note that BW increases as the depletion width gets longer (reducing the junction capacitance) till a maximum where the transit time starts to govern the BW. A similar plot of responsivity can be found by using Eq. (8.8) at 1600 nm.

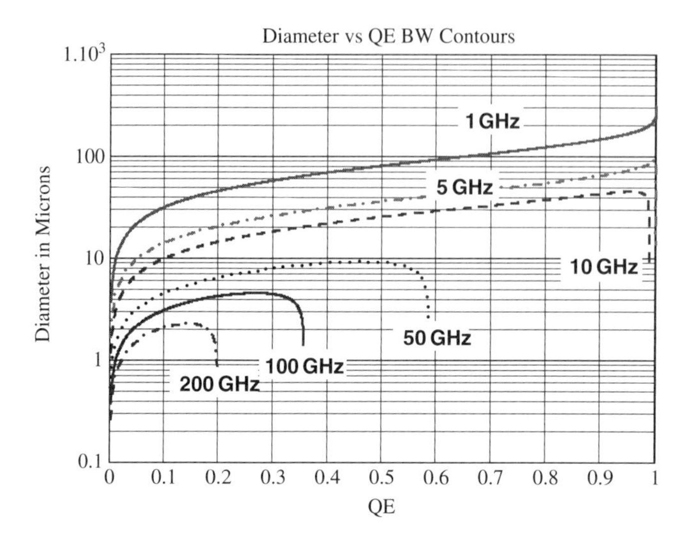


**Figure 8.8** Photodetector BW [GHz] contours as a function of PD diameter. Simulation parameters: Relative dielectric constant $\varepsilon_r = 12$, $V_S = 10^7$ cm/s, absorption factor $\alpha = 10^6\ \text{m}^{-1}$, which describes InGaAs at 1600 nm per Fig. 8.2. Series resistance $R_S = 5\ \Omega$, with 20 $\Omega$ TIA load. Note that as BW requirement increases, the QE goes lower. The reason for that is that the depletion region width becomes narrower to reduce the transit time effect over the combined BW. It can be observed from that graph that for larger BW contour, the photodetector diameter should go lower. For communication applications of 10 GBit/s and lower, a 0.7 QE and higher detectors are available with diameters starting from 30 to 40 μm. Responsivity at 1600 nm can be found by using Eq. (8.8) and its value is 1.2 mA/mW. Ultrafast photodetectors for BW of up to 50 GHz and data rates of 40 GB/s, would require 10–20 μm diameter and the responsivity in that case would be 0.6 mA/mW.

have low responsivity and vice versa. This is because the active depth is shorter for faster transit time, and the diameter is smaller to minimize the junction capacitance. As a result, less minority carriers are generated, since the active volume is smaller; hence, very wide-band photodetectors have low responsivity $r$. Moreover, since these photodetectors have a narrowed $I$ region, they will have higher distortions. Large diameter detectors would have a larger depletion capacitor and therefore would have a lower BW. This can be improved a bit by a larger reversed bias. For fiber to the home curb building business premises (FTTx) community access television (CATV), the photodetector diameter is 40 μm, while in the digital section of such FTTx integrated triplexer (ITR) the diameter is 20 μm.

Using Eqs. (8.30), (8.41), and (8.42), the depletion depth can be expressed and optimized as a function of QE and the absorption factor $\alpha$. Useful quantities of contour plots showing the trade-offs between depletion depth to QE, transit time BW, and junction time constant RC, as well as the QE to PD diameter with frequency as a parameter are provided in Figs. 8.7 and 8.8.

### 8.2.3 Small signal modeling of a PIN photodetector

The RF model of the photodetector is a current source sensitive to light and with reactive output impedance.[6,8,10,14]

Since the photodetector is reverse biased, the photodetector junction resistance $R_j$ is high. The Fermi-levels are bent, creating a high PIN potential barrier. The DC current is low and there is leakage current only, which is the dark current of the photodetector when there is no light. $C_j$, the depletion capacitor, is at its minimum value. $R_s$ is a low value representing the bulk contact resistance, generally about 5 Ω. There are additional parasitics due to packaging and bonding. The bonding inductances are at the order of 0.5 nHy max, and the lead is up to 1 nHy in the worst case, depending on the lead length. Hence, it is preferred to connect the photodetector module in case of a high digital data rate with flexible controlled-impedance transmission lines. In case of a CATV analog design, generally a pickup inductor is added in series with the photodetector in order to compensate $C_j$ and extend the PD BW. The equivalent RF circuit of a photodetector showing its high frequency parasitics is given by Fig. 8.9.

From that scheme, the RF nature of a PD is near to an RF open at the Smith chart or high impedance load. As the frequency increases, the PD impedance shows a capacitive trajectory at the Smith chart until it reaches its first serial resonance point where it shows a short impedance. Hence, PD would appear on the Smith chart as high impedance at DC (see Figs. 8.10 and 8.11). The photodetector RF and noise currents are given by Eq. (8.43), where $H(j\omega)$ is the photodetector, RF, the current transfer function, and $r$ is the photodetector responsivity:

$$I_{\text{photo}} = P_{\text{opt}} \cdot \text{OMI} \cdot r \cdot H(j\omega) + i_{\text{shot\_dark}} + i_{\text{noise}}. \tag{8.43}$$

Due to its high impedance and capacitive nature, the photodetector has a low-pass response. For this reason, one of the ways to extend the PD BW is by connecting the series inductor at its output. Figures 8.10, 8.11, and 8.12 show simulation results for inductive matching and BW improvement using a pickup inductor for a 40 μm InGaAs PIN PD reverse biased at 15 V. Simulation parameters were $C_j = 0.6$ pF, $R_j = 1000\ \Omega$, $R_s = 5\ \Omega$, $L_{\text{bond}} = 0.5$ nHy, and Cp = 0.1 pF; and $L_1$ varies between the lead inductance of 1 nHy and the max value of the pickup inductance, which is 35 nHy. Four cases were examined:

(1) Case 1 describes the photodetector connected to a 50 Ω load without any additional matching. From Figs. 8.10 and 8.11 it can be seen that the reflection factor is poor, and the photodetector represents an open at low frequency. On the other hand, the BW of the photodiode is large.

(2) Case 2 shows a match with a pickup inductor of 30 nHy and a 50 Ω load. The perfect matching occurs at the crossing from the capacitive region to the inductive region in the Smith chart. This point is the resonance point. The S11 notch depth and width depends on the quality factor $Q$ of the circuit. $Q$ is affected by the value of the resistive part of the matching network.

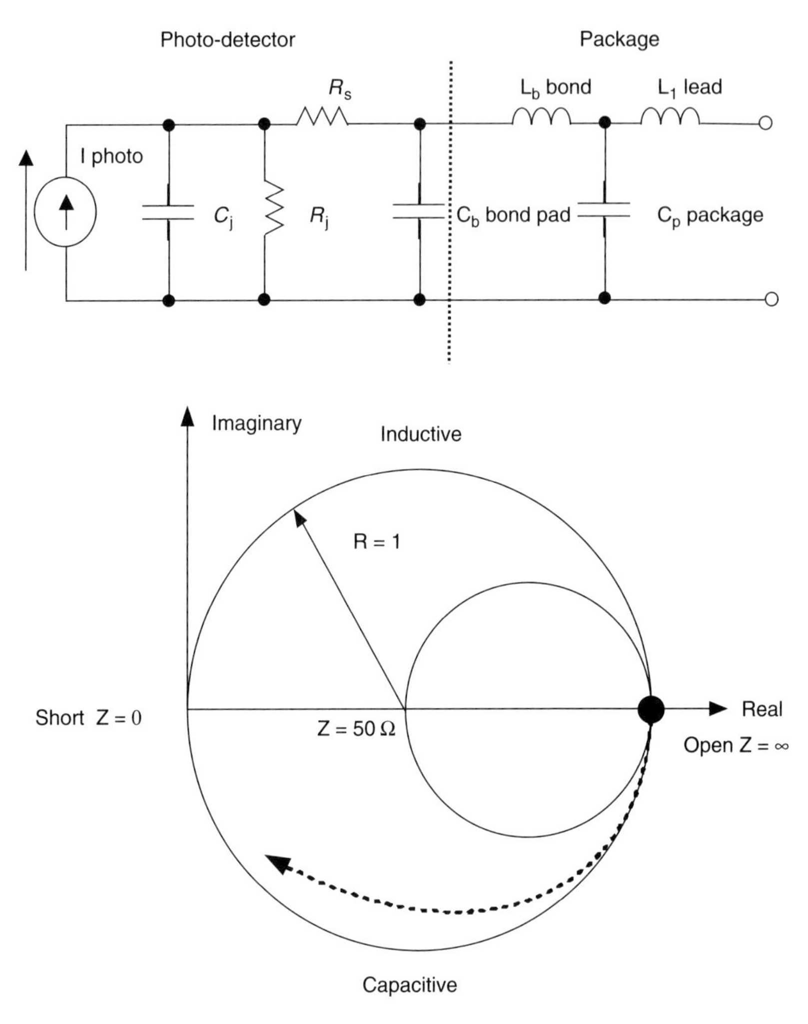


**Figure 8.9** Photodetector equivalent model, including package and Smith chart of S11 response vs. frequency.

(3) Case 3 displays a simulation result with a 2:1 transformer and without a pickup inductor. At this point, the output of the detector is matched to 200 Ω because of the 4:1 impedance ratio created by the 2:1 transformer. The BW is reduced, and the gain is increased, since the load value is much closer to the current source output impedance. However, the BW is reduced because the RC time constant is larger.

(4) Case 4 is the last simulation case that combines the pickup inductor and 2:1 transformer. In this case, S11 improves and S21 frequency response becomes higher with some overshoot due to the resonance of the photodetector capacitance by the pickup inductor. More discussion of the RF front-end design approach appears in Chapter 13.[32]

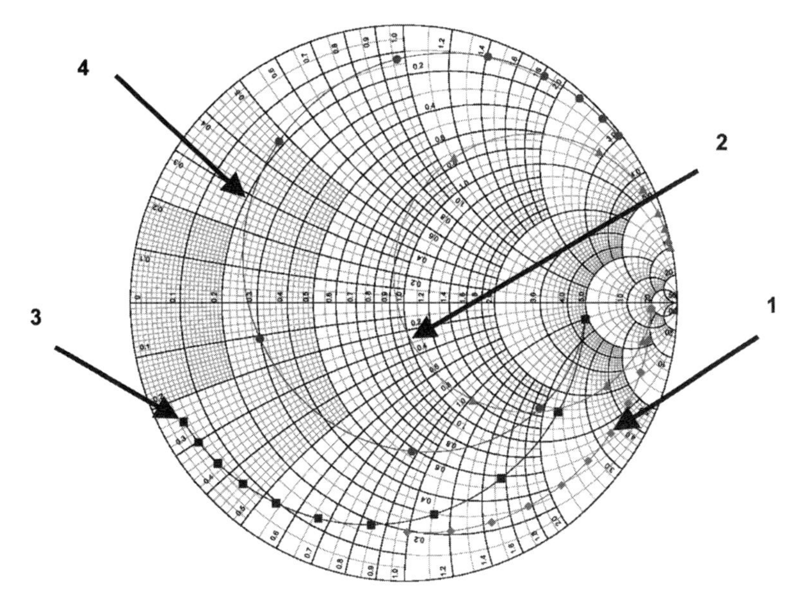


**Figure 8.10** Photodetector S11 response vs. frequency over Smith chart.

### 8.2.4 RF detection and modulation loading

RF detection by the photodetector is proportional to the optical power level impinging on the photodetector active surface, responsivity, signal OMI, and RF load. Hence, the RF current is given by

$$I_{\mathrm{RF}} = P_{\mathrm{OPT}} \cdot \mathrm{OMI}_{\mathrm{ch}} \cdot r, \tag{8.44}$$

where $r$ is the responsivity of the photodetector. The detected RF current is the peak current. However, the RF power refers to the rms current through the RF load $Z_{\mathrm{L}}$ connected to the photodetector ports. Hence, the RF power is given by

$$P_{\mathrm{RF\_ch}} = \left(\frac{P_{\mathrm{OPT}} \cdot \mathrm{OMI}_{\mathrm{ch}} \cdot r}{\sqrt{2}}\right)^2 \cdot Z_{\mathrm{L}}. \tag{8.45}$$

Equation (8.45) describes the photodetector output power for a single CATV channel loading.

In case of multichannel loading with $N$ channels with equal OMI per channel, the overall RF power is given by

$$P_{\mathrm{RF\_tot}} = \sum_{i=1}^{N} \left(\frac{P_{\mathrm{OPT}} \cdot \mathrm{OMI}_{\mathrm{ch}} \cdot r}{\sqrt{2}}\right)^2 \cdot Z_{\mathrm{L}} = N \cdot \mathrm{OMI}_{\mathrm{ch}}^2 \left(\frac{P_{\mathrm{OPT}} \cdot r}{\sqrt{2}}\right)^2 \cdot Z_{\mathrm{L}}. \tag{8.46}$$

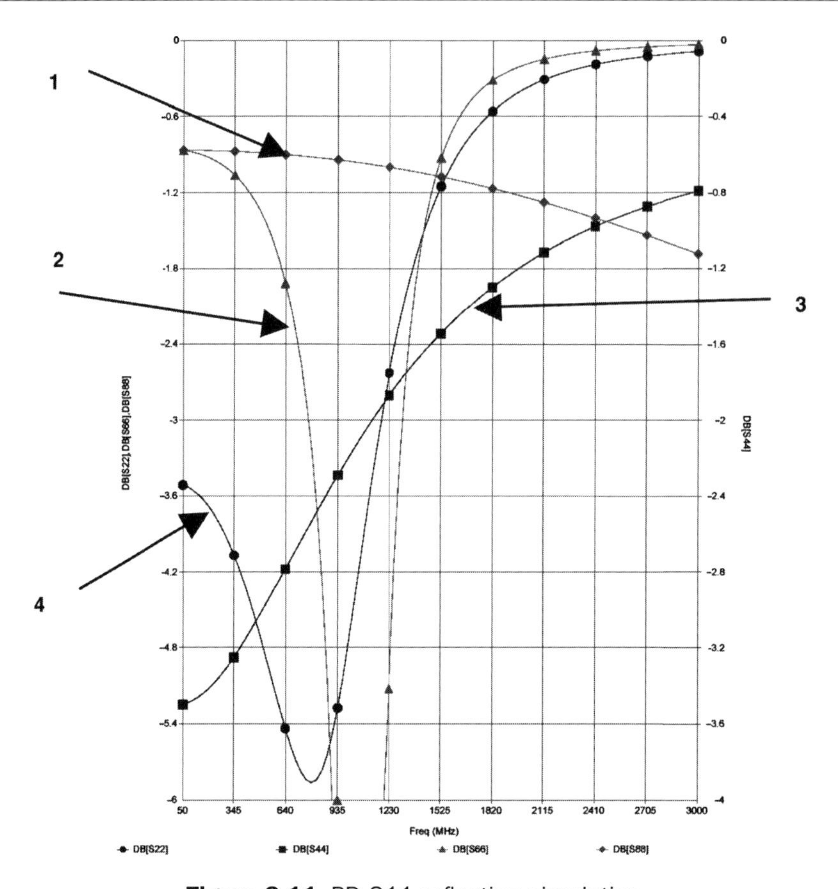


**Figure 8.11** PD S11 reflection simulation.

Assuming an equivalent tone with equivalent OMI, $\mathrm{OMI_{Eq}}$, results in the same RF power as

$$P_{\mathrm{RF\_tot}} = P_{\mathrm{RF\_Eq}} = \mathrm{OMI}_{\mathrm{Eq}}^2 \left( \frac{P_{\mathrm{OPT}} \cdot r}{\sqrt{2}} \right)^2 \cdot Z_{\mathrm{L}}, \tag{8.47}$$

which results in the equivalent OMI value as

$$\mathrm{OMI}_{\mathrm{Eq}} = \mathrm{OMI}_{\mathrm{ch}} \cdot \sqrt{N}. \tag{8.48}$$

The 110-channel CATV frequency plan has two OMI levels, $\mathrm{OMI_{ch\text{-}A}}$ for the first 79 channels between 50 and 550 MHz and a lower OMI, $\mathrm{OMI_{ch\text{-}B}}$, generally for the remaining channels between 550 and 870 MHz. In the same way, the overall RF power detected by the PD is

$$P_{\mathrm{RF\_tot}} = \sum_{i=1}^{N} \left( \frac{P_{\mathrm{OPT}} \cdot \mathrm{OMI}_{\mathrm{ch\text{-}A}} \cdot r}{\sqrt{2}} \right)^2 \cdot Z_{\mathrm{L}} + \sum_{i=N+1}^{M} \left( \frac{P_{\mathrm{OPT}} \cdot \mathrm{OMI}_{\mathrm{ch\text{-}B}} \cdot r}{\sqrt{2}} \right)^2 \cdot Z_{\mathrm{L}}. \tag{8.49}$$

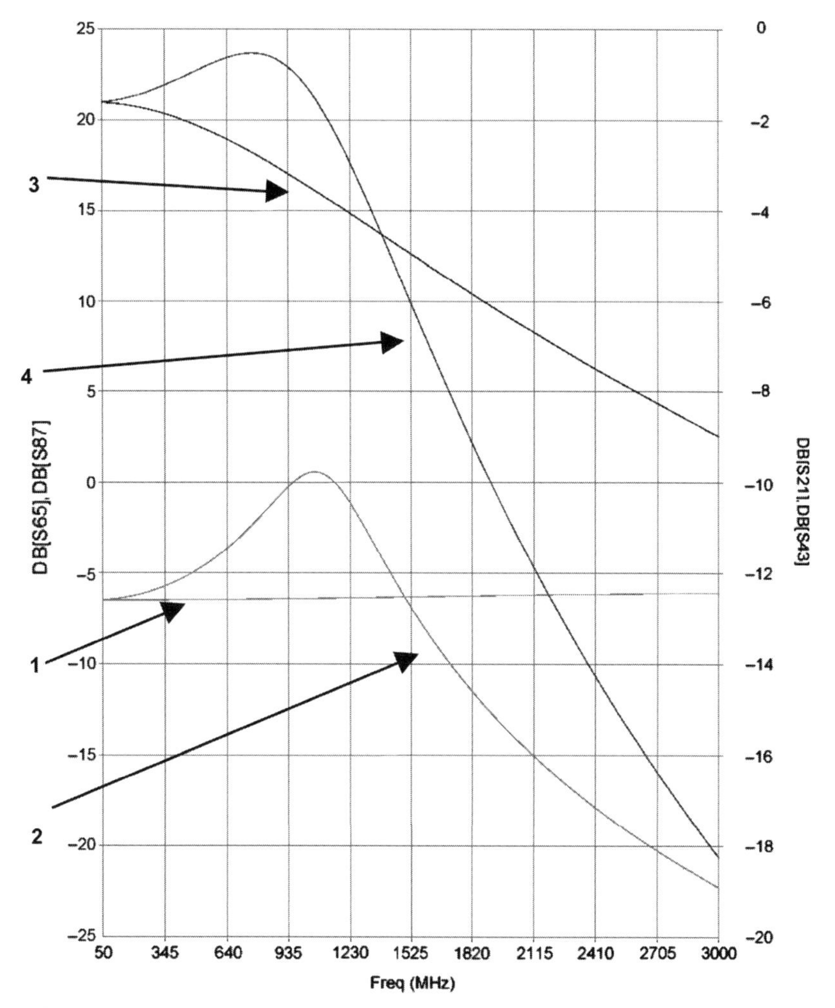


**Figure 8.12** PD S21 simulation with and without a BW extending pickup inductor.

Hence, the total RF power is given by Eq. (8.50), where $M$ is the total number of channels, and $N$ is the total number of channels with $\mathrm{OMI}_{\text{ch-A}}$ level:

$$P_{\text{RF\_tot}} = N \cdot \mathrm{OMI}_{\text{ch-A}}^2 \left(\frac{P_{\text{OPT}} \cdot r}{\sqrt{2}}\right)^2 \cdot Z_{\text{L}} + (M - N) \times \mathrm{OMI}_{\text{ch-B}}^2 \left(\frac{\mathrm{P}_{\text{OPT}} \cdot r}{\sqrt{2}}\right)^2 \cdot Z_{\text{L}}. \tag{8.50}$$

Defining the ratio between $\mathrm{OMI}_{\text{ch-B}}$ and $\mathrm{OMI}_{\text{ch-A}}$ as k, and defining an equivalent power tone with equivalent OMI, $\mathrm{OMI}_{\text{Eq}}$, as in Eq. (8.48), the equivalent OMI in this case is given by Eq. (8.51)

$$\mathrm{OMI}_{\text{Eq}} = \mathrm{OMI}_{\text{ch-A}} \cdot \sqrt{N + k^2(M - N)}. \tag{8.51}$$

These relations are important when analyzing the receiver for two-tone performance and CNR. Further discussion is provided in Chapter 10 and setups in Chapter 21.

### 8.2.5 Nonlinear response of PIN photodetector

Optical telecommunication systems using a 1.3 or 1.55 μm wavelength are commonly used due to the high transmission capacity. Analog video transmission using fiber optical semiconductors laser diode (L.D) and PDs have been specially developed for CATV.

Humphreys and Lobbet[12] reported on optoelectronic nonlinear (NL) mixing effect between photocurrents, and modulated bias voltage is a back-illuminated InGaAs/InP PD for high frequencies between 1.2 and 15 GHz. Dentan[7] reported numerical simulation of a back-illuminated InGaAs/InP PIN PD under high illumination power of 0 dBm. A simplified calculation method was presented by Hayer and Persechini.[3] Williams[5] also measured harmonic distortions and numerically modeled nonlinearity in PIN PD with a 0.95 μm long intrinsic region. NL behavior in a PD can be an important limiting factor in high-fidelity a analog and digital communication systems. The CATV IMD standard requires optical receivers to have CSO and CTB lower than −60 dBc. Analog fiber links operate at high average optical power levels on which a small RF modulation is imposed. Thus, the resulting photocurrent causes the carrier densities to become large enough to pinch off the applied field, which is a space-charged saturation. Eventually the device becomes NL, resulting in distortions that limit the dynamic range of the link. Today, receivers are designed with an IMD range of −60 to −80 dBc, depending on the topology of the receiver. As discussed in Chapters 10 and 11, distortions created by a PD cannot be removed by a push-pull receiver or any other receiver concept. There are several mechanisms creating distortions inside a PIN PD and methods to optimize the PD to improve IMD performance:

- generation of highly doped absorbing regions where the electric field is low;
- effects of space charge that induce changes in carrier velocities;
- nonzero load resistance; and
- intrinsic region length.

These are the semiconductor limitations. There are several methods being used to overcome distortions created in the PD such as:

- High reverse voltage, which increases the electric field, carrier velocity, and *I*-region (depletion) thickness, and reduces the capacitance.

- Spot defocusing, which creates a homogeneous carrier flux with a minimal lateral diffusion current, compared to a narrow spot illumination of the PD.

### 8.2.6 Photodetector chip structure

Figure 8.13 shows a cross sectional view of an InGaAs/InP PIN PD chip. The typical structure of a PIN photodetector is composed of three epitaxial layers grown on an S-doped InP substrate carrier. The carrier thickness is 350 μm, and the doping concentration is $6 \times 10^{18}\ \text{cm}^{-3}$. The first layer is an N-InP buffer layer with a thickness of 2.5 μm. The second layer is an N-InGaAs absorption layer with a thickness of 3.5 μm and a doping concentration of $1 \times 10^{15}\ \text{cm}^{-3}$. The absorption layer thickness is 3.5 μm in order to obtain high responsivity at 1.3 and 1.55 μm wavelengths. This layer is the *I* region of the PIN diode. The third-layer is n-InP, which is the window layer where light hits its surface, generating the photocurrent minority carriers. The third-layer thickness is 1.5 μm and the doping concentration is 2 to $3 \times 10^{15}\ \text{cm}^{-3}$. The active area was used to create a planar PN junction with the intrinsic absorption layer by selective diffusion of Zn through the InP window layer (the active area). This area defines the diameter of the PD. An additional peripheral P layer was made to create a PN junction on the perimeter of the PD. This prevents photocarriers that generated out of the active window periphery from slowly diffusing into the depletion layer and creating a photocurrent with delay in time with respect to the main generated current. That is one of the ways to prevent distortions due to time delay, which creates phase lag.

### 8.2.7 Semiconductor PIN detector distortion sources

The pertinent basic equation governing carrier transport is given by Poisson's and Gauss's laws given in Eq. (8.3). The continuity equations of a junction[1,5] for holes

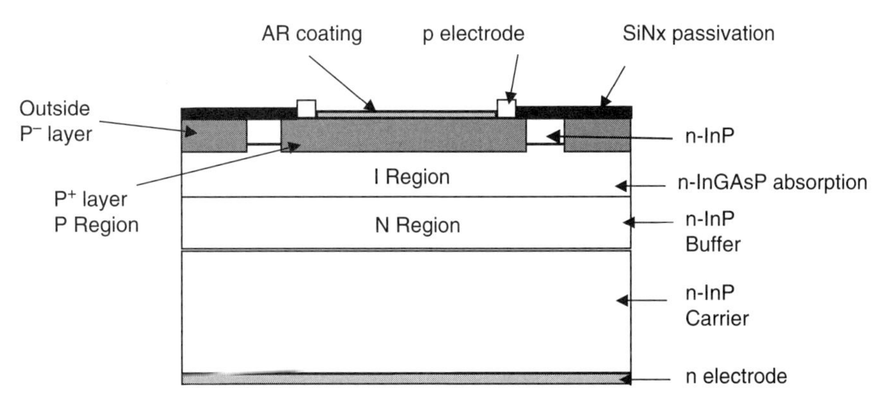


**Figure 8.13** Cross sectional view of an InGaAs/InP PIN PD chip.[2] (Reprinted with permission from the *Journal of Lightwave Technology* © IEEE, 1997.)

and electrons control the conservation of carriers, of both drift and diffusion, in any volume and are given by

$$\frac{\partial p}{\partial t} = G - R - \frac{1}{q}\nabla \cdot \left(J_{\text{p-drift}} + J_{\text{p-diff}}\right), \tag{8.52}$$

$$\frac{\partial n}{\partial t} = G - R + \frac{1}{q}\nabla \cdot \left(J_{\text{n-drift}} + J_{\text{n-diff}}\right), \tag{8.53}$$

where $G$ and $R$ are generation and recombination rates, respectively. $J$ represents current densities while the P and N signs and index represent the holes and electrons, respectively. The indices drift and diff represent drift and diffusion. These are a coupled with the partial differential equation solved by the variable separation method.[11] The total PD current is the sum of the hole and electron currents with the addition of the displacement current. Thus, there is a need to integrate the hole and electron current densities over the photodiode length:[5]

$$J = \frac{\varepsilon}{W} \cdot \frac{\partial V_{PD}}{\partial t} + \frac{1}{(x_2 - x_1)} \int_{x_1}^{x_2} \left(J_n + J_p\right) dx, \tag{8.54}$$

where $x_1 = 0$ represents the anode to the P side contact, and $x_2 = W$ is the entire PIN length, where the N contact of the cathode is shown by Fig. 8.9. $V_{PD}$ is reverse voltage value applied on the PD contacts. At first approximation, the displacement current is zero since $V_{PD}$ is constant.

The solution of Eqs. (8.3), (8.52), and (8.53) provides the drift current resulting from the $E$ field for both holes p and electrons n:

$$J_{\text{p-drift}} = qpv_p(E), \tag{8.55}$$

$$J_{\text{n-drift}} = -qnv_n(E), \tag{8.56}$$

where $v_p(E)$ and $v_n(E)$ are the electric field-dependent hole and electron drift velocities, respectively. At first observation, it looks as if both current densities are linearly dependent on the hole and electron concentration, in case the drift velocities are independent of N and P carrier densities. However, the real case is not that the velocities through the PIN are related to the $E$ field, as in Eqs. (8.57) and (8.58.), they are affected by the P and N concentrations:[5]

$$v_n(E) = \frac{E\left(\mu_n + v_{nhf}\beta|E|\right)}{1 + \beta E^2}, \tag{8.57}$$

$$v_p(E) = \frac{\mu_p v_{nhf} E}{\left(v_{nhf}^{\gamma} + \mu_p^{\gamma} E^{\gamma}\right)^{1/\gamma}}, \tag{8.58}$$

where $\mu_n$ and $\mu_p$ are the electron and hole low-field mobilities, respectively, $v_{nhf}$ is the high-field electron velocity, $\gamma$ is a temperature function, and $\beta$ is a fitting parameter. Figure 8.14 presents some experimental result plots of Eqs. (8.57) and (8.58).

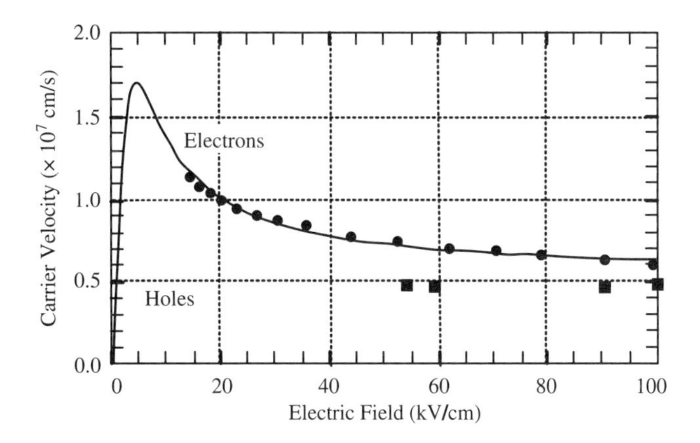


**Figure 8.14** Empirical expressions and measured results (*circles* and *squares*) for electron and hole velocities vs. electric field. Electron and hole low-field mobilities of 8000 and 300 $cm^2/V_s$.[5] (Reprinted with permission from the *Journal of Lightwave Technology* © IEEE, 1996.)

It is assumed that the light incident of the PD is radially Gaussian in power and enters through an opening in the n side antireflection-coated contact (an opposite case to Fig. 8.13) as described by

$$G(x,t) = G_0(t)\exp\left[-\alpha\left(w_p + w_i - x\right)\right], \tag{8.59}$$

where $G_0(t)$ is the time-dependent generation rate per volume, $\alpha$ is the absorption factor, and $w_p$ and $w_i$ are the P- and the *I*-region thicknesses. This equation can be written for the whole depletion layer length $L$ of the PIN detector while having a harmonic CW modulation signal with modulation index $m$:

$$G(x,t) = G_0(1 + m\sin\omega t)\exp[-\alpha(L - x)]. \tag{8.60}$$

Consequently, the resultant P and N will contain an NL harmonic that will result in distortions in the current densities of $J_p$ and $J_n$. Buckley[11] provided an analytic expressions for strong $\alpha \gg 1$ and weak $\alpha \ll 1$ absorption factors of $\alpha$ for both modulated and unmodulated $E$ fields by solving the continuity equations of a PIN. Short wavelength carrier absorption, $\alpha \gg 1$, corresponds to carrier generation at the edge of the depletion region:

$$J_0 = \left.\frac{9\varepsilon\mu V^2}{8L^3}\right|_{@\alpha\gg 1}, \tag{8.61}$$

$$J_0 = \left.\frac{4\varepsilon\mu_P\mu_n\left(\mu_P + \mu_n\right)^3 V^2}{\left(\mu_P^2 + \mu_n^2\right)^2 L^3}\right|_{@\alpha\ll 1}, \tag{8.62}$$

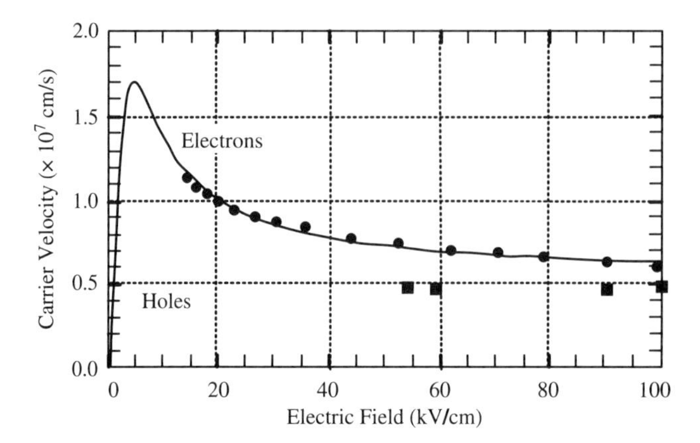


**Figure 8.15** Measured second harmonic power of 0.95 μm/region PD showing the regimes of importance for the two dominating NL mechanisms.[5] (Reprinted with permission from the *Journal of Lightwave Technology* © IEEE, 1996.)

where $V$ is the bias voltage and $L$ is the depletion layer length. For a strong absorption factor, the carrier transport current depends on the magnitude of the ratio of bias voltage mobility to the length of the depletion layer. For weak mobility, the carrier transport and pinch-off location depend on the relative magnitudes of the mobilities. NL behavior regimes of a PD are presented in Figs. 8.11 and 8.15). Figures 8.16 and 8.17 show a 40-μm diameter InGaAs PIN PD

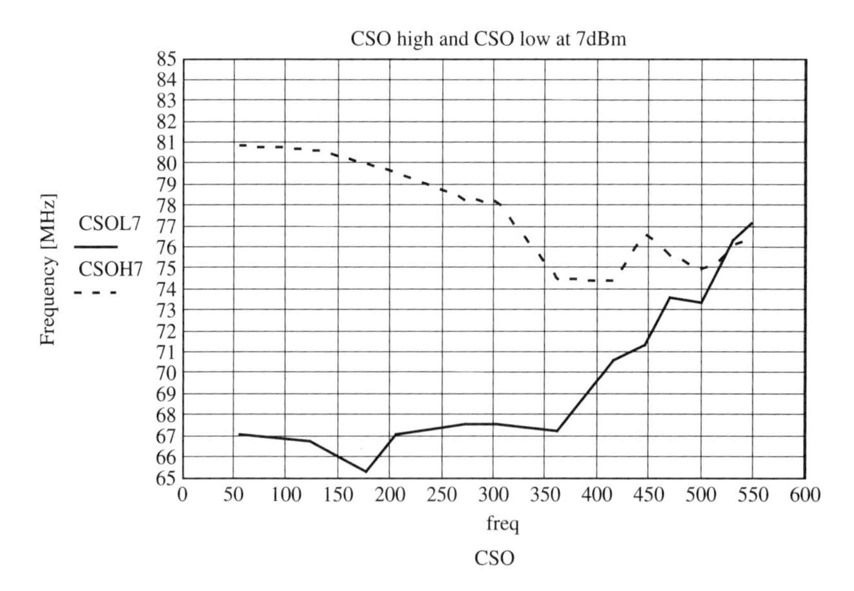


**Figure 8.16** CSO_H and CSO_L as functions of frequency and matching when the bias responsivity voltage sampling load equals 1 KΩ, the over drive protection is 1 KΩ, and the optical level is 7 dBm. Test conditions: 110 channels and 79 channels between 50 MHz and 550 MHz at 2.7% OMI; and 550 MHz and 870 MHz at 1.35% OMI with a reversed bias voltage of 13 V.

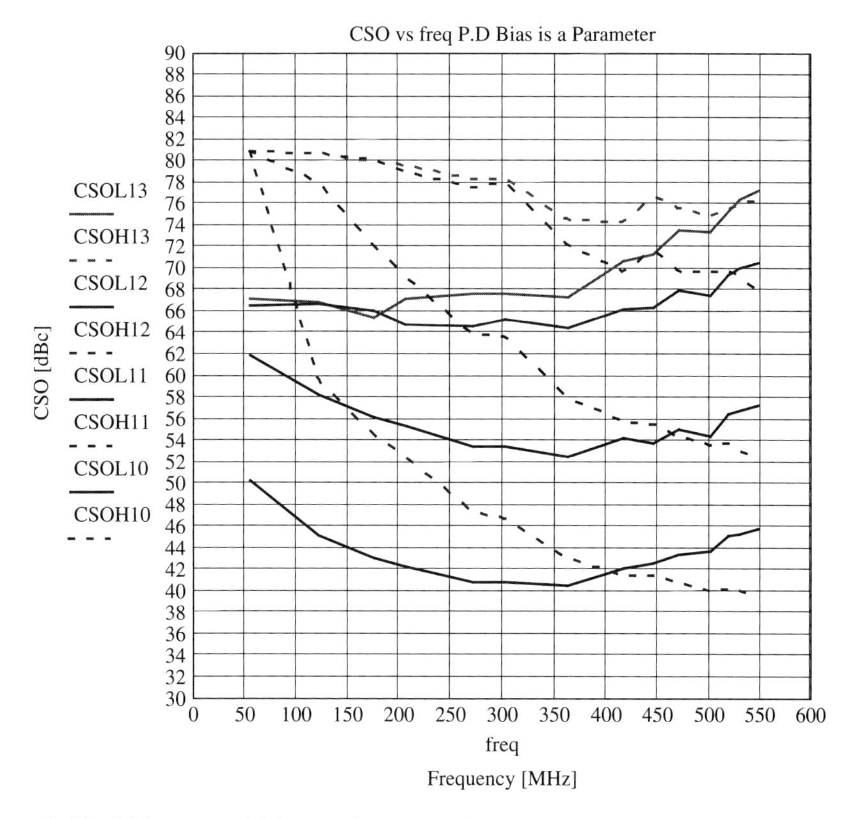


**Figure 8.17** CSO_H and CSO_L as function of reverse voltage and matching when the bias responsivity voltage sampling load equals 1 KΩ, overdrive protection is 1 KΩ, and the optical level is 7 dBm. Test conditions: 110 channels and 79 channels between 50 and 550 MHz at 2.7% OMI and 550 and 870 MHz at 1.35% OMI. It was tested at reversed bias voltages of 10 V, 11 V, 12 V, and 13 V.

composite second-order (CSO) performance under a 110-CATV-channel plan as a function of frequency and reverse voltage. CSO_L is measured at 1.25 MHz below the carrier and CSO_H is measured at 1.125 MHz above the carrier. Measurements were taken at 7 dBm when 79 channels were at 2.7% OMI, 31 channels were at 1.35% OMI as a standard CATV frequency, and the power plan was at 1550 nm. Optical power correction for 3.5% OMI is 1.127 dB. Thus, 7 dBm corresponds to 5.7 dB at 3.5% OMI. The test was performed by an Aeroflex-RDL multitone tester using a Rode-and-Schwartz high dynamic range spectrum analyzer (model ESCS30). A 2:1 matching transformer was used.

At first, the nonlinear distortions (NLDs) are strongly dependent on the bias voltage value. Up to 2 V second-order harmonics increase up to a peak at –2 V. Between 2 V and 10 V, NLDs decrease. In this region, the $E$ field is increased as the reverse voltage increases. Therefore, the carriers' velocity increases. That improves until a saturation point. At voltages above 10 volts, other mechanisms dominates such as P-region absorption dominates.[1]

More experiments show that the responsivity improves as the reverse bias increases. It was also explained that distortions are dependent on the load resistance $r$. Thus, the responsivity is expressed by

$$R = \alpha - \beta r I_{\mathrm{p}}, \tag{8.63}$$

where $R$ is the overall responsivity, $\alpha$ is the responsivity [mA/mW] at $V_{\mathrm{r}}$, $\beta$ [mA/mW/V] is the reverse voltage dependence of the responsivity, $r$ is the dc load resistance, and $I_{\mathrm{p}}$ is the photocurrent. The photocurrent is also expressed by the input optical power. Hence,

$$I_{\mathrm{p}} = R \cdot P_{\mathrm{in}} = \left(\alpha - \beta r I_{\mathrm{p}}\right) P_{\mathrm{in}}. \tag{8.64}$$

It can be seen that the factor $rI_{\mathrm{p}}$ is the voltage drop across the dc load resistance.

Rearranging Eq. (8.64) for $I_{\mathrm{p}}$, results in an equation that shows the second-order distortion related to the optical power level. Assuming $\beta r P_{\mathrm{in}} \leq 1$, the Taylor expansion provides the following relation for the photocurrent:

$$I_{\mathrm{p}} = \frac{\alpha P_{\mathrm{in}}}{1 + \beta r P_{\mathrm{in}}} \approx \alpha P_{\mathrm{in}} - \alpha \beta r P_{\mathrm{in}}^{2} + \alpha (\beta r)^{2} P_{\mathrm{in}}^{3}. \tag{8.65}$$

The first term of Eq. (8.65) refers to the fundamental frequency, the second term refers to the second-order distortion, and the third refers to the third-order distortion. Hence, the IMD level for a single test tone is as given by Eq. (8.66), where the OMI is the optical modulation index. Thus, the signal current and its harmonics are given by

$$I_{\mathrm{p-Sig}} \approx \alpha \mathrm{OMI} \cdot P_{\mathrm{in}} - \alpha \beta r \mathrm{OMI}^{2} \cdot P_{\mathrm{in}}^{2} + \alpha (\beta r)^{2} \mathrm{OMI}^{3} \cdot P_{\mathrm{in}}^{3}. \tag{8.66}$$

Therefore, the ratio between the fundamentals to the second order is given by

$$\begin{aligned} \mathrm{IMD}_2[dBc] &= 20 \log\left(\frac{\alpha \cdot \mathrm{OMI} \cdot P_{\mathrm{in}}}{\alpha \cdot \beta \cdot r \cdot (\mathrm{OMI} \cdot P_{\mathrm{in}})^{2}}\right) \\ &= 20 \log\left(\frac{1}{\beta \cdot r \cdot \mathrm{OMI} \cdot P_{\mathrm{in}}}\right). \end{aligned} \tag{8.67}$$

In the two-tone test, the optical power impinging on the photodetector is composed of dc and ac signal components, which is an AM on-off keying (OOK) expression. Hence, the optical powers of the two tones are given by $P_1 = P_0[1 + \mathrm{OMI} \cdot \cos(\omega_1 t)]$ and $P_2 = P_0[1 + \mathrm{OMI} \cdot \cos(\omega_2 t)]$, and the second-order distortion signal power would be

$$P_{\mathrm{IMD}} = \{\alpha \beta r \cdot \mathrm{OMI}^{2} \cdot P_0 \cos[(\omega_1 \pm \omega_2) t]\}^{2} Z_{\mathrm{L}}. \tag{8.68}$$

After some manipulations[2] IMD2 results in

$$\mathrm{IMD}_2 = 20\log\frac{\mathrm{OMI}\beta\cdot r\cdot P_0}{1-4\beta\cdot r\cdot P_0}, \tag{8.69}$$

where $Z_L$ is the RF load impedance ac coupled to the PD.

In conclusion, it is desirable to bias the PD with a high reverse voltage but with a low dc load. On the other hand, the goal of a dc load is to provide current protection for the PD at high-level optical powers. Generally, 1 K$\Omega$ is used as a responsivity sense for CATV designs, while the PD is biased to 10–15 V, separately from the signal processing circuitry.

As was reviewed in Sec. 8.3, the PD BW is a compromise between depletion length affecting the transit time and junction RC. Thus, modulation BW depends on those limitations. It is also desired to have equally distributed carrier velocities all over the $I$-region cross section. Therefore, the illumination spot of the PD should cover the maximum active area. Further discussion on those effects is provided in Sec. 8.6.3. CATV is a wide-band operating system, from 50 to 870 MHz. Thus, second-order distortions fall in the band of active channels. Several papers presented test results of the IP3 of PIN photodetectors as a function of frequency using microwave harmonic–balance analysis methods.[4,9] The key parameter for characterizing PD distortion is to have a highly linear setup. The setup should use the two two-tone lasers using the heterodyne method[4,5] or a highly linear predistortion laser transmitter which was used for testing web.

### 8.2.8 Optimization of photodetector packaging for low distortions

In previous sections, it was explained that the most important characteristics on an analog PD module are responsivity and second-order intermodulation, IMD2. Generally, the third-order, IMD3, is less than −90 dBc and is negligible in relation to IMD2. In Chapter 5, it was explained that the PD is coupled to the beam of light using a spherical lens, taking advantage of its aberration for increasing the spot size on the active area of the PD. Optimizing the illumination spot size on the photo was reported by Ref. [2]. A diode chip is soldered on an Alumina $Al_2O_3$ submount carrier, where the n electrode is soldered. The chip is sealed then by a metal cup with a spherical lens. Then a single-mode fiber (SMF) is inserted and aligned so the light spot is located at the active area of the chip. Figure 8.18 illustrates a cross section of the photodetector module with its pigtail fiber. As was explained in Chapter 5, there is an advantage for lens aberration, since it can be used for defocusing. In this case, a spherical lens is used, and the illumination spot size is optimized. This done by varying the pigtail distance from the lens, as shown in Fig. 8.19.

The PD responsivity and IMD are directly related to the distance $Z$ of the SMF from the spherical lens. The optimal distance $Z$ between SMF and the lens is tuned

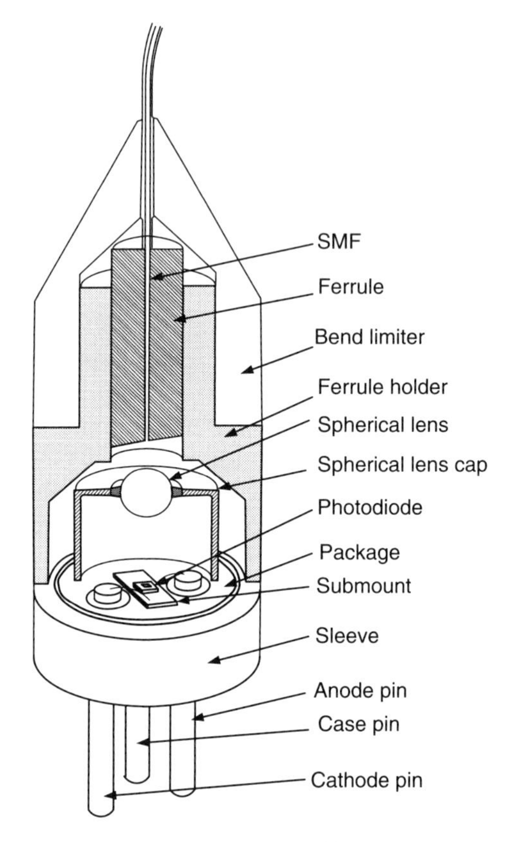


**Figure 8.18** Photodetector module cross section.[2] (Reprinted with permission from the *Journal of Lightwave Technology* © IEEE, 1997.)

for a high responsivity performance and IMD2. The technique uses two two-tone lasers. The sampling resistor for the photocurrent is relatively high at about 220 Ω. The measurement is done by a spectrum analyzer with a characteristic impedance of 50 Ω. Each laser is modulated to 40% OMI, which represents high CATV-channel loading.

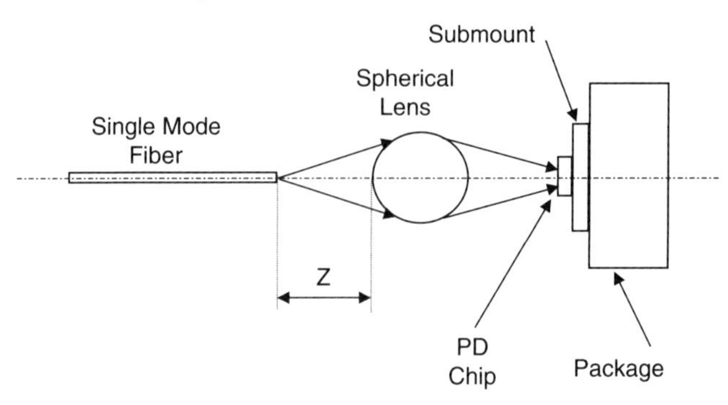


**Figure 8.19** Optical alignment for distortion and responsivity.[2] (Reprinted with permission from the *Journal of Lightwave Technology* © IEEE, 1997.)

The conclusion is that, for higher responsivity performance, the optimal distance $Z$ is achieved by moving the fiber farther from the lens. This increases the focusing of the light on the active area. Defocusing decreases the responsivity but improves IMD2 performance. For instance, at $1.0\ \text{mm} \leq Z \leq 1.7\ \text{mm}$, the responsivity is 0.95 mA/mW. In that case, IMD2 worst case at $f1 + f2$ at 824.5 MHz shows the highest distortions of −57.0 dBc. Defocusing by increasing $Z$ reduces responsivity to 0.88 mA/mW and improves IMD to −79.5 dBc. Moving the SMF closer to the lens improves responsivity and IMD as well—to 0.97 mA/mW and −80.5 dBc. Figure 8.20 presents the relationship between the distance $Z$ to the responsivity $R$ and IMD2. It is observed that IMD2 is very sensitive to the distance $Z$. According to the reports of Dentan and Williams, high optical input power creates harmonic distortions due to the electric field intensity modulation in the depletion region by space-charge effect. According to Ref. [2], it is related to the optical power density per unit area. This power density distribution is related to the focal distance $Z$. From Fig. 8.20 it can be seen that there is an optimum distance marked as $Z_a$ and $Z_b$, where the second-order IMD levels at the two-tone test are below −80 dBc, and the responsivity is above 0.95 mA/mW and 0.9 mA/mW, respectively. At $Z_p$, the IMD reaches its maximum value without any significant improvement in the responsivity.

Observing the power density distribution as given by Fig. 8.21, it can be seen that when the distance $Z$ equals 1500 μm, which is $Z_p$, all the optical power is concentrated at the center, illuminating a small spot diameter of 20 μm, while the detector at that experiment had a diameter of 80 μm. Hence, the responsivity is high, but the $IMD_2$ is low. That means that a lot of minority carriers are generated in a small area, and the photocurrent does not homogeneously spread through the PIN photodetector. Compromising on focal point spot illumination and setting

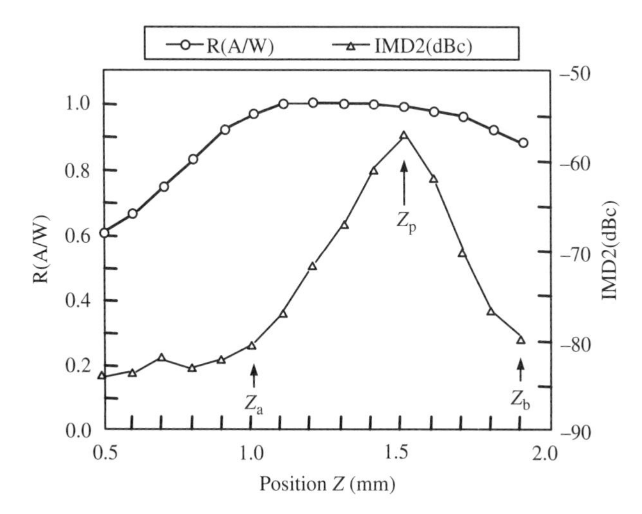


**Figure 8.20** IMD and responsivity performance of a photodetector module as a function of the distance Z.[2] (Reprinted with permission from the *Journal of Lightwave Technology* © IEEE, 1997.)

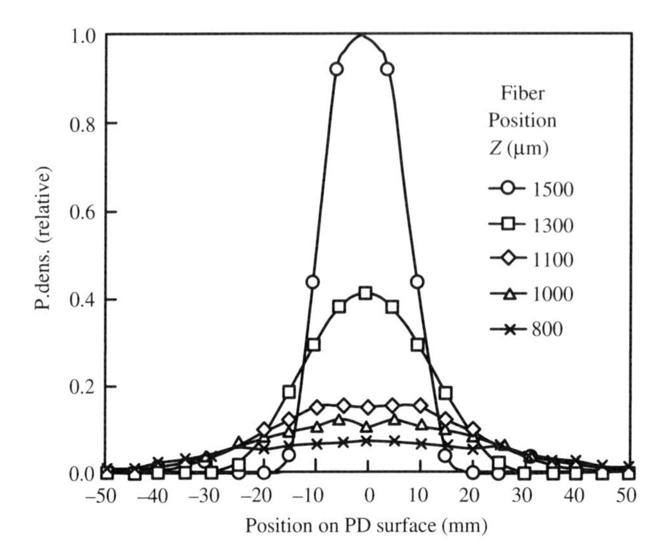


**Figure 8.21** Distribution of optical power density on the PD surface as a function of the distance from the PD photosensitive area.[2] (Reprinted with permission from the *Journal of Lightwave Technology* © IEEE, 1997.)

the distance $Z$ to be $Z_a$ or $Z_b$ would create a very flat, even power distribution. Hence, the responsivity is preserved, since the whole active area is illuminated, but the IMD performance is significantly improved.

## 8.3 Avalanche Photodetector

As an alternative to optical amplifiers and an optical front-end sensitivity increase by supporting electronics (such as post detection TIAs), Avalanche photodetectors (APDs) have been widely considered to be used in digital communication networks, analog applications,[23] wireless and space optical links,[17,27] and for measurement equipment.[25] The advantage of an APD over a PIN and PN photodetector is its gain. Thus, it can improve the optical front-end performance. This section reviews the APD structure and its performance envelope and limitations.

### 8.3.1 APD structure and concept

The common concept for PN and PIN detectors is reversed bias in order to have large depletion region and, hence, a high active region of photons absorption. Therefore, the photodetector would have higher responsivity, which is the main goal behind that. The other aspect of reverse bias is low dc current and, thus, low shot noise, since the diode has a wide depletion region and higher potential barrier. The third aspect of reversed bias is a strong electric field because of the high potential barrier across the depletion region. This results in a very high velocity for the holes and electrons. Furthermore, it can be observed that the field across a PN junction is much stronger than for PIN and peaks to a higher value. Due to the high concentration of charges and because the depletion region of the PN junction is thinner than one for PIN diode. Thus, a combination of PIN and PN may result in a photodetector with the advantages of both topologies and even more.

A simple APD structure is made of four sections, $P^+\pi PN^+$, where the $\pi$ region is a very lightly P-doped region is as shown in Fig. 8.22. This way it guarantees a long absorption region and a high responsivity. However, the electric field in that region is uniformly similar to a PIN *I* region. In this region, carriers gain speed. The other section of the APD is the $PN^+$, which is a narrow PN junction with a stronger field. Hence, at that region, the carriers' velocity is increased. Because of this, the carriers have higher kinetic energy than in the $\pi$ region. If the carriers' energy is high

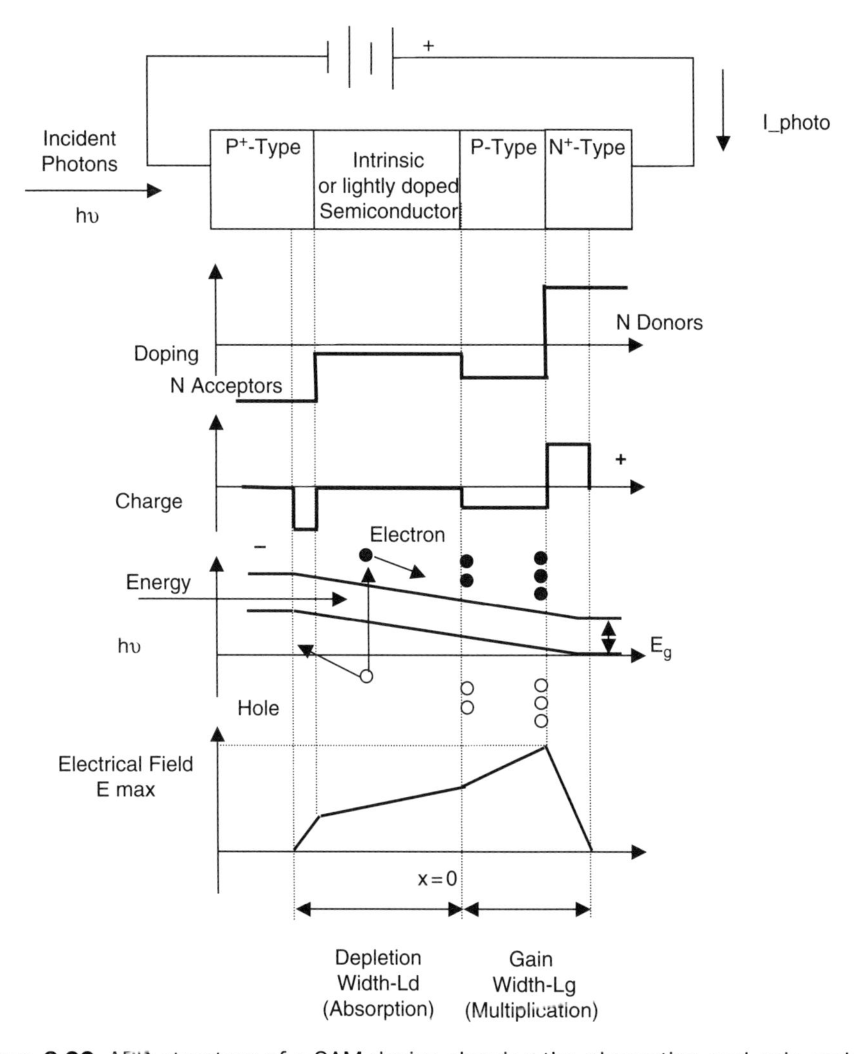


**Figure 8.22** APD structure of a SAM device showing the absorption and gain regions' doping profile, charge, multiplication, and field.

enough, they can kick new electrons or holes from the valence level to the conduction band. This is a process of gain creating an electron-hole pair, as demonstrated in Fig. 8.23. At the $PN^+$ region, the acceleration is higher than in the $\pi$ region because the field is stronger with a larger gradient as well; thus, the force is stronger, resulting in higher acceleration values, according to the second law of Newton. There is a limitation for the carrier velocity. It cannot accelerate to an unlimited value. Eventually, the carrier will reach saturation due to lattice scatterings.

Accordingly, the $PN^+$ region is called a multiplication region, whereas the $\pi$ region is the absorption region. The process of energy increase in the $PN^+$ region

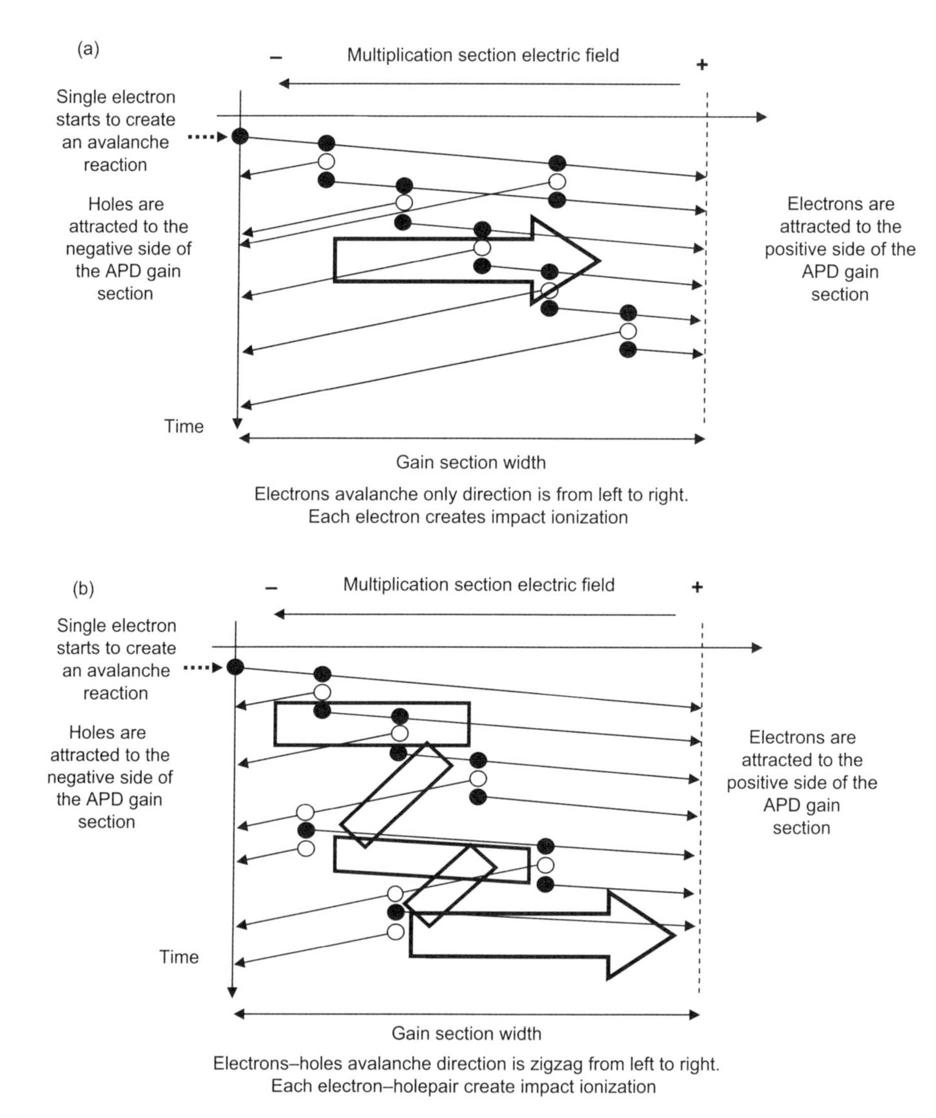


**Figure 8.23** (a) APD avalanche process created by electrons only (b) Both electrons and holes contribute to the avalanche.[14]

can create a chain reaction of ionization or feedback in which each hole and electron creates more electron and hole pairs. These new pairs may create new pairs by the same process. Thus, the higher the gain, the more time it takes to build the avalanche. Consequently, the BW is reduced or the speed of the APD becomes lower. The structure of APD described above is called a separate absorption and multiplication region (SAM). As a conclusion from the above description, APDs are biased at very high reverse voltage in order to achieve strong fields and high acceleration, resulting in high speed and momentum to kick electrons or holes to the conduction level.

The abilities of electrons and holes to impact and ionize are characterized by the ionization coefficients $\alpha_e$ and $\alpha_h$. These values represent the ionization probabilities per unit of length. The inverse coefficients, $1/\alpha_e$ and $1/\alpha_h$, represent the average distance between consecutive ionizations. These coefficients increase with the depletion electric field due to the acceleration. However, these factors decrease with temperature, since more collisions occur. Thus, the carriers do not gain enough speed and energy to ionize. Hence, the APD gain decreases versus temperature.

One of the factors characterizing the APD performance and defines the APD nature of ionization is the ionization ratio $k$:

$$k = \frac{\alpha_h}{\alpha_e}. \tag{8.70}$$

The value of $k$ defines the avalanche direction; an avalanche can be caused by holes or electrons. If $k < 1$, it means that the electrons generate most of the avalanche, and if $k > 1$, it is mainly due to the holes. Hence, for $k < 1$, the avalanche direction is from left to right. A special case is when $k = 1$. In this situation, both holes and electrons ionize, creating an ionization chain reaction, which is a positive feedback as described before. For this reason, the APD is designed for $k < 1$ or $k > 1$. The other disadvantages of $k = 1$ include noise created by this random process and the uncontrolled avalanche and generation of hole-electron pairs that lead to unstable gain and avalanche breakdown. APD gain is a multiplication process that creates noise. This noise can be reduced conceptually by use of a multilayer APD. One such structure is called a staircase.[18] The bandgap is compositionally graded from a low value $E_{g1}$ to a high value $E_{g2}$. Because of the material properties, the hole-induced ionizations are depressed. Thereby, the value of the ionization factor is reduced.

### 8.3.2 APD parameters

When considering selection of an APD, there are several parameters that are important in order to achieve desired performance. These parameters are frequency response, QE gain, gain-BW product, noise, and noise multiplication. All these parameters are related to each other, and the whole APD design and structure is a compromise process among all these performances. The goal however is to achieve them all.

The avalanche gain is also called the multiplication factor, denoted as $M$. The low-frequency avalanche gain is given by[15,20]

$$M=\left\{1-\int_0^W \alpha_n \exp\left[-\int_0^x (\alpha_n-\alpha_h)dx\right]dx\right\}^{-1}, \tag{8.71}$$

where $w$ is the depletion layer width (see Fig. 8.24).

The above equation is valid for pure electron injection from $x=0$ and is hard to solve, since the ionization coefficients depend on location and field values. For example, the electron ionization coefficient is given by[21,22]

$$\alpha(x)=A\exp\left\{-\left[\frac{E_0}{E(x)}\right]^m\right\}[1/\text{cm}], \tag{8.72}$$

where $E(x)$ is the applied field in the multiplication layer and the width-independent fitting parameters $E_0$, $A$, and $m$ are given in Table 8.1.[27] For uniform field profile position-independent ionization coefficients, as in PIN, the multiplication electrons are injected into the high field region at $x=0$, and the $M$ expression is reduced to a simpler one:[15,20]

$$M=\frac{\left(1-\frac{\alpha_h}{\alpha_n}\right)\exp\left[\alpha_n W\left(1-\frac{\alpha_h}{\alpha_n}\right)\right]}{1-\frac{\alpha_h}{\alpha_n}\exp\left[\alpha_n A\left(1-\frac{\alpha_h}{\alpha_n}\right)\right]}. \tag{8.73}$$

For equal ionization coefficients $1/\alpha_e=1/\alpha_h$ or $k=1$, the multiplication factor $M$ given in Eq. (8.71), which becomes

$$M=\frac{1}{1-\alpha_n W}. \tag{8.74}$$

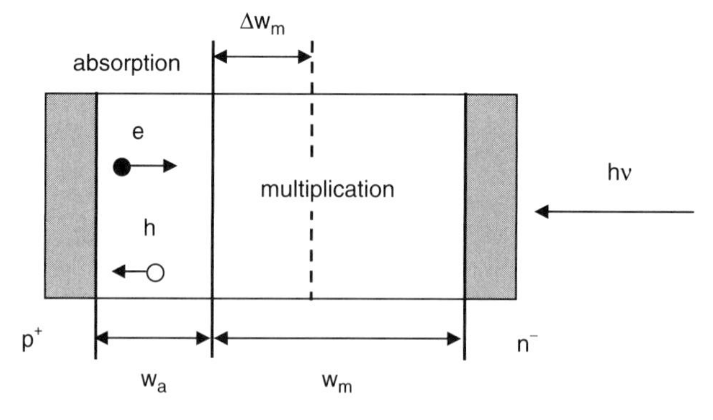


**Figure 8.24** 1D model of a SAM APD.[20] (Reprinted with permission from the *Journal of Lightwave Technology* © IEEE, 1996.)

From that equation, it is obvious that oscillations and instability are at $W\alpha_e = 1$. This point corresponds to the breakdown voltage. Thus, one of the design considerations makes the term $W\alpha_e \neq 1$. Conventional InGaAs APDs have limited gain-BW product and poor noise figure because of the small difference between the hole and electron ionization coefficient.[20,28,29] Si APDs are well known for low excess noise and high gain-BW product, the due to the large difference between the hole and the electron ionization coefficient.[20] However, the QE of Si APDs is negligible at 1.3–1.55 μm; thus, cannot be used for fiber communication systems. To take advantage of the Si low excess noise of the multiplication process and high gain-BW product, and to obtain high QE, a structure based on fusing silicon to InGaAs has been suggested and demonstrated.[30] Si is used as the multiplication material, and InGaAs is used as the absorption material, since its bandgap energy is excited by long wavelengths. This APD structure is called a silicon heterointerface photodetector (SHIP). The other parameter to be optimized in an APD for high data rate is the gain-BW product. This is accomplished by growing a narrow multiplication region,[20,22] where SHIP structures have 500 GHz of gain-BW product. As in other types of photodetection diodes, the BW limitations are due to the inherent capacitance and resistance of the detector, the transit time, and the average avalanche multiplication build-up time, $\tau_m$.[15,18,20] Photons detected by a PD generate an electron-hole pair. These drift by the electric field in a saturation velocity. Hence, the overall response time is given by the history of absorption time at the edge of the intrinsic region's depletion and multiplication time:

$$\tau = \frac{W_d}{v_e} + \frac{W_d}{v_h} + \tau_m. \tag{8.75}$$

In most APDs, the gain region is thinner compared to the absorption region and can be ignored. An approximation for the build-up time and ionization coefficients is given by[18,20]

$$\tau_m = k_1 M \frac{\alpha_h}{\alpha_e} \cdot \frac{W_g}{v_e} + \frac{W_g}{v_h}, \tag{8.76}$$

where $k$ is a correction factor for the ionization factor ratio and slowly varies with the ratio change. $W_g$ is the gain-region width. For equal ionization coefficients and $M \rightarrow \infty$, the current gain-BW product is given by[15]

$$\text{GBW} = \frac{3}{2\pi\tau_{ave}}, \tag{8.77}$$

where $\tau_{ave}$ is the average sum of the electron and hole transit time. An approximation for APD gain versus frequency is provided by[13,14]

$$M(\omega) = \frac{M_0}{\sqrt{1 + \left(\frac{\omega}{\omega_c}\right)^2}}, \tag{8.78}$$

where $M_0$ is the dc gain of the APD, and $\omega_c$ is the multiplication cutoff frequency denoted by Eq. (8.75). In practical APDs, the maximum achievable dc multiplications at high light intensities are limited by the series resistance and the space charge effect. All these factors are combined into an empirical relationship:[15]

$$M_0 = \frac{I - I_{\mathrm{MD}}}{I_{\mathrm{P}} - I_{\mathrm{D}}} = \frac{1}{\left[1 - \left(\frac{V_{\mathrm{R}} - IR}{V_{\mathrm{B}}}\right)^n\right]}, \tag{8.79}$$

where $I$ is the total multiplied current, $I_{\mathrm{P}}$ is the primary unmultiplied current, and $I_{\mathrm{D}}$ and $I_{\mathrm{MD}}$ are the primary and multiplied dark currents, respectively. $V_{\mathrm{R}}$ is the reversed biased voltage and $V_{\mathrm{B}}$ is the breakdown voltage. The exponent $n$ is a constant depending on the semiconductor material and doping profile. For high optical intensities, $I_{\mathrm{P}} \gg I_{\mathrm{D}}$ and $V_{\mathrm{B}} \gg I \times R$; thus, $V_{\mathrm{R}}$ approaches the value of $V_{\mathrm{B}}$. Hence, Eq. (8.79) becomes

$$M_{0\_\mathrm{MAX}} = \frac{V_{\mathrm{B}}}{nIR} \text{ or } M_{0\_\mathrm{MAX}} = \sqrt{\frac{V_B}{nI_P R}}. \tag{8.80}$$

In practical transceiver designs, in order to avoid breakdown and damage to the photodetector, a current limit protection circuit is provided in series with the APD. This way the voltage across the APD drops to a lower value at high optical levels.

### 8.3.3 APD-equivalent circuit

The APD-equivalent circuit is similar somewhat to a regular PIN or PN photodetector. Both these circuits exhibit the same parasitic concepts of junction capacitance and junction resistance, which is high due to the potential barrier, series resistance, and other packaging interface inductances and capacitance. However, in APD, the equivalent small signal circuit is divided into three main blocks as shown in Fig. 8.25. The first section is the photodetection region. This refers to the absorption process at the lightly doped or intrinsic region of the APD.

Thus, the equivalent transconductance of the current source is given by Eq. (8.15) of the primary current. After detection, the carriers reach the gain section or the so-called multiplication section. The output of this process is current multiplication. Thus, a transformer can be used to symbolize that. The turn ratio of such a transformer is given by the gain Eq. (8.74), showing the ratio between primary and multiplied currents.

These two models provided in Figs. 8.9 and 8.25 can be expended for noise modeling, which will be discussed later on in Chapter 11. Moreover, these small signal concepts are very useful in simulation tools for analog transport design. Distortions analysis using harmonic balance methods can be used here as well by separating the circuit into linear and nonlinear networks. It is clear that the current source impedance values, $C_{\mathrm{j}}$ and $R_{\mathrm{j}}$, are functions of biasing point and

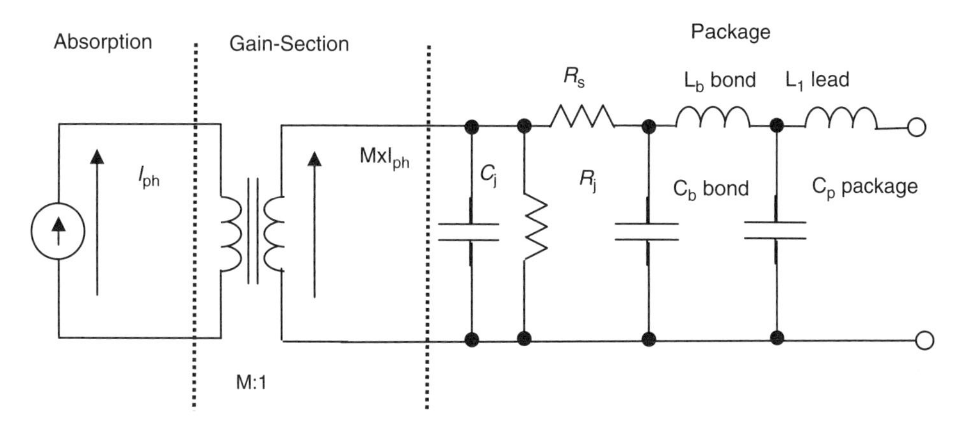


**Figure 8.25** APD-equivalent circuit showing the photonic current-gain process without the effects of noise.

signal. A different way to explain distortions is by observing the effects of a strong signal on these parameters. The same observation can be made on the gain factor $M$.

## 8.4 Bias Considerations

Optoelectronics PDs are reverse biased and thus operating at the third quadrant of the IV chart in Fig. 8.6. The current across the junction at a given voltage $V$ is the sum of two currents. The first one is the photocurrent $I_{ph}$, generated by the photons incident to the PD, and the other is the reversed bias leakage current $I_0$. Thus, the PD current at reversed bias is given by[1,16]

$$I = I_0\left[\exp\left(\frac{qV}{nKT}\right) - 1\right] - I_{ph}. \tag{8.81}$$

At reverse bias $V < 0$, the dc current is the leakage dark current of the PD $I_0$. It is as much a minority current as it is a reverse current. Hence, the PD under reversed bias at the third quadrant provides the total current of

$$I = -I_0 - I_{ph}. \tag{8.82}$$

Increasing $V$ to zero operates the PD at a short region, canceling the dark current $I_0$. However, the PD junction RC increases, reducing its BW. The dark current of the photodetector defines the shot-noise limit of a receiver as explained in Chapter 11. Both reversed bias and zero bias are called photoconductive modes. The other mode of operation is as a solar cell with $I = 0$. This mode is called photovoltaic. The load resistance goes to infinite since

$I = 0$. Hence, the detected voltage is proportional to the light. In that case, Eq. (8.81) is modified into

$$V = \frac{\mathrm{n}kT}{q}\ln\left(\frac{I_{\mathrm{ph}}}{I_0} + 1\right). \tag{8.83}$$

## 8.5 Photodetection and Coherent Detection

From Eqs. (8.1) and (8.15), it is clear that the photodetector reacts to the square of the electric field. In other words, it is a square law device. Hence, using Eqs. (8.1) and (8.15) with some trigonometric identities, the dc photocurrent of the unmodulated beam of light is given by

$$i_{\mathrm{ph}} = \frac{\eta q}{h\nu} P_{\mathrm{s}}[1 + \cos 2(\omega_{\mathrm{c}}t + \varphi)]. \tag{8.84}$$

Since the nature of the photodetector is a low pass, the high-frequency component is filtered out and only the dc argument remains. This characteristic of photodetection is useful for heterodyne in a coherent detection and optical modulation schemes. Assume two lasers emitting two wavelengths denoted as local oscillator (LO) and signal (S):

$$E_{\mathrm{LO}}(t) = \sqrt{2P_{\text{LO-Ave}}Z_0}\cos(2\pi\nu_{\mathrm{LO}}t), \tag{8.85}$$

$$E_{\mathrm{S}}(t) = \sqrt{2P_{\text{S-Ave}}Z_0}\cos(2\pi\nu_{\mathrm{S}}t + \phi). \tag{8.86}$$

Then, in coherent detection, maximum correlation occurs when both frequencies are in the same phase and are coherent. In other words, both signals are locked on each other as detection criteria. Assume now that the mixing system contains a beam splitter with reflection index $\Gamma$ and a PD with QE $\eta$ as drawn in Fig. 8.26. Thus, the incident field at the PD surface is given by

$$E(t) = \sqrt{2P_{\text{LO-Ave}}Z_0}\cos(2\pi\nu_{\mathrm{LO}}t)\Gamma + \sqrt{2P_{\text{S-Ave}}Z_0}\cos(2\pi\nu_{\mathrm{S}}t + \phi) \times (1 - \Gamma). \tag{8.87}$$

Since the detector responds to the field square, i.e., photons flux, the detected optical current is given by

$$i(t) = \frac{\eta q}{h\nu}\cdot\frac{1}{Z_0}\Big[\sqrt{2P_{\text{LO-Ave}}Z_0}\cos(2\pi\nu_{\mathrm{LO}}t)\Gamma + \sqrt{2P_{\text{S-Ave}}Z_0}\cos(2\pi\nu_{\mathrm{S}}t + \phi)(1 - \Gamma)\Big]^2. \tag{8.88}$$

The idea here is to calculate the heterodyne low frequency and dc current components. The other square values result in a frequency multiplication that

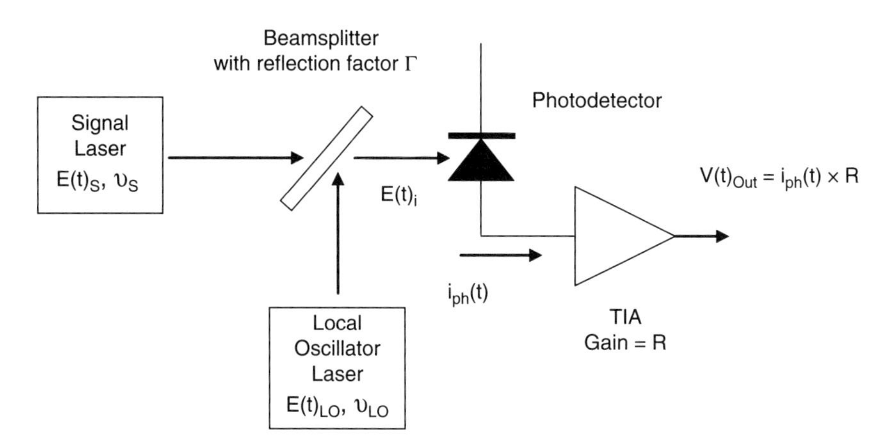


**Figure 8.26** Coherent detection concept. The front end contains a beamsplitter with reflection factor Γ, LO laser, and TIA with transimpedance gain $R$ [Ω].

is above the PD BW. Thus, these values are omitted. Hence, after using some trigonometric identities, it becomes

$$i(t) = \frac{nq}{h\upsilon}\Big[P_{\text{LO-Ave}}\Gamma + P_{\text{s-Ave}}(1-\Gamma) + 2\sqrt{P_{\text{LO-Ave}}P_{\text{S-Ave}}\Gamma(1-\Gamma)}\cos[2\pi(\nu_s - \nu_{\text{LO}})t + \phi]\Big]. \tag{8.89}$$

The first two terms in Eq. (8.89) are the dc components. The second dc term resulting from the signal received is relatively low compared to LO dc and thus can be neglected. The third term is an ac component that is the intermediate frequency (IF). Hence, $i$(t) is composed from a dc level, and an RF level is the IF frequency. It is obvious that $i(t)$ is at its maximum value when $\phi = 2k\pi$, where $k = 0, 1, 2 \ldots, n$ and when both frequencies of the LO and signal are equal. As a conclusion, coherent detection contains information about the phase and frequency of the received signal. These parameters, which were reviewed in previous sections, are not detected directly. One of the major advantages of coherent detection is better CNR, which will be reviewed later on in Chapter 11.

## 8.6 Main Points of this Chapter

1. A fiber optical communication wavelength of 800–1600 nm is within is the infrared region.

2. The most popular photodetectors for optical communication wavelength are PIN and APD.

3. Photodetection is a process where a photon excites a hole and electrons are elevated from their valence band to the conduction band.

4. The cutoff wavelength is defined by the bandgap energy depth.

5. Flux is the amount of photons per second.

6. Power flux is the amount of photon energy per unit of time per unit of area.

7. QE, probability function $\eta$ is defined as the probability that a single photon incident to the device would generate a photocarrier, pair of hole and electron, which contributes to the photocurrent.

8. Since QE is a probability function, it is limited to $0 \leq \eta \leq 1$.

9. The complementary probability of failing to generate hole and electron is $1 - \eta$.

10. QE depends on the reflection coefficient of the detector surface, absorption coefficient $\alpha$, and absorption region length. See Eq. (8.4).

11. Photodetector responsivity is a measure that defines the amount of photocurrent generated in a PD by a given optical power.

12. Responsivity units are in A/W, mA/mW, or any other factor of current-to-power ratio that has the same order of magnitude for both power and current. See Eq. (8.7).

13. Responsivity of a PD linearly depends on the wavelength $\lambda_0$. The wavelength should satisfy the bandgap range. See Eq. (8.8).

14. Photodetection is a process in which the semiconductor conductivity increases as a result of the carriers' concentration increment due to light. Thus, the current through the PD increases. It is a "light-controlled register."

15. A photodetector frequency response indicates ability to change its conductivity with respect to the modulated light incident to the semiconductor surface.

16. Photodetector noise results from several reasons: the random process of detection, which is called photon noise; photoelectron noise resulting from failure to create an electron-hole pair, which creates a phonon and vibration in the lattice; dark current shot noise resulting from the junction leakage current; gain noise resulting from the multiplication and impact ionization process as in APD; and, finally, the added receiver's electronics noise. (This is not an inherent noise of the PD but referred to as a limitation affecting the carrier-to-noise ratio, known as CNR.)

17. Transit time is the time it takes the hole or electron to travel through the semiconductor until it recombines. In PN and PIN, it refers to the time to pass the depletion region.

18. Gain process of a photodetector is defined as the ratio between carrier lifetime and its transit time until recombination.

19. Drift current results from the electric field attraction of a charge applying a force, providing acceleration to the charge. The charge reaches its saturation velocity, and the drift current is given by $q \times v$, where $q$ is the charge and $v$ is the saturated velocity.

20. A junction photodetector is made of P- and N-type materials or Schottkey metal semiconductor.

21. There are several types of junction detectors: PN, PIN, MSM, and APD.

22. The reason for creation of the depletion region in a PN junction is the doping concentration gradient between P and N. This results in diffusion and creation of an internal potential barrier on the junction and thus an electric field that generates a depletion region.

23. The thickness of the depletion region and the junction cross section defines the junction capacitance.

24. Einstein's law claims that the thermodynamic potential equals the ratio between diffusion coefficients and mobility factors. This ratio is constant for the hole and the electron, indicating thermodynamic equilibrium.

25. The junction build in potential depends on the donors' and acceptors' doping. This doping is a multiplication factor of the thermodynamic voltage. See Eqs. (8.21) and (8.22).

26. Depletion depth will be maximal at the region of low-doping concentration.

27. The absorption region of a photodetector is in its depletion region. The longer it is, the higher the QE of the detector.

28. PIN detector has wider depletion region than an ordinary abrupt PN junction.

29. The advantage of a PIN over a PN detector is its lower junction capacitance and higher QE.

30. BW limitation and trade-off in a photodetector results from the transit time of the depletion region and junction capacitance.

31. The reason for reverse biasing a photodetector is to increase its depletion region, increase the electric field across the junction, and reduce the shot noise by operating the PD at dc leakage current.

32. Small signal model of a photodetector composed of a current source represents the photocurrent, junction capacitance, resistance $C_j$ and $R_j$

parallel to the current source, series contact resistance $R_s$, package and bond capacitances, and bond and lead inductances.

33. The RF *S* parameters of a photodetector at low frequency are at the left side of the Smith chart, representing high impedance.

34. The NLD of a PD results from electron-hole traps, space charge, absorption, velocity fluctuations affecting the drift current, and responsivity fluctuations.

35. One of the methods to improve NLD performance without reducing the responsivity is by defocusing and using the aberration properties of a spherical lens. This results in a larger spot on the active area, making more uniform absorption and photocurrent, less lateral diffusion, and much more homogeneous drift velocity.

36. An APD is a photodetector exhibit gain.

37. APD gain results from the impact ionization due to high velocities of the carriers.

38. An APD is composed of two sections absorption section, which is a depletion region with low doping. This region defines the QE of the APD and gain section that is a PN junction with a much stronger field than the absorption region. This structure is called SAM.

39. The electric field in a PN junction is stronger than the one in PIN, since it has a thinner depletion region between the P and N sides and larger charge a concentrations.

40. APD BW limitations result from the gain build-up process and depletion transit time.

41. The ability of holes or electrons to create impact ionization is called ionization factor $\alpha_e$ and $\alpha_h$. The ratio between these values, $k = \alpha_h/\alpha_e$, defines whether the electron is ionizing, the hole is ionizing, or both are ionizing. The ionization depends on the semiconductor.

42. APD gain buildup is a feedback process.

43. By using the fusing grow method, an APD can be fabricated for long wavelengths. This structure of an APD is called silicon hetero-SHIP.

44. Coherent detection is done by heterodyne, the input signal with an LO signal, and by detecting the IF frequency.

45. A PD is a square-law device, which reacts to optical power, i.e., the square value of the electric field. Thus, a photodetector can be used as a mixer.

46. A PD detects dc and the low frequencies of optical wavelength beats. Higher-order harmonics are filtered out due to the LPF characteristics of the photodetector.

47. The main advantage of the coherent detection method is better CNR compared to direct detection.

## References


1. Bar-Lev, A., *Semiconductors and Electronic Devices*. Prentice Hall (1979).
2. Kuhara, Y., Y. Fujimura, N. Nishiyama, Y. Michituji, H. Terauchi, and N. Yamabayashi, "Characterization and theoretical analysis of second-order intermodulation distortion of InGaAs/InP PIN photodiode modules for fiber-optic CATV." *IEEE Journal of Lightwave Technology*, Vol. 15 No. 4, pp. 636–641 (1997).
3. Hayer, R.R., and D. Persechini, "Nonlinearity of PIN photodetectors." *IEEE Photonics Technology Letters*, Vol. 5 No. 1, pp. 70–72 (1993).
4. Scott, D.C., T.A. Vang, J. Elliot, D. Forbes, J. Lacey, K. Everett, F. Alvarez, R. Johnson, A. Krispin, J. Brock, L. Lembo, H. Jiang, D.S. Shin, J.T. Zhu, and P.K.L. Yu, "Measurement of IP3 in PIN photodetectors and proposed performance requirements for RF fiber-optic links." *IEEE Photonics Technology Letters*, Vol. 12 No. 4, pp. 422–424 (2000).
5. Williams, K.J., R.D. Esman, and M. Dagenais, "Nonlinearities in PIN microwave photodetectors." *IEEE Journal of Lightwave Technology*, Vol. 14 No. 1, pp. 84–96 (1996).
6. Sjöström, J., "Measurement and modeling of photodetectors for monolithic optoelectronic receivers for bit rates up to 40GP/sec." Master of Science Thesis in Photonics and Microwave Engineering. Royal Institute of Technology, Department of Electronics Stockholm (1998).
7. Dentan, M., and B.D. Cremoux, "Numerical simulation of the nonlinear response of a PIN photodiode under high illumination." *IEEE Journal of Lightwave Technology*, Vol. 8 No. 8, pp. 1137–1144 (1990).
8. Goldsmith, C.L., and B. Kanack, "Broadband microwave matching of high-speed photodiodes." *Microwave Symposium Digest*, Vol. 1, pp. 233–236 (1993).
9. Jiang, H., D.S. Shin, G.L. Li, T.A. Vang, D.C. Scott, and P.K.L. Yu, "The frequency behavior of the third-order intercept point in a waveguide photodiode." *IEEE Photonic Technology Letters*, Vol. 12 No. 5, pp. 540–542 (2000).
10. Chizh, A.I., and S.A. Malyshev, "Modeling and characterization of microwave PIN photodiode." *Proceedings of IEEE International Euroconference on Advanced Semiconductor Devices and Microsystems*, pp. 239–242 (2000).

11. Buckley, R.H., "Analytic solutions for distortion products in PIN photodiodes." *Proceedings of IEEE Microwave Photonics Conference*, pp. 157–160 (2002).
12. Humphreys, D.A., and R.A. Lobbet, "Investigation of an optoelectronic non-linear effect in a GaAs photodiode, and its application in a coherent optical communication system." *Proceedings of IEE, J.* Vol. 135 No.1, pp. 45–51 (1988).
13. Agrawal, G.P., *Fiber-Optic Communications Systems.* John Wiley and Sons (2004).
14. Alexander, S.B., *Optical Communication Receiver Design.* SPIE Press, Bellingham, WA (1997).
15. Sze, S.M., *Physics of Semiconductors Devices.* Willey Interscience (1981).
16. Donati, S., *Photodetectors: Devices, Circuits and Applications.* Prentice Hall (1999).
17. Li, J., and M. Uysal, "Optical wireless communication: system model, capacity and coding." *Proceedings of IEEE Vehicular Technolgy Conference*, Vol. 1, pp. 168–172 (2003).
18. Saleh, B.E.A., and M.C. Teich, *Fundamentals of Photonics.* Wiley Interscience (1991).
19. Susnjar, Z., Z. Djuric, M. Smiljanic, and Z. Lazic, "Numerical calculation of photodetector response time using Ramo's theorem." *Proceedings of IEEE International Conference on Microelectronics*, Vol. 2, pp. 717–720 (1995).
20. Wu, W., A.R. Hawkins, and J.E. Bowers, "Frequency response of avalanche photodetectors with separate absorption and multiplication layers." *IEEE Journal of Lightwave Technology*, Vol. 14 No. 12, pp. 2778–2785 (1996).
21. Hayat, M.M., O.H. Kwon, S. Wang, J.C. Campbell, B.E.A. Saleh, and M.C. Teich, "Boundary effects on multiplication noise in thin heterostructure avalanche photodiodes theory and experiment." *IEEE Transactions on Electron Devices*, Vol. 49 No. 12, pp. 2114–2123 (2002).
22. Hayat, M.M., O.H. Kwon, Y. Pan, P. Sotirelis, J.C. Campbell, B.E.A. Saleh, and M.C. Teich, "Gain-bandwidth characteristics of thin avalanche photodiodes." *IEEE Transactions on Electron Devices*, Vol. 49 No. 5, pp. 770–781 (2002).
23. Kang, Y., P. Mages, A.R. Clawson, A. Pauchard, S. Hummel, M. Bitter, Z. Pan, Y. H. Lo, and P.K.L. Yu, "Nonlinear distortions and excess noise behavior of fused InGaAs/Si avalanche photodiode." *Proceedings of IEEE International Topical Meeting on Microwave Photonics*, pp. 153–156 (2002).
24. Wang, S., J.B. Hurst, F. Ma, R. Sidhu, X. Sun, X.G. Zheng, A.L. Holmes Jr., A. Huntington, L.A. Coldern, and J.C. Campbell, "Low-noise impact-ionization-engineered avalanche photodiodes grown on InP substrates." *IEEE Photonics Technology Letters*, Vol. 14 No. 12, pp. 1722–1724 (2002).
25. Zheng, X.G., J.S. Hsu, J.B. Hurst, X. Li, S. Wang, X. Sun, A.L. Holmes Jr., J.C. Campbell, A.S. Huntington, and L.A. Coldern, "Long-wavelength InGaAs-InAlAs large-area avalanche photodiodes and arrays." *IEEE Journal of Quantum Electronics*, Vol. 40 No. 8, pp. 1068–1073 (2004).

26. Bowers, J., and A. Hawkins, "Gain sensitivity of InGaAs/Si avalanche photodetectors." Final report 1996-97 for MICRO project 96-041 (1997).
27. Srinivasan, M., J. Hamkins, B. Madden-Woods, A. Biswas, and J. Beebe, "Laboratory characterization of silicon avalanche photodiodes (APDs) for pulse-position modulation (PPM) detection." IPN Progress Report 42–146 (2001).
28. Pearsall, T.P., "Impact ionization rates for elecrons and holes in $Ga_{0.47}In_{0.53}As$." *Applied Physics Letters*, Vol. 36 No. 3, pp. 218–220 (1980).
29. Cook, L.W., G.E. Bulman, and G.E. Stillman, "Electron and hole impact ionization coefficients in GaAs." *Applied Physics Letters*, Vol. 66 No. 25, pp. 3507–3509 (1995).
30. Hawkins, A.R., T.E. Reynolds, D.R. England, D.I. Babic, M.J. Mondry, K. Streubel, and J.E. Bowers, "Silicon hetero-interface photodetector." *Applied Physics Letters*, Vol. 68 No. 26, pp. 3692–3694 (1996).
31. Dosunmu, O.I., D.D. Cannon, M.K. Emsley, B. Ghyselen, J. Liu, L.C. Kimerling, and M.S. Unlu, "Resonant cavity enhanced Ge photodetectors for 1550 nm operation on reflecting Si substrates." *IEEE Journal of Selected Topics in Quantum Electronics*, Vol. 10 No. 4, pp. 694–701 (2004).
32. Blauvelt, H., I. Ury, D.B. Huff, and H.L. Loboda, "Broadband receiver with passiner tuning network." U.S. Patent 5179461 (1993).

# Part 3

# RF and Control Concepts

# Chapter 9

# Basic RF Definitions and IMD Effects on TV Picture

## 9.1 Distortions and Dynamic Range

Any active network can be described by the power series. This term is correct in the first approximation, when the system has no memory. Memory effects of an active system are caused by a time-varying phase response, which is, in turn, manifested in the frequency response. This phase response is referred to as the amplitude modulation (AM) to phase modulation (PM) characteristic. This phenomenon describes how the phase response is affected by input power. Under these terms of memory, the system is described by a Volterra series.[15–18] The power series presentation here considers AM- to-PM effects only and the associated calculation is scalar only, which means a simple polynomial with real coefficients.

Assume an RF device such that its output voltage performance is described by power series at the operating frequency ω as follows:[6,4]

$$A(x)|_{@\omega} = a_0 + a_1 x + a_2 x^2 + a_3 x^3 + \cdots + a_n x^n, \qquad (9.1)$$

where $a_0$ describes the dc component, (in ac coupling, this can be neglected) $a_1 x$ describes the linear gain, $a_2 x^2$ describes the second-order distortion, $a_3 x^3$ describes the third-order distortion, and $a_n x^n$ describes the $n$th-order distortion.

Since the transfer function is not linear, intermodulation interference occurs, In fact, this power series can be referred to as a sum of transfer functions, where each one generates distortions of a different order. Presenting these arguments in a log scale would result in a linear function where its slope is the distortion order (the $Y$ axis represents power in dBm, since 20 log of the voltage is measured, and the $X$ axis is the input power). Thus, it can be written as

$$Y_1(\omega) = 20 \log (a_1) + 20 \log x = A_1 + X. \qquad (9.2)$$

The variable $A_1$ is the linear gain $G$.

In the same manner, the other terms of the power series are derived:

$$Y_2(\omega) = 20\log(a_2) + 20\log x^2 = 20\log(a_2) + 2\cdot 20\log x = A_2 + 2X, \quad (9.3)$$

$$Y_3(\omega) = 20\log(a_3) + 20\log x^3 = 20\log(a_3) + 3\cdot 20\log x = A_3 + 3X, \quad (9.4)$$

$$Y_n(\omega) = 20\log(a_n) + 20\log x^n = 20\log(a_n) + n\cdot 20\log x = A_n + nX. \quad (9.5)$$

System performance for linearity and dynamic range (DR) is measured by the two-tone test predicting the second-order intercept point (IP2) and third-order intercept point (IP3). The intercept point (IP) is a virtual point at which the distortion order line intersects the linear gain line, as shown in Fig. 9.1. Hence, IP2 is the virtual point where the second distortion line intersects the first-order linear gain line. In the same way, IP3 and $\mathrm{IP}_n$ are defined. Spectrumwise, this means that at the IP, the power level of the spurious intermodulations (IMDs) are equal to the fundamental signal power.

In community access television (CATV), third-order IMDs are problematic, since they are at the same frequency of the carrier. In a narrow-band system, they fall within the intermediate frequency (IF) band. As for a second-order IMD, in a narrow-band system, they are filtered out; but in CATV system, they are within the band, with an offset of $\pm 1.25$ MHz from the channel carrier. The composite second order (CSO) at $+1.25$ MHz is called CSO_H, and the one at $-1.25$ MHz is CSO_L. Observing the NTSC frequency plan given in Sec. 3.4, it is understood that CSO distortions are denoted as within the CATV band. When the bandwidth (BW) of the system is less than octave, second-order distortions can be filtered out; in fact, even each second-order intermodulation product can be filtered out in some cases. Generally, intermodulations and spurious-free dynamic range (SFDR) are tested by two tones, while IP2 and IP3 are evaluated by two tones as well.

The goal is to find the general expression for intercept point $\mathrm{IP}_n$, as described in Fig. 9.1. From Fig. 9.1, it is observed that in any system design, the goal is to have the maximum input power at which its IMDs are within the system noise floor. The SFDR of a system is defined by the maximum allowable power limit that generates IMDs that are within the noise floor of the system. For that condition of SFDR, there is a need to solve the two sets of linear line equations describing the gain and the IMD, to relate the equations' solution to the IP point, and then to constrain the IMD level to be within the noise floor:

$$Y = A_1 + X, \quad (9.6)$$

$$Y = A_n + nX. \quad (9.7)$$

At the $\mathrm{IP}_n$ point, both terms are equal to the linear power given by Eq. (9.6) and the distortions given by Eq. (9.7); therefore, the solution for $X$ is

$$X = \frac{A_n - A_1}{1-n} = \frac{A_1 - A_n}{n-1} = P_{\mathrm{IN}}. \quad (9.8)$$

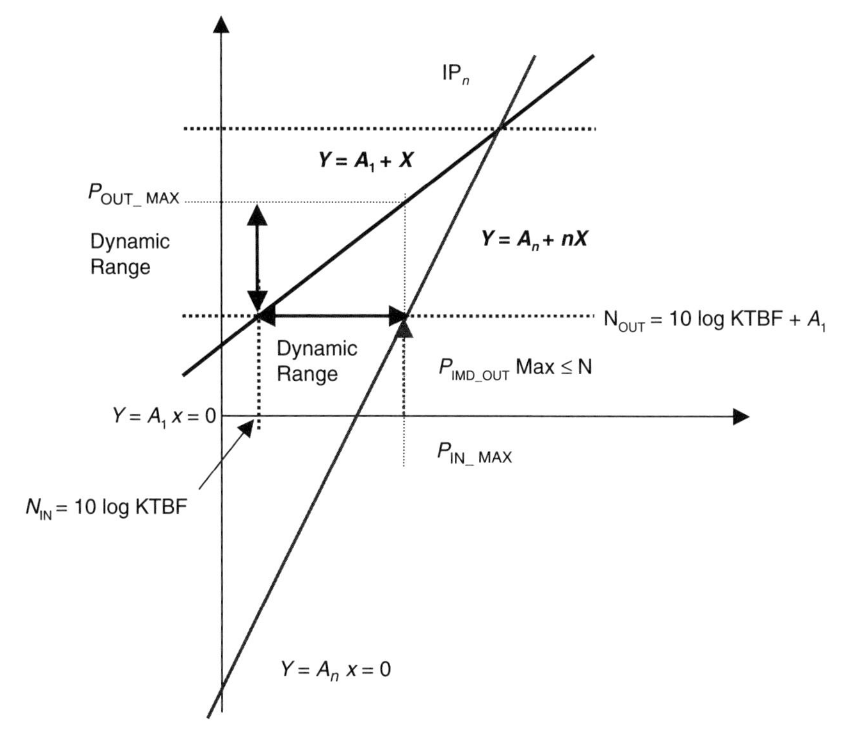


**Figure 9.1** Distortion line vs. linear gain and dynamic range.

Hence, the maximum input power at the IMD intercept equals

$$\mathrm{IP}_n = A_1 + \frac{A_1 - A_n}{n-1}. \tag{9.9}$$

For easier identification, the notation will be as follows for fundamental tone and IMD, respectively:

$$P_{\mathrm{OUT}} = A_1 + P_{\mathrm{IN}}, \tag{9.10}$$

$$P_{\mathrm{IMD}} = A_n + nP_{\mathrm{IN}}. \tag{9.11}$$

Thus, the power difference $\Delta$ between a fundamental signal and the $n$th-order IMD is defined as

$$P_{\mathrm{OUT}} - P_{\mathrm{IMD}} = \Delta = A_1 - A_n + (1-n)P_{\mathrm{IN}}. \tag{9.12}$$

Solving Eq. (9.12) for $P_{\mathrm{IN}}$ with respect to $\Delta$ and then substituting it in Eq. (9.6) results in $Y = P_{\mathrm{OUT}}$. Equation (9.13) provides a ratio between $P_{\mathrm{OUT}}$ and the power series AM-to-AM curve coefficients, $A_1$ and $A_n$:

$$P_{\mathrm{OUT}} = A_1 + \frac{\Delta - (A_1 - A_n)}{1-n} = A_1 + \frac{\Delta}{1-n} - \frac{(A_1 - A_n)}{1-n}$$
$$= A_1 - \frac{\Delta}{n-1} + \frac{(A_1 - A_n)}{n-1}. \tag{9.13}$$

From the $\mathrm{IP}_n$ expression given in Eq. (9.9), $P_{\mathrm{OUT}}$ at IP equals $\mathrm{IP}_n$ when $P_{\mathrm{IN}}$ satisfies the conditions given by Eqs. (9.8) and (9.10). Hence, after some algebra,

$$\mathrm{IP}_n = P_{\mathrm{OUT}} + \frac{\Delta}{n-1}. \tag{9.14}$$

Therefore, second $\mathrm{IP}_2$ and $\mathrm{IP}_3$ are given by

$$\mathrm{IP}_2 = P_{\mathrm{OUT}} + \Delta, \tag{9.15}$$

$$\mathrm{IP}_3 = P_{\mathrm{OUT}} + \frac{\Delta}{2}. \tag{9.16}$$

Now, when the relations between distortions and signal are known, DR terms can be calculated as shown in Fig. 9.1. At maximum input power case, the IMD level constraint is that it should be within the noise floor at a given system BW. Hence, the following terms are given when the IMD should be equal to the system's noise:

$$P_{\mathrm{OUT}} = A_1 + P_{\mathrm{IN\_MAX}}, \tag{9.17}$$
$$P_{\mathrm{IMD}} = N_{\mathrm{OUT}} = 10\log \mathrm{KTBF} + A_1 = A_n + nP_{\mathrm{IN\_MAX}}. \tag{9.18}$$

Therefore, the DR is given by

$$\mathrm{DR} = P_{\mathrm{OUT\_MAX}} - N_{\mathrm{OUT}} = A_1 - A_n + (1-n)P_{\mathrm{IN\_MAX}}, \tag{9.19}$$

$$\frac{\mathrm{DR} - (A_1 - A_n)}{1-n} = P_{\mathrm{IN\_MAX}}. \tag{9.20}$$

But, from the noise constraint in Eqs. (9.18) and (9.20), the result is

$$\frac{\mathrm{DR} - (A_1 - A_n)}{1-n} = \frac{N - A_n}{n} = \frac{(10\log \mathrm{KTBF} + A_1) - A_n}{n}. \tag{9.21}$$

Hence,

$$\mathrm{DR} = (A_1 - A_n) + (1-n)\frac{(10\log \mathrm{KTBF} + A_1) - A_n}{n}$$
$$= (A_1 - A_n) - (1-n)\frac{A_n}{n} + (1-n)\frac{(10\log \mathrm{KTBF} + A_1)}{n}. \tag{9.22}$$

Rearranging the expression above results in

$$\begin{aligned}\mathrm{DR} &= A_1 + \frac{-nA_n - (1-n)A_n}{n} + (1-n)\frac{(10\log \mathrm{KTBF} + A_1)}{n} \\ &= A_1 + \frac{(n-1)A_n - nA_n}{n} + (1-n)\frac{\mathrm{KTBF}}{n},\end{aligned} \tag{9.23}$$

$$\begin{aligned}\mathrm{DR} &= A_1 - \frac{A_n}{n} + (1-n)\frac{(10\log \mathrm{KTBF} + A_1)}{n} = \frac{nA_1 - A_n}{n} \\ &+ (1-n)\frac{(10\log \mathrm{KTBF} + A_1)}{n} = \frac{(n-1)A_1}{n} + \frac{A_1 - A_n}{n} \\ &+ (1-n)\frac{(10\log \mathrm{KTBF} + A_1)}{n},\end{aligned} \tag{9.24}$$

$$\begin{aligned}\mathrm{DR} &= \frac{(n-1)A_1}{n} + \frac{A_1 - A_n}{n} \cdot \frac{n-1}{n-1} + (1-n)\frac{(10\log \mathrm{KTBF} + A_1)}{n} \\ &= \frac{n-1}{n}\left(A_1 + \frac{A_1 - A_n}{n-1}\right) + (1-n)\frac{(10\log \mathrm{KTBF} + A_1)}{n}.\end{aligned} \tag{9.25}$$

Replacing the expression in the parenthesis by an expression for $\mathrm{IP}_n$ given in Eq. (9.9) results in

$$\mathrm{DR}_n = \frac{n-1}{n}[\mathrm{IP}_n - (10\log \mathrm{KTBF} + A_1)]. \tag{9.26}$$

Hence, the DR for second order and third order are given by

$$\mathrm{DR}_2 = \frac{1}{2}[\mathrm{IP}_2 - (10\log \mathrm{KTBF} + A_1)], \tag{9.27}$$

$$\mathrm{DR}_3 = \frac{2}{3}[\mathrm{IP}_3 - (10\log \mathrm{KTBF} + A_1)]. \tag{9.28}$$

As a general conclusion, the higher the IP and the lower the noise floor, the higher the DR.

From the plot at Fig. 9.2, it is observed that any system is linear up to a point where the gain starts to fall off its linear gain. This drop is marked as a compression point where the gain drops by 1 dB from its linear line. The reason for that is seen at the polynomial description in Eq. (9.1). In fact, the higher the signal power is, the higher is its swing. Hence, it reaches a point of clipping and rectifying. Additionally, the signals start to affect the dc bias operating point. Hence, for highly linear system designs, such optical video receivers, a class A amplifier bias approach is preferred.

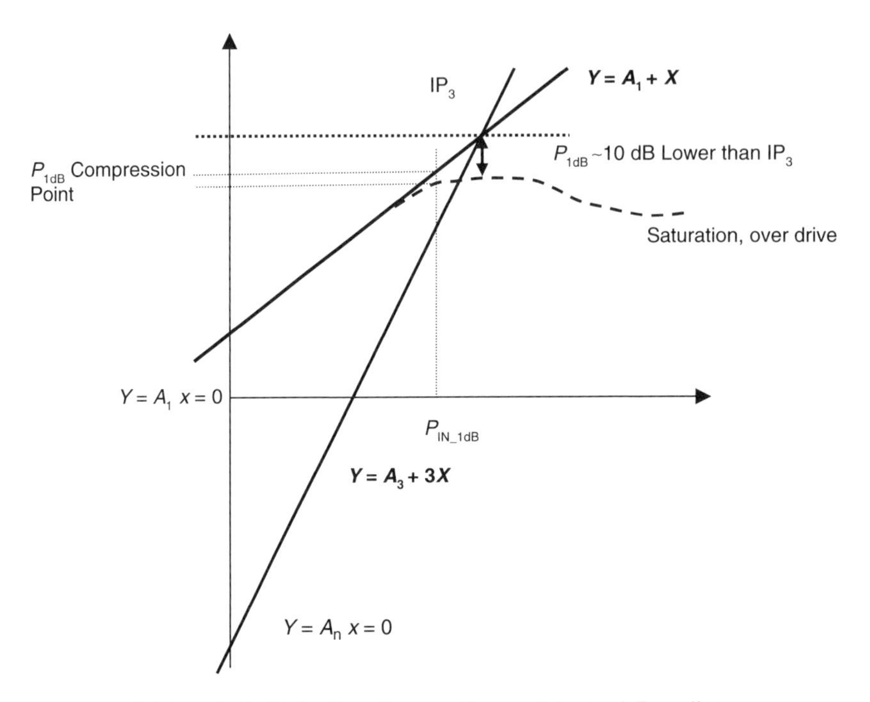


**Figure 9.2** Distortion line vs. linear gain and $P_{1dB}$ line.

## 9.2 1-dB Compression Point and IP3 Relations

The distortion created by the third-order term in Eq. (9.1) is interesting since it relates to the composite triple beat (CTB) in a multitone loading and creates the 1-dB gain compression or reduction.[8] Assuming a harmonic signal,

$$V(\omega) = A \cdot \cos \omega t, \tag{9.29}$$

its third-order value after some trigonometric manipulations is given by

$$a_3 V^3(\omega) = a_3(A \cdot \cos \omega t)^3 = a_3 A^3 \cdot \left(\frac{3}{4}\cos \omega t + \frac{1}{4}\cos 3\omega t\right). \tag{9.30}$$

If $a_3$ is a negative, then the third-order contribution at the fundamental frequency decreases the signal level and the gain is compressed. The reason is given by observing the curve $I_{OUT}$ versus $V_{IN}$ of the device; a large signal creates clipping. The energy used to create the harmonics is taken from the given constraint of the dc bias point. Hence, the energy of the fundamental tone is reduced due to the generation of harmonics. Ignoring the high-order harmonics, the power-compression point can be defined by the following equation:

$$20 \log \frac{a_1 V(\omega) - \frac{3}{4} a_3 V(\omega)}{a_1 V(\omega)} = -1 \text{ dB}. \tag{9.31}$$

The reason for looking at the fundamental frequency term in Eq. (9.30) and substituting it in Eq. (9.31) is that its frequency equals the fundamental level and its amplitude is the distortion level. That distortion product creates the compression. It can be seen that the third-order distortion is within the signal BW. Thus, it is hard to filter it out. Equation (9.31) can be written as the ratio of the AM-to-AM polynomial argument coefficients according to Eq. (9.32) where the coefficients are given in Eq. (9.1). In this manner, the 1-dB compression point can be evaluated analytically:

$$V(\omega)^2_{1\mathrm{dB}} = \frac{4a_1(1 - 10^{-1/20})}{3a_3}. \quad (9.32)$$

In the two-tone test, the input signal is given by

$$V(\omega) = A \cdot (\cos\omega_1 t + \cos\omega_2 t). \quad (9.33)$$

Both tones are equal in amplitude; when substituting this value in Eq. (9.1) and solving the third-order term using some trigonometric identities, a fundamental tone is generated by the third-order term with an amplitude of $2.25a_3A^3$. The third harmonic generated for each frequency is with an amplitude of $0.25a_3A^3$, as in the case of a single tone. Additional terms of $2f_1 - f_2$ and $2f_2 - f_1$ are generated with an amplitude of $0.75a_3A^3$. At the IP, the third-order IMD amplitudes of $2f_1 - f_2$ and $2f_2 - f_1$ are equal to the fundamental level. That means that $V_{\mathrm{IP3}}$ satisfies both the linear and the third-order IMD terms of Eq. (9.1). Thus, the following condition is valid:

$$a_1 V(\omega)_{\mathrm{IP3}} = \frac{3}{4} a_3 V(\omega)^3_{\mathrm{IP3}}. \quad (9.34)$$

Hence, the relation between $a_1$ and $a_3$ is given by

$$V(\omega)^2_{\mathrm{IP3}} = \frac{4\,a_1}{3\,a_3}. \quad (9.35)$$

Substituting the relation given by Eq. (9.35) in Eq. (9.32) would provide the power ratio between the IP3 and the 1 dB compression point:

$$\frac{V(\omega)^2_{\mathrm{IP3}}}{V(\omega)^2_{1\mathrm{dB}}} = \frac{1}{1 - 10^{-1/20}} = 9.6 \Longrightarrow 10\log 9.6 = 9.636\ \mathrm{dB}. \quad (9.36)$$

Hence, the relation between IP3 compression point and 1 dB is ~10 dB at first approximation.

For HBT devices, this difference ranges between 15 dB and 20 dB.

## 9.3 Amplifier Gain Reduction Due to Third-Order Nonlinearity

The purpose of this section is to provide a method for predicting the gain changes due to a strong signal interferer caused by third-order phenomena (CTB).[13] The analysis will refer to the two-tone case. Using a Taylor series from Eq. (9.1), the output voltage $V_0(t)$ of the gain stage as function of input voltage $V_i(t)$ is expressed as in Eq. (9.1). For two-tone inputs, the input voltage is expressed by

$$x = V_i(t) = A\cos(\omega_1 t) + B\cos(\omega_2 t). \tag{9.37}$$

Using the cubic formula of $(A+B)^3 = A^3 + B^3 + 3A^2B + 3B^2A$ and substituting Eq. (9.37) in Eq. (9.1) yields

$$\begin{aligned} V_0(t)_{\text{third order}} &= a_3[A\cos(\omega_1 t) + B\cos(\omega_2 t)]^3 \\ &= k_3[A^3\cos^3(\omega_1 t) + B^3\cos^3(\omega_2 t) \\ &\quad + 3A^2B\cos^2(\omega_1 t)\cos(\omega_2 t) + 3AB^2\cos(\omega_1 t)\cos^2(\omega_2 t)]. \end{aligned} \tag{9.38}$$

Using the trigonometric identities for $\cos^3(\omega t) = \frac{1}{4}\cos(3\omega t) + \frac{3}{4}\cos(\omega t)$[14] and for

$$3\cos^2(\omega_1 t)\cos(\omega_2 t) = \frac{3}{2}\cos(\omega_2 t) + \frac{3}{4}\cos(2\omega_1 t + \omega_2 t) + \frac{3}{4}\cos(2\omega_1 t - \omega_2 t),$$

the output voltage is expressed by

$$\begin{aligned} V_0(t)_{\text{third order}} = a_3\Bigg[&\frac{3}{4}A^3\cos(\omega_1 t) + \frac{3}{4}B^3\cos(\omega_2 t) + \frac{3}{2}A^2B\cos(\omega_2 t) \\ &+ \frac{3}{2}B^2A\cos(\omega_1 t) + \frac{3}{4}A^2B\cos(2\omega_1 t - \omega_2 t) \\ &+ \frac{3}{4}B^2A\cos(2\omega_2 t - \omega_1 t)\Bigg]. \end{aligned} \tag{9.39}$$

Cos(3ωt) is a high frequency that filtered out due to the CATV high-band limit of operation, but it falls within the band when $f \leq 1/3\, f_{\max}$. From the power series given by Eq. (9.1), the gain for the first-order term is $a_1$. In the case of $V_{\text{IN}} = A\cos(\omega_1 t)$, $V_{\text{OUT}} = k_1 A\cos(\omega_1 t)$. Hence, the linear voltage gain can be written as

$$\text{Gv} = \frac{V_{\text{OUT}}}{V_{\text{IN}}} = \frac{a_1 A\cos(\omega_1 t)}{A\cos(\omega_1 t)} = a_1, \tag{9.40}$$

where $\frac{3}{4}A^2B\cos(2\omega_1 t - \omega_2 t)$ and $\frac{3}{4}B^2A\cos(2\omega_2 t - \omega_1 t)$ are the IIP3 (input $\text{IP}_3$) products. According to the IP3 definition as a virtual IP at which the fundamental signal is equal to the third-order product, and assuming equal tones, where $A = B$, it can be written

$$\frac{3}{4}A^2B\cos(2\omega_1-\omega_2)t = \frac{3}{4}A^3\cos(2\omega_1-\omega_2)t = \frac{3}{4}A^3\cos(2\omega_2-\omega_1)t. \tag{9.41}$$

The goal is to find the small signal gain compression due to a strong signal $A \neq B$ and $A < B$. It is known that the output voltage at the IP for equal tone levels is given by

$$V_{OUT} = \mathrm{Gv}\cdot V_{IN} = a_1\cdot A = a_3\cdot\frac{3}{4}A^2B. \tag{9.42}$$

The attention is on the low IMD product, since $A < B$. At the IP level, the input power for the intercept level is given by

$$P_{IN} = \frac{V_{RMS}^2}{R_{IN}} = \mathrm{IIP3}. \tag{9.43}$$

Thus, for the strong sinusoidal signal B, the input $IP_3$ power is given by

$$\mathrm{IIP3} = \left(\frac{B}{\sqrt{2}}\right)^2\cdot\frac{1}{R_{IN}} = \frac{B^2}{2R_{IN}}\ [\mathrm{watt}], \tag{9.44}$$

where $B^2 = 2\cdot R_{IN}\cdot P_{IN}$. Thereby, it yields

$$V_{OUT} = \mathrm{Gv}\cdot B = a_3\cdot\frac{3}{4}\cdot AB^2, \tag{9.45}$$

where $\mathrm{Gv} = a_3\cdot\frac{3}{4}B^2$. Knowing that $a_3$ is always negative, $a_3$ can be written as

$$a_3 = -\frac{\mathrm{Gv}}{B^2}\cdot\frac{4}{3} = -\frac{\mathrm{Gv}}{2\cdot R_{IN}\cdot\mathrm{IIP3}}\cdot\frac{4}{3} = -\frac{\mathrm{Gv}}{R_{IN}\cdot\mathrm{IIP3}}\cdot\frac{2}{3}. \tag{9.46}$$

Assume $A \ll B$, which is the case of two highly unequal signals at the amplifier input. Then, from Eq. (9.38), the fundamental frequency component of the smaller signal, of $V_0(t)$, arising from the third order term is $a_3\frac{3}{2}B^2A\cos(\omega_1 t)$. Recalling that $a_3 < 0$, it can be said that the small signal $A$ is compressed by the strong signal $B$ relative to the linear gain $a_1$ as denoted by

$$\mathrm{Gv}_A = \frac{V_{OUT}}{V_{IN}} = \frac{a_1A\cos(\omega_1 t) + a_3\cdot\frac{3}{2}B^2A\cos(\omega_1 t)}{A\cos(\omega_1 t)} = a_1 + a_3\frac{3}{2}B^2. \tag{9.47}$$

This is an important conclusion, showing that signal gain is affected by fluctuations of a strong interfering signal. Substituting Eqs. (9.44) and (9.46) for Eq. (9.47) results in

$$\mathrm{Gv}_A = \mathrm{Gv} - \frac{2}{3}\frac{\mathrm{Gv}}{R_{\mathrm{IN}} \cdot \mathrm{IIP3}} \cdot \frac{3}{2} \cdot 2 \cdot R_{\mathrm{IN}} \cdot P_{\mathrm{IN}} = \mathrm{Gv} - \mathrm{Gv} \cdot 2\frac{P_{\mathrm{IN}}}{\mathrm{IIP3}}. \tag{9.48}$$

Thus,

$$\mathrm{Gv}_A = \mathrm{Gv}\left(1 - 2\frac{P_{\mathrm{IN}}}{\mathrm{IIP3}}\right). \tag{9.49}$$

The power gain is given by

$$\mathrm{Gp} = \mathrm{Gv}^2 \cdot \frac{R_{\mathrm{OUT}}}{R_{\mathrm{IN}}}. \tag{9.50}$$

Assuming $R_{\mathrm{IN}} = R_{\mathrm{OUT}}$, then Eq. (9.49) becomes

$$\mathrm{Gp}_A = \mathrm{Gv}^2\left(1 - 2\frac{P_{\mathrm{in}}}{\mathrm{IIP3}}\right)^2 = \mathrm{Gp}\left(1 - 2\frac{P_{\mathrm{in}}}{\mathrm{IIP3}}\right)^2. \tag{9.51}$$

Thus, the small signal compression can be written as

$$\mathrm{Gp}_A(\mathrm{dB}) = 10\log\mathrm{Gp} + 20\log\left(1 - 2\frac{P_{\mathrm{in}}}{\mathrm{IIP3}}\right). \tag{9.52}$$

In the same manner, strong signal compression is evaluated. Observing Eq. (9.39), the term $\frac{3}{4}B^3\cos(\omega_2 t)$ is the compression of strong signal. Hence, it can be written as

$$\mathrm{Gv}_B = \frac{V_{\mathrm{OUT}}}{V_{\mathrm{IN}}} = \frac{a_1 B\cos(\omega_2 t) + a_3 \cdot \frac{3}{4}B^3\cos(\omega_2 t)}{B\cos(\omega_2 t)} = a_1 + a_3\frac{3}{4}B^2. \tag{9.53}$$

Using the relation from Eq. (9.44) for IP3 and the identities from Eq. (9.44), and substituting it in Eq. (9.53) results in

$$\begin{aligned}\mathrm{Gv}_B &= \mathrm{Gv} - \frac{\mathrm{Gv}}{R_{\mathrm{in}} \cdot \mathrm{IIP3}} \cdot \frac{2}{3} \cdot \frac{3}{4} \cdot 2R_{\mathrm{IN}} \cdot P_{\mathrm{IN}} = \mathrm{Gv} - \mathrm{Gv}\frac{P_{\mathrm{IN}}}{\mathrm{IIP3}} \\ &= \mathrm{Gv}\left(1 - \frac{\mathrm{P}_{\mathrm{IN}}}{\mathrm{IIP3}}\right).\end{aligned} \tag{9.54}$$

Using the same relation for Gp, given by Eq. (9.50) while assuming $R_{\mathrm{IN}} = R_{\mathrm{OUT}}$, results in

$$\mathrm{Gp}_B = \mathrm{Gv}2\left(1 - \frac{P_{\mathrm{IN}}}{\mathrm{IIP3}}\right)^2. \tag{9.55}$$

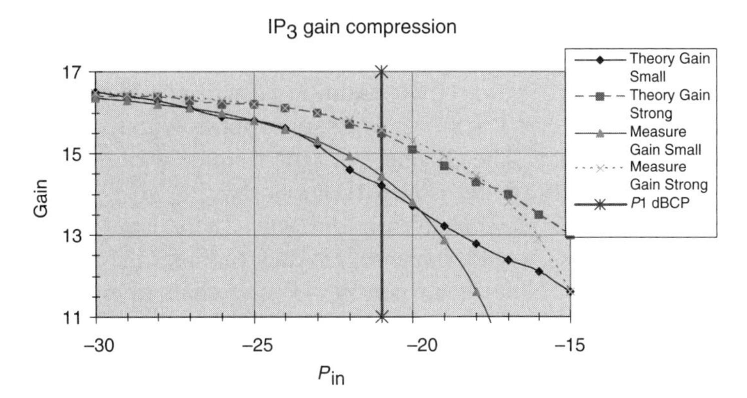


**Figure 9.3** Test results vs. theory for $P_{1dB}$ compression affects small and large signal gain.

Hence, the strong signal compression with respect to IP3 is given by

$$\mathrm{Gp}_B(\mathrm{dB}) = 10\log \mathrm{Gp} + 20\log\left(1 - \frac{P_{\mathrm{IN}}}{\mathrm{IIP3}}\right). \tag{9.56}$$

Here, it can be observed that compression and 1 dB reduction results from the $a_3$ coefficient. Figure 9.3 demonstrates this phenomena.

## 9.4 Cross Modulation Effects

One of the major distortion effects in a multichannel CATV transport is the cross-modulation (X-mod) effect that occurs as the AM-vestigial side band (VSB) tones are AM modulated for the image. All tones are assumed to be equally modulated and equal in power. X-mod results from a third-order distortion level. It is a phenomenon in which the modulation of a strong carrier, in this case all are of the same power, is transferred to a nearby interfering signal. For the sake of argument and simplicity of analysis, assume the case of two CW tones and an AM modulated tone. Assume the following scenario of input tones: desired CW signal given by $V_1 \cos \omega_1 t$, near by tone $V_2 \cos \omega_2 t$, and an AM modulated tone $V_3[1 + m\cos(\omega_m t)]\cos \omega_3 t$. The goal is to see how the X-mod falls in the band of $\omega_1$. Observing the third-order component under this assumption, it can be written

$$V_{\mathrm{OUT}} \cong V_1 a_1 \cos(\omega_1 t) + 3a_3 V_2 V_3^2 [1 + m\cos(\omega_m t)]^2 \cos^2(\omega_3 t)\cos(\omega_2 t). \tag{9.57}$$

After using some trigonometric manipulations, the result is as follows:

$$V_{\mathrm{OUT}} \cong V_1 a_1 \cos(\omega_1 t) + 3a_3 V_2 V_3^2 \left[1 + \frac{m^2}{2} + 2\,m\cos\omega_m t + \frac{m^2}{2}\cos 2\omega_m t\right]$$
$$\times \left[\frac{1}{2}\cos\omega_2 t + \frac{1}{4}\cos(2\omega_3 - \omega_2)t + \frac{1}{4}\cos(2\omega_3 + \omega_2)t\right]. \tag{9.58}$$

Assume that the delta is $(2\omega_3 - \omega_2) \approx \omega_1$. Then, this third-order IMD product will transfer the modulating tone is side bands into the channel under test and, thereby, reduces its carrier to noise ratio (CNR). Additionally, it would be transferred as a CW multiplied by the factor $1 + m^2/2$. The same applies with $(2\omega_3 + \omega_2) \approx \omega_1$ in the case of high frequency. The channel with the frequency of $\omega_2$ will have CNR reduction by the quantity of the CW IMD given by $1 + m^2/2$ and would have modulated distortions from $2m\cos\omega_m t$ and $\left(m^2/2\right)\cos\omega_m t$. Furthermore, the channel adjacent to $\omega_2$ would have $m^2/2$ due to $2\omega_m$ in case the channel spacing is at that value. Note that when $m = 0$, the channel $\omega_1$ suffers the beat note $(2\omega_3 - \omega_2)$ only and $\omega_2$ is affected by CW IMD.

## 9.5 AM-to-PM Effects

When taking the IP3 two-tone test, sometimes it can be observed that the IMD distortion levels are not equal in amplitude as shown in Fig. 9.4.

The reason behind this phenomenon is the AM-to-PM conversion resulting from memory effects and fifth-order terms beats. As a consequence, the IMD products are not added in phase versus frequency. Thus, the characterization of the device has to be both, by measuring its AM-to-AM curve and by AM-to-PM, curve showing the amount of change in the transfer phase of the device as a function of power. Hence, the power series representing such a case at the operating frequency $\omega$ is given by

$$A(x,\varphi)|_{@\omega} = a_0\exp(\varphi_0) + a_1 x\exp(\varphi_1) + a_2x^2\exp(\varphi_2) + a_3x^3\exp(\varphi_3) + \cdots + a_nx^n\exp(\varphi_n). \tag{9.59}$$

The above relation affects the equality of second-order IMDs as well. Hence, when dealing with the two-tone test, it can appear that for a given channel, there are differences between its discrete second-order high (DSO), which is above the high-frequency test signal carrier and its DSO low, which is below the low-frequency test signal carrier at a given frequency test, when the input power is varied. This analysis type related to memory effects is called Volterra series analysis.

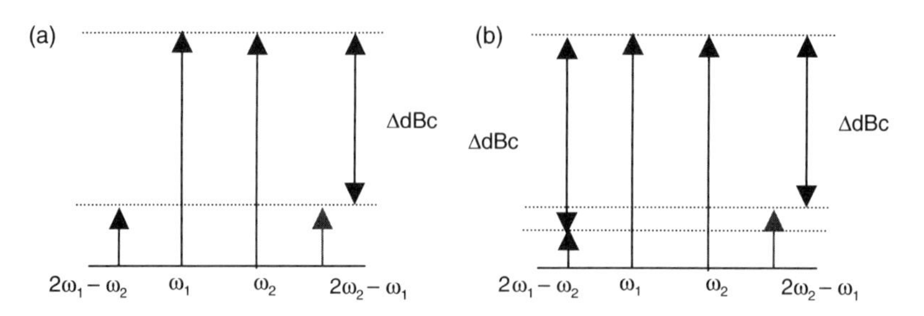


**Figure 9.4** AM to PM affects IP3 IMDs; (a) no AM to PM effects, (b) with AM to PM effects.

## 9.6 Multitone CTB Relations

When more than two carriers are present in a channel, third-order interference can be created by the multiplication of three fundamental carriers. In CATV systems, it is generally used to measure the distortions with more than two signals. Hence, IP3 corresponds to CTB, and IP2 corresponds to CSO. These CTB spurious signals are generally 6 dB higher than the regular two-tone test products defined as discrete two-tone, third-order beat (DTB). The level of CTB is further enhanced by the fact that multiple CTB signals can occur in the same frequency band. The number of CTB signals being superimposed on any particular channel is related to the number of desired carriers present. Statistics shows that more CTB interference occurs in the central band.

It is assumed that the system is fed by three tones with the following amplitudes and frequencies, respectively: $E_1 \sin \omega_1 t$, $E_2 \sin \omega_2 t$, and $E_3 \sin \omega_3 t$, while $E_1 = E_2 = E_3 = E$.

Using the polynomial transfer function, the condition in Eq. (9.1), and the condition for the input signal per Eq. (9.60), the input signal is[2,3,5]

$$x = E_1 \cos \omega_1 t + E_2 \cos \omega_2 t + E_3 \cos \omega_3 t. \tag{9.60}$$

Solving the third-order condition would result in the following terms:

$$\text{Third IMD\_CTB} = A_3 E^3$$

$$\times \begin{pmatrix} [\cos(\omega_1 t)]^3 + [\cos(\omega_2 t)]^3 + [\cos(\omega_3 t)]^3 + \frac{3}{2}[2\cos(\omega_1 t) + 2\cos(\omega_2 t) + 2\cos(\omega_3 t)] + \\ \frac{3}{2}\left\{ \begin{array}{l} \frac{1}{2}\cos[(2\omega_1 - \omega_2)t] + \frac{1}{2}\cos[(2\omega_1 + \omega_2)t] \\ +\frac{1}{2}\cos[(2\omega_1 - \omega_3)t] + \frac{1}{2}\cos[(2\omega_1 + \omega_3)t] + \\ \frac{1}{2}\cos[(2\omega_2 - \omega_1)t] + \frac{1}{2}\cos[(2\omega_2 + \omega_1)t] \\ +\frac{1}{2}\cos[(2\omega_2 - \omega_3)t] + \frac{1}{2}\cos[(2\omega_2 + \omega_3)t] + \\ \frac{1}{2}\cos[(2\omega_3 - \omega_2)t] + \frac{1}{2}\cos[(2\omega_3 + \omega_2)t] \\ +\frac{1}{2}\cos[(2\omega_3 - \omega_1)t] + \frac{1}{2}\cos[(2\omega_3 + \omega_1)t] + \end{array} \right\} + \\ \frac{6}{4}\left\{ \begin{array}{l} \cos[(\omega_1 - \omega_2 + \omega_3)t] + \cos[(\omega_1 - \omega_2 - \omega_3)t] + \\ \cos[(\omega_1 + \omega_2 - \omega_3)t] \end{array} \right\} \end{pmatrix}. \tag{9.61}$$

The resultant frequencies are third order and CTB with relative amplitudes $3/4E^3$ and $6/4E^3$, respectively. The result shows that the CTB triple beat terms $\omega_1 \pm \omega_2 \pm \omega_3$ are two times higher than the two-signal, third-order terms $2\omega_1 \pm \omega_2$ and $2\omega_2 \pm \omega_1$, as given in Eq. (9.62).

Hence, the triple beat is 6 dB higher than the two-tone, third-order intermodulations.

$$\text{Third IMD\_IP}_3 = A_3E^3\left(\begin{array}{l}[\cos(\omega_1 t)]^3 + [\cos(\omega_2 t)]^3 + \frac{3}{2}[\cos(\omega_1 t) + \cos(\omega_2 t)] + \\ \frac{3}{2}\left\{\begin{array}{l}\frac{1}{2}\cos[(2\omega_1 - \omega_2)t] + \frac{1}{2}\cos[(2\omega_1 + \omega_2)t] \\ + \frac{1}{2}\cos[(2\omega_2 - \omega_1)t] + \frac{1}{2}\cos[(2\omega_2 + \omega_1)t]\end{array}\right\}\end{array}\right) \tag{9.62}$$

The cubic arguments generate gain compression. The resultant frequencies are a third harmonic signal and an additional fundamental frequency:

$$[\cos(\omega_1 t)]^3 = \frac{3}{4}\cos(\omega_1 t) + \frac{1}{4}\cos(3\omega_1 t). \tag{9.63}$$

In both cases, the gain compression argument generates the third harmonic and contributes an equal level of the fundamental frequency. As a result, the relation between the triple beat and $IP_3$ is given by Eq. (9.64), where $\Delta$ is the difference in dBc between the test-tone output and the IMD levels:

$$2(\text{IP}_3 - P_{\text{OUT}}) - 6 = \Delta - 6 = \text{CTB}. \tag{9.64}$$

That means CTB can be measured indirectly by the two-tone test. First, $IP_3$ is evaluated as described by Eq. (9.16). Then, the IMD of the two-tone test is measured in reference to the test tones. The CTB is the two-tone, third-order IMD plus 6 dB. Figure 9.5 illustrates the two-tone, third-order test, and the carrier triple-beat test. Figure 9.5 describes CSO and CTB products of a CATV link bearing 110 channels.

In reality, there are more than three channels, and CTB is evaluated under full CATV OMI loading, which consists of 79 AM-VSB tones with an OMI of 3.5–4%

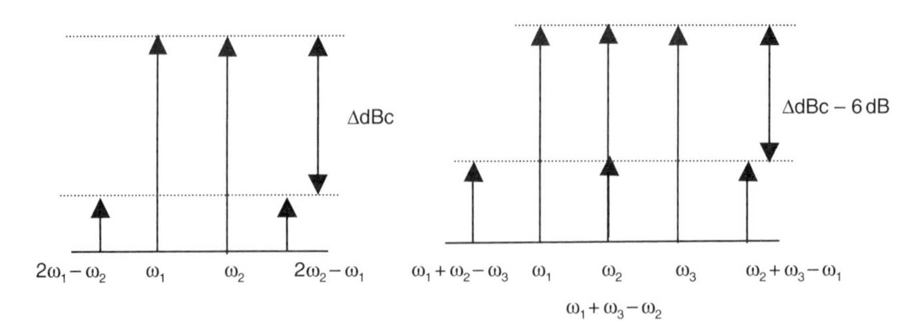


**Figure 9.5** Two-tone $IP_3$ test vs. triple-beat CTB test.

and 31 tones for QAM digital channels with an OMI of 1.75–2%, respectively, and thus 110 tones.[12] Eventually, nonlinear distortions (NLD) are generated by the laser due to clipping as well as by the receiver. Unlike third-order intermodulation interference, CTB signals can overlap each other and add none coherently. This considerably increases the overall spuriousness in any given channel. The total spurious interference is related to the number of carriers and the position of the carrier within the BW of operations. From Eq. (9.61), it is clear that there are three groups of CTB: $A+B+C$, $A-B+C$, and $-A+B+C$, generally presented by $A \pm B \pm C$. Thus, at any given CATV channel, there are different product counts or beat counts. That means carriers that are at the end of the BW have less interference products than those at the midband. The expression in Eq. (9.65) can be used to find the number of beats in any channel with very low error.[9] The full analysis is provided by Fig. 9.6 illustrates the CSO and CTB bean count versus frequency:

$$\text{beats} = \frac{N^2}{4} + \frac{(N-M)(M-1)}{2}, \tag{9.65}$$

where "beats" is the number of interference carriers in the measured channel, $N$ is the total number of channels, and $M$ is the channel under test and being measured, $1 \leq M \leq N$.

Performing the derivative of $(\text{dbeats}/\text{d}M) = 0$ results in the max beats being at $M = N/2 + 1$. The conclusion is that maximum number of interference occurs at the midband.

In case $N >> 1$, then $M \approx N/2$. The maximum number of beat notes is

$$\text{beat}_{\text{max}} = \frac{3N^2}{8}. \tag{9.66}$$

At the BW edges where $M = 1$ or $N$, the minimum beat note is

$$\text{beat}_{\text{min}} = \frac{N^2}{4}. \tag{9.67}$$

In the same manner, two-tone tests an be used to estimate the midband CTB of multitone as

$$\text{CTB}_{\text{max}} = 2(\text{IP}_3 - P_{\text{OUT}}) - 6 - 10\log\frac{3N^2}{8}. \tag{9.68}$$

For BW edges, CTB is given by

$$\text{CTB}_{\text{BW\_edge}} = 2(\text{IP}_3 - P_{\text{OUT}}) - 6 - 10\log\frac{N^2}{4}. \tag{9.69}$$

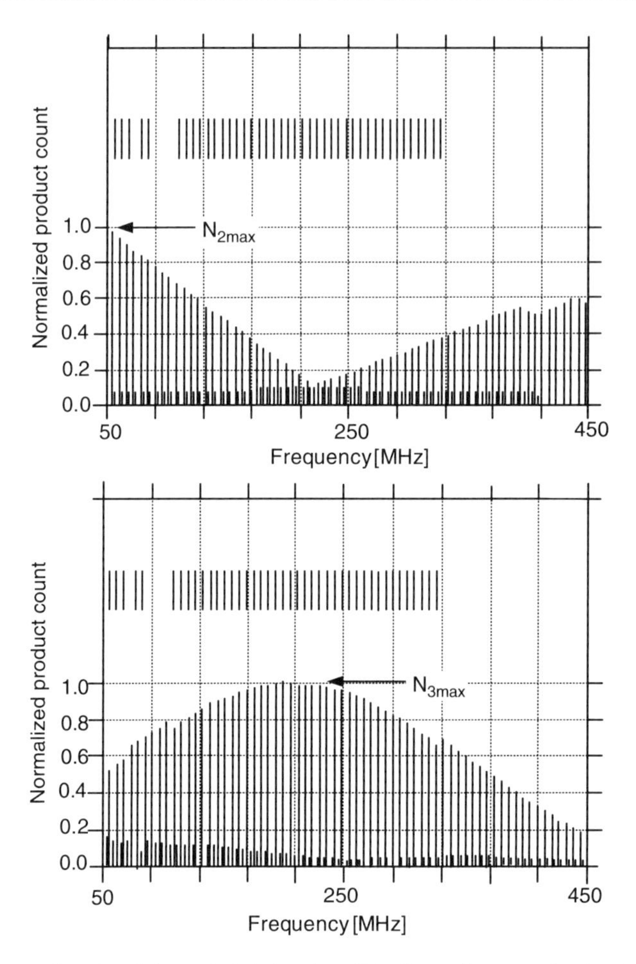


**Figure 9.6** Second-order (a) and third-order (b) distortion products for 42 channel AM-VSB. The maximum number of second-order products occurs at the lowest frequency channel, where 30 products contribute to CSO. The maximum number of third-order products occurs near the center channel where 530 products contribute to CTB.

In the previous discussion, CSO and CTB calculation was presented in time domain. The concept is straightforward, applying the sum of cosines $\sum_{k=1}^{n} \cos(\omega_k t)$ through a nonlinear transfer function. This transfer function is the polynomial characterization of the receiver's amplifier and other nonlinear building blocks. It is clear that second-order distortion is a result of the second-order term in the polynomial with the weight of the coefficient $a_2$. However, a more intuitive explanation for CSO spectral shape is by observing the frequency domain. Second-order distortion and polynomial calculation in the time domain is transformed the to convolution of the Fourier transform (FT) of the cosine frequencies' sum, which becomes discrete convolution of a train of $\sum_{k=1}^{n} \delta(\omega - \omega_k)$ in the frequency domain by itself. Thus, the CSO spectrum can be written as

$$\left[\sum_{k=1}^{n} \delta(\omega - \omega_k) \otimes \sum_{k=1}^{n} \delta(\omega - \omega_k)\right] \cdot a_2. \tag{9.70}$$

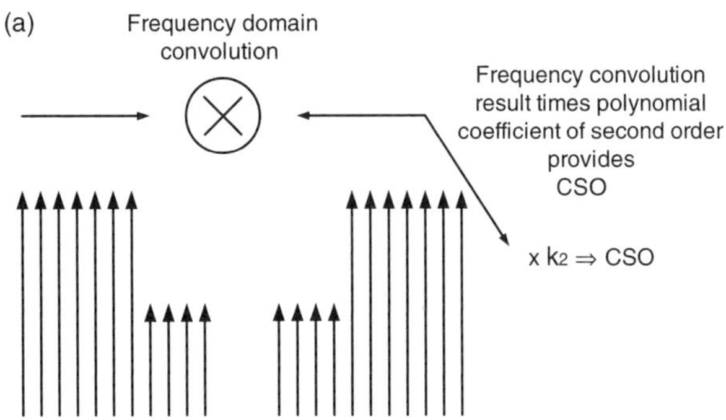


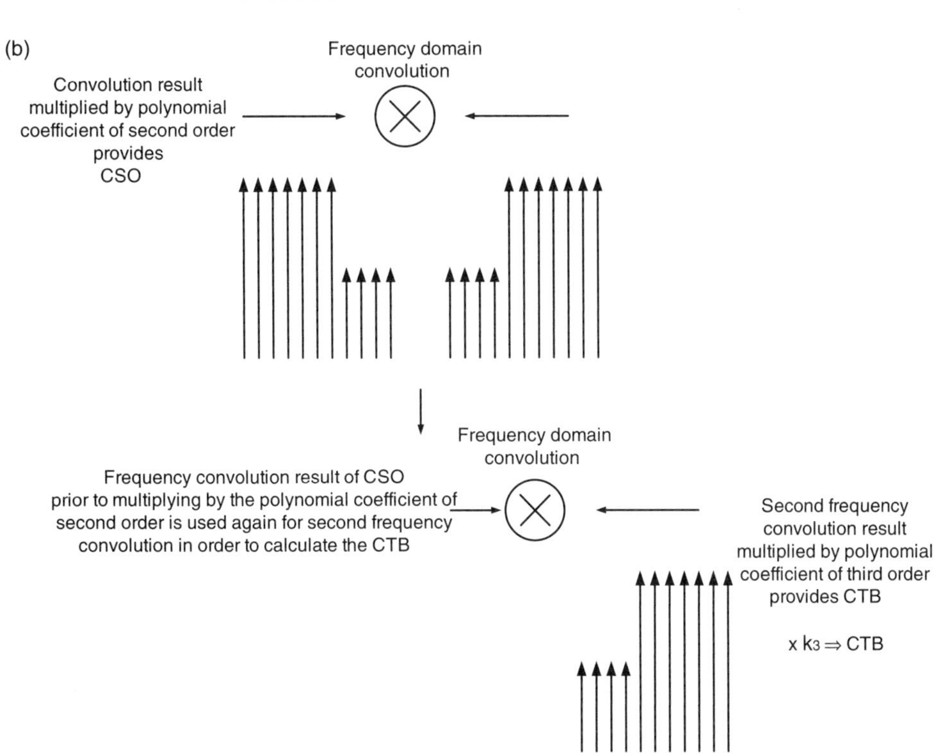


**Figure 9.7** CSO and CTB calculation using frequency domain convolution.

For the same reason, the CTB shape can be evaluated as

$$\left\{\left[\sum_{k=1}^{n}\delta(\omega-\omega_k)\otimes\sum_{k=1}^{n}\delta(\omega-\omega_k)\right]\otimes\sum_{k=1}^{n}\delta(\omega-\omega_k)\right\}\cdot a_3. \quad (9.71)$$

Figure 9.7 graphically illustrates this calculation approach. It can be said that the second-order distortion is convolution of the signal by itself, and third-order distortion is the second-order convolution result convolved again with the input signal.

## 9.7 RF Lineups, NF Calculations, and Considerations

RF chain lineup is an important stage when designing any kind of RF or analog optical receiver. The correct lineup will provide the desired noise figure 10 log($F$) (NF) for the system and thus satisfies CNR performance, gain, and low distortion, which create low IMD levels that define the system DR to be high. When designing an RF lineup, there is a need to know the input power range to the RF chain, the required output power from the receiver at its minimum input power, and the CNR at that point. The CNR requirement defines the RF chain max NF. In the case of an optical receiver, as will be discussed later on, there is a need to know the OMI and the optical power range, the responsivity of the photodetector (PD), as well as the input impedance at the PD output port in order to derive the input power.

Generally, when designing any kind of receiver, the first stage at the RF front end (RFFE) should be a low-noise amplifier (LNA), with high-output P and IP3. These requirements contradict each other, and a compromise between NF and P and $IP_3$ is made. The first stage in the RF chain defines the RF chain NF $P_{1dB}$ and $IP_3$, which results in the receiver's DR. A typical lineup of a CATV receiver with an automatic gain control (AGC) attenuator, voltage variable attenuator (VVA), or digital controlled attenuator (DCA) is given in Fig. 9.8.

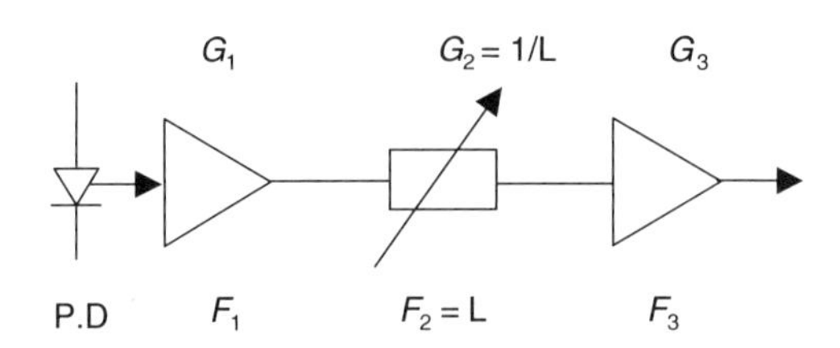


**Figure 9.8** Receiver lineup with AGC.

The noise figure of the RF chain is given by[7]

$$F = F_1 + \frac{F_2 - 1}{G_1} + \frac{F_3 - 1}{G_1 G_2}. \tag{9.72}$$

However, the second stage is a loss (see Fig. 9.8) or attenuation; hence, Eq. (9.72) becomes

$$F = F_1 + \frac{L - 1}{G_1} + \frac{L(F_3 - 1)}{G_1}. \tag{9.73}$$

From Eq. (9.73), it can be observed that the attenuation affects the system NF by multiplying all the consecutive stages of NF by the value of its attenuation state. Furthermore, Eq. (9.73) shows that the first-stage gain compensates for the rest of the RF chain NF and losses. However, there is a limit that is defined by the value of the attenuation $L$. At the instance that the second NF term in Eq. (9.73) equals the first-stage noise factor, $F_1$, as given by Eq. (9.74), the system NF will degrade by 3 dB:

$$F_1 = \frac{L - 1}{G_1} \Rightarrow L_{\max} < G_1 F_1 + 1. \tag{9.74}$$

That condition is valid in case $L(F_3 - 1) < L - 1$

For high gain $G_1$ and a low loss-over-gain ratio, as in Eq. (9.75), the system NF will be approximately equal to the first-stage LNA noise factor $F_1$, with a slight degradation of 0.4 dB:

$$\frac{L - 1}{G_1} \le 0.1 \cdot F_1. \tag{9.75}$$

Assuming an RF lineup with $N$ identical blocks, i.e., equal noise factors and gains in cascaded stages, the overall RF chain NF is given by

$$F_{\mathrm{TOT}} = F_1 + \frac{F_2 - 1}{G_1} + \frac{F_3 - 1}{G_1 G_2} + \cdots + \frac{F_N - 1}{G_1 G_2 \cdots G_{N-1}}. \tag{9.76}$$

Using a geometric sum, Eq. (9.76) becomes

$$\begin{aligned} F_{\mathrm{TOT}} &= F + (F - 1)\frac{1}{G} + \frac{1}{G^2} + \cdots + \frac{1}{G^{N-1}} = F + (F - 1)\sum_{K=1}^{N-1} \frac{1}{G^K} \\ &= F + (F - 1)\frac{1}{G}\left[\frac{1 - \left(\frac{1}{G}\right)^K}{1 - \left(\frac{1}{G}\right)}\right]. \end{aligned} \tag{9.77}$$

Hence, the NF of $N$ cascaded identical amplifiers is given by

$$\mathrm{NF_{TOT}[dB]} = 10\log\left\{F + (F-1)\left[\frac{1-\left(\frac{1}{G}\right)^{K}}{G-1}\right]\right\}. \tag{9.78}$$

For high gain and $K \to \infty$, the noise figure limit is given by

$$\mathrm{NF_{TOT}[dB]} = 10\log\left(F + \frac{F-1}{G-1}\right), \tag{9.79}$$

and the noise factor is given by

$$F_{\mathrm{TOT}} = F + \frac{F-1}{G-1}. \tag{9.80}$$

The CNR at the output of the first stage is given by

$$\mathrm{CNR}_1 = \frac{P_{\mathrm{IN}} \cdot G}{\mathrm{KTBFG}} = \frac{P_{\mathrm{IN}}}{\mathrm{KTBF}}. \tag{9.81}$$

The CNR at the output of the $N$ stage is given by Eq. (9.82):

$$\mathrm{CNR}_N|_{N\to\infty} = \frac{P_{\mathrm{IN}} \cdot G^N}{\mathrm{KTB}\left\{F + (F-1)\left[\frac{1-\left(\frac{1}{G}\right)^{N}}{G-1}\right]\right\}G^N} \approx \frac{P_{\mathrm{IN}}}{\mathrm{KTB}\left(F + \frac{F-1}{G-1}\right)}. \tag{9.82}$$

Hence, the CNR degradation limit referenced to the first-gain stage is given by

$$\frac{\mathrm{CNR}_N}{\mathrm{CNR}_1} = \frac{F}{F + \frac{F-1}{G-1}} < 1. \tag{9.83}$$

When the CNR is referenced to the input, the first-stage noise factor is not included.

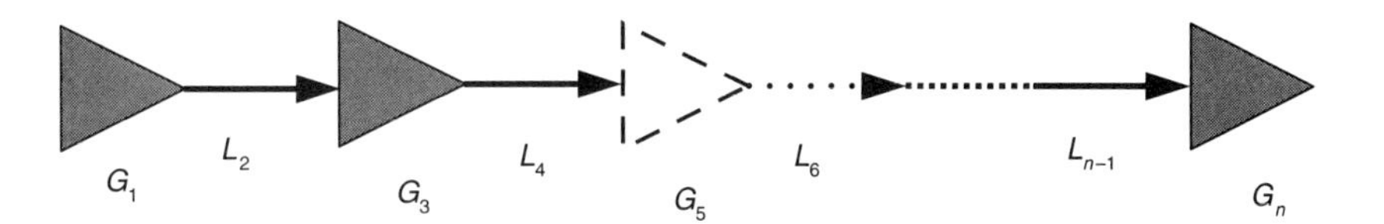


**Figure 9.9** Typical trunk of RF amplifiers and coax cables.

Hence, the CNR degradation limit referenced to the input is given by

$$\frac{\text{CNR}_N}{\text{CNR}_i} = \frac{1}{F + \dfrac{F-1}{G-1}} < 1. \tag{9.84}$$

As expected, the CNR performance is reduced throughout the RF chain.

Assume an RF chain that consists of RF amplifiers with a gain of $G$, noise factor $F$, and coax cables with loss $L$ equal to $1/G$ in alternate lineup, as given in Fig. 9.9.

The value of $L < 1$, since it is attenuation. Hence, the noise factor for an attenuator is $1/L$. As a consequence, the noise figure of the RF chain from Fig. 9.9 would be

$$F = F_1 + \frac{1/L_2 - 1}{G_1} + \frac{(F_3 - 1)}{G_1 L_2} + \frac{1/L_2 - 1}{G_1 L_2 G_3} + \cdots + \frac{(F_i - 1)}{\prod_{i=1}^{n} G_i L_{i-1}}. \tag{9.85}$$

Since $1/L \gg 1$ and the $GL = 1$, Eq. (9.85) becomes

$$F = F_1 + 1 + (F_3 - 1) + 1 + (F_5 - 1) + \cdots + 1 + (F_n - 1) \approx nF. \tag{9.86}$$

Therefore, the cascaded noise factor at the $n$th stage output equals to $nF$. This result indicates that the CNR after $n$ stages degrades by 10 log $n$. As a conclusion, the CNR of RF chain with long enough cables, with losses on the order of the RF amplification, is given by

$$\text{CNR}_n \approx \text{CNR}_1 - 10\log(n). \tag{9.87}$$

The noise figure is approximately given by

$$\text{NF}_n \approx 10\log(n) + \text{NF}_1. \tag{9.88}$$

The aforementioned analysis provides the system constraints with respect to CNR, CSO, and CTB products. When deploying a system in a cascade, each active device adds small distortions and noise. Even passive devices contribute noise. The distortions and noise compound are such that with each additional device, the signal is degraded and becomes less perfect. As mentioned in Sec. 1.2.2, any bimetallic junctions cause distortions.

In real scenarios, hybrid fiber coax (HFC) deployment of the coaxial distributions requires cascaded amplifiers and coax cables. The amplifiers boost the signal power and build gain in the line. The coax cable adds attenuation and loss. The amplifiers' gain reduces the channel DR since buildup of a gain increases distortions. On the other hand, amplifiers' NF and gain compensate the coax loss. Hence, there is a maximum number of trunk amplifiers that can be cascaded against the specifications of CNR and CTB.[19]

Figure 9.10 illustrates the noise floor and distortions ceiling as a function of amplifiers in cascade for a system at 300 MHz. The plot shows that a cascade of 46 trunk amplifiers is possible while realizing a 46-dB CNR and a 53-dB CTB. Nevertheless, other operating realities dictate that substantially more headroom for both distortion and noise be incorporated into the design. The factors to be considered are change in cable attenuation and noise with temperature, AGC for increasing DR as mentioned in Chapter 12, automatic slope control (ASC) tolerances, system frequency response, accuracy of field test equipment, design anticipation, and maintenance probabilities. In this example, allowing a tolerance of for AGC/ASC, a 3-dB peak to valley, and 2 dB of test equipment results in a 6 dB tolerance. Limiting distortions, CTB implies that the cascade should only be half the length predicted in the plot or 23 amplifiers.

The above discussion is applicable only to a trunk portion of the system. As signal levels are increased in distribution sections, additional allowances must be made in the system design. As a rule of thumb, CNR is determined primarily by the conditions of trunk operation and signal-to-distortion ratio (SDR) is determined primarily by the conditions of distribution operation.

The other factor mentioned above is channel loading; the higher it is, the greater the distortions.

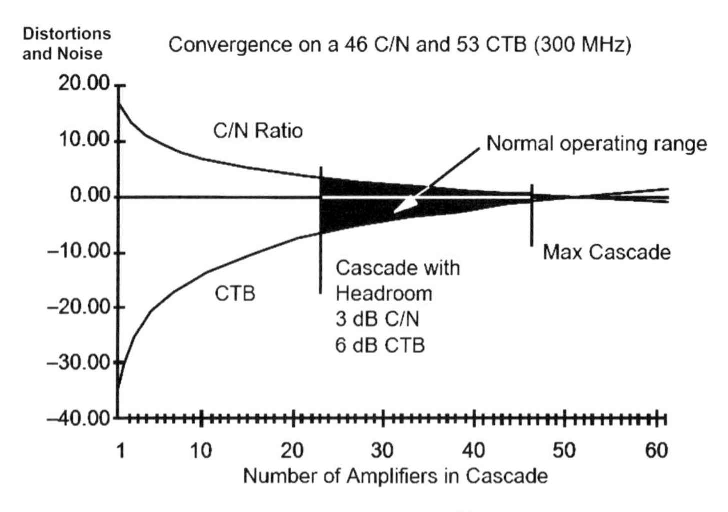


**Figure 9.10** CTB distortions and CNR in a cascade.[19] (Reprinted with permission from *CableLabs* © CableLabs, 1995.)

## 9.8 RF Lineups, $P_{1dB}$ and IP$_3$ Calculations, and Considerations

The DR of an amplifier, receiver or, more generally, any chain of cascaded two-port networks extends between noise level and the nonlinearity threshold. This is defined by compression point and IP as was analyzed in Sec. 9.1 and shown in Figs. 9.1 and 9.2. The noise figure of the system is mostly affected by the performance of first amplifying stage and by the performance of the negative gain two-port networks preceding it. Similarly, the gain compression level is affected mostly by the last amplifying stage and by the negative gain of the two ports following it. As was explained in Secs. 9.1 and 9.6, noise and DR are optimized under noise and distortion considerations. The goal is to find an expression for IP and compression point of an RF cascade. Consider the RF chain provided in Fig. 9.11.

In Sec. 9.1, an expression of the $n$th order IP was derived by Eq. (9.14). Therefore, the IMD power of the $n$th order is given by Eq. (9.89):[1]

$$P_{\mathrm{IMD}-n,k} = nP_{\mathrm{OUT},k} - (n-1)\mathrm{IP}_{n,k}, \tag{9.89}$$

where $k$ designates the two-port network considered in the cascade and n, the order of distortion. In order to combine nonlinear effects of different two-port networks, it is convenient to refer the intermodulation products created by each of them to a unique point of the chain, such as the chain input. This is done by replacing the power of these products, $P_{\mathrm{IMD}-n,k}$, created by each two-port network with an equivalent input signal power at the intermodulation frequencies, $\tilde{P}_{\mathrm{IMD}-n,k}$, which is amplified by the consecutive networks 1, 2, ..., $k$, producing the same effect:

$$\tilde{P}_{\mathrm{IMD}-n,k} = P_{\mathrm{IMD}-n,k} - \sum_{j=1}^{k} \mathrm{GL}_j, \tag{9.90}$$

where GL is the linearized gain for each two port. A similar transformation is applied to the $n$-order IP and signal power by

$$\tilde{\mathrm{IP}}_{n,k} = \mathrm{IP}_{n,k} - \sum_{j=1}^{k} \mathrm{GL}_j \tag{9.91}$$

$$\tilde{P}_{\mathrm{IN}} = P_{\mathrm{OUT},k} - \sum_{j=1}^{k} \mathrm{GL}_j. \tag{9.92}$$

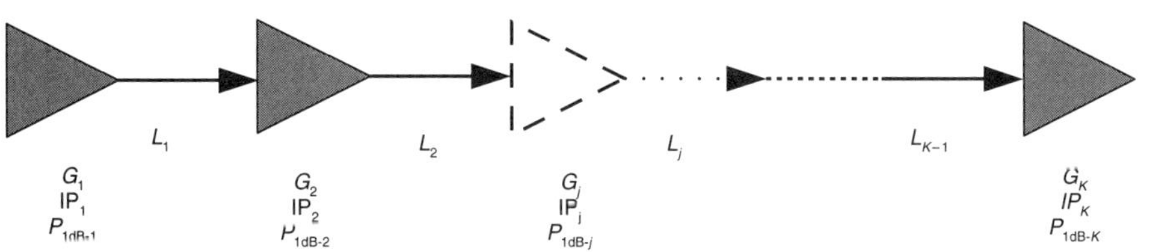


**Figure 9.11** Cascade connection of $n$ two-port networks.

With these transformation equations, similar expressions to Eq. (9.89) can be written for the equivalent input signals:

$$\tilde{P}_{\mathrm{IMD}-n,k} = n\tilde{P}_{\mathrm{OUT},k} - (n-1)\tilde{\mathrm{I}}\mathrm{P}_{n,k}. \tag{9.93}$$

Assuming the worst case, the voltage amplitudes $\tilde{e}_{\mathrm{IMD}-n,k}$ of these spurious signals can be added in order to obtain the total $\tilde{e}_{\mathrm{IMD}-n}$. Hence, the voltage can be written using Eq. (9.89) as

$$\tilde{e}_{\mathrm{IMD}-n,k} = e_0 \times 10^{P_{\mathrm{IMD}-n,k}/20} = e_0 \times 10^{n\tilde{P}_{\mathrm{in}}-(n-1)\tilde{I}P_{n,k}/20}. \tag{9.94}$$

Using the relations given by Eqs. (9.90) to (9.94), the equivalent voltage of all intermodulations created by all of the RF chain two ports is given by

$$\tilde{e}_{\mathrm{IMD}} = \sum_{j=1}^{k} \tilde{e}_{\mathrm{IMD}-n,k} = e_0 \times 10^{n\tilde{P}_{\mathrm{in}}/20} \times \sum_{j=1}^{k} 10^{-(n-1)\tilde{I}P_{n,k}/20}. \tag{9.95}$$

However, because of the transformation to the chain input showing the IP result at the input as given by Eq. (9.93), Eq. (9.95) should be equal to the condition given by

$$\tilde{e}_{\mathrm{IMD}-n} = e_0 \times 10^{n\tilde{P}_{\mathrm{in}}-(n-1)\tilde{\mathrm{I}}\mathrm{P}_{n,k}/20}. \tag{9.96}$$

Hence,

$$10^{-(n-1)\tilde{\mathrm{I}}\mathrm{P}_{n,k}/20} = \sum_{j=1}^{k} 10^{-(n-1)\tilde{\mathrm{I}}\mathrm{P}_{n,k}/20.} \tag{9.97}$$

Since the relation between the power in dBm and the power in mW is

$$P[\mathrm{mW}] = 10^{0.1P\mathrm{dBm}}. \tag{9.98}$$

Then the cascade $n$-order $\mathrm{IP}_n$ is given by

$$\frac{1}{\tilde{\mathrm{IP}}_n[\mathrm{mW}]^{n-1/2}} = \sum_{j=1}^{k} 1/\tilde{\mathrm{IP}}_{n,k}[mW]^{n-1/2}. \tag{9.99}$$

Transferring the IP to the output results in the equivalent output IP, including the gain values of the cascade as

$$\mathrm{IP}_{\mathrm{OUT}-n}[mW] = \frac{1}{\dfrac{1}{\left(\mathrm{IP}_{n,1}[mW]\prod_{j=2}^{k} g_j\right)^{(n-1)/2}} + \dfrac{1}{\left(\mathrm{IP}_{n,2}[mW]\prod_{j=3}^{k} g_j\right)^{(n-1)/2}} + \cdots + \dfrac{1}{\left(\mathrm{IP}_{n,k}[mW]\right)^{(n-1)/2}}}. \tag{9.100}$$

In the case of second order, $n = 2$; and $n = 3$ in the third order. A similar expression holds for compression power:

$$P1\text{dB}_{\text{OUT}}[mW] = \frac{1}{\dfrac{1}{P1\text{dB}_{\text{OUT},1}[mW]\prod_{j=2}^{k} g_j} + \dfrac{1}{P1\text{dB}_{\text{OUT},2}[mW]\prod_{j=3}^{k} g_j} + \cdots + \dfrac{1}{P1\text{dB}_{\text{OUT},k}[mW]}} \tag{9.101}$$

In the case of a cascade (as shown by Fig. 9.9), where $g = 1/L$, and assuming all gain stages have the same compression point and IP, the compression point and IP are degraded by $1/k$. In the case of a cascade with $k$ identical consecutive gain blocks with no loss, the overall equivalent IP and compression point are calculated by a geometric series:

$$P1\ \text{dB}[mW]_{\text{OUT}} = P1\ \text{dB}[mW]\frac{1 - 1/g}{1 - 1/g^k}. \tag{9.102}$$

When $k$ goes to infinity, it can be seen that the compression point is reduced by $1/g$, where $g = \text{antilog}\ G$.

## 9.9 Mismatch Effects

Mismatch loss is one of the ripple contributors. Having a device or network with a low return loss or reflection coefficient does not imply low ripple or mismatch loss. Assume the following case of a low $\Gamma_L$ or perfectly matched load device to 50 Ω. Also, assume a source with an unperfected reflection factor $\Gamma_S$. The mismatch loss is not zero, even though the load is perfectly matched to 50 Ω. The system should be transformed to the source plane. Thus, a new reflection coefficient is defined for the load $\Gamma_{SL}$:

$$\Gamma_{SL} = \left|\frac{Z_S - Z_L}{Z_S + Z_L}\right|. \tag{9.103}$$

This value will be used in Chapter 11 for noise factor and noise parameter analysis.

It is well known that the mismatch loss is the average of the peak-to-peak ripple generated by the voltage standing wave ratio (VSWR) mismatch between the source and load. Peak-to-peak ripple as of mismatch loss is given by

$$\text{MM}_{\text{loss-ripple}} = 1 \perp |\Gamma_{SL}|^2. \tag{9.104}$$

This notation shows destructive and constrictive interference. Hence, the peak-to-peak ripple is given by

$$\mathrm{MM\_PTP}_{\mathrm{loss-ripple}} = 10\log\left(\frac{1+|\Gamma_{\mathrm{SL}}|^2}{1-|\Gamma_{\mathrm{SL}}|^2}\right) > 0. \tag{9.105}$$

Note that return loss is given by

$$\mathrm{RL[dB]} = 20\log\Gamma = 20\log\frac{\mathrm{VSWR}-1}{\mathrm{VSWR}+1}, \tag{9.106}$$

and the impedance is given by

$$Z = \frac{1-\Gamma}{1+\Gamma}Z_0. \tag{9.107}$$

The following plots in Figs. 9.12 and 9.13 describe ripple created between source and load as a function of the source VSWR, while load VSWR mismatch is a parameter. When both source and load VSWR are equal in magnitude and phase, the ripple is at a minimum in the $\Gamma_{\mathrm{SL}}$ plane.

The next question describes the ripple of a cascade. There are two ways to calculate it. The first is an rms decibel ripple assuming, low probability of coherent ripple addition, generally related to narrow band systems. The second approach is the worst case where, at a certain frequency, all ripples are added in phase. This

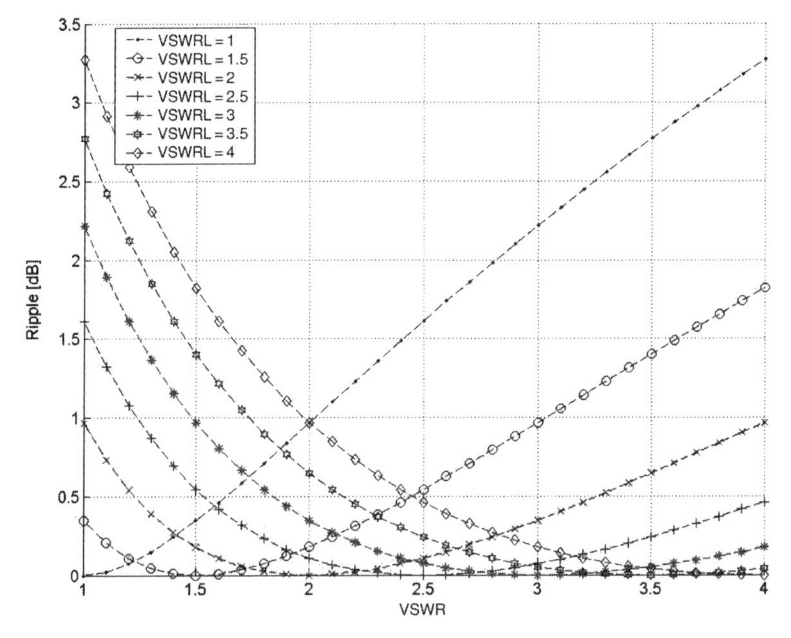


**Figure 9.12** Ripple as a function at source load plane of reflection coefficient with respect to the source impedance at 50 Ω system, while the load VSWR at 50 Ω system is a parameter. Note that the minimum ripple minima refer to the VSWRs; resolution 0.5.

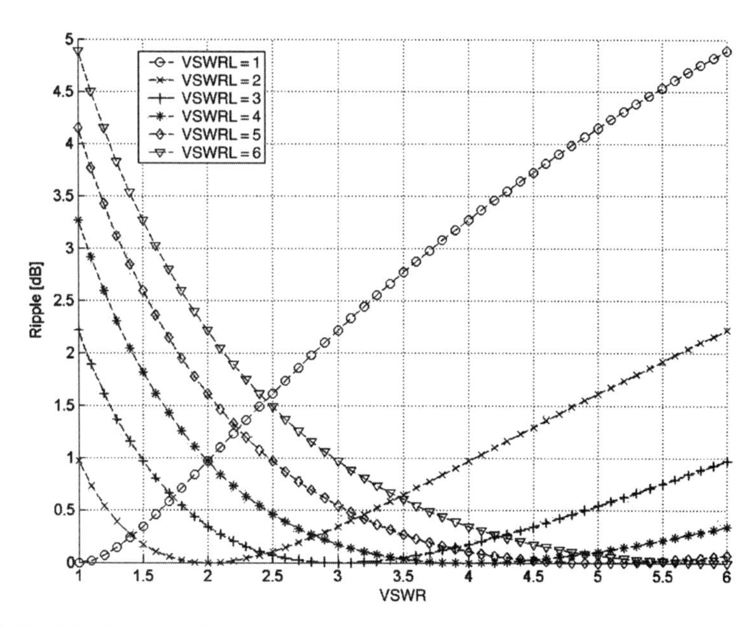


**Figure 9.13** Ripple as a function at source load plane of reflection coefficient with respect to the source impedance at 50 Ω system while the load VSWR at 50 Ω system is a parameter. Note that the minimum ripple minima refer to the VSWRs; resolution-1 for a higher mismatch.

refers to very wide band systems and provides an upper limit of performance. Note that losses between stages minimize the effects of mismatch loss due to the final load-to-source reflections attenuation. Equation (9.108) provides the decibel rms ripple:

$$\mathrm{rip}_{\mathrm{rms}} = \sqrt{\sum_{i=1}^{n} (\mathrm{ripple}_i)^2}. \tag{9.108}$$

## 9.10 CSO and CTB Distortion Effects on TV Picture

Section 3.2.1 provides, by Eqs. (3.2)–(3.4), the basic concept of scanning, which is bidimensional Fourier transform (FT) presented by Fig. 3.6. In the aforementioned sections, it was mentioned that the CATV transport induces NLD beats. The goal is to understand how the NLD beat affects the image on the screen.[11] When the RF signal is received, it is demodulated and converted into a two-dimensional screen by a raster scanner. The screen is scanned horizontally by a triangle signal, as demonstrated in Figs. 3.4 and 3.5.

When a TV receiver receives an AM-VSB RF signal, it is composed of the desired signal [called the victim signal $f(t)$] and an interferer denoted as $A\cos(\omega_{\mathrm{IMD}}t + \phi')$. IMDs such as CSO and CTB can be added coherently or randomly, which is the most likely case. For further analysis, the RF received signal is marked $f'(t)$, and the demodulated signal as $f(t)$.

The triangle signal scan time will be marked as $T_P$, and the flyback to the next scan row is marked as $\Delta T$. Thus, one triangle period equals $T = T_P + \Delta T$. As was explained in Sec. 3.2.1, in NTSC, the frame rate is 30 frames per second, where each frame corresponds to $L = 525$ horizontal lines. Since each triangle corresponds to a row, the period $T$ is given by

$$T = T_P + \Delta T = \frac{1}{30L}. \tag{9.109}$$

Recalling Sec. 3.2.1, only the $\Delta T$ part results in the active part of the demodulated RF on the screen. That is the only time the signal $f(x,y)$ in Eqs. (3.2)–(3.4) active. Assume the screen consists of a horizontal active dimension marked as $X$, an additional $\Delta X$ for the flyback time, the same for the vertical marked as $Y$, and an additional $\Delta Y$. Thus, the screen is larger and marked as $X'$ and $Y'$. Figure 9.14 illustrates this model. Based on those assumptions, it can be written that the excessive length in $X$ is proportional to the flyback according to

$$\frac{\Delta X}{X} = \frac{\Delta T}{T_P}. \tag{9.110}$$

In the same manner, based on previous data, it can be noted that the vertical ratio provides

$$\frac{\Delta Y}{Y} = \frac{\text{lines corresponding to the vertical retrace time}}{\text{lines appearing in } f(x,y)} = \frac{41}{484}. \tag{9.111}$$

The value of $Y'$ is given by

$$Y' = Y + \Delta Y = 525\Delta y = L\Delta y, \tag{9.112}$$

where $\Delta y$ is the distance between any two horizontal lines. Due to interlacing, the odd lines appear first on the screen, followed by the even lines. Note that only the first half of line $L$ appears on the screen as well as the second half of line zero, as shown in Fig. 9.14. It is also assumed that the retrace time from any line to the one that follows it is zero; recall that a frame consists of two fields, odd and even. For ease of analysis, 30 frame-lines' time coverage on screen is defined as $S$. This parameter describes the screen area scan progress and is given by

$$S = (30X'L)t. \tag{9.113}$$

Hence, the demodulated signal $f_t(t)$ in the time domain becomes

$$f_t(S) = f_t\left(\frac{S}{30X'L}\right). \tag{9.114}$$

Since the frame rate is 30 frames per second, $t$ satisfies $0 \leq t \leq 1/30$. The $S$ parameter corresponds to where the scanning beam is on the screen. For odd field $(x,y)$,

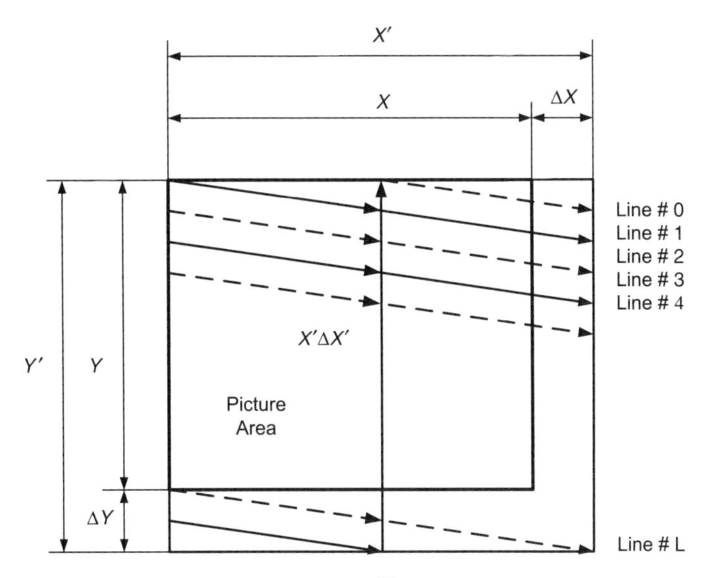


**Figure 9.14** NTSC screen scanning process.[11] (Reprinted with permission from *IEEE Transactions on Broadcasting* © IEEE, 1992.)

$S$ is given by

$$S = x + \frac{X'(y - \Delta y)}{2\Delta y}. \tag{9.115}$$

And in case $(x,y)$ lies at the even field, $S$ is given by

$$S = x + \frac{X'y}{2\Delta y} + \frac{X'(L-1)}{2}. \tag{9.116}$$

Hence, the sampled version of $f(x,y)$ for the received signal $f_t(t)$ is given by the FT of Eq. (9.117), which describes the sampled picture for even and odd fields:

$$\begin{aligned} f(x,y) = {} & f_t\left[x + \frac{X'(y-\Delta y)}{2\Delta y}\right]\mathrm{rect}_{0,X}(x)\mathrm{rect}_{0,Y}(y) \\ & \times \left[\sum_{k,j} \delta(x - k\Delta x)(y - 2j\Delta y - \Delta y)\right] \\ & + f_t\left(x + \frac{X'y}{2\Delta y} + \frac{X'(L-1)}{2}\right)\mathrm{rect}_{0,X}(x)\mathrm{rect}_{0,Y}(y) \\ & \times \left[\sum_{k,j} \delta(x - k\Delta x)(y - 2j\Delta y - \Delta y)\right], \end{aligned} \tag{9.117}$$

where $y = 2j\Delta y$ and the following rect sub functions describing the illumination scan beam within the $X$ $Y$ active screen area are defined as follows:

$$\mathrm{rect}_{0,X}(x)\mathrm{rect}_{0,Y}(y) = \begin{cases} 1 & \forall \quad 0 \le x \le X \quad \text{and} \quad 0 \le y \le Y \\ 0 & \text{otherwise.} \end{cases} \tag{9.118}$$

So far, the relationship between the picture $f(x,y)$ and the received signal were established. The next step is to find the effect of the distortion signal, $A\cos(\omega_{\mathrm{IMD}}t + \phi)$, on the victim signal, $f(t)$. For that, the variable change would have to be made from $t$ to $S$ based on the same strategy mentioned above. Hence, $\omega_{\mathrm{IMD}}t$ becomes

$$\omega_s = \frac{\omega_{\mathrm{IMD}}}{30X'L}. \tag{9.119}$$

Using this notation and replacing $f_t(s)$ by A $\cos(\omega_s t + \phi)$ in Eq. (9.117) and performing the FT results in[11]

$$F(u,v) = \frac{8\pi^4}{2\Delta x\Delta y}XY\left(\begin{array}{l}\left\{\begin{array}{l}\sum_{m,n}\exp\left[-j\left(u-\omega_s-\frac{2\pi n}{\Delta x}\right)\frac{X}{2}-j\left(v-\frac{X'}{2\Delta y}\omega_s-\frac{\pi m}{\Delta y}\right)\right]x \\ \sin c\left(u-\omega_s-\frac{2\pi n}{\Delta x}\right)\frac{X}{2}\sin c \quad \times\left(v-\frac{X'}{2\Delta y}\omega_s-\frac{\pi m}{\Delta y}\right)x \\ \frac{Y}{2}\left[\exp\left(-j\omega_s\frac{X'}{2}-j\pi m\right)+\exp\left(jX'\frac{L-1}{2}\omega_s\right)\right]\end{array}\right\} \\ + \\ \left\{\begin{array}{l}\sum_{m,n}\exp\left[-j\left(u+\omega_s+\frac{2\pi n}{\Delta x}\right)\frac{X}{2}-j\left(v+\frac{X'}{2\Delta y}\omega_s+\frac{\pi m}{\Delta y}\right)\frac{Y}{2}\right]x \\ \sin c\left(u+\omega_s+\frac{2\pi n}{\Delta x}\right)\frac{X}{2} \quad \times\sin c\left(v+\frac{X'}{2\Delta y}\omega_s+\frac{\pi m}{\Delta y}\right)x \\ \frac{Y}{2}\left[\exp\left(j\omega_s\frac{X'}{2}-j\pi m\right)+\exp\left(-jX'\frac{L-1}{2}\omega_s\right)\right]\end{array}\right\}\end{array}\right). \tag{9.120}$$

The peaks resulting from the IMD are defined by the four-sin $c$ functions within Eq. (9.120). Investigating the zero conditions of those sincs would provide the impulse locations for the IMD in the $U$, $V$ plane. For instance at $n$, $m$ equal to zero $u = \pm\omega_s$, the peak magnitude is calculated from $|F(u,v)|$. Each pair of peaks is symmetrical around the (0, 0) point. Moreover, the vertical distance, $v$ axis, between any two vertically adjacent peaks is $\pi/\Delta y$. When the discrete Fourier transform (DFT) of size $N \times N$ is used, $N = X/\Delta x = Y/\Delta y$. There would be four peaks where the vertical distance between the two vertical peaks equals $N/2$. Figures 9.15 and 9.16 describe the peak locations for solving the sin $c$ function in Eq. (9.120).

Because $2\pi/\Delta x$ in the FT at Fig. 9.15 corresponds to $N$ in the DFT as it appears in Fig. 9.16, the horizontal distance $D$ between the peaks in Fig. 9.16 satisfies

$$\frac{D}{2\omega_s} = \frac{N\Delta x}{2\pi} \Longrightarrow D = \frac{N\Delta x}{\pi}\omega_s = \frac{X}{\pi}\omega_s. \tag{9.121}$$

Hence, if the peak location is found, $\omega_s$ is determined. Since the values of the DFT are proportional to the sampled values of the FT, the location of the peaks cannot be precisely determined unless they happen to lie exactly on the points where the samples are taken. The calculation goal is to find out in what resolution the location of the peaks can be determined.

The rectangular function (9.122) (also known as the rectangle function, rect function, unit pulse, or the normalized boxcar function) is defined as,

$$\text{rect}\,(t) = \sqcap(t) = \begin{cases} 0 \;\; if|t| > \frac{1}{2} \\ \frac{1}{2} \;\; if|t| = \frac{1}{2} \\ 1 \;\; if|t| < \frac{1}{2} \end{cases} \tag{9.122}$$

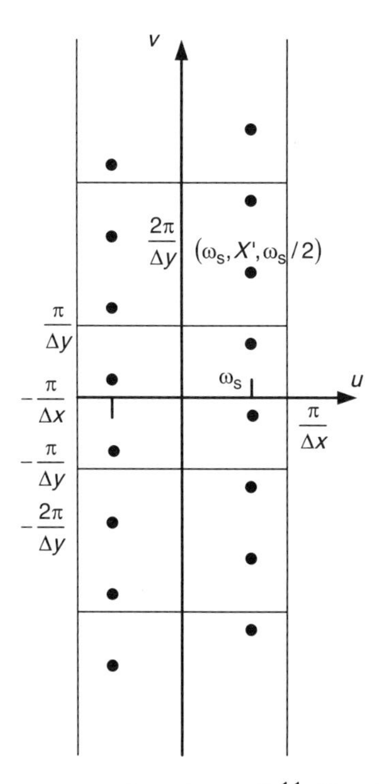


**Figure 9.15** Peak location $n = m = 0$ and $n \neq 0$.[11] (Reprinted with permission from *IEEE Transactions on Broadcasting* © IEEE, 1992.)

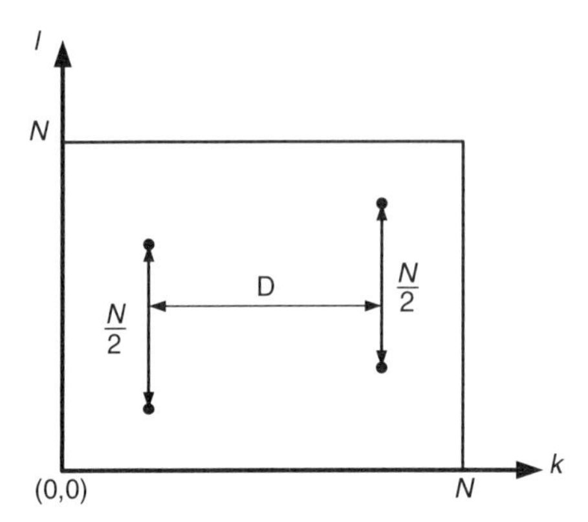


**Figure 9.16** $n = m = 0$ $N \times N$ DFT shows four peaks.[11] (Reprinted with permission from *IEEE Transaction on Broadcasting* © IEEE, 1992.)

Because the accuracy of measuring the location of the point in the DFT is half the index, this corresponds to $\omega_s = 2\pi/2X$. In turn, this corresponds to $\omega_t = (30X'L/2X)2\pi = 9025$ Hz, which is the resolution at which $\omega_t$ can be determined. Assume a peak $n = m = 0$, where $u = \omega_s$ and $v = \omega_s X'/2\Delta y$ with the locus of this peak being $\omega_s$, it increases from the value of zero in the line $v = uX'/2\Delta y$. For $n = 0, m = \pm 1$, the loci are the lines $v = uX'/2\Delta y \pm \pi/\Delta y$. Thereby, the horizontal distance between any two adjacent loci lines is $2\pi/X'$, which in the DFT corresponds to $X/X' = 53/64 = 0.83$. In the DFT, these loci will be parallel lines as shown in Fig. 9.17. The slope of any of these loci lines is $N'/2$, where $N'/N = X'/X$, and the horizontal distance between any two lines equals 0.83. In Fig. 9.17, **a** is the point $(N/2, N/2)$, **b** is $(N/2 + 0.83, N/2)$, etc.

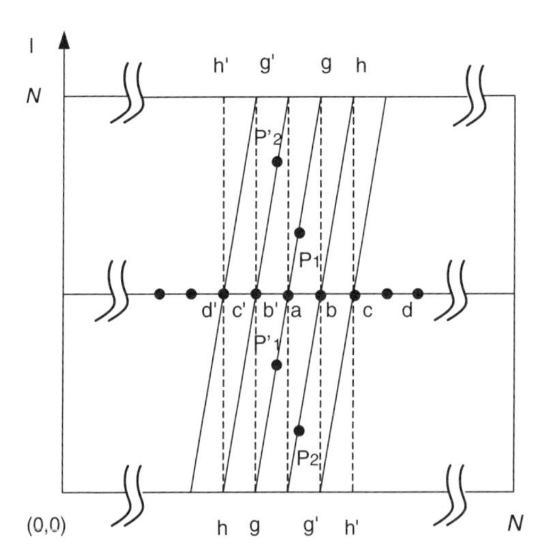


**Figure 9.17** Effect of IMD on screen.[11] (Reprinted with permission from *IEEE Transactions on Broadcasting* © IEEE, 1992.)

Since each pair of pulses are symmetrical, if a pulse occurs at **b**, the corresponding will be at **b**′ and so on. In this case, the other pair will be at points ($\mathbf{g} - \varepsilon$) and ($\mathbf{g} + \varepsilon$). It can be shown that if a pair of pulses occurs on the middle horizontal line passing through point **a**, then their amplitudes will take the maximum value possible. In this case, the pulses of the corresponding pair will have amplitudes equal to zero.

At $\omega_s = 0$, **p1** and **p1**′ are both at point **a**. As $\omega_s$ increases, **p1** travels toward **g**. When **p1** is at **g**, **p2** is at **b**. As the pulse **p1** goes to $\mathbf{g} + \varepsilon$, its alias will appear at $\mathbf{g}' + \varepsilon$ and so on. **p1** reaches **g** at $\omega_S = 2\pi/X$, which corresponds to $\omega_t = 2\pi 30L = 15.750$ KHz. At this instant, **p2** reaches **b**. As a consequence, IMD will create tilted lines on screen, while their slope determined by the IMD frequency.

## 9.11 Main Points of this Chapter

1. Any active device can be described by a AM to AM polynomial, which approximates its linear and high-order distortion term.
2. The second-order term in the polynomial describes second-order distortions, which results in CSO.
3. The third-order term in the AM to AM polynomial describes third-order distortions, which results in CTB.
4. System linearity is measured by three ways: $P_{1\text{dB}}$, $\text{IP}_2$, and $\text{IP}_3$.
5. $\text{IP}_3$ is a virtual point at which the third-order distortion terms are equal to the fundamental tone.
6. $\text{IP}_2$ is a virtual point at which the second-order distortion terms are equal to the fundamental tone.
7. $P_{1\text{dB}}$ is defined as a compression point at which the gain drops by 1 dB where a clipping event occurs at a large input signal swing. $P_{1\text{dB}}$ is affected by a bias point.
8. A third order distortion in a multitone environment falls in the band of the fundamental tone.
9. Composite triple beat (CTB) is a result of three tones compared to two test tones for $\text{IP}_3$, and it is higher by 6 dB than the $\text{IP}_3$ products.
10. System DR is traditionally defined as the maximum input power allowed, at which the resultant distortion products are within the noise. Thus, a higher $\text{IP}_3$ and a low noise figure guarantee high DR.
11. Noise factor $F$ is a figure of merit of CNR degradation caused by the system and is expressed as the ratio between input CNR and output CNR. $NF = 10 \log F$.
12. When cascading RF blocks, the NF performance of the cascade degrades. Thus, it is important to have at the head of the cascade LNA with a low NF and a high $P_{1\text{dB}}$ to have high a high enough DR and low-noise figure.
13. When having an amplifier with a gain of G followed by a loss L equal to the amplifier gain as $L = 1/G$, the noise figure is not compensated for during the next stage and is added directly.

14. A high-gain cascaded amplifier would have a low equivalent $IP_3$ and $P_{1dB}$.
15. A high-power interferer tone would increase the system noise figure.
16. There are three types of CTB: $A + B + C$, $A - B + C$, and $-A + B + C$ (generally presented by $A \pm B \pm C$). Each CATV channel has its own beat-counts statistics.
17. When an LNA receives a CATV channel, all other channels that are equal to an equivalent tone compress the weak channel gain; thus, CNR is reduced at that channel.
18. Compression results at the first approximation from the negative coefficient $a_3$ of the third-order term.
19. Cross modulation results when an AM tone accompanied by CW tones exhibits third-order distortion.
20. A cross-modulation signal transfers its AM to the channel under test by the third-order products. Additionally, it adds CW tone proportional to the square of the AM index $m$.
21. When describing cascade mismatch, it is important to transfer the reflection coefficient of the cascaded stage $\Gamma_L$ to the source impedance plane $\Gamma_S$. This reflection coefficient is marked as $\Gamma_{SL}$.
22. Mismatch in VSWR results in loss and ripple.
23. The case of a two-tone IP3 test showing two unequal third-order IMD is due to AM-to-PM effect of the RF.
24. AM-to-PM effect of RF chain is due to memory effect. The regular polynomial describes scalar approximation. Memory effect can be described by the Volterra series.
25. In the case of cascading RF amplifiers with a gain of G and losses equal to the gain of the amplifier, i.e., $GL = 1$, the overall NF is degraded by $10 \log N$, where $N$ is the number of staggered stages of gain and loss.
26. CTB and CSO result in distortions appearing on the CATV screen by tilted lines. Analysis is done by DFT.
27. CSO is a second-order distortion, which in time domain multiplies the signal by itself. In a frequency domain, using (FT) second-order distortions is a convolution of the signal on itself.
28. CTB is a third-order distortion and is a cubic term. In frequency domain, it is the convolution of the signal on itself twice. Hence, the CSO convolution result is convolved again on the signal to provide the CTB shape.
29. The polynomial coefficients, $a_2$ and $a_3$, are scalars that multiply the second-order convolution result and third-order convolution result, respectively. The product of it is the CSO and CTB voltage levels.

## References


1. Sorger, G.U., "The 1 dB gain compression point for cascade two-port networks." *Microwave Journal*, pp. 136–142 (1988).

2. Laffel, M., "Intermodulation distortion in a multi-tone signal environment." *RF Design*, pp. 78–84, June (1995).
3. Hausman, H., "Topics in communications system design: carrier triple beats." *Microwave Journal*, No. 1, p. 20 (2002).
4. Turlington, T.R., *Behavioral Modeling of Nonlinerar RF and Microwave Devices*. Artech House (1999).
5. Smith, J.L., "A method to predict the level of intermodulation products in broadband power amplifiers." *Microwave Journal*, Vol. 46 No. 2, pp. 62–78 (2003).
6. Ha, T.T., *Solid State Microwave Amplifier Design*. Krieger Publishing Company (1991).
7. Vendelin, G.D., A.M. Pavio, and U.L. Rohde, "Microwave circuit design." Wiley Interscience.
8. Abrie, P.L.D., *Design of RF and Microwave Amplifiers and Oscillators*. Artech House (1999).
9. Ciciora, W., J. Farmer, and D. Large, *Modern Cable Television Technology: Video, Voice and Data Communications*. Morgan Kaufman Publishers (1999).
10. Kos, T., B. Zokov-Cihlar, and S. Gric, "An algorithm for counting of intermodulation products in multicarrier broadband systems." *Proceedings of IEEE International Symposium on Industrial Electronics*, Vol. 1, pp. 95–98 (1999).
11. Ward, R.K., and Q. Zhang, "Automatic identification of impairments caused by intermodulation distortion in cable television pictures." *IEEE Transactions on Broadcasting*, Vol. 38 No. 1, pp. 60–68 (1992).
12. Rainal, A.J., "Laser clipping distortions in analog and digital channels." *IEEE Journal of Lightwave Technology*, Vol. 15 No. 10, pp. 1805–1807 (1997).
13. Domino, W., N. Vakilian, and D. Agahi, "Polynomial model of blocker effects on LNA mixer devices." *Applied Microwave and Wireless*, pp. 30–44 (June, 2001).
14. Spiegel, M.R., *Mathematical Handbook of Formulas and Tables, 2nd Ed.* McGraw-Hill (1998).
15. Cripps, S.C., *Advanced Techniques in RF Power Amplifiers Design*. Artech House (2002).
16. Cripps, S.C., *RF Power Amplifiers for Wireless Communications*. Artech House (1999).
17. Zhang, L.X., E. Larson, and P.M. Asbeck, *Design of Linear RF Outphasing Power Amplifiers*. Artech House (2003).
18. Pedro, J.C., and N.N. Carvalho, *Intermodulation Distortion in Microwave and Wireless Circuits*. Artech House (2003).
19. Ciciora, W.S., *Cable Television in The United States An Overview, 2nd Edition*, Cable Labs (1995).

## Chapter 10

# Introduction to Receiver Front-End Noise Modeling

The goal of all communications systems is the same: to produce a signal that was applied at the transmitter as input, suffered the link losses and distortions (such as cable attenuation, dispersion, fading, and reflections), and then was recovered by the demodulator or detector in the receiver section. When the signal reaches the detection stage, it is accompanied by an additional signal voltage that varies in time in an entirely unpredictable manner. This unpredictable voltage waveform is a random process called *noise*. All signals accompanied by such a waveform are described as contaminated or corrupted by noise. Therefore, one of the parameters that needs to be kept in mind when designing any kind of receiver is its noise performance. The noise performance of a system is defined by the system's noise figure; 10 log(F) = (NF). This affects the receiver's C/N (carrier to noise), or $E_b/n_o$ bit energy per noise, which directly affects the detection quality carrier-to-noise ratio (CNR) and bit error rate (BER), respectively.

In fiber optics receiver design, there are several noise sources that affect the overall C/N result of a link and receiver performance, affecting the receiver's NF and system CNR. Since there is no way to access the optical receiver RF front end (RFFE) due to its connection and matching to a photodetector, other methods to measure noise performance and other definitions are used, which will be reviewed below.

There are several circuits noise models; once these models are established, the noise expression for noise density can be determined, enabling us to analyze the receiver's performance in the presence of noise. Moreover, the modern-communications-theory models can be used to predict the system's performance in case of random signal and random noise inputs. In addition, there are models that are used in random signal analysis to analyze and predict system performance in the presence of random distortions and different noise mechanisms, such as white noise. This type of process has an equal power versus frequency spectrum; in other words, its density is a frequency-independent function.

Noise densities of the type $1/f^n$ are sometimes defined as colored noise because of their frequency dependency, or as a phase noise component in the case of oscillators and synthesizers. Distortions, however, are not referred to as noise and have their own statistical modeling. Other electrical sources of noise can be due to the bias current of a device or optical current generated by the photodetector. This kind of noise is defined as shot noise.

Additional noise sources that should be attended to are mechanical and acoustical, which are generated by mechanical resonance due to the vibrations created by a cooling fan or any other environmental condition that creates vibrations. Assume the following case of a vibration resonance, which vibrates an radio frequency choke (RFC). The RFC coil would have therefore variable impedance, which is dependent on the mechanical vibration frequency. This would result in small fluctuations over the signal amplitude, inducing amplitude modulation (AM) at the mechanical resonance frequency. The modulation depth is directly related to the vibration energy. The worst vibrations are random vibrations with a wide spectrum.

Another example are acoustical effects over an SAW filter. Optical modules such as OTP (optical three port) can be affected by modulating the fiber alignment distance to a detector surface by creating reflections and changing the optical power hitting the detector surface; this would create undesired residual AM over the input signal. This type of noise sometimes is referred to as *microphonics*. One more noise source can be X-talk due to fast-switching logic or desensitization of one receive channel by another receive channel. This case is well investigated in Chapters 17 and 20.

Noise sources in a receiver can be divided into two main categories. The first mechanism is intrinsic noise, which arises from the physical structure of the semiconductor. This category includes thermal noise, shot-noise, $1/f^n$ noise, and quantum noise from the photodetector. The other group includes noise mechanisms that are coupled into the receiver from the surrounding environment. This includes X-talk, microphonics, switched power supplies, solar noise in case of free space, adjacent-channel interference, and laser relative intensity noise (RIN). Laser RIN is an intrinsic noise source of a laser device, and it is not a part of noise that is generated in the receiver. Hence, it is considered as coupled noise source to the link. When referring a link budget calculation and CNR estimation, RIN is treated as the link intrinsic noise.

The following paragraphs provide a basic introduction to different noise source mechanisms and modeling related to the first group, which are commonly used to analyze optical receivers performance and system link performance. They also serve as a brief introduction to random signal processes.

## 10.1 Noise Analysis Basics

Noise is a random process variable. As such, it cannot be presented and predicted by an exact deterministic function. Since the process is random and is statistical in value, it does not matter how many observations were made on the random variable. Another method for a good estimation is required, using the aid of probability theory and random signals. Assuming the random noise power or voltage as an event, it can be measured by mean and variance, or characterized by the energy integral over its density function. This way, the average power and rms voltage of the event can be evaluated during the event period of occurrence. Assume a random noise voltage process with a density function $n(t)$. Additionally, assume that the noise density function is a continuous-wave-shape function and the event is sampled within the time interval $0 \leq t \leq T$. Therefore, the observed noise process would have the following average energy:

$$\overline{\langle n(t)^2 \rangle} = \frac{1}{T}\int_0^T n^2(t)\mathrm{d}t, \tag{10.1}$$

where the horizontal bar symbolizes average value and the $\langle \ \rangle$ parenthesis indicates a statistic value average. One conclusion is that when dealing with random signals, the average power is investigated. The average power is proportional to the voltage square. Therefore, the rms noise voltage can be evaluated by

$$Vn_{\mathrm{rms}} = \sqrt{\overline{\langle n(t)^2 \rangle}} = \sqrt{\frac{1}{T}\int_0^T n^2(t)\mathrm{d}t}. \tag{10.2}$$

The random variable $n(t)$ is a function that satisfies the condition $t \in T$ at the event domain in such a way that for any real value of $V$, $\{t\text{: } n(t) \leq V\}$is an event. The total average energy of random processes is given by

$$\begin{aligned} N_{\mathrm{e}} &= \overline{\langle n_{\mathrm{Tot}}^2(t) \rangle} = \overline{\left[\langle n(t)_1^2 \rangle + \langle n(t)_2^2 \rangle\right]^2} = \overline{\langle n_1^2(t) \rangle} \\ &+ 2\gamma\overline{\langle n(t)_1 \rangle \langle n(t)_2 \rangle} + \overline{\langle n_2^2(t) \rangle} = \overline{\langle n_1^2(t) \rangle} \\ &+ 2\gamma\sqrt{\overline{\langle n_1^2(t) \rangle}}\sqrt{\overline{\langle n_2^2(t) \rangle}} + \overline{\langle n_2^2(t) \rangle}. \end{aligned} \tag{10.3}$$

The amount of statistical event correlation or independency is defined by the correlation factor $\gamma$. In case of two independent events or processes, $\gamma = 0$. This means that the two processes are orthogonal or statistically independent. Noise can be analyzed in the frequency domain as well. Generally, this is the preferred method to analyze systems. In this case, the random variable $n(\omega)$ is defined by the condition $\omega \in \Omega$ and the event domain $\{\omega\text{: } n(\omega) \leq V\}$. In this case, the noise process $n(\omega)$ is described by its power spectral density (PSD). The total noise power is given by

$$\begin{aligned} N(\omega)_{\mathrm{e}}[W/Hz] &= \left|\overline{\langle n_{\mathrm{Tot}}^2(\omega) \rangle}\right| = \left|\overline{\langle n_1^2(\omega) \rangle}\right| \\ &+ 2\gamma(\omega)\sqrt{\left|\overline{\langle n_1^2(\omega) \rangle}\right|}\sqrt{\left|\overline{\langle n_2^2(\omega) \rangle}\right|} + \left|\overline{\langle n_2^2(\omega) \rangle}\right| \end{aligned} \tag{10.4}$$

The notation in Eq. (10.4) is commonly used when analyzing two-port networks such as transistor noise performance, noise impedance, and noise matching.

The method used involves creating an ideal quiet device whose intrinsic noise sources are represented by thermal noise voltage and noise current. Hence, $\langle \overline{n_1^2(\omega)} \rangle$ represents the noise voltage $\overline{e_n^2}$, and the second term $\langle \overline{n_2^2(\omega)} \rangle$ represents $\overline{i_n^2}$.

It is obvious that when the two noise sources are fully correlated, they would produce the maximum correlated power, since their covariance is the geometric average of each term's variance multiplication result. This leads to the definition of the correlation term $\gamma$ given by

$$\gamma(\omega) = \frac{\sqrt{\left|\langle \overline{n_1^2(\omega)} \rangle \langle \overline{n_2^2(\omega)} \rangle\right|}}{\sqrt{\left|\langle \overline{n_1^2(\omega)} \rangle\right|}\sqrt{\left|\langle \overline{n_2^2(\omega)} \rangle\right|}} = \frac{\sqrt{\left|\langle \overline{n_1^2(\omega)} \rangle \langle \overline{n_2^2(\omega)} \rangle\right|}}{\sqrt{\left|\langle \overline{n_1^2(\omega)} \rangle\right| \left|\langle \overline{n_2^2(\omega)} \rangle\right|}} = \frac{\overline{e_n \cdot i_n^*}}{\sqrt{\left|\overline{i_n^2}\right| \cdot \left|\overline{e_n^2}\right|}}. \tag{10.5}$$

The value range of the correlation factor is limited between $-1 \leq \gamma(\omega) \leq 1$. In noise analysis, it is common to use the terms "correlation impedance" and "admittance," which are described by

$$\begin{aligned} Z_c(\omega) &= \gamma \sqrt{\frac{\left|\overline{e_n^2}\right|}{\left|\overline{i_n^2}\right|}} = \frac{\overline{e_n i_n^*}}{\left|\overline{i_n^2}\right|}, \\ Y_c^*(\omega) &= \gamma \sqrt{\frac{\left|\overline{I_n^2}\right|}{\left|\overline{e_n^2}\right|}} = \frac{\overline{e_n i_n^*}}{\left|\overline{e_n^2}\right|}. \end{aligned} \tag{10.6}$$

The correlation definitions are used for investigating the optimum-matching transformer at the RFFE and extracting the minimum noise factor–matching expression as presented in Secs. 10.9 and 11.2. An intuitive way to understand the correlation factor is by the diagram given in Fig. 10.1.

In the case of linear dependency between the two random variables, the scalar multiplication is not a zero value. If both events are fully correlated, this means the two vectors are parallel and are linear combinations of each other, and are therefore defined as fully linear dependent vectors. In the case of no correlation, the vectors are orthogonal and the covariance of the two events is zero. Two events that have no statistical dependency require that they are not linear combinations of each other. The reverse claim is not always correct. Another way to observe the correlation factor is as the value of cos $\theta$ between the two phasors.

In Fig. 10.1, the current vector $\mathbf{i}_n$, composed of two components, one in full correlation with the thermal voltage $e_n$, marked as $\mathbf{i}_c$, and one orthogonal, is out of correlation with $e_n$, marked as $\mathbf{i}_u$; hence, the following relation is applicable in a vector notation:

$$\mathbf{i}_c = \mathbf{i}_n - \mathbf{i}_u. \tag{10.7}$$

The relation between the noise voltage $e_n$ and the noise current components provides the correlation impedance, none-correlated impedance, and the optimal impedance is

$$\mathbf{R}_c = \frac{e_n}{\mathbf{i}_c},\quad \mathbf{R}_u = \frac{e_n}{\mathbf{i}_u},\quad \mathbf{R}_{opt} = \frac{e_n}{\mathbf{i}_n}. \tag{10.8}$$

In the same manner, correlated, none-correlated, and optimal noise conductance are defined as

$$\mathbf{G}_c = \frac{\mathbf{i}_c}{e_n},\quad \mathbf{G}_u = \frac{\mathbf{i}_u}{e_n},\quad \mathbf{G}_{opt} = \frac{\mathbf{i}_n}{e_n}. \tag{10.9}$$

These variables will be used later in Chapter 11 and when analyzing minimum noise figure matching. However, it is obvious that $\mathbf{G}_c$ is orthogonal to $\mathbf{G}_u$ and $\mathbf{R}_c$ is orthogonal to $\mathbf{G}_u$. Hence, the following relations are applicable:

$$\begin{aligned} \mathbf{G}_c \times \mathbf{G}_u &= 0,\\ \mathbf{R}_c \times \mathbf{R}_u &= 0,\\ e_n \times \mathbf{i}_u &= 0,\\ \mathbf{i}_c \times \mathbf{i}_u &= 0,\\ \mathbf{n}_c &= e_n \times \mathbf{i}_c,\\ \mathbf{i}_c &= e_n \times \mathbf{G}_c. \end{aligned} \tag{10.10}$$

In some cases, $\mathbf{G}_c$ is noted as $\mathbf{Y}_c$. Since the symbol represents admittance, it consists of a real part and an imaginary part; hence, $\mathbf{Y}_c = \mathbf{G}_c + \mathrm{j}\mathbf{B}$.

Each network is characterized by a transfer function $H(f)$. In case there is power gain, then the transfer function is $G(f)$. Therefore, in case of white Gaussian noise, the output noise would be according to the network shape factor:

$$N = n_{in} \times \frac{1}{G_0}\int_0^{\infty} G(f)\mathrm{d}f = n_{in} \times \mathrm{NEB}, \tag{10.11}$$

where $G_0$ is the network gain at dc or the peak gain, $G(f)$ is the network gain versus frequency, and NEB is the noise equivalent bandwidth, which is the integral result of Eq. (10.11) in units of Hz.

A different definition would be through observing the LPF response $H(j\omega)$. This transfer function is a voltage or current transfer versus frequency. Since power is proportional to the amplitude square power, the following notation is applicable for power transfer:

$$|H(j\omega)|^2 = |H(j\omega)| \times |H(j\omega)|^* = |H(j\omega)| \times |H(-j\omega)|, \tag{10.12}$$

where the sign * marks the complex conjugate and $j\omega = j2\pi f$. Assume a single pole LPF function of the type

$$H(j\omega) = \frac{1}{1 + j(f/f_0)}, \tag{10.13}$$

where $f_0$ is the 3 dB BW and $H(0) = 1$. Within the same approach, the equivalent noise BW at the 3 dB frequency can be defined as

$$\mathrm{NEB} = \frac{1}{2}\int_{-\infty}^{\infty} |H(f)|^2 \mathrm{d}f = \frac{f_0}{2}\int_{-\infty}^{\infty} \frac{\mathrm{d}f/f_0}{1 + (f/f_0)^2} = \frac{\pi}{2} f_0. \tag{10.14}$$

Several interesting conclusions can be made from this discussion. The NEB of a single pole LPF with 20 dB per decade roll-off is 1.57 times greater than the 3 dB BW (an increase of 1.95 dB in noise power), i.e., $\mathrm{NEB} = 1.57 f_0$. In order to equalize between NEB and 3 dB BW, there is a need to get the multipole LPF network closer to the ideal shape of a rectangular LPF or use elliptic multipole response. It can be observed that any colored noise with spectral density $S(f)$ is generated by white noise and transfer function $H(f) = \sqrt{S(f)}$. In addition, it can be proved by the Paley-Wiener criteria[16] that, if the condition in Eq. (10.15) exists,

$$\int_{-\infty}^{\infty} \frac{|\log S(f)|}{1 + f^2} \mathrm{d}f < \infty, \tag{10.15}$$

then there is a causal transfer function $H(j\omega)$ that provides at its output the spectral density $S(f)$ while exhibiting a white noise at its input. Figure 10.2 illustrates the

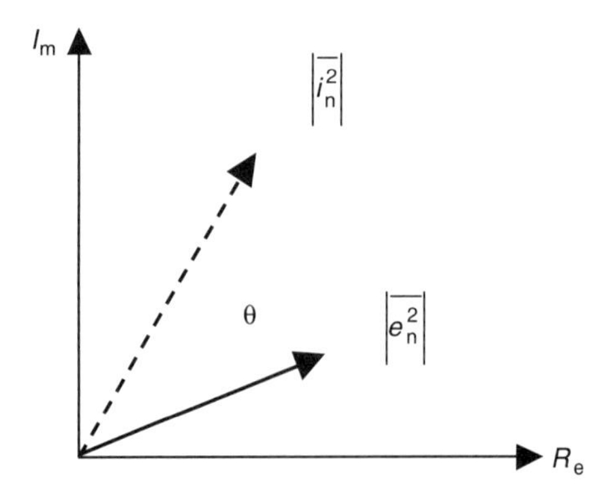


**Figure 10.1** Random noise sources phasors in the complex plan.

equivalent noise BW in the case of LPF and BPF. It can be observed that the actual filter is not as sharp as the ideal filter. Its shape factor has no infinite rejection. Hence, due to equality of filter energy integrals, both marked areas are equal.

## 10.2 Noise Sources in an Optical Receiver

When designing an optical receiver or a gain block, it is common to separate the receiver from its intrinsic noise and coupled noise, and refer to the receiver as an ideal one. Optical modules such as triplexers [optical triple port (OTP)], duplexers [optical dual port (ODP)], and detectors have wide wavelength responses. For instance, in a 1550 nm receiver, its detector would respond to 1310 nm. Hence, in the vicinity of several wavelengths, the amount of isolation defines the optical X-talk and optical background interfering noise to the desired wavelength. The photodiode bias current or the gain-block devices contribute shot noise. In a free space link, there is background noise too. Figure 10.3 illustrates the main noise sources of an optical receiver.

As was explained previously, some of the noise mechanisms are not correlated; hence, the overall power would be the sum of the squares of the noise voltages and currents.

## 10.3 Thermal Noise (Johnson Noise)

The Johnson noise (named after Johnson and Nyquist's 1928 work on a metallic resistor; it is also called resistance noise or white thermal noise, due to its equal power density over the spectrum[17]) represents resistive losses generated by the active device's internal resistance or by the passive components' losses. A random cloud of electrons, which move in all directions without the applying voltage, generates Johnson noise. When voltage is applied, the average random motion is in the direction of the $E$ field, which is generated by the voltage; hence, the random electron cloud moves toward the positive potential. The energy of the electron is temperature dependent.

Hence, when a metallic resistance $R$ is at a temperature $T$, the random electron motion produces a noise voltage $e_n(T)$ at the open circuit terminals. Consistent with the central limit theory, $e_n(T)$ has a Gaussian distribution with zero mean and variance.

The average power is related to the voltage square; thus, any component can be described as a noise-free device with a noise voltage source. From quantum mechanics:

$$\frac{d\langle e_n e_n^* \rangle}{d\nu} = 4R\left[\frac{h\nu}{2} + \frac{h\nu}{\exp(h\nu/kT) - 1}\right], \tag{10.16}$$

$$\frac{d\langle i_n i_n^* \rangle}{d\nu} = \frac{4}{R}\left[\frac{h\nu}{2} + \frac{h\nu}{\exp(h\nu/kT) - 1}\right], \tag{10.17}$$

where $k$ is Boltzmann's constant, $1.38 \times 10^{-23}$ J/K, $T$ is temperature in Kelvin, $h$ is Planck's constant, $6.62 \times 10^{-34}$ J sec, $R$ is the resistance, and $\nu$ is the optical frequency. Sometimes this symbol is replaced by $f$ to represent electrical signal frequency.

For the case of $hf \ll kT$ using Taylor expansion of exp($x$) or the L'Hopital limit for $h\nu \to 0$, Eqs. (10.16) and (10.17) become:

$$\frac{d\langle e_n e_n^* \rangle}{d\nu} = 4kTR, \tag{10.18}$$

$$\frac{d\langle i_n i_n^* \rangle}{d\nu} = \frac{4kT}{R}, \tag{10.19}$$

where $k$ is the Boltzmann's constant, $1.38 \times 10^{-23}$ Joule/K, $T$ is temperature in Kelvin, $h$ is Planck's constant, $6.62 \times 10^{-34}$ J sec, $R$ is resistance, $\nu$ is the optical frequency. Sometimes this symbol is replaced by $f$ to represent electrical signal frequency.

Thus, any thermal noise can be represented by a network and equivalent noise voltage, or noise current and noise resistance, as shown in Fig. 10.4.

Quantum mechanics shows that the spectral density of thermal noise is given by Eq. (10.20) when $hf \ll kT$:

$$n^2(f) = \frac{2Rh|f|}{\exp(h|f|/kT) - 1}. \tag{10.20}$$

Since the system is causal, only the positive frequency is taken under account; hence, the factor of four was changed to two, since only half spectrum is observed. At optical frequencies, Eq. (10.20) becomes

$$n^2(\nu) = \frac{h\nu}{\exp(h\nu/kT) - 1} + h\nu. \tag{10.21}$$

Equation (10.21) defines the noise limits for optical receivers compared to electrical signal receivers. In optical receivers, $f = \nu$ and $h\nu \gg kT$, and this governs the noise density; thus, at high frequencies, $n^2(\nu)$ becomes a linear line, and Eq. (10.21) becomes quantum energy noise:

$$n^2(\nu) = h\nu. \tag{10.22}$$

The first term in Eq. (10.21) under an optical frequency regime goes to zero; hence, at high frequencies, the noise energy is not equally distributed and becomes frequency dependent.

From Eq. (10.18), the noise voltage square or power in a given system BW is given by Eq. (10.23):

$$\bar{e}_n^2 = 4kTBR, \tag{10.23}$$

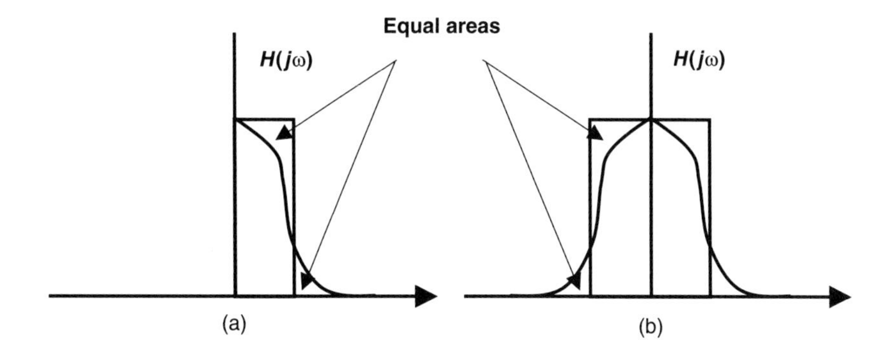


**Figure 10.2** Noise equivalent bandwidth plots: (a) low-pass, (b) band-pass.

where $k$ is the Boltzmann constant, $1.38 \times 10^{-23}$ J/K, $T$ is temperature in Kelvin, $B$ is bandwidth, and $R$ is resistance. Equation (10.24) provides the noise voltage source:

$$\bar{e}_{\mathrm{n}} = \sqrt{4kTBR}. \tag{10.24}$$

For a system impedance $Z_0$ and load impedance $Z_{\mathrm{L}}$, the equivalent matched noise source is described by Fig. 10.5.

The maximum power and noise transfer occurs when $R_0 = R_{\mathrm{L}} = R$. Hence, using voltage divider relations and power relation results in

$$P_{\mathrm{ave}} = \frac{\left[(R_0/(R_0 + R_{\mathrm{L}})\right]\sqrt{4KTR_0}\big)^2}{R_{\mathrm{L}}} = \frac{\left(\frac{1}{2}\sqrt{4KTR}\right)^2}{R} = \frac{KTR}{R} = KT. \tag{10.25}$$

This is a fundamental relationship. All resistors have equal available noise power. Any component under thermal equilibrium, meaning no bias or current, satisfies the above condition. The noise power density in dBm/Hz units is given by:

$$10 \log KT + 30\,\mathrm{dB} = -174[\mathrm{dBm/Hz}]. \tag{10.26}$$

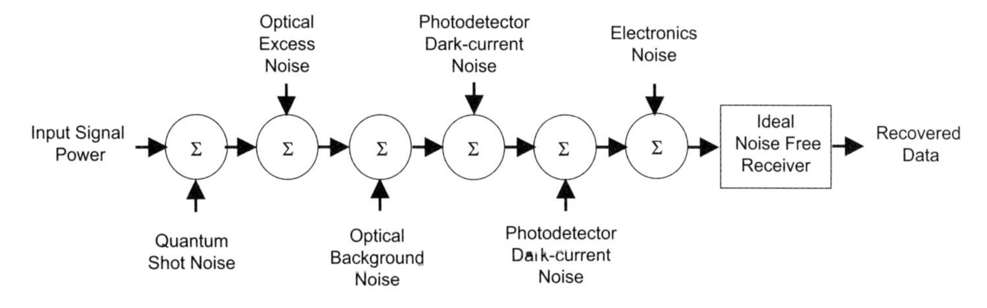


**Figure 10.3** Noise sources in an optical receiver.

In conclusion, it is a good engineering practice to increase the impedance at the receiver input in order to minimize white-noise amplification and improve CNR; however, it affects distortion performance. In the case of optical receiver front ends, increasing the input impedance improves gain as well, as was explained in Sec. 8.2.3. Moreover, in this scenario, any component under thermal equilibrium satisfies the following conditions:

$$\frac{\mathrm{d}\langle e_{\mathrm{n}} e_{\mathrm{n}}^{*}\rangle}{\mathrm{d}f} = 4kT\,\mathrm{Re}(Z), \tag{10.27}$$

$$\frac{\mathrm{d}\langle i_{\mathrm{n}} i_{\mathrm{n}}^{*}\rangle}{\mathrm{d}f} = 4kT\,\mathrm{Re}(Y), \tag{10.28}$$

$$\frac{\mathrm{d}\langle P_{\mathrm{ave-noise}}\rangle}{\mathrm{d}f} = kT. \tag{10.29}$$

These relations are useful for modeling and analyzing noise in electrical networks, such as modeling of transistors and matching networks. For instance, consider the following problem to find noise on a capacitor, per Fig. 10.6.

Using the definition given by Eq. (10.27), and finding the voltage across the capacitor using voltage divider calculation, we get,

$$\begin{aligned}\frac{\mathrm{d}\langle v_{\mathrm{c\_n}} v_{\mathrm{c\_n}}^{*}\rangle}{\mathrm{d}f} &= \frac{1}{1+j2\pi fRC}\,\frac{1}{1-j2\pi fRC}\,\frac{\mathrm{d}\langle v_{\mathrm{n}} v_{\mathrm{n}}^{*}\rangle}{\mathrm{d}f}\\ &= \frac{1}{1+f^{2}(2\pi RC)^{2}}\,\frac{\mathrm{d}\langle v_{\mathrm{n}} v_{\mathrm{n}}^{*}\rangle}{\mathrm{d}f}.\end{aligned} \tag{10.30}$$

Thus, the energy stored on the capacitor is

$$\int_{0}^{\infty} \frac{1}{1+4\pi^{2}(RC)^{2}f^{2}}\,\frac{\mathrm{d}\langle v_{\mathrm{n}} v_{\mathrm{n}}^{*}\rangle}{\mathrm{d}f}\,\mathrm{d}f = \frac{1}{2}CV_{\mathrm{C}}^{2} = \frac{1}{2}kT. \tag{10.31}$$

Note that the second term in the integral is the noise density in the circuit, which is given by Eq. (10.27). The same solution approach goes for resistor inductor (RL) and resistor inductor capacitor (RLC) circuits.

## 10.4 Shot Noise

Shot noise is sometimes called Schottky noise. The flow of an electrical current across a semiconductor junction is not smooth and continuous. The current is composed of charge carriers with $q$ Coulombs flowing across the circuit. The current flow across the junction's potential barrier is determined by a statistical function that describes the noise process associated with that process. If the current flow across the barrier follows a periodic, predictable way, the current flow is

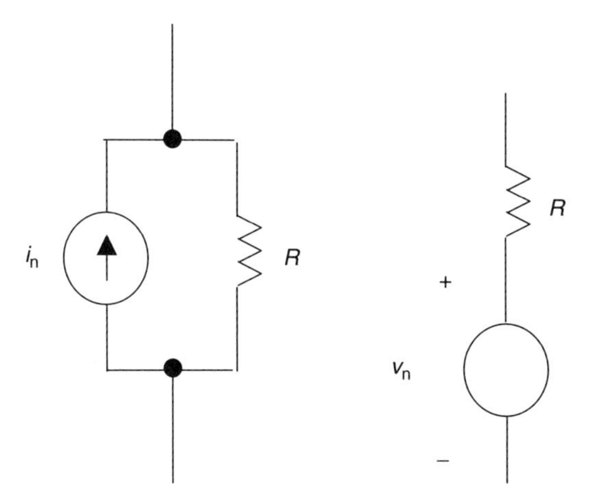


**Figure 10.4** Noise current and noise voltage with noise resistance $R$.

uniform, and noise is not generated. That kind of noise is called shot noise. It is a Poisson distribution, since the arrival of each electron is statistically independent of every other electron. When large observations are made, it becomes a Gaussian process. Shot noise is generated by bias current fluctuations $I_O$ in a semiconductor's junction. It is represented by a current source parallel to the device (Fig. 10.7):

$$\bar{i}_n^2[\mathrm{W/Hz}] = 2qI_O B \tag{10.32}$$

or

$$\frac{\mathrm{d}\langle I_n I_n^* \rangle}{\mathrm{d}f} = 2qI. \tag{10.33}$$

In conclusion, lower bias minimizes the noise; hence, transistors' $S$ parameters for $\Gamma_{opt}$ for noise matching are provided at low current. At higher currents, the noise performance of the device degrades. The noise density current is given by

$$\bar{i}_n[\mathrm{W}/\sqrt{\mathrm{Hz}}] = \sqrt{2qI_O B}. \tag{10.34}$$

## 10.5 1/f Noise

$1/f$ noise has been observed in many physical systems. Low-frequency fluctuations in the resistance of a semiconductor create $1/f$ noise. In vacuum tubes, this type of noise was described as flicker noise. In a resistor, it is called excess noise. One of the variants of $1/f$ noise is a component of phase noise in oscillators. Further analysis will be given in Chapters 14 and 15 while discussing jitter analysis.

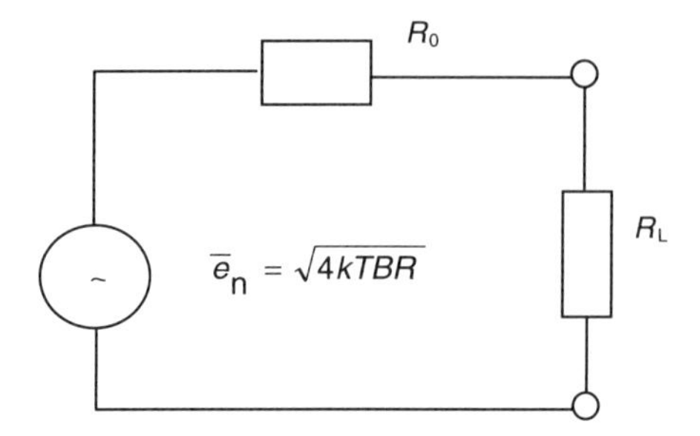


**Figure 10.5** Noise free load $R_L = R$ and noise source.

The $1/f$ power spectral density (PSD), known also as flicker noise, is given by

$$n(f) = \frac{a}{f^b}[\mathrm{W/Hz}], \tag{10.35}$$

where $a$ is a constant defining the absolute level of $n(f)$, and $b$ is an exponent generally close to one. In many semiconductors, the $1/f$ noise is related to the bias current flow in the device; hence, Eq. (10.35) is modified to

$$n(f) = I_{\mathrm{DC}}^{d} \cdot \frac{a \cdot f_{\mathrm{c}}}{f^b}[\mathrm{W/Hz}], \tag{10.36}$$

where $d$ is a constant that sets the relation between the dc current and the $1/f$ noise level, and $f_c$ is the corner frequency. For classic $1/f$ noise, $b = 1$. It can be seen that the $1/f$ noise rolls off versus frequency with a slope of –10 dB/decade. The total $1/f$ noise power at a given BW is given by

$$N = \int_{f_1}^{f_2} I_{\mathrm{DC}} \frac{a}{f} \mathrm{d}f = I_{\mathrm{DC}} \cdot a \cdot \ln\left(\frac{f_{\mathrm{c}}}{f_1}\right)[\mathrm{watts}]. \tag{10.37}$$

The overall noise near the carrier and the noise measured with some frequency offset from the signal carrier is given by the sum of two noise sources. Since both noises are not correlated, it is a sum of squares of the currents:

$$N = i_{\mathrm{n}}^2 + i_{\mathrm{n}}^2\left(\frac{f_{\mathrm{c}}}{f_1}\right)[A^2\mathrm{Hz}]. \tag{10.38}$$

The first term has a constant power versus frequency, i.e., equal power over the spectrum. The $1/f$ gets higher near the carrier, adding jitter and rolls off at a rate of –10 dB/decade, until it gets below the constant-density-region noise; therefore, $1/f$ is called colored noise. The noise presentation of $N$ versus frequency is given by Fig. 10.8.

$1/f$ noise becomes a concern in optical receiver design when the low cut-off frequency becomes a few tenths of MHz. This case is more relevant to digital transport, where the data occupies the low-spectrum frequencies as well. In

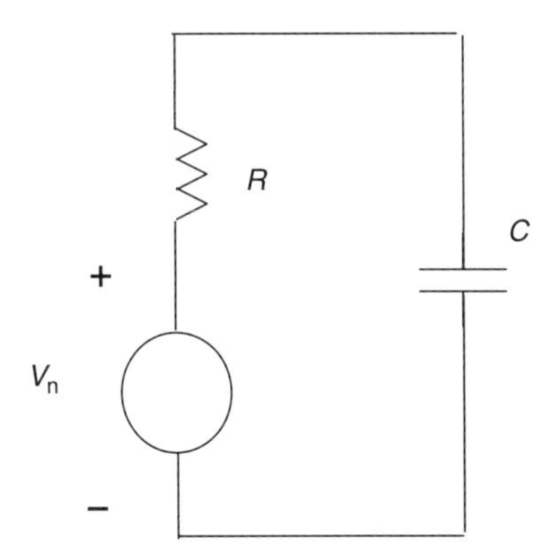


**Figure 10.6** Noisy RC network.

analog return path transport, the BW is 5 to 50 MHz, and in CATV, it is between 47 and 870 MHz; hence, in digital designs, TIA (transimpedance amplifier) choice is vital for proper receiver BER performance. In silicon devices, $1/f$ corner frequency is tens of kHz, while in gallium arsenide (GaAs) metal Schottky field effect transistors (MESFETs), it can reach 100 MHz.

## 10.6 Carrier to Noise Ratio

### 10.6.1 CNR and noise figure

The ratio between $(C/N)_{\text{IN}}$ to $(C/N)_{\text{OUT}}$ or $\text{CNR}_{\text{IN}}$ to $\text{CNR}_{\text{OUT}}$ is called $F$ or noise factor. The logarithmic value of this figure of merit is called noise figure, $NF$[dB].

Any receiver would degrade $C/N$; hence, $F > 1$. Assuming all the parameters are normalized to the input, the following expressions are applicable:

$$F = \frac{(C/N)_{\text{IN}}}{(C/N)_{\text{OUT}}} = \frac{C/KTB}{CG/KTBFG} = \frac{C/KTB}{C/KTBF} = \frac{N_{\text{OUT}}}{N_{\text{IN}}} > 1, \tag{10.39}$$

$$NF[\text{dB}] = 10\log F = \text{CNR}_{\text{IN}}[\text{dB}] - \text{CNR}_{\text{OUT}}[\text{dB}]. \tag{10.40}$$

A different notation is to define the system as a block with excess noise $N_e$; thus, the output CNR to input CNR ratio would be

$$F = \frac{C/N_0}{C/(N_0 + N_e)} = 1 + \frac{N_e}{N_0}. \tag{10.41}$$

In conclusion, $F$ is a figure of merit showing the system additive noise with

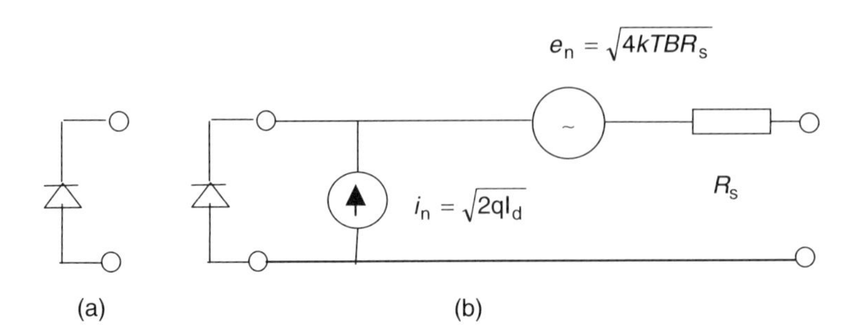


**Figure 10.7** Shot noise and thermal noise equivalent circuit. (a) Ideal diode. (b) With shot noise current source and thermal noise voltage.

respect to the input reference noise. Assume a simple case of shot noise:

$$\frac{C}{N_{\mathrm{OUT}}} = \frac{C}{N_{\mathrm{shot}} + N_0}. \tag{10.42}$$

Hence,

$$\frac{1}{N/C_{\mathrm{OUT}}} = \frac{1}{(N_{\mathrm{shot}} + N_0)C}. \tag{10.43}$$

Thus, based upon the above, the total CNR of a CATV optical receiver or any type of optical receiver is given by

$$\frac{1}{1/\mathrm{CNR}_{\mathrm{tot}}} = \frac{1}{1/\mathrm{CNR}_{\mathrm{shot}}} + \frac{1}{1/\mathrm{CNR}_{\mathrm{Johnson}}} + \frac{1}{1/\mathrm{CNR}_{\mathrm{shot\_det}}}. \tag{10.44}$$

Equation (10.43) represents the receiver CNR when the photodetector is not illuminated; i.e., the receiver-intrinsic noise appears only when the shot noise stands for the photodetector dark current. A general conceptual block diagram of an analog receiver is given in Fig. 10.9.

The noise figure and CNR tests are done when the receiver is in its maximum gain state. This means the automatic gain control (AGC) is in its minimum attenuation state or at its threshold state; thus, the RF chain produces its minimum NF value, as was explained in Eq. (9.71) in Chapter 9. Noise figure measurement refers to Johnson noise; therefore, the Johnson CNR term in Eq. (10.44) will govern the overall CNR at high attenuation levels. As the AGC loss increases, the NF and Johnson CNR becomes dominant and determines the system's total CNR. When analyzing the system noise figure or CNR, tests are done at dark conditions; thus, the shot noise originating from the photodetector relates to the leakage of photodetector dark current. Illuminating the fiber would increase the photodetector shot noise, and the overall system CNR may degrade by 1 dB to 2 dB with respect to the dark CNR state.

### 10.6.2 Quantum CNR limit

Based upon Eqs. (8.7) and (8.44), the dc level and RF level of a received signal at the output of the photodetector is given by

$$I_{\mathrm{O}} = P_{\mathrm{opt\_in}} \cdot r, \tag{10.45}$$

$$I_{\mathrm{RF\_peak}} = \mathrm{OMI} \cdot P_{\mathrm{opt\_in}} \cdot r, \tag{10.46}$$

$$P_{\mathrm{RF}} = \left( \frac{\mathrm{OMI} \cdot P_{\mathrm{opt\_in}} \cdot r}{\sqrt{2}} \right)^2 \cdot R_{\mathrm{in}}, \tag{10.47}$$

where $r$ is the photodetector responsivity.

However, on the other hand, the shot noise current is given by

$$\bar{i}_{\mathrm{n}}^2 = 2qI_{\mathrm{O}}B. \tag{10.48}$$

Thus, $\mathrm{CNR_{shot}}$ is given by

$$\mathrm{CNR_{shot}} = \frac{\left(I_{\mathrm{RF\_peak}}^2 \cdot R_{\mathrm{in}}\right)/2}{2qI_{\mathrm{O}}B \cdot R_{\mathrm{in}}} = \frac{\left(I_{\mathrm{RF\_peak}}^2\right)/2}{2qI_{\mathrm{O}}B} = \frac{I_{\mathrm{RF\_peak}}^2}{4qI_{\mathrm{O}}B}, \tag{10.49}$$

but from OMI expression is

$$\mathrm{OMI} = \frac{I_{\mathrm{RF\_peak}}}{I_{\mathrm{O}}}. \tag{10.50}$$

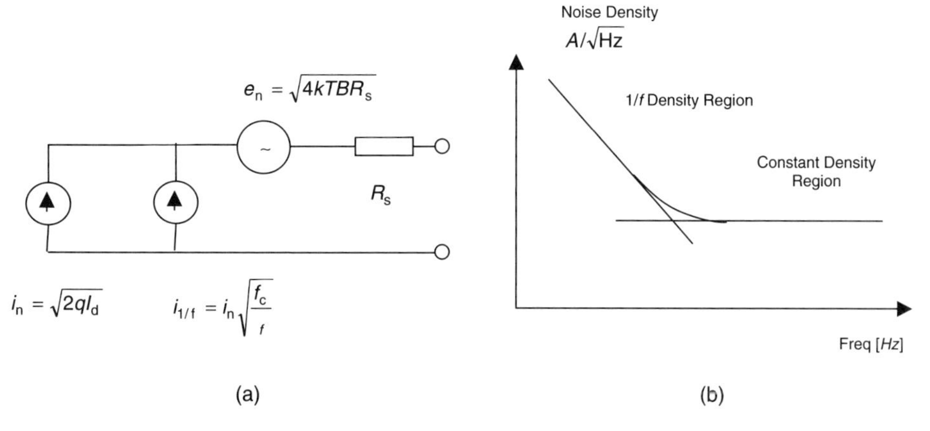


**Figure 10.8** Equal density noise sources and $1/f$ noise source. (a) Noise equivalent circuit. (b) Noise characteristic vs. frequency note that the thermal noise can be presented as a current source using Norton theory.

As a result, the quantum CNR limit is given by

$$\mathrm{CNR_{shot}} = \frac{\mathrm{OMI}^2 \mathrm{I_O}}{4qB} \tag{10.51}$$

or

$$\mathrm{CNR_{shot}} = \frac{\mathrm{OMI}^2 r}{4qB} P_{\mathrm{opt\_in}}. \tag{10.52}$$

In conclusion, it is recommended to have a high-responsivity photodetector at the receiver front end. This way, the photodetector noise contribution to the receiver's overall CNR is minimized [see Eq. (10.52)]. Obviously, CNR is highly dependent on the OMI modulation index. Moreover, it is linearly improving when power is increased. This describes a system CNR test where there is light at the detector input.

The receiver noise performance is evaluated when there is no illumination on the photodetector. In this case, the AGC is set to the maximum gain, the receiver noise figure is at its minimum. The photodetector shot noise in this case is due to dark current and reverse bias leakage current. It is very easy to derive the dark CNR formulation at the AGC threshold state. $I_O$ in Eq. (10.48) is replaced by the dark leakage current, while the signal level is evaluated per the optical level at the AGC threshold state as in Eq. (10.46); hence, the overall noise power at the receiver input is the RMS sum of all the noise sources. Knowing what the CNR results are under its optical state at its sensitivity level, and what the photodetector's dark leakage current and responsivity are, and controlling the OMI allows the extraction of illumination noise and the calculation of the dark CNR from test results are done by using Eq. (10.44). It is important to emphasize that the AGC should be at threshold state, which will be explained in Chapter 12.

### 10.6.3 Equivalent input noise current density

When measuring the receiver front-end noise performance, it is typical to present its noise performance by an equivalent input noise current, EiNC. The receiver output-measured noise can be transformed into the input and represented by two noise mechanisms: Johnson and shot noise. The rms summation of the Johnson noise power and shot noise power, transformed to the receiver input noise, can be represented as an equivalent density noise as given by Eq. (10.53). The reference here to $N_{\mathrm{shot}}$ is to the photodetector dark current:

$$N_{\mathrm{eq}}[\mathrm{W/Hz}] = kTF = \sqrt{N_{\mathrm{shot}} + N_{\mathrm{Johnson}}}. \tag{10.53}$$

Assuming that the receiver's input impedance $Z_0$ is known, then the noise can be presented as an equivalent input noise current density, creating the noise power at the receiver's input:

$$N_{\mathrm{eq}}[\mathrm{W/Hz}] = kTF = i^2_{\mathrm{n\_rms\_eq}} \cdot Z_0. \tag{10.54}$$

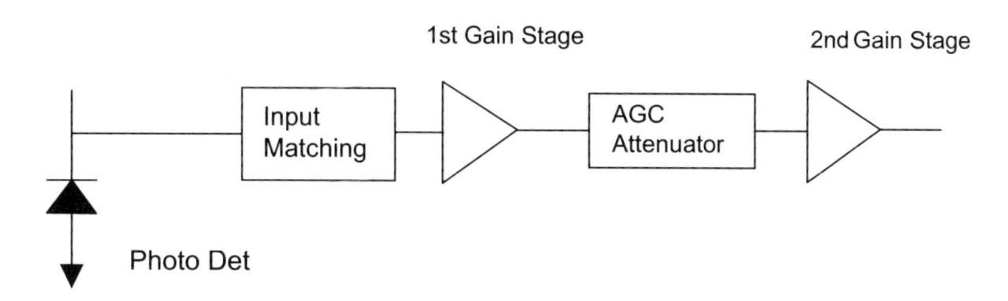


**Figure 10.9** Top level block diagram of an analog CATV receiver.

Therefore, the equivalent input noise current density is given by

$$i_{\text{n\_rms\_eq}}[A/\sqrt{\text{Hz}}] = \sqrt{\frac{kTF}{Z_0}}. \tag{10.55}$$

The value $kT$ equals $-174$ dBm/Hz or $-204$ dBW/Hz; hence, Eq. (10.55) can be written with reference to an input noise current density in $[pA/\sqrt{\text{Hz}}]$:

$$i_{\text{n\_rms\_eq}}[pA/\sqrt{\text{Hz}}] = \frac{10^{-204NF/20}\cdot 10^{12}}{\sqrt{Z_0}} = 63.095\frac{10^{0.1NF}}{\sqrt{Z_0}}. \tag{10.56}$$

Assuming an impedance matching transformer with a turn ratio of $N$:1, then the equivalent noise current density transformed to the photodetector output ports is given by

$$i_{\text{EiNC}}[pA/\sqrt{\text{Hz}}] = \frac{10^{-204NF/20}\cdot 10^{12}}{N\sqrt{Z_0}} = 63.095\frac{10^{0.1NF}}{N\sqrt{Z_0}}. \tag{10.57}$$

In conclusion, the higher the input impedance of the receiver's first stage, the lower the noise current. Matching the transformer with a high turn ratio would increase the receiver's input impedance even further. However, there is a limit and optimum value. Too high an input impedance results in too high a signal and reduces the distortion performance. Moreover, since the input impedance observed from the photodiode is higher, this limits the front end BW, since the $RC$ time constant is higher, where $C$ is the photodiode's stray capacitance. Figure 10.10 depicts the receiver's input noise.

### 10.6.4 Equivalent input noise current density derivation from CNR test

As shown by Figs. 10.9 and 10.10, it is almost impossible to directly measure the optical receiver noise figure using a noise figure meter. There is no way to access the RF input. Additionally, there is no way to directly measure the RF-to-RF gain.

Hence, a different approach is used to evaluate the receiver noise figure: measuring CNR at the output when there is no illumination at the photodetector input. This test provides information about the receive section performance. It is not affected by additional parameters of the whole link, such as RIN, modal noise, etc.; however, it provides some envelope limits for the receiver.

Input noise density to a CATV receiver is measured indirectly by measuring CNR. CNR, or (carrier-to-noise ratio) is the ratio between the carrier signal and the noise at a given BW optical level, and modulation index. The receiver, input CNR is given by:

$$\mathrm{CNR} = \frac{P_{\mathrm{IN}}}{kTBF + N_{\mathrm{shot}}}, \tag{10.58}$$

where $k$ is Boltzmann's constant, $T$ is the temperature in Kelvin (generally 290 K), and $B$ is the bandwidth of interest (generally the last IF section BW; in our case 4 MHz). $F$ is the overall equivalent noise figure of the receiver. $N_{\mathrm{shot}}$ is the photodetector shot noise at dark current. Referring to the output, and assuming the receiver gain is $G$, gives

$$\mathrm{CNR} = \frac{P_{\mathrm{IN}}G}{(KTBF + N_{\mathrm{shot}})G} = \frac{P_{\mathrm{IN}}}{KTBF + N_{\mathrm{shot}}}. \tag{10.59}$$

As was expected, CNR is not dependent on the gain of the device. Both noise mechanisms can be represented by an equivalent noise current density $I_{\mathrm{EiCN}}$ for all noise sources per BW of 1 Hz. Since the current square is proportional to power, and the noise power is dependent on the BW of operation, the following notation is applicable:

$$N = I_{\mathrm{EiNC}}^2 \cdot Z_0, \tag{10.60}$$

where $N$ is the overall noise power per 1 Hz, and $Z_0$ is characteristic impedance. Using the Norton theorem thermal voltage, $\bar{e}_{\mathrm{n}}$ is converted to a current source. In this case, the goal is to find the overall current density $I_{\mathrm{EiNC}}$. The overall noise power density is $N_{\mathrm{shot}} + N_{\mathrm{Johnson}}$; hence, the equivalent current density will be

$$I_{\mathrm{EiNC}}[pA/\sqrt{\mathrm{Hz}}] = \sqrt{I_{\mathrm{shot}}^2 + I_{\mathrm{Johnson}}^2}. \tag{10.61}$$

Hence, the CNR term at 1 Hz BW is given by

$$\begin{aligned} \mathrm{CNR@1\,Hz}\frac{P_{\mathrm{RFOUT}}}{N_{\mathrm{OUT1\,Hz}}} &= \frac{P_{\mathrm{RF-IN}}}{N_{\mathrm{IN-1\,Hz}}} = \frac{P_{\mathrm{RF-IN}}}{I_{\mathrm{EiNC}}^2 \cdot Z_0} \\ &= \frac{(\mathrm{OMI} \cdot r \cdot P)^2 \cdot Z_0}{I_{\mathrm{EiNC}}^2 \cdot Z_0 \cdot 2} = \frac{(\mathrm{OMI} \cdot r \cdot P)^2}{I_{\mathrm{EiNC}}^2 \cdot 2}, \end{aligned} \tag{10.62}$$

where $P$ is the optical power, $r$ is the responsivity, and OMI is the optical modulation index.

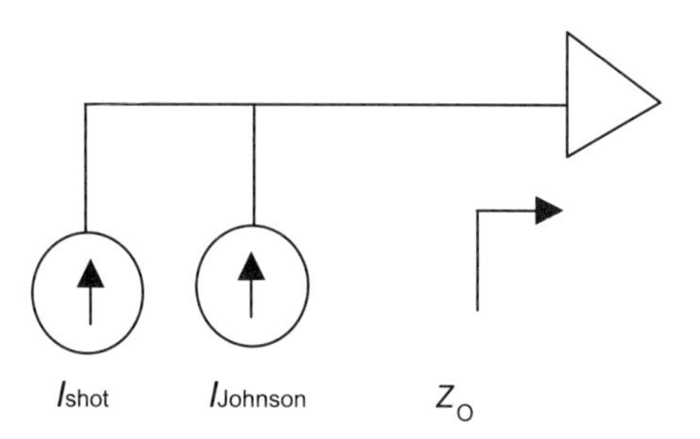


**Figure 10.10** Input noise at the receiver front end.

Moreover, it can be observed that $Z_0$ does not affect the CNR result. From the above notation, noise current density is evaluated in $pA/\sqrt{Hz}$, as shown in Fig. 10.11.

In CATV, the BW per channel is 4 MHz, and the OMI is 3.5%/Ch. The ultimate CNR test is done at the receiver minimum noise level: the AGC is at the threshold level, which means minimum noise figure and maximum gain. These test conditions provide information regarding the noise density current at the receiver input. In this way, the system noise figure can be estimated. According to the NCTA standard, CNR is measured at 4 MHz BW. Equation (10.62) is modified to

$$\mathrm{CNR@BW[MHz]} = \frac{P_{\mathrm{RFOUT}}}{N_{\mathrm{OUT1\,Hz}}} = \frac{P_{\mathrm{RF-IN}}}{N_{\mathrm{IN-1\,Hz}}} = \frac{P_{\mathrm{RF-IN}}}{I_n^2 \cdot Z_0 \cdot \mathrm{BW}} = \frac{\left(\mathrm{OMI} \cdot r \cdot P \cdot 1/\sqrt{2}\right)^2}{I_n^2 \cdot \mathrm{BW_{MHz}} \cdot 10^6}. \tag{10.63}$$

$$\mathrm{CNR@BW} = 4\,\mathrm{MHz} = \frac{P_{\mathrm{RFOUT}}}{N_{\mathrm{OUT1\,Hz}}} = \frac{P_{\mathrm{RF-IN}}}{N_{\mathrm{IN-1\,Hz}}} = \frac{P_{\mathrm{RF-IN}}}{I_n^2 \cdot Z_0 \cdot \mathrm{BW}} = \frac{\left(\mathrm{OMI} \cdot r \cdot P \cdot 1/\sqrt{2}\right)^2}{I_n^2 \cdot 4 \cdot 10^6}. \tag{10.64}$$

In case the noise current is in pA units and optical power in dBm, Eq. (10.64) becomes

$$\mathrm{CNR[dB]@4\ MHzBW} = 20\log P_{\mathrm{OPTICAL}}[\mathrm{mW}] + 20\log\frac{\mathrm{OMI}}{\sqrt{2}} + 20\log r - 20\log I_n + 114. \tag{10.65}$$

Using Eq. (10.64), at −4 dBm and −6 dBm optical input power, OMI = 3.5% (0.035); CNR is calculated for different photodetector responsivities $r$ shown as parameter $r = 0.7$ mW/mA, $r = 0.8$ mW/mA, $r = 0.9$ mW/mA, and

$r = 1$ mW/mA in Figs. 10.11 to 10.14. Hence, noise current in $pA/\sqrt{\text{Hz}}$ can be estimated from the CNR test at maximum gain. Noise level should be measured at dark without light and signal.

Signal level and noise density should be tested at AGC threshold, meaning high gain and minimum attenuation. The assumption of the calculated plots is that $-4$ dBm and $-6$ dBm are the minimum optical level, generating the minimum RF level that brings the AGC to its threshold point. In case this test is done under illumination, the CNR should be corrected by subtracting the shot noise contribution using Eq. (10.61). It is important to know the photodetector responsivity in order to predict the correct input noise current. These tests are controlled measurements where OMI is calibrated, the optical level is monitored, and the photodiode responsivity is a known parameter. A typical FTTX CATV receiver's performance is 5.5 $pA/\sqrt{\text{Hz}}$. For this purpose, the receiver's front end utilizes PHEMT technology or HBT devices. The matching techniques vary from transformer matching to resonance matching methods, as will be explained in Chapter 11. By measuring the equivalent input noise current, it is possible to estimate the receiver's noise figure using Eqs. (10.55) and (10.56). It is assumed that $Z_0$ is known and is a fixed value.

However, to have the correct estimation of the noise figure, there is a need to map S11, which is the receiver's input reflection parameter, and then extract the value of $Z_0$ versus frequency; this can be done by microwave simulation tools. The next step would be CNR measurements at several frequencies. In this manner, the noise figure of an optical receiver can be estimated.

An additional parameter used to define the thermal noise level threshold for $\text{CNR} = 1$ is NEP (noise equivalent power). Using the CNR definitions, $\text{NEP} = (1/r) \times i_n$, where $i_n^2$ is given by Eq. (10.19), and $r$ is responsivity. This figure of merit defines the signal threshold for a given receiver noise factor $F$, meaning for $\text{CNR} > 1$, it should satisfy $i_n = \sqrt{4kTF/R} < i_{\text{sig}}$, where $i_{\text{sig}}$ is the signal current given by $i_{\text{sig}} = \text{OMI} \times P_{\text{opt}} \times r/\sqrt{2}$.

### 10.6.5 Laser relative intensity noise

In system testing, the link performance is examined. The system CNR test is different from the receiver CNR test, since it takes into account noise generated by illuminating the photodetector of the receiver. As a result, additional parameters are added to the CNR budget equation:

$$\frac{1}{\text{CNR}_{\text{tot}}} = \frac{1}{\text{CNR}_{\text{RIN}}} + \frac{1}{\text{CNR}_{\text{shot\_p.d.}}} + \frac{1}{\text{CNR}_{\text{receiver}}}. \tag{10.66}$$

The output amplitude of the transmitter's laser is not perfectly constant. It fluctuates randomly around its average level (analog to oscillator AM noise), which was explained in Chapter 7. The reason for this phenomenon is temperature, acoustic disturbances, and quantum mechanics considerations. In the long term, the power level is the average, but in the short term, it is a sum of the constant term and the time-varying term. This was explained in detail in Sec. 7.4.4. Thus, the instantaneous power $P(t)$ is written by

$$P(t) = P_{\mathrm{O}} + \Delta P(t). \tag{10.67}$$

The optical current signal is described by Eq. (8.7) and can be written as

$$i_{\mathrm{photo}}(t) = rP(t) = rP_{\mathrm{O}} + r\Delta P(t) = I_{\mathrm{dc}} + i_{\mathrm{ac}}(t). \tag{10.68}$$

This means that the photocurrent, when no modulation is applied on the laser, consists of dc and ac terms; hence, this noise source would have spectral density and an average rms value. RIN is defined as the ratio between the average fluctuation and the average power:

$$\mathrm{RIN}(\omega) = \frac{\overline{|\Delta P(\omega)|^2}}{\overline{P}^2}. \tag{10.69}$$

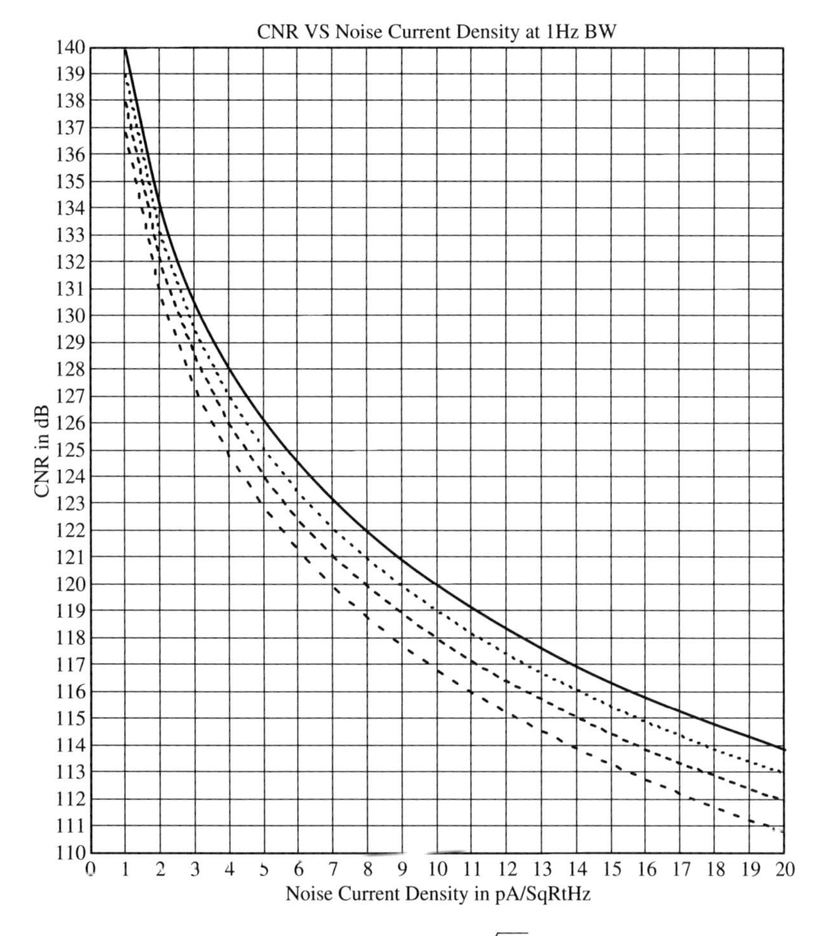


**Figure 10.11** CNR[dB] vs. noise density in $pA/\sqrt{\mathrm{Hz}}$ at $-4$ dBm optical signal and 1 Hz BW OMI = 3.5%. Responsivity codes: solid-1, dotted-0.9, dashed-0.8, dash-dot-0.7.

In terms of measured photocurrent, RIN is expressed by

$$\mathrm{RIN}(\omega) = \frac{i_{\mathrm{n}}^2(\omega)}{I_{\mathrm{dc}}^2}, \tag{10.70}$$

where $i_{\mathrm{n}}^2(\omega)$ is the measured noise current PSD in amps$^2$/Hz that is associated with the average dc photocurrent $I_{\mathrm{dc}}$. Usually RIN is expressed in terms of [dB/Hz];

$$\mathrm{RIN}(\omega)[\mathrm{dB/Hz}] = 10\log\left(\frac{i_{\mathrm{n}}^2(\omega)}{I_{\mathrm{dc}}^2}\right). \tag{10.71}$$

Since an inherent quantum shot noise is associated with the generation of

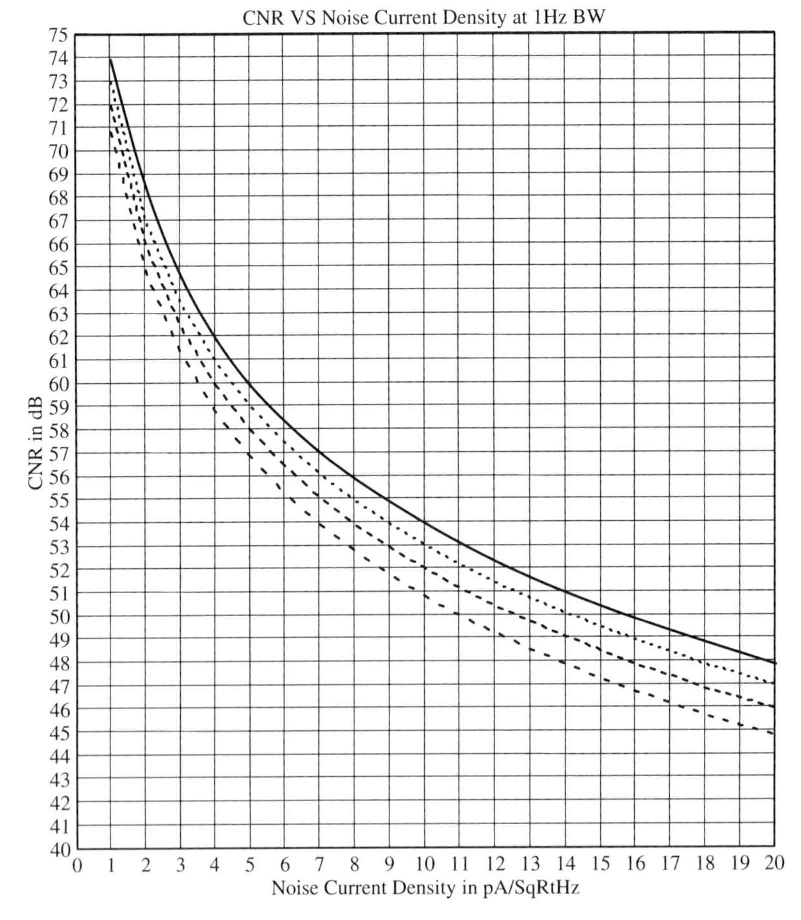


**Figure 10.12** CNR[dB] vs. noise density in $pA/\sqrt{\mathrm{Hz}}$ at −4 dBm optical signal and 4 MHz BW OMI = 3.5%. Responsivity codes: solid-1, dotted-0.9, dashed-0.8, dash-dot-0.7.

photocurrent, the quantum shot noise will be the lowest RIN limit:

$$\mathrm{RIN_{QL}}\left[\frac{1}{\mathrm{Hz}}\right] = \frac{i_n^2(\omega)}{I_{dc}^2} = \frac{2qi_{dc}}{I_{dc}^2} = \frac{2q}{I_{dc}} = \frac{3.2 \cdot 10^{-19}}{I_{dc}}. \tag{10.72}$$

At a photocurrent of 0.1 mA, the RIN limit is −145 dB/Hz. Using Eq. (10.72) and the responsivity equation given in Sec. 8.15, assuming an ideal detector with quantum efficiency $\gamma = 1$ provides the RIN quantum limit as a function of wavelength and optical power:

$$\begin{aligned}\mathrm{RIN_{QL}} &= \frac{2q}{I_{DC}} = \frac{2q}{rP_{OPT}} = \frac{2q}{(\gamma q/h\nu)P_{OPT}} = \frac{2\,h\nu}{P_{OPT}} \\ &= \frac{C}{\lambda}\cdot\frac{2\,h}{P_{OPT}} = \frac{3.96\cdot 10^{-16}}{\lambda(\mu\mathrm{m})\cdot P_{OPT}(\mathrm{mW})}.\end{aligned} \tag{10.73}$$

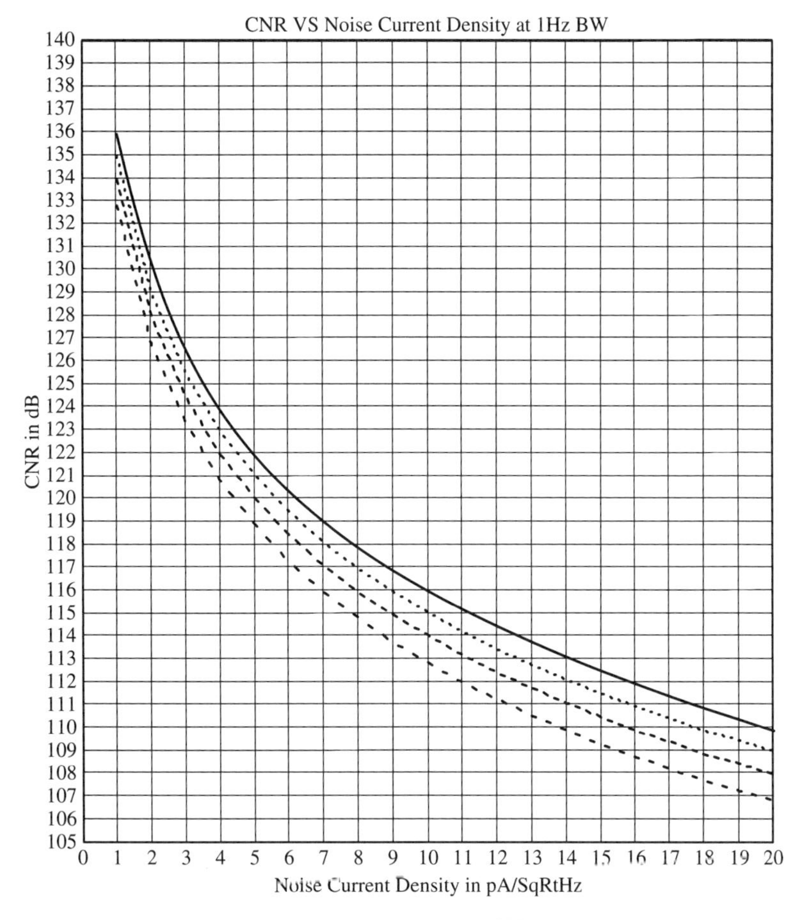


**Figure 10.13** CNR[dB] vs. noise density in $pA/\sqrt{\mathrm{Hz}}$ at −6 dBm optical signal and 1 Hz BW OMI = 3.5%. Responsivity codes: solid-1, dotted-0.9, dashed-0.8, dash-dot-0.7.

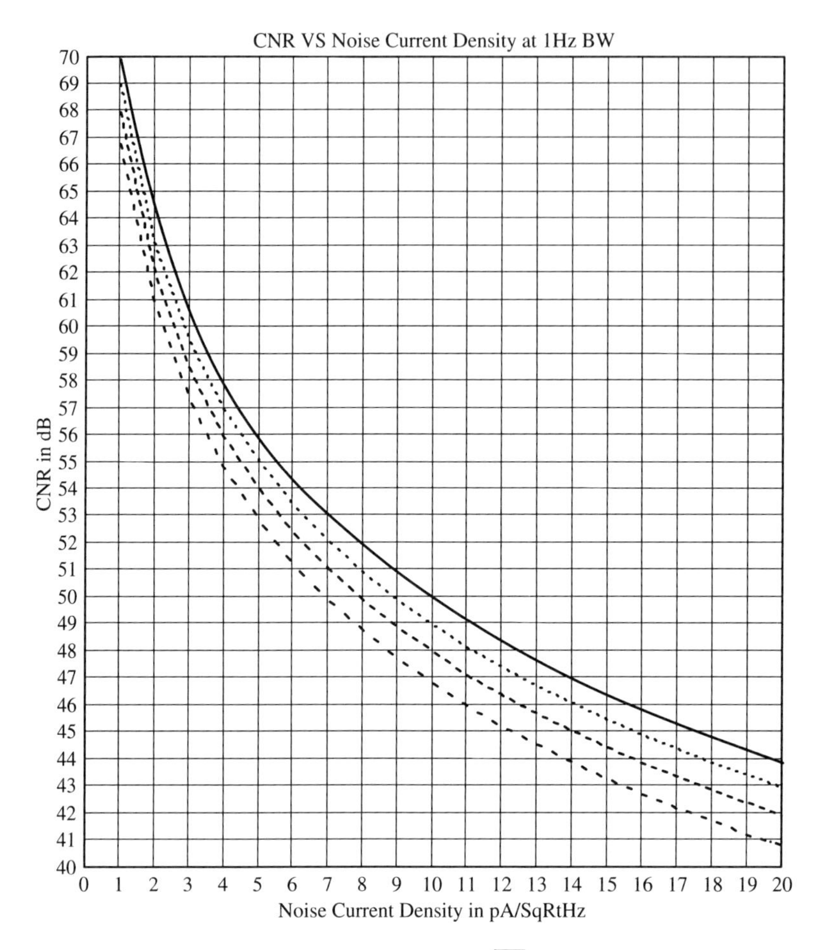


**Figure 10.14** CNR[dB] vs. noise density in $pA/\sqrt{\mathrm{Hz}}$ at $-6$ dBm optical signal and 4 MHz BW OMI = 3.5%. Responsivity codes: solid-1, dotted- 0.9, dashed-0.8, dash-dot-0.7.

Hence, for a 1550-nm laser at 1 mW (0 dBm), the RIN is $-156$ dB/Hz. The higher the power of the laser, the higher the RIN. From Eq. (10.70), RIN noise is derived for a given BW:

$$i_{\mathrm{n}}^2(\omega)\cdot \mathrm{BW} = \mathrm{RIN}\cdot I_{\mathrm{dc}}^2\cdot \mathrm{BW}. \tag{10.74}$$

Using Eq. (8.45) for the rms RF power, RIN CNR is calculated. If it is assumed that there are no other effects, then the system $\mathrm{CNR}_{\mathrm{RIN}}$ upper limit is given by

$$\mathrm{CNR}_{\mathrm{RIN}} = \frac{P_{\mathrm{RF}}}{P_{\mathrm{RIN}}} = \frac{\left((\mathrm{OMI}^2\cdot P_{\mathrm{opt}}^2\cdot r^2)/2\right)Z_{\mathrm{L}}}{\mathrm{RIN}\cdot P_{\mathrm{opt}}^2\cdot r^2\cdot \mathrm{BW}\cdot Z_{\mathrm{L}}} = \frac{\mathrm{OMI}^2}{2\cdot \mathrm{RIN}\cdot \mathrm{BW}}. \tag{10.75}$$

It can be observed that the value is not dependent on the current but is related to

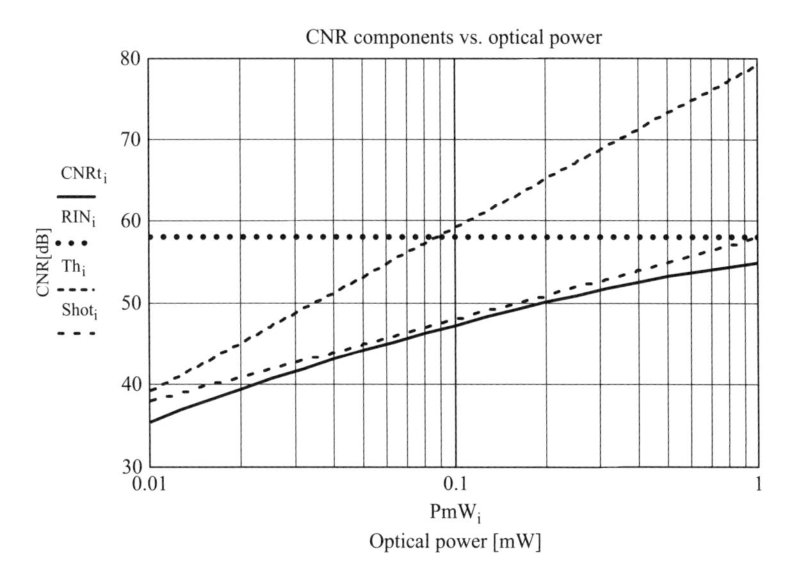


**Figure 10.15** Analog receiver CNR limits of shot noise, RIN and Johnson (Th) noise as function of optical power. Responsivity is $1A/W$, transformer turn ratio is 2:1, $Z_0 = 50\,\Omega$, dark current is 5 $pA$. RIN = z−155 $dB$/Hz, OMI = 4% and BW = 4 MHz.

modulation depth; hence, the higher the OMI, the better the CNR performance. The impact of RIN on the receiver noise is accompanied with shot noise; thus, the latter should be subtracted from the overall noise:

$$i^2_{\text{RIN}}(\omega) = \text{RIN}(\omega) \times I_{\text{P}}^2 - 2q \times I_{\text{P}}. \tag{10.76}$$

From Eq. (10.76), it can be seen that RIN is a signal-dependent noise: the higher the signal, the higher the noise; hence, it reduces the receiver's CNR. In digital designs, the RIN limit for quality of service is −120 dB/Hz, up to −100 dB/Hz. It is more forgiving than the analog design.

### 10.6.6 System CNR limits

The link CNR budget is described by Eq. (10.77). Substituting the above relations from Eqs. (10.19), (10.49), and (10.74) provides a measure to estimate the system CNR performance and limits:

$$\text{CNR} = \frac{(\text{OMI} \cdot I_{\text{dc}})^2}{2BW\left(I_{\text{dc}}^2 \cdot \text{RIN} + 2q(I_{\text{dc}} + I_{\text{dark}}) + \frac{4kTF}{R_{\text{L}}}\right)}. \tag{10.77}$$

Figure 10.15 provides the CNR limits for an analog CATV receiver taking under account thermal noise (Th), RIN, and shot noise.

## 10.7 PIN Photodetector Noise Modeling

In Chapter 8, the PIN photodiode was reviewed, and a small-signal equivalent model circuit was provided in Fig. 8.9. This schematic should be modified to include noise sources of the photodiode. There are several noise sources that are inherent to the photodiode.[10–12,14] The first is the dark current noise source, which is the dc reverse current leakage noise. This noise source is parallel to the signal source. The second one is the shot noise current source. This current source value varies with optical power and is in parallel to the signal current source. This shot noise is also called Schottky noise. These two current sources are referred to as shot noise sources. The next two noise sources are thermal noise sources of the photodiode resulting from ohm resistance build-up in the photodetector. These resistances are the junction resistance and the bond contact resistance. These thermal noises are called Johnson noise sources. The fifth noise source results from the carriers' recombination process. This noise source is called generation recombination (GR) noise. This noise statistics is similar to thermal noise rather than shot noise. Despite this similarity, it is a totally different noise mechanism. In GR noise, the conductance of the material fluctuates as the number of carriers randomly changes. On the other hand, in thermal noise, the instantaneous velocity of the carrier changes, but the number of carriers is relatively constant. Hence GR noise is AM of the semiconductor conductance.

For a photodiode operating at constant bias voltage point and an average dc current of photocurrent plus dark current, the random fluctuations in conductance due to recombination and trapping of carriers will induce random fluctuations on the dc current. In Chapter 8, it was explained that a photodetector is a photoconductive detector. This noise adds somewhat to the average AM noise. This can be modeled as an additional parallel current source whose value is given by:[14]

$$i_{\mathrm{GR}}^2 = 4qI_{\mathrm{DC}}\left(\frac{\tau_{\mathrm{carrier}}}{\tau_{\mathrm{transit}}}\right)\frac{1}{1+(\omega/\omega_c)^2} \cong 4q\left|G(\omega)\right|^2 RP_{\mathrm{OP-Rx}}, \tag{10.78}$$

where, from Chapter 8, the cut-off frequency is $\omega_c = 1/\tau_{\mathrm{carrier}}$, the optical gain is the ratio between the carrier lifetime to its transit time $G(\omega) = \tau_{\mathrm{carrier}}/\tau_{\mathrm{transit}}$, $R$ is responsivity, $P_{\mathrm{OP-Rx}}$ is the received optical power, and the dc current is the sum of the dark current and the photocurrent. Remember that the photodiode is in reverse bias.

The intrinsic photodiode equivalent circuit after taking these noise sources into account in shown in Fig. 10.16.

From this schematic, the CNR can be calculated as the ratio between the photocurrent delivered to the load RL and the sum of all noise power delivered to the load.

## 10.8 APD Photodetector Noise Modeling

In Chapter 8, it was explained that APD gain is a result of ionization process due to the impact of carriers accelerated in the absorption and gain regions. The ionization process depends on the electron ionization coefficient factor. However, this process is random; thus, the APD gain is the root mean square value $\langle M^2 \rangle$[11] and is a statistical variance. Since this is a random process, there is a mean gain value as well, denoted as $M = \overline{M}$, that is given by Eq. (8.74) and is a statistical average. The power detected by an APD is related to the average gain square. The shot noise is related to the APD gain variance; hence, the power and noise can be written as:

$$P_{\mathrm{sig}} = i_{\mathrm{DC}}^2 \overline{M}^2. \tag{10.79}$$

$$P_{\mathrm{noise}} = 2qi_{\mathrm{dc}} \langle M \rangle^2 \mathrm{BW}. \tag{10.80}$$

Hence, the SNR for an APD is given by the ratio of Eq. (10.79) to Eq. (10.80):

$$\mathrm{SNR} = \frac{i_{\mathrm{DC}}^2 \overline{M}^2}{2qi_{\mathrm{DC}} \langle M \rangle^2 \mathrm{BW}}. \tag{10.81}$$

The ratio of the gain average to its variance in Eq. (10.80) is called excess noise factor $F(M)$.

In a deterministic process, the expected value of the mean is the same as the expected value of the variance. This means that both the noise and the signal are multiplied by the same factor. In other words, the SNR at the input of the APD is the same as at the output of the APD. Because of this gain fluctuation, there is an excess noise factor $F(M)$ that is related to the mean gain and the ionization ratio $k$, which is discussed in Sec. 8.7.2. Hence, the shot noise in Eq. (10.80) can be descsribed by the excess noise factor that describes the SNR degradation due to the APD gain process:

$$P_{\mathrm{noise}} = 2qi_{\mathrm{DC}} M^2 F(M). \tag{10.82}$$

Here, it is clear that $F(M) \geq 1$. Sometimes the approximation of $F(M) = M^x$ is used for Eq. (10.82), where $0.1 < x < 1$. The excess factor can be represented with respect to the effective ionization factor $k_{\mathrm{eff}}$ and the mean gain $M$ by

$$F(M) = kM + (1 - k)\left(2 - \frac{1}{M}\right). \tag{10.83}$$

Equation (10.83) is valid when the electrons are injected at the edge of the depletion layer. This expression describes a process of noise gain due to both holes and electrons. In the case of only the holes generating the impact multiplication, $k$ is replaced by $1/k$. The noise gain per $F(M)$ [Eq. (10.83)] for holes is minimized by injecting carriers with a higher ionization coefficient $k$ so that $1/k$ goes lower, and fabricating a structure with lowest $k$ value, if electrons are injected. Moreover, Eq. (10.83) is claimed to be valid even for single carrier–initiated double multiplication. This is since both types of carriers have the ability to ionize. In cases where both types of carriers are injected simultaneously, electrons and holes, then the overall response is the sum of both partial results per Eq. (10.83). Equation (10.83) is plotted in Fig. 10.17 for electrons, where $k$ is a parameter between 0 and 1. It can be seen that for $k = 0$ and a high value of $M$, $F(M)$ reaches the value of 2. For $k = 1$, the excess noise increases linearly. Therefore, the optimum $k$ value for minimum $F(M) = 2$ is $k = 0$.

From the above discussion, it is possible to estimate the SNR of APD front end. The noise at the APD front end is composed of shot noise enhanced by the excess factor $F(M)$ as well as thermal noise. Therefore, APD SNR is given by[11,12,14]

$$\mathrm{SNR} = \frac{\left(\mathrm{MRP_{opt}}\right)^2}{2qM^2F(M)\left(RP_{\mathrm{opt}} + I_{\mathrm{dark}}\right)\mathrm{BW} + 4\left(\frac{kT}{R_{\mathrm{L}}}\right)F_{\mathrm{n}}BW}, \tag{10.84}$$

where $R_{\mathrm{L}}$ is the load resistance, $k$ is the Boltzmann constant, $R$ is responsivity, and $q$ is the electron charge. APD gain can be optimized to achieve maximal SNR by setting $\partial\mathrm{SNR}/\partial M = 0$.[12] Thus, the optimal APD gain for maximal SNR is given by

$$kM_{\mathrm{opt}}^3 + (1-k)M_{\mathrm{opt}} = \frac{4\,k_{\mathrm{B}}TF_{\mathrm{n}}}{qR_{\mathrm{L}}\left(RP_{\mathrm{opt}} + I_{\mathrm{dark}}\right)}, \tag{10.85}$$

where $k$ is the ionization coefficient, $k_{\mathrm{B}}$ is the Boltzmann constant, and $F_{\mathrm{n}}$ is the noise figure of the electronics following the APD. This expression can be approximated for large $M$ values, and for $M > 10$ it is approximated to

$$M_{\mathrm{opt}} \approx \left(\frac{4\,k_{\mathrm{B}}TF_{\mathrm{n}}}{kqR_{\mathrm{L}}\left(RP_{\mathrm{opt}} + I_{\mathrm{dark}}\right)}\right)^{1/3}. \tag{10.86}$$

It can be seen that the optimal gain is sensitive to $k$.

Now there is enough information to construct the APD noise equivalent circuit with respect to the intrinsic model provided in Fig. 8.25. However, not all shot

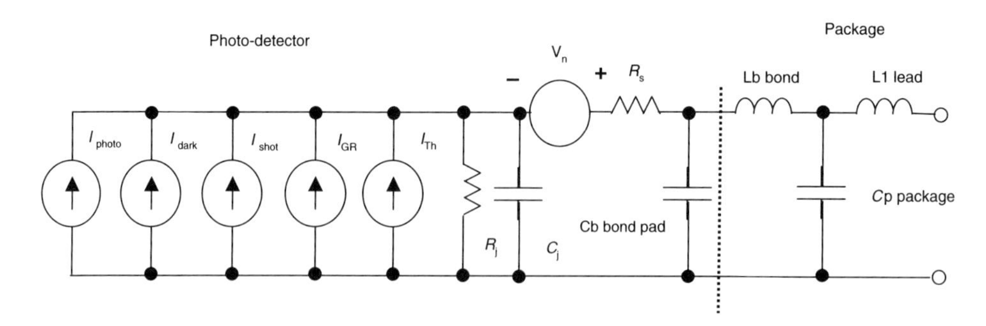


**Figure 10.16** Equivalent photodetector circuit including thermal noises, shot noises and generation recombination noise. The symbols in Fig. 10.6 are as follows; I photo is the received signal $I_{\mathrm{photo}} = RP_{\mathrm{OP-Rx}}$. Dark current shot noise is $I_{\mathrm{dark}} = \sqrt{2qI_{\mathrm{d}}}$. Shot noise current is $I_{\mathrm{shot}} = \sqrt{2qI_{\mathrm{photo}}}$ thermal noise current is $I_{\mathrm{Th}} = \sqrt{\frac{4kT}{R_{\mathrm{j}}}}$. Thermal noise voltage $V_{\mathrm{n}} = \sqrt{4kTR_{\mathrm{S}}}$. The generation recombination current, $I_{\mathrm{GR}}$, is given by Eq. (10.78).

noise currents are multiplied; therefore, the noise currents consist of a multiplied current source and a nonmultiplied current source; the current has two leakage mechanisms. The first is from the surface, which is not multiplied ($i_{dnm}$), and the other comes from the bulk and is multiplied ($i_{dm}$). The signal shot noise is fully multiplied. Similar to a PIN diode, there is thermal noise related to the APD ohmic losses. Therefore, the following shot noise relations with respect to the APD output are valid:[14]

$$i_{sig-out} = M(\omega)RP_{opt}(\omega) = M(\omega)\frac{\eta q}{h\nu}P_{opt}(\omega), \tag{10.87}$$

$$i_{dark-out} = i_{dnm} + i_{dm}M, \tag{10.88}$$

$$i^2_{n-out}(\omega) = 2q\left[i_{dnm} + i_{dm}M^2(\omega)F(M)\right] + i_{shot}M^2(\omega)F(M), \tag{10.89}$$

$$i_{shot} = M(\omega)\sqrt{2qRP_{opt}}, \tag{10.90}$$

$$i_{shot-dm} = M(\omega)\sqrt{2qi_d} = M(\omega)i_{dm}. \tag{10.91}$$

There are several important conclusions from these noise relations. First, the shot noise is amplified by $M^2$ due to the relation given by Eq. (10.90). The amplified noise is increased also by the excess factor $F(M)$. The shot noise of the multiplied dark current is given by Eq. (10.91), where $i_d$ is the dark current.

In conclusion, the APD noise equivalent circuit in Fig. 10.18 is presented with the multiplied noise sources at the primary of the transformer, while the nonmultiplied dark current noise sources are at the secondary. The transformer represents the gain $M$ and excess noise $F(M)$. Normalizing Eqs. (10.87) to (10.91) by $M(\omega)$ would provide the values for the noise sources with respect to the input.

The upper section of Fig. 10.18 shows the coupling of the multiplied and nonmultiplied shot noise. This shot noise is made of dark current, which has two contributors: the multiplied and unmultiplied leakage. In addition, there is a shot noise current source related to the received signal. The noise-multiplying gain is made of the APD current gain represented by a transformer of $M$:1, and the excess noise factor gain factor is represented by the cascaded transformer with a turn ratio of $F(M)$:1. These noises are in parallel to the signal gain transformer at the lower section of this circuit. At the secondary of these transformers, there is a thermal noise voltage source related to $R_s$, and in the upper part of this schematic, there is a thermal current noise source related to $R_j$. For CAD simulation, in order to have directivity on the coupling of noise and signal with high reverse isolation, it is useful to have a unity operational amplifier at the secondary of the signal gain transformer and the output of the $F(M)$ transformer.

## 10.9 Receiver Front-End Design Considerations

Generally speaking, the design considerations of an optical CATV receiver can be divided into two parts:

A. The interface matching between the PIN diode and the first stage of the receiver

B. RF stages, inter-stage matching, and AGC (RF lineup architecture)

The first part is the most critical one in aspects of NF. High NF affects the noise floor, which defines CNR. The second part of the process affects the receiver dynamic range and ripple. Step A involves several considerations:

1. Minimizing the total stray capacitance to GND from any element in the interface circuitry, and any input capacitance of the first stage that is translated back by the transformer across the PIN diode output ports. The result

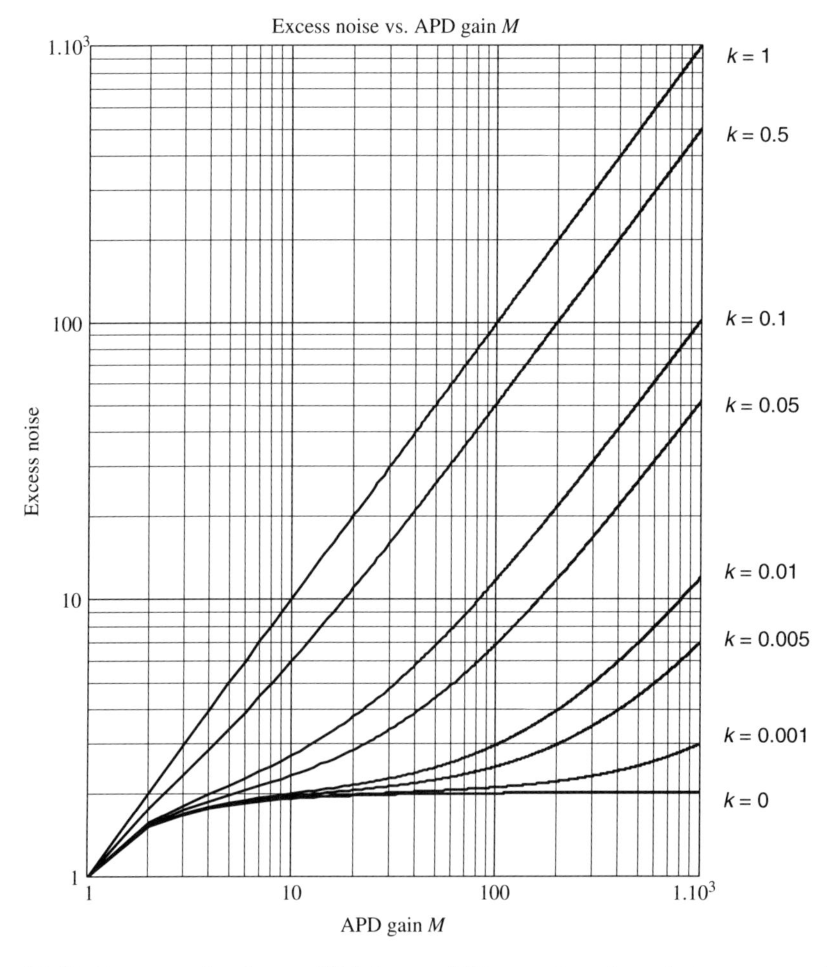


**Figure 10.17** Excess noise factor $F(M)$ for an APD as a function of the mean gain M under electron injection, while $k$ is a parameter.

of this matching is to increase the PIN photodetector frequency response. Sec. 8.2.3 provides some insight into matching.

2. Minimizing white noise current by increasing noise resistance at the photo detector port output (receiver's input). This achieves two goals: noise and impedance matching.

3. Improving dynamic range by canceling distortions.

Usually an impedance transformer with a 2:1 turn ratio is used[18] (with some matching between the photodetector and the transformer) as the interface circuit between the first stage and the photodetector. This approach is efficient since it has to operate within a BW of 50 MHz to 1 GHz, and conventional RF transformers are available.[3] A higher-turn-ratio transformer will improve noise performance, since the impedance at the photodetector port would be higher; thus, Johnson, noise would be lower and power transfer would be optimal. However, it would result in BW degradation due to the transformer stray capacitance and the photodiode output capacitance.

Figure 10.19 illustrates several typical receiver front ends. Figs. 10.19(a) and (b) present a tapping connection. The input impedance of the first gain stage, which is an LNA, is reflected in the transformer primary. Assuming the first stage is perfectly matched to the characteristic impedance $Z_0$, the reflected impedance at the primary is $N^2Z_0$, denoted as $R$.

The bearing signal photocurrent is divided between the bias resistors, marked as $R_3$ in Fig. 10.19(a) or $R_1$ in Fig. 10.19(b), to the primary impedance $R$, according to the current divider. For simplicity $R_3$ and $R_1$ are denoted as dc bias load RL. Thus, for maximum RF power transfer, it is desired to have RF load $RL\_RF = R = N^2Z_0 = R_o$, where $R_o$ is the photodetector output impedance. However, the disadvantage in this method is the loss in the dc bias load parallel to the RF load $RL\_RF$. This is because the photodetector output resistance $R_o$ is in parallel to both R and RL.

The optimized solution is provided in Fig. 10.19(c) by using a 2:1 autotransformer. In this way the reflected load, $N^2Z_0$, is in parallel to the photodetector output resistance $R_o$ without having an additional dc bias load loss. Moreover, this would improve noise performance, since the thermal noise current $4kTB/RL$, contribution of the dc bias resistor RL is omitted. It looks as if increasing the transformation ratio would improve performance and power transfer. However, the penalty would be a decrease in BW, which is demonstrated in Sec. 8.2.3; this is because of the photodetector's stray output capacitance and the nonideal transformer impairments, such as parasitic stray capacitance and increased losses versus frequency. Assuming an ideal transformer and photodetector with $R_o \to \infty$ and $Z_0 = 75\ \Omega$, then the RFFE (RF front end) BW proportion limits at first approximation are provided by Eq. (10.92) for Figs. 10.19(a) and (b), while Eq. (10.93) provides the limit

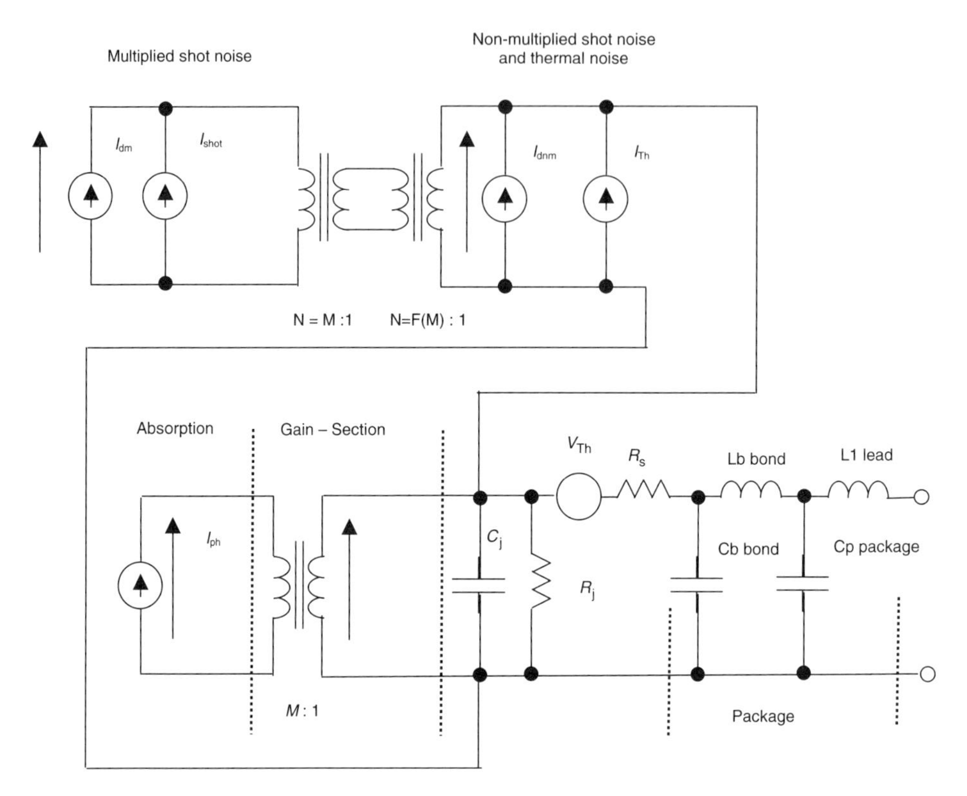


**Figure 10.18** Equivalent noise circuit for APD showing excess noise coupling for shot noise.

for the autotransformer in Fig. 10.19(c):

$$BW \propto \frac{1}{2\pi \frac{R_L \cdot N^2 75}{R_L + N^2 75} C}, \tag{10.92}$$

$$BW \propto \frac{1}{2\pi N^2 \cdot 75 \cdot C}, \tag{10.93}$$

where $C$ represents the photodetector depletion capacitance and other stray capacitance of the front end. The approximation assumption in this case is an impedance level of 75 Ω at the receiver input stage, but one should remember that the general purpose or medium power MMICs available are 50 Ω matched and dedicated CATV power stages are 75 Ω matched. From Eqs. (10.92) and (10.93), it can be seen that the configurations in Figs. 10.19(a) and 10.19(b) may have wider BW, since the equivalent resistance at the primary is lower compared to the direct current sampling in Fig. 10.19(c). Nevertheless, the thermal noise would be higher due to RL.

Increasing the value of RL in order to reduce the thermal noise and improve power transfer in the case of tapping configurations of Figs. 10.19(a) and 10.19(b) would increase distortions. As was explained previously using Eqs. (8.63) and (8.64) in Sec. 8.2.7, as the optical power increases, the dc voltage drop over RL increases. As a consequence, the photodetector reverse bias decreases, and its stray capacitance and distortions eventually increase. Hence, the role of RL is dc protection of the photodetector against optical current overdrive. Adding an inductor marked as L1 in series to the photodetector may resonate the photodetector capacitance, which was reviewed in Sec. 8.2.3.

### 10.9.1 Optimum matching with optimal transformer

As was explained in Sec. 10.7, there are several noise and current sources at the receiver front end. The design goal is to define the optimal receiver matching for gain and a minimum noise figure. Minimum noise figure is minimum noise energy and better CNR. This section will discuss the matching between the photodetector and the receiver's front-end LNA.

Photodiode noise sources are provided in Fig. 10.20, which describes its equivalent circuit, including noise sources.

The photodiode equivalent circuit contains a signal current source and noise sources of a photodiode composed of shot and thermal noise generators, which are also marked as current sources as per the following equations. The RF signal photocurrent, $i_{\mathrm{L}}$, resulting in the photodetector is given by

$$i_{\mathrm{L}} = \frac{I_{\mathrm{L}} \cdot \mathrm{OMI} \cdot r}{\sqrt{2}}. \tag{10.94}$$

The noise generators in a photodetector are described below.

1. Shot noise current, $i_{\mathrm{SL}}$, as a result of the photoelectric dc current $I_{\mathrm{L}}$:

$$\overline{i_{\mathrm{SL}}^2} = 2qI_{\mathrm{L}}B. \tag{10.95}$$

2. The dark current of photodetector, $i_{\mathrm{D}}$. This is the reverse leakage current at no illumination:

$$i_{\mathrm{D}} = I_{\mathrm{S}} \exp(-qv/kT). \tag{10.96}$$

3. The dark current shot noise, $i_{\mathrm{sd}}$. This noise current results from the leakage current, and the CNR dark test is also referred to this noise level:

$$\overline{i_{\mathrm{Sd}}^2} = 2qi_{\mathrm{D}}B. \tag{10.97}$$

4. The Johnson noise resulting from $R_{\mathrm{P}}$, where $R_{\mathrm{P}}$ is the photodetector output resistance. The contact resistance noise voltage of the photodetector is omitted in this analysis and is assumed to be very low:

$$\overline{i_{\mathrm{JP}}^2} = 4kTB/R_{\mathrm{P}}. \tag{10.98}$$

To measure the dark CNR value of an optical receiver when the receiver input has no light, the equivalent noise current $i_{\mathrm{eq}}$ is measured. This value represents the

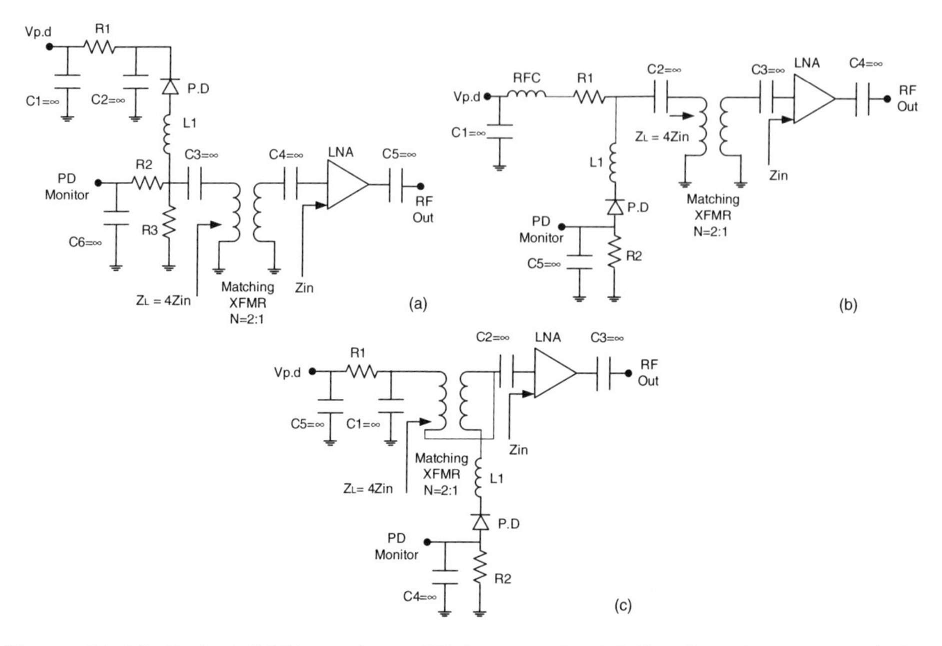


**Figure 10.19** Typical CATV receivers RF front ends. (a) Tapping the anode of the photodetector, (b) tapping the Cathode of the photodetector, (c) direct Serial Current Sampling.

receiver's intrinsic noise. Thus, the relevant noise sources are the dark current shot noise $i_{sd}$, the white noise $i_{JP}$, and the two equivalent noise sources of the front-end input amplifier, which is the LNA. The front-end LNA noise components $e_n$ and $i_n$ are the uncorrelated noise sources. The assumption is that noncorrelated noise sources do not reduce the general presentation of the problem but simplify the calculations to some extent. That assumption is not far from reality in the case of low-frequency amplifiers.

The CNR is calculated by the method of superposition while shorting the output of the secondary for calculating the short noise current at the input as per Fig. 10.21. The turn ratio of the RFFE transformer is $N$:1. The signal $S$ is given by:

$$\overline{i_L^2} \equiv S = \left(\frac{I_L \cdot \mathrm{OMI} \cdot r}{\sqrt{2}}\right)^2 \cdot N^2. \tag{10.99}$$

It is clear that the signal is independent of $R_P$, the photodetector equivalent resistance, since the load at the secondary is short. This is due to checking the photodetector as a current source of signal and noise by using the superposition method. The noise $N_e$ at the secondary is given by

$$N_e = \overline{i_{sd}^2} \cdot N^2 + \overline{i_{JP}^2} \cdot N^2 + \overline{e_n^2} \cdot \frac{N^4}{R_P^2} + i_n^2. \tag{10.100}$$

Note that the noise voltage develops a noise current across the imaged $R_P$ at the secondary in Eq. (10.100), while outputs are not shorted; hence, the CNR observing the primary is given by

$$\frac{S}{N} = \frac{\overline{i_L^2}}{\overline{i_{sd}^2} + \overline{i_{JP}^2} + \overline{e_n^2} \cdot \left(N^2/R_P^2\right) + \left(\overline{i_n^2}/N^2\right)} \equiv \frac{\overline{i_L^2}}{M}. \tag{10.101}$$

The maximum value of CNR ($S/N$) occurs when $M$ is at its minimum value; hence, there is an optimum value of turns ratio for $N$:

$$\frac{\partial M}{\partial N} = 2N\frac{\overline{e_n^2}}{R_P^2} - 2N^{-3} \cdot \overline{i_n^2} = 0, \tag{10.102}$$

which results in

$$N_{opt}^4 = \frac{\overline{i_n^2}}{\overline{e_n^2}} R_P^2. \tag{10.103}$$

But the ratio $\overline{e_n^2}/\overline{i_n^2}$ is the optimal noise resistance $R_{opt}$; therefore, the optimal matching turn ratio is given by

$$N_{opt} = \sqrt{\frac{R_P}{R_{opt}}}. \tag{10.104}$$

In conclusion, the photodetector noise sources $i_{sd}$ and $i_{JP}$ do not affect and define the transformer turns ratio in order to achieve maximum CNR value. The transformer impedance transformation guarantees that the RFFE LNA would have the optimal noise impedance $R_{opt}$ at its input port. If we neglect the photodetector noise sources, the maximum CNR is given by

$$\begin{aligned}\frac{S}{N} &= \frac{\overline{i_L^2}}{\overline{i_{sd}^2} + \overline{i_{JP}^2} + \overline{e_n^2} \cdot \left[1/(R_P \cdot R_{opt})\right] + \left(\overline{i_n^2} \cdot R_{opt}/R_P\right)} \\ &\approx \frac{\overline{i_L^2}}{2 \cdot \left(R_{opt}/R_P\right) \cdot \overline{i_n^2}}.\end{aligned} \tag{10.105}$$

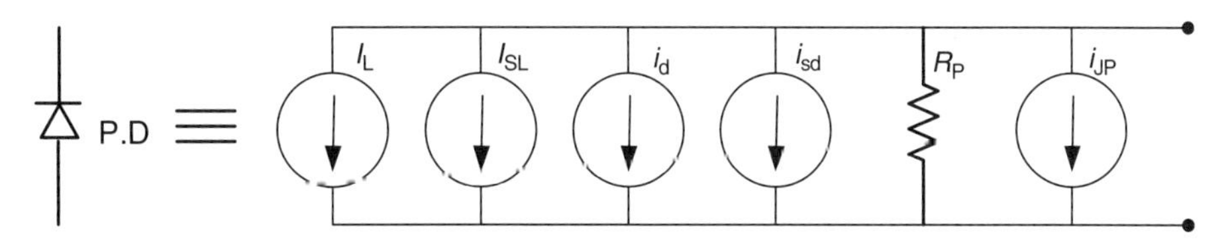


**Figure 10.20** Noise sources of the photodetector.

The denominator value is called the equivalent noise density:

$$i_{eq}\left[pA/\sqrt{\text{Hz}}\right] = \sqrt{2 \cdot \frac{R_{opt}}{R_P} \cdot \overline{i_n^2} \cdot 10^{12}}. \tag{10.106}$$

## 10.9.2 Numerical example

Consider a commercial MOSFET or transistor used as an amplifier with the following parameters:

$$F_{min} = 3 \text{ dB noise figure}, \tag{10.107}$$

$$e_n, \text{ equivalent impedance}, R_n = 25\ \Omega, \tag{10.108}$$

$$\overline{e_n^2} = 4kTBR_n, \tag{10.109}$$

$$i_n, \text{ equivalent conductance}, G_u = 0.01[1/\Omega], \tag{10.110}$$

$$\overline{i_n^2} = 4kTBG_u. \tag{10.111}$$

From the above data and previous relations, the optimal noise resistance is given by

$$R_{opt} = \sqrt{\frac{\overline{e_n^2}}{\overline{i_n^2}}} = \sqrt{\frac{R_n}{G_u}} = \sqrt{\frac{25}{0.01}} = 50\ \Omega. \tag{10.112}$$

The output parallel resistance of the photodetector is $R_P \gg R_{opt}$. For the sake of discussion, it is assumed to be $R_{opt} = 10\ \text{K}\Omega$. The equivalent noise current in this case would be

$$i_{eq}\left[pA/\sqrt{\text{Hz}}\right] = \sqrt{2 \cdot \frac{R_{opt}}{R_P} \cdot \overline{i_n^2} \cdot 10^{12}} = \sqrt{2 \cdot \frac{50}{10^4} \cdot 4kTG_u} \approx 1.26\ pA/\sqrt{\text{Hz}}. \tag{10.113}$$

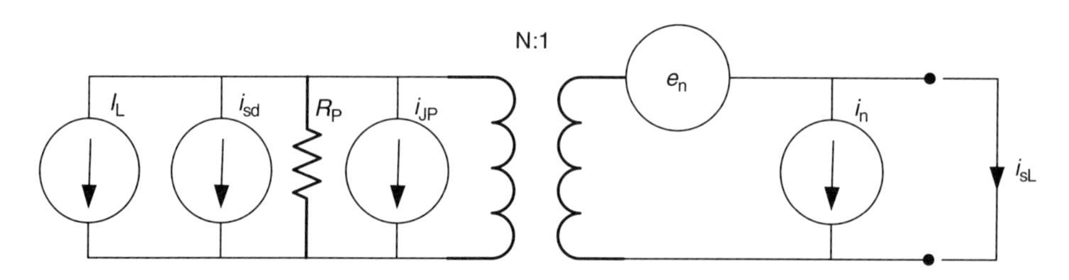


**Figure 10.21** Noise sources of an optical receiver.

The result shows a very low noise current due to a high transformation ratio:

$$N_{\text{opt}} = \sqrt{\frac{R_{\text{P}}}{R_{\text{opt}}}} = \sqrt{\frac{10^4}{50}} = \sqrt{200} \approx 14. \tag{10.114}$$

Practically, such a ratio is impossible to achieve due to the photodetector's parasitic capacitance $C$ as well as the transformer's limitation over BW, such as parasitic capacitance and losses.

### 10.9.3 Photodetector shunt capacitance

Assume a stray capacitance $C$ at the photodetector output parallel to $R_{\text{P}}$. The solution to this problem follows the method presented in Sec. 10.9.1. Equation (10.115) is evaluated under the same approximation taken for Eq. (10.105), which neglects the photodiode noise sources. Figure 10.22 illustrates the equivalent circuit in this discussion. The signal level is given by Eq. (10.99). Under these conditions, the noise power at the secondary is given by

$$\begin{aligned} N_{\text{e}} &= \overline{i_{\text{sd}}^2} \cdot N^2 + \overline{i_{\text{JP}}^2} \cdot N^2 + \overline{e_{\text{n}}^2} \cdot \frac{N^4}{\left(R_{\text{P}}^2 \cdot \dfrac{1}{\omega^2 c^2} \Big/ R_{\text{P}}^2 + \dfrac{1}{\omega^2 c^2}\right)} + \overline{i_{\text{n}}^2} \\ &\approx \frac{(1 + \omega^2 c^2 R_{\text{P}}^2)\overline{e_{\text{n}}^2} \cdot N^4}{R_{\text{P}}^2} + \overline{i_{\text{n}}^2}. \end{aligned} \tag{10.115}$$

Assuming the quality factor satisfies $Q = \omega c R_{\text{P}} \gg 1$, the CNR is given by

$$\frac{S}{N} = \frac{\overline{i_{\text{L}}^2}}{\overline{e_{\text{n}}^2} \cdot \dfrac{N^2}{1/\omega^2 c^2} + \dfrac{\overline{i_{\text{n}}^2}}{N^2}} \equiv \frac{\overline{i_{\text{L}}^2}}{M}. \tag{10.116}$$

This result is similar to Eq. (10.100), and $1/\omega^2 c^2$ replaces $R_{\text{P}}^2$. The optimal transformer is evaluated as for Eq. (10.102) and is given by

$$N_{\text{opt}} = \sqrt{\frac{1}{R_{\text{opt}} \cdot \omega c}}. \tag{10.117}$$

The equivalent noise current is given by

$$i_{\text{eq}}\left[pA/\sqrt{\text{Hz}}\right] = \sqrt{2 \cdot \frac{\overline{i_{\text{n}}^2}}{N^2} \cdot 10^{12}} = \sqrt{2\omega c R_{\text{opt}} \overline{i_{\text{n}}^2}}. \tag{10.118}$$

Using the above numerical example, and assuming a photodetector with shut capacitance of 1 pf, and a receiver with a maximum operating frequency of

860 MHz, the optimal transformer would be calculated at the maximum frequency as

$$N_{\text{opt}} = \sqrt{\frac{1}{2\pi \cdot 860 \cdot 10^6 \cdot 10^{-12} \cdot 50}} \approx 1.924. \tag{10.119}$$

Hence, the equivalent noise density is given by

$$i_{\text{eq}}\left[pA/\sqrt{\text{Hz}}\right] = \sqrt{2 \cdot \frac{\overline{i_{\text{n}}^2}}{N^2} \cdot 10^{12}} = \sqrt{\frac{2}{1.924^2} 4kTG_{\text{u}}} = 9.3\ pA/\sqrt{\text{Hz}}. \tag{10.120}$$

The noise result is a bit higher and the transformer is ideal. This noise current can be reduced by using a series inductor between the photodetector and the transformer. This kind of matching inductor is called a peaking inductor. Interestingly, the transformer turn ratio is commonly used in practical receivers.

### 10.9.4 The matching inductor for noise reduction

The next step in this front-end transformer analysis is adding inductive matching, such as the pickup inductor L illustrated in Fig. 10.23. The circuit can be analyzed in two ways, by using the Norton equivalent circuit, as in Fig. 10.24, or the Norton identities, shown in Fig. 10.25.

By using the Norton equivalent, the current source at the primary, shown in Fig. 10.24, can be converted into a current source with the output reactance $Y_{\text{e}}$ and equivalent current $i_{\text{e}}$. The equivalent reactance of the photodetector is solved by looking from the inductor port shown in Fig. 10.24 leftside into the photodiode series inductor and parallel capacitor and resistor pair. So there is an inductor in series to the parallel pair of $C$ and $R_{\text{P}}$, after calculating the impedance Z seen from the port the reactance is the 1/2 value given by Eq. (10.121).

$$Y_{\text{e}} = \frac{\dfrac{1}{j\omega L}\left(\dfrac{1}{R_{\text{P}}} + j\omega C\right)}{\dfrac{1}{j\omega L} + \dfrac{1}{R_{\text{P}}} + j\omega C} = \frac{1 + j\omega C R_{\text{P}}}{R_{\text{P}}(1 - \omega^2 LC) + j\omega L}. \tag{10.121}$$

The equivalent current is calculated by current divider when the output of the photodetector is shorted. This is because the short current test is performed while the secondary of the transformer is shorted, as was explained for Figs. 10.21 to 10.23; hence, ac-wise, it is reflected at the transformer primary as a short:

$$i_{\text{e}} = i_{\text{L}} \cdot \frac{1/j\omega L}{(1/j\omega L) + (1/R_{\text{P}}) + j\omega C} = \frac{i_L}{1 - \omega^2 LC + j(\omega L/R_{\text{P}})}. \tag{10.122}$$

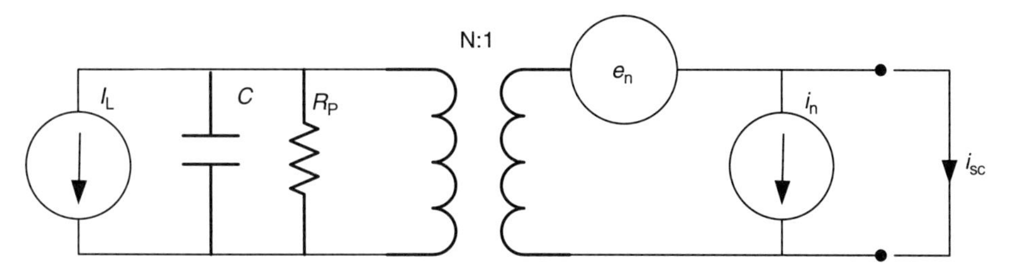


**Figure 10.22** Photodetector with shunt capacitor equivalent circuit.

Hence, the signal current square at the primary, which is proportional to power, would be

$$S = \overline{i_e^2} = \frac{\overline{i_L^2}}{(1 - \omega^2 LC)^2 + (\omega L/R_P)^2}. \tag{10.123}$$

The noise power reflected to the primary would be

$$N_e = \frac{\overline{i_n^2}}{N^2} + N^2 \overline{e_n^2} \frac{1 + \omega^2 C^2 R_P^2}{R_P^2(1 - \omega^2 LC) + \omega^2 L^2}. \tag{10.124}$$

Thus, the signal-to-noise ratio is given by

$$\frac{S}{N_e} = \frac{\overline{i_L^2}}{\frac{\overline{i_n^2}}{N^2}\left[(1 - \omega^2 LC) + \frac{\omega^2 L^2}{R_P^2}\right] + \frac{N^2 \overline{e_n^2}}{R_P^2}\left(1 + \omega^2 C^2 R_P^2\right)} = \frac{\overline{i_L^2}}{M}. \tag{10.125}$$

The CNR is maximal when $M$ is minimal. For this reason, the optimal $L$ is calculated by

$$\frac{\partial M}{\partial L} = \frac{\overline{i_n^2}}{N^2}\left[-2\omega LC(1 - \omega L^2 C) + 2\frac{\omega^2 L}{R_P^2}\right] = 0. \tag{10.126}$$

The result for the optimal $L$ is given by

$$L_{OPT} = \frac{R_P^2}{1 + \omega^2 c^2 R_P^2}. \tag{10.127}$$

For a high $Q$, the optimum $L$ value is given by

$$L_{OPT} = \frac{1}{\omega^2 c}. \tag{10.128}$$

At high $Q$ case, the CNR is given by

$$\frac{S}{N_{\mathrm{e}}} \approx \frac{\overline{i_{\mathrm{L}}^2}}{\frac{\overline{i_{\mathrm{n}}^2}}{N^2}\frac{\omega^2 L^2}{R_{\mathrm{P}}^2} + N^2\overline{e_{\mathrm{n}}^2}\omega^2 C^2} = \frac{\overline{i_{\mathrm{L}}^2}}{M}. \tag{10.129}$$

In the same case, $S/N_{\mathrm{e}}$ is maximal when $M$ is minimal; thus, the optimal turns-ratio $N$ is defined by

$$\frac{\partial M}{\partial N} = -2\frac{\overline{i_{\mathrm{n}}^2}}{N^3}\frac{\omega^2 L}{R_{\mathrm{P}}^2} + 2N\overline{e_{\mathrm{n}}^2}\omega^2 C^2 = 0. \tag{10.130}$$

Knowing that the ratio $\overline{e_{\mathrm{n}}^2}/\overline{i_{\mathrm{n}}^2} = R_{\mathrm{opt}}^2$, and by substituting the value of $L_{\mathrm{OPT}}$, the optimal transformation ratio is given by

$$N_{\mathrm{OPT}} = \frac{1}{\omega C\sqrt{R_{\mathrm{p}}R_{\mathrm{opt}}}}. \tag{10.131}$$

Substituting all those optimal values in the CNR relation results in the

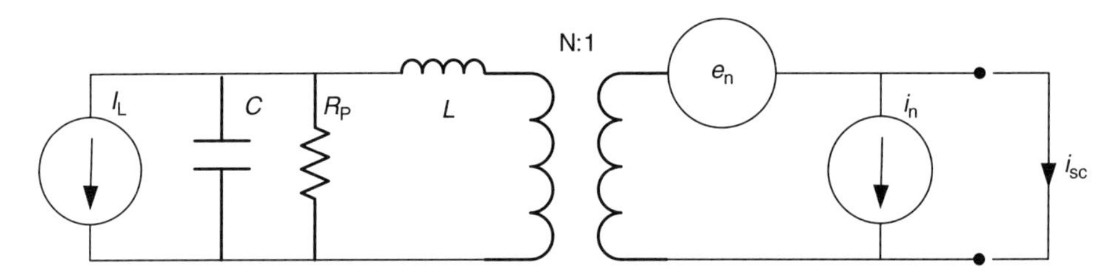


**Figure 10.23** Photodetector noise modeling with peaking inductor compensation.

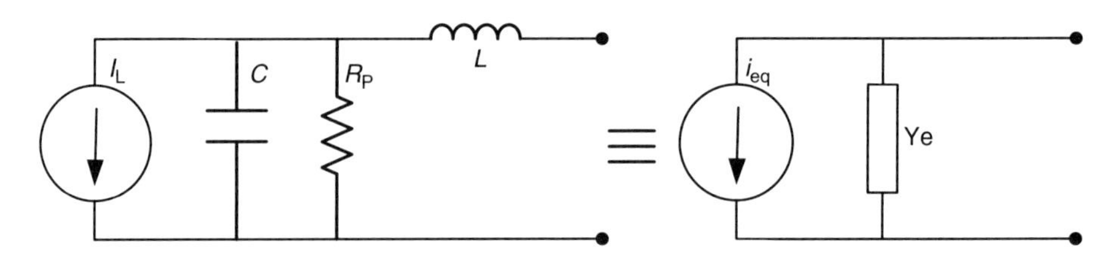


**Figure 10.24** Photodetector with shunt capacitor and matching inductor Norton equivalent circuit.

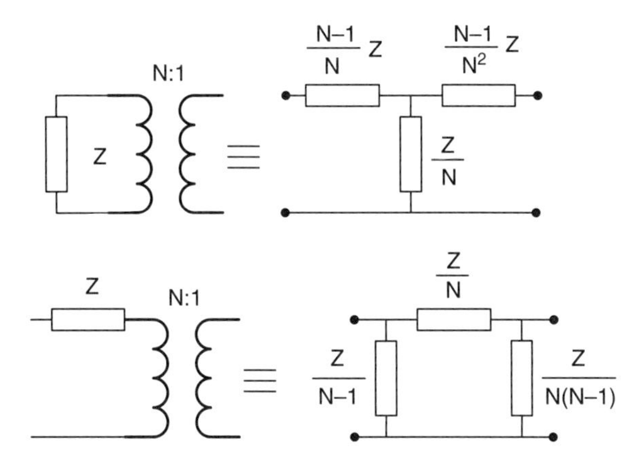


**Figure 10.25** Norton identities, pay attention that the equivalent network always contains an element with negative impedance.

optimum CNR expression:

$$\frac{S}{N} \approx \frac{\overline{i_{\mathrm{L}}^2}}{\left(\overline{i_{\mathrm{n}}^2}/R_{\mathrm{P}}^2\right)\left[\omega^4 L^2 C^2 R_{\mathrm{P}} R_{\mathrm{opt}} + R_{\mathrm{P}} R_{\mathrm{opt}}\right]}$$

$$= \frac{\overline{i_{\mathrm{L}}^2}}{2\left(\overline{i_{\mathrm{n}}^2}/R_{\mathrm{P}}^2\right) R_{\mathrm{P}} R_{\mathrm{opt}}} = \frac{\overline{i_{\mathrm{L}}^2}}{2\left(R_{\mathrm{opt}}/R_{\mathrm{P}}\right)\overline{i_{\mathrm{n}}^2}}. \tag{10.132}$$

Going back to the numerical example and to examining the results, let us calculate the optimal $N$ value at 860 MHz:

$$N_{\mathrm{OPT}} = \frac{1}{\omega C \sqrt{R_{\mathrm{P}} R_{\mathrm{opt}}}} = \frac{1}{2\pi \cdot 860 \cdot 10^6 \cdot 10^{12} \sqrt{10^4 50}} = 0.26. \tag{10.133}$$

This result looks a bit strange, $N < 1$, but there is a simple explanation. Each parallel RC has an equivalent series value of RC, at a discrete frequency as shown in Fig. 10.26. Observing the parallel RC versus series RC impedance,

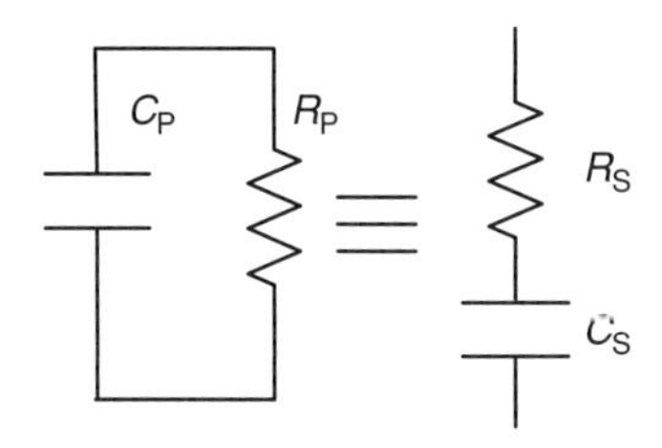


**Figure 10.26** Series and parallel modeling of a photodetector impedance.

and comparing coefficients at high $Q$, results in the following formulas:

$$\frac{R_P(1/j\omega C_P)}{R_P + (1/j\omega C_P)} = R_S + \frac{1}{j\omega C_S}, \tag{10.134}$$

which yields

$$\frac{R_P - j\omega C_P R_P^2}{1 + \omega^2 C_P^2 R_P^2} = R_S - j\frac{1}{\omega C_S}. \tag{10.135}$$

After a comparison of coefficients, it becomes

$$R_S \approx \frac{1}{\omega^2 C_P^2 R_P}, \qquad C_P \approx C_S. \tag{10.136}$$

After resonating the capacitor, there is a very small residual resistance, $R_S$. An explanation for that can be achieved by using the Smith chart. The series-resonating LC is equal to a $\lambda/4$ transformer; hence, there are two impedance transformers ratios:

$$\frac{R_P}{R_S} \approx \omega^2 C_P^2 R_P^2, \qquad N^2 \approx \frac{1}{\omega^2 C^2 R_P R_{opt}}. \tag{10.137}$$

Hence,

$$\frac{R_P}{R_S} N^2 = \frac{\omega^2 C_P^2 R_P^2}{\omega^2 C^2 R_P R_{opt}} = \frac{R_P}{R_{opt}}. \tag{10.138}$$

It can be observed that both transformers match the input amplifier to its optimal noise impedance. Therefore, the same $S/N$ results, as in the first example. The question is: Is it practical? The answer is no. First, what is the logic behind doing two transformations opposing each other? Meaning, one ratio to increase the impedance, and one to reduce it. That would increase the losses due to the finite $Q$ value of the transformer. Additionally, this reflects a singular matching at a single frequency, and we are dealing with wideband matching. Moreover, when dealing with such a high $Q$ impedance value, it is important to consider the transformer parasitic values, such as ferrite losses, capacitance, etc. It is difficult, if not impossible, to find transformer vendors who provide a wide band equivalent circuit to their transformers.

The best way is to characterize the transformer is by using a vector network analyzer (VNA) and a deembedding technique to derive the equivalent transformer circuit after getting the scattering parameters of the transformer. This way, the optimum inductor value $L$ can be estimated.

## 10.10 Main Points of this Chapter

1. Noise is a random process of an interfering signal accompanying the information signal.

2. There are several noise statistics: thermal noise (known as Gaussian white noise or Johnson noise) shot noise (or Schottky noise) related to the dc current across a semiconductor junction, generation recombination noise related to a carrier's generation recombination process, and the laser's relative intensity noise (known as RIN).

3. Thermodynamic equilibrium is a condition of no current.

4. In semiconductors, there are two noise sources: commonly analyzed noise voltage and noise current. The relation between the two is measured by the correlation factor.

5. Another noise mechanism is colored noise, which is also known as $1/f^n$ noise or flicker noise, where $n = 1, 2$.

6. The concept of colored noise is that any white noise can be transformed to a frequency-dependent noise when passing through a frequency dependent transfer function, that shapes the noise such as a filter. This concept is known as Paley-Wiener.

7. Noise equivalent bandwidth (NEB) is the noise energy ratio between the realistic network energy passing through to the half-power BW.

8. CNR or SNR is the ratio of the signal to the corrupting noise power at a given receivers state.

9. Noise factor $F$ is the ratio of the input CNR and the output CNR.

10. Any series element in a network will reduce CNR.

11. One of the methods to measure noise is to estimate the equivalent noise current density at the front-end of a receiver in $pA/\sqrt{\text{Hz}}$.

12. CNR curves are useful in estimating the equivalent noise current density.

13. Reduction in photodiode responsivity reduces CNR by $20\log(R_1/R_2)$, where $R_1$ and $R_2$ are responsivities; thus, a 10% responsivity reduction is a 1 dB loss in CNR.

14. In order to reduce the noise density, it is better to have high impedance at the receiver's front end.

15. Higher impedance at the front end will reduce BW because of the RC increase. However, since there is better matching, the power transfer is maximized and the noise current is minimized.

16. Dark current and photocurrent generate shot noise; thus, it is important to have a low-leakage current at the photodetector, which is the dark current.

17. In analog systems, the RIN requirement is in the range of −155 dB/Hz. In digital systems, which are less demanding, the RIN is between −100 and −120 dB/Hz.

18. Thermal noise in a photodiode results from ohmic losses. Such losses are contact resistance and junction resistance.

19. In an APD, the dark current shot noise is composed of multiplied and unmultiplied leakage currents.

20. Shot noise in APD is multiplied by $M^2$.

21. The excess noise factor is a measure of noise penalty during amplification. It results from the random process of noise multiplication and is dependent on the ionization factor $k$.

22. The minimum excess noise is $k = 0$. See Fig. 10.7.

23. In an APD and PIN, photodetector noise source currents are parallel to the signal current source; thus, it is much more clear that any additional noise mechanism reduces CNR.

24. RIN is an external coupled noise generated by the laser. This value defines the input CNR and therefore defines the best-case input CNR. See Fig. 10.5.

25. The higher the OMI, the better the CNR, since OMI determines the signal current, while the dc photocurrent defines the shot noise.

26. An additional mechanism of noise and IMD is acoustical noise, which results in AM residual modulation. This noise is known as microphonics.

27. NEP is a figure of merit defining the thermal noise level for CNR = 1 under a known receiver photodetector responsivity.

## References

1. Rothe, H., and W. Dahlke, "Theory of noisy fourpoles." *Proceedings of the IRE*, Vol. 44 No. 6, pp. 811–818 (1956).
2. "IRE standards on methods of measuring noise in linear twoports, 1959." *Proceedings of the IRE*, Vol. 48 No. 1, pp. 60–68 (1960).
3. Haus, H.A., W.R. Atkinson, G.M. Branch, W.B. Davenport, W.H. Fonger, W.A. Harris, S.W. Harrison, W.W. McLeod, W.K. Stodola, and

T.E. Talpey, "Representation of noise in linear twoports." *Proceedings of the IRE*, Vol. 48 No. 1, pp. 69–74 (1960).
4. Carlson, A.B., P.B. Crilly, and J. Rutledge, *Communications Systems, 4th Ed.* McGraw-Hill (2001).
5. Taub, H., and D.L. Schilling, *Principles of Communication Systems, 2nd ed.* McGraw-Hill (1986).
6. Davenport, W.B., and W.L. Root, *An Introduction to the Theory of Random Signals and Noise*. Wiley-IEEE Press (1987).
7. Zverev, A.I., *Handbook of Filters Synthesis*. John Wiley & Sons (2005).
8. Ha., T.T., *Solid State Microwave Amplifier Design*. Wiley Interscience (1981).
9. Schwartz, M., *Information Transmission Modulation and Noise: A Unified Approach to Communication Systems.* McGraw-Hill (1980).
10. Sze, S.M., *Physics of Semiconductor Devices, 2nd Ed.* Wiley Interscience (1981).
11. Saleh, B.E.A., and M.C. Teich, *Fundamentals of Photonics.* Wiley-Interscience (1991).
12. Agrawal, G.P., *Fiber-Optic Communication Systems.* John Wiley & Sons (1997).
13. Ovadia, S., *Broadband Cable TV Access Networks.* Prentice Hall PTR (2001).
14. Alexander, S.B., *Optical Communication Receiver Design.* SPIE Press, Bellingham, WA (1997).
15. Bass, M., J.M. Enoch, E.W. Van Stryland, and W.L. Wolf, *Handbook of Optics Volume IV*, McGraw Hill (2000).
16. Papoulis, A., *Signal Analysis.* McGraw-Hill (1977).
17. Wood, D., *Optoelectronic Semiconductor Devices.* Prentice–Hall (1994).
18. Blauvelt, H., I. Ury, D.B. Huff, and H.L. Loboda, "Broadband receiver with passiner tuning network." U.S. Patent 5179461 (1993).

# Chapter 11

# Amplifier Analysis and Design Concepts

In the previous chapters, it was explained that the RF front end (RFFE) plays a major role in defining the receiver performance. It was demonstrated that the system carrier-to-noise ratio (CNR) performance is significantly dependent on the front-end input low-noise amplifier's (LNA's) NF. Furthermore, it was seen that there is an optimum matching for NF by an input matching transformer. It was demonstrated in Chapter 9 that the receiver's dynamic range is highly dependent on the RF chain line-up and the location of the automatic gain control (AGC) attenuator. (Further AGC analysis will be provided in Chapter 12.) Those factors affect the overall system performance and the link budget. Additionally, the topology of the video receiver defines its linearity and its distortion-free dynamic range. This section will focus on the design and tradeoffs of gain blocks and equalizers used as building blocks of a receiver. It will explain how to achieve good noise performances of gain and ripple. Gain block design is essential to achieving a receiver with high performance. Emphasis on noise analysis will be provided in Sec. 11.11 for various design cases and CMOS RF cases.

## 11.1 Noise Parameters of a Two-Port Device

Noise analysis of such two-port networks was investigated by several authors, and the results of those analyses are provided in Refs. [1–5,17]. Any linear noise hybrid network can be represented by a quiet network and two equivalent noise sources, $e_n$ and $i_n$, at its input, as described in Fig. 11.1. The input signal is represented by the current source $I_s$ and its output impedance $Y_s = G_s + jB_s$. The dashed line is a virtual short for the calculation of the signal short current and noise short current in order to derive the signal-to-noise ratio (SNR). The method of calculation is by superposition, since this is a linear network. In superposition, current sources are left open, and voltage sources are shorted. The only remaining connected source to the network is the one being calculated. With the contributions of every source, they are summed as a full result. This is how to calculate the noise factor $F$ for an active device, which is explained later.

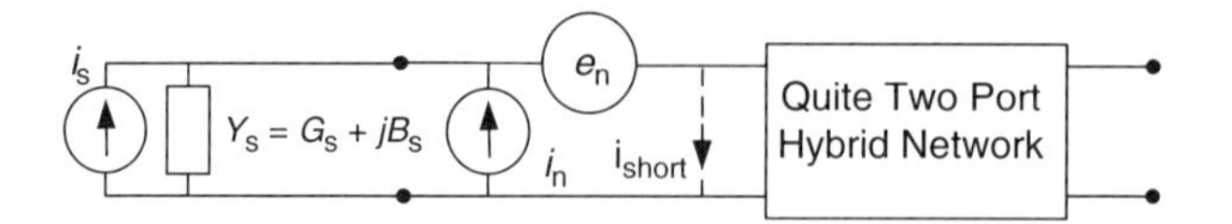


**Figure 11.1** Hybrid network representation with noise.

The two independent noise sources, $e_n$ and $i_n$, are characterized by four values:

- $\overline{e_n^2}$ and $\overline{i_n^2}$, which are real numbers that represent the average noise power density of each noise source.
- $\overline{e_n \cdot i_n^*}$, which represents the correlation between $e_n$ and $i_n$; it is a complex number (it is not the correlation factor since it is not normalized).
- $\overline{e_n^2}$, which is represented by an equivalent noise resistance $R_n$, according to the relation presented in Chapter 10 and provided by:

$$\overline{e_n^2} = 4kT_0\Delta f R_n, \tag{11.1}$$

where, $K$ is Boltzman's constant, $T_0$ is the absolute Kelvin temperature at room temperature, and $\Delta f$ is the bandwidth (BW), sometimes noted as B.

The noise current $i_n$ is composed of two components:

First is $I_u$, which is the noncorrelated part and orthogonal to $e_n$ and represented by an equivalent noise conductance $G_u$, according to

$$\overline{i_u^2} = 4kT_0\Delta f G_u. \tag{11.2}$$

The second current component is $i_n - i_u$, which is correlated with $e_n$ and is represented by the correlation admittance $Y_c$:

$$Y_c = G_c + jB_c. \tag{11.3}$$

The correlated current value is given by

$$i_c = i_n - i_u = Y_c e_n. \tag{11.4}$$

This is a complex vector equation in the complex numbers domain. Additionally, it can be noted that the correlation coefficient between the noise voltage $e_n$ and noise current $i_n$ is given by

$$\gamma = \frac{\overline{e_n \cdot i_n^*}}{\sqrt{\left|\overline{i_n^2}\right| \cdot \left|\overline{e_n^2}\right|}}, \tag{11.5}$$

where the sign $*$ marks a complex conjugate.

Using the correlation ratio from Eq. (11.5) and using Eq. (11.4), the correlation admittance can be found by

$$Y_c^* = \gamma\sqrt{\frac{\overline{|I_n^2|}}{\overline{|e_n^2|}}} = \frac{\overline{e_n \cdot i_n^*}}{\overline{|e_n^2|}}. \tag{11.6}$$

Hence, by using Eqs. (11.1) and (11.4), it can be noted that

$$\overline{e_n \cdot i_n^*} = \overline{e_n^2} Y_c^* = 4kT_0 \Delta f R_n Y_c^*. \tag{11.7}$$

From Eqs. (11.2) and (11.4), and due to the orthogonal condition between correlated and noncorrelated current, it can be noted that the total noise current is given by

$$i_n^2 = i_u^2 + i_c^2 = 4kT_0 \Delta f \left(G_u + R_n |Y_c|^2\right). \tag{11.8}$$

The noise factor $F$ is defined as the ratio of the input CNR to output CNR. Since the signal for the input and output CNR is the same but the amount of noise power contributed by the hybrid network is different, $F > 1$. The input noise power is the signal source noise. The output noise referenced to the device input is the sum of the device noise current power, the noise voltage power, and the source noise power densities. Hence, the following definition applies for $F$:

$$F = \frac{\overline{i_s^2} + \overline{i_{n1}^2}}{\overline{i_s^2}} = 1 + \frac{\overline{i_{n1}^2}}{\overline{i_s^2}}, \tag{11.9}$$

where $\overline{i_s^2}$ is the source noise power, and $\overline{i_{n1}^2}$ is the device total noise power at its input. Using the virtual short in Fig. 11.1, it can be noted that the total noise current due to noise voltage source and noise current source is given by

$$F = 1 + \frac{|i_n + Y_s e_n|^2}{\overline{i_s^2}}. \tag{11.10}$$

Rearranging Eq. (11.10) would result in

$$F = 1 + \frac{(i_n + Y_s e_n)(i_n + Y_s e_n)^*}{\overline{i_s^2}} = 1 + \frac{\overline{i_n^2} + \overline{Y_s^2 e_n^2} + Y_s e_n i_n^* + Y_n^* e_n^* i_n}{\overline{i_s^2}}. \tag{11.11}$$

Using the correlation and correlation impedance definitions in Eqs. (11.5) and (11.6), Eq. (11.11) can be written as

$$F = 1 + \frac{(i_n + Y_s e_n)(i_n + Y_s e_n)^*}{\overline{i_s^2}} = 1 + \frac{\overline{i_n^2} + \overline{Y_s^2 e_n^2} + Y_s e_n^2 Y_c^* + Y_s^* e_n^2 Y_c}{\overline{i_s^2}}. \tag{11.12}$$

The source impedance $Y_s$ is given by

$$Y_s = G_s + jB_s. \tag{11.13}$$

The correlation impedance $Y_c$ is given by

$$Y_c = G_c + jB_c. \tag{11.14}$$

The term $\overline{i_s^2}$ represents the signal source–thermal noise power, which is given by

$$\overline{i_s^2} = 4kT_sG_s\Delta f \quad \text{if} \quad T_s = T_0. \tag{11.15}$$

Taking Eqs. (11.1), (11.8), (11.13), (11.14), and (11.15) into Eq. (11.12) results the following expression for $F$:

$$F = 1 + \frac{G_u + R_n\left|G_c + jB_c\right|^2 + R_n\left|G_s + jB_s\right|^2 + R_n[(G_s + jB_s)(G_c - jB_c) + (G_s - jB_s)(G_c + jB_c)]}{G_s} \tag{11.16}$$

After some manipulations, it becomes

$$F = 1 + \frac{G_u}{G_s} + R_n\frac{(G_c + G_s)^2 + (B_c + B_s)^2}{G_s}. \tag{11.17}$$

The partial derivatives of Eq. (11.17) per $G_s$ and $B_s$ provide the optimal load condition for a minimum noise figure:

$$\frac{\partial F(G_s, B_s)}{\partial G_s} = 0, \quad \frac{\partial F(G_s, B_s)}{\partial B_s} = 0. \tag{11.18}$$

Let us assume that $Y_{opt}$ is given by

$$Y_{opt} = G_{opt} + jB_{opt}. \tag{11.19}$$

The solution for the load condition, given by the Eq. (11.18), results in values for the real and imaginary part of the optimal impedance $Y_{opt}$, which is given by Eqs. (11.20) and (11.21), respectively:

$$G_{opt} = \sqrt{G_c^2 + \frac{G_u}{R_n}}, \tag{11.20}$$

$$B_{opt} = -B_c. \tag{11.21}$$

An expression for minimum noise factor $F$ can be derived by substituting the results of Eqs. (11.20) and (11.21) into Eq. (11.17), after it has been manipulated to a convenient set of identities:

$$F = 1 + \frac{R_n}{G_s}\left[\frac{G_u}{R_n} + G_c^2\right] + 2R_nG_c + R_nG_s. \tag{11.22}$$

Knowing that at optimal matching, $G_s$ equals to $G_{opt}$, given by Eqs. (11.20) and (11.22) becomes

$$F_{opt} = 1 + 2R_n(G_{opt} + G_c). \tag{11.23}$$

Hence, the general noise factor given in Eq. (11.17) can be represented with respect to the optimal noise factor using the relations founded for the optimal load noise factor:

$$F = F_{opt} + \frac{R_n}{G_s} + R_n\left[\left(G_s - G_{opt}\right)^2 + \left(B_s - B_{opt}\right)^2\right]. \tag{11.24}$$

The remarkable part of this analysis is shown by the conditions in Eqs. (11.20) and (11.21). Each device has its noise impedance as a given constraint. The goal of matching to minimum noise is identical to matching maximum power transfer to the real part of the noise impedance of the device. By matching the source signal impedance under this constraint, the output CNR is improved. The condition of Eq. (11.21) defines that in order to achieve the maximum power transfer to a load, the reactive part of the load should be neutralized; i.e., the signal phasor is parallel to the resistive part of the noise impedance. Since the resistive part of the noise impedance is composed of correlated and noncorrelated resistive components, which are orthogonal to each other, the equivalent is derived by a Pythagorean summation of the two resistive vectors. The condition in Eq. (11.20) defines that in order to transfer maximum power to the resistive value of the noise impedance, the signal source impedance should be equal to the value given by Eq. (11.20), which is the equivalent noise conductance. Figure 11.2 provides an illustrative explanation for the above argument in the phasor domain.

Observing the noise factor function at the 3D axis system, $F(G_s, B_s)$, $G_s$, and $B_s$, it can be seen that the noise factor depends on the loading conditions of the signal source (Fig. 11.3).

On the $G_S$ versus $B_S$ plan, it can be seen that equal noise-factor load-circles are generated, and $F_{min}$ is at the center of the load circles. On the $F(G_S, B_S)$ versus $G_S$ plan and $F(G_S, B_S)$ versus $B_S$ plan, the noise factor function is a parabolic function with a minimum at $F_{min}$. In reality, there are additional constrains for matching such as minimum return loss. Return loss scattering parameters are different from noise-matching parameters and a compromise is made between the two.

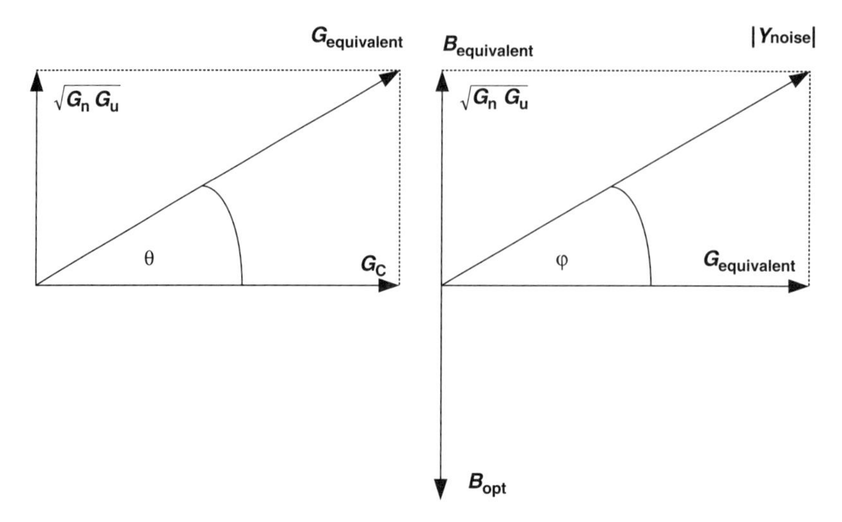


**Figure 11.2** Noise impedances, left equivalent noise conductance, and right neutralization of the noise reactance imaginary part. The equivalent noise conductance $G_{equivalent}$ is the $G_{opt}$ value required as an output impedance from the signal current source.

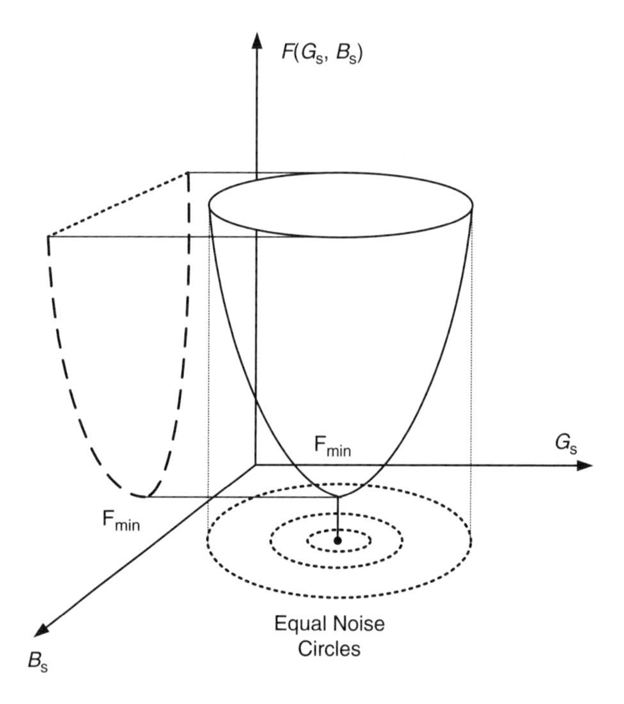


**Figure 11.3** Noise factor function $F(G_s B_s)$.

A more practical expression for noise factor is given by the noise-scattering parameters:

$$F = F_{\text{opt}} + \frac{R_n Y_o |\Gamma_s - \Gamma_{\text{opt}}|^2}{|1 + \Gamma_{\text{opt}}|^2 (1 + |\Gamma_s|^2)}, \tag{11.25}$$

where, $Y_o$ is the characteristic admittance of the transmission line, generally $1/50$ $\Omega$. $R_n$ is the noise resistance and $\Gamma_{\text{opt}}$ and $\Gamma_s$ are the scattering parameters for the minimum noise figure and the source impedance, respectively, as

$$\Gamma_{\text{opt}} = \frac{Y_o - Y_{\text{opt}}}{Y_o + Y_{\text{opt}}}. \tag{11.26}$$

$$\Gamma_s = \frac{Y_o - Y_s}{Y_o + Y_s}. \tag{11.27}$$

Since $\Gamma_{\text{opt}}$ provides the minimum value for noise factor $F$, it is sometimes called $\Gamma_m$ and it has the same meaning $\Gamma_{\text{opt}}$. Henceforth, all the "opt" indexing will be replaced with "m" indexing.

The parameters $F_m$, $\Gamma_m$, $R_n$, and $G_m$ characterize the device noise performance and are provided by the transistor manufacturers in the data sheets for different frequencies.

The noise resistance $R_n$, as mentioned by Eqs. (11.23) and (11.24), does not necessarily reflect the noise factor dependency on the input source admittance. The fact that a certain network or device has a low or high $R_n$ value does not imply strong or weak effects on the noise factor $F$. New transformation is required to calculate $F$ with respect to the normalized source impedance reflection coefficient $\Gamma_s$. This value is $\Gamma_{sm}$, and is sometimes called the source-scattering parameter. A preliminary definition of such transformation is provided in Chapter 9. For transformation purpose, Eq. (11.25) is modified and called the proportion factor $k$ and is defined as

$$k = \frac{Y_o |\Gamma_s - \Gamma_m|^2}{|1 + \Gamma_m|^2 (1 + |\Gamma_s|^2)} = \frac{F - F_m}{R_n}. \tag{11.28}$$

The next step is to transfer the reflection factors $\Gamma_m$ and $\Gamma_s$ with respect to the optimal load-plane impedance, which is the optimal impedance of the source, mentioned before. Hence, the following relation is defined:

$$\Gamma_{sm} = \frac{Y_m - Y_s}{Y_m^* + Y_s}. \tag{11.29}$$

Using Eqs. (11.24) and (11.25) the normalized value of source impedance reflection factor $\Gamma_{sm}$ can be obtained by

$$|\Gamma_{sm}|^2 = \frac{k}{4G_m + k}. \tag{11.30}$$

Note that according to Eq. (11.27), $\Gamma_s = 0$ for $Z_s = Z_o$ source impedance, while $\Gamma_{sm} \neq 0$.

Using the value of $k$ and Eq. (11.24), the expression for the noise factor $F$ refers to the normalized source impedance:

$$F = F_m + 4R_n G_m \frac{|\Gamma_{sm}|^2}{1 - |\Gamma_{sm}|^2}, \tag{11.31}$$

which provides a clear relation between the noise factor $F$ to the normalized noise resistance $R_n$, the minimum noise conductance $G_m$, and to the normalized source impedance reflection factor $\Gamma_{sm}$. The conclusion is that the minimum noise factor $F_m$ occurs when the difference between $\Gamma_m$ and $\Gamma_s$ is 0. The magnitude of $\Gamma_{sm}$ represents the distance. This can also be seen from Eq. (11.28), since at that point $k = 0$.

## 11.2 Two-Port Network Matching to Minimum Noise

Assume that a two-port network is marked as network $N$ with equivalent noise sources $i_n$ and $e_n$ and noise parameters $F_m$, $G_m$, $B_m$, and $R_n$. When the source impedance, generally $Y_o$, is matched to the optimal impedance $Y_m$, it requires a new network definition for the matching network and the network $N$ is marked as $N_1$ network. The new network $N_1$ is characterized by new noise sources and noise parameters $i_1$ and $e_n$, the noise parameters $F_1$, $G_1$, $B_1$, and $R_1$. Figure 11.4 describes the two-port matched network. Since the matching network is for minimum noise factor $F_m$, it can be concluded that the new noise parameters of $N_1$ network are as follows:

$$F_1 = F_m. \tag{11.32}$$

Assuming the matching network matches the characteristic admittance $Y_o$ to the optimal admittance $Y_m$, the matching network input admittance is satisfying

$$G_1 + jB_1 = Y_o + j \cdot 0. \tag{11.33}$$

The value of the input noise resistance $R_1$ can be calculated from Eq. (11.31), for the case network, $N_1$ is fed by source admittance $Y_{s1}$ and the results are given as

$$F = F_m + 4R_1 Y_o \frac{|\Gamma_s 1|^2}{1 - |\Gamma_{s1}|^2}, \tag{11.34}$$

where

$$\Gamma_{s1} = \frac{Y_{s1} - Y_o}{Y_{s1} + Y_o}. \tag{11.35}$$

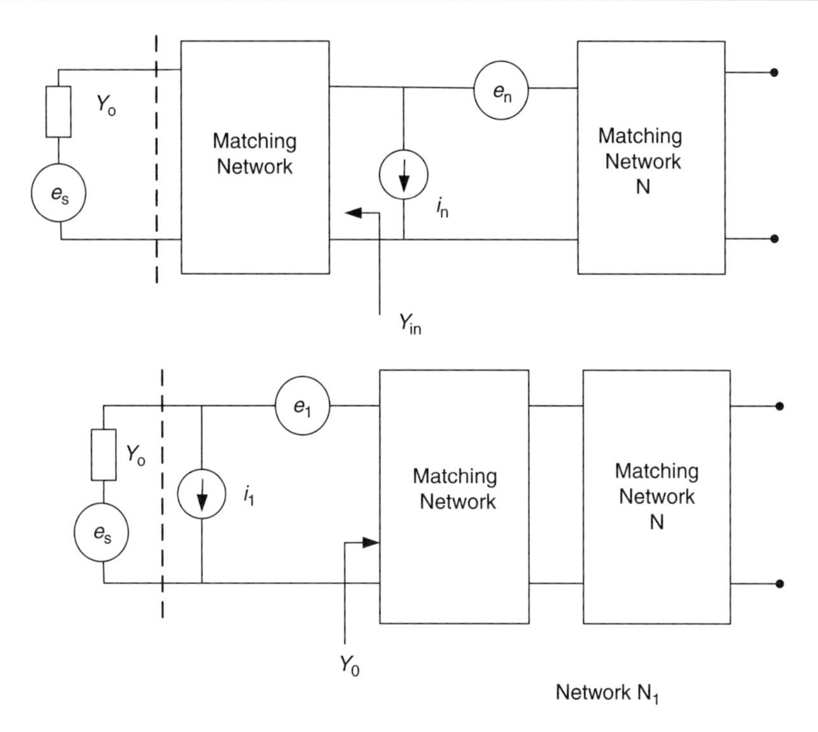


**Figure 11.4** Equivalent noise parameters of two-port matched network.

On the other hand, from Eq. (11.29), it is known that

$$|\Gamma_{sm}| = |\Gamma_{s1}| = \left|\frac{Y_m - Y_s}{Y_m^* + Y_s}\right|, \tag{11.36}$$

where $Y_s$ is the source impedance of the network $N$. By using the equality between Eqs. (11.31) and (11.34), $R_1$ can be solved by

$$R_1 = \frac{G_m}{Y_o} R_n. \tag{11.37}$$

Substituting the value of $R_1$ into Eq. (11.34) results in a general expression for the noise factor of the network $N_1$:

$$F = F_m + 4R_n G_m \frac{|\Gamma_{s1}|^2}{1 - |\Gamma_{s1}|^2}. \tag{11.38}$$

Equation (11.38) can be presented in terms of VSWR:

$$F = F_m + 4R_n G_m \frac{|\rho - 1|^2}{\rho}, \tag{11.39}$$

where the networks' VSWR, ρ, is given by

$$\rho = \frac{1 + |\Gamma_{s1}|}{1 - |\Gamma_{s1}|}. \tag{11.40}$$

The last expression for noise factor is very useful; the values $F_m G_m R_n$ are known from the noise parameters of the device. The noise factor dependency on the normalized noise resistance $G_m R_n$ and the source VSWR, ρ, is presented here. Moreover, the optimal admittance of the matched gain stage is a real value that is equal to the characteristic admittance. As a result, the correlation admittance provided in Eq. (11.21) is a real value. Thus, the matched stage would have dual expressions to those in Eqs. (11.8) and (11.20) for the total noise current and the optimal conductance:

$$i_1^2 = 4kT_o\Delta f\left(G_{u1} + R_1 G_{c1}^2\right), \tag{11.41}$$

$$Y_o = \sqrt{G_{c1}^2 + \frac{G_{u1}}{R_1}}. \tag{11.42}$$

Taking the last two expressions and solving the noise current a simple expression for the matched stage noise current energy can be obtained:

$$i_1^2 = 4kT_o\Delta f Y_o^2 R_1. \tag{11.43}$$

## 11.3 Noise Modeling of MESFET

One of the building blocks for analog receivers is the MESFET (metal semiconductor field effect transistor) device. These transistors are manufactured in GaAs or AlGaN/gAN PHEMT (pseudo heterojunction transistor) technology. The advantage of these devices is their high cut-off frequency, high gain of about 15–18 dB, determined by $g_m$, the transconductance value of the device, (high gain and $f_T$ are resulting due to the high mobility of the semiconductor), high IP3, high compression point P1 dB, and low noise figure typically of 1.0–1.5 dB. All these parameters make PHEMT devices suitable for analog front end LNA. A field effect transistor (FET) is a voltage-controlled current source. Any voltage change at the input port, called a "gate," affects the drain source current. The FET schematics are given in Fig. 11.5.

Assume now the channel current is corrupted by noise. As a result, noise voltage is induced on the gate in correlation to the channel noise. This is the correlated portion of the total noise current density, as explained in the previous section. The correlation between these two noise parameters increases with frequency. An FET device is a white-noise mechanism. This is because the current through the channel is a drift current and not a diffusion current across a PN junction. However, there is a small shot-noise component caused by the gate-leakage current from the gate bias to the source. The ohmic losses that are

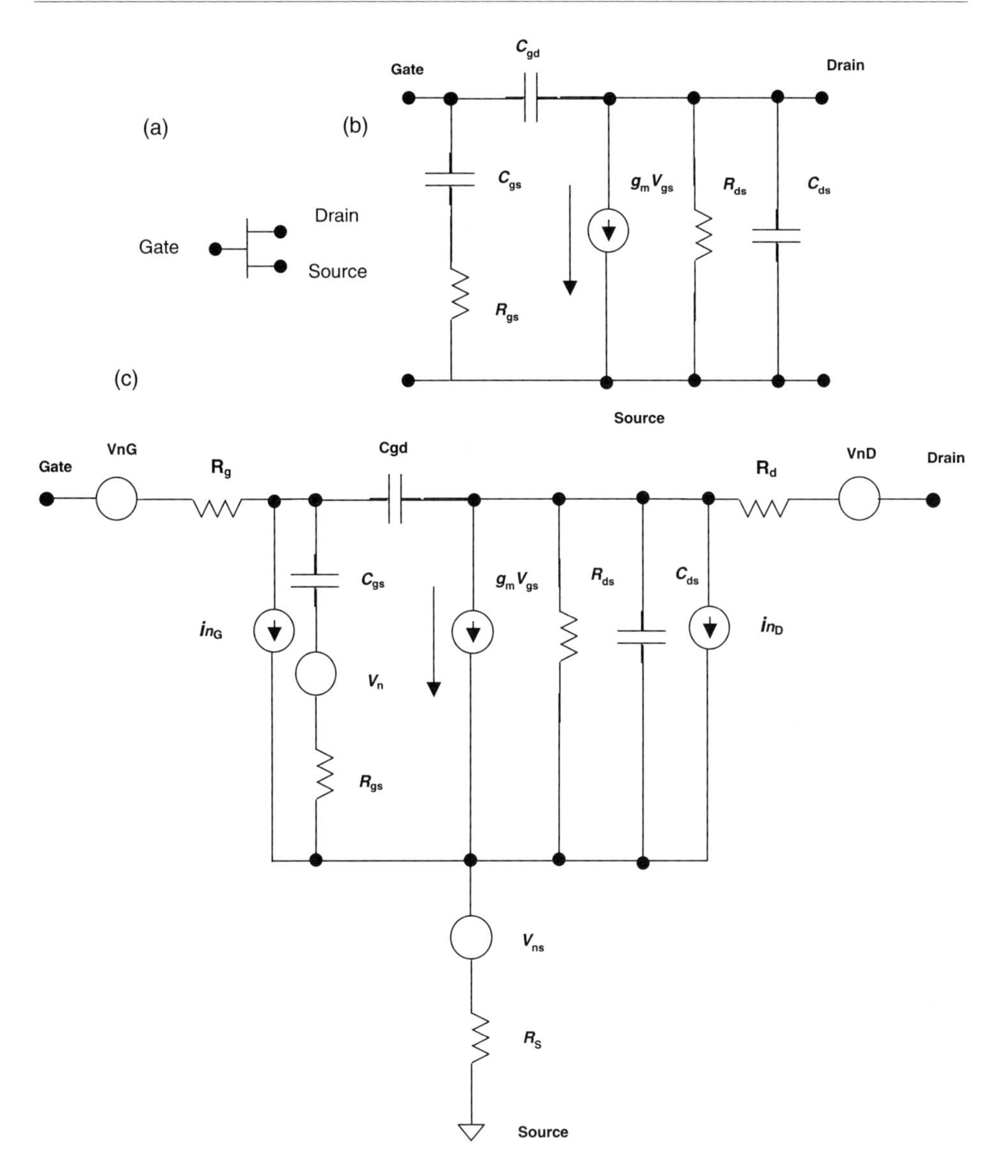


**Figure 11.5** (a) A basic FET; (b) a small-signal equivalent circuit; (c) noise source equivalents in an FET model. The resistances $R_g$, $R_S$, and $R_d$ are neglected in some of FET modeling. Note that the current source $i_{nG}$ is the gate source leakage shot noise.

internal to the device and are resistive result in thermal noise or Johnson noise. The noise nature of the FET channel is thermal and the gate-leakage current is shot noise. Any model of FET involving parasite junctions would add shot noise. The analysis here is a basic one.

Thus, the small-signal equivalent circuit is modified to the equivalent noisy-circuit model by adding those noise sources on the basis of the rules previously presented in Fig. 11.5(c).

Microwave FET noise models were investigated and approximated by several researchers and text books[6–13] and their results are provided by the analysis below. The common topology used in FET devices is common source, which this model refers to.

As previously explained, any noisy network can be presented by a noiseless network and with noise currents and voltages per Fig. 11.5(c). FET noise sources are thermal at the channel and internal resistances. Shot current in a FET results from the gate source leakage current. The FET channel is a thermal-noise source since there is no junction barrier that characterizes shot noise. The short-circuit current from the channel noise source generated in the drain source path is given by[11]

$$\overline{i_{nD}}^2 = 4kTBg_mP, \tag{11.44}$$

where $P$ is a factor depending on the device geometry and the dc bias. For zero-drain voltage, $i_{nD}$ characterizes the thermal noise generated by the drain output resistance or conductance $1/R_{ds}$. Thus, $P$, in this case, satisfies $P = (1/R_{ds}) \times g_m$. For positive-drain voltages, the noise generated in the channel is larger than the thermal noise that is generated by $R_{ds}$ because of the following reasons: first, the thermal noise voltage generated locally in the channel modulates the conductive cross-section of the channel and results in an amplified noise voltage at the drain. Second, the electrons are accelerated in the electric field, then scattered in all directions due to interaction with lattice phonons. This velocity increases with the $E$ field. Third, in GaAs, carriers undergo field-dependent transitions from central valley in the conduction band to satellite valleys and vice versa. Fourth, in high-drain voltages, the carriers' velocity is saturated on the drain side of the channel. Thus, the field has no influence on the drift velocity; thus this channel section cannot be referred to as ohmic conductance resulting in a thermal noise like $R_{ds}$ may create.

As previously explained, a noise voltage generated locally in the channel causes a fluctuation in the depletion layer width. The resulting charge fluctuation in the depletion layer, in turn, induces a compensating charge variation on the gate electrode. The total induced gate-charge fluctuation is described in Fig. 11.5 by the $i_{nG}$ noise generator at the gate terminal and is given by

$$\overline{i_{nG}}^2 = 4kTB\frac{(2\pi C_{gs})^2 f^2}{g_m}R, \tag{11.45}$$

where $R$ is a factor dependent on the FET geometry and bias.

The two noise currents $i_{nG}$ and $i_{nD}$ result from the same noise voltages in the channel. Hence, they are partially correlated. Thereby, a correlation factor $C$ is given by

$$jC = \frac{\overline{i_{nG}^* i_{nD}}}{\sqrt{\overline{i_{nG}^2} \cdot \overline{i_{nG}^2}}}. \tag{11.46}$$

The correlation factor is purely imaginary since $i_{nG}$ is caused by capacitive coupling of the gate circuit to the noise sources in the drain circuit.

Other thermal noise sources of FET are due to its resistances at the gate drain and source:

$$\overline{e_g^2} = 4kTR_g, \quad (11.47)$$

$$\overline{e_d^2} = 4kTR_d, \quad (11.48)$$

$$\overline{e_s^2} = 4kTR_s. \quad (11.49)$$

The shot noise results from the dc-leakage current on the gate to the channel, which is a junction in reverse bias and is given by

$$\overline{i_{shot}^2} = 2qI_g. \quad (11.50)$$

In a small-signal low-noise FET, the shot noise values are negligible because of the low value of the gate leakage current $I_g$. Thus, the dominant thermal noise generators come mainly from the gate and the drain. This was presented by Van Der Ziel in a series of classic papers.[14–16] This may simplify the calculation and therefore the thermal voltage $v_n$ is given by

$$v_n^2(f)[\mathrm{V}^2/\mathrm{Hz}] \cong 4kTr_n\left(1 + \frac{f_c}{f}\right), \quad (11.51)$$

where $f_c$ is the corner frequency. The effective noise resistance of the FET, $r_n$, is given by the work of Pucel et al.:[10]

$$r_n = (R_s + R_g)\frac{T_d}{T_o} + K_r\left(\frac{1 + (\omega C_{gs}R_{gs})^2}{g_m}\right), \quad (11.52)$$

where $T_d$ is the temperature of the FET, $K_r$ is a numerical noise coefficient that represents the properties of the intrinsic noise generators $i_{nG}$ and $i_{nD}$ and their correlation factors. The thermal noise current with the gate-leakage current causing the shot noise is given by

$$i_n^2(f)[\mathrm{A}^2/\mathrm{Hz}] \cong 2qI_g + 4kTr_n(2\pi C_{gs})^2 f^2. \quad (11.53)$$

The correlation noise current is given by

$$\begin{aligned} i_{n-corr}^2(f) &= v_n^2(f)Y_{corr}(f)Y_s(f) \\ &= 16kTr_nC_{gs}\pi f\, Y_s(f), \end{aligned} \quad (11.54)$$

where the value $Y_s(f)$ is the source admittance given by

$$Y_s(f) = 2\pi f C_{ds}, \tag{11.55}$$

and finally, the noise current energy, including shot noise is given by

$$i_n^2(f) = 2qI_g + 4kTr_n\left\{\left[2\pi f_c(C_{gs} + C_{ds})\right]^2 f + \left[2\pi(C_{gs} + C_{ds})\right]^2 f^2\right\}. \tag{11.56}$$

The first term in this equation is the shot noise due to the gate-leakage current, the second term is $1/f$ noise, and the third is the thermal noise of the channel coupled into the gate, as explained previously. Using these relations, the noise factor of an FET device is given by[10]

$$F = 1 + \frac{1}{R_{rs}}\left(r_n + g_n|Z_s + Z_c|^2\right), \tag{11.57}$$

where $R_{rs}$ is the real part of the source impedance, $Z_s$ is the signal source impedance, $Z_c$ is the correlation impedance, and $g_n$ is a noise property of the FET given by

$$g_n = K_g\frac{(\omega C_{gs})^2}{g_m}. \tag{11.58}$$

The correlation impedance is given by[10]

$$Z_c = R_{rs} + R_g + \frac{K_c}{Y_{11}}, \tag{11.59}$$

where $K_g$ and $K_c$ are numerical noise coefficients, which represent the properties of the intrinsic noise generators $i_{nG}$ and $i_{nD}$ and correlation factor and $1/Y_{11}$ is given by

$$\frac{1}{Y_{11}} = R_{gs} + \frac{1}{j\omega C_{gs}}. \tag{11.60}$$

Substituting these relations in Eq. (11.57) and applying the conditions for minimum noise-figure matching given by Eqs. (11.20) and (11.21) results in an expression for $F_{min}$:[10]

$$F_{min}(\omega) = 1 + 2\left(\frac{\omega C_{gs}}{g_m}\right)\sqrt{K_g[K_r + g_m(R_{rs} + R_g)]} + \left(\frac{\omega C_{gs}}{g_m}\right)^2 K_g g_m(R_{rs} + R_g + K_c R_{gs}). \tag{11.61}$$

There are several important conclusions from the $F_{min}$ expression of an FET. Since $F$ is a ratio of input CNR to output CNR, the first term is unity because this is the ratio between the input noises to itself. The second term in this expression represents $1/f$ noise contribution and the third represents the channel thermal noise contribution coupled to the gate. At high frequencies, the noise is proportional to the ratio $C_{gs}/g_m$. Therefore, when selecting an FET, it desirable to have high transconductance. On the other hand, it is an advantage to have a low gate source capacitance in order to minimize the stored noise energy or coupling of noise. When designing a circuit, it is desired to have a low source resistance value of $R_{rs}$, since its contribution is amplified by the $g_m$. The same applies to the gate series resistance $R_g$. All three coefficients $K_g$, $K_c$, and $K_r$ depend on the gate length.[7] $K_r$ is an order of magnitude lower than $K_c$ and $K_g$. Generally, in FETs, the bias is expressed in a normalized value of $I_{ds}/I_{dss}$, where $I_{dss}$ is the saturated current for $V_{gs} = 0$, and $I_{ds}$ is the bias current. $K_g$ is a strong function of the $I_{ds}$ current; hence, in RFFE, it is required to use a lower current for low noise performances. Note that the device $f_T$ is the cut-off frequency and is given by

$$f_T = \frac{g_m}{2\pi C_{gs}}. \tag{11.62}$$

## 11.4 Noise Modeling of Bipolar Junction Transistors

In the recent two decades, monolithic microwave circuit (MMIC) technology has advanced in several directions; PHEMT, which is related to GaAs, and diverse MESFETs, and other semiconductors such as silicon, InGaP, and SiGe. The last two semiconductors are used for heterojunction bipolar transistor (HBT) gain blocks, which belong to the bipolar junction transistor (BJT) family. These MMICs are commonly used for RF and CATV transport RF blocks. Si bipolar are used also for digital transports as building blocks of laser drivers for FP and DFBs, where high current is required, unlike CMOS, which is used for VCSEL lasers, where low modulation current is needed.

BJT transistors are described by the traditional hybrid $\pi$ small-signal approximation circuit.[18] This is an approximation of the Ebers—Moll BJT modeling.[19] Many research works were conducted on BJT noise modeling.[20–22] Some of these works were presented by Van Der Ziel and coworkers.[23,24] The intrinsic and noise modeling of a bipolar transistor are given in Fig. 11.6.

Unlike a FET device, which is a drift-current mechanism and the governing noise is thermal Johnson noise, a BJT device is a diffusion-current device across semiconductor PN junctions; therefore, the governing noise is shot noise. The losses across the internal resistor $r_x$ of the base port is thermal noise. This resistor is sometimes marked as $r_b$. The resistances $r_\pi$, $r_\mu$, and $r_o$ are dynamic resistances that do not dissipate energy, where the Miller feedback resistance is marked as $r_\mu$. The common topology used in BJT devices is a common emitter (CE), which this model refers to.

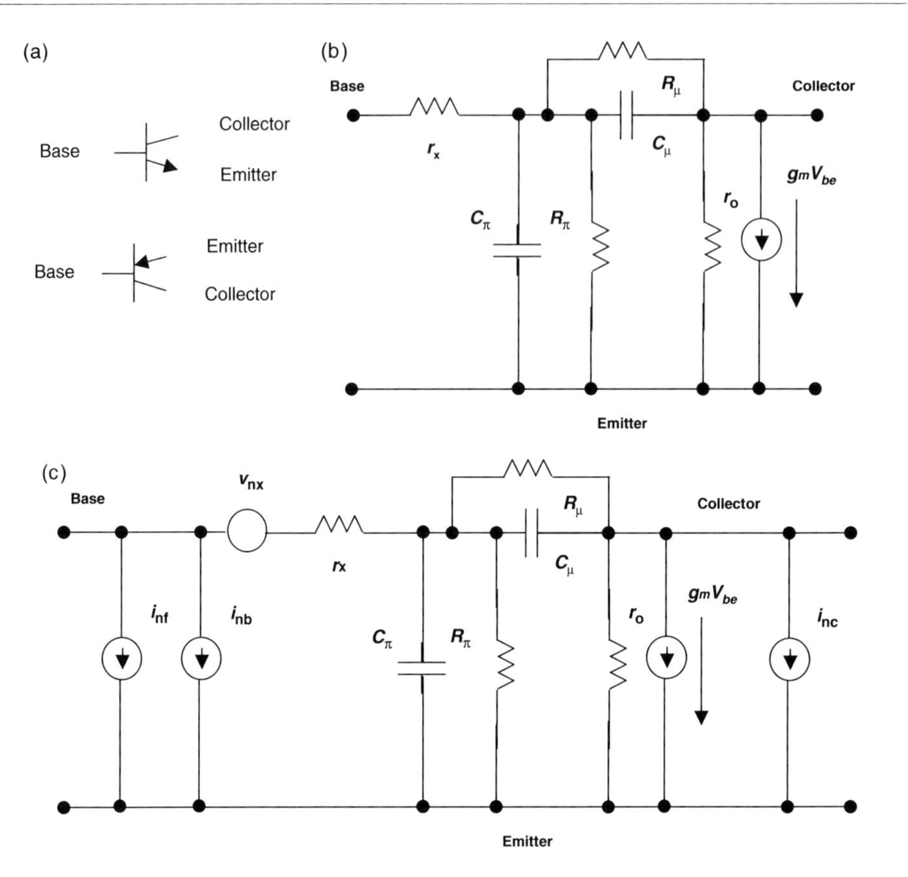


**Figure 11.6** Hybrid $\pi$ small signal modeling and nose modeling. (a) Basic, (b) small signal, and (c) noise modeling.

From Refs. [25] and [26], neglecting the Miller feedback effects of $r_\mu$, $C_\mu$, the following equations describe the noise voltage and current in a BJT device:

$$v_n^2(f)[\mathrm{V}^2/\mathrm{Hz}] \cong \frac{1}{2} r_x q I_B \left(\frac{f}{f_c}\right) + 4kTr_x + 2qI_C r_e^2, \tag{11.63}$$

$$i_n^2(f)\left[\mathrm{A}^2/\mathrm{Hz}\right] \cong 2qI_B \left(\frac{f_c}{f}\right) + 2qI_B + 2qI_C \left(\frac{f}{f_T}\right)^2, \tag{11.64}$$

where $I_B$, and $I_C$ are the base and collector bias currents, respectively. The value $r_e = 1/g_m$ and $g_m = (q/kT) \times I_C$; note that $kT/q$ is the thermal voltage equal to 0.025 V at $T = 290$ K; $f_c$ is the "$1/f$" corner frequency, and the cut-off frequency is given by $f_T = g_m/2\pi C_\pi$.

Equation (11.63) contains $1/f$ noise in the first term and the second term is thermal noise, both at the base. The last term is shot noise voltage due to the transistor transconductance. Equation (11.64) describes the noise current terms. The first term is $1/f$ noise current on the base, the second term is shot noise due to the base current, and the last term is collector shot noise. Three topologies are used to build BJT amplifiers,[18] CEs, common collectors, and common bases. The most commonly used is the CE configuration due to its high-gain properties.

The noise voltage and current can be reflected to the transistor input, which is its base port. The stray capacitance at that port is the PD capacitance $C_d$. Thus, the voltage noise develops a noise current on that capacitance, assuming that admittance is $Y = j\omega C_d$, $|Y|^2 = (\omega C_d)$.$^2$ Thus, the total noise current without $1/f$ noise is the sum of currents from the noise voltage and current source:

$$i_n^2(f)[\mathrm{A}^2/\mathrm{Hz}] = 2qI_B + 4kTr_x(2\pi f C_d)^2 + 2qI_C r_e^2[2\pi f(C_\pi + C_d)]^2. \quad (11.65)$$

Noise wise, it is an advantage to have a low capacitance PD at the input of the receiver. Note that the transconductance is increasing as the bias current is increasing. Since $1/r_e = g_m$, $r_e$ is decreasing. Thus, observing Eq. (11.65), the first term, which is shot current, is not dependent on the collector current or, to be precise, it is weakly dependent on it. Note that $I_B\beta = I_C$; thus noise current increases as bias point $I_C$ increases. The second term is the thermal voltage that results from the base resistance $r_x$ and has no dependency on the bias point. This thermal voltage creates noise current on the PD stray capacitance. The last term is decreasing since, as gain or transconductance increases, the dynamic resistance decreases. This way, it can be seen that transistor gain compensates for noise. Note that $hf_e = \beta = g_m \times r_\pi = r_\pi/r_e$;[18] this way, Eq. (11.65) can be simplified with respect to the base current as follows:

$$i_n^2(f)[\mathrm{A}^2/\mathrm{Hz}] = 2qI_B + 4kTr_x(2\pi f C_d)^2 + \frac{2q\beta I_B}{g_m^2}[2\pi f(C_\pi + C_d)]^2, \quad (11.66)$$

or

$$i_n^2(f)[\mathrm{A}^2/Hz] = 2qI_B\left\{1 + \frac{\beta}{g_m^2}[2\pi f(C_\pi + C_d)]^2\right\} + 4kTr_x(2\pi f C_d)^2. \quad (11.67)$$

This means that the input shot noise is composed of the base current as well as the amplification coupling of the collector current with cut-off frequency $f_T$ that is given by

$$f_T = \frac{g_m}{2\pi(C_\pi + C_d)}. \quad (11.68)$$

Observing Eqs. (11.63) and (11.64), one can see that there are two regions of noise. The first is when the transistor is operating below the corner frequency $f_c$, where the $1/f$ noise is dominant. Up to the corner frequency, the noise rolls off at a rate of 20 dB/dec. The second region is above the corner frequency, the noise floor is determined by the shot noise and thermal noise dictated by the base parasitic resistor $r_x$. As the transistor operates above its cut-off, the noise climbs up by 40 dB/dec. From Eq. (11.67), it is desirable to minimize the input capacitance by two methods: low PD stray capacitance and by selecting a high-frequency performance transistor with high $f_T$. More detailed analysis for BJT noise figure was provided by Hawkins.[27]

## 11.5 MESFET Feedback Amplifier

The idea of a feedback regenerative circuit was introduced first in 1913 by Alexander Meissner, when he was granted a feedback oscillator patent by the German patent office. In the same year, Edwin H. Armstrong presented a paper on regenerative circuits. A year later, Lee DeForest filed a patent on regenerative circuits. Since then, a great number of new concepts for negative feedback have emerged. One of the features of great importance is the use of negative feedback to control gain, gain BW, and input and output impedance of an amplifier. Hence, it is not surprising that the negative-feedback principle began to be used to control gain BW flatness and return–loss of GaAs MESFET amplifiers.[28–30] One of their advantages is the distortion correction due to the parallel feedback.[31]

### 11.5.1 Basic feedback amplifier circuit concept

The GaAs MESFET parasitic elements given in Figs. 11.5 and 11.7 restrict the amplifier BW. Minimization of these parasitics was a major goal in the development of microwave devices. Recently, the technology had achieved high-performance PHEMT devices with a high cut-off frequency $f_T$, low-noise figure performance, flatness, and a high 1-dB compression point $P_{1dB}$. However, there is still a need to extend the BW performance while maintaining a high compression point, a low IMD for high power performance, and a decent noise figure. As it can be observed from Figs. 11.5 to 11.7, the MESFET BW is limited by the gate-time constant, $\tau_g = R_i C_{gs}$, and the drain-time constant, $\tau_d = R_{ds} C_{ds}$. Additionally, the Miller capacitance $C_{dg}$, between the drain to the gate, limits the isolation between the output of the device at the drain port to the input at the gate port. By using MESFET with a high cut-off frequency $f_T$ of approximately 6 GHz or higher and optimizing the performance between 50 and 1000 MHz at first approximation, those parasitics are omitted. Hence, the feedback model for low frequency is given by [Fig. 11.7(b)], assuming a unilateral FET.[32]

### 11.5.2 Low-frequency model

The use of low-frequency model results in several simple expressions for the feedback resistor $R_{FB}$ and matching conditions. Assuming the contact resistances at the gate and the drain $R_g$ and $R_d$ are very small compared to the feedback resistor $R_{FB}$

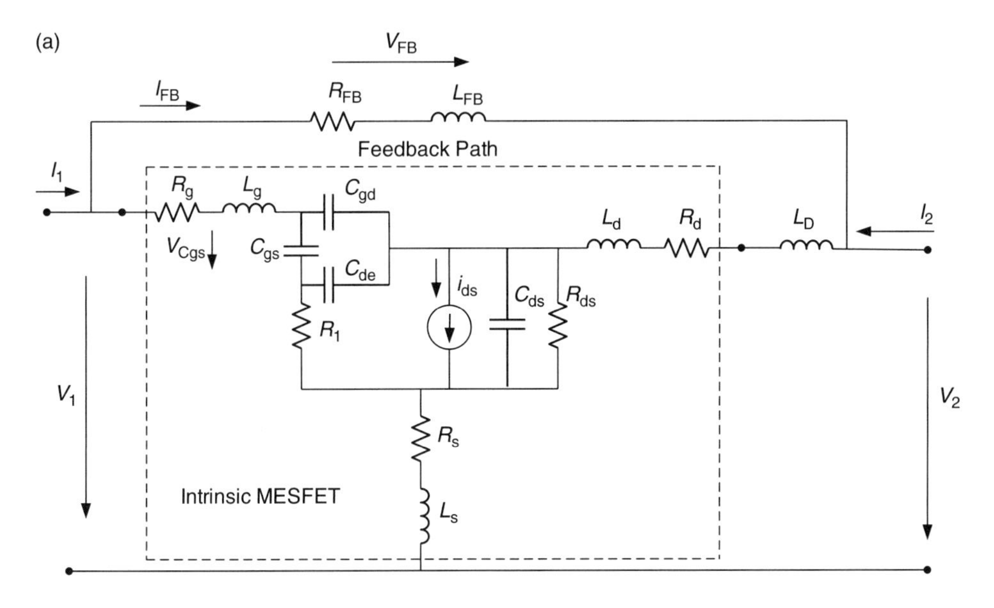


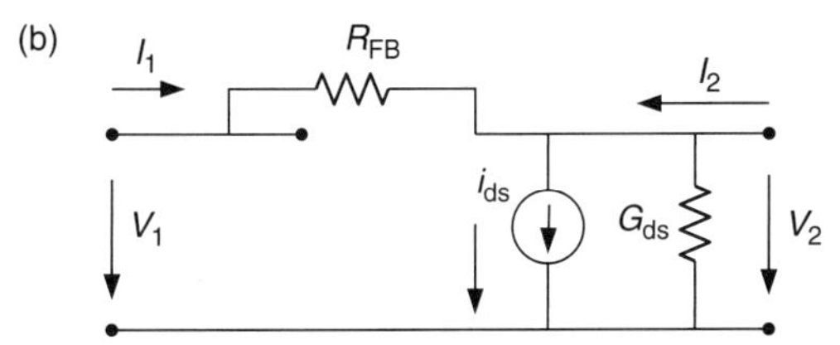


**Figure 11.7** Feedback-FET amplifier small-signal model (a) for high frequency (b) approximation for low frequency.[32] (Reprinted with permission from *IEEE Microwave Theory and Techniques* © IEEE, 1980.)

and assuming a load resistor satisfies $R_L = Z_o$, the currents are described by the following conductance matrix:[32]

$$\begin{bmatrix} I_1 \\ I_2 \end{bmatrix} = \begin{bmatrix} G_{FB} & -G_{FB} \\ g_m - G_{FB} & G_{FB} - G_{ds} \end{bmatrix} \begin{bmatrix} V_1 \\ V_2 \end{bmatrix}, \tag{11.69}$$

where

$$G_{FB} = R_{FB}^{-1}, \tag{11.70}$$

$$G_{ds} - R_{ds}^{-1}, \tag{11.71}$$

and

$$i_{gs} = g_m V_{gs}. \tag{11.72}$$

Using the relation between $Y$ parameters and $S$ parameters, the scattering parameters matrix is obtained

$$S_{i,j} = \begin{bmatrix} S_{11} & S_{12} \\ S_{21} & S_{22} \end{bmatrix}. \tag{11.73}$$

The resulting scattering parameters are

$$S_{11} = \frac{1}{\sum}\left[\frac{R_{\rm FB}}{Z_0}(1 + G_{\rm ds}Z_0) - (g_{\rm m} + G_{\rm ds})Z_0\right], \tag{11.74}$$

$$S_{12} = \frac{2}{\sum}, \tag{11.75}$$

$$S_{21} = \frac{-2}{\sum}[g_{\rm m}R_{\rm FB} - 1], \tag{11.76}$$

$$S_{22} = \frac{1}{\sum}\left[\frac{R_{\rm FB}}{Z_0}(1 - G_{\rm ds}Z_0) - (g_{\rm m} + G_{\rm ds})Z_0\right], \tag{11.77}$$

where

$$\sum = 2 + (g_{\rm m} + G_{\rm ds})Z_0 + \frac{R_{\rm FB}}{Z_0}(1 + G_{\rm ds}Z_0). \tag{11.78}$$

Additional assumptions are that the output resistance of the FET $R_{\rm ds} \gg Z_0$; i.e., all the output current from the FET is driven into the matched load $Z_0$. Additionally, it is assumed that the transconductance of the FET $g_{\rm m} \gg G_{\rm ds}$, i.e., the output impedance of the device does not affect the gain performance. With these assumptions and the requirement for perfect matching of $S_{11} = S_{22} = 0$, the following conditions are derived:

$$S_{11} = \frac{1}{\sum}\left[\frac{R_{\rm FB}}{Z_0} - g_{\rm m}Z_0\right], \tag{11.79}$$

$$\frac{R_{\rm FB}}{Z_0} = g_{\rm m}Z_0, \tag{11.80}$$

$$\sum = 2 + g_{\rm m}Z_0 + \frac{R_{\rm FB}}{Z_0}, \tag{11.81}$$

and by using Eq. (11.80),

$$\sum = 2 + 2\,g_{\rm m}Z_0. \tag{11.82}$$

Hence,

$$S_{12} = \frac{1}{1 + g_m Z_0}, \qquad (11.83)$$

$$S_{21} = (g_m Z_0 - 1). \qquad (11.84)$$

As a consequence, the condition for an ideal match is given by

$$R_{FB} = g_m Z_0^2, \qquad (11.85)$$

and the gain of a feedback GaAs MESFET or PHEMT feedback amplifier is given by

$$G[\mathrm{dB}] = 20 \log (g_m Z_0 - 1). \qquad (11.86)$$

Figure 11.8 provides a graphic design tradeoff tool between gain, voltage-standing-wave ratio (VSWR), and feedback value for designing a feedback FET amplifier.

### 11.5.3 Noise performance of the feedback amplifier

Several papers were published on the subject of equivalent noise parameters and figures of parallel feedback amplifiers.[33,34] Niclas[34] describes an analytical approach for deriving NF of a parallel feedback amplifier, and Vendelin[33] describes the computer-analysis results for amplifiers employing parallel feedback. With today's RF solvers, parallel feedback amplifiers can be easily analyzed for all performance parameters. The circuit diagram of the basic feedback amplifier is provided in Fig. 11.9(a). In order to analyze the noise behavior of the amplifier, the diagram in Fig. 11.9(a) was converted to the equivalent circuit of Fig. 11.9(b). The latter contains the voltage noise source $v_T$ and the current noise source $i_T$ of the MESFET as well as the thermal noise of the feedback resistor $v_{FB}$. The feedback noise voltage $v_{FB}$ is not correlated to the transistor's noise source. The MESFET itself is represented by a $T$ network with one voltage source. For simplifying the calculations, the analysis was divided into two parts. The first calculation is of the noise expressions for the intrinsic MESFET and then adding the feedback with its noise source. Figures 11.9(b), 11.10(a) and (b) show the MESFET equivalent circuit with and without the feedback's noise effects. The pure MESFET without the feedback, as shown in [Fig. 11.10(a)],

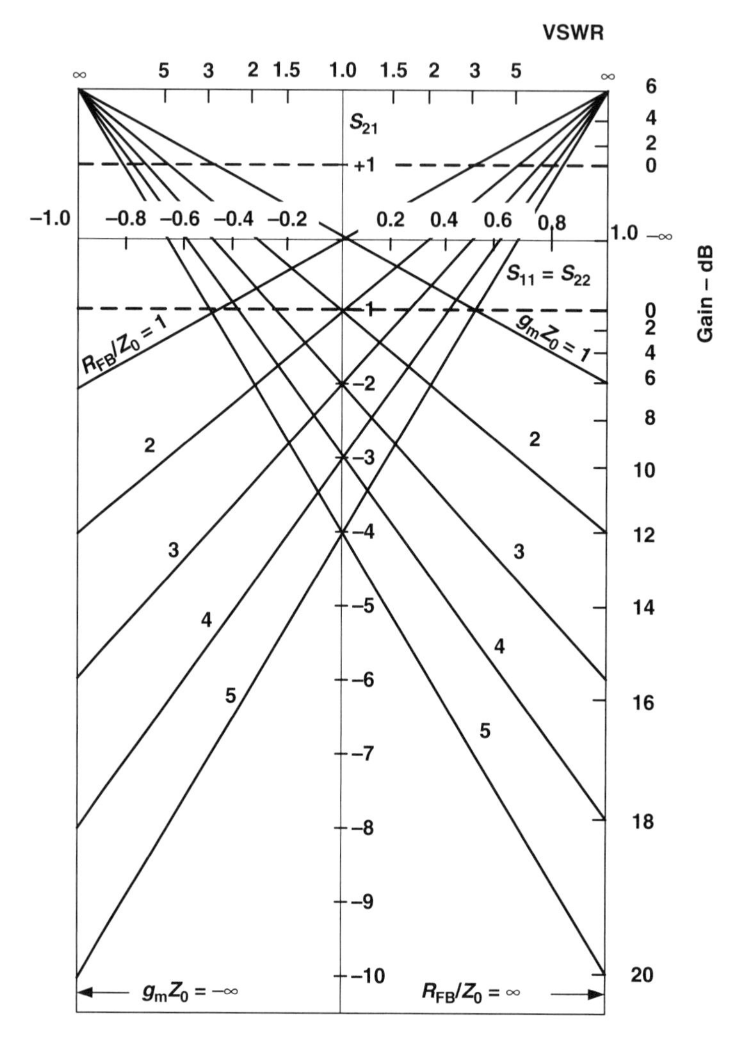


**Figure 11.8** Tradeoffs between feedback value to gain and VSWR at low frequency. Conditions are $G_{ds} = 0$ and $R_{FB} = g_m \cdot Z_0^2$.[32] (Reprinted with permission from *IEEE Microwaves Theory and Techniques* © IEEE, 1980.)

has the following noise parameters:

$$R_n = \frac{\overline{|v_1|^2}}{4kT_0\Delta f} = R_n^T\left[1 + jX_G Y_{cor}^T\right]^2 + G_n^T X_G^2, \tag{11.87}$$

$$G_n = \frac{\overline{|i_n|^2}}{4kT_0\Delta f} = \frac{G_n^T R_n^T}{R_n^T\left[1 + jX_G Y_{cor}^T\right]^2 + G_n^T X_G^2}, \tag{11.88}$$

$$Y_{cor} = \frac{\overline{i_1 v_1^*}}{\overline{|v_1|^2}} = \frac{R_n^T\left(Y_{cor}^T + jX_G\left|Y_{cor}^T\right|^2\right) - jX_G R_n^T}{R_n^T\left[1 + jX_G Y_{cor}^T\right]^2 + G_n^T X_G^2}, \tag{11.89}$$

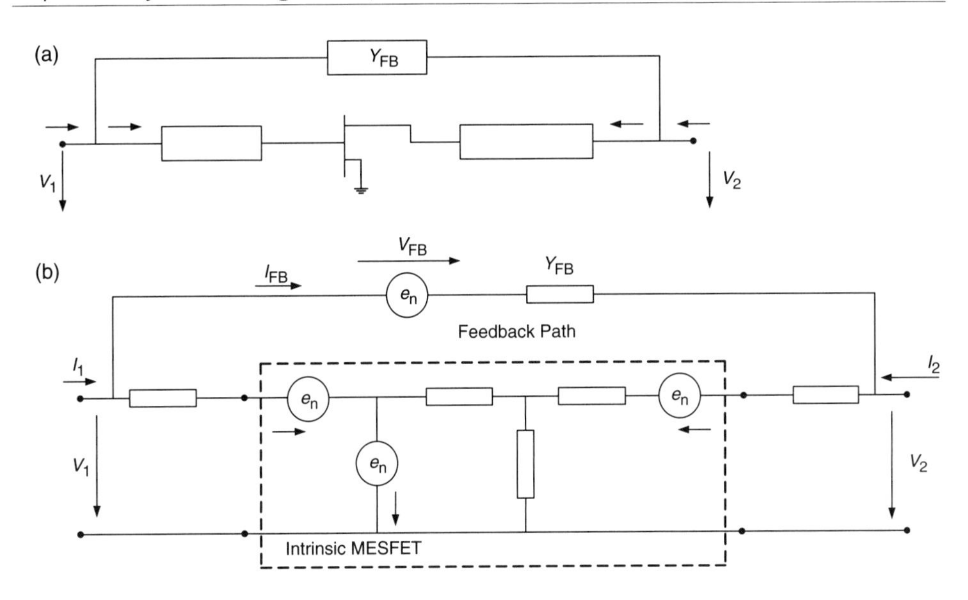


**Figure 11.9** The basic feedback amplifier. (a) Circuit diagram; (b) equivalent circuit with noise voltage and current sources.[34] (Reprinted with permission from *IEEE Microwaves Theory and Techniques* © IEEE, 1982.)

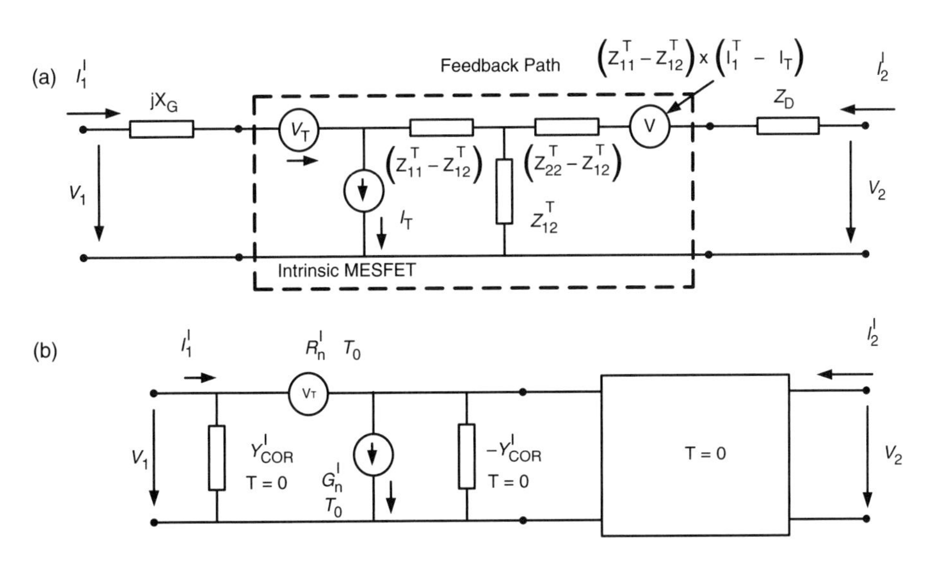


**Figure 11.10** Equivalent circuit of the amplifier with an open feedback loop (a) A two port representation, including the transistor external noise $v_T$, $i_T$. (b) A two-port with a correlation admittance $Y_{corr}$ and uncorrelated noise-source impedances $R_n$, $G_n$.[34] (Reprinted with permission from *IEEE Microwaves Theory and Techniques* © IEEE, 1982.)

where $k$ is Boltzman's constant, $\Delta_f$ is BW, sometimes known as B, $T_0$ is the absolute temperature in K and $T$ is the MESFET parameters.

The optimal noise factor matching conditions of a device are given in Eqs. (11.20), (11.21), and (11.23). Substitution of Eqs. (11.87), (11.8), and (11.89) into Eqs. (11.20), (11.21), and (11.23) demonstrates the well-known fact that the minimum noise figure of a device remains unchanged when it is preceded by a lossless passive two port. Now the parallel feedback is added as shown by Fig. 11.11. The following relationship exists between the equivalent voltages and currents shown in Fig. 11.10(a) and the MESFET's noise parameters:

$$v_1 = v_T + jX_G i_T, \tag{11.90}$$

$$i_1 = i_T = i_n^T + Y_{corr}^T v_T, \tag{11.91}$$

$$\overline{|v_1|^2} = \left|1 + jX_G Y_{corr}^T\right|^2 \overline{|v_T|^2} + X_G^2 \overline{\left|i_n^T\right|^2}, \tag{11.92}$$

$$\overline{|i_1|^2} = \overline{\left|i_n^T\right|^2} + \left|Y_{corr}^T\right|^2 \overline{|v_T|^2}, \tag{11.93}$$

where $i_n^T$ is the uncorrelated portion of the MESFET's equivalent noise current.

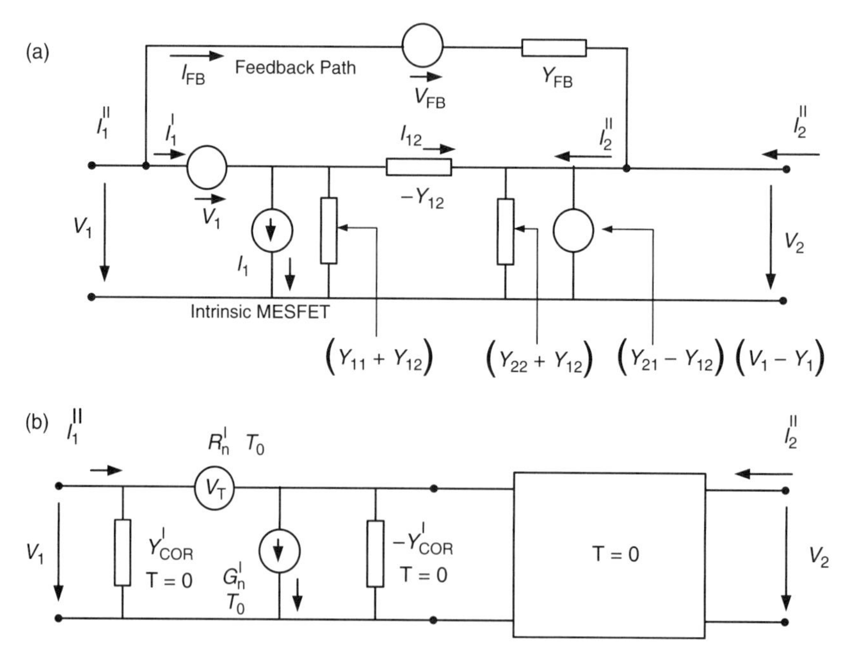


**Figure 11.11** The equivalent circuit of a feedback amplifier. (a) A two-port representation including the external noise sources $v_1$, $i_1$ of the open-loop amplifier. (b) A two-port with a correlation admittance $Y_{corr\text{-}1}$ and uncorrelated noise-source impedances $R_n^1$, $G_n^1$.[34] (Reprinted with permission from *IEEE Microwaves Theory and Techniques* © IEEE, 1982.)

The resulting noise parameters are located outside the noiseless two-port as shown in Fig. 11.11:

$$R_n^1 = \frac{|v_1^1|}{4KT_0\Delta f} = \left|\frac{Y_{21}}{Y_{21} - Y_{FB}}\right|^2 \left(R_n + \left|\frac{Y_{FB}}{Y21}\right|^2 R_{FB}\right), \tag{11.94}$$

$$G_n^1 = \frac{|i_1^1|}{4KT_0\Delta f} = G_n + |Y_{11} + Y_{21} + Y_{corr}|^2 \frac{|Y_{FB}/Y_{21}|^2 R_n R_{FB}}{R_n + |Y_{FB}/Y_{21}|^2 R_{FB}}, \tag{11.95}$$

$$Y_{corr}^1 = \frac{\overline{i_1^1 (v_1^1)^*}}{|v_1^1|^2} = Y_{corr} + (Y_{11} + Y_{21} - Y_{corr}) \times \frac{(Y_{FB}/Y_{21})R_n + |Y_{FB}/Y_{21}|^2 R_{FB}}{R_n + |Y_{FB}/Y_{21}|^2 R_{FB}}. \tag{11.96}$$

Using the relation of a distributed transmission line, impedance transformation values of $Y_{21}$ and $Y_{21} + Y_{11}$ are provided:

$$Y_{21} = \frac{Z_{21}^T}{(Z_{11}^T Z_{22}^T - Z_{12}^T Z_{21}^T)\cos\theta + j(Z_{11}^T + jX_G)Z_D \sin\theta}, \tag{11.97}$$

$$Y_{11} + Y_{21} = \frac{Z_{21}^T + Z_{22}^T \cos\theta + jZ_D \sin\theta}{(Z_{11}^T Z_{22}^T - Z_{12}^T Z_{21}^T)\cos\theta + j(Z_{11}^T + jX_G)Z_D \sin\theta}, \tag{11.98}$$

where $D$ stands for MESFET drain, $G$ for the gate, and $\theta$ for the electrical length of the drain transmission line, or phase change in case of an inductor.

Several important conclusions are derived from the above set of equations. From Eq. (11.94), it is observed that $R_n^1$ can be reduced by proper selection of feedback inductance. In addition, since $Y_{FB} \gg Y_{21}$, pure inductive feedback means $R_{FB} = 0\ \Omega$, which result in

$$R_n^1 = \left|\frac{Y_{21}}{Y_{21} - Y_{FB}}\right|^2 R_n, \tag{11.99}$$

$$G_n^1 = G_n, \tag{11.100}$$

$$Y_{corr}^1 = Y_{corr}\left(1 - \frac{Y_{FB}}{Y_{21}}\right) + Y_{FB}\left(1 - \frac{Y_{11}}{Y_{21}}\right). \tag{11.101}$$

The influence of a finite parallel-feedback resistor is always shown by increasing the value of $R_n^1$. When using reactive feedback or resistive and reactive feedback, proper attention should be paid to stability criterion. The impedance level of the drain transmission line $Z_D$ and its length or phase response $\theta$ have significant impact on the noise figure VSWR (return loss) and gain flatness. This phase response corresponds to the inductor value in a lumped-element design.

## 11.6 Distributed Amplifier

The concept of distributed amplifiers is old and was introduced in the 1940s for very broadband design of vacuum tube amplifiers. However, with the availability and advanced GaAs and PHEMT technology, distributed amplifiers were implemented in MIC technology and later on in monolithic microwave integrated circuit (MMIC) technology, which made it much more reproducible. The relatively wide BW gain of this topology made these amplifiers suitable TIA for high-data rate transport such as 9.92 Gb/s OC-192[35] and 40 GB/s.[36] One of the reasons for distributed amplifier usage in high data rate traffic is the recent ability to have BW that covers the low harmonics of the digital data stream. Assuming PRBS of 2^31-1, the number of spectral lines of 9.92 Gb/s NRZ code is 2147483647 with frequency spacing of approximately 4.62 Hz. Hence, in order to recover the data properly for having a high-performance eye pattern, the amplifier should have a very low start frequency. This is achieved by proper design of the distributed amplifier transmission line for the input and output signals as well as getting as close as possible to dc-coupling performance in both of the amplifier's input and output ports. That way, it is assured that the signal harmonics are sufficient to recover the eye diagram. Additionally, due to the same reason a of very wide digital signal, biasing should have wide-band isolation; hence, the RFC coil should provide high impedance at low frequencies as well. In MMIC technology, RFC is realized by an active RFC circuitry based on FETs.

### 11.6.1 Concept of operation

The distributed amplifier is composed of two main transmission lines: a gate transmission line path for the input signal and drain transmission line path for the output signal (see Fig. 11.12). Each transmission line is terminated by a load of approximately 50 Ω. The idea is to create the path of the gate transmission line by using the gate source $C_{gs}$ capacitance and connect the gate with series inductance $L_{gs}$. Similarly, the output transmission line is made by using the drain source capacitance $C_{ds}$ and connecting the consecutive FETs with the serial inductor $L_{ds}$ that creates the drain transmission line path.[37–39] On the basis of these concepts, the gate cut-off frequency is defined as

$$\omega_g = \frac{1}{R_i C_{gs}}. \tag{11.102}$$

In the same way, the drain cut-off frequency is defined by

$$\omega_d = \frac{1}{R_{ds} C_{ds}}. \tag{11.103}$$

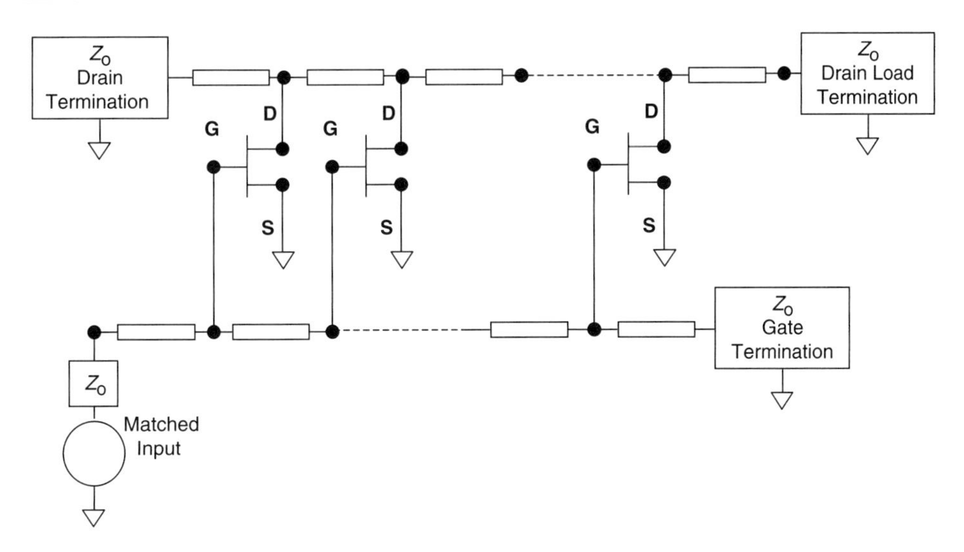


**Figure 11.12** Distributed FET amplifier concept.

The other definition refers to the *k*-derived transmission-line ladder networks that represent the artificial gate and drain transmission lines (Fig. 11.13). Both transmission lines should have the same propagation velocity, or in other words, the phase shift versus frequency between each gate-line section should be equal to the phase shift between each drain-line section. This way, it is

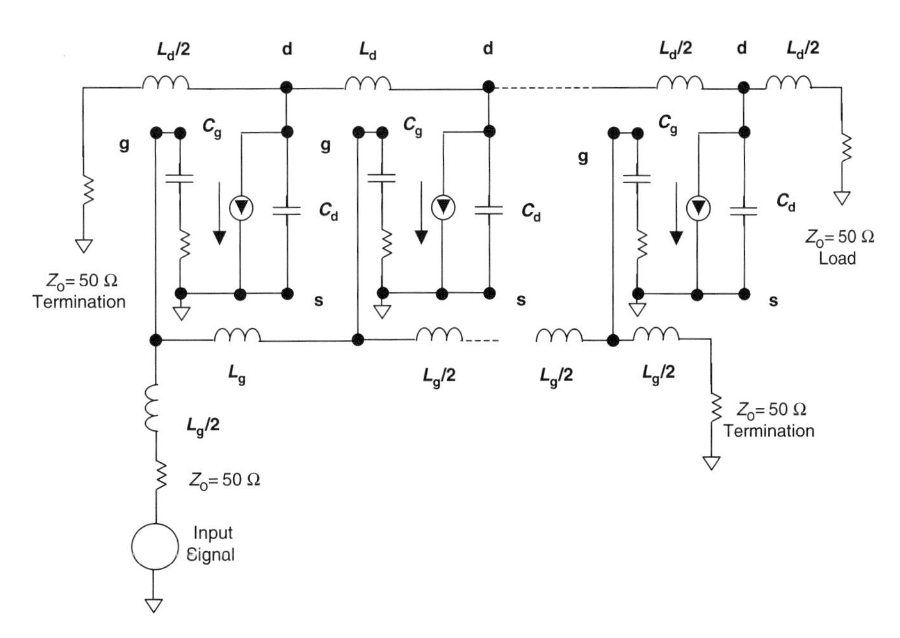


**Figure 11.13** Simplified FET amplifier circuit model showing $C_g$, $C_d$, and gate resistance $R_i$ sometimes called $R_g$.

assured that the power summation of the traveling wave would be in phase: Hence, the following constraint is applicable:[40]

$$\omega_c = \frac{2}{\sqrt{L_g C_{gs}}} = \frac{2}{\sqrt{L_d C_{ds}}}. \tag{11.104}$$

Additionally, both transmission lines should have the same characteristic impedance:

$$Z_{og} = Z_{od}, \tag{11.105}$$

hence,

$$Z_o = \sqrt{\frac{L_g}{C_{gs}}} = \sqrt{\frac{L_d}{C_{ds}}}. \tag{11.106}$$

The losses of the transmission line are mainly due to the active device input resistance at the gate and output resistance at the drain (Fig. 11.14). In fact, the dominant loss is due to the gate resistance $R_i$. For that reason, the following normalized frequency term is defined:

$$k_g = \frac{2R_i C_{gs}}{\sqrt{L_g C_{gs}}} = \frac{\omega_c}{\omega_g}, \tag{11.107}$$

$$k_d = \sqrt{\frac{L_d C_{ds}}{2R_{ds} C_{ds}}} = \frac{\omega_d}{\omega_c}. \tag{11.108}$$

The gate transmission line losses are defined by

$$A_g = \frac{k_g X_k^2}{\left[1 - (1 - k_g^2) X_k^2\right]^{\frac{1}{2}}}. \tag{11.109}$$

The drain transmission losses are defined by

$$A_d = \frac{k_d}{(1 - X_k^2)^{\frac{1}{2}}}, \tag{11.110}$$

where

$$X_K = \frac{\omega_c}{\omega}. \tag{11.111}$$

Figure 11.15 illustrates the gate transmission line losses versus normalized frequency $X_k$, where $k_g$ is a parameter. Figure 11.16 illustrates the drain transmission line losses versus normalized frequency $X_k$, where $k_d$ is a parameter. It can be observed that the drain path has a higher attenuation for the same $k$ parameter.

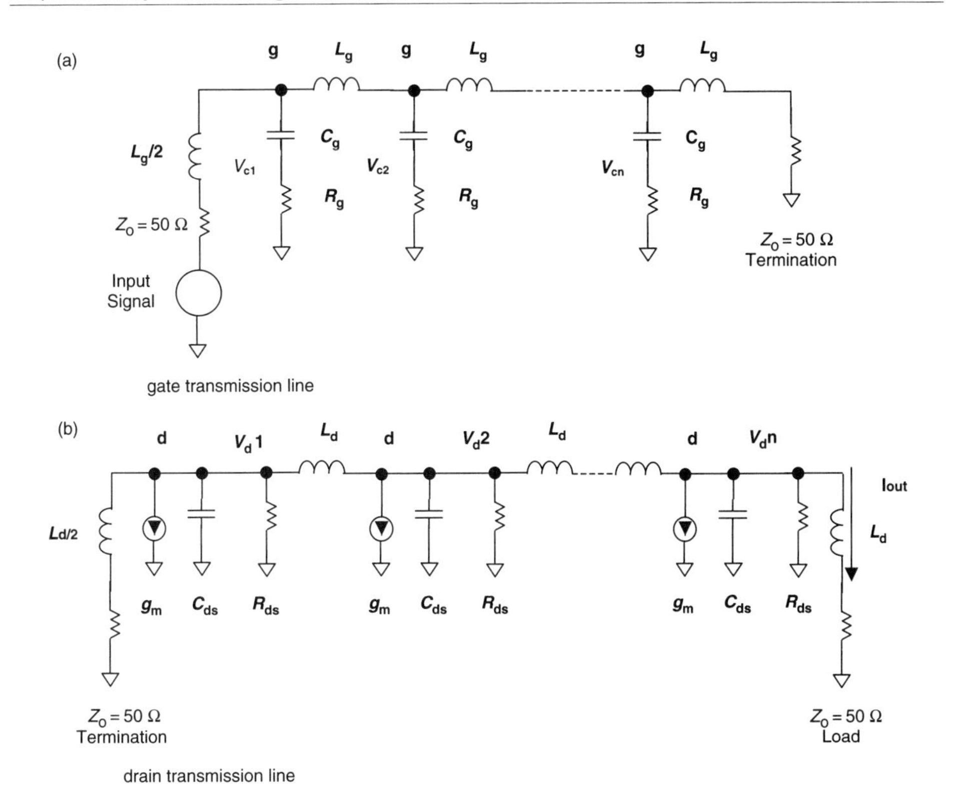


**Figure 11.14** (a) A distributed FET amplifier gate transmission line; (b) distributed FET amplifier drain transmission line.

It is evident from the figures that the gate-line attenuation is more sensitive to frequency than drain-line attenuation. Further, unlike attenuation in the gate line, the drain-line attenuation does not vanish in the low-frequency limit. Therefore, the frequency response of the amplifier can be expected to be predominantly controlled by the attenuation on the gate line as can the dc gain by the attenuation on the drain line.

The design goal is to minimize the value of $k_g$ and $k_d$, which means to extend the gate transmission line frequency response by minimizing the gate input capacitance; this can be done either by a series capacitor to the gate or by an input inductor to the gate. The disadvantage is a voltage divider network that reduces the input signal amplitude to the FET. The advantage, in the case of a high-power driver design, is that a larger FET can be used. The same resolution applies to the drain, where the drain is tapped by a parallel capacitor. However, this reduces the drain cut-off frequency and the drain transmission-line-frequency response as well. The gain of the $k$-derived transmission-line distributed amplifier is given by

$$G = \frac{g_m(Z_{og}Z_{od})^{\frac{1}{2}} \sin h[n/2(A_d - A_g)] \exp(-n(A_d - A_g)/2)}{2[1 + (\omega/\omega_g)^2]^{\frac{1}{2}}[1 - (\omega/\omega_c)^2]^{\frac{1}{2}} \sin h[(A_d - A_g)/2]}, \tag{11.112}$$

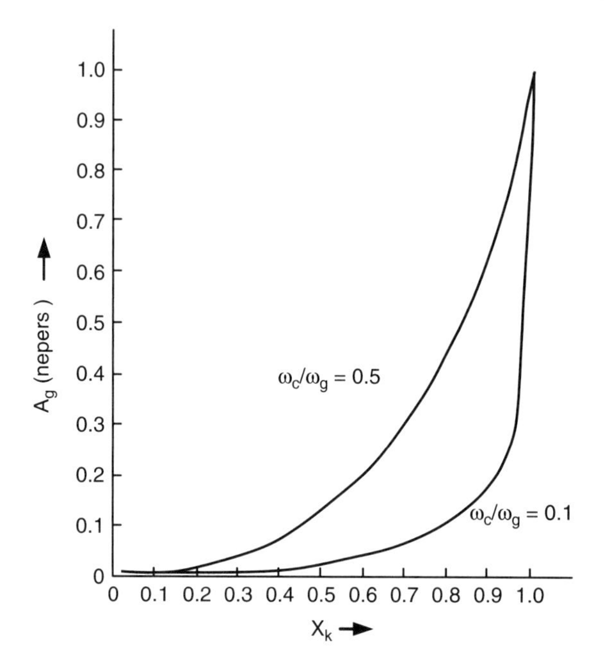


**Figure 11.15** Gate path transmission line losses vs normalized frequency $k_g$ is a parameter.[41] (Reprinted with permission from *IEEE Microwaves Theory and Techniques* © IEEE, 1984.)

where $n$ is the number of gain cells and $Z_{og}$ and $Z_{od}$ are the gate and the drain transmission-line characteristic impedances, respectively. The optimum number of cells for a gain performance at a given frequency is

$$\left.\frac{\partial G}{\partial n}\right|_{@\omega=\omega o} = 0 \Rightarrow N_{opt} = \frac{\ln(A_d/A_g)}{A_d - A_g}. \quad (11.113)$$

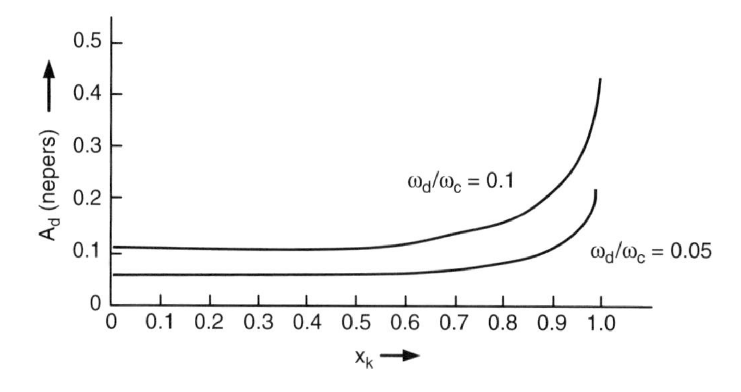


**Figure 11.16** Drain path transmission line losses vs. normalized frequency $k_d$ is a parameter.[41] (Reprinted with permission from *IEEE Microwaves Theory and Techniques* © IEEE, 1984.)

Practically, there is a limit for $n$. This is because very little gain improvement is obtained due to the increase in the losses of the gate and the drain transmission line paths. Generally, the maximum number of gain cells is not greater than eight. Different network topology, called constant $R$, can be used to synthesize distributed amplifiers. The advantage of constant $R$ over the $k$-derived network is that it can achieve approximately $\sqrt{2}$ times larger BW. The noise figure of a $k$-derived distributed amplifier was calculated by Aitchinson and Niclas[43] and is given by

$$F = 1 + \left[\frac{\sin(n\beta)}{n \cdot \sin(\beta)}\right]^2 + \frac{4}{n^2 g_m Z_{og} Z_{od}} + \frac{Z_{og}\omega^2 C_{gs}^2 R \sum_{r=1}^{n} f(r,\beta)}{n^2\, g_m} + \frac{4P}{n g_m Z_{og}}, \tag{11.114}$$

where $R$ and $P$ are numerical factors, given by Ref. [44]. The characteristic impedance of the gate or drain artificial transmission line versus frequency is given by

$$Z_\pi = \sqrt{\frac{L/C}{1 - (\omega/\omega_c)^2}}, \tag{11.115}$$

$$\beta = 2\sin^{-1}\frac{\omega}{\omega_c}. \tag{11.116}$$

Equations (11.104) and (11.116) are the phase constant per gain section. The design goal is to have equality between the gate transmission-line propagation factor and the drain propagation factor, i.e., $\beta = \beta_g = \beta_d$. This constraint defines that both the artificial gate and drain transmission lines would have the same cut-off frequency:

$$\omega_c = \frac{4}{LC}, \tag{11.117}$$

where $g_m$ is the FET transconductance, $R$ is the FET noise resistance, $P$ is the FET noise parameter $f(r, \beta) \approx n^3/3$.

The noise figure in Fig. 11.17 describes a distributed amplifier with the following MESFET parameters and resulted calculation values of $L_g$, $Z_{od}$, $L_d$, and cut-off frequency $f_c$. The MESFET parameters are given in Table 11.1. The fact that this is not a minimum noise-figure matching example should be kept in mind.

## 11.7 Operational Transimpedance Amplifiers PIN TIA

Operational PIN TIA amplifiers are commonly used in digital receivers up to 10 Gb/s and even 40 GHz using a multipole BW enhancement.[45,46] In this

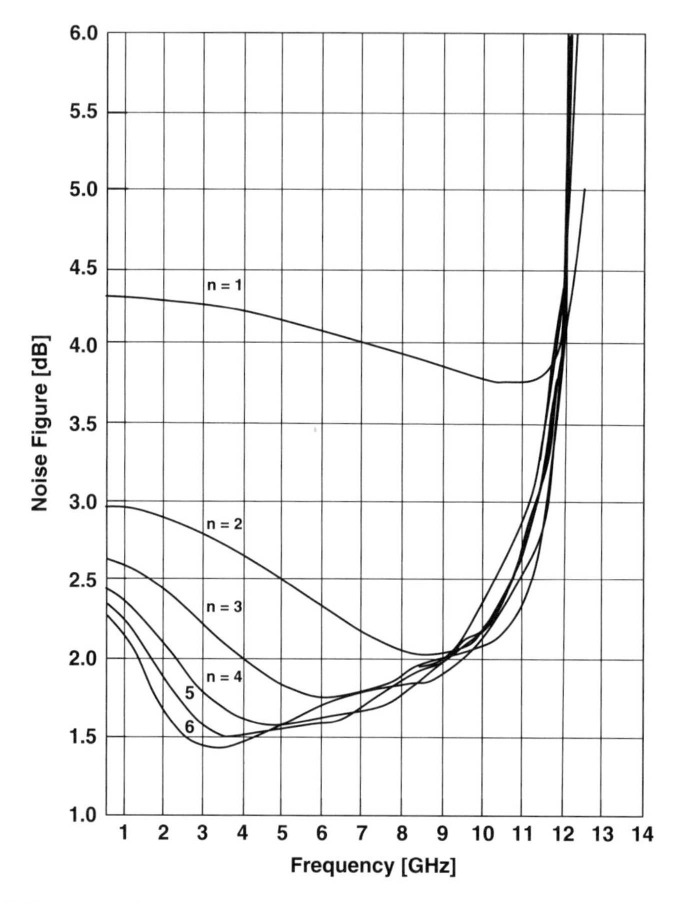


**Figure 11.17** Noise figure of a distributed amplifier vs frequency and number of gain cells *n* as a parameter.[42] (Reprinted with permission from *IEEE Microwaves Theory and Techniques* © IEEE, 1985.)

**Table 11.1** Distributed amplifier parameters for min NF.

| | Value |
|---|---|
| MESFET parameter symbol | |
| $Z_{og}$ | 50 Ω |
| $C_{gs}$ | 0.5 pF |
| $C_{ds}$ | 0.2 pF |
| $g_m$ | 30 ms |
| $R$ | 0.2 |
| $P$ | 0.3 |
| MESFET calculated | |
| $L_g$ | 1.25 nHy |
| $Z_{od}$ | 125 Ω |
| $L_d$ | 3.13 nHy |
| $f_c$ | 12.7 GHz |

technique, several passive networks are added. Each network is designed independently to control transfer function and improve the frequency response. Parasitic capacitors between stages are isolated and absorbed into the passive networks, as will be elaborated later on. For higher data rates, the PIN TIA approach involves the use of distributed amplifiers and high-power distributed amplifiers as laser drivers. In analog applications, operational amplifiers (OP-amps) are used for return-path receiving and transmitting, which occupies the BW of 5–50 MHz. With recent technology, the high-dynamic range and high $IP_3$ as well as wide BW operation are available. Since the input signal to the amplifier is the signal photocurrent, and since the output is signal voltage, the gain ratio of output to input is transimpedance.

The transimpedance gain is determined by the shunt feedback, connected between the OP-amp output and the noninverting input of the OP-amp. Since a transimpedance amplifier uses feedback, it is important to pay attention to the stability criteria of phase margin and gain margin. As a rule of thumb, phase margin should be 50 deg and gain margin should be 10 dB at a minimum.[18] Where pure resistive feedback occurs, there is no phase shift between output and input. The analysis of a regular OP-amp with reactive feedback will be straightforward.

### 11.7.1 Analysis and synthesis methods

A typical front-end configuration using an OP-amp PIN TIA is illustrated by Fig. 11.18. The photodetector PIN diode is considered to be a current source with output impedance $Z_{PD}(\omega)$. As was explained previously in Sec. 8.2.3, the PIN photodetector output impedance is governed by a parallel load of the output capacitance and resistance. The input impedance of an ideal OP-amp is $Z_{in}(\omega) \to \infty$. The parallel feedback connection between the output and the inverting input is $Z_{FB}(\omega)$. The voltage gain of the OP-amp is given by the term $A_V(\omega)$. Assuming an inversion gain, the transimpedance gain is given by solving the schematic case in Fig. 11.18(b) and finding the transimpedance gain $Z_{TIA}(\omega)$ in Eq. (11.118).[17,47] TIA can be based on either bipolar or FET and CMOS devices[46,48,49] or regular MSM (metal semiconductor metal)/HEMT PIN TIA synthesized in RF methods, and does not have to be a differential amplifier:[50]

$$Z_{TIA}(\omega) = \frac{V_{OUT}(\omega)}{I_{IN}(\omega)} = \frac{-Z_{FB}(\omega)}{1 + \frac{1}{A_V(\omega)}\left(1 + \{[Z_{FB}(\omega)/Z_{PD}(\omega)]\}\right)}. \tag{11.118}$$

The ratio $Z_{FB}(\omega)/Z_{PD}(\omega)$ is a loss ratio that shows the shunt current dissipated on the photodiode output impedance. Hence not all the photodetection current flows into the TIA input port. The ideal case is similar to an ideal current source, where $Z_{PD}(\omega) \gg Z_{FB}(\omega)$ and $Z_{PD}(\omega) \to \infty$. In case of $A_V(\omega) \to \infty$, an ideal OP-amp, the transimpedance gain becomes $-Z_{FB}(\omega)$. This is because the input is a current source to an ideal inverting OP-amp.

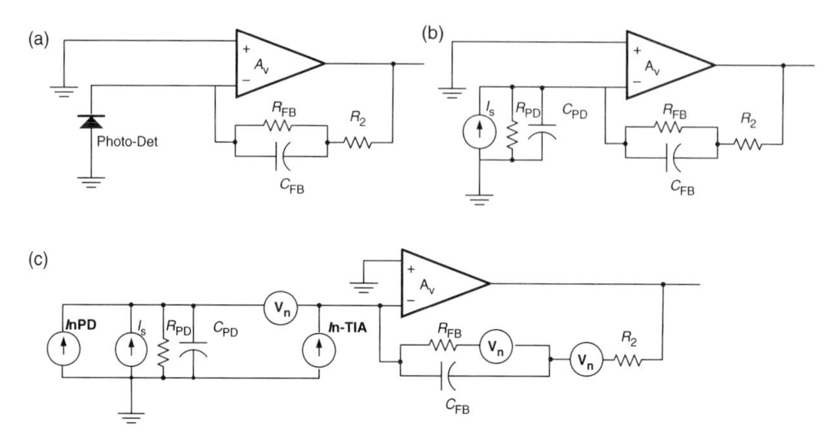


**Figure 11.18** Operational TIA front-end. (a) Typical application; (b) equivalent circuit; (c) noise modeling thermal noise due to PD output resistance is ignored but can be added using Norton identity as current source in parallel to the shot noise of the PD, as was explained in Chapter 10.

The next step is to find the input impedance of the above-mentioned TIA, including the shunt output impedance of the photodetector. Since the discussion here is about a $Z$-matrix transfer function, where the excitation is current and the result is voltage, the output of the TIA will be left open. Solving the ratio of $V_{\mathrm{IN}}(\omega)/I_{\mathrm{S}}(\omega)$ provides the value of $Z_{\mathrm{IN}}(\omega)$ for the circuit in Fig. 11.18(b):

$$\begin{aligned} Z_{\mathrm{IN}} &= \frac{1}{[1/Z_{\mathrm{PD}}(\omega)] + [1 + A_{\mathrm{V}}(\omega)/Z_{\mathrm{FB}}(\omega)]} \\ &= \frac{Z_{\mathrm{PD}}(\omega)Z_{\mathrm{FB}}(\omega)}{Z_{\mathrm{FB}}(\omega) + [1 + A_{\mathrm{V}}(\omega)]Z_{\mathrm{PD}}(\omega)} \\ &= \frac{Z_{\mathrm{PD}}(\omega)\{Z_{\mathrm{FB}}(\omega)/[1 + A_{\mathrm{V}}(\omega)]\}}{Z_{\mathrm{PD}}(\omega) + \{Z_{\mathrm{FB}}(\omega)/[1 + A_{\mathrm{V}}(\omega)]\}} \\ \Rightarrow Z_{\mathrm{IN}} &= \frac{Z_{\mathrm{FB}}(\omega)}{[1 + A_{\mathrm{V}}(\omega)]} \Big| Z_{\mathrm{PD}}(\omega) \end{aligned} \tag{11.119}$$

It is observed that for high gain, the input impedance is $Z_{\mathrm{IN}}(\omega) \approx 0$ since the current feedback reduces the input impedance of the TIA. This method of matching the photodetector, which is a high-output impedance, to a low-input impedance TIA is called "cascading" by impedance mismatch. In this manner, the whole photodetector current is loaded by the TIA input impedance. Having low-input impedance TIA at the front end minimizes the influence of the photodiode stray capacitance on the overall BW of the system by having a large BW starting from dc to its designated BW. On the other hand, having a high-feedback resistance may increase the transimpedance gain, thus improving the dynamic range due to a high SNR, consequently resulting a dynamic range that can reach 30 dB. Implementation of designs using avalanche photo detector (APD) and AGC may exceed a dynamic range on the order of 40 dB.[25] Observing $Z_{TIA}(\omega)$ and modeling $Z_{FB}(\omega)$ as a pure resistor with residual stray capacitance $C_{FB}$, the TIA gain as function of frequency is given by

$$Z_{TIA}(\omega) = \frac{V_{OUT}(\omega)}{I_{IN}(\omega)} = \frac{\{[-R_{FB}(1/j\omega C_{FB})/R_{FB} + (1/j\omega C_{FB})]\}}{1 + [1/A_V(\omega)](1 + \{[R_{FB}(1/j\omega C_{FB})/R_{FB} + (1/j\omega C_{FB})]/(1/j\omega C_{PD})\})}. \quad (11.120)$$

After some simplification of the above expression, the transimpedance gain is noted as

$$Z_{TIA}(\omega) = \frac{-R_{FB}}{1 + \dfrac{1}{A_V(\omega)} + j\omega R_{FB}\{C_{FB} + [(C_{FB} + C_{PD})/A_V(\omega)]\}}, \quad (11.121)$$

or

$$Z_{TIA}(\omega) = \frac{V_{OUT}(\omega)}{I_{IN}(\omega)} = \frac{-R_{FB}}{1 + j(\omega/\omega_p)}, \quad (11.122)$$

where

$$\omega_p = \left. \frac{1 + [1/A_V(\omega)]}{R_{FB}\{C_{FB} + [(C_{PD} + C_{FB})/A_V(\omega)]\}} \right|_{@A_V(\omega)\to\infty} \approx \frac{1}{C_{FB}R_{FB}} \quad (11.123)$$

In conclusion, at a high gain of $A_V(\omega) \to \infty$, the TIA BW is independent of the photodiode shunt capacitance and is related only to the feedback impedance. This is clear because the TIA input impedance with a parallel current feedback tends to zero. However, in real case, there are more parasites that start to affect high frequencies and add additional frequency rolloff to the transimpedance gain. For instance, the PIN diode is internal series resistance $R_s$ adds a pole of $\tau = C_{PD}R_s$; however it would be insignificant in case of low $R_s$ value. Moreover, in order to minimize the value of $C_{PD}$, the PIN photodiode is reverse biased with high voltage, where the depletion capacitance $C_{PD}$ is minimized even further. A typical reversed bias 40-μm PIN photodetector has a capacitance at the range of about $0.5\text{ pF} \le C_{PD} \le 0.6\text{ pF}$.

The more commonly used network is a $T$ feedback network in order to compensate the rolloff created by $C_{FB}$ (Fig. 11.19). This method is called zero-pole

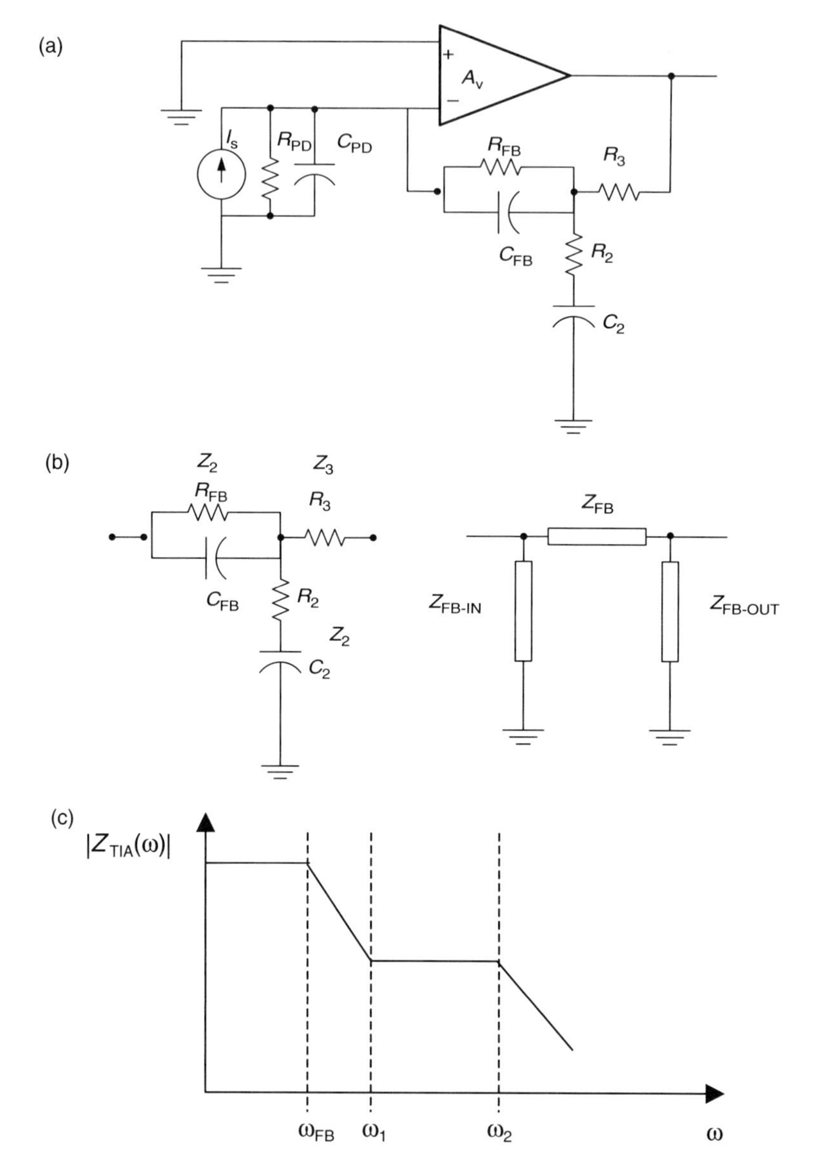


**Figure 11.19** Zero-pole cancellation $T$ network. (a) T concept, (b) $Y$ to $\Delta$ transformation, (c) frequency response.

cancellation. The $T$ network is not so convenient to analyze and the network is simplified by using $Y$ to $\Delta$ transformation:[51]

$$Z_{FB}(\omega) = \frac{Z_1(\omega)Z_2(\omega) + Z_2(\omega)Z_3(\omega) + Z_3(\omega)Z_1(\omega)}{Z_2(\omega).}. \tag{11.124}$$

Using the same technique, the two other impedance values are calculated, resulting in a $\pi$ network in the feedback path:

$$Z_{\text{FB-in}}(\omega) = \frac{Z_1(\omega)Z_2(\omega) + Z_2(\omega)Z_3(\omega) + Z_3(\omega)Z_1(\omega)}{Z_3(\omega),}, \tag{11.125}$$

$$Z_{\text{FB-out}}(\omega) = \frac{Z_1(\omega)Z_2(\omega) + Z_2(\omega)Z_3(\omega) + Z_3(\omega)Z_1(\omega)}{Z_1(\omega)}. \tag{11.126}$$

Hence, the first impedance $Z_{\text{FB-in}}(\omega)$ loads the input of the TIA. This load is in parallel with the photodetector shunt capacitance $C_{\text{PD}}$. The second impedance is the feedback $Z_{\text{FB}}(\omega)$, which is given by the $Y$ to $\Delta$ transformation and the third impedance is $Z_{\text{FB-out}}(\omega)$, which loads the output of the TIA and merges with the load in parallel. The effective transimpedance value is $Z_{\text{FB}}(\omega)$, given by Eq. (11.125).

Kirchhoff's mesh equations, or direct solution of voltage divider, can be obtained from Fig. 11.19(b)—an equivalent circuit. Assuming following time constraints

$$R_{\text{FB}} \gg R_3 \gg R_2 \quad \text{and} \quad C_2 \gg C_{\text{FB}}, \tag{11.127}$$

the following approximated transfer impedance function is given by

$$Z_{\text{TIA}}(\omega) \cong -R_{\text{FB}} \frac{\left(1 + j\dfrac{\omega}{\omega_1}\right)}{\left(1 + j\dfrac{\omega}{\omega_{\text{FB}}}\right)\left(1 + j\dfrac{\omega}{\omega_2}\right)}, \tag{11.128}$$

where

$$\omega_{\text{FB}} = \frac{1}{R_{\text{FB}}C_{\text{FB}}}, \quad \omega_2 = \frac{1}{R_2C_2}, \quad \omega_1 = \frac{1}{R_3C_2} \quad \Rightarrow \omega_2 > \omega_1 > \omega_{\text{FB}}. \tag{11.129}$$

Setting the zero frequency to $\omega_1 = \omega_{\text{FB}}$ would achieve an ideal zero-pole cancellation. This, of course, requires a manual alignment of $R_3$ by setting its value, which in turn requires a calibration method similar to an oscilloscopic probe calibration for minimum overshoot or undershoot.[52] Such an approach is problematic in a large-scale production line.

In the above equations the OP-amp-gain approximation is as given by

$$A_{\text{V}}(\omega) = \frac{A_0}{1 + j(\omega/\omega_\alpha)}, \tag{11.130}$$

where $\omega_\alpha$ is the pole frequency of the OP-amp, which results in an open-loop-gain rolloff of 20 dB/dec. Substituting Eq. (11.130) into Eq. (11.123) results in a second-order transfer function response for the OP-amp:[17]

$$Z_{\mathrm{TIA}} = -R_{\mathrm{TIA}} \frac{1}{1 + j(\omega/Q\omega_0) - (\omega/\omega_0)^2}, \tag{11.131}$$

and the input impedance is given by

$$Z_{\mathrm{IN}} = R_{\mathrm{IN}} \frac{1 + j(\omega/\omega_\alpha)}{1 + j(\omega/Q\omega_0) - (\omega/\omega_0)^2}, \tag{11.132}$$

where, after comparing coefficients[53]

$$R_{\mathrm{TIA}} = R_{\mathrm{FB}} \frac{A_0 R_{\mathrm{PD}}}{(A_0 + 1)R_{\mathrm{PD}} + R_{\mathrm{FB}}}, \tag{11.133}$$

$$R_{\mathrm{IN}} = \frac{[R_{\mathrm{FB}}/(A_0 + 1)]R_{\mathrm{PD}}}{R_{\mathrm{PD}} + [R_{\mathrm{FB}}/(A_0 + 1)]}. \tag{11.134}$$

[See Eq. (11.118).]

$$Q = \sqrt{\frac{R_{\mathrm{PD}} R_{\mathrm{FB}} (C_{\mathrm{FB}} + C_{\mathrm{PD}})}{\omega_\alpha [(A_0 + 1)R_{\mathrm{PD}} + R_{\mathrm{FB}}]}} \times \frac{\omega_\alpha [(A_0 + 1)R_{\mathrm{PD}} + R_{\mathrm{FB}}]}{R_{\mathrm{PD}} + R_{\mathrm{FB}} + \omega_\alpha R_{\mathrm{PD}} R_{\mathrm{FB}} [C_{\mathrm{PD}} + C_{\mathrm{FB}}(A_0 + 1)]}, \tag{11.135}$$

$$\omega_0 = \sqrt{\frac{\omega_\alpha [(A_0 + 1)R_{\mathrm{PD}} + R_{\mathrm{FB}}]}{R_{\mathrm{PD}} R_{\mathrm{FB}} (C_{\mathrm{FB}} + C_{\mathrm{PD}})}}, \tag{11.136}$$

$$A_0 = A_{\mathrm{V}}(\omega)|_{@\omega=0} \tag{11.137}$$

As can be observed, a second-order response is characterized by its resonance frequency and quality factor $Q$ and damping factor $\xi$.[47,54,55] The TIA frequency response is a low-pass response with overshoot at the high-end frequency. When the quality factor is high the overshoot peak is high and narrow; this means the circuit has a low damping factor value. The rolloff is steeper than in a low $Q$ design with high damping factor and low overshoot response.[18] The damping factor is given by

$$\xi = \frac{1}{2Q}. \tag{11.138}$$

Proper selection of $Q$-factor values or damping values allows the design of several transfer functions such as maximally flat amplitude response

Butterworth,[53] equal ripple Chebyshev, or maximally flat group delay Bessel. In applications where a monotonic step response is required, $Q = 0.5$ is recommended; however, with $Q = 0.707$ a maximally flat frequency response is obtained at the expense of only 4.3% overshoot in the step response of a two-pole Butterworth design. Given the desired transimpedance BW $\omega_0$ and the values for the total input capacitance, feedback resistance, and feedback capacitance, Eq. (11.135) can be solved. Due to the complexity of the relation between $Q$ and $\omega_0$, CAD tools are used to produce design curves.[47,53] Moreover, controlling, individually, the frequency response of each gain section of the TIA can help to extend the TIA GBW (gain BW product).[45] An inspection of the mathematical expression describing the transfer characteristic of an amplifier with lumped components as coupling elements will reveal that the gain and the BW cannot be simultaneously increased beyond a certain limit. Bode showed that there exists a maximum GBW regardless of the network used. Fano and Youla further generalized the theory for a larger class of impedances. The GBW of such an amplifier was given by Wong:[39]

$$\mathrm{GBW}_{\mathrm{MAX}} = \frac{g_{\mathrm{m}}}{\pi C}, \tag{11.139}$$

where $g_{\mathrm{m}}$ is the device transconductance and the quantity $C$ is defined as

$$C = \lim_{\omega \to \infty} \left( \frac{1}{j\omega Z} \right). \tag{11.140}$$

However, the condition in Eq. (11.139) is not valid for amplifiers where the load does not satisfy the condition of impedance function in Eq. (11.140). An impedance function in the complex frequency domain is a ratio between two polynomials with real coefficients. With no singularities in the interior of the right half of the plane, the numerator degree is higher than the denominator by, at most, one degree. In other words, the overall transfer function of an amplifier is

$$A_{\mathrm{V}}(j\omega) = g_{\mathrm{m}} Z(\omega). \tag{11.141}$$

Since $Z(\omega)$ is not an impedance function, the Bode–Fano limit in Eq. (11.140) is not valid.

The role of multipole BW enhancement[45] is to arrange a ladder LC passive network between the stages to control the transfer-function shape. This method results in a transfer function that does not satisfy the Bode–Fano impedance function condition of Eq. (11.140). Figures 11.20(a) and (b) show an equivalent small-signal-circuit model for a transistor used in a cascaded configuration.

Hence for such an approach, the transistor voltage gain, the cascaded network gain, and the cascaded load impedance are given by

$$A_{\mathrm{Veq}}(j\omega) = G_{\mathrm{m}} Z_{21}(\omega) \tag{11.142}$$

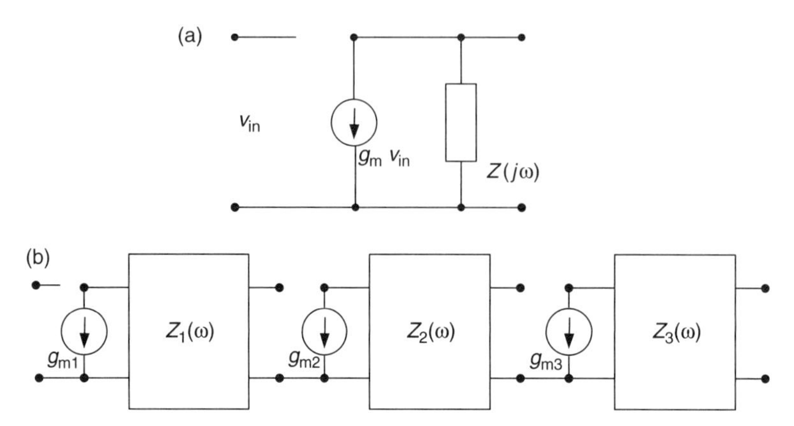


**Figure 11.20** General small signal model of an amplifier with arbitrary interstage passive matching. (a) Gain block model; (b) cascaded network.[45] (Reprinted with permission from *IEEE Solid State Circuits* © IEEE, 2004.)

$$G_m = g_{m1} \cdot g_{m2} \cdot g_{m3} \tag{11.143}$$

$$Z_{21}(j\omega) = Z_{21,1}(j\omega) \cdot Z_{21,2}(j\omega) \cdot Z_{21,3}(j\omega). \tag{11.144}$$

It is clear that $Z_{21}(j\omega)$ is the product of the three interstage network impedances.

Further, each network can be optimized individually. $Z_{21}(j\omega)$ is a rational function; however it can have a numerator one degree higher than the denominator. Hence, Eq. (11.142) is of the form of Eq. (11.141), but not limited by the condition of Eq. (11.139).

One design approach is the stagger design. Each low pass ladder network is optimized to different response (Fig. 11.21). The first ladder rolloff is

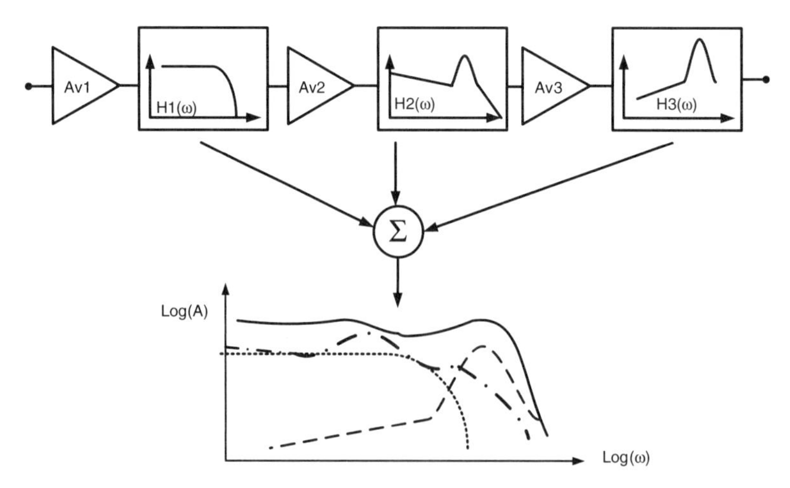


**Figure 11.21** Stagger design tuning for designing wide band amplifiers.[45] (Reprinted with permission from *IEEE Solid State Circuits* © IEEE, 2004.)

compensated by the second ladder peaking, and the third ladder resonates as a band pass to compensate the rolloff of the second LPF, which extends the BW even further. Moreover, Eq. (11.144) demonstrates that the entire transfer function of the TIA is wholly dependent upon the passive network $Z_{21}(j\omega)$ characteristics.

The design approach is based on the LPF response-matching-network technique. In this method, parasitic components of the transistors, such as output capacitance and input capacitance, are absorbed as a part of the matching. Parasitic components absorption is used in distributed amplifiers as well.[8] The filter element transformer is designed by standard network synthesis methods.[56] The network order $n$ is a free-design parameter that may improve the network response at a higher order; however, one has to remember that there is a limit after which losses begin to affect the performance. As a design example, a Butterworth maximally flat third order ($n = 3$) is provided:

$$C_1 = \frac{1}{R_1(1-\delta)\omega_C}, \tag{11.145}$$

$$L_2 = \frac{2}{\left(1-\delta+\delta^2\right)\cdot\omega_C^2\cdot C_1}, \tag{11.146}$$

$$C_3 = \frac{1}{R_2(1+\delta)\omega_C}, \tag{11.147}$$

where $\delta$ is the impedance transformation ratio between $R_1$ and $R_2$ defined as

$$\delta = n\sqrt{\frac{R_1-R_2}{R_1+R_2}}, \tag{11.148}$$

and $\omega_C$ is the 3-dB cut-off frequency of the network. The resulting network is a $\pi$ configuration starting with parallel capacitor $C_1$, which absorbs the output capacitance of the first-stage amplifier and ending with what absorbs $C_3$, the input capacitance of the next stage amplifier. Note that $R_1$ and $R_2$ are the output and the input resistance of the first transistor and the second amplifier, respectively. The general case of an $n$-order ladder network is presented in Fig. 11.22.

The BW enhancement ratio (BWER) is defined as the ratio between the 3-dB BW of the amplifier prior to inserting the passive network and the extended BW of the amplifier with the interstage passive-matching network:

$$\text{BWER} = \frac{\omega_{\text{C-After}}}{\omega_{\text{C-Before}}} = \frac{1}{1-\delta}\cdot\frac{R_2}{R_1+R_2}\cdot\frac{C_1+C_3}{C_1}. \tag{11.149}$$

A special case refers to the input stage impedance and the photodiode output impedance, where there is a large $\delta$ factor due to the reverse bias high impedance of the photodiode. The photodiode impedance composed from a high-value resistance and low-value stray capacitance.

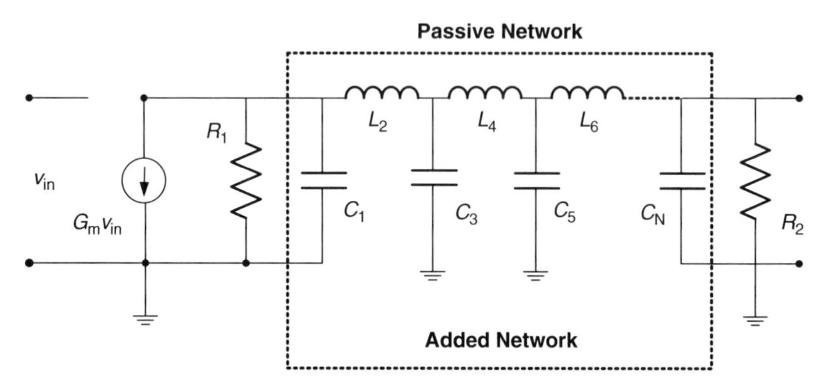


**Figure 11.22** Interstage ladder LPF-type matching network.[45] (Reprinted with permission from *IEEE Solid State Circuits* © IEEE 2004.)

Since $R_1 \gg R_2$ then the BWER expression is modified to

$$\text{BWER} = \frac{\omega_{\text{C-After}}}{\omega_{\text{C-Before}}} = \frac{1}{1-\delta} \cdot \frac{R_2}{R_1} \cdot \frac{C_1 + C_3}{C_1}. \tag{11.150}$$

BWER expressions are used as guidelines for amplifier design and network synthesis by using CAD tools. Figure 11.23 demonstrates a simplified design approach of a CMOS TIA. Figure 11.24 depicts the transimpedance multipole performance.

Using a multipole BW enhancement technique enables designs of CMOS and BiCMOS TIAs with data rates of up to OC-192 and linear-phase response.

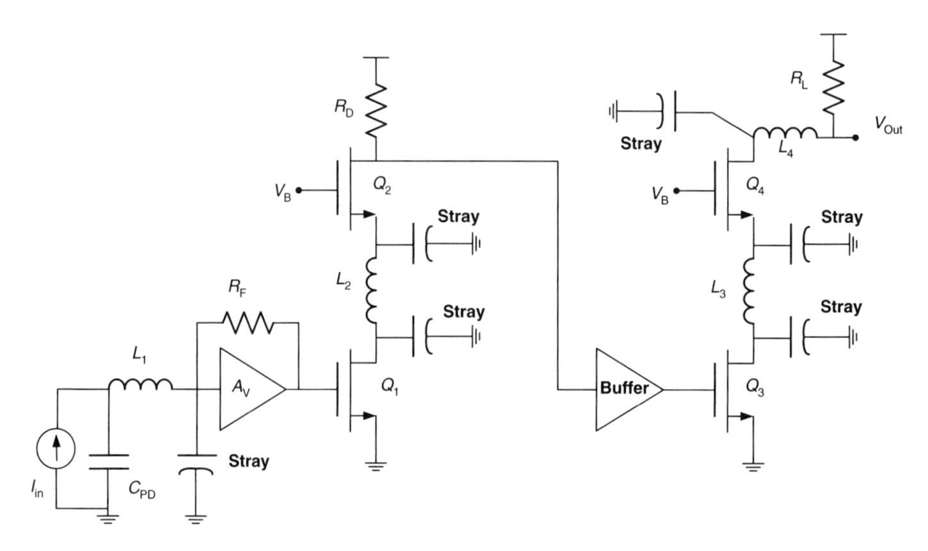


**Figure 11.23** Simplified schematic of TIA with stray parasitic-capacitance absorption at the front end and at the interstage matching.[45] (Reprinted with permission from *IEEE Solid State Circuits* © IEEE, 2004.)

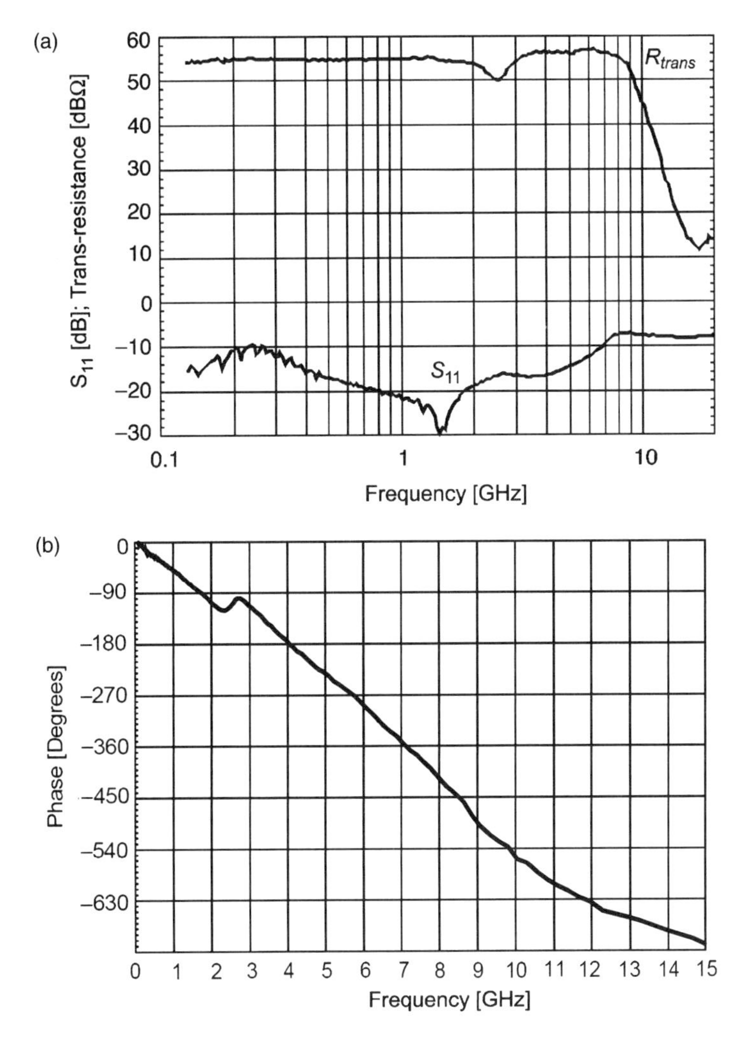


**Figure 11.24** (a) Measured transimpedance gain of multipole BW enhancement TIA with 0.5 pF photodetector. (b) Phase response of multipole BW enhancement TIA 0.5 pF photodetector.[45] (Reprinted with permission from *IEEE Solid State Circuits* © *IEEE*, 2004.)

Linear phase response is an essential prerequisite for minimum group delay, as a condition for minimum deterministic jitter.

### 11.7.2 Noise analysis of operational transimpedance amplifiers PIN TIA

In Secs. 11.1 and 11.2, noise analysis was demonstrated using methods of superposition and introducing the noise correlation factor. TIA can be implemented as a common source[57] amplifier or OP-amp. High-speed OP-amp TIAs are used for high-speed data and are realized in CMOS RF methods.[49,58,59] The overall noise density at the output of an operational TIA is the sum of the squares of all noise variances at the TIA input. These variances are composed of thermal noise sources (Johnson white noise is mentioned in Secs. 10.6.3 and 10.6.3), resulting from the resistive parts of the feedback and the PD output impedance, as well as shot noise currents due to the PD photo current, dark current, and TIA bias. The output noise voltage is given by[47]

$$V^2_{\text{n-out}} = (I^2_S + I^2_D + I^2_{FB} + I^2_{PD} + I^2_{NA})A^2_Z + e^2_{NA}A^2_V, \tag{11.151}$$

where

- $I^2_S$, is the shot noise square associated with the PD signal level;
- $I^2_D$, the dark current shot noise square associated with the PD leakage current;
- $I^2_{FB}$, the thermal noise square resulting from the resistive part of the feedback impedance;
- $I^2_{PD}$, thermal noise square resulting from the PD resistive part of its output impedance;
- $I^2_{NA}$, the noise current square resulting from the OP-amp;
- $e^2_{NA}$, the noise voltage square resulting from the OP-amp;
- $A^2_Z$, the transimpedance gain square; and
- $A^2_V$, the voltage gain square.

The voltage gain for a noninverting case is given by

$$A_V = \frac{A(Z_{PD} + Z_{FB})}{(A-1)Z_{PD} + Z_{FB}} \approx \frac{A(Z_{PD} + Z_{FB})}{AZ_{PD} + Z_{FB}}. \tag{11.152}$$

The voltage gain in the inverting case is given by

$$A_V = -\frac{AZ_{FB}}{Z_{PD}(A+1) + Z_{FB}} \approx -\frac{AZ_{FB}}{Z_{PD}A + Z_{FB}}. \tag{11.153}$$

The transimpedance gain at the inversion is provided by Eq. (11.121) in Sec. 11.7.1. In the case of a noninverting mode, the transimpedance gain would be approximately[47]

$$A_Z = \frac{Z_{FB}Z_{PD}A}{Z_{PD}A + Z_{FB}}, \tag{11.154}$$

where $A$ is the OP-amp open-loop gain function given by Eq. (11.129). Substituting Eqs. (11.130) into (11.154) provides a second-order transimpedance transfer function. The second-order function can be written as[47]

$$H(\omega) = H_0 \frac{1}{(j\omega)^2 + 2j\xi\omega\omega_p + \omega_p^2}, \quad (11.155)$$

where $\omega_p$ is the natural frequency of the loop and $\xi$ is the damping factor.

These parameters are solved by comparing coefficients between Eqs. (11.154) and (11.155). In order to determine the noise dominance, it is a good engineering practice to examine the ratio of the transimpedance gain to the voltage gain.[47]

For the noninverting gain configuration, the ratio is given by

$$\frac{A_Z}{A_V} = \frac{Z_{FB}Z_{PD}}{Z_{FB} + Z_{PD}} = Z_{FB}\|Z_{PD}, \quad (11.156)$$

and for the inverting case, the ratio would be

$$\frac{A_Z}{A_V} = \frac{ssZ_{PD}(Z_{PD}A + Z_{FB})}{Z_{PD} + Z_{PD}A + Z_{FB}} = Z_{PD}\|(Z_{PD}A + Z_{FB}). \quad (11.157)$$

A comparison factor $k$ is then determined to examine which of the TIA noise mechanisms is dominant:[47]

$$k(\omega) = \frac{A_Z(\omega) \times I_{NA}}{A_V(\omega) \times e_{NA}}. \quad (11.158)$$

In the case of a noninverting amplifier configuration, the cut-off frequency is determined by

$$\omega_k = \frac{1}{[R_{FB}R_{PD}/(R_{FB} + R_{PD})](C_{PD} + C_{FB})}, \quad (11.159)$$

where, as before, $C_{FB}$ symbolizes the residual parasitic capacitance of the transimpedance feedback resistor. The function $k(\omega)$ is used as a ballpark value to check which of the noise mechanisms is dominant and which can be ignored. In the case of $|k| > 1$ ($|k|\text{dB} > 0$), the noise current is dominant and visa versa. Hence, at frequencies above $\omega > \omega_k$, the noise voltage generated by the TIA becomes more important. Furthermore, as the capacitances $C_{FB}$ and $C_{PD}$ become higher, the noise voltage becomes more significant and starts to be affected at lower frequencies. The $k$-factor function is also used to determine dominance of all other noise currents with respect to the TIA noise voltage.

## 11.8 Biasing Methods

There are several ways to bias GaAs MESFET or PHEMT. The methods of biasing are shown in Fig. 11.25 and are based on finding the optimal bias point.[9,60,61]

The dc-bias point is wholly dependent on the application. Figure 11.25 depicts the bias considerations based upon the required application. The main rule is that before selecting a MESFET device for power, the saturated current, $I_{DSS}$, should be

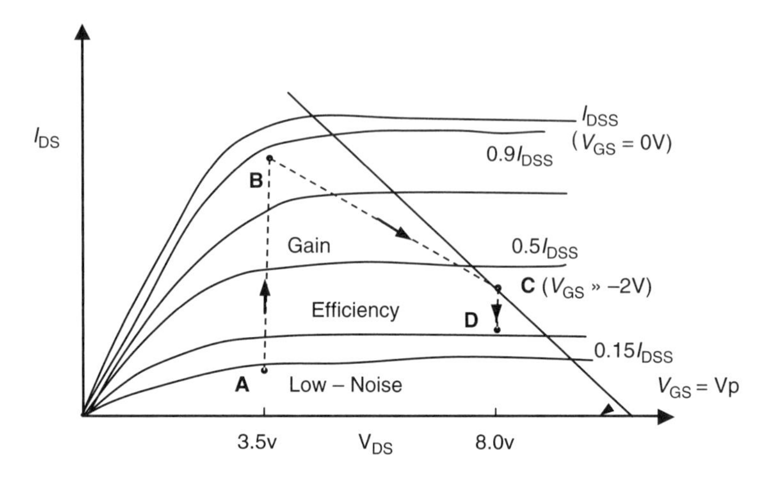


**Figure 11.25** Four basic biasing operating points after Ref. [25]. (Reprinted with permission from *Microwaves and RF* © 1978.)

sufficient for handling the power requirements. The $I_{DSS}$ rating is directly related to the gate periphery. For applications below 20 dBm, it would be enough to have a gate periphery of 500 μm. For a 27-dBm-application, a gate periphery or width of 1500 μm is sufficient.[60] Higher power per gate width unit implies risks for a higher channel temperature breakdown and a lower mean time between failures or poor functional integrity testing occurs. For low-noise operation, the FET would be biased to a low dc operating point. For higher gain and intermediate noise, the current would be higher. For high power and low distortions, the device would be biased at 50% of $I_{DSS}$; this way the RF swing is not rectified and clipped. The optimal load is characterized by power setup and phase shifter to reach the maximum power at a given bias point and frequency. This is called the load-pull test.[62] Moreover, the RF load line at the $I_{DS}/V_{DS}$ curve is not a straight line but an elliptical one as shown in Fig. 11.26.[62] The reason for that phenomenon is due to the FET output impedance nature, which is complex, consisting of real and reactive parts. Distortions are observed when current swing exceeds $I_{DSS}/2$ and $V_{DS}$, respectively. For a large signal, which is full CATV-channel loading and high OMI of about 40%, the MESFET output impedance, composed of $C_P$ and $R_P$, is signal dependent.

In general, $C_P$ is an increasing function of power, but decreasing function of bias voltage, $V_{DS}$. The variations of $R_P$ are the opposite. Such behaviors are not surprising; they are related to the extended depletion of the gate toward the drain—hence a decrease in $C_P$ and an increase in $R_P$ due to the increase of the channel constriction. The MESFET output impedance components are in parallel. Hence, the MESFET output power at the fundamental frequency delivered to the load is given by the empirical formula[62]

$$P_S = \frac{1}{2}\Delta I_D \Delta V_{DS} \cos\psi, \tag{11.160}$$

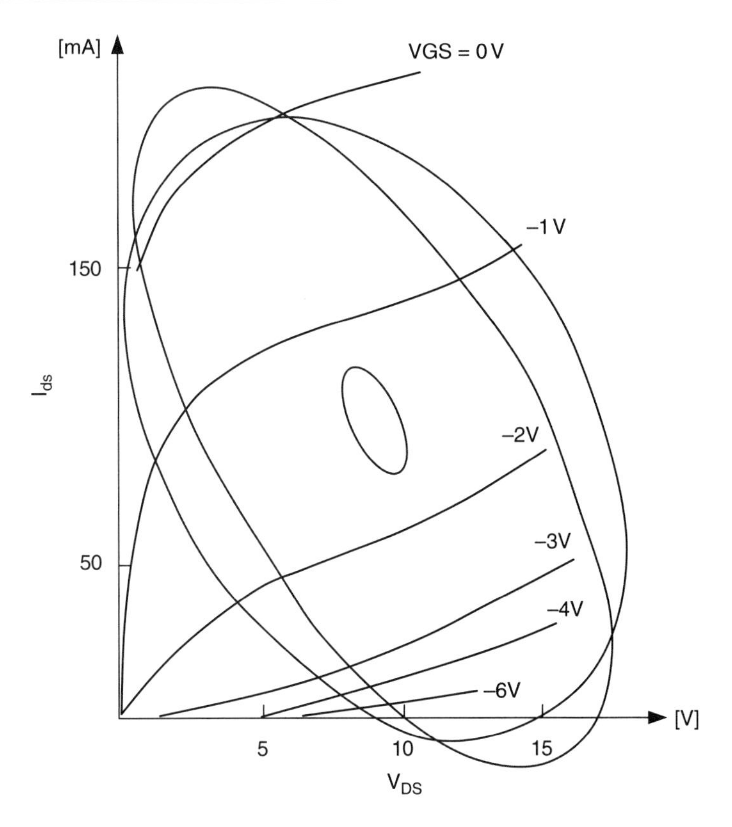


**Figure 11.26** Three signal levels load lines of NE 900 at 10 GHz, small signal class-A, class-A limit, and 3-dB compression.[62] (Reprinted with permission from *IEE Proceedings* © IEE, 1987.)

where $I_D$ and $V_D$ are the maximum amplitudes of drain current $I_D$ and drain source voltage $V_{DS}$, respectively. $\psi$ is the phase angle between $I_{DS}$ and $V_{DS}$ due to that complex load. Assuming this matches the conjugate load impedance of the FET, $\psi$ is expressed as

$$\cos\psi = \frac{1}{\sqrt{\left(1 + \omega^2 C_P^2 R_P^2\right)}}. \tag{11.161}$$

In the load line, it was observed that in the linear region, the load ellipse does not cross the $I/V$ diagram axis boundaries,[62] as shown in Fig. 11.26, the ellipse is less steep and wider.

Modern PHEMT devices provide a very low noise figure of 1.8–2.2 dB and a high 1-dB compression point of about 21 dBm at 50% $I_{DSS}$. In CATV FTTx receiving modules, the bias method is a self-bias class-A, single positive voltage supply for cost effective design and simplicity. This method avoids a second negative voltage to be applied on the MESFET gate, with a voltage sequence circuit applying the negative voltage on the gate first and the drain voltage at the second, when turned on, and turning the negative voltage off after the positive at the amplifier.

In the case of a push–pull design single ended topology or cascaded design, a stack-biasing method is used to prevent a large voltage drop on a dc resistor and large power dissipation.

The stack bias method is useful when the *V*dd supply voltage is high. Hence, the high dc supply voltage is used as an advantage to bias two FETs or more in series dc configuration as shown in Figs. 11.27 and 11.28. This method was used in MMICs by AVANNTEK.[8] The bottom FET, *Q*2, in this configuration is a current source. The stack method is also a self-bias method. The stack-bias method suffers from disadvantages when the GaAs FETs are not identical in properties. For instance, different GaAs FETs power compression specifications, different $I_{DSS}$ due to process fluctuations, or when two types of FETs are used in a stack, such as LNA for *Q*1 and power for *Q*2. These differences affect the circuit performance for distortions or NF since the GaAs FETs are used as a current source, *Q*2 drive the upper FET, and *Q*1, under the same $I_{DS}$, which results in a nonoptimal bias operating point. To solve this problem, a shunt load, R3, is placed in parallel with the upper GaAs FET, and *Q*1, to "bleed" current and drive the upper GaAs FETs to an optimal current, resulting in an optimal

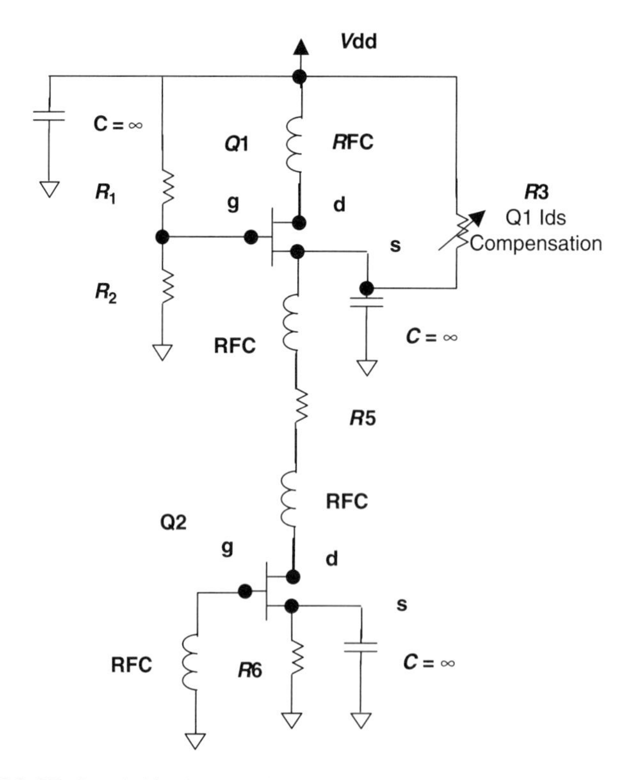


**Figure 11.27** Stack-biasing method with passive current compensation of Q1.

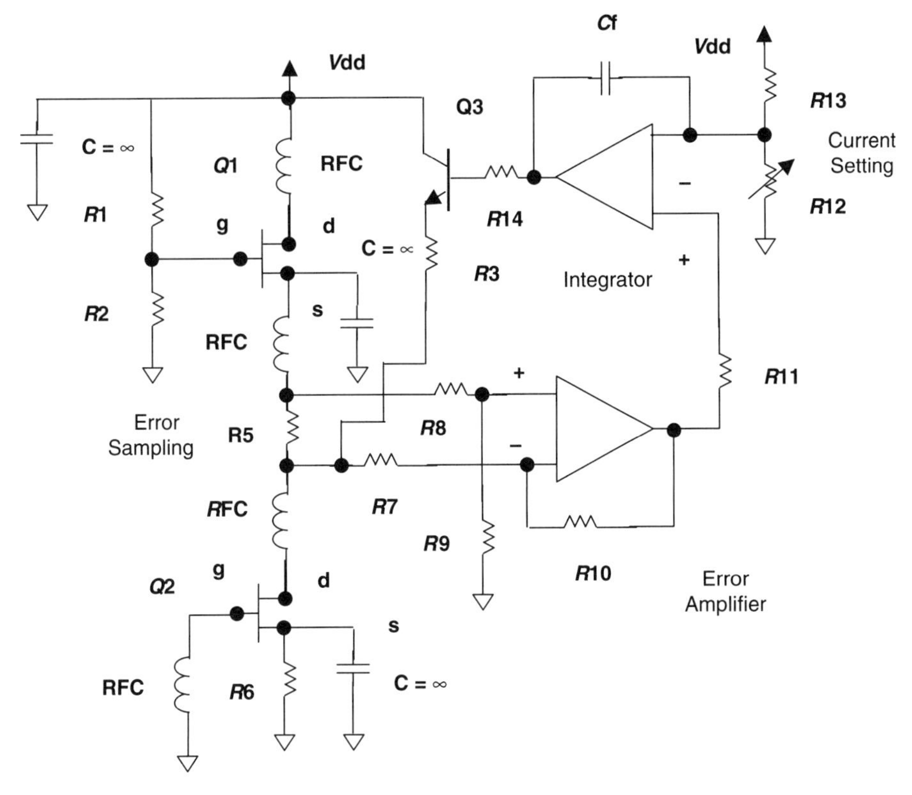


**Figure 11.28** The stack-biasing method and current compensation with active current locking of *Q*1, known also as current starving.

biasing point, as shown in Fig. 11.27. Since this is dc compensation, it is isolated from the RF path by RFC, and the source capacitor of *Q*1 is used as an RF GND path. In order to achieve current accuracy for *Q*1 and a stable operating point over temperature, a feedback correction circuit is used, as shown in Fig. 11.28. The feedback should have a narrow BW to prevent fluctuations on the $I_{DS}$ current that would result in residual AM of the signal. A wide band bias-control circuit will reduce the CSO CTB performance by adding spurious AM or reduce BER performance in QAM transport since the current is modulated during the correction process. The concept of automatic current-control (ACC) operation, also known as a current-starving circuit, is accomplished by sampling the *Q*1 current over R5. The error voltage is amplified by the error amplifier with an error gain of $R10/R7 = R9/R8$. This error voltage is compared to the reference voltage set by *R*13 and *R*12 dividers. This error fluctuation is integrated with an ideal integrator, having a time constant of $\tau = C_{fx}R11$. The integrator and the error amplifier are noninverting and the negative feedback is created by the NPN transistor *Q*3. If the $I_{DS}$ current through *Q*1 is too high, the error voltage across *R*5 is increased. Thus, the error gain of the error amplifier is higher and the error voltage increases. This may result in a higher integration result of voltage to the base of *Q*3. As a consequence, $Q3$ increases its conductivity and bleeds more current from $Q1$. Therefore, the current through $Q1$ would be reduced and the error voltage across $R5$ and the feedback locks the current of $Q1$. Note that the current's sum of $Q1$ and $Q3$ is constant and equal to the current defined by $Q2$, which is a current source. Reference voltage in an advanced design is a programmed microcontroller by setting a binary level to a digital-to-analog converter (DAC) in order to adjust a band-gap reference-regulated voltage.

Biasing of MESFET is crucial when there is a distortion specification to comply with, especially, when dealing with multiple-carrier telecommunication system such as CATV. For this purpose, one may use a method called bias compensation. This method is based on experimental results showing that for a given fixed $V_{DS}$ voltage, intermodulation levels at $2\omega_1$-$\omega_2$ and $2\omega_2$-$\omega_1$ are reduced when the gate voltage becomes more negative.[63] The gate voltage that produces the minimum intermodulation products is called $V_{GSM}$. The idea here is a feed-forward biasing by sampling the RF at the input of the MESFET, using an RF detector and correcting the gate voltage versus RF power levels. It is well known that the more negative the gate voltage is, the greater the decrease in the gain of the device. However, the distortions are reduced more than the gain reduction; hence, the net improvement in distortion is achieved. In case the voltage $V_{GS}$ is lower than $V_{GSM}$, the MESFET is near its pinch-off voltage. At that point, the intermodulation products would increase since the device is operating at a nonlinear point. Figure 11.29 demonstrates that concept.

MMIC devices of today are based on HBT using a Darlington pair with reactive-resonance matching networks to the photodetectors.[63] BJT MMICs are current bias amplifiers and require a current-source biasing method. Figure 11.30 shows an example of MMIC bias. The current-source bias circuit is temperature compensated and provides an accurate bias current to the device over temperature. Some other MMICs are based on PHMT feedback amplifiers with a common source as an input stage and common gate as a buffer. Those MMICs also can be biased in the stack method.

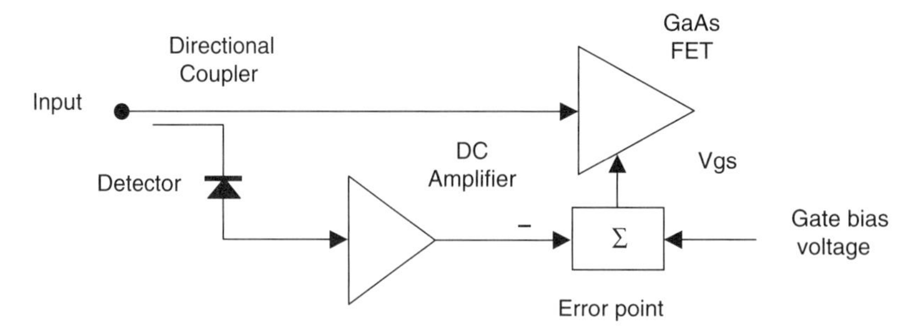


**Figure 11.29** Gate bias compensation method in GaAs MESFET.[63] (Reprinted with permission from T. T. Ha, *Solid State Microwave Amplifiers Design*, Krieger Publishing Company © 1991.)

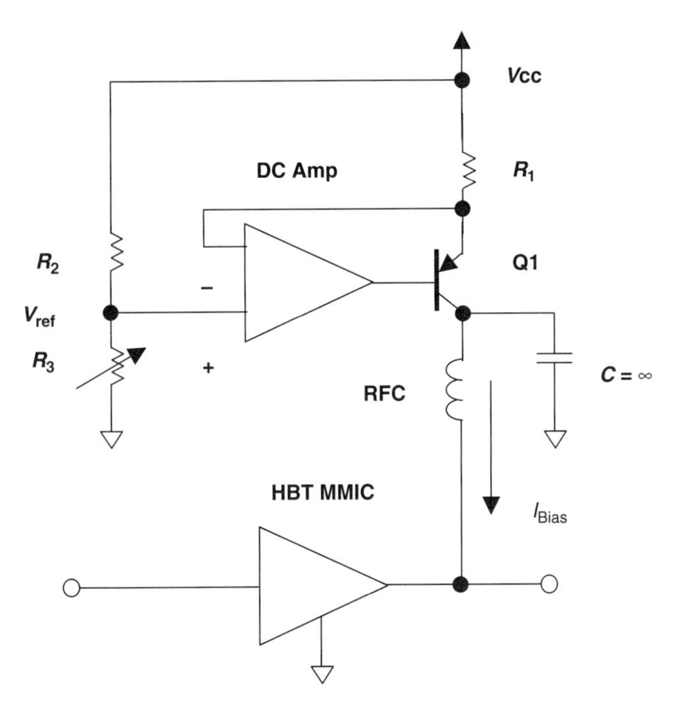


**Figure 11.30** Typical MMIC biasing. Bias current value is $I_{\text{Bias}} = V_{\text{cc}}\{1 - [R_3/(R_2 + R_3)]\}(1/R_1)$.

## 11.9 Equalizers and Optimum Placement of Equalizers

Equalizers are essential networks for improving the frequency response of a receiver.

The system requirements are to compensate the losses of the coax cables, which are increasing versus frequency and have a downward tilt. One method to compensate these cable losses is by designing a receiver with an upward tilt versus frequency. Equalizers are generally implemented after the first input LNA stage. This way the RF chain NF performance is not affected and the equalizer losses are compensated by the RF-gain stage. A zero-pole cancellation network realizes such an equalizer. Figure 11.31 illustrates two ways of realizing a zero-pole cancellation equalizer.

Referring to Fig. 11.31(a) and assuming the use of Tevenin identity and that the signal source stage output impedance is $R_s$, the equalizer resistor and capacitor are $R_{\text{eq}}$ and $C_{\text{eq}}$, respectively, and the load value or next stage input impedance is $R_{\text{L}}$, the equalizer frequency response is given by

$$H(s) = \frac{R_{\text{L}}\left(SC_{\text{eq}}R_{\text{eq}} + 1\right)}{\left(R_{\text{L}} + R_{\text{S}} + R_{\text{eq}}\right) \times \left[\left(R_{\text{S}} + R_{\text{L}}/R_{\text{L}} + R_{\text{S}} + R_{\text{eq}}\right)SC_{\text{eq}}R_{\text{eq}} + 1\right]}. \quad (11.162)$$

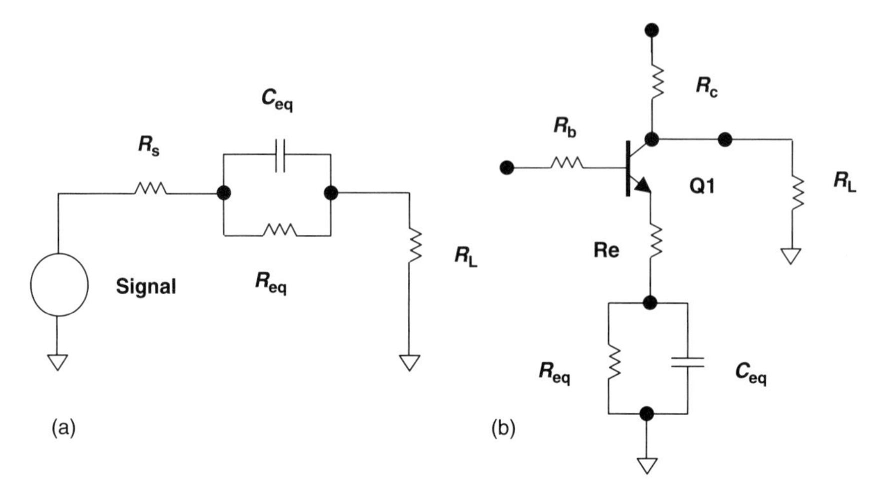


**Figure 11.31** Up-tilt equalizer circuits: (a) passive, (b) active used in return path channel.

The equalizer losses at low frequency are determined by

$$H\,[\mathrm{dB}] = 20\,\log \frac{R_\mathrm{L}}{\left(R_\mathrm{L} + R_\mathrm{S} + R_\mathrm{eq}\right)}. \tag{11.163}$$

The equalizer losses at high frequency are determined by

$$H\,[\mathrm{dB}] = 20\,\log \frac{R_\mathrm{L}}{(R_\mathrm{L} + R_\mathrm{S})}. \tag{11.164}$$

The zero frequency $\omega_\mathrm{Z}$ and pole frequency $\omega_\mathrm{P}$ (Fig. 11.32) are given by

$$\omega_\mathrm{P} = \frac{R_\mathrm{L} + R_\mathrm{S} + R_\mathrm{eq}}{R_\mathrm{S} + R_\mathrm{L}} \cdot \frac{1}{C_\mathrm{eq} R_\mathrm{eq}}, \tag{11.165}$$

$$\omega_\mathrm{Z} = \frac{1}{C_\mathrm{eq} R_\mathrm{eq}}. \tag{11.166}$$

For typical CATV RF chain with a $50\,\Omega$ characteristic impedance, and $R_\mathrm{eq} = 43\,\Omega$ and $C_\mathrm{eq} = 33\,\mathrm{pF}$, the equalizer insertion loss is 9 dB. At a high frequency, the equalizer losses are 6 dB. From this example, the equalizer's zero-to-pole attenuation range is determined by the value of $R_\mathrm{eq}$. The equalization depth in decibels is the ratio between Eqs. (11.163) and (11.164), which is given by

$$H\,[\mathrm{dB}] = 20\,\log \frac{R_\mathrm{L} + R_\mathrm{S} + R_\mathrm{eq}}{R_L + R_\mathrm{S}}. \tag{11.167}$$

The equalizer frequency range is determined by the ratio between $\omega_\mathrm{P}$ and $\omega_\mathrm{Z}$. In order to have a wideband response, the ratio $\omega_\mathrm{P}/\omega_\mathrm{Z}$, which is the ratio between Eqs. (11.164) and (11.165), respectively, should be high. The equalizer BW factor,

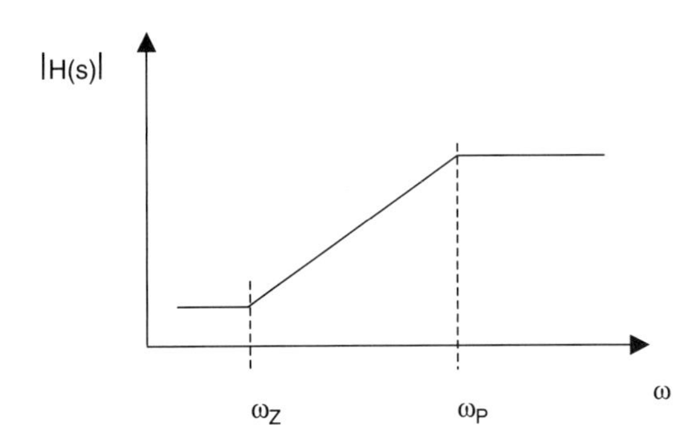


**Figure 11.32** A typical equalizer frequency response in a log scale Bode-plot.

therefore, is determined by

$$\text{BW\_factor} = \frac{R_L + R_S + R_{eq}}{R_L + R_S}. \tag{11.168}$$

Increasing $R_{eq}$ would increase the equalizer BW factor but will increase its losses at low frequency as well as its equalization depth. As was explained before in RF-lineup considerations, such a high equalization depth loss may affect the gain and NF performance as a result of concentrating too much equalization loss at one point of the RF chain. Hence, it is a common engineering practice to distribute the up-tilt equalizers throughout the RF chain.

Resonance equalizers are used to compensate flatness for better frequency response. A common network is a parallel RLC band-stop-resonance circuit. It can be shunt coupled or connected in series as depicted by Fig. 11.33.

Assume the following parallel RLC equalizer with shunt coupling as shown in [Fig. 11.33(b)]. The quality factor of parallel RLC network is defined by Desoer and Kuh[54] as

$$Q = \frac{\omega_0}{2\Delta\omega} = \omega_0 RC = \frac{R}{\omega_0 L} = \frac{R}{\sqrt{L/C}}, \tag{11.169}$$

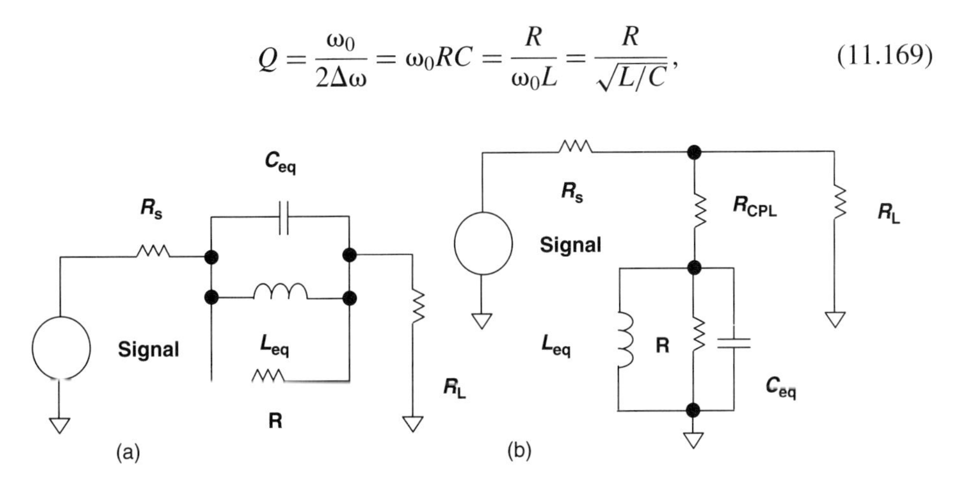


**Figure 11.33** Series and shunt parallel RLC equalizer.

where $2\Delta\omega$ is the RLC resonator's angular BW. Its transfer function is given by

$$H(j\omega) = \frac{1}{1 + jQ[(\omega/\omega_0) - (\omega_0/\omega)]}, \tag{11.170}$$

and its admittance is given by

$$Y(j\omega) = \frac{1}{R \times H(j\omega)}. \tag{11.171}$$

Assume that the equalizer is coupled to the network by a coupling resistor $R_{\mathrm{CPL}}$. The load is marked as $R_{\mathrm{L}}$ and the signal source impedance is $R_{\mathrm{S}}$. Additionally, the equalizer is loaded by a parallel de-$Q$ load $R$. That loading is in order to have a wider BW response. Hence, the port of the equalizer would be loaded by $R'_{\mathrm{L}}$, which is given by

$$R'_{\mathrm{L}} = R_{\mathrm{CPL}} + (R_{\mathrm{S}} \| R_{\mathrm{L}}). \tag{11.172}$$

Thereby, the equalizer overall equivalent load would be

$$R_{\mathrm{Leq}} = [R_{\mathrm{CPL}} + (R_{\mathrm{S}} \| R_{\mathrm{L}})] \| R = R'_{L} R \|, \tag{11.173}$$

where the symbol $\|$ means parallel components. The unloaded quality factor $Q_0$ is defined as

$$Q_0 = \frac{R}{\omega_0 L}. \tag{11.174}$$

The loaded quality factor $Q_{\mathrm{L}}$ is given by

$$Q_{\mathrm{L}} = \frac{R'_{\mathrm{L}}}{\omega_0 L}, \tag{11.175}$$

and the equivalent load quality factor $Q_{\mathrm{Leq}}$ is given by

$$Q_{\mathrm{Leq}} = \frac{R_{\mathrm{Leq}}}{\omega_0 L} = \frac{Q_0 Q_{\mathrm{L}}}{Q_0 + Q_{\mathrm{L}}}. \tag{11.176}$$

Let us define the coupling factor $\beta$ as the ratio between the equalizer resistance $R$ and the load reflected to the equalizer via the coupling resistor $R_{\mathrm{CPL}}$:

$$\beta = \frac{R}{R_{\mathrm{S}} \| R_{\mathrm{L}} + R_{\mathrm{CPL}}} = \frac{R}{R'_{\mathrm{L}}} = \frac{Q_0}{Q_{\mathrm{L}}}. \tag{11.177}$$

Hence, after substituting Eq. (11.177) into Eq. (11.176), the $Q_{Leq}$ is given by

$$Q_{Leq} = \frac{Q_0}{1+\beta}. \quad (11.178)$$

Observing Fig. 11.32, the overall power transferred from the source to the load and the equalizer is given by Eqs. (11.179) and (11.180), respectively:

$$P_L = \frac{V_O^2}{R_L} = i_L^2 R_L = \frac{i_S^2}{R_L}\left[\frac{R_L(R_{CPL}+R)}{R_L+R_{CPL}+R}\right]^2, \quad (11.179)$$

$$P_{Resonator} = \frac{V_R^2}{R} = i_R^2 R = R\left\{\frac{i_S}{R+R_{CPL}}\left[\frac{R_L(R_{CPL}+R)}{R_L+R_{CPL}+R}\right]\right\}^2 \quad (11.180)$$

The calculation did not take into account the source impedance loss but observed the losses ongoing from the source's ports. The overall transferred power is given by

$$P_{tot} = \frac{V_O^2}{R_{eq}} = i_S^2 R_{eq} = i_S^2\left[\frac{R_L(R_{CPL}+R)}{R_L+R_{CPL}+R}\right]. \quad (11.181)$$

The power coupled to the equalizer's resonator is given by

$$\frac{P_{resonator}}{P_{tot}} = \frac{R\left\{\frac{i_S}{R+R_{CPL}}\left[\frac{R_L(R_{CPL}+R)}{R_L+R_{CPL}+R}\right]\right\}^2}{i_S^2\left[\frac{R_L(R_{CPL}+R)}{R_L+R_{CPL}+R}\right]}$$

$$= \frac{R}{(R+R_{CPL})^2}\left[\frac{R_L(R_{CPL}+R)}{R_L+R_{CPL}+R}\right] \quad (11.182)$$

The power losses due to the equalizer coupling are given by

$$\frac{P_L}{P_{Tot}} = \frac{1}{R_L}\left[\frac{R_L(R_{CPL}+R)}{R_L+R_{CPL}+R}\right]. \quad (11.183)$$

A different way of compensating flatness response is by using a parallel RLC equalizer in series, as described in Fig. 11.33(a). In this case, $R_{CPL} = 0$. Hence, the equivalent $Q$ loading of the equalizer is the sum of $R_L$ and $R_S$. Therefore, the coupling factor $\beta$ is given by

$$\beta = \frac{R}{R_L+R_S} = \frac{R}{R_L'} = \frac{Q_0}{Q_L}, \quad (11.184)$$

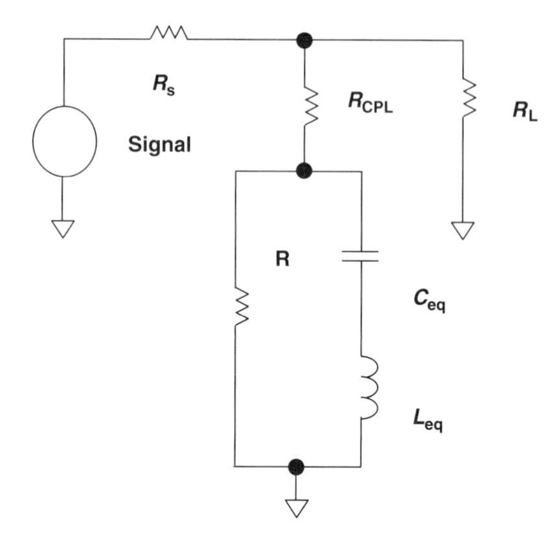


**Figure 11.34** Series RLC equalizer connected by a shunt.

and the equivalent loading of the equalizer is given by

$$R_{\mathrm{Leq}} = \frac{R(R_{\mathrm{L}} + R_{\mathrm{S}})}{R_{\mathrm{L}} + R_{\mathrm{S}} + R}. \tag{11.185}$$

The current from the source into the load is a simple calculation of resistors in series at the resonance point:

$$i_{\mathrm{L}} = \frac{v_{\mathrm{S}}}{R_{\mathrm{L}} + R_{\mathrm{S}} + R}. \tag{11.186}$$

The insertion loss in this case is given by

$$IL\,[\mathrm{dB}] = 20\ \log \frac{R}{R_{\mathrm{L}} + R_{\mathrm{S}} + R_{\mathrm{L}}}. \tag{11.187}$$

In some cases, a shunt equalizer using a series RLC circuit is used to leak some undesired power at a given frequency to the GND. In some applications, the bias RFC network and the GND decoupling capacitor are used as equalizers (Fig. 11.36).

In order to achieve high CNR performance as well as a high dynamic range and flatness, it is important to install the equalizer networks as interstate and output networks. Equalizers should be distributed throughout the RF chain. Placing an equalization network at the RFFE would reduce the CNR performance by the amount of losses the equalizer adds. Proper RF matching and optimization using CAD tools and RF solvers would minimize the number of equalizers in a design. The smaller the number of equalizers used, the lower the interstate and

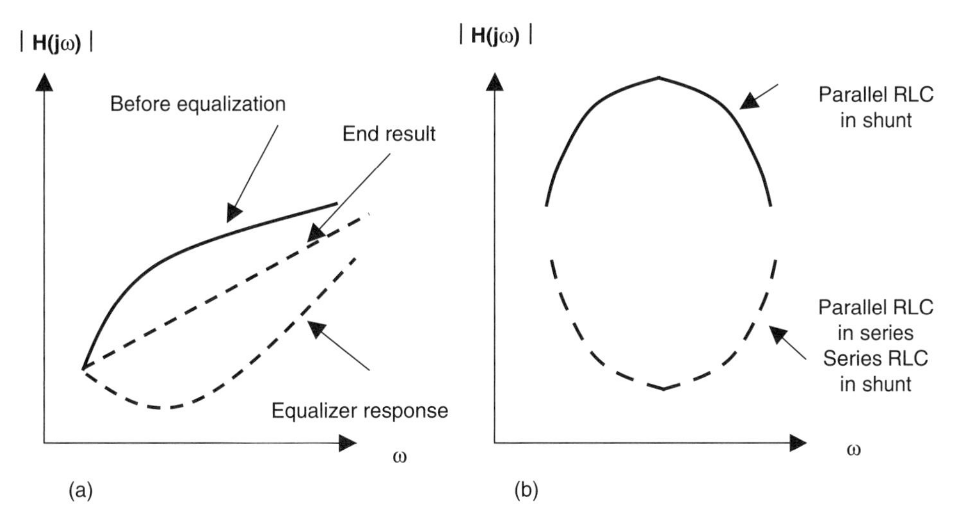


**Figure 11.35** RLC equalizer frequency response correction. (a) Correction using parallel RLC connected in series as in Fig. 11.33(a) or using series RLC connected in a shunt as in Fig. 11.34. (b) RLC equalizer response, solid line–parallel RLC connected in a shunt as in Fig. 11.33(b). Dashed line–parallel RLC connected in series as in Fig. 11.33(a) or series RLC connected in shunt as in Fig. 11.34. Up tilt is accomplished by a series inductor to the PD or RC equalizer as shown in Fig. 11.31.

output losses are. Hence, the higher output level can be reached with minimum gain blocks and current consumption. Figures 11.34 and 11.35 show equalizer frequency response correction and optimal RF lineup locations for equalizers.

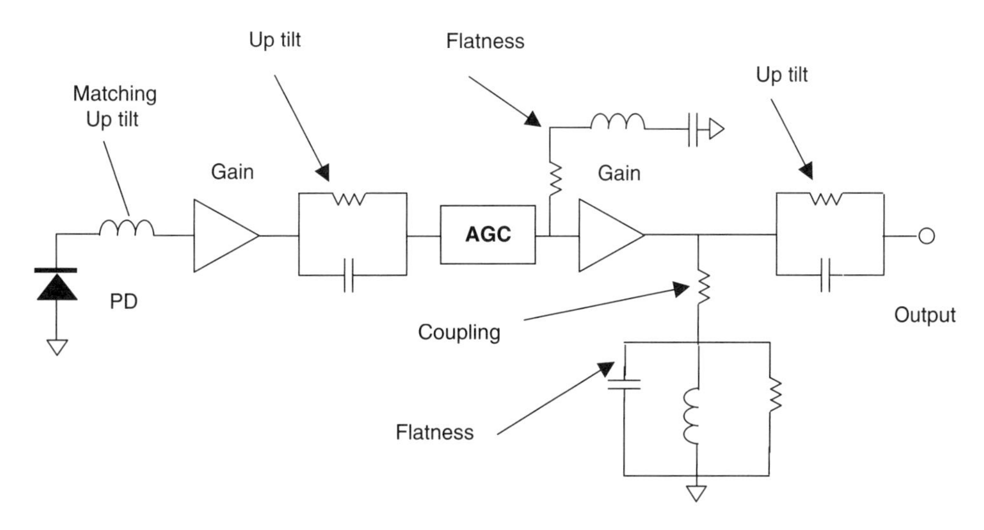


**Figure 11.36** Example of equalizer placements in RF lineup. Other options are available per design.

## 11.10 Matched PIN VVA

As will be explained in Chapter 12, analog AGC or microcontrolled AGC require a voltage variable attenuator (VVA). VVA advantage is due to its linearity and simple control circuit. However, in order to maintain low gain fluctuations versus frequency, the requirement is to have a matched PIN VVA (Fig. 11.37). The match should be held throughout the entire attenuation range of the VVA. One of the best configurations to control matching conditions is by using an L-type PIN diode VVA.

The idea is to have low intermodulation performance as well as good matching conditions. In order to reduce intermodulations, there should be a series resistor with $Z_0 \approx 50\ \Omega$ to the shunt diode. The shunt diode is marked as $R_2$; the series diode is marked as $R_1$. The series diode is connected to a matched transistor $Q_2$ with $Z_0 \approx 50\ \Omega$. The shunt resistor load would also help to achieve partial matching conditions. The VVA is not symmetrical; hence, the matching is available only in one direction. Assume that $Z_0 \approx 50\ \Omega$, then the conditions for the first gain stage, $Q_1$, to have a value of 50 Ω are given by

$$\frac{(Z_0 + R_1)(Z_0 + R_2)}{Z_0 + R_1 + Z_0 + R_2} = Z_0 = 50. \tag{11.188}$$

This equation yields the following matching condition:

$$R_1 R_2 = Z_0 = 50^2. \tag{11.189}$$

An additional condition is required in order to maintain matched impedance at any attenuation level. This condition can be controlled by monitoring the diode impedance versus current. It is well known that the diode resistance as a function

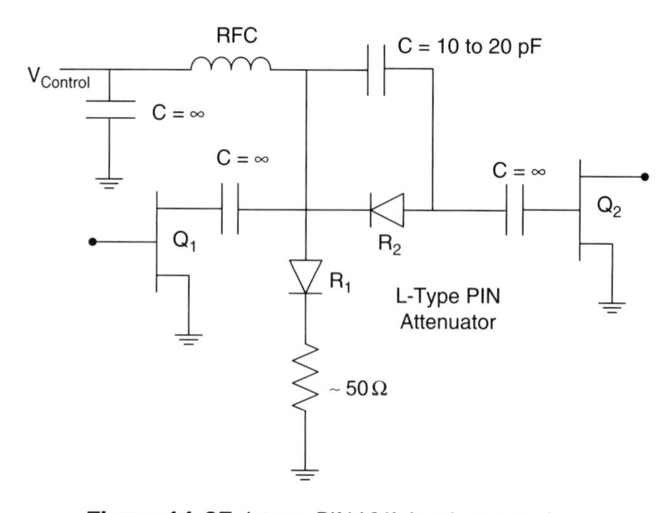


**Figure 11.37** L-type PIN VVA implementation.

of bias current is given by Eq. (11.190). This relation is accurate for a relatively wide range of currents:

$$R = K \cdot I^{-\gamma}. \tag{11.190}$$

where $\gamma$ is the constant, generally its value is about 0.9; $K$ is the constant, which depends on the diode PN junction cross section; and $I$ is the diode dc-bias current.

The relation between the voltage across the diode and the diode current is given by

$$I = I_0\left(e^{\frac{qv}{kT}} - 1\right). \tag{11.191}$$

In forward voltage conditions, Eq. (11.190) becomes

$$I \approx I_0 e^{\frac{qv}{kT}}. \tag{11.192}$$

Taking Eq. (11.192) into Eq. (11.190) and substituting into the relation of the matching conditions given by Eq. (11.189) would result in the following matching condition vs. voltage for the L diodes:

$$K^2 I_0^{-2\gamma} \exp\left(-\frac{\gamma q}{kT}(v_1 + v_2)\right) = Z_0^2 = 50^2, \tag{11.193}$$

thus

$$v_1 + v_2 = -\frac{kT}{\gamma q} \ln\left(\frac{Z_0^2}{K^2 I_0^{-2\gamma}}\right) = \text{constant}. \tag{11.194}$$

Hence, the matching condition requires that the voltage across the L diode have a constant value. That condition saves the circuit complexity with respect to control voltages and linearizers. Consequently, the control voltage would be at the anode–cathode common port of the diodes and the diodes would have a fixed shunt voltage as shown in Fig. 11.38.

In the case of low current operation, the "—" voltage polarity can be connected directly to ground since the voltage drop across the resistor is negligible. This would reduce the matching conditions but not to a large extent. This matching condition applies to a constant temperature; however, the voltage sum across the L-section diodes is strongly temperature dependent. In order to compensate for these temperature changes, the reference voltage should track these temperature changes. One of the ways to realize this reference is by a $V_{BE}$ amplifier as shown in Fig. 11.39. Assuming that the current through $R_A$ and $R_B$ is much

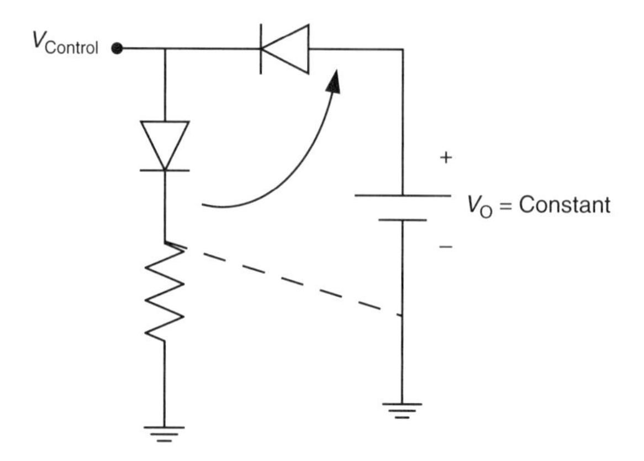


**Figure 11.38** The matching condition of the L section requires constant voltage across the L diodes.

higher than the base current of the transistor, the following bias conditions are applicable:

$$V_{BE} = V_O \frac{R_B}{R_A + R_B} \tag{11.195}$$

or

$$V_O = \left(1 + \frac{R_A}{R_B}\right) V_{BE}. \tag{11.196}$$

In this case, it is required that $V_O \approx 2V_{BE}$ since it should be equal to twice the diode PN junction voltage. The control voltage in this case is the AGC loop control voltage.

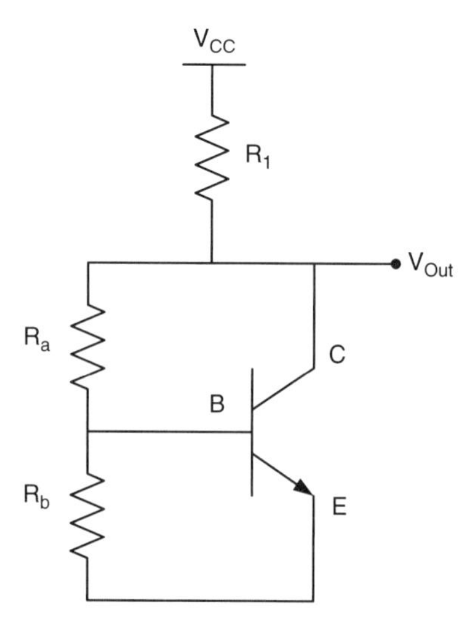


**Figure 11.39** Reference voltage with temperature compensation.

The above analysis had shown the advantage of an L-section VVA. A symmetrical VVA can be realized by using a bridged T attenuator. It has similar matching conditions and it is a symmetrical VVA. Hence, it provides both stages of matched impedance conditions.

## 11.11 Examples for Noise Analysis CMOS and BJT RF IC Designs

In previous sections, classical design approaches using discrete components were presented. Traditionally, RF designs are based on GaAs PHEMT design InGaAs and SiGe. Early TIAs were based on PHEMTs; however, high-speed low-current digital and RF are moving toward CMOS RF.[13,65]

The dominant noise sources for MOSFET transistors are flicker and thermal noise. This is similar to GaAs and is modeled as a voltage source with respect to the gate:

$$V_{\mathrm{g}}^2(f) = \frac{K}{WxLC_{\mathrm{ox}}f}, \tag{11.197}$$

where $K$ is a constant dependent on the device characteristics and varies for different devices. $WxL$ and $C_{\mathrm{ox}}$ represent the FET width, length, and gate capacitance, respectively. The $1/f$ is inversely proportional to $WxL$. In other words a larger device has a lower $1/f$. This noise is dominant in low frequencies and affects phase noise and jitter in CDR (clock data recovery) applications.

The MOS thermal noise is due to its $R_{\mathrm{ds}}$ resistance and is given by

$$I_{\mathrm{d}}^2(f) = \frac{4kT}{R_{\mathrm{ds}}}. \tag{11.198}$$

However, when the MOS is active, the channel cannot be considered homogeneous thereby over small portions of the channel. The result is given as

$$I_{\mathrm{d}}^2(f) = 4kT\frac{2}{3}g_{\mathrm{m}}, \tag{11.199}$$

when $V_{\mathrm{DS}} = V_{\mathrm{GS}} - V_{\mathrm{T}}$.

FET is a current-controlled voltage source, Hence, $I_{\mathrm{d}}(f) = g_{\mathrm{m}}V_{\mathrm{i}}(f)$ is the noise current and the square current is given by $I_{\mathrm{d}}^2(f) = [g_{\mathrm{m}}V_{\mathrm{i}}(f)]^2$, which results in total input noise voltage of

$$V_{\mathrm{i}}^2(f) = 4kT\left(\frac{2}{3}\right)\frac{1}{g_{\mathrm{m}}} + \frac{K}{WLC_{\mathrm{ox}}f}. \tag{11.200}$$

## 11.12 Main Points of this Chapter

1. In any two-port network, there are two noise sources: current and voltage.
2. The ratio between the noise voltage and current is measured by the correlation factor or coefficient.
3. Thermal voltage is associated with Johnson white noise that represent internal losses within the network or semiconductor.
4. Each transistor is characterized by noise impedance that represent its noise generators.
5. The noise current is composed of two vectors: correlated and uncorrelated with respect to the noise voltage.
6. The ratio between the noise current and the noise voltage is the noise admittance, which is composed of correlated and uncorrelated vectors.
7. The noise source at the device input is composed of the signal impedance noise generator and the device's noise sources, which are the noise voltage and the noise current.
8. The noise figure is calculated as the ratio between the noise current of the device and the noise of the signal source to the noise of the signal source.
9. The rules for noisy electrical network calculations applied for noise voltage and noise current contributions to total noise power are the same as in a regular network, calculated by using KVL (Kirchhoff voltage law), and KCL (Kirchhoff current law), as well as superposition for linear circuits.
10. The noise-power equivalent current is calculated as an rms sum, assuming uncorrelated noise sources.
11. The method of calculating the noise figure with respect to noise impedances is by substituting the ratios between thermal voltage, thermal current, and noise impedance.
12. The minimum noise-figure conditions are evaluated by taking the derivative of the expression for $F$ with respect to the source conductance $G_s$ and admittance $Y_s$; the result provides $Y_{opt}$ and $G_{opt}$.
13. Matching to a minimum noise factor means matching the signal source for maximum power transfer to the device's noise impedance.
14. Matching for a minimum noise figure involves canceling the reactive part of the device's noise impedance, and the real part of the signal impedance

should be equal to the equivalent real part of the noise source, which is composed of the correlated and uncorrelated values.

15. The noise factor is a 3D sphere whose projection on the source impedance plane provides the equal noise-factor circles. The center of those noise circles is the minimum noise-factor point.

16. Minimum noise reflection coefficients are calculated with respect to $Y_o$. The same approach is for the source-reflection coefficient. Thus, it is natural to change the frame of reference to the source impedance and calculate the minimum noise reflection coefficient with respect to the signal source.

17. The FET transistor is characterized by thermal noise. Shot noise in a FET device is due to the gate-leakage current.

18. The correlation factor in a FET device results from the channel current fluctuations that modulate the depletion layer of the device and thus the gate charge is affected and the correlation factor is imaginary.

19. A BJT transistor is an injection and diffusion device, while an FET is a drift device. Hence, BJT has a large component of shot noise. The thermal noise results from the inner resistance of the device $R_{\mu}$, the Miller resistance between the base, the collector, and $R_X$ between the base and the emitter. Those resistances represent the junction resistance.

20. In a single-ended RF amplifier based on a FET device, drain gate feedback is used.

21. The feedback-design approach in RF combats return loss and improves distortions performance. However, it adds thermal noise.

22. One of the methods of improving BW and tilt in FET-feedback RF amplifiers is by adding inductance in series to the FET.

23. In order to reach a wide BW, early high-data-rate amplifiers were based on distributed amplifiers, which are active transmission lines.

24. A distributed amplifier consists of two transmission lines, which use the FET stray capacitance as part of the transmission line. Those are the gate- and the drain-transmission lines.

25. In order to have a coherent power addition and thereby gain, the phase velocity of the gate transmission line should be equal to that of the drain transmission line.

26. The losses of the transmission line are mainly due to active device input resistance, mainly the gate.

27. The maximum number of distributed amplifier gain cells is eight due to noise figure considerations.

28. Advances in technology enable the design of OP-amps TIA up to 10 GHz. Early Op-amps used PHEMT devices and the advanced design uses RFIC technologies and CMOS RF designs.

29. The method to enhance OP-amp BW is by interstaged low-pass networks.

30. In order to minimize distortions in RF amplifiers, the bias point should be in class A.

31. The FET output impedance varies since its $C_p$ and $R_p$ vary with power and bias point, and, hence, draw an ellipse at the $I_D$ versus $V_{DS}$ curve.

32. The self-bias method commonly used in RF FET biasing.

33. The method of series biasing of FETs is called "stack," where the bottom FET operates as the current source. In this manner, large dc supply voltage is taken advantage of.

34. Generally in stack, the upper FET serves as the input stage, LNA, and the current source FET serves as an output stage.

35. The method of operating the upper FET at different biases with respect to the bottom FET is by current starving, where a negative feedback current-sensing circuit locks the upper FET current by bleeding its dc current to a parallel transistor.

36. An additional method for combating distortions is by RF sensing, which uses an RF detector, and adjusting the FET bias point that corresponds to the RF power.

37. Darlington MMICs are current-driven biases requiring a current source as a biasing method.

38. Equalizers in an RF chain are used for up-tilting and ripple optimization.

39. RC equalizers generate upward-tilt response. Parallel LC equalizers in a shunt operate as BPF and improve response at the $f_o$, which is the LC resonance frequency.

40. BW of LC equalizers is increased by a de-Q resistor in parallel to the LC resonator.

41. A series inductor to the photodetector is called a pickup inductor and compensates the photodetector stray capacitance, thus increasing BW and generating an up-tilt response.

42. In order to have a matched impedance PIN VVA (PIN-diode variable attenuator), an L-type configuration is used and voltage across the two diodes is kept constant.

43. CMOS flicker noise is inversely proportional to the WL area; thus larger CMOS FETs have lower flicker noise.

44. Typically, P channel transistors have less noise than the N channel since the majority of carriers and holes are less likely to be trapped.

## References


1. Rothe, H., and W. Dahlke, "Theory of noisy fourpoles." *Proceedings of the IRE*, Vol. 44 No. 6, pp. 811–818 (1956).
2. "IRE standards on methods of measuring noise in linear twoports, 1959." *Proceedings of the IRE*, Vol. 48 No. 1, pp. 60–68 (1960).
3. "IRE subcommittee 7.9 on noise: representation of noise in linear two-ports." *Proceedings of the IRE*, Vol. 48, pp. 69–74 (1960).
4. Mitama, M., and H. Katoh, "An improved computational method for noise parameter measurement." *IEEE Transact Microwave Theory Technology*, Vol. 27 No. 6, pp. 612–615 (1979).
5. Biran, Y., "The noise resistance of a balanced amplifier." *Proceedings of IEEE Conference of Electrical and Electronics Engineers in Israel.* pp. 1–3 (1989).
6. Fukui, H., "Design of microwave GaAs MESFET's for broadband low noise amplifiers." *IEEE Transact Microwave Theory Technology*, Vol. 29 No. 10, pp. 1119 (1981).
7. Fukui, H., "Optimal noise-figure of microwave GaAs MESFET's." *IEEE Transactions on Electron Devices.* Vol. 26 No. 7, pp. 1032–1037 (1979).
8. Vendelin, G.D., A.M. Pavio, and U.L. Rohde, *Microwaves Circuit Design Using Linear and Nonlinear Techniques.* John Wiley & Sons Inc. (1992).
9. Ha, T.T., *Solid State Microwave Amplifier Design.* Krieger Publishing Company (1991).
10. Pucel, R.A., D.J. Masse, and C.F. Krumm, "Noise performance of gallium arsenide field effect transistors." *IEEE Journal of Solid State Circuits*, Vol. 11 No. 2, pp. 243–255 (1976).
11. Liechti, C.A., "Microwave field effect transistors." *IEEE Transactions on Microwave Theory Technology*, Vol. 24, pp. 279–300 (1976).
12. Lee, S., K.J. Webb, V. Tialk, and L.F. Eastman, "Intrinsic noise equivalent circuit parameters for AlGaN/GaN HEMTs." *IEEE Transactions on Microwave Theory Techology*, Vol. 51 No. 5, pp. 1567–1577 (2003).
13. Lee, T.H., *The Design of CMOS Radio-Frequency Integrated Circuits.* Cambridge University Press (1998).

14. Van Der Ziel, A., and J. W. Ero, "Small signal high-frequency theory of field effect transistors." *IEEE Transactions on Electronic Devices*, Vol. 11 No. 4, pp. 128–135 (1964).
15. Van Der Ziel, A., "Thermal noise in field effect transistor." *Proceedings of the Institute of Radio Engineers*, Vol. 50 No. 8, pp. 1808–1812 (1962).
16. Van Der Ziel, A., "Gate noise in field effect transistors at moderately high frequencies." *Proceedings of IEEE*, Vol. 51 No. 3, pp. 461–467 (1963).
17. Holt, C.A., *Frequency and Transient Response of Feedback Amplifiers in Electronic Circuits: Digital and Analog*, John Wiley and Sons (1978).
18. Millman, J., and C.C. Halkias, *Integrated Electronics: Stability and Oscillators*. McGraw-Hill, New York, NY (1972).
19. Bar-Lev, A., *Semiconductors and Electronic Devices*. Prentice Hall Europe (1979).
20. Cook, H.F., "Microwave transistor: theory and design." *Proceedings of IEEE*, Vol. 59 No. 8, pp. 1163–1181 (1971).
21. Fukui, H., "The noise performance of microwave transistor." *IEEE Transactions on Electronic Devices*, Vol. 13 No. 3, pp. 329–341 (1966).
22. Hawkins, R.J., "Limitations of Nielsen's and related noise equations applied to microwave bipolar transistors, and a new expression for the frequency and current dependent noise figure." *Solid State Electronics*, Vol. 20, pp. 191–196 (1977).
23. Malaviya, S.D., and A. Van Der Ziel, "A simplified approach to noise in microwave transistors." *Solid State Electronics*, Vol. 13, pp. 1511–1518 (1970).
24. Van Der Ziel, A., "More accurate expression for the noise figure of transistors." *Solid State Electronics*, Vol. 19, pp. 149–151 (1976).
25. Muoi, T.V., "Receiver design for optical-fiber systems." *Journal of Lightwave Technology*, Vol. 2 No. 3, pp. 243–265 (1984).
26. Smith, R.G., and S.D. Personick, "Receiver design for optical fiber communication systems." *Semiconductor Devices for Optical Communications*, Vol. 39, pp. 89–160 (1979).
27. Hawkins, R.J., "Limitations of Nielsen's and related noise equations applied to microwave bipolar transistors and a new expression for the frequency and the current dependent noise figure." *IEEE Solid State Electronics*, Vol. 20, 191–196 (1977).
28. Ulrich, E., "Use of negative feedback to slash wideband VSWR." Avantek Inc. (1978).
29. Sweet, A.A., and C.A. Palo-Alto, "A parametric study of the behavior of 1Micron GaAs FET amplifiers." *Microwave Journal*, pp. 155–168 (1985).
30. Niclas, K.B., "Compact multi-stage single-ended amplifiers for S-C band operation." *Microwave Symposium Digest*, Vol. 81, pp. 132–134 (1981).
31. Cripps, S., *Advanced Techniques in RF Power Amplifier Design*. Artech-House, Boston, MA (1992).
32. Niclas, K.B., W.T. Wilser, R.B. Gold, and W.R. Hitchens, "The matched feedback amplifier: ultrawide-band microwave amplification with GaAs MESFET's." *IEEE Transactions on Microwave Theory Technology*, Vol. 28, pp. 285–294 (1980).

33. Vendelin, G.D., "Feedback effects on the noise figure performance of GaAs MESFETs." *IEEE Microwave Symposium Digest*, Vol. 75 No. 1, pp. 324–326 (1975).
34. Niclas, K.B., "Noise in broadband GaAs MESFET amplifiers with parallel feedback." *IEEE Transactions on Microwave Theory Technology*, Vol. 30 No. 1, pp. 63–70 (1982).
35. Liang, J.Y., and C.S. Aitchison, "Signal to noise performance of the optical receiver using a distributed amplifier and PIN photodiode combination." *IEEE Transactions on Microwave Theory Technology*, Vol. 43 No. 9, pp. 2342–2350 (1985).
36. Liang, J.Y., and C.S. Aitchison, "An optical photodetector using a distributed amplifier and PIN diode photodiode combination." *IEEE Microwave Symposium Digest*, Vol. 2, pp. 1101–1104 (1964).
37. Wong, T.T.Y., *Fundamentals of Distributed Amplification*. Artech-House, Boston, MA (1993).
38. Abrie, P.L.D., *Design of RF and Microwave Amplifiers and Oscillators*. Artech-House, Boston, MA (1999).
39. Wong, T.T.Y., *Fundamentals of Distributed Amplification*. Artech House, Boston, MA (1993).
40. Biran, Y., "2–22 GHz low noise distributed amplifier design." *Proceedings of IEEE Electrical and Electronics Engineers in Israel*, pp. 1–4 (1989).
41. Beyer, J.B., S.N. Prasad, R.C. Becker, J.E. Norman, and G.K. Hohenwarter, "MESFET distributed amplifier design guidelines." *IEEE Transactions of Microwave Theory Technology,* Vol. 32 No. 3, pp. 268–75 (1984).
42. Aitchison, C.S., "The intrinsic Noise Figure of MESFET Distributed Amplifier." *IEEE Transactions of Microwave Theory Technology*, Vol. 33 No. 6, pp. 460–466 (1985).
43. Niclas, K.B., and B. A. Tucker, "On noise in distributed amplifiers at microwave frequencies." *IEEE Transactions of Microwave Theory Technology*, Vol. 31 No. 8, pp. 661–668 (1983).
44. Brewitt-Taylor, C.R., P.N. Robson, and J.E. Sitch, "Noise figure of MESFETS." *IEE Proceedings, Part 1: Solid State and Electron Devices*, Vol. 127 No. 1, pp. 1–8 (1980).
45. Analui, B., and A. Hajirmiri, "Bandwidth enhancement technique for transimpedance amplifiers." *IEEE Journal of Solid State Circuits*, Vol. 39 No. 8, pp. 1263–1270 (2004).
46. Sanduleanu, M.A.T, and P. Manteman, "Low noise wide dynamic range transimpedance amplifier with automatic gain control for SDH/SONET (STM16/OC48) in a 30 GHz fT BiCMOS process." *Solid-State Circuits Conference, 2001. ESSCIRC 2001.* Philips Research Eindhoven, The Netherlands.
47. Barros, M.A.M., "Low-noise InSb photodetector preamp for the infrared." *IEEE Journal of Solid-State Circuits*, Vol. 17 No. 4, pp. 761–6 (1982).
48. Cole et al., "Fiber optic receiver and amplifier," US Patent 5095286, March 10 (1992).

49. Chen, W.Z., Y.L. Cheng, and D.S. Lin, "A 1.8 V 10 Gb/s fully integrated CMOS optical receiver analog front end." *IEEE Journal of Solid State Circuits*. Vol. 40 No. 6, pp. 1388–1396 (2005).
50. Fay, P., Caneau, and I. Adesia, "High speed MSM/HEMT and PIN/HEMT monolithic photoreceivers." *IEEE Transactions on Microwave Theory Technology*, Vol. 50 No. 1, pp. 62–66 (2002).
51. Edminister, J.A., *Electric Circuits Theory and Problems*. New York: Schaum Publishing Co. (1965).
52. Alexander, S., *Optical Communication Receiver Design*, SPIE Press, Bellingham, WA (1997).
53. Abraham, M., "Design of Butterworth-type transimpedance and bootstrap-transimpedance preamplifiers for fiber-optic receivers." *IEEE Transactions on Circuit Systems*, Vol. 29 No. 6, pp. 375–382 (1982).
54. Desoer, C.A., and E.S. Kuh, *Basic Circuit Theory*, McGraw-Hill, New York, N.Y. (1969).
55. Edminister, J.A., *Electric Circuits Theory and Problems*. New York: Schaum Publishing Co. (1965).
56. Matthaei, G.L., "Tables of Chebyshev impedance transforming networks of low pass filter form." *Proceedings of the IEEE*, Vol. 52 No. 8, pp. 939–963 (1964).
57. Park, S.M., and C. Papavassiliou, "On the design of low-noise giga-preamplifiers for optical receiver application." *Proceedings of the IEEE*, Vol. 2, pp. 785–788 (1999).
58. Shaeffer, D.K., and T.H. Lee, "A 1.5 V, 1.5 GHz CMOS low noise amplifier." *IEEE Journal Solid State Circuits*, Vol. 32 No. 5, pp. 745–59 (1997).
59. Razavi, B., *Design of Analog CMOS Integrated Circuits*. McGraw-Hill, New York, N.Y. (2001).
60. Millman, J., and C.C. Halkias, *Integrated Electronics: Analog and Digital Circuits and Systems*. McGraw-Hill (1972).
61. Vendelin, G.D., "Five basic bias designs for GaAs FET amplifiers." *MicroWaves*, Vol. 17 No. 2, pp. 40–42 (1978).
62. Crosneir, Y., H. Gerard, and G. Salmer, "Analysis and understanding of GaAs MESFET behavior in power amplification." *Proceedings of the IEEE*, Vol. 134 No. 1, pp. 7–16 (1987).
63. Ha, T.T., *Solid State Microwave Amplifier Design*. Krieger Publishing Company, pp. 246–247 (1981).
64. Moreira P.M.R.S., I.Z. Darwazeh, and J.J. O'Reilly, "Design and optimization of a fully integrated GaAs tuned receiver preamplifier MMIC for optical SCM applications." *IEE Proceedings-Optoelectronics*, Vol. 140 No. 6, pp. 411–415 (1993).
65. Razavi, B., *Professional Design of Integrated Circuits for Optical Communications*. McGraw-Hill, New York, N.Y. (2002).
66. Hammad, H.F., A.P. Freundorfer, and Y.M.M. Antar, "Unconditional stabilization of common source and common gate MESFET transistor." Queen's University, and Royal Military College of Canada, Kingston, ON Canada.

67. Faulkner, D.W., "A wide-band limiting amplifier for optical fiber repeaters." *IEEE Journal of Solid State Circuits*, Vol. 1B No. 3, pp. 333–339 (1983).
68. Cooke, M.I.P., G.W. Sumerling, T.V. Moui, and A.C. Carter, "Integrated circuits for a 200-Mbit/s fiber-optic link." *IEEE Journal of Solid-State Circuits*, Vol. 21 No. 6, pp. 909–915 (1986).
69. Caruth, D., S.C. Shen, D. Chan, M. Feng, and J.S. Aine, "A 40 Gb/s integrated differential PIN-TIA with dc offset control using InP SHBT technology." *IEEE GaAs Digest*, pp. 59–62 (2002).
70. Streit, D., R. Lai, A. Gutierrez-Aitken, M. Siddiqui, B. Allen, A. Chau, W. Beale, and A. Oki. "InP and GaAs components for 40 GBPS applications." *IEEE GaAs Digest*, pp. 247–250 (2001).
71. Johns, D.A., and K. Martin, *Analog Integrated Circuit Design*. John Wiley & Sons (1997).

# Chapter 12

# AGC Topologies and Concepts

This chapter provides the reader with an introduction to the theory of feed forward (FF) and feedback automatic gain control (AGC) systems, followed by design examples, noise analysis, and design approaches for practical FTTx an analog receivers. This is a key chapter and has a major importance when designing any amplitude control system for hybrid fiber coax (HFC) receivers.

The purpose of AGC is to increase the receiver's dynamic range when large signal levels are applied at the receiver input. Additionally, AGC should preserve and track input signal-to-noise ratio (SNR) as the signal gets into higher levels. That means that if the input signal at the receiver front end gets higher, the carrier-to-noise ratio (CNR) (same definition as SNR) is getting better. Hence, the requirement is to preserve, with minimum degradation, the input CNR. Since any signal processing system adds noise, the output CNR is reduced and the noise figure increases. The AGC role is to preserve a constant output when the input signal level is above the AGC threshold. Hence, it can be referred to as a kind of limiting circuit. AGC attenuator for that purpose can be an analog voltage-variable attenuator (VVA), a digital-controlled attenuator (DCA), or a variable-gain amplifier (VGA).

There are two main concepts to realizing and implementing AGC in a CATV receiver, feedback AGC and FF AGC. In both cases, the system constraints are the same. The design approach and design limitations described below:

1. The AGC response should be slow, meaning slower by the order of magnitude than the slowest transient of the signals handled by the system. It should not respond to fast changes in power and should not track them. Generally, optical power changes are very slow. Thus, the resulting effects on RF signal are slow as well.

2. In the case where the AGC tracks fast changes of the optical power, that is tracking fast amplitude changes in the RF signal, it would add undesirable residual amplitude-modulation (AM) fluctuation on the data. This is even more critical when the video data is a high-order quadrature amplitude modulator (QAM), for instance, a 256 QAM. That kind of residual AM

affects BER and would result in lower BER performance as will be explained in Chapters 14 and 15.

3. The AGC should have a sufficient dynamic range in order to prevent the output stage from being driven into the nonlinear operation mode region.

4. The AGC system should be independent of channel loading, modulation index, and modulation type.

5. As the input signal gets higher, the AGC system should increase the dynamic range of the receiver by increasing the attenuation without degrading output SNR. SNR reduction results because of the increased attenuation that produces a higher noise figure $10\log(F) = NF[dB]$. (This will be explored in Sec. 12.2.6.) In order to fulfill these requirements, the gain is generally distributed throughout the receiver's RF chain. The AGC attenuator is placed in the middle of the RF chain, taking into account noise-figure and compression-point.

## 12.1 Feed Forward (FF) Automatic Gain Control (AGC)

The FF AGC approach is commonly used in CATV receivers. In fact, an FF AGC is an open-loop gain-control system. Hence, it is much more sensitive to parameter variations of the circuit and less accurate than a regular feedback AGC. However, it is not sensitive to channel loading and modulation indexing since it senses the dc level of optical power at the photodetector (PD) monitor.

There are several ways to realize FF gain control in a CATV receiver. One approach is by using DCA and the second is by using VVA as an attenuation element.

The first method senses the dc monitor optical voltage generated by the PD, using an analog-to-digital converter (ADC) microprocessor and creating a lookup table for the required attenuation bits required for the DCA per the PD voltage level. Other methods are to use comparators and tune them to different levels of attenuation states to provide the attenuation bits for the DCA. A third way is to use a PIN VVA and an operational amplifier (OP amp) so that its output voltage gain is aligned to the PIN diode curve. This is a much more complex approach but less expensive. A fourth way is to use a microcontroller lookup table. The lookup table bits are used to define the DAC, which provides a control voltage to the PIN VVA. This is a cost-effective method commonly used in CATV receivers.

The power quantization (flatness or ripple accuracy) of the digital FF AGC is directly dependent and related to the number of attenuation bits or AGC steps. Since an FF-gain-control system is not a continuous tracking system, it would have a power step (jump) between transitions; this step size is related to the quantization value. This is true for DCA realization and also for VVA realization since the VVA control voltage is directly related to the DAC resolution. Moreover, in coarse DCA/VVA control realization, the RF output power would be increased if the optical power level impinging on the PD gets higher or decreased if the light level impinging on the photodetector gets lower until the next AGC-state step. Hence, the more attenuation bits the FF gain control has, the smaller the power step and power variation at the output. As a consequence, the RF power versus light response would be much flatter. However, the higher the resolution of the system, it is much more sensitive to fluctuations over the dc level sampled from the PD. As a result, it would generate undesired residual AM on the bearing information signal.

There are several ways to overcome this FF AGC toggling problem. The first approach is to increase the hysteresis of the transitioning point, i.e., between attenuation states. This would sacrifice dynamic range when the power goes up, and would degrade sensitivity and reduce CNR when the power goes down since the gain control state does not update the per signal level but operates on programmed dc thresholds. The second way is to add some delay (time constant) in the software to reduce the low side band (LSB) rapid changes (toggling). This is equivalent to the integration time of the optical current fluctuations as it appears on the dc monitor voltage-sampling resistor as it appears in Fig. 12.1. Another method of resolution increase, when realizing AGC with a VVA, which includes integration as well, is to use pulse-width modulation (PWM). At the out-of-lock state, the error pulse is wider and there is a larger current flowing through the integrator, which is the loop filter. As the system goes toward the locking state, the pulse becomes narrower and the system maintains the AGC state. Hence, for an analog FF gain control, a large, time-constant narrow bandwidth (BW) integrator is needed. Another method is to use microcontroller and VVA while sampling the PD voltage with an ADC and drive the VVA with a high-resolution DAC. Here, the strategy is to create an interpolated lookup table and add a sampling routine that measures the derivatives of the dc photovoltage between each sampling. This way the algorithm can be detected if it is an upward or downward power trend, or if this is just a interference created by the mechanical vibration of the optical connector as an example. For that purpose, averaging of voltage and derivative is performed to the evaluated trend. Derivative averages are compared to the previous state and voltage readings for the FF AGC would operate as a continuous power control system with high resolution and accuracy.

The FF AGC threshold and the first trigger point setting is done by calibrating the lowest optical voltage value at the minimum optical power level specified as the input value to the receiver. The dc voltage-sampling resistor is a potentiometer. Hence, it is used to compensate the PD responsivity factor and set the reference voltage. At low-level responsivity, it can be tuned at a higher resistance to produce the desired trigger voltage, and vice versa since the PD is a light-controlled current source. This way the AGC trigger-point voltage versus optical level is preserved. This calibration would guarantee a fixed RF level at the receiver's output. Fig. 12.1 illustrates the FF gain control diagram. In Fig. 12.1 (a) FF AGC realization using a DCA and a microcontroller is presented. In Fig. 12.1 (b), the FF AGC is realized by using a VVA microcontroller and a DAC to drive the VVA. Figure 12.1 (c) describes a simple coarse FF AGC system with four-step

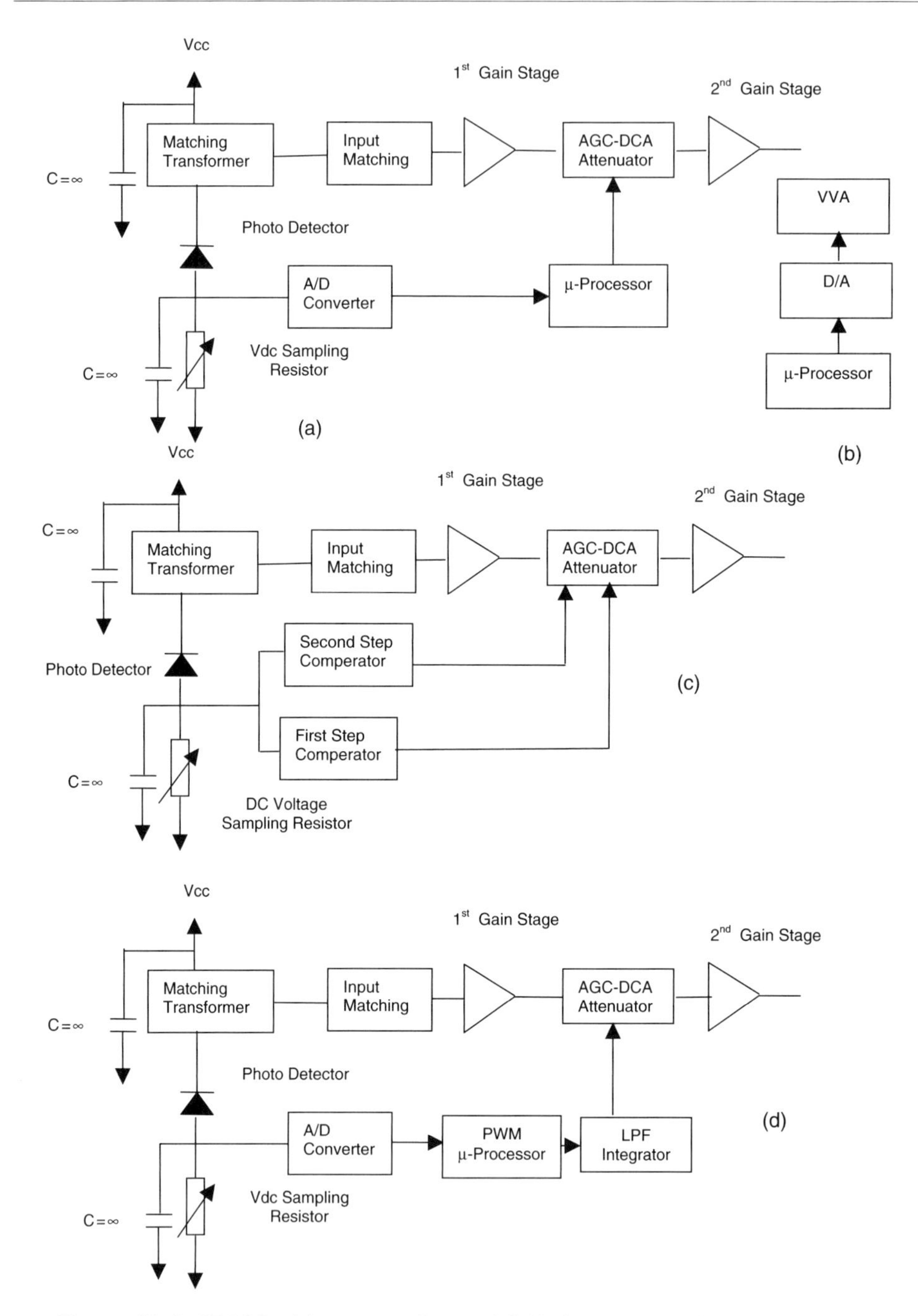


**Figure 12.1** FF AGC with a controller and 2-bit four steps using comparators.

DCA, and Fig. 12.1 (d) describes a modern PWM FF AGC system using a VVA. In this case, the output of the microcontroller is a PWM pulse that is integrated into a control voltage. Integration and filtering in this case is essential and important for two reasons. First, it creates the control voltage relative to the pulse width of the

PWM controller. Second, the filtering should be sharp enough in order to reduce any harmonics that might create an undesired residual AM through the VVA control voltage. This residual AM can be observed as two AM sidebands, while transmitting a CW signal. The AM sidebands' frequency is determined by the PWM pulse frequency. The minimum specification for any of these residual AM sidebands is driven by the composite second-order (CSO) and composite triple-beat (CTB) specifications. Residual AM sidebands may reduce QAM bit-error-rate (BER) performance as explained in Chapters 14 and 15.

## 12.2 Feedback AGC

Feedback amplitude leveling system configuration is widely used in communication systems to maintain a constant-signal power-control system.[18] A typical AGC loop includes a PIN-VVA, an amplitude rms detector, and a loop filter. The leveling is done by comparing the detected power voltage to an accurate reference voltage that represents the desired power. A feedback AGC system is an accurate tracking system that is not affected by the gain chain components' tolerances. Due to the nature of power variations considered to be slow to change versus time, AGC loop is a narrow BW feedback system. For a CATV receiver, the BW should be below 1 Hz. That way the AGC loop does not track the rapid power changes due to frame sync, blanking, and cross modulation. However, there are several details that should be paid attention to when realizing such a system in a CATV receiver. The rms power detector samples many channels; therefore, the detected power is directly related to the number of channels and their optical modulation index (OMI). Moreover, not all the channels have the same modulation methods; there is a variation in the OMI as well as the peak to average of the signal (sometimes this value is called peak to rms). As a consequence, the system design should refer this sensitivity to a modulation scheme OMI and peak to rms values. Additionally, when using an AGC with PIN VVA, it should be highly linear at a high attenuation mode as well as in low attenuation. This requirement is valid for FF AGC as well.

A way to overcome VVA distortions is by using a PIN diode with a wide intrinsic I region. That way the VVA would generate fewer and lower level distortions at a high attenuation mode when attenuating high-RF power levels as well. The only disadvantage of a wide PIN VVA is its insertion loss at a zero level attenuation state, which is a bit higher by 2–3 dB compared to a narrow I-region PIN diode VVA. Moreover, since the AGC is a slow narrow BW tracking system and since a wide PIN diode has slow switching time, it does not affect the AGC performance since the PIN diode recovery time constants are still faster than those of the AGC loop.[17]

To avoid the AGC detector dependency on the number of channels, a portion of the energy is sampled by the rms detector through a directional coupler and then a low pass filter (LPF). Generally, the low-frequency channels are sampled and good engineering practice calls for 30 channels LPF BW, i.e., the pass band of the LPF transfers only 30 channels. Those 30 channels are equal to a single

AGC calibration pilot tone such that its OMI is calculated by Eqs. (6.84) and (8.48). These channels are analog NTSC channels that have lower peak to rms compared to QAM schemes as will be explained in Sec. 12.2.8 and in Chapter 14. Therefore, this solution is similar to a conventional AGC approach using an out-of-band leveling pilot tone so that its OMI is equal to 30 tones and the equivalent OMI is per Eqs. (6.84) and (8.48). As was explained in Chapter 3, the analog NTSC channels have a higher OMI compared to the QAM channels. Therefore, fewer NTSC channels are needed to create the AGC power that is equivalent to the AGC pilot tone. In new frequency-advised plans, all the channels will be QAM. Since the QAM scheme operates at a 50% lower OMI compared to the NTSC analog plan, the required number of channels to be sampled for the detector according to Eqs. (6.84) and (8.48) would be greater by a factor of $2\sqrt{30}$. Thus, the sampling LPF is of a wider BW, with a larger number of channels, almost four times as much, in order to sustain the same pilot-equivalent OMI level and compensate the power difference due to the difference in OMI. The main drawback of this method is the higher peak-to-rms error of 256 QAM compared to the analog channels. Section 12.2.8 provides a detailed elaboration on that matter.

The AGC system is a bit more complex to analyze compared to the phase-locked loop (PLL).[4,19–24] Common to most closed-loop RF leveling circuits are two nonlinear elements: the power detector and the PIN VVA. These nonlinearities often affect the closed-loop stability and various methods are commonly used to ensure stability within the entire range of attenuation. Since loop stability depends on the operating point, stable operation at one point does not guarantee total stability over the entire AGC dynamic range. A way to understand this is to observe the VVA constant definition $K_{\mathrm{VVA}} = \mathrm{d}[\mathrm{att}(V)]/\mathrm{d}V = f(V) \neq \text{constant}$; att($V$) is the VVA attenuation versus the control voltage. The same applies for the detector $K_{\mathrm{det}} = \mathrm{d}V/\mathrm{d}P = f(P) \neq \text{constant}$. Here, the voltage detected at the detector output is a function of input RF power and detector type, i.e., an envelop rms averaging detector as a power meter or peak. Therefore, the AGC system is a nonlinear system. However, at the operating point when the AGC acquires the power level it was set for, it operates under small signal fluctuations around its set point. Hence, linearization approximation is used for small signal analysis to set loop parameters $K_{\mathrm{VVA}}$ and $K_{\mathrm{det}}$.

Feedback AGC can be realized by using a microcontroller. For that purpose, the AGC software should have the VVA and the detector curves. In some implementations, the VVA can be replaced by a DCA. Today, DCAs available in the market have high accuracy with a resolution step size of 0.5 dB with an attenuation depth of up to 32 dB. However, the DCAs should have high IP2 and IP3 levels in order to maintain stringent CSO CTB specifications. Figure 12.2 illustrates typical CATV AGC system approaches with VVA and DCA.

From Fig. 12.2 it can be observed that the power detector would operate under the same operating point for different input levels at the receiver's RF front end (RFFE). This is due to the fact that the feedback corrects the VVA attenuation state to have a constant RF output power. Hence, at different optical levels, the VVA operates under different control voltages, whereas the detector remains at

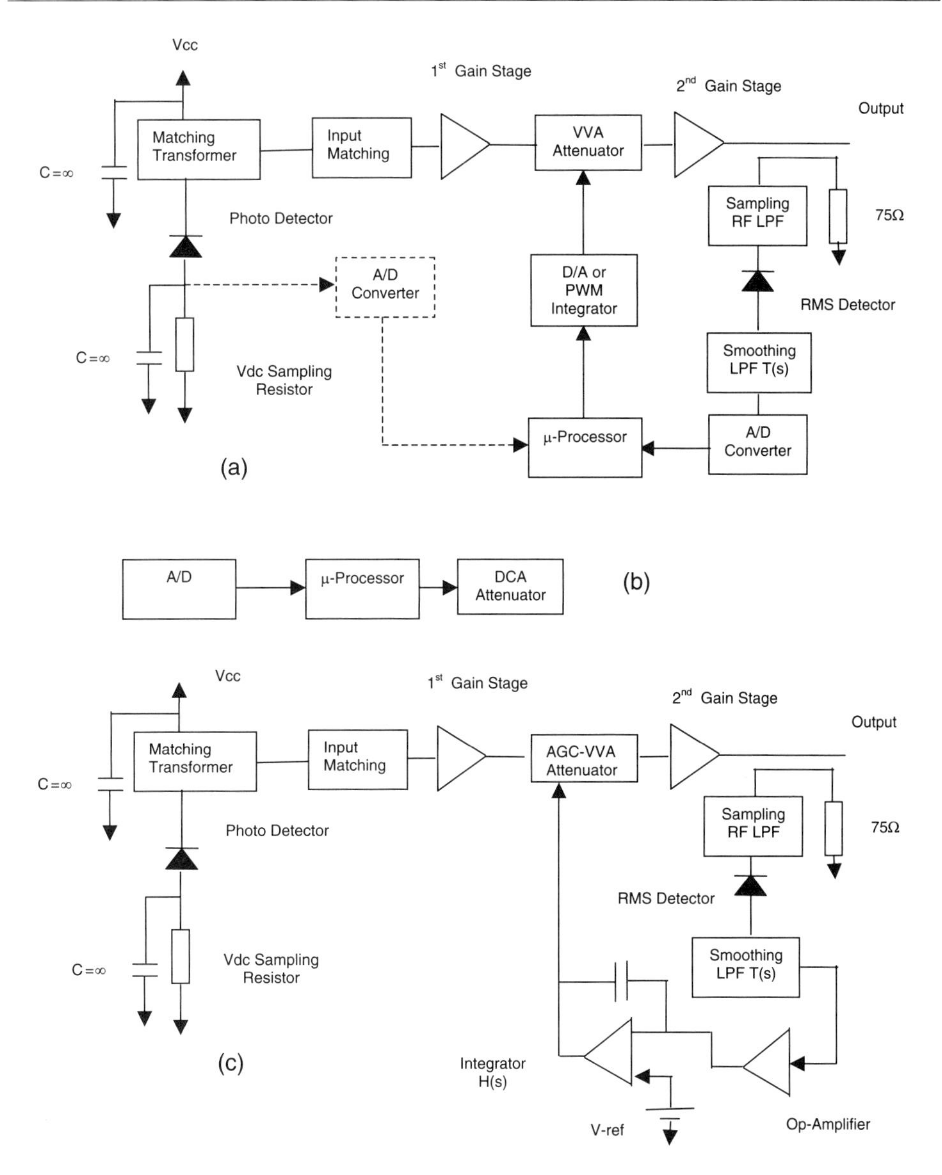


**Figure 12.2** Feedback AGC with a controller and DCA and with analog circuit using VVA. Note that VVA can be used with a controller as well.

the same RF operating point and senses the same RF level at the steady-state locked condition. Therefore, under locking at a fixed output RF level, $K_{\text{det}} = \mathrm{d}V/\mathrm{d}P = \text{constant}$, whereas $K_{\text{VVA}} = \mathrm{d}[\text{att}(V)]/\mathrm{d}V = f(V) \neq \text{constant}$ since it varies from one RF level to another. During a large magnitude input power change to the receiver, such as a step response, the system gets out of lock and both coefficients $K_{\text{det}}$ and $K_{\text{VVA}}$ vary until the system stabilizes to its new set point. This input power step response requires large signal analysis. If the only change were in input power, $K_{\text{det}}$ would return to its previous value since RF power is locked to a fixed level and $K_{VVA}$ would stabilize to its new value depending on the attenuation level. For both RF and output leveling change, the mean input step and required RF level at the receiver output is programmed to a new power state, then both constants change to new values. Those phenomena can drive the AGC system through nonstable zones causing oscillations since both constants vary until they stabilize to the steady-state values. Since the AGC is a feedback system, it may go through the unstable state with a low-phase and gain margin. If the AGC goes through a stable zone until it locks, it would be similar to cycle slips in PLL until the system is locked again. A detailed analysis about AGC time stabilization is provided by Refs. [24] and [25] As will be demonstrated in Sec. 12.2.5, it is important to check stability in transition from one level to the other for changes in both the input power and in the level locking. For small input-power changes, the AGC system would operate as a tracking system. In case of a sudden abrupt step change of the input power, the AGC would operate as a nonlinear system until it acquires its new power-leveling state.

Figure 12.2 summarizes several design approaches for feedback AGC. In Fig. 12.2(a), a PWM controller implementation is presented. At a large power-level error, the PWM generates a wide-current pulse integrated to large VVA voltage as the loop error goes down and the power level is locked, the pulse becomes very narrow and the system is in tracking mode. This is due to the fact that the signal is corrupted by noise and there are small amplitude fluctuations around the equilibrium point. Advanced popular controllers have several ADC ports; their architecture can support both FF and feedback concepts. A flag-bit level can activate either feedback or FF software algorithms and in that manner, the receiver can have both the AGC mechanisms of coarse FF and fine feedback leveling. The VVA can be driven by DAC with high resolution as shown by Fig. 12.2(a). In this case, the VVA control voltage resolution is determined by the DAC. Feedback structure can support DCA as shown in Fig. 12.2(b). The attenuation resolution is determined by the number of control bits as was elaborated for FF AGC. Traditional straightforward analog realization is presented in Fig. 12.2(c).

A different approach is used in current FTTx deployments called the "flat AGC sensing." In this approach, there is no sampling LPF filter at the coupled arm. All tones are routed to the RF detector. Calibration is made for all tones. This enables flexibility in the frequency plan since providers do not have to use all low-band channels as a pilot. This elevates the additional requirement of flatness in order to have accurate sensing by the detector throughout the entire band from 50 to 870 MHz. A typical specification is $\pm 0.5$ dB and an equalization network is inserted between the RF detector and the coupler. Changing the number of channels may affect the AGC and a correction for RF power is done by actively changing the reference voltage to lock the AGC. That functionality is done via the diagnostic channel through the digital link of the integrated triplexer (ITR). A dedicated controller of the AGC will update the reference voltage for the AGC in order to lock on a new reference value for preserving the same RF level. Otherwise, for reducing the channel count, equal to reducing OMI, AGC locks on higher power and for increasing the channel count, equal to increasing OMI, the AGC locks on lower power level.

### 12.2.1 AGC loop analysis

Unlike a phase-locked loop transfer function,[4] AGC transfer function is a bit more complex to analyze. The reason is that the AGC loop has more degrees of freedom that are needed to be determined.[2,19–24]

AGC locking requires solving the block diagram as described in Fig. 12.3, showing an envelop detection of a noncoherent AGC loop.[2] The voltage-controlled amplifier is marked as $G[V(t)]$ in time domain or $G[V(s)]$ in Laplace domain, where the Laplace variable is denoted by "$s$." The steady-state voltage would be marked

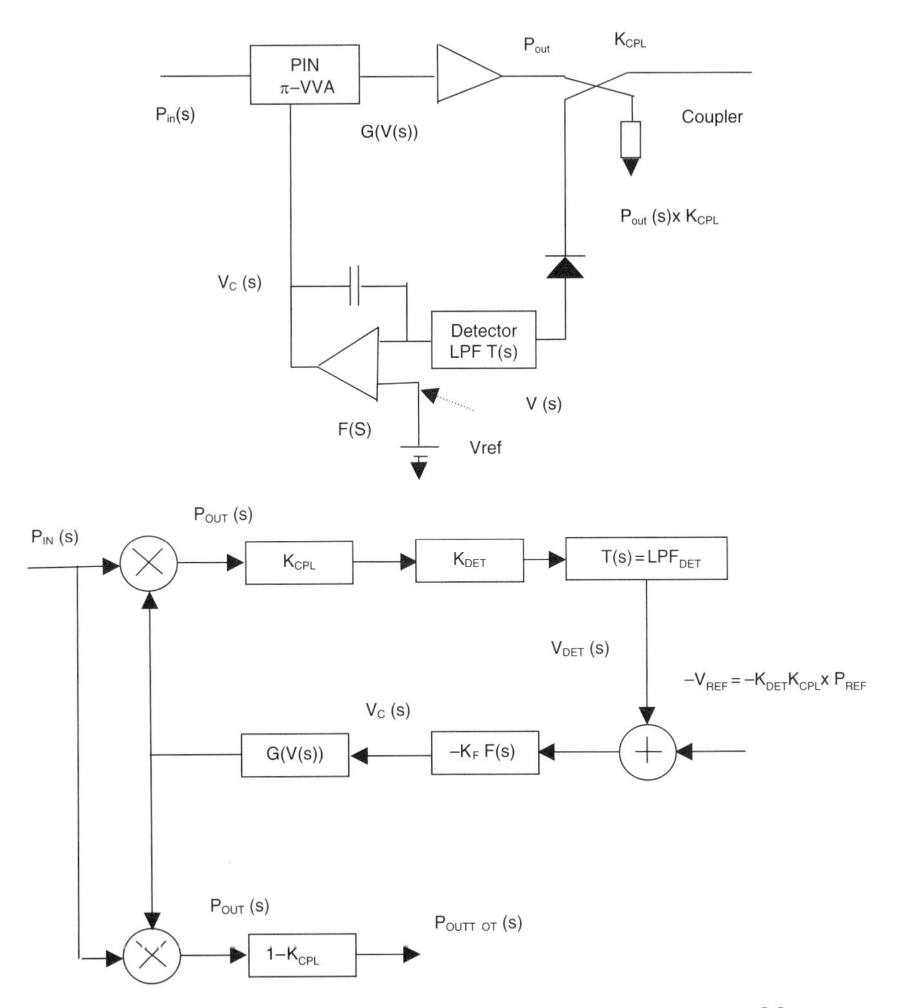


**Figure 12.3** AGC block diagram parameters and signal flow diagram.[2,3] (Reprinted with permission from John Wiley and Sons © 1997.)

as $V_{cs}$ at the integrator output and $V_{in\_cs}$ at the integrator input. The model includes the $\pi$ PIN attenuator multiplied by the output stage gain. Hence, the voltage variable gain is marked by $G[V(s)]$. The loop amplifier or integrator is denoted as $F(s)$. The minus sign is due to the inverting integrator. The reference power is represented by a precession-controlled bandgap voltage regulator $V_{REF}$. Since this block diagram refers to linear units and not logarithmic units in dB, it shows multiplication rather than summation. The first RF-input stage is not a part of the loop and is referred to as multiplying the transfer function by a constant. Hence, the first stage is omitted in the following analysis.

The analysis presented here refers to narrow-band fluctuations over the time of $P_{IN}(s)$. The BW of the RF components is wider and is transparent to the AGC control loop BW. In the steady state, the reference power is approximately equal to the RF power. Strictly speaking, there is no steady state since the AGC continuously tracks the signal due to signal corruption by noise. Assuming that the detector's smoothing filter $T(s)$ in Fig. 12.3 is much wider than the loop BW, $T(s) = 1$ within the loop BW. That filter is generally used to remove high-order harmonics and distortions from the OP-amp-integration circuit $F(s)$. Any additional gain of the OP-amp in Fig. 12.2 is absorbed by $K_F$, meaning, $K_F$ is multiplied by a constant. In the steady state, the integrator input is almost zero, the error voltage is extremely low, and the first approximation goes to zero. The AGC loop gain is at its maximum and therefore the integrator gain is $F(s) = \infty$, $s = 0$. Thus, the following condition applies:

$$G(V_{CS})P_{IN\_SS} - P_{REF} = 0. \tag{12.1}$$

$P_{IN\_SS}$ is the steady-state input power and $P_{REF}$ is the steady-state RF reference power for locking the output of the RF amplifier. As a conclusion, the steady-state VVA control voltage can be written by Eq. (12.2), where $V_{CS}$ is the AGC steady-state operating point voltage:

$$V_{CS} = -K_F K_D[G(V_{CS})P_{IN\_SS} - P_{REF}]. \tag{12.2}$$

This equation is a recursive equation and $G(V_{cs})$ is an exponent function. Rearranging Eq. (12.2), provides a graphic solution for the AGC-operating point value of $V_{CS}$ as described by Fig 12.4:

$$P_{IN} \cdot G(V_{CS}) = P_{REF} - \frac{V_{CS}}{K_F K_D}. \tag{12.3}$$

It is found that $V_{CS}$ is given by the intersection of the curve $P_{IN}G(V_c)$ with the straight line,

$$P_{REF} - \frac{V_C}{K_F K_D},$$

where $V_C$ at the intersection point equals the VVA control voltage $V_{CS}$.

From that plot it can be observed that $V_{CS}$ equals the reference voltage $V_{C0}$, which occurs when the gain of the loop is very high and goes to infinity, i.e., the slope of the straight line $-1/K_F\ K_D \to 0$ (the straight line is parallel to the

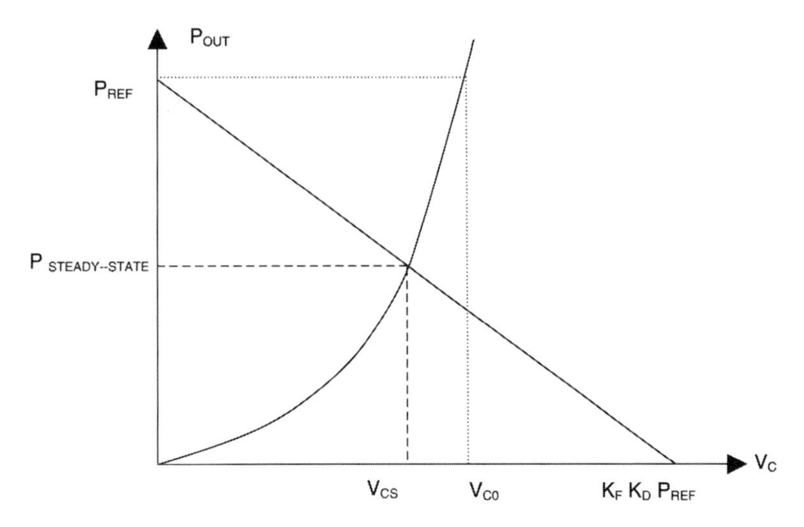


**Figure 12.4** Output power $P_{OUT}$ as a function of gain control voltage $V_C$ and $V_{CO}$ is the reference voltage.[2,3] (Reprinted with permission from John Wiley & Sons © 1997.)

voltage axis), and $P_{REF}$ almost merges with $P_{STEADY\text{-}STATE}$. Thus, high AGC loop gain minimizes its error and improves its limiting factor $P_{OUT}/P_{REF}$. If the $V_{CS}$ condition of having a close value to the reference voltage $V_{C0}$ is satisfied, linearization approximation is applicable around $V_{C0}$. That condition means that the loop had stabilized and all nonlinear transitions and acquisition states are completed and $P_{OUT}/P_{REF} \cong 1$.

### 12.2.2 Linear approximation

When the gain-control voltage exhibits only small fluctuations around its steady-state voltage $V_{CS}$, a linear approximation using a Taylor series expansion of $G[V_C(t)]$ is applicable. These control voltage fluctuations result from power fluctuations around its nominal locking value. The goal in this analysis is to find the ratio between the control voltage fluctuations and the power fluctuations in the Laplace domain.[2] This ratio is the AGC closed-loop transfer function at a steady state under a linear approximation. When the fluctuations are considered to be small enough, the second-order terms and higher-order terms in the power series expansion are omitted and the following linearization is valid:

$$G[V_C(t)] \approx G[V_{CS}(t)] + [V_C(t) - V_{CS}] \cdot g,$$

where

$$g - \left.\frac{\partial G[V_c(t)]}{\partial V_c}\right|_{V_c = V_{cs}}, \tag{12.4}$$

where $V_{CS}(t)$ is the steady-state locking voltage and $V_C(t)$ is the fluctuated voltage around the steady-state voltage value $V_{CS}(t)$.

Substituting Eq. (12.4) in to Eq. (12.5) results in the approximated term given in (12.6):

$$V_C(t) = -K_F K_D \{G[V_C(t)] \cdot P_{IN}(t) - P_{REF}\} * f(t), \tag{12.5}$$

$$V_C(t) = -K_F K_D \left( \left\{ G(V_{CS}) + [V_C(t) - V_{CS}] \cdot g \right\} \cdot P_{IN}(t) - P_{REF} \right) * f(t), \tag{12.6}$$

where $f(t)$ is the loop-filter (integrator) impulse response for $\delta(t)$ and $^*$ marks convolution.

The goal now is to manipulate Eq. (12.6) into an equation of fluctuating voltage versus power in order to find the AGC transfer function. Knowing that $P_{REF} = G(V_{CS}) \times P_{REF\text{-}IN}$, Eq. (12.6) becomes

$$V_C(t) = -K_F K_D \left\{ G(V_{CS}) \cdot [P_{IN}(t) - P_{REF-IN}] + [V_C(t) - V_{CS}] \cdot g \cdot P_{IN}(t) \right\} * f(t). \tag{12.7}$$

The coefficient $K_F$ is negative since the loop integrator is inverting. The above notation of Eq. (12.7) can be written as

$$V_C(t) = -K_F K_D \left\{ \begin{array}{l} G[V_{CS}][P_{IN}(t) - P_{REF-IN}] + [V_C(t) - V_{CS}] \cdot \\ g \cdot P_{IN}(t) - [V_C(t) - V_{CS}] \cdot g \cdot P_{REF-IN} \\ + [V_C(t) - V_{CS}] \cdot g \cdot P_{REF-IN} \end{array} \right\} * f(t). \tag{12.8}$$

To redefine the variables as voltage and power fluctuations

$$\Delta v(t) = V_C(t) - V_{CS} \ \text{and} \ \Delta P_{IN}(t) = P_{IN}(t) - P_{IN-REF}, \tag{12.9}$$

Eq. (12.8) becomes

$$V_C(t) = -K_F K_D \Big\{ G[v_{CS}][\Delta P_{IN}(t)] + [\Delta v_C(t)] \cdot [\Delta P_{IN}(t)] \cdot g + [\Delta v_C(t)] \cdot g \cdot P_{REF-IN} \Big\} * f(t). \tag{12.10}$$

After rearranging the variables, Eq. (12.10) becomes

$$V_C(t) = -K_F K_D \cdot g \cdot P_{REF-IN} \cdot \left\{ [\Delta v_C(t)] * f(t) \right\} - K_F K_D G[v_{CS}] \cdot \left\{ [\Delta P_{IN}(t)] \left( 1 + \frac{[\Delta v_C(t)] \cdot g}{G[v_{CS}]} \right) \right\} * f(t). \tag{12.11}$$

This is still a nonlinear equation, but when the fluctuations around the steady state are small as was postulated,

$$\frac{[\Delta v_C(t)] \cdot g}{G[v_{CS}]} << 1.$$

Hence, this term is negligible. Additionally, the voltage $V_C(t)$ is a fluctuation around a dc and the final linear differential equation of the AGC is obtained by

$$\Delta V_C(t) = -K_F K_D \cdot g \cdot P_{REF-IN} \cdot \left\{[\Delta v_C(t)] * f(t)\right\} - K_F K_D G[v_{CS}] \cdot \left\{[\Delta P_{IN}(t)] * f(t)\right\}. \quad (12.12)$$

This expression is converted into a Laplace domain

$$\Delta V_C(s) = -K_F K_D \cdot g \cdot P_{REF-IN} \cdot \left\{[\Delta v_C(s)] \cdot F(s)\right\} - K_F K_D G[v_{CS}] \cdot \left\{[\Delta P_{IN}(s)] \cdot F(S)\right\}. \quad (12.13)$$

The transfer function is defined as the ratio $\Delta V_C(s)/\Delta P_{IN}(s)$. Therefore, the AGC transfer function is given by

$$H(s)_{AGC} = \frac{\Delta v_c(s)}{\Delta P_{IN}(s)} = \frac{-K_F K_D \cdot G(v_{cs}) \cdot F(s)}{1 + K_F K_D P_{REF-IN} \cdot g \cdot F(s)}. \quad (12.14)$$

However, in order to have the transfer function by way of open-loop gain and feedback gain, both sides of Eq. (12.14) are multiplied by a factor:

$$H(s)_{AGC} \cdot \frac{g \cdot P_{REF-IN}}{G(v_{cs})} = \frac{\Delta v_c(s)}{\Delta P_{IN}(s)} \cdot \frac{g \cdot P_{REF-IN}}{G(v_{cs})} = \frac{-K_F K_D \cdot g \cdot P_{REF-IN} \cdot F(s)}{1 + K_F K_D P_{REF-IN} \cdot g \cdot F(s)}. \quad (12.15)$$

Redefining the AGC transfer function as the right-hand side of Eq. 12.15 (with positive sign) results in

$$H(s)'_{AGC} = \frac{K_F K_D \cdot g \cdot P_{REF-IN} \cdot F(s)}{1 + K_F K_D P_{REF-IN} \cdot g \cdot F(s)}. \quad (12.16)$$

Thus, the total AGC loop response equals

$$H(s) = \frac{-G(v_{cs})}{g \cdot P_{REF-IN}} \cdot H(s)'_{AGC}. \quad (12.17)$$

Hence, it is identical to Eq. (12.14).

The main point behind this notation is expressed in Eq. (12.18). It shows how the relative power fluctuations around the reference level are filtered by the loop transfer function and are equal to relative control voltage fluctuations:

$$\Delta v(s) \approx \frac{-G(v_{cs}) \cdot \Delta P_{IN}(s)}{g \cdot P_{REF-IN}} \cdot H(s)'_{AGC}. \quad (12.18)$$

Moreover, that factor has no effect on the loop stability and time constants since it is a real number factoring the transfer function. But when designing an AGC loop and calculating its stability, there is a need to know the reference power and the AGC operating point.

A different point of view is by looking at Eq. (12.14) as

$$H(s) = \frac{A}{1+\beta A} \text{ or } \beta \cdot H(s) = \beta \frac{A}{1+\beta A}. \tag{12.19}$$

The main idea behind this notation is that the relative changes around the reference point at the output of a feedback system are sampled by the feedback and generate an error signal. In the AGC system, power changes are translated to VVA control voltage. These control voltage fluctuations are filtered out by $H(s)$ until they converge into the desired equilibrium. Moreover, the AGC is transparent to fast input signal changes with a frequency response higher than $H(s)$ 3-dB corner frequency. That means that the AGC feedback system is not tracking these changes because of its low pass nature.[3,5] In FTTx designs for CATV transport, these fast signals are the blanking and the frame synchronization as was mentioned in Chapter 3. Taking the ratio between the denominator expression without the unity added and the numerator of Eq. (12.14) results in the value of the feedback coefficient $\beta$:

$$\beta = \frac{K_F K_D P_{REF-IN} \cdot g \cdot F(s)}{-K_F K_D \cdot G(v_{cs}) \cdot F(s)} = -\frac{P_{REF-IN} \cdot g}{G(v_{cs})}. \tag{12.20}$$

Hence, at a large forward $A$ value gain as in Eq. (12.19), the loop transfer equals $1/\beta$. Thus,

$$\frac{1}{\beta} = -\frac{G(v_{cs})}{P_{REF-IN} \cdot g}. \tag{12.21}$$

This means that there is a full tracking of voltage fluctuations versus power fluctuations. The feedback coefficient units are [V/W] and $g$ is in units of [1/$V$]. Hence, AGC can be considered an AM detector when the fluctuation frequencies are within the loop BW since the control voltage will start tracking them. This is an analog to PLL that is used as an FM detector.

Additional coefficients should be added to the loop calculations. These values are the coupling factor $K_{CPL}$ and the detector smoothing filter $T(s)$. Thus, the final general expression of an AGC loop is given by

$$H(s)_{AGC} = \frac{K_F K_D K_{CPL} \cdot g \cdot P_{REF-IN} \cdot F(s)T(s)}{1 + K_F K_D K_{CPL} P_{REF-IN} \cdot g \cdot F(s)T(s)}. \tag{12.22}$$

### 12.2.3 Transfer function analysis of exponential gain AGC with PIN VVA

For large dynamic range gain control, AGC systems are realized with an exponential gain control characteristic:[2,17]

$$G(v_c) = G_0 \exp(G_1 v_c), \tag{12.23}$$

where $G_0$ and $G_1$ are design constants. Such exponential behavior refers to a diode. The PIN VVA attenuation depends on its junction series resistance $R_s$. That resistance depends on the drive current. In feedback AGC, the drive current is related to the output voltage from the AGC integrator, the series resistance between the integrator and the VVA, and the voltage drop across the VVA.[6] Hence, $G_1$ has units of $1/\Omega$ and $G_0$ is the system gain. $G_0$ can be the minimum gain. As a result, the gain function approximation would be similar to a diode current function:

$$G(v_c) = G_0[\exp(G_1 v_c) - 1]. \tag{12.24}$$

It can be seen that as the control voltage gets higher, $G(v_c)$ increases. However, there is some limit that is defined by the system maximum gain. Maximum-gain state occurs at the AGC threshold. At that point, the current drive of the VVA and the AGC control voltage are at their maximum value. If the RF-input power is below the threshold point, for any 1-dB decrement in optical input power, the RF output would be reduced by 2 dB. On the other hand, when the signal is too strong, the control voltage reaches its minimum value. Hence, the current to the VVA is minimal and the attenuation is at the maximum value. That point is called the AGC saturation point. If the RF input power is above the saturation point, then for any 1-dB increment in input optical power, the RF output will rise by 2 dB until it reaches the compression point. The difference between the AGC-saturation point and the AGC-threshold point is defined as the AGC dynamic range.

Since the AGC system has a maximum gain and a minimum gain, as was explained above, Eq. (12.25) expresses the gain versus voltage within the dynamic range of the AGC:

$$G(v_c) = [G_{max}^{-1} + G_0^{-1} \exp(G_1 v_c)]^{-1}. \tag{12.25}$$

When designing an optical CATV-receiver-link budget, it is important to know the maximum gain and power-leveling requirements. Power leveling by AGC is defined as power limiting. These two points define the AGC-threshold point and its dynamic range.

Assuming that AGC is roughly a limiting mechanism, several design limitations can be observed from a graphical design approach presented by the plot in Fig. 12.5.[6] Figure 12.5 shows two straight lines describing RF power versus optical input power. The first dashed line describes the RF output from

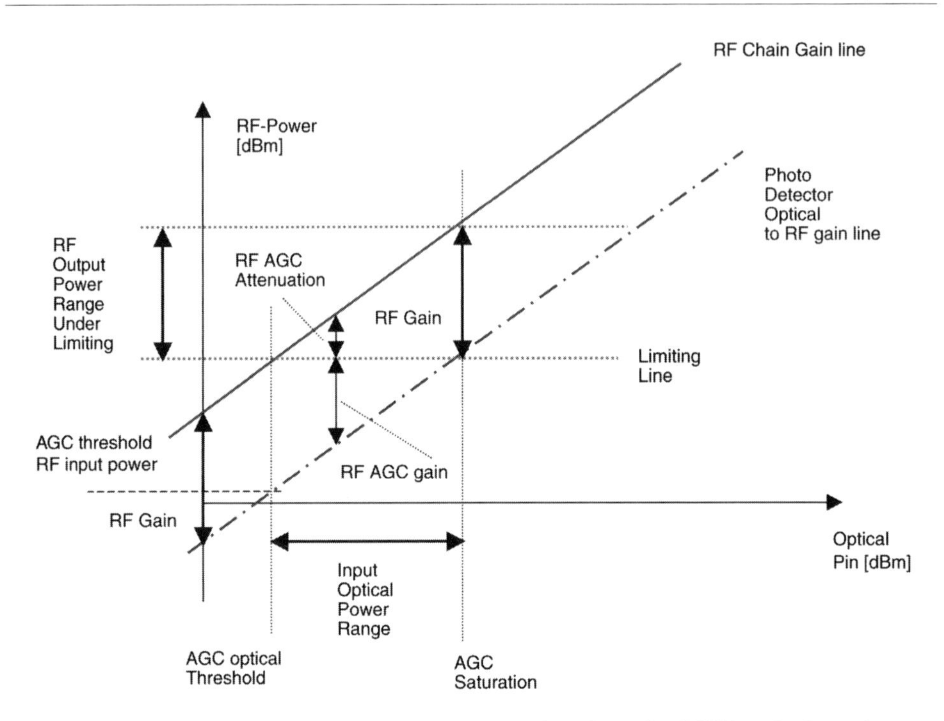


**Figure 12.5** AGC dynamic-range design considerations for CATV optical receiver.

the photodetector, and the second solid line shows the RF output from the RF-gain chain when the receiver is at its maximum-gain state, i.e., the VVA is at its minimum attenuation and maximum AGC-control-voltage value is applied. Figure 12.17 depicts the VVA attenuation versus AGC control voltage. The power difference between the two lines is constant and is the maximum RF-gain chain. The AGC optical-threshold level defines the RF-limiting level set by the AGC at the receiver's output for a given maximum gain. In other words, the optical-threshold level is limited by the maximum gain of the RF chain, which defines the minimum RF input that the AGC can lock. Hence, if a higher RF-limiting level is required at the receiver's output for a given RF-gain chain, a higher optical-threshold level is needed. In the ideal case, the saturation point is defined at the point where the RF-gain chain equals 0 dB. That means that the VVA absolute-attenuation value at that point equals the maximum net gain. Hence, the maximum dynamic range of the AGC equals the RF-gain chain. As a consequence, the maximum optical power range equals half the RF-chain gain. Practically, the AGC dynamic range is lower than the RF net gain because of noise considerations as well as distortions that reduce the CNR and C/I (carrier to intermodulation ratio).

Since the VVA PIN diode nature is of exponential gain, the linear approximation of the gain coefficient $g$ is given by Eq. (12.26):

$$g = g_0 = \frac{\partial}{\partial v_c} G_0 \exp(G_1 v_c) = G_0 G_1 \exp(G_1 v_c) = G_1 G(v_{c,o}). \tag{12.26}$$

Assuming the AGC control voltage is produced from an ideal integrator with a single pole at the origin as described by Eq. (12.27), the detector's smoothing LPF filter is described by Eq. (12.28). Then the AGC transfer function can be written by Eq. (12.29). Figure 12.6 illustrates several kinds of loop filters of integrators:

$$F(s) = -\frac{1}{sR_1C_1} = -\frac{1}{s\tau_1}, \tag{12.27}$$

$$T(s) = \frac{1}{1 + sR_2C_2} = \frac{1}{1 + s\tau_2}, \tag{12.28}$$

$$H(s)_{\mathrm{AGC}} = \frac{1}{1 + \dfrac{s\tau_1(1 + s\tau_2)}{K_F K_D K_{\mathrm{CPL}} P_{\mathrm{REF-IN}} \cdot g}} = \frac{\dfrac{K_F K_D K_{\mathrm{CPL}} P_{\mathrm{REF-IN}} \cdot g}{s\tau_1(1 + s\tau_2)}}{1 + \dfrac{K_F K_D K_{\mathrm{CPL}} P_{\mathrm{REF-IN}} \cdot g}{s\tau_1(1 + s\tau_2)}}. \tag{12.29}$$

This loop shows the pole in the origin $\tau 1$ due to the integrator and the dominant pole in $\tau 2$.

In conclusion, the AGC loop BW is defined by the ratio between the integrator time constant and the loop-transfer-function parameters. The solution is checking when the magnitude of Eq. (12.29) is $|H(j\omega)| = 1/2$. Moreover, for stability

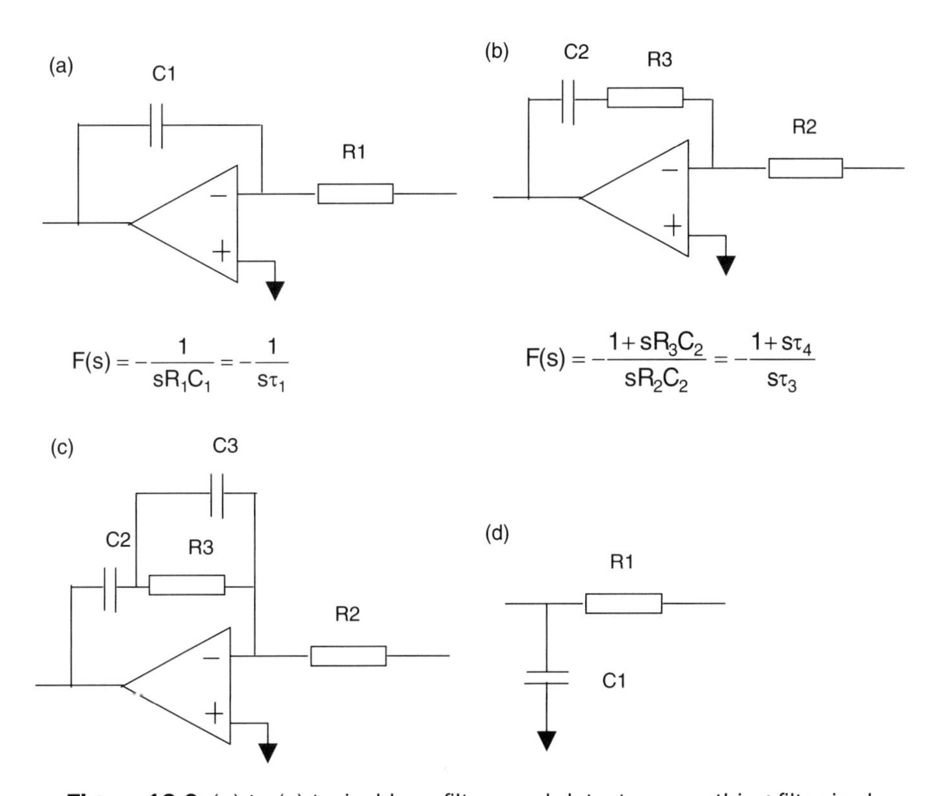


**Figure 12.6** (a) to (c) typical loop filters and detector smoothing filter in d.

considerations, the second pole created by the detector's smoothing filter should be at least one decade above the dominant pole. (Section 12.2.4 and Fig. 12.8 discusses this further in detail.) This condition would satisfy the phase-margin and gain-margin requirements.[1,14–16] [Note that the—sign of $F(s)$ was previously absorbed by $K_F$, see Eq. (12.7).] A way to make the AGC system even more stable is to use a zero-pole cancellation integrator cascaded to the ideal integrator output (see Fig. 12.9). That approach would provide better gain and phase margins. Figure 12.6 illustrates several loop filters and detector smoothing filter functions described by

$$F_1(s) = -\frac{1 + sR_3C_2}{sR_2C_2} = -\frac{1 + s\tau_4}{s\tau_3}, \tag{12.30}$$

$$H(s)_{AGC} = \frac{1}{1 + \dfrac{s\tau_1 s\tau_3(1 + s\tau_2)}{K_F K_D K_{CPL} P_{REF-IN} \cdot g \cdot (1 + s\tau_4)}}, \tag{12.31}$$

$$F(s) = -\frac{s(R_3C_2 + R_3C_3) + 1}{R_2C_2(1 + sR_3C_3)} = -\frac{1 + s\tau_5}{s\tau_3(1 + s\tau_4)}, \tag{12.32}$$

$$H(s)_{AGC} = \frac{1}{1 + \dfrac{s\tau_1 s\tau_3(1 + s\tau_4)(1 + s\tau_2)}{K_F K_D K_{CPL} P_{REF-IN} \cdot g(1 + s\tau_5)}}. \tag{12.33}$$

### 12.2.4 Stability analysis of exponential gain AGC with PIN VVA

Feedback system transfer function in the Laplace domain is described by

$$H(s) = \frac{\mathrm{LT}(s)}{1 + \mathrm{LT}(s)}, \tag{12.34}$$

where $\mathrm{LT}(s)$ is the open-loop transmission function. One of the conditions for $H(s)$ stability is the Nyquist stability criterion when drawing the Bode plot of $\mathrm{LT}(s)$ while $s = j\omega$.[1,7] The simplified Nyquist stability criterion states that a system is stable if the phase margin is positive. The phase margin $\phi_R$ is obtained by subtracting $-180$ deg from the argument, $\arg\{\mathrm{LT}(\omega_C)\}$, of the open-loop gain transfer function, where $\omega_C$ is the angular frequency at which $|\mathrm{LT}(\omega_C)| = 1$. Hence, the phase margin can be written as

$$\phi_R = \arg\{\mathrm{LT}(\omega_C)\} + 180. \tag{12.35}$$

A way to understand it is by observing Eq. (12.34) and looking for that condition where the denominator is smaller than 1 but $\neq 0$ whereas the real value of $\mathrm{LT}(\omega_C)$ is given by $|\mathrm{LT}(\omega_C)|\cos\phi_R$, which is eventually $\cos\phi_R$ since $|\mathrm{LT}(\omega_C)| = 1$ when intersecting the unity circle. The Nyquist polar diagram in

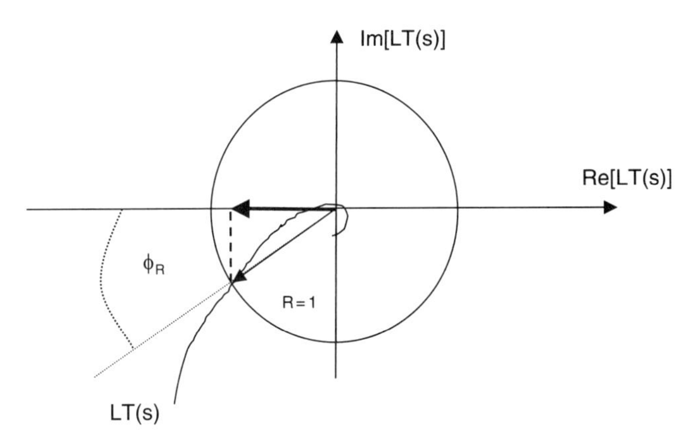


**Figure 12.7** LT(s) trajectory on the Nyquist polar diagram.

Fig. 12.7 provides the trajectory of LT($s$) versus frequency. The point at which LT($s$) intersects the unity radius circle defines the gain and phase margins. Observing the trajectory of LT($s$) after intersecting the unity circle, the magnitude of LT($s$) is lower than unity but it is still not a real number. At the instance $\arg\{\mathrm{LT}(\omega)\} = 0$, LT($s$) is a real number.

Hence, the gain margin is given by $1 - \mathrm{LT}(\omega) < 0$. That condition defines $\omega$ and relatively defines the value of the denominator of $H(s)$. The closer it is to unity, the better. In other words, it is desirable to have $0 \gg 20\log[\mathrm{LT}(\omega) < 0]$.

When examining the phase margin, the higher the $\phi_R$, the lower the real value of $|\mathrm{LT}(\omega_C)|$ and the closed-loop transfer function denominator is farther from 0 value. The two conditions of phase margin and gain margin define relative stability. That relative stability condition is known also as Berkhousen oscillation condition when $|\mathrm{LT}(\omega_C)| = 1$. Hence, for a stable system, the closed-loop gain is not infinite, which is the condition of oscillations. Figures 12.8 and 12.9 show the asymptotic plot of two different kinds of AGC loop filters. Note that each pole adds $-90$ deg lag and each zero adds $+90$ deg lead.

The outcome from the above discussion is categorizing the feedback system per the polynomial order and poles, for which the commonly used definition is order and type. The order of the transfer function is defined by the highest ranked polynomial. Since AGC is a feedback system, it has the nature of LPF. Therefore, the highest ranked polynomial is the denominator's polynomial. The transfer function type is defined by the number of ideal integrators it has, which contribute a pole in the origin. Therefore the system type is defined by the number of poles in the origin. Hence, Eq. (12.29) and Fig. 12.8 describe a second-order, first-type control loop.

The zero crossing of LT($s$) that is given by Eq. (12.29) occurs at $1/\tau_c$. The value of $1/\tau_c$ is given by

$$1/\tau_c = \frac{K_F K_D K_{CPL} P_{REF-IN} \cdot g}{\tau_1}. \tag{12.36}$$

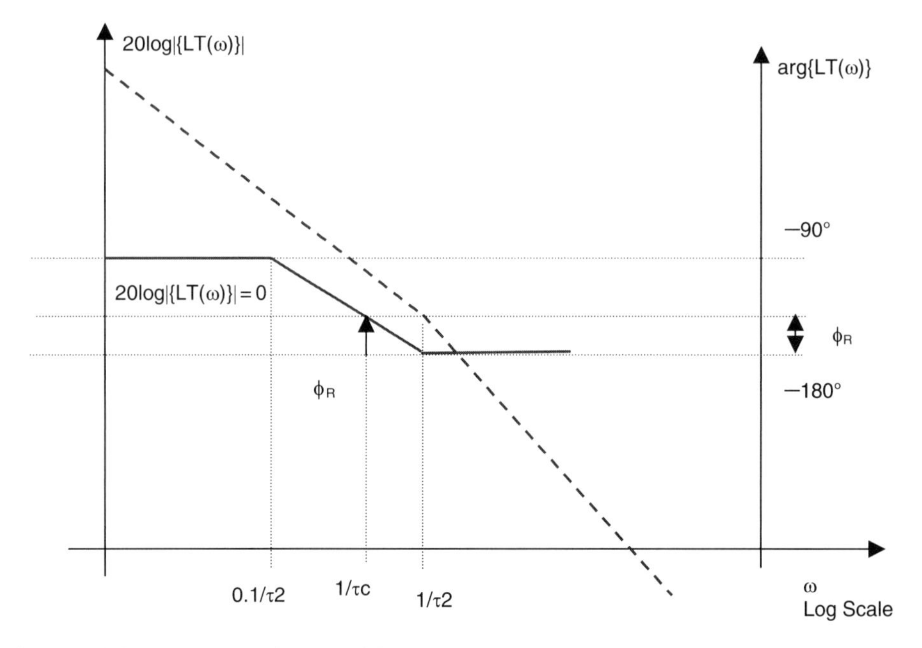


**Figure 12.8** A Bode plot for an AGC loop with an ideal integrator that has a pole at the origin and a detector smoothing LPF as in Figs. 12.6(a) and (d).

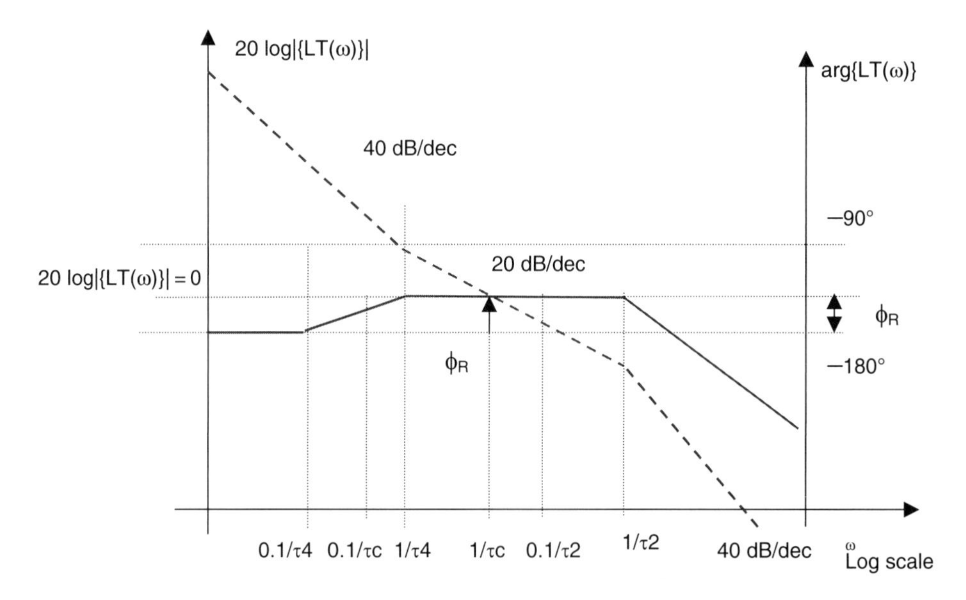


**Figure 12.9** A Bode plot for an AGC loop with an integrator that has a zero-pole cancellation integrator as in Fig. 12.6(b) cascaded to the ideal integrator as in Fig. 12.6(a), and detector smoothing LPF as in Fig. 12.6(d).

Several conclusions are driven from the above equation. The time constant $\tau_1$ is the ideal pole constant for the AGC loop integrator. Hence, to have a slow loop, for instance below 0.5 Hz, would result in the following requirement:

$$\frac{\tau_1}{K_F K_D K_{CPL} P_{REF-IN} \cdot g} > 1. \tag{12.37}$$

That means that the AGC loop BW is limited by the integrator's time constant. Additionally, $1/\tau_{c3\text{ dB}}$ is the closed-loop 3-dB BW frequency. Stabilitywise, it is required that the detector's LPF pole frequency $1/\tau_2$ or time constant $\tau_2$ would be at least 10 times higher than $1/\tau c$. This way, at the zero crossing point, a decent phase margin of $\phi_R$ is maintained.

The advantage of an AGC loop with zero-pole cancellation integrator as described by Fig. 12.9 is accomplished by cascading Fig. 12.6(b) to the integrator in Fig. 12.6(a) as shown in Fig. 12.10. In that manner, it provides a wider BW for the phase margin. This kind of loop with two poles at the origin is defined as the second type.[4] Since the "$s$" polynomial is at a power of 3, it is defined as a third-order, second-type AGC loop. The zero of the open-loop transfer function using zero-pole cancellation provides a wider BW response for the phase margin. If we observe the Bode plot of the simple first-type second-order AGC loop with an ideal integrator, we can see that the phase margin changes versus frequency at the rate of $-90$ deg/Dec. Fig. 12.8 illustrates the LT($\omega$) Bode plot for first-type, second-order AGC with an integrator as given in Fig. 12.6(b). Thus, adding a zero to the loop integrator provides a flat response for $\phi_R$ around the zero crossing frequency $1/\tau c$. Then, other poles such those of the smoothing filter take over and the phase rate goes in a down slope.

It can be seen from Fig. 12.9 that the dual pole at the origin creates $-40$ dB/dec as roll off for LT($s$). This dual pole is a result of $\tau_1$ of the ideal integrator in Fig. 12.6(a) and $\tau_3$ of the zero-pole cancellation filter in Fig. 12.6(b). At a certain instance, the slope of LT($s$) is reduced to $-20$ dB/dec due to the zero of the zero-pole cancellation filter. Observing the asymptotic plots of the phase in Fig. 12.9, it can be seen that the dual pole adds $-180$ deg lag to LT($s$). The effect of the zero frequency $1/\tau_4$ starts a decade before the

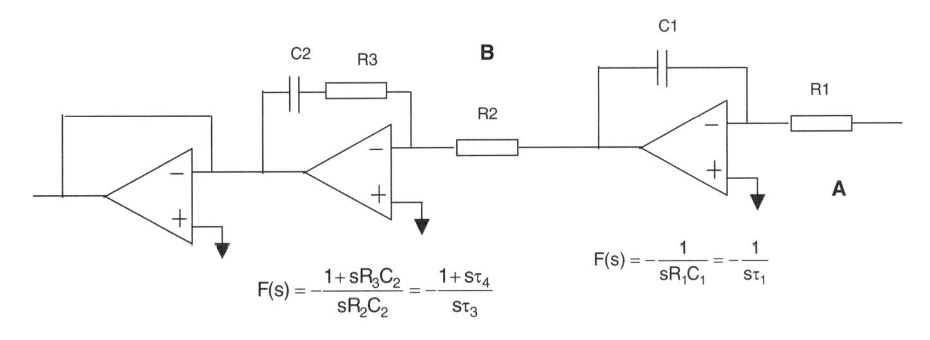


**Figure 12.10** Zero-pole cancellation loop integrator with dual pole at the origin.

zero frequency, i.e., $0.1/\tau_4$. Hence, the LT($s$) phase adds a lead and the phase slope is positive. However, it can be seen that the pole of the zero crossing, which is at $1/\tau c$, starts to affect the phase of LT($s$) at $0.1(1/\tau c)$, which is between 10% of the zero frequency value $0.1(1/\tau_4)$ and the zero frequency $1/\tau_4$. Hence, it adds a phase lag preventing LT($s$) from reaching a phase of $-90$ deg. Thus, the phase is flattened at $\phi_R$. The pole of the detector-smoothing filter starts to affect approximately at $0.1(1/\tau_2)$. Hence, the phase margin $\phi_R$ starts to roll off at a rate of 90 deg/dec. From Eq. (12.31) (which describes AGC function for the filter illustrated in Fig. 12.10), several conditions are required to satisfy the stability for the response in Fig. 12.9. The detector-smoothing LPF pole frequency $1/\tau_2$ should be at least one decade above the zero frequency $1/\tau_4$. The zero frequency should start before the AGC zero crossing frequency $1/\tau c$. This way it is guaranteed that the phase response at the zero crossing is flat, providing a wide-BW phase margin.

The zero-crossing frequency value, assuming the affects of the smoothing filter are in higher frequency than zero, i.e., $1/\tau_2 \gg 1/\tau_4$, is given by

$$\omega_c = \frac{-K\tau_4 \pm \sqrt{(K\tau_4)^2 - 4K\tau_1\tau_3}}{2\tau_1\tau_3}, \tag{12.38}$$

where $K$ is given by

$$K = K_F K_D K_{CPL} P_{REF-IN} \cdot g. \tag{12.39}$$

This example demonstrates the advantage of a third-order, second-type AGC loop over a second-order, first-type loop.

Figure 12.10 displays the AGC integrator with a zero-pole cancellation that has dual poles at the origin. The loop type is determined by the number of poles at the origin it has.

Hence, the Bode plot given in Fig. 12.8 represents the first-type loop whereas the Bode plot in Fig. 12.9 represents the second-type of loop. Figure 12.10 demonstrates the AGC filter concept with dual poles at the origin discussed above.

The inverting buffer at the output is used to correct the phase of the integrator as an inverting integrator.

### 12.2.5 Numerical analysis of exponential gain AGC with PIN VVA

The AGC transfer function is given by a linear approximation at the operating point. The main assumption is that the amplitude fluctuations around the operating point are considered to be a small signal. However, when locking the AGC at various power levels, the detector gain parameter, $K_D$, and the VVA gain parameter, $g$, change their values per each power level.[26] This is due to the fact that both parameters are linear approximations of nonlinear functions. $K_D$ is the derivative of the detector transfer function $V_D(P)$, when $P$ equals the RF reference level at steady state. The derivative of the VVA gain function $G_{VVA}(V_C)$ $g$ is when the AGC control voltage level equals the required voltage value for preserving the output power for the given reference value (the reference RF power value is represented by a reference voltage as in Fig. 12.2). Thus, the following notations are valid:

$$K_{\mathrm{D}} = \frac{\partial V_{\mathrm{D}}(P)}{\partial P}, \tag{12.40}$$

$$g = \frac{\partial G_{\mathrm{VVA}}(V_{\mathrm{C}})}{\partial V_{\mathrm{C}}}. \tag{12.41}$$

The dependency of both values on the AGC-loop-operating point affects the open-loop gain and, as a result, might affect the closed-loop frequency response and stability.

Assume the following design problem of a second-order first-type loop with a limiting level of 14-dB mv/ch and topology as in Fig. 12.2. The specification is as listed below:

- Number of channels: 79
- Optical power input range: $-6$d$-1$ dBm
- Frequency range: 47–550 MHz

Additional assumptions are:

- A typical RF-stage of $G_1$ and $G_2$ is 13 dB $\pm$ 1 dB
- The $\pi$ PIN-diode VVA with a $I$ region thickness is 175 $\mu$
- The VVA zero attenuation loss is 3 dB
- A loss due to an RF-chain mismatch is 3 dB
- Power sampling is done with a 10-dB directional coupler (coupler $=$ 10)
- The AGC sampling BW of 180 MHz is from 50–230 MHz
- Channel spacing is 6 MHz
- The OMI is 3.5%/ch
- The characteristic impedance is $Z_0 = 75\ \Omega$
- The photodetector matching transformer-turn ratio is $N = 2{:}1$[11]
- The photodetector worst-case responsivity is $r = 0.8$ mA/mW
- The photodetector max responsivity is $r = 1.0$ mA/mW

The number of channels $N_C$ sampled by the AGC is calculated by

$$\mathrm{N_C} = \frac{\mathrm{BW}}{\text{channel\_spacing}} = \frac{180\,\mathrm{MHz}}{6} = 30\,\mathrm{ch}. \tag{12.42}$$

From the above technical information and $N_C$, the reference power to lock the AGC is given by

$$P[\mathrm{dBm}] = 14\,\mathrm{dBmV} + 10\log 30 - 10\mathrm{dB} - 48.7\mathrm{dB} = -29.9\mathrm{dBm}. \tag{12.43}$$

The RF power budget of the receiver that defines the sampled power by the AGC is defined by

$$\begin{aligned} P_{\mathrm{ERR}}[\mathrm{dBm}] = {} & 10\log Nc + 20\log\left(\frac{\mathrm{OMI}\cdot r\cdot P}{\sqrt{2}}\right) + 10\log\left(N^2\cdot Z_0\right) \\ & + G_1 - \mathrm{VVA}_{\mathrm{LOSS\_0dB\_att}} - \text{mismatch\_loss} + G_2 - \mathrm{CPLR} + 30. \end{aligned} \tag{12.44}$$

During acquisition time, the minimum RF power sampled into the detector occurs when the AGC system is at its minimum attenuation, the receiver is at its maximum gain, and the input optical level is at its minimum value as described by Fig. 12.5. The maximum RF power sampled into the detector is when the AGC is at minimum attenuation and maximum gain and the input optical level is at its maximum value. This phenomenon is due to the nature of AGC loops to idle at max gain and converge to the limiting level. This is called fast-attack slow-release or slow decay.

At the quiescent point, the AGC system would maintain a constant value for $K_{\mathrm{det}}$ and would change $K_{\mathrm{VVA}}$ as the input power changes from one level to the other, as was explained previously. Hence, $K_{\mathrm{det}}$ should be calculated at the limiting level after stabilization and also the two extreme conditions during the acquisition period to verify that the system dynamics passes through a stable trajectory. The same applies to $K_{\mathrm{VVA}}$, which should be calculated by referring to the input level, which defines its desired attenuation state as well as the two extremes.

The method of AGC evaluation and calculation is straightforward. For worst-case envelope performance, the minimum power is calculated for a low-responsivity detector value $r = 0.8$ mA/mW, a low optical level, and a low RF gain. This defines the sensitivity requirements of the RF detector. On the other hand, maximum RF power is calculated for high RF gain and high photodetector responsivity. This defines the VVA attenuation margin. The difference between the two extreme RF power points defines the AGC RF dynamic range.

Additionally, those two extreme RF powers define the constants $g$ and $K_D$ during the extreme acquisition states. The typical operating point will be at −3.5-dBm optical input. At that point, the calculations would be made by the typical values of $g$ or as denoted by $K_{VVA}$ and $K_{det}$ at the limiting level. Therefore, the typical operating point would provide the constants $g$ and $K_D$ at the typical power leveling with typical loop values.

From the above conditions and Eq. (12.44), the RF detector would sense approximately the following RF levels: −24.5 dBm at minimum optical input level, −20.5 dBm at typical level, and −10.5 dBm max. This power range describes the AGC-attack state when the VVA is at its minimum attenuation. It can be seen from the following data that the output range of the receiver is between −14.5 and −0.5 dBm. That range is the power sum of 30 channels. Thus, the RF dynamic range of the AGC is 14 dB. This result defines the VVA attenuation range to be at least 14 dB in order to maintain the reference level indicated by Eq. (12.43). Additionally, the detector sensitivity should be below −34.5 dBm to provide at least a 10-dB detection margin for the low power level and low responsivity state. A decent choice for this design example would be a high dynamic-range detector log-video amplifier (DLVA) such as AD8313 or AD8314 from analog devices. The disadvantage of a DLVA is that it is not a temperature compensated detector and it is not a true power meter detector. Thus, it will be much more sensitive to the peak-to-average ratio (PAR) for modulated signal. This issue is discussed in Sec. 12.2.8. Such an extremely sensitive detector is not essentially needed, that is, because of the AGC-attack mode of high output RF power delivered to the detector. Thus, the detector will always sense a sufficient RF level that is above its tangential sensitivity level $T_{SS}$ and correct the loop. However, it is a good enough example.

Using MathCAD[13] and numerical methods,[12] the approximation function of the AD8313 DLVA is given in Eq. (12.45). The reason for selecting exponent approximation function rather a polynomial is due to DLVA $\log_e$ characteristic, which provides higher accuracy:

$$V_D(P) = S_0 \cdot e^{\sqrt{P}} + S_1 \cdot e^{\sqrt[3]{P}} + S_2 \cdot e^{\sqrt[4]{P}} + S_3 \cdot e^{\sqrt[5]{P}} + S_4 \cdot e^{\sqrt[6]{P}} + S_5 \cdot e^{\sqrt[7]{P}} + S_6 \cdot e^{P}. \tag{12.45}$$

The coefficients of the exponential approximation function are given by the **S** vector shown by Eq. (12.46), while $P$ is the input power to the detector in watts and $V_D$ is the output voltage in volts:

$$\mathrm{S} = \begin{pmatrix} 822.043 \\ -4.443 \times 10^3 \\ 1.457 \times 10^4 \\ -2.493 \times 10^4 \\ 2.066 \times 10^4 \\ -6.569 \times 10^3 \\ -118.715 \end{pmatrix}. \tag{12.46}$$

Figure 12.11 shows the approximation function results versus the detector's measured results.

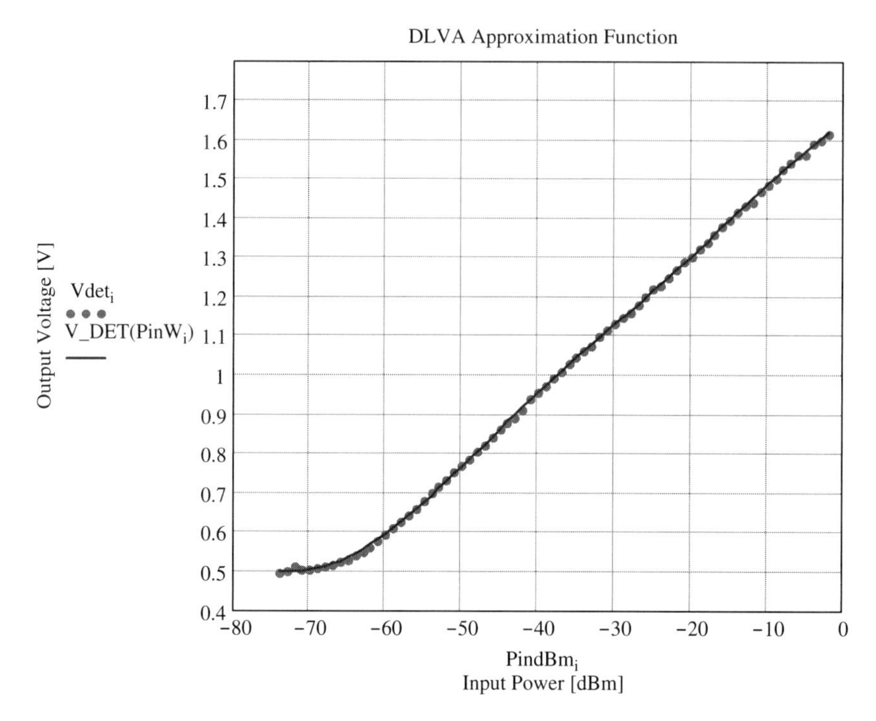


**Figure 12.11** DLVA approximation function $V_{out}$ vs. $P_{in}$[dBm] (*solid*) vs. measured (*dotted*).

From the plot in Fig. 12.12, it is observed as expected that the detector constant $K_D$ (Fig. 12.14) would be a downward function since the derivative values of $K_D$ decrease versus input power; that means $K_D$ values decrease at high-input optical-power levels.

The plot in Fig. 12.13 describes $K_D$ values versus sampled power to the detector in a log scale. It is typical for a DLVA to have a linear $K_D$. Figure 12.14 shows the detector $K_D$ function in a linear scale. It can be seen that $K_D$ values depend on the locking level, i.e., the AGC operating point.

In the same manner as was demonstrated previously, the VVA attenuation approximation of the exponential function $G_{VVA}(V_C)$ as a function of the control voltage $V_C$ was evaluated. The result is given by

$$G_{VVA}(V_C) = A_0 \cdot e^{\sqrt{V_C}} + A_1 \cdot e^{\sqrt[3]{V_C}} + A_2 \cdot e^{\sqrt[4]{V_C}} + A_3 \cdot e^{\sqrt[5]{V_C}} + A_4 \cdot e^{\sqrt[6]{V_C}} + A_5 \cdot e^{\sqrt[7]{V_C}} + A_6 \cdot e^{V_C}. \quad (12.47)$$

Again, an exponent approximation function was preferred over a polynomial due to approximation accuracy because of the nature of PIN-diode junction as an

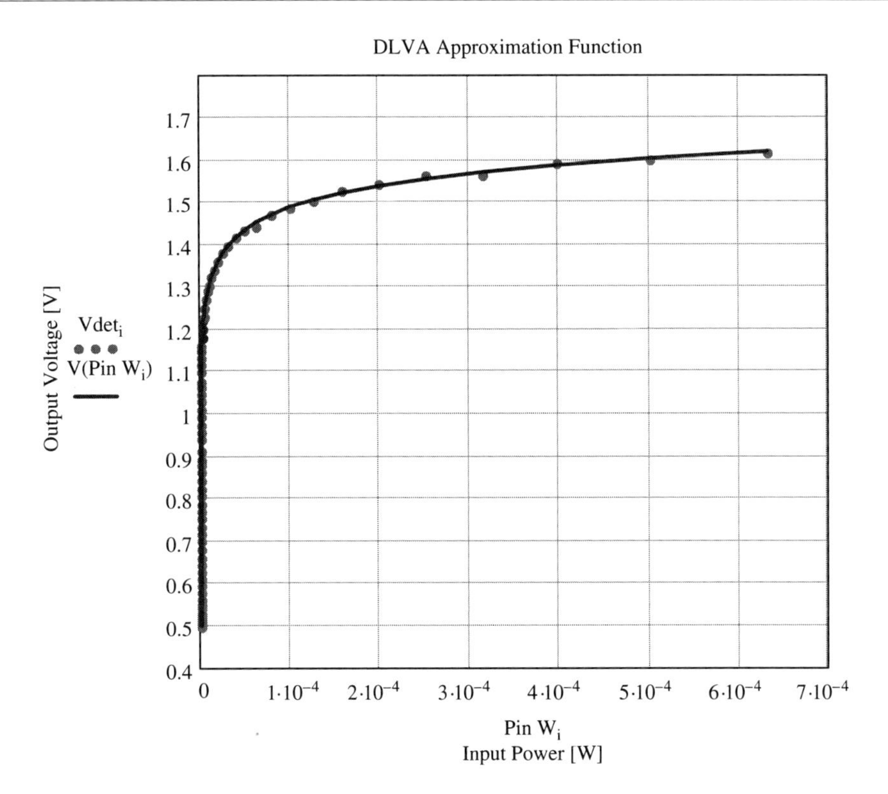


**Figure 12.12** DLVA approximation function $V_{out}$ vs. $P_{in}$[W], (*solid*) vs. measured (*dotted*), on a linear power scale. The solid-line function is used to calculate $K_D$.

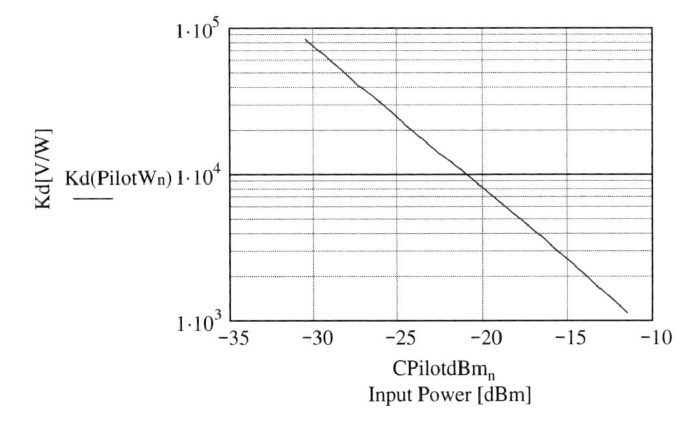


**Figure 12.13** RF power detector gain coefficient $K_D$ as a function of input power to the detector in dBm. Note that the following ratio between the minimum value and maximum value of $K_D$ coefficients is $K_{Dmin} \approx 15\, K_{Dmax}$ for RF sampling between −24.5 and −10.5 dBm.

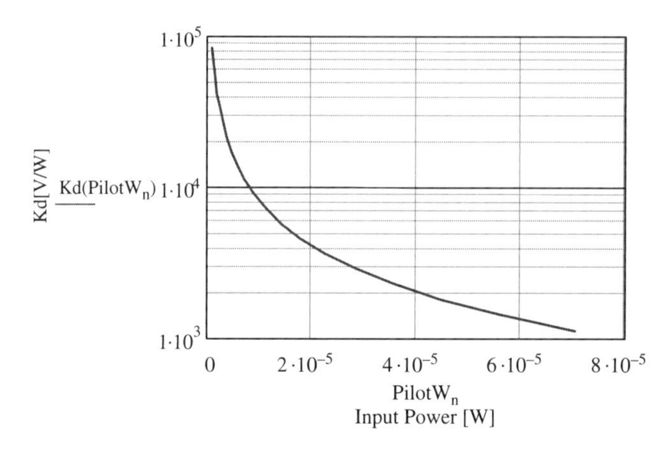


**Figure 12.14** RF Power detector gain factor $K_D$ as a function of input power in Watts on a linear scale.

exponent function. The coefficients of the above approximation are given in the vector described as

$$\mathbf{A} = \begin{pmatrix} 0.039 \\ 1.701 \\ -55.113 \\ 251.534 \\ -368.562 \\ 170.402 \\ -1.376 \times 10^{-8} \end{pmatrix}. \tag{12.48}$$

In the approximation process, the zero attenuation loss was taken out. That fixed value is absorbed in Eq. (12.44). The plot in Fig. 12.15 shows the VVA attenuation transfer function versus control voltage in a linear scale, and in log scale in Fig. 12.17. The plot in Fig. 12.16 shows the VVA derivative of the approximated curve presented in Fig. 12.15. This derivative function is $g(V)$, which is the VVA gain constant, sometimes denoted as $K_{\mathrm{VVA}}$, at different values of AGC control voltage in linear scales. Note the VVA subthreshold point of 0.5 V to the knee at 1.5 V.

At this point, the series of PIN diodes at the $\pi$ configuration start the conduction state. Therefore, the derivative at the knee is maximal and will go to lower values as the control voltage gets higher since the attenuation rate goes lower. The high attenuation state at subthreshold drops abruptly to a lower state at conduction; after that, the attenuation rate is moderate. Hence, at the transition point from subthreshold to conduction, attenuation rate drops rapidly between the off state and the on state. For that reason, the derivative is an upgoing function until it reaches the on state, and a downgoing function at conduction (Fig. 12.16).

The rapid increase of $g$ between 1 V and 2 V is because of the transition from subthreshold to conduction of the VVA, where the voltage increases until the diode

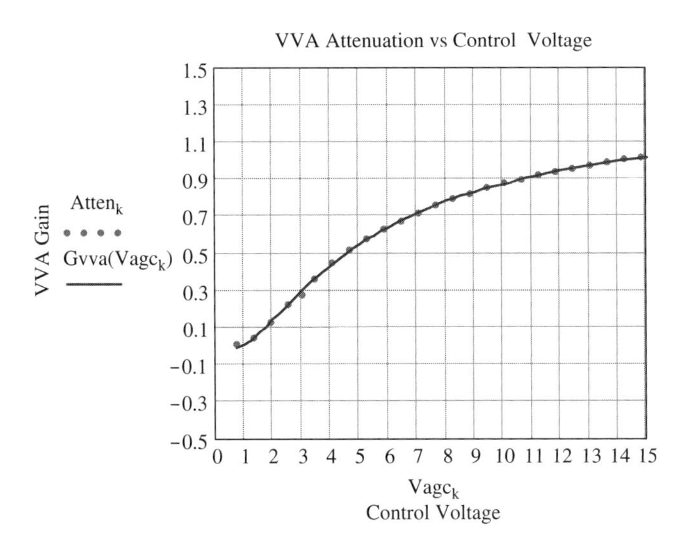


**Figure 12.15** VVA gain function $G(V)$ vs. control voltage; solid: calculated, dotted: measured.

fully conducts as shown by Fig. 12.15. Hence, at that point, attenuation drops rapidly from off state to on state.

Note the following ratio from the minimum value to maximum value of $K_{\text{VVA}}$ coefficients is $K_{\text{VVAminpower}}$ $(g_{\text{LOW}}) \approx K_{\text{VVAmaxpower}}/4$, i.e., $(g_{\text{HIGH}}/4)$ for RF sampling between $-24.5$ and $-10.5$ dBm and control voltage between 12 and 2 V, respectively.

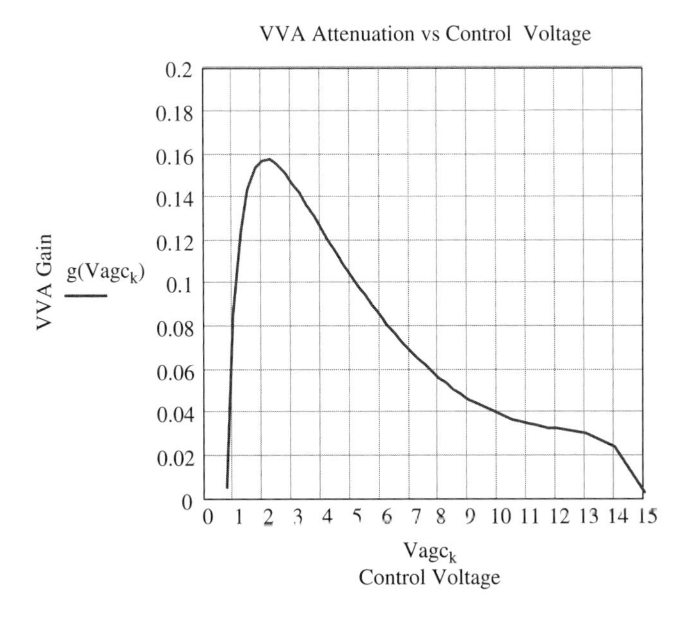


**Figure 12.16** VVA gain constant $g(V)$ function vs. control voltage.

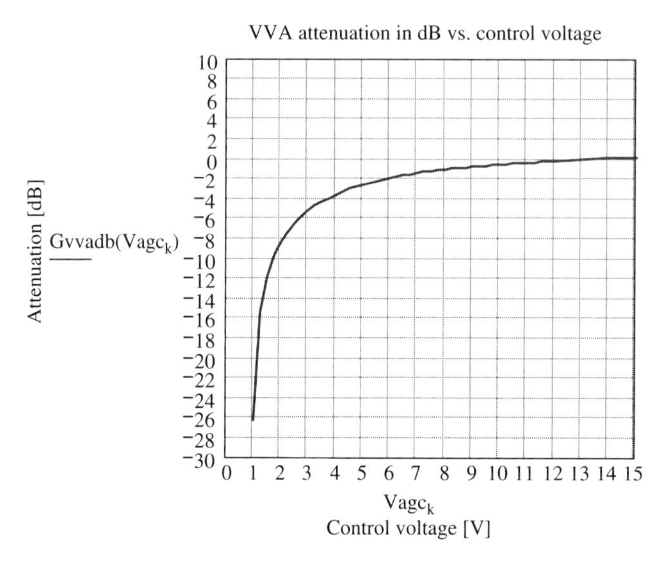


**Figure 12.17** VVA attenuation function vs. the control voltage dB scale.

Several interesting conclusions can be derived from the above plots and the open-loop transfer function of the AGC marked as LT($s$). The VVA gain constant $g$ is a downgoing function since the attenuation rate decreases versus the control voltage increment as observed in Fig. 12.5. The ratio between the maximal value of $g$ and the minimal value of $g$ within the control voltage range of 2 to 12 V is approximately 4:1 as it appears in Fig. 12.16. This means that while the RF power decreases, $g$ decreases because the attenuation rate gets lower and the control voltage gets higher. On the other hand, as the RF power increases within the receiver dynamic range of $-24.5$ to $-10.5$ dBm, $K_D$ decreases by the ratio of 1:15, as it appears in the plots at Figs. 12.13 and 12.14. Hence, it can be noted that the VVA coefficient satisfies $4\,g_{\text{LOW-}P} \approx g_{\text{HIGH-}P}$ and the RF detector satisfies $K_{\text{D-LOW-}P} \approx 15K_{\text{D-HIGH-}P}$. Recalling the open-loop transfer-function equation, and assuming $T(s) = 1$ within the AGC loop-pass band, for a realization case of loop filter $F(s)$ as in Fig. 12.6(a), LT($s$) can be written as

$$\begin{aligned}\mathrm{LT(s)} &= \mathrm{F(s)T(s)P_{REF\text{-}IN}K_{DAV}K_{CPL}VVA_{LOSS}G_2 \cdot mismatchloss \cdot gK_D}\\ &= \mathrm{C}\frac{\mathrm{gK_D}}{\mathrm{s}\tau_1},\end{aligned} \tag{12.49}$$

where $F(s)$ is the loop-filter transfer function, $T(s)$ is the detector-smoothing filter transfer function, $K_{\text{DAV}}$ is the detector dc gain amplifier, $K_{\text{D}}$ is the detector coefficient gain in [V/W], $K_{\text{CPL}}$ is the directional coupler coupling factor, $P_{\text{REF-IN}}$ is the reference power to lock the AGC in W, $G_2$ is the RF gain in the loop, $\text{VVA}_{\text{LOSS}}$ is the VVA insertion loss at zero state attenuation, mismatchloss is the additional loss in the RF chain, and $g$ is the VVA gain factor in [1/V].

Hence, based on the above $K_D$ and $g$ ratios versus optical power, LT($s$) varies during the acquisition state identified as attack according to

$$C\frac{15}{4}\frac{g_{\text{HIGH-P}} \cdot K_{D\text{-HIGH-P}}}{s\tau_1} = C\frac{g_{\text{LOW-}P} \cdot K_{\text{D-LOW-}P}}{s\tau_1}. \tag{12.50}$$

Observing the closed-loop Eq. (12.29) and the zero-crossing frequency equation when solving the denominator of Eq. (12.29) or (12.36), it can be seen that the AGC BW varies between its maximum BW at minimum power and its minimum BW at maximum power. Equation (12.51) shows the AGC time constant variation with respect to maximum power:

$$\frac{4}{15}\frac{\tau_1}{K} \leq \tau_c \leq \frac{\tau_1}{K}. \tag{12.51}$$

That means that the closed-loop BW during the low-power input acquisition state is 3.75 times wider than the loop BW during acquisition at a high-power state. This phenomenon is demonstrated for the following design scenario.

Selecting R1 = 100 KΩ and C1 = 2.2 μF for the loop filter as in Fig. 12.6(a) and R2 = 4.4 KΩ and C2 = 0.022 μF for the detector-smoothing filter results in the following closed-loop and open-loop responses as given by Figs. 12.18 to 12.21 during steady-state locking. The closed-loop response shown by Fig. 12.18 versus Fig. 12.19 demonstrates that the closed-loop BW in −6 dBm is wider by almost 1.5 times the BW at high-optical power in Fig. 12.19.

Figures 12.20 and 12.21 displays the open-loop poles location function of design values.

### 12.2.6 Noise and dynamic range analysis of feedback AGC

When dealing with AGC and noise, the question is how to define the AGC dynamic range. This question has two aspects:

1. What is the maximum input power the AGC can handle without driving the system into distortions?
2. What is the maximum attenuation the AGC can handle without limiting the CNR? In other words, where is the turning point at which the AGC stops to show output CNR improvement at the output as the input signal increases.

The second question is a bit strange but can be understood by observing the AGC effects on system performance. Observing Figs. 12.22 and 12.23 and calculating noise figure and output noise density at different AGC states would provide a simple explanation.

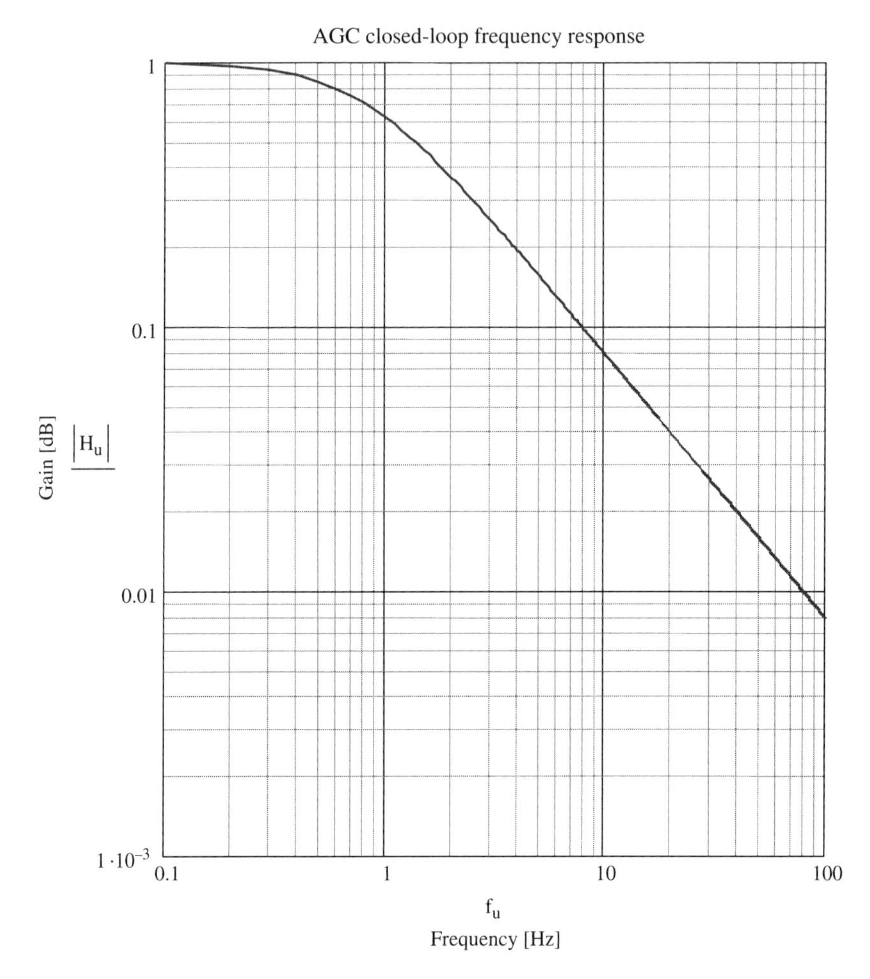


**Figure 12.18** Closed-loop response at −6 dBm; 3-dB BW is 0.8 Hz.

At very low input power, the AGC is at its maximum gain state and its minimum attenuation. Hence, the system noise figure is at the minimum. The system noise figure is given by

$$F = F_1 + \frac{L^{-1} - 1}{G_1} + \frac{L^{-1}(F_2 - 1)}{G_1}, \tag{12.52}$$

where $L$ is the AGC attenuation, $G_1$, $G_2$, $F_1$, and $F_2$ are the gain values and noise factors of the first and the second stages, respectively. Remembering that the attenuator gain is given by $G_{\text{att}} = 1/L$, and knowing that $0 < L \leq 1$, the output-stage noise-figure contribution increases by the amount of attenuation value $1/L$. At the AGC threshold state, the output-noise density is at the maximum value even though $F$ is minimal. This is because the overall RF gain is at its maximum state since the RF signal is weak. Thus, the input CNR is low and

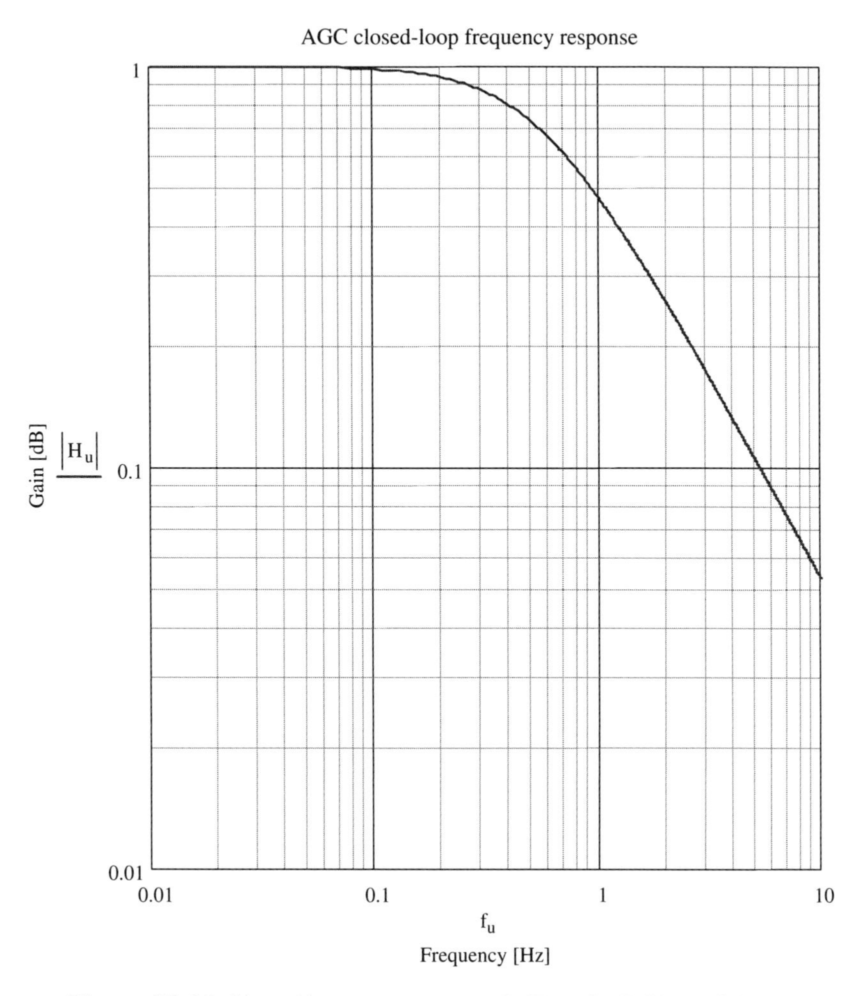


**Figure 12.19** Closed-loop response at 0 dBm; 3-dB BW is 0.5 Hz.

minimal. However, when the input signal is increased, the AGC preserves the output power at the same level as was in the threshold. On the other hand, the gain is decreased and attenuated by the AGC attenuator $L$. As a result, $F$ slightly increases, but its increment is less than the AGC attenuation setting of $L$ due to the first-stage gain, which compensates for the rest of the chain NF. The result is that the output-noise density gets lower. Hence, output CNR gets better. Note that the input CNR improves as well, the signal increases when the intrinsic noise is the thermal noise KTB. The receiver's thermal noise, referred to the input, is degraded by $F$ and equals KTBF. The factor KTBF is the noise density referred to the receiver's input, which degrades the output CNR by $F$, and the AGC role is to preserve the output CNR with minimum degradation compared to the intrinsic input CNR, which is $S/\text{kTB}$, where $S$ is the signal level. The AGC noise suppression process would take place until the attenuation value $L$ would be at the order of the RF gain

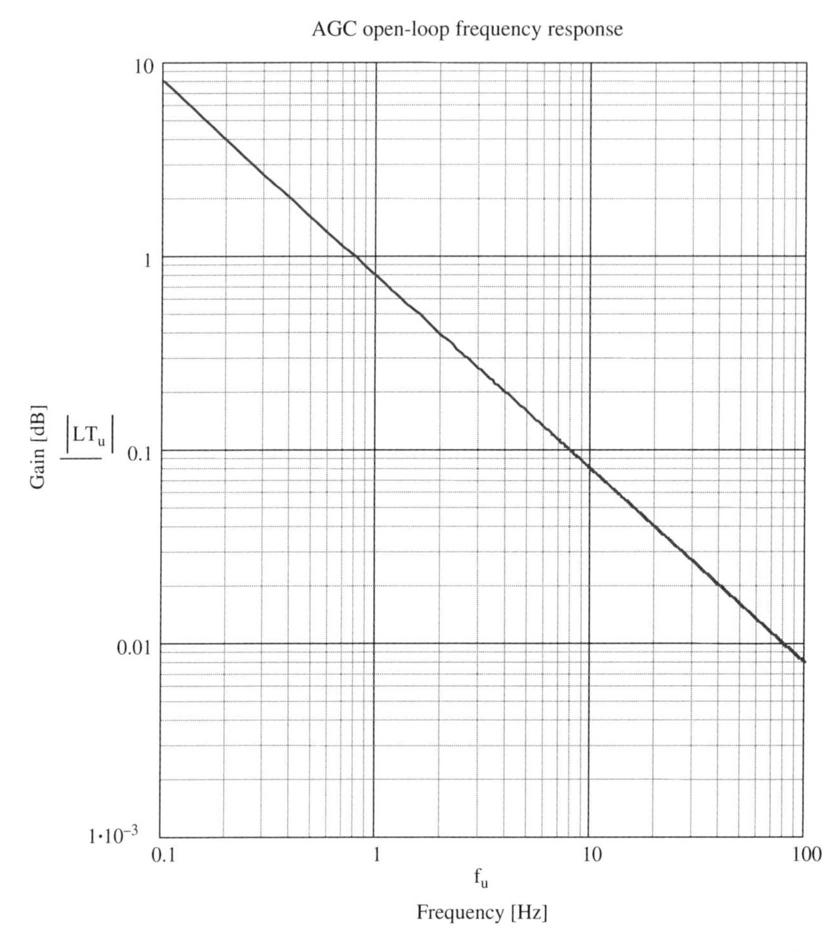


**Figure 12.20** Open-loop response at −6 dBm; 3-dB BW is 0.8 Hz. The LT(s) zero crossing is at 0.8 Hz. The detector pole is at a high frequency and has no effect on the phase margin. Increasing the value of C2, the detector's filter capacitor to 22 μF from 0.022 μF would reduce stability and bring the second pole one decade above the dominant pole at 0.8 Hz. Figure 12.21 provides the plot for that case.

and as a result, of course, at the first-gain stage level. From that point on, the noise density at the output would remain flat and the CNR would decrease. The reason is as follows: at the instance the AGC attenuator $L$ is at the order of the first-gain stage, the rest of the RF chain is not compensated for the first-gain stage. Equation (12.52) provides the reason this. It can be seen from Fig. 12.23 that NF increases decibel per decibel of RF or decibel per half decibel of the optical power increment. This is approximately a straight line. The decibel per decibel increment of NF results due to the VVA attenuation of decibel-per-decibel RF. Since at that point NF is governed by the VVA, any increment of the VVA attenuation by a decibel increases NF by a decibel but reduces the output-noise density by the same amount. As a consequence, the output-noise density remains flat, whereas

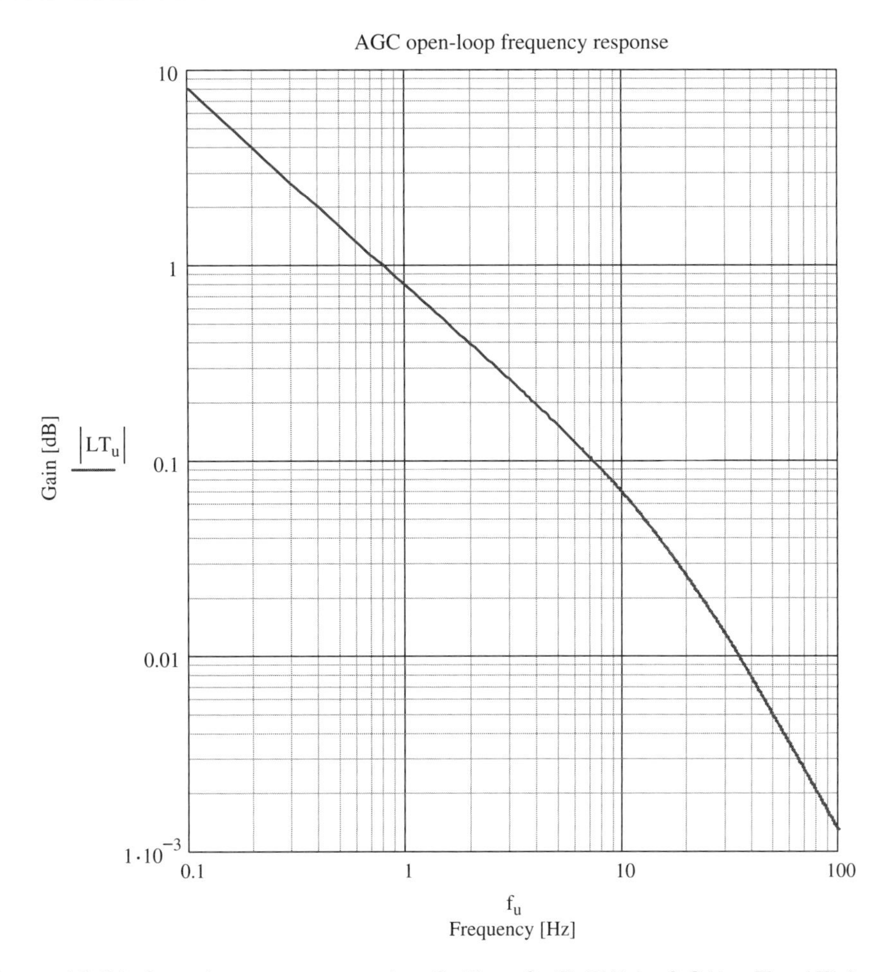


**Figure 12.21** Open-loop response at −6 dBm; 3-dB BW is 0.8 Hz. The LT(s) zero crossing is at 0.8 Hz. C2 capacitor value of the detector filter was increased to 22 μF from 0.022 μF. Second pole is at 10 – 11 Hz.

the input noise density increases decibel per decibel for signal increment. This point is the maximal CNR that the AGC can support. This phenomenon threshold starts at the instant the VVA attenuation is equal to the LNA gain, which is the first-gain stage $G1$-gain value in Eq. (12.52).

At the instant the output-noise density equals the input-noise density, because $F$ is strongly affected by $L$, the AGC reaches its limit of attenuating high input power levels and does not track input CNR. That can be understood as follows. The first case is of a finite VVA attenuation range. The AGC reached its limit and ability to attenuate. The loop integrator provided the VVA its lowest control voltage. The noise figure is at its maximum value. Thus, the output noise density remains constant. There is no further attenuation of noise. If the VVA dynamic range equals the RF gain, the overall system's gain equals zero. Another case is theoretical, where the VVA has an infinite attenuation range, making a 1-dB increase in NF for any 1-dB increase in power thus the CNR remains constant. This scenario is also valid for a finite VVA range where the VVA attenuation is sufficiently high to affect the cascaded noise figure as given by Eq. (12.52). This is well demonstrated in Fig. 12.23. In very high-gain receivers, the AGC is distributed through the RF chain. The AGC order of attenuation moves from the back end to the front end. Attenuation the second AGC stage. This stage is located after the second-gain or third-gain stage in the receiver chain. For very high-power signals, the first-stage AGC starts to operate, preventing the input stage from being driven into compression.

The following plots describe simulation and measured results of a FTTx CATV receiver with feedback AGC:

- Figure 12.22 shows the AGC attenuation state versus the optical level.
- Figure 12.23 shows the input-and output-noise densities as a function of the AGC state.

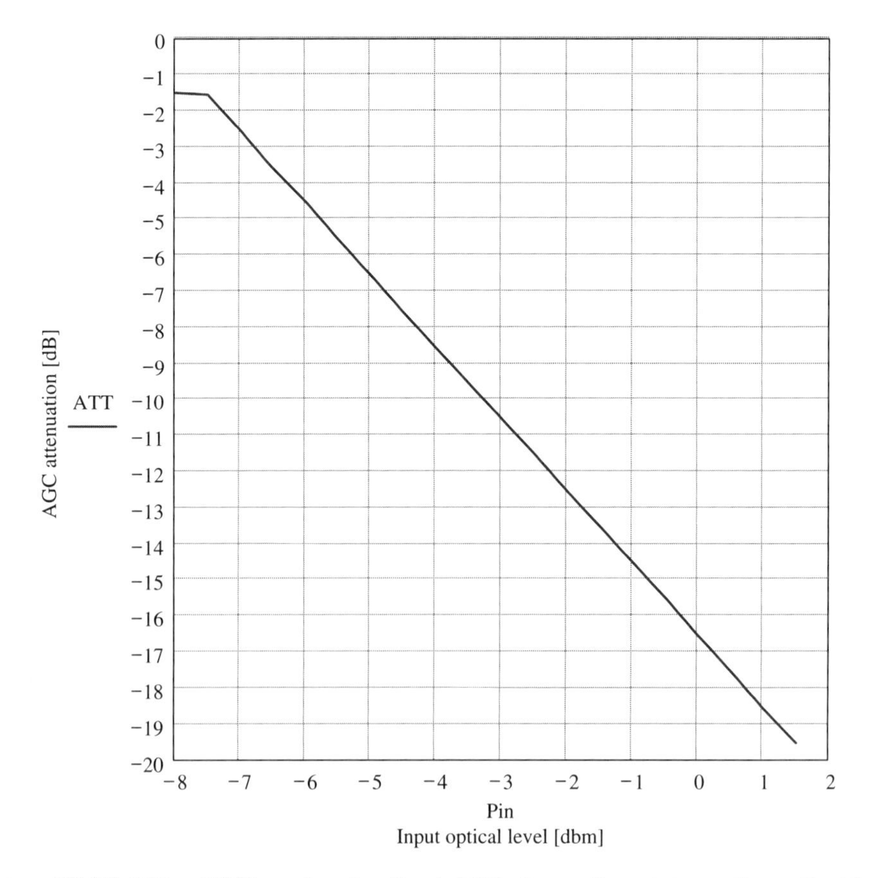


**Figure 12.22** FTTx – CATV receiver-feedback AGC-attenuation state vs. the optical level 4% OMI, responsivity 0.8 mA/mW.

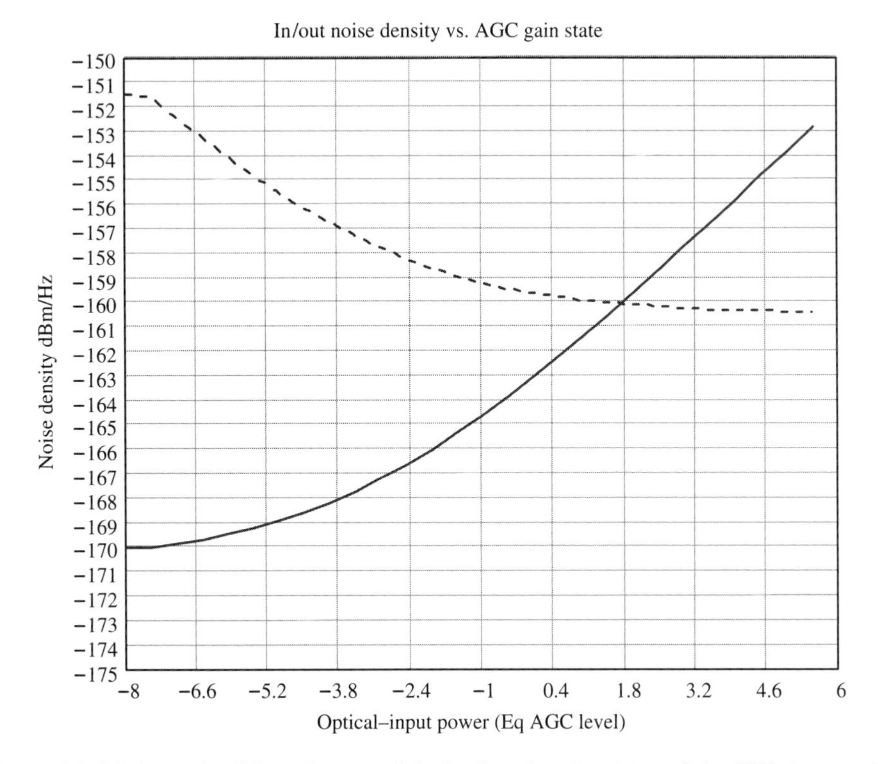


**Figure 12.23** Input (*solid*) and output (*dashed*) noise densities of the FTTx transceiver as a function of AGC attenuation vs. optical level. Note that above 1.8-dBm, the output noise density becomes flat. Thus, AGC does not improve CNR and starts to operate as a CNR limiter. The optical level refers to the AGC state where 1-dB light is equal 2-dB RF. Responsivity $r = 0.8$ mA/mW, OMI = 4% as it appears in Fig. 12.22.

- Figure 12.24 shows the measured output-noise density versus the AGC state.
- Figure 12.25 shows simulation of the CNR versus the AGC attenuation stage.
- Figure 12.26 shows the measured CNR of an FTTx receiver.
- Figure 12.27 shows the NF degradation as a function of AGC attenuation increments.

The assumptions and constraints for those plots are as follows: NF per each gain stage is 3.5 dB and gain is 13 dB. The VVA zero-state insertion loss is 3 dB and output-matching losses are an additional 3 dB. OMI per CATV tone equals 4% and the characteristic impedance of 75 Ω. The AGC power leveling at the output is 14 dBmV/tone. The plots show very good agreement between calculated and measured results taken on an integrated FTTx triplexer.

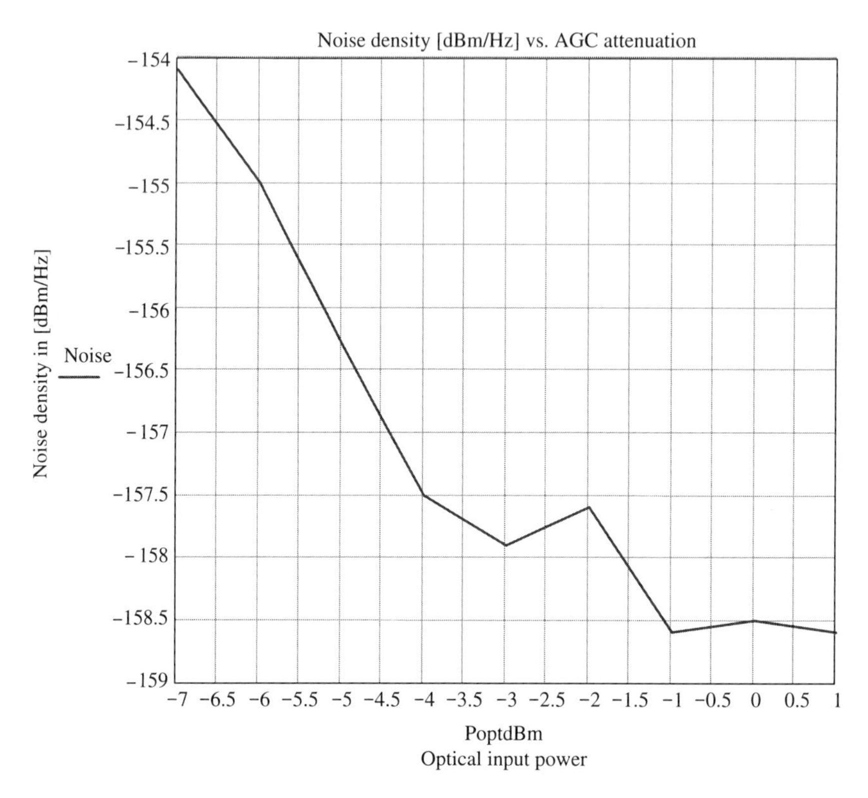


**Figure 12.24** Measured noise density at output of the FTTx CATV receiver as a function of AGC attenuation vs. the optical level. Optical level refers to the AGC state where 1-dB of light is equal to 2-dB RF. Responsivity is a bit better, $r = 0.9$ mA/mW, OMI = 4% as appears in Fig. 12.22.

The reason for improved performance at low-power levels in the measured results is the higher responsivity of the photodetector compared to the simulation's worst case. Hence, the CNR penalty or gain improvement due to responsivity is given by

$$\chi = 20\log\frac{r_2}{r_1} = 20\log\frac{0.9}{0.8} \approx 1\text{ dB}, \tag{12.53}$$

where $r$ represents responsivity.

That explains a little about the difference between the simulated results in Fig. 12.25 against the measured results in Fig. 12.26. Additionally, higher-gain and lower matching losses at low optical levels add more RF gain; thus, a higher signal power results as well as an improvement in the RF-chain NF. As a consequence, the AGC threshold is measured at a lower optical power and the noise attenuation is higher at a given optical level or AGC state. The

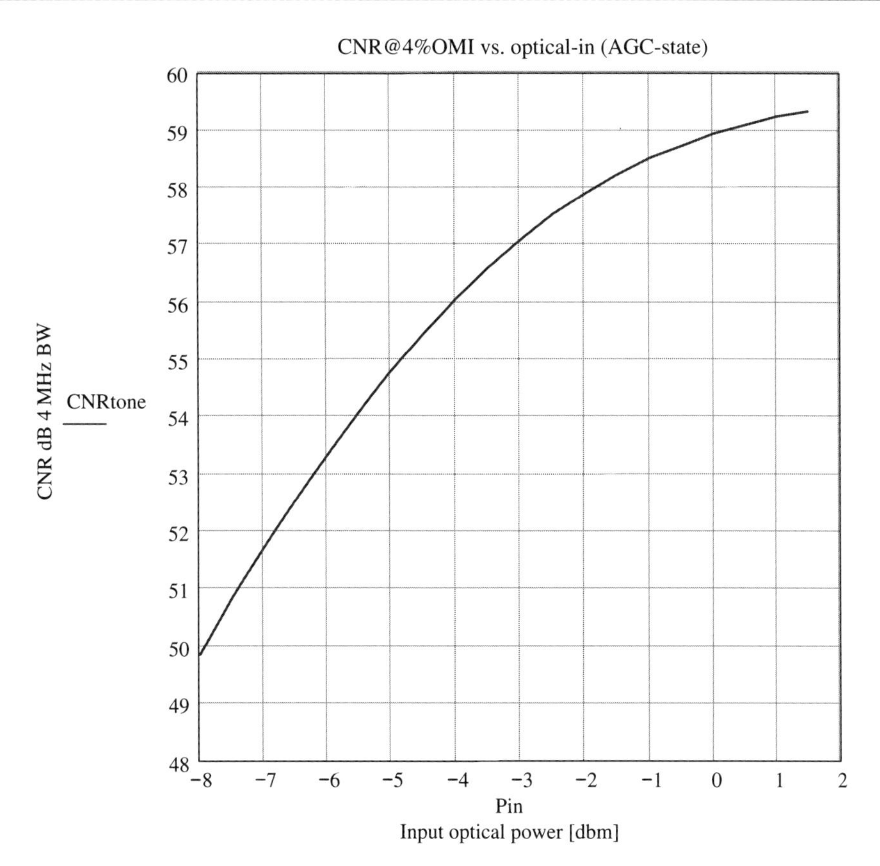


**Figure 12.25** Simulation of the worst-case CNR vs. the AGC state as a function of the optical level at the FTTx transceiver input. The optical level refers to the AGC state where 1-dB light is equal to 2-dB RF. Responsivity $r = 0.8$ mA/mW, OMI = 4% as appears in Fig. 12.22.

mismatch loss due to VVA attenuation at high power adds more losses, increases the NF, and thus the measured CNR is lower by 1.5 dB compared to the simulated CNR.

The measurement was done with an open AGC loop and controlling the VVA signal leveling with a power supply. The noise was measured without illuminating the photodetector (CNR = dark) and the same VVA-driving-voltage conditions were preserved as when the carrier level was tested. The AGC signal attenuation corresponds to the input-optical-power level. Hence, the open-loop test simulated a limited closed loop, which is the role of the AGC.

Note that Fig. 12.22 demonstrates that at a −2-dB optical level, the AGC attenuation level equals 13 dB, which is the first-stage LNA gain. Hence, Fig. 12.23 shows about a −2-dBm optical NF increase of decibel per 0.5-dB optical increment. This phenomenon is described by the NF plot in Fig. 12.27.

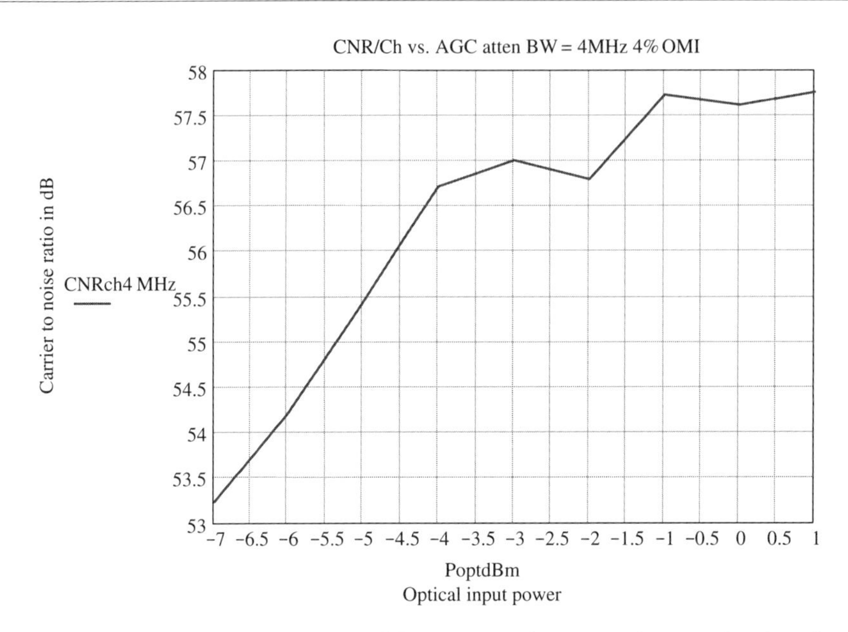


**Figure 12.26** Measured CNR vs. the AGC state as a function of optical level at the FTTx transceiver input. Optical level refers to AGC state where 1-dB light is equal to 2-dB RF. Responsivity $r = 0.9$ mA/mW, OMI $= 4\%$ as appears in Fig. 12.22.

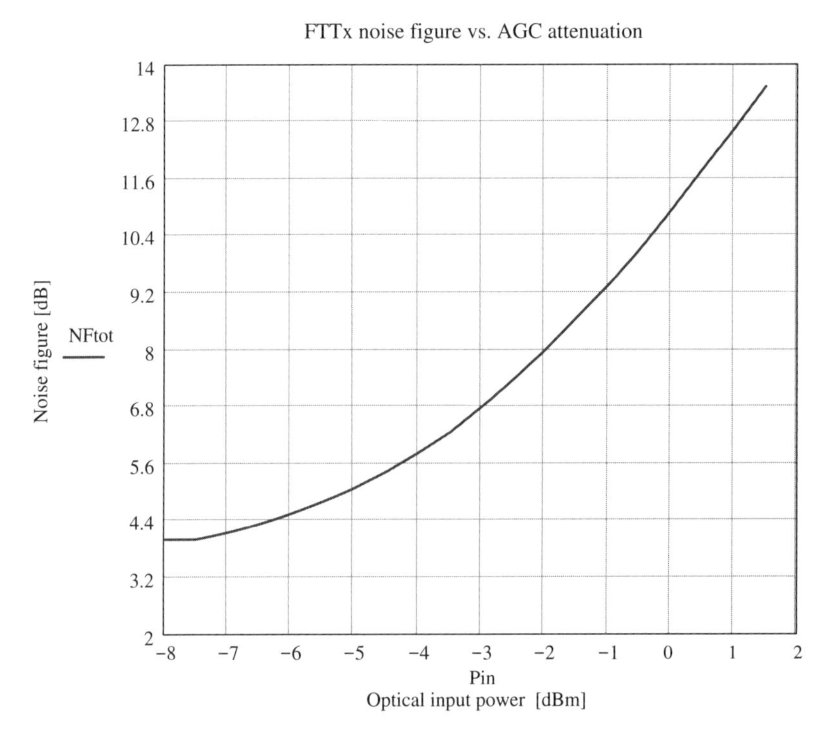


**Figure 12.27** Simulation of noise figure degradation as a function of the AGC state. Optical level refers to the AGC state where 1-dB light is equal to 2-dB RF. Responsivity $r = 0.8$ mA/mW, OMI $= 4\%$ as appears in Fig. 12.22.

### 12.2.7 Noise and dynamic range analysis of feed forward AGC

Unlike feedback AGC the FF AGC power-leveling system with DCA or VVA operates in an open loop. When dealing with coarse-step attenuators, the system operates as a regular amplifier between steps; there is no continuous power tracking and level locking. Hence, between attenuation steps, the signal is amplified linearly. The mechanism of noise suppression, while at a constant signal level, occurs only at the transitions.

As was explained previously, the AGC increases the dynamic range of a system and preserves the input CNR as the signal gets stronger, with a slight degradation compared to the input CNR caused by NF. In the FF system, where the gain is fixed at a given AGC step state, the result is a fixed value of NF. Hence, throughout the input-signal power increment, the AGC attenuation level remains constant until the next AGC state. Thus, there is no change in the output-noise density. The gain and NF remain the same, which is why the CNR improves when the carrier gets higher in coarse FF. Obviously, the same happens in a feedback system. The input CNR increases as the signal at the input increases. Hence, the AGC preserves the CNR with a slight degradation due to incremental NF as the VVA loss gets higher. That is another way to observe the AGC–CNR tracking. To demonstrate, a two-step FF coarse-AGC system was simulated. It was observed that the higher the number of steps of an FF-AGC system, the fewer power fluctuations it had at the output. For a coarse AGC system, it would exhibit larger power steps between attenuation states, as appears in Fig. 12.28.

Proper system gain and attenuation distribution prevents a large NF degradation step change, illustrated in Fig. 12.31. Thus, overall CNR and a dark CNR curve is smooth, as is shown in Fig. 12.30. The AGC system trigger point is adjusted by the sampling resistor that converts the dc photocurrent to a sampling voltage, as described by the block diagram in Fig. 12.1. The photovoltage developed on a 1-K sampling resistor is directly related to responsivity as shown by Fig. 12.29. Photodetector responsivity distributions would affect the AGC trigger point, as appears in Fig. 12.28.

The simulation plot in Fig. 12.28 describes the FF AGC RF output as a function of optical input power and responsivity as a parameter. It has hysteresis of approximately 0.15–0.2 dB, which equal 0.3–0.4 dB RF. As was explained previously, this method has no feedback and tracking. Hence, trigger points and RF levels depend on the alignment of the AGC circuit, and it is susceptible to the circuit parameters' distribution. The plot shows four different curves at four different trigger points due to four different photodetector responsivities values for $r$. The highest responsivity the system has, the earlier the trigger point. This is due to the fact that the dc voltage sampled from the photodetector sampling resistor in Fig. 12.1 reaches its trigger value at a lower optical power, as appears in Fig. 12.29. Hence, to compensate for early trigger point, the sampling resistor value should be decreased in order to provide a higher optical level for the AGC trigger point. The optical power difference between triggering points relates to the ratio of the responsivities log according to

$$\Delta\mathrm{dB} = 10\log\left(\frac{r_1}{r_2}\right). \tag{12.54}$$

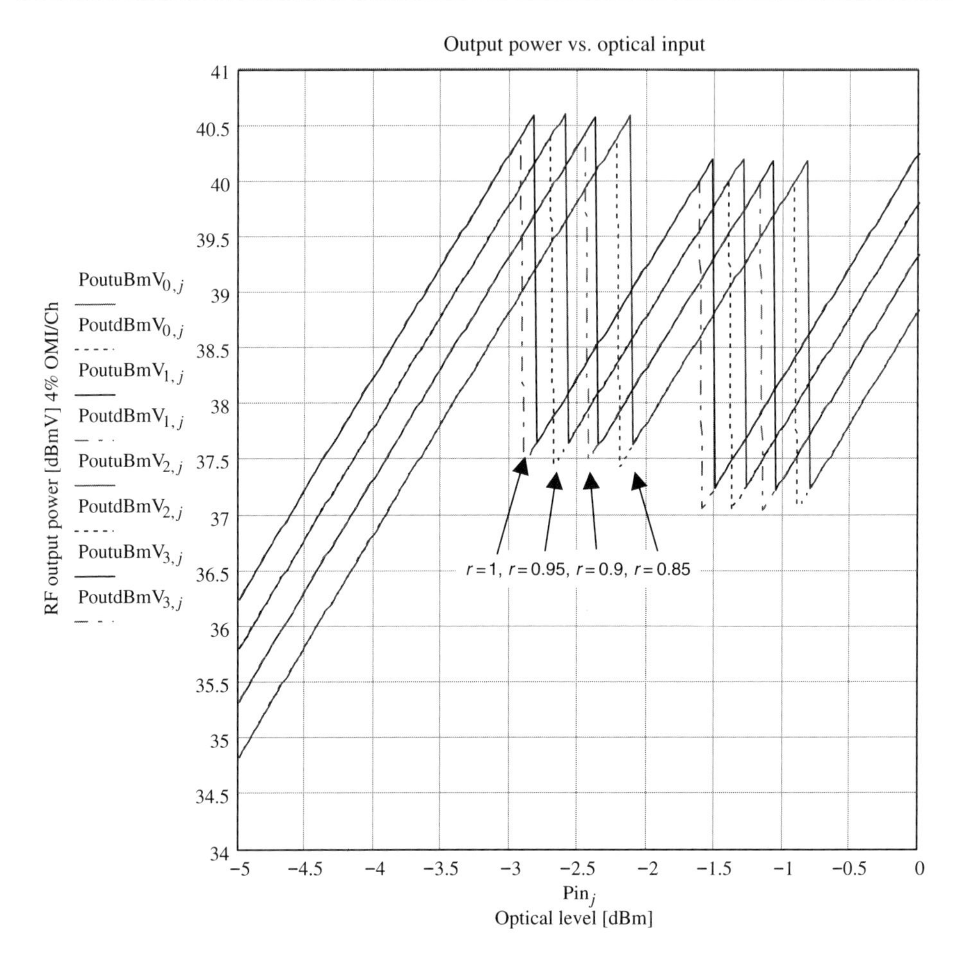


**Figure 12.28** Simulation of two-step coarse FF AGC RF output power vs. the optical level, while the responsivity $r$ is a parameter.

For instance, the ratio between $r_1 = 1$ mA/mW and $r_2 = 0.95$ mA/mW results approximately in a 0.22 dB difference between trigger points as can be observed in the plot of Fig. 12.28. The RF delta between the trigger points would be twice as much as the optical difference.

The above simulation in Fig. 12.30 demonstrates the output CNR components as a function of the optical level in a coarse two-step AGC. In this example, the responsivity is the worst-case value of $r = 0.85$ mA/mW.

As the optical power gets higher, the AGC sampling voltage gets higher until the first trigger point; there, the first AGC step is 3 dB, and the output power is attenuated. However, this attenuates the input-noise powers of RIN shot-noise and dark-current shot noise at the same level as the RF signal. Therefore, the corresponding CNR is preserved. As for thermal noise (white noise), when attenuation is incremental the noise figure gets higher. Figure 12.31 shows the step response in the noise figure caused by the incremental AGC attenuation. However, the change in noise figure due to the 3-dB AGC step is about 0.1 dB at the first AGC stage and

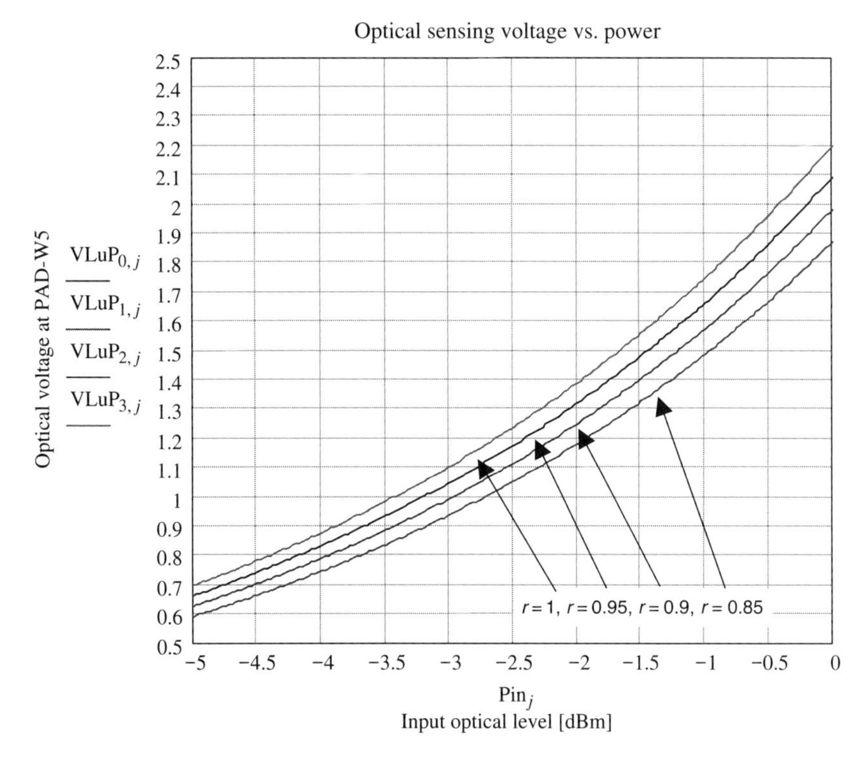


**Figure 12.29** The FF AGC DC sensing-voltage simulation from the photodetector as a function of power and responsivity *r*.

about an additional 0.16 dB in the second step. This is the reason that in the thermal-noise CNR curve, there is no observed degradation of CNR at the trigger point.

The AGC chain is designed in such a way that the first amplifier gain is high enough to compensate for the DCA attenuation. Thus, at first approximation, it is accurate to say that the thermal noise is more or less in a fixed level. Therefore, the thermal output CNR marked as Ne in Fig. 12.30 keeps improving as the optical level increases. This is because the RF increases proportionally to the OMI square as explained in Secs. 8.2.4 and 10.9. However, as the input RF signal increases versus optical level increment and AGC attenuation occurs, both RF power and thermal noise power are attenuated at the same level. Hence, the thermal noise CNR improves versus power. One more important conclusion is that the dominant noise variance is the laser RIN, and for an ideal laser, the shot noise will dominate as optical power increases. Furthermore, it is observed that the dark current is by order of magnitude lower than the RIN, shot, and thermal noise CNRs. Hence, the equivalent link CNR is dominated by the laser RIN.

Note the different trigger points of the AGC as a function of responsivity. Noise figure is not affected by responsivity, only the trigger point is, and NF changes are related to different responsivity values. The NF value is determined by the RF-chain lineup only.

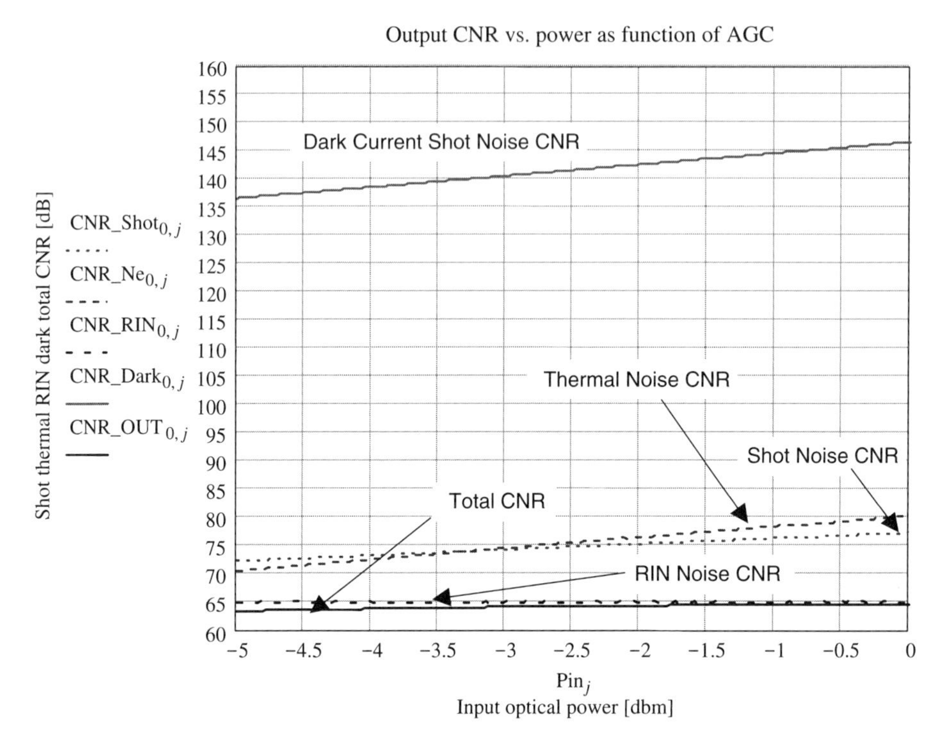


**Figure 12.30** FF-AGC-CNR contributor simulation as a function of the optical level at responsivity: $r = 0.85$ mA/mW, transformer turn ratio: $N = 2$, laser RIN: 145 dB/Hz, transformer loss: 1.5 dB, amplifier gain: 19 dB in two stages, NF: 3.5 dB, AGC step size: 3 dB, losses: 3 dB, and OMI 4%.

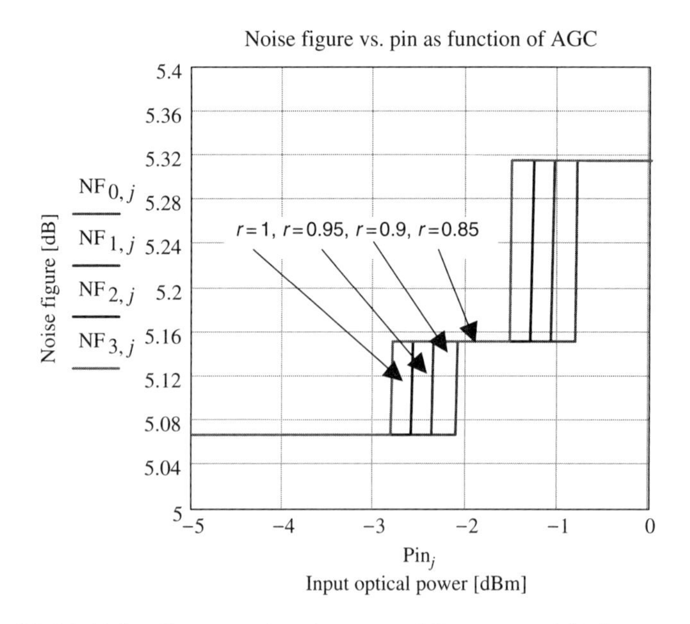


**Figure 12.31** Noise figure vs. input power while responsivity is a parameter.

### 12.2.8 Feedback AGC in the presence of modulated CATV signal

The specifications of community access television (CATV) receivers define the peak power level per CATV channel. However, since the CATV channels are modulated signals, the AGC detector would sense the average power level. Moreover, although a CATV receiver is tested and evaluated for CSO, CTB, CNR performances, its AGC is calibrated for CW nonmodulated carriers.

In a CW-testing mode, the peak power of the signal equals the rms. Hence, the AGC leveling locks to the correct signal-leveling requirement. In a real case of modulated tones, there is a ratio between peak and the RMS power levels. Hence, if the AGC was tuned to be locked for the desired spec level in CW mode, it will create an error. The error results because the AGC RF detector is a power estimator that measures the rms level. There are several types of detectors: a peak detector that is not used for AGC, a true rms power detector, and a fast DLVA detector. The DLVA is more sensitive to PAR. As a consequence, when using a high-dynamic range DLVA, the AGC will lock the modulated signal to the rms value as it was set by the CW tones. Therefore, the peak-signal level of the modulated carrier would be higher by the peak-to-rms power ratio known also as PAR. So, there is a need to compensate for the difference of the peak-to-rms ratio. Since the AGC samples are for locking the lower channels in the frequency plane, which are NTSC AM–VSB, the peak-to-rms analysis would refer to the AM, which carries the video picture (image) as part of the NTSC modulation scheme. Note that sound (aural) is the FM part of composite video and is a constant envelope, hence, it theoretically has no PAR. The peak-to-rms compensation applies to feedback AGC topology only. Since FF AGC is driven by the PD dc-monitor voltage and does not sense the RF signal. QAM suffers from higher PAR compared to NTSC AM–VSB as explained by Chapter 14. Further elaboration on QAM peak to average appears in Sec. 14.3.

#### 12.2.8.1 Peak-to-rms analysis

NTSC power is characterized by its peak power as measured during the horizontal and vertical sync intervals. On the other hand the peak power relates the average power of a CW and not the instantaneous peak power of a CW signal. That means that when observing time domain $A\sin(\omega t)$ or $A\cos(\omega t)$ the voltage measured for power calculations is the CW RMS. Hence $P_{\text{peak}} \equiv P_{\text{ave-CW}} = V_{\text{rms}}^2/R$ and not and not instantaneous peak power $P_{\text{peak}} = V_{\text{peak}}^2(t)/R$. This value is formally called the peak envelope power (PEP)[27]. This definition states that continuous sine-wave-voltage peaks are equal to the voltage peaks of the RF carrier during the sync interval. CW peak equals its average, thus PEP $= 0$ dB. The NTSC peak is very consistent since sync signals are always at the same height; its average power is not. Instead, its average power is highly dependent on the video signal content, which means the average luminance level and therefore is not used to describe NTSC RF signals. The analysis is as below.

NTSC video signal modulation is composed of two types of modulation: AM for the image picture on the screen, and FM for the audio (aural). The image signal's modulation depth defines the color, the blanking pulse, which blanks the scan beam, and the horizontal synchronization pulse, which is 5-μs wide. The blanking pulse width is 10 μs and the composed pulse repetition period is 53.5 μs (see Fig. 12.32).[8–10]

Simple AM is defined as given by Eq. (12.54).[8,9] Equation (12.54) describes dual sideband plus carrier. The NTSC signal is AM VSB + C (amplitude modulation vestigial side band plus carrier). The clusters around the carrier result due to the horizontal repetition rate thus each 15.8 KHz.

$$\varphi_{AM}(t) = m(t)\cos\omega_c t + A\cos\omega_c t = [A + m(t)]\cos\omega_c t, \tag{12.54}$$

where $m(t)$ is the modulating-signal amplitude versus time shown in Fig. 12.32. The condition for demodulation by an envelope detector is

$$A + m(t) > 0 \quad \text{for all } t. \tag{12.55}$$

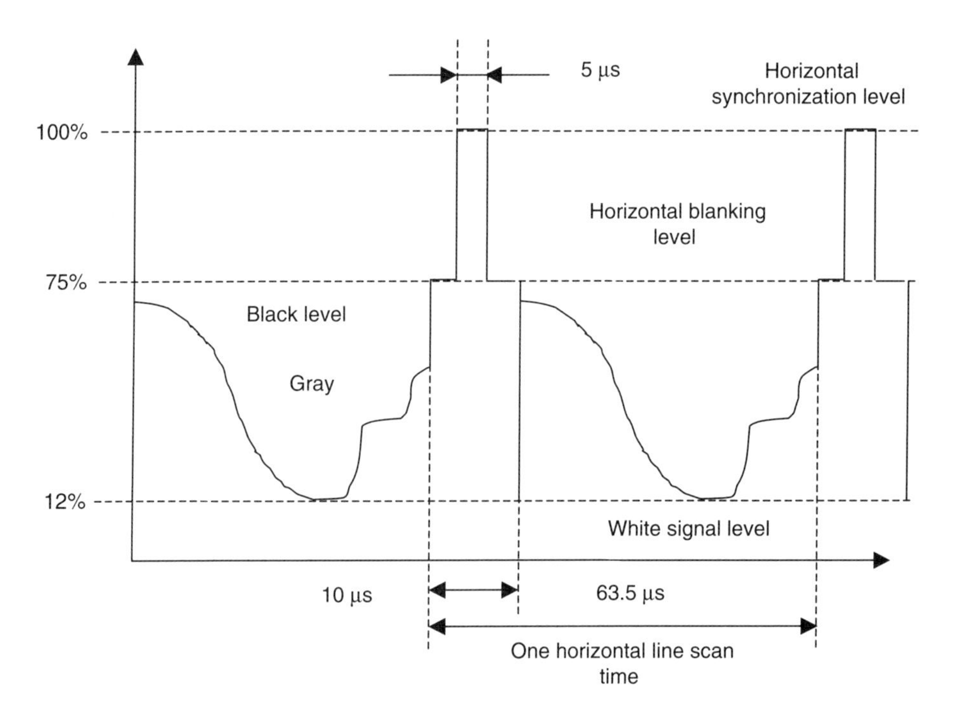


**Figure 12.32** Composite monochrome television signal levels.

The modulation index is defined as

$$\mu = \frac{-m(t)_{\min}}{A}. \tag{12.56}$$

In order to have proper modulation, the following condition is required as a result from the definition in Eqs. (12.55) and (12.56):

$$\mu \leq 1. \tag{12.57}$$

It can be seen that $\mu$ defines how much the carrier amplitude fluctuates by the modulating tone $m(t)$. Moreover from Eq. (12.56), $m(t)$ indicates that there are instances where the carrier (RF) is at its maximum level, $|\mu| = 1$, and instances where the carrier is at its minimum. Since the modulating signal $m(t)$ is described by Fig 12.32 the carrier peak occurs at horizontal synchronization level.

The average RF level would be related to the average amplitude of the modulating carrier $m(t)$ over a single scan period as given by

$$<m(T)>= \sqrt{\frac{1}{T}\int_0^T [V(t)]^2 \mathrm{d}t}. \tag{12.58}$$

Thus, the power refers to

$$<m(T)^2>= \frac{1}{T}\int_0^T [V(t)]^2 \mathrm{d}t. \tag{12.59}$$

The integration in Eq. (12.59) would be per the normalized voltage-square value, as shown in Fig. 12.32. In fact Fig. 12.32 represents $|\mu|$. Assuming a Gaussian signal statistics with zero mean for the image modulation, then by observing Fig. 12.32 it can be said that its average is $(75\% + 12\%)/2 = 43.5\%$. Using this result and taking the square area of the blanking level, the synchronization square area, and the video square area sum as given by

$$\begin{aligned} <m(T)^2>&= \frac{1}{T}\int_0^T [V(t)]^2 \mathrm{d}t \\ &= \frac{1}{63.5}\left[2\left(0.75^2 \cdot 2.5\right) + 1 \cdot 5 + 0.43^2 \cdot 53.5\right]\frac{A^2}{2} = 0.1394A^2. \end{aligned} \tag{12.60}$$

where $T = 63.5$ μs the normalized rms value is 0.379 or 37.9% of the peak voltage amplitude, if the peak represents 100%.

Equation (12.60) demonstrates the approach for calculating the NTSC average power. It is like a "mask shape" applied on the CW carrier. It defines the percentage of the CW power or voltage swing used when observing the CW modulated envelope. Hence, when calibrating the AGC for CW there is a need to compensate the peak to average compared to a real NTSC signal. The CW power is $A^2/2$; therefore the correction is $10 \log(0.5/0.1394) \approx 5.55$ dB.

This correction is implemented in AGC. However, since the FTTx AGC is fed by 30 NTSC signals, a more accurate statistical approach should be made for analyzing PAR of 30 uncorrelated NTSC signals. Those signals may have a Gaussian distribution and very similar statistics as additive white Gaussian noise (AWGN). This is made by evaluating average and peak signals. The AGC calibration method is measuring 30 CW tones and compensating them for the expected PAR of a single NTSC tone.

The FCC rules specify that power measurements of transmitted NTSC RF signals are to be referred to peak sync during NTSC broadcast transmitter calibration. For that, special NTSC signal is used that consists of composite syncs and blanks only. AGC PAR calibration can be made by NTSC generator such as Fluke 54200 and Rhode–Schwartz SFE and SFU.

When discussing QAM channels the signal nature is statistical and numeric analysis is used. Pseudorandom code is created for sending a QAM pattern. For having an accurate result it is desired to have a long random series and measuring process. If large enough samples are taken, an envelope of the HDTV QAM signal could be measured and collected into "power bins." This may result in an accurate statistical histogram. This histogram is based on a statistical tabulation of the frequency of occurrence of a random power variable within adjacent ranges or bins. If the sampling is large enough and the bins are narrow enough this sampling histogram is normalized by dividing each bin number sample by the total number of samples. This generates a probability density function (PDF). The goal is to find the percentage of time in which the HDTV RF signal rises above a particular power level. Hence, after finding the PDF, a form of cumulative distribution function (CDF) can be obtained by integrating overall power levels, starting from the maximum power level and working down to zero power. As each bin's frequency of occurrence is added, a percentage number is obtained for how often the HDTV signal envelope power exceeded that power bin level. As more and more bins are added and numerically integrated, the number rises toward 100% at the zero-power level. Meaning, 100% of the measured RF envelope power samples are greater than the zero-power level. It should be noted that the histogram could have been integrated from zero power to maximum power, creating a standard CDF. Then the value found in each bin would represent the percentage of time the HDTV RF signal is below that particular level.

In order to create the peak-to-average statistics, the above process is modified, where each bin's power value is divided by the average power of the HDTV sample signal. This power is the power measured within the 6-MHz HDTV BW of operation.

As was described above, the main interest is to determine the frequency of occurrence for and level of each bin and create a useful plot to work with that can show PAR at the $x$ axis and the probability of occurrence in $y$ axis. For that there is a need to plot the standard CDF marked as $F(x)$. This is a statistical function bounded by $0 \leq F(x) \leq 1$, which indicates power-sample occurrence frequency for values below a certain level. It is cumulative since it adds up all occurrences of all the signal power samples that are below a given power value, in this case this is the average power. Hence, the case of interest is to evaluate the complementary case of $1-F(x)$. Figure 12.33 depicts the flow chart concept for analyzing PAR. It is important to be careful when producing histograms and statistics that relate to sampling. If the random vector is large enough, there are sufficient symbols to produce a PDF, and a single iteration is made, then the PDF is given by normalizing the histogram to a random vector length. An alternative method is to have several iterations on a shorter random signal. In that case histogram is normalized to number of iterations, where each iteration provides a peak. The histogram bins collect peaks for either single runs or loops. Additionally, there is a significant difference between producing the amplitude distribution of an ensemble of symbols in a complex signal. These statistics provide the instantaneous value of signal voltage. It can be above or below the average value since the statistical sample is not for the peak, and mapping into histogram bins is for the signal amplitude distribution. Therefore, when calculating statistics of samples per average, the ratio may show negative values as well, however, peak to average will always be positive. Moreover, the normalization of the histogram in this case is per the signal vector length. The statistics for an event of amplitude peaks and null to occur compared to PAR analysis are different since the process of producing a data stream distribution of amplitudes is not by searching for peaks but by observing the signal's instantaneous nature. Having this in mind, PAR calculation and signal statistics can be produced.

Before the statistical calculation of PAR, there is a need to define the constellation size that is used. A constellation is a map of symbols on the I-Q axis system, where the distance between symbols in each direction is $2a$, and the distance from the I and Q axes is $a$. Equation (14.10) in Sec. 14.3 is used to evaluate the voltage distance between symbols and calculate $a$. It is important to note that the constellation used should be square, that is, QAM 16, 64 and so on. QAM 32 is not a square constellation since it has no square root.

The energy per symbol $E_s$ equals the number of bits used to represent a symbol. For instance in 16 QAM, 2 bits are used for I and 2 for the Q. Hence, there are 4 states for each axis and 16 states overall. Therefore, $E_s = 4\,E_b$. In the same manner $E_s = 6\,E_b$ for 64 QAM and $E_s = 8\,E_b$ for 256 QAM.

Note that the constellation axis refers to voltage, and voltage is related to the square root of power; therefore, the following relation is valid:

$$\frac{E_s}{T_s} = P_s. \tag{12.61}$$

The voltage of each symbol versus time composed from I voltage and Q voltage. For the sake of simplicity, let's assume a 16 QAM scheme. The mesh-grid for such constellation would be $\pm 1$ and $\pm 3$. These mesh grid are marked as $k_I$ and $k_Q$ in Eq. (12.61). Thus the highest energy symbol is at the following coordinates; $(3a, 3a)$, $(-3a, 3a)$, $(-3a, -3a)$, $(3a, -3a)$. These coordinates are the projections of the carrier phasor amplitude on the I and Q axes. The AM process controlls the phase of a sine wave carrier signal versus time by a sequence of bits, analog or digital. In a QAM two orthogonal carriers are used to create a

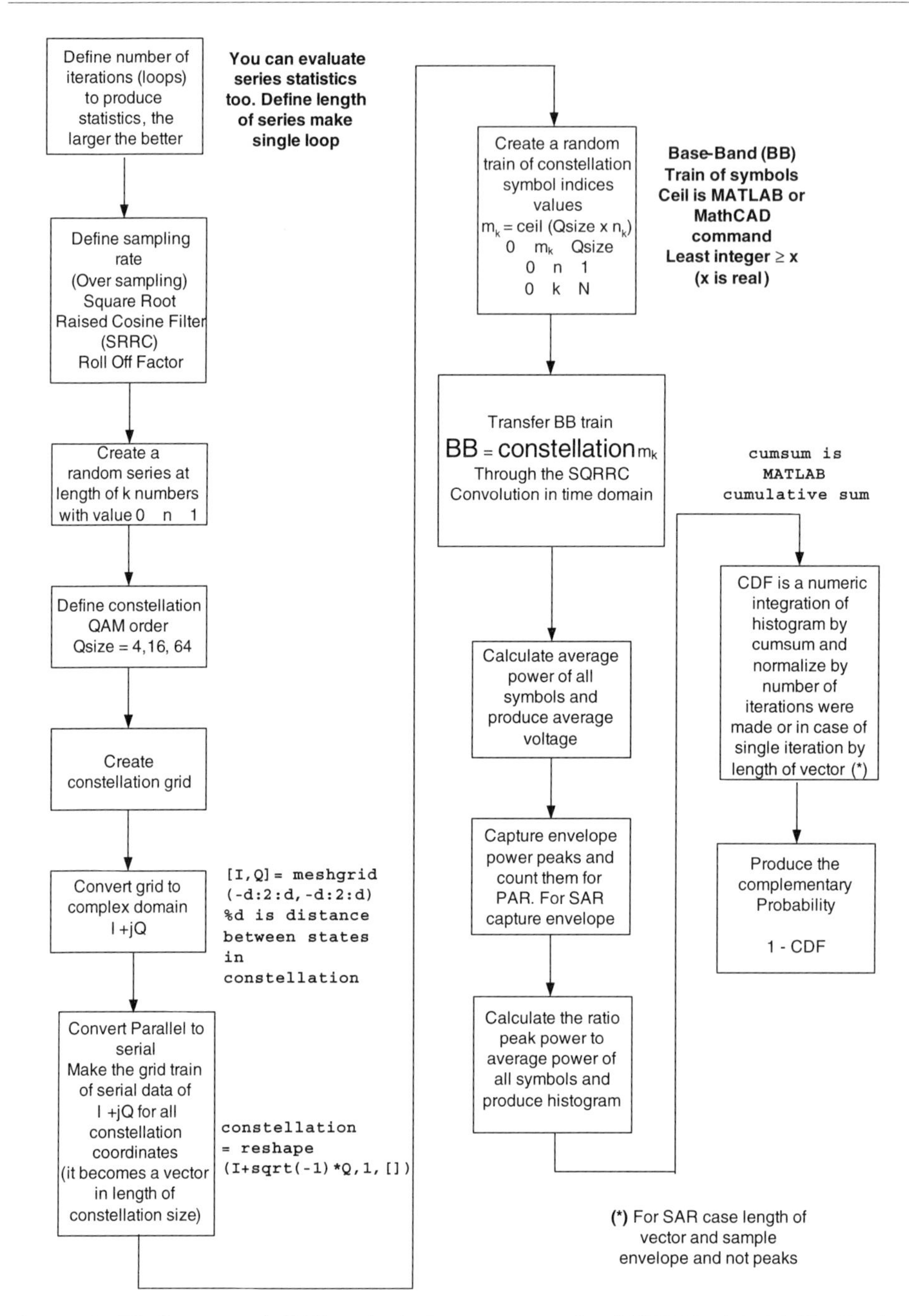


**Figure 12.33** Suggested PAR calculation process for MATLAB, MathCAD Further reading about QAM is in Chapter 14 and 15.

constellation that produces a two dimensional scheme in the I and Q plan. If the sinusoidal carrier signal has an amplitude of A, then its power is $P_s = 1/2 \cdot A^2$, so that $A = \sqrt{2 \cdot P_s}$. With this information the constellation

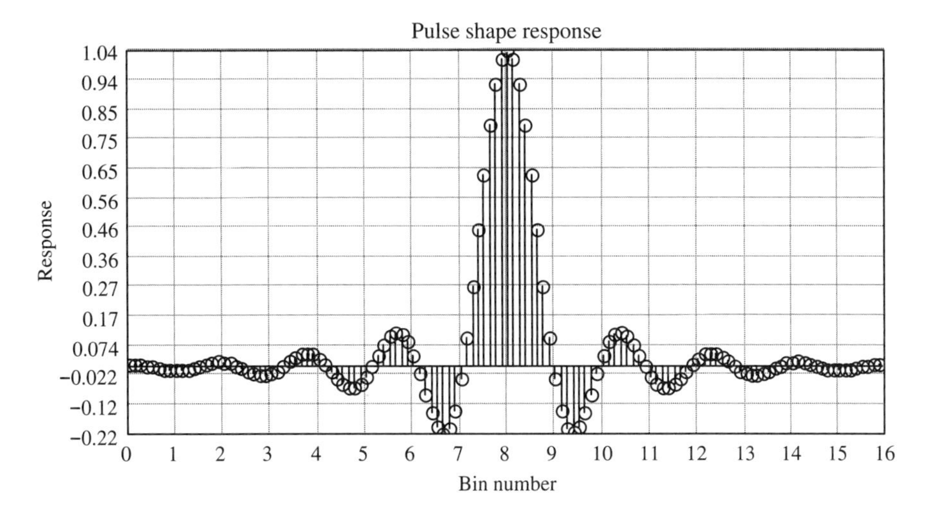


**Figure 12.34** SRRC the square-root raised cosine pulse shape sampled at N-8 samples/symbol with $\alpha = 0.15$ and truncated to span 16 symbols. Note that bin 8 is $t$/Tsymbol = 0, bin 0 is $t$/Tsymbol = −8 and bin 16 is $t$/Tsymbol = 8 where $t$ = nT_samp and Ts is Tsymbol and $n$ is integer $n = 1, 2, 3$.

voltage versus time can be plotted by using the following equations where $a$ is analyzed in Eq. (14.10) in Sec. 14.3:

$$V_{\text{QAM}}(t) = k_I \cdot a \cdot \sqrt{2 \cdot P_s} \cos(\omega_0 \cdot t) + k_Q \cdot a \cdot \sqrt{2 \cdot P_s} \sin(\omega_0 \cdot t). \tag{12.62}$$

With a sampling periodicity of $T_s$ and a signal period of $T$, Eq. (12.62) is modified to

$$V_{\text{QAM}}(n) = k_I \cdot a \cdot \sqrt{2 \cdot P_s} \cos\left(\frac{2 \cdot \pi_0}{T} \cdot nT_s\right) + k_Q \cdot a \cdot \sqrt{2 \cdot P_s} \sin\left(\frac{2 \cdot \pi_0}{T} \cdot nT_s\right). \tag{12.63}$$

The coefficient $a$ determines the symbol's average energy depending on the quadrature order. The coefficients $k_I$ and $k_Q$ in Eq. (12.63) determine the random location of the symbol in the constellation, per Eq. (14.10) in Sec. 14.3, and the sine and cosine functions represent the carrier. The local oscillator (LO) feeding the I/Q modulator is sometimes called the QPSK modulator. However, for base-band (BB) analysis there is no need for the LO, but there is a need to generate the I and Q in the complex domain. Hence, Eq. (12.60) turns into

$$V_{\text{QAM}}(n) = \left[k_I\left(\frac{nT_s}{T}\right)a + k_Q\left(\frac{nT_s}{T}\right)a\right] \otimes h\left(\frac{nT_s}{T}\right), \tag{12.64}$$

where $\otimes$ marks the convolution in the time domain, and $h(t)$ is the impulse response of a square-root-raised cosine (SRRC) filter provided in Eq. (12.64).[28]

The recipe for PAR calculation is provided in Fig. 12.34 and the following preparations should be made. First, define the QAM order M. The array dimension must be at the power of 2. Next, define data rate D [bits/sec]. Define number of symbols for simulation, which must be $2^m$, for instance, 2048. Define number of samples per symbol to be N_samp. Determine the SRRC shape filter roll-off factor $\alpha$. Define the symbol rate according to

$$S\_\text{rate} = \frac{\text{D\_rate}}{\log_2 \text{M}} = \log_{10} 2 \cdot \frac{\text{D\_rate}}{\log_{10} \text{M}}, \tag{12.65}$$

$$h(t) = \frac{\sin\left[\frac{\pi\cdot(1-\alpha)}{T\text{.symb}}\cdot(t+\Delta t)\right] + \frac{4\cdot\alpha\cdot(t+\Delta t)}{T\text{.symb}}\cdot\cos\left[\frac{\pi\cdot(1+\alpha)\cdot(t+\Delta t)}{T\text{.symb}}\right]}{\left[\frac{\pi\cdot(t+\Delta t)}{T\text{.symb}}\right]\cdot\left[1-\left[\frac{4\cdot\alpha\cdot(t+\Delta t)}{T\text{.symb}}\right]^2\right]}, \tag{12.66}$$

where $\Delta t$ is a very small number used to prevent the simulation error of real numbers divided by zero, and $T$_symb is the symbol period. Since this is a numeric analysis there is a need to create vectors for the $h(t)$, which is a finite impulse response filter (FIR), and the signal I + jQ. Assume that $T$_symb = 1, then the sampling period $T$_samp = $T$_symb/$N$_samp, and the sampling frequency is $F$_samp = 1/$T$_samp. Therefore the number of wave form samples is $S$_samp = $N$_samp × $N$_symb, which is the total number of samples. This is the length of the **I + jQ** vector samples. The next step is to define the length of the FIR that provides the shaping pulse, let's assume L1 = 16 × N_samp. That means that the FIR length is 16 symbols. Hence, the **t** vector range is given by $t$ = $T$_samp × (0 to L1-1). Because SRRC($t$) is noncausal, it must be truncated, and the pulse shaping filter are typically implemented for $\pm 6Ts$ to $\pm 8Ts$ about the $t = 0$. Observe that $h(t)$ in Fig. 12.34 has zero crossings at $\pm nTs$ but the center is located in bin number 8. It is a finite filter in the time domain. The result of truncation is the presence of nonzero side lobes in the frequency domain, the spectrum is no longer zero for $|\,f\,| > (1+\alpha)/2Ts$. Since $h(t)$ is a noncasual function, it was shifted to the right by 8 symbols, therefore its center is bin 8. There is no negative time in casual system, and in this example, $h(t)$ symmetry prior to the time shift is $\pm 8Ts$ about the $t = 0$ point for each symbol. After $\pm 8Ts$ the response is flat: equal to zero. Figure 12.34 illustrates the SRRC bins for 16 symbols. After having the two vectors there are two options to calculate the output: the first, as was presented, is by time-domain convolution and the second is to use fast Fourier transform (FFT) for the SRRC. The signal performs the frequency domain calculation and uses inverse fast Fourier transform (IFFT) to evaluate the output. A hamming window is used in FFT to increase accuracy. Boundary cases of SRRC are cut and ignored. The statistics for PAR is estimated and plot is produced.

There are two kinds of PAR charts. Power and voltage where PAR relates to 20 log($V/V_{rms}$). When calculating RF back-off against compression power, PAR is used. For ADC voltage PAR is used to examine margins against clipping. Since in the RF channel both I and Q are transferred, PAR is estimated for the complex signal. In analog base–band, (ABB) where there is a channel for I and Q, each component is examined for PAR. The sum of them would produce the overall PAR. Both I and Q have the same likelihood; therefore, the difference between the $I^2$ and $Q^2$ to the $I^2 + Q^2$ is 3 dB.

The question asked is what statistics to use: PAR or the instantaneous signal sampling statistic sample-to-average ratio (SAR), which affects back-off and probability results are different. Generally the design tradeoffs refer to signal statistics and the likelihood of a peak to happen within a stream of symbols. Therefore the MATLAB file of sample to average is used in orthogonal frequency division multiplexing (OFDM), where a system is designed for a back off within a given probability that is since QAM OFM PAR is stronger than regular QAM where PAR is used.

Figure 12.35 provides the probabilities for SAR signal statistic amplitudes to average. The MATLAB code in this case observes in a snapshot the statistics of the same random signal used for PAR. This analysis is called SAR for which the highest positive SAR value is the PAR value of this stream. The difference in simulation is that the histogram is normalized to the QAM symbol trail length and histogram bins collect data of instantaneous amplitude to average value. Thus there are bins showing negative values as well. In PAR the statistics run several times on the stream and collects peaks for each run. The histogram is normalized to the number of runs. In other words, the MATLAB code observes the abs value in SAR rather than the peak of abs value within a stream in PAR analysis.

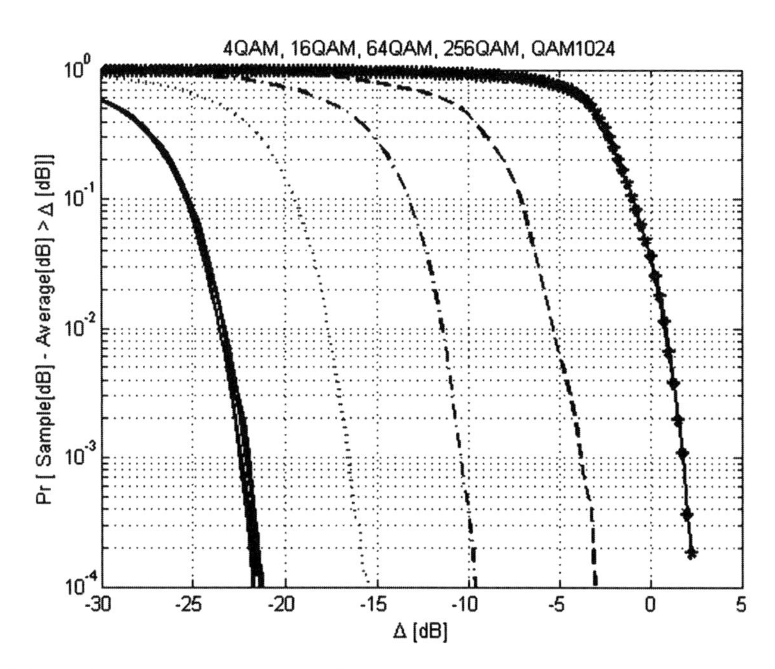


**Figure 12.35** SAR probability vs. SAR value. Left to right 4QAM, 16QAM, 64QAM, 256QAM, and 1024QAM. Simulation parameters are SRRC roll-off factor $\alpha$ is 15%, the series length is 2048 symbols, and number of iterations is 1 per each modulation scheme.

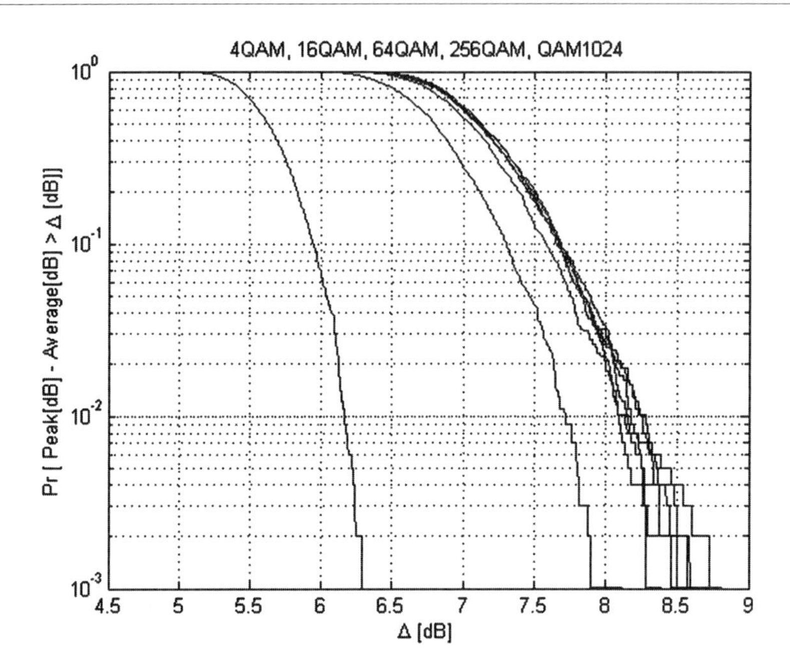


**Figure 12.36** PAR probability vs. PAR value left to right, 4QAM, 16QAM, 64QAM, 256QAM, and 1024QAM. Simulation parameters are SRRC roll-off factor $\alpha$ is 15%, the series length is 2048 symbols, and number of iterations is 1000 per each modulation scheme.

This simulation tool can be expanded to check ratios of multiple QAM channels transmitted at the same time. A value of SAR and PAR can be produced for the AGC correction of peak to average.

Observing Fig. 12.36, it can be seen that as the constellation order increases the PAR likelihood to occur. Moreover, it can be seen that 64 QAM curve is very close to 256QAM and 1024QAM. That is because it is the nature of QAM to be more Gaussian at high orders. The MATLAB code for producing PAR statistic is listed below.

## MATLAB Code for PAR Estimation

```
%% QAM PAR statistics written by Avi Brillant 6 of October
   2007
% This program evaluates PAR for 4QAM, 16QAM, 64QAM, 256QAM
  and 1024QAM

function [xbbpdf, proba] = peakToAverage(runs, N_symb, osr,
constTp, constSz);
```

```
%Running parameters

if nargin == 0              % Data Control and Monitoring

   alpha  = 0.15            % Roll Off Factor of SRRC FIR shaping
                              filter
   runs   = 1e3;            % Number of runs to gather stats. The
                              larger the better (and the
                              longer...)
   D_Rate = 10e8            % Data Rate Bits/Sec
   N_symb = 2048;           % Number of symbols in this simulation
   %constTp= 'qam';         % Constellation type ('qam' or 'psk')
                              USE this for manual run and dissable
                              Constellation_Size case
   %constSz = 256;          % Constellation size USE this for
                              manual run and dissable
                              Constellation_Size case
   T_symb = 1;              % Symbol period
   osr    = 8;              % oversampling rate for BB signal
end

%% Check all QAM types

% Modulation index selector case

Constellation_Size=
str2mat('4_QAM','16_QAM','64_QAM','256_QAM','1024_QAM');
Const_Index=length(Constellation_Size);

for Const_Counter =1:Const_Index;

     switch Const_Counter;
     case 1
         constTp = 'qam'
         constSz = 4;
     case 2
         constTp = 'qam'
         constSz = 16;
     case 3
         constTp = 'qam'
         constSz = 64;
     case 4
         constTp = 'qam'
         constSz = 256;
```

```
      case 5
          constTp = 'qam'
          constSz = 1024;

          otherwise

          disp('Select Valid QAM Constellation Satisfies
CostSize = X^2, 32QAM type is not valid')

    end;

% baseband samplig parameters
    T_samp     = T_symb/osr;
%% Samplig period
    F_samp     = 1/T_samp;
%% Samplig frequency
    S_samp     = osr*N_symb;
 %% Overall number of samples

% Simulation parameters. Don't touch.

bbEpf =[];BB_I=[];BB_Q=[]; %% bbEpf base band Statistics
Nosr  = N_symb*osr;        %% Nosr is total number of
                              samples
S_rate= D_Rate*(log10(2)/  %% Symbol rate [Hz] is the
  log10(constSz));            data rate divided by
                              number of bits per symbol


%% Making call to procedures and create vectors

SRRC = FIR(T_symb, alpha,T_samp,osr); %% Square Root Raised
Cosine Filter Call function per sampling parameter: symbol
time is T_symb,Roll off factor is alpha, sampling period
T_samp,osr

constellation = BuildConst(constTp, constSz); %% Build-
Const Function Call Constellation Parameters are: constTp
(Type) and constSz (Size)

for k=1:runs; % number of statistical calculation iterations
                per modulation scheme
```

```
% signal train of I + jQ

    Symbols             =constellation(ceil(constSz*rand(1,
                         N_symb)))

                         %% random choice of constellation
                            elements by telling their
                            position as an index for the
                            signal vector

    I_BB                =real(Symbols);
    Q_BB                =imag(Symbols);

%% Over sampling baseband of signal train of I +jQ

   for i=1:S_samp; % over sampled vector BB_I

      BB_I(i)     =0;
      BB_Q(i)     =0;

end;

for n=1:N_symb; % over sampling of vector BB_I

    k=n*osr;

    BB_I(k)     = I_BB(n);
    BB_Q(k)     = Q_BB(n);

end;

for j=1:S_samp;          % over sampled complex vector made of
                          real and image of BB_I

baseband(j)              =complex(BB_I(j),BB_Q(j));
end;

%% Making convolution of SRRC with train of impulses and
   getting physical signal

%% Convolution of I+jQ with SRRC
BBsignal    = conv(baseband,SRRC);
```

```
%% Normalized VOLTAGE
BBsignal     = BBsignal/sqrt(mean(abs(BBsignal).^2));
RMS          = sqrt(mean(abs(BBsignal).^2));

%This is PAR
bbEpf        = [ bbEpf     10* log10(max(abs(BBsignal).^2)/
                (RMS^2))];

end


%You can use BBsignal to plot physical BB constellation
I=real(BBsignal);
Q=imag(BBsignal);

%You can use Symbols to plot ideal BB constellation
I_BB         =real(Symbols);
Q_BB         =imag(Symbols);

%histogram on peaks found per run so we have peaks distri-
 bution over 1000 runs
[bbPdf, xbbpdf] = hist(bbEpf, runs);

% integration
bbcf            = cumsum(bbPdf);

%normalizing by number of runs taking the probability of
 being larger
proba           = 1-bbcf/runs;


figure(1)

   semilogy(xbbpdf, proba);
grid on; xlabel('\Delta [ dB] '), ylabel('Pr [ Peak [dB] -
Average[dB] > \Delta[dB] ] ');
%title([ constTp'' num2str(constSz) ' - Peak To Average']);
title([ '4QAM, 16QAM, 64QAM, 256QAM, QAM1024']);
hold on;

end


%% help function CREATING CONSTELLATION IN TIME DOMAIN
function constellation = BuildConst(constTp, constSz);
```

```
switch constTp
    case('psk')
       constellation = exp(j*(2*pi/constSz*(1:constSz)+pi/
         constSz));
    case('qam')
       d              = sqrt(constSz)-1;
       if mod(d,1), error('constellation size in QAM should
be a square'); end

       [I,Q] = meshgrid(-d:2:d,-d:2:d);
       constellation = reshape(I + sqrt(-1)*Q, 1,[]);
    otherwise
     error('illegal modulation technique (choose qam or psk)')
end
return

%% help function CREATINGIN FIR SRRC IN TIME DOMAIN

function SRRC = FIR(T_symb,alpha,T_samp,osr)

%L1 is the length of the SRRC FIR filter that provides the
shaping pulse
L1=16*osr;

L=0:L1-1;

%SRRC FIR Equation Parameters

t=L*T_samp;

A=sin(pi*(1-alpha)*(t-8*T_symb+1e-9)/T_symb);

B=(4*alpha*(t-8*T_symb+1e-9)/T_symb);

C=cos((pi*(1+alpha)*(t-8*T_symb+1e-9))/T_symb);

D=(pi*(t-8*T_symb+1e-9)/T_symb);

E= 1-((4*alpha*(t-8*T_symb+1e-9)/T_symb).^2);

%SRRC Function

[ SRRC_FIR] = (A+B.*C)./(D.*E);

SRRC=SRRC_FIR;

return
```

**MATLAB Code for Sample to Average Ratio (SAR) Estimation**

```
%% QAM SAR Sampling to Average Statistics written by Avi Brillant
%% This program evaluates SAR for 4QAM, 16QAM, 64QAM, 256QAM and
%% 1024QAM

function [ ] = SAR_QAM_4_64_256_1024_()

%Clear the MATLAB workspace + close all previous windows
clear all, close all

%Running parameters

if nargin == 0              % Data Control and Monitoring

   alpha = 0.15             % Roll Off Factor of SRRC FIR shaping
                              filter
   D_Rate = 10e8            % Data Rate Bits/Sec
   N_symb = 2048;           % Number of symbols in this simulation
                              the longer
                           % the better
   %constTp = 'qam'; % Constellation type ('qam' or
     'psk')USE this for manual run and dissable
     Constellation_Size case
   %constSz = 256;          % Constellation size USE this for
     manual run and dissable Constellation_Size case
   T_symb = 1;              % Symbol period
   osr    = 8;              % oversampling rate for BB signal
end

%% Check all QAM types

% Modulation index selector case

Constellation_Size=
str2mat('4_QAM','16_QAM','64_QAM','256_QAM','1024_QAM');
Const_Index=length(Constellation_Size);

for Const_Counter =1:Const_Index; %% Main Loop Check all
Constellation Types

   switch Const_Counter;   %% Constellation Types Menu
   case 1
      constTp = 'qam'
      constSz = 4;
```

```
    case 2
        constTp = 'qam'
        constSz = 16;
    case 3
        constTp = 'qam'
        constSz = 64;
    case 4
        constTp = 'qam'
        constSz = 256;
    case 5
        constTp = 'qam'
        constSz = 1024;

        otherwise
        disp('Select Valid QAM Constellation Satisfies Cost-
Size = X^2, 32 QAM type is not valid')
    end;
% baseband samplig parameters
    T_samp = T_symb/osr;
%% Samplig period
    F_samp = 1/T_samp;
%% Samplig frequency
    S_samp = osr*N_symb;
%% Overall number of samples

% Simulation parameters. Don't touch.

bbEpf =[];BB_I =[];BB_Q =[]; %% bbEpf base band Statistics

Nosr  = N_symb*osr;             %% Nosr is total number
                                   of sampls


%% Symbol rate[Hz] is the data rate divided by number of bits
per symbol
S_rate= D_Rate*(log10(2)/log10(constSz));

%% Making call to procedures and create vectors

%% Square Root Raised Cosine Filter Call function per
sampling parameter: symbol time is T_symb,Roll off factor
is alpha, sampling period T_samp,osr
```

```
SRRC = FIR(T_symb,alpha,T_samp,osr);

%% BuildConst Function Call Constellation Parameters are:
constTp (Type) and constSz (Size)
constellation = BuildConst(constTp, constSz);

%% signal train of I +jQ

%% random choice of constellation elements by telling their
position as
%% an index for the signal vector

   Symbols =constellation(ceil(constSz* rand(1,N_symb)));

   I_BB     =real(Symbols);
   Q_BB     =imag(Symbols);

  % Over sampling baseband of signal train of I +jQ

   for i=1:S_samp;

   BB_I(i)          =0;
   BB_Q(i)          =0;

   end;

   for n=1:N_symb;
       k=n* osr;

   BB_I(k)          =I_BB(n);
   BB_Q(k)          =Q_BB(n);

   end;

   for j=1:S_samp;

   baseband(j)      =complex(BB_I(j),BB_Q(j));

   end;

%% Making convolution of SRRC with train of impulses and
   getting
%% physical signal
```

```
%% Convolution of I+jQ with SRRC

   BBsignal = conv(baseband,SRRC);

   RMS      = sqrt(mean(abs(BBsignal).^2));

%% Normalized VOLTAGE
   BBsignal = BBsignal/(RMS);

%% This is SAR
   SAR      = 10*log10((abs(BBsignal) + eps).^2/(RMS^2));

%%calculate CCDF

%% SAR bins range -30 to 5 dB and increments are 0.25dB

   bin = -30:0.25:5;

%% SAR bbPDF is histogram normalized to length of symbols
   series

   bbPDF = hist(SAR,bin)/length(SAR);

%% integration and producing complementary probability of
   being larger
%% than SAR

   proba = 1 - cumsum(bbPDF) + eps;


%%Output and Plotting For more info see following site %%
%%%%%%%%%%%%%%%%%%%%%%%%%%%%%%%%%%%%%%%%%%%%%
http://www.mathworks.com/access/helpdesk/help/techdoc/
index.html?/access/helpdesk/help/techdoc/ref/plot.html
&http://www.google.com/search?hl=en&q=matlab+plot+
lines+

figure(1)

%% QAM Plot Menu

switch constSz
          case 4
              semilogy(bin, proba, '-b*'); %Asterisk +
                                            Solid line 4QAM
```

```
          case 16
             semilogy(bin, proba, '--b'); %Dashed line
                                          16QAM
          case 64
             semilogy(bin, proba, '-.b'); %Dash-dot line
                                          64QAM
          case 256
             semilogy(bin, proba, '.:b'); %Dotted line
                                          256QAM
          case 1024
             semilogy(bin, proba, 'b');   %Solid line
                                          1024QAM
          otherwise
             disp('Unknown method.')
       end

ylim([1e-4 1]); %Range of Probability from 1 to 0.0001

grid on;
xlabel('\Delta [dB] '), ylabel('Pr [Sample[dB] - Average[dB]
> \Delta[dB]] ');
%title([constTp ' ' num2str(constSz) ' - Sample To
Average']);
title(['' 4QAM, 16QAM, 64QAM, 256QAM, QAM1024']);
hold on;

end

%% Help function CREATING CONSTELLATION IN TIME DOMAIN
function constellation = BuildConst(constTp, constSz);

switch constTp
     case('psk')
        constellation = exp(j* (2*pi/constSz* (1:constSz)
                        +pi/constSz));
     case('qam')
        d             = sqrt(constSz)-1;
        if mod(d,1), error('constellation size in QAM
        should be a
         square'); end

        [I,Q]         = meshgrid(-d:2:d,-d:2:d);
        constellation = reshape(I + sqrt(-1)*Q, 1, []);
     otherwise
        error('illegal modulation technique (choose qam or
        psk) ')
```

```
    end
    return


%% Help function CREATINGIN FIR SRRC IN TIME DOMAIN

function SRRC = FIR(T_symb,alpha,T_samp,osr)

%L1 is the length of the SRRC FIR filter that provides the
shaping
%pulse

L1=16*osr;

L=0:L1-1;

%SRRC FIR Equation Parameters

t=L*T_samp;

A=sin(pi*(1-alpha)*(t-8*T_symb+1e-9)/T_symb);

B=(4*alpha*(t-8*T_symb+1e-9)/T_symb);

C=cos((pi*(1+alpha)*(t-8*T_symb+1e-9))/T_symb);

D=(pi*(t-8*T_symb+1e-9)/T_symb);

E= 1-((4*alpha*(t-8*T_symb+1e-9)/T_symb).^2) ;

%SRRC Function

[SRRC_FIR] = (A+B.*C)./(D.*E);

SRRC=SRRC_FIR;

return
```

#### 12.2.8.2 Peak-to-rms compensation

The difference between peak and rms requires the compensation of the AGC leveling lock. For example, if the AGC goal is to maintain peak leveling of 14 dBmV per CATV channel, then the CW calibration should be 9.4 dBmV. That means that if the AGC is calibrated to a 14 dBmV power level at CW, it would result in a 18 dBmV peak level. This is problematic because it complicates the testing of a

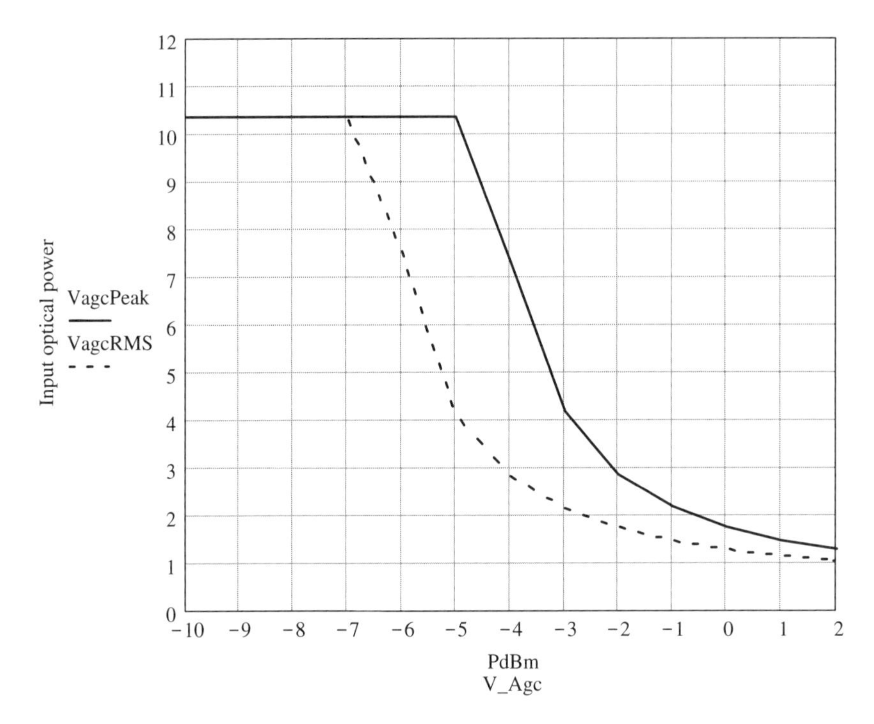


**Figure 12.37** VVA control voltage for AGC vs. input optical power before and after PAR compensation. AGC peak-to-RMS compensation is done when using 110-CW channels. Test conditions: 79 channels between 50 and 550 MHz at 2.7% OMI; 31 Channels between 550 and 870 MHz at 1.35% OMI; RF AGC pilot compensation: 4 dB; and the RF detector type is DLVA. A true RMS detector would require lower PAR corrections.

receiver with feedback AGC. The system should be tested for CSO CTB linearity and CNR sensitivity performance when the AGC is calibrated to the peak level under CW, in our example, 14 dBmV. After the evaluation is completed, the AGC reference is reduced by 4.6 dB to lock on 9.4 dBmV CW. This reference correction of the AGC shifts the AGC threshold to start at a lower optical-input power. On the optical-power scale, it means the threshold point is shifted down by half of the RF reference value correction, in our case, 2.3 dB lower in power. Figure 12.37 illustrates the curve of AGC control voltage versus optical level before and after the reference level correction of 4 dB.

The threshold point shifted from −5 dBm optically to −7 dBm. Saturation occurs at AGC voltage of 1.5 V and optical saturation point for AGC was shifted from 1 to −1 dBm. The AGC dynamic range is preserved at 6 dB optical or 12 dB RF.

### 12.2.9 Adaptive feedback AGC DSP algorithm

AGC realization can be done by using a digital signal processing (DSP) method. In this case, the integration is done by means of software. The advantage of using a DSP for AGC is its leveling program flexibility. Furthermore, the DSP can perform any type of integrator, and the peak-to-rms problem can be solved by a DSP algorithm. The power detector can be an envelope detector, and the AGC algorithm can estimate the peak power and average power (rms). That compensation can be implemented on a true rms detector too, and accuracy is increased. Hence, if the AGC is set to a specific CW level, it can be an adaptive leveling. In a CW mode, both estimations of peak and rms are equal. Hence, the AGC correction of peak to rms would be zero. For modulated signal leveling, the peak and rms estimators would result in different readings. Hence, the AGC would automatically lock to the rms level to keep the peak level estimates equal to CW leveling at which the AGC was programmed. The AGC BW calculation is determined as was explained in previous paragraphs. However, the integration time constant is determined by the DSP algorithm. In FTTx ITR, this can be a simple microcontroller. Knowing the peak-to-average ratio is

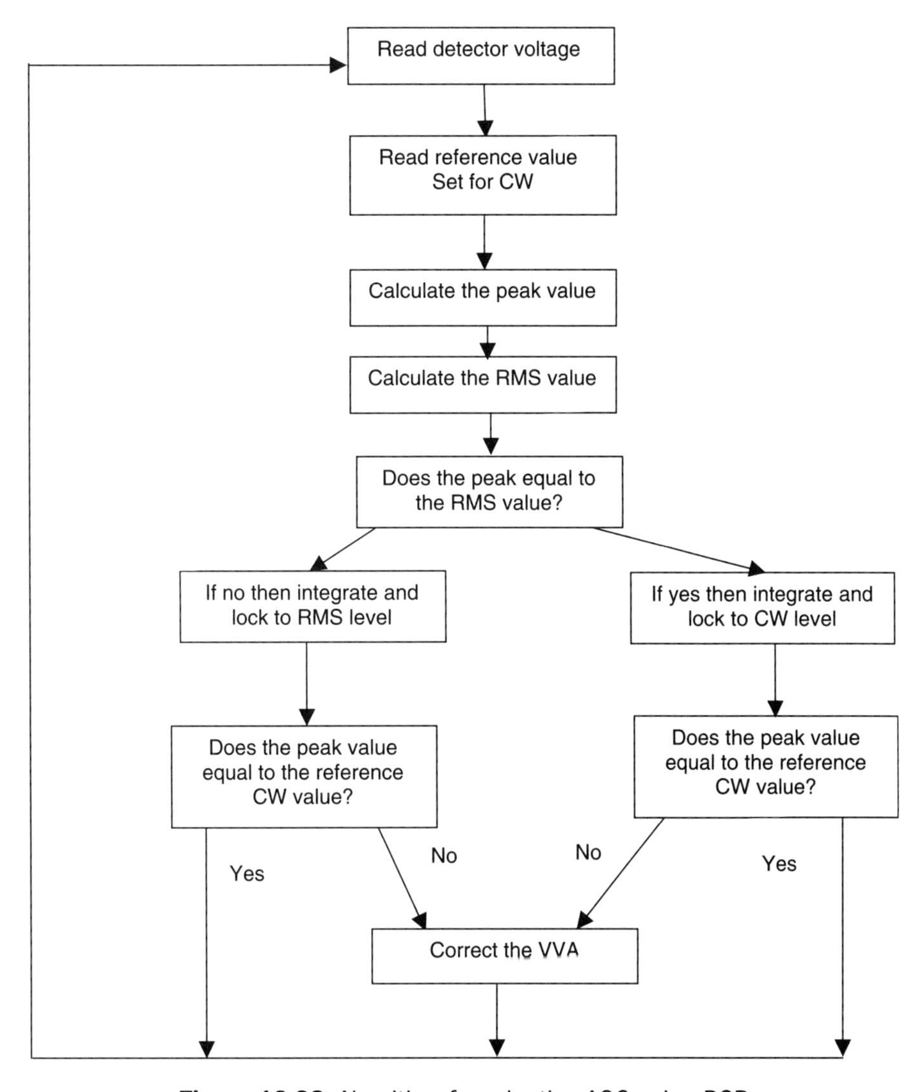


**Figure 12.38** Algorithm for adaptive AGC using DSP.

useful for laser transmitters in order to improve RF amplifiers' linearity by having an adaptive bias. Using a true power detector rather than a DLVA would minimize PAR error due to the true RMS detector structure compared to DLVA structure. Figure 12.38 provides a suggestion for adaptive PAR compensation. The PAR compensation is a learning process and is frozen after it sensed the PAR factor. It should not be dynamic since it might add undesired residual AM to the signal. Hence this process compensates the difference between CW leveling to modulated level.

## 12.3 Main Points of this Chapter

1. The following differences and advantages per each AGC method are given below:

2. The AGC closed loop is given by

$$H(s)_{\mathrm{AGC}} = \frac{K_{\mathrm{F}}K_{\mathrm{D}}K_{\mathrm{CPL}} \cdot g \cdot P_{\mathrm{REF-IN}} \cdot F(s)T(s)}{1 + K_{\mathrm{F}}K_{\mathrm{D}}K_{\mathrm{CPL}}P_{\mathrm{REF-IN}} \cdot g \cdot F(s)T(s)},$$

where $F(s)$ is the loop filter, $T(s)$ is the detector smoothing filter, $K_{\mathrm{F}}$ is the loop filter, dc gain generally is 1, $K_{\mathrm{D}}$ is the detector factor, $K_{\mathrm{CPL}}$ is the coupling factor, $g$ is the VVA gain factor, and $P_{\mathrm{REF\text{-}IN}}$ is the locking level.

3. The AGC-loop BW should be 0.5–1 Hz at a maximum. It should be a slow system. The reason is to prevent the AGC circuit from becoming an envelope detector of the vertical blanking and sync signal. The pulse repetition frequency (PRF) of the vertical sync-blank is 50–60 Hz. Disobeying this would add residual AM.

4. The peak-to-rms correction is approximately 4.6 dB for NTSC plan when using DLVA detector type.

5. The AGC type is defined by the number of ideal integrators and poles it has at the origin.

6. The AGC order is defined by the order of the poles' polynomial.

7. As the input signal becomes higher, the CNR at the AGC-controlled system improves with a slight degradation caused by noise figure increase.

8. As the input signal becomes higher, the system NF degrades by the following relation:

$$F = F_1 + \frac{L^{-1} - 1}{G_1} + \frac{L^{-1}(F_2 - 1)}{G_1},$$

where $L$ is the attenuation, $G$ is gain, and $F$ is noise factor.

| Requirement | Feedback | FF |
|---|---|---|
| To provide constant power vs. optical input with high accuracy | Yes, it is a tracking system | No, it requires alignment |
| To increase dynamic range | Yes, in case of VVA | Yes, but with limitation on minimum number of steps |
| To be independent of OMI | No, low CATV band sampling as a Pilot | Yes |
| To be independent of channel loading | No, low CATV band sampling as a pilot Number of channels should be fixed | Yes |
| To be independent from the receiver components' gain | Yes | No, alignment is needed |
| To be slow enough so the AGC would not track slow optical power changes (may create undesired residual AM) | Yes, narrow AGC BW of 0.5 Hz | No, hence, hysteresis should be wider, thus loosing some dynamic range. Large integration constant time can be used to solve this problem in software |
| To be independent of modulation | No, AGC detects average power. Peak to RMS offset should be included in reference calibration | Yes |
| Simplicity and cost-effective solution | Yes | Yes |

9. The AGC dynamic range is determined by the system gain-attenuation range and noise.

10. The AGC threshold is determined by the system maximum gain and locking level.

11. The AGC ability to preserve and track CNR as the input signal increases is limited by NF. At the instant the suppressed output noise equals the noise referenced to the input, it means that the AGC is out of its noise suppression range.

12. The more steps an FF AGC has, it would have a better power leveling and power flatness response. However, it would be susceptible to residual AM due to the LSB toggling.

13. To resolve fine FF AGC toggling, one option is to use a coarse step with a wide hysteresis, the second option is to use integration and averaging of the dc-sampled photovoltage by software.

14. The FF-AGC trigger point is set by the dc sampling resistor of the PD.

15. The role of the directional coupler in an AGC system is to sample the RF power and isolate the feedback path from the output-load return-loss and reflections. In that manner, it prevents an error in sampling due to VSWR.

16. When a feedback AGC is locked, the voltage level at the detector output is constant throughout the entire AGC dynamic range.

17. The peak-to-rms ratio locking error occurs because of the RF-detector type used for the AGC. A true rms detector is the rms power meter. Comparing CW mode versus modulated mode, the peak equals rms with a slight error rate compared to the modulated scenario. For a DLVA envelope detector, or DLVA with larger integration time to average the signal, the error of PAR is higher compared to a true rms detector.

18. The AGC RF locking is done by sampling portions of the lower band channel that are analog NTSC channels with high OMI and lower peak-to-rms compared to QAM channels.

19. Future frequency plan design will use all channels as QAM. Therefore, the OMI would be lower. Hence, larger number of channels should be sampled to the AGC RF detector.

20. Using QAM signals as sample bands to lock on would require a larger peak to rms correction.

21. Feedback AGC operates as fast-attack slow-release. Thus, the receiver with no signal will wait at maximum-gain highest-sensitivity. That is the nature of feedback systems. As soon as a signal is detected, the AGC will converge to the required level with a slow decay according to the integrator time constant.

22. The nature of feedback AGC, fast attack slow release, can create a peak power hit to the TV when initially connected, or when optical power exists and modulation stops and starts again or starts later. This can be solved by using a software integrator that sets the AGC to minimum gain without signal presence. This does not exist in the case of FF AGC.

23. The AGC loop BW is affected by the operating point and the coefficients of the VVA and RF detector. The loop BW extremes can be evaluated at the two extremes of high and low input power and the extreme conditions of the VVA and detector.

24. The CW peak power definition refers to its average power and not its instantaneous peak: CW $P_{\text{peak}} \equiv P_{\text{ave-CW}} = V_{\text{rms}}^2/R$ and not $P_{\text{peak}} = V_{\text{peak}}^2(t)/R$.

25. FCC rules require the NTSC peak power to occur during NTSC sync peak.

26. The NTSC signal shape defines the average power of modulated RF.

27. PAR of QAM data is a statistical process where a PDF function is created and a numerical integration process of cumulative sum is processed on the PDF.

28. PAR PDF density function is Gaussian and its integral is bounded by one. The PAR is a complementary calculation of the PDF, resulting in the following property of probability $\mathrm{PAR} = 1 - \int_{-\infty}^{x} \mathrm{PDF}(v) \cdot dv = 1 - \mathrm{CDF} = \int_{x}^{\infty} \mathrm{PDF}(v) \cdot dv$. The CDF is numerical integral.

29. In receivers where analog BB is sampled by ADC the AGC role is to protect ADC from clipping zone. Hence ADC margin design takes under consideration PAR as a back-off. Meaning of this is that the ADC full scale voltage range should handle the signal peaks as well as jamming of leaking adjacent and alternate tones (called blockers and jammers) that reach the analog base-band (ABB) Chapter 15.

30. In modems, energy estimation for AGC control is made by evaluating $I^2 + Q^2$ at digital BB. In RF case energy estimation is made by RF detector.

31. The peak-to-average value defines the RF back-off that should be taken in order to prevent gain compression. This protection is implemented by the AGC as a limiting circuit. Sometimes this is called automatic power control (APC) in RF transmitters.

32. The process of a digital loop integrator algorithm that observes history with a weight coefficient in order to avoid an error state change due to abrupt RF level change, as was explained in AGC topology with a microprocessor, is called infinite impulse response (IIR) DSP filter.

33. There are two kinds of statistics used to evaluate a signal, in this case QAM was explained. PAR, which is peak to average and SAR signal sampling statistics.

34. PAR statistic histogram bins are positive only when peaks are above average. SAR statistic histogram bins are negative and positive since voltage amplitudes and power can be above or below average value.

35. PAR and SAR statistics for a certain PAR value to occur are different.

36. The highest positive value in the SAR vector is the PAR value of the symbol ensemble stream that was evaluated.

## References

1. Thaler, J.T., and R.G. Brown, *Analysis and Design of Feedback Control Systems. Second Edition.* McGraw-Hill (1960).
2. Meyr, H., and G. Ascheid, *Synchronization in Digital Communications, Volume 1. Phase-Frequency-Locked Loops and Amplitude Control.* John Wiley & Sons, New York, N.Y., Chapter 7 Amplitude Control (1990).
3. Meyr, H., and G. Ascheid, *Synchronization in Digital Communications, Volume 1. Phase-Frequency-Locked Loops and Amplitude Control.* John Wiley & Sons, New York, N.Y., Chapter 5 Aided Acquisition (1990).
4. Gardner, F.M., *Phaselock Techniques. Second Edition.* John Wiley & Sons, New York, N.Y. Chapter 2 Loop Fundamentals (1979).
5. Gardner, F.M., *Phaselock Techniques. Second Edition.* John Wiley & Sons, New York, N.Y., Chapter 4 Tracking, and Chapter 5 Acquisition (1979).
6. Even, B., "Graphical analysis of high speed nonlinear RF leveling loops." *Microwave Journal*, pp. 67–80 (1990).
7. DiStefano, J.J., A.R. Stubberud, and I.J. Williams, *Feedback and Control Systems, Theory and Problems.* Schaum Publishing Co. New York (1967).
8. Lathi, B.P., *Modern Digital and Analog Communication Systems: International Edition.* Holt Rinehart and Winston, Inc., pp. 259–293 (1983).
9. Lathi, B.P., *Modern Digital and Analog Communication Systems: International Edition.* Holt Rinehart and Winston, Inc., pp. 234–251 (1983).
10. Ciciora, W., J. Farmer, and D. Large. *Modern Cable Television Technology.* Morgan Kaufman Publishers Inc., pp. 45–101 (1999).
11. Martin, G.H., "Analysis and Design of a CATV Inductive Directional Coupler." *Applied Microwave & Wireless*, pp. 70–78, May/June (1997).
12. Press, W. H., S.A. Teukolsky, W.T. Vetterling, and B.P. Flannery, *Numerical Recipes in C.* Cambridge University Press (2002).
13. MathCad Professional 2001, *User's Guide with Reference Manual*, MathSoft (2001).
14. Hodisan, A., and Z. Hellman, "Methods optimize performance of phase-locked loops." *Microwaves & RF*, pp. 87–96 (1994).
15. Hodisan, A., and Z. Hellman, "CAE Software Predicts PLL Phase Noise." *Microwaves & RF*, pp. 95–102 (1994).
16. Brillant, A., "Understanding phase-locked DRO design aspects." *Microwave Journal*, pp. 22–42 (1999).
17. Skyworks Solutions, Inc., "Wideband general purpose PIN diode attenuator," ALPHA Application-Notes 1003 (2006).
18. Sinyanskiy, V., J. Cukier, A. Davidson, and T. Poon, "Font-end of a digital atv receiver." *IEEE Transactions on Consumer Electronics*, Vol. 44 No. 3, pp. 817–822 (1998).
19. Eberle, W., G. Vandersteen, P. Wambacq, S. Donnay, G. Gielen, and H. De Man, "Behavioral modeling and simulation of a mixed analog/digital automatic gain control loop in a 5 GHz WLAN receiver." *IEEE Computers*

*Society Proceedings of the Design, Automation and Test in Europe Conference and Exhibition*, pp. 642–647 (2003).
20. Banta, E.D., "Analysis of an automatic gain control (AGC)." *IEEE Transactions on Automatic Control*, Vol. 9 No. 2, pp. 181–182 (1964).
21. Jiang, H., A. Primatic, P. Wilford, and L. Wu, "Analysis of the tuner AGC in a VSB demodulator IC." *IEEE Transactions on Consumer Electronics*, Vol. 46 No. 4, pp. 986–991 (2000).
22. Cai, L.Y., W.T. Song, H.W. Luo, and Z.H. Fang, "AGC and IF amplifier circuits design." *IEEE Topical Conference on Wireless Communication Technology*, pp. 42–46 (2003).
23. Bertran, E., and J.M. Palacin, "Control theory applied to the design of AGC circuits." *IEEE: Electrotechnical Conference, Proceedings, 6th Mediterranean LJubljana, Slovenia*, pp. 66–70, 22–24 (1991).
24. Khoury, J.M., "On the design of constant settling time AGC circuits." *IEEE Transactions on Circuits and Systems—II: Analog and Digital Signal Processing*, Vol. 45 No. 3, pp. 283–294 (1998).
25. Ohlson, J.E., "Exact dynamics of automatic gain control." *IEEE Transactions on Communications*, Vol. 22 No. 1, pp. 72–75 (1974).
26. Douglas, N., "Global stability analysis of automatic gain control circuits." *IEEE Transactions on Circuits and Systems*, Vol. CAS-30 No. 2, pp. 78–83 (1983).
27. Sgrignoli, G., "Measuring peak/average power ratio of the zenith/AT&T DSC–HDTV signal with a vector signal analyzer." *IEEE Transactions on Broadcasting*, Vol. 39 No. 2, pp. 255–264 (1993).
28. S. Chennakeshu, and G. J. Saulnier, "Differential detection of $\pi/4$ shifted DQPSK for digital cellular radio," *IEEE Trans on Vehicular Technology*, Vol. 42, pp. 46–57 (1993).

# Chapter 13

# Laser Power and Temperature Control Loops

Chapter 12 provided an extensive explanation about automatic gain control (AGC) topologies for RF leveling in both feedforward (FF) and traditional feedback methods. It was explained that amplitude control using the feedback approach is a linear approximation around the operating point to which the leveling system is locked. Amplitude control is used in signal-leveling systems such as receiver systems, whereas AGC is used in transmitter systems where automatic level control (ALC) is used. Additional amplitude leveling systems that use feedback optically control power to stabilize the emitted power from a laser. This system is called automatic-power-control (APC) loop. An additional control system utilizing feedback is for leveling the amount of laser cooling, using a thermoelectric cooler (TEC) to achieve tight wavelength control. This is required for DWDM systems per the international telecommunications union (ITU) grid specifications as provided in Chapter 1.[6] In the TEC loop, the temperature is controlled for wavelength stabilization, as well as for optimizing the laser's environmental conditions. Temperature control is done in order to deliver the required optical power in community access television (CATV) transmitters and WDM under extreme ambient conditions. The RF power leveling used in AGC is dual to a TEC loop, where RF power is equivalent to the heat capacity denoted as $Q$. In APC, RF power is dual to the optical level. In this short chapter, analogies to AGC models are provided as well as an introduction to APC and TEC loop concepts. Many of these TECs and APC commercial controllers are available. Both TEC and APC loops in current advanced small-form pluggable (SFP) and analog designs are realized by using microcontrollers. TEC controllers and APC controllers are part of laser drivers, and loop filter implementations are done in software. However, it is beneficial to understand the design considerations of these kinds of loops. This chapter is intentionally placed after Chapter 12 to provide a review about APC and TEC loops used for both analog and digital applications.

## 13.1 Automatic-Power-Control Loop

Laser power control in digital transmission is composed of two power control loops. The first loop controls the average power transmitted from the laser. The second loop controls the peak power level. As a consequence, by applying the two loops simultaneously, the peak power is maintained at a constant level as well as the average. In this manner, power stability and extinction ratio (ER) stability are accomplished (see Sec. 6.9 for definitions of ER). Note that since the average power level is the half value of the ER, setting the average and the peak power level defines the ER. Generally, the optical power monitoring of a laser is done by the back-facet monitoring photodiode. The back-facet monitor samples a portion of the average optical power and provides a feedback current to the control loop. Power monitor systems such as AGC, APC, and even TEC loops are slow systems, since the power changes are slow and vary over time, which is the reason for their narrow bandwidth (BW) loop of a few hertz. Therefore, the back-facet monitoring photodetector has a relatively large diameter since power control loops are slow systems. Figure 13.1 illustrates a typical concept of APC loop and peak power loop.

The peak-power-control loop is a faster loop that measures the optical-peak level using RF-peak detection. The RF-peak detection controls the modulation depth, which is the ER. The modulation depth is determined by the drive current provided to the laser by the laser driver.

There are several key design considerations related to the APC loop. These design considerations are related to the data pattern and APC loop BW. Assume digital transmission of long patterns with the same consecutive logic levels of either high level when the laser illuminates or low levels where the laser is dark and biased to its threshold. The APC loop should keep the same optical average power level without being driven to overcompensation. Hence, the loop BW or step response should be slower than the equivalent 50% duty cycle signal equal to those long consecutive bits that have the same logic level. For instance, a gigabit $2^{23}-1$ pattern with 23 consecutive bits at "0" state and 23 consecutive bits at "1" state creates an equivalent 50% duty-cycle signal at the frequency of 1 GHz/46, which is 21.7 MHz. The problem is that long patterns start to affect APC loops at lower data rates. For instance, a 154.44 MB/s OC-3 with the same pattern generates an equivalent signal of 3.35 MHz. As a result, the wideband APC loop starts to follow the optical envelope rather than averaging. This problem is similar to the AGC system applied on a modulated signal. Thus, the APC loop should be a slow narrow BW feedback-control system. A good engineering practice is to design the APC BW to 10% of the lowest optical data rate, which results in an equivalent 50% duty cycle created by the train pattern of identical consecutive bits.

In both cases of AGC and APC, the control loops should be transparent to the AM-modulated signal. In other words, the rule of thumb is that the lowest AM rate should be higher by at least 10 times (orders of magnitude) than the AGC and APC loop 3-dB cutoff frequency.

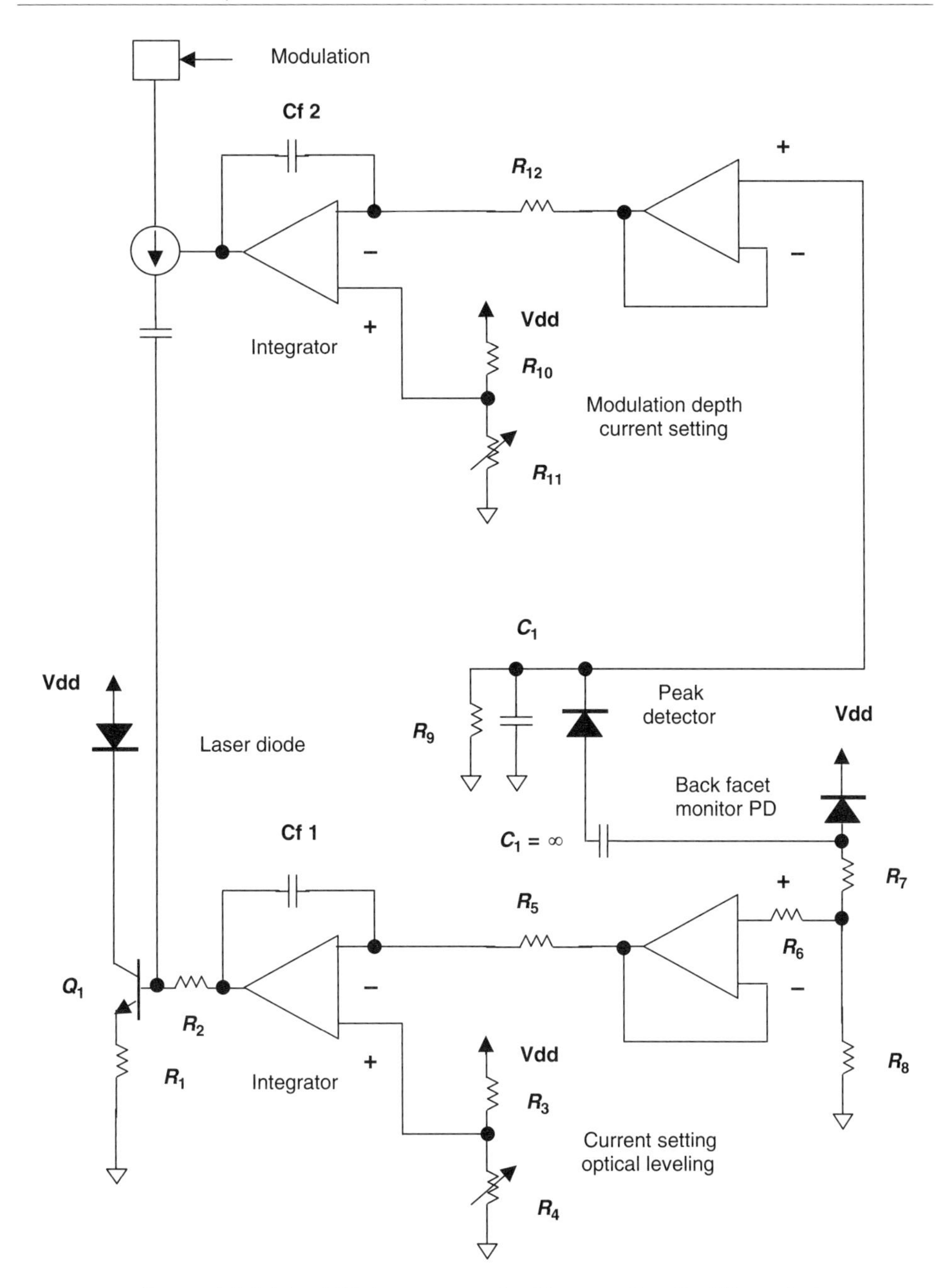


**Figure 13.1** The APC loop concept with an option for ER control.

In order to calculate the APC transfer function, there is a need to find the ratio between the laser output-power fluctuations over time and the control-current or bias-current fluctuations. The calculation and analysis processes are similar to the feedback AGC analysis presented in Chapter 12. The analog relations between the variables are given in Table 13.1.

**Table 13.1** Similarities between AGC loop and APC loop.

| AGC | APC |
|---|---|
| $G[V_C(s)]$ | $G[i_{cs}(s)] = \eta[i_{cs}(s)] = \frac{\partial P[i_{cs}(s)]}{\partial i_{cs}(s)}$<br>In the linear case it is slope efficiency |
| $g = \left.\frac{\partial G[V_C(t)]}{\partial V_C(t)}\right\|_{V_C=V_{CS}}$ | $g = \left.\frac{\partial G[i_C(t)]}{\partial i_C(t)}\right\|_{V_C=V_{CS}}$ |
| $K_D = \frac{\partial V_D(P)}{\partial P}$ | $r[mA/mW] = K_{PD} = \frac{\partial I_{PD}(P_{OPT})}{\partial P_{OPT}}$<br>Responsivity |
| $P_{out}(s) = P_{in}(s) \cdot G[V_C(s)]$ | $P_{out}(s) = i_{in}(s) \cdot G[i_C(s)]$ |
| $H(s)$ Loop filter<br>$T(s)$ Smoothing filter | $H(s)$ Loop filter<br>$T(s)$ Smoothing filter |

Figure 13.1 illustrates a typical APC loop. The laser bias current $I_C$ is controlled by the base current $I_B$ of $Q_1$. The transistor $Q_1$ is in its linear region bias point. Therefore, the collector current is given by

$$I_C(s) = I_B(s) \cdot (1 + \beta), \tag{13.1}$$

where $s$ refers to the Laplace domain and $\beta$ refers to the transistor dc hfe. The back-facet monitoring photodetector samples the laser's optical power and generates an error current. The error current generates an error voltage across a load resistor. That voltage is sampled via a buffering unity-gain amplifier and compared to a reference level on the noninverting port of the loop integrator operational amplifier. The result is an error voltage that is converted into an error current by the base resistor of $Q_1$. Sometimes a parallel feedback resistor is placed on the negative feedback of the operational amplifier of the integrator circuit in order to reduce the dc gain. However, that would reduce the loop accuracy since the finite gain of the integrator would create a residual error voltage between the inverting and noninverting ports of the integrator.

## 13.2 Thermoelectric Cooler (TEC)

TEC circuits are essential for having accurate wavelength transmissions without wavelength drifts to adjacent channels. Traditionally, coarse wavelength division multiplexing (CWDM) solutions are preferred to DWDM due to cost simplicity, power consumption, and equipment density. However, DWDM solutions provide maximum scalability in both channel count and channel distance. C band DWDM currently employs 44 channels that can be simultaneously amplified by a low-cost erbium-doped fiber amplifier (EDFA), enabling the distance and BW required for extended metro area networks.[5] Recently, multisource agreement (MSA)-based transceivers, and especially pluggable MSAs, have become the de facto choice in system design. The most popular MSA is the SFP transceiver.[7] Recently, there was a report about the first subwatt hot pluggable DWDM SFP transceiver with wavelength stabilization to use an optimized programmable TEC loop for an ITU grid of 100 GHz (see Fig. 13.9).[8]

As ITU wavelength grid spacing goes to 50 GHz, DWDM becomes denser. Thus, the TEC loop is essential in DWDM applications. TEC loops should be accurate and cheap. A more expensive way is to have wavelength locking by using a tunable laser or mode locking. This method requires larger-layout "real-estate." Wavelength control using TEC is similar to oven crystal oscillators (OCXO) where the oscillator is heated in order to maintain an accurate frequency.

### 13.2.1 TEC physics

The TEC module is a small solid-state device that operates as a heat pump. The principle was discovered in 1843 by Peltier.[4] The concept is that when an electrical current passes through a junction of two different types of conductors, a temperature gradient is created. Therefore, the Peltier phenomenon requires that the semiconductor be an excellent electrical conductor and poor heat conductor. For instance, Bismuth-Telluride N and P types are used as the semiconductor in a TEC. The TEC consists of P-type and N-type pairs connected electrically in series and sandwiched between two ceramic plates, which prevents a shorting of the laser by the TEC.[1,4] Extremely low temperatures can be achieved by cascading several TECs in series.[2]

Note that the semiconductor elements are connected in series electrically but are parallel thermally.

The plates of the TEC on the hot and the cold sides are excellent thermal conductors. When the TEC is driven by a dc supply, the current creates a cold side and a hot side on the TEC. The cold side is exposed to the laser diode and the hot side to the heat sink, which transfers the heat to the environment. The dissipated heat is delivered to the hot side marked by $Q_H$. The heat removed from the laser is marked as $Q_C$. Therefore, the power equation is

$$Q_H = Q_C + I_{TEC}^2 R, \tag{13.2}$$

where $R$ is the TEC total resistance in ohms of the serial P-type and N-type elements, and $I$ is the TEC dc drive current. TEC coolers are used in butterfly laser packages and recently were used in small form factor (SFF) and SFP transceivers. Figure 13.2 shows a typical TEC structure. TEC operation can be analyzed by differential equations:[9,10,16]

$$\begin{cases} q_p = \alpha_p I \cdot T - k_p A \dfrac{dT}{dx}, \\ q_n = -\alpha_n I \cdot T - k_n A \dfrac{dT}{dx}, \end{cases} \tag{13.3}$$

where $\alpha_p$ and $\alpha_n$ are the Seebeck coefficients in $WK^{-1}A^{-1}$ of the P and N materials, $k_p$ and $k_n$ are their thermal conductivities in $Wm^{-1}K^{-1}$, $I$ is the TEC

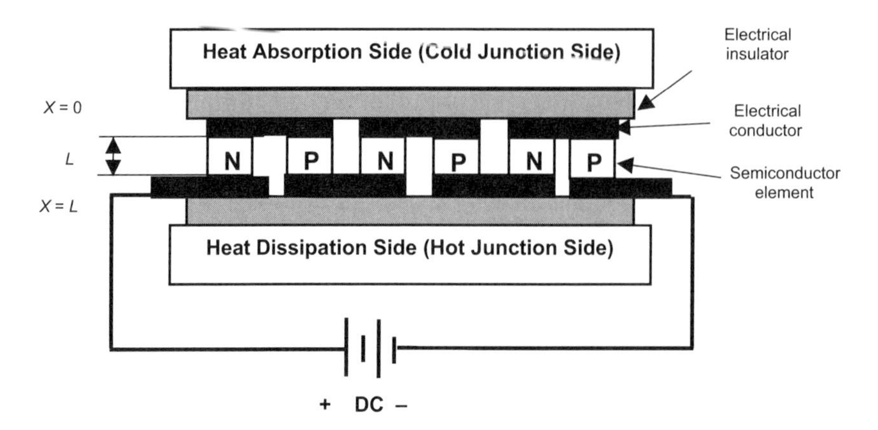


**Figure 13.2** Typical cross section of a TEC module.

current, $A$ is the semiconductor pellet is cross sectional area, and $T$ is temperature in Kelvin. The coefficient $\alpha_n$ is a negative quantity and the thermoelectric heat flow from the source through adjacent P and N branches, as shown in Fig. 13.2, is positive and opposed to the thermal conduction.

Within the branches, the rate of heat generation per unit length ($Js^{-1}m^{-1}$) from the Joule effect is given by

$$\begin{cases} -k_pA\dfrac{d^2T}{dx^2} = \dfrac{I^2\rho_p}{A}, \\ -k_nA\dfrac{d^2T}{dx^2} = \dfrac{I^2\rho_n}{A}, \end{cases} \tag{13.4}$$

where $\rho_p$ and $\rho_n$ are the electrical conductivities ($\Omega$.m) of P and N, respectively. Assume that the branches length is $L$; then, the boundary conditions are $T = T_C$ at $x = 0$, and $T = T_H$ at $x = L$, which is the heat sink. Solving the pair of differential eq. given in Eq. (13.4) results in[9]

$$\begin{cases} q_p = \alpha_p IT_C - k_pA\dfrac{k_pA(T_H - T_C)}{L} - \dfrac{I^2\rho_pL}{A}, \\ q_p = \alpha_n IT_C - k_pA\dfrac{k_nA(T_H - T_C)}{L} - \dfrac{I^2\rho_nL}{A}. \end{cases} \tag{13.5}$$

The cooling power $Q_C$ at the sink is the sum of the two equations and is given by

$$Q_C = \alpha IT_C - K(T_H - T_C) - \frac{1}{2}I^2R, \tag{13.6}$$

where $\alpha = \alpha_p - \alpha_n$ is known as the differential Seebeck coefficient of the unit. The thermal conductance $K$ of the two parallel branches can be written as

$$K = \frac{k_pA}{L} + \frac{k_nA}{L}, \tag{13.7}$$

and the electrical resistance is the series resistance on the N and P materials is

$$R = \frac{\rho_p L}{A} + \frac{\rho_n L}{A}. \quad (13.8)$$

Equation (13.6) demonstrates the cooling process by stating that the net cooling is the difference between the reversible cooling process, which is the first term, and the irreversible terms identified as the thermal potential on the branches and ohm losses, which are the second and the third terms in Eq. (13.6). Using Eq. (13.8), the electrical power consumed in the P and the N branches can be written as

$$\begin{cases} P_p = \alpha_P I(T_H - T_C) + \dfrac{I^2 \rho_p L}{A}, \\ P_n = -\alpha_n I(T_H - T_C) + \dfrac{I^2 \rho_n L}{A}. \end{cases} \quad (13.9)$$

The total electric power is the sum of the power at P and N resulting in

$$P = \alpha I(T_H - T_C) + I \cdot R^2. \quad (13.10)$$

According to the first law of thermodynamics, the heat dissipated at the heat sink is the sum of Eqs. (13.6) and (13.10), which results in

$$Q_H = \alpha I T_H - K(T_H - T_C) + \frac{1}{2} I^2 R. \quad (13.11)$$

Equation (13.11) is the same as Eq. (13.2), when substituting Eq. (13.6). Having all the thermal and power relations, the system efficiency is examined by the refrigeration coefficient of performance (COP):

$$\mathrm{COP} = \frac{Q_C}{P} = \frac{\alpha I T_C - K(T_H - T_C) - \frac{1}{2} I^2 R}{\alpha I \cdot (T_H - T_C) + \frac{1}{2} I^2 R} > 1. \quad (13.12)$$

Maximum COP is accomplished at $\partial(\mathrm{COP})/\partial I = 0$, which results in optimum values for COP current and provides the TEC figure of merit:

$$\mathrm{COP}_{max} = \frac{T_C\left[\sqrt{1 + ZT_m} - T_H/T_m\right]}{(T_H - T_C)\left[\sqrt{1 + ZT_m} + 1\right]}, \quad (13.13)$$

where the figure merit of TEC is given by

$$Z = \frac{\alpha^2}{KR}, \quad (13.14)$$

and $T_m$ is the average temperature, which is given by

$$T_m = \frac{T_H + T_C}{2}. \quad (13.15)$$

### 13.2.2 TEC control types

Basically, there are two types of temperature controls: thermostatic and steady state. With thermostatic control, the thermal load is maintained between two temperature limits, for instance, 27–30°C. The system would continually vary between the two limits. The difference between the two limits is defined as the system hysteresis. This is done by switching the current to the TEC module.

Whenever a system must be maintained within tight limits, for example, a laser that should not drift above 0.05% of its wavelength, a continual temperature control should be considered. The system is locked on a set-point temperature with very little variation around it. That means a feedback-control loop is required by using a TEC controller. If the steady-state condition is suddenly disrupted by a sudden change in the ambient conditions, the control circuit will correct the system and bring it back to its steady state by minimizing the error voltage created at the error amplifier or loop integrator. Figure 13.3 describes a system plot that is locked to 15°C by a proportional TEC controller. The proportional controller creates a correction voltage to power the TEC module, proportional to the error between the set point and the temperature sensor. This can be described in the Laplace domain as the output of the error amplifier by

$$[V_{\text{sens}}(s) - V_{\text{set}}]K_A = V_{\text{TEC}}(s). \tag{13.16}$$

It can be observed that there is a certain time to converge to the minimum required error voltage. In case the ambient temperature is higher than the set point and is increasing, the error voltage would increase. That would reduce the system's accuracy. Generally, such controllers are not suitable for wavelength control and their temperature accuracy is within tenths of a degree. The conclusion from the above mentioned description of operation, is that a proportional controller hardly amplifies

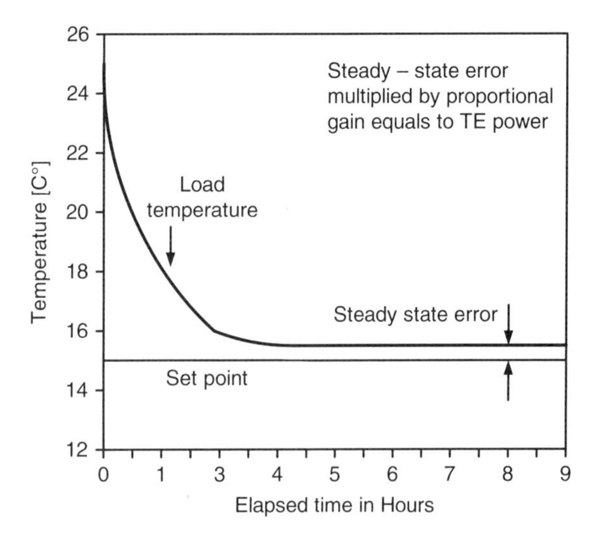


**Figure 13.3** Steady-state error.

the error voltage is between the set point and the temperature sensor. Hence, there is a need to increase the error amplifier gain, which is the loop integrator. Increasing the error amplifier gain minimizes the error voltage at its inputs and increases the TEC loop accuracy. That, of course, creates overshoots and undershoots, causing oscillations around the temperature set point as demonstrated in Fig. 13.4.

As was explained in Chapter 12, power control and AGC feedback control system stability should be optimized to have sufficient phase and gain margins. A TEC controller that uses a proportional integrator (PI) amplifier is known as a "PI" controller. Unfortunately, such a feedback loop is not stable enough since it has a single pole at the origin; therefore, its phase margin response at the zero crossing is not flat. (An elaborate stability analysis theory was given in Sec. 12.2.4.) Moreover, the PI TEC control system is useful only at the steady state, but it cannot stabilize the environmental fluctuations well enough. This disadvantage would result in fluctuations in the steady-state error voltage, as shown in Fig. 13.5. To overcome stability problems in a single integrator loop, a derivative amplifier is often employed to bring the feedback system parameters to a more optimal stable point.[12] The proportional amplifier at such a system controls the amount of change and the derivative amplifier controls the rate of change. This way, the overshoot response is prevented. The derivative amplifier therefore operates as a charge pump. In case there is a large error due to a sudden thermal load change, the derivative amplifier would increase the current through the TEC. As the TEC gets closer to its set point, the derivative amplifier would reduce its current to the TEC, preventing overshoot and ringing. The higher the gain of the derivative amplifier, the greater its instantaneous response to change. There is a certain amount of maximum gain utilized by the derivative amplifier. It should compensate for sudden disruptions but not overcompensate for the system and

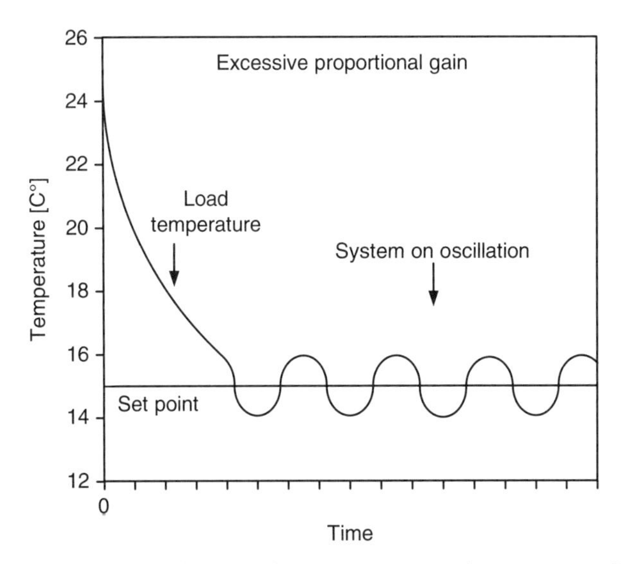


**Figure 13.4** TEC oscillation of a PI TEC control system due to nonstability caused by very high gain of the error amplifier.

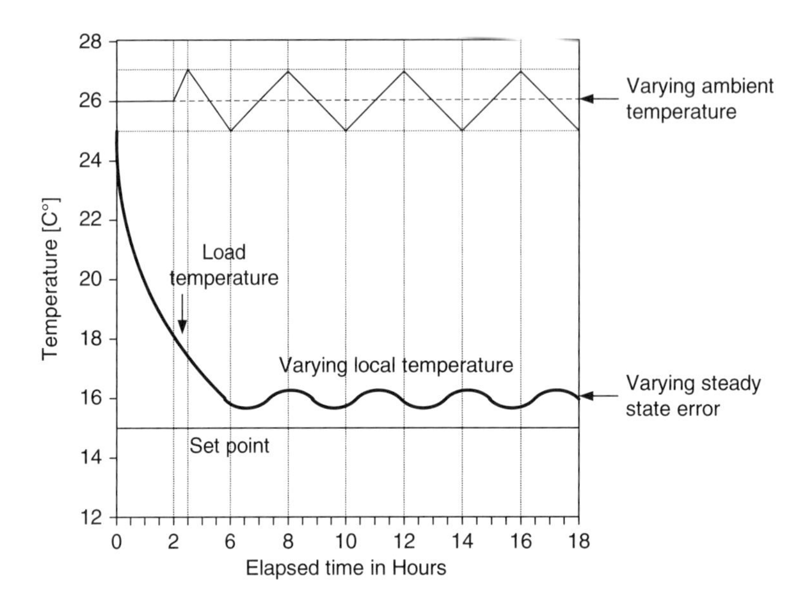


**Figure 13.5** Ambient temperature fluctuations in a proportional integral TEC circuit are minimized over the thermal controlled load, but some variations remain.

bring it to oscillations. A controller that uses both a proportional integral amplifier and a derivative amplifier is called a proportional integral derivative (PID). This is the most common steady-state temperature controller.

There are two approaches to realize a PID control loop to monitor the amount of dc power delivered to the TEC:

1. Pulse width modulation (PWM), which appears in Fig. 13.6. In this method, the pulse width is proportional to the temperature difference between the set point and the actual laser temperature. The higher the error voltage, the wider the current pulse driving the TEC module. As the temperature stabilizes and the error voltage is reduced, the pulse width becomes narrower. The disadvantage of this method is EMI, emissions, and large inductors required for the MOSFET's current.

2. Linear control using classic negative feedback. This method is relatively simple but generates a lot of heat on the current drive transistors for the TEC module. PWM's control circuit operation is straightforward (Fig. 13.6). When the laser temperature is too high, the negative thermal coefficient (NTC) resistance sensor generates an error voltage at the temperature-sensing circuit. Generally, a bridge circuit is used (Fig. 13.7), so the reference set point is the point of the balanced bridge where the error voltage is zero. The advantage of this sampling method is that the error voltage polarity indicates the temperature error state; i.e., the device temperature is above or below the reference. Since the bridge is unbalanced, an error voltage is created at the error amplifier or integrator. That error voltage

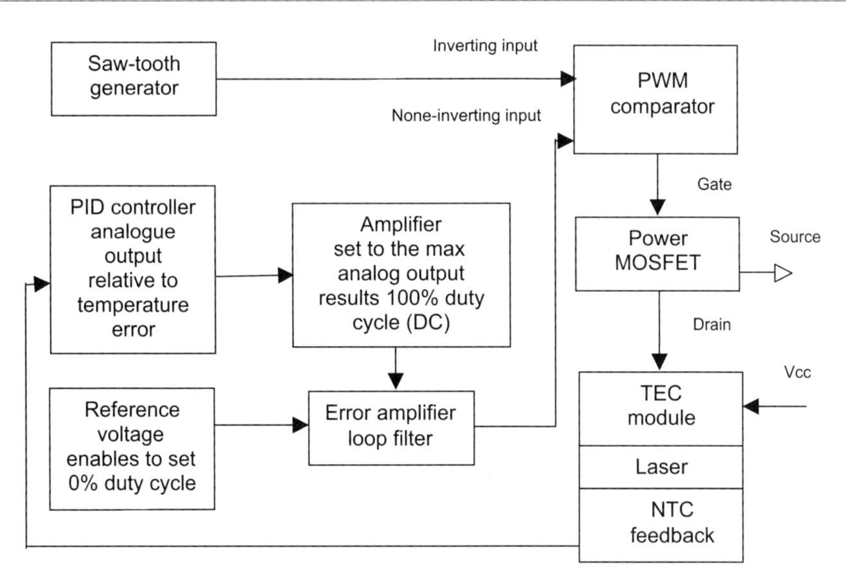


**Figure 13.6** PWM-TEC-control circuit. Note that the NTC is the feedback, and the loop filter can be the same as Fig. 13.7(a), whereas the error-correction signal is the pulse controlled by the integrator's voltage level. This replaces Q1 in Fig. 13.7.

level serves as a reference at a level comparision of the pulse width in the following manner. The feeding saw-tooth wave is the level comparator–inverting input. As long as the sweep voltage is below the error level, the output of the comparator is high level. The instant the sweep saw-tooth level goes above the error voltage, the output becomes zero. As a consequence, when the temperature difference to the set point is the highest, the duty cycle would be 100%. This means that the TEC module is continually driven by a dc level. As the error voltage decreases due to temperature correction, the duty cycle is reduced and a pulse train is created at the input of the error amplifier loop filter. As the error is minimised (strictly speaking, there is no zero error owing to the ambient temperature fluctuations), the reference level at the comparator is minimum. As a result, the pulse is narrow or the duty cycle is low. The dc pulses created by the comparator are fed into a power MOSFET gate, which controls the current fed into the TEC by switching the VCC of the TEC according to the duty cycle created by the comparator. Today's controllers are digital and programmed for the desired temperature lock via a $I^2C$-bus interface. This way a DAC output voltage is set to determine the reference level for target temperature.

A TEC-loop analysis is similar to the AGC analysis. In both cases, it is a power-control loop. A typical AGC controls the RF level and the detection is done by an RF detector. In APC, the loop controls the amount of heat over the laser by monitoring the temperature with NTC. In AGC, the power control is done by a voltage variable attenuator (VVA) or digital controlled attenuator

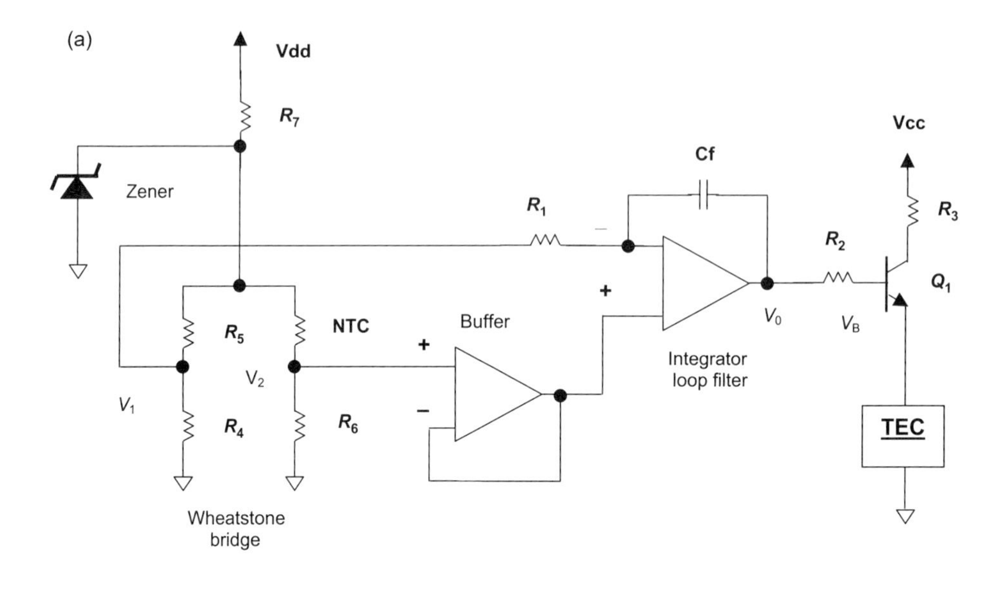


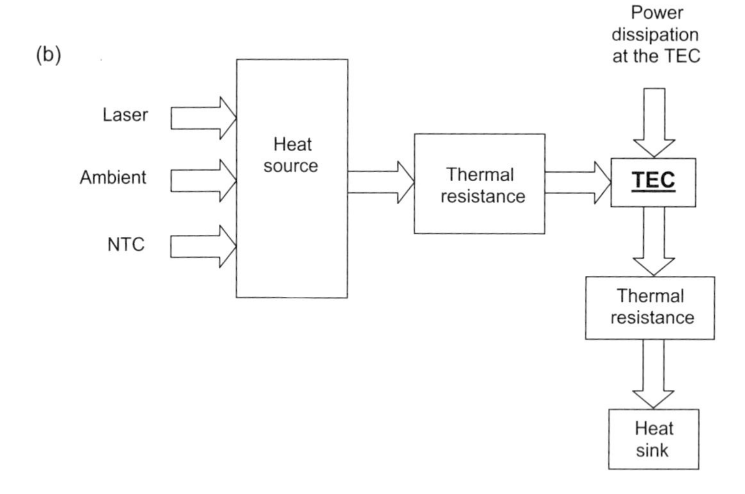


**Figure 13.7** (a) A negative-feedback TEC loop using the balanced-bridge method, (b) thermal balance.

(DCA), which, in a TEC-loop analogy, is the TEC. The heat load over the laser is given by

$$Q = mCT = \Gamma V_{\text{bias}} \cdot I_{\text{dc-bias}} = \Gamma I^2_{\text{dc-bias}} R_{\text{ON}}, \tag{13.17}$$

where $Q$ is the heat energy in calories or Joules, $m$ is the thermal mass in kilograms, $C$ is the heat coefficient of the object in cal/kg $\times$ $K$, $T$ is temperature in Kelvin, $V_{\text{BIAS}}$ is the laser voltage at the on state, $I_{\text{CD-BIAS}}$ is the laser current at the bias point, $R_{\text{ON}}$ is the laser resistance at the on state typically, and 4–6 $\Omega$ $\Gamma$ is a conversion constant from watts to thermal units. The amount of heat needed to be removed from the laser is given by

$$\Delta Q(s) = mC(T_2 - T_1), \tag{13.18}$$

where $T_2$ and $T_1$ are the final and initial temperatures in Kelvin. For slow fluctuations around the equilibrium setting point, $T$, Eq. (13.17) can be written as

$$Q(s) = mC[T + \Delta T(s)]. \tag{13.19}$$

Equation (13.18) describes the potential difference between the laser and the TEC at the steady state where $T(s)$ and $Q(s)$ are the temperature and heat fluctuations in the Laplace domain, respectively. Moreover, the presentation of $Q$ as a function of mass provides a ballpark figure for the amount of heat.

As was explained in Chapter 12, the most accurate leveling method is done by feedback, where the feedback can be made by an NTC resistor. NTC characteristics are given by

$$R_T = R_0 \exp\left[\beta\left(\frac{1}{T_2} - \frac{1}{T_1}\right)\right], \tag{13.20}$$

where $\beta$ is the temperature coefficient, $R_0$ is the room temperature resistance of the NTC, $T_1$ is the initial state temperature, and $T_2$ is final state temperature in Kelvin. Such a temperature control system is described in Fig. 13.7. The NTC is part of a balanced Wheatstone bridge. When the temperature starts to drift from the set point, the bridge goes out of balance, an error voltage at the loop integrator brings back the TEC to its temperature, and the bridge-error voltage is minimised.[3]

The analog of an AGC RF detector and its power sensing coefficient is the NTC thermistor constant, which is given by $\partial R_T/\partial T$. The transfer function of temperature changes to resistance change and error voltage of the bridge is identical to the RF detector RF power to voltage curve. Thus, $\partial V_{RF}/\partial P \Leftrightarrow \partial V_T/\partial T$. In this manner, the bridge-error voltage as a function of temperature can be calculated. An additional parameter required is the thermal resistance between the cold-side heat sink and the NTC, which is equivalent to heat coupling or temperature sampling, since in AGC, there is an RF coupler. With these factors, a similar analysis can be performed for a TEC loop using the AGC modeling provided in Chapter 12. The TEC-control-loop block diagram is shown in Fig. 13.8, where "$s$" indicates the Laplace domain. The goal in solving a TEC-control loop is to find the transfer function between the fluctuations of $Q(s)$ and the fluctuations of the control current $I(s)$.[15] This can be solved in the same manner as was done for AGC in Chapter 12, by using the thermal relations presented here.

From Fig. 13.7, the error voltage is given by $V_{ERR} = (V_2 - V_1)$. The reference voltage is determined by the voltage divider created by $R_5$ and $R_4$, denoted as $V_1$. The error or temperature monitoring is created by the $V_2$ created by the voltage divider of $R_6$ and $R_{NTC}$. In case of overheating, the NTC decreases its resistance, $V_2$ goes lower, the error voltage has a positive sign, the noninverting integrator

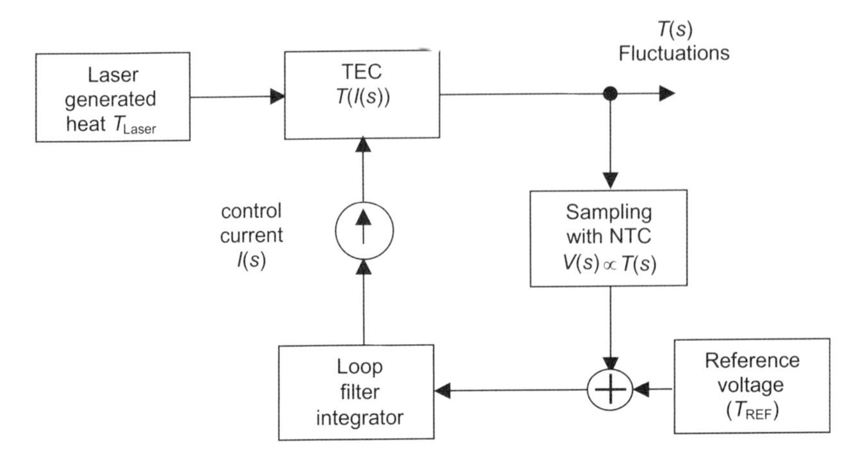


**Figure 13.8** Negative feedback small signal TEC loop block diagram.

increases the base current to $Q_1$, and the temperature goes down. An overcooling operation has an opposite behavior. The zener diode is used to stabilize the bridge voltage.

In order to isolate the error voltage, an operational amplifier is used as a buffer, while the other operational amplifier serves as an integrator.

The output voltage from the integrator at Fig. 13.7(a) is given by

$$V_0(S) = V_2(S) + \frac{1}{SC_f R_1}[V_2(S) - V_1]. \tag{13.21}$$

The TEC drive current is given by

$$I_E(s) = \beta \frac{V_0(s) - V_B(s)}{R_2}, \tag{13.22}$$

where

$$V_B(s) = I_E(s) R_{TEC} + V_{BE}. \tag{13.23}$$

The thermal voltage in the Laplace domain is given by

$$V_2[T(s)] = V_Z\left(\frac{R_6}{R_6 + R_{NTC}[T(s)]}\right). \tag{13.24}$$

The temperature detection constant $K_T$ at the operating point $T_{CS}$ is given by

$$K_T = \frac{\partial V_2(T)}{\partial T}\Big|_{@T_{CS}}. \tag{13.25}$$

Note that the TEC dynamic range is determined by the difference between the laser temperature and the target temperature. The noncooled laser temperature denoted as "laser temperature" is not $T_H$, it is the heat dissipating to the heat sink. Also note that at steady state the laser temperature is kept low; however, the TEC hot side $T_H$ behaves according to the first law of thermodynamics, per Eqs. (13.2) and (13.11).

The TEC control constant $K_{TEC}$ at the operating point $I_{E\_CS}$ is given by

$$K_{TEC} = \frac{\partial TEC_2(I_E)}{\partial I_E}|_{@I_{E\_CS}}. \tag{13.26}$$

The thermal loads in watts are given by

$$Q_{laser} = I_{bias}^2 \cdot R_{ON}, \tag{13.27}$$

$$Q_{NTC} = \frac{V_2^2}{R_{NTC}}, \tag{13.28}$$

and the ambient radiation due to the temperature gradient between the laser and the ambient temperature is given by[13]

$$Q_{RAD} = F \cdot e \cdot s \cdot A_{TEC}\left(T_{amb}^4 - T_C^4\right), \tag{13.29}$$

where $T_{amb}$ is the ambient temperature, $T_C$ is the TEC cold temperature, $A_{TEC}$ is the area of the TEC cold surface, $s$ is the Stephan–Boltzmann constant of $5.667 \times 10^{-8}$ W/m$^2$K$^4$, $e$ is emissivity (for the worst case value $e = 1$), and $F$ is shape factor (for theworst case value $F = 1$). Other heat loads, such as the NTC cold-side radiation as well as air conduction between the hot and cold side TEC plates can be added too.

The TEC hot side accumulative heat should be cooled by a heat sink, with a surface area $A_{HS}$ and should be transferred by radiation where SFP devices as well as other small package components are used. This value is given by

$$Q_{HS} = F \cdot e \cdot s \cdot A_{HS}\left(T_H^4 - T_{amb}^4\right). \tag{13.30}$$

Direct contact heat conduction is given by

$$Q_{cond} = \Delta T \frac{\Theta A}{L}, \tag{13.31}$$

where $\Theta$ is the thermal conductivity in W/mC, $A$ is the cross-sectional area of the material in square meters, and $L$ is the length path in meters. The thermal resistance $R_\Theta$ is given by

$$R_\Theta = \frac{\Theta A}{L}. \tag{13.32}$$

### 13.2.3 TEC models for simulation

Using the thermal equations given above, simulation program with integrated circuit emphasis (SPICE) modeling can be derived to simulate heat conduction.[14]

**Table 13.2** Analogy of thermal to electrical.

| Thermal quantities | Units | Analogous electrical quantities | Units |
|---|---|---|---|
| Heat $Q$ or $q$ | W | Current $I$ | A |
| Temperature $T$ | K | Voltage $V$ | V |
| Thermal resistance $\Theta$ | K/W | Resistance $R$ | $\Omega$ |
| Heat capacity $C$ | J/K | Capacity $C$ | F |
| Absolute zero temperature | 0 K | Ground | 0 V |

This model is useful for whole TEC-control-loop simulation in time and frequency domain. As was explained previously, there are five energy processes that take place in a thermoelectric module (TEM). They are conductive heat transfer, Joule heating, Peltier cooling/heating, Seebeck power generation, and the Thompson phenomenon. All these processes account for the interrelations between thermal and electrical energies. In order to proceed and create the SPICE modeling, thermal to electrical analogies should be established. Table 13.2 provides the conversion.

The analogy is simple: heat flows from higher temperature to lower temperature through the material. Each material has its characteristic heat conduction. Heat is accumulated in the material and its accumulation depends upon the mass and the specific heat coefficient of the material. Therefore, the following analogies per Table 13.2 are applicable. Temperature is potential and the temperature difference becomes voltage across the thermal conductor. Heat flows through the material; thus, it is similar to current. Heat accumulation is the charge and the capacity to accumulate heat is capacitance. Thus it has the same phenomenon as a capacitor, i.e., decay of charge. The ability to transfer heat is thermal resistance, which is similar to a resistor. The absolute 0 K is zero potential, similar to zero volts. Hence, it is denoted as ground (GND).

The next step is to establish a SPICE model based upon the relations in Sec. 13.2.1 with a slightly different notation. According to the first law of thermodynamics, the energy equilibrium for the heat-absorbing side, denoted with index $a$, is given by

$$q_{\mathrm{a}} = \frac{\Delta T}{\Theta_{\mathrm{m}}} + \alpha_{\mathrm{m}} T_{\mathrm{a}} I - \frac{I^2 R_{\mathrm{m}}}{2}. \tag{13.33}$$

For the emitting side,

$$q_{\mathrm{e}} = \frac{\Delta T}{\Theta_{\mathrm{m}}} + \alpha_{\mathrm{m}} T_{\mathrm{e}} I + \frac{I^2 R_{\mathrm{m}}}{2}. \tag{13.34}$$

Note that the absorbing side is the cooling side, whereas the emitting side is the warm side that releases heat to the environment through a heat sink. The factors in

Eqs. (13.33) and (13.34) are provided by

$$\alpha_m = \alpha N, \tag{13.35}$$

$$R_m = RN, \tag{13.36}$$

$$\Theta_m = \Theta/N, \tag{13.37}$$

where $N$ is the number of couplers between the cold and hot side, per Fig. 13.2, $T_a$ and $T_e$ are temperatures of absorbing and emitting sides in Kelvin, and $\Theta$ is thermal resistance of a single coupler. There are $N$ couplers in parallel that create an equivalent thermal resistance $\Theta_m$. $R$ is the electrical resistance of the TEC, $\alpha$ is Seebeck coefficient, and $\Delta T = (T_e - T_a)$. It is conventional to leave out the Thompson effect as it is negligibly small. The electrical part of the module is described as an electrical resistance and a potential difference $V$:

$$V = \alpha_m T_e - \alpha_m T_a = \alpha_m \Delta T. \tag{13.38}$$

Using the above, TEC parameters can be driven as well as their 1D equivalent circuit based on the Cauer network C–Θm–C, which is presented in Fig. 13.10.[14,17,18] The current sources show Joule heating of the TEC, $q_j$, Peltier cooling on the heat-absorbing side TEC, $q_{pa}$, and Peltier heating on the heat-emitting side of the TEC, $q_{pe}$. The electrical part consists of a voltage source $V_s$ and electrical resistance $R_m$. All capacitors have an initial charge of IC = $T_{amb}$, which defines the start conditions of the ambient temperature potential.

A modified circuit is provided in Fig. 13.10.[14] It is based on the previous suggestion, but instead of three dependent current sources, there are only two. In addition, it is a lumped model rather than a distributed one, thus, a simpler model that provides an intuitive understanding of TEC.

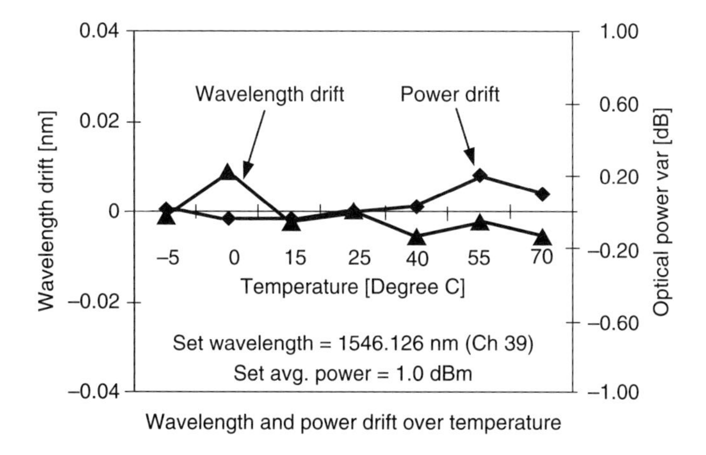


**Figure 13.9** Wavelength and power drift over temperature in an SFP unit with TEC and APC loops.[8] (Reprinted with permission from the *Optical Communications* © IEEE, 2005.)

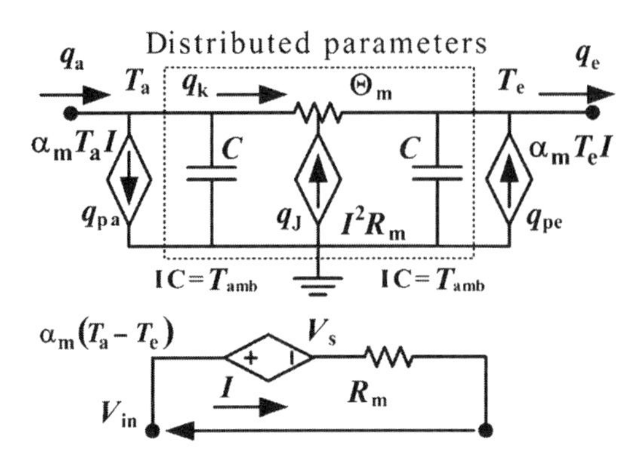


**Figure 13.10** TEC-equivalent circuit based on a Cauer-type network for describing heat transfer in a solid with internal heat sources ($q_j$), $q_{pa}$, and $q_{pe}$ (Peltier cooling and heating). The $V_s$ voltage source describes Seebeck power generation. IC is the initial temperature of the device, $T_{amb}$.[14] (Reprinted with permission from the *Power Electronics Specialists* © IEEE, 2005.)

Manufacturers of TEC use the following parameters to specify their product:

- $\Delta T_{max}$ is the largest temperature differential in Kelvin that can be obtained between the hot and the cold ceramic plates of the TEC for a given $T_h$, the temperature of the hot side.
- $I_{max}$ is the input current in $A$, that will produce the maximum possible $\Delta T$ across the TEM.
- $V_{max}$ is the dc voltage, in V, that will deliver maximum possible $\Delta T$ at the supplied $I_{max}$.
- $Q_{max}$ is the maximum amount of heat, in W, that can be absorbed at the TEC's cold plate with $I_{max}$ and $\Delta T = 0$. It is important to emphasize that $Q_{max}$ in not the maximum possible amount of heat that can be handled by the TEC; rather, it is the heat flow that corresponds to the current $I_{max}$.
- $Q_{opt}$ is the maximum amount of heat that can be absorbed at the TEC's cold plate for $\Delta T = 0$. $Q_{opt}$ is larger than $Q_{max}$. Some manufacturers apply the notation $Q_{max}$ instead of $Q_{opt}$. Hence, one should pay attention to the data sheet descriptions.

Using the aforementioned equations and Sec. 13.2.1, the characteristic parameter equations can be derived:

$$\Delta T_{max} = T_h + \frac{\left(1 - \sqrt{1 + 2T_h Z}\right)}{Z}, \tag{13.39}$$

$$I_{\mathrm{max}} = \frac{\sqrt{1 + 2T_{\mathrm{h}}Z} - 1}{\alpha_{\mathrm{m}}\Theta_{\mathrm{m}}}, \tag{13.40}$$

$$V_{\mathrm{max}} = \alpha_{\mathrm{m}}T_{\mathrm{h}}, \tag{13.41}$$

$$Q_{\mathrm{max}} = \frac{\sqrt{1 + 2T_{\mathrm{h}}Z}\left(\sqrt{1 + 2T_{\mathrm{h}}Z} - 1\right)^2}{2\Theta_{\mathrm{m}}Z}, \tag{13.42}$$

$$I_{\mathrm{opt}} = \frac{\alpha_{\mathrm{m}}T_{\mathrm{h}}}{R_{\mathrm{m}}}, \tag{13.43}$$

$$Q_{\mathrm{opt}} = \frac{\alpha_{\mathrm{m}}^2 T_{\mathrm{h}}^2}{2R_{\mathrm{m}}}, \tag{13.44}$$

where $Z$ is the TEC's figure of merit given by

$$Z = \frac{\alpha_{\mathrm{m}}\Theta_{\mathrm{m}}}{R_{\mathrm{m}}}. \tag{13.45}$$

Applying Eqs. (13.39)–(13.45), the data set of the SPICE model in Fig. 13.10 is derived:

$$R_{\mathrm{m}} = \frac{V_{\mathrm{max}}}{I_{\mathrm{max}}} \frac{(T_{\mathrm{h}} - \Delta T_{\mathrm{max}})}{T_{\mathrm{h}}} [\Omega], \tag{13.46}$$

$$\Theta_{\mathrm{m}} = \frac{\Delta T_{\mathrm{max}}}{I_{\mathrm{max}}V_{\mathrm{max}}} \frac{2T_{\mathrm{h}}}{(T_{\mathrm{h}} - \Delta T_{\mathrm{max}})} \left[\frac{K}{W}\right], \tag{13.47}$$

$$\alpha_{\mathrm{m}} = \frac{V_{\mathrm{max}}}{T_{\mathrm{h}}} \left[\frac{V}{K}\right]. \tag{13.48}$$

Like AGC, the TEC loop is a nonlinear system[15] whose transfer function is affected by each bias point. The SPICE model is a small signal approximation at the operating point. Thus, it can be useful for determining the frequency response of the transfer function. The detailed TEC loop analysis is given in Ref. [15].

Figure 13.11 describes the experimental system of a TEC with a thermal load of two massive aluminum plates. The system is thermally insulated. There are two thermocouplers inserted into the thermal load for temperature measurement. The set-up permits response measurement of the system to a sine wave voltage input. Sweeping the frequency provides the frequency response. A simulation tool can be OrCAD or personal computer SPICE (PC-SPICE), using the suggested TEC equivalent circuit provided in Fig. 13.10. The gain and phase of the TEC system can be evaluated by simulation. Simulation results and model accuracy are provided in Fig. 13.12.

### 13.2.4 Wavelength locking by a wavelength-selective photodetector

In previous sections, wavelength control using a TEC loop was done with NTC thermistors and comparing the temperature-related voltage to a reference

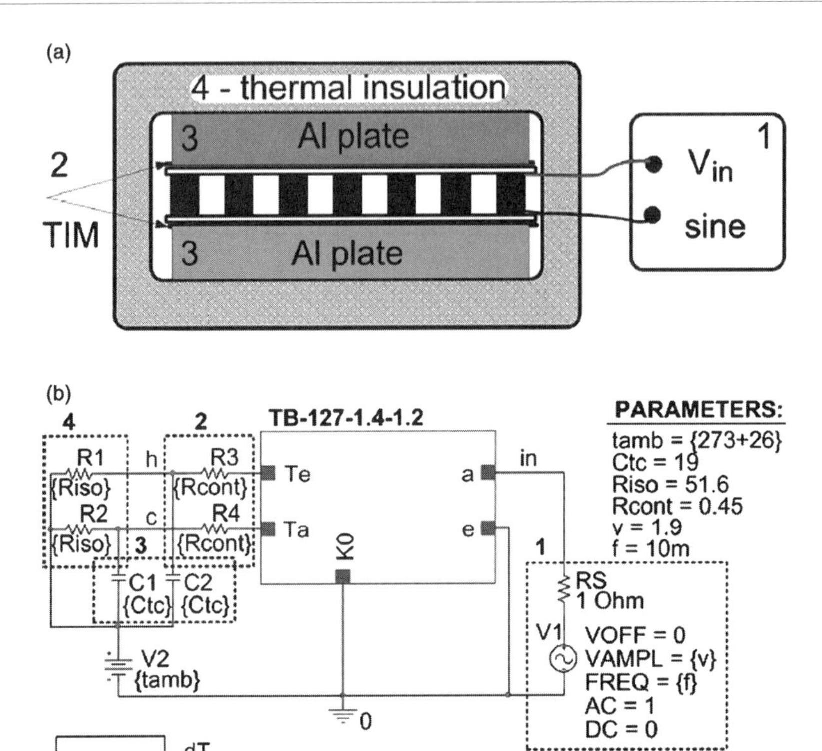


**Figure 13.11** Measurement of the small-signal transfer function of the system: TEC sandwiched between two massive aluminum plates with built-in thermocouples. (a) Experimental setup. (b) PSPICE/OrCAD simulation scheme. This model of the TEC has the following parameters: $\alpha_m = 0.053$ V/K, $R_m = 1.6\ \Omega$, $\Theta_m = 1.5$ K/W calculated above, 1: sine voltage source, 2: thermal interface material (TIM), 3: aluminum plates.[14] (Reprinted with permission from the *Power Electronics Specialists* © IEEE, 2005).

voltage that guarantees the desired wavelength. A different approach for wavelength stabilization may be considered by using a voltage-tunable wavelength-selective photodetector (VT-WSPD).[19] In this case, the wavelength monitoring is done by a wavelength detector rather than by indirect wavelength stabilization using temperature control with NTC.

A wavelength-selective photodetector (WSP) structure is made of back-to-back coupled PIN and NIP photodetectors, where the NIP is an upside down PIN. The photodiodes exhibit distinct but adjacent spectral responses. The latter characterization can be obtained by employing materials with distinct absorption spectra, i.e., different quaternary III–V compounds in the intrinsic layer. Because the diodes are connected back-to-back, the overall photocurrent is the difference between the individually generated carrier flows from each of the two diodes. Note that in such a structure, as illustrated in Fig. 13.13, the upper diode is a reversed bias PIN and the lower diode is a forward biased NIP. Chapter 8 explains that PIN PD is reversed biased for improving RF BW and reducing distortions.

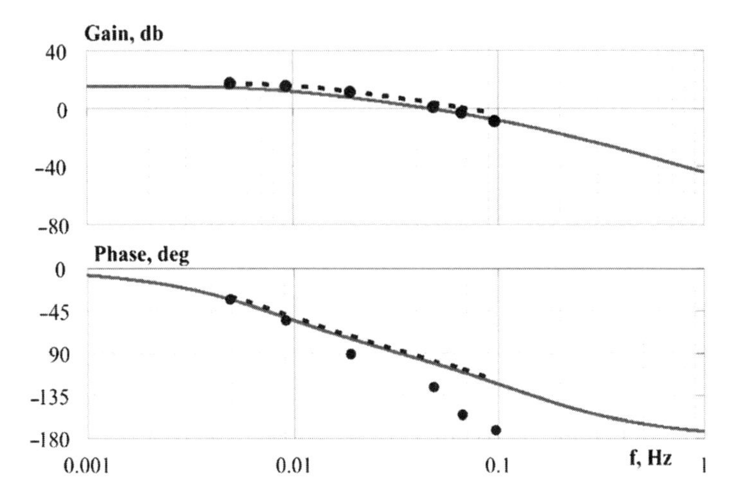


**Figure 13.12** Bode plot of PSPICE/OrCAD simulation circuit. Simulation parameters: $R_m = 1.6\ \Omega$, $\Theta_m = 1.5$ K/W (K stands for Kelvin), and $\alpha_m = 0.053$ V/K. Dashed lines are results of cycle-by-cycle transient simulation with sine voltage amplitude 1.3 V. The solid line is simulation and points are measured experimental results taken on test set-up in Fig. 13.11.[14] (Reprinted with permission from the *Power Electronics Specialists* © IEEE, 2004.)

The design goal of a WSP is to exhibit a linear photocurrent versus wavelength chart; i.e., $I/\lambda$ = constant. It is also desirable that the straight-line chart will have zero crossing at the desirable wavelength $\lambda_0$. Due to the double-junction structure of the WSP, the wavelength is voltage selectable by an external bias, making the WSP a VT-WSPD.

The double photodiode structure is shown in Fig. 13.13 on the right-hand side. The layers are arranged in "n–i–p–i–n" sequence, corresponding to the NIP and

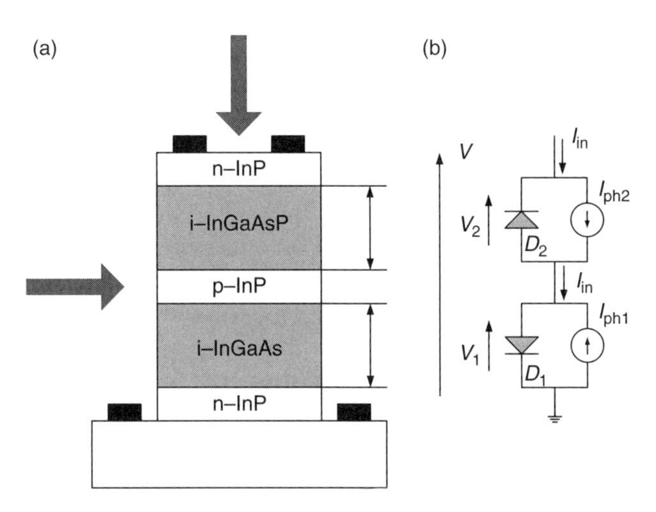


**Figure 13.13** WSP structure. (a) Device structure and incident scenarios showing normal and lateral incidences. (b) An equivalent circuit.[19] (Reprinted with permission from the *Journal of Lightwave Technology* © IEEE, 2003.)

the PIN denoted as D1 and D2, respectively. The material of choice for operating around 1550 nm is the III–V quaternary InGaAsP, which can be epitaxially grown on InP substrates. The heavily doped films are realized in p– or n–InP (transparent near infrared), while the two intrinsic layers consist of $In_{1-x}Ga_xAs$ and $In_{1-x}Ga_xAs_yP_{1-y}$, respectively, which exhibit different spectral coverages. Specifically, $In_{0.53}Ga_{0.47}As$ is lattice matched to InP and employed in D1 to obtain the largest cutoff at $\lambda_{gap} = 1680$ nm. $In_{1-x}Ga_xAs_yP_{1-y}$ is used for D2 and provides a wavelength-working range of 950 nm $\leq \lambda_{gap} \leq$ 1680 nm. Notice that the lattice match with InP can be obtained for any $y$ $(0 < y < 1)$, provided $x \approx 0.47y$.

Analysis of the circuit in Fig. 13.13(b) is done by using Kirchhoff's voltage and current laws (KVL) and (KCL), and by using the diode current equation.[20,21] The basic equations are:

$$V = V_1 + V_2, \tag{13.49}$$

$$I_{D1} = I_{01} \cdot \left[1 + \exp\left(\frac{V_1}{V_T}\right)\right], \tag{13.50}$$

$$I_{D2} = I_{02} \cdot \left[1 + \exp\left(\frac{-V_2}{V_T}\right)\right], \tag{13.51}$$

$$I_{in} = I_{ph2} + I_{D2} = -I_{ph1} + I_{D1}, \tag{13.52}$$

where $I_{ph1}$ and $I_{ph2}$ are D1 and D2 photocurrents, $I_{01}$ and $I_{02}$ are D1 and D2 reverse leakage currents, and $V_T$ is the thermal Boltzman voltage $(Kt)/q$. The solutions to these equations provide the input current to the WSP:

$$\begin{aligned} I_{in} = {} & I_{ph2}\left[1 + \frac{I_{02}}{I_{01}}\exp\left(\frac{-V}{V_T}\right)\right]^{-1} - I_{ph1}\left[1 + \frac{I_{01}}{I_{02}}\exp\left(\frac{V}{V_T}\right)\right]^{-1} \\ & + I_{02}\left[1 + \frac{I_{02}}{I_{01}}\exp\left(\frac{-V}{V_T}\right)\right]^{-1} \cdot \left[1 - \exp\left(\frac{-V}{V_T}\right)\right]. \end{aligned} \tag{13.53}$$

The leakage currents are dominated by the generation of space–charge layers with thicknesses $d_1$ and $d_2$, and are evaluated by[19]

$$\begin{cases} I_{01} = \dfrac{q \cdot n_{i1}(T) \cdot d_1 \cdot A_1}{\tau_1}, \\ I_{02} = \dfrac{q \cdot n_{i2}(T) \cdot d_2 \cdot A_2}{\tau_2}, \end{cases} \tag{13.54}$$

where $A_1$ and $A_2$ are the device areas, and the carrier lifetimes $\tau_1$ and $\tau_2$ are assumed to be equal and are within the magnitude of μs. The intrinsic carrier concentrations $n_{i1}$ and $n_{i2}$ are calculated taking into account the temperature dependence:

$$n_i(T) = k_n \cdot T^{\frac{3}{2}} \cdot \exp\left[\frac{-E_g(T)}{2 \cdot k \cdot T}\right], \tag{13.55}$$

where $k_n$ is the material constant and $E_g$ is the fundamental band-gap energy. The gap of InGaAs is given by

$$E_g^{\mathrm{InGaAs}}(T) = 0.714 - 0.47 \cdot 10^{-4} \cdot \left(\frac{5.8}{T+300} - \frac{4.19}{T+271}\right) - 4.19 \cdot 10^{-4} \cdot \left(\frac{T^2}{T+271}\right). \tag{13.56}$$

The temperature and composition dependence of InGaAsP are given by

$$\begin{cases} E_g^{\mathrm{InGaAs}}(T) = E_g(0) - 4.3 \cdot 10^{-4} \dfrac{T^2}{T+224}, \\ E_g^{\mathrm{InGaAs}}(y) = 1.344 - 0.738y + 0.138y. \end{cases} \tag{13.57}$$

These are also employed to evaluate the optical absorption factor $\alpha$ versus wavelength and temperature per Eq. (13.58):

$$\alpha(\lambda, T) = k_a \cdot \sqrt{\left[\frac{h \cdot c}{\lambda} - E_g(T)\right]}, \tag{13.58}$$

where $k_a$ is an experimental constant whose data is provided in Ref. [22], $c$ is the speed of light in vacuum, $h$ is the Planck's constant, and $\lambda$ is wavelength.

Having all these formulations, the photocurrent expressions can be derived with respect to wavelength, temperature, absorption coefficient, and physical dimensions. There are two cases for analyzing the photocurrent, as depicted in Fig. 13.13. The first is when light impinges the WSP from the top. Under this condition, the photocurrents are given by

$$\begin{cases} I_{ph2} = \dfrac{q \cdot \lambda}{h \cdot c}\{1 - \exp[-\alpha_2(\lambda, T) \cdot d_2]\} \cdot P \cdot R, \\ I_{ph1} = \dfrac{q \cdot \lambda}{h \cdot c}\{1 - \exp[-\alpha_1(\lambda, T) \cdot d_1]\} \cdot P \cdot R \cdot \exp[-\alpha_2(\lambda, T) \cdot d_2], \end{cases} \tag{13.59}$$

where $P$ is the incident optical power, the optical frequency $\upsilon = c/\lambda$, and $R$ is the surface reflectivity; multiple reflections are neglected. Note that the second exponential for $I_{ph1}$ accounts for the light absorbed in the first top layer.

The second case is sideways illumination where light impinges on lateral incidence. The optical powers coupled to the active layers $l_1$ and $l_2$ are $P_1$ and $P_2$, respectively. Under this scenario, the photocurrents are given by

$$\begin{cases} I_{\text{ph2}} = \dfrac{q \cdot \lambda}{h \cdot c} \{1 - \exp[-\alpha_2(\lambda, T) \cdot l_2]\} \cdot P_2, \\[2ex] I_{\text{ph1}} = \dfrac{q \cdot \lambda}{h \cdot c} \{1 - \exp[-\alpha_1(\lambda, T) \cdot l_1]\} \cdot P_1. \end{cases} \tag{13.60}$$

The above equations can be solved numerically and provide useful charts for design parameters. First, the zero current for the reference wavelength, defined as zero crossing, can be found. Second, the calculation of photocurrent versus bias voltage $V$ and temperature can be evaluated for the design of the TEC control loop.

Figure 13.14 displays the spectral photoresponse of the WSP diodes, D1 and D2. These plots provide guidelines to locate a crossover wavelength within the appropriate spectral window. The current equations were presented in the aforementioned discussion.

Figure 13.15 presents the WSP operation as VT-WSPD, showing the good linearity in the vicinity of the zero-crossing wavelength $\lambda_0$.

Figure 13.16 presents the VT-WSPD tunability. Because the two diodes are connected back-to-back, changes in the external voltage provide reversed bias and result in a shift of $\lambda_0$. In WDM, the zero-crossing wavelength $\lambda_0$ is crucial. The wave has to be selected corresponding to a specific carrier within the spectral window of operation. Since this is the main requirement of such a light-sensitive element, a detailed analysis of $\lambda_0$ sensitivity should be made.

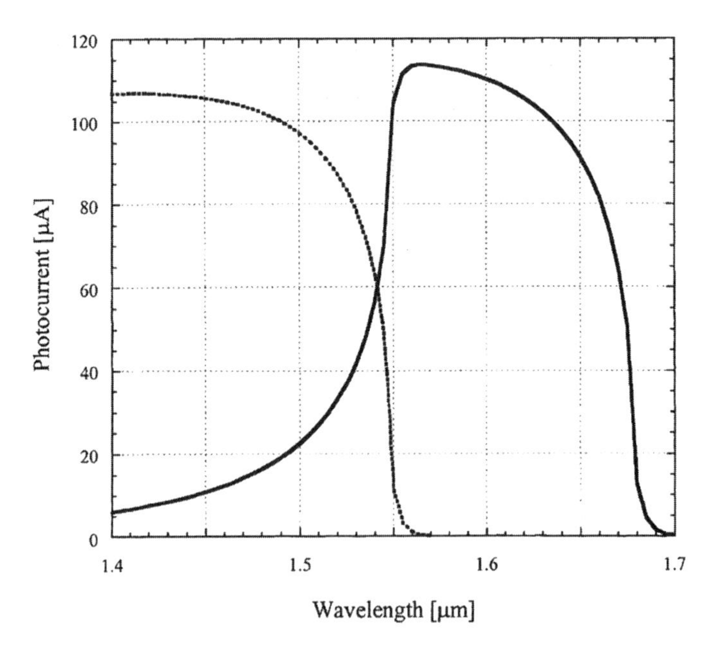


**Figure 13.14** WSP spectral photoresponses for the WSP diode pair D1 (solid line) and D2 (dots).[19] (Reprinted with permission from the *Journal of Lightwave Technology* © IEEE, 2003.)

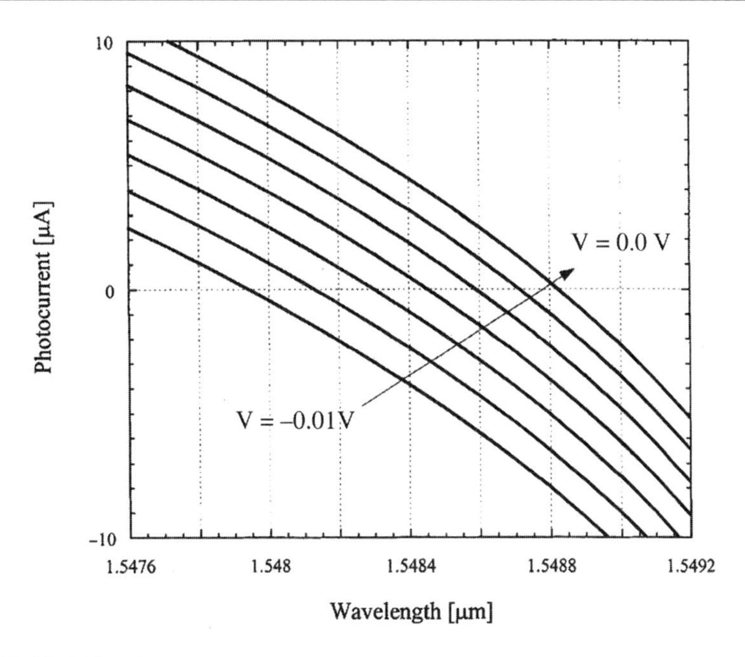


**Figure 13.15** WSP is total current vs. incident wavelength, whereas bias is a parameter within the range of 0.01 - 0.0 V in increments of 1.5 mV. The zero-crossing portion generates a bipolar error signal.[19] (Reprinted with permission from the *Journal of Lightwave Technology* © IEEE, 2003.)

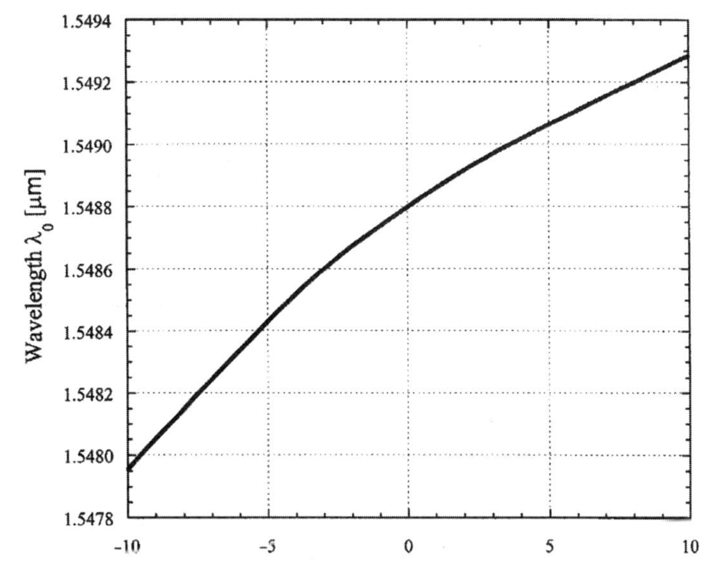


**Figure 13.16** VT-WSPD zero-crossing wavelength $\lambda_0$ vs. bias voltage $V$.[19] (Reprinted with permission from the *Journal of Lightwave Technology* © IEEE, 2003.)

Device evaluation and analysis should include parameter sensitivity for predicting the trends of $\lambda_0$. These parameters are linear approximations for photo-response sensitivity to wavelength variations $\Delta I/\Delta\lambda$, tenability and sensitivity of $\lambda_0$ to bias changes, and $\Delta\lambda_0/\Delta V$ and sensitivity of $\lambda_0$ to input light power. Additional parameters are fabrication tolerances in geometric factors such as layer thicknesses and areas. These parameters are important since they generate error signals suitable to the control system. Hence, the PD should have a high sensitivity to wavelength variation. Mathematically, the first-order derivative of the photocurrent with respect to wavelength should be of high value. Figure 13.17 provides these values. This is similar to the AGC analysis presented in Chapter 12, where the RF detector constant and VVA constant were evaluated. This value defines the system's open-loop gain. Moreover, since the photodetector operates as a wavelength error detector, it defines the error signal sensitivity to wavelength deviation.

Lastly, the VT-WSPD tenability is examined. This can be obtained by acting on voltage bias or by modifying the composition of the active layer in $D_2$, InGaAsP. Figure 13.18 demonstrates the voltage-tuning curves for various thicknesses, $d_1$ and $d_2$. In both cases of normal and lateral incidences, the wavelength tenability exceeds 20 nm. This is sufficient to stabilize 25 channels on a 100-GHz grid. However, it should be emphasized that there are a couple of drawbacks in spanning the whole operating range in $V$, mainly with the normal

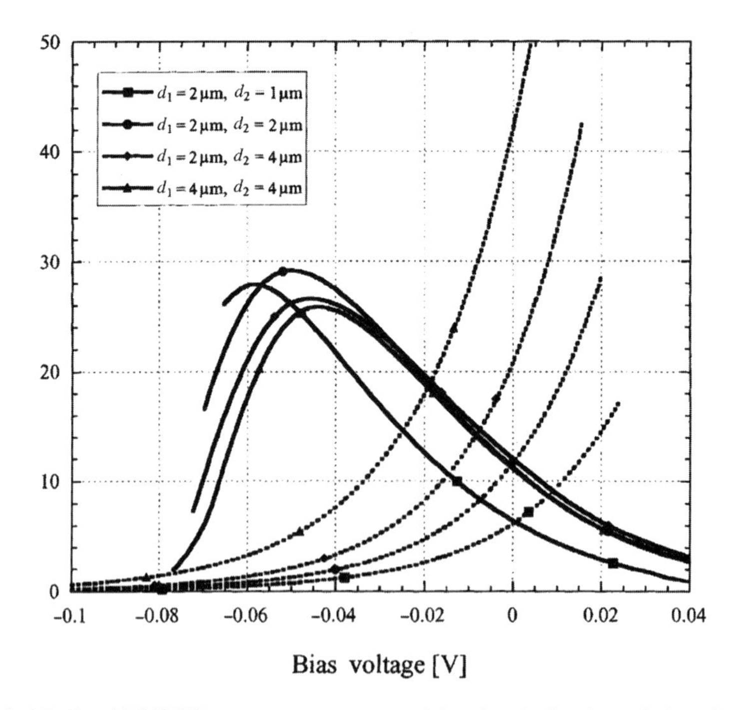


**Figure 13.17** The VT-WSPD constant presented by the derivative of the photocurrent with respect to wavelength $\Delta I/\Delta\lambda$. The chart was evaluated for various thickness combinations, $d_1$ and $d_2$, for each PD.[19] (Reprinted with permission from the *Journal of Lightwave Technology* © IEEE, 2003.)

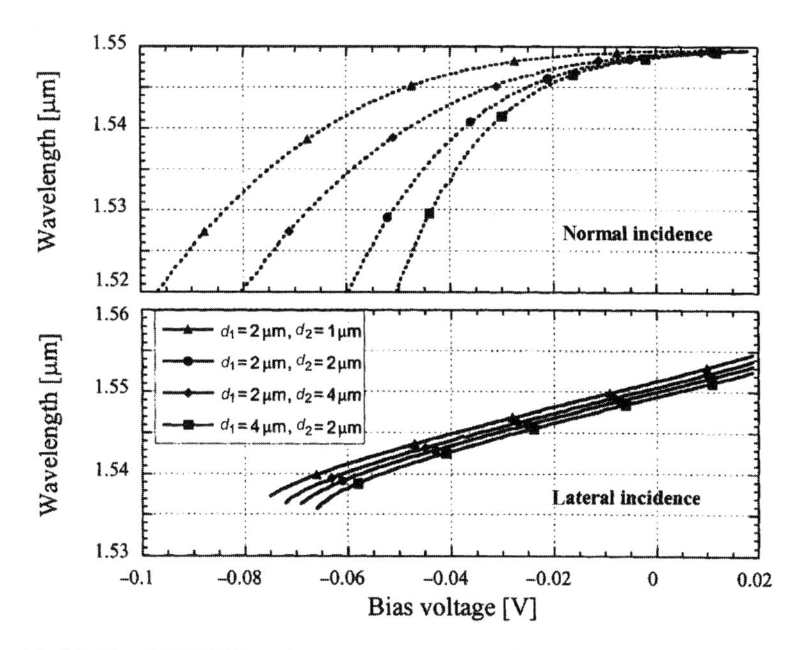


**Figure 13.18** The VT-WSPD voltage-tuning curves for several $d_1$ and $d_2$ thickness combinations.[19] (Reprinted with permission from the *Journal of Lightwave Technology* © IEEE, 2003.)

incidence of the device. The first issue is lack of linearity in the tuning curve, which may require the adoption of more complex wavelength selection circuitry. The second stems from $\lambda_0$ sensitivity to voltage $\Delta\lambda_0/\Delta V$, which affects the accuracy in bias. This parameter, according to simulations, varies from 25 to 750 pm/mV for normal incidence, whereas it is nearly constant for lateral incidence having 180 pm/mV.

Overcoming this problem requires modifying the InGaAsP composition in the $d_2$ absorption layer.[19] As was previously explained, $x \approx 0.47y$ to guarantee lattice matching to InP. The λ gap of $In_{1-x}Ga_xAs_yP_{1-y}$ can be tailored from 950 to 1680 nm. Such an adjustment can only be set during growth, but if the accuracy required in $y$ is tight, voltage control can also be adopted. Figure 13.19 displays a family of curves with $y$ changing stepwise from 0.81 to 0.91, in voltage intervals suitable for 20-nm fine wavelength tuning. The InGaAsP composition can be modified to operate in S, C, or L bands. C and L bands can be covered by varying the concentration only, whereas working in the S band requires changes of the active layer thicknesses in $d_1$ and $d_2$.

Figure 13.20 displays the $\lambda_0$ shift versus the power error in P1 and P2 with respect to their nominal percentage values. Note that to fulfill the specifications of ±3 GHz and ±1 GHz, the imbalance should be limited to 0.6% and 0.2%, respectively. This is because the latter would imply an unrealistically tight alignment. A calibration in tuning voltage can compensate for the wavelength error associated with the misalignment.

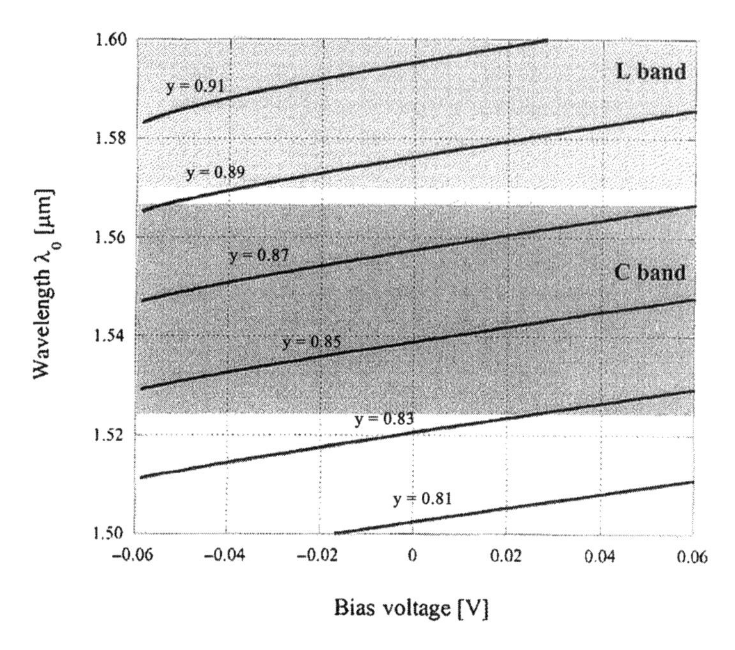


**Figure 13.19** Tuning curves obtained with combination of composition; *y* are changed from 0.81 to 0.91 in the InGaAsP composition and voltage ranges from −60 to 60 mv in case of lateral incidence.[19] (Reprinted with permission from the *Journal of Lightwave Technology* © IEEE, 2003.)

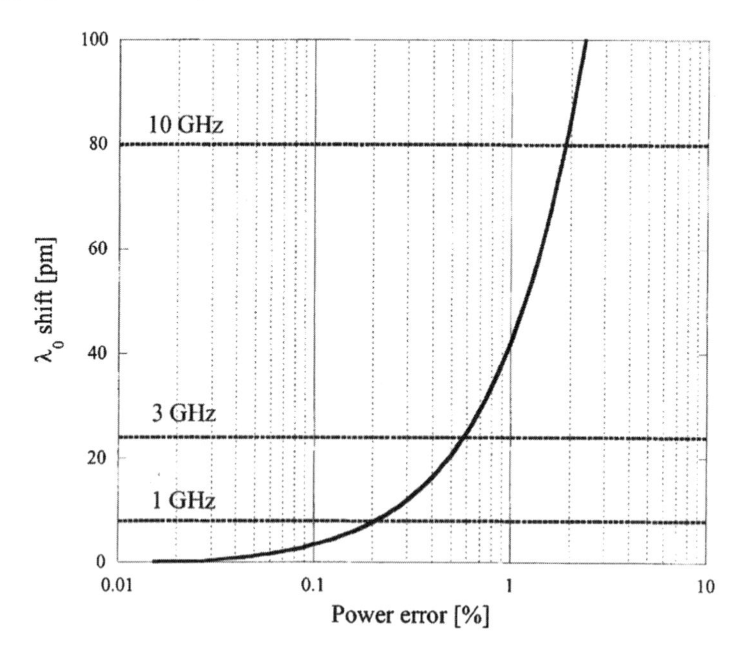


**Figure 13.20** Wavelength sensitivity to power error (percentage $P_1$ or $P_2$ with respect to the nominal value) for lateral incidence. The horizontal dashed lines help in identifying the target accuracy of 10, 3, and 1 GHz, respectively.[19] (Reprinted with permission from the *Journal of Lightwave Technology* © IEEE, 2003.)

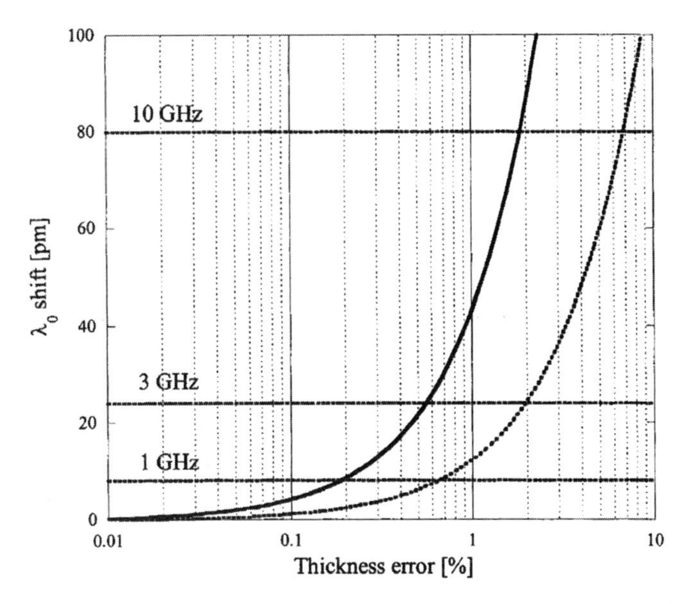


**Figure 13.21** Wavelength sensitivity to thickness error (percentage $d_1$ or $d_2$ with respect to the nominal value) for both lateral incidence (*solid line*) and normal incidence (*dashed*) geometries. The horizontal dashed lines help in identifying the target accuracy of 10, 3, and 1 GHz, respectively.[19] (Reprinted with permission from the *Journal of Lightwave Technology* © IEEE, 2003.)

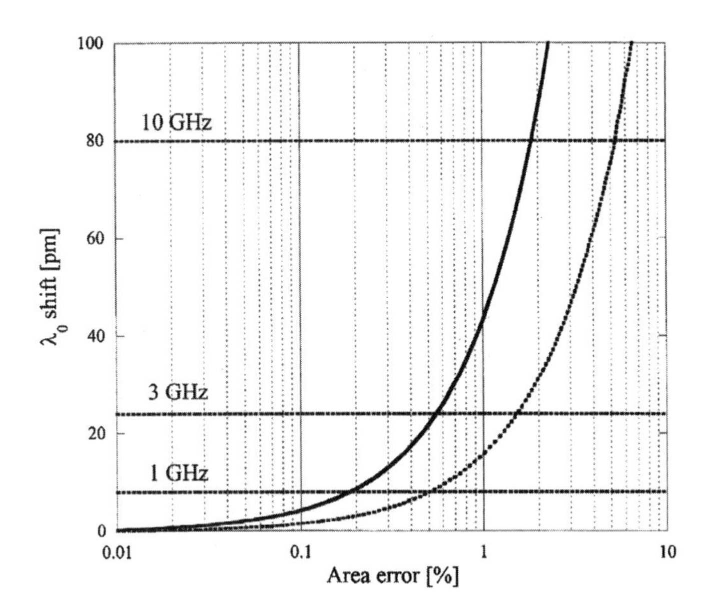


**Figure 13.22** Wavelength sensitivity to area error (percentage $A_1$ or $A_2$ with respect to the nominal valuc) for both lateral Incidence (solid line) and normal incidence (dashed) geometries. The horizontal dashed lines help in identifying the target accuracy of 10, 3, and 1 GHz, respectively.[19] (Reprinted with permission from the *Journal of Lightwave Technology* © IEEE, 2003.)

Figure 13.21 presents $\lambda_0$ shift versus thickness error with respect to a nominal value of $d = 2$ μm. The shift can be kept within ±3 GHz (±1 GHz), with tolerances of 2% (0.6) and 0.5% (0.2%) for normal incidence and lateral incidence, respectively.

Figure 13.22 is a plot of sensitivity to area errors; i.e., $\lambda_0$ shift versus error with respect to the nominal value of $A = 10^{-4}$ cm$^2$. The wavelength drifts remain within ±3 GHz (±1 GHz) if tolerances of 1.5 percent (0.5 percnet) and 0.5 percent (0.2 percent) are met for normal incidence and lateral incidence, respectively. Based on the abovementioned, normal incidence appears slightly more convenient.

Using the above analysis and design parametrical charts, a wavelength stabilization and locking system can be realized using VT-WSPD. Such a system is comprised of a laser source and beamsplitter that routes the laser power to the fiber and samples a portion of it to the feedback path. The feedback contains the VT-WSPD, a current to voltage converter, such as TIA, a controller, power amplifier to drive the TEC, and the Peltier heat sink and sources. The block diagram of

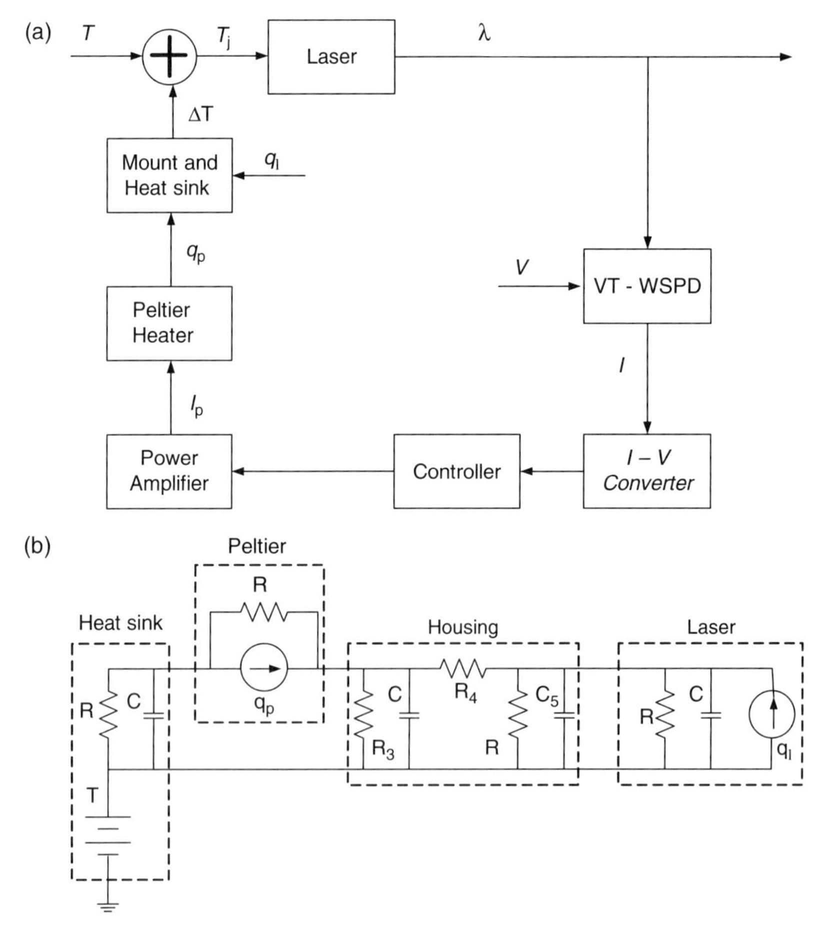


**Figure 13.23** TEC control loop: (a) block diagram, (b) equivalent circuit for SPICE analysis.[19] (Reprinted with permission from the *Journal of Lightwave Technology* © IEEE, 2003.)

the control system and equivalent circuit for the SPICE model are provided in Fig. 13.23. As it is explained in Chapter 12 and in the aforementioned sections, the analysis of a control system is based on linear approximations, SPICE is a small signal modeling for the TEC control loop.

The laser source wavelength emission is approximated by

$$\lambda_{\text{laser}} = \lambda^{0}_{\text{laser}} + m_{\text{laser}} T_{\text{j}}, \tag{13.61}$$

where $T_{\text{j}}$ is the junction temperature, $\lambda^{0}_{\text{laser}}$ is the wavelength at 0°C, and $m_{\text{laser}}$ is the temperature coefficient that is evaluated experimentally.

The photodetector is the VT-WSPD whose linear approximation equations are provided by Eq. (13.62). This equation provides the total current versus wavelength at voltage $V$ and temperature $T$. The parameters $m_{\text{PD}}$, $a$, $b$, $c$, and $d$ are driven from the previous discussion and aforementioned equations:

$$\begin{cases} I(\lambda, T, V) = m_{\text{PD}}\lambda + i_{\text{PD}}(T, V), \\ I_{\text{PD}} I(T, V) = a + bT + cV + \text{d}V^2. \end{cases} \tag{13.62}$$

The thermal stage takes into account two heat sources: laser and Peltier cell, the mounting hardware, and the heat sink. Heat flow is emulated to current; hence, a linear approximation is given by

$$q_{\text{p}} = I_{\text{p}} + m_{\text{p}} \cdot I_{\text{p}}. \tag{13.63}$$

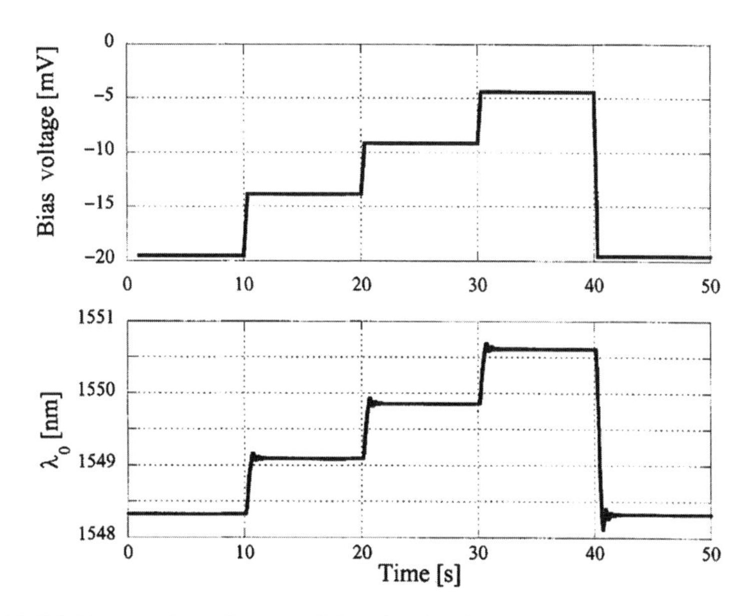


**Figure 13.24** Temporal evolution of $\lambda_0$ in the bottom in response to four-level voltage ramp at the top of the chart. This is designed to scan over four 100-GHz spaced channel grid.[19] (Reprinted with permission from the *Journal of Lightwave Technology* © IEEE, 2003.)

The equation defines the heat flow versus the supplied current $I_p$. The constants $I_p$ and $m_p$ and the resistance $R_2$ are defined by cell manufacture. The heat sink is represented by a parallel resistor and capacitor, marked as RC, in series to the voltage source that corresponds to room temperature. The metal housing between Peltier cell and laser is described by a $\pi$ circuit with $R_3 = R_5$ equal to half of the convection resistance, $R_4$ accounting for internal heat conductivity, and $C_3 = C_5$ equal to one-half of the thermal capacitance. The housing parameters can be calculated for the material and geometry. The thermal properties of the laser are modeled by a parallel RC network. This network emulates the thermal resistance and capacitance that are in parallel with the heat source $q_I$, i.e., the heat is generated while the laser emits light. Using superposition, Eq. (13.64) is obtained, which yields the dependency of the laser junction temperature $T_j$ on room temperature $T$. It also provides the dependency of $T_j$ on heater and laser heat flows $q_p$ and $q_I$, respectively, where $G_1$ and $G_2$ are the appropriate transfer functions:

$$T_j = T + G_1 q_p + G_2 q_I, \tag{13.64}$$

which can be used for a controller algorithm that can be designed at a specific settling time and absolute maximum steady state error. A PID analog controller can be used.

The wavelength control loop performance can be examined in several aspects. The system dynamics of wavelength locking versus time by acting on the voltage is applied to the VT-WSPD. After the $\lambda_0 - V$ relationship is obtained, the $\lambda_0$ sweep is readily accomplished by biasing the device with a voltage. Figure 13.24 displays the temporal evolution of $\lambda_0$ on its bottom, corresponding to the four-level voltage ramp on the top of this chart. The voltage ramp is designed to sweep four 100 GHz–spaced channels.

## 13.3 Main Points of this Chapter

1. APC and TEC are feedback control loops.

2. APC loop and TEC loop are narrow-band control loops because of the response to slow time-varying events.

3. A long train of same-level consecutive bits creates the equivalent of a lower rate signal that may affect the APC power locking. Thus, the loop BW should be narrower by at least ten times than the lowest equal rate of the long train signal.

4. The BW loop of APC should support a long train of consecutive bits at the same level, such as PRBS $2^{23-1}$, without making a power error due to AM detection.

5. ER can be controlled by a faster APC loop, which locks on the peak energy of the modulated laser feedback from the back-facet monitor using a peak detector.

6. TEC current control can be done by the traditional linear method as shown in Fig. 13.7, or by a more efficient method using a PWM controller.
7. When the TEC loop is locked by PWM, the current control pulses are narrow. When the TEC integrates to the desired temperature, the current pulses are wider to minimise the error.
8. In both the regular linear loop and the PWM, there is a loop filter, which is the integrator. The integrator determines the loop dynamics and BW.
9. In order to accomplish fast and accurate locking, PID controllers are used.
10. The PI loop is a proportional integral regular feedback system.
11. TEC loops are used to stabilize the wavelength and enable the laser to emit higher power.
12. Prior to any TEC design, heat balance is made. This balance contains heat to be removed, invested energy heat, which is the TEC current, and the total heat to the heat sink.
13. The COP is a measure of cooling performance. COP $> 1$ and describes the ratio of $Q_C/P_{TEC}$ in Eq. (13.12). The TEC design should maximize the COP as much as possible.
14. The TEC can be represented as an equivalent linear electrical circuit.
15. Analogies of thermal to electrical SPICE-modeling derivation are made per Table 13.2.
16. Temperature is potential and temperature difference is voltage across the thermal conductor. Heat flows through the material and thus is similar to current. Heat accumulation is charge and the capacity to accumulate heat is capacitance. Heat generates a potential since it is measured by temperature.
17. PSPICE OrCAD modeling is a small-signal approximation of TEC at its operating point. The TEC loop is affected by its operating point. Thus, the TEC loop is a small-signal approximation of the nonlinear control loop.
18. Linear approximation of a control system is done at the operating point, where it is locked.
19. VT-WSPD is a voltage-tuned wavelength selective photodiode composed of PIN and NIP photodetectors, where NIP is an upside down PIN.
20. The VT-WSPD is composed of two PDs with adjacent responsivities. Hence, a specific wavelength $\lambda_0$ can be optimized to produce a zero photocurrent at the desired wavelength $\lambda_0$. (see Fig. 13.15).

**Table 13.3** Analogies between AGC, TEC and APC cntrol loops.

| AGC | TEC | APC |
|---|---|---|
| RF detector | NTC | Back facet |
| VVA | TEC | Current control |
| $RF_{in}$ | Laser dc power load | Laser unlocked bias |
| $RF_{out}$ | Cooling level | Power bias point |
| Loop filter $H(s)$ | Loop filter $H(s)$ | Loop filter $H$(s) |
| Smoothing filter $T(s)$ | Smoothing filter $T(s)$ | Smoothing filter T(s) |
| RF detector constant $K_D$ $K_D = \frac{\partial V_D}{\partial P_{RF}}$ | NTC bridge constant $K_D$ $K_{NTC} = \frac{\partial V_{NTC-BRIDGE}}{\partial T}$ | Back facet monitor constant responsivity $K_{PD}$ $r = K_{PD} = \frac{\partial I_{PD}}{\partial P_{OPT}}$ |
| VVA constant $g$ $g = \left.\frac{\partial G[V_C(t)]}{\partial V_C(t)}\right\vert_{V_C=V_{CS}}$ | TEC constant $g$ $g = \left.\frac{\partial T[I_C(t)]}{\partial I_C(t)}\right\vert_{V_C=V_{CS}}$ | Laser constant (slope efficiency) $g$ $g = \left.\frac{\partial G[i_C(t)]}{\partial i_C(t)}\right\vert_{V_C=V_{CS}}$ |
| $P_{out}(s) = P_{in}(s) \cdot G[V_C(s)]$ | $T_C(s) = I_{in}(s) \cdot G[I_C(s)]$ | $P_{out}(s) = I_{in}(s) \cdot G[I_C(s)]$ |
| RF coupler | Thermal coupling | Back facet free space |

21. At the target wavelength there is a zero photocurrent, which is due to the structure of two the VT-WSPD, which is a back-to-back structure of two PIN photodetectors. This structure creates the PIN–NIP stack. Two back-to-back PDs create photocurrents in opposite directions. Wavelength selection is done by treating the external bias voltage as a tuning voltage.

22. Wavelength discrimination for two responsivities is accomplished by doping recipes for each of the back-to-back PD pairs.

23. Analogies to AGC loop analysis are provided in Chapter 12.

## References


1. Godfrey, S., "An introduction to thermoelectric coolers." *Electronic Cooling*, Vol. 2 No. 3 (1996).
2. Yang, R., G. Chen, G.J. Snyder, and J.P. Fleurial, "Multistage thermoelectric micro coolers." *IEEE Eighth Intersociety Conference on Thermal and Thermomechanical Phenomena in Electronic Systems*, pp. 323–329 (2002).
3. Rafaat, T.F., W.S. Luck Jr, and R. DeYoung, "Temperature control of avalanche photodiode using thermoelectric cooler." NASA/TM-1999-209680. Langley Research Center, Hampton, VA (1999).
4. Rowe, D.M., *CRC Handbook of Thermoelectrics*. CRC Press (1995).
5. Gupta, G.C., et al. "3.2 Terabit/sec(40ch/80GB/s) transmission over 1000 Km with 100 km span and 0.8 bit/s/Hz spectral efficiency." OFC Tutorial (2002).

6. "Optical interfaces for equipments and systems relating to the synchronous digital hierarchy." ITU-T Recommendation G.957 (1995).
7. "Small form factor pluggable (SFP) multi source agreement." www.schelto.com/SFP/ (2000).
8. Priyadarshi, S., S.Z. Zhang, M. Tokhmakhian, C. Martinez, N. Margalit, and A. Black, "First sub watt hot pluggable DWDM SFP transceiver." *IEEE Optical Communications*, pp. 529-531 (2005).
9. Goldsmid, H.J., "Electronic refrigeration." *European Conference on Thermophysical Properties*. Pion Ltd. (1986).
10. Kapitulnik, A., "Thermoelectric colling at very low temperature." Applied Physics Letters, p. 60 (1992).
11. "NTC Thermistors" Bowthorpe Thermometrics Application notes, www.thermometrics.com/assets/images/ntcnotes.pdf (2007).
12. Galan, P., "Enhanced temperature controller is both fast and precise." EDN, p. 111 (March, 2001).
13. "Thermoelectric cooling systems design guide." Marlow Industries Inc. www.marlow.com/AboutMarlow/pdf/MRL1002G.pdf
14. Lineykin, S., and S. Ben-Ya'akov, "PSPICE compatible equivalent circuit of thermoelectric coolers." *Proceedings of IEEE 36th Conference on Power Electronics Specialists*, pp. 608-612 (2005).
15. Huang, B.J., and C.L. Duang, "System dynamic model and temperature control of a thermoelectric cooler." *International Journal of Refrigeration*, Vol. 23, pp. 197-207 (2000).
16. Simons, R.E., and R.C. Chu, "Application of thermoelectric cooling to electrical equipment." *Proceedings of IEEE Sixteenth Semi-Therm Symposium*, pp. 1-9 (2000).
17. Chavez, J.A., J.A. Ortega, J. Salazar, A. Turo, and M.J. Garcia, "SPICE model of thermoelectric elements including thermal effects." *Proceedings of IEEE Instrumentation and Measurement Technology Conference*, pp. 1019-1023 (2000).
18. Bagnoli, P.E., C. Casarosa, M. Ciampi, and E. Dallago, "Thermal resistance analysis by induced transient (TRAIT) method for power electronic devices thermal characterization—Part I: Fundamentals and theory." *IEEE Transactions on Power Electronics*, Vol. 13 No. 6, pp. 1208-1219 (1998).
19. Colace, L., G. Masini, and G. Assanto, "Wavelength stabilizer for telecommunication lasers: design and optimization." *IEEE Journal of Lightwave Technology*, Vol. 21 No. 8, pp. 1749-1757 (2003).
20. Millman, J., and C.C. Halkias, *Integrated Electronics*. McGraw-Hill, New York, N.Y. (1972).
21. Desoer, C.A., and E.S. Kuh, *Basic Circuit Theory*. McGraw-Hill, New York, N.Y. (1969).
22. Levinstine, M., S. Rumyantsev, and M. Shur, *Handbook Series on Semiconductor Parameters*. Singapore World Scientific, Vol. 2, p. 65 (1999).

Part 4

# Introduction to CATV MODEM and Transmitters

# Chapter 14

# Quadrature Amplitude Modulation (QAM) in CATV Optical Transmitters

The goal of this chapter is to provide an introduction to community access television (CATV) quadrature phase-shift keying (QPSK) modulators. QPSK modulators are used as building blocks for *I* and *Q* modulation schemes in modulator–demodulators (modems). The quadrature amplitude modulation (QAM) Modem baseband (BB) output is up-converted into RF and transmitted over fiber by linearized transmitters, as will be reviewed in Chapter 16, to the subscriber's home. By reviewing the basics of QAM and modulation impairments, the motivation behind modem calibration methods, RF and optics linearization methods, and implementations would be much clearer. System requirements from a fiber-to-coax CATV receiver converter will be discussed and their effects on the data quality will be examined. This chapter presents a review of a QPSK modulator, and then Chapter 15 provides a review of QAM modems, including a review of MPEG-framing Reed–Solomon coding M-ary modulators, equalizers, and BER (of an M-ary QAM). Chapter 16 provides a review of predistortion methods and linearization methods. Chapters 15 and 16 also provide the final structure and block diagram of the fiber CATV transmission side. The receiving side is explored, and the general concept of CATV receive modems and timing recovery is provided. Using this, an HFC FTTx receiver designer is able to identify and rectify any degradation on QAM transport, resulting from an RF-circuitry problem. Based on Chapter 9, RF receiver relations are further explored.

## 14.1 Quadrature Modulators

Quadrature modulators have been used for single-sideband transmission. In recent years, QAMs and other digital modulation schemes have been used in digital radios. In the old IS54 dual-mode cellular phone radios, the modem is designed to use one modulator for both digital and analog mode transmissions. Modern cellular phones utilize QPSK modulators and demodulators for 2.5G GSM Extended global system for mobile communications (EGSM) general packet

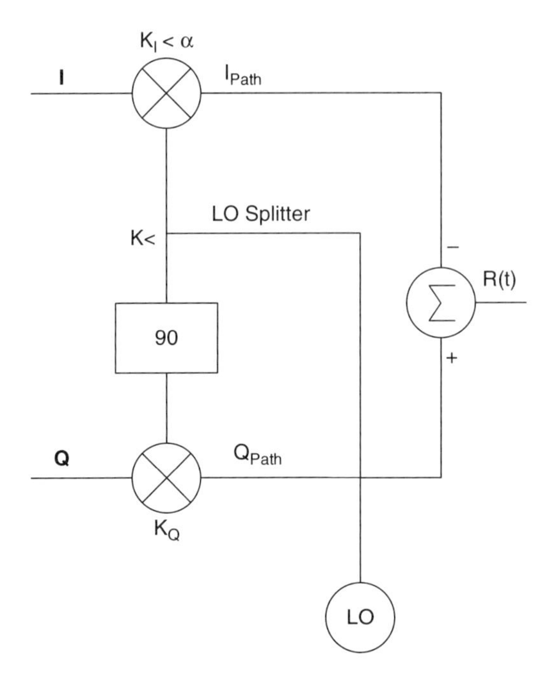


**Figure 14.1** Quadrature modulator and its impairments.

radio service (GPRS) as well as 3G wideband code division multiple access (WCDMA) universal mobile telecommunications system (UMTS) modulation schemes. In CATV transmission, it is used to transfer the QAM digital channels at the frequency range of 550 to 870 MHz.

The block diagram of a quadrature modulator given in Fig. 14.1 consists of two mixers and a summation circuit to combine the mixers' output. The input to the mixers are the information bearing BB (low pass) signals $i(t)$ and $q(t)$. A quadrature carrier of the local oscillator (LO) signal $\cos(\omega_c t)$ mixes with the $I$ input, and $\sin(\omega_c t)$ mixes with the $Q$ input. Hence, the transmitted up-converted signal is given by a Cartesian presentation in the complex domain,[4]

$$r(t) = i(t)\cos(\omega_c t) + jq(t)\sin(\omega_c t). \tag{14.1}$$

After some trigonometric manipulations, it becomes

$$r(t) = \sqrt{i^2(t) + q^2(t)} \cdot \exp j\left\{\omega_c t + \tan^{-1}\left[\frac{q(t)}{i(t)}\right]\right\} = A(t)\exp j[\omega_c t + \theta(t)], \tag{14.2}$$

where $r(t)$ is a vector in the $I$–$Q$ plan, which can be presented in Cartesian or polar coordinates:

$$A(t) \cdot e^{j\theta(t)} = i(t) + jq(t). \tag{14.3}$$

The later presentation of the QPSK modulator output signal implies that any complex modulation can be generated by properly chosen amplitudes and phases of $I$ and $Q$. Moreover, since the $I$ and $Q$ signals have low-pass characteristics, the low-frequency modulation (FM) circuitry is usually easier to implement at the required accuracy. This is a major reason for the popularity of quadrature modulators.

## 14.2 Generating FM using QPSK Modulators

FM is used in many applications due to its amplitude noise immunity. In TV transmissions, the sound is an FM signal. To establish FM, the carrier is held constant and the phase is varied:[4]

$$r(t) = A(t)\cos[\phi(t)] = A(t)\cos[\omega_c t + \theta(t)]. \tag{14.4}$$

Using some trigonometric identities and using the fact $A(t) = A$, Eq. (14.4) becomes

$$r(t) = A\{\cos(\omega_c t)\cos[\theta(t)] - \sin(\omega_c t)\sin[\theta(t)]\}. \tag{14.5}$$

The BB signals at the $I$ and $Q$ inputs are $\cos[\theta(t)]$ and $\sin[\theta(t)]$, respectively. The instantaneous carrier frequency $\omega_c(t)$ is raised or lowered around the nominal carrier frequency $\omega_c$, according to the information bearing signal $m(t)$:

$$\omega_c(t) = \omega_c + \Delta\omega m(t), \tag{14.6}$$

where $\Delta\omega$ is the peak radian deviation, and the peak of $m(t)$ equals 1. The phase $\theta(t)$ is given by

$$\theta(t) = \int_{-\infty}^{t} \Delta\omega m(\tau)\mathrm{d}\tau = \Delta\omega \int_{-\infty}^{t} m(\tau)\mathrm{d}\tau. \tag{14.7}$$

Hence, the FM BB signal is given by

$$\begin{aligned} A \cdot e^{j\theta(t)} &= i(t) + jq(t) = A\exp\left[j\Delta\omega \int_{-\infty}^{t} m(\tau)\mathrm{d}\tau\right] \\ &= A\left\{\cos\left[\Delta\omega \int_{-\infty}^{t} m(\tau)\mathrm{d}\tau\right] + j\sin\left[\Delta\omega \int_{-\infty}^{t} m(\tau)\mathrm{d}\tau\right]\right\}. \end{aligned} \tag{14.8}$$

The advantage of this FM method is the accuracy of the deviation since it is independent of linearity of the VCO constant $\mathrm{d}f/\mathrm{d}v$, where $f$ is the frequency and $v$ is the tuning voltage. It can be used in the analog channels to produce the sound (aural) for the video signal.

## 14.3 Digital QAM

In digital channel bands, QAM techniques are used. The most common are QAM 64 and QAM 256 constellations, $M = 64$, $M = 256$. However using QAM 256 or a large-symbol constellation requires a low-jitter LO, and highly linear amplifier with low-amplitude modulation (AM) to PM and high compression point, or AM-to-AM characteristics. Moreover, there is a need to know how to estimate the normalized symbol energy. Figure 14.2 illustrates the QAM 256 quadrant. Each symbol of the constellation is represented by vertical and horizontal voltage coordinates. Hence, the symbol energy is related to its coordinate's voltage square.[8,11,14] Since all symbols have equal likelihoods, and due to the constellation symmetry, the average energy per symbol $E_s$ is determined for a single quadrant. Therefore, the quadrant average energy per symbol is the sum of all symbol energies divided by the total number of symbols per quadrant, as shown in Fig. 14.2. The horizontal or vertical voltage distance between two symbols would be defined as $2a$. Thus the QAM 256 symbol normalized energy $E_s$ is given by Eq. (14.9). It can be said that $2^p$ states' QAM is bidimensional pulse-amplitude modulated (PAM), where each axis consists of $k = \sqrt{M}$ states. Each state is defined by different permutations of bits, where there are $\sqrt{M}$ bit levels for each PAM axis. As a consequence, $k$ and $M$ are both even and have a root as appears in this QAM 256 example. Moreover, $p_1 = \log_2 \sqrt{M}$ is the number of bits required to accommodate each axis level or energy state. For example, in QAM 256, $p_1 = 4$, thus there are 4 bits to describe 16 states for the $I$ axis and 16 states for the $Q$ axis. Thereby, each QAM symbol state is described by 8 bits. Due to QAM symmetry, the calculation is done by observing a single quadrant. In this case, each axis is described by 3 bits for 8 states; as a result, each quadrant has 64 states. This is a simplified case

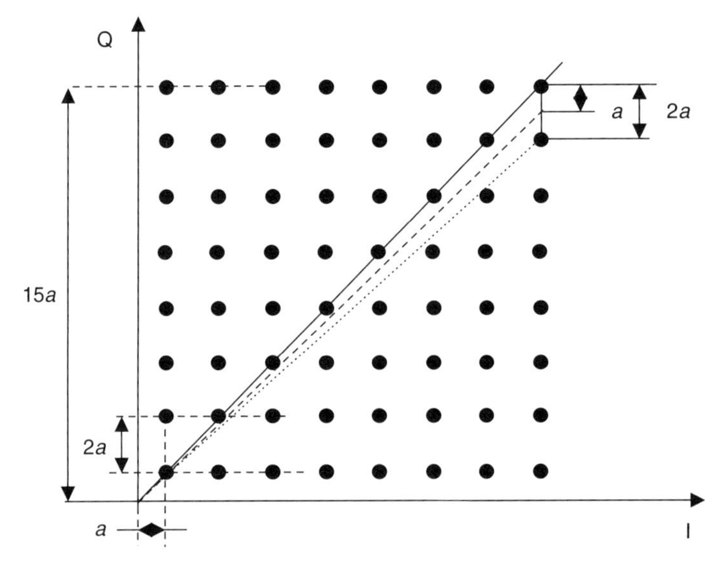


**Figure 14.2** First-quadrant QAM constellation symbols map.

where $M = 64$. The conclusion is that QAM 256 can be used to transfer 4 blocks of 64 QAM; thus Eq. (14.9) is proved. Note that QAM 32 does not apply to this condition since $M = 5$:

$$E_s = \frac{a^2}{64}\sum_{q=1}^{8}\sum_{i=q+1}^{8}\left[(2i-1)^2 + (2q-1)^2\right]. \tag{14.9}$$

The unit distance $a$ between the symbols can be calculated from Eq. (14.9) and is given by

$$a = \sqrt{\frac{64E_s}{\sum_{q=1}^{8}\sum_{i=q+1}^{8}\left[(2i-1)^2 + (2q-1)^2\right]}}. \tag{14.10}$$

The distance between the symbols is defined as $d = 2a$. Each symbol in this modulation scheme represents 8 bits; if the normalized energy per bit is $E_b$, then the normalized symbol energy satisfies $E_s = 8E_b$; therefore,

$$a = \sqrt{\frac{E_s}{170}}, \tag{14.11}$$

$$a = \sqrt{\frac{E_b}{21.25}}. \tag{14.12}$$

It can be observed that the distances between the symbols are significantly reduced as the constellation order gets higher. One immediate requirement is to have a low-jitter (phase noise) or LO-integrated phase noise. The extreme requirement for minimum jitter appears between the symbols at the coordinates $I = 15a$, $Q = 15a$, and $I = 15a$, $Q = 13a$ as depicted in Fig. 14.2. As a rule of thumb, jitter should below 10% of the angle pointing to the half Euclidian distance, which is $a$, between two adjacent symbols. The angle calculation is approximately given by $1/2[\text{arctangent}(15/15) - \text{arctangent}(13/15)] = 1/2(45\quad\text{deg} - 40.9\quad\text{deg}) \approx 1.5$ deg, which is about 0.15 deg. This means that the LO jitter should be below 0.15 deg.[4]

Digital modulation schemes exhibit higher voltage and power peaks than the average power which is the result of power meter reading. This issue is affecting AGC calibration explained in Sec. 12.2.8. Figure 14.3 illustrates the peak value for QAM 256. The first quadrant is good enough due to the symmetry in constellation calculation and QAM 256 structure of four 64 quadrants of symbol.

The symbols' energy levels are determined by the geometry of the constellation. The number of symbols in each quadrant of QAM 256 is 64. The average power is defined as the sum of the squares of the magnitudes of all symbols divided by the number of symbols in the quadrant. This is due to the power being proportional to the voltage square. The average voltage is defined as the sum of all the voltage magnitude values divided by the number of symbols. The peak energy and voltage relates to the symbol with the highest $I$ and $Q$ coordinates. In this case, it is the symbol at coordinates $15a$,$15a$. The peak to rms ratio, also known as peak

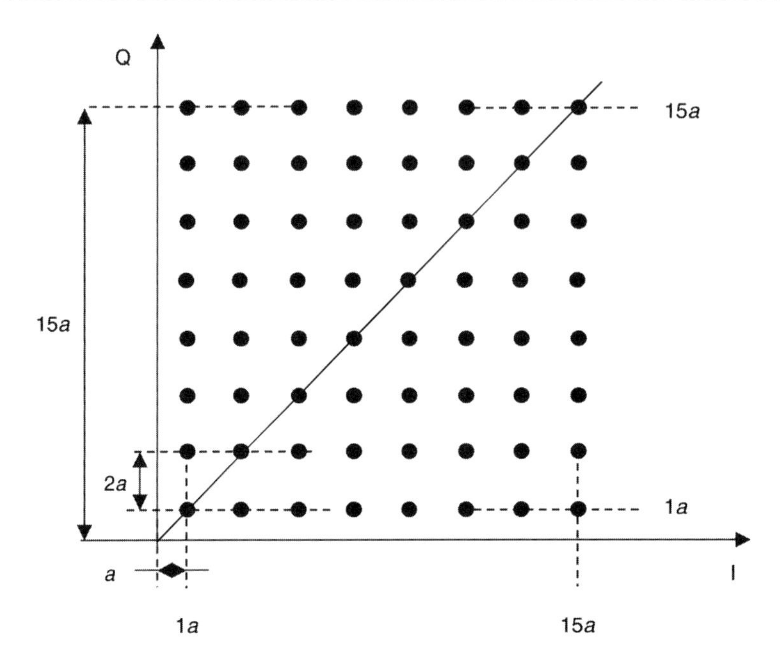


**Figure 14.3** Constellation-energy-level points and coordinate states.

to average ratio (PAR), is calculated by taking the power level of the symbol located at the coordinate $15a, 15a$ and dividing it by the average power value of a symbol. In the same way, the peak to rms voltage ratio is defined. One observation is obvious: the higher the constellation level is, the higher is the peak-to-rms value, as shown in Fig. 14.4. For instance, in a QPSK modulation scheme, the peak-to-rms value is 1 or 0 dB. The peak to rms value for QAM 256 power is provided by

$$P_{\mathrm{PEAK}}/P_{\mathrm{rms}} = \frac{2a^2 15^2}{(a^2/64)\sum_{q=1}^{8}\sum_{i=q+1}^{8}\left[(2i-1)^2+(2q-1)^2\right]},$$

$$P_{\mathrm{PEAK}}/P_{\mathrm{rms}} = \frac{128 \cdot 15^2}{\sum_{q=1}^{8}\sum_{i=q+1}^{8}\left[(2i-1)^2+(2q-1)^2\right]}. \tag{14.13}$$

The voltage peak-to-rms value is defined by

$$V_{\mathrm{PEAK}}/V_{\mathrm{rms}} = \frac{\sqrt{2a^2 15^2}}{(a/64)\sum_{q=1}^{8}\sum_{i=q+1}^{8}\left[(2i-1)^2+(2q-1)^2\right]},$$

$$V_{\mathrm{PEAK}}/V_{\mathrm{rms}} = \frac{64\sqrt{2} \cdot 15^2}{\sum_{q=1}^{8}\sum_{i=q+1}^{8}\left[(2i-1)^2+(2q-1)^2\right]}. \tag{14.14}$$

The above calculation shows the geometrical relationships between the peak symbol energy in the constellation and the average values of the constellation's symbols. However, additional impairments should be taken into account. For

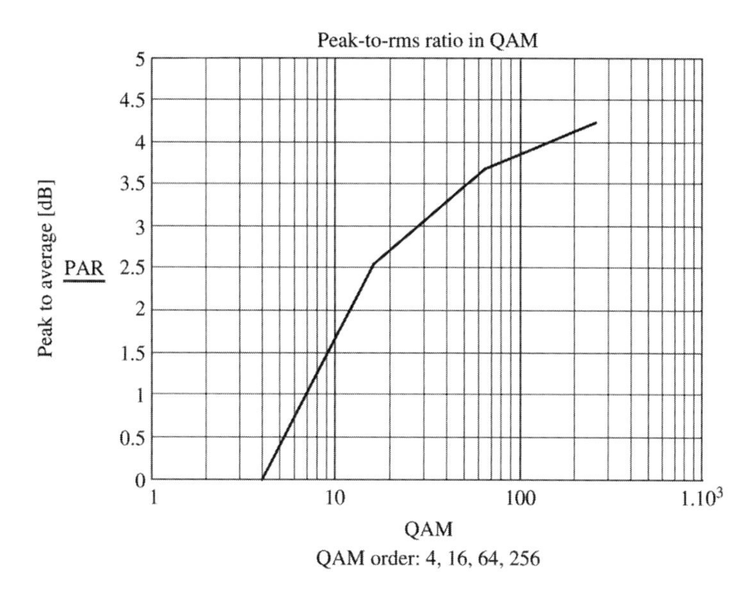


**Figure 14.4** Geometrical peak-to-average-power ratio for QAM 4 (QPSK), 16, 64, 256. Peak-to-rms values in decibles are: QPSK, 0; 16 QAM, 2.55; 64 QAM, 3.68; 2 and 56 QAM, 4.23. ISI effects due to the raised cosine filter are not included.

instance, the inter-symbol interference (ISI) defeat filter has a raised cosine shape with a roll-off factor value $\alpha$ measured in percentage. The more moderate the filter, i.e., the higher the value of the roll-off factor $\alpha$, the greater the ringing the filter would have in its impulse response. This means that additional amplitudes are added to the signal, and therefore to the symbol coordinates. As a consequence, the peak-to-rms ratio would be even greater. An ideal brick-wall filter with $\alpha = 0$ is represented by a rectangular filter that adds no ISI; hence, geometrical calculation is applicable. However, when the SQRT (square root raised cosine) filter is used, where $\alpha \neq 0$, peak to average (PAR) is not per the geometrical calculation and is significantly increased. This is demonstrated in section 12.2.8 using numerical analysis methods. In a multicarrier transport environment, the amplitude distribution has a probability distribution close to that of random noise. This function has a finite probability with a high value in a short period of time. As a result, the peak to rms reaches even greater values. This is one more reason, as was explained in Chapter 12, why AGC sampling is on the low-channel frequencies with analog modulation, thereby avoiding this problem. Furthermore, this is the reason, among many others, that CATV receivers use feedforward AGC schemes. In case of sampling QAM, the peak-to-rms fluctuations result in higher errors in power tracking.

## 14.4 Signal Impairments and Effects

Impairment types can be divided into two main categories:[4]

1. Linear impairments related to an $I$ and $Q$ amplitude imbalance that generates constellation deformation, an $I$ and $Q$ arms-phase imbalance that

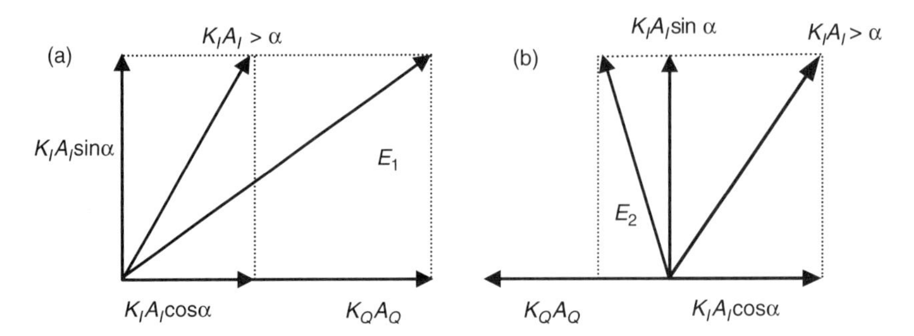


**Figure 14.5** Sideband phasors: (a) upper sideband, and (b) lower sideband.

generates constellation deformation, and $I$ and $Q$ dc offset that creates carrier leakage, which affects the constellation origin's offset. Figures 14.5 and 14.6 illustrate these effects.

2. Nonlinear impairments related to phase noise, which affects the constellation symbol coordinates' accuracy, by adding $\pm\theta$ rms jitter around the symbol target. AM to AM and AM to PM, which are generated by the transmitter or receiver compression, affect the symbol coordinates' accuracy, and create amplitude and phase error. IMD, which is mixed with the signal, creates a rotating vector around the symbol target point. Additive white Gaussian noise (AWGN) adds a random vector around the symbol target, thus generating a clouded symbol.

These impairments create a displacement vector, which is a measure of symbol-placement error. This value of error is known as the error vector magnitude (EVM). The exact formulation will be provided in Sec. 14.8. and Fig. 14.6.

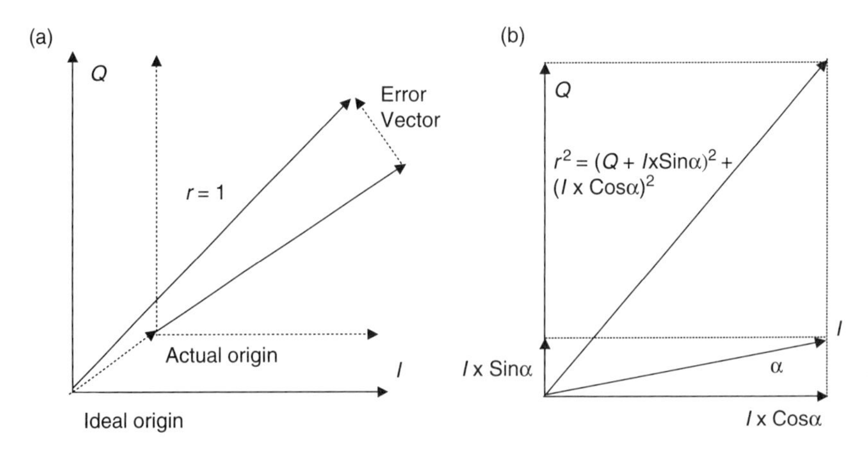


**Figure 14.6** Linear impairments in QPSK modulator. (a) Carrier leakage. (b) $I$ and $Q$ amplitude imbalance effects on quadrature shape.

### 14.4.1 *I* and *Q* amplitude imbalance and carrier leakage

A way to observe and analyze QPSK modulator impairments and calibrate the constellation is by applying orthogonal signals at the $I$ and $Q$ inputs, for instance $A_I \cos(\omega t)$ and $A_Q \sin(\omega t)$ respectively. For the modulator in Fig. 14.1, the LO is given by $\cos(\omega_c t)$; then after using some trigonometric identities, the upper sideband voltage $E_1$ is

$$E_1 = \frac{K_I e^{j\alpha} A_I + K_Q A_Q}{2} \cos(\omega_c + \omega)t. \tag{14.15}$$

The lower sideband voltage would be

$$E_2 = \frac{K_I e^{j\alpha} A_I - K_Q A_Q}{2} \cos(\omega_c - \omega)t. \tag{14.16}$$

The above expressions describe the upper sideband and lower sideband phasors, which rotate at $(\omega_c + \omega)t$ and $(\omega_c - \omega)t$, respectively. Figure 14.5 describes the two sideband vectors.

Hence the upper sideband power and lower sideband power are given by

$$E_1^2 = \left(\frac{K_I A_I}{2}\right)^2 + \left(\frac{K_Q A_Q}{2}\right)^2 + 2\left(\frac{K_I A_I K_Q A_Q}{4}\right)\cos(\alpha), \tag{14.17}$$

$$E_2^2 = \left(\frac{K_I A_I}{2}\right)^2 + \left(\frac{K_Q A_Q}{2}\right)^2 - 2\left(\frac{K_I A_I K_Q A_Q}{4}\right)\cos(\alpha), \tag{14.18}$$

where $K_I$ is the $I$-path-mixer-transfer factor, $K_Q$ is the $Q$-path-mixer transfer factor, and $\alpha$ is the $I$-path-phase imbalance with respect to $Q$ section.

Using the definition of the relative amplitude balance factor:

$$\delta = \left|\frac{K_I A_I}{K_Q A_Q}\right|. \tag{14.19}$$

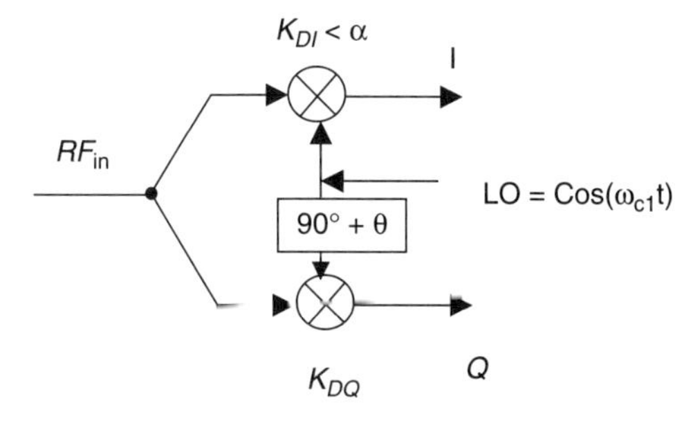


**Figure 14.7** A QPSK demodulator with its amplitude and phase impairments.

Using Eq. (14.19) and taking the ratio between Eqs. (14.17) and (14.18), then the sideband suppression is defined by

$$S[\mathrm{dBc}] = 10\log\left[\frac{\delta^2 + 2\delta\cos\alpha + 1}{\delta^2 - 2\delta\cos\alpha + 1}\right]. \tag{14.20}$$

As a result, the QAM constellation quadrature can be optimized and calibrated. Feeding the modulator with orthogonal tones and optimizing their quadrature amplitudes and phases in order to achieve the maximum sideband suppression accomplishes quadrature calibration. This process is done at the BB level by the cable MODEM calibration algorithm. Equation (14.20) for the case of $\cos(\alpha) < 0$ means spectrum inversion; i.e., the upper sideband is suppressed with respect to the lower sideband. Optimizing the constellation quadrature would result in a high-performance link with low BER, due to the fact that each symbol-decision section is optimal. Hence the symbol coordinates are well inside the symbol decision zone. Figure 14.8 provides useful plots for QAM sideband suppression calibration as a function of amplitude balancing between the in-phase arm $I$ and the quadrature arm $Q$, and as a function of phase balancing between the two arms. This calculation assumes an ideal LO splitter. However, that additional impairment can be solved in the same manner presented above.

The EVM includes the error generated by the $I$ and $Q$ imbalances. The smaller the value of the EVM, the higher the quality of the constellation, as can be observed in Fig. 14.6.

The second required calibration is the dc offset between $I$ and Q. This parameter affects the QPSK (quadrature) modulator carrier leakage. A carrier signal leaking from the modulator generates a constellation origin offset at the receiver. This results in BER degradation, since the QAM constellation center is shifted. Assume a QPSK modulator, as given in Fig. 14.1, with the following BB signals and LO impairments:

| | |
|---|---|
| $\alpha$ | Phase error between $I$ and $Q$ signals |
| $\beta$ | Phase balance of the LO splitter within the modulator |
| $\delta$ | Amplitude balance between $I$ and Q |
| $dc_1$ | dc offset of the LO at the $I$ input |
| $dc_2$ | dc offset of the LO at the $Q$ input |
| $dm_1$ | dc offset error at the $I$ input |
| $dm_2$ | dc offset error at the $Q$ input |
| $K_I < \alpha$ | $I$-path-mixer-transfer function, including the phase error $\alpha$ |
| $K_Q$ | $Q$-path-mixer transfer function with zero phase error |
| $K$ | Amplitude balance of the LO splitter within the modulator |

The modulator is assumed to be fed by two orthogonal BB signals and LO:

$$I = A_1 \sin\omega_1 t + dm_1, \tag{14.21}$$

$$Q = A_2 \cos\omega_1 t + dm_2, \tag{14.22}$$

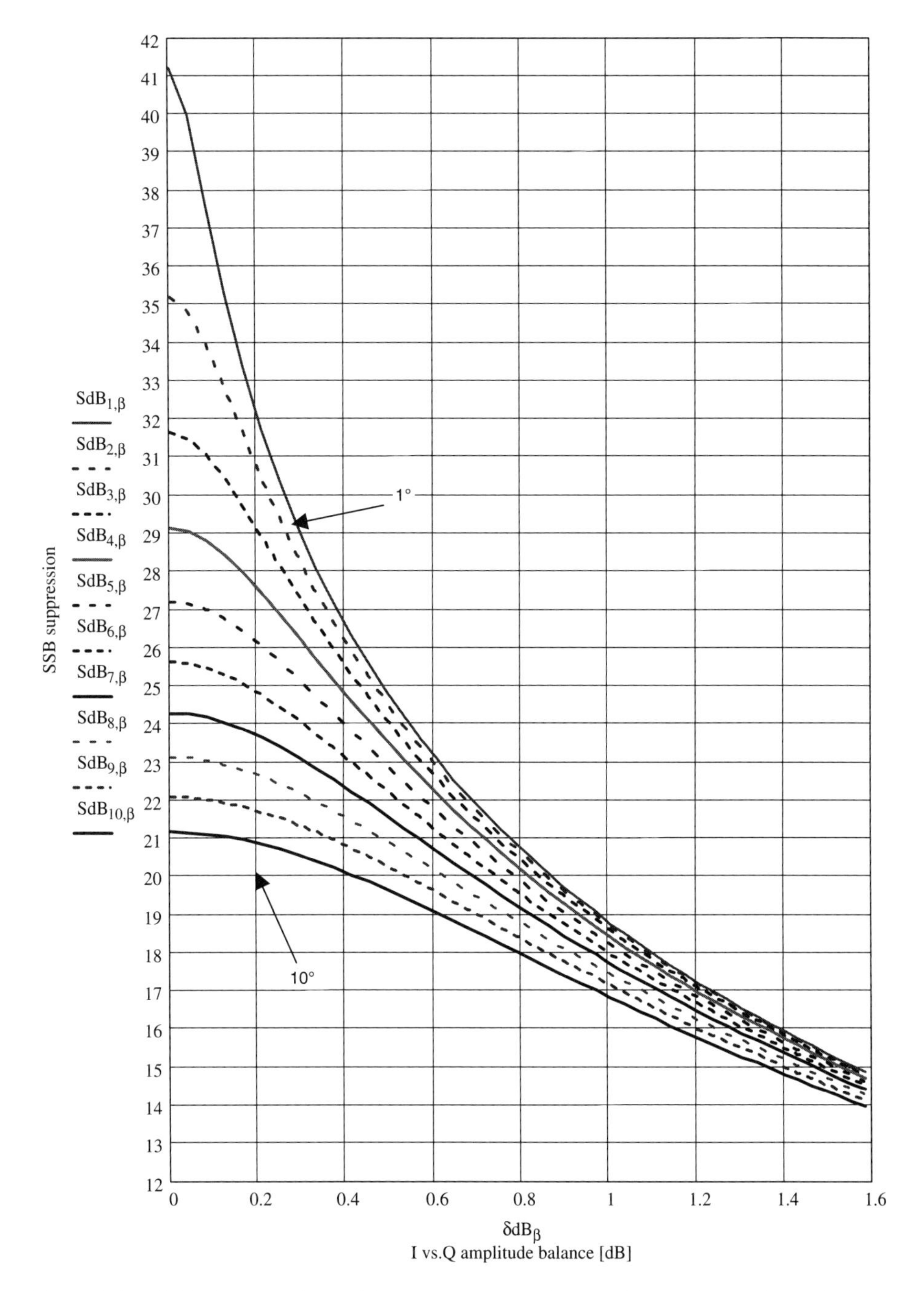


**Figure 14.8** QPSK calibration. Sideband suppression vs. *I* and *Q* amplitude balance, while the phase error is a parameter varying from 1 deg to 10 deg.

$$LO_I = A_I \sin(\omega_c t + \beta) + dc_1, \tag{14.23}$$

$$LO_Q = A_Q \cos \omega_c t + dc_2, \tag{14.24}$$

where $\omega_1$ is the BB frequency, $\omega_c$ is the LO frequency, $A_1$ is the in-phase BB signal amplitude, $A_2$ is the quadrature BB signal amplitude, $A_I$ is the in-phase LO signal amplitude, and $A_Q$ is the quadrature LO signal amplitude. The $I$ and $Q$ arms' outputs are

$$r_I(t) = K_I A_I A_1[\sin(\omega_1 t + \alpha) + dm_1] \cdot [\sin(\omega_c t + \beta) + dc_1], \tag{14.25}$$

$$r_Q(t) = K_Q A_Q A_2[\cos \omega_1 t + dm_2] \cdot [\cos \omega_c t + dc_2]. \tag{14.26}$$

Hence, the modulator output would be the summation of the above signals. Normalizing the result by $|K_Q A_Q A_2|$ provides the normalized magnitude value of $r'(t)$:

$$\begin{aligned} r'(t) = K\delta[\sin(\omega_1 t + \alpha) + dm_1] \cdot [\sin(\omega_c t + \beta) + dc_1] \\ + [\cos \omega_1 t + dm_2] \cdot [\cos \omega_c t + dc_2]. \end{aligned} \tag{14.27}$$

After some trigonometric manipulations as well as using Pythagorean identities, this results in the following signal amplitudes:

$$V_{\mathrm{LSB}}^2 = \frac{1}{4}\left[1 + 2K\delta \cos(\beta - \alpha) + (K\delta)^2\right], \tag{14.28}$$

which is the lower sideband power.

$$V_{\mathrm{USB}}^2 = \frac{1}{4}\left[1 - 2K\delta \cos(\beta + \alpha) + (K\delta)^2\right], \tag{14.29}$$

which is the suppressed upper sideband power, and the carrier power is given by

$$V_{\mathrm{C}}^2 = dm_2^2 + 2K\delta dm_1 dm_2 \sin \beta + (K\delta dm_1)^2. \tag{14.30}$$

Using the above results, the carrier suppression, or carrier leakage with respect to nonsuppressed low sideband is

$$S_{\mathrm{C}}[\mathrm{dBc}] = 10 \log \frac{\frac{1}{4}\left[1 + 2K\delta \cos(\beta - \alpha) + (K\delta)^2\right]}{dm_2^2 + 2K\delta dm_1 dm_2 \sin \beta + (K\delta dm_1)^2}. \tag{14.31}$$

The sideband suppression (residual sideband) is

$$S_{SSB}[\mathrm{dBc}] = 10\log\frac{1 + 2K\delta\cos(\beta - \alpha) + (K\delta)^2}{1 - 2K\delta\cos(\beta + \alpha) + (K\delta)^2}. \tag{14.32}$$

It can be seen that the sideband suppression limits for an ideal amplitude balance of $K = \delta = 1$ are defined by the phase error $\beta$ at the LO splitter and the phase imbalance $\alpha$ between $I$ and Q. The ability to suppress the carrier below the desired sideband is related to the digital to analog (D/A) resolution. A higher D/A resolution provides finer voltage steps that enable a better adjustment for the carrier leakage. Another interesting conclusion is that equality between $\alpha$ and $\beta$ results in an optimum for both conditions of carrier leakage and sideband suppression. Figure 14.9 provides the carrier leakage values for $I$ and $Q$ dc offset imbalance, while the phase balance of the LO splitter is a parameter. It can be seen that the dc calibration has significant importance.

### 14.4.2 Demodulation and QPSK demodulator

QPSK demodulators, shown in Fig. 14.7, have a structure similar to QPSK modulators. Hence, the same impairment-analysis method is useful to analyze the demodulation side of the quadrature demodulator used for QAM transfer. Figure 14.7 illustrates a QPSK demodulator with its impairments as follows:

| | |
|---|---|
| $\alpha$ | Phase error between $I$ and $Q$ signals |
| $\beta$ | Phase imbalance of the LO splitter within the demodulator |
| $\delta$ | Amplitude imbalance ratio between $I$ and Q |
| $dc_1$ | dc offset of the LO at the $I$ input |
| $dc_2$ | dc offset of the LO at the $Q$ input |
| $dm_1$ | dc offset error at the $I$ input |
| $dm_2$ | dc offset error at the $Q$ input |
| $K_{DI} < \alpha$ | $I$-path-mixer transfer function, including the phase error $\alpha$ |
| $K_{DQ}$ | $Q$-path-mixer transfer function with zero phase error, with respect to the $I$ path |
| $K$ | Amplitude imbalance factor of the LO splitter within the demodulator |

Assume an input RF signal of a complex modulation, which is composed of two vectors with completely balanced quadratures, but with a carrier leakage and intermodulation (IMD) or spuriousness at a frequency of $\omega_{\mathrm{IMD}}$. It is also assumed that the IMD frequency is within the BB IF frequency and the carrier leakage value and

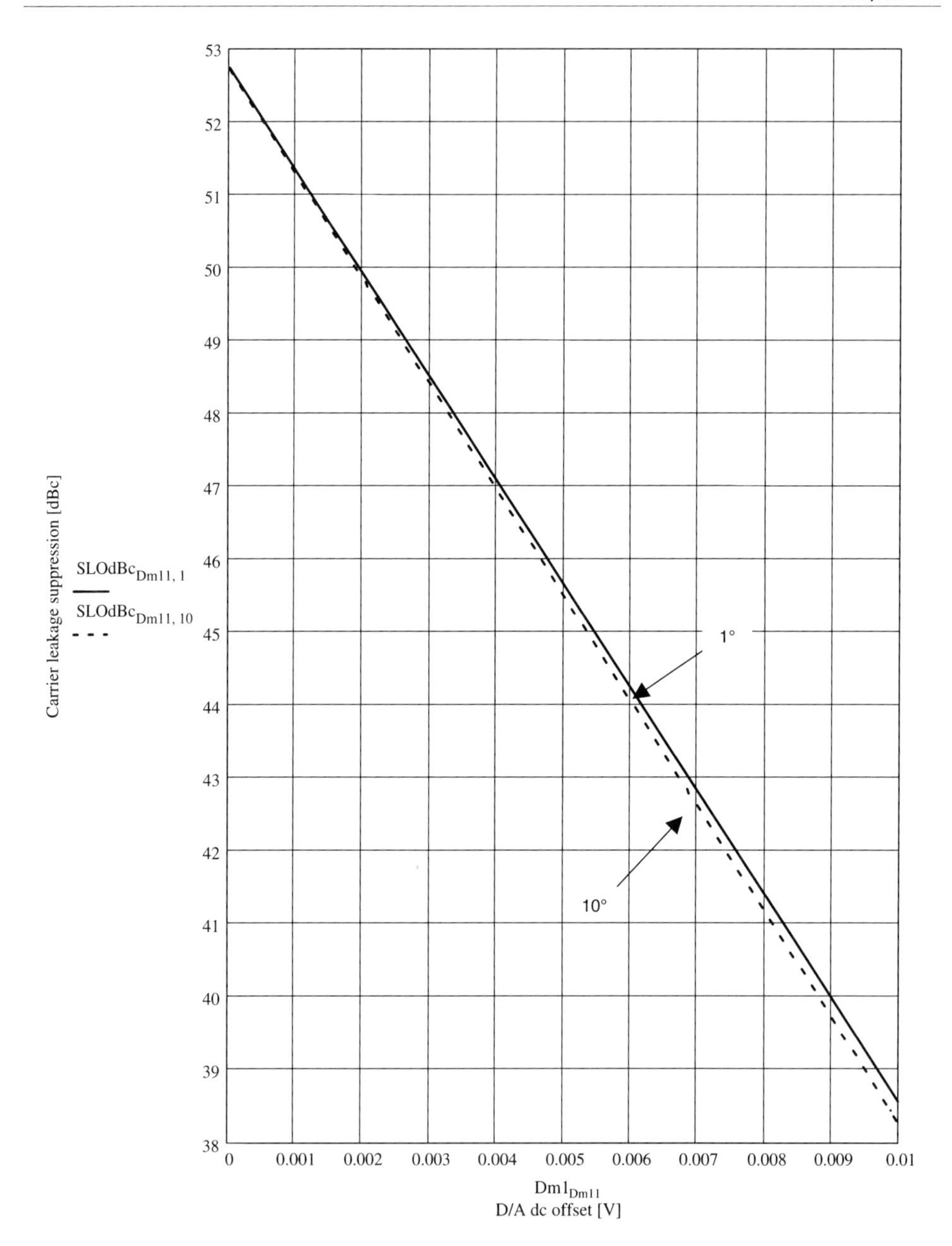


**Figure 14.9** LO carrier leakage suppression below nonsuppressed sideband vs. *Q* BB D/A dc offset error in volts. *I* and *Q* phase error is 1 deg, LO splitter error is a parameter of 1 deg and 9 deg, and *I* and *Q* and LO splitter amplitude balance ratio is 1 or 0 dB.

IMD level values below the signal are given by

$$C[\text{dBc}] = 20\log\frac{\sqrt{A_I^2 + A_Q^2}}{\sqrt{A_{LO-I}^2 + A_{LO-Q}^2}} = 20\log\frac{A}{A_{LO}} = 20\log N_{\text{LO}}, \tag{14.33}$$

and

$$\text{IMD}[\text{dBc}] = 20\log\frac{\sqrt{A_I^2 + A_Q^2}}{\sqrt{A_{\text{IMD}-I}^2 + A_{\text{IMD}-Q}^2}} = 20\log\frac{A}{A_{\text{IMD}}} = 20\log N_{\text{IMD}}. \tag{14.34}$$

Assume that the AWGN projections provide the CNR:

$$\text{CNR}[\text{dBc}] = 20\log\frac{\sqrt{A_I^2 + A_Q^2}}{\sqrt{A_{\text{AWGN}-I}^2 + A_{\text{AWGN}-Q}^2}} = 20\log\frac{A}{A_{\text{AWGN}}}. \tag{14.35}$$

These aforementioned expressions show three mechanisms for generating the error vector, which are shown in Figs. 14.6 and 14.10. The impairment residual power defines the radius of error around the desired constellating-symbol placement. In real-time measurements, all impairments generate an equivalent CNR. The calculation approach for estimating the radius of error sector is provided in Sec. 14.12 in Eq. (14.97). This equivalent radius is related to the total CNR value. The dc error generates complex dc components in both $I$ and $Q$ directions. Hence there is a shift of the decision boundary when determining a symbol. This kind of dc shift is deterministic; thus, it can be somewhat claimed that the probability of error is diverted from zero mean. Section 14.9 briefly discusses the QAM probability of error.

Assume that the receiver selects only the upper sideband during calibration, and the IMD frequency can be above or below the data frequency. Therefore,

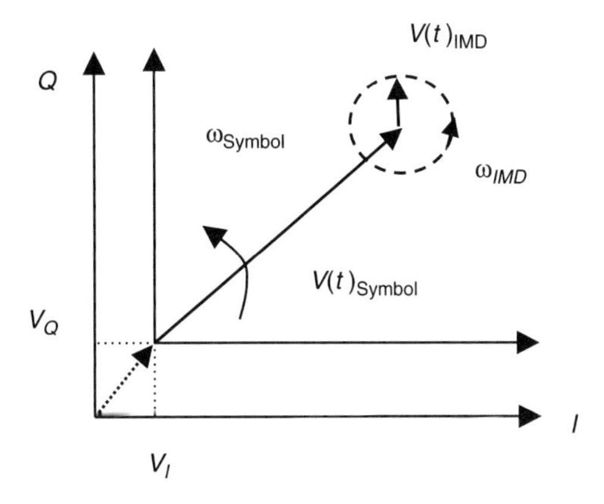


**Figure 14.10** QAM constellation errors due to impairments.

the input voltage to the demodulator would be

$$V_{\mathrm{RF}} = \frac{\sqrt{A_I^2 + A_Q^2}}{2}\cos[(\omega_c + \omega_D)t + \varphi] + \frac{\sqrt{A_{\mathrm{IMD}-I}^2 + A_{\mathrm{IMD}-Q}^2}}{2}\cos[(\omega_c + \omega_{\mathrm{IMD}})t + \psi] + \frac{\sqrt{A_{\mathrm{LO}-I}^2 + A_{\mathrm{LO}-Q}^2}}{2}\cos(\omega_c t + \gamma), \tag{14.36}$$

where the first term describes the data, the second describes the IMD products, and the third describes the carrier leakage. If the interference voltages are referred to the signal level, then using Eqs. (14.33) to (14.45), Eq. (14.36) can be written as

$$V_{\mathrm{RF}} = \frac{\sqrt{A_I^2 + A_Q^2}}{2}\left\{\cos[(\omega_c + \omega_D)t + \varphi] + \frac{1}{N_{\mathrm{IMD}}}\cos[(\omega_c + \omega_{\mathrm{IMD}})t + \psi]. + \frac{1}{N_{\mathrm{LO}}}\cos(\omega_c t + \gamma)\right\}. \tag{14.37}$$

Therefore, by using the image in Fig. 14.7, the demodulator BB in-phase and quadrature output are given by Eqs. (14.38) and (14.39), respectively:

$$V_I = \frac{\sqrt{A_I^2 + A_Q^2}}{4}\left[\cos(\omega_D t + \varphi) + \frac{1}{N_{\mathrm{IMD}}}\cos(\omega_{\mathrm{IMD}} t + \psi) + \frac{1}{N_{\mathrm{LO}}}\cos(\gamma)\right], \tag{14.38}$$

$$V_Q = -\frac{\sqrt{A_I^2 + A_Q^2}}{4}\left[\sin(\omega_D t + \varphi) + \frac{1}{N_{\mathrm{IMD}}}\sin(\omega_{\mathrm{IMD}} t + \psi) + \frac{1}{N_{\mathrm{LO}}}\sin(\gamma)\right]. \tag{14.39}$$

The above results lead to the following conclusions on the QAM constellation picture and the symbol point at a given sampling time. The carrier leakage creates an origin dc offset in both directions $I$ and $Q$, as described by Eq. (14.40). The origin of the dc offset is directly related to LO leakage below the carrier in dBc. The problem becomes more severe when having zero IF (ZIF) topology. In this case, an analog BB (ABB) should have the means to compensate offsets, such as dc injection, with proper polarity using digital-to-analog converters (DAC).

Other root causes for dc offset are second-order distortions that may occur in both $I$ and $Q$ paths. In case of sampled IF topology, ac coupling solves the problem. However, for a ZIF design approach, the ABB should have the means to nullify those dc effects as previously mentioned.

Equation (14.40) demonstrates that there is a scenario that dc offset does not have to be equal in both $I$ and $Q$ arms. This depends on the balance between the $I$ and $Q$ arms:

$$V_i = \frac{\sqrt{A_I^2 + A_Q^2}}{4} \frac{1}{N_{\text{LO}}} \cos(\gamma),$$
$$V_Q = \frac{\sqrt{A_I^2 + A_Q^2}}{4} \frac{1}{N_{\text{LO}}} \sin(\gamma). \tag{14.40}$$

The IMD signal would create a circle around the symbol-target coordinates. The IMD phasor radius and rotation versus time is given by Eq. (14.37); the IMD circle radius is a direct function of the IMD level below the carrier in dBc; Eqs. (14.38) and (14.39) provide the IMD phasor projections on both $I$ and $Q$ axes, respectively. Note that the IMD frequency defies the rotation direction. If the IMD frequency is above the LO or below, the rotations will be clockwise or counter-clockwise, respectively. The IMD phasor can be described by

$$V_{\text{IMD}}(t) = \frac{\sqrt{A_I^2 + A_Q^2}}{4} \frac{1}{N_{\text{IMD}}} \exp(j\omega_{\text{IMD}}) \tag{14.41}$$

as well.

The symbol position at a given sampling instant is given by a phasor:

$$V_{\text{symbol}}(t) = \frac{\sqrt{A(t)_I^2 + A(t)_Q^2}}{4} \exp(j\varphi)\exp(j\omega_D t). \tag{14.42}$$

Since the transmitted signal beats in time to different levels of the in-phase and quadrature voltages, the origin offset is defined by a correction factor, taking into account the average power of the signal. The same applies to the IMD level below the signal. Consequently, the origin offset and the IMD radius around the symbol coordinate are statistical values. Figure 14.10 illustrates a symbol in a constellation diagram with the above impairments.

## 14.5 Jitter-to-Phase-Noise Relationship

Jitter is a significant parameter for characterizing short-term stability of clocks, crystal oscillators, and LOs in the time domain. The relation between the Allen variance (time domain) and phase noise (frequency domain) is described in numerous papers.[3,5,7,13,15] Theoretically, jitter is defined as short-term, noncumulative variations of the significant instants of a digital signal from their ideal positions.

In practice, when evaluating a synchronous optical network (SONET), this is done by a digital storage oscilloscope or time-interval analyzer.

In QAM, there is a need to define the symbol position variance in the constellation map. This is due to phase variation $\Delta\theta_{rms}$ caused by the LO phase noise, as provided by Fig. 14.11. Phase noise is translated into time domain jitter $\Delta t_{rms}$ as illustrated in Fig. 14.12 which provides a MathCAD file to calculate jitter from phase noise data.[6,7]

The accumulated jitter for a number of clock periods $N$ is typically defined as the rms deviation of $N$ periods from average value. The one-period jitter most frequently used for the characterization of clock sources used as references in phase locked loop (PLL) systems corresponds to $N = 1$.

Consider an oscillator model with the absence of AM noise and only phase noise as described by

$$V(t) = V \sin[2\pi f_0 t + \varphi(t)] \tag{14.43}$$

where $f_0$ is the oscillator nominal frequency and $\varphi(t)$ is the oscillator phase noise. Hence Eq. (14.43) describes FM tone. Jitter measurements consist of measuring the time between the zero-crossing value of the voltage $V(t)$ described by Eq. (14.44). This is why phase noise is referred to as FM noise and jitter, since the variance of the signal frequency is around its mean value. In case of an accumulated jitter measurement over $N$ clock periods, the following condition is applicable:

$$V(t_1) = V(t_2) = 0. \tag{14.44}$$

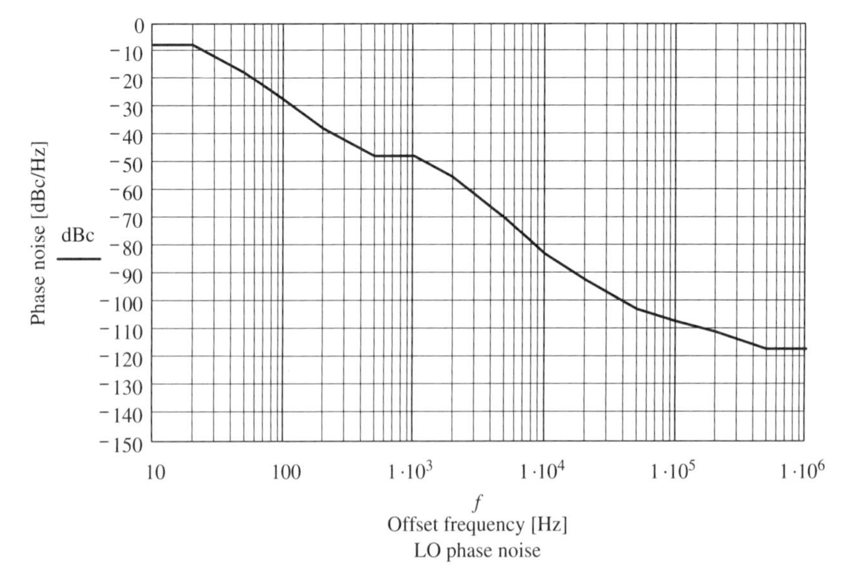


**Figure 14.11** Typical QAM LO phase noise used in RF-tuner frequency converters.

$$\mathrm{dBc} := \begin{pmatrix} -8 \\ -8 \\ -18 \\ -28 \\ -38 \\ -48 \\ -48 \\ -55 \\ -70 \\ -83 \\ -92 \\ -103 \\ -107 \\ -111 \\ -117 \\ -117 \end{pmatrix} \qquad f := \begin{pmatrix} 10 \\ 20 \\ 50 \\ 100 \\ 200 \\ 500 \\ 1000 \\ 2000 \\ 5000 \\ 10000 \\ 20000 \\ 50000 \\ 100000 \\ 200000 \\ 500000 \\ 1000000 \end{pmatrix}$$

Data vectors of phase noise [dBc/Hz] versus offset frequency [Hz]:

$i := 0..\ \mathrm{last}(\mathrm{dBc}) - 1$

$$a_i := \frac{\mathrm{dBc}_i - \mathrm{dBc}_{i+1}}{\log(f_i) - \log(f_{i+1})}$$

$$b_i := \mathrm{dBc}_i - \left[a_i \cdot \log(f_i)\right]$$

$$\Delta\theta\mathrm{rms\_deg} := \frac{180\cdot\sqrt{2}}{\pi}\cdot\sqrt{\sum_{i=0}^{\mathrm{last}(\mathrm{dBc})-1}\int_{f_i}^{f_{i+1}} 10^{\frac{b_i}{10}}\cdot\left[(X)^{0.1\cdot a_i}\right]dX}$$

$$\Delta\theta\mathrm{rms\_deg} = 151.957$$

$$\Delta\theta\mathrm{rms} := \frac{\pi}{180}\cdot\Delta\theta\mathrm{rms\_deg}$$

$$\Delta\theta\mathrm{rms} = 2.652$$

**Figure 14.12** MathCAD program to calculate the integrated phase-noise jitter given in Fig. 14.11. The calculation is based on the narrow FM approximation method.[15]

Hence

$$\begin{aligned} 2\pi f_0 t_1 + \varphi(t_1) &= 0, \\ 2\pi f_0 t_2 + \varphi(t_2) &= 2\pi N. \end{aligned} \tag{14.45}$$

As a result of subtracting the two conditions in Eq. (14.45), the jitter is given by

$$2\pi f_0(t_2 - t_1) + \varphi(t_2) - \varphi(t_1) = 2\pi N. \tag{14.46}$$

However, by definition of a jitter at zero–crossing, the sampling interval is given by

$$t_2 - t_1 = NT_0 + \Delta t, \tag{14.47}$$

where $T_0 = 1/f_0$ and $\Delta t$ is the jitter accumulation during $N$ clock periods. Substituting Eq. (14.47) into Eq. (14.46) provides the relation between $\Delta\theta_{\mathrm{rms}}$ and $\Delta t_{\mathrm{rms}}$:

$$\Delta t = \frac{T_0}{2\pi}[(\varphi(t_1) - \varphi(t_2))]. \tag{14.48}$$

Equation (14.48) describes a stochastic process with the random variables $\varphi(t_1)$, $\varphi(t_2)$, and $\Delta t$. Therefore the average result of the square rms value is given by

$$\langle \Delta t^2 \rangle = \frac{T_0^2}{4\pi^2}\left[\langle \varphi(t_1)^2 \rangle - 2\langle \varphi(t_1) \times \varphi(t_2) \rangle + \langle \varphi(t_2)^2 \rangle\right]. \tag{14.49}$$

The above equation describes an integration over the frequency domain of the power spectral density $S_\varphi(f)$ as described by

$$\langle \varphi(t)^2 \rangle = \int_0^\infty S_\varphi(f)\mathrm{d}f \tag{14.50}$$

and the autocorrelation is given by

$$\langle \varphi(t_1) \times \varphi(t_2) \rangle = R_\varphi(t_2 - t_1) = R_\varphi(\tau) = \int_0^\infty S_\varphi(f)\cos(2\pi f\tau)\mathrm{d}f. \tag{14.51}$$

Hence, the overall jitter is given by substituting Eqs. (14.50) and (14.51) into Eq. (14.44) resulting in

$$\Delta t_{\mathrm{rms}} = \sqrt{\frac{T_0^2}{\pi^2}\int_0^\infty S_\varphi(f) \times \sin^2(\pi f\tau)\mathrm{d}f}. \tag{14.52}$$

There are several definitions for oscillator phase noise. The common one is single-sideband (SSB) phase noise, which is marked as $\mathfrak{I}(f)$ and its units are measured in dBc/Hz. The value $\mathfrak{I}(f)$ is evaluated using narrow FM approximation[15] as integrated phase noise $\Delta\theta_{\mathrm{rms}}$. Assume the oscillator is described per

Eq. (14.43). This can be modified into

$$S(t) = A(t)\cos\left(\omega t + \frac{\Delta f}{f_{\mathrm{m}}}\sin\omega_{\mathrm{m}} t\right), \tag{14.53}$$

where $f_{\mathrm{m}}$ is the modulation frequency and $\Delta f$ is the frequency deviation. Using trigonometry identities, Eq. (14.53) is modified to

$$\begin{aligned} S(t) &= V\cos(\omega t + \beta\sin\omega_{\mathrm{m}} t) \\ &= V[\cos(\omega t)\cos(\beta\sin\omega_{\mathrm{m}} t) - \sin(\omega t)\sin(\beta\sin\omega_{\mathrm{m}} t)]. \end{aligned} \tag{14.54}$$

For a small deviation, $\beta \ll 1$, $\cos(\beta\sin\omega_{\mathrm{m}} t) \approx 1$, and $\sin(\beta\sin\omega_{\mathrm{m}} t) \approx \beta\sin\omega_{\mathrm{m}} t$, Eq. (14.54) is approximated to

$$S(t) = V[\cos\omega t - \sin\omega t(\beta\sin\omega_{\mathrm{m}} t)], \tag{14.55}$$

which becomes

$$S(t) = V\left\{\cos\omega t - \frac{\beta}{2}[\cos(\omega + \omega_{\mathrm{m}})t - \cos(\omega - \omega_{\mathrm{m}})t]\right\}. \tag{14.56}$$

The function $S(t)$ represents a narrow FM oscillator carrier and its two sidebands, where each sideband's amplitude coefficient is $\beta/2$. The sideband energy at a given 1-Hz BW is denoted as SSB phase noise $L(f_{\mathrm{m}})$. Note that the first argument in Eq. (14.56) represents the carrier voltage and the other two terms are the sideband voltages at 1-Hz BW. It is known that the power is proportional to the voltage square and the power ratio in decibels below the carrier is the ratio of the noise voltage square to the carrier voltage square. Thus the phase noise at 1 Hz, known as dBc/Hz, for a single sideband is

$$L(f_{\mathrm{m}}) = \left(\frac{v_n}{V}\right)^2 = \frac{\beta^2}{4} = \frac{\Delta\theta_{\mathrm{rms}}^2}{2}. \tag{14.57}$$

The double sideband phase noise is given by

$$2L(f_{\mathrm{m}}) = \left(\frac{v_n}{V}\right)^2 = 2\frac{\beta^2}{4} = 2\frac{\Delta\theta_{\mathrm{rms}}^2}{2} = \Delta\theta_{\mathrm{rms}}^2. \tag{14.58}$$

Since $v_{\mathrm{n}}$ is noise density at 1 Hz and the ratio of $(v_{\mathrm{n}}/V)^2$ is the phase noise value at a given offset, the integrated phase noise is given by the integral of the noise density over the band of interest.

The next question is what are the boundaries of integration? Referring to Eq. (14.52) as a case study, assume a clock of 250 KHz, and hence a period of 4 μs. Assume a time division multiple access (TDMA) slot period of 5 KHz. It can be the measurement interval drift rate or any other time-slot event. Assume that the symbol rate is 1 MHz over a period of 1 μs. Thus the lower boundary of integration is 5 KHz and the upper bound is 1 MHz. $T_0 = 4$ μs in which we have $N$ cycles of cumulative jitter; the same applies to $\tau$, and the same approach holds for Eq. (14.58).

A practical example is given below that demonstrates how to implement Eq. (14.58).

Assume the following synthesized LO is for a QAM RF converter. The total jitter of this LO is 152 deg and the MathCAD integral calculation is provided below in Eq. (14.59).[1,10] However, the excessive jitter is reduced by the synchronization circuits that have a high-pass response. More details are provided in the following paragraphs that deal with clock and carrier-recovery circuits and algorithms. As was previously explained, the jitter requirement is 0.15 deg at QAM 256. Therefore, the maximum phase noise should not exceed that high value of 152 deg. This requirement results in a stringent spec of noise rejection by the carrier recovery PLL, which has a response of a high-pass filter.[2]

The next question is how much the EVM and modulation error ratio (MER) are affected. Using Fig. 14.16, the phase error generated by the phase noise defines the EVM error vector, while assuming no other degradations. Using geometrical approximation of small angles where $\theta \approx \sin\theta$, the following can be stated. CNR is the ratio of the ideal symbol vector to the error vector:

$$\text{err\_vector\_power} \propto \sin^2\theta \approx \theta^2. \tag{14.59}$$

## 14.6 Residual AM Effects

One of the problems introduced in QAM transport is undesired residual AM. Such an impairment may result from amplitude fluctuation in a receiving or transmitting channel. Assume that a receiver case has a fluctuating AGC. As was mentioned in Chapter 12, AGC should be a slow system to avoid residual AM. However, in case of an improper AGC design it may oscillate or have a fast response to optical fluctuations. Thus, both $I$ and Q, which pass through the voltage variable attenuator (VVA), may exhibit the same amplitude fluctuations. Hence, both vectors are correlated with their amplitude fluctuations. Equation (14.1) can be modified to contain the modulation information with the residual AM. Consider weak sinusoidal amplitude fluctuations represented by a modulating signal,

$$K(t) = k \cdot [1 + M\cos(\omega t)], \tag{14.60}$$

where $k$ is the voltage variable attenuator's (VVA's) average attenuation, $M$ is modulation index, $0 < M < 1$ or coupling to the VVA and $K(t)$ is the VVA swing around its average. Then $r(t)$ is given by

$$r_{\text{resAM}}(t) = K(t) \cdot i(t)\cos(\omega_c t) + jK(t) \cdot q(t)\sin(\omega_c t). \tag{14.61}$$

Substituting Eq. (14.60) into Eq. (14.61) results in

$$r_{\text{resAM}}(t) = k \cdot i(t) \cdot [1 + M\cos(\omega t)]\cos(\omega_c t) + jk \cdot q(t) \cdot [1 + M\cos(\omega t)] \times \sin(\omega_c t). \tag{14.62}$$

Rewriting Eq. (14.62) in a polar presentation as in Eq. (14.2) results in

$$\begin{aligned} r_{\text{resAM}}(t) &= \sqrt{\{i(t) \cdot k[1 + M\cos(\omega t)]\}^2 + \{q(t) \cdot k[1 + M\cos(\omega t)]\}^2} \\ &\quad \times \exp\left[j\left((\omega_c t) + \tan^{-1}\left\{\frac{q(t) \cdot k \cdot [1 + M\cos(\omega t)]}{i(t) \cdot k \cdot [1 + M\cos(\omega t)]}\right\}\right)\right] \\ &= k[1 + M\cos(\omega t)]\sqrt{i(t)^2 + q(t)^2} \cdot \exp\left(j\left\{(\omega_c t) + \tan^{-1}\left[\frac{q(t)}{i(t)}\right]\right\}\right). \end{aligned} \tag{14.63}$$

This result shows that the symbol target point draws a diagonal line toward the constellation center as shown in Fig. 14.13. For an FTTx ITR, such residual AM may be also due to strong X-talk coupling into the VVA control voltage, resulting in residual AM. The amount of AM in $I$ and $Q$ is described in Fig. 14.13 and is

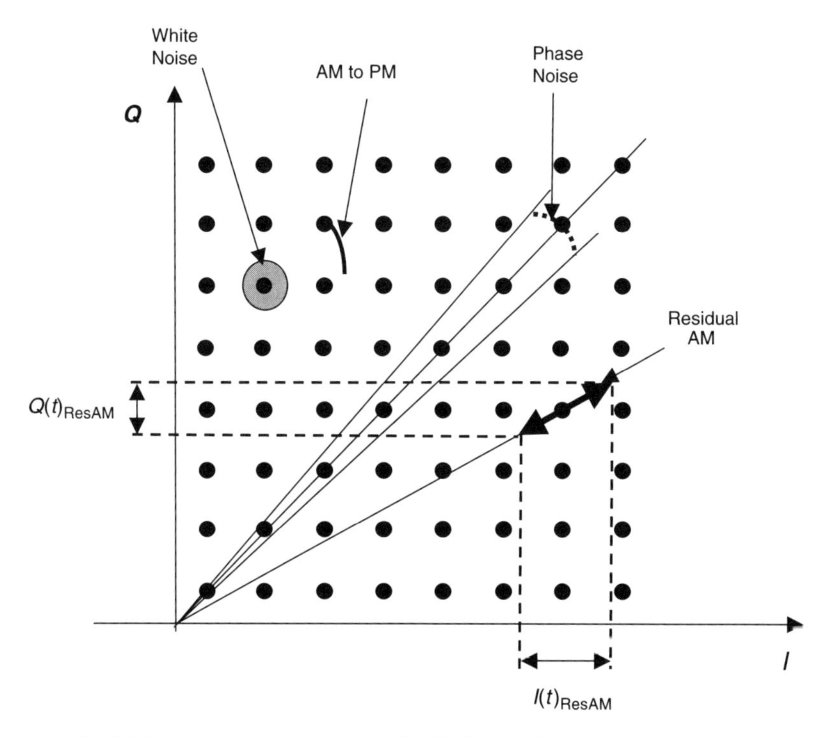


**Figure 14.13** QAM impairment effects of AWGN, AM to PM, phase noise, and residual AM.

given by Eq. (14.64):

$$I(t) = k \cdot i(t)[1 + M\cos(\omega t)],$$
$$Q(t) = k \cdot q(t)[1 + M\cos(\omega t)]. \quad (14.64)$$

Note that Eq. (14.63) describes the quiescent location of $r(t)$ projections on $I$ and $Q$ axes and the deviation from the desired point by the $M\cos(\omega t)$, where $-1 \leq \cos(\omega t) \leq 1$.

## 14.7 Nonlinear Effects

Three other RF imperilments that may affect the constellation are AM to PM, AM to AM, and phase noise. Again the analysis is similar to the previous one, since both $I$ and $Q$ pass through the same RF channel they are correlated.

Assume in the first case that the integrated phase noise after the carrier recovery loop is $\pm\Delta\theta_{\mathrm{rms}}$. This quantity will modify Eq. (14.2) into[4]

$$r(t) = \sqrt{i^2(t) + q^2(t)} \cdot \cos\left\{\omega_c t + \tan^{-1}\left[\frac{q(t)}{i(t)}\right] \pm \Delta\theta_{\mathrm{rms}}\right\}$$
$$= A(t)\cos(\omega_c t + \theta(t) \pm \Delta\theta_{\mathrm{rms}}). \quad (14.65)$$

Note that $\pm\Delta\theta_{\mathrm{rms}}$ is the result of FM noise, which is a constant envelope modulation; thus, there is no change in magnitude. The radius vector of that symbol will draw a banana shape around the target. The sector opening will be the angle value of $2|\Delta\theta_{\mathrm{rms}}|$ with respect to the constellation center, as shown in Fig. 14.13.

Using the same approaches used for residual AM and phase noise will help in understanding AM-to-AM and AM-to-PM compression with memory effects, which introduces amplitude and phase changes due to these memory effects. At high-energy modulation levels, i.e., high $I$ and $Q$ voltages, the instantaneous power through the RF channel reaches its peak level. This may result in compression and AM to PM. Thus for high-energy symbols in the constellation, one would expect to see larger deviations from the desired target symbol compared to lower-energy symbols. Since AM to PM involves both amplitude compression and phase change, the symbol draws a curved shape toward the constellation center. The compression effect reduces both the $I$ and the $Q$ amplitudes, and rotates the resultant $I$ and $Q$ radii vectors because of the phase change. This impairment is described in Fig. 14.13.

Another parameter affecting the constellation symbol is AWGN. Since white noise is a random signal process with zero mean, it has an rms value. The noise signal randomly changes its amplitude and frequency. Thus it can be described as an additive vector with variable amplitude and frequency. Following the same process results in the demonstration in Fig. 14.10. AWGN would create a cloud around the symbol target. The diameter of the noise cloud is related to the $C/N$ or $E_b/N_0$, and the noise energy is defined by the system's IF bandwidth.

In addition to these parameters, there are other parameters such as ISI as well as those that result due to poor $I$ and $Q$ calibration, as mentioned in Sec. 14.4.1, and sampling misalignment.

## 14.8 EVM and MER

To quantify the figure of merit of QAM transport that can be measured, the EVM and MER values were established. The MER definition is commonly used for the QAM constellation quality evaluation and is given by

$$\mathrm{MER[dB]} = -10\log\left[\frac{\sum_{j=1}^{N}\left(\delta I_j^2 + \delta Q_j^2\right)}{\sum_{j=1}^{N}\left(I_j^2 + Q_j^2\right)}\right]. \quad (14.66)$$

The denominator of Eq. (14.66) describes the ideal constellation data points. Looking back at Sec. 14.3, we find that it also describes the ideal constellation sampled energy. The nominator describes the error energy between the target ideal symbol point to the actual symbol coordinate. In other words, it is the sum of all energy errors, which is the sum of all EVM squares. Hence MER information can provide a measure of the difference between the ideal constellation energy and the actual energy due to transmitter imperfections as well as additional channel imperfections.

An additional parameter used to evaluate the constellation is EVM, which is defined by

$$\mathrm{EVM}[\%] = 100\sqrt{\frac{(1/N)\sum_{j=1}^{N}\left(\delta I_j^2 + \delta Q_j^2\right)}{S_{\max}^2}}, \quad (14.67)$$

where $S_{\max}^2$ describes the peak energy of the constellation, as elaborated in Sec. 14.3. The nominator describes the average energy error between the ideal constellation and the actual sampled constellation. Thus, the EVM describes the average EVM normalized to the constellation peak radius or peak energy. The EVM value and MER value are related by

$$\frac{\mathrm{EVM}^2}{\mathrm{MER}} = \frac{(1/N^2)\sum_{j=1}^{N}\left(I_j^2 + Q_j^2\right)}{S_{\max}^2} = \frac{P_{\mathrm{rms}}}{P_{\mathrm{peak}}} = \frac{1}{\mathrm{PAR}}, \quad (14.68)$$

where PAR is the peak-to-rms value described in Eq. (14.13).

## 14.9 BER of M-ary QAM

One of the methods to evaluate data transport is by measuring probability of error. In previous sections, several impairments were examined. These impairments affect the location of the symbol in the constellation map and, therefore, may introduce an error. In this section, a basic bit error rate (BER) analysis of M-ary QAM as a function of SNR is presented. For this purpose, it is necessary to first find the bit-to-noise energy ratio, and then to derive the probability-of-error model. The concept of symbol energy and bit energy calculation is given in Sec. 14.3. For $M$ states, the following condition describes the number of bits per symbol: $p = \log_2 M$. Note that $M = 2^p$, and $M$ is the number of states that $p$ bits can provide.

According to Proakis,[12] QAM-modulated BER can be analyzed due to the similarity of energies as in PAM with even $k$. M-ary PAM is a 1D modulation scheme organized as described in Fig. 14.14. Note that the symbol coding is Gray encoded, where the symbols differ by a bit digit.

In PAM, if, for instance, $M = 64$, then there are 6 bits to describe 64 states. In a 64 QAM case, it is a 2D modulation; thus, 3 bits are needed for the $I$ and 3 for the Q. Hence it can be said that QAM is bidimensional PAM, where each axis consists of $k = \sqrt{M}$ states, where each state is defined by a different bits-permutation $p = \log_2 \sqrt{M}$ and there are $2^p$ levels for each PAM axis. Accordingly, each quadrant consists of $4 \times 4$ states, with a total of 64 states for all four quadrants. Therefore, it is clear that $k$ is an even and rational number. In this way, the PAM BER probability can be converted to QAM BER.

The Euclidian distance between two adjacent symbols' amplitudes is denoted as $2d$. This distance represents voltage and is square energy; the distance between the symmetry $Y$-axis to the nearest symbol is $d$. The decision threshold between two adjacent symbols is then half of the Euclidian distance. Hence the amplitude of each state $A_m$ is given by

$$A_m = (2m - 1 - M)d, \tag{14.69}$$

where $m = 1, 2, \ldots, M$. The energy of each symbol is marked as $E_g$ and the average energy of each pulse is marked as $E_g/2$. Therefore the voltage of each

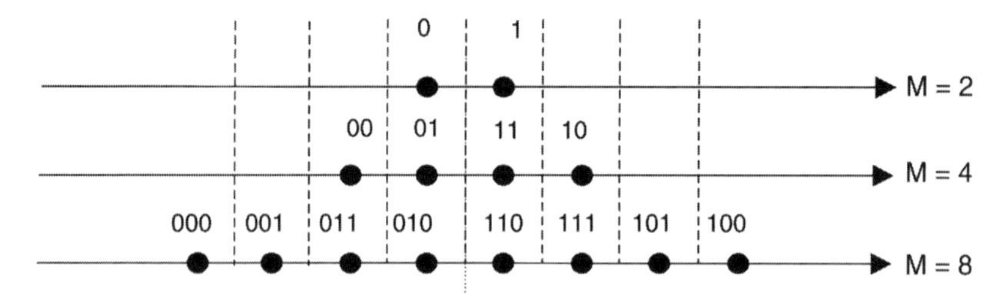


**Figure 14.14** Signal space diagram for digital PAM-modulated signals.[11] (Reprinted with permission from Digital Communications © McGraw Hill.)

symbol is related to its coordinate and is given by

$$V_m = A_m\sqrt{\frac{E_g}{2}}. \qquad (14.70)$$

Applying Eq. (16.69) into Eq. (16.70) results in the Euclidian distance between two adjacent signals:

$$V_{m+1} - V_m = 2\sqrt{\frac{E_g}{2}} = \sqrt{2E_g}. \qquad (14.71)$$

Assuming equally probable signals, the average energy is given by

$$\begin{aligned} E_{ave} &= \frac{1}{M}\sum_{m=1}^{M} E_m = \frac{1}{M}\sum_{m=1}^{M}(A_m V_m)^2 \\ &= \frac{d^2 E_g}{2M}\sum_{m=1}^{M}(2m-1-M)^2 = \frac{1}{6}d^2 E_g\left(M^2-1\right). \end{aligned} \qquad (14.72)$$

Normalizing to $E_g/2$ would result in the geometrical symbol average energy

$$E_s = \frac{\left(M^2-1\right)}{3}d^2. \qquad (14.73)$$

Signals are characterized by their average power, and thus the power of a symbol throughout its duration is given by $E_s/T$. The average probability of error for M-ary PAM can be determined from a decision-rule set by the threshold boundary between two adjacent symbols at the midpoint between the symbols. The detector compares the demodulator output $r$ with a set of $M-1$ thresholds, as shown in Fig. 14.14 by the dashed lines. In this manner, a decision is made in favor of the amplitude level that is closest to $r$.

In the presence of AWGN noise with zero mean and variance $\sigma_n^2 = \frac{1}{2}N_0$ denoted by the variable $n$, the output of the demodulator is

$$r = V_m + n = A_m\sqrt{\frac{E_g}{2}} + n. \qquad (14.74)$$

Assuming all amplitude levels are equally likely *a priori*, the average probability of a symbol error (SER) is when the noise is exceeding the amplitude of one-half distance between levels. Thus the error can be in both directions. However, when the two outside symbols are transmitted, $\pm(M-1)$, the error is in one direction only. Therefore, the probability of error occurs when the symbol displacement distance between the desired point to the actual point is larger than the threshold value. Hence the probability of error is given by

$$P_M = \frac{M-1}{M} P\left( |r - V_m| > A_m \sqrt{\frac{E_g}{2}} \right) = \frac{M-1}{M} \frac{2}{\sqrt{\pi N_0}} \int_{d\sqrt{E_g/2}}^{\infty} \exp\left(-\frac{x^2}{N_0}\right) dx$$

$$= \frac{M-1}{M} \frac{2}{\sqrt{2\pi}} \int_{\sqrt{E_g d^2/N_0}}^{\infty} \exp\left(-\frac{x^2}{2}\right) dx = \frac{2(M-1)}{M} Q \sqrt{\frac{E_g d^2}{N_0}}, \tag{14.75}$$

where $Q$ is the complementary error function denoted as $\mathrm{erfc}(x)$:

$$Q(x) = \frac{1}{\sqrt{2\pi}} \int_{x}^{\infty} \exp\left(\frac{-u^2}{2}\right) du = \frac{1}{2} \mathrm{erfc}\left(\frac{x}{\sqrt{2}}\right). \tag{14.76}$$

Equation (14.76) can be modified by applying the expression for $E_g$ by Eq. (14.71) where

$$E_g = \frac{6E_{ave}}{d^2(M^2-1)}. \tag{14.77}$$

Therefore Eq. (14.75) becomes

$$P_M = \frac{2(M-1)}{M} Q \sqrt{\frac{6E_{ave}}{N_0(M^2-1)}}. \tag{14.78}$$

However, $E_{ave}$ is the average energy of the symbol. Assuming Gray coding, each symbol state is represented by $k$ bits, as per $k = \log_2 M$. Thus, the average symbol energy is $k$ times the average energy of a bit $E_b$. Thus the SER probability of error is given by:

$$P_M = \frac{2(M-1)}{M} Q \sqrt{\frac{(6 \log_2 M)}{(M^2-1)} \frac{E_b}{N_0}}. \tag{14.79}$$

Hence the BER of PAM is given by

$$P_M = \frac{2}{\log_2 M} \left(1 - \frac{1}{M}\right) Q \sqrt{\frac{(6 \log_2 M)}{(M^2-1)} \frac{E_b}{N_0}}. \tag{14.80}$$

Recall that QAM consists of two orthogonal PAM; then, when analyzing M-ary PAM, it is equivalent to two orthogonal $\sqrt{\text{M}}$-ary PAM.[11] By using the same procedure in Sec. 14.3, the PAM probability of error can be modified into QAM.[11] Since the case here is of two orthogonal PAM, the overall probability of QAM is the square of PAM probability. It is important to emphasize the fact that error propabilities in most of the equations are symbol error probabilities. The above analysis is for the occurrence of an error; hence the $\sqrt{M}$ PAM complementary probability is for a correct decision and its square related to M-ary QAM.

$$P_{\text{M\_QAM}} = \left(1 - P_{\sqrt{\text{M}}\text{\_PAM}}\right)^2. \tag{14.81}$$

Therefore the probability of error is given by the complementary:

$$\overline{P}_{\text{M-QAM}} = 1 - \left(1 - P_{\sqrt{\text{M}}\text{\_PAM}}\right)^2. \tag{14.82}$$

Since QAM is made of two orthogonal PAM, each quadrature would have half of the energy. Thus Eq. (14.78) is modified to

$$P_{\sqrt{M}} = 2\left(1 - \frac{1}{\sqrt{M}}\right) Q\sqrt{\frac{3E_{\text{ave}}}{N_0(M-1)}}. \tag{14.83}$$

Assuming $P_{\sqrt{M}} \ll 1$, then by using Taylor series,[9] Eq. (14.82) is approximated to QAM SER:

$$\overline{P}_{\text{M-QAM}} \approx 2P_{\sqrt{M}} = 4\left(1 - \frac{1}{\sqrt{M}}\right) Q\sqrt{\frac{3E_{\text{ave}}}{N_0(M-1)}}. \tag{14.84}$$

Hence, QAM BER is given by

$$\overline{P}_{\text{M-QAMB}} = \frac{4}{\log_2 M}\left(1 - \frac{1}{\sqrt{M}}\right) Q\sqrt{\frac{\left(3\log_2\sqrt{M}\right)}{(M-1)} \cdot \frac{E_b}{N_0}}. \tag{14.85}$$

The Euclid energy distance $d$ for M-ary QAM can be drived from

$$E_{\text{ave}} = \left(\log_2 M\right) E_b = \frac{d^2}{M} \sum_{i=1}^{\sqrt{M}} \sum_{q=1}^{\sqrt{M}} \left(A_i^2 + A_q^2\right) = \left[\frac{2(M-1)}{3}\right] d^2, \tag{14.86}$$

resulting in[15]

$$d = \sqrt{\frac{3(\log_2 M)E_b}{2(M-1)}}. \tag{14.87}$$

Figure 14.15 illustrates QAM BER versus $E_b/N_0$ calculation for 16 QAM, 64 QAM, and 256 QAM using MathCAD.[10]

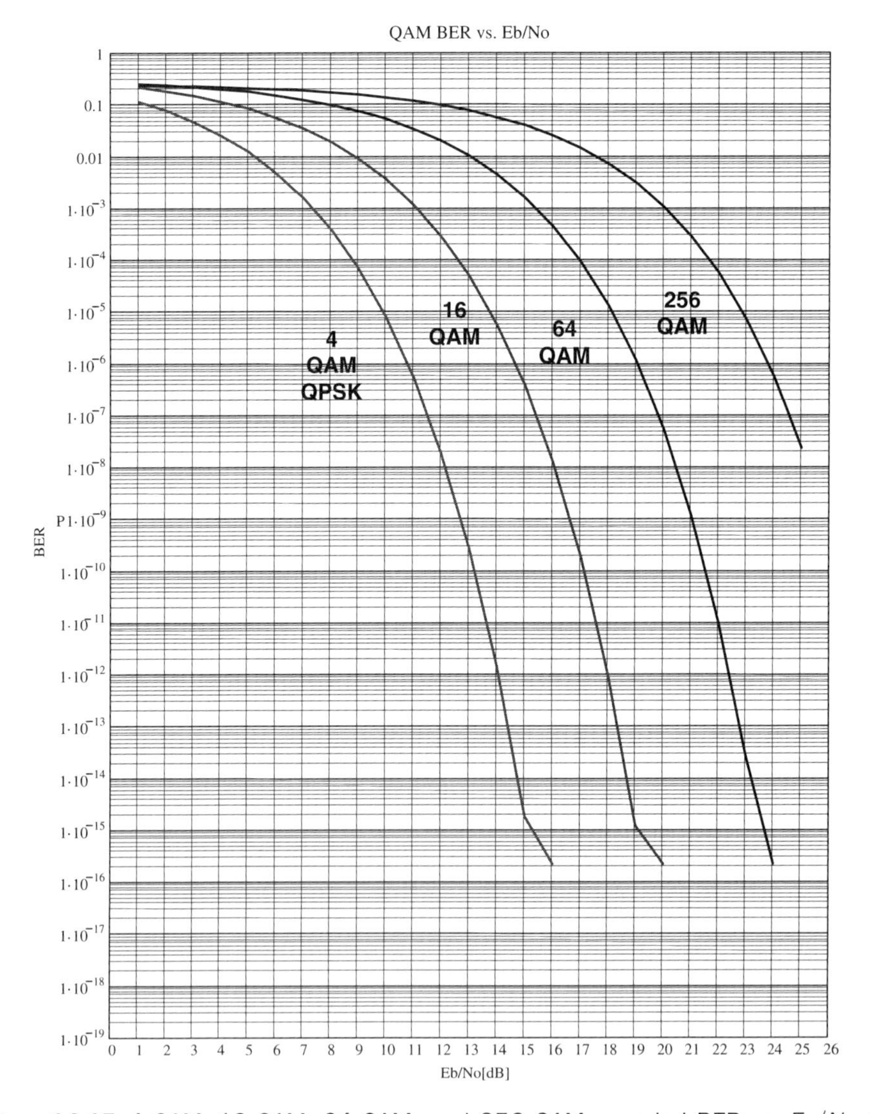


**Figure 14.15** 4 QAM, 16 QAM, 64 QAM, and 256 QAM uncoded BER vs. $E_b/N_0$ per Eq. (14.81).

## 14.10 Relationship between $E_b/N_0$ and $C/N$

In digital transport, it is convenient to use the ratio of the required bit energy $E_b$ to noise density $N_0$, denoted as $E_b/N_0$. However, in practical measurements, it is more common to measure the average carrier to average noise, $C/N$. The conversion from these two quantities is as follows:[13,14]

$$E_b = CT_b = C\left(\frac{1}{f_b}\right), \tag{14.88}$$

$$N_0 = \frac{N}{B_W}, \tag{14.89}$$

$$\frac{E_b}{N_0} = \frac{CT_b}{N/B_W} = \frac{CB_W}{Nf_b}. \tag{14.90}$$

Thus, the transformation is

$$\frac{E_b}{N_0} = \frac{C}{N} \cdot \frac{B_W}{f_b}, \tag{14.91}$$

where $B_W$ is the receiver noise BW, $f_b$ is the bit rate, $C$ is the carrier level, and $N$ is the noise level. Thus CNR times the ratio of receiver BW to bit rate provides $E_b/N_0$.

Using the above-mentioned relations between $E_b/N_0$ and SNR or CNR, the M-ary QAM BER is given by the following approximation:[14]

$$P_{M\text{-QAM}} = 2\left(1 - \frac{1}{\sqrt{M}}\right)\text{erfc}\left[\sqrt{\frac{3\text{CNR}(\text{BW}/R_s)}{2(M-1]}}\right] \times \left\{1 - \frac{1}{2}\left(1 - \frac{1}{\sqrt{M}}\right)\text{erfc}\left[\sqrt{\frac{3\text{CNR}(\text{BW}/R_s)}{2(M-1)}}\right]\right\}, \tag{14.92}$$

where BW is the channel BW, typically 6 MHz, $R_s$ is the symbol rate, typically 5 Mbps at 64 QAM, $M$ is the modulation order, and CNR, sometimes denoted as SNR, is carrier-to-noise ratio. Note that $E_b/N_0$ is given by

$$\frac{E_b}{N_0} = \text{CNR}\frac{\text{BW}}{R_s}. \tag{14.93}$$

## 14.11 BER versus $E_b/N_0$ to $C/N$ Performance Limits

The BER versus $E_b/N_0$ chart displays the performance of QAM with AWGN without any additional DSP gain processing. This is accomplished by forward-error-correction (FEC) encoding. In Sec. 3.3.6, an introduction to MPEG and

TV signal digital coding was provided. Gain processing improves $E_b/N_0$, and thus, the entire chart shifts to the left and becomes much sharper. However, it cannot be improved beyond a limit. These BER versus $E_b/N_0$ limits may rise from additional link impairments as were mentioned above. One of them is phase noise that directly transforms into jitter. The jitter is commonly described by the integrated phase noise $\Delta\theta_{rms}$ and $\Delta t_{rms}$ mentioned in Sec. 14.5. Additional parameters such as NLD due to signal clipping as well as constellation calibration impairments would limit BER performance. An intuitive explanation is by the MER and EVM values describing the offset errors due to symbol displacement in the constellation. The next chapter provides a simple analytical approach for jitter effects on the QAM scheme after providing the structural concept behind a CATV MODEM.[1,3]

## 14.12 EVM Relations to $C/N$

EVM is an error voltage between the actual symbol placement on the constellation diagram and its desired target point. The displacement magnitude is the error vector value. In previous sections, various errors were presented. Those errors affecting the overall $C/N$, and thus the rms sum of all channel impairments such as AWGN, integrated phase noise, IMD products, and spuriousness construct an equivalent vector. This vector magnitude is the symbol displacement error in the constellation. In other words the EVM represents the equivalent noise voltage. From the definitions for MER in Sec. 14.8, it is clear that 1/MER represents the average $C/N$. In the same way, the inverse of EVM$^2$ value (1/EVM$^2$) represents $C/N$ between the peak power and the average noise. Geometrically, 1/EVM represents the error angle in a constellation. For the sake of simplicity, assume an 8 PSK or $\pi/4D$ QPSK constellation. From the definition of EVM in Eq. (14.67), the approximated phase error is derived by assuming a right-angle triangle. Figure 14.16 illustrates the ideal signal voltage, the overall error vector, and the actual vector. Note that the actual vector is the sum of the ideal vector and signal power value, with noise equivalent, which is the error vector.

From these assumptions, assuming $\theta << 1$, Fig. 14.16 describes a right angle triangle in the first approximation and the CNR is given by

$$\text{CNR} = \text{SNR} = \frac{C}{N} = tg^2\theta. \tag{14.94}$$

Thus by knowing MER, the average CNR is the actual CNR given by

$$\text{CNR}_{\text{ave}} = \text{CNR} = \frac{1}{\text{MER}}. \tag{14.95}$$

Using this approach, peak CNR is given by

$$\text{CNR}_{\text{peak}} = \left(\frac{1}{\text{EVM}}\right)^2. \tag{14.96}$$

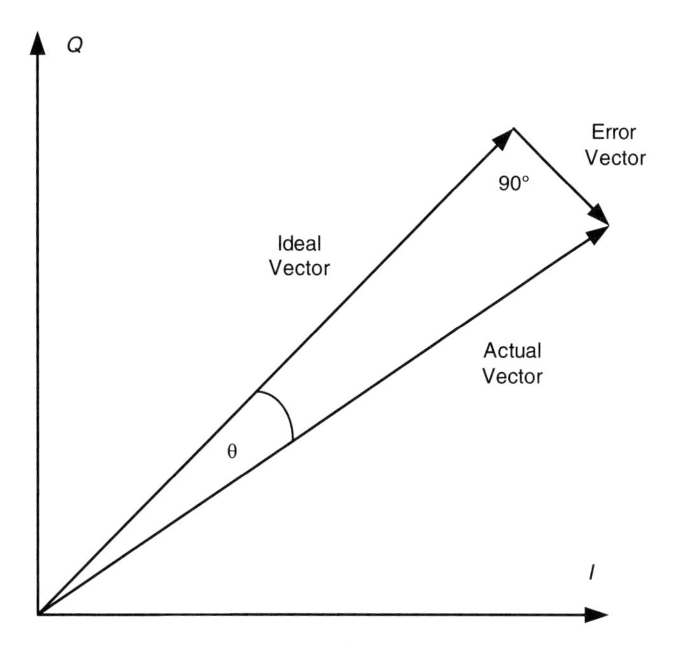


**Figure 14.16** EVM presentation as $C/N$, right-angle triangle approximation.

The next step is to evaluate the error vector as a design ballpark number. In the previous section, noise effects on symbol displacement were examined. Error vector approximation is given by the rms sum of all error power contributors:

$$V_{\text{error}} = \sqrt{Z_0 \cdot (P_{\text{AWGN}} + P_{\text{integrated phase noise}} + P_{\text{res-am}} + P_{\text{AM-PM}} + P_{\text{dc-offstet}} + P_{\text{CSO}} + P_{\text{CTB}})}. \tag{14.97}$$

The actual voltage vector is the signal power without any impairments. Hence it can be noted as

$$V_{\text{signal}} = \sqrt{Z_0 \cdot P_{\text{signal}}}. \tag{14.98}$$

If we neglect phase noise, the residual AM, AM–PM, and dc-offset and their effects in FTTx receiver are negligible or do not exist, and the MER at first approximation is given by

$$\text{MER} = \frac{1}{\text{CNR}} = \left(\frac{V_{\text{error}}}{V_{\text{signal}}}\right)^2 = \frac{10^{0.1 \cdot P_{\text{AWGN}}} + 10^{0.1 \cdot P_{\text{CSO}}} + 10^{0.1 \cdot P_{\text{CTB}}}}{10^{0.1 \cdot P_{\text{signal}}}}. \tag{14.99}$$

From this notation, a more general MER approximation can be written:

$$\mathrm{MER} = \sum_{i=1}^{n} \mathrm{MER}_i, \tag{14.100}$$

where $i$ represents the MER due to each noise source as sum of all 1/CNR. Knowing PAR provides an estimate for EVM as given by Eq. (14.68).

## 14.13 Main Points of this Chapter

1. The QAM modulator is a device that creates two orthogonal voltages denoted as $I$ for in-phase and $Q$ for quadrature. In fact, one can look at it as an image reject mixer (IRM) fed by orthogonal signals.

2. By changing the voltages of both $I$ and $Q$ according to a specific modulation code, and sequence, symbol coordinates that define its location in the $I$ and $Q$ plane are provided.

3. An M-ary QAM describes $M$ symbols in the $I$ and $Q$ plane.

4. Each axis, $I$ or Q, provides $\sqrt{M}$ coordinates; thus $M = \sqrt{M} \times \sqrt{M}$ and each axis requires $p_1 = \log_2 \sqrt{M}$ bits.

5. Each axis provides $\sqrt{M} = 2^{p1}$ combinations; thus $2^{2p1}$ bits describe all positions of the $M$ constellation's symbols.

6. There are two presentations of symbol locations in the $I$-$Q$ plane versus time $t$: Cartesian, given by $V(t)_{\mathrm{I}}$ and $V(t)_{\mathrm{Q}}$, and polar phasor, by $R(t) = \sqrt{V(t)_I^2 + V(t)_Q^2} < \mathrm{arctg}[V(t)_Q/V(t)_I]$.

7. The sum of squares of $I$ voltages and $Q$ voltages of all the constellation symbols divided by the total number of symbols $M$ provides the average power or energy per symbol.

8. The highest energy symbol in a QAM constellation is the one with the highest coordinate or largest radius value $R$.

9. The ratio of the largest energy symbol to the average energy of a constellation is called peak to rms or PAR.

10. The denser the QAM is, the higher the PAR will be.

11. Constellations whose symbols are located at a fixed $R$ but varying at phase-per-point 6, have a peak-to-rms ratio of 1 or 0 dB. Such a constellation, for example, may be QPSK, $\pi/4$ DQPSK.

12. QAM impairments can be divided into two sources: calibration of the quadrature and channel impairments.

13. The calibration processes of a constellation provide the proper dc offset between the $I$ and the $Q$ to minimize carrier leakage, amplitude offset between $I$ and $Q$ in order to assure proper quadrature, and phase calibration between $I$ and $Q$ to make sure both $I$ and $Q$ are orthogonal.

14. LO or carrier leakage results in the constellation origin offset with respect to its ideal origin.

15. By feeding the $I$ and $Q$ with two orthogonal signals with equal amplitudes $Vi \cos(\omega t)$ and $Vi \sin(\omega t)$ the quadrature is calibrated. The lowest value of the undesired sideband accomplished means that $I$ and $Q$ are closer to be orthogonal.

16. The $I$ and $Q$ calibration process consists of amplitude and phase adjustment.

17. Carrier leakage calibration consists of dc offset calibration between the $I$ and $Q$ ports. The lower the carrier leakage, the lower the origin offset is.

18. Additional constellation impairments result from phase noise (jitter), residual AM, AM to PM, AM to AM (compression), AWGN, and NLD in band spuriousness.

19. Constellation impairments result in the following effects:

   - Phase noise creates a banana-shaped symbol with an angle of $\Delta\theta_{rms}$.
   - White noise generates a cloud-shaped target, with the cloud radius defined by the CNR.
   - NLD spuriousness creates a circle-shaped symbol, with the circle diameter related to $C/I$ carrier-to-IMD ratio. The IMD rotation depends upon the delta between the IMD frequency and the LO, i.e., if the IMD is above or below LO.
   - Residual AM would create a diagonal-shaped symbol toward the constellation's center.
   - AM to PM combines phase and amplitude shapes toward the center.
   - AM to AM distorts high-energy symbols by reducing their radius.
   - $I$ and $Q$ calibration error distorts QAM quadrature into a trapezoid shape.
   - Carrier leakage shifts the constellation in both $I$ and $Q$ directions.

20. The integrated phase noise is reduced to match the maximum allowable jitter by the carrier recovery loop.

21. The figure of merit to measure the constellation quadrature is EVM and MER.

22. EVM is defined as the sum of all error vectors between the desired point of a symbol and its actual placement divided by the peak energy level.

23. MER is defined as the sum of all error vector energies between the desired point of a symbol to its actual placement divided by the ideal constellation energy level.

24. The ratio $EVM^2/MER$ equals the inverse peak-to-rms value.

25. QAM constellation BER function is derived by analyzing PAM with $M$ symbols and then presenting the QAM as two orthogonal $\sqrt{M}$-ary PAM schemes.

26. The denser the QAM, the more stringent the phase noise spec.

27. To improve the BER of a QAM channel, DSP gain processing is added by encoding methods such as FEC code.

28. Impairments reduce $E_b/N_0$, thus reducing BER performance.

29. $MER = 1/CNR$.

## References

1. Cheah, J.Y.C., *Practical Wireless Data Modem Design.* Artech House (1999).
2. Gardner, F.M., "Synchronization of Data Receivers." Seminar Presented in the S. Neaman Institute at the Technion, Haifa, Israel May 25–26 (1998).
3. Brillant, A., and D. Pezo, "Modulation Imperfections in IS54 Dual Mode Cellular Radio." *The 27th European Microwave Conference*, Jerusalem Israel, Vol. 1, pp. 347–353 (1997).
4. Brillant, A., and D. Pezo, "Modulation Imperfections in IS54/136 Dual Mode Cellular Radio." *Microwave Journal*, pp. 300–312 (2000).
5. Hodisan, A., and Z. Hellman. "CAE Software Predicts PLL Phase Noise." Brillant MTI Technology and Engineering Ltd. *Microwaves & RF*, pp. 95–102 (1994).
6. Brillant, A., "Understanding phase-locked DRO design aspects." *Microwave Journal*, pp. 22–42 (1999).
7. Drakhlis, B., "Calculate oscillator jitter by using phase-noise analysis." *Microwaves & RF*, pp. 82–90 (2001).
8. Taub, H., and D.L. Schilling, *Principles of Communication Systems.* McGraw-Hill (1986).

9. Spiegel, M.R., *Mathematical Handbook of Formulas and Tables*. McGraw-Hill Professional (1968).
10. Schwartz, M., *Information Transmission Modulation and Noise, Fourth Edition*. McGraw-Hill College (1990).
11. Proakis, J.G., *Digital Communications, Fourth Edition*. McGraw-Hill (2000).
12. Feher, K., *Digital Communications: Satellite/Earth Station Engineering*. Nobel Publishing Corp. Atlanta (1997).
13. Cociora, W., J. Farmer, and D. Large, *Modern Cable Television Technology*. Morgan Kaufman Publishers, Inc. (2004).
14. Yang, L.L., and L. Hanzo, "A recursive algorithm for the error probability evaluation of M–QAM." *IEEE Communications Letters*, Vol. 4 No. 10, pp. 304–306 (2000).
15. Rohde, U.L., *Microwave and Wireless Synthesizers Theory and Design*. Wiley & Sons (1997).

# Chapter 15

# Introduction to CATV MODEM

## 15.1 QAM MODEM Block Diagram

As was previously reviewed in Sec. 3.3.6, the CATV MODEM is a key building block in hybrid fiber coax (HFC) systems. There are two types of MODEMs, 8 VSB (vestigial side band) shown in Fig. 3.21, and QAM. The basic link block diagram of a QAM HFC MODEM is given in Fig. 15.1 with more details on its block functionality and its operation.

The first block in a QAM or VSB MODEM after Moving Picture Experts Group (MPEG) frame synchronization is the data randomizer, which is also called a scrambler. Its role is to achieve dc balance and eliminate long sequences of consecutive zeros to ensure accurate timing recovery on the receiver side. Next is the Reed–Solomon (RS) encoder. The RS encoder provides block encoding for block identification. After the RS encoder is the interleaver block that disperses the symbols in time. This way, it prevents a burst-error situation from occurring at the RS coder. The interleaver spreads the error burst so that the RS encoder refers to them as if they were random errors. This way, errors are corrected and the RS encoder is not overloaded. After the interleaver, there is an additional convolution encoding done by Trellis coding. Then sync signals are inserted. In case of 8 VSB, a pilot is inserted and the nonreturn-to-zero (NRZ) serial-beat stream has to feed the $I$ and $Q$ inputs of the QPSK modulator mentioned in Chapter 14. The data stream goes through a serial-to-parallel (SP) converter. The demultiplexed data is mapped by a Gray code mapper, differential encoder, a digital-to-analog converter (DAC), an amplitude equalizer $1/\text{sinc}(x)$ to equalize the $\text{sinc}(x)$ shape of the NRZ spectral shape, and a square-root–raised cosine (SRRC) filter at each input arm of the QPSK modulator to reduce intersymbol interference (ISI). Note that the receiving side also has an SRRC filter so that the whole response is cosine. The resultant base-band (BB) signal is up-converted by an RF superheterodyne section with synthesized local oscillators (LOs). The RF signal feeds the CATV transmitter RF signal is modulating a linearized laser which is externally modulated. Linearization methods will be reviewed in Chapter 16. The QAM transport

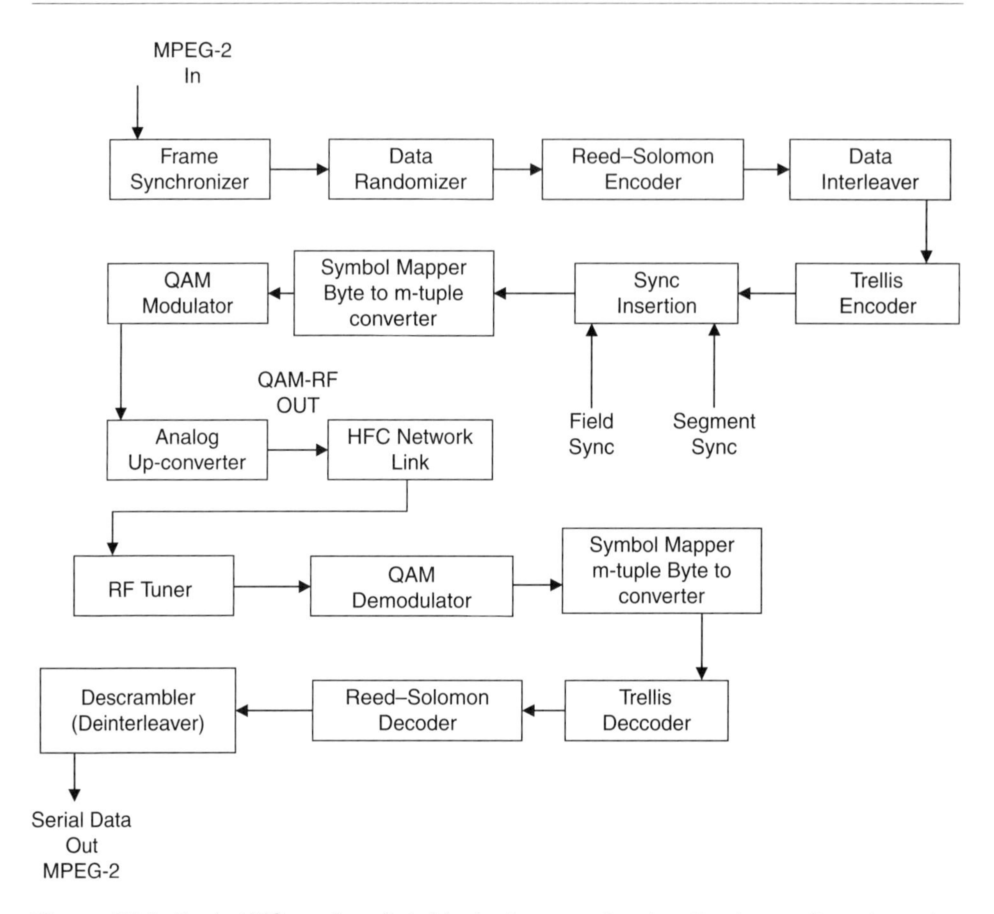


**Figure 15.1** Basic HFC modem link block diagram showing the transmit and receive chain.

is received by the subscriber's FTTx integrated triplexer (ITR) optical receiver, and converted into an electric RF signal. The CATV transport from the optical transmitter is composed of 110 channels (see Chapter 3). Thus, the RF signal enters the RF tuner and is down-converted by a super-heterodyne receiver into a BB signal. The conversion to BB is done by the QPSK demodulator described in Sec. 14.4.2. Both the $I$ and $Q$ outputs pass through the SRRC filter, then through a symbol demapper, convolutional deinterleaver, RS decoder, and data descrambler (derandomizer). This process has been standardized by DAVIC[8] (digital audio video council) for predominantly digital video broadcasting (DVB), ITU J.83 annex A[9] (based on European DVB), ITU J.83 annex B (based on General Instrument's DigiCipher II), IEEE 802.14[10] (based on ITU J.83 annex A and B), MCNS DOCSIS[11] (Multimedia Cable Network System, Data Over Cable Service Interface Specifications) (based on ITU J.83 annex B), and ITU J.83 annex D.[9] Almost all of these were reviewed in Chapter 3. This chapter will discuss implementation and realization.

## 15.2 MPEG Scrambler/Descrambler

As presented in Fig. 15.1, the first block of the modem data transmitter is the scrambler and the last block of the modem data receiver is the descrambler. The scrambler is a machine that converts the data sequence into a channel sequence. It achieves two basic goals, both of which improve the distribution of the output signal. It increases the number of transitions in the transmitted signal and thus ensures accurate timing recovery, hence, eliminating dc imbalance and expanding the period of input signals. The basic scrambler/descrambler consists of a linear sequential filter with feedback/feedforward paths.[12] Figure 15.2 depicts their different elements. Constants in the figure are binary, which means that the tap is present when the constant $b_k$ takes the value "1" in Eq. (15.1). These constants are selected such that the scrambler increases the period of the signal in the desired way. From Eq. (15.1), it has been shown that the period of the output signal, when the input period of the signal is $s$, is the least common multiple (LCM) of $s$ and $2^m - 1$, where $m$ is the number of delay elements. This is always correct except for one initial state of the scrambler: when the signal period remains unchanged. The self-synchronizing properties of this machine are easily verified by observing their block diagrams in Fig. 15.1. The scrambler is defined to be synchronized with the descrambler if its register contents are the same as those of the descrambler. This is because the same scrambler output passes through the descrambler, and synchronization is automatically achieved.

There are two forms of scrambling: self-synchronizing and frame synchronizing. Both these scramblers use maximum-length-shift register sequences, which are generated by the pseudorandom binary sequence (PRBS) generator shown in Fig. 15.2(a). The output of the PRBS generator $h(X)$ is given in Eq. (15.1) by a primitive polynomial with a degree of $m$ representing the number of delay elements:

$$h(X) = 1 + \sum_{k=1}^{m} b_k X^k, \tag{15.1}$$

where

$$b_k = \begin{cases} 1, & k = 0, m, \\ 0,1, & k = 1, 2, \ldots, m-1. \end{cases} \tag{15.2}$$

This is called the characteristic polynomial of the scrambler/descrambler or the tap polynomial. Similarly a sequence $C$ can be denoted:

$$C(X) = \sum_{i=0}^{\infty} c_i X^i, \tag{15.3}$$

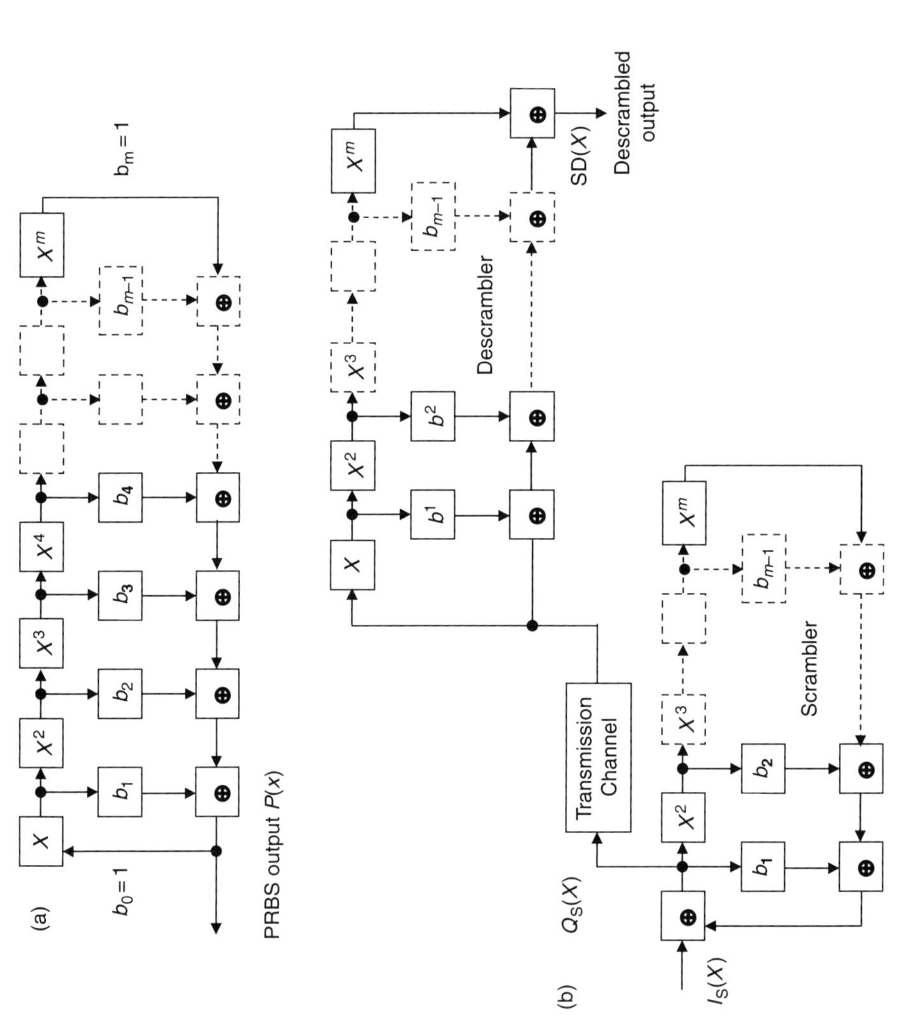


**Figure 15.2** (a) PRBS generator. (b) Self-synchronized scramble/descramble link. The summation is modulo-2.

where $X$ is the delay of one bit and $i$ is its order. As can be seen from Fig. 15.1, a PRBS generator is made of feedback shift register (FSR) with $m$ stages and modulo-2 adders. The selection of $m$ determines the period of the sequence, which eventually equals $2^m - 1$. The condition of all zero states should be excluded since it prevents any further changes in the register states.

Assume a case of self-synchronization as per Fig. 15.1(b). From Fig. 15.1(b), it can be deduced that the scrambler divides the input sequence $I_S(X)$ by its characteristic polynomial. The scrambler output $Q_S(X)$ is the input of the descrambler and the descrambler output is $S_D(X)$:

$$\begin{aligned} Q_S(X) &= I_S(X) + \left[Q_S(X)b_m X^m + Q_S(X)b_{m-1}X^{m-1} + \cdots + Q_S(X)b_1X^1\right] \\ &= I_S(X) + Q_S(X)\sum_{k=1}^{m} b_k X^k. \end{aligned} \tag{15.4}$$

Hence, the scrambler output $Q_S(X)$ is given by

$$I_S(X) = Q_S(X)\left(1 - \sum_{k=1}^{m} b_k X^k\right) \Rightarrow Q_S(X) = \frac{I_S(X)}{\left(1 - \sum_{k=1}^{m} b_k X^k\right)}. \tag{15.5}$$

In the same manner, the descrambler can be calculated and it can be shown that the descrambler multiplies the input by its characteristic polynomial:

$$\begin{aligned} S_D(X) &= Q_S(X) + \left[Q_S(X)b_m X^m + Q_S(X)b_{m-1}X^{m-1} + \cdots + Q_S(X)b_1X^1\right] \\ &= Q_S(X)\left(1 + \sum_{k=1}^{m} b_k X^k\right). \end{aligned} \tag{15.6}$$

Substituting Eq. (15.5) for $Q_S(X)$ results in the transfer function of scrambling/descrambling, while Eq. (15.5) provides the scrambler transfer function, Eq. (15.7) provides the descrambler transfer function:

$$S_D(X) = \frac{I_S(X)}{\left(1 - \sum_{k=1}^{m} b_k X^k\right)}\left(1 + \sum_{k=1}^{m} b_k X^k\right) = Q_S \cdot \left(1 + \sum_{k=1}^{m} b_k X^k\right), \tag{15.7}$$

which can be written as

$$S_D(X) = \frac{I_S(X)}{(A - B)}(A + B). \tag{15.8}$$

Since in binary system, $A - B = A + B$, it can be proved that both sequences are the same and $I_S(X) = S_D(X)$. However, there is one drawback with this method: it is not error-proof. Assume that the descrambler receives $Q_S(X) + P_{SE}(X)$, where $P_{SE}(X)$ is an error polynomial created by the transmission channel. Thus the descramble output contains an additional undesired string, since it was not normalized back, as shown in Eq. (15.8), and because this random string was created between the scrambler and the descrambler. The additional string is directly related to the descrambler's characteristic polynomial. Thus, the error can be noted as in Eq. (15.9). In conclusion, the longer the shift register is and the more nonzero taps it has, i.e., $b_k \neq 0$ it would have more errors. In fact the number of errors are equal to the number of nonzero taps in the descrambler:

$$S_{\text{D-ERR}}(X) = \frac{I_S(X)}{\left(1 - \sum_{k=1}^{m} b_k X^k\right)}\left(1 + \sum_{k=1}^{m} b_k X^k\right) + P_{SE}\left(1 + \sum_{k=1}^{m} b_k X^k\right). \quad (15.9)$$

One of the simple ways to overcome error propagation is to design a scrambler managed by a primitive polynomial with the minimum number of nonzero taps: three. However, to prevent the error-propagation problem, the self-synchronization scheme is modified to frame a synchronization mechanism. This method is nonself-synchronized. The input bit stream $I_S(X)$ is summed with the PRBS generator by XOR (modulo-2) to create the scrambler output. The recovery is accomplished by the same process. The scrambler output is again passed through an XOR (modulo-2) summation with an identically coded PRBS generator, as shown in Fig. 15.3(a). The PRBS generator is an FSR with a length of 15 bits. From Fig. 15.3(b), it is clear that the scrambler characteristic polynomial is given by

$$h(X) = 1 + X^{14} + X^{15}. \quad (15.10)$$

The recovery is accomplished because of the XOR $\oplus$ function since $h(X) \oplus h(X) = 0$; thus $I_S(X) \oplus h(X) \oplus h(X) = I_S(X)$. Parallel loading the initial sequence of 100101010000000 into the shift register in Fig. 15.2(b), per the ITU J.83 annex A and DAVIC, PRBS is initiated by the same value at the beginning of every eight transport packets. The spectrum of the PRBS generator spreads the spectrum of the original signal with spectral increment of $1/[(2^m - 1)T_S]$, where $T_S$ is the symbol period. Thus, the period of the scrambled output was increased by $(2^m - 1)T_S$.[13]

## 15.3 Codes Concept

The theory of correcting codes has presented numerous constructions with corresponding decoding algorithms.[57] However, for applications where very strong error-correcting capabilities are required, these constructions all result in far too complex of decoder solutions. The method to combat this is to use concatenated coding where two or more constituent codes are used in sequence or in parallel, usually with some kind of interleaving. The constituent codes are decoded with

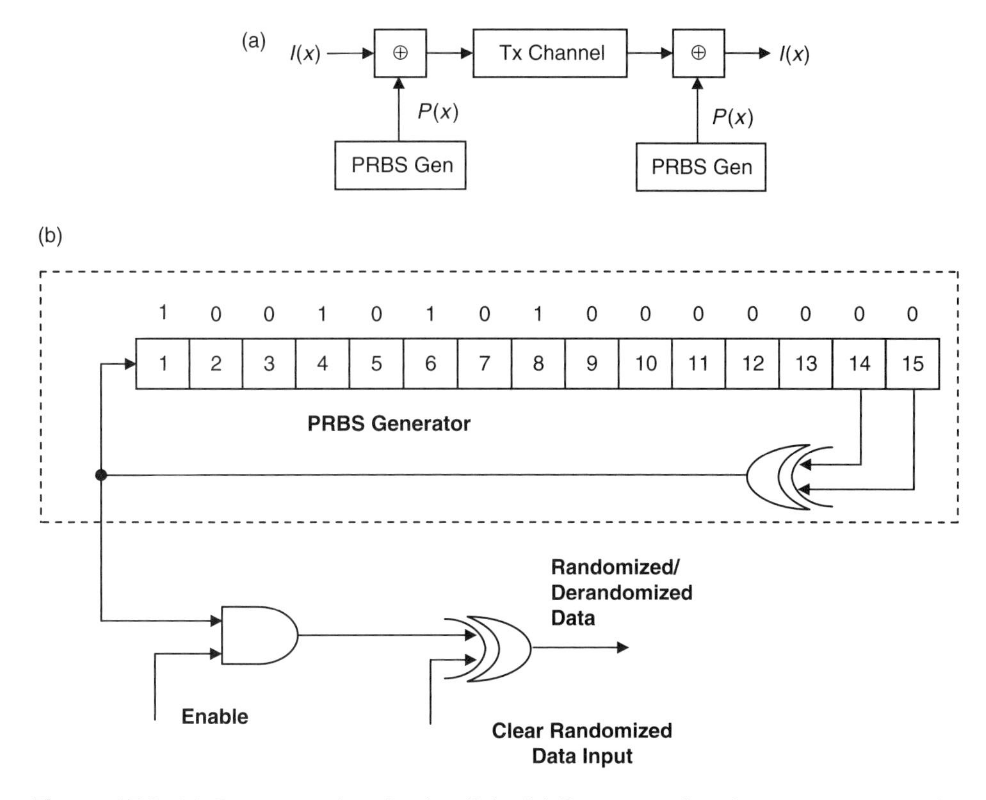


**Figure 15.3** (a) Frame synchronization link. (b) Frame synchronizer structure per the ITU J.83 annex A.

their respective decoders, but the final decoded result is suboptimal. This means that better results may be accomplished with more complicated decoding algorithms, such as the brute force approach of trying all coding options. However, concatenated coding offers an acceptable trade-off between error-correction capabilities and decoding complexity.

The concept of concatenated coding is given in Fig. 15.4. The information frame is illustrated as a square. Assuming block interleaving, this block can be monitored by horizontal and vertical parities. This way, any error can be located

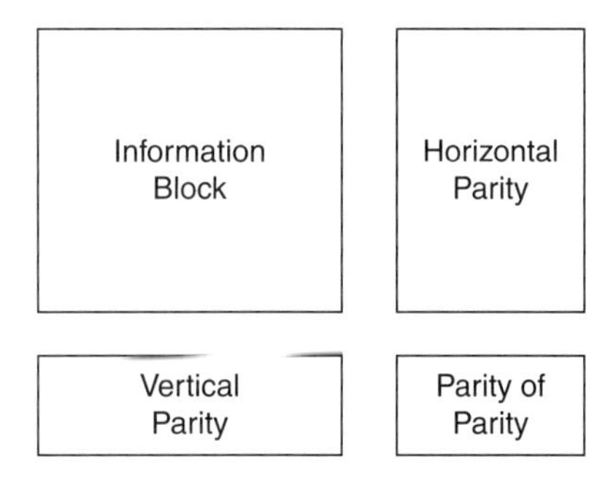


**Figure 15.4** The bidimensional concatenated error correcting and encoding concept.

and corrected. For serial concatenation, the parity bits from one of the constituent codes are encoded with a second code, resulting in parity of the parity. If the codes are operating in parallel, then there is no need for this additional parity.

## 15.4 Reed–Solomon Codes

The Reed–Solomon paper "Polynomial codes over certain finite fields" appeared in 1960 in the *Journal of the Society for Industrial and Applied Mathematics.*[14] The authors were then members of the MIT-Lincoln Laboratory. Their code maps from a vector space of dimension $m$ over a finite field $\mathbf{K}$ denoted $\mathbf{V}_m(\mathbf{K})$. $\mathbf{K}$ is usually taken to be the field of two elements $\mathbf{Z}_2$, in which it is a mapping of $m$-tuples of binary digits (bits) into $n$-tuples of binary digits.

Let $\mathbf{K}$ be the field-of-degree $n$ over the field of two elements $\mathbf{Z}_2$. $\mathbf{K}$ contains $2^n$ elements. Its multiplicative group is cyclic and generated by powers of $\alpha$, where $\alpha$ is the root of a suitable irreducible polynomial over $\mathbf{Z}_2$. The discussion here is about a code $E$ that maps $m$-tuples of $\mathbf{K}$ into $2^n$-tuples of $\mathbf{K}$.

In other words, starting from a "message" $(a_0, a_1, \ldots, a_{m-1})$, where each $a_k$ is an element of the field $\mathbf{K}$, an RS code produces $[P(0), P(g), P(g^2), \ldots, P(g^{N-1})]$, where $N$ is the number of elements in $\mathbf{K}$, $g$ is a generator of the cyclic group of nonzero elements in $\mathbf{K}$, and $P(x)$ is the polynomial $a_0 + a_1x + \cdots + a_{m-1}x^{m-1}$. If $N$ is greater than $m$, then the values of $P$ overdetermine the polynomial, and the properties of finite fields guarantee that the coefficients of $P$, i.e., the original message, can be recovered from any $m$ of the values.

Conceptually, the RS code specifies a polynomial by "plotting" a large number of points. And just as the eye can recognize and correct for a couple of "bad" points in what is otherwise a smooth parabola, the RS code can spot incorrect values of $P$ and still recover the original message. A modicum of combinatorial reasoning and a bit of linear algebra establishes that this approach can cope with up to $s$ errors, as long as $m$, the message length, is strictly less than $N - 2s$.

In today's byte-sized world, for example, it might make sense to let $K$ be the field of degree 8 over $\mathbf{Z}_2$, so that each element of $K$ corresponds to a single byte (in computers, there are four bits to a nibble and two nibbles to a byte). In that case, $N = 2^8 = 256$, and hence messages up to 251 bytes long can be recovered even if two errors occur in transmitting the values $P(0), P(g), \ldots, P(g^{255})$.

The error-correcting ability of any RS code is determined by $n - k$, the measure of redundancy in the block. If the locations of the errors are not known in advance, then an RS code can correct up to $(n - k)/2$ error symbols; i.e., it can correct half as many errors as there are redundant symbols added to the block. Sometimes error locations are known in advance (e.g., "side information" in demodulator SNRs: these are called erasures. An RS code is twice as powerful at erasure correction than at error correction, and any combination of errors and erasures can be corrected as long as the equation $2E + S \leq (n - k)$, where $E$ is the number of errors and $S$ is the number of erasures in the block is satisfied.

For the above-mentioned reasons, RS codes are commonly used in linear block codes[17] for multilevel signals such as M-QAM and N-VSB as was mentioned in

Chapter 3. This is a nonbinary code that is well matched for $M$-phase modulation such as the M-ary QAM with $M$ symbols created by $2^m$ bits, where $k$ is the size of the bit packet per symbol, which is one byte.

RS codes are created in the number space called the Galois field denoted as GF($2^m$). In the RS, there are $n-k$ parity check symbols at the end of $k$ information-symbol blocks. Thus the RS coded block is at the length of $n$ and is characterized by a pair of bits $(n, k)$. Thus the $k$ data bits are tagged by $n-k$ suffix bits (see Sec. 3.3.6 for more description). Observing the field-generator's primitive polynomial $g$ of the cyclic group of nonzero elements over GF(2), per ITU J.83 annex ADC and DAVIC, the extended field GF($2^8$) generator is

$$P(X) = X^8 + X^4 + X^3 + X^2 + 1. \tag{15.11}$$

Hence, for $g = X^8$ and nonbinary (255, 239) RS, GF(256) $m = 8$ would produce $P(g^{239}) = (2^{8\times 239})$ massage vectors out of $(2^{8\times 255})$ code vectors compared to regular binary code that yields $2^{239}$ massage vectors out of $2^{255}$ code vectors. As was explained, the RS code has $k$ information symbols (rather than bits), and $r$ parity symbols that satisfies $r = 2T$. Hence, the total length of the code word of $n$ is $k + 2T$ symbols. Furthermore, it is characteristic that the number of symbols in the code word is arranged to be $n = 2^m - 1$. Thus, the RS is able to correct $T$ symbols where $T = r/2$.

As an example, assume $m = 8$. Then, $n = 2^8 - 1 = 255$ symbols in the code word. Assume further that if $T = 16$, then $r = 2T = 32$ and $k = n - 2T$. Therefore, there are $k = 255 - 32 = 223$ information symbols per code word. The code rate $R_c$ teaches about coding efficiency. $R_c = k/n = 223/255 \approx 7/8$. The total number of bits in the data of the code word is $255 \times 8 = 2040$ bits. Since the RS code in this example can correct 16 symbols, $T = 16$, it can correct bursts of $16\times 8$ consecutive bit errors.

This leads to another definition of coding called $d_{\text{min}}$"minimum distance." Minimum distance is defined as possible words become code words; it is the measure of the distance between $T$ code words in an $(n, k)$ block code, which is the number of corresponding nonbinary symbols by which the code words differ.[58] Coming back again to the above example, the RS code has a smaller fraction of possible words, becoming code words compared to regular binary coding. Using this RS alphabet of $2^m$ symbols, then for $T$ error-correcting code words, the minimum distance is given by[16]

$$\begin{aligned} n &= 2^m - 1, \\ k &= 2^m - 1 - 2T, \\ d_{\text{min}} &= n - k + 1 = 2T + 1 \\ T &= \frac{d_{\text{min}} - 1}{2}. \end{aligned} \tag{15.12}$$

Per Ref. [15], for MPEG-2 transport, RS uses (128, 122) code over GF(128). This code is capable of correcting up to $T = 3$ symbol errors per RS block.

The same RS block is used for both 64 QAM and 256 QAM. However, the FEC frame format is different for each modulation type as is explained later on. The primitive polynomial used to form that field over GF(128) is given by Eq. (15.11), where the generator polynomial is given by

$$\begin{aligned} g(X) &= (x+\alpha)(x+\alpha^2)(x+\alpha^3)(x+\alpha^4)(x+\alpha^5) \\ &= X^5 + \alpha^{52}X^4 + \alpha^{116}X^3 + \alpha^{119}X^2 + \alpha^{61}X + \alpha^{15}, \end{aligned} \tag{15.13}$$

where the generator polynomial roots $\alpha$ satisfy $P(\alpha) = 0$.

The message polynomial input to the encoder consists of 122,7 bit symbols:

$$m(X) = m_{121}X^{121} + m_{120}X^{120} + \cdots + m_1X + m_0. \tag{15.14}$$

This message polynomial is multiplied first by $X^5$ and then divided by the generator polynomial $g(X)$ to form a reminder given by

$$r(X) = r_4X^4 + m_3X^3 + m_2X^2 + r_1X + r_0. \tag{15.15}$$

This reminder constitutes five parity symbols that are then added to the message polynomial to form a 127-symbol code word that is an even multiple of the generator polynomial. The output is described by the following polynomial:

$$c(X) = m_{121}X^{126} + m_{120}X^{125} + \cdots + m_4X^4 + r_3X^3 + r_2X^2 + r_1X + r_0. \tag{15.16}$$

A valid code word will have roots at the first through the fifth power of $\alpha$, as per Eq. (15.13).

The last symbol transmitted by the RS block is used to form an extended parity symbol to be the sixth power of $\alpha$:

$$\widehat{c} = c(\alpha^6). \tag{15.17}$$

This extended symbol is used to form the last symbol transmitted by the RS block. Thus, the extended code word appears as

$$\begin{aligned} \bar{c} = X \cdot c(X) + \widehat{c} &= m_{121}X^{127} + m_{120}X^{126} + \cdots + m_1X^7 + m_0X^6 \\ &+ r_4X^5 + r_3X^4 + r_2X^3 + r_1X^2 + r_0X + \widehat{c}\,. \end{aligned} \tag{15.18}$$

The structure of the RS block, which illustrates the order of transmitted symbols output from the RS encoder, is

$$m_{121}m_{120}m_{119} \cdots m_1\, m_0r_4r_3r_2r_1r_0\, \widehat{c}, \tag{15.19}$$

where the order of the pattern transmission is from left to right.

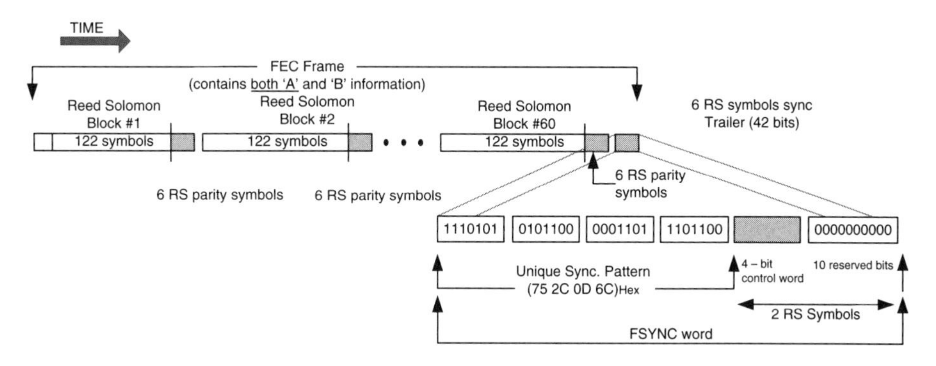


**Figure 15.5** 64 QAM frame packet format.[15] (Reprinted with permission of *SCTE* © 2000.)

For 64 QAM, the FEC frame consists of a 42-bit sync trailer, which is appended to the end of the 60-RS blocks, where each RS block contains 128 symbols. Each RS symbol consists of 7 bits; thus, there are a total of 53,760 data bits and 42 frame sync trailer bits in this FEC. The first four 7-bit symbols of the frame sync trailer contain the 28-bit unique synchronization pattern $(75\ 2C\ 0D\ 6C)_{HEX}$. The remaining two symbols of 14 bits are used as follows: the first 4 bits are for interleaver mode control, and 10 bits are reserved and set to zero. The frame sync trailer is inserted by the encoder and detected at the decoder. The decoder circuits search for this pattern and determine the location of frame boundary and interleaver depth mode when found. Figure 15.5 provides information on a QAM 64-frame packet format.

For QAM 256, the FEC consists of a 40-bit trailer, which is appended at the end of 88 RS blocks. Each RS symbol consists of 7 bits; thus, there are a total of 78,848 data bits and 40 frame sync trailer bits in this FEC. Here, the 40 bits are divided as follows: 32 bits form the unique synchronization pattern of $(71\ E8\ 4D\ D4)_{HEX}$, the first 4 bits are for interleaver mode control, and 4 bits are reserved and set to zero.

## 15.5 Interleaver/Deinterleaver

In the previous section, it was explained that FEC is limited by the amount of errors it can handle. For a burst of errors, it means that in case there is one bit of error, there is a large likelihood that the successor and predecessor bits are errors as well. This can cause a burst of error bits more than the FEC can handle.[57,58] For the RS previously reviewed, it means more than $T = 8$ (i.e. more parity symbols since $T = r/2$) that RS(204, 188) $n - k = 16$ can correct. Thus, there is a need for a commutation mechanism to spread the large burst of errors into a smaller quantity of bits, which can be corrected by the RS FEC. Therefore the interleaver disperses the burst of error bits so that it looks like random errors. Therefore, the interleaving for the R–S code is done to spread out the symbol error so that the total number of symbol errors in each block is smaller.

There are several types of interleavers.[18] One type of interleavers increases the minimum Euclidean distance of FEC, which can be seen from Eq. (15.12). According to the ITU J.83[9] and IEEE 802.14,[10] a convolution interleaver is used for QAM transport. From Fig. 15.5, it is obvious that the interleaver is characterized by two indices: the depth $I$ and delay increment per register. Hence, the interleaver is a matrix of the size $\mathbf{J} \times \mathbf{I}$. The $I$ index refers to the commutating switch state and $J$ refers to the shift register length or number of symbol delay periods. So the interleaver is a switched-shift-register-bank matrix with $1 \leq I \leq 128$ and $0 \leq J \leq 127J$. Therefore, the largest delay is of $127J$ symbol delay periods and the shortest is zero, where $J$ is a basic delay unit per symbol. The size of each $J$ cell size is $J = n/I$, where $n$ is the RS code-protected frame length. Note that $J = 1$ is no delay and $I = 128$. Hence $J$ is at the size of one symbol of 7 bits. In other words, each RS block is spread through the interleaver and shuffled with symbols over the span of the interleaving depth when encoded. The operation of the interleaver and deinterleaver is synchronized, meaning the states of the encoder, commutation switch, and decoder are synchronized. The shift registers operate on an FIFO (first in first out) basis. Again, from Fig. 15.6 (interleave matrix), it is clear that the end-to-end delay (latency) of the interleaver/deinterleaver is given by

$$L[\mu\,\text{sec}] = I \times J(I-1)T_{\text{RS}}, \tag{15.20}$$

where $T_{\text{RS}}$ is the period of one RS symbol. Since each section of the interleaver/deinterleaver is a triangle matrix, the memory size of each is given by the number of RS symbol cells $J$ in each matrix:

$$\text{mem} = \frac{I \times J(I-1)RS_{\text{symbol}}}{2}. \tag{15.21}$$

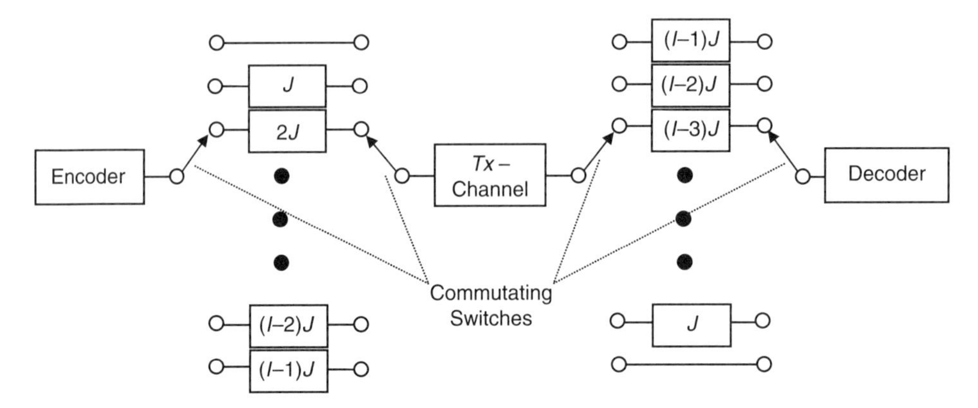


**Figure 15.6** Convolutional interleaver and deinterleaver conceptual block diagram.[15] (Reprinted with permission from *SCTE* © 2000.)

Since error bits are noise, the maximum noise burst (NB) length that can be handled by the RS coder is given by

$$\mathrm{NB} = (I \times J + 1)\frac{T \times I}{T_{\mathrm{RS}}}, \tag{15.22}$$

where $T$ is given in Eq. (15.12). Thus, as was claimed, $T$ is increased and so too is $d_{\min}$.

## 15.6 Trellis-Coded Modulation

Section 15.3 discussed RS coding, which is a block coding, and Sec. 3.3.6 briefly explained the difference between trellis and RS coding. Trellis-coded modulation (TCM) is convolutional (or recurrent) coding that evolves with the data. Convolution codes, such as TCM, differ from block codes as follows. In block code, the block of $n$-code digits generated by the encoder in any particular time unit depends only on the block of $k$-input data digits within that time unit. In convolution code, the block of $n$-code digits, generated by the encoder in a particular time unit does not depend only on the block of $k$-message digits within that time unit, but also on the block of data digits within the previous span of $N-1$ time units where $N > 1$. For convolution codes, $k$ and $n$ are usually small. These codes can be used for random errors, burst errors, or both. RS is utilized as external coding, while TCM is utilized as an internal-coding scheme. The description of these two schemes is mentioned in Sec. 3.3.6. A TCM modulator is located after the RS encoder, as shown in Fig. 15.1 and described in Sec. 3.3.6.

### 15.6.1 Convolutional coding

The advantage of TCM coding for QAM schemes is its ability to introduce redundancy and gain processing to improve SNR or BER versus $E_b/N_0$. (This is analyzed in detail in Refs. [19, 20]). It is efficient, since it does not need extra BW. The TCM coding method provides considerably larger free distance or minimum Euclidean distance compared to an uncoded modulation signal bearing the same information rate, BW, and power.

The convolutional encoder is based on a shift register with $M$ stages. The number of stages is defined by the number of bits, $n$, and the constraint length of the code $N$. Thus, the total stages of the shift register is $M = n \times N$, which reflects changes generated by a single bit.

At any given time, a gate signal enables $n$ bits to enter the shift register. The shift register is FIFO, so the content of the last $n$-bit packet is shifted out. The shift register outputs define the characteristic polynomials of the encoder by XOR with previous states, as shown in Fig. 15.6.[22] It can be observed from the shift register in Fig. 15.7 that the constraint length $N$ equals the number of storage cells plus one for the input, hence $N = 1 + 6 = 7$. Since this encoder has no feedback

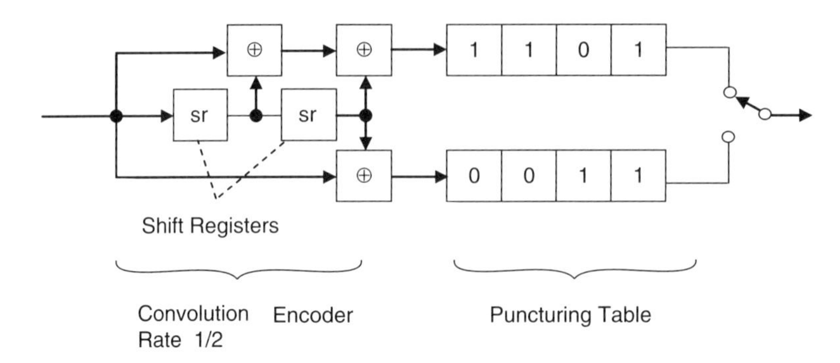


**Figure 15.7** The Trellis convolutional encoder with a 4/5 rate puncturing block.

(not recursive), it is considered a feedforward type. The next step is to define the shift register rate. Assume that each memory cell $Z^{-1}$ contains only one bit. First, there is a need to define the memory length as the number of information blocks of the past that are used to create the present state. This is a different way of presenting the encoder's constraint length by $k = N - 1$. Coming back to the example, the information blocks consist only of one bit. Thus it is noted as

$$U_r = (u_{r,1}), \qquad U_{r-1} = (u_{r-1,1}) \Rightarrow n = 1. \tag{15.23}$$

Only one information block of the past $(u_{r-1,1})$ is used; thus $k = 1$ or 6 for Fig. 15.6. But the code block consists of two bits since two outputs are used:

$$a_r = (a_{r,1}), \qquad a_{r-1} = (a_{r,2}) \Rightarrow R_C = \frac{1}{2}. \tag{15.24}$$

Thus, the code rate $R_C$ equals $\frac{1}{2}$, and the conclusion is that the rate definition indicates the ratio between the number of input bits to memory cell block $Z^{-1}$ to the number of output bits at one encoding cycle. Coming back to the example in Fig. 15.7, it is clear that the encoder rate is $\frac{1}{2}$.

The next step is to be able to create a selectable-rate convolutional encoder. For this purpose, a puncture unit is used.[23] The puncture block consists of a table with 0 and 1, where 0 means that the channel symbol is to be deleted, and 1 is to be transmitted. The puncture table ends by a commutating switch, which serializes the outputs into one bit stream. Then, in case of one cycle of operation and a puncturing rule as in Fig. 15.7, for any 4-bit input to each branch, e.g., the total input is 8, the output is 5. Thus the puncture block rate is defined as $n/Q$, where $Q$ is the number of 1 programmed in the puncturing table. Thus, from Fig. 15.5, if each 4-RS symbol for a total of 28 bits occupies one row of the puncturing rule, with a rate of 14/15 per the ITU J.83–B for QAM 64, then each 1 cell represents two bits, which translates into a total of 30 bits, which are 5 symbols of 64 QAM. Each symbol is made of 6 bits, 3 for the $I$ coordinate and 3 for the $Q$ coordinate, which translates to 8 states for each, or a total of 64 states. This means that the rule should be modified from Fig. 15.7. However, this can be

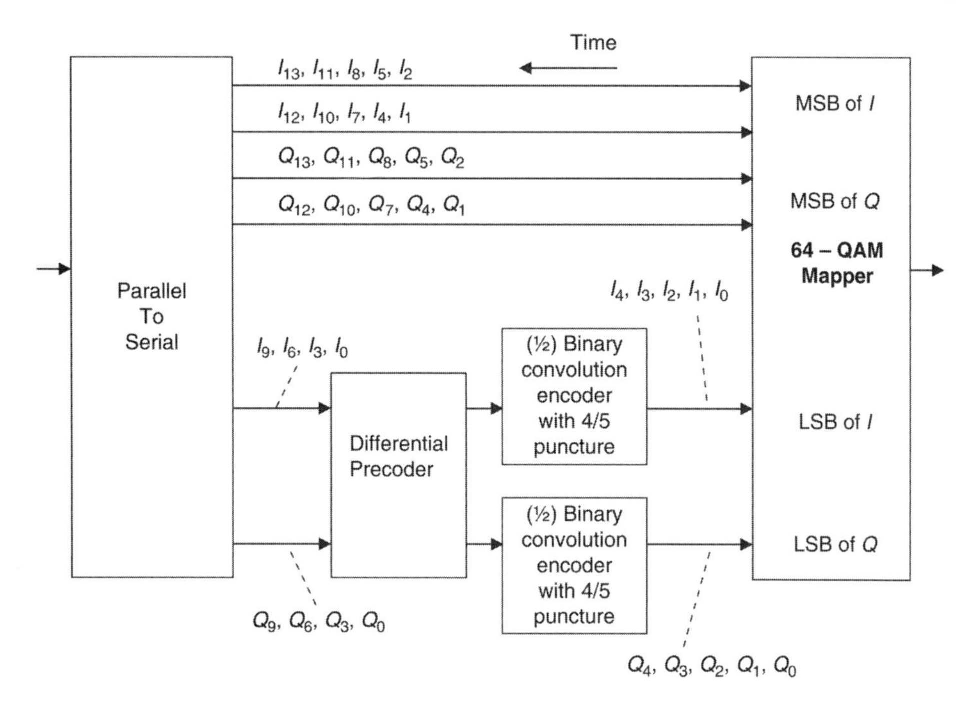


**Figure 15.8** Trellis convolutional encoder with 14/15 rate using 4/5 puncturing block per the ITU J.83 - B for 64 - QAM.[15] (Reprinted with permission from *SCTE* © 2000.)

done by a different method[20,21] and ITU J.83–B[9] by coding only 4 $I$ and 4 $Q$ bits from the RS train, as shown in Fig. 15.8, while letting the remained RS to be not coded. Thus, by using a 4/5 structure, for each 4 $I$ bits there are 5 output bits. Observing Figure 15.8, there are 20 uncoded bits: 10 for the $I$, 10 for the $Q$ and 4 are coded and transformed into 5 bits as follows: 5 bits for the $I$ and 5 for the $Q$. So for the input of 28 bits, the output is 30 bits, which results in a 14/15 rate as shown in Fig. 15.8. This result is due to the following; 10 uncoded bits (Q or I), plus 4 punctured bits transformed into 5. This results in a total of 15, hence for input of 14 bits there are 15 going out; therefore, the rate is 14/15.

## 15.6.2 Viterbi decoding

One of the trellis-decoding algorithms is Viterbi decoding. For understanding the Viterbi algorithm concept, it is necessary to first introduce the trellis diagram for describing a set of signal sequences. The distance properties of the TCM scheme can be studied through its trellis in the same manner as for convolutional codes. The optimum decoding is the search for the most likely path through the trellis once the received sequence has been observed at the channel output. However, due to noise corruption of signal, the path chosen may not coincide with the correct path. Meaning, the path will be traced by the sequence of source symbols, but will sporadically deviate from it and merge again at a later time.

The trellis decoder can be described in several ways commonly used in logic design, such as: state table, state diagram, which are presenting input, the actual state of output, and the next state. The third method to describe trellis decoding is by the trellis diagram, previously described. This diagram describes too the state of the encoder and decoder, but with a temporal approach. There are several key rules for drawing the trellis diagram:

- The nodes of the trellis distinguish the encoder state.
- Solid and dashed lines represent outputs created by "0" and "1," respectively.
- The starting state always is the zero state.
- The reaction of the encoder, output, and next state, at the actual state to every possible input is determined.
- The next state is considered and again the reaction of the encoder at the actual state to every possible input is determined, and so on . . . .

A major parameter of a convolutional code is its free distance denoted as $d_{\mathrm{FREE}}$. Since it specifies the decoding capability of a code under a maximum likelihood (ML) decoding, this encourages the search for convolutional codes with a specified rate and degree, or constraint length, which have a maximum free distance $d_{\mathrm{FREE}}$. Hence the key rules here are:

- Starting at zero state and ending at zero state, without being in zero state in between.
- Note all zero states that are away.
- The path has to be chosen so that the number of 1 at the output is minimized.
- Every different path that starts at zero and ends at zero, except an all zero-state path would have a higher number of 1 at the output.
- The free distance parameter is the number of 1 at the output using a fundamental way, so that it is the distance to the all zero sequence.

Free distance is a measure of the capability to correct errors. The higher the free distance, the higher the error-correction capability.

The Viterbi algorithm as a general technique for decoding the convolutional codes is also used in the TCM decoder. Because of one-to-one correspondence between the signal sequences and the paths traversing the trellis, the ML decoding consists of searching for the trellis path with the minimum Euclidean distance to the received signal sequence. In case of a sequence length of $K$ being transmitted, and the sequence $r_0, r_1, \ldots, r_{K-1}$ corrupted by additive white Gaussian noise (AWGN) being observed at the channel output, the ML receiver looks for the sequence $x_0, x_1, \ldots, x_{K-1}$ that minimizes $\sum_{i=0}^{K-1} |r_i - x_i|^2$. This search is done by the Viterbi algorithm for any constellation dimension.[24] The branch metrics to be used are obtained as follows. The branch in the trellis used for coding is labeled by a signal $s$; if there are no parallel transitions, then at a discrete time $i$, the metric associated with that branch is $|r_i - x|^2$. If a pair of nodes is connected by parallel transitions, and the branches have labels $x', x'', \ldots$ in the $x$ set, then the trellis used for decoding the same pair of nodes is connected by a signal branch, whose metric is $\min_{x^i \in x} \left| r_i - x^i \right|^2$. This is in the presence of parallel transitions; the decoder first selects the signal among $x', x'', \ldots$ with the minimum distance from $r_i$, which is a demodulation operation, then creates the metric based on the selected signal.

### 15.6.3 Viterbi decoding implementation process example

In order to understand the coding–decoding process according to Viterbi's algorithm and the use of the trellis diagram, assume the following problem of a state machine. Consider the convolution encoder as depicted in Fig. 15.9. This encoder has a single input and two outputs; hence, the code rate is $r = 1/2$. The system number of states is depending upon number of flip flops (FF). In this case, there are two denoted as $D$. Thus, there are $2^2 = 4$ states, $S = 4$. For a system with $k$ delay units the number of states is $2^k$. The encoder outputs are marked as $G_0(n)$ and $G_1(n)$. The system block diagram is expressed with the following state equations:

$$G_0(n) = x(n) + x(n-1) + x(n-2) \tag{15.25}$$

$$G_1(n) = x(n) + x(n-2) \tag{15.26}$$

The state diagram is provided in Fig. 15.10 and the states are defined as $x(n-1)$, $x(n-2)$ pairs and the state transitions are defined as $G_0(n)$, $G_1(n)/x(n)$. The transitioning from state $G_0(n)$,$G_1(n)$ to the next state is for a specific $x(n)$

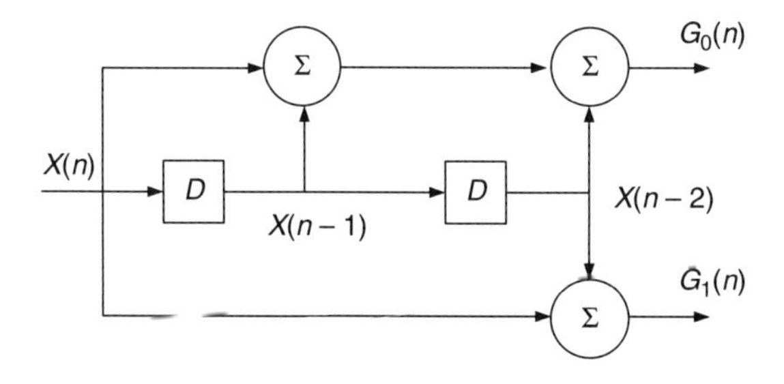


**Figure 15.9** Convolution-encoder block-diagram code rate $r = 1/2$.

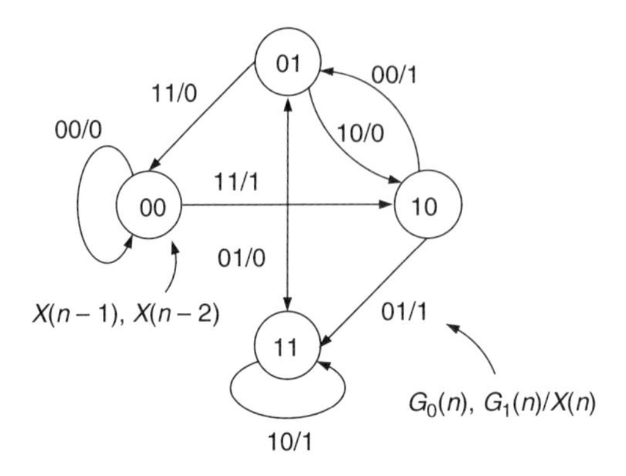


**Figure 15.10** Encoder state machine diagram

input. This is a graphical notation. The states are indicative system memory. The state transition provides the path and the outputs associated with each possible input. Since the state transition is associated with each possible input, the number of state transitions depends on the number of possible inputs. If there are $m$ inputs then there are $2^m$ transitions. The total number of state transitions at a point in time is the product of the number of state transitions and the number of states; hence it becomes $2^{m+k}$.

The next step is to describe the trellis diagram, but first there is a need to define the system states per FF states as described in the Table 15.1.

The state diagram offers a complete description of the system as a state machine. However, it shows only the instantaneous transitions. It does not illustrate how the states change in time. To include time in state transitions, a trellis diagram is used as shown in Fig. 15.11. It is a different presentation of the state machine bubble diagram presented in Fig.15.10. Note that the number of rows in Fig. 15.11 is identical to the number of states in Fig. 15.10. Each node in the trellis diagram denotes a state at a point in time. The branches connecting the nodes denote the state transitions. Notice that the inputs are not labeled on the state transitions compared to the bubble diagram in Fig. 15.10; they are implied since the input, $x(n)$, is included in the destination states, $x(n)$, $x(n-1)$. Note that the state transitions are defined by the definition of the source, $x(n-1)$, $x(n-2)$, and the destination states $x(n)$, $x(n-1)$. The state transitions are independent of the system equations in this illustration. This is a consequence of including

**Table 15.1** Encoder FF states.

| State | $FF_1$ | $FF_0$ |
|---|---|---|
| $s_0$ | 0 | 0 |
| $s_1$ | 0 | 1 |
| $s_2$ | 1 | 0 |
| $s_3$ | 1 | 1 |

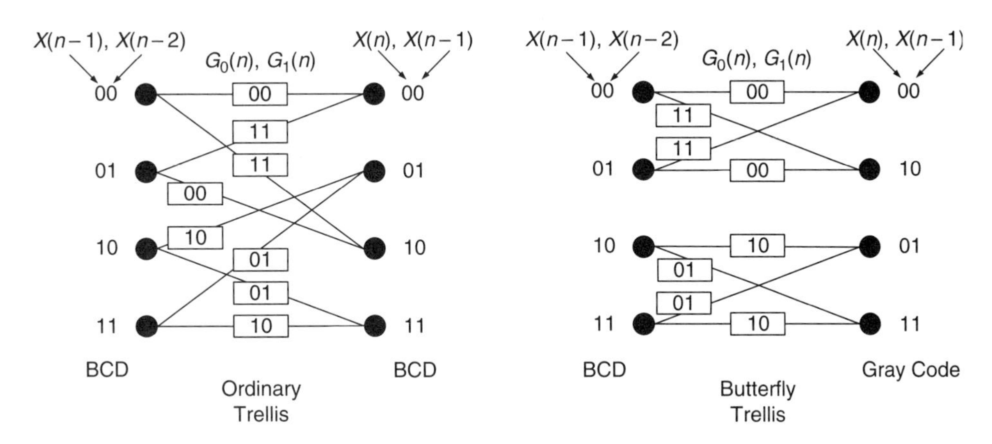


**Figure 15.11** Conversion of state diagram to Trellis diagram. Left: BCD coded, and right: Butterfly with Gray code.

the input $x(n)$ in the state definition. In certain systems, the definition of states might be more complicated and may require special considerations. An explanation of such systems (such as the radix 4 trellis) is outside the scope in this example.

Figure 15.11 displays two trellis diagrams. The one on the left is ordered by binary-coded decimal (BCD), while the other on the right side is ordered by Gray code, as in a Karnaugh map. This obtains higher computational efficiency.

Figure 15.11 on the right side is called the "butterfly" due to its physical resemblance to a butterfly. It is also referred to as the radix-2 trellis structure since each group contains two source and two destination states, and since $\log_2 4 = 2$. The Texas Instruments TMS320C54x DSP has particular hardware that is dedicated to compute such parameters associated with a butterfly. Note, again, that the structure of the reordered trellis is fixed depending of how the states are defined. Not all trellis diagrams can be reordered into butterflies. Fortunately, a number of commonly used convolution coders have this nice property. An important feature of this convolution code is that in each butterfly, there are only two (out of four possible) distinct outputs: 00 and 11; or 01 and 10. Notice that the Hamming distance between each output pairs is 2, where for binary strings $a$ and $b$ the Hamming distance is equal to the number of ones in $a$ OR $b$. This feature is available to a very small subset of all possible convolution codes. The usefulness of this feature will be noticeable later on when Viterbi decoding is described.

Figure 15.12 illustrates the entire process of encoding an input pattern for quantizing the received signal and decoding the quantized signal. The sample input pattern used is 1011010100. For the sake of this example, this input pattern is sufficiently long; however, in practice and for proper error correction, the input pattern needs to be 10 times longer than the number of delay units plus one, meaning the pattern length is equal to $10(k+1)$, where $(k+1)$ is often referred as the constraint of pattern length. Sometimes marked as constraint length, $K = k + 1$. In this example there are two delay units implemented by flip

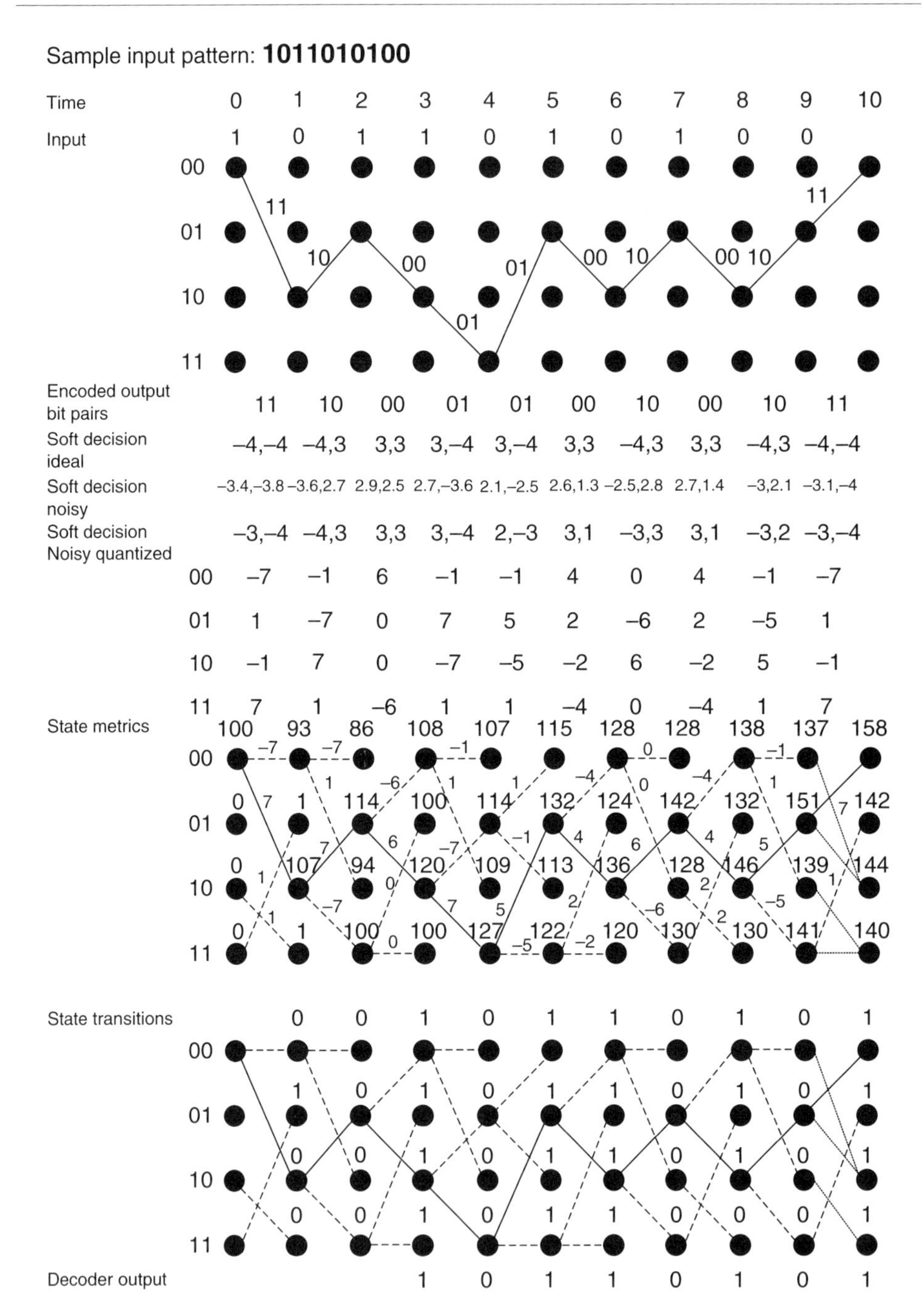


**Figure 15.12** Trellis diagram showing an analysis of convolution encoding and Viterbi decoding algorithm process.

flops, hence $k = 2$, which results in 30 symbols at the input that need to be received prior to the decoding process to improve the bit error rate. This is due to the fact of filling up the decoder and avoiding boundary conditions similar to DSP filters.

The first trellis diagram, provided in Fig. 15.12, illustrates the state transitions associated with each input. The trellis diagram is used here to illustrate how the decoder operates as a state machine. Figure 15.11 is used to verify the state transitions provided by the state machine diagram in Fig. 15.10 and to check the outputs of the encoder, which is shown in Fig. 15.9. Note that encoder outputs do not have to be generated using the trellis diagram. They can be produced by using the system equations directly. The encoded outputs are transmitted as signed antipodal analog signals (i.e., 0 is transmitted with a positive voltage and 1 is transmitted with a negative voltage). They are received at the decoder and quantized with a 3-bit quantizer. The quantized number is represented in complementary twos, providing a range of −4 to 3. The process of quantizing a binary analog signal with a multibit quantizer is called a *soft decision.* In contrast, a *hard decision* quantizes the binary analog signal using a single bit quantizer (i.e., the quantized signal is either a 0 or 1). A soft decision offers better performance results since it provides a better estimate of the noise (i.e., less quantization noise is introduced). In most circumstances, the noise is strong enough just to tip the signal over the decision boundary. If a hard decision is used, a significant amount of quantization noise will be introduced. The quantized soft decision values are used to calculate the parameters for the Viterbi decoder.

At this point, no Viterbi decoding has been performed and the concept of a state-machine-diagram conversion into a Trellis diagram is presented. Viterbi decoding process starts after a certain number of encoded symbols have been received. This length is $10(k + 1)$, as was previously explained, and is application dependent. In this example, 20 encoded symbols are received prior to being decoded. In some other applications, Viterbi decoding is operated on a frame of received symbols and is independent of the neighboring frames. It is also possible to perform Viterbi decoding on a sliding window in which a block of decoded bits at the beginning of the window are error free and the windows will have to overlap. With frame-by-frame decoding, a number of trailing zeros equal to the number of states are added. This forces the last state of the frame to be zero, which provides a starting point for trace back (trace-back memory for storing the survival path data of Viterbi decoded data). With a sliding-window decoding, the starting point for trace back is the state with the optimal accumulated metric. This starting state may be erroneous; however, the trace-back path converges to the correct states before reaching the block of error-free bits.

Viterbi decoding can be broken down into two major operations, metric update and trace back. In metric update, two operations are done for each symbol interval (in this example one symbol equals a single input bit, which results in two encoded bits, i.e., the code rate is $1/2$). First the accumulated (state) metric is calculated for each state and the optimal incoming path associated with each state is determined. Second, trace back uses this information to derive an optimal path through the trellis. Referring to the second trellis in Figure 15.12, it can be seen that a state at any time instance has exactly one incoming path, but the number of outgoing paths may vary. This fact results in a unique path tracing backward. The question though is what are state metrics and how the optimal incoming paths are determined? To understand how the metrics are used in Viterbi decoding process, consider the received encoded bit pairs at the output of $G_0(n)$, and $G_1(n)$ at the encoder provided in Fig. 15.9. Assume that 11 are transmitted, hence it is expected to see (1,1) as received pair with a soft decision pair of (−4,−4). However, channel noise may corrupt the transmitted signal such that the soft decision pair might be (−0.4, −3.3) resulting in a quantized value of (0,−3) as shown in Fig. 15.13. Without having any a priory information on the encoded bit stream, the detector would decide that (3,−4) or (0,1) is transmitted, which results in a bit error. This is called symbol-by-symbol detection.

The Viterbi algorithm exploits the structure of the convolution code and makes its decision based on previously received data, i.e., tracking is kept by using a "states" table as provided in Fig. 15.12. Observing Fig. 15.11, it can be seen that an encoded bit pair is associated with only two possible originating states, $x(n-1)$, $x(n-2)$. For instance, if 11 is transmitted, then the originating state must be either state 00 or state 01, this can be seen from Fig. 15.10 of the state machine as well. If the original state is known to be 00, it would be impossible for the detected encoded bit pairs to be 01; it could only be 11. The Viterbi decoder would then have to decide between the two possible encoded bit pairs, 00 or 11 as can be seen from Figs. 15.10, 15.11, and 15.12. This decision depends on how far away the received bit pairs are from the two possible transmitted bit pairs of constellation symbol-by-symbol detection, as shown in Fig. 15.14. The metric used to measure this provides the Euclidean distance. It

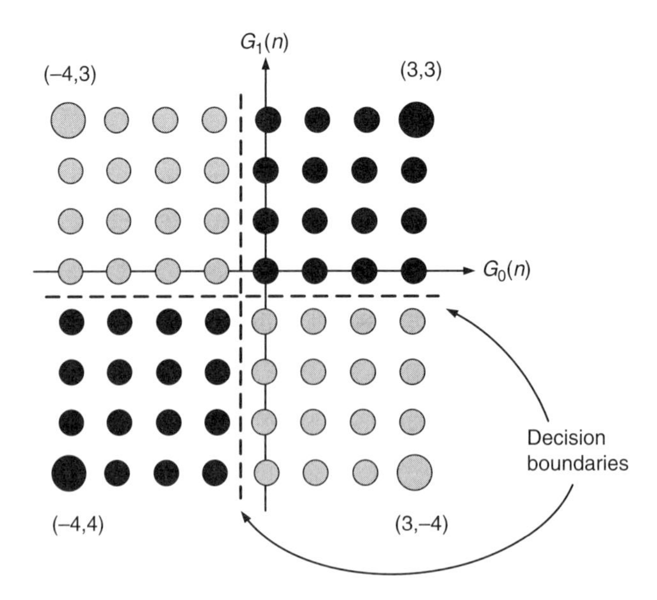


**Figure 15.13** Signal constellation used for symbol to symbol detection.

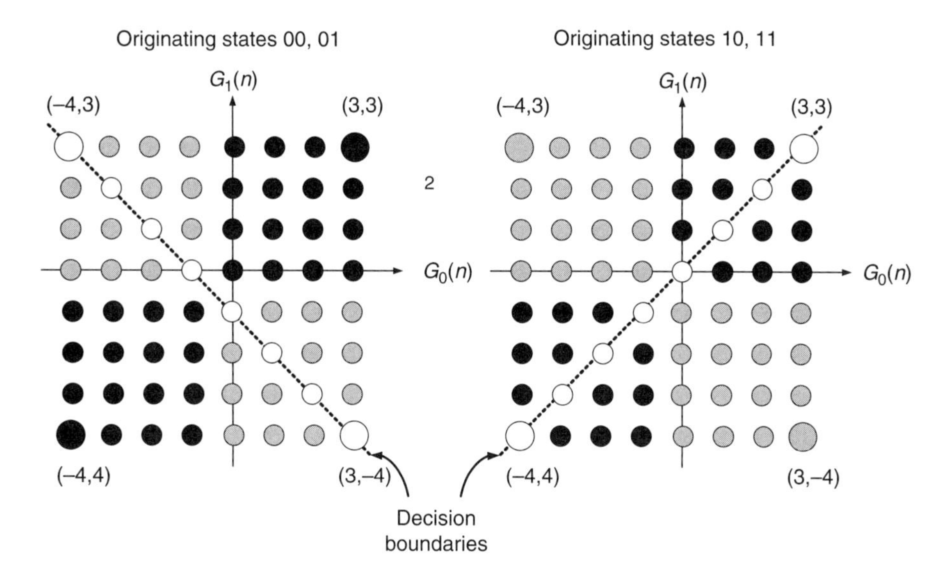


**Figure 15.14** Signal constellation for Viterbi decoding.

is also referred to as the local distance *Ld* calculation provided by

$$Ld(n,i) = \sum_{all\_j} [S_j(n) - G_j(n)]^2 \quad j \in \{\text{all bits associated with a given input}\}, \tag{15.27}$$

where $n$ denotes the time instance and $j$ denotes the path calculated for. Using the law of $(a - b)^2 = a^2 - 2ab + b^2$, Eq. (15.27) can be written as

$$Ld(n,i) = \sum_{all\_j} [S_j^2(n) - 2 \cdot S_j(n) \cdot G_j(n) + G_j^2(n)] \tag{15.28}$$

The values $\sum_{all\_j} S_j^2(n)$ and $\sum_{all\_j} G_j^2(n)$ are constants in a given symbol period, hence, they can be ignored as the concern is to find the minimum local distance. Hence the value $\sum_{all\_j} S_j(n) \cdot G_j(n)$ defines the minimum local distance. Therefore, Eq. (15.28) becomes

$$Ld(n,i) = \sum_{all\_j} S_j(n) \cdot G_j(n). \tag{15.29}$$

Thus, instead of finding the path with the minimum local distance, we would look for the path with the maximum. Observing Fig. 15.12, the branch metrics can now be calculated using Eq. (15.29). As an example, the branch metrics for times 0 and 1 are calculated here. The received quantized soft decision pair is (−3,−4). The branch metric corresponding to is equal to $[G_0(n),G_1(n)] = (0,0)$ which equals to $-3(1)-4(1) = -7$. For the case of $[G_0(n),G_1(n)] = (0,1)$, it is $-3(1)-4(-1) = 1$, for $[G_0(n),G_1(n)] = (1,0)$, it is $-3(-1)-4(1) = -1$ and for $[G_0(n),G_1(n)] = (1,1)$, it is $3(1)+4(1) = 7$, where values are derived from the signal constellation provided in Fig. 15.14. Observe that the value of $G_j(n)$ used in the local distance calculation are +1 and −1, rather than +3 and −4, respectively. There are three reasons for that. First, using unities makes computation very simple. The branch metrics can be calculated by adding and subtracting, and subtracting the soft decision values. Second, 3 and −4 can be scaled to approximately 1 and −1, respectively. As mentioned in the derivation of local distance, constant scaling can be ignored in determining the maxima and the minima. Finally, note that there are only two unique metric magnitudes: 1 and 7. Therefore, to compute the branch metric, the only thing that needs to be done is to 1 add, 1 subtract, and 2 negation.

Once the branch metrics are calculated, the state metrics and the best incoming paths for each destination state can be determined. The state metrics at time equal to 0 are initialized to 0 except for state 00, which takes on the value of 100. This value is arbitrarily chosen and is large enough so that the other initial states cannot contribute to the best path. This basically forces the trace back to converge on state 00 at time 0. Then the state metrics are updated in two steps. First, for each of the two incoming paths, the corresponding branch metrics are added to the state metrics of the originating state. The two sums are compared and the larger one is stored as the new state metric, and the corresponding path is stored as the best path. For example, observe state number 10 at a time instance of 2, as appears in Fig. 15.12. The two paths coming in to this state originate from states 00 and 01, appearing in Fig. 15.11. It can be seen at the second trellis that state 00 has a value of 93 and state 01 has a value of 1. The branch metric of the top path, which connects state 00 at time 1 and state 10 at time 2, is 1. The branch metric of the bottom path, between states 01 and 10, is −1. The top path yields a sum of $93 + 1 = 94$ and the bottom path yields a sum of $1 + (-1) = 0$. As a result, 94 is stored as the state metric for state 10 at time 2, and the top path is stored as the best path in the transition buffer.

The transition buffer stores the best incoming path for each state. For a radix 2 trellis, only 1 bit is needed to indicate the chosen path. A value of 0 indicates that the top incoming path of the given state is chosen as the best path, whereas a value of 1 indicates that the bottom path is chosen. The transition bits for this example are illustrated in Fig. 15.12 in the third trellis. Trace back begins after completing the metric update of the last symbol in the frame. For frame-by-frame Viterbi decoding, all that is needed by the trace-back algorithm are the state transitions; the starting state is 00. For sliding-window decoding, the starting state is the state with the largest state metric. In this example, the starting state for both types of decoding is 00 since it has the largest state metric, 158. Looking at the state transition for state 00, it can be observed that the bottom path is optimal where the state transition equals 1 implies that it is a bottom path. Two operations are happening at that stage. First, the transition bit 1 is sent to the decoder output.

**Table 15.2** Coding gains for various trellis coded M-ary QAM schemes.[61]

| Number of states | Code rate | $G$(16QAM / 8QAM) [dB] | $G$(32QAM/ 16QAM) [dB] | $G$(64QAM/ 32QAM) [dB] | Asymptotic Coding gain [dB] | $N_{\text{fed}}$ |
|---|---|---|---|---|---|---|
| 4 | 1/2 | 3.01 | 3.01 | 2.88 | 3.01 | 4 |
| 8 | 2/3 | 3.98 | 3.98 | 3.77 | 3.98 | 16 |
| 16 | 2/3 | 4.77 | 4.77 | 4.56 | 4.77 | 56 |
| 32 | 2/3 | 4.77 | 4.77 | 4.56 | 4.77 | 16 |
| 64 | 2/3 | 5.44 | 5.44 | 4.23 | 5.44 | 56 |
| 128 | 2/3 | 6.02 | 6.02 | 5.81 | 6.02 | 344 |
| 256 | 2/3 | 6.02 | 6.02 | 5.81 | 6.02 | 44 |

Second, the originating state 01 of this optimal path is determined. This process is repeated until time 2, which corresponds to the first input bit that was transmitted. The decoded output is shown in Fig. 15.12 in the last row. Notice that the order in which the decoded output is generated is reversed, i.e., decoded output = 10101101, whereas the sample input pattern is 10110101. An additional code is required to be added to reverse the ordering of the decoded output in order to exactly reconstruct the sample input.

The advantage of trellis coded modulation (TCM) is by the improving system performance compared to uncoded modulation. This term is called processing gain $G$ and is

$$G = \frac{\left[\dfrac{d_{\min}^2}{P_{\text{ave}}}\right]_{\text{coded}}}{\left[\dfrac{d_{\min}^2}{P_{\text{ave}}}\right]_{\text{uncoded}}}, \tag{15.30}$$

where $P_{\text{ave}}$ is the average power of the modulation and $d_{\min}$ is the minimum Euclidian distance demonstrated in Eq. (15.27).

Table 15.2 illustrates the coding gains for various constraint lengths of trellis coded M-ary QAM formats.[61] $N_{\text{fed}}$ represents number of signal sequences with the minimum distance, which diverge at any state and after one or more transitions remerges at that state. The asymptotic coding gain represents the maximum possible coding gain for given M-ary QAM over uncoded modulation. For $K = 17$ meaning 16 cells in a shift register since $k = 16$ the coding gain of 64 QAM is 4.56 dB. Viterbi decoding is more tolerant than sequential decoding to the metric table and a receiver's AGC errors. The natural parallelism of the Viterbi decoder makes it easy to implement in hardware. This is an important consideration for high-speed decoding, which is an additional design parameter of a Viterbi decoder and plays important role in the path memory length. An ideal Viterbi decoder would keep every possible path in the memory and delays its final decision about the first bits of a packet pending to the very end. On the other hand, simulations have shown that several constraint lengths back, the paths usually merge into a single ML (maximum likelihood) candidate. Hence, the realization typically retains 4–5 constraint lengths in order to achieve the same uncorrected error rate as in an ideal decoder. For a constraint length of $K = 7$, a 32-bit path memory is a good match for a long word, where a word is 16 bits, a long word is 32 bits, and a byte is 8s bit. This is easy to implement in a 32 bit DSP.

## 15.7 M-ary QAM Transmitter Design

The QAM scheme was selected as the HFC standard format by ITU, DAVIC, MCNS, and IEEE802.14 for its spectrum efficiency. From Chapter 14, it is understood that QAM is composed of both FM to PM and AM when presented in its polar notation. Thus, when a larger number of bits represent a symbol, the frequency deviation is relatively small. The same applies to the AM component. The BW of QAM can be expressed as follows:

$$\mathrm{BW} = \frac{R_0}{\log_2 M} \approx 0.3\frac{R_0}{\log M}, \tag{15.31}$$

where $R_0$ is the bit rate. The $\log_2 M$ provides the information for the required number of bits per symbol. The QAM data stream that feeds the complex modulator is a series BB NRZ data that already has been encoded for FEC. This data is demultiplexed by SP into two outputs of $I$ and $Q$. The generated bits are mapped by the Gary code mapper, then by a differential encoder. The digital digits, after the differential encoding, contain both amplitude and phase information about which quadrant of the $I$ and $Q$ plan the $I$ and $Q$ voltages should point to. Such a Gray-coded symbol QAM constellation is presented in Figure 15.16. The digital-coded voltage is converted into analog voltage by a DAC. For generating 64 QAM, 4 bits are needed for the $I$ and 4 for the $Q$. Each of the $I$ and $Q$ analog BB signals then pass through an amplitude equalizer of $1/\mathrm{sinc}(x)$ to equalize the NRX spectral shape of $\mathrm{sinc}(x)$ and finally through an SRRC filter as a smoothing filter to reject ISI. Note that the receiver section has the other half of the SRRS, so that the overall response is raised cosine. The transmitting BB contains two SRRC one for the $I$ port and the other for the $Q$ port of the QPSK modulator. The transmitting QPSK modulator is calibrated for $Z$ and $Q$ dc offset and $Z$ and $Q$ mismatch per the procedure described in Chapter 14. Each arm of the modulator transfers $\sqrt{M}$ PAM and both are creating the QAM scheme. In older technologies, the QPSK modulation was done by discrete components. However, higher levels of integration and current advanced CMOS technology for cellular and RF, the intermediate frequency (IF) section can be done digitally by means of DSP.[27] This of course would achieve greater accuracy for the QAM quadrature because of higher tolerance control. From Eq. (14.1), the BB is sampled 4 times faster, according to $2\pi f_{\mathrm{BB}} = 2\pi f_{\mathrm{LO}}/4 = \pi f_{\mathrm{LO}}/2$. Thus, $1/T_{\mathrm{LO}} = 4/T_{\mathrm{BB}}$ and the LO clock (CLK) sampling in the discrete time domain is $\cos(n\pi/2)$ and $\sin(n\pi/2)$, which results in $1, 0, -1, 0, \ldots$ and $0, 1, 0, -1, \ldots$, respectively. Forcing the IF frequency to be $1/T$, which is the symbol rate, is advantageous. As a consequence, it eliminates the need for high-speed digital multipliers and adders to implement the mixing functions in both the modulator and demodulator. Moreover, since half of the cosine and sine oscillator samples are zeros, only a single 40-tap interpolate-by-4 transmitter filter can be used to process the data in both the $I$ and $Q$ rails of the modulator. In the same way, a single 40-tap decimate-by-2 receiver filter is needed to process the data in both $I$ and $Q$ rails of the demodulator. Even though the LO CLK sampling rate of the modulation and demodulation is $4/T$, 90% of the modulator still requires the $1/T$ CLK rate, and for demodulation, $2/T$ CLK rate is required, which reduces heating problems, circuit design, and chip area. Signal quantization and finesse is accomplished by a 12-bit high-resolution D/A rather than the 6-bit used for the analog QPSK modulator solution.[28]

### 15.7.1 Differential encoder/decoder

In Chapter 14, it was explained that the M-ary QAM constellation has four quadrants; thus, it has 90-deg symmetry where $I$ and $Q$ are orthogonal. Hence, in the demodulator portion, there is a phase ambiguity. The carrier-phase detector might lock the carrier for every 90-deg of phase error.

To overcome the phase ambiguity, a differential encoding technique is used. Under this method, the next symbol is encoded based on the input symbol and the previous state of the encoded symbol. The differential encoder is defined per the ITU J.83-A in Table 15.3. Phase-transition in encoding is a feedback system, but on the receiver side it is feedforward. Figure 15.15 provides a conceptual

**Table 15.3** Differential encoding law per the ITU J.83-A.

| $I_n$ | $Q_n$ | $C_{n-1}$ | $D_{n-1}$ | Phase change from the previous QPSK symbol | $C_n$ | $D_n$ |
|---|---|---|---|---|---|---|
| 0 | 0 | 0 | 0 | 0 | 0 | 0 |
| 0 | 0 | 0 | 1 | 0 | 0 | 1 |
| 0 | 0 | 1 | 1 | 0 | 1 | 1 |
| 0 | 0 | 1 | 0 | 0 | 1 | 0 |
| 0 | 1 | 0 | 0 | $\pi/2$ | 0 | 1 |
| 0 | 1 | 0 | 1 | $\pi/2$ | 1 | 1 |
| 0 | 1 | 1 | 1 | $\pi/2$ | 1 | 0 |
| 0 | 1 | 1 | 0 | $\pi/2$ | 0 | 0 |
| 1 | 1 | 0 | 0 | $-\pi$ 2 | 1 | 1 |
| 1 | 1 | 0 | 1 | $-\pi/2$ | 1 | 0 |
| 1 | 1 | 1 | 1 | $-\pi/2$ | 0 | 0 |
| 1 | 1 | 1 | 0 | $-\pi/2$ | 0 | 1 |
| 1 | 0 | 0 | 0 | $\pi$ | 1 | 0 |
| 1 | 0 | 0 | 1 | $\pi$ | 0 | 0 |
| 1 | 0 | 1 | 1 | $\pi$ | 0 | 1 |
| 1 | 0 | 1 | 0 | $\pi$ | 1 | 1 |

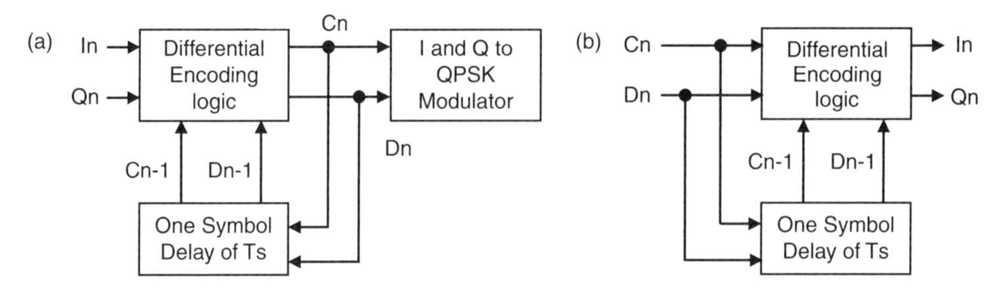


**Figure 15.15** (a) QAM differential encoder. (b) QAM decoder concept.

structure for the encoder and decoder. Since the QAM signal appears with an initial phase $\phi_0$, the encoding table defines the next quadrant the symbol will appear in. The use of this table is as follows. If the previous encoded state was $C_{n-1}D_{n-1} = 00$ with initial phase $\phi_0$, then the next state $C_nD_n$ is defined by $C_{n-1}D_{n-1}$ and the inputs $I_nQ_n$. For instance if $I_nQ_n$ is 01, respectively, then as per the fifth row, the phase change would be $\pi/2$ and the next state will be $C_n D_n = 01$ or $01 < \pi/2$. This state becomes the previous state for the next input 1 0 of $I_nQ_n$. Thus for $C_{n-1}D_{n-1} = 01 < \pi/2$ and $I_nQ_n$ 10, per row 14, $C_nD_n = 00$, which also directs the additional phase of $\pi$, and this state becomes $00 < 3\pi/2$, since it is an accumulated phase and so on.

Sometimes there is a tendency for differential encoding, coherent detection, and BER to mix. Per Fig. 15.8, it is clear that the decoding process is based on two consecutive symbols, $C_{n-1}D_{n-1}$ and $C_nD_n$, to produce $I_nQ_n$. Thus the symbol error (SER) may be in pairs; the average probability of SER for differential encoding QPSK that uses coherent detection is about twice that for nondifferential coherent QPSK. However, as was briefly demonstrated in Chapter 8, coherent detection improves CNR; the same applies to any kind of channel, light, or electric signal. Therefore, noncoherent detection of QPSK signals lose sensitivity by 2.3 dB in $E_b/N_0$, which means a shift of BER versus $E_b/N_0$ to the right. Furthermore, coherent detection with automatic frequency control (AFC) locking and carrier recovery reduces jitter because the source transmitter jitter is CLK jitter and no additional CLK is involved. A Gray-coded-symbol QAM constellation is presented in Fig. 15.16.

Further remarks regarding the QAM quadrature calibration, mentioned previously in Chapter 14: an uncalibrated quadrature would reduce the QAM BER performance with no relation to detection type, whether it is coherent, noncoherent, differential, etc. The explanation for that is provided in the BER modeling and requirement for symbol placement in the constellation as reviewed in Chapter 14.

### 15.7.2 SRRC filter

After the signal passes FEC and is encoded into the Gray-code format, it is sampled by an ADC. Thus, the digitized signal at a given time of sampling $T$ may be affected from the previous sampled signal ringing. In other words,

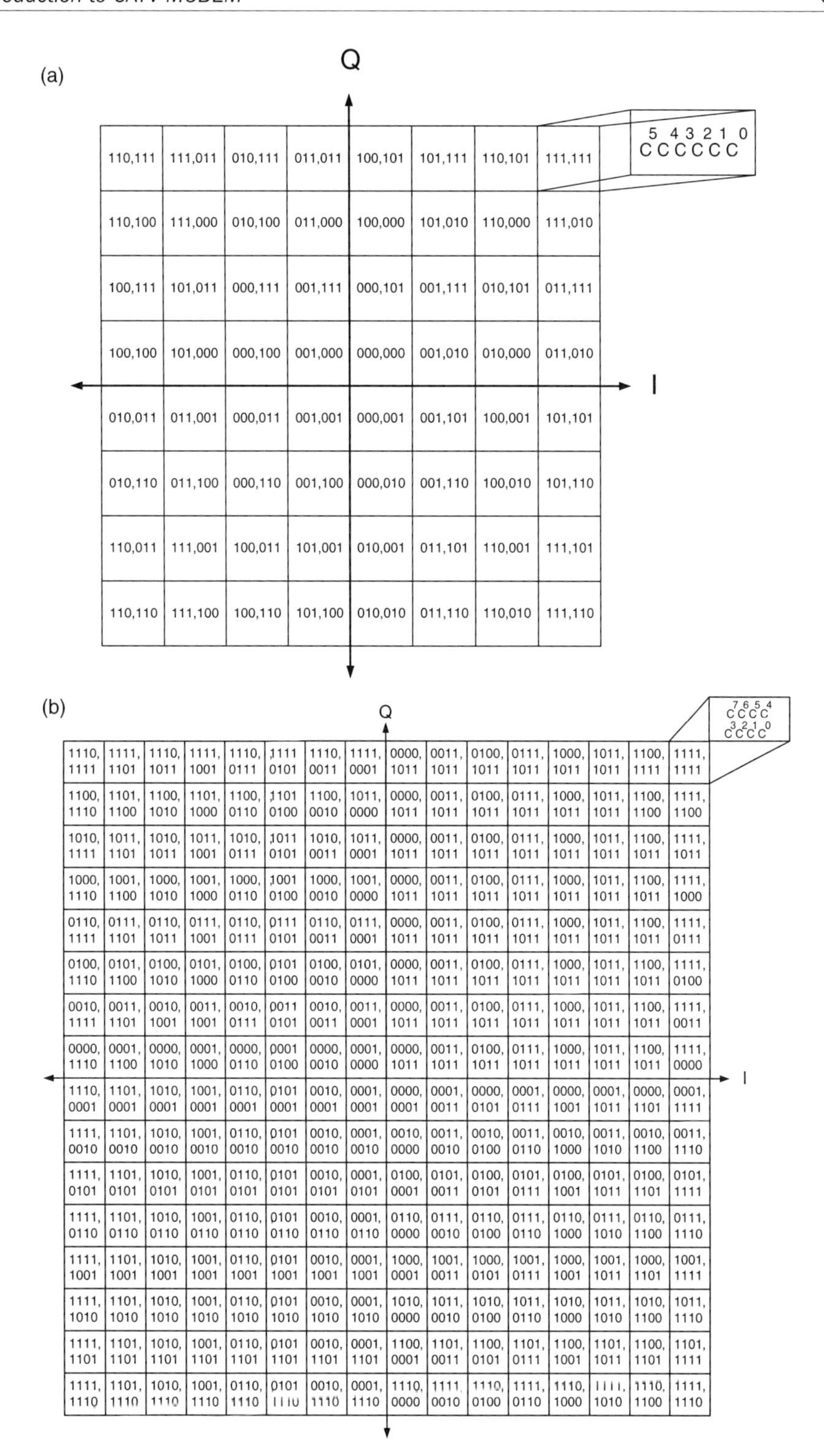


**Figure 15.16** (a) 64-QAM and (b) 256-QAM constellations per the ITU-83.J. Symbol coding is as per the Gray code.[15] (Reprinted with permission from *SCTE* © 2000.)

oscillating trails of the previous or the following pulses may interfere with the current pulse. This is called ISI. To overcome this problem a rectangular brick wall filter that has a BW of half the data rate $1/(2Ts)$ is used. However, such a rectangular shape means an infinite number of poles, which is an infinite delay time. The inverse Fourier transform of such an ideal filter is $\mathrm{sinc}(t/T_S)$. Thus to overcome ISI, the compromise is a finite impulse response (FIR), called Nyquist raised cosine (NRC), is a DSP FIR filter. This type of a design is called "matched filter design." The full response of the receiver and transmitter is raised cosine, and each one of them has an SRRC. The SRRC frequency domain response is given by[3,4]

$$H(f) = \begin{cases} 1 & \forall \quad 0 < f < \dfrac{(1-\alpha)}{2} f_s \\ \sqrt{\dfrac{1}{2}\left[1 - \sin\dfrac{\pi}{2\alpha f_s}(f - f_s)\right]} & \forall \quad \dfrac{(1-\alpha)}{2} f_s < f < \dfrac{(1+\alpha)}{2} f_s, \\ 0 & \forall \quad \dfrac{(1+\alpha)}{2} f_s < f \end{cases} \tag{15.32}$$

where $f_s$ can be written as $1/T_S$, which is the symbol interval, and $\alpha$ is the roll-off factor defining the percentage of excess BW relative to the 3-dB BW. This filter response can be written in the time domain as:

$$h(t) = \frac{\sin(\pi t/T_S)}{\pi t/T_S} \frac{\cos(\pi\alpha t/T_S)}{1 - (2\pi\alpha t/T_S)^2} = \frac{1}{T_S} \mathrm{sinc}(t/T_S) \frac{\cos(\pi\alpha t/T_S)}{1 - (2\pi\alpha t/T_S)^2}. \tag{15.33}$$

Taking the root square of $h(t)$ provides the Nyquist SRRC (NSRRC) response:[3]

$$h(t)_{\mathrm{NSRRC}} = \sqrt{\frac{1}{T_S}} \frac{\sin[\pi(1/T_S)(1-\alpha)t] + [4\alpha(t/T_S)]\cos[\pi(1/T_S)(1+\alpha)t]}{\pi(t/T_S)\left[1 - (4\alpha t/T_S)^2\right]}. \tag{15.34}$$

Figure 15.17 provides a graphical presentation of the SRRC presented by Eqs. (15.32) and (15.34) for the frequency and time domain respectively, using MathCAD.[4] It can be observed from the time domain impulse response that at each sampling time, the SRRC is at zero with the exception of $t = 0$, where the sinc function is at its peak. Thus the ringing of previous pulses are null and ISI is cancelled. The ITU J.83 defines the shape factor or rejection of the raised cosine (RC) filter given in Eq. (15.32) such that its out-of-band rejection will be better than 43 dB and its band ripple lower than 0.4 dB. Note that the symbol $\mathrm{SRRC} = \sqrt{\mathrm{RC}}$, or square root raised cosine. Other design trade-offs are made by giving up on rejection of SRRC and the compensation is made by a sharp surface acoustic wave (SAW) filter in the IF section. Additional spectral purity parameter is the adjacent channel interference (ACI) defined by the modem's inherent spectral mask, which results from the band limitation of the SRRC.

(a) Time domain impulse response.

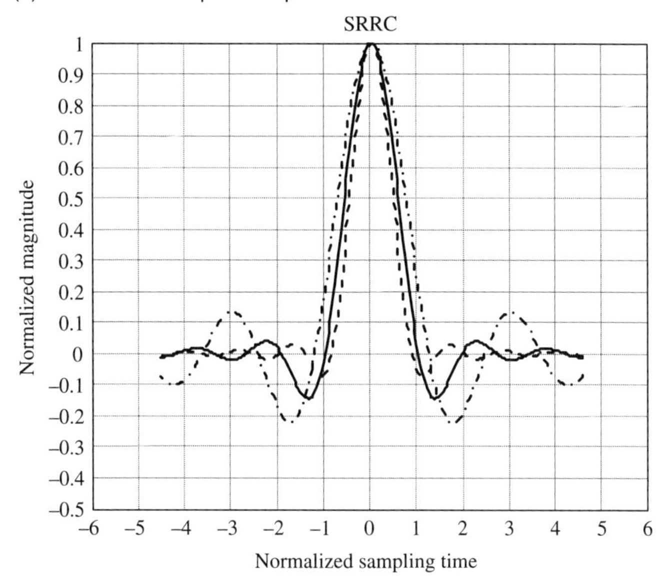


(b) Frequency response.

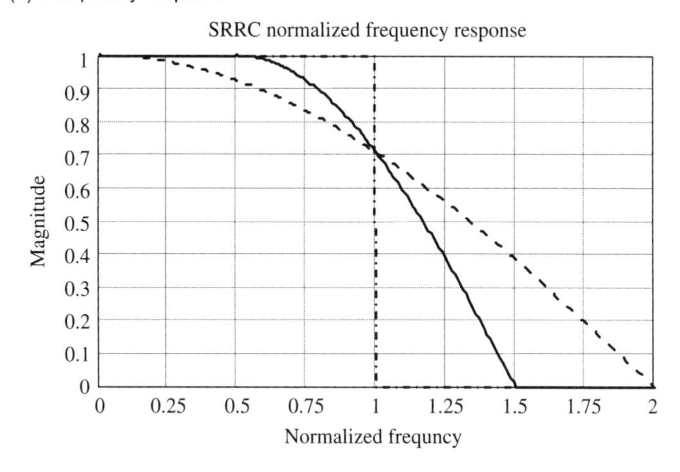


**Figure 15.17** Nyquist SRRC filter: solid line: $\alpha = \frac{1}{2}$; dashed line: $\alpha = 1$; dash - dotted line: $\alpha = 0$.

This spectral regrowth is worsened due to the nonlinearities of the RF chain by orders of magnitude. Hence, a sharp SAW in the IF may provide a solution to that problem.

## 15.7.3 IF to RF up-converter

Up-conversion is a process where the BB frequency is converted into IF by the QPSK modulator as shown in Fig. 15.18. Then it passes through an SAW filter

(a) RF up converter

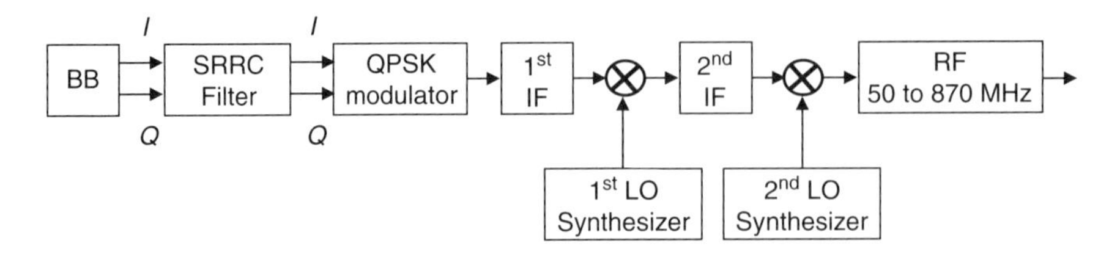


(b) Base – Band to IF analog QPSK modulator

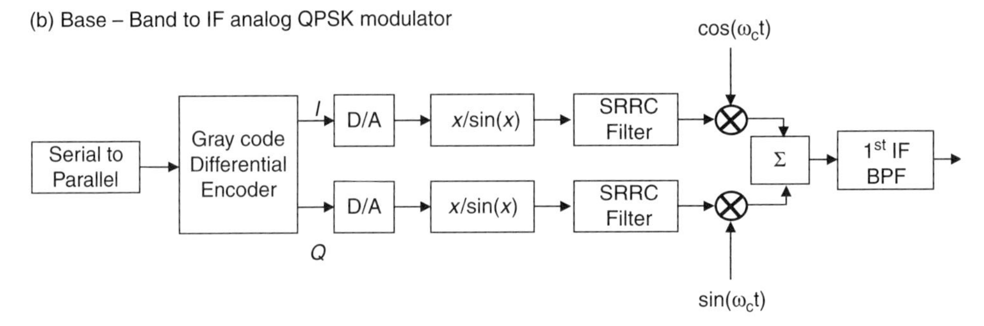


(c) Base – Band to IF integrated solution fully digital

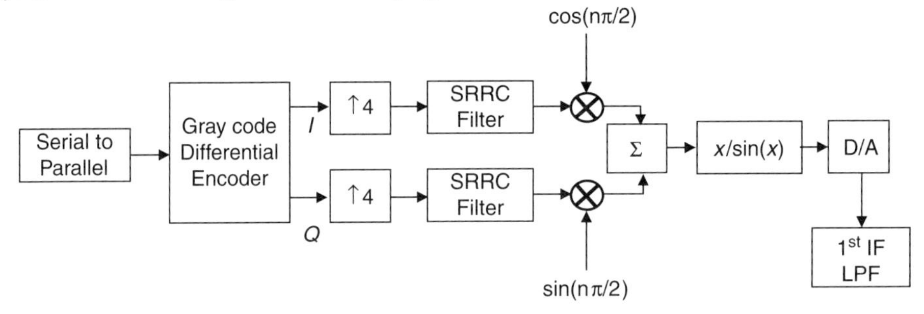


**Figure 15.18** Conceptual block diagrams for BB to RF up-conversion.

to reduce undesired spectrum regrowth and to minimize reciprocal modulation by the LO when up-converted to the final RF level. An additional factor when defining the architecture of an up-converter involves the selecting of LO frequencies and IF. This is done in such a manner that the up-converted signal is free from all spurious IF LO intermodulation combinations created by the mixing process according to $|\pm\text{mLO} \ \pm\text{nBB}| \neq \text{IF}$, where $m = \pm 1, \ \pm 2, \ldots$ and $n = \pm 1, \ \pm 2, \ldots,$ LO is the LO frequency, BB is the baseband frequency, and IF is the second IF. Generally, an RF up-converter involves dual conversion, where the output of the QPSK modulator is converted to a second IF and then converted to the final RF band between 50 to 750 MHz. The conversion to the higher IF is done with a high frequency LO with respect to the QPSK modulator output so that if the IF has an inverted spectrum, the final up-conversion to RF is done in the same manner with a higher LO frequency so the RF is not an inverted spectrum because it was twice inverted. In both cases, the frequencies are selected in such a way that the RF output is free of spurious intermodulation products, and IMD are below the maximum level allowed for composite second order (CSO) and composite triple beat (CTB) specifications requirements. The same requirements apply to the spurious LO that results from the prescalars and division CLKs of the synthesizer.

These spurious LO intermodulations may result in undesirable reciprocal mixing and noise folding into the IF band. Today's new technologies such as $\sum\Delta$ enable a decent phase noise and low spuriousness from LO synthesizers to be acheived. Figure 15.18 provides a conceptual block diagram of a BB to RF up-converter.

## 15.8 M-ary QAM Receiver Design

This section provides a review of CATV signal processing from the RF level to the BB level. The design aspects of RF chain, carrier recovery, timing recovery, equalizers realizations, and encoding are presented. The phase noise limits for BER are explained and provides the hybrid fiber coax (HFC) receiver's designer some more background on the data transport through the channel. Some of this analysis is useful for understanding digital transceivers with data clock recovery (DCR) as well as jitter transfer functions.

### 15.8.1 RF to IF down-converter and BB processing

The receiving side of a CATV transport is done by a digital set top box (STB). The STB is a super-heterodyne receiver that consists of two main blocks: an RF down-converter or tuner and a digital section, which performs the decoding carrier and timing recovery and then provides the video BB.

The RF down-converter is a super-heterodyne tuner. The RF front end (RFFE) consists of a low-pass filter between dc to 750 MHz to provide image rejection and noise folding due to reciprocal mixing. The first down-conversion is done by a preselector. The preselector can be a monolithic integrated circuit (MMIC) that consists of a preamplifier, mixer, amplifier, and first IF filter.

The LO (local oscillator) that drives the mixer is a variable-frequency synthesizer that selects and converts the desired RF channel into the first IF band. The LO frequency is above the RF and the IF is at high frequency. There are several reasons for this topology. First, the IF band-pass filter (BPF) frequency bandwidth should be 6 MHz: it is easier to realize a filter with a relatively narrow BW than a wider one; a 700-MHz BPF would have a relative BW of 8.5% ($\approx$6 MHz/700 MHz) $\times$ 100. The IF section defines the receiver's selectivity, which is the ability of the receiver to separate between channels. Hence a narrow filter with a relatively low BW would have a sharper shape factor and thus a higher selectivity IF. The second consideration is driven by the requirement of spurious-free IF per $| \pm m\mathrm{LO} \pm n\mathrm{RF}| \neq \mathrm{IF}$, where $m = \pm 1, \pm 2, \ldots$ and $n = \pm 1, \pm 2, \ldots$, LO is the LO frequency, RF is the RF frequency, and IF is the second IF. Thus in both cases of up- and down-conversion, mixers are selected by their IP2, IP3, and LO power drive that defines their IP2 and IP3. Since the preselector LO is above the RF input, the first IF spectrum is inverted.

The first IF is converted to the second IF by the fixed LO, which is below the first IF; hence the spectrum remains inverted. The second IF is additionally filtered by an SAW filter, and thus the selectivity of the tuner is increased even further. Since SAW filters have sharp shape factors, the adjacent channel interference and alternate channel interference are suppressed even further. The second IF is a standard 43.75 MHz. The second IF is down converted to the last IF, which is centered to the data rate frequency. In this manner, the QPSK modulator operates at the symbol rate and hence the carrier recovery is obtained for free, saving the carrier recovery loop. The other parameter of concern in the realization of a tuner is the LO phase noise. It is well known that the higher the LO frequency the more its phase noise will degrade. This is because of the synthesizer's frequency resolution, which defines its step size and is determined by the division ratio of the output frequency, $f_{\text{out}}$, to the phase detector frequency, $f_{\text{ref}}$. The dividers phase noise determines the PLL phase noise floor within the PLL loop bandwidth. In this case, the preselector, marked as LO $1^{\text{st}}$, step size is 6 MHz. To overcome this problem, a fractional $N$ technique is used as well as $\sum\Delta$ frequency synthesizers. First, LO phase noise values are $-85$ to 90 dBc/Hz at 10 KHZ offset from the carrier. The analog STB is more forgiving in phase noise for about 10 to 15 dB worse. This will be important when we deal later with carrier recovery and maximum jitter allowed for a QAM. Figure 15.19 provides a conceptual structure for a digital STB. IF to BB conversion is done by the QPSK demodulator, whose basics were reviewed in Chapter 14. However, in order to prevent distortions

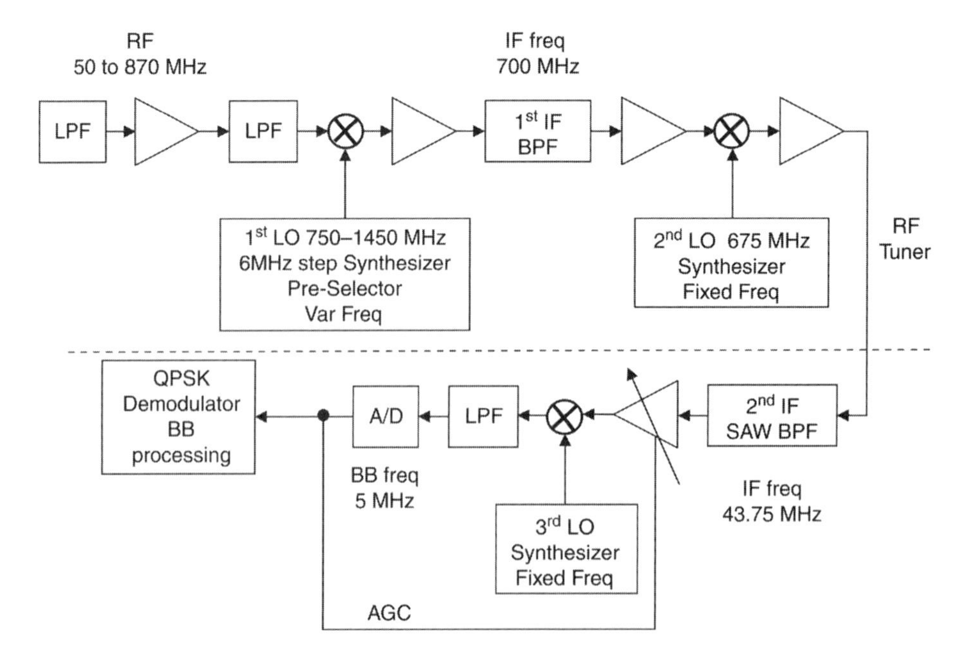


**Figure 15.19** Conceptual block diagram for super-heterodyne STB receiver.

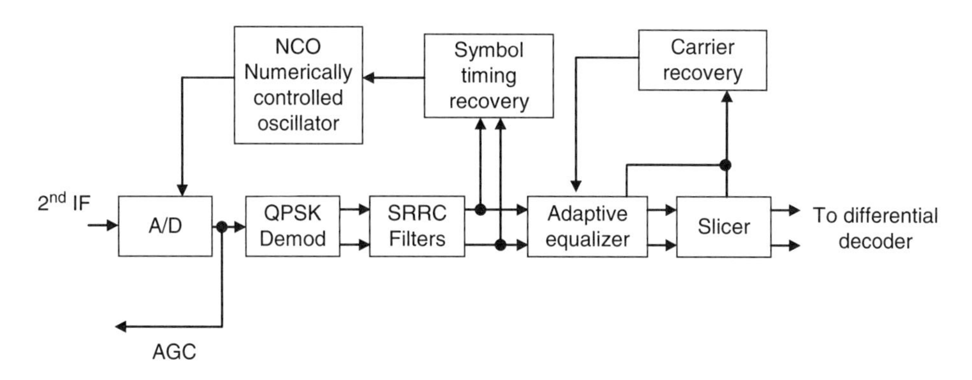


**Figure 15.20** Conceptual block diagram for BB QAM all digital signal processing.

and over driving of the ADC prior to QPSK digital processing, the automatic gain control (AGC) loop is closed between the ADC output and the IF amplifier. In this manner, the third mixer and its IMD products are kept under a constant power level. Following the ADC, the receiver is able to fully process the digital signal.

A fully digital signal-processing-receiver section consists of the following blocks: an *I*-*Q* QPSK demodulator, a SRRC filter for both *I* and *Q*, an *I* and *Q* adaptive equalizer, and a slicer. The *I* and *Q* outputs of the SRRC filter are sampled by the symbol timing recovery loop to feed a numerically controlled oscillator (NCO) that serves as the ADC sampling CLK. This way, sampling alignment is accomplished by the ADC. In order to accelerate the acquisition process and save training series time and preamble, the adaptive equalization and carrier recovery are peformed together. Figure 15.20 provides a conceptual block diagram of the digital BB processing.

### 15.8.2 Adaptive equalizer concepts and design aspects

The role of the equalizer filter is to remove linear channel distortion effects, such as cable reflections that create echoes in terrestrial receiving to combat ghost images[25] and SAW group delay at its BW frequency edges, where the phase derivative is not a constant, e.g., $d\phi/d\omega \neq$ constant, its role is to remove dispersion effects. Hence, an equalizer can be considered as an adaptive delay line. One of the basic equalizers is the linear transversal equalizer (LTE). This equalizer is constructed by using a tapped delay line and series of tap weights. Figure 15.21 illustrates the structure of LTE.[25,26]

Using these properties of equalization, the LTE impulse response in time domain and its Fourier transform to frequency domain can be written as

$$h_{E_k}(t) = \sum_{n=0}^{N} W_{nk}\delta(t - nT), \tag{15.35}$$

$$HE_{kT}(f) = \sum_{n=0}^{N} W_{nk} \exp(-jn\omega T), \tag{15.36}$$

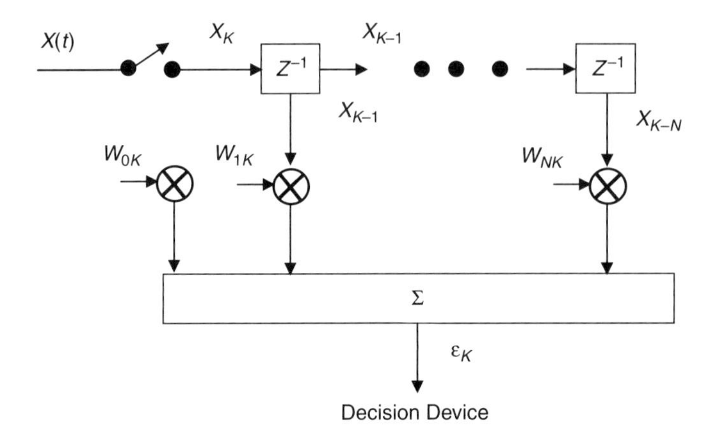


**Figure 15.21** LTE equalizer concept.

where $k$ is the $k^{\text{th}}$ sampling time, $W_{nk}$ is the weight for each tap, $t$ is the sampling interval, and δ is Dirac's delta function.

The channel input signal to the equalizer is a superposition of the desired signal and ISI caused by the channel delays and dispersion impulse responses. Assuming a ADC sampling at $t_0$, the input signal can be written as

$$x(t_0 + kT) = \sum_m a_m h(t_0 + kT - mT). \tag{15.37}$$

Therefore the desired sampling instant is where $m = k$, providing $h(t_0)$. As a consequence, the other terms are ISI and echoes. Replacing $kT$ by $t$, then $h(t_0 + t)$, the aliased spectrum of the sampled impulse, can be written as

$$H_{\text{samsig}}(f) = \sum_i H\left(f - \frac{i}{T}\right) \exp\left[\pm j2\pi \cdot \left(f - \frac{i}{T}\right) t_0\right]. \tag{15.38}$$

It is well known from the properties of the Fourier transform that time domain convolution becomes multiplication in frequency domain, and time delay becomes phase delay response. This equation represents $m \neq$ k. The exponent term in Eq. (15.38) represents timing jitter created by the sampler instance $t_0$. The $\pm$ sign represents both spectral side bands.

One of the roles of the equalizer is to reduce the signal amplitude ripple that results due to sampling jitter. Thus the overall transfer function result after equalization becomes constant, as denoted by Eq. (15.39). Furthermore, it becomes clear that while sampling the desired signal and satisfying the Nyquist sampling criteria, this is not sufficient for the ISI signal, whose interference frequency is higher. Thus, the overall result is equal to the sampling below the Nyquist frequency resulting in aliasing:

$$HE_{kT}(f) H_{\text{samsig}} = \text{constant}. \tag{15.39}$$

Hence, the equalizer weights can be optimized to flatten the notches created at the band edges. This may increase the noise at that region, which increases the sampling frequency. Figure 15.22 illustrates under-sampling and over-sampling spectrums for LTE and fractionally spaced equalizers (FSEs), respectively.

Assume that the sampling rate is twice as much; instead of $T$, the interval is $T/2$, thus the sampling frequency $f = 2/T$ is twice as fast compared to $f = 1/T$. This new sampling interval is denoted as $T'$. Hence Eq. (15.39) will be modified to[26]

$$Y_{T'}(f) = E_{kT'} \sum_i H\left(f - \frac{i}{T'}\right) \exp\left[\pm j2\pi \cdot \left(f - \frac{i}{T'}\right) t_0\right]. \quad (15.40)$$

In fact, the method here divides the sampling process into two stages. At first, the sampling is done at twice the rate. An antialiasing LPF is placed before the first $T'$ sampler; then the sampling process takes place at twice the rate of $2/T = 1/T'$. The desired information data is passed through a matched filter centered to the sampling CLK frequency. Since the sampling is twice as much, the Nyquist

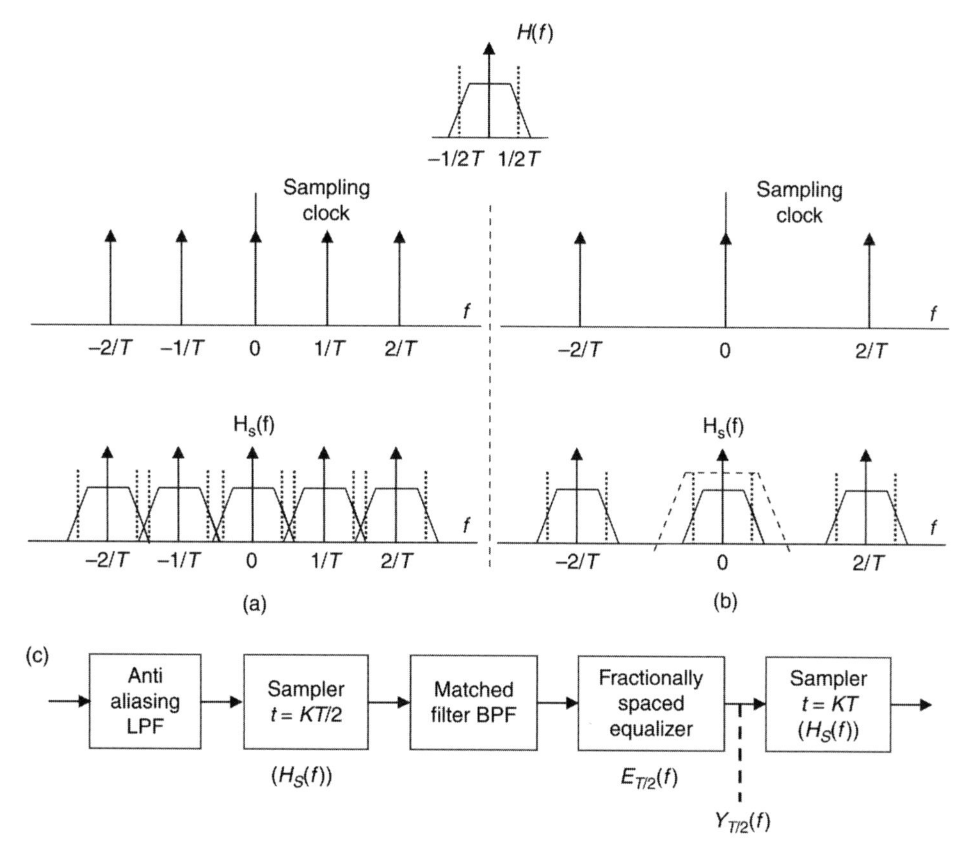


**Figure 15.22** (a) Sampling spectrum with an aliasing problem for the LTE equalizer, (b) solution by over-sampling for FSE equalizer, (c) and block diagram.

BW of the previous spectral components are filtered out without any aliasing. Thus, the equalizer operates on the desired sample without any ISI distortions and produces the output $Y_{T'}$ without aliasing. As a result, Eq. (15.41) after the matched filter can be written as

$$Y_{T/2}(f) = E_{kT/2}H(f)\exp(j2\pi \cdot f \cdot t_0). \tag{15.41}$$

This process allows the equalizer to operate on a pure signal without any ISI folding in it. The isolated sample $Y_{T/2}(f)$ is now ready to be sampled at the desired $T$ interval to create a periodic spectrum. Therefore, the equalizer output after the second sampler can be written as

$$Y_T(f) = \sum_i Y_{T/2}\left(f - \frac{i}{T}\right) = \sum_i E_{kT/2}\left(f - \frac{i}{T}\right)H\left(f - \frac{i}{T}\right) \times \exp\left[j2\pi \cdot \left(f - \frac{i}{T}\right) \cdot t_0\right]. \tag{15.42}$$

No equalization is done after the sampling at the desired rate, and there is no increment of ISI noise as in the previous suggested solution of LTE. In this manner, the equalizer can compensate for any jitter that results from the phase argument $\exp(j2\pi f t_0)$.

Returning to the equalizer output, it can be written as the sum of all tap output where equalization takes place after the matched filter, as was explained for the LTE method:

$$y_k = \sum_{n=0}^{N} W_{nk}x_{k-n}, \tag{15.43}$$

where $k$ denotes the $k^{\text{th}}$ sampling instance and $W_{nk}$ denotes the LTE weights ($n = 0, 1, \ldots, N$). The weight can be selected in a way that all the samples of the combined channel and equalizer impulse are zero except for one of the NT spaced instants in the span of the equalizer. In other words, only the desired instant sample would not be zero, while all other samples are forced by the weight function to be zero. Hence, an equalizer can be considered to be a curve-fitting device that smoothes the shape of a pulse by taking out undesirable ringing. This is called a zero-forcing equalizer (ZFE). Figure 15.23 provides the block diagram of an FSE equalizer.

However, this method suffers also from its inability to overcome sudden-signal notches and nulls. When a sudden-signal null occurs due to reflection and Rayleigh fading, the equalizer tap will try to compensate for it by additional gain at the weight parameter. This of course emphasizes the noise and consequently reduces the SNR of a channel being received. Hence one of the solutions for this problem in LTE is to apply the last mean square (LMS) adaptive equalizer. This solution is also used in fiber optics systems to remove chromatic dispersion

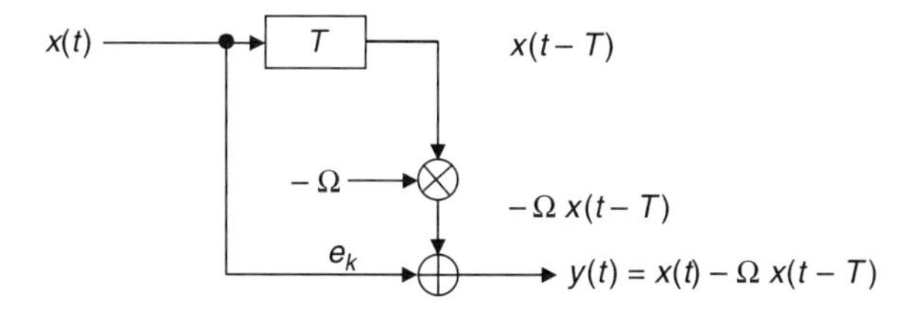


**Figure 15.23** ZFE tap concept.

effects at high data rates of 10 GB[29] by realizing it at the chip level. It is not the ideal solution, but it can adjust the weight coefficients in such a manner that the signal is optimized. This way, the equalizer becomes an adaptive equalizer. This adaptation algorithm operates in the following manner. The equalizer has to set its weight coefficients while receiving a signal. However the nature and sequence of the signal is unknown. As a consequence, the initial acquisition may result in large error and time. In conclusion, the equalizer has to be initiated and guided to optimize its tap weights; this way, it is trained and prepared for the signal-time variations. This is accomplished by a known series called the training series or preamble. The method of training series initiation is as follows. Prior to ordinary data transmission, the transmitter sends a training signal sequence known to the receiver. There the same sequence is regenerated locally by a local training signal generator. This local preamble generator serves as a feedback to correct the weight coefficient errors and adjust their values. The structure of a preamble adaptive equalizer is given in Fig. 15.24. It is clear from the block diagram that the error $e_k$ can be written as[30]

$$e_k = d_k - \sum_{n=0}^{N} W_{nk} x_{k-n}. \tag{15.44}$$

The LMS algorithm searches for and specifies the steepest successor of adaptation of the weight coefficient with respect to the square error. Hence the next weight coefficient can be written as

$$W_{n(k+1)} = W_{nk} - \frac{\mu}{2} \cdot \frac{\partial e_k^2}{\partial W_{nk}}, \tag{15.45}$$

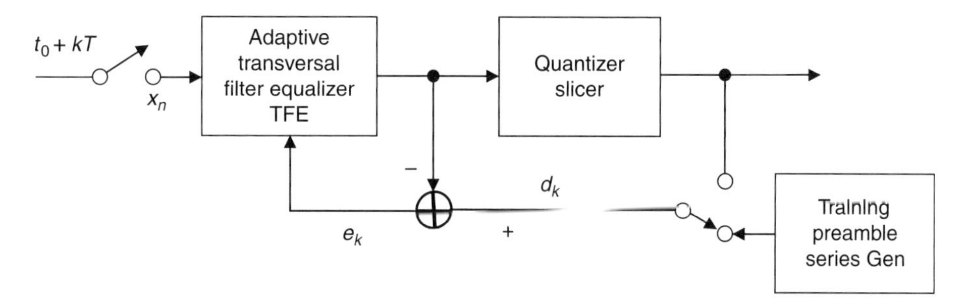


**Figure 15.24** Adaptive equalizer with preamble training series.

where $\mu$ is the adaptation initialization step size that regulates the speed and stability of the adaptation, and $n = 0, 1, \ldots, N$.

For continuing the discussion, it is clear that any error signal introduces noise, and any error signal is referred to as noise introduced into the system. Thus Eq. (15.45) describes the noise squared-error gradient and it is not the square of the error mean. Because of the feedback associated, the error is minimized in time. Hence this feedback system operates as a low-pass filter on the noise while emphasizing the signal. In this manner, the equalizer improves SNR. If we substitute Eq. (15.44) into Eq. (15.45), the coefficients of the LMS are updated according to

$$W_{n(k+1)} = W_{nk} + \mu e_k x_{k-n}, \quad \text{where } n = 0, 1, \ldots, N. \tag{15.46}$$

Since QAM is a complex modulation, Eq. (15.46) can be written as

$$W_{n(k+1)} = W_{nk} + \mu e_k x^*_{k-n}, \quad \text{where } n = 0, 1, \ldots, N, \tag{15.47}$$

where * denotes the complex conjugate.

Adaptive equalization can also be used on ZFEs. In such cases, Japanese television broadcasters employ an impulse-like training signal, and thus many of the video deghosters in Japan employ ZFE. However, to improve SNR for the receiver training signal, these systems average the training signal over several training intervals. To further improve SNR, the U.S. broadcast had selected a chirplike training signal, which has inherently higher energy. This signal was transmitted during a vertical blanking interval, allowing suitably equipped receivers to automatically synthesize filters to alleviate the effects of multipath and reflection interferences that appear as a visible ghost.

In summary, the concept of an LMS is a more general approach to automatic synthesis. Instead of solving a set of $N$ simultaneous equations as in the ZFE, the coefficients are gradually adjusted to converge into a filter that minimizes the error between the equalized signal and the stored reference. The filter convergence is based on approximations to a gradient calculation of the quadratic Eq. (15.45). The beauty is that only one parameter is adjusted and it is the adaptation parameter step size $\mu$. Through an iterative process, all filter tap weights are adjusted during each sample period at the training sequence. In this manner, the filter reaches a configuration that minimizes mean square error between the equalized signal and the stored reference. As it could be figured out, there is a trade-off between rapid convergence and the magnitude of the residual error. A too large setting of the coefficient $\mu$ would result in a fast converging system that, in a steady state, chops around the optimal coefficients. The LMS equalizer has a better noise performance compared to ZFE. Heuristically, the ZFE calculates coefficients based on samples received from the original training signal. However, the captured data will always contain some noise, and thus the calculated coefficients will be noisy or, by terminology, noise-in-noise-out. On the other hand, the LMS algorithm is an adaptive feedback process. If the noise is zero mean and it is averaged over time, its effect is minimized.

After the initialization and preamble adaptation are completed, the equalizer's mode changes to decision directed (DD). The feedback switch changes from the training-signal input generator to the slicer (quantizer) output while in steady state operation. The adaptive equalizer continuously corrects the weight coefficients using the LMS algorithm.

The use of a training sequence is sometimes not motivated because it consumes BW. This is because the preamble must be sent prior to any random data transport. So the frame of data is sent by a training block followed by data each time. There are equalization schemes that do not require training sequences, the so-called blind equalization algorithms.[36] A practical, commonly used algorithm is the constant modulus algorithm (CMA).[31–36] Blind equalization is a way to estimate unknown sources of symbols; however, the known information is the constellation alphabet, which is the key to solving the equalizer problem. The concept here is similar in its approach to the trained LMS. The equalizer update algorithm uses the stochastic gradient descent (SGD) method. The SGD algorithm uses the steepest descent method, meaning, it minimizes moving in the least-squared-error direction. In this case, the feedback replaces $d_k$ from Eq. (15.44), and there is no switch in the feedback for a training signal as appears in Fig. 15.24.

$$e'_k = \left[ |y_k|^2 - \frac{E\left(|y_k|^4\right)}{E\left(|y_k|^2\right)} \right] \cdot y_k. \tag{15.48}$$

In the same manner, equalizer weights are estimated by

$$W_{n(k+1)} = W_{nk} + \mu e'_k x_{k-n}, \quad \text{where } n = 0, 1, \ldots, N, \tag{15.49}$$

$$W_{n(k+1)} = W_{nk} + \mu e'_k x^*_{k-n}, \quad \text{where } n = 0, 1, \ldots, N. \tag{15.50}$$

which is a DD structure. Hence, an LMS algorithm is used for SGD as well as part of its feedback.

The above-mentioned algorithms can be implemented into the structures of different equalizers such as linear feedforward equalizers (LFE) and nonlinear decision feedback equalizers (DFE). Both equalizers are used in a QAM receiver scheme as shown in Fig. 15.25. The feedforward equalizer concept (FFE) is provided in Fig. 15.26 and decision-feedback equalizer is given in Fig. 15.27.

### 15.8.3 Synchronization

Etymologically, the word "synchronization" refers to the process of making two or more events occur at the same time, or, by duality, at the same frequency. In a digital communication context, various levels of synchronization must be established before data decoding can take place, including carrier synchronization, symbol synchronization, and frame synchronization.

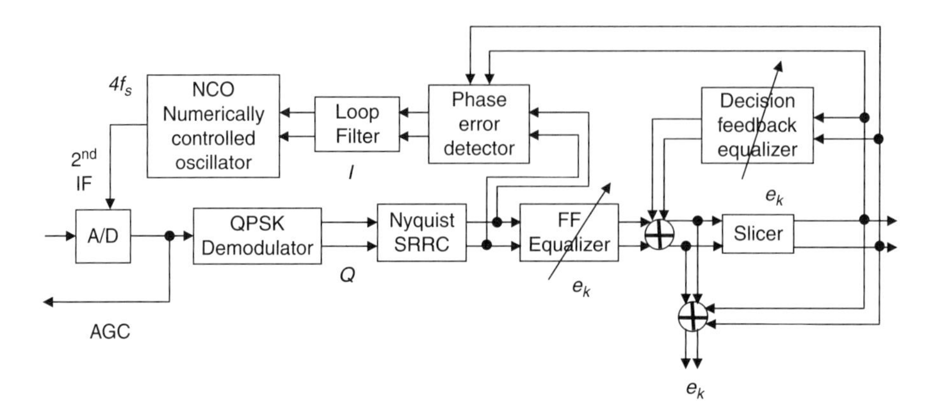


**Figure 15.25** QAM BB processing and demodulation showing all equalization.

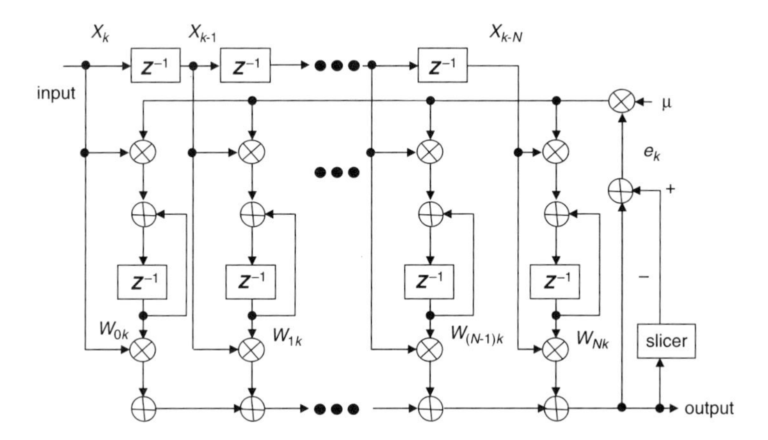


**Figure 15.26** An FFE structure.

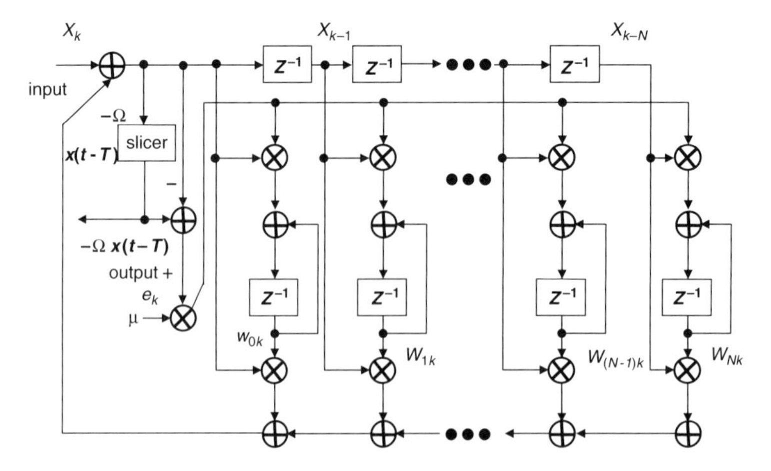


**Figure 15.27** A decision feedback equalizer.

In CATV transport, which is part of radio communications technology, "carrier synchronization" refers to the process of generating a sinusoidal signal that closely tracks the *phase* and *frequency* of a received noisy carrier transmitted possibly by a distant transmitter. Hence, carrier synchronization refers to both frequency, and phase acquisition and tracking. The most difficult process of the two, i.e., frequency and phase acquisition and tracking, is extracting the phase since it is a faster varying process than a frequency drift.

Communication systems that have available, or somehow extract and make use of good, theoretically perfect, carrier frequency and phase information are known as coherent systems, in contrast to incoherent systems that neglect the carrier phase. Systems that attempt to acquire phase information, but do not manage to get it done perfectly, are categorized as partially coherent. Coherent systems perform better due to the improvement in CNR but are more expensive due to the complexity required for carrier acquisition.

Symbol synchronization is a process of deriving at the receiver timing signals that indicate where in time the transmitted symbols are located. The receiver subsequently uses this information to decide what the symbols are. As with carrier synchronization, data available to the receiver for making timing estimates are noisy. Thus perfect timing information cannot be obtained in practice even though practical systems come close. Once symbol synchronization is achieved, the next highest synchronization level is the frame synchronization. So there are three levels of synchronization.

There are two general methodologies for achieving the various synchronization levels needed by a receiver. One way is to provide at the receiver side an unmodulated carrier, or modulated carrier by a known sequence, which cannot be used solely for the purpose of synchronization. It is advantageous since there is no need to detect data and synchronize; however, its drawback is wasted power for transmission of no information. The second approach, which is often preferred, is to derive synchronization from a data-modulated carrier. An excellent general treatment of this method can be found in the classic texts of Refs. [1–3, 6, 37–39].

### 15.8.4 Carrier recovery

The objective of a carrier-frequency synchronization in a stationary system consists of estimating and compensating the carrier frequency offset that may be induced at the receiver by oscillator instabilities. According to the degree of knowledge of the transmitted symbols, carrier frequencies are classified into three main categories: data-aided (DA), DD, and nondata aided (NDA) or blind methods. DA methods assume perfect knowledge of the transmitted symbols whereas NDA do not require any aid. Therefore NDA are spectrally efficient and are well suited for burst mode applications. An intermediate category between the DA and NDA is the DD method, which relies on symbols obtained at the output of a symbol-by-symbol decoder.

According to the magnitude of the carrier frequency offset that they can cope with, carrier frequency synchronizers may be classified into two classes: coarse and fine.

- Coarse: A carrier frequency synchronizer that can compensate for large frequency offsets of the order of the symbol rate $1/T$.
- Fine: A carrier frequency synchronizer that can compensate for frequency offsets much smaller than the symbol rate $1/T$; in general, less than 10% of the symbol rate.

Thus, during initial carrier frequency acquisition, the coarse synchronizer is employed. In this manner, large carrier frequency offsets are reduced to a small percentage of the symbol rate. This can facilitate other carrier or symbol synchronization operations. Compensation of large frequency offsets with magnitudes at the range of 100% of the symbol rate can be performed using either a closed loop, involving feedback, or an open loop, feedforward, frequency synchronizer.

Within the class of closed loop frequency synchronizers, the most commonly used carrier recovery system is the approximate ML (maximum likelihood) frequency error detectors,[5,40] quadricorrelators,[41–43] and dual filter detectors,[44] which are equivalent. Open-loop frequency recovery schemes[45,46] are referred to as delay-and-multiply methods, hence, they have a shorter acquisition time and are simpler to implement than closed-loop frequency recovery methods. They also exhibit performance that is comparable to the approximate ML frequency error detectors.[40] Due to these features, open-loop carrier recovery schemes are used in spontaneous packet transmissions, where the frequency synchronization must be performed within a fixed interval.

Once the compensation of the large carrier-frequency offsets has been accomplished and the receiver operates under steady-state conditions, carrier-recovery systems that can track and compensate carrier-frequency offsets of magnitudes less than the symbol are usually employed. Compensation of frequency offsets of less than 10% of the symbol's range can be done by DA methods.

Digital QAM receivers should also be synchronized. Receiver synchronization tasks are:

- Adjust the receiver frequency to agree with the carrier frequency of the signal (often incorporated within the phase adjustment).
- Adjust the receiver phase to agree with carrier-phase of the signal.
- Adjust the receiver timing to agree with the timing of the signal symbols.

If the carrier is out of lock and receiver is not synchronized, the constellation will rotate about its origin. The rotation rate would be as per the difference between the local timing and the transmitter's CLK. Thus, AFC circuits are used and the above-mentioned carrier recovery techniques are implemented. In addition, a carrier-recovery circuit rejects the close phase noise and thus minimizes the jitter error in the constellation.

The hardware utilized for M-ary QAM carrier recovery can be classified into DD and multiplying. The multiplying hardware can be classified into the feedforward power of $N$ and the Costas feedback loop, developed by John Costas in 1956. From theory of phase-locked loops (PLLs), it is known that PLL's phase detector is a nonlinear device;[1] also in feedforward, the frequency multiplier is a nonlinear element.[5] The power-of-$N$-feedforward carrier-recovery multiplies each phase as well as frequency in the constellation by $N$. As a result, points of the QAM constellation are rotated to the same location. For example, in the QAM, 256 points on the 45-deg diagonals are rotated to the same phase and are arranged on the $I$ axis where $N = 4$. Hence, a nonlinear device of at least the fourth order removes modulation from the signal, leaving discrete-carrier-plus-self noise. However, raising the signal to the fourth power produces a discrete line in the spectrum at 4.fo, but with a phase uncertainty of 90 deg, 180 deg, 270 deg. Therefore, the higher the number of modulation phases, the higher $N$ is. Figure 15.28 illustrates the structure of a feedforward $N$-order power analog solution. The drawback of this method is that it is an open-loop carrier-recovery circuit without phase control.

The second method is using Costas loops. Costas loops are useful synchronizers for a modulated carrier. These are closed-loop synchronizers composed of two orthogonal PLLs. The Costas loop is a means to implement ML phase estimation in addition to carrier recovery. Figure 15.29 provides the block diagram of a Costas loop.

The regular Costas loop is a square law and thus the output of the voltage-controlled oscillator (VCO) contains a phase ambiguity of 180 deg, necessitating the

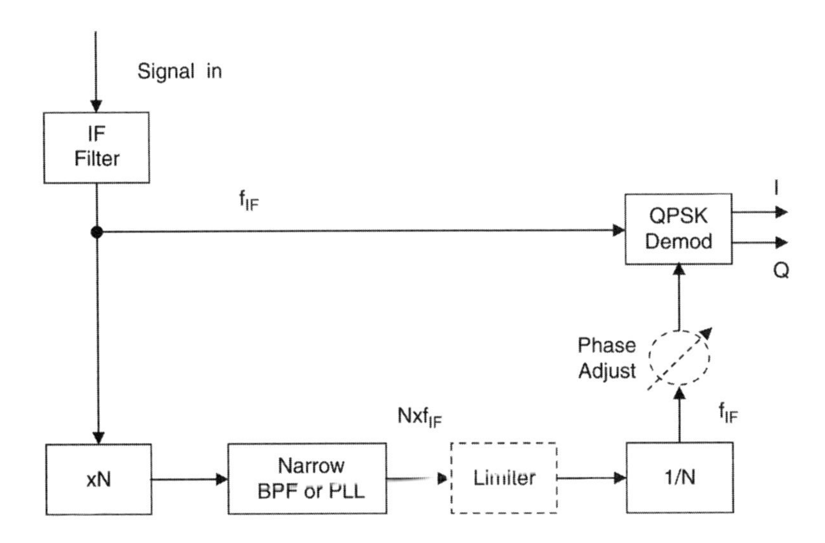


**Figure 15.28** Feedforward $N$-power synchronizer.

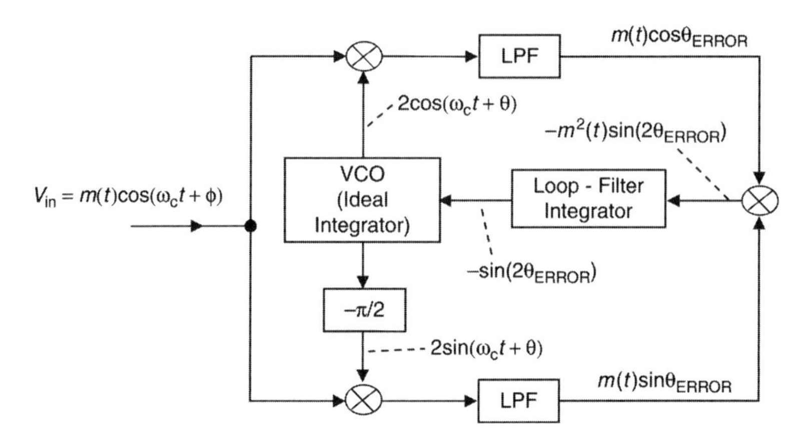


**Figure 15.29** ML carrier recovery and ML phase tracking in a Costas loop synchronizer.

need for differential encoding of data prior to transmission and differential decoding at the demodulator.

Assume an input modulated signal:

$$V_{\text{in}} = m(t)\cos(\omega_{\text{c}}t + \phi). \tag{15.51}$$

The carrier feed in phase $I$ and the quadrature $Q$ mixers (phase detectors) are driven by a VCO as denoted by

$$V_I = 2\cos(\omega_{\text{c}}t + \theta), \tag{15.52}$$

and

$$V_Q = 2\sin(\omega_{\text{c}}t + \theta). \tag{15.53}$$

The output of the $I$-path mixer and $Q$-path mixer go through a low-pass filter removing the carrier signal and second harmonics, leaving the BB only. Using some trigonometric identities, the $I$ and $Q$ outputs are $m(t)\cos(\phi - \theta)$ and $m(t)\sin(\phi - \theta)$, respectively. These two $I$ and $Q$ outputs contain a demodulated bit stream. The output of the mixed $I$ and $Q$ would be

$$V_{I,Q} = \frac{m^2(t)}{2}\sin 2(\phi - \theta). \tag{15.54}$$

Thus, the tuning voltage to the VCO after the loop filter (integrator) is proportional to the phase error between the input to the VCO, which is $\sin 2(\phi - \theta)$. In the same way, the square-law loop output of the $I$ path represents the phase error. Therefore, this, with respect to the input, operates as a high-pass filter that rejects the close-to-carrier phase noise from the LOs. As a consequence, the order of the loop filter defines the steepness of the close-to-carrier phase noise rejection. This is an important parameter for QAM detection, since the integrated phase noise defines the jitter, sometimes also denoted as $\Delta\theta_{\text{rms}}$.

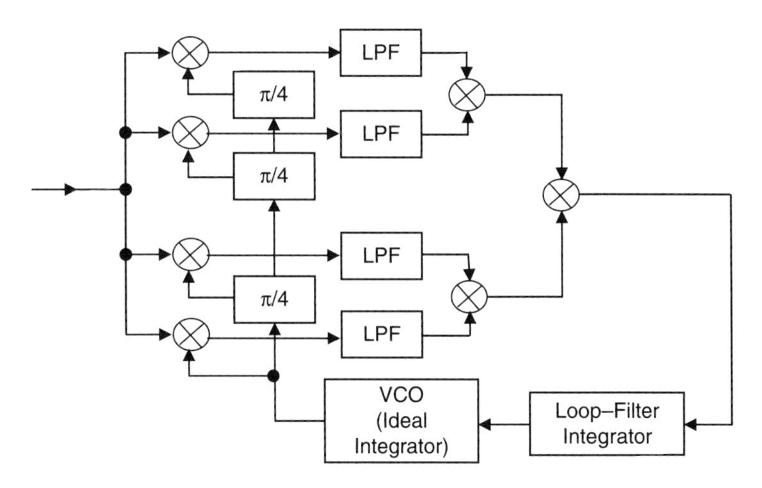


**Figure 15.30** Fourth-order Costas-loop synchronizer for M-ary QAM.

For QPSK modems, the fourth-order polarity-type Costas loop shown in Fig. 15.30 is used. It consists of four VCO phasors feeding each one of the phase detectors. Each phase detector is at 45 deg to its successor and predecessor. Hence the input signal splits into four branches and is multiplied by a VCO with the following phases: 0 deg, 45 deg, 90 deg, and 135 deg. Again, the LPF branches are used to remove high-order frequencies and pass the BB only. The LPF prior to the VCO is the loop filter, which defines the loop's dynamics. Using the block diagram in Fig. 15.29 with some trigonometric identities,[47] it can be determined that the VCO tuning voltage is proportional to sin $4(\phi - \theta)$, where the phase error is denoted by $(\phi - \theta) = \theta_{\mathrm{ERROR}}$.

So far, Costas loop concepts of carrier synchronization were presented in an analog way. However, almost all Costas loops are realized by DSP. Assume modulated data $r(t)$ accompanied with random noise $n(t)$ as follows:

$$r(t) = s(t,d,\phi) + n(t), \tag{15.55}$$

where $d$ is a sequence of $N$ modulation symbols. The additional assumption is a binary antipodal signal, in which case, each component $d_K$ of $d$ is either 1 or $-1$. If the band rate is $1/T$ symbols/s, then the signal component expressed in Eq. (15.55) can be written as

$$s(t,d,\phi) = A \sum_{K=0}^{N-1} d_K \cos(2\pi f_c t) p(t - kT), \tag{15.56}$$

where $p(t)$ is the BB pulse that determines to a large extent the spectral content of the transmitted signal. It is also assumed that a data window of $N$ exists. To simplify the calculation, it is assumed that $p(t)$ is a unit high rectangular pulse, however, results can be generalized to any pulse shape. It is also assumed that the modulation sequence $d$ is known. Thus the following likelihood function is applicable:

$$L(\phi) = \exp\left[\frac{2}{N}\int_0^{NT} r(t)s(t,d,\phi)\mathrm{d}t\right]$$
$$= \exp\left[\frac{2A}{N}\sum_{K=0}^{N-1} d_K \int_{KT}^{(K+1)T} r(t)\cos(2\pi f_c t + \phi)\mathrm{d}t\right]. \quad (15.57)$$

All that is needed now is to take the expectation of the conditional likelihood function in Eq. (15.57) with respect to a random modulation sequence. Assume that each binary symbol does not depend on past and future symbols and thus its probability of level is $1/2$. This way the log-likelihood function is obtained after dropping terms that are not functions of $\phi$ and taking the logarithm of the resultant expression:

$$\Lambda(\phi) = \sum_{K=0}^{N-1} \ln\left\{\cosh\left[\frac{2A}{N_0}\int_{kT}^{(k+1)T} r(t)\cos(2\pi f_c t + \phi)\mathrm{d}t\right]\right\}. \quad (15.58)$$

The ML estimator maximizes the above expression with respect to $\phi$. To simplify the calculation, the Taylor expansion of $\ln[\cosh(x)] \propto x^2$ and $\ln[\cosh(x)] \propto |x|$ is used for small and large SNR, respectively. In small SNR, the log-likelihood comes under the above approximation:

$$\ell(\phi) = \sum\left[\int_{kT}^{(k+1)T} r(t)\cos(2\pi f_c t + \phi)\mathrm{d}t\right]^2. \quad (15.59)$$

Taking the derivative of Eq. (15.59) with respect to $\phi$, the following necessary condition for the ML estimate $\hat{\phi}$ of $\phi$ is obtained:

$$\sum_{k=0}^{N-1}\int_{kT}^{(k+1)T} r(t)\cos\left(2\pi f_c t + \hat{\phi}\right)\mathrm{d}t \times \int_{kT}^{(k+1)T} r(t)\sin\left(2\pi f_c t + \hat{\phi}\right)\mathrm{d}t = 0. \quad (15.60)$$

Such a loop that dynamically forces that condition is provided in Fig. 15.31, which is again a Costas loop.

Using DSP techniques for current CATV modems, it is possible to combine the $I$ and $Q$ demodulator ADC and carrier recovery after blind equalization as a digital block, as demonstrated in Fig. 15.32. Figure 15.33 shows a digital receiver with a complex derotator located after the adaptive equalizer. The $I$ and $Q$ demodulator marked as QPSK demodulator, is driven by a free-running oscillator. The corrections are done on the symbols after equalization by a derotator, and adding them and to the phase by the operator $\exp(-j\phi_k)$. The adaptive equalizer generates a good estimate of the rotated symbols and the carrier recovery loop rotates them back after the carrier recovery by the operator $\exp(+j\phi_k)$.

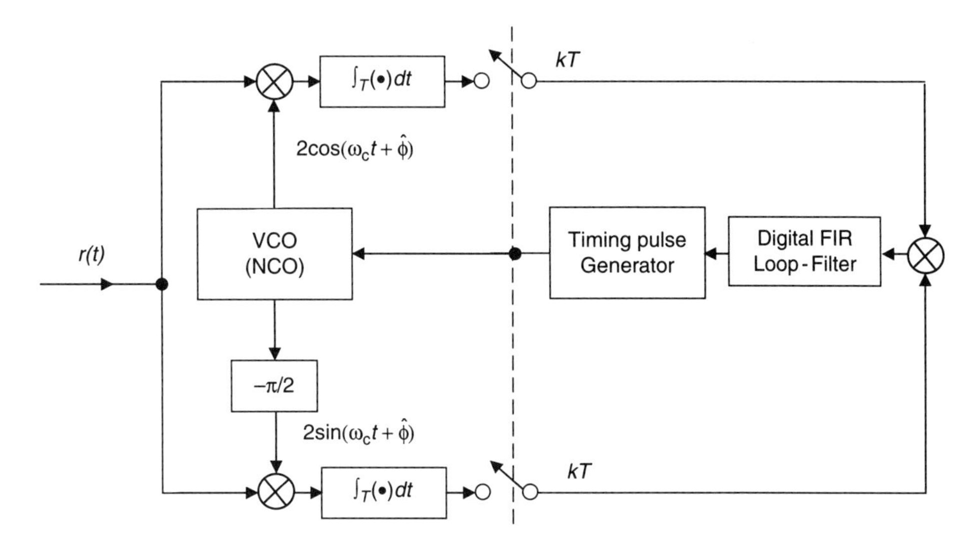


**Figure 15.31** Carrier synchronization from modulated data.

### 15.8.5 Timing recovery

Timing recovery is the next level of the synchronization process. The main role of timing-recovery loops is to generate a CLK signal at the symbol rate, $1/T_S$ from the received signal. In QAM transport, the CLK is extracted from the data. The data encodes the CLK, for instance, by using an NRZ format.

Timing recovery can be classified into two methods of synchronization. The first is forward-acting,[5,48] and the second uses feedback methods.[5,495] In the

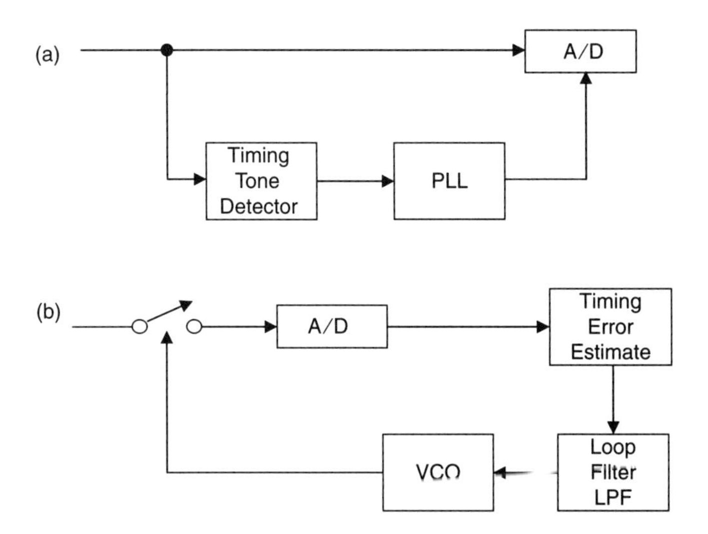


**Figure 15.32** The (a) spectral-line timing-recovery loop after blind equalization. (b) The minimum mean-square-error (MMSE) feedback timing-recovery loop.

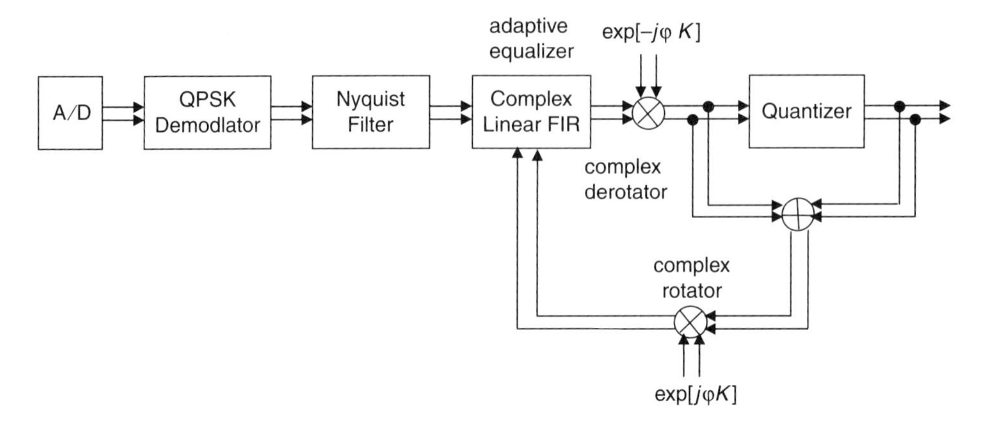


**Figure 15.33** A DD BB receiver, where the *I* and *Q* demodulator is free running and a phase-error correction is done after equalization by a complex derotator recovery loop.

forward-acting approach, a timing tone is extracted directly from the received signal. The timing tone, which has an average frequency identical to the symbol rate, is used to synchronize the receiver to the detected digital signal. To remove the resultant jitter of the timing tone, a PLL is used. An example of PLL phase noise analysis is provided in Refs. [50] and [51]. In feedback timing-recovery techniques, a feedback loop is used in combination with the PLL to tune the sampled phase in an iterative way and drive the VCO in order to reduce the timing error.

One of the most common methods of timing recovery in the forward-acting class is the spectral line method. In this approach, the timing error $\tau_k$ is determined by the zero crossing of the timing tone and the PLL. The modulated signal goes through a timing tone detector such as an envelope detector, LPF, and a PLL, where the resultant signal is utilized to drive a VCO. So an AM detector provides a BB that the PLL is locked on. The VCO frequency will then be the CLK.

The feedback timing recovery method is also divided into two methods. The first is classified as minimum mean squared error (MMSE) and the second is the band-rate timing recovery. Figure 15.32 illustrates MMSE-feedback timing. A timing error estimator creates a phase error voltage that drives a VCO. The VCO is aligned to drive the sampler in exact timing, thus minimizing the timing error. The VCO is tuned to sample every symbol period of *T*. This can be implemented by using the LMS or using the stochastic-gradient method algorithm. This approach is similar to the one used for the adaptive equalizer.

Another timing method is called early–late,[6] shown in Fig. 15.34. In this method, the BB is squared to make all peaks positive, and then sampled at two different times: once before and once after the sampling instance at the same timing shift $\pm\delta$. The sampling instants are adjusted by the VCO until both are equal. In this method, it is assumed that the received waveform peaks are at the correct sampling points and the waveform around the peak is symmetric. The average timing error is utilized to correct the VCO. The main drawback of this

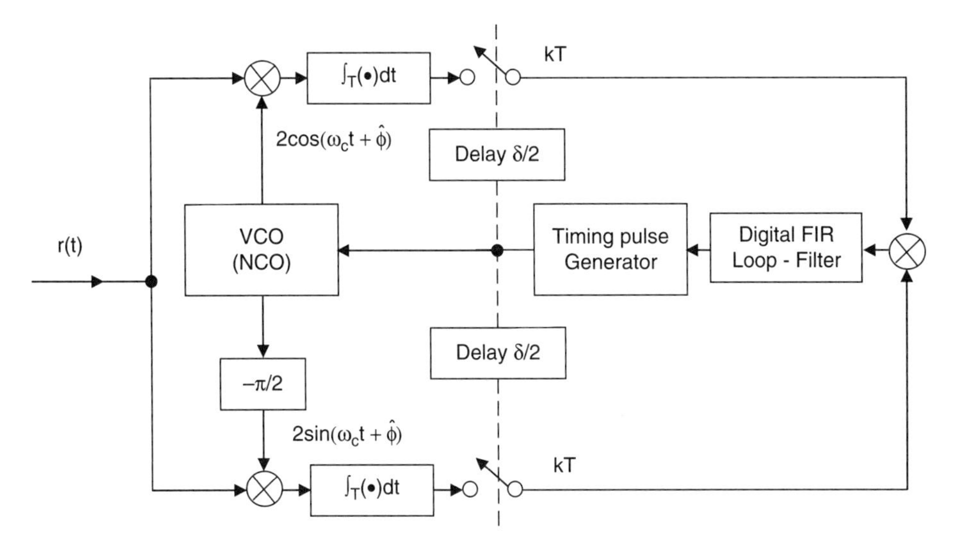


**Figure 15.34** Early–late timing-recovery loop.

concept is that the waveform peaks of the received signal do not take place at every sampling period of QAM.

In advanced designs, timing and phase recovery can be processed in a joint recovery process,[5,52] as shown in Fig. 15.35.

## 15.8.6 Timing-jitter phase noise and BER

One of the parameters affecting QAM BER performance is integrated phase noise. The phase noise of the recovered carrier is used as LO for the QAM demodulator.

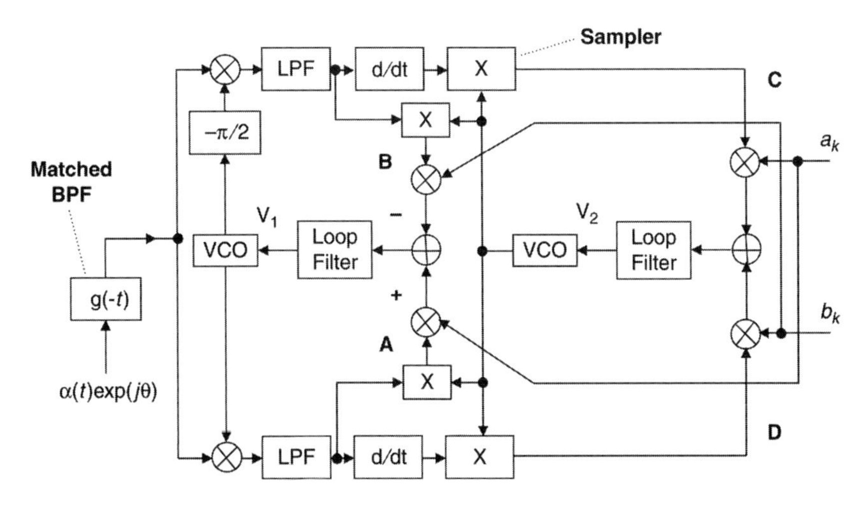


**Figure 15.35** Joint recovery of the carrier phase and symbol timing (DA) for QAM.

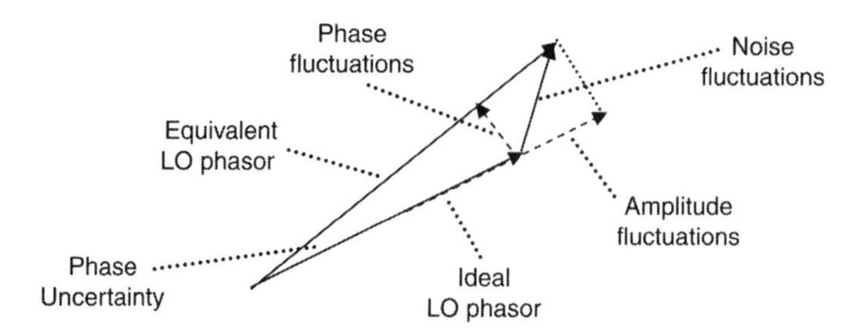


**Figure 15.36** Noise phasors showing amplitude and phase fluctuations.

The jitter of the recovered carrier and timing CLK is used for the symbol-sampling process. An oscillator is characterized by a phasor, where the noise can be described as a rotating phasor, as described in Fig. 15.36.

The noise of the LO is composed of two components: amplitude, which is a parallel vector to the LO phasor, and phase, which is orthogonal to the carrier phasor. The amplitude fluctuations can be removed by a limiter; however, the normal component is harder to remove. For this purpose, a PLL is used:

Timing jitter is a characteristic of the CLK. In the ideal case, the decisive instant is determined by the CLK during sampling. Figure 15.37 describes a digital sampling circuit and its equivalents. The ideal sampling instant is deterministic. In real time, the sampling CLK transitions fluctuate from bit to bit by $\pm\Delta T$. The changes of $\pm\Delta T$ are random and are related to frequency fluctuations of the CLK around its average value $f_0$. Hence, stochastic jitter of the sampling CLK is related to the CLK FM noise. This defines the phase noise of the CLK. The sampling process is related to the multiplication of the sampling CLK with the

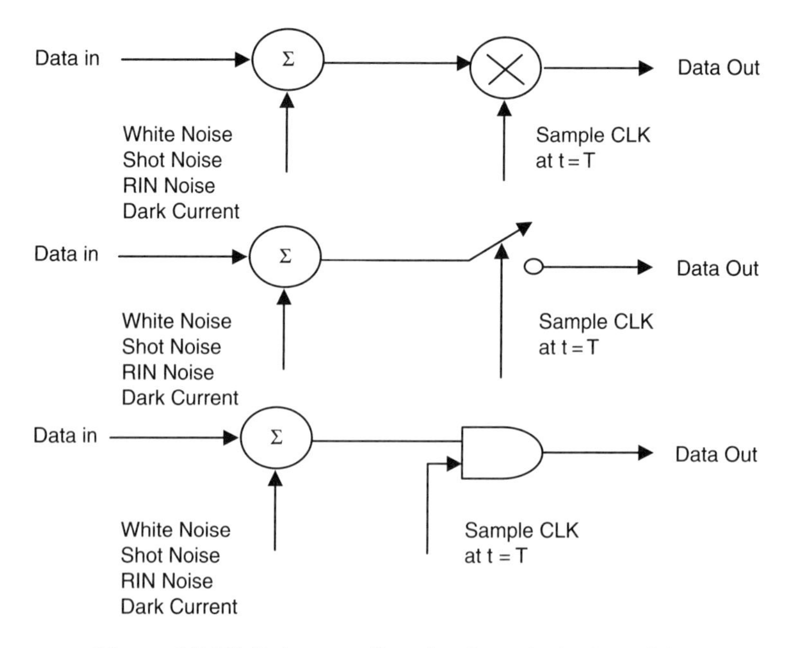


**Figure 15.37** Data sampling circuit equivalent models.

data at the sampling instant $T$. However, the sampling time $T$ is not a deterministic process, due to the excessive jitter of the CLK, which is a random process.

As a result, during the sampling period, an error may occur and the wrong bit level might be sampled if the jitter is high enough. Fig. 15.38 shows that the sampling range is twice the swing of the rms jitter.

For the sake of simplicity in analyzing BER degradation as a function of phase noise, assume a QPSK modulation for which the sampling instance $T_0$ is uncertain and it is assumed that it has a Gaussian distribution with zero mean and variance of $\Delta T$. Also assume that the period of the CLK signal is $T$. Therefore, the sampling-signal phase error with respect to the time error can be written as

$$\begin{cases} T \Longleftrightarrow 2\pi \text{ or } 360\,\text{deg}, \\ \Delta T \Longleftrightarrow \Delta\theta_{\text{rms}}, \\ \Delta T = \dfrac{\Delta\theta_{\text{rms}}}{2\pi}\dfrac{1}{f_{\text{CLK}}} = \dfrac{\Delta\theta_{\text{rms}}}{2\pi} T, \\ \Delta\theta_{\text{rms}} = \dfrac{\Delta T \cdot 2\pi}{T}. \end{cases} \tag{15.61}$$

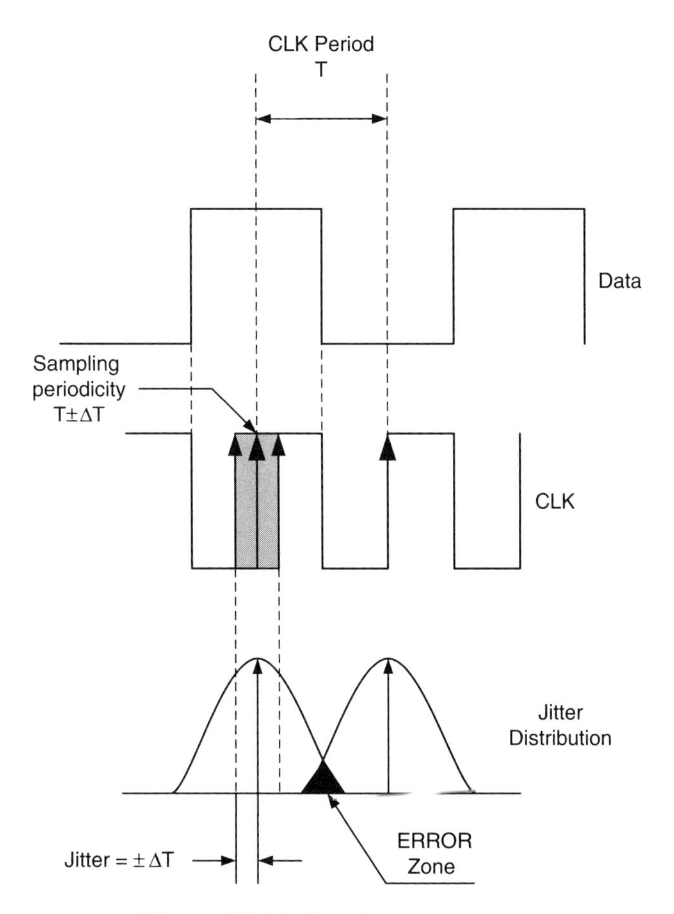


**Figure 15.38** Jitter in time domain.

Hence, the carrier can be written as

$$V(t) = A\cos(\omega_c t + \Delta\theta_{rms}). \tag{15.62}$$

If the jitter is stochastic and not deterministic due to group-delay, and if the CLK's fundamental harmonic is cosine, as in Eq. (15.62), and if the data are a toggling signal at the CLK frequency, and the sampling is at the bit center at half of its period $T$, the following sampling relations are applicable for a 1D modulation:

$$\begin{aligned} V_{out}(t) &= A\cos(\omega t)\cos(\omega t + \Delta\theta_{rms}) \\ &= A\cos(\omega t)[\cos(\omega t)\cos(\Delta\theta_{rms}) - \sin(\omega t)\sin(\Delta\theta_{rms})]. \end{aligned} \tag{15.63}$$

The output of the sampler is rms voltage or power. Hence, during the period $T$, the sampling average is

$$\begin{aligned} \langle V_{out}(T)\rangle &= \frac{A}{T}\int_0^T \cos(\omega t)[\cos(\omega t)\cos(\Delta\theta_{rms}) - \sin(\omega t)\sin(\Delta\theta_{rms})]\mathrm{d}t \\ &= \frac{A}{2T}\cos(\Delta\theta_{rms}), \end{aligned} \tag{15.64}$$

which shows that the output-voltage amplitude of the signal is reduced by $\cos(\Delta\theta_{rms})$. In other words, the CNR is reduced by $\cos(\Delta\theta_{rms})^2$ since the power is related to the square of the voltage. The conclusion is that jitter reduces the CNR by $\cos(\Delta\theta_{rms})^2$, assuming the SNR is given now by[56]

$$\mathrm{SNR} = \frac{E_b}{N_0}\cos^2\Delta\theta_{rms}. \tag{15.65}$$

The conclusion drawn from Eq. (15.65) is that $(E_b/N_0)(1 - \cos^2\Delta\theta_{rms}) = (E_b/N_0)\sin^2\Delta\theta_{rms}$, which is the CNR delta. In Chapter 14 [Eq. (14.58)], it was explained that the phase-noise modulation-error ratio (MER) is proportional to $\sin^2\Delta\theta_{rms}$. The degradation of CNR due to phase noise is the penalty in MER and the phase-noise error vector magnitude is $\sqrt{(E_b/N_0)\sin^2\Delta\theta_{rms}}$.

The conditioned probability of the error per bit for QPSK with phase-noise effects is given by the Bayes rule:[4]

$$P_b\left(\frac{E_b}{N_0}, \Delta\theta_{rms}\right) = \int_{-\infty}^{\infty} P_b\left(\frac{E_b}{N_0}\middle|\theta.\right)\cdot P(\theta)\mathrm{d}\theta, \tag{15.66}$$

which results in

$$P_b = \frac{1}{\pi\cdot\sigma_\theta}\int_0^{\infty}\exp\left(-\frac{\theta^2}{2\sigma_\theta}\right)\int_{\sqrt{2(E_b/N_0)}\cos\theta}^{\infty}\exp\left(-\frac{y^2}{2}\right)\mathrm{d}y\mathrm{d}\theta. \tag{15.67}$$

Now assume coherent detection from a QPSK modulator, as illustrated in Fig. 15.39. It is assumed that the pulse shape $p(t)$ is appropriately chosen such that the matched filter $h(t)$ outputs are free of ISI. Additionally, if the data-symbol timing is known exactly, then the recovered bits of information can be written as

$$\begin{aligned} I_k &= a_k\cos(\Delta\omega t + \Delta\theta_{rms}) + b_k\sin(\Delta\omega t + \Delta\theta_{rms}), \\ Q_k &= b_k\cos(\Delta\omega t + \Delta\theta_{rms}) - a_k\sin(\Delta\omega t + \Delta\theta_{rms}), \end{aligned} \tag{15.68}$$

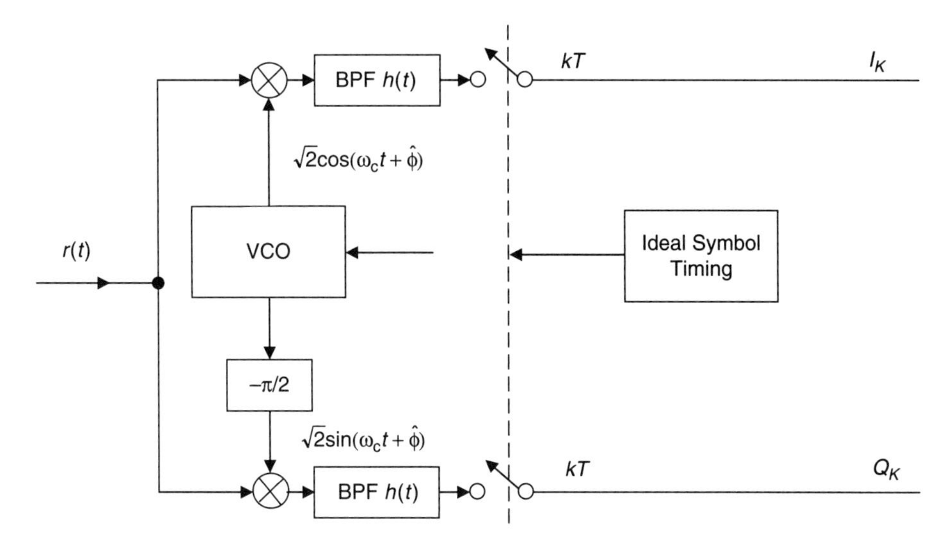


**Figure 15.39** QPSK coherent receiver concept.

where $\Delta\omega$ is the error frequency between the LO and the input data. If $\Delta\omega = 0$, then Eq. (15.68) becomes

$$I_k = a_k \cos(\Delta\theta_{\text{rms}}) + b_k \sin(\Delta\theta_{\text{rms}}),$$
$$Q_k = b_k \cos(\Delta\theta_{\text{rms}}) - a_k \sin(\Delta\theta_{\text{rms}}). \tag{15.69}$$

Any phase error due to jitter results in:

- A loss of correlation, which is represented by cosine terms.
- An undesirable coupling between the $I$ and $Q$ channels, which is represented by sine terms. This latter effect is particularly damaging to QAM systems.
- A reduction in $E_b/N_0$ as was shown in Eq. (15.65).

Then the conditioned probability of error per bit for QPSK is given by:[53]

$$P_{bQPSK}\left(\frac{E_b}{N_0}, \Delta\theta_{\text{rms}}\right) = \frac{1}{2}\int_{-\pi}^{\pi} Q_{QPSK}\left[\sqrt{\frac{2E_b}{N_0}}(\cos\theta + \sin\theta)\right] \cdot P(\theta)d\theta$$
$$+ \frac{1}{2}\int_{-\pi}^{\pi} Q_{QPSK}\left[\sqrt{\frac{2E_b}{N_0}}(\cos\theta - \sin\theta)\right] \cdot P(\theta)d\theta, \tag{15.70}$$

where $Q$ for QPSK modulation is given by:[56]

$$Q(x) = \int_{x}^{\infty} \frac{\exp(-u^2/2)}{\sqrt{2\pi}} du \Rightarrow Q_{\mathrm{QPSK}}\left(\sqrt{\frac{2E_{\mathrm{b}}}{N_0}}\right) = \int_{\sqrt{2E_{\mathrm{b}}/N_0}}^{\infty} \frac{\exp(-u^2/2)}{\sqrt{2\pi}} du. \quad (15.71)$$

A synthesizer's phase-noise performance introduces phase error, denoted as $\Delta\theta_{\mathrm{rms}}$, which is assumed to remain constant over individual symbol intervals. This phase error is assumed to have a Tikhonov probability distribution function:[6,54,55]

$$P(\theta) = \frac{\exp(\alpha \cos \theta)}{2\pi \cdot I_0(\alpha)}, \quad (15.72)$$

where $I_0(\bullet)$ is the zero order–modified Bessel function and $\alpha$ is the SNR for carrier recovery PLL. For large $\alpha$ arguments, a numerical evaluation of the Bessel function can become numerically ill conditioned, and it is better to use the asymptotic approximation for the density function in a frequency synthesizer, where the total integrated phase noise is $\sigma_\theta^2 \Rightarrow \alpha = \sigma_\theta^{-2}$.

The next square constellation above QPSK is 16 QAM. An ideal 16 QAM with Gray coding is shown in Fig. 15.40.

The Gray coding assigns the specified $I$- and $Q$-bit patterns to the constellation points in such a way that any nearest-neighbor reception results only in a single-bit error. This coding results in the BER system performance being very nearly equivalent to a more simply computed symbol error rate of the system. When there is no phase noise, the 16-QAM symbol-error rate is given by

$$P_{\mathrm{S}} = 1 - \left[\frac{4}{16}P(C|I) + \frac{8}{16}P(C|II) + \frac{4}{16}P(C|III)\right], \quad (15.73)$$

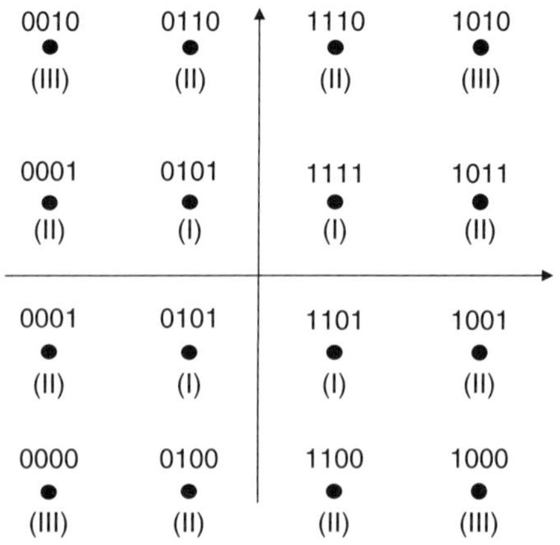

**Figure 15.40** 16-QAM Gray-coded constellation and symmetry indices.[53] (Reproduced with permission from *Frequency, Synthesizer Design Handbook*, © Artech House, Inc., 1994.)

where the $P(|)$ functions represent probabilities of symbol error for regions of symmetry per Fig. 15.40. These quantities are written as

$$P(C|I) = P(C|III) = \left[1 - Q(\gamma_0)\right]^2, \tag{15.74}$$

$$P(C|II) = \left[1 - 2Q(\gamma_0)\right] \cdot \left[1 - Q(\gamma_0)\right], \tag{15.75}$$

where

$$\gamma_0 = \sqrt{\frac{2a^2}{N_0}} = \sqrt{\frac{4}{5}\frac{\bar{E}_b}{N_0}}, \tag{15.76}$$

where $a$ is defined in Fig. 14.2 (Chapter 14). When phase noise is introduced, and using the symmetry conditions in Eqs. (15.74) and (15.75), Eq. (15.76) is modified to

$$P_S(\alpha,\gamma_0) \leq \sqrt{\frac{\alpha}{2\pi}} \int_{-\pi}^{\pi} 1 - \sum_{K=1}^{12} Qm_k(\gamma_0,\theta) \times \exp\{\alpha[\cos(\theta) - 1]\} d\theta, \tag{15.77}$$

where

$$\begin{cases} Qm_1(\gamma_0,\theta) = Q\{\gamma_0[\cos(\theta) - \sin(\theta)]\}, \\ Qm_2(\gamma_0,\theta) = Q\{\gamma_0[2 - \cos(\theta) + \sin(\theta)]\}, \\ Qm_3(\gamma_0,\theta) = Q\{\gamma_0[\cos(\theta) + \sin(\theta)]\}, \\ Qm_4(\gamma_0,\theta) = Q\{\gamma_0[2 - \cos(\theta) - \sin(\theta)]\}, \\ Qm_5(\gamma_0,\theta) = Q\{\gamma_0[3\cos(\theta) - \sin(\theta) - 2]\}, \\ Qm_6(\gamma_0,\theta) = Q\{\gamma_0[\cos(\theta) + 3\sin(\theta)]\}, \\ Qm_7(\gamma_0,\theta) = Q\{\gamma_0[2 - \cos(\theta) - 3\sin(\theta)]\}, \\ Qm_8(\gamma_0,\theta) = Q\{\gamma_0[3\cos(\theta) + \sin(\theta) - 2]\}, \\ Qm_9(\gamma_0,\theta) = Q\{\gamma_0[2 - \cos(\theta) + 3\sin(\theta)]\}, \\ Qm_{10}(\gamma_0,\theta) = Q\{\gamma_0[\cos(\theta) - 3\sin(\theta)]\}, \\ Qm_{11}(\gamma_0,\theta) = Q\{\gamma_0[3\cos(\theta) - 3\sin(\theta) - 2]\}, \\ Qm_{12}(\gamma_0,\theta) = Q\{\gamma_0[3\cos(\theta) + 3\sin(\theta) - 2]\}, \end{cases} \tag{15.78}$$

and

$$Q(x) = \begin{bmatrix} \dfrac{\exp(-x^2/2)}{\sqrt{2\pi}} \displaystyle\sum_{i=1}^{5} b_i t^i & \text{for} \quad x \geq 0 \\ 1 - \dfrac{\exp(-x^2/2)}{\sqrt{2\pi}} \displaystyle\sum_{i=1}^{5} b_i t^i & \text{for} \quad x < 0 \end{bmatrix}, \tag{15.79}$$

where

$$t = \frac{1}{1 + p|x|}, \tag{15.80}$$

and the constants are

$$\begin{cases} p = 0.2316419, \\ b_1 = 0.319381530, \\ b_2 = -0.356563782, \\ b_3 = 1.781477937, \\ b_4 = -1.821255978, \\ b_5 = 1.330274429. \end{cases} \tag{15.81}$$

As the constellation becomes of a higher order, the calculation becomes more complicated and numerical methods are used. Fig. 15.41 provides QAM-64 performance as a function of $E_b/N_0$. Note that the phase-noise limits the BER performance even though $E_b/N_0$ is improving.

A more straightforward approximation to analyze QAM BER versus phase-noise jitter is by analyzing the phase noise as a narrow FM. Assume a signal $X(t)$ with residual FM $\phi_n(t)$ and amplitude $A_c$:

$$X(t) = A_c \cos[\omega_c t + \phi_n(t)]. \tag{15.82}$$

This is dual to

$$X(t) = X_I(t)\cos(\omega_c t) + X_Q(t)\sin(\omega_c t), \tag{15.83}$$

where

$$X_I(t) = A_c \cos\phi_n(t) = A_c\left[1 + \frac{1}{2!}\phi_n(t)^2 + \cdots\right], \tag{15.84}$$

$$X_I(t) = A_c \sin\phi_n(t) = A_c\left[1 - \frac{1}{3!}\phi_n(t)^3 + \cdots\right], \tag{15.85}$$

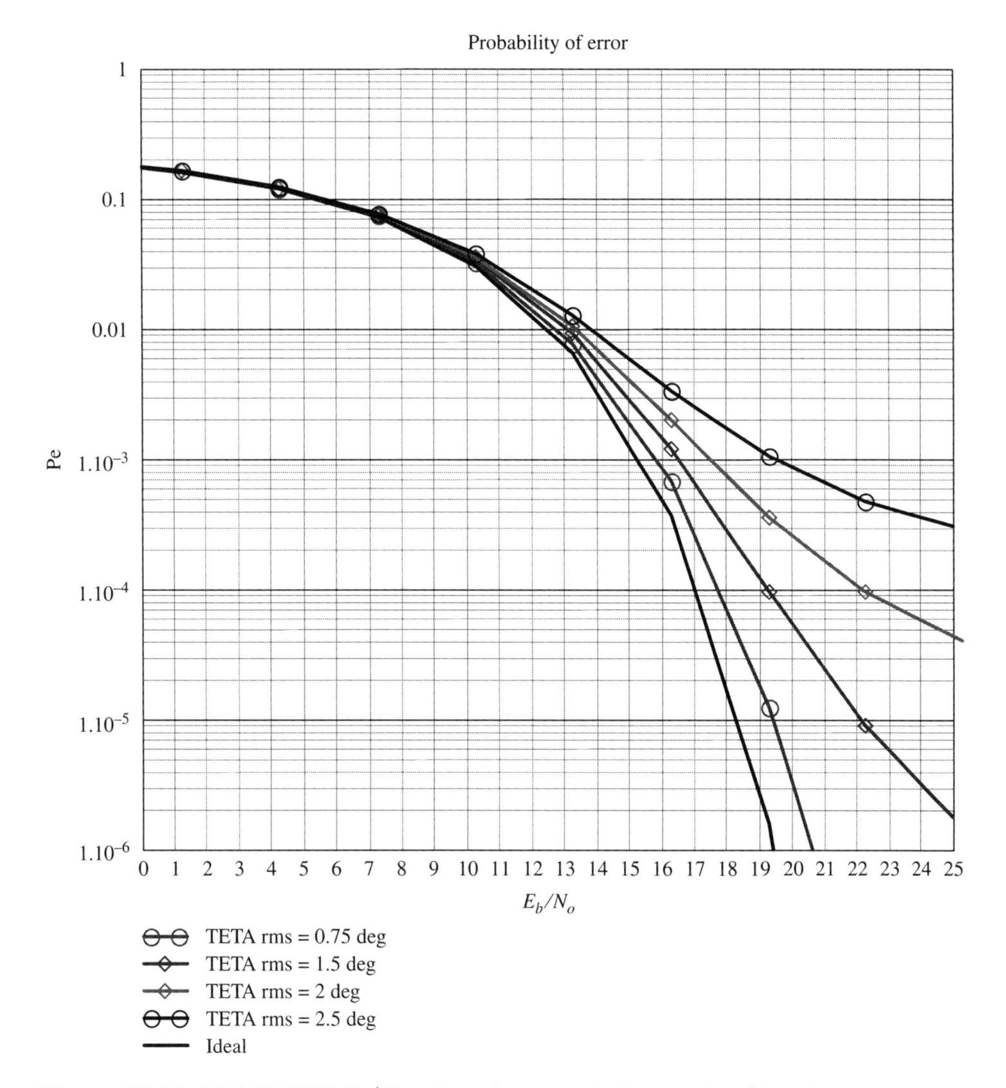


**Figure 15.41** 64-QAM BER $E_b/N_0$, where integrated phase noise $\Delta\theta_{rms}$ is a parameter. Note that the decision angle limit for 16 QAM is 13.25 deg; hence, a 0.6625-deg-BER is reduced by an order of one magnitude and a 1.325-deg-BER is reduced by an order of two magnitudes. The rule of thumb for $\Delta\theta_{rms}$ is more or less to be within 0.1 for low-order QAM and 0.05 for high-order QAM. See Chapter 14.

for $|\phi_n(t)| \ll 1$ rad; then $X_I(t) \approx A_c$ and $X_Q(t) \approx A_c\phi_n(t)$, which is the term of leakage in the QPSK modulator shown by Eq. (15.69). Thus the signal can be approximated as

$$X(t) \approx \underbrace{A_c(t)\cos(\omega_c t)}_{\text{desired signal}} + \underbrace{A_c(t)\phi_n(t)\sin(\omega_c t)}_{\text{noise}}. \tag{15.86}$$

The phase noise $\phi_n(t)$ is a random noise with standard deviation of $\Delta\theta_{rms}$ and zero mean. The Fourier transform of $X(t)$ is given by

$$X(f) \approx \frac{1}{\sqrt{2}} A_c \delta(f - f_c) + \frac{j}{\sqrt{2}} A_c \phi_n \delta(f - f_c), \tag{15.87}$$

and the spectral power density can be written as

$$|X(f)|^2 \approx \tfrac{1}{2} A_c^2 \delta(f - f_c) + \tfrac{1}{2} A_c^2 \phi_n^2 \delta(f - f_c), \tag{15.88}$$

If we observe the signal $X(t)$ with AWGN, then Eq. (15.86) can be written as

$$X(t) \approx \underbrace{A_c(t)\cos(\omega_c t)}_{\text{desired signal}} + \underbrace{A_c(t)\phi_n(t)\sin(\omega_c t)}_{\text{phase-noise}} + \underbrace{n(t)}_{\text{AWGN}}. \tag{15.89}$$

Hence, CNR in the presence of phase noise and AWGN can be written as

$$\begin{aligned} \text{CNR} &= \frac{A_c^2/2}{(A_c^2/2)\Delta\theta_{rms} + N} = \frac{C}{C\Delta\theta_{rms} + N} \\ &= \frac{1}{\Delta\theta_{rms} + (N/C)} = \frac{1}{\Delta\theta_{rms} + (1/CNR)}. \end{aligned} \tag{15.90}$$

Equation (15.90) shows that the weighted CNR is limited by $\Delta\theta_{rms}$. Therefore, BER cannot be improved beyond the limit set by $\Delta\theta_{rms}$ and becomes flat for high values of $E_b/N_0$. This is since the denominator term $1/\text{CNR}$ goes to zero and it is dominated by $\Delta\theta_{rms}$.

This approximation method is useful for analyzing high-order QAM-BER performances, where phase-noise jitter is a parameter.

### 15.8.7 Quantization noise

The outputs of RF receiver section of in a set-top box (STB) are I and Q signals from a quadrature demodulator. Those signals are sampled by an analog to digital converter (ADC). Such a device samples an continuous analog signal at certain instances determined by the sampling rate $T_s$ defines the sampling CLK. The ADC sampling process adds noise. This is due to the quantization process of the continuous analog signal. Assume a continuous signal $x(t)$ with a sampling CLK at a rate denoted as $T_s$ as depicted in Fig. 15.42. The sampled signal $x(t)$ has a BW B[$x(t)$]. Therefore, according to Nyquist criterion B[$x(t)$] $\leq 2/T_s$. The next step when designing an ABB (analog base band) interface to an ADC is to define the full scale of voltage sampling $V_{FS} = V_{max} - V_{min}$ of the ADC (See Fig. 15.50). Assume now that the ADC has $N$ bits; hence the ADC has $2^N$ state or levels marked as NOL (number of levels) and $2^N - 1$ voltage increments marked as number of steps (NOS). This value defines the resolution of the ADC

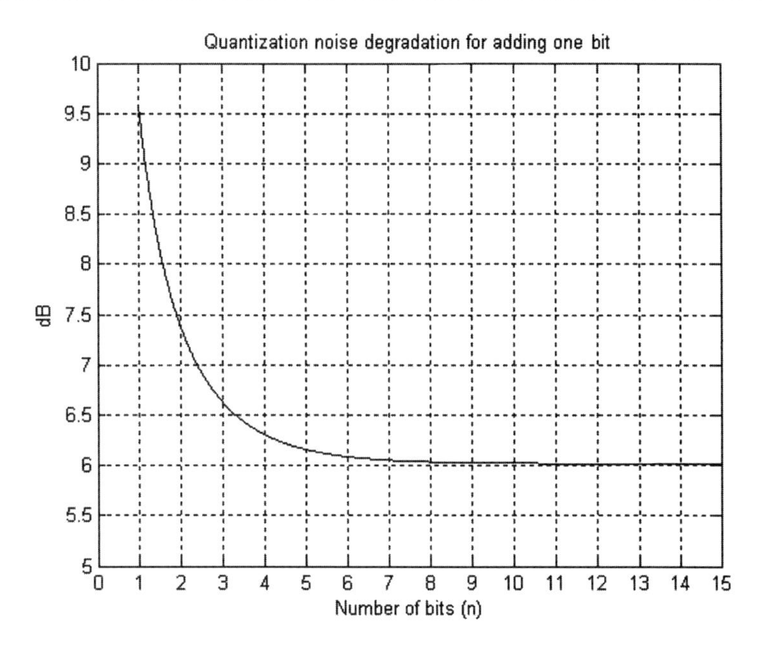


**Figure 15.42** Quantization noise as a function of the ADC number of bits.

marked as $\Delta$ and is given by

$$\Delta = \frac{V_{FS}}{2^N - 1} \tag{15.91}$$

Figure 15.43a illustrates the quantization of a continuous signal represented by a straight line. The X axis represents the input signal to the ADC and the Y axis represents the quantized signal. The quantization is represented by a staircase that generates an error voltage $\pm e$. If the slope of the analog signal $x$ that produces the analog signal $y$ is $a = 1$, it can be written that the error signal $e$ between the analog signal and the quantized signal is given by

$$e = y - x. \tag{15.92}$$

Recall that $y = ax$ is the analog signal. Then it is clear that Fig. 15.43b presents the ADC error transfer function.

The next stage of the analysis is to define the noise statistics and relation to the sampling rate. It is assumed that the error is random, it has a uniform PDF (probability density function). Recall the resolution definitions previously mentioned, the number of output levels is defined by:

$$\text{NOL} = 2^N, \tag{15.93}$$

where $N$ is number of bits. The number of steps between the output levels is

$$\text{NOS} = \text{NOL} - 1 = 2^N - 1. \tag{15.94}$$

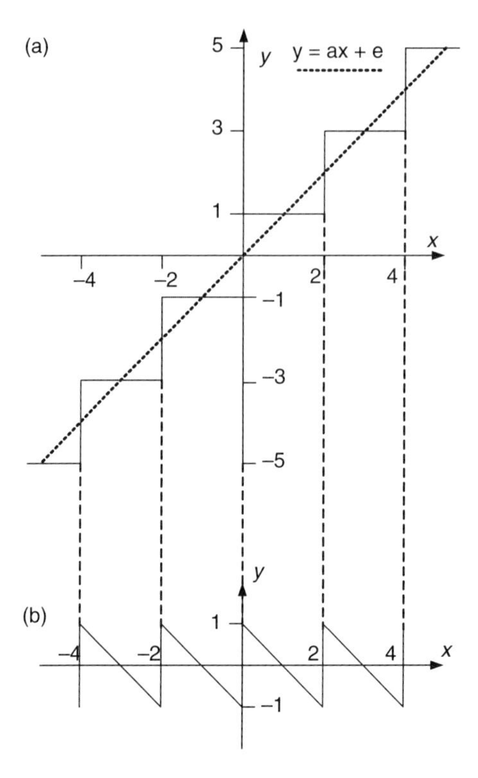


**Figure 15.43** (a) Quantization process of a continuous signal $y = ax$. (b) Quantization error transfer function.

The size of each step is defined by Eq. (15.91).

If we assume that the quantization error has an equal probability of lying anywhere in the range of $\pm\frac{\Delta}{2}$, its mean square value is

$$e_{\text{rms}}^2 = \frac{1}{\Delta}\int_{-\frac{\Delta}{2}}^{\frac{\Delta}{2}} e^2 de = \frac{1}{\Delta}\left(\frac{e^3}{3}\Bigg|_{-\frac{\Delta}{2}}^{\frac{\Delta}{2}}\right) = \frac{1}{\Delta}\left[\frac{\left(\frac{\Delta}{2}\right)^3}{3} - \frac{\left(-\frac{\Delta}{2}\right)^3}{3}\right] = \frac{\Delta^2}{12}. \quad (15.95)$$

The sampled signal and the noise BW are illustrated by Fig. 15.44.

The noise power spectral density (PSD) of the sampled signal at the sampling frequency $f_s = 1/T_s$ is illustrated by Fig. 15.45. This is a band-limited uniform PSD between $\pm f_s/2$, which can be written as

$$\text{PSD} = \frac{e_{\text{rms}}^2}{f_s}. \quad (15.96)$$

Observing a single sided band all the noise is folded into the positive side. Hence, as the PSD is written.

$$\text{PSD} = \frac{e_{\text{rms}}^2}{\frac{f_s}{2}} = \frac{2 \cdot e_{\text{rms}}^2}{f_s} \quad (15.97)$$

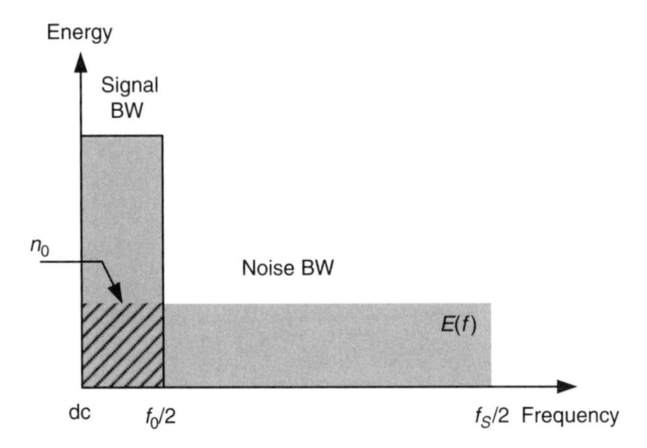


**Figure 15.44** Signal and quantization noise BW. Note that the noise density is *E(f)* and the energy within the signal BW is $n_0$.

Sampling noise voltage spectral density is described by

$$E(f) = \sqrt{PSD} = e_{\text{rms}}\sqrt{\frac{2}{f_s}} \tag{15.98}$$

Observing Fig. 15.44, the PSD power within the signal BW is given by Eq. (15.99).

$$n_0^2 = \int_0^B PSD \cdot df = e_{\text{rms}}^2 \frac{2}{f_s} \int_0^{f_0} df = e_{\text{rms}}^2 \frac{2f_0}{f_s} \tag{15.99}$$

The value $f_s/2f_0$ is defined as over sampling ratio (OSR) and describes the over sampling with respect to Nyquist criterion claiming stating that $f_s$ should be higher twice the sampled signal frequency. Hence OSR of 1 describes $f_s = 2f_0$. Hence, the noise power density can be written as

$$n_0^2 = e_{\text{rms}}^2 \frac{1}{\text{OSR}}. \tag{15.100}$$

As a conclusion, the higher the sampling rate the lower the quantization noise. However, it is a uniform level noise. The advanced technique of $\Sigma\Delta$ modulator enables to shape the noise as will be explained later on.

The next step of this analysis is to examine the effect of number of bits on quantization noise and SNR. Assume two cases. Case1—the number of bits is $n$, case2—the number of bits is $N+1$. Therefore, the step size of each case is described by $\Delta_1$ and $\Delta_2$ respectively:

$$\Delta_1 = \frac{V_{FS}}{2^N - 1}$$
$$\Delta_2 = \frac{V_{FS}}{2^{N+1} - 1} \tag{15.101}$$

The ratio between the quantization errors for each case is:

$$\frac{e_{1_{\text{rms}}}}{e_{2_{\text{rms}}}} = \frac{\frac{\Delta_1}{\sqrt{12}}}{\frac{\Delta_2}{\sqrt{12}}} = \frac{\Delta_1}{\Delta_2} = \frac{\frac{V_{FS}}{2^N-1}}{\frac{V_{FS}}{2^{N+1}-1}} = \frac{2^{N+1} - 1}{2^N - 1} = \frac{2 \cdot 2^N - 1}{2^N - 1} \tag{15.102}$$

The result is not constant but a function of $N$, when $N$ goes to infinity the result goes to 2 and when $N$ goes to 0 the result goes to infinity.

$$\lim_{N \to \infty} \left( \frac{2 \cdot 2^N - 1}{2^N - 1} \right) = 2$$
$$\lim_{N \to 0+} \left( \frac{2 \cdot 2^N - 1}{2^N - 1} \right) = \infty \tag{15.103}$$

It can be seen that if the rms (voltage) ratio is improved by $\sim 2$, the power ratio is improved by $\sim 4$:

$$\frac{e_{1_{\text{rms}}}}{e_{2_{\text{rms}}}} \approx 2 \Rightarrow \frac{e^2_{1_{\text{rms}}}}{e^2_{2_{\text{rms}}}} \approx 4 \Rightarrow 10 \cdot \log_{10}(4) \approx 6\,\text{dB (for high number of bits)}. \tag{15.104}$$

Only if the number of bits is high we can say that the quantization noise improved by 6 dB to bit (See Fig. 15.42).

Based on above definitions, and Eqs. (15.91) and (15.95) SNR can be evaluated.

$$V^2_{FS\text{rms}} = \left[ \frac{(2^N - 1) \cdot \Delta}{\sqrt{2}} \right]^2 \tag{15.105}$$

When observing a sine wave, the full scale ADC's input voltage is a full sine swing, meaning peak to peak. Hence, the peak voltage is $V_{FS}/2$. Since power is measured with respect to rms voltage, the sine wave rms amplitude is $V_{FS}/(2\sqrt{2})$. For a large $N$, the full scale voltage can be written as

$$V_{FS} = \Delta \cdot (2^N - 1) \approx \Delta \cdot 2^N. \tag{15.106}$$

Using the sine case and Eqs. (15.95), (15.100) and the above definitions, SNR can be evaluated by

$$\mathrm{SNR} = \frac{\left(\frac{V_{FS}}{2\sqrt{2}}\right)^2}{n_0^2} = \frac{\left(\frac{e_{\mathrm{rms}}\cdot\sqrt{12}\cdot 2^N}{2\sqrt{2}}\right)^2}{e_{\mathrm{rms}}^2\cdot\frac{1}{\mathrm{OSR}}} = \frac{3}{2}\cdot 2^N\cdot\mathrm{OSR}. \tag{15.107}$$

In dB log scale SNR is given by

$$\mathrm{SNR} = 1.76 + 6.02\cdot N + 10\cdot\log(\mathrm{OSR}). \tag{15.108}$$

## 15.8.8 Sigma delta ADC noise analysis

The sigma delta modulator concept is used in ADC for noise shaping in order to improve SNR. The concept of analyzing is NTF (noise transfer function) and STF (signal transfer function) is based on control theory in the discrete domain. Since this is a signal processing method using sampling theory, the $Z$ transform is used rather than the Laplace transform. Therefore, the following relations are used to convert from continuous time domain to discrete:

$$\frac{1}{s} = \frac{1}{1 - z^{-1}}. \tag{15.109}$$

The delay unit is a flip flop (FF) and is $z^{-1}$:

$$z^{-1} = \exp(j\varphi), \tag{15.110}$$

where $\varphi$ is the angular frequency related to the sampling rate. These equations will be used to analyze the sigma–delta NTF in the discrete and continuous domains.

A sigma delta PCM (pulse code modulator) was first introduced by Inose, Yasuda and Marakami in 1962.[59] Figure 15.45a illustrates the concept for PCM modulator and de-modulator. The PCM is a feedback system that translates an analog signal into a train of pulses. The input of a PCM is an analog signal passing through a LPF antialiasing filter. The LPF output is connected to a comparator. The second input of the comparator is fed by the feedback path. The comparator's output feeds a delay block implemented by a D-FF. That manner the comparator output is released according to the CLK rate feeding the FF. This process samples the comparator output according to the CLK rate. The D–FF output is a train of pulses, which are routed to a feedback path and integrated by an ideal integrator that creates a triangle wave. This triangle wave feeds the comparator and serves as a variable reference for the comparator. The comparator modulates the analog signal and generates a train of pulses adequately describing the analog signal. The analog signal is quantized and the variable threshold levels of the comparator create the modulation. The decoding process of the pulses is

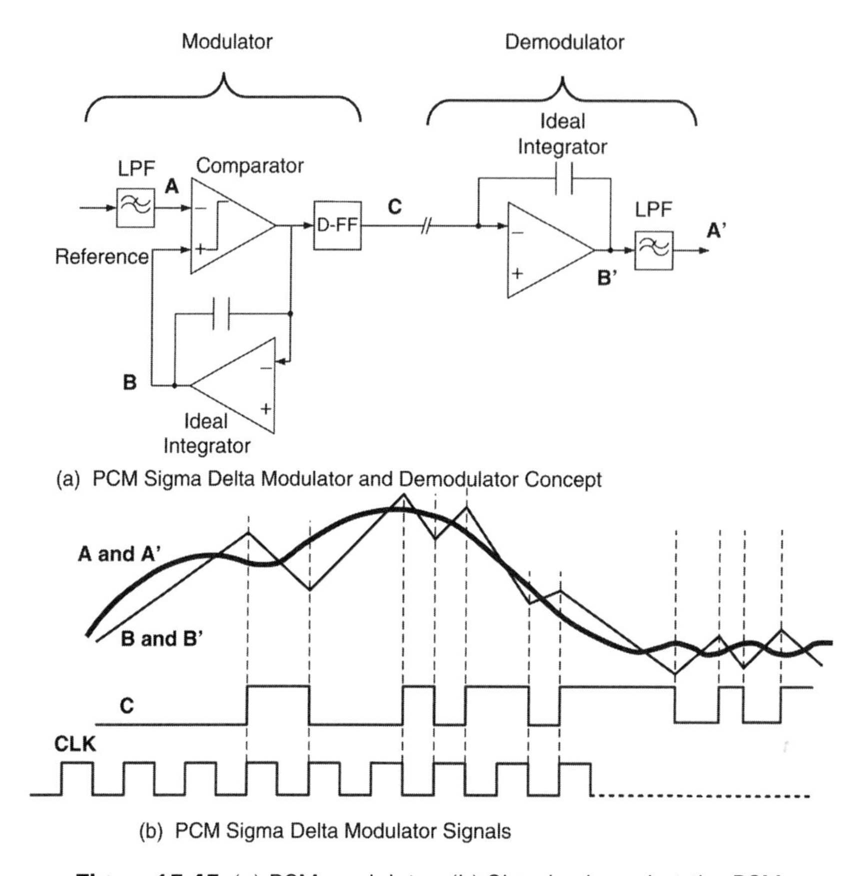


**Figure 15.45** (a) PCM modulator. (b) Signals shaped at the PCM.

accomplished by an ideal integrator and smoothing LPF. Figure 15.45b describes the analog signal and the integrated pulse routed by a feedback loop to the comparator and the output bit stream train of pulses representing the analog signal.

The above process can be described by a first-order feedback system in $z$ domain as in Fig. 15.46. Quantization error voltage is inserted at the comparator's output. This block diagram is solved according to control theory per Eqs. (15.111)–(15.113).

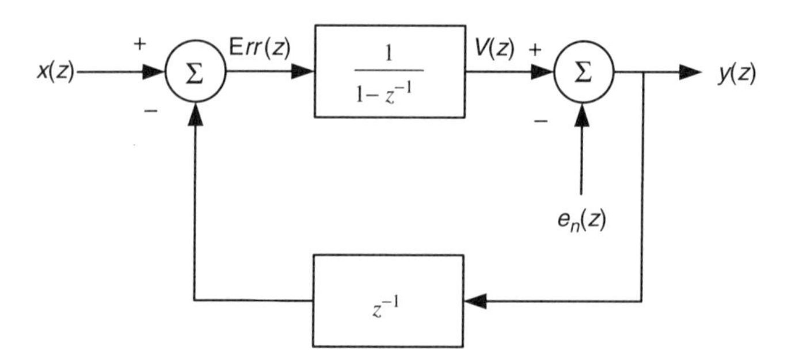


**Figure 15.46** PCM modulator block diagram in $z$ domain.

The block diagram in Fig. 15.46 is solved by a set of equations that provide a solution for STF and NTF.

$$\mathrm{Err}(z) = x(z) - z^{-1} \cdot y(z), \tag{15.111}$$

$$V(z) = \mathrm{Err}(z) \cdot \frac{1}{1 - z^{-1}}, \tag{15.112}$$

$$y(z) = V(z) + e_n(z). \tag{15.113}$$

It can be seen that the noise is passing a high pass response and the system tracks the input signal. This is very similar to PLL STF and NTF results. The next step is to understand the noise shaping process of the sigma delta modulator. Figure 15.47 illustrates a first order sigma delta block diagram containing two feedback paths.

The solution of Eqs. (15.111)–(15.113) provides the NTF and STF:

$$y(z) = x(z) + \left(1 + z^{-1}\right) \cdot e_n. \tag{15.114}$$

It can be seen that the noise voltage $e_n$ is passing a HPF response while the input tracks the output. In case of no delay block as in Fig. 15.47 results in

$$\frac{y(z)}{x(z)} = \frac{1}{1 + (1 - z^{-1})} \quad \Leftrightarrow \quad \frac{y(s)}{x(s)} = \frac{1}{1 + s}. \tag{15.115}$$

The result is similar to PLL's phase noise response of the VCO.

Figure 15.47 illustrates a block diagram of first order sigma-delta modulator with two feedback loops. Solving the signal and noise transfer functions provides the following results.[60]

$$y_i = y_{i-1} + (e_i - e_{i-1}). \tag{15.116}$$

The noise delta is expressed by $n_i = e_i - e_{i-1}$, where the second term is delayed. Hence, the noise component in Eq. (15.116) is modified to noise spectral density $N(\omega)$, where the delay in discrete domain is converted to $\exp(j\omega T)$:

$$N(\omega) = E(t) \cdot [1 - \exp(-j\omega T)]. \tag{15.117}$$

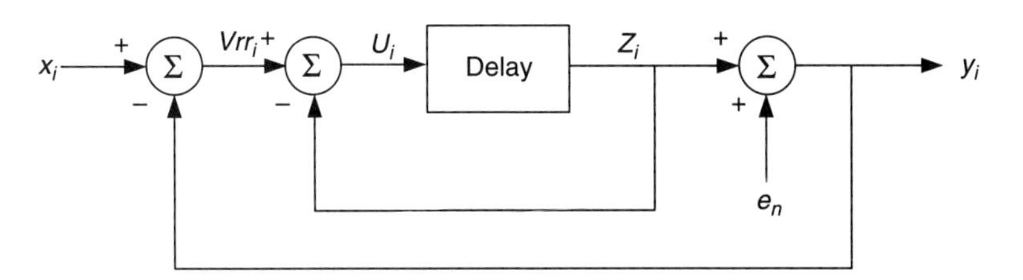


**Figure 15.47** First order sigma delta block diagram in discrete domain.

Using the identity

$$1-\exp(-j\omega T)=2\cdot j\cdot \exp\left(-\frac{j\omega T}{2}\right)\cdot\frac{\exp\left(\frac{j\omega T}{2}\right)-\exp\left(-\frac{j\omega T}{2}\right)}{2j}$$
$$=2\cdot j\cdot\exp\left(\frac{-j\omega T}{2}\right)\cdot\sin\left(\frac{\omega T}{2}\right), \tag{15.118}$$

it becomes

$$\left|[1-\exp(-j\omega T)]\right|=2\cdot\sin\left(\frac{\omega T}{2}\right) \tag{15.119}$$

Using Eqs. (15.98), (15.117) and (15.119), the noise voltage spectral density can be written as

$$\left|N(f)\right|=2\cdot\sqrt{2T}\cdot e_{\mathrm{rms}}\cdot\sin\left(\frac{\omega T}{2}\right), \tag{15.120}$$

where $T$ is the sampling rate $T_s$ and $\omega=2\cdot\pi\cdot f_0$ and $f_0<f_s$. The noise-power density is proportional to the noise-voltage square. Therefore, the noise power within the signal BW is the integration of Eq. (15.120). This can be simplified by using the fact that $T_s/T_0\ll 1$; therefore, the approximation of $\sin(x)\approx x$ is used:

$$n_0^2=\int_0^{f_0}\left|N(f)\right|^2\cdot df=\int_0^{f_0}\left[2\cdot\sqrt{2T}\cdot e_{\mathrm{rms}}\cdot\sin\left(\frac{\omega T}{2}\right)\right]^2\cdot df$$
$$\approx 8\cdot T\cdot e_{\mathrm{rms}}^2\cdot\int_0^{f_0}\left(\frac{\omega T}{2}\right)^2\cdot df. \tag{15.121}$$

This integration results in the following formulas, which describe the noise shaping of first-order sigma delta modulator:

$$n_0^2\left[V^2/Hz\right]=(2\cdot f_0\cdot T)^3\cdot\frac{\pi^2}{3}\cdot e_{\mathrm{rms}}^2=\frac{\pi^2}{3}\cdot e_{\mathrm{rms}}^2\left(\frac{1}{\mathrm{OSR}}\right)^3; \tag{15.122}$$

therefore, the noise voltage value is given by

$$n_0\left[V/\sqrt{Hz}\right]=e_{\mathrm{rms}}\cdot\frac{\pi}{\sqrt{3}}\cdot\left(\frac{1}{\mathrm{OSR}}\right)^{3/2} \tag{15.123}$$

Equations (15.122) and (15.123) demonstrate that the sigma-delta modulator utilized in ADC reduces the in-band noise energy when comparing it to ordinary ADC NTF as in Eq. (15.100). This can be seen when observing the increased OSR noise rejection rate of 10 dB/dec against 30 dB/dec. The result is due to the effect

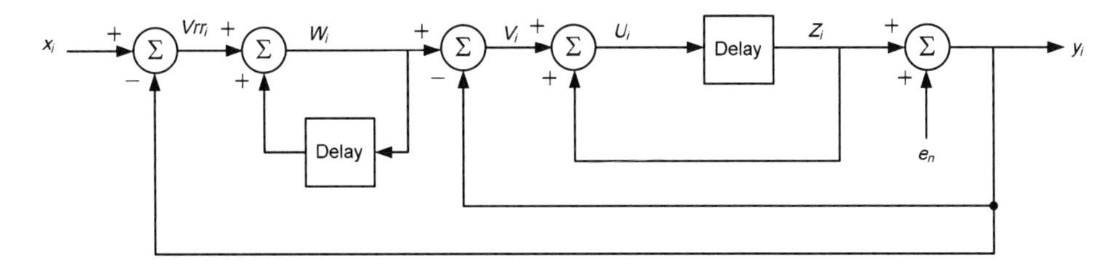


**Figure 15.48** Second-order sigma-delta block diagram.

of the feedback system because first-order ΣΔ has two loops. The next step is to examine higher-order ΣΔ noise shaping. Figure 15.48 depicts a second-order ΣΔ block diagram. The concept for solving it is straightforward and was demonstrated previously for first-order ΣΔ. Hence, the following equations can be written:

$$V\text{err}_i = x_i - y_i, \qquad (15.124)$$

$$W_i = V\text{err}_i + W_{i-1}, \qquad (15.125)$$

$$V_i = W_i - y_i, \qquad (15.126)$$

$$U_i = V_i + Z_i, \qquad (15.127)$$

$$y_i = e_i + Z_i. \qquad (15.128)$$

The solution of Eqs. (15.124)–(15.128) results in the following in-to-out equations:

$$y_i = x_{i-1} + (e_i - 2 \cdot e_{i-1} + e_{i-2}). \qquad (15.129)$$

The parenthetical noise expression can be written using a delay argument as used in Eq. (15.117). Using the relation $a^2 + 2 \cdot a \cdot b + b^2 = (a+b)^2$ the noise density term $|N(f)|$ is given by

$$|N(f)| = 4 \cdot e_{\text{rms}} \cdot \sqrt{2 \cdot T} \cdot \sin\left(\frac{\omega T}{2}\right)^2 \qquad (15.130)$$

Integration of Eq. (15.130) between $f = 0$ to $f = f_0$ provides the noise power within the pass-band:

$$n_0^2[V/Hz] = \frac{\pi^4}{5} \cdot e_{\text{rms}}^2 \cdot \left(\frac{2 \cdot f_0}{F_s}\right)^5 = \frac{\pi^4}{5} \cdot e_{\text{rms}}^2 \cdot \left(\frac{1}{\text{OSR}}\right)^5, \tag{15.131}$$

$$n_0\left[V/\sqrt{Hz}\right] = \frac{\pi^2}{\sqrt{5}} \cdot e_{\text{rms}} \cdot \left(\frac{2 \cdot f_0}{F_s}\right)^{5/2} = \frac{\pi^2}{\sqrt{5}} \cdot e_{\text{rms}} \cdot \left(\frac{1}{\text{OSR}}\right)^{5/2}. \tag{15.132}$$

Using induction, a general formulation is driven by

$$\begin{aligned} n_0\left[V/\sqrt{Hz}\right] &= \frac{\pi^L}{\sqrt{2 \cdot L + 1}} \cdot e_{\text{rms}} \cdot \left(\frac{2 \cdot f_0}{F_s}\right)^{(2 \cdot L+1)/2} \\ &= \frac{\pi^L}{\sqrt{2 \cdot L + 1}} \cdot e_{\text{rms}} \cdot \left(\frac{1}{\text{OSR}}\right)^{(2 \cdot L+1)/2}, \end{aligned} \tag{15.133}$$

where $L$ is the $\Sigma\Delta$ out-to-in number of loops.

The conclusion is that the higher-order $\Sigma\Delta$ loop rejects the quantizer in band noise sharper. This is similar to PLL rejection of VCO noise by the loop filter. Figure 15.49 demonstrates the NTF for various $\Sigma\Delta$ orders. It can be seen that

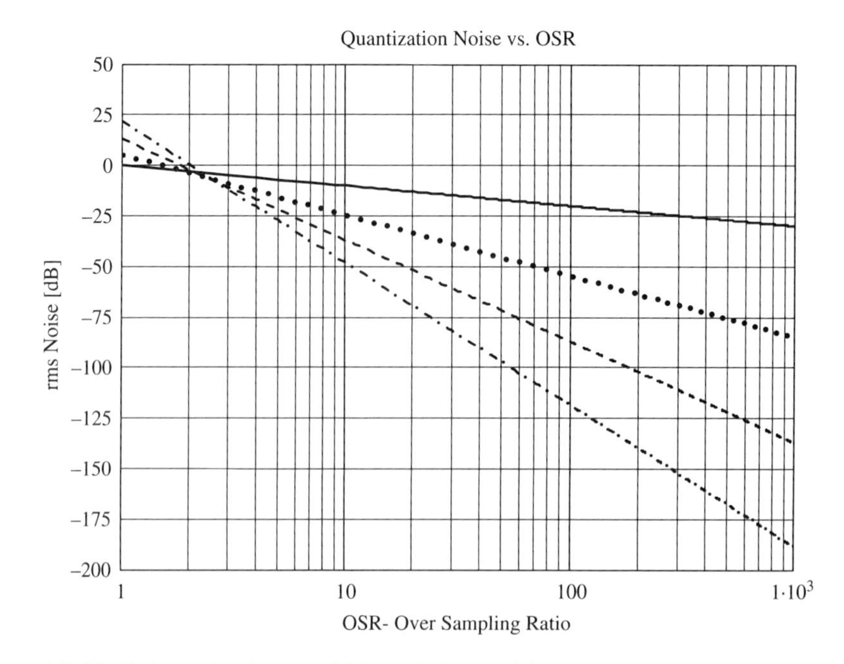


**Figure 15.49** Noise rejection vs. OSR and sigma delta order: $e_n = 1$, solid line $L = 0$ ordinary ADC no feedback; dotted line $L = 1$, dashed Line $L = 2$, dash dotted $L = 3$.

the noise suppression rate is increased by $3(2L-1)$ for every doubling of the sampling rate $F_s$ or OSR.

The next step is defining noise over gain (NOG), which is a measure that shows how the RF-to-ABB (analog base band) converter has sufficient gain so AGC noise floor will not affect over all CNR. Hence, the noise density at the ADC input should be at least 10 dB below the noise density of the RF-to-ABB converter output. It is a similar design compared to the LNA that compensates the RF chain NF.

NOG is defined by

$$\mathrm{NOG} = \frac{N_{\mathrm{ADC}} \cdot + k \cdot T \cdot G \cdot F}{k \cdot T \cdot G \cdot F}, \qquad (15.134)$$

where $N_{\mathrm{ADC}}$ is the ADC noise-density analyzed above, $k$ is Boltzman constant, $T$ is room temperature in Kelvin, $G$ is the RF-to-ABB conversion gain, and $F$ is noise figure. There are two degrees of freedom in this design expression in order to meet CNR performance at the ADC input: the gain and the noise figure. Increasing OSR reduces $N_{\mathrm{ADC}}$, which means the NOG is less affected by quantization noise. This is a similar definition to noise factor.

Another definition is take over gain (TOG):

$$\mathrm{TOG} = \frac{N_{\mathrm{ADC}} \cdot + k \cdot T \cdot G \cdot F}{N_{\mathrm{ADC}}}. \qquad (15.135)$$

Note that when $k \cdot T \cdot G \cdot F \rightarrow \infty$, NOG goes to unity, similar to the LNA noise figure definition. On the other hand, TOG goes to infinity, meaning $k \cdot T \cdot G \cdot F$ is dominant and overall CNR is not affected by ADC quantization noise. When designing the RF down converter, AGC has to be given special attention. It is not a continuous AGC but DSP controlled based on a DSP energy estimation of $P = I^2 + Q^2$.

The designer should also pay attention to the gain-switch point based on SINAD. SINAD is determined by

$$\mathrm{SINAD} = \frac{C}{N + N_{\mathrm{ADC}} + \mathrm{IMR}_3 + \mathrm{IMR}_2}, \qquad (15.136)$$

where $N$ is AWGN, $N_{\mathrm{ADC}}$ is ADC quantization noise, $\mathrm{IMR}_3$ is third-order intermodulations or CTB products, and $\mathrm{IMR}_2$ is second-order intermodulations or CSO products. Note that the intermodulation products formulation is given in Section 9.1. The design goal is to keep linearity and prevent saturation of ADC caused by too strong of a signal. This tradeoff is measured in units of decibels full scale (dBFS) compared to the input power of the RF-to-ABB down converter. The dBFS is the ratio between the input voltage to the ADC at a given power, normalized to the ADC full scale. Figure 15.50 demonstrates the AGC upper margin, which is the ADC full scale and the CNR lower margin defined by SINAD.

(a)
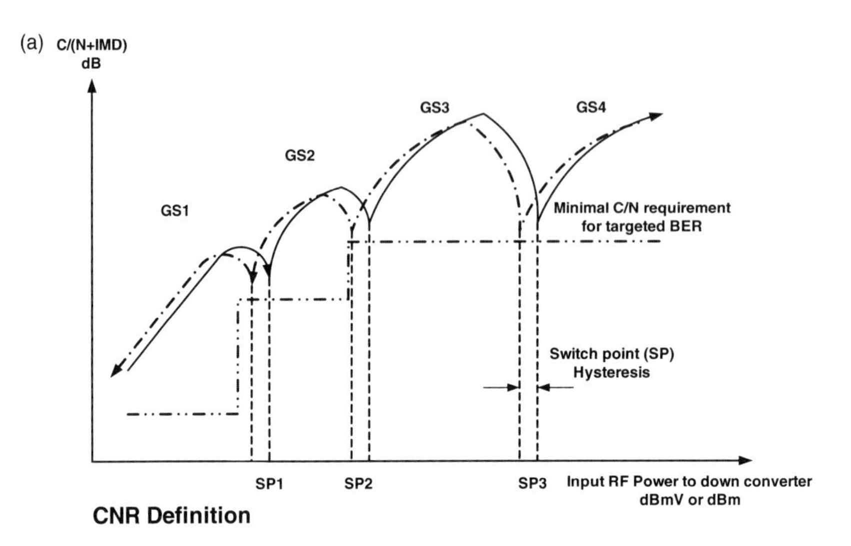


(b)
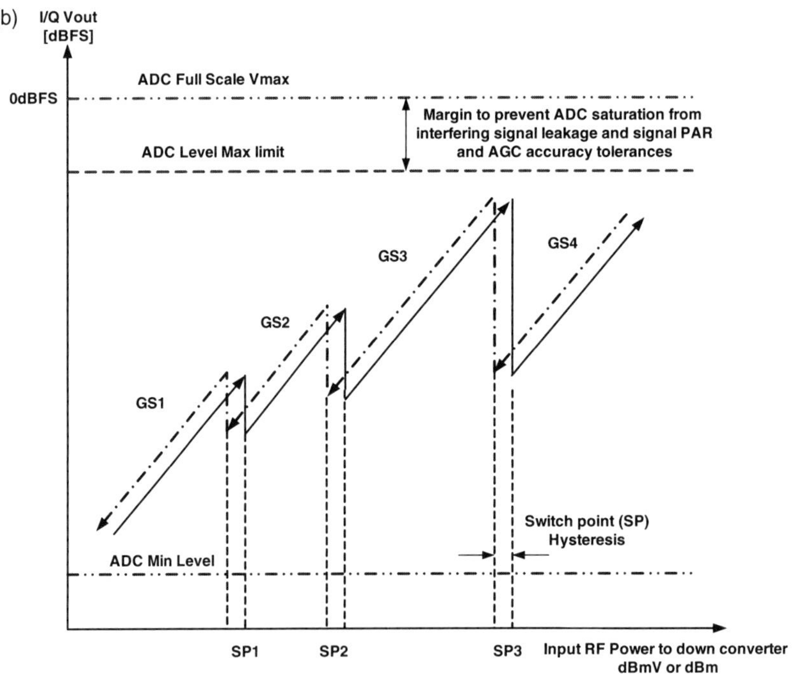


**Figure 15.50** AGC plot for the RF to ABB (analog base band) or IF I/Q down converter. a) The up going power increases C/N until it is affected by IMR products and compression which are causing the curvature in C/N plot. This is the AGC switch point (SP) to next gain state (GS). $\mathrm{dBFS} = 20\log(V/V_{Full-Scale})$. b) The criterion for AGC planning is not to saturate the ADC and not degrade the C/N. Hysteresis is applied to prevent toggling at the switch point. The take over gain defines the C/N margin above the min C/N. Note that

$$\mathrm{CNR[dB]} = 10\log\left[\frac{1}{(Noise/\mathrm{C}) + (IMR/\mathrm{C})}\right]$$

## 15.9 Main Points of this Chapter

1. QAM MODEM has layers: MPEG framing, forward error correction (FEC) encoding/decoding QAM/demodulation.

2. FEC has four layers. The order of the FEC process is as follows:

   RS encoder (RS layer)

   Interleaver (interleaver layer)

   Trellis encoding (Trellis layer)

   Link (Trellis layer)

   Trellis decoder (Trellis layer)

   Derandomizer (interleaver layer)

   Deinterleaver (interleaver layer)

   RS decoder (RS layer)

3. The main role of a data randomizer is to prevent long sequences of "0" or "1," thus helping the CLK, and to create a balanced spectrum.

4. There are two kinds of randomizers in use. The first is self-synchronized and the other is frame synchronized.

5. The drawback of a self-synchronized scrambler/descrambler is that error bits can propagate through it during detection. The number of errors will correspond to the number of nonzero taps in the decoding side.

6. A frame-synchronized encoder/decoder scrambles the data by a known pattern on both receiving and transmitting sides.

7. The role of a forward-error-correction code is to improve the BER curve. The change in the BER performance curve might become steeper and add gain processing, so for the same $E_b/N_0$, BER would improve. In other words, less CNR is needed for the same BER result.

8. Error-correcting codes are 2D at the minimum, which can measure the bits of the data matrix parity both horizontally and vertically. This way, the coordinate of the fault bit is known and the bit is corrected.

9. Reed–Solomon (RS) is a block-error-correction code.

10. The purpose of an interleaver and a deinterleaver is to randomize bursts of errors, making them shorter. This way, the RS decoder will not crash under an overload of errors.

11. The Trellis encoding process is convolutional coding.

12. The Viterbi algorithm is a general technique for decoding Trellis codes by checking the path with a minimum Euclidean distance to receive a signal.

13. The advantage of QAM schemes is their spectrum efficiency [see Eq. (15.25)]. Each symbol in the constellation map represents $\sqrt{M}$ bits for $I$ and $Q$.

14. In QAM, there is a 90-deg ambiguity for carrier locking. Thus, a phase encoder is used to prevent false lock.

15. One of the difficult problems in digital transport is ISI. ISI results from previous signal trails falling in the sample signal.

16. To overcome ISI, a Nyquist raised cosine (NRC) filter is used; it is an approximation to an ideal square filter.

17. NRC is implemented on both transmitter and receiver sides. Hence, each side has the square root filter known as SRRC.

18. SRRC is a finite impulse response filter (FIR) realized in DSP. It receives its maximum at $T = 0$ and goes to zero at $\pm nT$ sampling periods. Thus ISI, resulting from previous signal trails, is cancelled.

19. An RF up-converter in the QAM transmitter converts the BB into IF frequency and then to a final RF frequency.

20. A QAM receiver consists of two main blocks: an RF super-heterodyne down-converter and a BB demodulator known as a modem. The purpose of a double-conversion super-heterodyne RF unit is to increase selectivity, and adjacent and alternate channel interference.

21. The purpose of equalizers is to remove linear channel distortion effects such as cable reflections, ghost images, and any depressive interference. An equalizer is a matched filter realized in DSP.

22. The equalizer structure consists of a storage register that stores the data samples and weight function to adjust the signal to the desired shape. Equalization is an iterative process.

23. There are two main methods to adjust the equalizer weight parameters: preamble, using a training series, and blind equalization.

24. In the preamble method, the equalizer weight coefficients are adjusted by a known training sequence, which is transmitted prior to the data string. After the equalizer is ready, it is set to be decision directed (DD) operated. This is done by feedback to the weight coefficients.

25. Preamble-guided equalizers are inefficient in BW due to the time-consuming training series.

26. The blind equalization technique uses feedback and it is a DD method. The feedback uses an algorithm to check and minimize the stochastic-error gradient.

27. There are several types of equalization techniques: a linear-transversal equalizer (LTE), zero-forcing equalizer (ZFE), and fractionally spaced equalizers (FSE). These concepts are hardware methods to realize equalization.

28. One of the important roles of equalizers is to improve SNR. For this reason, in each equalizer, there are some drawbacks, since while increasing gain to compensate for a notch in signal, it increases noise as well, which is one of the main challenges when designing an equalizer.The main drawback of an LTE equalizer is its sampling rate, which causes spectrum aliasing of consecutive samples. Thus, for a signal notch at the BW edge of the corrected signal, a gain increase causes an increase in aliasing noise and reduces SNR.

29. To overcome the LTE sampling problem, sampling is done in two stages. This technique is utilized in an FSE. In FSE, the first step is sampling the signal at twice the data rate. The signal is passed through a matched filter to isolate it from the remaining periodic spectrum. The pulse is equalized and then sampled at the symbol rate.

30. The concept of a ZFE equalizer is to compensate for the signal but not the noise; hence an adaptive feedback algorithm is used to minimize the noise error. These algorithms are called LMS (least mean square) and are mainly used for trained (using preamble series) equalizers. The other kind of equalizers, which are a modification of LMS, are called modified error signal (MES) or constant modulus algorithm (CMA), which are used for blind QAM equalizers.

31. DD equalizers are feedback equalizers that adapt their states per the symbol outputs.

32. The rate of a Trellis convolutional encoder without a puncturing table is determined as the ratio of number of inputs to number of outputs. For example, the rate for one input and two outputs is 1:2.

33. The purpose of a puncturing table is to accomplish a variable rate at the output of a Trellis convolutional encoder.

34. In order to detect and perform carrier recovery, modulation effects should be removed. This is done by a nonlinear operation that brings all modulation phases to one phase point. There are two kinds of nonlinear carrier

recovery methods: power of $N$ and Costas loop.

35. A Costas loop is composed of two orthogonal PLLs with a square-law detector, or fourth-order detector for QAM.

36. Since a Costas loop is a PLL, it can recover phase and timing. Since it is a feedback system, it is decision directed (DD) and is an ML algorithm realization for phase-error minimization.

37. Phase-noise jitter reduces CNR and therefore affects BER versus CNR by limiting BER performance. This means that if CNR is improved, BER remains the same.

38. The integrated phase noise variance is $\Delta\theta_{rms}$ and its mean is zero. The phase noise is a colored noise of $1/f_n$ type.

39. The intuitive way to explain phase-noise jitter is by introducing a sampling error; thus, the CNR is degraded.

40. One of the roles of a carrier recovery loop is to reject the close-to-carrier phase noise. If the jitter transfer function of a PLL is LPF. Then the carrier recovery loop transfer function is HPF. Thus, it rejects the phase noise. Note that $1 - \text{LPF} = \text{HPF}$, where the LPF is the PLL low-pass characteristic transfer function.

41. Phase noise is FM noise that results from a residual FM modulation caused by noise.

42. Quantization noise results from ADC and is related to the voltage step of the AGC and number of bits. For a high number of bits, the quantization noise improves for any additional bit by 6 dB.

43. ADC quantization noise is improved by increasing over sampling ratio OSR.

44. ADC quantization noise can be shaped by using the sigma delta modulator.

45. Sigma delta is a feedback system and the higher the order of it, it would reject stronger the quantization noise within the signal BW.

46. Quantizer noise transfer function is similar to the rejection of a VCO phase noise in PLL, where the loop filter rejects the VCO phase noise leakage into the PLL band. The order of the loop filter and its steepness define the rejection rate.

47. The quantization noise rejection NTF is a high pass. This process is very similar to PLL rejection of a VCO's noise.

48. Take over gain (TOG) and noise over gain are design measurements to

evaluate the RF down converter SNR immunity against the nose contribution of ADC.

49. NOG is similar to the nose figure and measures the ratio between the SNR of the down converter output compared to the SNR, including the ADC noise. Both are normalized to the input of the down converter. The ideal case is when NOG $\cong 1$, so ADC has a minimum contribution of nose. It is desirable that ADC noise will beat least 10 to 15 dB below the down converter noise at the I/Q output.

50. To accomplish optimal NOG, TOG parameter should be high.

## References


1. Gardner, F.M., *Phaselock Techniques*. John Wiley & Sons (1979).
2. Meyr, H., and G. Ascheid, *Phase Frequency Locked Loops and Amplitude Control*. John Wiley & Sons (1990).
3. Meyr, H., G. Ascheid, and S.A. Fechtel, *Digital Communication Receivers: Synchronization Channel Estimation and Signal Processing*. John Wiley & Sons (1998).
4. Cheanh, J.Y.C., *Practical Wireless Data Modem Design*. Artech House (1999).
5. Gardner, F.M., "Synchronization of data receivers." Seminar presented May 25–26, 1998 in the S. Neaman Institute at the Technion, Haifa, Israel.
6. Proakis, J.G., *Digital Communications Fourth Edition*. McGraw Hill (2000).
7. Belouchrani, A., and W. Ren, "Blind carrier phase tracking with guaranteed global convergence." *IEEE Transactions on Signal Processing*, Vol. 45 No. 7, pp. 1889–1894 (1997).
8. DAVIC 1.1 Digital Audio Visual Council. Specifications http:// www.davic.org/DOWN1.html
9. ITU-T Recommendation J.83, "Digital multi-programme systems for television sound and data services for cable distribution." (October 1995/1997).
10. IEEE 802.14 Physical Layer Specifications V1.0 http://walkingdog.com/catv/index.html.
11. MCNS Specifications, Data over cable interface specifications, radio frequency interface specifications SP-RFII02-9701008. http://www.cablemodem.com
12. Mouine, J., *A Simple Algebraic and Analogical Approach to a Scrambler/Descrambler*. RF Design, pp. 45–52 (1992).
13. Feher, K., et al., *Telecommunications Measurements, Analysis, and Instrumentation*. Nobel Publishing, pp. 14–21 (1987).
14. Reed, I.S., and G. Solomon, "Polynomial codes over certain finite fields." *J. SIAM*, Vol. 8, pp. 300–304 (1960).
15. SCTE—Society of Cable Telecommunications Engineers, Digital Video

Subcommittee. *Digital Video Transmission Standard for Cable Television*, ANSI/STCE 07 (2000).
16. Sklar, B., *Digital Communications: Fundamentals and Applications*. Prentice Hall (2001).
17. Lin, S., and D.J. Catello, *Error Control Coding*. Prentice Hall (2004).
18. Simon, M.K., S.M. Hinedi, and W.C. Lindsey, *Digital Communication Techniques*. Prentice Hall (1995).
19. Ungerboeck, G., "Trellis coded modulation with redundant signal sets, Part I: Introduction." *IEEE Communication Magazine*, Vol. 25, pp. 5–11 (1987).
20. Ungerboeck, G., "Trellis coded modulation with redundant signal sets, Part II: State of the art." *IEEE Communication Magazine*, Vol. 25, pp. 12–21 (1987).
21. Benedetto, S., and Biglieri, E., *Principles of Digital Transmission with Wireless Application*. Kluwer Academic/Plenum Publisher (1999).
22. Shah, S.S., S. Yaqub, and F. Suleman, "Self-correcting codes conquer noise, Part I: Viterbi codecs." *EDN*, pp. 131–140 (2001).
23. Wolf, H.K., and E. Zehavi, "P2 codes parametric trellis codes utilizing punctured convolutional codes." *IEEE Communication Magazine*, Vol. 33, pp. 94–99 (1995).
24. Wei, L.F., "Trellis coded modulation with multidimensional constellations." *IEEE Transactions on Information Theory*, Vol. IT-33 No. 4, pp. 483–501 (1987).
25. Rohde, L.U., and C. Jerry, *Communications Receivers: Principles and Design, Whitaker 3rd. Edition*, McGraw-Hill, New York, N.Y. (2000).
26. Lee, E.A., and D.G. Messerschmitt, *Digital Communications*. Kluwer Academic Publisher (1994).
27. Cho, K.H., and H. Samueli, "A frequency agile single chip qam modulator with beam forming diversity." *IEEE Journal of Solid State Circuits*, Vol. 36 No. 3 (2001).
28. Wong, B.C., and H. Samueli, "A 200 MHz all digital QAM modulator and demodulator in 1.2 mm CMOS for digital radio applications." *IEEE Journal of Solid State Circuits*, Vol. 26 No. 12, pp. 1970–1980 (1991).
29. Wu, H., J.A. Tierno, P. Pepeljugoski, J. Schaub, S. Gowada, J.A. Kashe, and A. Hajimiri, "Integrated transversal equalizers in high-speed fiber-optic systems." *IEEE Journal of Solid-State Circuits*, Vol. 38 No. 12, pp. 2131–2137 (2003).
30. Widrow, B., and S.D. Stearns, *Adaptive Signal Processing*. Prentice Hall (1985).
31. Johnson, C.R., et al., "Blind equalization using the constant modulus criterion." *IEEE Proceedings* (1998).
32. Treichler, J.R., M.G. Larimore, and J.C. Harp, "Practical blind demodulators for high-order QAM signals." *Proceedings of the IEEE Special Issue on Blind System Identification and Estimation*, Vol. 86 No. 10, pp. 1907–1926 (1998).
33. Godard, D.N., "Self recovering equalizer and carrier tracking in two dimen-

sional data communications system." *IEEE Transactions Communications*, Vol. 28, pp. 1876–1875 (1980).
34. Sato, Y., "Methods of self-recovering equalization for multilevel amplitude modulation systems." *IEEE Transactions Communications*, Vol. 23, pp. 679–682 (1975).
35. Picchi, G., and G. Prati, "Blind equalization and carrier recovery using a stop and go decision directed algorithm." *IEEE Transactions Communications*, Vol. 35, pp. 877–887 (1987).
36. Haykin, S., *Unsupervised Adaptive Filtering*. John Wiley N.Y. (1999).
37. Stiffler, J.J., *Theory of Synchronous Communications*. Prentice Hall (1971).
38. Lindsey, W., and M. Simon, *Telecommunication System Engineering*. Prentice Hall (1973).
39. Mengali, U., and A.N. D'Andrea, *Synchronization Techniques for Digital Receivers*. Plenum Press N.Y. (1997).
40. D'Andrea, A.N., and U. Mengali, "Noise performance of two frequency—error detectors derived from maximum likelihood estimation methods." *IEEE Transactions on Communications*, Vol. 42, pp. 793–802 (1994).
41. D'Andrea, A.N., and U. Mengali, "Design of quadricorrelators for automatic frequency control systems." *IEEE Transactions on Communications*, Vol. 41, pp. 988–997 (1993).
42. D'Andrea, A.N., and U. Mengali, "Performance of quadricorrelators driven by modulated signals." *IEEE Transactions on Communications*, Vol. 38, pp. 1952–1957 (1990).
43. Gardner, F.M., "Properties of frequency difference detectors." *IEEE Transactions on Communications*, Vol. 33, pp. 131–138 (1985).
44. Alberty, T., and V. Hespelt, "New pattern jitter free frequency error detector." *IEEE Transactions on Communications*, Vol. 37, pp. 159–163 (1989).
45. Classen, F., H. Meyr, and P. Sehier, "Maximum likelihood open loop carrier synchronizer for digital radio." *ICC' 93*, pp. 493–497 (1993).
46. Classen, F., and H. Meyr, "Two frequency estimation schemes operating independently of timing information." *Globalcom'93 Conference*, Houston, TX, pp. 1996–2000 (1993).
47. Spiegel, M.R., *Mathematical Handbook of Formulas and Tables. Schaum's Outline Series*. McGraw-Hill (1998).
48. Bingham, J.A.C., *The Theory and Practice of MODEM Design*. John Wiley & Sons (1988).
49. Franks, L.E., "Carrier and bit synchronization in data communication a tutorial review." *IEEE Transactions on Communications*, Vol. COM–28 No. 8, pp. 1107–1121 (1980).
50. Brillant, A., "Understanding phase-locked DRO design aspects." *Optomic-Microwaves Ltd Microwave Journal*, pp. 22–42 (1999).
51. Kroupa, V.F., "Noise properties of PLL systems." *IEEE Transactions on Communications*, Vol. COM-30, pp. 2244–2252 (1982).
52. Feher, K., *Digital Communications Satellite Earth Station Engineering*. Nobel Publishing Corp. (1983).
53. Crawford, J.A., *Frequency Synthesizer Design Handbook*. Artech-House

(1994).
54. Gilmore, R., "Specifying local oscillator phase noise performance: how good is enough." RF Expo (1991).
55. Zimmer, R.E., and R.L. Peterson, *Digital Communication and Spread Spectrum Systems*. Macmillan N.Y. (1985).
56. Taub, H., and D.L. Schilling, *Principles of Communication Systems*. McGraw-Hill (1986).
57. Lathi, B.P., *Modern Digital and Analog Communication System*. Oxford University Press (1998).
58. Taub, H., and D.L. Schilling, *Principles of Communications Systems, 2nd Edition*. McGraw-Hill (1986).
59. Norsworthy, S.R., R. Schreier, and G. C. Temes, *Delta-Sigma Converters: Theory, Design and Simulation*. Wiley-IEEE Press (1996).
60. Inose, H., Y. Yasuda, and J. Marakami, "A telemetering system by code modulation, delta-sigma modulation," *IRE Trans. on Space, Electronics and Telemetry*, SET-8, pp. 204–209 (1962).
61. Ungerboeck, G., "Trellis coded modulation with redundant signal sets, Part II, State of the art." *IEEE Communication Magazine*, Vol. 25, pp. 12–21 (1987).

# Chapter 16

# Linearization Techniques

From the previous chapters, it is known that any linear analog signal is accompanied by an interfering signal. Interference is composed of noise, which is a random process, and distortions. Distortions result from the nonlinear (NL) response of a network and can be divided into two categories: distortions created by the electronic circuit and those created by the optics. Nonlinear distortions (NLDs) may result in undesired tones, which may interfere with the desired signal detection process, thus reducing the system's dynamic range. NLD is a stochastic random process as well.[30] There are two definitions for that purpose, signal-to-noise ratio (SNR) or carrier-to-noise ratio (CNR), in analog transport that measure the ratios of the signal-to-noise level, and signal-to-noise and distortion (SINAD). In digital modulation schemes, CNR is replaced by $E_b/N_0$, which is the bit energy per noise. However, in both cases, noise and distortions may reduce the signal quality of service by affecting the bit error rate (BER) in digital transport, CNR, and dynamic range. To overcome this problem, linearization techniques are used. Linearization can be done in the optics and in the electronic circuit that drives the optics.

In wideband systems such as community access television (CATV) transport, the goals are to reduce both second-order distortions and third-order distortions, since due to the enormous bandwidth (BW) of 50 to 870 MHz, both composite second order (CSO) and composite triple beat (CTB) distortions fall within the channel's receiver band. Thus, these linearizers should be wideband. In narrow-band systems such as cellular and radio over fiber (ROF), second-order distortions are out of band and are filtered by the transmitter and receiver. However, third-order distortions are in band and pass through the system's intermediate frequency (IF) as a legal signal, or in direct conversion, zero IF creates dc offsets due to IP2 and IP3. Therefore, when designing a narrow-band system, high IP3 is a major consideration, while in wideband systems, both IP2 and IP3 are important. In other words, linearization circuits improve the system's IP2 and IP3, thus the system's dynamic range is improved.

The transmitter's linearization circuits are divided into two categories of predistortion circuits. One creates a transfer function that compensates for the NL products of a device,[7,13] and the other is a feedforward (FF) linearizer that creates distortions with the opposite phase of the system distortions. Therefore, when summed with the output, the NLD is minimizes.

In the FTTx-receiving portion, a simpler and cheaper method is required. This is done by using push–pull configuration to minimize the receiver electrical circuit second-order distortions, or parallel feedback in a single-ended case. A third method combines both push–pull and parallel feedback.

This chapter reviews electrical and optical linearization methods for transmitting and receiving.

## 16.1 Electronic Linearization Methods in CATV Transmitters

### 16.1.1 Feed forward (FF) linearization

FF has been extensively used with a solid state power amplifier (SSPA) and in cellular base station applications.[12] The FF linearization method does not use pre-distortion circuit. This technique processes the distortions created by the main amplifier [power amplifier (PA)] in such a manner that they are subtracted and minimized at its output. No correction is done prior to PA or on the PA itself. In the theoretical and ideal case of FF, the distortions are totally cancelled. Figure 16.1 illustrates the basic concept of operation. The system consists of two loops. The first loop is the distortions isolation loop. The loop subtracts

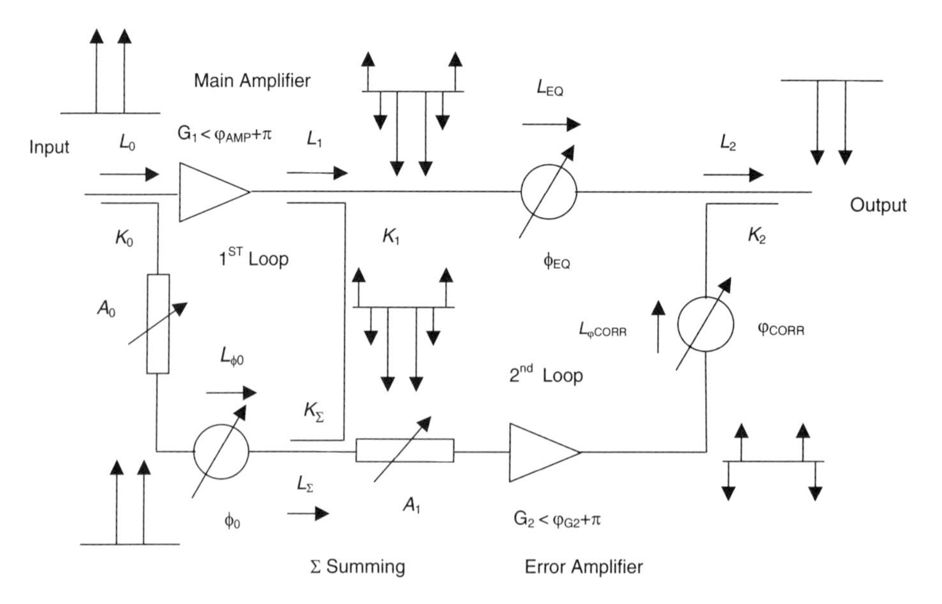


**Figure 16.1** FF linearization concept showing signal and distortions with phase inversions. Note that third order are odd frequency and are inverted like the fundamental. The second order are even and thus are not inverted.

samples of the input signal ($S_{in}$), which is distortion free, from the output signal ($S_{out}$), which contains the main amplifier distortions and fundamental tones, to produce a correction sample of the main amplifier distortion. This correction sample is the PA's distortions, which feed the second loop. The second loop is the inter-modulation (IMD) correction, alignment, and cancellation loop. $S_{out}$ is the main amplifier output. It consists of the amplified input signal, fundamental tones, plus any distortion introduced by the main amplifier.[7] A similar approach is used for optics linearization,[10] as will be described later on. Using phasor notation, the following equation is applicable for the main amplifier marked as G1 in Fig. 16.1:

$$S_{out} = G_1 S_{in} \exp(j\varphi_{AMP}) + \text{IMD}, \tag{16.1}$$

where $G_1$ is the linear gain and $\varphi_{AMP}$ is the phase shift introduced by the main amplifier. The samples of the input and output signals of the main amplifier are given by

$$\begin{cases} S_{\text{in-cplr}} = K_0 S_{in}, \\ S_{\text{out-cplr}} = K_1 S_{out}, \end{cases} \tag{16.2}$$

where $K_0$ and $K_1$ are the coupling coefficients of the directional coupler (DIR) used for sampling the input and output signals of the main amplifier. It is clear from Fig. 16.1 that the input signal is distortion free and the output signal is accompanied with distortions. Thus, if the input signal matches the output signal's amplitude and phase, the fundamental tones of the two are identical. Hence, if the input signal sampling path is adjusted with an additional phase lag placed in the first loop to compensate for the main amplifier delay and the amplifier signal has a 180-deg phase inversion, then both the input signal and main PA output of fundamental frequencies are identical in amplitude but opposite in phase. As a consequence, the output of the first loop produces the main amplifier distortions $S_{ERR}$ as written by Eq. (16.3) and shown in Fig. 16.1:

$$\begin{aligned} S_{ERR} = S_{IMD\_1} = S_{in}\Big\{&K_0 A_0 L_{\phi 0} L_{\Sigma} \exp\big(j\phi_0\big) \\ &+ L_0 G_1 K_1 K_{\Sigma} \exp\big[j(\varphi_{AMP} + \pi)\big]\Big\} + K_1 K_{\Sigma}\text{IMD}, \end{aligned} \tag{16.3}$$

where $L_0$ is the insertion loss of the input signal sampling coupler, $L_1$ is the insertion loss of the PA output signal sampling coupler, $L_{\phi 0}$ is the insertion loss of the first loop phase shifter (PS), and $K_{\Sigma}$ is the coupling factor of the sum point combiner that sums the input sampling and the PA output sampling. The output of the combiner, $\sum$, feeds the second FF loop. $L_{\Sigma}$ is the insertion loss of the sum point and $\phi_0$ is the first loop's PS phase response or state. It is assumed that the main amplifier is an inverting amplifier. In addition, it is assumed that the coupling coefficient $K_1$ has enough margin to provide fine amplitude tuning of the input signal sampling. Thus, the sampled input signal amplitude is adjusted using an attenuator $A_0$ and PS $\phi_0$ to match the main amplifier output signal sampling in both amplitude and phase $\varphi_{AMP}$. In other words, since full alignment is needed for fundamental signal cancellation in the first loop, a PS is added to the input-sampled signal path to compensate for the main amplifier phase. This yields the phase and amplitude alignment condition for the fundamental signal given by Eq. (16.4). Note that the linear signal results from the linear gain coefficient of the main amplifier power series denoted as $G_1$. The gain values of the second- and third-order distortions are driven from the second and third coefficients of the power series mentioned in Chapter 9:

$$0 = K_0 A_0 L_{\phi 0} L_{\Sigma} \exp\left(j\phi_0\right) + L_0 G_1 K_1 K_{\Sigma} \exp\left[j(\varphi_{AMP} + \pi)\right], \tag{16.4}$$

which provides phase and amplitude conditions. The results are the amplitude condition $K_0 A_0 L_{\phi 0} L_{\Sigma} = L_0 G_1 K_1 K_{\Sigma}$ and the phase condition $\varphi_{AMP} + \pi - \phi_0 = \pi$, so $\varphi_{AMP} = \phi_0$. Assuming an ideal first loop, the error-distorting signal produced by the first loop is amplified by the second loop's RF-error amplifier $G_2$, which is an inverting amplifier as well. The output of the second loop's error amplifier is aligned in phase, by a $\varphi_{CORR}$ PS, and amplitude to the main amplifier's output is summed with the main amplifier's output signal that contains distortions. In order to achieve amplitude alignment between the second loop's distortion output and the main amplifier there is an additional attenuator $A_1$ at the input of the second loop's error amplifier. However, an additional PS $\phi_{EQ}$ is needed after the main amplifier's output-signal coupler to compensate for the delay of the second loop's error amplifier $\varphi_{G2}$. The correction signal from the second loop is coupled to the main amplifier's-signal path by a DIR $K_2$. This structure results in the following condition for linearization:

$$\begin{aligned} 0 = \mathrm{IMD}\{&L_1 L_{EQ} L_2 \exp\left(j\phi_{EQ}\right) + K_1 K_2 K_{\Sigma} G_2 A_1 L_{CORR} \\ &\times \exp\left[j(\varphi_{G2} + \pi)\right] \exp(j\varphi_{CORR})\}. \end{aligned} \tag{16.5}$$

The output level and phase of the desired linear signal from the main amplifier is given by

$$S_{out} = S_{in} L_0 L_1 L_{EQ} L_2 G_1 \exp\left[j\left(\varphi_{AMP} + \phi_{EQ} + \pi\right)\right], \tag{16.6}$$

where $\phi_{EQ}$ is the equalization PS phase, $L_{EQ}$ is the insertion loss of the equalization PS, $L_2$ is the insertion loss of the IMD's subtracting coupler, $L_{CORR}$ is the insertion loss of the second loop's PS at the IMD frequency, $\varphi_{CORR}$ is the second loop's PS phase at the IMD frequency, $K_2$ is the coupling coefficient of the output coupler for the second loop, and $\varphi_{G2}$ is the error amplifier phase at the IMD's frequency.

In real time, the alignment conditions given by Eqs. (16.4) and (16.5) are not ideal and are finite. Thus the fundamental signal has a finite value at the output of the first loop given by

$$S_{\text{in}}K\exp(j\psi) = S_{\text{in}}\{K_0A_0L_{\phi 0}L_{\Sigma}\exp(j\phi_0) + L_0G_1K_1K_{\Sigma}\exp[j(\varphi_{\text{AMP}} + \pi)]\}. \tag{16.7}$$

Since the second loop's amplifier is assumed to be inverting, the leaked signal to the main amplifier path is given by

$$S_{\text{sigcplr}} = S_{\text{in}}K\exp(j\psi)A_1G_2K_2L'_{\text{CORR}}\exp[j(\varphi'_{G2} + \pi + \varphi'_{\text{CORR}})], \tag{16.8}$$

where the $'$sign symbolizes phase shift values of the second loop's amplifier and the PS at the fundamental frequency as well as the insertion losses of the second loop's PS at the fundamental frequency. It is assumed that there is equal gain attenuation and coupling for both the fundamental frequency and the IMD.

It can be observed that the fundamental signal coupled from the second loop to the main output is 180 deg plus the additional phase with respect to the fundamental signal coming out from the main amplifier. The overall fundamental signal power is given by the summation of the phasors presented by Eqs. (16.6) and (16.8).

Using the same considerations, the IMD value in the real case of imperfect alignment is presented by the following phasor:

$$\begin{aligned}\text{IMD} \times M\exp(j\theta) = \text{IMD}\{&L_1L_{\text{EQ}}L_2\exp(j\phi_{\text{EQ}}) + K_1K_2K_{\Sigma}G_2A_1L_{\text{CORR}} \\ &\times \exp[j(\varphi_{G_2} + \pi + \varphi_{\text{CORR}})]\}.\end{aligned} \tag{16.9}$$

The value IMD $\times$ $M$ is equivalent to the IMD magnitude. Thus the C/I ratio (carrier to IMD) is given by

$$\text{C/I[dB]} = 20\log\left|\frac{S_{\text{out}} + S_{\text{sigcplr}}}{\text{IMD} \times M}\right|. \tag{16.10}$$

The reason for 20 log is because the phasors are voltages. Note that Eq. (16.10) describes the vector sum of magnitudes. The disadvantage of the FF method is its narrow BW response. As a consequence, FF is useful for cellular applications rather than CATV, which is a wideband application, and the realization of wideband PSs is more complicated. One of the methods to realize PSs with a group delay similar to that of amplifiers is by a low-pass filter tuned by a voltage-variable capacitance diode (VARICAP). Additionally, it requires a complex calibration process. Figure 16.1 provides a conceptual block diagram of an FF linearizer. This technique cancels both CSOs and CTBs but has limited BW. In the area of optical frequencies, this concept becomes very attractive for laser linearization because at the optical domain, the correction and wavelength deviations due to distortion is considered narrow BW. The reader can use the concept provided in this section together with the RF identities in Chapter 9 to calculate and optimize an FF linearizer for narrow-band RF applications.

### 16.1.2 Distortions of unlinearized modulators

This section reviews and derives the CSO and CTB intermodulation distortions of an unlinearized modulator in order to determine the condition to null the CSO and establish the specification for the CTB.[9,10,15] The overmodulation limit of an externally modulated system is more severe than the directly modulated laser system used for amplitude modulation (AM), vestigial side band (VSB), and CATV systems.[17]

The normalized optical output power of the modulator is given by the normalized delta between the ac power component $P$ to the dc power $\langle P\rangle$:

$$p = \frac{P - \langle P\rangle}{\langle P\rangle} = \sin(\theta + \phi_b), \tag{16.11}$$

where $\theta$ is given by $\theta(t) = \pi V(t)/V_{\pi}$ (see Sec. 7.5.1) and $\phi_b$ is the bias phase. Assume a multichannel CATV modulating signal consisting of $N$ unmodulated RF carrier signals plus a dc bias term $V_B$:

$$V(t) = \sum_{q=1}^{N} V_0 \cos(\omega_q t + \psi_q) + V_B. \tag{16.12}$$

The voltage described by Eq. (16.12) is the modulating voltage applied on the modulator plus the dc bias voltage $V_B$. If $\psi_q$ is a random phase of each carrier voltage, they are not coherent. Back to Sec. 7.5.1: $\beta$ defines the modulation index, $\beta = \pi V_0/V_{\pi}$. Thus, the total modulation phase, including the bias, equals $\phi_T = \beta + \phi_b$. Using these notations, $\theta(t)$ is given by

$$\theta(t) = \beta \sum_{q=1}^{N} \cos(\omega_q t + \psi_q). \tag{16.13}$$

Using the Taylor power series expansion for $\sin(\theta)$ and $\cos(\theta)$, and using the trigonometric identity of $\sin(\theta + \alpha)$ and $\cos(\theta + \alpha)$, as well as Eqs. (16.11) and (16.12), Eq. (16.11) is modified to:

$$\begin{aligned} p = {} & \cos(\phi_T)\left\{\beta \sum_{q=1}^{N} \cos(\omega_q t + \psi_q) - \frac{\beta^3}{6}\left[\sum_{q=1}^{N} \cos(\omega_q t + \psi_q)\right]^3\right\} \\ & + \sin(\phi_T)\left\{1 - \frac{\beta^2}{2}\left[\sum_{q=1}^{N} \cos(\omega_q t + \psi_q)\right]^2\right\}, \end{aligned} \tag{16.14}$$

where the modulation index $\beta$ represents the linear signal, $\beta^2$ represents the CSO, and $\beta^3$ is the third-order distortion (CTB). From Eq. (16.14), it can be seen that for $\phi_T = 0$,[16] second-order distortions are nullified while the fundamental linear signal is at its maximum value. It can be seen that the fundamental and the third-order terms are orthogonal to the second-order term.

The intrinsic bias point $\phi_b$ of the device is determined by the geometry, e.g., whether it is an Mach-Zehnder (MZ) or a balanced-bridge interferometer (BBI) as well as by the imbalance of the two interferometer arms. Temperature variations and stress may cause the intrinsic bias to change a small function of $V_\pi$. Such drifts are slow with large time constants of minutes and even hours. Observing the condition $\phi_T = \beta + \phi_b$ and setting $\phi_T = 0$ to meet the quadrature bias point in Fig. 16.2, set the phase drifts that change the bias point:

$$V_B = -\frac{\phi_b(t)V_\pi}{\pi}. \tag{16.15}$$

To maintain bias stability requires a special parametric feedback loop for locking the bias point against temperature changes. This is established by taking into account the property presented by Eq. (16.15) and nullifying CSOs. Doing so and stabilizing the optimal bias point, also known as the quadrature point, results in only one requirement for a third-order predistortion circuit.

Setting the bias to the optimal point modifies Eq. (16.14) to contain only the fundamental and the cubic terms, which can be expanded to

$$p = \beta \sum_{q=1}^{n} \cos(\omega_q t + \psi_q) - \frac{\beta^3}{6}\left[\sum_{q1=1}^{N}\sum_{q2=1}^{N}\sum_{q3=1}^{N} \cos(\omega_{q1} + \psi_{q1})\cos(\omega_{q2} + \psi_{q2})\cos(\omega_{q3} + \psi_{q3})\right]. \tag{16.16}$$

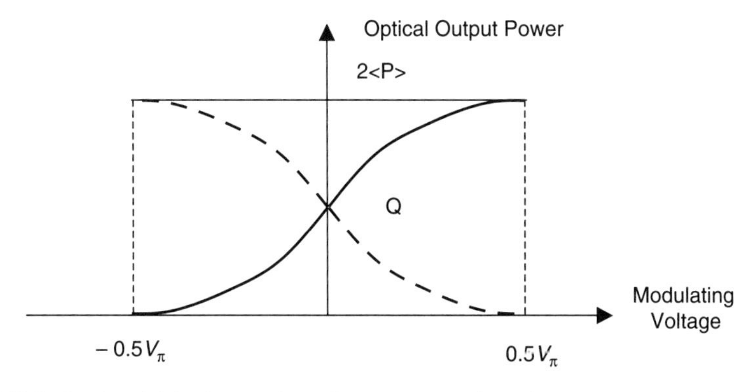


**Figure 16.2** Optimal biasing point of a two-port interferometer modulator showing output power vs. modulating voltage.

This expression can be further expanded using a combination, resulting in

$$p^3 = -\frac{\beta^3}{24}\sum_{\pm}\sum_{q1=1}^{N}\sum_{q2=1}^{N}\sum_{q3=1}^{N}\cos\left[\left(\omega_{q1} \pm \omega_{q2} \pm \omega_{q3}\right)t + \psi_{q1} + \psi_{q2} + \psi_{q3}\right], \tag{16.17}$$

when making the following convention for negative indices: $\omega - q = -\omega q$. As was explained in Chapter 3, CATV channel plans usually contain large subsets of commensurate channel frequencies arranged on a grid of equidistant frequency values. These values are the channel spacing of 6 MHz, satisfying channel frequency $f_N$ [MHz] $= 1.25$ MHz $+ 6N$. Thus many mixing products can fall on the same beat frequency $\omega_b$, which is a valid channel frequency, as was explained in Chapter 9. Triple beat occurs whenever three channel frequencies satisfy the condition

$$\omega_b = \omega_{q1} + \omega_{q2} + \omega_{q3}. \tag{16.18}$$

If the beats are added coherently, meaning that phases coincide, the beats' phasor phases satisfy

$$\psi_b = \psi_{q1} + \psi_{q2} + \psi_{q3}. \tag{16.19}$$

If there are three indices, then there are 3! permutations, which is 6. Thus, if there is a particular triplet $\{q1, q2, q3\}$, the other five permutations correspond to beats that are mutually coherent, meaning that they all have equal phases. As a result, the total contribution associated with a given combination of three distinct indices is six times larger than any of the individual beats associated with any particular permutation. Similarly for a given triplet $\{q1, q1, q2\}$ with only two distinct indices, there are only three permutations. Therefore, the total contribution associated with a given combination of two distinct indices is three times larger than any of the individual beats associated with any particular permutation. These considerations lead to the following expression for the CTB:

$$\begin{aligned} p^3 = \sum_{\pm}\Bigg\{ & 6\sum_{q1=1}^{N-2}\sum_{q2=q1+1}^{N-1}\sum_{q3=q2+1}^{N} B(q_1 \pm q_2 \pm q_3) \\ & + 3\sum_{q1=1}^{N-1}\sum_{q2=q1+1}^{N}\left[B(q_1 \pm q_2 \pm q_2) + B(q_2 \pm q_1 \pm q_1)\right] \\ & + \sum_{q=1}^{N} B(q, \ \pm q, \ \pm q)\Bigg\}, \end{aligned} \tag{16.20}$$

where $B$ is given by

$$B = -\frac{\beta^3}{24}\cos\left[\left(\omega_{q1} + \omega_{q2} + \omega_{q3}\right)t + \psi_{q1} + \psi_{q2} + \psi_{q3}\right]. \quad (16.21)$$

From a combinatory point of view, Eq. (16.20) provides all possible combinations with repetition as generated by the summation. It can be solved efficiently by a computer program.

### 16.1.3 Predistortion linearization fundamentals

A state-of-the-art design approach to broadband linearization is based on the mathematical analysis given in the previous section. This method takes advantage of the fact that external modulators have a consistent and stable transfer characteristic that is generally unaffected by optical-power temperature and aging. The linearizer is installed between the transmitter output and the modulator input. The mathematical presentation uses a memoryless model as was explained in Chapter 9, followed by a finite memory system with transfer function $G(\omega)$. A phase conversion factor $\pi/V_\pi$ to the modulator phase retardation follows and then the modulator frequency response becomes $M(\omega) = \sin(\theta)$, as shown in Figs. 16.3 and 16.4.

The main idea behind this predistortion concept is to create an arcsine circuit. Based on the previous analysis, the shape of the NL transfer

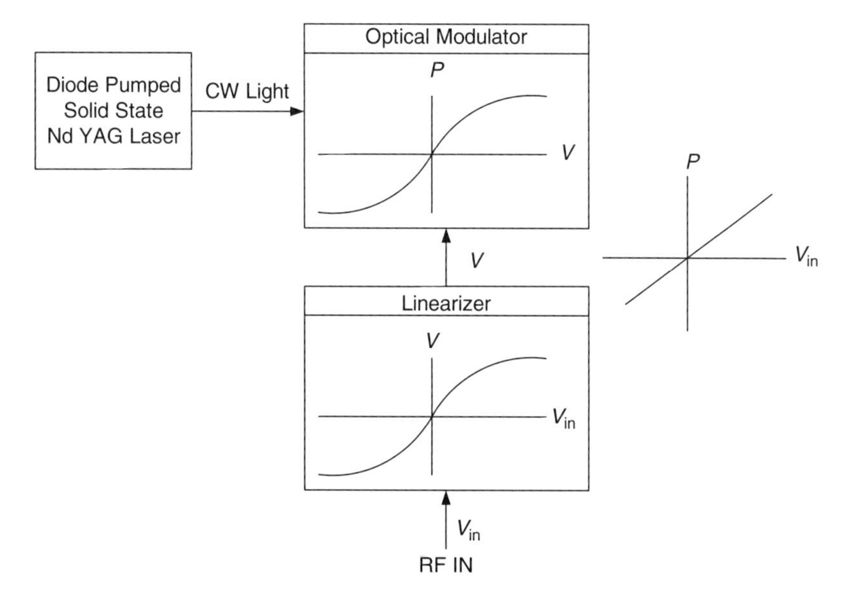


**Figure 16.3** Predistortion linearization concept of external modulator.[10] (Reprinted with permission from the *Journal of Lightwave Technology* © IEEE, 1993.)

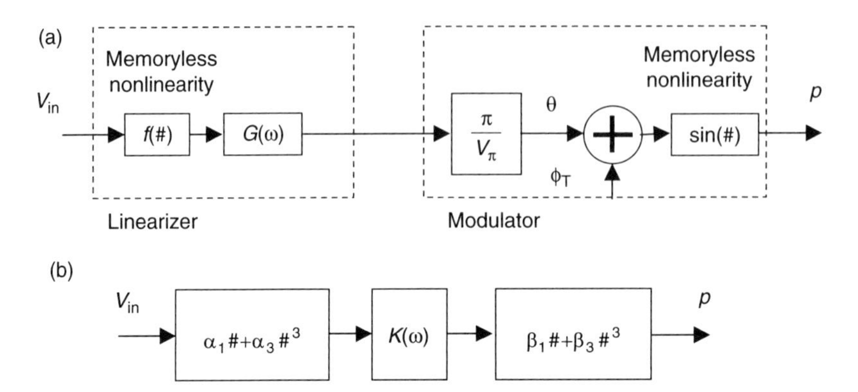


**Figure 16.4** Predistortion general model: (a) simplified model; (b) NL model with memory.[10] (Reprinted with permission from the *Journal from the Lightwave Technology* © IEEE, 1993.)

characteristic should be

$$f(V_{\text{in}}) = K^{-1}\arcsin(K\alpha_1 v_{\text{in}}), \tag{16.22}$$

where $K$ is given by

$$K = \frac{\pi}{V_\pi}G \tag{16.23}$$

and

$$p = \sin[Kf(V_{\text{in}})]. \tag{16.24}$$

Hence, for a small signal approximation and using the Taylor series expansion for the first linear argument, the linearizer produces

$$p = K\alpha_1 V_{\text{in}}, \tag{16.25}$$

where $\alpha_1$ is the coefficient of the linear term in the power series. Assume now that the modulating voltage is expressed by an odd power series:

$$V_{\text{in}} = \alpha_1 V_{\text{in}} + \alpha_3 V_{\text{in}}^3 + \alpha_5 V_{\text{in}}^5 + \cdots. \tag{16.26}$$

Then the normalized optical power $p$ is given by

$$p = \sin[Kf(V_{\text{in}})] = \sin\big[K\big(\alpha_1 V_{\text{in}} + \alpha_3 V_{\text{in}}^3 + \alpha_5 V_{\text{in}}^5 + \cdots\big)\big]. \tag{16.27}$$

Again, using the Taylor series expansion while omitting higher orders than the power of 3 results in

$$p = K\alpha_1 V_{in} + \left(K\alpha_3 - \frac{K^3\alpha_1}{6}\right)V_{in}^3. \tag{16.28}$$

From this equation, it becomes clear that the condition to nullify the third-order term is given by

$$K = \sqrt{\frac{6\alpha_3,}{\alpha_1}}, \tag{16.29}$$

which means that careful tuning is required for the linearization gain $K$. Moreover, higher-orders terms should be examined, though they are omitted here, to make sure their residual contribution does not affect performance.

So far, this analysis did not take into account the frequency response. Consider the following model, as provided by Fig. 16.4(a):

$$K(\omega) = G(\omega)\frac{\pi}{V_\pi}M(\omega). \tag{16.30}$$

Using the Taylor approximation of $\sin\theta$ up to the cubic term and taking the voltage as described by Eq. (16.26), the problem is reduced to analyzing a system that consists of a linear filter $K(\omega)$ inserted between two memoryless nonlinearities, as shown in Fig. 16.4(b). The first is the distortion transfer function of the modulating voltage and the other is the modulator's NL transfer function. Taking the complex notation of $\exp(\pm j\omega_1 t)$, $\exp(\pm j\omega_2 t)$, $\exp(\pm j\omega_3 t)$, where the three frequencies may be or may not be distinct, then the linearizer output $V$ contains the intermodulation term $\alpha_3 \exp[j(\pm\omega_1 \pm \omega_2 \pm \omega_3)t]$ and the fundamental terms with the coefficient $\alpha_1$. In the frequency domain, these terms at the filter output are multiplied by the complex transfer function of the filter $K(\omega)$. Thus the intermodulation products generated by the linearizer at the filter output are

$$\alpha_3 K(\pm\omega_1 \pm \omega_2 \pm \omega_3)\exp[j(\pm\omega_1 \pm \omega_2 \pm \omega_3)t] \tag{16.31}$$

and the fundamentals are given by

$$\alpha_1 K(\pm\omega_i)\exp[j(\pm\omega_i)t], \tag{16.32}$$

where the index $i = 1, 2, 3$.

Finally both the intermodulations and the fundamental terms pass the modulator transfer function. These products are further distorted. The third-order term generates higher orders than the cubic orders that are omitted. Additionally, the third-order term transparently propagates through the modulator's linear slope, resulting in

$$\beta_1\alpha_3 K(\pm\omega_1 \pm \omega_2 \pm \omega_3)\exp[j(\pm\omega_1 \pm \omega_2 \pm \omega_3)t]. \tag{16.33}$$

Equation (16.33) shows the linear transfer of the input-distorted signal. The fundamental term is distorted by the cubic argument of the modulator resulting in

$$\beta_3\alpha_1^3 K(\pm\omega_1)K(\pm\omega_2)K(\pm\omega_3)\exp[j(\pm\omega_1 \pm \omega_2 \pm \omega_3)t]. \tag{16.34}$$

In addition, the fundamental signal passes the linear transfer function of the modulator, resulting in the desired signal.

The sum of Eqs. (16.33) and (16.34) presents the conditions for linearity when the sum result equals zero:

$$\beta_3\alpha_1^3 K(\pm\omega_1)K(\pm\omega_2)K(\pm\omega_3) + \beta_1\alpha_3 K(\pm\omega_1 \pm \omega_2 \pm \omega_3) = 0. \tag{16.35}$$

Writing $K(\omega)$ as a phasor with magnitude and phase, which is frequency dependent,

$$K(\omega) = k(\omega)\exp[j\phi(\omega)] \tag{16.36}$$

modifies Eq. (16.35). Hence, in order to nullify the third-order terms, the phase versus frequency should be linear. In other words, the phase derivative should be a constant value, which states a constant group delay over the frequency:

$$\begin{cases} \phi(\omega) = T_g\omega, \\ \dfrac{d\phi(\omega)}{d\omega} = T_g. \end{cases} \tag{16.37}$$

This condition results in linear properties,

$$\phi(\pm\omega_1 \pm \omega_2 \pm \omega_3) = \phi(\pm\omega_1) + \phi(\pm\omega_3) + \phi(\pm\omega_3), \tag{16.38}$$

which simplify Eq. (16.35) to

$$\beta_3\alpha_1^3\, k(\pm\omega_1)k(\pm\omega_2)k(\pm\omega_3) + \beta_1\alpha_3\, k(\pm\omega_1 \pm \omega_2 \pm \omega_3) = 0, \tag{16.39}$$

which can be reduced to

$$\beta_3\alpha_1^3 k^2 + \beta_1\alpha_3 = 0, \tag{16.40}$$

where $k$ is a constant equivalent to the condition of Eq. (16.29). The outcome of this analysis is that the prerequisite for wideband predistortion is to have a constant group delay, and amplitude versus frequency in order to have a third-order cancellation. Such ripples may come from the RF chain of amplifiers and

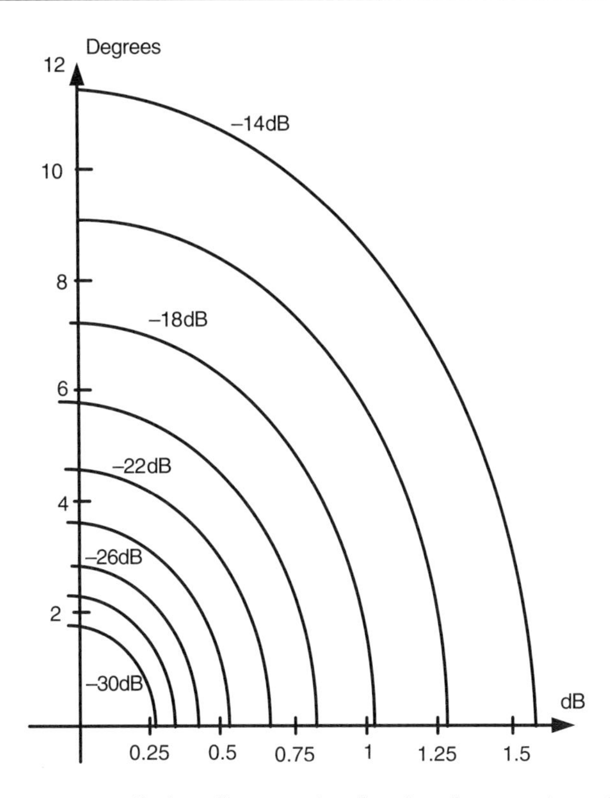


**Figure 16.5** Phasor-cancellation diagram showing the phase and amplitude mismatch while the suppression is a parameter. The suppression is the ratio of the resultant signal to the average power of the two IMD signals.[10] (Reprinted with permission from the *Journal of Lightwave Technology* © IEEE, 1993.)

RF components, modulator velocity mismatch between the optical and the electrical fields, and the modulator piezoelectric-related ripple. Thus, for an optimized frequency response, a traveling wave modulator is preferred. In the real case, phase and amplitude alignment would define the amount of IMD suppression as described by Eq. (16.35) from which it is clear that the ratio between the two phasors' magnitudes should be equal to unity and the phase between the two phasors is 180 deg. A parametric chart of amplitude and phase mismatch can provide the amount of suppression parameter for a contour. Figure 16.5 describes a phasor-cancellation diagram, where the required amplitude mismatch and phase mismatch defines the amount of IMD cancellation. The contours create an elliptical shape since the phase and amplitude errors can be $\pm$. The abscissas describes the amplitude mismatch and the ordinates describe the phase.

### 16.1.4 Predistortions linearization circuits

There are several methods to realize the desired arcsine function in an electronic circuit such as a back-to-back diode expansive circuit[10] or a compressive one.

A distorted sine wave signal can be classified into two categories: odd mode harmonics–distorted signal and even mode–distorted signal. The odd mode–distorted sine wave is a signal in which both the top peak and the bottom peak are distorted in the same manner. For instance, both peaks are clipped at the same magnitude. The even mode–distorted sine is a half-cycle sine containing only the positive peaks. This explanation is important for understanding predistortion topologies. Predistortion techniques using a Schottky barrier for their lower forward voltage were developed in the early 1970s for satellite communications.[1] Diodes and a combination of a bipolar junction transistor (BJT) differential amplifiers with diodes were used for CATV linearization circuits.[10] Linearizers became more advanced when integrated complementary MOS (CMOS) predistortion circuits were developed.[8,14]

The first approach presents a back-to-back expansive circuit made of diodes, as shown in Fig. 16.6. The dc current source biases the $N$ diodes in a branch up to the point that the dc block capacitors are charged to a fixed dc value. This voltage is presented as a bias voltage source. The diodes are in forward bias and the current of a diode at that point is given by

$$I = I_S \exp(V/V_T), \tag{16.41}$$

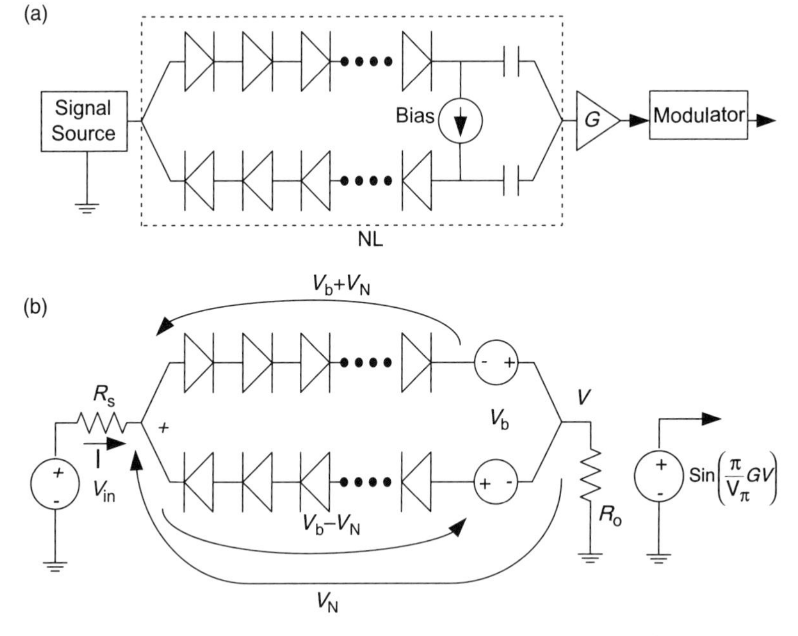


**Figure 16.6** (a) Linearizer circuit and (b) its equivalent circuit.[10,15] (Reprinted with permission from the *Journal of Lightwave Technology* © IEEE, 1993 and from the *Journal on Selected Areas on Communications* © IEEE, 1990.)

where $V_T = kT/q$. Figures 16.6(a) and 16.6(b) describe dc and ac conditions. In the dc state, the dc bias current source is tapped differentially. Thus the upper capacitor is charged to negative while the lower capacitor is in a positive dc charge. These two capacitors at first approximation are at a fixed dc voltage marked as $V_b$. Hence these capacitors are two opposing voltages, as shown in Fig. 16.16(b). When the ac scheme is used during the positive cycle of $V_{in}$, the upper branch has a higher conductivity because the chain of diodes has a higher forward voltage while the lower branch has a lower voltage, thus its conductivity is lower. In the second half of the cycle this is reversed. Note that $V_b$ remains unchanged in its polarity throughout the entire cycle. So the ac model describes the ac signal on a dc. Therefore, the following notation is applicable while $V_N$ is defined in Fig. 16.16(b):

$$\begin{aligned} I &= I_S \exp[(V_b + V_N)/NV_T] - I_S \exp[(V_b - V_N)/NV_T] \\ &= 2I_S \exp[V_b/NV_T] \sinh \frac{V_N}{NV_T} = 2I_B \sinh \frac{V_N}{NV_T}, \end{aligned} \tag{16.42}$$

where $I_B$ is the current flowing through the diodes without the ac signal and $N$ is the number of diodes in each branch. Equation (16.42) describes the diode's current fluctuations around the dc point on the $I-V$ curve. Thus the $V-I$ impedance characteristic is given by the inverse of Eq. (16.42):

$$V_N = NV_T a \sinh\left(\frac{I}{2 \cdot I_B}\right), \tag{16.43}$$

$$R_N(I) = \frac{\partial V_N}{\partial I}. \tag{16.44}$$

Observe the load resistance $R_0$ in Fig. 16(b): the voltage rate developed on it is faster than the voltage changes at the input because of the nature of the sinh function of the down-going impedance of the series diodes, creating the NL block. This change compensates for the compressive sine-shaped $L-V$ characteristic of the modulator, resulting in a fairly linear overall $L-I$ characteristic.

Assume that the varying voltage $V_N$ given by Eq. (16.43) is expressed as a Taylor power series:

$$V_N = r_1 I + r_3 I^3, \tag{16.45}$$

where

$$\begin{cases} r_1 = \frac{NV_T}{2I_B}. \\ r_3 = -\frac{NV_T}{48 I_B^3}. \end{cases} \tag{16.46}$$

From Fig. 16.16(b), using Kirchhoff's voltage law (KVL) in a loop results in

$$V_{in} = R_S I + \left(r_1 I + r_3 I^3\right) + R_0 I = (R_S + R_0 + r_1)I + r_3 I^3. \tag{16.47}$$

Inverting Eq. (16.47) results in

$$I = g_1 V_{in} + g_3 V_{in}^3, \tag{16.48}$$

where

$$\begin{cases} g_1 = \dfrac{1}{R_S + R_0 + r_1}, \\ g_3 = -\dfrac{r_3}{(R_S + R_0 + r_1)^4}. \end{cases} \tag{16.49}$$

Hence the linearizer output voltage is given by

$$V = R_0 I = R_0 g_1 V_{in} + R_0 g_3 V_{in}^3. \tag{16.50}$$

From this analysis, the coefficients mentioned in Eq. (16.28) are given by

$$\begin{cases} \alpha_1 = R_0 g_1, \\ \alpha_3 = R_0 g_3. \end{cases} \tag{16.51}$$

These coefficients' dependency over the bias point and number of diodes is available by substituting Eq. (16.45) into Eq. (16.49) and then into Eq. (16.51). These coefficients provide the linearization gain when applied into Eq. (16.22). Optimization and tuning of coefficients is made by controlling the diodes bias.

The CMOS linearization chip circuit in an adaptive predistortion linearizer that was, presented in Refs. [8] and [14]. This technique has two nonlinear gain (NLG) paths generating second-order and third-order predistortions. In CMOS integrated circuits, the differential amplifier with variable transconductance is a widely used architecture for implementing a variable gain amplifier (VGA). Since $g_m$ of the MOS device varies as the square root of the bias current, the voltage gain $A_V$ has the same nature:

$$\begin{cases} g_m = \sqrt{\dfrac{2\mu C_{OX} W I_{bias}}{L}}, \\ A_V = g_m R_{out}, \end{cases} \tag{16.52}$$

where $W$ and $L$ are the channel width and length, $C_{ox}$ is the gate oxide capacitance, $A_v$ is the voltage gain, $g_m$ is the transconductance, and $\mu$ is mobility. The second-order terms are generated because of the $I-V$ square-law relationship of the CMOS device given by

$$I_D = \frac{\mu C_{OX} W (V_{GS} - V_t)^2}{2L}, \tag{16.53}$$

where $V_t$ is the field-effect-transistor (FET) threshold voltage, and $V_{GS}$ is the gate source voltage. If $V_{GS}$ is composed from dc and ac terms, $V_B$ and $V_{AC}$, Eq. (16.53) becomes

$$I_D = K(V_B - V_t)^2 + KV_{AC}^2 + 2K(V_B - V_t)V_{AC}, \tag{16.54}$$

where $K$ is given by

$$K = \frac{\mu C_{OX} W}{2L}. \tag{16.55}$$

The goal is to amplify $KV_{ac}^2$ and suppress the linear current term $2K(V_B - V_t)V_{ac}$. This is done by a differential transistor pair as shown in Fig. 16.7.

The two transistors are perfectly matched and the input signal is perfectly balanced around the bias voltage. The output load resistor $R_1$ converts the current into voltage. This way a square-law voltage is created. When the two ac phasors feed the differential pair, they are at 180 deg of each other, and thus the first order ac terms' sum is null and the second-order terms are at

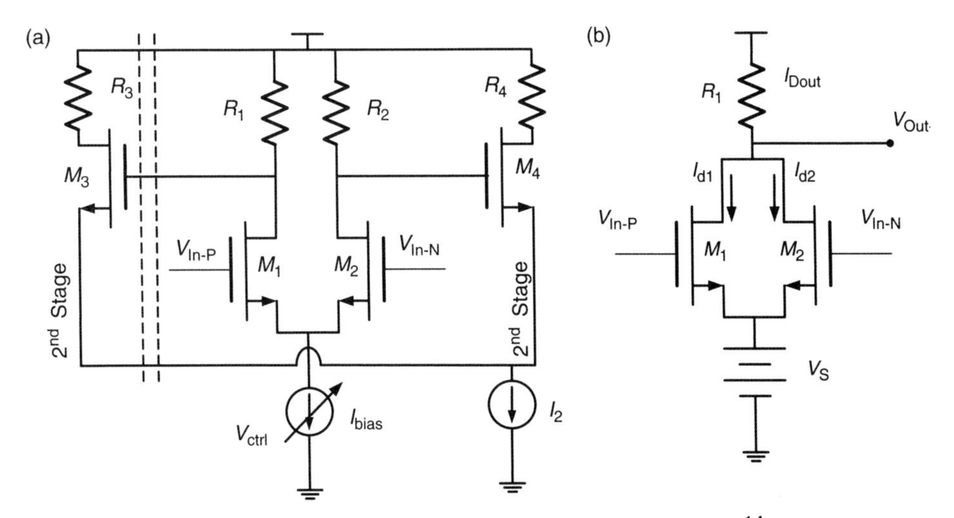


**Figure 16.7** (a) VGA architecture. (b) Two-transistor squaring circuit.[14] (Reprinted with permission from the *Journal of Lightwave Technology* © IEEE, 2003.)

phases of 0 deg and 360 deg; thus, they sum in phase. Hence the second-order term is given by

$$I_{D\_2f} = I_{D1} + I_{D2} = 2K(V_B - V_S - V_t)^2 + 2KV_{ac}^2, \tag{16.56}$$

where $V_s$ is the source bias shown in Fig. 16.7(b). The amount of rejection can be calculated in the same manner as was presented in Sec. 16.1.3.

Third-order distortion is created by a cross-coupled differential pair (CCDP). An ordinary CCDP is used to cancel odd-order nonlinearities in the differential amplifier, but it can be used as an effective architecture to synthesize a cubic transfer function as shown in Fig. 16.8.

A common-source biased differential pair with $I_B$, the differential output current $I_D$, and input signal voltage $V_D$ are related by

$$I_D = \sqrt{\frac{\mu C_{OX} W I_B}{L}} V_D - \frac{1}{8}\sqrt{\frac{1}{I_B}}\left(\frac{\mu C_{OX} W}{L}\right) V_D^3. \tag{16.57}$$

Observing Fig. 16.8, the differential output current from the CCDP is $I_{out} = I_{O\_P} - I_{O\_N} = I_{out\text{-}1} - I_{out\text{-}2}$ where $I_{out\text{-}1} = I_{O\_P1} - I_{O\_N1}$ and $I_{out\text{-}2} = I_{O\_P2} - I_{O\_N2}$. Since $W/L$ and the bias current are different for both differential

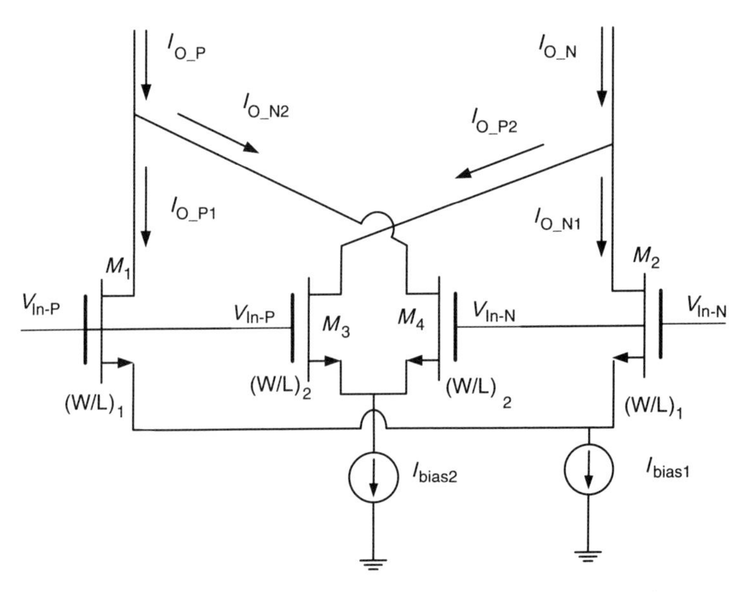


**Figure 16.8** CCDP cell for generating a cubic transfer function.[14] (Reprinted with permission from the *Journal of Lightwave Technology* © IEEE, 2003.)

pairs, the following condition guarantees that the linear component of the output current is cancelled:

$$\left(\frac{W}{L}\right)_1 I_{\text{B-1}} = \left(\frac{W}{L}\right)_2 I_{\text{B-2}}. \tag{16.58}$$

This results in the cubic transfer function,

$$I_{\text{out}} = -\sqrt{\frac{(\mu C_{\text{OX}})^3}{64K_{\text{O}}}}\left[\left(\frac{W}{L}\right)_1^2 - \left(\frac{W}{L}\right)_2^2\right]V_{\text{D}}^3. \tag{16.59}$$

Such a predistortion chip-block diagram is given in Fig. 16.9. This chip contains a PS to compensate for the phase between the primary signal path and the NL block. The phase control is done by a 5-stage polyphase filter using a resistor–capacitor mesh. This filter generates broadband quadrature signals $I$ and $Q$, which are added with a variable gain factor to generate two outputs. These outputs have a tunable relative phase angle and zero relative amplitude variation between the NL block path and the primary signal path. The second control is of gain to compensate amplitude variances between the NL block and the primary path.[14]

### 16.1.5 Parametric feedback for CSO and CTB control

In previous sections, distortion mechanisms and cancellation methods were reviewed. It was explained that CSO orders can be cancelled by controlling the phase argument of the modulator, which is related to the bias point. The long-term CSO and CTB performance of a predistortion-linearized transmitter can be maintained by a closed loop. Since the dc change and signal are slow and are quasi-dc, the loop is a narrow BW feedback system. Such loops are called parametric loops, since the feedback quantity of intermodulation beats detected by the optical output feedback detector vary slowly due to the slow drifts mentioned. A faster loop creates more distortions as it would track the instantaneous change rather than the average. This is similar to the automatic gain control (AGC) requirements previously explained in Chapter 12; recall that fast AGC would generate a residual AM on the signal, since it follows and detects the AM envelope of the signal.

The goal of a CSO parametric control loop is to maintain the modulator at a fixed phase retardation. The terminology retardation locked loop (RLL) is analogous to a phase locked loop (PLL) in that as a PLL maintains the phase of the voltage controlled oscillator (VCO) at a fixed value and locks its frequency, the RLL maintains the phase retardation of the optical modulator at a fixed value of $\phi_{\text{T}} = 0$ for canceling CSOs. Hence this system has to be locked on a reference IMD that is analogous to a PLL reference frequency.

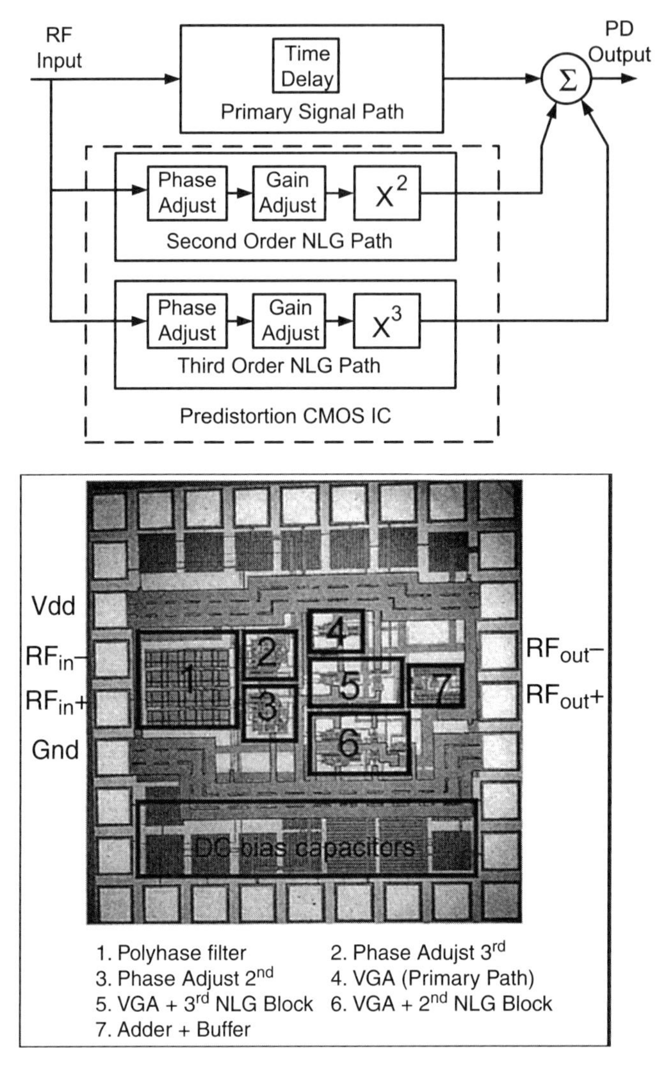


**Figure 16.9** CMOS predistorter chip-block diagram and actual chip.[14] (Reprinted with permission from the *Journal of Lightwave Technology* © IEEE, 2003.)

For this purpose, there is a need for a known IMD that can be used as a reference to lock on. This is accomplished by inserting two tones as pilots and using the resultant second-order IMD or the second harmonic as a reference signal to lock on. The reference IMD is produced by a an NL element. Since the NL element produces many combinations from the two pilot tones, the desired IMD is selected by a band-pass filter (BPF), as shown in Fig. 16.10. This way, the RLL is locked on a reference IMD at $f_{\mathrm{REF}} = f_2 - f_1$. The RLL is tuned to a bias point that sets the CSO level to $-70$ dBc or better. This is below the

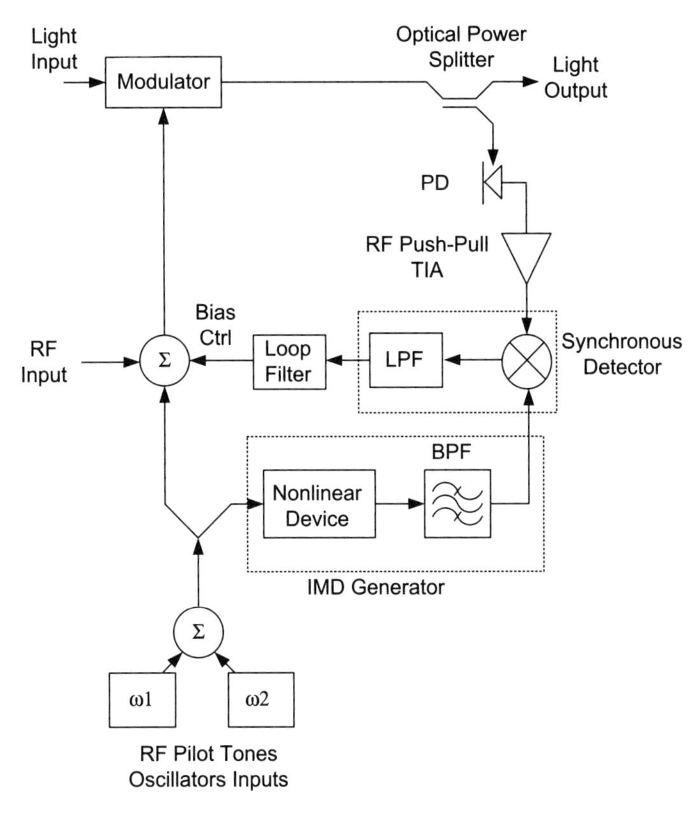


**Figure 16.10** RLL for biasing a Mach-Zehnder interferometer (MZI) modulator at the inflexion point of its *L*-*V* characteristic in order to null CSO.[10] (Reprinted with permission from the *Journal of Lightwave Technology* © IEEE, 1993.)

FTTx receiver CSO limit that mainly results from the photodetector's (PD's) residual nonlinearities.

Assume two pilot carriers, per Eq. (16.13) (where $i = 2$), injected into the CATV channel path going into the modulator. These pilots' frequencies are selected in such a manner that they are at unused CATV frequencies and so are the IMD products. In addition, their power level is way below the signal, so that the RLL does not interfere with the RF functioning of the transmitter. The second-order products are then given by

$$p^{(2)} = \beta \cos(\phi_T) \sum_{i=1}^{2} \cos(\omega_i t + \psi_i) - \frac{\beta^2 \sin(\phi_T)}{2} \times \left\{ \sum_{\pm} \cos[(\omega_1 \pm \omega_2)t + \psi_1 \pm \psi_2] + \frac{1}{2} \sum_{i=1}^{2} \cos(\omega_i t + \psi_i) \right\} \qquad (16.60)$$

If a small value of $\phi_T$ is assumed, the amplitudes of the intermodulations and harmonics are given by

$$\begin{cases} m[\omega_1 \pm \omega_2] \approx \dfrac{\beta^2 \phi_T}{2}, \\ m[2\omega_i] \approx \dfrac{\beta^2 \phi_T}{4}. \end{cases} \tag{16.61}$$

Equation (16.61) describes the reference IMD parameter that the RLL is locked to. It is convenient to lock on the difference between the two tones, so that the RF circuit becomes simpler to realize. From this, it is easy to derive the IMD locking level specification. Knowing that $\beta$ is referred to as the modulator's modulation index, the carrier-to-IMD ratio is given by:

$$\frac{C}{2\mathrm{IMD}_{2\mathrm{TONE}}} = \frac{1}{\beta^2}\left(\frac{\beta^2 \phi_T}{2}\right)^2 = \left(\frac{\beta \phi_T}{2}\right)^2. \tag{16.62}$$

Note that if $\beta$ refers to the optical modulation index (OMI), then power refers to its square. For a standard CATV OMI of 3.6% and a desired two-tone second-order IMD of $-87$ dBc, it follows from Eq. (16.62) that $\phi_T = 0.15$ deg.

The input of the loop is the reference phase $\phi_{REF} = 0$. This is the desired phase of the modulator in order to null the CSO. The modulator output is the actual phase $\phi_T$. The sampled phase is converted into the second-order IMD signal given in Eq. (16.61), which produces the NL device $\beta^2/2$. This modulated quantity is converted into the optical amplitude by factor $\langle P \rangle$ and then sampled via a DIR with coupling coefficient $k$ to the RLL loop. The optical receiver responsivity $r$ transforms it into a current and the trans-impedance amplifier (TIA) converts this current plus noise into voltage. This low-frequency fluctuating error voltage passes through the loop filter $H(s)$ and is converted into a control voltage that locks the phase of the modulator, according to the conversion $\theta(s) = V(s)(\pi/V_\pi)$, where $s$ is the Laplace domain. The resultant error angle is the sum of $\theta$ and $\phi_B$, which is the bias drift phase error. At the instance $\theta + \phi_B$ equal to zero, the loop is locked to zero CSO. The loop feedback is given by

$$B = \frac{\beta^2}{2}\langle P \rangle kr. \tag{16.63}$$

Similar to the modeling of the AGC signal flow chart in Chapter 12, the parametric loop signal flow chart is given in Fig. 16.11.[10] The parametric loop gain is given by

$$A(s) = ZH(s)\frac{\pi}{V_\pi}, \tag{16.64}$$

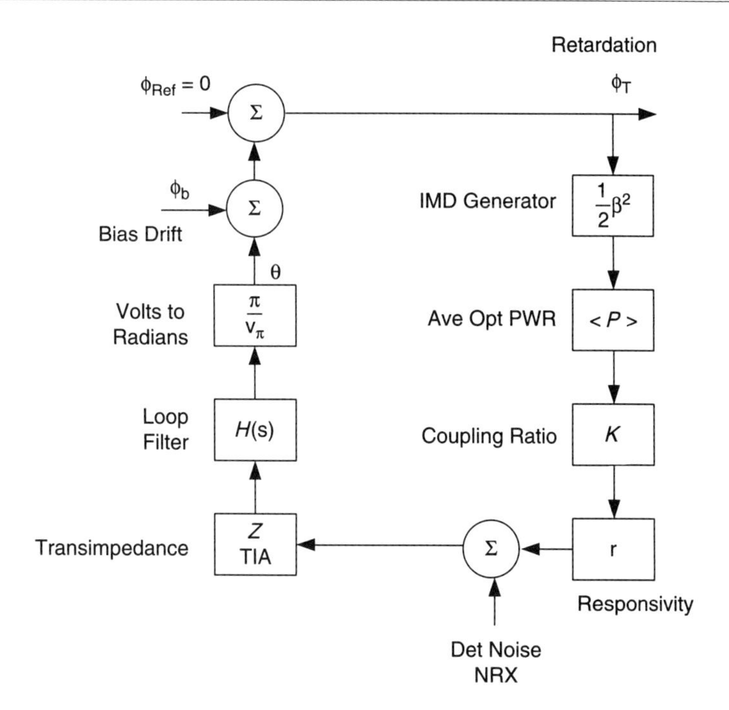


**Figure 16.11** Signal flow chart for RLL in order to nullify CSO.[10] (Reprinted with permission from the *Journal of Lightwave Technology* © IEEE, 1993.)

where $Z$ is the PIN TIA, $H(s)$ is the loop filter response and the last term is the volts-to-radians conversion factor. The closed loop transfer function is given by

$$F(s) = \frac{A(s)}{1 + BA(s)}. \tag{16.65}$$

Similarly a parametric loop can be applied on the predistortion linearizer, locking it to its optimal bias point and granting the desired coefficients for minimum CTB levels. It is possible to use the pilot tones for both the CSO and CTB parametric loops. This architecture is called an adaptive predistortion linearizer.[4,8,14]

The technique of predistortion is also adopted for ROF systems. These systems are mainly used in buildings and serve as transponders for cellular operators to create a distributed antenna system. Such a system is called a radio access point (RAP). These systems must be low cost; thus the laser is directly modulated. However, it is also a multichannel RF transport, which requires the linearization of both second and third order.[2,3,6] Due to the multichannel load, the ROF system exhibits similar multicarrier problems as in a CATV system,[5] affecting its dynamic range. The concept of such a predistorter is given in Fig. 16.12.

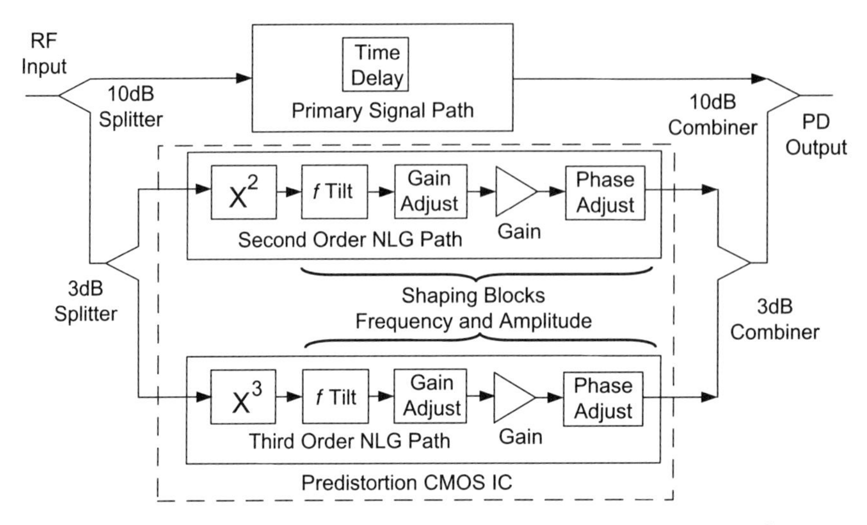


**Figure 16.12** Second-order and third-order predistortion system for ROF.[2] (Reprinted with permission from the *Journal of Lightwave Technology* © IEEE, 2003.)

## 16.2 Push–Pull

The push–pull concept is used for both receivers with a high-performance low-CSO level and for a driving and amplification chain in the CATV transmitter. It is a relatively simple, effective method to cancel even distortions generated by the RF amplifiers. This section provides an elaborate analysis and design approach. The push–pull concept is not a linearizer or predistortion, but a symmetric design that cancels out even distortions.

### 16.2.1 Distortions analysis

The role of the first-stage amplification block is to convert the RF photocurrent into a voltage with low noise and low NLD. The first stage can be, for instance, a BJT amplifier, GaAs MESFET, PHEMT amplifier, or an heterojunction bipolar transistor (HBT) monolithic microwave integrated circuit (MMIC). The first stage should be matched to provide a low-noise figure. Without a proper front end, matching the first stage would enhance the NLD from the multivideo channels.

The solution for the problem is by using differential sampling of the PD in a push–pull configuration. This solution requires identical amplifiers at each sampling arm. In order to have a highly linear performance, each amplifier should be biased in class A. Special-amplitude and phase-balancing networks are required to have equally matched amplitude and phase arms in order to cancel the distortions generated by the gain stages.[28]

This method does not cancel distortions that are generated by the PD. The configuration only cancels even harmonics (CSO) that are produced by the amplification chain. Figure 16.13 illustrates the PD differential sampling principle. Consider two arms in a push–pull configured receiver, as described in Fig. 16.13. Both arms have identical gain, frequency response (flatness), and phase response. The upper arm has a linear gain coefficient and phase response, marked as $A_1 x < \alpha_1$. The lower arm response is $A_1' x < \alpha_1'$, where $x$ is the input signal. In the same way higher-distortion orders are described: $A_2 x^2 < \alpha_2$, $A_2' x^2 < \alpha_2'$ is described for second order, and $A_3 x^3 < \alpha_3$, $A_3' x^3 < \alpha_3'$ for third order. It is important to remember that the unit of $A_1$ is a pure number, $A_2$ is 1/volt, and $A_3$ is $1/\text{volt}^2$. This is because the polynomial sum is in voltages, which contains the fundamental and its higher-distortions order. With that in mind, the next stage is to evaluate suppression of the second-order harmonics, while applying a single tone represented as phasor sampled from both sides of the PD.

The input signal sampled from the upper arm is marked as $V_1 = E \cdot e^{j\omega t + \phi} \cdot e^{j\pi}$. The input signal sampled from the upper arm is marked as $V_2 = E \cdot e^{j\omega t + \phi}$. Both voltages and currents are equal in amplitude but opposite in phase. The output voltage from the upper arm and the lower arm is given by the following set of equations:

$$\begin{cases} V_{\text{OUT\_UP\_ARM}} = A_1 E \cdot e^{j\omega t+\phi} e^{j\pi} e^{j\alpha_1} \\ \qquad + A_2 e^{j\alpha_2} \left(E \cdot e^{j\omega t+\phi} e^{j\pi}\right)^2 + A_3 e^{j\alpha_3} \left(E \cdot e^{j\omega t+\phi} e^{j\pi}\right)^3, \\ V_{\text{OUT\_LOW\_ARM}} = A_1' E \cdot e^{j\omega t+\phi} e^{j\alpha_1'} \\ \qquad + A_2' e^{j\alpha_2'} \left(E \cdot e^{j\omega t+\phi}\right)^2 + A_3' e^{j\alpha_3'} \left(E \cdot e^{j\omega t+\phi}\right)^3. \end{cases} \tag{16.66}$$

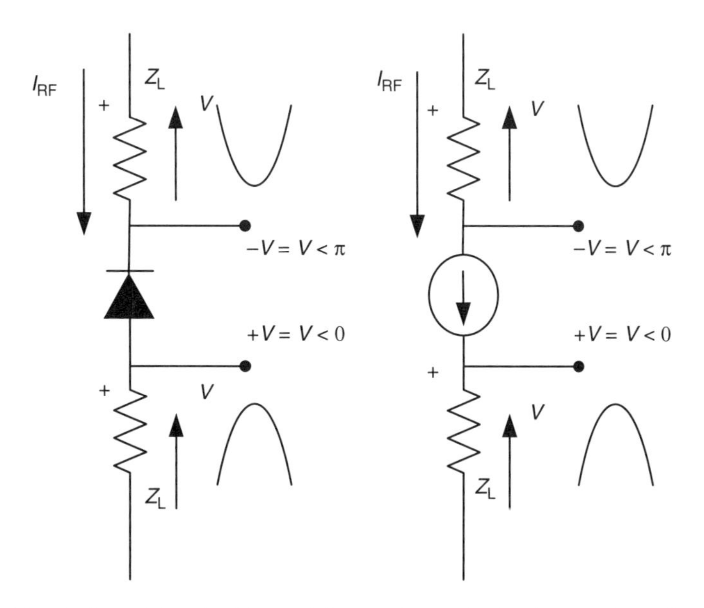


**Figure 16.13** Principle of differential sampling of a PD for push–pull receiver.

The output combiner balanced to unbalanced (BALUN) transformer sums both arms out of phase. Hence the lower arm has an additional 180 deg. The output voltage is given by

$$V_{OUT} = V_{OUT\_UP\_ARM} + V_{OUT\_LOW\_ARM} \cdot e^{j\pi}. \tag{16.67}$$

Replacing the variables with the actual phasors shows that the push–pull configuration cancels even harmonics (note these are not the two-tone discrete second-order DSO). Hence, even distortions such as CSOs are nullified; DSOs are cancelled for the same reasons:

$$\begin{aligned} V_{OUT} &= A_1 E \cdot e^{j\omega t+\phi} e^{j\pi} e^{j\alpha_1} + A_1' E \cdot e^{j\omega t+\phi} e^{j\alpha_1'} \cdot e^{j\pi} \\ &\quad + A_2 e^{j\alpha_2}\left(E \cdot e^{j\omega t+\phi} e^{j\pi}\right)^2 + A_2' e^{j\alpha_2'}\left(E \cdot e^{j\omega t+\phi}\right)^2 \cdot e^{j\pi} \\ &\quad + A_3 e^{j\alpha_3}\left(E \cdot e^{j\omega t+\phi} e^{j\pi}\right)^3 + A_3' e^{j\alpha_3'}\left(E \cdot e^{j\omega t+\phi}\right)^3 \cdot e^{j\pi} \\ &= A_1 E \cdot e^{j\omega t+\phi} e^{j\pi} e^{j\alpha_1} + A_1' E \cdot e^{j\omega t+\phi} e^{j\alpha_1'} \cdot e^{j\pi} \\ &\quad + A_3 e^{j\alpha_3}\left(E \cdot e^{j\omega t+\phi} e^{j\pi}\right)^3 + A_3' e^{j\alpha_3'}\left(E \cdot e^{j\omega t+\phi}\right)^3 \cdot e^{j\pi}. \end{aligned} \tag{16.68}$$

From Eq. (16.68), it can be observed that the push–pull configuration suppresses even modes. Third-order IMD is an odd harmonic signal, by which “push-pull” does not cancel odd NLD products, as demonstrated by Eq. (16.68). The push–pull method is useful for improving the linearity of the RF chain since any distortion created by the PD is a valid RF signal passing through the RF amplification chain of the receiver. Only distortions that are created by the RF amplification chain are cancelled by the “push–pull” topology.

Furthermore, the suppression of the second-order IMD and harmonics is directly dependent on the balance between the sampled amplitudes at both sides of the detector, as well as the gain balance between stages and AM-to-AM coefficients $A_1$, $A_2$, $A_3$, versus $A_1'$, $A_2'$, $A_3'$. Additionally, phase alignment between the branches is essential for efficient 2nd-order IMD suppression. The 2nd-order harmonic IMD suppression relates to the expression in Eqs. (16.69) and (16.70), describing the equivalent of the $X$ and $Y$ distortion projections versus the fundamental. This expression represents second-order two-tones IMD known as DSOs or second-order harmonics and their voltage projections are described by

$$\text{IMD[dBm]} = 10\log\left(\frac{E_{IMD\_X}^2 + E_{IMD\_Y}^2}{2Z_0}\right) + 30. \tag{16.69}$$

$$\text{CSO[dB]} = 10\log\frac{P_{OUT}}{P_{IMD}} = 10\log\frac{E_{\_X}^2 + E_{\_Y}^2}{E_{IMD\_X}^2 + E_{IMD\_Y}^2}. \tag{16.70}$$

also see Figs. 16.14 and 16.15.

If we define distortions’ amplitude balance between the branches as $\delta$ and phase balance between the branches as $\psi$, then the following notation is applicable; note

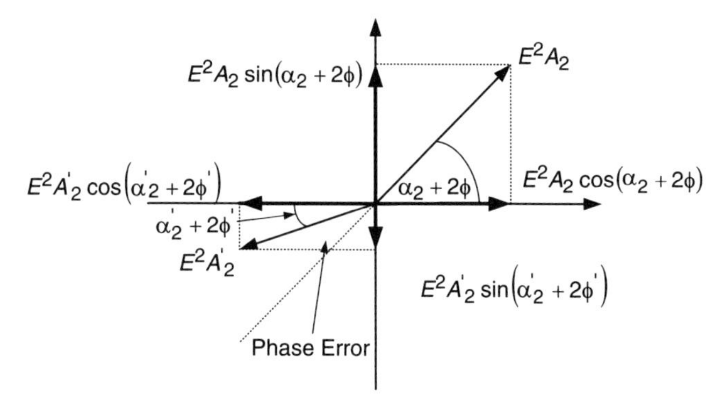


**Figure 16.14** Second-order distortion voltages.

that $\phi$, the phase of the 2nd harmonic or IMD tone, is twice as much compared to the fundamental frequency:

$$\delta_{\mathrm{IMD}} = \frac{A'_2}{A_2}, \tag{16.71}$$

$$\psi_{\mathrm{IMD}} = (2\phi + \alpha_2) - (2\phi' + \alpha'_2). \tag{16.72}$$

In the same way, phase and amplitude balance are defined for the fundamental frequency:

$$\delta = \frac{A'_1}{A_1}, \tag{16.73}$$

$$\psi = (\phi + \alpha_1) - (\phi' + \alpha'_1). \tag{16.74}$$

Calculation of the equivalent vectors for the fundamental frequency and second harmonic voltages is done as per Pythagorean identity. Referring to the equivalent

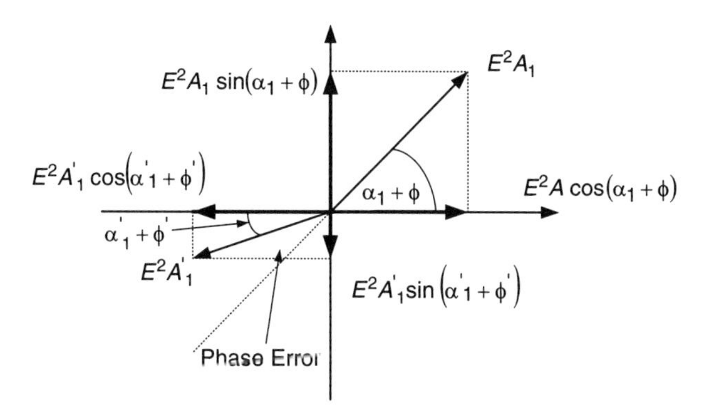


**Figure 16.15** Fundamental signal voltages.

voltage square to the power, where $Z_0$ is the characteristic impedance, results in the following relations for the fundamental frequency in Eq. (16.75) and IMD second harmonic in Eq. (16.76):

$$P_{\text{OUT}} = \frac{E^2 A_1^2}{2 Z_0}\left(1 + \delta^2 + 2\delta \cos \psi\right), \tag{16.75}$$

$$P_{\text{IMD}} = \frac{E^4 A_2^2}{4 Z_0}\left(1 + \delta_{\text{IMD}}^2 - 2\delta_{\text{IMD}} \cos \psi_{\text{IMD}}\right). \tag{16.76}$$

The last two equations show that when both conditions of phase balance and amplitude balance are optimized, the second-order harmonic distortions are zero and the output power is at its maximum and equals twice the power of a single arm or twice the voltage square.

There are some configurations that use push–pull for the first-stage low-noise amplifier (LNA) only. However, by using a topology of two identical arms, including the AGC and the output stage, the CSO level is lowered and the second-order CSO dynamic range is increased. This is important when designing a fiber to the curb where the output stage should drive a long lossy coax and splitters because the receiver's output power should be high enough to provide a high signal with minimum IMD and decent CNR. Hence, the PA dynamic range is increased as well. The push–pull configuration typically reduces second-order distortions by 20 dB. The disadvantage is size, power consumption, complexity, and cost. In a digital design, which has a large signal, a differential approach has better immunity against common-mode interference. That kind of interference is cancelled in video push–pull as well. A traditional push–pull design uses an input-matching transformer and output BALUN that has a wideband response for CATV applications.[26,27]

## 16.2.2 Numerical example

Assume the following design problem of a single gain stage in a push–pull configuration. The active device would be a WJ FP1189 PHEMT in a configuration of drain-source parallel feedback. The biasing conditions are 50% $I_{\text{dss}}$, $V_{\text{ds}} = 6$ V. This device has a 27-dBm compression point and 2-dB NF. The goal is to calculate and predict the second-order harmonic distortion levels below the carrier as a function of phase and amplitude balancing between the two arms of the push–pull receiver. The receiver structure is given in Fig. 16.16.

At first approximation, to simplify the calculation and evaluation of the device, only an AM-to-AM test is done to characterize the device. Additionally, it is assumed that the only phase imbalance between the two arms is due to the arm phase response differences. It is assumed that due to the high-compression-point specification of the device, the AM-to-PM phase changes are very small at this operating point. Thus, the differences between the phase response of the second-order term of the device are equal to those of the first-order term and are absorbed

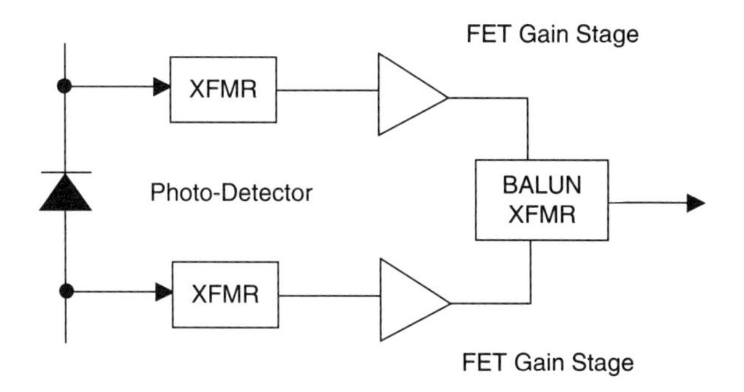


**Figure 16.16** High dynamic range push - pull single-stage optical receiver.

within the arm phase differences. Note that second-order phase delay is multiplied by a factor of two. We shall assume that the phasor initial sampling phase $\phi = 0$. In addition, it is assumed that the amplitude misbalancing between the two arms for the first-order terms is equal to the second-order terms, and there is no frequency dependency. Because of these approximations, the following relations are applicable for predicting the envelope of performance for the receiver:

$$P_{\mathrm{OUT}} = \frac{E^2 A_1^2}{2 Z_0} \left(1 + \delta^2 + 2\delta \cos \psi\right), \tag{16.77}$$

$$P_{\mathrm{IMD}} = \frac{E^4 A_2^2}{4 Z_0} \left(1 + \delta_{\mathrm{IMD}}^2 - 2\delta_{\mathrm{IMD}} \cos \psi_{\mathrm{IMD}}\right), \tag{16.78}$$

$$\psi = (\phi + \alpha_1) - \left(\phi' + \alpha_1'\right) = \alpha_1 - \alpha_1', \tag{16.79}$$

$$\psi_{\mathrm{IMD}} = (2\phi + \alpha_2) - \left(2\phi' + \alpha_2'\right) = \alpha_2 - \alpha_2' = 2\left(\alpha_1 - \alpha_1'\right). \tag{16.80}$$

Bear in mind that the measured AM-to-AM polynomial is a power approximation curve and not a voltage-curve approximation; hence $a_n = A_n^2$ and $x^n = E^{2n}$. The AM-to-AM approximation was taken at 550 MHz and results in the following polynomial coefficients,[25] using MathCAD for $a_n$:

$$P_{\mathrm{out}} = \sum_{n=1}^{7} a_n x^n. \tag{16.81}$$

$$a_n = \begin{pmatrix} 20.156 \\ 785.08 \\ 1.874 \times 10^4 \\ -3.024 \times 10^7 \\ 2.325 \times 10^9 \\ -5.095 \times 10^{10} \\ -3.177 \times 10^9 \end{pmatrix}. \tag{16.82}$$

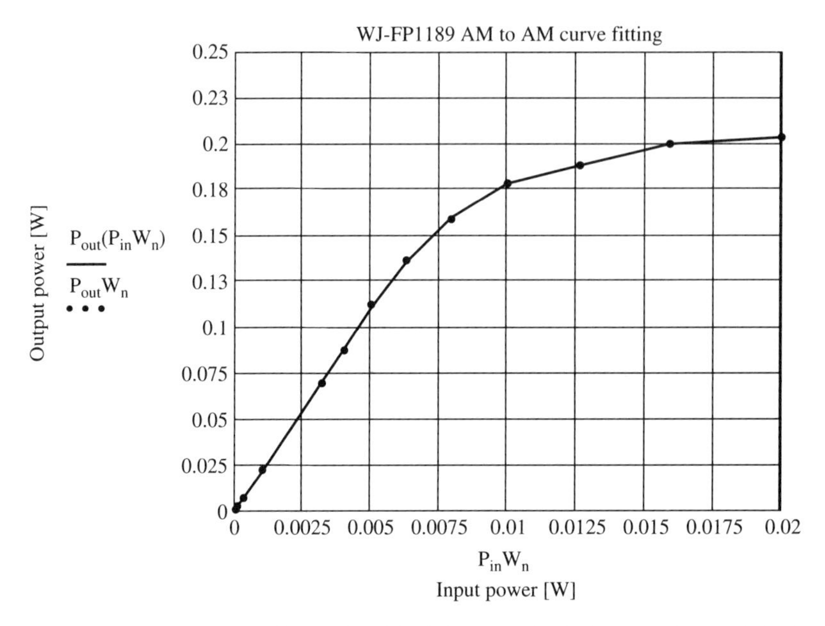


**Figure 16.17** AM-to-AM curve fitting for WJ-FP1189 at 50% $I_{dss}$ and $V_{ds} = 6$ V. The dots are measured results; solid line is the approximated AM-to-AM curve.

Figure 16.17 shows the curve fitting between the measured AM-to-AM results and approximated polynomial for the WJ-FP1189.

The next step is to evaluate the RF loading of the receiver for a given PD with a responsivity of 1 mA/mW. The standard OMI per channel plan is 3.5% OMI for 79 channels between 50 and 550 MHz and 1.75% for the upper 31 channels between 550 and 870 MHz. Using the Eq. (8.51) for the equivalent OMI at full channel loading, the equivalent modulation index is given by:

$$\mathrm{OMI}_{\mathrm{eq}} = \mathrm{OMI}_{\text{ch-A}} \cdot \sqrt{N + k^2(M-N)} \Rightarrow \mathrm{OMI}_{\mathrm{eq}} = 3.5 \cdot \sqrt{79 + \left(\frac{1}{2}\right)^2 (110-79)} \Rightarrow \mathrm{OMI}_{\mathrm{eq}} = 32.6\%. \tag{16.83}$$

Using the expression for RF power given in Eq. (8.47) and using a 2:1 RF front end (RFFE) matching transformer, the equivalent input RF power at optical power of 1 dBm is given by

$$P_{\mathrm{RF\_tot}} = P_{\mathrm{RF\_eq}} = \mathrm{OMI}_{\mathrm{eq}}^2 \left(\frac{P_{\mathrm{OPT}} \cdot r}{\sqrt{2}}\right)^2 \cdot Z_{\mathrm{L}} \Rightarrow P_{\mathrm{RF\_tot}} = 0.326_{\mathrm{eq}}^2 \left(\frac{10^{0.1-3} \cdot 1}{\sqrt{2}}\right)^2 \cdot 4 \cdot 50 = 1.684 \cdot 10^{-5}\,\mathrm{W}. \tag{16.84}$$

Hence the input power at maximum load is −17.7 dBm or 30.96 dBmV at the input to each arm.

The second-order suppression versus amplitude imbalance, assuming a perfect PD without any distortions, is given by Fig. 16.18.

It can be observed that phase alignment between the two arms of the receiver is critical while the amplitude imbalance is below 0.5 dB. At that region, the phase misbalancing is dominated. Above 1.5-dB amplitude imbalance, the asymptotic performance converges to 40 dBc. It is important to remember that distortions generated by the PD are not cancelled out by the receiver and are added to the distortions generated by the receiver. Hence the second-order performance is limited by the PD and by the receiver's arms' alignment.

From Fig. 16.19, it can be seen that phase imbalance is dominant and can improve performance up to 0.5 dB amplitude imbalance. At the amplitude imbalance of 1.5 dB between the two arms of the receiver, the second-order suppression is limited to 50 dBc and is not affected by the phase alignment. Hence, it is recommended as a design goal to have a phase error less than 2 deg and amplitude

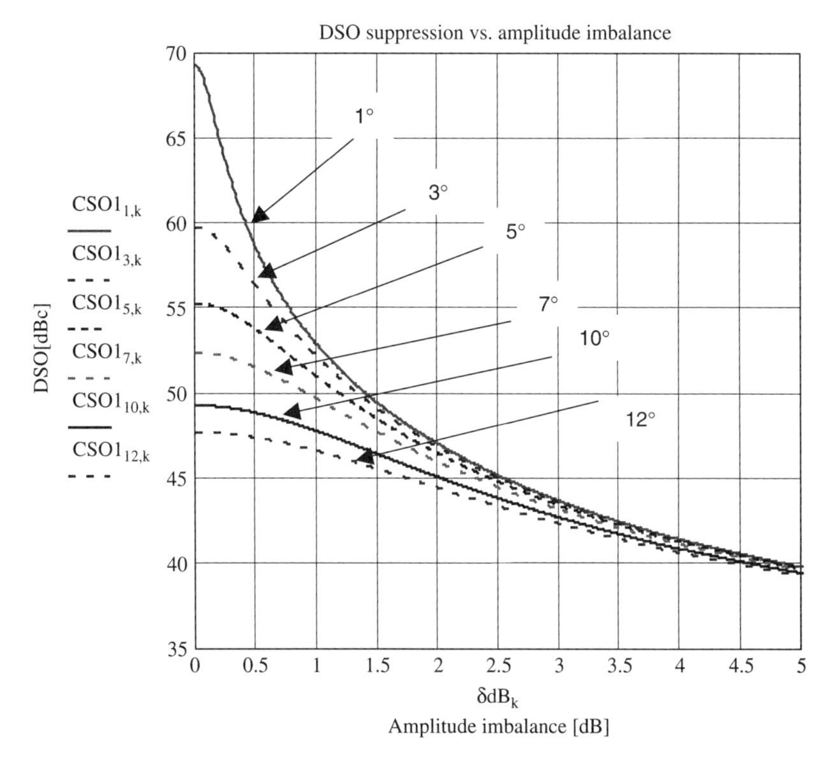


**Figure 16.18** Second-order harmonic interference suppression as a function of amplitude imbalance while phase imbalance is a parameter. Input-power conditions are 79 channels with 3.5% OMI and 31 channels with 1.75% OMI. The optical power level is 1 dBm; PD responsivity is $r = 1$ mA/mW; input-transformer-turn ratio is 2:1, and amplifier-input impedance is 50 Ω. FET device WJ-FP1189 at 50% $I_{dss}$ and $V_{ds} = 6$ V dc operating point and 23-dBm compression point.

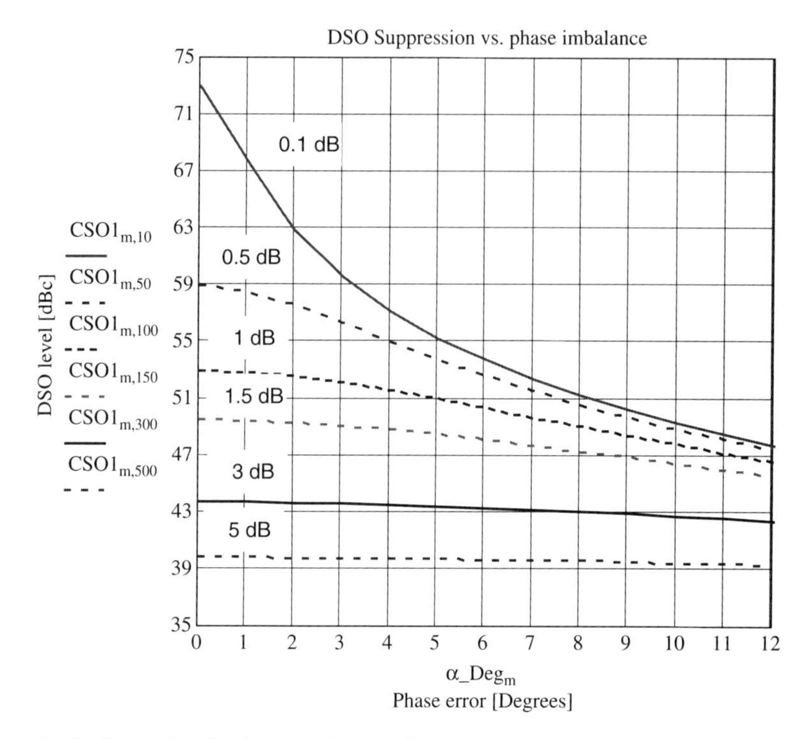


**Figure 16.19** Second-order harmonic interference suppression as a function of phase imbalance while amplitude imbalance is a parameter. Input power conditions are 79 channels with 3.5% OMI and 31 channels with 1.75% OMI. The optical power level is 1 dBm; PD responsivity is $r = 1$ mA/mW; input-transformer-turn ratio is 2:1, and amplifier-input impedance is 50 Ω. FET device WJ-FP1189 at 50% $I_{dss}$ and $V_{ds} = 6$ V dc operating point and 23-dBm compression point.

error less than 0.5 dB. Selecting a higher compression point device or biasing the FET for higher $I_{ds}$ current would improve the CSO performance at higher optical level inputs. In AGC receivers, as discussed in Chapter 12, it is important to select the first stage LNA device for a low NF in one hand and for a high compression point on the other. The compression point AM-to-AM curve of the device defines the receiver's performance for CSO and CTB. In AGC receivers, the gain stage placed after the power leveling attenuator exhibits a constant power level. At high optical levels where the AGC reaches its highest attenuations, it is the input LNA stage that receives the highest RF levels at its input port. In conclusion, it is a good engineering practice to have a high dynamic range input LNA, even in a push–pull configuration where high gain and high power are required.

### 16.2.3 Push–pull structure

As was explained above, it is important to have in a push–pull receiver, design features for tuning and making phase and amplitude alignments in order to have acceptable performance for CSO suppression. Phase alignment can be done for each arm while observing the spectrum analyzer two-tone test for the second order. This method of tuning each arm is not effective and requires a lot of touch time to tune. It can be used as a fine tune, but not as a fast main tune to converge performance. One of the ways to overcome this is by using a differential phase and amplitude alignment as shown in Fig. 16.20. The idea is to simultaneously tune both arms for phase and amplitude alignment in reference to a common virtual ground for the phase and a common ground for the amplitude. Hence, the alignment tuning networks are composed from a resistive network for amplitude alignment and a reactive network for phase adjustments. The phase adjustment network generally is a trimmer capacitor, and has a low-pass response. Hence it can be considered a phase equalizer. The amplitude alignment network is a potentiometer.

In a short RF lineup, where a single stage RF amplifier is used, the phase and amplitude alignment networks are less critical. However, for a high-gain receiver with AGC, the tolerances of the components and layout transmission lines add up, increasing the overall phase and amplitude error. Hence there is a need to use phase and amplitude compensation networks as shown above. It is important to place the trimming components before the gain stages in order to align the phase and the amplitude of the fundamental frequency from which the CSO is produced. A post tune network at the receiver output would not improve the receiver performance and would not cancel out the second-order distortions with the same efficiency as input networks. The alignment process should be done in several

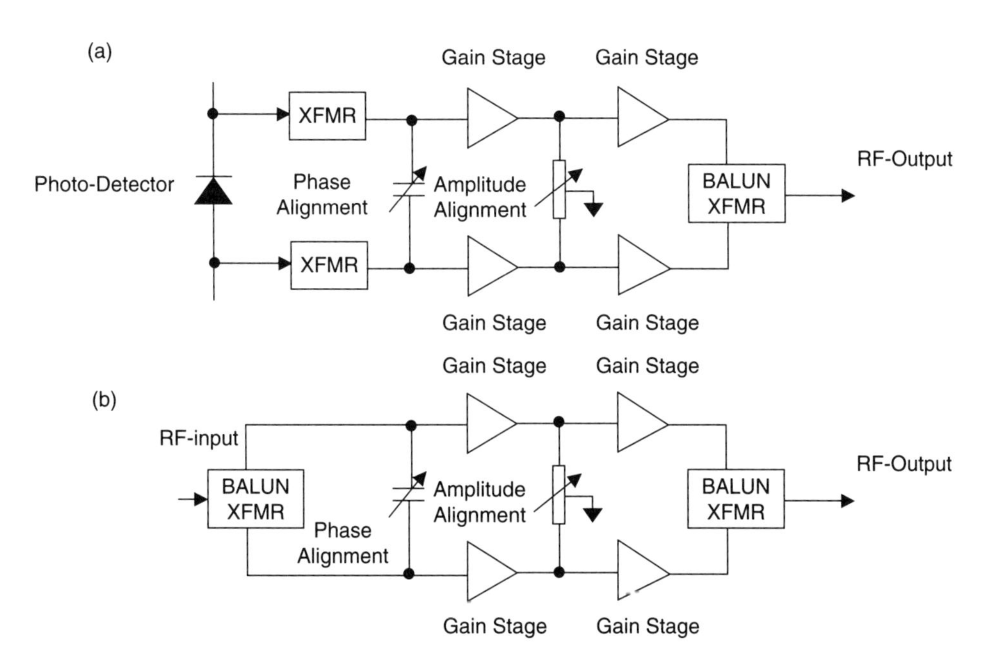


**Figure 16.20** (a) Phase and amplitude balancing in a push-pull CATV receiver. (b) Booster drive in CATV transmitter.

frequencies since the passive components have different responses at different frequencies, and due to the fact that the active components have different AM-to-AM and AM-to-PM response curves at different frequencies. The solution is that a compromise value is set as a minimum acceptable standard for all frequencies. Performance is evaluated by a two-tone two-lasers test for DSO, or by a multitonetest using a high-quality predistorted laser transmitter for the CATV standard.

## 16.3 Optical Linearization Methods in CATV Transmitters

In the previous section, linearization methods reviewed were implemented on the supporting electronics around the optical modulator, laser diode, and PD. However, linearization is feasible in optics as well. Optical linearization can be divided into four categories: FF, which is similar to RF FF, cascaded linearized modulator, dual parallel and feedback. These three methods are implemented on systems with external modulation. The last method is used for low-cost systems with direct modulated lasers and involves feedback from a PD, which feeds the laser with IMDs in a phase opposite to those created in the laser.

### 16.3.1 Feed-forward (FF) linearization

The optical concept of FF linearization is similar to RF FF explained in Sec. 16.1.1. Figure 16.21 shows an optical FF concept.[10] The RF signal is

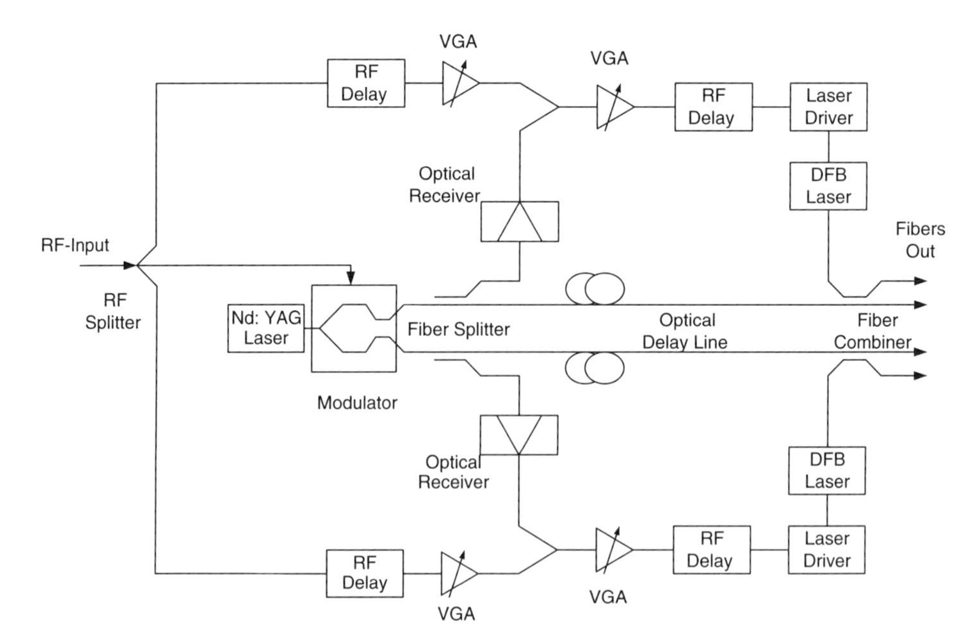


**Figure 16.21** Optical FF linearization block diagram.[10] (Reprinted with permission from *Journal of Lightwave Technology* © IEEE, 1993.)

split into three branches prior to modulation. The main route feeds the balanced bridge MZI. The other two RF signals pass through the RF delay to compensate for the modulator group delay. Then the signal passes through a VGA. The first loop is closed similarly, as in Sec. 16.1.1, with the difference that the optical signal is sampled by an optical coupler and then detected by a PIN photodiode. Both the FF RF signal and sampled signal, which is detected by the PIN PD, are summed by the RF combiner. The RF combiner can be a wideband transformer that subtracts the fundamental. The remaining second-order and third-order signals coming out from the primary loop feeds the secondary loop. The secondary loop consists of a VGA, which is an inverting amplifier, and an additional RF delay to compensate for the auxiliary laser and laser-driver amplifier response. Thus, both CSOs and CTBs are out of phase with respect to the main laser's IMD products. The output of the laser driver modulates a DFB laser with a similar wavelength as that of the Nd-YAG main laser. However the auxiliary laser's wavelength must be detuned to prevent coherent interference. The optical path coming out of the modulator has a fiber delay line for signal alignment, which is required for compensating the single-mode fiber's (SMF's) chromatic dispersion. This requirement is because of the wavelength differences between the main and auxiliary lasers. The second loop is closed by an optical combiner with a power ratio of 20/80%. The auxiliary laser does not have to be a high-power laser since it only has to drive IMD products, which are $-40$ dBc. Additional parametric locking of the MZI can be done as well to further cancel the CSO products. The main drawback of such a design is the fiber delay, which is related to the deployment link length, as well as the ripple and phase frequency response of the RF components.

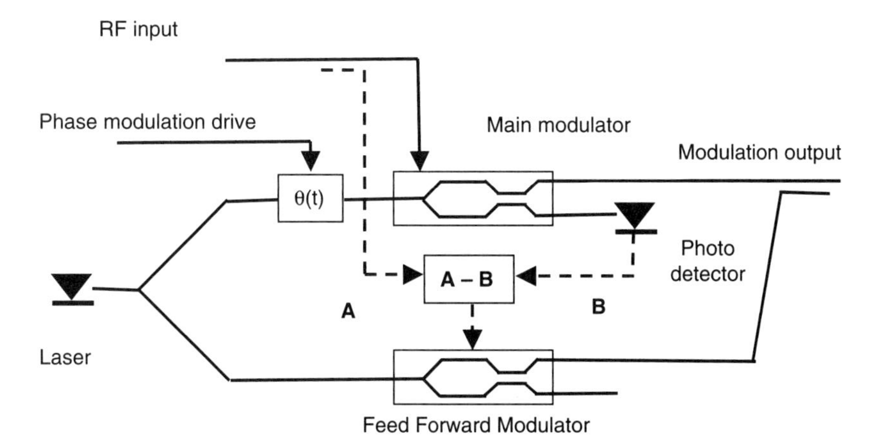


**Figure 16.22** Novel concept of optical FF linearization block diagram. This method uses a phase modulator to achieve stable quasi-coherent addition at the output coupler, thus eliminating phase control and noise problems.[24] (Reprinted with permission from *Proceedings of OFC* (© OSA, 1996.)

An improved version of FF using a phase modulator for achieving a quasi-coherent addition between the main MZI modulator and the FF loop modulator is described in Fig. 16.22, per Ref. [24].

### 16.3.2 Dual parallel optical linearization

The concept of dual parallel linearization was presented in Refs. [20] and [23]. The functional representation of the optical circuit is illustrated in Fig. 16.23. Two modulators designated $M_A$ and $M_B$ are fed from a single optical source. It is assumed that the modulators produce pure AM without any additional chirp[29] that results in phase modulation due to the refractive index change of the modulator. The optical power-splitting ratio of the taps is $T_1$ and $T_2$, which are design parameters. In addition, an optical phase shift is introduced between the two modulators by the PS element. The role of the PS block is to maintain the lower modulator, $M_B$, with $\pi$ radians with respect to the upper modulator $M_A$, so that the output power is as described by Eq. (7.95) in Chapter 7. It is also assumed that both modulators are locked in to the optimal bias point Q, and thus the second-order IMDs are minimal. Additionally, the RF power-splitting taps that feed the modulator are asymmetrically marked as $1{:}\alpha^2$, or for RF voltage the split ratio is $1{:}\alpha$, where RF is coming from the same RF source for both modulators. Since the optical power at each modulator is not at the same value because the optical splitting ratio is $T_1$ and $T_2$, in favor for the upper modulator, $M_A$, the lower modulator, $M_B$, operates at a higher OMI and produces more distortions. By providing higher optical power to the primary modulator, the distortions created by the lower modulator may cancel the upper modulator distortions. The optical power

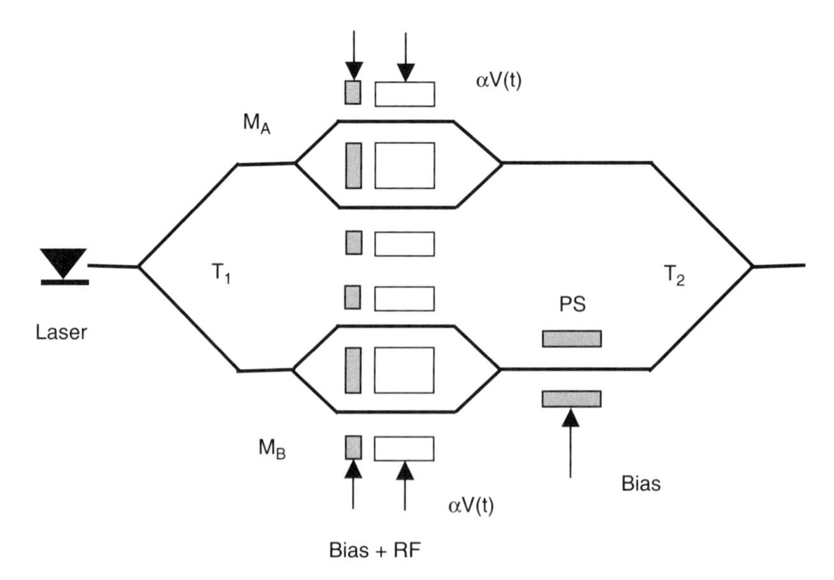


**Figure 16.23** The dual parallel linearization concept.[23] (Reprinted with permission from the *Journal of Lightwave Technology* © IEEE, 1993.)

incident on the upper modulator is marked as $P_{PR\text{-}in}$ primary, and the lower as $P_{SR\text{-}in}$ secondary. As a result, using Eq. (7.95) and the Taylor series expansion, the upper modulator optical power $P_{PR\text{-}out}$ and the lower $P_{SR\text{-}out}$ modulator are given by[23]

$$P_{PR\text{-}out}(t) = \frac{P_{PR\text{-}in}}{2}\left\{1 - \sin\left[\frac{\pi V_{RF}(t)}{V_\pi}\right]\right\}$$
$$\approx \frac{P_{PR\text{-}in}}{2}\left\{1 - \left[\frac{\pi V_{RF}(t)}{V_\pi}\right] + \frac{1}{6}\left[\frac{\pi V_{RF}(t)}{V_\pi}\right]^3\right\}, \qquad (16.85)$$

$$P_{SR\text{-}out}(t) = \frac{P_{SR\text{-}in}}{2}\left\{1 + \sin\left[\frac{\alpha\pi V_{RF}(t)}{V_\pi}\right]\right\}$$
$$\approx \frac{P_{SR\text{-}in}}{2}\left\{1 + \left[\frac{\alpha\pi V_{RF}(t)}{V_\pi}\right] - \frac{1}{6}\left[\frac{\alpha\pi V_{RF}(t)}{V_\pi}\right]^3\right\}, \qquad (16.86)$$

where $V_{RF}(t)$ is the RF voltage at the modulator electrodes. Note again that the RF power and voltage split between the two modulators is $1{:}\alpha^2$ and $1{:}\alpha$ respectively, and the optical power is not evenly split, so that the lower modulator operates with higher OMI and distortions. If the optical sources are incoherent, the combined output from the modulators is given by the sum of Eqs. (16.85) and (16.86):

$$P_{tot}(t) \approx \frac{1}{2}\left\{P_{PR\text{-}in} + P_{SR\text{-}in} - \frac{\pi V(t)}{V_\pi}(P_{PR\text{-}in} - \alpha P_{SR\text{-}in}) + \frac{1}{6}\left[\frac{\pi V(t)}{V_\pi}\right]^3\left(P_{PR\text{-}in} - \alpha^3 P_{SR\text{-}in}\right)\right\}. \qquad (16.87)$$

In conclusion, in order to cancel the third order, the following condition must be satisfied:

$$\frac{P_{PR-in}}{P_{SR-in}} = \alpha^3. \qquad (16.88)$$

Taking the result of Eq. (16.88) and applying to Eq. (16.87) provides the optimum conditions with respect to the primary modulator. The modulated optical power and the optical dc power are provided by

$$P_{tot\text{-}ac} = \frac{\pi|V(t)|}{V_\pi}\langle P_{PR\text{-}in}\rangle\left[1 - \frac{1}{\alpha^2}\right], \qquad (16.89)$$

$$P_{tot\text{-}dc} = \langle P_{PR\text{-}in}\rangle\left[1 + \frac{1}{\alpha^3}\right]. \qquad (16.90)$$

where the symbol $\langle P\rangle = P/2$ refers to the optical average power.

The OMI after linearization is the ratio of the modulated optical signal to the dc optical power, as was defined in Chapter 6. Thus the OMI is given by the ratio of Eq. (16.89) to Eq. (16.90):

$$\mathrm{OMI} = \frac{1-(1/\alpha^2)}{1-(1/\alpha^3)} = \frac{\alpha(\alpha-1)}{1-\alpha+\alpha^2}. \tag{16.91}$$

If we assume imperfect optical phase alignment between the primary and secondary modulators, with a phase error value of $\phi$, then by using phasor calculation and assuming an optical power ratio of $\alpha^3$, an expression for the fundamental, second harmonic, and third tones are given by

$$\hat{P}_{\mathrm{F}} = \frac{1-\alpha^2}{2\alpha^3} + \frac{\alpha-1}{2\sqrt{\alpha^3}}\cos\phi, \tag{16.92}$$

$$\hat{P}_{2\mathrm{F}} = \left[\frac{1+\alpha}{\alpha}\right]^2 + \frac{1}{4\sqrt{\alpha^3}}\cos\phi, \tag{16.93}$$

$$\hat{P}_{3\mathrm{F}} = \left[\frac{\alpha-1}{\alpha}\right]^3 + \frac{1}{12\sqrt{\alpha^3}}\cos\phi. \tag{16.94}$$

Note that the power is normalized to the primary path marked by^.

For the second harmonic to be −65 dBc with $\alpha = 3.3$, the relative phase tolerance is 89.94 deg $< \phi <$ 90.06 deg. For the second harmonic to be −70 dBc, the relative phase tolerance is 89.46 deg $< \phi <$ 90.54 deg; these are extremely tight tolerances to achieve. Hence, this method is better for the incoherent approach than for the coherent one. This method resembles the FF concept in the sense of creating out-of-phase distortions and using forward combining to cancel them out.

### 16.3.3 Dual-cascade optical linearization

In the previous section, the dual parallel linearization method was introduced. As was briefly explained, this technique is similar in its concept to the FF method. A different approach was introduced by having cascaded modulators in order to meet the CATV CSO and CTB requirements,[11,18,19,21] as shown in Fig. 16.24. An additional method was introduced as a combination of the dual parallel and the cascaded linearization techniques,[22] as shown in Fig. 16.25.

Referring to Fig. 16.24, the MZI modulator after the $Y$ branch splitter is described by its electric field $\vec{E}$ excitation data of phase and amplitude. These phasor values of $\vec{E}_1$ and $\vec{E}_2$ feed the MZI upper path and lower path:

$$\vec{E}_{\mathrm{IN}} = \begin{bmatrix} \vec{E}_1 \\ \vec{E}_2 \end{bmatrix}. \tag{16.95}$$

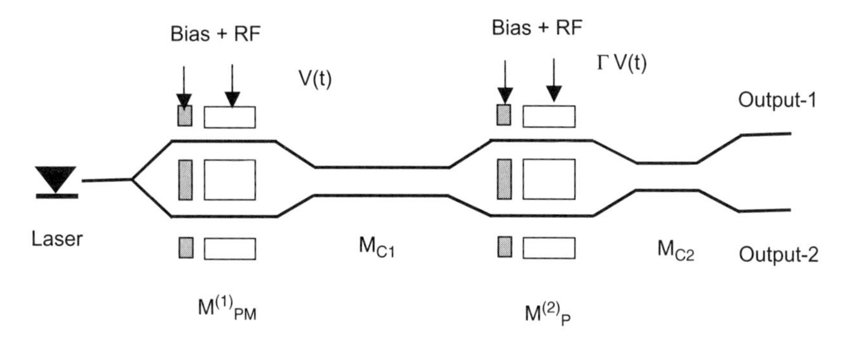


**Figure 16.24** Schematic representation of cascaded linearized modulators.[19] (Reprinted with permission from *Photonics Technology Letters* © IEEE, 1995.)

The cascaded modulators are described by the first and the second modulator transfer function matrixes:

$$M_{\mathrm{pm}}^{1}=\begin{bmatrix}\exp\left(\dfrac{j\pi V_{\mathrm{RF}}}{2V\pi}\right) & 0\\ 0 & \exp\left(-\dfrac{j\pi V_{\mathrm{RF}}}{2V\pi}\right)\end{bmatrix}, \tag{16.96}$$

$$M_{\mathrm{pm}}^{2}=\begin{bmatrix}\exp\left(\dfrac{j\pi\Gamma V_{\mathrm{RF}}}{2V\pi}\right) & 0\\ 0 & \exp\left(-\dfrac{j\pi\Gamma V_{\mathrm{RF}}}{2V\pi}\right)\end{bmatrix}. \tag{16.97}$$

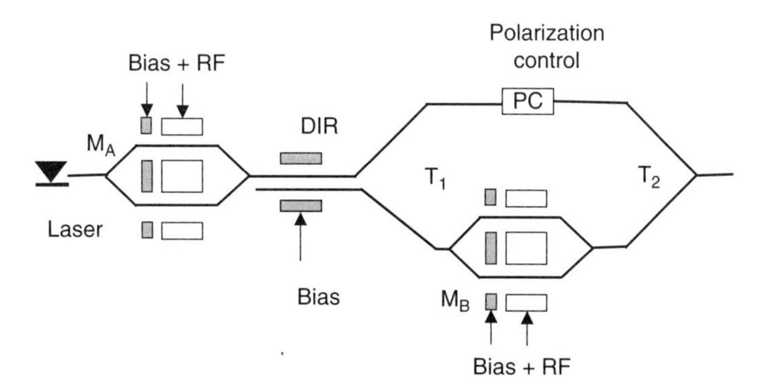


**Figure 16.25** A schematic representation of cascaded dual MZI linearized modulators. The primary modulator, $M_A$, feeds a dc bias-controlled DIR that controls the power split ratio to the auxiliary coupler $M_B$. Compensation is made by a delay block marked as PC. 6This method combines both dual parallel and dual series linearization.[22] (Reprinted with permission from the *Journal of Lightwave Technology* © IEEE, 1992.)

The parameter $\Gamma$ is the amplitude coupling control of the second modulator's modulating voltage $V_{RF}$, which is coupled from the first modulator. These two dual-output modulators are cascaded, while having in between and at the output, an optical DIR. This optical DIR has transfer function $M_C$, given by Eq. (16.98), and coupling angle $\gamma = kL$, where $L$ is the coupling length and $k$ is the coupling coefficient between the two waveguides:

$$M_C = \begin{bmatrix} \cos\gamma & -j\sin\gamma \\ -j\sin\gamma & \cos\gamma \end{bmatrix}. \tag{16.98}$$

Since both couplers are not identical, for the sake of general tuning, the overall transfer function and electric field phasor outputs are given by the matrixes' multiplication:

$$\vec{E}_{Out} = \begin{bmatrix} \vec{E}_{out\text{-}1} \\ \vec{E}_{out\text{-}2} \end{bmatrix} = M_{C2} M_{pm}^{(2)} M_{C1} M_{pm}^{(1)} \begin{bmatrix} 1/\sqrt{2} \\ 1/\sqrt{2} \end{bmatrix}. \tag{16.99}$$

The reason for the $1/\sqrt{2}$ is for equal splitting of the electric field whose amplitude is normalized to unity. The output field magnitude is given by its power density equivalent:

$$P = \frac{1}{2}\begin{bmatrix} |E_{out\text{-}1}|^2 \\ |E_{out\text{-}2}|^2 \end{bmatrix}. \tag{16.100}$$

Thus the polar presentation of the output $E$ field phasors is given by

$$\vec{P} = \frac{1}{\sqrt{2}}\begin{bmatrix} |E_{out\text{-}1}|\exp(j\phi_{(1)}) \\ |E_{out\text{-}2}|\exp(j\phi_{(2)}) \end{bmatrix}, \tag{16.101}$$

where the indices (1) and (2) refer to the output port, and $\phi$ corresponds to the port's output phases. From Eq. (16.99) and using phasor algebra, we get the following result for the upper branch:

$$\begin{aligned} F(V_{RF}) &= \left|\frac{E_{out\text{-}1}}{E_{in\text{-}1}}\right|^2 \\ &= \left\{ \frac{1}{2} \begin{matrix} 1 - \sin 2\gamma_1 \cos 2\gamma_2 \sin\left(\dfrac{\pi V_{RF}}{V_\pi} + \phi_1\right) \\ + \sin 2\gamma_2 \begin{bmatrix} -\cos 2\gamma_1 \sin\left(\dfrac{\pi V_{RF}}{V_\pi} + \phi_1\right)\cos\left(\dfrac{\pi\Gamma V_{RF}}{V_\pi} + \phi_2\right) \\ -\cos\left(\dfrac{\pi V_{RF}}{V_\pi} + \phi_1\right)\sin\left(\dfrac{\pi\Gamma V_{RF}}{V_\pi} + \phi_2\right) \end{bmatrix} \end{matrix} \right\}. \end{aligned} \tag{16.102}$$

In case of equal DIRs, Eq. (16.102) becomes

$$F(V_{\mathrm{RF}})=\left|\frac{E_{\text{out-1}}}{E_{\text{in-1}}}\right|^2 =\frac{1}{2}\left\{\begin{array}{l}1-\frac{1}{2}\sin 4\gamma\sin\left(\frac{\pi V_{\mathrm{RF}}}{V_\pi}+\phi_1\right)\left[1+\cos\left(\frac{\pi\Gamma V_{\mathrm{RF}}}{V_\pi}+\phi_2\right)\right]\\ -\sin 2\gamma\cos\left(\frac{\pi V_{\mathrm{RF}}}{V_\pi}+\phi_1\right)\sin\left(\frac{\pi\Gamma V_{\mathrm{RF}}}{V_\pi}+\phi_2\right)\end{array}\right\}. \quad (16.103)$$

It is assumed that the modulator is biased to the optimal $Q$-bias point. Then by using Taylor expansion, it results in the condition of third-order cancellation, which dictates the value of optimal $\Gamma$. Therefore the condition of $0=\partial F^3(V_{\mathrm{RF}})/\partial V_{\mathrm{RF}}^3$ provides the condition of distortion optimization, and $\partial F(V_{\mathrm{RF}})/\partial\Gamma$, $\partial F(V_{\mathrm{RF}})/\partial\gamma$ provides the sensitivity to design parameters. These kind of chips showed optimal coupling values in the range of 58 deg to 68 deg. In case the coefficient $\Gamma=0$, then there is no linearization. The optimum value of $\Gamma$ can be calculated by requiring the maximum RF voltage by selecting $\partial F(V_{\mathrm{RF}})/\partial\Gamma=0$ and $\partial F^3(V_{\mathrm{RF}})/\partial V_{\mathrm{RF}}^3\partial\Gamma=0$ for minimum value of the third order. The main drawback of this method is the insertion loss introduced by cascading the two modulators.

To provide a summary of this method: linearization is accomplished by having the second modulator with low OMI to nullify out the primary modulator distortions. The second modulator creates out-of-phase IMD products with respect to the primary modulator because of the mutual coupling between its two arms, which is accomplished by the coupler installed between the two modulators. There is no need for high OMI at the second modulator since the IMD products that result from the first modulator are low.

A similar approach from this section, together with the dual parallel, is used to analyze the structure provided in Fig. 16.25.[22] The structure is to create a parallel path to the second modulator, which produces an orthogonal field. This is done by a polarization control (PC) block in Fig. 16.25. The input modulator's output is split by a 3-dB coupler or a polarization splitter. This configuration has some advantages. First, the changeable RF modulation level ratio marked as $\gamma$ can be used to optimize the optical power. This is in contrast to a single MZ modulator with a fixed $\gamma$ value. Second, there is the option of applying the same modulation levels for both MZ modulators, and hence no electrical degeneration is experienced.

Figure 16.25 describes the structure of such a linearizer. A dc-biased DIR cascades two MZI. A fixed DIR can be utilized as well. The amplitude of the electrical field at the PC output is given by:

$$E^{\mathrm{PC}}=\sqrt{P_{\mathrm{in}}}\cos\left[\frac{\phi_1}{2}+\frac{\phi_{\mathrm{m1}}(t)}{2}\right]\cdot\alpha, \quad (16.104)$$

where $P_{in}$ is the input optical power, $\phi_1$ is the constant phase shift by dc bias, $\phi_{m1}(t)$ is RF modulation on the first MZI, marked as $M_A$, and $\alpha^2$ is the power-splitting ratio to the PC output.

In the same way, the output field of the second modulator, $M_B$, is given by:

$$E^{M_B} = \sqrt{P_{in}}\left\{\cos\left[\frac{\phi_1}{2} + \frac{\phi_{m1}(t)}{2} + \frac{\phi_2}{2} + \frac{\phi_{m2}(t)}{2}\right] + \cos\left[\frac{\phi_1}{2} + \frac{\phi_{m1}(t)}{2} - \frac{\phi_2}{2} - \frac{\phi_{m2}(t)}{2}\right]\right\} \cdot \beta, \tag{16.105}$$

where $\phi_2$ is a constant phase shift, $\phi_{m2}(t)$ is RF modulation applied on the second MZI, which is linearly related to $\phi_{m1}(t)$ by a $\gamma$ factor, $\gamma$ is a constant, and $\beta^2 = 1 - \alpha^2$ for lossless power split of both MZI and DIR.

By using the PC, the fields from the outputs of the PC and $M_B$ modulator can be tuned for orthogonality so $E^{PC} \perp E^{M_B}$. Hence the total output power is given by the sum of the fields' squares:

$$P_{out} = \frac{1}{2}\left(\left|E^{PC}\right|^2 + \left|E^{M_B}\right|^2\right). \tag{16.106}$$

From Eq. (16.104),

$$\left|E^{PC}\right|^2 = P_{in}\cos^2\left[\frac{\phi_1}{2} + \frac{\phi_{m1}(t)}{2}\right] \cdot \alpha^2. \tag{16.107}$$

To reduce IMD, $\phi_1$ and $\phi_2$ should have opposite signs, if modulation $\phi_{m1}(t)$ and $\phi_{m2}(t)$ have the same polarity. Additionally, the lesser the difference between the RF modulation levels $\phi_{m1}(t)$ and $\phi_{m2}(t)$, the lesser the electrical RF modulation power lost. In fact, for the two-modulator scheme for parallel modulators, there is an optimum:

$$\phi_{m2}(t) = 2\phi_{m1}(t). \tag{16.108}$$

For a cascaded case, this optimum is given by

$$\phi_{m2}(t) = \phi_{m1}(t). \tag{16.109}$$

The bias voltage can be adjusted to give

$$\phi_1 - \phi_2 = \pi. \tag{16.110}$$

This minimizes the dc term in the optical power of $E^{M_B}$ so that the last terms in Eq. (16.105) vanishes, resulting in

$$|E^{M_B}|^2 = P_{in} \cdot \left\{ \cos\left[ \frac{\phi_1}{2} + \frac{\phi_{m1}(t)}{2} + \frac{\phi_2}{2} + \frac{\phi_{m2}(t)}{2} \right] \right\}^2 \cdot \beta^2. \tag{16.111}$$

The constant phase difference between the PC channel and $M_B$ does not have to be taken into account for moderately high modulation frequencies due to the fact that an optical power addition is used here rather than an optical fields addition. Note that power is scalar while the field is a vector.

An RF IMD analysis is done by applying the two-tones modulation test. Here $\phi_m(t)$ is given by

$$\phi_m(t) = \phi_m(\sin\omega_1 t + \sin\omega_2 t), \tag{16.112}$$

where $\phi_m$ is the driven signal amplitude.

Assume that $\phi_{m2}(t) = \gamma\phi_{m1}(t)$; this may result in the following modulating signals:

$$\begin{aligned} \phi_{m1}(t) &= \phi_m \frac{\sin\omega_1 t + \sin\omega_2 t}{1+\gamma}, \\ \phi_{m1}(t) &= \gamma \cdot \phi_m \frac{\sin\omega_1 t + \sin\omega_2 t}{1+\gamma}. \end{aligned} \tag{16.113}$$

Substituting Eqs. (16.107), (16.111), (16.113), and (16.114) into Eq. (16.106) and expressing $P_{out}$ as a series results in $C_1$, which describes the fundamental terms $\sin\omega_1 t + \sin\omega_2 t$:

$$C_1 = \frac{1}{2} \left\{ \begin{aligned} &\alpha^2 \cdot P_{in} \cdot J_0[\phi_m/(1+\gamma)] \cdot J_1[\phi_m/(1+\gamma)] \cdot \sin(\phi_1) \\ &+ \beta^2 \cdot P_{in} \cdot J_0[\phi_m/(1+\gamma)] \cdot J_1[\phi_m/(1+\gamma)] \cdot \sin(\phi_1) \end{aligned} \right\}, \tag{16.114}$$

where $J_0$ and $J_1$ are Bessel functions.

For linearizing the modulator, the coefficient $C_3$ for the third order has to be equal to or near zero. Assuming modulating amplitude $\phi_m \ll 1$, the result is

$$J_1(\phi_m) \cdot J_2(\phi_m) \approx -\frac{\phi_m^3}{16}. \tag{16.115}$$

Substituting this result into Eqs. (16.113) and (16.114) would provide a condition for nullifying $C_3$:

$$-\alpha^2 \cdot \sin(\phi_1) = \beta^2 \cdot (1+\gamma)^3 \cdot \sin(\phi_1 + \phi_2). \tag{16.116}$$

Using $\beta^2 = 1 - \alpha^2$ for the ideal case, $\alpha^2$ and $\beta^2$ can be expressed as functions of $\phi_1$, $\phi_2$, and $\gamma$; thus, the following condition applies, which provides no severe degradation of optical output:

$$\frac{\sin(\phi_1)}{\sin(\phi_1 + \phi_2)} = -1. \tag{16.117}$$

From Eqs. (16.117) and (16.110), the condition is given by

$$\begin{cases} \phi_1 = \dfrac{\pi}{3}, \\ \phi_2 = -\dfrac{2\pi}{3}. \end{cases} \tag{16.118}$$

From Eqs. (16.116) and (16.118) and the relation between $\alpha$ and $\beta$, formal expressions for the last two can be provided

$$\alpha^2 = \frac{(1+\gamma)^3}{1+(1+\gamma)^3}. \tag{16.119}$$

$$\beta^2 = \frac{1}{1+(1+\gamma)^3}. \tag{16.120}$$

The values of $\phi_1$ and $\phi_2$, $\alpha^2$ and $\beta^2$, do not have to be very precisely defined. Detuning one of the parameters can be compensated by adjusting other parameters. That means the degree of freedom and tolerances are relatively large. By substituting $\gamma = 1$ and $\phi_1 - \phi_2 = \pi$ into Eq. (16.116) results in

$$\frac{\alpha^2}{2 \cdot \beta^2 \cdot \cos\phi_1} = 8. \tag{16.121}$$

Equation (16.121) provides the condition $C_3 = 0$ under $\phi_m \ll 1$. However, for an arbitrary $\phi_m$, the condition of $C/\mathrm{IMD}_3 > 90$ dBc for third order may not be satisfied. Assume that the left-hand side of Eq. (16.121) equals $x$, where $x$ can lie between $x_1$ and $x_2$. The optical splitting ratio $\alpha^2/\beta^2$ and the dc bias of the first electrode, which provides the phase offset $\phi_1$ and thus $\phi_2$, would have a tolerance range. There is a need to determine the tolerance ranges for $\phi_1$ with respect to the optical-splitting ratio $\alpha^2/\beta^2$. Rewriting Eq. (16.121) would set a range:

$$\frac{\alpha^2}{2 \cdot \beta^2 \cdot x_1} \leq \cos\phi_1 \leq \frac{\alpha^2}{2 \cdot \beta^2 \cdot x_2}. \tag{16.122}$$

The condition of $C_3$ satisfying $C/\mathrm{IMD}_3 > 90$ dBc would show values for $\phi_1$ as a function of splitting ratio $\alpha^2/\beta^2$. Forbidden zones are defined as areas where $C/\mathrm{IMD}_3 < 90$ dBc.

### 16.3.4 Feedback linearization

In some applications such as return path or cellular where the laser is directly modulated, there is a need for low cost and simple linearization. The feedback linearization is reported by Ref. [30], but requires some modifications to implement it for laser modulation scheme. This is accomplished by a feedback method. The outcoming modulated light from the laser is sampled by a PD. In order to prevent the PD from being driven into saturation and distortion, a beamsplitter of 20/80% may be used as power coupler. The structure of such an optical block is similar to a duplexer and is shown in Sec. 5.2. The modulated light and distortion sampled by the PD is fed back to the laser input by a wideband feedback to minimize delays. The sampled signal and distortions coming from the feedback are summed with the input signal to cancel the fundamental frequency. Thus the requirement is that the input signal and the feedback would be at 180 deg with respect to each other.

The correction IMD signal and the input tones feed the laser. In this way, IMD products are cancelled. The cancellation is not absolute since the feedback group delay results in a residual phase with the correction signal. Thus the two phasors are not opposite in phase and the subtraction of the IMDs is not complete. Figure 16.26 provides a conceptual block diagram of this linearization technique. In low-BW systems, the feedback can be done by the back-facet monitor owing to its limitation of a large-diameter PD, and thus a large capacitance is not crucial. The drawback of that option for high-frequency wideband data transport is the large diameter of the photodiode, which limits the BW, as was explained in Chapter 8. Thus a return path up to 50 MHz can be realized with the back-facet PD.

In narrowband systems, IMD cancellation can be further optimized by adding PSs and better alignment is achieved; Fig. 16.27 illustrates such a system. An input RF signal is split into two paths by a power splitter, marked as PS1. The upper route is the laser modulation and correction path. The upper arm of PS1 is connected to a second power combiner/splitter, marked as PS2. At this point, the IMD correction signal is added to the fundamental tones. The IMD correction signals are opposite in phase to the IMD that results from the modulated laser. A driving linear RF modulates the laser. The RF amplifier's high linearity is accomplished by its push–pull configuration. The modulated laser output is sampled by a directional optical coupler and then detected by a PD. The linearity of the receiver is accomplished by a differential sampling of the PD, resulting in a push–pull receiver configuration. Amplitude control is added by a variable attenuator, marked as ATT2. Phase inversion of 180 deg is done by the output BALUN of the PD receiver. This output is connected to the RF power combiner/splitter, marked as PS3. A PS marked as $\phi_1$ is inserted after the lower arm of the input RF power splitter PS1 for optimizing cancellation of the fundamental tones at the sum point, which is the output of PS3. The output of PS3 is the correction

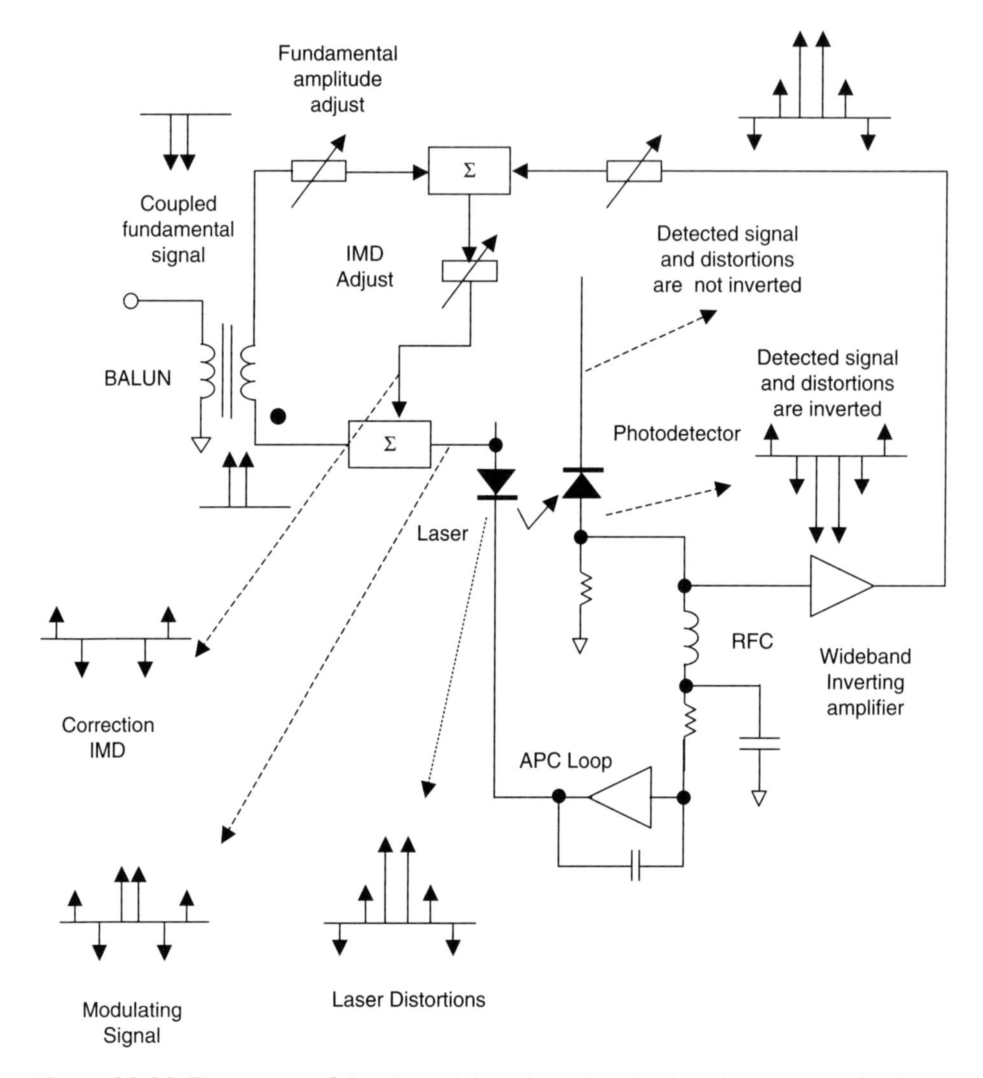


**Figure 16.26** The concept of directly modulated laser linearization with wideband feedback.

feedback containing only IMD products. This path contains an RF amplifier, amplitude alignment using a variable attenuator marked as ATT1, and PS $\phi_2$ for optimizing the IMD nullification.

The next step in such a design is to define constraints and outputs. The laser output power defines the dc current. The modulation index, OMI [%], defines the RF current and therefore the driving RF power applied at the laser input.

Consider a design problem where a laser with a known slope efficiency of $\eta$, CSO, and CTB parameters are provided for a two-tone test with known OMI conditions and the PD responsivity is $r$. The design requirements are for specific optical power $P_{\text{opt1}}$ and CATV modulation index $\text{OMI}_{\text{CATV}}$. It is known that the PD is coupled via an optical coupler with a coupling value of CPL[dB]. From

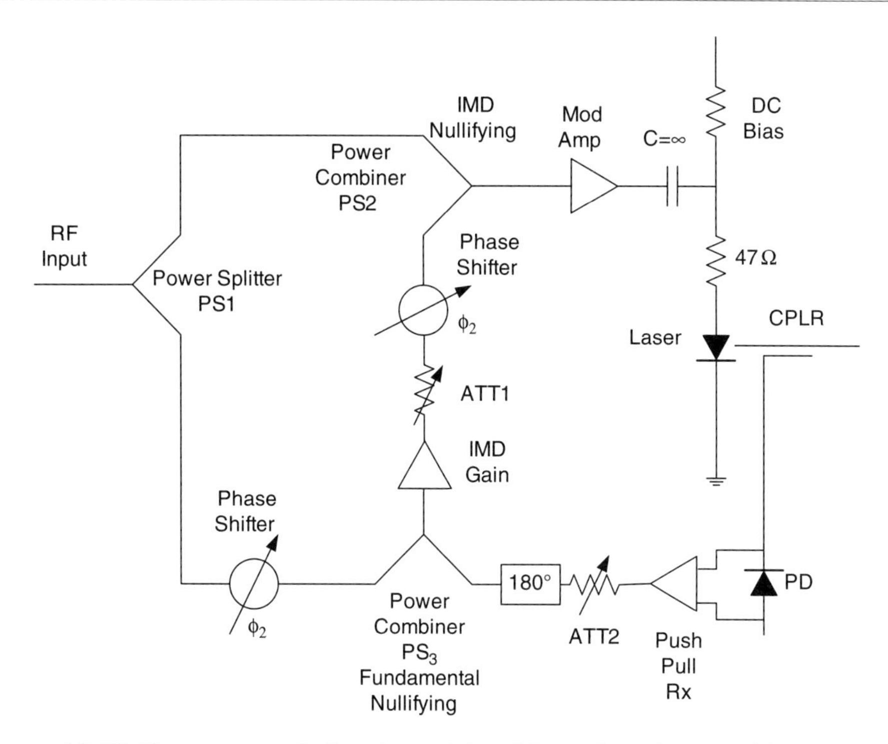


**Figure 16.27** The concept of directly modulated laser linearization with phase- and amplitude-correction feedback. Note that PSs are narrow-BW devices and thus this concept is narrowband. Phase locking to the IMD level using a pilot tone can be done. This may require RF BPF to isolate IMD, RF detector, and loop filter. The leveling method is similar to AGC.

these parameters, the following are concluded. The laser bias current $I_{dc}$ is defined to satisfy the required optical power. Hence, the RF current for laser is driven by Eq. (16.123) and RF power at the laser input is evaluated by Eq. (16.124), using the equations provided in Chapter 6.

$$i_{RF} = \frac{OMI}{100} \cdot I_{dc}, \tag{16.123}$$

$$P_{RF} = Z_0 \cdot \left(\frac{i_{RF}}{\sqrt{2}}\right)^2 = Z_0 \cdot \left(\frac{I_{dc} \cdot OMI}{\sqrt{2} \cdot 100}\right)^2. \tag{16.124}$$

The modulated optical power that creates the RF current at the PD is given by

$$i_{RF\_PD} = I_{dc} \cdot \frac{OMI}{100} \cdot \eta \cdot 10^{0.1 \cdot CPL} \cdot r. \tag{16.125}$$

Hence the RF current phasor detected by the PD is provided by

$$i_{RF\_PD}(t) = \left(I_{dc} \cdot \frac{OMI}{100} \cdot \eta \cdot 10^{0.1 \cdot CPL} \cdot r\right) \cdot \cos \omega_{RF} \cdot t. \tag{16.126}$$

Assume a laser diode such as a Mitsubishi FU450SFD. The linearity test for two tones is defined at OMI = 10%/tone, including a 15-km SMF to add dispersion effects, due to laser chirp. The data sheet presents the IMD result in decibels relative to the carrier power. Hence the next stage is to evaluate the laser polynomial. Assume that the model for a modulated laser has a linear slope efficiency as well as two additional slope efficiencies for the second- and third-order DSO and DTO two-tone test. Thus it can be written as follows:

$$P_{\mathrm{opt}} = \eta \cdot x + b \cdot x^2 + c \cdot x^3. \tag{16.127}$$

The units of the coefficients of the linear and NL slope efficiencies are $\eta$ [W/A] or [mW/mA], $b$ [W/A$^2$] or [mW/mA$^2$], and $c$ [W/A$^3$] or [mW/mA$^3$]. Such tests are performed by vendors using a spectrum analyzer as described in Fig. 16.28. The two tones applied on Eq. (16.127) result in the following:

$$P_{\mathrm{opt}} = \begin{cases} \eta \cdot \left[ I_{\mathrm{dc}} \cdot \dfrac{\mathrm{OMI}}{100} \cdot (\cos\omega_{\mathrm{RF1}} \cdot t + \cos\omega_{\mathrm{RF2}} \cdot t) \right] \\ + b \cdot \left[ I_{\mathrm{dc}} \cdot \dfrac{\mathrm{OMI}}{100} \cdot (\cos\omega_{\mathrm{RF1}} \cdot t + \cos\omega_{\mathrm{RF2}} \cdot t) \right]^2 \\ + c \cdot \left[ I_{\mathrm{dc}} \cdot \dfrac{\mathrm{OMI}}{100} \cdot (\cos\omega_{\mathrm{RF1}} \cdot t + \cos\omega_{\mathrm{RF2}} \cdot t) \right]^3. \end{cases} \tag{16.128}$$

The next stage is to evaluate $b$ and $c$ from the CSO and CTB data. These are provided by using trigonometric identities and selecting the relevant frequencies. The third-order intermodulation (TOI) currents are given by

$$i_{\mathrm{RF\text{-}TOI}} = r \cdot \left( I_{\mathrm{dc}} \cdot \frac{\mathrm{OMI}}{100} \right)^3 \cdot c \times \frac{3}{4} [\cos(2\omega_{\mathrm{RF1}} - \omega_{\mathrm{RF2}}) + \cos(2\omega_{\mathrm{RF2}} - \omega_{\mathrm{RF1}})], \tag{16.129}$$

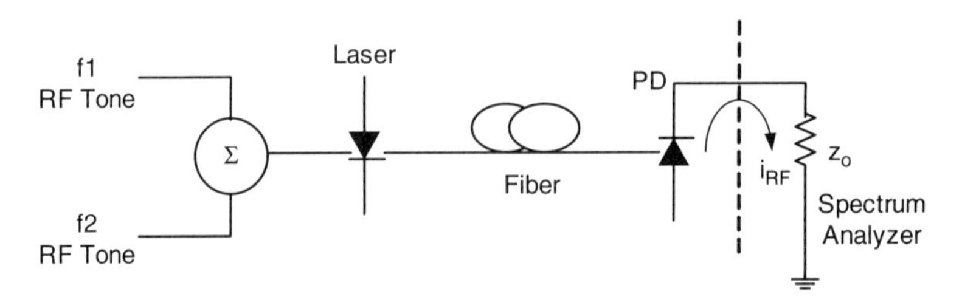


**Figure 16.28** Laser IMD test in a spectrum analyzer.

where $r$ is PD responsivity. The second-order IMD (SOI) currents are given by

$$i_{\text{RF-SOI}} = r \cdot \left(I_{\text{dc}} \cdot \frac{\text{OMI}}{100}\right)^2 \cdot b \times [\cos(\omega_{\text{RF1}} + \omega_{\text{RF2}}) + \cos(\omega_{\text{RF2}} - \omega_{\text{RF1}})]. \tag{16.130}$$

The IMD products are measured by a spectrum analyzer with input impedance of $Z_0$. The power is related to rms currents. Hence, the TOI result in dBc would be given by:

$$dBc_{\text{TOI}} = 10\log\frac{\left[r \cdot \left(I_{\text{dc}} \cdot \frac{\text{OMI}}{100}\right)^3 \cdot c \cdot \frac{3}{4}\right]^2 \cdot \frac{1}{2} \cdot Z_0}{\left[r \cdot \left(I_{\text{dc}} \cdot \frac{\text{OMI}}{100}\right) \cdot \eta\right]^2 \cdot \frac{1}{2} \cdot Z_0} = 10\log\frac{c^2 \cdot \left(I_{\text{dc}} \cdot \frac{\text{OMI}}{100}\right)^4 \cdot \left(\frac{3}{4}\right)^2}{\eta^2}. \tag{16.131}$$

Consequently the $c$ coefficient is given by:

$$c\left[\frac{W}{A^3}\right] = \frac{4 \cdot \sqrt{10^{0.1 \cdot \text{dBc}}} \cdot \eta \cdot 10^4}{3 \cdot I_{\text{dc}}^2 \cdot \text{OMI}^2}. \tag{16.132}$$

Using the same approach, $b$ is given by:

$$\text{dBc}_{\text{SOI}} = 10\log\frac{\left[r \cdot \left(I_{\text{dc}} \cdot \frac{\text{OMI}}{100}\right)^2 \cdot b\right]^2 \cdot \frac{1}{2} \cdot Z_0}{\left[r \cdot \left(I_{\text{dc}} \cdot \frac{\text{OMI}}{100}\right) \cdot \eta\right]^2 \cdot \frac{1}{2} \cdot Z_0} = 10\log\frac{b^2 \cdot \left(I_{\text{dc}} \cdot \frac{\text{OMI}}{100}\right)^2}{\eta^2}. \tag{16.133}$$

Hence the $b$ coefficient is given by:

$$b\left[\frac{W}{A^2}\right] = \frac{\sqrt{10^{0.1 \cdot \text{dBc}}} \cdot \eta \cdot 10^2}{I_{\text{DC}} \cdot \text{OMI}}. \tag{16.134}$$

The next step is to estimate the fundamental cancellation by amplitude- and phase-alignment tolerances of the PS $\phi_1$ and amplitude control ATT2. For that purpose, the voltage summation at the PS3 output should be observed. Assume that the detected voltages at the feedback path are given by

$$V = V_2 \cdot [\cos(\omega_{\mathrm{RF1}} t + \theta + \pi) + \cos(\omega_{\mathrm{RF2}} t + \theta + \pi)]. \tag{16.135}$$

The voltages, resulting from the push–pull analog TIA, have phase inversion and phase shift $\theta$ due to the laser, detector, and RF delays. Recall that phase inversion is made by the BALUN for subtraction at PS3:

The fundamental tones feeding PS3 after passing through $\phi_1$ are

$$V = V1 \cdot [\cos(\omega_{\mathrm{RF1}} t + \varphi) + \cos(\omega_{\mathrm{RF2}} t + \varphi)]. \tag{16.136}$$

From the phasor diagram, in Fig. 16.29, the calculation of the $Vx$ and $Vy$ phasor projections would be

$$\begin{aligned} Vx &= V_1 \cdot \cos\varphi - V_2 \cdot \cos\theta, \\ Vy &= V_1 \cdot \sin\varphi - V_2 \cdot \sin\theta. \end{aligned} \tag{16.137}$$

The equivalent fundamental phasor magnitude is given by

$$V = \sqrt{V_X^2 + V_Y^2}. \tag{16.138}$$

Defining the voltage-amplitude balancing coefficient as the ratio of the laser fundamental frequency feedback and the fundamental input, $\delta = V_2/V_1$, would

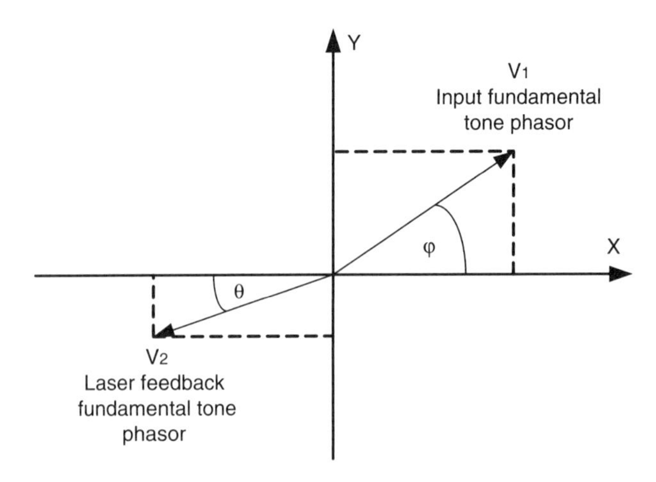


**Figure 16.29** Fundamental tone cancellation as a phasor diagram.

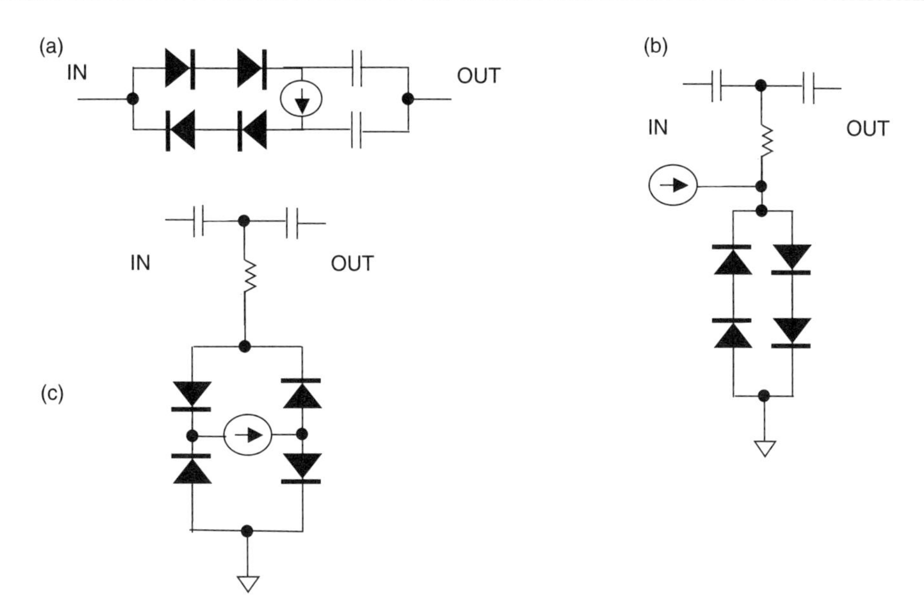


**Figure 16.30** Back-to-back linearizer concepts: (a) expansive hyperbolic arcsin (TRL); (b) compressive (RFL); (c) hyperbolic arctangent (RFL) based on comsat technical review [1].

provide the equivalent phasor nullification tolerances of phase and amplitude mismatch:

$$V = V_1 \cdot \sqrt{1 + \delta^2 - 2 \cdot \cos(\varphi - \theta)}. \tag{16.139}$$

where the phasor phase is $\xi = \tan(Vy/Vx)$.

In the same way, IMD nullifying is evaluated for the laser output by observing full roundtrip–inverted IMDs against newly generated IMDs created by the laser. For this calculation, the laser polynomial model is used. For the calculation on the current domain, assume PD has responsivity $r = 1$ and zero delay. For this iterative calculation, the fundamental tones would generate the current OMI according to the amount of the fundamental nullification ability. Note that the upper branch used to modulate the laser sums both antipodal IMD and residues of the fundamental tones. These fundamental residues are summed with the main RF tones used to modulate the laser. The fundamental tones modulating the laser produce IMDs according to Eq. (16.127), where the $c$ and $b$ coefficients are given by Eqs. (16.132) and (16.134). The full roundtrip IMDs pass the laser as linear tones per Eq. (16.127), where the linear slope efficiency coefficient is $\eta$. The total IMD would be the phasor equivalent of the full roundtrip with the newly generated IMDs.

The drawback of this linearization technique is that it has a narrow BW. The ability to cancel distortions depends on the phase response of the RF devices such as amplifiers. Therefore, for narrow BW, RF fundamentals and distortions may have the same phase, therefore the IMD and fundamentals are canceled according to the suggested scheme in Fig 16.27. However, for a wide-band span, each fundamental RF tone and IMD product accumulate different phase from the RF chain, and the phase difference between them is large. As a consequence, the phasors are not aligned and IMD products are not canceled.

## 16.4 CATV Transmitter Structure

In previous sections, several linearization techniques for state-of-the-art CATV transmitters were reviewed. In Chapters 12 and 13, the AGC, APC, and TEC loops were described. All of these concepts are part of modern fiber-optical

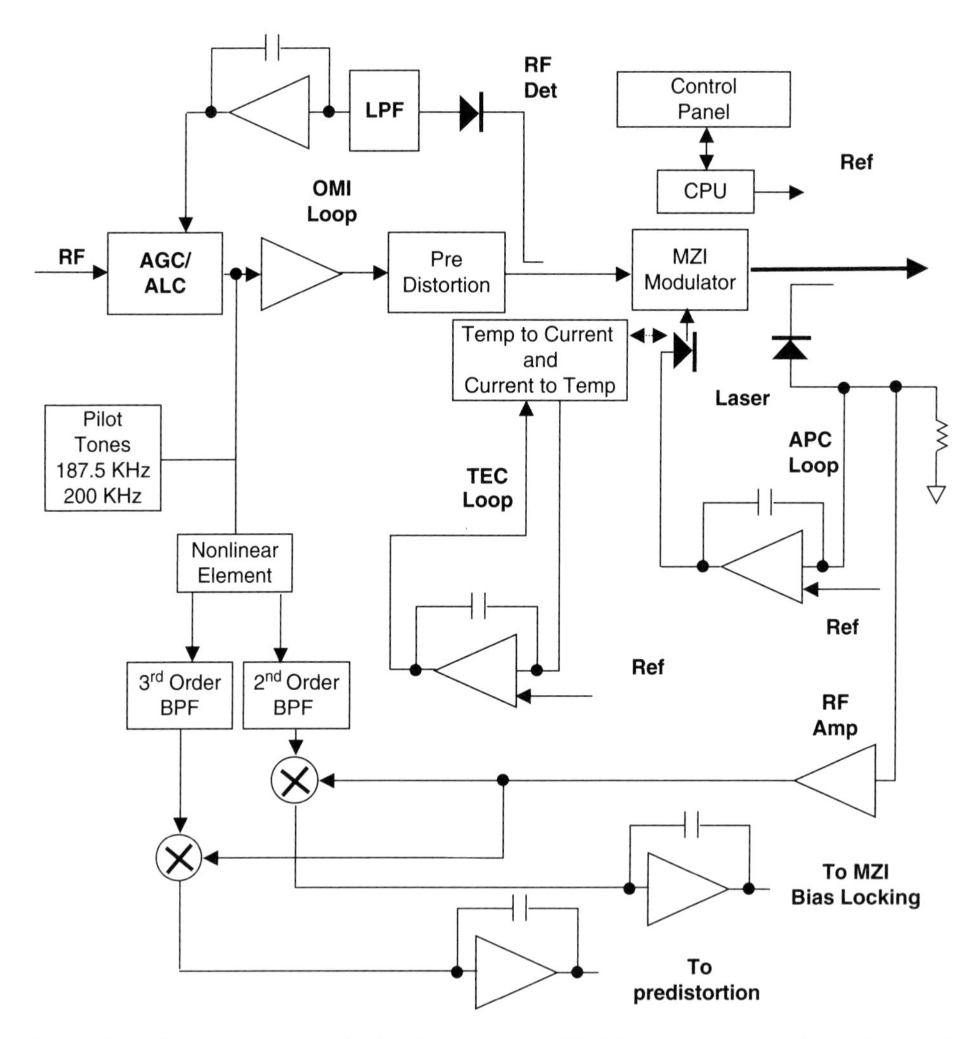


**Figure 16.31** A conceptual block diagram of a CATV transmitter showing all control loops and parametric loops.

CATV transmitter architectures. In order to maintain a fixed-level OMI and avoid overmodulation, the RF signal driving the external modulator is locked to a constant RF level by automatic level control (ALC), which is the AGC concept for transmitters. The second control loop is the TEC, which sets the laser diode to a constant temperature and thus stabilizes its wavelength. The last loop is the APC loop controlling the laser power. In addition to these loops, the modulator is introduced by a predistortion circuit. For that purpose, two additional loops are used. The first locks the modulator bias point using a parametric feedback, hence canceling the CSO. The second loop is a parametric loop to lock the optimal point of the third-order predistortion circuit. This loop is also parametric feedback; both feedbacks use the same pilot tones.[10] Figure 16.31 provides a block diagram of such a CATV transmitter.

RF lineups in CATV transmitters are configured in push–pull to minimize second-order distortions. In addition to modulator linearization, an RF chain is linearized as well, using the same concept of back-to-back diodes.[1,13] Phase alignment is accomplished by VARICAP diodes.

Figure 16.30 provides some linearizer concepts such as reflective transmission line (RFL) and transmission line type (TRL).[1]

The overall system performance is measured by carrier-to-intermodulation $(C/I)$ and calculated by

$$\left(\frac{C}{N}\right)_{\mathrm{T}}=\frac{1}{\dfrac{1}{\left(\frac{C}{I}\right)_{\mathrm{IMD}}}+\dfrac{1}{\left(\frac{C}{N}\right)_{\mathrm{RIN}}}+\dfrac{1}{\left(\frac{C}{N}\right)_{\mathrm{shot}}}\dfrac{1}{\left(\frac{C}{N}\right)_{\mathrm{dark}}}+\dfrac{1}{\left(\frac{C}{N}\right)_{\mathrm{thermal}}}}. \tag{16.140}$$

It is clear that the lower the intermodulations are, the higher the dynamic range of the entire link.

## 16.5 Main Points of this Chapter

1. Analog system or link performance is measured by carrier to intermodulation $(C/I)$ and noise, which defines the system's dynamic range.

2. In digital transport such as QAM, $C/I$ affects the performance of BER versus $E_{\mathrm{b}}/N_0$.

3. Analog system performance and dynamic range is evaluated by IP2 and IP3, which are correlated to the CSO and CTB.

4. To overcome NLD and reduce distortions, there is a need for linearization circuits.

5. Linearization can be done on the modulating electronic circuit or on the optics.

6. Linearization in electronic circuits is categorized into two main configurations: FF linearizer, which is made of linear elements that sample distortions, and the signal created in the RF path, which is subtracted. The remaining distortions are inverted and feed the output with inverted phase, and phase and amplitude alignment, so the output IMD are cancelled.

7. The FF technique is implemented both in optics and RF.

8. The main drawback of FF for CATV applications is its alignment as well as its limited BW, because of the PS's BW limitations.

9. RF FF is useful for cellular applications.

10. In a CATV, the modulator's distortions result from the $\sin(x)$ transfer function of the modulator.

11. By selecting a proper bias point for the optical modulator, second-order distortions theoretically can be cancelled.

12. The predistortion circuit's role is to cancel the CTB of the modulator.

13. NLD analysis is based on the Taylor power series with no memory. This means amplitude changes with no phase changes due to AM-to-PM.

14. The MZI predistorter generates an arcsinh function that linearizes the MZI $\sin(x)$ transfer function.

15. One of the methods to have an arcsinh function is by a back-to-back diode linearizer, as shown in Fig. 16.6.

16. In order to lock the linearizer to its optimum operating bias and to optimize the MZI bias point for minimum CSOs, a special feedback loop is used and is identified as parametric feedback.

17. A parametric feedback loop is similar to a PLL since it is a PLL locked on reference IMD. Because of this, it is called a RLL.

18. An RLL loop requires two pilot tones to produce IMDs. This way, it can be locked on the second-order IMD and third-order IMD for bias optimization of the MZI and the bias of the predistorter, respectively.

19. The pilot tones, used in RLL to create distortions, are selected to be out of the CATV frequency band in order to prevent interference. As a result, the IMD products are out of the CATV band as well. This is the frequency design concept of RLL pilot tones.

20. To create a reference IMD, the two pilot tones pass through the NL device.

21. An IMD generated by the MZI are sampled back, then pass a phase detector, and create an error voltage. The error voltage is integrated by the loop filter, creating a correction voltage.

22. RLL for CSO and CTB consist of two loops; such a system is called also an adaptive predistorter.

23. Predistorters are available for directly modulated lasers as well. Such predistorters cancel second- and third-order IMDs.

24. CMOS technology is also used for linearization since the FET is a square-law device.

25. The push–pull method is a linearization circuit that cancels second-order distortions created by the RF amplifiers. It does not cancel the PD distortions.

26. Push–pull is useful for both CATV receivers and RF drive in CATV transmitters.

27. Push–pull performance is sensitive to the phase and amplitude alignment between its two arms.

28. There are three methods used for modulators: linearization, dual parallel, and dual cascade linearization, a combination of the previous two.

29. The main idea behind the three methods mentioned in point 28 is to have a main modulation path and auxiliary modulation path, which provides out-of-phase distortions. These distortions are summed with the output of the main modulator and thus the distortions are cancelled.

30. Another optical linearization method is wideband fast feedback of the modulated optical signal by sampling it with a PD. The sampled signal is fed back with inverted IMDs to cancel the laser's IMDs. The delay of the feedback defines the phase lag and sets the limit of linearization.

31. A CATV transmitter has five control loops: (1) an OMI loop for constant RF modulation depth; (2) an APC loop for optics power; (3) a TEC loop for a fixed temperature of the transmitter's laser in order to stabilize the wavelength; (4) A dc bias parametric loop to cancel second-order CSO of the modulator by locking its bias to the optimal bias point; and (5) a dc bias parametric loop to optimize third-order CTB of the modulator by locking the predistorter to the optimal bias point.

32. Parametric loop BW is narrow since the bias drifts are slow with time constants of hours to days. A wideband fast loop would track undesirable amplitude changes and would generate residual AM, consequently making higher IMDs.

## References


1. Cahana, D., J.R. Potukuchi, R.G. Marshalek, and D.K. Paul, "Linearized transponder technology for satellite communications; Part-1 Linearizer circuit development and experiment characterization." *COMSAT Technical Review*, Vol. 15 No. 2A, pp. 277–307 (1985).
2. Rosly, L., V. Borgioni, F. Zepparelli, F. Ambrosi, M. Comez, P. Faccin, and A. Casini, "Analog laser predistortion for multiservice radio over fiber systems." *IEEE Journal of Lightwave Technology*, Vol. 21 No. 5, pp. 1121–1223 (2003).
3. Fernando, X.N., and A.B. Sesay, "Higher order adaptive filter based predistortion for nonlinear distortion compensation of radio over fiber links." *Proceedings of IEEE* (2000).
4. Sadhwani, R., J. Basak, and B. Jalali, "Adaptive electronic linearization of fiber optic links." Paper ThG7. *Proceedings of OFC*, Vol. 2, pp. 477–497 (2003).
5. Salgado, H.M., and J.J. O'Reilly. "Accurate performance modeling of subcarrier multiplexed fiber/radio systems: Implications of laser nonlinear distortions and wide dynamic range." *IEEE Transactions on Communications*, Vol. 44 No. 8, pp. 988–996 (1996).
6. Roselli, L., V. Borgioni, F. Zepparelli, M. Comez, P. Faccin, and A. Casini, Predistortion, "Circuit design for I and III order simultaneous linearization in multiservice telecommunications apparatuses." Paper TH2F-7, *IEEE MTTS Digest*, pp. 1711–1714 (2002).
7. Katz, A., "Linearization reducing distortion in power amplifiers." *IEEE Microwave Magazine*, Vol. 37, pp. 37–49 (2001).
8. Chiu, Y., B. Jalali, S. Garner, and W. Steier, "Broad band electronic linearizer for externally modulated analog fiber optic links." *IEEE Photonics Technology Letters*, Vol. 11 No. 1, pp. 48–50 (1999).
9. Wilson, G.C., T.H. Wood, and U. Koren, "Integrated electroabsorption modulator/DBR laser linearized by RF current modulation." *IEEE Photonics Technology Letters*, Vol. 7 No. 10, pp. 1154–1156 (1995).
10. Nazarathy, M., J. Berger, A.J. Ley, I.M. Levi, and Y. Kagan, "Progress in externally modulated AM CATV transmission systems." *Journal of Lightwave Technology*, Vol. 11, pp. 82–105 (1993).
11. Jackson, M.K., V.M. Smith, W.J. Hallam, and J.C. Maycock, "Optically linearized modulators: chirp control for low-distortion analog transmission." *IEEE Journal of Lightwave Technology*, Vol. 15 No. 8, pp. 1538–1545 (1997).
12. Katz, A., "SSPA Linearization." *Microwave Journal*, Vol. 42 No. 4, pp. 22–73 (1999).
13. Cripps, S., *Advanced Techniques in RF Power Amplifier Design.* Artech House (1999).
14. Sadhwani, R., and B. Jalaly, "Adaptive CMOS predistortion linearizer for fiber optic links." *IEEE Journal of Lightwave Technology*, Vol. 21 No. 12 (2003).

15. Childs, R.B., and V.A. O'Brian, "Multichannel AM video transmission using a high-power Nd. YAG laser and linearized external modulator." *IEEE Journal on Selected Areas in Communications*, Vol. 8 No. 7, pp. 1369–1376 (1990).
16. Wilson, G.C., "Optimized predistortion of overmodulated mach-zehnder modulators with multicarrier input." *IEEE Photonics Technology Letters*, Vol. 9 No. 11, pp. 1535–1537 (1997).
17. Atlas, D.A., "On the overmodulation limit on externally modulated lightwave AM–VSB CATV systems." *IEEE Photonics Technology Letters*, Vol. 8 No. 5, pp. 697–699 (1996).
18. LaViolette, K.D., "CTB performance of cascaded externally modulated and directly modulated CATV transmitters." *IEEE Photonics Letters*, Vol. 8 No. 2, pp. 281–283 (1996).
19. Sabido, D.J.M., M. Tabara, T.K. Fong, C.L. Lu, and L.G. Kazovsky, "Improving the dynamic range of a coherent AM analog optical link using a cascaded linearized modulator." *IEEE Photonics Technology Letter*, Vol. 7 No. 7, pp. 813–815 (1995).
20. Korotky, K., and R. M. DeRidder, "Dual parallel modulation schemes for low-distortion analog optical transmission." *IEEE Journal on Selected Areas in Communications*, Vol. 8 No. 7, pp. 1377–1381 (1990).
21. Burns, W.K., "Linearized optical modulator with fifth order correction." *IEEE Journal of Lightwave Technology*, Vol. 13 No. 8, pp. 1724–1727 (1995).
22. Boulic, Y.W., "A linearized optical modulator for reducing third order intermodulation distortion." *IEEE Journal of Lightwave Technology*, Vol. 10 No. 8, pp. 1066–1070 (1992).
23. Brooks, J.L., G.S. Maurer, and R.A. Becker, "Implementation and evaluation of a dual parallel linearization system for AM-SCM video transmission." *IEEE Journal of Lightwave Technology*, Vol. 11 No. 1, pp. 34–41 (1993).
24. Farina, J.D., B.R. Higgins and J.P. Farina, New Linearization Technique for Analog Fiber Optic Links. OFC 96 Technical Digest. Paper ThR6. pp. 283–284.
25. *MathCad Professional 2001 User Guide Book.* Parametric Technology Corporation (2001).
26. Transmission Lines Accurately Model Autotransformers. By K. B. Nicals and R.R Pereira and A.P. Chang. Microwaves and RF. November 1992 pp. 67–76.
27. Walston, J.A., and J.R. Miller, *Transistor Circuit Design.* Texas Instruments Inc. and McGraw-Hill (1963).
28. Hagensen, M., "Influence of imbalance on distortion in optical push-pull front ends." *Journal of Lightwave Technology*, Vol. 13 No. 4, pp. 650–657 (1995).
29. Koyama, F., and K. Iga, "Frequency chirping in external modulators." *IEEE Journal of Lightwave Technology*, Vol. 6 No. 1, pp. 87–93 (1988).
30. Qiang, L., Z.Z. Ying, and G. Wei, "Design of a feedback predistortion linear power amplifier." *Microwave Journal*, Vol. 48 No.5, pp. 232–241 (2005).

# Chapter 17

# System Link Budget Calculation and Impairment Aspects

The previous chapters provided a detailed review of an optical analog link at both ends. In the central downlink, there is a predistorted multichannel community access television (CATV) transmitter. On the user's end, an analog receiver employs simple linearization techniques to combat internally created distortions. However, when considering the entire system, there are several parameters that should be taken into account to measure the quality of service. Quality of service is measured either by carrier-to-noise ratio (CNR) or bit error rate (BER) performance. Quality-of-service parameters such as BER and CNR are affected by optical-link building blocks and their inherent limitations. Those limitations include reflections, dispersion, and fiber-related nonlinearities. Additionally, the CATV channel is affected by external modulation and clipping effects of the laser used in the transmitter in case of DML. Moreover the whole ling is affected system wise by the transmitter quality, RF circuits and laser as well as the receiver's photo detector and the receivers RF circuit performance. An analog receiver produces NLD (nonlinear distortions) such as composite second orders (CSO) and composite triple beats (CTB), which can be countered with proper electrical designing. A photodetector (PD) produces NLDs such as CSO and CTB that are minimized by defocusing the spot and the PD active area as well as using a high reverse bias. The inherent receiver noise floor is determined by the PD shot noise as well as its noise figure performance. The CATV laser transmitter is optimized and predistortion techniques are utilized to combat NLD products. Laser transmitters are divided into four categories related to wavelengths and modulation scheme, such as a direct modulated laser (DML) or an externally modulated laser (EML):

- DML transmitter operating at 1310 nm
- EML transmitter operating at 1319 nm
- DML transmitter operating at 1550 nm
- EML transmitter operating at 1550 nm

An EML is used to reduce distortions and chirp effects, which also increases the chromatic dispersion.

An additional performance-degradation source is an erbium-doped fiber amplifier (EDFA).[4] This section focuses on link degradations without getting into the internal structure of EDFAs, which is beyond the scope of this book. Further information on this subject can be found in Refs. [80] and [81].

In conclusion, a link budget is affected by both electronic RF-related distortions, reviewed in Chapter 9, and optical lightwave-degradation mechanisms. The optical mechanisms are as follows:

- clipping-induced impulse noise,
- bursty NLDs,
- dispersion-induced distortions,
- stimulated Brillouin scattering (SBS),
- self-phase modulation (SPM),
- stimulated Raman scattering (SRS),
- cross-phase modulation (XPM),
- polarization-dependent distortions (PDDs), and
- coexistence between nonreturn-to-zero (NRZ) and CATV signals that occur at the digital fiber to the home, curb, building, business premises (FTTx) and video links services.

In addition, there is a need to have a fast method to approximate the link budget in order to have a ballpark estimate of the link constraints. These effects are analyzed in detail in the following sections.

## 17.1 Link Design Calculations

One of the first steps in optical link design is to estimate the link budget at first-order approximation. The link budget calculation defines the required gain, which is affected by optical loss, resistive matching loss, and the voltage standing-wave ratio (VSWR) mismatch. Link bandwidth (BW) affects the system threshold since it determines the total noise power of the link and its spurious-free dynamic range (SFDR), as explained in Chapter 9. With these parameters, one can select the proper optical components to build a link.

### 17.1.1 System gain

As was previously explained, in optical fibers, one cannot define the gain of an optical device since there is no access to it. On one side is an optical pigtail, and on the other is an RF sub miniature version B connector (SMB) or sub miniature version A connector (SMA) output connector. The same applies to a transmitter; thus, alternative definitions can be used.[5,56] Gain can be defined with respect to RF currents and resistances on transmitter and receiver sides. This way, the optical link gain can be defined as the ratio of RF-output power to RF-input power:

$$G_{\text{link}} = \left(\frac{I_{\text{out}}}{I_{\text{in}}}\right)^2 \frac{R_{\text{out}}}{R_{\text{in}}}, \tag{17.1}$$

where $I_{\text{out}}$ is the receiver's RF-output current to the output load $R_{\text{out}}$, $I_{\text{in}}$ is the RF current to the laser transmitter's input resistance $R_{\text{in}}$. This notation can be modified in terms of link characteristics according to the transmitter's RF-to-optical conversion efficiency $\eta_{\text{TxRF}}$ [mA/mW], the receiver's optical-to-RF conversion efficiency $\eta_{\text{RxRF}}$ [mW/mA], and optical fiber loss $L_{\text{opt}}$:

$$\frac{I_{\text{out}}}{I_{\text{in}}} = \frac{\eta_{\text{TxRF}} \times \eta_{\text{RxRF}}}{L_{\text{opt}}}. \tag{17.2}$$

The optical fiber losses are defined according to the net point-to-point optical power budget:

$$G_{\text{link}} = \left(\frac{\eta_{\text{TxRF}} \times \eta_{\text{RxRF}}}{L_{\text{opt}}}\right)^2 \frac{R_{\text{out}}}{R_{\text{in}}}. \tag{17.3}$$

Substituting Eq. (17.2) into (17.1) results in a useful link-gain-budget equation. Taking efficiencies as unity, one can correct the plot in Fig. 17.1 by adjusting the link gain with the efficiencies' correction factor; hence it can be written as

$$G_{\text{link}}[\text{dB}] = 20\log(\eta_{\text{TxRF}}) + 20\log(\eta_{\text{RxRF}}) - 20\log(L_{\text{opt}}) + 10\log\left(\frac{R_{\text{out}}}{R_{\text{in}}}\right). \tag{17.4}$$

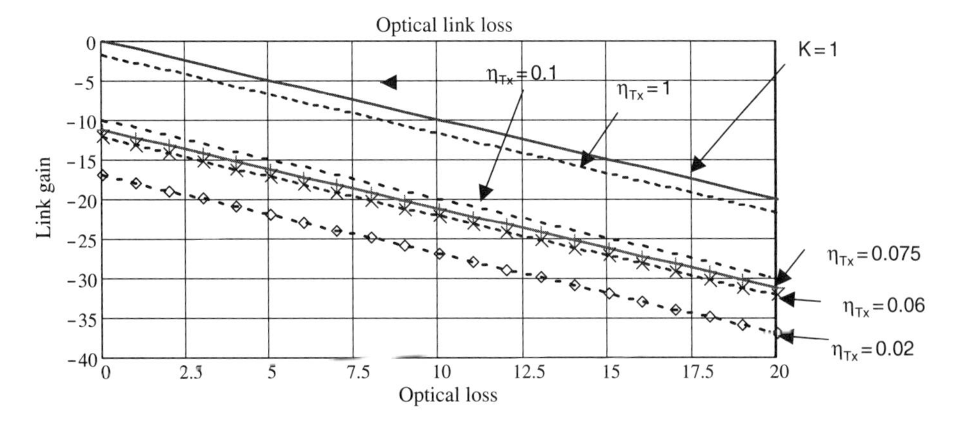


**Figure 17.1** Total link gain for three cases of impedance ratios $K = R_{\text{out}}/R_{\text{in}}$: (a) $K = 1$; (b) $K = 2/3$ such as 50/75 Ω for various transmitter efficiencies, $\eta_{\text{Tx}}$, and 100% receiver efficiency, which is $\eta_{\text{Rx}} = 1$.

### 17.1.2 Efficiency versus matching methods

There are two kinds of matching: the resistive method, which is used in simple DML, as shown in Sec. 6.9.3, and reactive matching, which is commonly used in optical receivers' front ends, as described in Sec. 10.9. As explained, the photodiode is a current source with a shunt-output impedance. Thus the transformer-matching concept improves power-transfer efficiency when the impedance reflected to the transformer primary equals the PD-output impedance. The current-transfer efficiency for an optical receiver is defined as a responsivity correction factor because not all of the photocurrent is transferred to the RF impedance. Therefore, the correction factor is the current divider between the PD and the RF load. Consequently, the overall receiving efficiency is given by Eq. (17.5). It is clearly observed that the higher the PD-output resistance is, the higher the efficiency is. An ideal PD would not have any shunt resistance. Power-matching considerations would reduce the link gain by 6 dB, as both PD and RF resistances are equal:

$$\eta_{\mathrm{RxRF}} = \frac{R_{\mathrm{PD}} R_{\mathrm{out}}}{R_{\mathrm{PD}} + R_{\mathrm{out}}} \cdot \frac{1}{R_{\mathrm{out}}} \cdot r = \frac{R_{\mathrm{PD}}}{R_{\mathrm{PD}} + R_{\mathrm{out}}}. \tag{17.5}$$

The transmitter side does not experience such a drop since the RF-matching resistor is in series; hence, ideally, neglecting ac parasitic effects, the RF efficiency is almost equal to the dc-modulation OMI gain of the laser. From Eq. (17.3), it is clear that for high-link gain, it is desirable to have $R_{\mathrm{out}}/R_{\mathrm{in}} > 1$, and a responsivity of $r \geq 1$. Therefore, the transformer-matching technique to a PD is the preferred method for high-link gain. Figure 17.2 illustrates the concept of a resistive matched link.

### 17.1.3 Link noise

When calculating the quality of service of a link, there is a need to quantify its inherent noise at the input. Chapters 9–11 analyzed noise constraints, where the concept was to have an ideal noise-free link and an equivalent noise source at the input. This noise is amplified and measured at the output of an optical link.

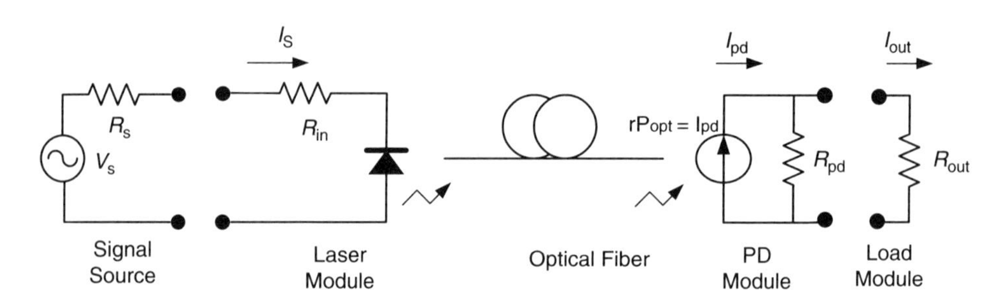


**Figure 17.2** A resistive matched link.

Chapter 12 demonstrates this method of referencing noise to the input when analyzing automatic gain control (AGC) CNR performance. The same concept is useful for link calculations. The equivalent input noise (EIN) density in milliwatts per Hertz or decibels relative to milliwatt per Hertz reflected to the input is defined by:

$$\mathrm{EIN}_{\mathrm{link\_mW/Hz}} = \frac{\mathrm{noise_{out}}}{G_{\mathrm{link}}}. \tag{17.6}$$

The EIN can be converted into a link-noise figure by

$$\mathrm{NF[dB]} = \mathrm{EIN}_{\mathrm{link\_dBm/Hz}} + 174[\mathrm{dBm/Hz}]. \tag{17.7}$$

Another presentation of a link-noise figure, $10\log(F)$ (NF), is the method adopted in satellite communications, by defining noise equivalent temperature $T_e$.

$$T_e = 10^{0.1\times\mathrm{EIN}_{\mathrm{link\_dBm/Hz}}} - 10\cdot\log T_0 - 10\cdot\log(k), \tag{17.8}$$

where $k$ is the Boltzman constant, $T_0$ is 290 K, and $T_e$ results from excess noise, whose sources were reviewed in previous chapters. These noise sources are:

- lasers
- photodiodes
- amplifiers in transmitter electronics
- amplifiers in receiver electronics

Using the definition for relative intensity noise (RIN) provided in Eq. (10.65), laser EIN is given by

$$\mathrm{EIN_{RIN}} = \mathrm{RIN}\cdot(I_{\mathrm{dc}} - I_{\mathrm{th}})^2\cdot R_{\mathrm{in}}\cdot\left(\frac{\mathrm{OMI_{dc}}}{\eta_{\mathrm{TxRF}}}\right)^2, \tag{17.9}$$

where $\mathrm{RIN} = \langle P^2\rangle/\langle P_0^2\rangle$ is the laser noise measured directly at the transmitter RIN, so named because it is the ratio of the mean-square amplitude of the noise fluctuations per unit BW, $\langle P^2\rangle$, to the square of the dc optical power, $P_0$; $I_{\mathrm{dc}}$ is the laser diode (LD) dc bias current, $\mathrm{OMI_{dc}}$ (not in %) is the laser slope efficiency called laser dc modulation gain, which is a regular OMI of 100% multiplied by the slope efficiency of the laser, generally denoted as $\eta$; thus $\mathrm{OMI_{dc}} = \eta$, $\eta_{\mathrm{T\times RF}}$ [mA/mW], the receiver's optical-to-RF conversion efficiency, $I_{\mathrm{th}}$ is the laser threshold, $R_{\mathrm{in}}$ is the laser on dc-input resistance, and OMI is the modulation index as explained in Sec. 6.9. An additional parameter is the receiver EIN reflected to the input. Knowing the link gain per Eq. (17.3) and PD efficiency as defined by Eq. (17.5), and the fiber optic attenuation, $L = P_{\text{optlaser}}/P_{\text{optPD}}$, EIN reflected to the input is given by Eq. (17.10) after performing all substitutions:

$$\text{EIN}_{\text{shot}} = 2q \frac{P_{\text{optlaser}}}{\left(\eta_{\text{T}\times\text{RF}}\right)^2 r} L, \tag{17.10}$$

where $q$ is the electron charge. Lastly there is a need to calculate the thermal EIN as it is reflected to the input. This can be done using Eq. (17.6) or (17.7) normalized to the link gain. Having all these parameters, the overall link EIN is calculated by summing all EIN densities. Having this value and knowing the channel BW, one can calculate the dynamic range of the entire link:

$$\text{EIN}_{\text{link}}[\text{mW/Hz}] = \text{EIN}_{\text{shot}} + \text{EIN}_{\text{RIN}} + \text{EIN}_{\text{thermal}}. \tag{17.11}$$

Note that the link NF is referred to as the overall EIN result in Eq. (17.11).

### 17.1.4 Link dynamic range versus P1 dB and IP3

Equation (9.28) provided the SFDR full analysis. Thus, the noise value is given by $N = \text{EIN} \times \text{BW}$. Another definition for the dynamic range is with respect to P1 dB as given by

$$\text{DR}_{1\text{dB}} = P_{1\text{dB}}[\text{dBm}] - \text{EIN}_{\text{link}}[\text{dBm/Hz}] + 10\log \text{BW}[\text{Hz}]. \tag{17.12}$$

## 17.2 Clipping-Induced Nonlinear Distortions

The sections above described a distortion-free link; however, there have been many studies showing amplitude modulation vestigial side band (AM-VSB) quadrature amplitude modulation (QAM) transport BER degradation due to the "clipping" nature of the laser transmitter.[2] Such a phenomenon results due to the high modulation index, OMI, as there is a high channel load and it is a DML. Research has been conducted on the NLD effects on BER.[6] Section 6.9 provides the basics of DML and the definition of OMI with respect to the laser $L$–$I$ curve. However, for a large number of channels, the OMI depth increases, and in some instances, which are statistically related, the magnitude of the modulating tones is added in phase. This results in a clipping effect, since the signal is rectified. The rectification process induces noise, which is referred to as "clipping noise." In practice, the multitone loading is per the frequency plan described in Chapter 3. The most adverse effect of clipping is on QAM. Thus to observe the clipping noise of a QAM signal, the spectrum analyzer should be set to the QAM channel's center frequency. Spectrum analyzer parameters are: resolution BW and video BW (VBW) set to 3 MHz, and the spectrum analyzer set to "zero-span." In "zero-span" mode, the spectrum operates as an oscilloscope.[9,36] The spectrum is used as a band pass on the channel and the video signal impulses are detected and displayed in time domain. Time division is controlled by the sweep time setting. The spectrum analyzer's video output cable can be used to display the signal on an oscilloscope.

### 17.2.1 Clipping statistical modeling

The theory of DML clipping distortion is provided in Refs. [1–3] and [14]. Numerical analysis presented in Ref. [29] shows that DML has an advantage over EM and CTB are better by 6 dB. Assume DML channel loading of $N$ uncorrelated channels:

$$I(t) = I_a + \sum_{n=1}^{N} a_n \cos(n\omega t + \theta_n), \tag{17.13}$$

where $a_n$ is the current amplitude of the frequency modulation (FM)–multiplexed channel, $I_a$ is the laser bias current, $I(t)$ is the modulating current versus time, and $\theta_n$ is the phase of each channel. It is easier to observe a single channel in order to derive the amplitude probability density function $f$:

$$f[I(t) - I_a] = \left\{ \begin{array}{ll} \dfrac{1}{\pi\sqrt{A^2 - [I(t) - I_a]^2}} & |I(t) - I_a| < A \\ 0 & \text{else} \end{array} \right\}, \tag{17.14}$$

where $A$ is the maximal swing, which is $a_n$ for $n = 1$, and the difference between $A$ and expression in the parenthesis provides the error between the max swing and actual swing. The "$\pi$ value" is a normalization factor to comply the probability density function of "$f$" integration is 1. When $N$ is sufficiently high, the variance $\sigma$ is given by the sum of all average powers of all the channels (note that channel power is related to the amplitude square, i.e., $p_n \propto a_n^2$):

$$\sigma^2 = \frac{1}{2}\sum_{1}^{N} a_n^2. \tag{17.15}$$

Assuming all amplitudes are equal to $A$, the variance of those harmonic channels is given by

$$\sigma = A\sqrt{\frac{N}{2}}. \tag{17.16}$$

The variance is similar to the OMI expression of $N$ coherent tones in Eq. (6.84), since $A$ defines the OMI of a single tone. Using the power average of a harmonic function as a sine or cosine makes the OMI expression identical to the variance expression, as was expected. Modifying Eq. (17.16) by substituting $A$ with $m$ per Eq. (17.17) leads to

$$m = \frac{A}{I_a - I_{\text{th}}}. \tag{17.17}$$

Thus:

$$\sigma = m\sqrt{\frac{N}{2}}(I_a - I_{\text{th}}), \tag{17.18}$$

where $I_{th}$ is the threshold current. Note that when $m\sqrt{N}$ is held constant, the variance is held constant. This parameter defines distortions, since it is a different way of representing multitone OMI. Figure 17.3(a) shows the amplitude probability densities as functions of a number of channels $N$. Figure 17.3(b) shows the effective transfer function of a laser for the input signal plus Gaussian noise from other carriers and the output clipping statistics.[2]

For a large number of channels $N$, the clipping probability is approximately normal, with zero mean and variance, as denoted by Eq. (17.18). Hence $P(|y| > R)$, where $R = N \times A/m = I_a - I_{th}$ is given by:

$$P(|y| > R) = 1 - P(|y| < R)$$

$$= 1 - \frac{1}{\sqrt{2\pi}\sigma}\int_{-R}^{R}\exp\left(\frac{-y^2}{2\sigma^2}\right)dy = \operatorname{erfc}\left(\frac{R}{A\sqrt{N}}\right). \qquad (17.19)$$

Several papers were published on the CSO and CTB calculations,[2,10,13] BER statistics as of NLD,[11,14,15,17] asymptotic bounds,[3] signal-to-noise ratio (SNR),[20] and turn-on delay[16] as causes of clipping and the result of clipping effects. The next step is to use clipping statistics to predict NLD and estimate QAM BER.

Assume that the laser is biased at $I_0$ and the threshold current is $I_{th}$. In case of a modulation current $I$ fluctuating around the bias point, the modulation current is written by:

$$I = I_0 - I_{th} + \Delta I \approx I_0 + \Delta I. \qquad (17.20)$$

Hence, $I$ can be presented as a modification of Eq. (17.13):

$$\Delta I = \sum_{i=1}^{N} a_i \cos(\omega_i t + \theta_i). \qquad (17.21)$$

Using the same process of Gaussian distribution and variance definition from Eq. (17.18), the over-modulation square value of current can be written as:

$$\langle I_{CH}^2\rangle = \frac{1}{\sqrt{2\pi}\sigma}\left\{\int_{-\infty}^{0} I^2\exp\left[\frac{-(I-I_0)^2}{2\sigma^2}\right]dI + \int_{2I_0}^{\infty} I^2\exp\left[\frac{-(I-I_0)^2}{2\sigma^2}\right]dI.\right\} \qquad (17.22)$$

The first term is $\langle I_C^2\rangle$, the clipping part, and the second term is $\langle I_H^2\rangle$, the hard limiting part. This is explained by Fig. 17.4. The idea now is that the variance is the delta modulation current:

$$I = I_0 + \Delta I = I_0 + \sqrt{2}\sigma\mu, \qquad (17.23)$$

where $\mu$ is given by

$$\mu = \frac{I - I_0}{\sqrt{2}\sigma}, \qquad (17.24)$$

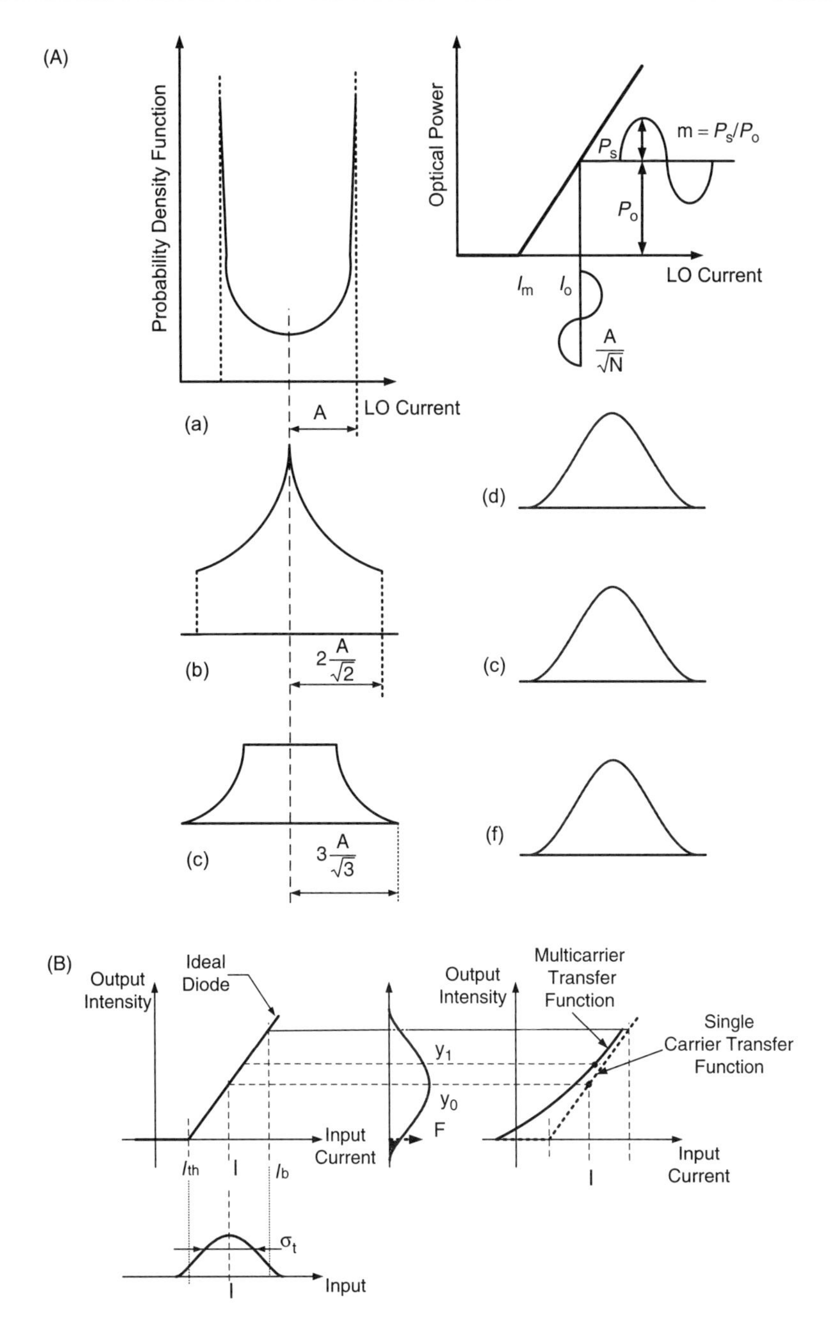


**Figure 17.3** (A) Amplitude probability density function vs. the number of transmitted channels $N$: (a) $N=1$; (b) $N=2$; (c) $N=3$; (d) $N=5$; (e) $N=10$; (f) $N=30$.[1] Reprinted with permission from *IEEE Selected Areas on Communications* © IEEE, 1990. (B) Development of the effective transfer function. Input signal plus Gaussian noise from other carriers (*left*) is clipped, resulting in expected output ($y_1$) greater than the output if there were no clipping ($y_0$). The set of all such inputs results in an effective transfer curve (*solid curve right*).[2] (Reprinted with permission from the *Journal of Lightwave Technology* © IEEE, 1993.)

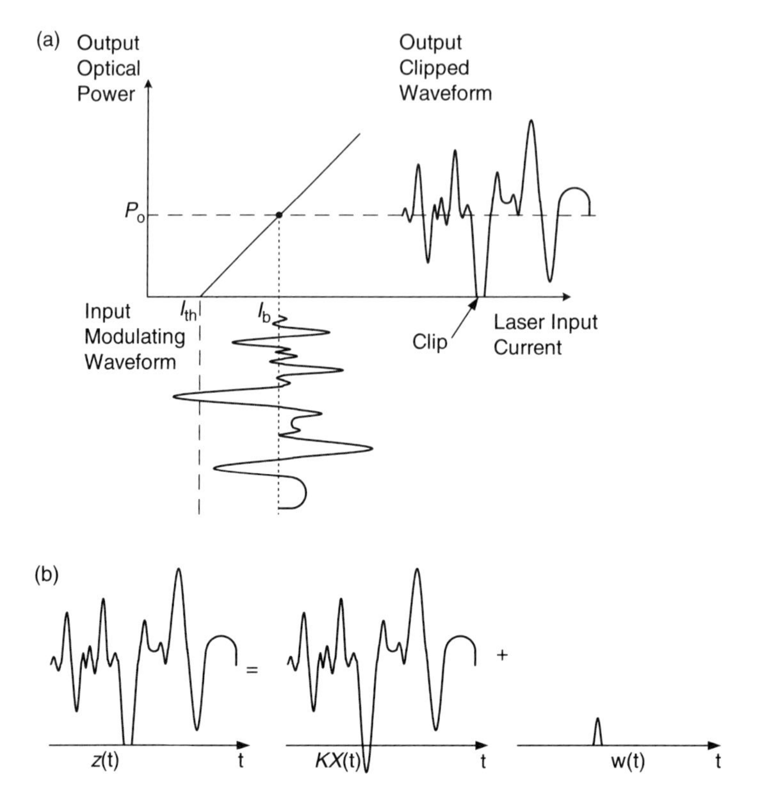


**Figure 17.4** Clipping and strong limiting effects when modulating a laser: (a) laser clipping; (b) a clipped signal as a sum of a linear signal plus a rectified portion.[21] (Reprinted with permission from the *Journal of Lightwave Technology* © IEEE, 1997.)

Hence, the integration is on the fluctuating current. We can use the approximation,

$$\int_x^{\infty} \exp(-\mu^2)\mathrm{d}\mu \approx \frac{1}{2x}\exp(x^2) - J\frac{1}{x^3}\exp(x^2), \tag{17.25}$$

where $J$ is a constant coefficient. Thus $I_{\mathrm{ch}}$ can be written as

$$\langle I_{\mathrm{C}}^2 \rangle = \frac{1}{\sqrt{2\pi}} I_0^2 \mu^3 (1 - \Delta - \Delta\mu^2) \exp\left(\frac{-1}{2\mu^2}\right), \tag{17.26}$$

where $\Delta = g(J)$, which is the mean-square contribution due to the clipping effect. Using the same process employed for the clipping effect, the mean square of the hard limiting current is given by

$$\langle I_{\mathrm{H}}^2 \rangle = \left[\frac{4}{\sqrt{2\pi}} I_0^2 \mu + \frac{1}{\sqrt{2\pi}} I_0^2 \mu^3 (1 - \Delta - \Delta\mu^2)\right] \exp\left(\frac{-1}{2\mu^2}\right). \tag{17.27}$$

The input current is given by

$$\langle I_{\mathrm{I}}^2 \rangle = I_0^2 \frac{N \cdot m^2}{2} = I_0^2 \mu^2. \tag{17.28}$$

Thus, carrier to interference (CIR) is given by

$$\mathrm{CIR} \equiv \frac{\langle I_{\mathrm{in}}^2 \rangle}{\langle I_{\mathrm{C}}^2 \rangle + \langle I_{\mathrm{H}}^2 \rangle} = \frac{\langle I_{\mathrm{in}}^2 \rangle}{\langle I_{\mathrm{ch}}^2 \rangle} = \sqrt{\frac{2}{\pi}} \mu \frac{1}{2 + \mu^2(1 - \Delta - \Delta\mu^2)} \exp\left(-\frac{1}{2\mu^2}\right). \tag{17.29}$$

## 17.2.2 Asymptotic statistical properties of clipping-induced noise

After the above statistical concept, it is clear that the next step is to find the asymptotic clipping of band-limited channel response. Assuming that the clipped signal is composed of the sum of the linear signals and the rectified part, as shown in Fig. 17.4, the following notation is applicable:[3,21]

$$z(t) = Kx(t) + w(t), \tag{17.30}$$

where $K$ is an attenuation factor, $x(t)$ is the original modulation signal, and $w(t)$ is an additive noise that represents the clipping. In Chapter 3, it was explained that the frequency plan has an analog AM-VSB band with high OMI, and digital channels with lower OMI. Hence, the signal is represented by

$$x(t) = \sum_{i=1}^{N} m \cos(2\pi f_i t + \theta_i), \tag{17.31}$$

where $N$ is the number of AM-VSB channels, $m$ is the peak-modulation index, assumed to be common to all channels, $f_i$ is the frequency of the $i$th carrier, and $\theta_i$ is a random phase assumed to be independently and uniformly distributed over the interval $(0-\pi)$. For a large number of channels, $N > 10$. It is assumed that a Gaussian approximation is a random process. Thus, the asymptotic statistical properties of the threshold crossing of a Gaussian process can be used to model $w(t)$, assuming a large difference between the bias current and the threshold current of the laser, as shown in Fig. 17.4.

There are three particularly useful asymptotic properties for $w(t)$. First, the clipping events in $w(t)$ are represented by a Poisson sequence of pulses that occur at an average rate of[18]

$$\lambda = \sqrt{\frac{f_b^3 - f_a^3}{3(f_b - f_a)}} \exp\left(\frac{-1}{2\mu^2}\right), \tag{17.32}$$

where $\mu = m\sqrt{N/2}$ is the rms OMI presented previously for the channels, and $x(t)$ is assumed to be a stationary Gaussian process with a flat power spectrum over the frequency band $(f_a, f_b)$. Second, the probability density of the duration for each clipping event, i.e., time interval between down-crossing and subsequent up-crossing of the laser threshold current by $w(t)$, follows an asymptotic Rayleigh distribution given by

$$p(\tau_p) = \frac{\pi}{2} \cdot \frac{\tau_p}{\overline{\tau_p}^2} \exp\left[ -\frac{\pi}{4} \left( \frac{\tau_p}{\overline{\tau_p}} \right)^2 \right], \quad \tau_p \geq 0 \tag{17.33}$$

where $\overline{\tau}_p$ is the average time duration and is related to the clipping rate $\lambda$ according to

$$\overline{\tau_p} = \frac{1}{2\lambda} \operatorname{erfc}\left( \sqrt{\frac{1}{2\mu^2}} \right). \tag{17.34}$$

Lastly, the shape of each pulse in $w(t)$ is approximated by a parabolic arc:[3]

$$y(t) = \left\{ \begin{array}{ll} \frac{1}{2}\ddot{g}t^2 - \frac{\ddot{g}\tau^2}{8}, & \text{if} \quad |t| < \frac{\tau}{2} \\ 0, & \text{otherwise} \end{array} \right\}. \tag{17.35}$$

The peak of the parabolic arc occurs at $t_k$, and is described by $y(t - t_k)$, or one that starts at $t_k$ and is denoted by $y(t - t_k - \tau_k/2)$, where $\tau_k$ is a support time that replaces $\tau$ in Eq. (17.35). The second derivative of $g$ is approximated by:

$$\ddot{g} \cong -\left( \frac{\dot{\sigma}}{\sigma} \right)^2 M, \tag{17.36}$$

where $M$ refers to the threshold level in Fig. 17.4, and the expression of variance ratio in Eq. (17.36) is denoted by:

$$\left( \frac{\dot{\sigma}}{\sigma} \right)^2 = (2\pi)^2 \frac{f_b^3 - f_a^3}{3(f_b - f_a)}. \tag{17.37}$$

It can be shown that the area $A$ of the parabolic pulse has a random distribution of the following form:[21]

$$p(A) = \frac{\pi}{6C_1^{2/3} A^{1/3} \overline{\tau_p}^2} \exp\left[ -\frac{\pi}{4\overline{\tau_p}^2} \left( \frac{A}{C_1} \right)^{2/3} \right] = \frac{2}{3} \frac{C_2}{A^{1/3}} \exp\left( C_2 A C^{2/3} \right), \tag{17.38}$$

where $A > 0$ and $C_1$ and $C_2$ are given by:

$$C_1 = \frac{\pi}{4\overline{\tau_p}^2} \left( \frac{\overline{A^2}}{6} \right)^{2/9}, \tag{17.39}$$

$$C_2 = \left( \frac{6}{\overline{A^2}} \right)^{1/3}, \tag{17.40}$$

and $A^2$ is the mean square of $A$. The impulsive noise impairment in a band-limited channel can be characterized by examining the channel response to the clipped signal $z(t)$.

This can be simplified by studying Eq. (17.30). Since the receiver's channel filter suppresses all frequency components in $x(t)$ except the desired channel signal, it is clear that the clipping noise contribution comes entirely from the frequency of $w(t)$. The average pulse duration, $\tau_p$, in $w(t)$ for the BW in a 6 MHz CATV channel, with $N = 42$ channels and $m = 0.04$, is calculated by Eqs. (17.32) and (17.34), and is about $\approx$334 ps against 1/6 MHz, which is 167 ns approximately. The clipping noise train can be approximated by:

$$w(t) \approx \sum_{k=-\infty}^{\infty} A_k \delta(t - t_k), \tag{17.41}$$

where $\delta(t - t_k)$ is the impulse function and $t_k$ are the Poisson-distributed event times, and $A_k$ are random variables representing the parabola area of $w(t)$. The channel response to $w(t)$ is obtained by convolving Eq. (17.41) with the effective impulse response of the band-limited channel. Thus the resulting clipping-induced impulsive noise waveform is given by:

$$u(t) \approx \sum_{k=-\infty}^{\infty} A_k h_c(t - t_k), \tag{17.42}$$

where $h_c$ is the impulse response of the channel in the receiver.

### 17.2.3 BER of M-ary QAM channels' clipping noise due to DML OMI

The preliminary background to clipping noise given above can be used for deriving the BER performance of an M-ary QAM in the presence of asymptotic clipping impulse noise and additive white Gaussian noise (AWGN).[11,21] A typical AM-VSB CATV signal is shown in Fig. 17.5 and the modeling of link noise is depicted in Fig. 17.6. In Ref. [11], a study was made to provide bounds for OMI in analog-digital-hybrid subcarrier multiplexed (SCM) transmission. In that scheme, the transport employs 79 AM-VSB analog channels and 31 QAM digital TV channels. Chapter 8 provides an analytic expression for such a scheme in Eq. (8.49) to (8.51). The assumption made in Ref. [11] is that clipping affects BER in the same way as it does AWGN, with the following corrections:

$$P_b = \frac{2}{\log_2[\sqrt{M}]}\left(1 - \frac{1}{\sqrt{M}}\right)\exp(-A)\sum_{j=0}^{\infty}\frac{A^j}{j!}\mathrm{erfc}\left[\sqrt{\frac{3\mathrm{DUR}}{2(M-1)}\cdot\frac{1}{\sigma_J}}\right], \tag{17.43}$$

where $A$ is the impulse index, which is defined as the product of clipping density occurrences, which is the frequency of the frequency-multiplexed (FDM) signal amplitude crossing over the threshold level of the LD. The ratio of demodulation (DUR) is the ratio of the averaged M-QAM signal power to the undesirable signal power, which is the sum of the averaged power of the impulse noise and the

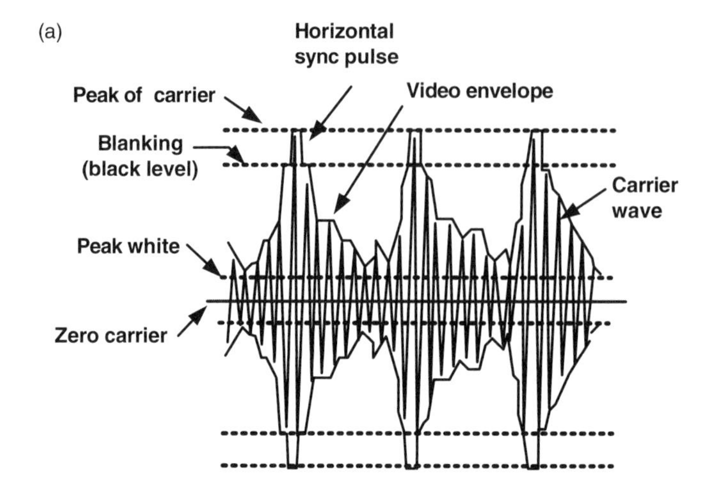


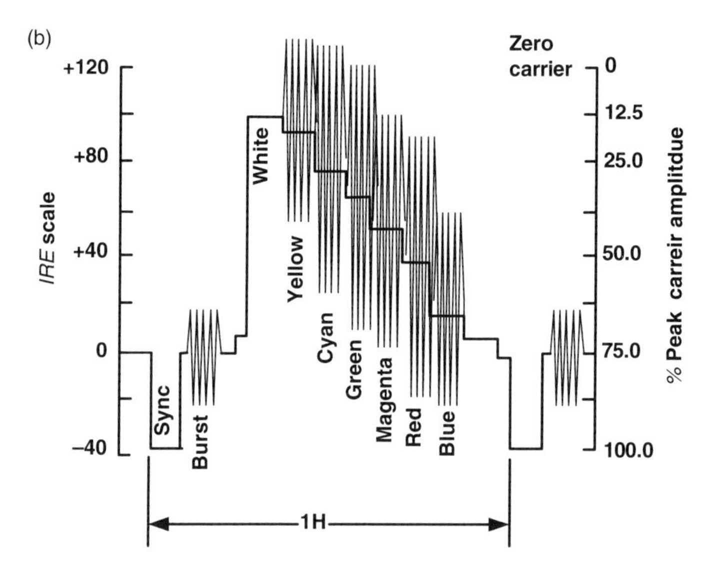


**Figure 17.5** AM-VSB video modulation profile. (a) The carrier signal and modulation envelope in time domain. (b) The standard color bar amplitude according to the IRE.[11] (Reprinted with permission from the *Journal of Lightwave Technology* © IEEE, 1999.)

Gaussian noise. The following relations describe the variables in Eq. (17.43):

$$\sigma_j^2 = \frac{(j/A) + \Gamma'}{1 + \Gamma'}, \tag{17.44}$$

$$\Gamma' = \frac{\text{CNLD}}{\text{CNR}_{\text{QAM}}}, \tag{17.45}$$

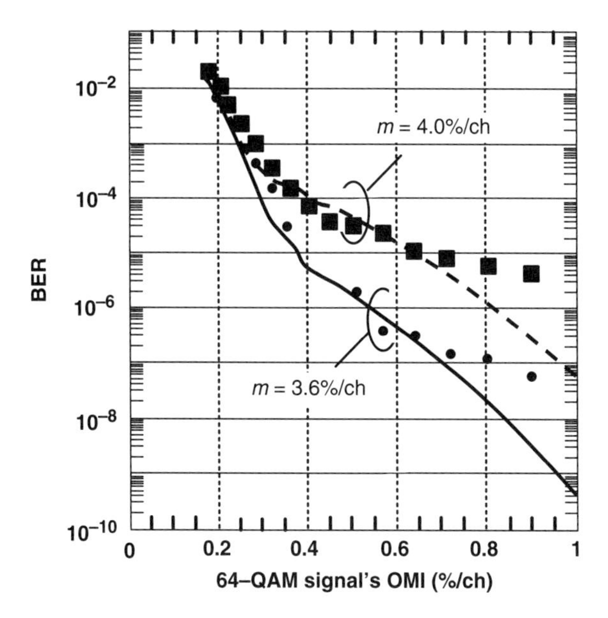


**Figure 17.6** A 64 QAM signal BER in an SCM system with 60 AM-VSB signals. The first AM-VSB case has $m = 4.0\%/\text{ch}$, and the second AM-VSB case has $m = 3.6\%/\text{ch}$. Square dots: experiments; solid line: calculated. QAM OMI; $h$ is varying.[11] (Reprinted with permission from the *Journal of Lightwave Technology* © IEEE, 1999.)

where $\text{CNR}_{\text{QAM}}$ is the M-QAM carrier to the AWGN SNR and CNLD is the carrier-to-impulse-noise power ratio. The impulse index $A$ is derived in the following manner. The multicarrier FDM signal is given by

$$x(t) = \sum_{i=1}^{N_A} m_i \cos(\omega_i t + \varphi_i + \Theta_i) + \sum_{k=1}^{N_Q} h_k \cos\left(\omega_k t + \phi_k + \Phi_k\right), \tag{17.46}$$

where the first part of the equation refers to the AM-VSB channels and the second to the QAM channels, $m$ is the OMI for the VSB, and $h$ is the OMI for the QAM; $N_A$ is the number of AM-VSB channels and $N_Q$ is the number of the QAM channels, $\varphi$ and $\phi$ are the phases of the carriers, and $\Phi$ and $\Theta$ are random variables with uniform distributions at intervals $[-\pi, \pi]$, which expresses phase changes that result from the VSB that and the fluctuation of phase resulting from frequency instability of the synthesizer used to generate the carrier.

The spectral power of $x(t)$ is calculated by the autocorrelation

$$R_{xx}(\tau) = \frac{1}{2}\sum_{i=1}^{N_A} R_{(m_i m_i)} \cos(\omega_i \tau) + \frac{1}{2}\sum_{k=1}^{N_Q} R_{(h_k h_k)} \cos(\omega_k \tau). \tag{17.47}$$

For $\tau = 0$, Eq. (17.47) becomes,

$$R[0] = \frac{1}{2}\sum_{i=1}^{N_A} R_{(m_i m_i)}[0] + \frac{1}{2}\sum_{k=1}^{N_Q} R_{(h_k h_k)}[0] = \frac{1}{2}\sum_{i=1}^{N_A} \overline{m_i}^2 + \frac{1}{2}\sum_{k=1}^{N_Q} \overline{h_k}^2, \tag{17.48}$$

where $\overline{m_i}$ and $\overline{h_k}$ are the averaged OMIs for each modulation signal. In the same manner, the derivative of $x(t)$, denoted as $x'(t)$, is calculated. This expression is the variance in the modulation index, or in other words, the instantaneous change in the signal during the clipping period of $\tau$. Thus,

$$\begin{aligned} R_{x'x'}(\tau) = &\frac{1}{2}\sum_{i=1}^{N_A} R_{(m_i' m_i')}[\tau]\cos(\omega_i\tau) + \omega_i^2 R_{(m_i m_i)}[\tau]\cos(\omega_i\tau) \\ &+ 2\omega_i R_{(m_i' m_i)}[\tau]\sin(\omega_i\tau) + \frac{1}{2}\sum_{k=1}^{N_Q} R_{(h_k' h_k')}[\tau]\cos(\omega_k\tau) \\ &+ \omega_k^2 R_{(h_k h_k)}[\tau]\cos(\omega_k\tau) + 2\omega_k R_{(h_k' h_k)}[\tau]\sin(\omega_k\tau). \end{aligned} \tag{17.49}$$

For $\tau = 0$, Eq. (17.49) becomes

$$\begin{aligned} R_{x'x'}(0) &= \frac{1}{2}\sum_{i=1}^{N_A} R_{(m_i' m_i')}[0] + \omega_i^2 R_{(m_i m_i)}[0] + \frac{1}{2}\sum_{k=1}^{N_Q} R_{(h_k' h_k')}[0] + \omega_k^2 R_{(h_k h_k)}[0] \\ &= \frac{1}{2}\sum_{i=1}^{N_A} \overline{m_i}'^2 + \omega_i^2 \overline{m_i}^2 + \frac{1}{2}\sum_{k=1}^{N_Q} \overline{h_k}'^2 + \omega_k^2 \overline{h_k}^2 \end{aligned} \tag{17.50}$$

where $\overline{m_i}'^2$ and $\overline{h_k}'^2$ are mean squares of the time differential values of each signal's OMI. Moreover, since $\overline{m_i}'^2$ and $\overline{h_k}'^2$ are much smaller than $\omega_i^2 \overline{m_i}^2$ and $\omega_k^2 \overline{h_k}^2$, it enables defining an approximation:

$$R''[0] = -R_{x'x'}(0) \approx -\frac{1}{2}\sum_{i=1}^{N_A} \omega_i^2 \overline{m_i}^2 - \frac{1}{2}\sum_{k=1}^{N_Q} \omega_k^2 \overline{h_k}^2. \tag{17.51}$$

The clipping occurrence density is defined by

$$\nu = \frac{1}{2\pi}\sqrt{\frac{-R''(0)}{R(0)}}\exp\left[\frac{-1}{2R(0)}\right]. \tag{17.52}$$

Using these approximations and substituting $\omega_i = 2\pi f_i$ and $\omega_k = 2\pi f_k$, Eq. (17.52) becomes

$$\nu \approx \sqrt{\frac{\sum_{i=1}^{N_A} f_i^2 \overline{m_i}^2 + \sum_{k=1}^{N_Q} f_k^2 \overline{h_k}^2}{\sum_{i=1}^{N_A} \overline{m_i}^2 + \sum_{k=1}^{N_Q} \overline{h_k}^2}} \exp\left[\frac{-1}{\sum_{i=1}^{N_A} \overline{m_i}^2 + \sum_{k=1}^{N_Q} \overline{h_k}^2}\right]. \tag{17.53}$$

The clipping impulse index is $\wedge = vT_s$. Assuming that the OMI average for AM-VSB and QAM satisfy $\overline{m_i} = m$ and $\overline{h_k} = h$, the impulse index is given by

$$\Lambda \approx T_s \sqrt{\frac{m^2 \sum_{i=1}^{N_A} f_i^2 + h^2 \sum_{k=1}^{N_Q} f_k^2}{m^2 N_A + h^2 N_Q}} \exp\left[\frac{-1}{m^2 N_A + h^2 N_Q}\right]. \tag{17.54}$$

Assume that all signal levels are as above, and that the timing of the horizontal scan synchronization pulse is random, as was mentioned in Chapter 3. Also assume that the average of OMI values for AM-VSB is given by $\overline{m_i} = \mathrm{PF}_A m$ and M-QAM is given by $\overline{h_k} = \mathrm{PF}_Q h$, where, $\mathrm{PF}_Q$ stands for peak factor, which is the index of the difference between the carrier level of the M-QAM signal and its average amplitude; for instance, a 64 QAM has $\mathrm{PF}_Q = 0.66$ ($-3.67$ dB). This issue was explained in Chapter 12 for dealing with AGC calibration of CW versus video. For the AM-VSB, $\mathrm{PF}_A$ is obtained from the video-signal structure in Chapter 3, as also displayed in Fig. 17.5. The AM signal employs negative modulation in such a way that it increases the transmission power and decreases the light intensity of the video signal. Hence, its average amplitude is smaller than that of the horizontal synchronization pulse that has the same level as the AM carrier. From Fig. 17.5, it can be seen that the peak-carrier amplitude appears at a synchronization pulse, which lasts only 8% of the horizontal scan interval, 1H in Fig. 17.5(b). Thus, this amplitude is negligible when estimating the averaged amplitude of the AM-VSB. As was explained in Chapter 12, Fig. 17.5, and Sec. 12.32, it is assumed that the averaged amplitude of practical AM is between 70% black and 12.5% white, with the peak carrier; averaging of $PF_{\mathrm{A}}$ value for the AM signal results in 0.4375 or $-7.7$ dB. Finally, Eq. (17.55) can be written as:

$$\Lambda \approx T_s \sqrt{\frac{(\mathrm{PF}_A m)^2 \sum_{i=1}^{N_A} f_i^2 + (\mathrm{PF}_Q h)^2 \sum_{k=1}^{N_Q} f_k^2}{(\mathrm{PF}_A m)^2 N_A + (\mathrm{PF}_Q h)^2 N_Q}} \exp\left[\frac{-1}{(\mathrm{PF}_A m)^2 N_A + (\mathrm{PF}_Q h)^2 N_Q}\right]. \tag{17.55}$$

The above discussion provides the tools for deciding on the OMI policy for various links that have AM-VSB and QAM against BER performance. An experiment was conducted to check how the AM-VSB distortion profile reduces 64 QAM. Forty analog channels between 91.25 and 445.25 MHz with 6-MHz channel spacing were transmitted with a QAM channel at 473 MHz. The optical power range was controlled by an attenuator. The optical power was $-1.0$ dBm, which provided a digital CNR of ~31 dB. The 64 QAM OMI, denoted as $h$, was 0.4%. The AM-VSB OMI was varied and increased, as shown in Fig. 17.6. In this case, 60 AM-VSB were launched. BER tended to expand as QAM OMI was increased. The example shows two test cases where the AM-VSB channels were held at 4.0% and 3.6% OMI per channel. It can be observed that the higher AM-VSB OMI provided a lower QAM BER performance and lower channel quality of service since the clipping effect that result from the AM-VSB affected the QAM, because the distortions were in the QAM band. In order to overcome that effect, QAM OMI is increased, which is equal to the improvement of $E_b/N_0$.

The next step is to find the OMI bounds for AM-VSB and QAM. Since those TV transports are FDMC (frequency division multiplexing) and affect each other. In fact the AM-VSB affects the BER performance of the QAM. There are two bounds: the lower bound for AM-VSB CNR and QAM BER, and the upper bound, which is affected by clipping and, as will be explained later on, by NLD such as CSO and CTB products. Hybrid SCM transmission requires 51-dB or better CNR at its threshold. This CNR depends on the received optical power and the receiver, as explained in Sec. 17.8 and in Chapter 10, as well as on its AGC strategy, reviewed in Chapter 12. For the case of AM-VSB signal at optical level $-1.0$ dBm and the data in Fig. 17.7 data, the AM-VSB opital modulation index (OMI) is given as $4\% \geq m \geq 3.6\%$. The upper bound is affected by CSO and CTB distortion products, which appear in distortion bursts.[19] Assuming that high-quality reception is for CSO $< -60$ dBc and CTB $< -65$ dBc, two limits are set. The first limit is OMI; the second is the maximum received optical power. The second constraint is solved in the receiver by AGC and PD biasing. The OMI affects the RF level through the channel and is controlled by modulation level in the transmitter, as explained in Chapter 16. Distortions are created by the transmitter and the receiver, and are affected by the number of channels.[22] In general, the approach is to observe the rms OMI average:[11,14]

$$\mu_{\text{OMI}} = \sqrt{\frac{m^2 N_A + h^2 N_Q}{2}} < 0.26, \tag{17.56}$$

which requires overall rms loading of less than 26%.

The lower bound for 64 GAM signal's OMI is analyzed as follows. The Digital Audio Visual Council (DAViC) specifications require BER $<10^{-12}$ after an error correction employing the Reed–Solomon error correction code (204,188,16) and a convolutional interleaver with a depth of 12. The OMI ranges for AM-VSB and 64 QAM are provided in Fig. 17.7.[11] In analog/digital hybrid SCM transmission,

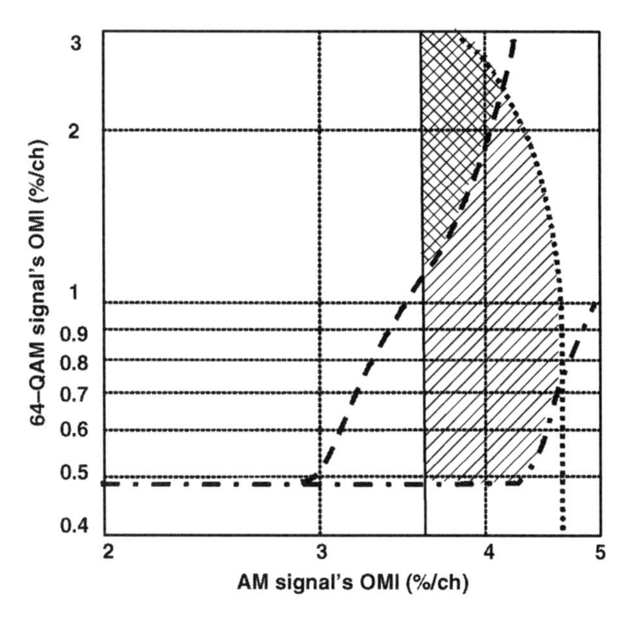


**Figure 17.7** OMI ranges when assuming carrier and AM signal of black picture for AM signal. Reticulated and hatched areas indicate OMI ranges with different AM conditions. Solid line: The lower bound of OMI given by assumed AM-CNR at received optical power of $-1$ dBm. Broken Line: The upper bound of OMI providing assumed distortions levels (CSO $< -60$ dBc, CTB $< -65$ dBc). Dashed line: The lower bound of OMI given by assumed 64 QAM for BER $<10^{-9}$ when employing carriers as AM signals. Dashed dotted line: The lower bound of OMI given by assumed 64 QAM BER when employing AM signal of a black picture as AM-VSB, under tills case modulation signals are wiih AGC.[11] (Reprinted with permission from the *Journal of Lightwave Technology* © IEEE 1999.)

carriers have generally been substituted for transmission signals such as AM-VSB and M-QAM, to evaluate transmission quality. In practical hybrid SCM, however, carriers are modulated by video or digital data. Hence, the amplitude of these signals is more compressed than that of carriers. This phenomenon causes a decrease in the frequency of clipping the multiplexed signal at the laser threshold. Consequently BER is lower than that of the experimental results for the same OMI.[30] Figure 17.7 superimposes the OMI bounds by assuming transmission signals, carriers, and AM-VSB signals with AGC have a 3.9 dB averaged penalty. The OMI's of the AM-VSB and 64 QAM are given by the carrier level. Note that these carriers are CW tones. Therefore, a peak-to-average phenomenon is observed. It can be seen that BER performance is restricted by both OMIs of the AM and the QAM signal. The reticulated area shows the OMI ranges for carriers, and the hatched area shows the OMI ranges for an AM signal with AGC and practical 64-QAM signals. For the carriers, the OMI range is restricted by the BER rather than the CSO and CTB within the analog CATV band. The curve giving the BER bounds rises sharply for the AM-VSB OMI. Hence, the OMI per AM-VSB channel is limited to about 4% in order to satisfy the required

BER prior to the error-correction process. The 64 QAM OMI has to be slightly higher than 2% for such AM-VSB OMI.

Consequently, the OMI ranges shrink and the OMI needs a larger 64-QAM signal to be compared with that for Gaussian noise. On the other hand, the OMI range is wider for modulation signals with AGC, as shown by the hatched area. This is because the clipping occurrence is reduced and the OMI restriction due to BER performance is relaxed. A typical CATV transmitter's AGC is shown in Chapter 16. The AGC protects against over-modulation of the laser; therefore, having a higher OMI range results in larger CNR margins for the AM-VSB signals. Furthermore, there are no limitations on the QAM OMI that results from BER degradation due to impulse noise for the practical SCM system. This is because the FDM (frequency division multiplexed) signal level becomes smaller than that, while assuming the AM signal and the occurrence frequency of clipping is within the value of having no influence on BER. The carriers create a condition under which it is too difficult to evaluate BER in hybrid SCM and the OMI's of the transmission are severely limited.

### 17.2.4 CSO and CTB statistical properties and effects on BER

So far, the analysis has focused on modulation depth effects on the statistics of BER. The next step is to evaluate how the clipping process affects the generation of CSO and CTB, and examine how those products degrade BER performance in an SCM channel. The analysis is based on Eqs. (17.13) and (17.46). The amount of CSO and CTB distortion products are given in Ref. [13]. It can be considered as a Gaussian random process, since the FDM current passes throughout the laser $I$–$P$ characteristics. The resulting distortions are found by the autocorrelation function of the laser output:

$$\mathrm{CSO} = N_{\mathrm{CSO}} \frac{m^2_{\text{AM-VSB}}}{8\pi\mu^2} \gamma^{-2} \exp\left(-\frac{1}{\mu^2}\right), \tag{17.57}$$

$$\mathrm{CTB} = N_{\mathrm{CTB}} \frac{m^4_{\text{AM-VSB}}}{32\pi\mu^6} \gamma^{-2} \exp\left(-\frac{1}{\mu^2}\right), \tag{17.58}$$

where

$$\gamma = \frac{1}{2}\left[1 + \mathrm{erf}\left(\frac{1}{\sqrt{2}\mu}\right)\right], \tag{17.59}$$

and $\mu$ is the rms modulation depth given by

$$\mu_{\mathrm{OMI}} = \frac{1}{2} m^2_{\text{AM-VSB}} \cdot N_{\text{AM-VSB}} + \frac{1}{2} m^2_{\mathrm{QAM}} \cdot N_{\mathrm{QAM}}, \tag{17.60}$$

which refers to the equivalent OMI of multicarrier transport mentioned in Eqs. (17.16) and (6.84). $N_{\mathrm{CSO}}$ and $N_{\mathrm{CTB}}$ are the relative power sums of the second and third intermodulation products falling at frequency $f$. $\mathrm{CSO}/N_{\mathrm{CSO}}$ and $\mathrm{CTB}/N_{\mathrm{CTB}}$ are the ratios of second- and third-order intermodulation products' power to carrier power, IMD2 and IMD3.

As was mentioned previously, the QAM depth is low, and thus its influence on CSO and CTB is low; hence, the AM-VSB has the more significant influence on the NLD products. When analyzing BER in the presence of clipping, there are two kinds of noise impaired digital QAM transmission. First is Gaussian noise, which includes PD shot noise, receiver's noise, and laser RIN. Clipping NLD is a second kind of non-Gaussian broadband noise, which is considered a random train of pulses passing through the band-pass filter (BPF) of the QAM channel. Thus the BER depends on the rate of pulses arriving at and the magnitude of pulses passing through the QAM BPF:

$$\mathrm{BER} = 4\frac{\sqrt{M}-1}{\sqrt{M}}P_{\mathrm{imp}}P\left(\tilde{i} > d/2\right), \tag{17.61}$$

where $M$ is the number of QAM symbols, $P_{\mathrm{imp}}$ is the impulse arrival probability, and $P\left(\tilde{i} > d/2\right)$ is the probability that the magnitude of filtered pulse is larger than the half distance between two QAM symbols. $P_{\mathrm{imp}}$ is given by the Poisson statistics of clipping noise, $P_{\mathrm{imp}} \approx \lambda T_{\mathrm{sym}}$, also known as the clipping index, and $\lambda$ is affected by $\mu_{\mathrm{OMI}} = \mu$ according to Eq. (17.32), and $T_{\mathrm{sym}}$ is equal to the period of symbols. The value of $P\left(\tilde{i} > d/2\right)$ is given by

$$P\left(\tilde{i} > d/2\right) = -\frac{3}{2\pi}E_i\left\{-\sqrt{\frac{9\pi}{16}\left(\frac{d/2}{4f_g k_{\mathrm{si}}\bar{A}}\right)^2} + \left(\frac{1}{2}-\frac{1}{\pi}\right)\right.$$
$$\left. \times \exp\left[-\sqrt{\frac{9\pi}{16}\left(\frac{d/2}{4f_g k_{\mathrm{si}}\bar{A}}\right)^2}\right]\right\}, \tag{17.62}$$

where $E_i()$ is the exponential integral function, $\bar{A}$ is mean area of the impulse whose statistics are given by Eq. (17.38), and $f_g$ and $k_{\mathrm{si}}$ are functions of the BPF of the receiver. Equation (17.61) provides the symbol error probability, which is equal to BER, assuming the use of Gray coding. Note that the distance between two adjacent states of an M-ary QAM constellation is given by

$$d = 2\sqrt{\frac{3T_{\mathrm{sym}}P_{\mathrm{ave}}}{M-1}}, \tag{17.63}$$

where $P_{\mathrm{ave}}$ is the average power per symbol.

### 17.2.5 Analysis of external modulator clipping statistics

An external modulator (EM) is used to combat chromatic dispersion due to laser wavelength chirp; however, it also suffers from clipping effects. Thus, linearization methods are used, as was elaborated in Chapter 16. The analysis of nonlinear EM $L$–$I$ curves is performed by treating them as an ideal two-level clipping event.[15] A direct modulated laser $L$–$I$ transfer function has been modeled as an ideal one-level clipping which is a special case of the two level clipping. Hence, under those assumptions, the output $y(t)$ is expressed in Eq. (17.30), where $w(t)$ is a noise sequence of Poisson impulses and $K$ is the attenuation factor. Equation (17.30) can be written as

$$y(t) = K_x(t) + \sum_{n=-\infty}^{\infty} A_{\tau_n}\delta(t - t_n) + \sum_{m=-\infty}^{\infty} B_{\tau_m}\delta(t - t_m), \tag{17.64}$$

where $t_n$ and $t_m$ are the Poisson clipping times, $\tau_n$ and $\tau_m$ are the clipping intervals corresponding to lower- and upper-level clippings, $L_B$ and $L_A$, respectively, and

$$A_{\tau_n} = (2\pi)^2 \frac{f_b^3 - f_a^3}{f_b - f_a}\frac{P_0\tau_n^3}{36}, \tag{17.65}$$

$$B_{\tau_m} = (2\pi)^2 \frac{f_b^3 - f_a^3}{f_b - f_a}\frac{P_0\tau_m^3}{36}\cdot\frac{L_B}{L_A}, \tag{17.66}$$

where $P_0$ is the output power and, as in Eq. (17.32), $(f_a, f_b)$ is the BW of interest. When such impulse sequences are input to an M-ary channel with matched filters, the output is a shot-noise process. Therefore, the first-order probability density function $f(z)$ of the combined clipping noise and Gaussian background noise is obtained:

$$f(z) \approx (1 - \gamma_A - \gamma_B)\phi^{(0)}\left(\frac{z}{\sigma_g}\right)\frac{1}{\sigma_g} + \gamma_A f_A(z) + \gamma_B f_B(z), \tag{17.67}$$

where

$$\begin{aligned} f_A(z) = {} & \phi^{(0)}\left(\frac{z}{\sigma_{10}}\right)\frac{1}{\sigma_{10}} + 4^{-3}\gamma_A C_4^A \phi^{(4)}\left(\frac{z}{\sigma_{20}}\right)\left(\frac{1}{\sigma_{20}}\right)^5 \\ & + 0.5\cdot 4^{-3}(3!)^{-2}\gamma_A^2 C_6^A \phi^{(6)}\left(\frac{z}{\sigma_{30}}\right)\left(\frac{1}{\sigma_{30}}\right)^7, \end{aligned} \tag{17.68}$$

$$\begin{aligned} f_B(z) = {} & \phi^{(0)}\left(\frac{z}{\sigma_{10}}\right)\frac{1}{\sigma_{10}} + 4^{-3}\gamma_B C_4^B \phi^{(4)}\left(\frac{z}{\sigma_{02}}\right)\left(\frac{1}{\sigma_{02}}\right)^5 \\ & + 0.5\cdot 4^{-3}(3!)^{-2}\gamma_B^2 C_6^B \phi^{(6)}\left(\frac{z}{\sigma_{03}}\right)\left(\frac{1}{\sigma_{03}}\right)^7, \end{aligned} \tag{17.69}$$

where

$$\phi^{(k)}(z) = (-1)^k H_k(z)\frac{1}{\sqrt{2\pi}}\exp\left(\frac{-z^2}{2}\right). \tag{17.70}$$

$H(z)$ is the Hermit polynomial, $\gamma_{A,B}$ are the clipping indices per symbol interval $T_s$, and

$$\sigma_{m,n}^2 = \sigma_g^2 + m\langle A_\tau^2\rangle \frac{1}{\gamma_A T_s} + \frac{n\langle B_\tau^2\rangle}{\gamma_B T_s}, \tag{17.71}$$

where $\sigma_g^2$ is the Gaussian noise variance, and $C_4^A$, $C_4^B$, $C_6^A$, and $C_6^B$ are given in Ref. [83] The variance of the clipping noise is given by:

$$\sigma_w^2 = \frac{\gamma_A\langle A_\tau^2\rangle + \gamma_B\langle B_\tau^2\rangle}{T_s} = \frac{16P_0^2}{3\pi^2}\frac{T_s}{\gamma_A\rho_A^3}$$
$$\times \exp{-\rho A)}\left\{1 - \frac{6}{\rho_A} + \frac{18}{\rho_A^2} + \frac{1}{\beta^2}\exp\left[\frac{(1-\beta)\rho_{\rho A}}{2}\right]\left[1 - \frac{6}{\beta\rho_A} + \frac{18}{\beta_A^2\rho_A^2}\right]\right\}, \tag{17.72}$$

where Eq. (17.73) was used and $\langle A_\tau^2\rangle$ and $\langle B_\tau^2\rangle$ were expended over $\tau$ using the Rayleigh distribution and $\beta = \rho_A/\rho_B$:

$$\frac{\gamma_{A,B}}{T_s} = \exp\left(-\frac{\rho_{A,B}}{2}\right)\sqrt{\frac{f_b^3 - f_a^3}{3(f_b - f_a)}} \tag{17.73}$$

The clipping noise power in the band of interest $(\pm f_a, \pm f_b)$ is given by

$$P_w = 2\sigma_w^2(f_b - f_a). \tag{17.74}$$

Now, this in-band noise power is compared to the total clipping noise power, $P_D$, which can be obtained from the transfer function;[15] i.e., the ratio, $\tau$, that is given by the condition $\rho_{A,B} \gg 1$ and $f_b \gg f_a$. Hence it can be written that

$$\Gamma = \frac{P_W}{P_D}$$
$$= 2.346\left(1 - 1.5\frac{f_a}{f_b}\right)\sqrt{\rho_A}\left\{1 - \frac{18}{\rho_A^2} + \frac{18}{\beta^2}\left(1 + \frac{1}{\beta^2}\right)\rho_A^2\exp\left[\frac{(1-\beta)\rho_A}{2}\right]\right\}. \tag{17.75}$$

If the input signal is comprised of $N$ consecutive channels, the fraction of the total clipping noise power distributed in each channel is $\Gamma_{in} = \Gamma/N$. In Table 17.1,[15] $\Gamma$ is compared to several values of $\rho_A$, where $f_b/f_a = 0.1$ and $\beta\infty$. $\Gamma_{in}$ values are listed for $N = 80$. It can be seen that $\Gamma =$ 40–50%, implying that more than half of the clipping noise is outside the band of interest. The clipping noise reduction, as per Table 17.1, is $10 \log N = 19$ dB.

Substituting Eqs. (17.67) to (17.69) in the integral of Eq. (17.76) provides the BER performance for M-ary transport using EM:[15]

$$P_e = C_M\int_d^\infty f(z)dz = (1 - \gamma_A - \gamma_B)C_M\text{erfc}\left[\frac{d}{\sqrt{2\sigma_g}}\right] + \gamma_A P_e^A + \gamma_B P_e^B, \tag{17.76}$$

**Table 17.1** Comparison between $\Gamma$ and $\Gamma_{in}$ vs. $\rho_A$.[15]

| $\rho_A$ | $\Gamma$ | $\Gamma_{in}$ |
|---|---|---|
| 10.82 | 0.525 | 0.0064 |
| 15.29 | 0.468 | 0.0059 |
| 22.37 | 0.398 | 0.0051 |

where $C_M$, $d$, $P_e^A$, and $P_e^B$ are given by

$$C_M = \frac{2\left[1 - (1/\sqrt{M})\right]}{\log_2 M}, \tag{17.77}$$

$$d = \sqrt{\frac{3T_s P_{ave}}{M-1}} \tag{17.78}$$

$$P_e^A = C_M \mathrm{erfc}\left(\frac{\Delta_{10}}{\sqrt{2}}\right) - \frac{9\gamma_A}{(2+\gamma_A G_A)^2}\phi^{(3)}\left(\frac{\Delta_{20}}{4}\right) - \frac{21\gamma_A^2}{(3+\gamma_A G_A)^3}\phi^{(5)}\left(\frac{\Delta_{30}}{4}\right), \tag{17.79}$$

$$P_e^B = C_M \mathrm{erfc}\left(\frac{\Delta_{01}}{\sqrt{2}}\right) - \frac{9\gamma_B}{(2+\gamma_B G_B)^2}\phi^{(3)}\left(\frac{\Delta_{02}}{4}\right) - \frac{21\gamma_B^2}{(3+\gamma_B G_B)^3}\phi^{(5)}\left(\frac{\Delta_{03}}{4}\right), \tag{17.80}$$

where $\Delta$ with $m$, $n$ indices is given by

$$\Delta_{mn} = \sqrt{\frac{d}{\sigma_g^2 + \frac{m\sigma_A^2}{\gamma_A} + \frac{n\sigma_B^2}{\gamma_B}}}, \tag{17.81}$$

and the $G$ parameters are given by

$$G_A = \frac{\sigma_g^2}{\sigma_A^2}, \tag{17.82}$$

$$G_B = \frac{\sigma_g^2}{\sigma_B^2}, \tag{17.83}$$

and the clipping and noise variances are respectively given by

$$\sigma_{\mathrm{w}}^2 = \sigma_A^2 + \sigma_B^2 \tag{17.84}$$

$$\sigma_g^2 = \mathrm{RIN}\left(rP_{\mathrm{opt}}\right)^2 + 2qP_{\mathrm{opt}} + i_n^2 + \mathrm{NF}, \tag{17.85}$$

where $r$ is the PD responsivity, $P_{\mathrm{opt}}$ is the received optical power, $i_n$ is the thermal EIN noise current, and NF is the digital noise figure. In lightwave transmission, the rms OMI is given by

$$\mu_A = \sqrt{\frac{1}{\rho_A}}. \tag{17.86}$$

For a multicarrier SCM system, the OMI is Eq. (17.56). Figure 17.8 provides data for 256-QAM BER versus OMI for an SCM system with two-level clipping at various $\beta$ values while the AM-VSB OMI is 4.1%.

Figure 17.9 shows the operating OMI range for 256 QAM at a given number of channels. An optimum OMI appears at ~4.5%. Below that, OMI clipping noise is less significant; however, AWGN governs and reduces BER performance due to low CNR. At higher OMI values, NLD affects BER performance. It also shows that BER performance is better for a single-clipping device, represented by the solid line, compared to the double-clipping device represented by the dotted line.

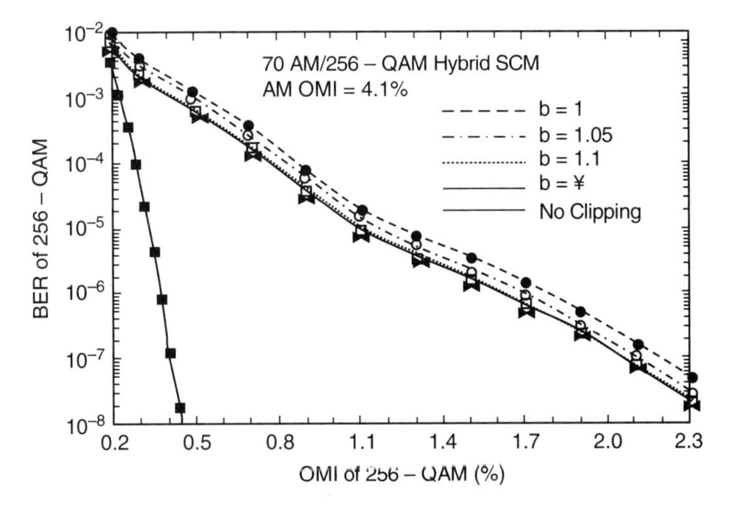


**Figure 17.8** 256 QAM BER vs. OMI for an HFC-SCM system with 70-AM VSB channels with two-level clipping at various $\beta$ values.[15] (Reprinted with permission from *Transactions on Communications* © IEEE, 1998.)

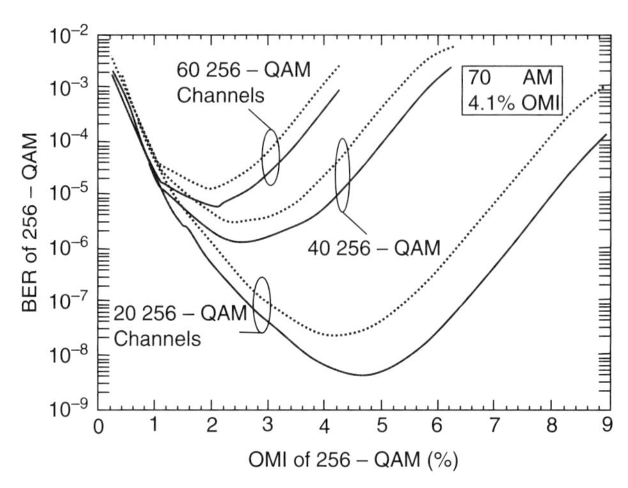


**Figure 17.9** Optimal OMI range for 256 QAM minimum BER. Solid line: single clipping; dotted line: double clipping.[15] (Reprinted with permission from *Transactions on Communications* © IEEE, 1998.)

### 17.2.6 Dynamic clipping-noise effects on BER

In previous sections, the LD was presented as an ideal clipping circuit, which has a linear limiting transfer function. The last section presented the EM as a dual limiting circuit. However, large-signal loading of an active device involves transitions of going in and out of the saturation state. Such situations need recovery time and result in a decay of charges from the previous large-signal state. Thus, there is a lag in the ability of the device to follow the input signal. Consequently, the stored energy decay affects the signal amplitude, because a portion of it is used for the decay of previous charges. This results in amplitude distortions, which yield to spectral distortions. This effect is called memory effect. Such an effect results in the turn-on delay of a semiconductor laser, which affects the laser clipping impulsive noise.[16]

The turn-on delay, $T_d$, of a semiconductor laser is important in high-rate direct modulation. It is the difference between the output and input signals. Typical values vary between 0.3 ns to 0.5 ns in addition to frequency oscillation. Figure 17.12 depicts a normalized time dependence of a laser (solid line) after being pulse modulated (dashed line). Recall that AM-VSB has a higher OMI than the resulted frequency dependent nonlinear distortions, which are dynamic rather than static clipping, and affect the QAM channel by reducing BER performance. Assume the carrier lifetime is $\tau_c$, the active volume is $V$, $q$ is the electron charge, level of threshold current is $I_{th}$, and signal dynamics is shown in Fig. 17.10.

Neglecting the stimulated recombination, since the clipping phenomenon occurs below the threshold, it is a limiting process; the carrier density is expressed by[16]

$$\frac{dn}{dt} \approx \frac{i(t)}{qV} - \frac{n}{\tau_c(n_{th})}. \tag{17.87}$$

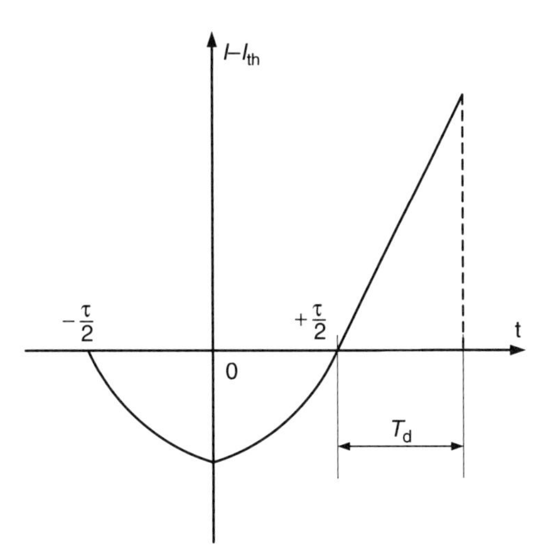


**Figure 17.10** Semiconductor laser model for clipping pulse of width $\tau$, considering a turn-on delay $T_d$.[16] (Reprinted with permission from *Photonics Technology Letters* © IEEE, 1997.)

It has been assumed that the carrier lifetime $\tau_c(n_{th})$ is the threshold level. The clipping pulse is a rare event of small amplitude and is short in time around the threshold value. Thus, changes of carrier lifetime $\tau_c(n_{th})$ against carrier density can be omitted. Further approximation of Eq. (17.87) yields a simpler expression, because the phenomenon of interest occurs around the laser threshold:

$$\frac{dn}{dt} \approx \frac{i(t)}{qV} - \frac{n}{\tau_c(n_{th})} = \frac{1}{qV}[i(t) - I_{th}]. \tag{17.88}$$

The statistics of $i(t)$ can be expressed by:[16]

$$i(t) = \begin{bmatrix} I_{th} - \frac{R''(0)}{R(0)} \cdot \frac{I_b}{2}\left(t^2 - \frac{\tau^2}{4}\right) & -\frac{\tau}{2} \le t \le \frac{\tau}{2} + T_d \\ 0 & \text{otherwise} \end{bmatrix}, \tag{17.89}$$

where $I_b$ is the laser bias current and $\tau$ is the clipping pulse width, which is asymptotically a Rayleigh random variable. $R''(0)/R(0)$ represents the ratio between the second derivative of the autocorrelation function and the autocorrelation function $R(\xi)$, which is a Gaussian process emitted by the semiconductor laser itself. Using these assumptions, integration of Eqs. (17.97) or (17.98) over the pulse width results in a linear dependency of $t_d = \tau/2$. However, the real dependency is exponential, as can be seen in Fig. 17.11.

Because the relationship for pulse width $\tau$ average and variance are per Eq. (17.90),[16] it turns out that the values of the clipping pulse width are at the

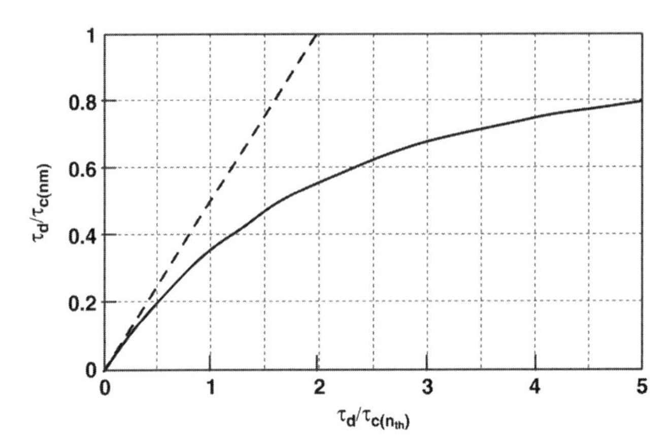


**Figure 17.11** Directly modulated laser turn-on delay $T_d$ vs. clipping pulse width $\tau$. Dashed line: upper bound, $T_d = \tau/2$. Both scales are normalized to the carrier lifetime, $\tau_c$ $n_{th}$, at threshold.[16] (Reprinted with permission from *Photonics Technology Letters* © IEEE, 1997.)

same order of magnitude as the turn-on delay $T_d$:

$$\left.\begin{aligned} \langle\tau\rangle &= \mu_{OMI}\sqrt{2\pi\frac{-R(0)}{R''(0)}} \\ \sigma_\tau^2 &= \langle\tau\rangle\left(\frac{4}{\pi} - 1\right) \end{aligned}\right\}, \tag{17.90}$$

where $\mu_{OMI}$ is the effective full-load OMI. For 60 AM-VSB channels with $OMI_{VSB} = 4\%/ch$ and 90 channels of 16 QAM with $OMI_{QAM} = 0.6\%$, $\langle\tau\rangle \approx 0.27$ ns, and $\sigma_\tau \approx 0.14$ ns. Furthermore, $\tau_c(n_{th}) \approx 0.4$ ns for a typical distributed feedback (DFB) laser. Hence, for the $T_{d-}-\tau$ relationship, a linear approximation can be assumed for values of $0 \leq \tau/\tau_c(n_{th}) \leq 1.5$, as can be observed in Figs. 17.11 and 17.12. As was previously explained, the clipping distortion is modeled as the difference between the sequence of random parabolic pulses and Rayleigh distribution for the pulse width of $\tau$. The difference is approximated by a Poisson sequence of those parabolic pulses. The rate of those clipping pulses is given by:

$$\lambda = \frac{1}{2\pi}\sqrt{-\frac{R''(0)}{R(0)}}\exp\left(\frac{-1}{2\mu_{OMI}^2}\right). \tag{17.91}$$

The spectral power of this process, $S(f)$, is expressed by:

$$S(f) = \lambda\langle|P(f)|^2\rangle + \frac{\pi}{2}\delta(f)\lambda\langle|P(f)|^2\rangle, \tag{17.92}$$

where $\delta(f)$ is the Dirac delta function, and $P(f)$ is the Fourier transform of the parabolic pulse in Fig. 17.10. The power spectral density is described in Fig. 17.13.

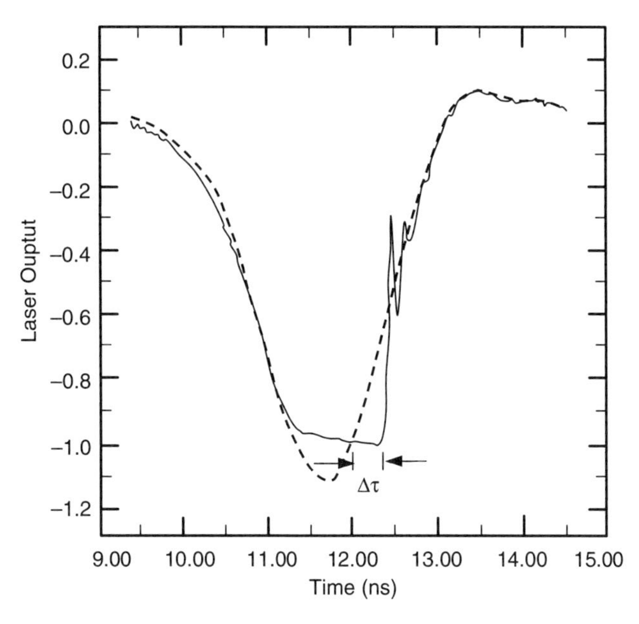


**Figure 17.12** Memory effect of turn-on delay in a laser. Dashed line: input pulse; solid line: output signal from the laser. The delay, denoted as $\Delta\tau$, is the delay in ns. Both curves are normalized to be at the same scale.[31] (Reprinted with permission from *Photonics Technology Letters* © IEEE, 1996.)

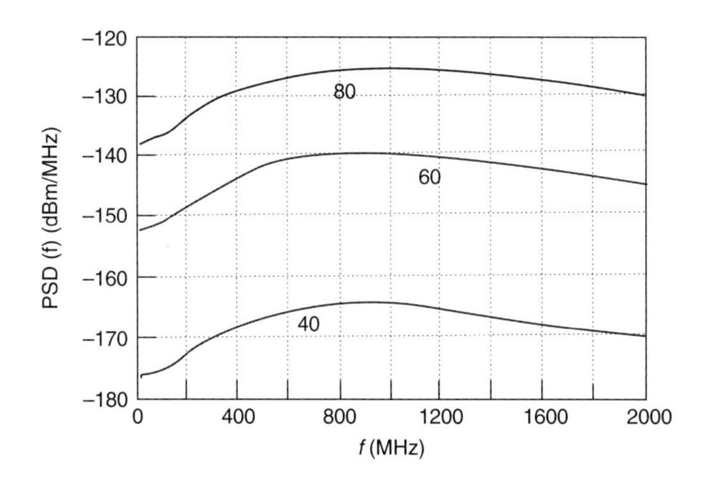


**Figure 17.13** The power spectral density of clipping noise, taking into account the turn-on delay effect. The parameters for this plot are PD responsivity $r = 1$, optical power receiving is 0 dBm, 90 channels of 16 QAM with OMI/ch of 0.6%, and BW of 6 MHz and 40 to 60 AM-VSB channels with 4% OMI/ch and BW = 8 MHz.[16] (Reprinted with permission from *Photonics Technology Letters* © IEEE, 1996.)

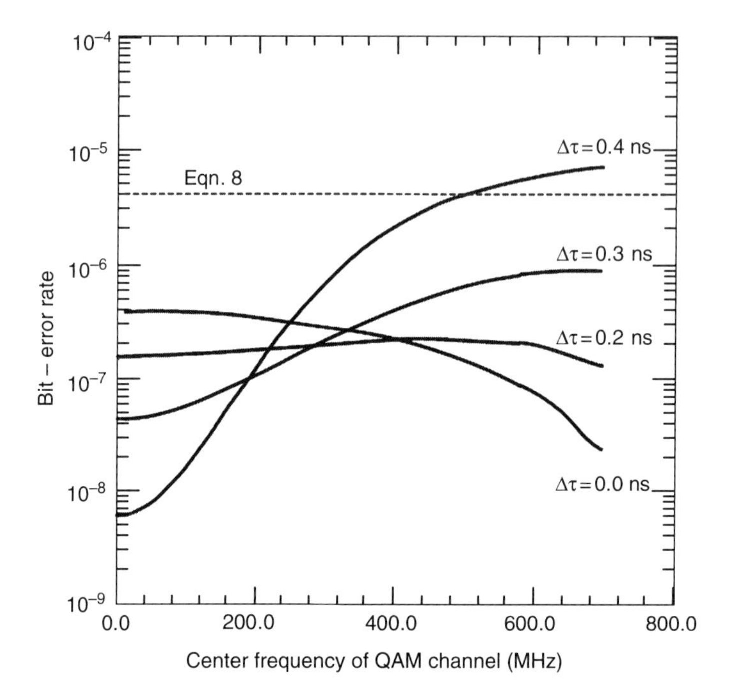


**Figure 17.14** BER vs. the 64-QAM center frequency for different values of the turn-on delay, $\Delta\tau$. Theoretical predictions assume 63 analog carriers (95.25 to 555.25 MHz) and a 64-QAM signal 10 dB below the analog carriers. Total rms modulation depth is assumed to be 25.3%. Dashed line: frequency-independent approximation.[31] (Reprinted with permission from *Photonics Technology Letters* © IEEE, 1996.)

It can be seen that turn-on delay enhances clipping impulsive noise at higher frequencies in hybrid AM-VSB QAM channeling.

From Ref. [31], it is observed that BER is affected by the turn-on delay as well as the QAM channel center frequency. Figure 17.14 illustrates BER performance for various delays.

### 17.2.7 Clipping-reduction methods

In previous sections, it was seen that the clipping effect is mainly dominated by the AM-VSB channels, since they have deeper OMI compared to the M-ary channels. Moreover, it was shown that the resulting clipping bursts induced by the AM-VSB fold into the QAM channels, thus degrading the BER performance of the channel. This subject was analyzed in several publications.[17,28,32–35] The main idea to overcome clipping is to create a predistorter[28] per Fig. 17.15 or have a preclipping circuit[17] as shown in Fig. 17.16.

Preclipping is used to avoid limiting a high OMI signal by the laser transfer function, which is the threshold point. Since the AM-VSB OMI is higher than

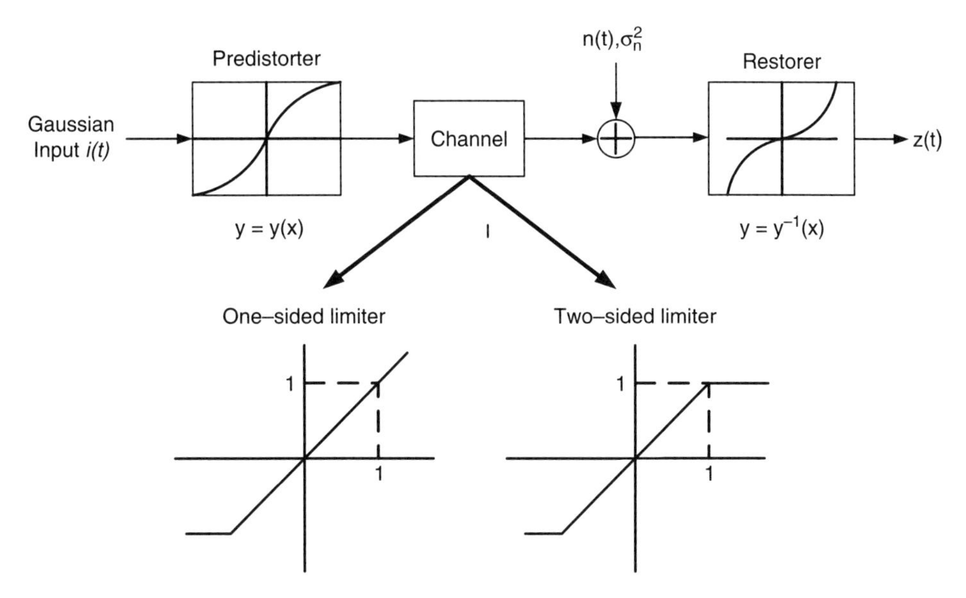


**Figure 17.15** A general model of predistortion and restoration for a one- or two-sided clipping channel with a Gaussian-distributed input signal. In the nonclipping region, the signal is normalized to have a transfer characteristic with a unit slope, unit clipping boundary, and zero intercept.[28] (Reprinted with permission from *Transactions on Communications* © IEEE, 1996.)

the QAM index, a slight clipping prior to the laser modulation will not harm the AM-VSB channels' OMI with respect to the QAM channels. The preclipping process is accomplished by an RF limiter. The limiter output is filtered to remove high-order distortions. The AM-VSB signal is then summed with the QAM transport and modulates the LD. Detection is done by an optical CATV receiver that separates the AM-VSB and the QAM by an RF-diplex filter.

The third approach to overcome clipping is to have an adaptive laser bias point. The circuit senses the incoming analog-modulated signal. The circuit is calibrated to a certain threshold level that defines the RF peak potential for clipping. A peak detector operates as a sampling block. The dc voltage created is filtered and the bias point of the laser is prepared for a large modulation swing, thus, it is momentarily increased. In order to align the RF signal with the bias set, the RF path is delayed by the time amount of the bias feedforward circuit's transfer-function delay.

Based on the previous bias feedforward and preclipping method, a threshold circuit detects the amount of the clipping signal, which is beyond the threshold level. This difference is amplified and then inverted. The inverted signal is added to the delayed SCM RF signal. In this method, the SCM RF peaks are cancelled, and the limiting process created by the laser threshold is minimized. As in the bias feedforward, the RF is delayed in order to align the RF signal

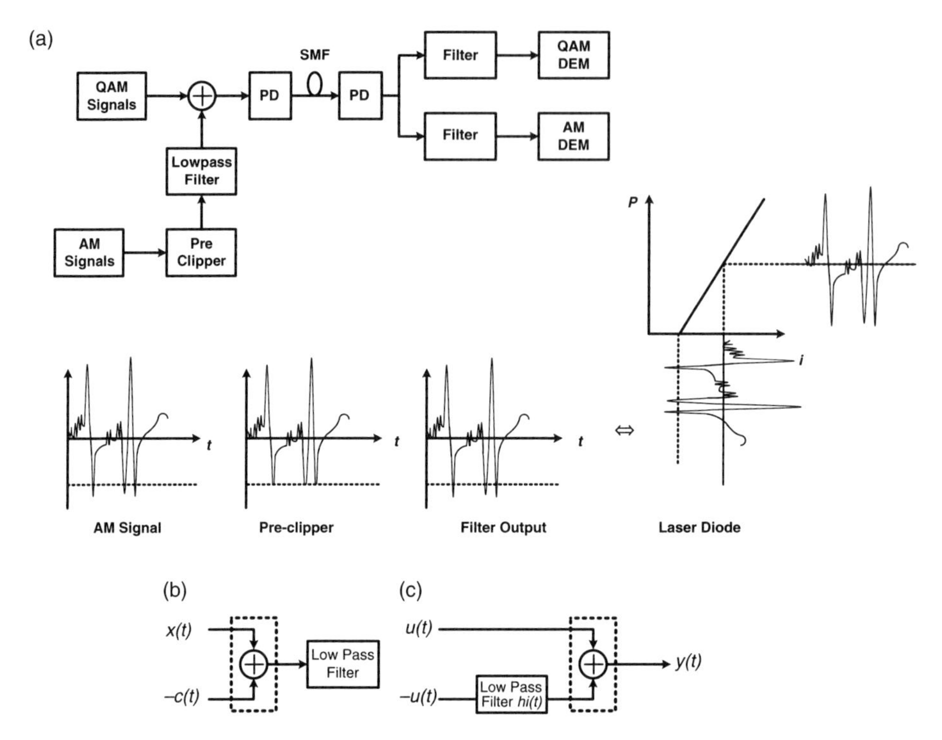


**Figure 17.16** (a) A schematic block diagram of a preclipped AM/QAM-hybrid system with signal illustrations. (b) An analysis model for the preclipper and low-pass filter. (c) An equivalent noise model.[17] (Reprinted with permission from the *Journal of Lightwave Technology* © IEEE, 1997.)

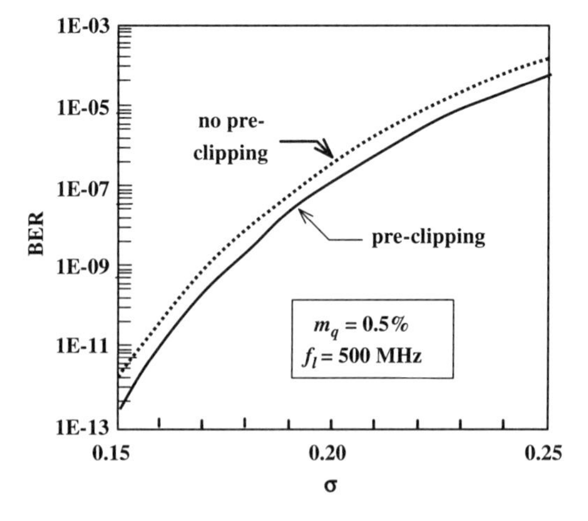


**Figure 17.17** A 64-QAM signal BER as a function of modulation index with and without preclipping circuits.[17] (Reprinted with permission from the *Journal of Lightwave Technology* © IEEE, 1997.)

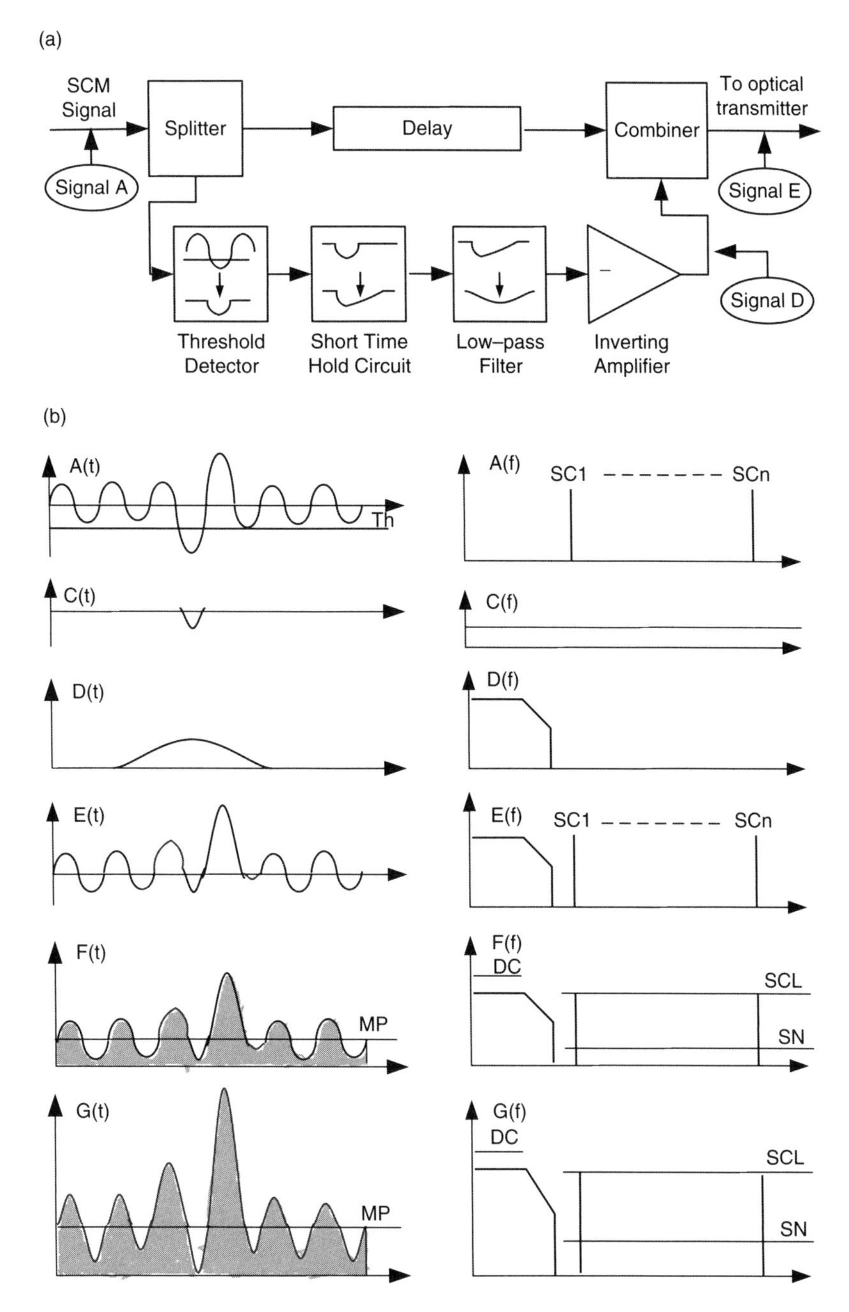


**Figure 17.18** (a) A block diagram of the dissymmetrization circuit; (b) time- and frequency-domain representations of signals relative to the proposed scheme.[35] (Reprinted with permission from *Photonics Technology Letters* © IEEE, 1994.)

with the inverted peaks generated by the processing circuit, which is known also as a dissymmetrization signal as shown in Fig. 17.18.

It has been shown here, and in Fig. 17.17, that preclipping techniques improve the link BER versus AM-VSB CNR by 20-log $OMI_A/OMI_B$.

## 17.3 Bursts of Nonlinear Distortions

In the previous sections, it was explained that the clipping process results in NLD[1–3] and creates CSO and CTB products. These distortion products are not continuous but appear in bursts. These bursts of NLD products are mainly generated by the AM-VSB channels and appear in the QAM digital-CATV band. Thus, QAM-BER performance is degraded.[26] The effect of the bursts can be measured by a spectrum analyzer, which is set to zero span. This way, the spectrum analyzer operates as an oscilloscope. Seventy-nine AM-VSB channels are combined with two 256 QAM. This experiment was previously presented when analyzing dopping effects on BER performance. The QAM channels are at 571.25 and 643.5 MHz, bearing 40.5 Mb/s and directly modulating a 1310-nm DFB laser.[26] The link was through a 10.6-km single-mode fiber (SMF) and was at 0 dBm when detected. A white-noise source with known excess noise ration (ENR) was coupled as well for BER versus CNR test purposes. The receive conditions were 51.8 dB CNR, average CSO −56.8 dBc, and CTB of −60 dBc. The modulation conditions were 3.5% OMI for the analog channel and clipping index of $\gamma \approx 4 \times 10^{-4}$. To observe a CSO burst, the zero-span testing is used. The spectrum analyzer is set to the center frequency of the channel under test; in this case, the QAM channel is at 571.25 MHz + 1.25 MHz offset, and thus it is set to 572.5 MHz for the CSO test (see Chapter 3). When doing zero-span, the trigger can be an external or internal video. In this case the trigger is internal and set to be 30 dB below the QAM level to start sweeping at the instance of

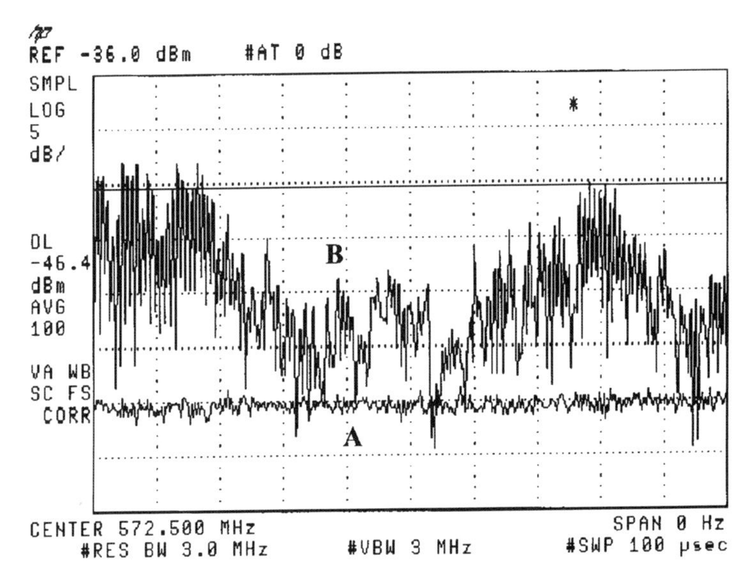


**Figure 17.19** A bursty NLD CSO test. Sweep time: 100 μs; center frequency: 572.5-MHz zero-span mode. Trace A is the average power sweep and the solid line is the single-sweep trigger level.[36] (Reprinted with permission from the *Journal of Lightwave Technology* © IEEE, 1998.)

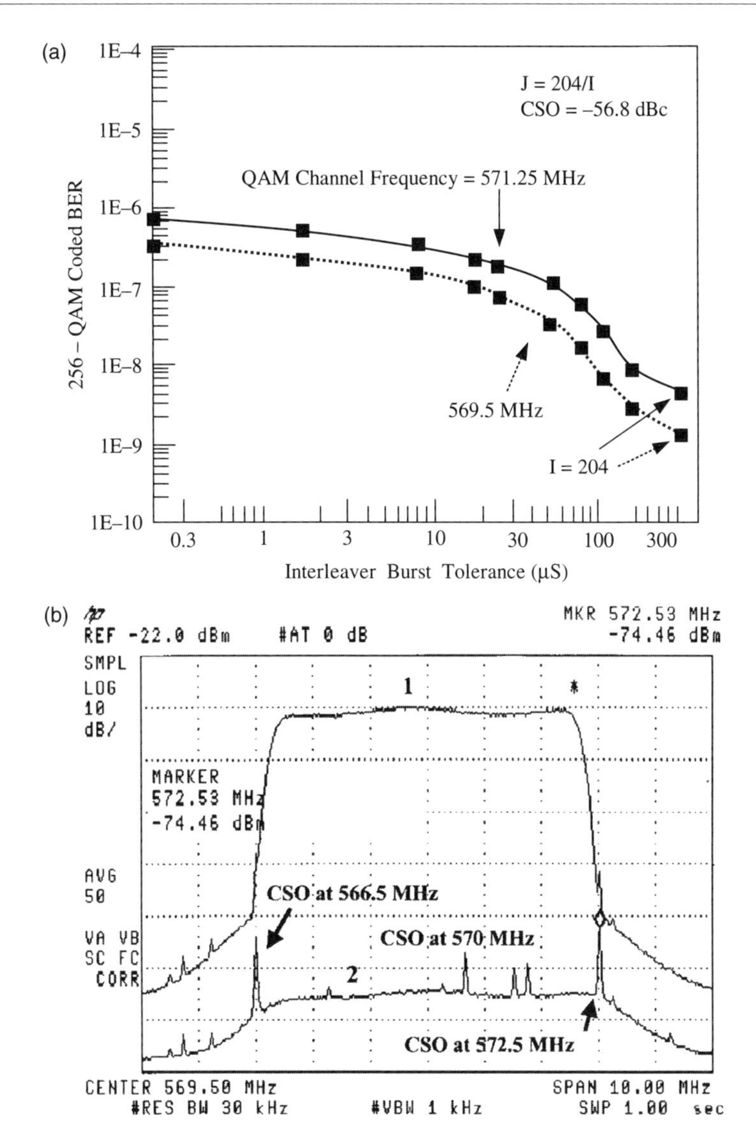


**Figure 17.20** (a) Code protected QAM BER vs. the interleaver EC length. QAM SNR equals 30 dB, and CSO distortion is −56.8 dBc. Two cases are shown: one where the CSO is in band, and another where it is out of band, which affects the BER. (b) Trace 1 shows an RF spectrum of a 256 QAM channel centered at 569.5 MHz and without; Trace 2 shows the NLD CSO products. Note that dominant CSO are at BW edges.[26] (Reprinted with permission from *Photonics Technology Letters* © IEEE, 1998.)

peak CSO distortion. Figure 17.19 illustrates the test, where Trace A is the average threshold level, and trace B shows the CSO peak. The horizontal solid line is the single sweep trigger level. The test of −30-dB NLD corresponds to −30-dB AWGN CNR for a coded BER = $1.5 \times 10^{-9}$; therefore any CSO above that line refers to BER degradation.

Chapters 12 and 21 explain that the AM-VSB peak-to-average can vary a lot, by up to 18 dB, depending on the picture content, and the sync and blank signals in special cases may result in a large X-mod effect. These changes affect CSO and CTB products as well. Hence, CW tones are not very accurate for examining NLD bursts. In previous sections, it was explained the laser RF AGC for modulation depth control fixes RF level. Thus OMI for CW tones is reduced, owing to the peak-to-average issue, as explained in Chapter 12. Hence in the CW case, NLD products are lowered; therefore, the CW test is not accepted.

The method to combat the generated burst errors is by a variable convolutional interleaver in QAM MODEM. The modem structure was briefly reviewed in Chapter 15. The interleaver depth is of up to 204 symbols, and $J = 204/I$ symbols are used with the RS code.[26] The interleaving and coding disperses the burst errors in time, so that there is no long error burst that the receiver cannot overcome. According to Ref. [26], the burst noise tolerance provided by the interleaver on the transmitter side and the deinterleaver on the receiver side is given by Eq. (17.93), which is also called erasure correction, EC [μs], which is equal to $\tau$:

$$\tau = \frac{(J \cdot I + 1)T}{R_{\mathrm{RS}}}\left(\frac{1}{N}\right), \tag{17.93}$$

where $I$ and $J$ are the interleaver parameters, and $N = 204$ symbols is the $R$–$S$ block size. $R_{\mathrm{RS}}$ is the transmitted $R$–$S$ symbol rate and $T = 8$ symbols. From Chapter 15, it is clear that the larger the end-to-end latency of the interleaver/deinterleaver, the larger the interleaver burst tolerance EC [μs]. Thus, the link is more protected against NLDs such as CSO bursts, since the NLD bursts are dispersed. An optimal interleaver would have $I = 205$ and $J = 1$. Figure 17.20 shows that as soon as the interleaver EC is larger than the NLD burst duration by about three times, BER improves significantly.[26] Additional optimization is to center the QAM band in such a way that NLDs are out of band, as can be seen in Fig. 17.20.

## 17.4 Multiple Optical Reflections

This chapter and previous sections demonstrated the effects on link-analog-CATV quality that results from modulation depth affecting the BER performance. The second impairment that affects an optical link is multiple optical reflections. Reflections are a result of fiber discontinuities in the refraction index and Rayleigh backscattering effect, which occurs in long cables due to microscopic inhomogeneities,[37,44] as was explained in Chapter 4. Chapter 8 explained that PDs respond to the $E$-field portion of a lightwave, where the optical light power is proportional to the $E$-field square. Since a light beam is an electromagnetic wave, discontinuities in its propagation path result in reflections. Such discontinuities are abrupt changes in refraction indices at connection points, which can be minimized by using angle-polished connectors, as explained in Sec. 4.2.

### 17.4.1 Double Rayleigh backscattering

Figure 17.21 illustrates the generation of double-Rayleigh backscattering (DRB). These reflections generate a beat noise. Assume $R_1$ is the reflection factor at point $r_1$, and $R_2$ is the reflection factor at point $r_2$, $P_{12}$ is the polarization coefficient between the two reflection points, and $\eta_{12}$ is the fiber loss between the two points. Hence at a given dual reflection combination, the equivalent reflectance is denoted by

$$R_{\mathrm{eq}}^2 = P_{12}\eta_{12}R_1R_2. \tag{17.94}$$

So far, this modeling is for a single reflection; however, assume the fiber length $L$ consists of $N$ identical separate microreflections. Hence it can be written that $L = N\Delta x$, while all reflections are incoherent, and $\Delta x$ is the differential fiber length. To analyze this problem, a microreflector cell is characterized. Each cell has a loss of $\eta = \exp(-\alpha\Delta x)$ and reflectivity $R = S\alpha_s\Delta x$, where $S$ is the fraction of scattered light that was captured by the fiber, and $\alpha_s$ is the proportion of the signal scattered per length unit. Having these definitions, all reflection combinations along the fiber can be calculated. Using an arithmetic-series sum for $N$ cells, there are $N(N-1)/2$ fabry perot (FP) etalon reflectors, resulting in a total equivalent reflectance coefficient given by

$$R_{\mathrm{eq}}^2 = \sum_{i=1}^{N-1}\sum_{j=i+1}^{N} P_{ij}R_iR_j\eta_{ij} = PR^2\frac{\eta}{(1-\eta)^2}\left[N(1-\eta) - \left(1-\eta^N\right)\right], \tag{17.95}$$

where $i$ and $j$ are the microreflector cell indices. To simplify Eq. (17.95), it is assumed that parameters are homogeneous for each cell, equal to $R$ and $P$, and that the spacing between adjacent microreflectors is equal. Interactions between fiber and discrete reflections Eq. (17.95) are written in a recursive way. Therefore,

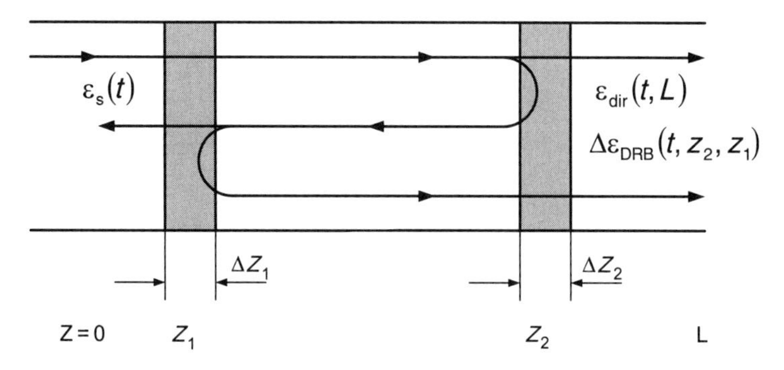


**Figurc 17.21** Interterence intensity noise that resuls from DRB and directly propagated light, which creates a beat noise.[37] (Reprinted with permission from the *Journal of Lightwave Technology* © IEEE, 1996.)

Eq. (17.96) can be written as

$$R_{eq}^2(N+1) = R_{eq}^2(N) + R_{N+1}R_{in}(N). \tag{17.96}$$

Equation (17.96) describes the increased noise due to interaction between the added reflector $R_{N+1}$ and the original group $R_{in}$ given by

$$R_{in}^2(N) = PR\left(\frac{1-\eta^N}{1-\eta}\right). \tag{17.97}$$

Using the above noted reflection cell properties, and letting $P = 0.5$, since Rayleigh scattering is randomly polarized, the insertion of a microreflector cell is given by:

$$R_{in}^2(L) = \frac{S\alpha_s}{4\alpha}[1 - \exp(-\alpha L)], \tag{17.98}$$

$$R_{eq}^2(L) = \frac{1}{2}\left(\frac{S\alpha_s}{\alpha}\right)^2 [\alpha L - 1 + \exp(-\alpha L)], \tag{17.99}$$

where using the Taylor series, $\eta$ is approximated by

$$\eta \approx 1 - \alpha \cdot \Delta x. \tag{17.100}$$

The parameters of $S$ and $\alpha$ and $\alpha_s$ are provided by fiber manufacturers and are measured for the wavelength of interest. For instance, at 1310 nm, $S = 0.0015$ and $\alpha = \alpha_s = 0.081\text{Np/km}$.

### 17.4.2 Double Rayleigh backscattering effects on laser RIN

One of the main concerns with reflections is their effect on laser performance. The reflected lightwave can parasitically modulate the transmitting laser. One of the ways to overcome this is to insert an isolator; however, the receiver is affected from those delayed reflections. An analysis of that phenomenon is provided in Refs. [27] and [37]. Assume an emitted, nonmodulated field according to:

$$E_i(t) = A\exp\{j[\omega_0 t + \varphi(t)]\}, \tag{17.101}$$

where $\omega_0$ is the center frequency of the laser and $\varphi(t)$ is the laser's random phase noise. Using Fig. 17.21, the receiver's PD is impinged by the same frequency fields delayed in time. This time delay results in phase difference and beat note between the delayed and undelayed signals. Thus, it can be written that between the two optical connectors, the effective forward loss has a reflection coefficient given by $\alpha_1\alpha_2$, and at the receiver's input is the directly propagating field given by

$E_i(t)\sqrt{\alpha_1\alpha_2}$ and the double reflection that has the same polarization, $E_i(t-\tau)\alpha_1\sqrt{\alpha_1\alpha_2R_1R_2}$, which is the worst-case scenario. The received signal for responsivity equal to unity, $r=1$, is given then by

$$I(t) = \left|E_i(t)\sqrt{\alpha_1\alpha_2} + E_i(t-\tau)\alpha_1\sqrt{\alpha_1\alpha_2R_1R_2}\right|^2, \tag{17.102}$$

where $\alpha_1$ is the first connector's forward insertion loss, $\alpha_2$ is forward loss for the second connector, $R_1$ is the first connector's reflection factor and $R_2$ is the reflection factor for the second connector, and $\tau$ is the associated delay with this traveling delay. Defining $R=\alpha_1\sqrt{R_1R_2}$ and $\alpha=\alpha_1\alpha_2\alpha_1$, $I(t)$ can be written as

$$I(t) = \left|E_i(t)\sqrt{\alpha} + E_i(t-\tau)R\sqrt{\alpha}\right|^2 \cong \alpha|A|^2[1+\rho(t,\tau)]. \tag{17.103}$$

The parameter $\rho(t,\tau)$, which contains the interference component affecting the intensity of $I(t)$ is given by

$$\rho(t,\tau) = 2R\cos[\omega_0\tau + \phi(t) - \phi(t-\tau)]. \tag{17.104}$$

Finding the autocorrelation of Eq. (17.104) results in

$$\begin{aligned} S_\rho(t,\tau) &= \langle \rho(t',\tau)\rho(t'+t,\tau)\rangle \\ &= 4R^2\langle \cos[\omega_0\tau + \theta(t',\tau)]\cos[\omega_0\tau + \theta(t+t',\tau)]\rangle, \end{aligned} \tag{17.105}$$

where the random zero crossing expressed by $\theta(t,\tau)=\phi(t)-\phi(t-\tau)$ is normally distributed with a variance of $\sigma^2 = 2\pi\tau\Delta\upsilon$, where $\Delta\upsilon$ is the 3-dB Lorentzian linewidth of the laser. According to Refs. [27], [36], [37], and [83], the autocorrelation function is given by

$$S_\rho = 2R^2\begin{Bmatrix} \exp(-2\pi\tau\Delta\nu)\cdot[1+\cos(2\omega_0\tau)], & |t|>\tau \\ \exp(-2\pi\Delta\nu|t|)\cdot\{1+\cos(2\omega_0\tau)\exp[-4\pi\Delta\nu(\tau-|t|)]\}, & |t|<\tau \end{Bmatrix}. \tag{17.106}$$

Taking the Fourier transform of Eq. (17.106) and removing out the dc terms provides the laser RIN($f$):

$$\begin{aligned} \text{RIN}(f) = 4R^2\left\{1+\left[\frac{-\sin(2\pi\tau f)}{\pi f}\right]\exp(-2\pi\tau\cdot\Delta\upsilon)\cdot\cos(2\omega_0\tau)\right\} + \frac{4R^2}{\pi}\left[\frac{\Delta\upsilon}{f^2+(\Delta\upsilon)^2}\right] \\ \times\begin{pmatrix} 1-\exp(-2\pi\tau\cdot\Delta\upsilon)\left[\cos(2\pi\tau f)-\dfrac{f}{\Delta\upsilon}\sin(2\pi\tau f)\right] \\ -\cos(2\omega_0\tau)\{\exp(-4\pi\tau\cdot\Delta\upsilon)-\exp(-2\pi\tau\cdot\Delta\upsilon) \\ \left[\cos(2\pi\tau f)+\dfrac{f}{\Delta\upsilon}\sin(2\pi\tau f)\right]\} \end{pmatrix}. \end{aligned} \tag{17.107}$$

Equation (17.107) shows that the maximum conversion of phase noise into RIN transpires when the main propagating beam of light is beats in quadrature with the doubly reflected beam of light. That condition is when $\omega_0\tau = (n \pm 1/2)\pi$. The expression for that case is provided by[38]

$$\mathrm{RIN}(f) = \frac{4R_1R_2}{\pi}\left[\frac{\Delta\nu}{f^2 + \Delta\nu^2}\right][1 + \exp(-4\pi\tau\cdot\Delta\nu) - 2\exp(-2\pi\tau\cdot\Delta\upsilon)\cos(2\pi f\tau)]. \tag{17.108}$$

Equation (17.108) can be further approximated; for $2\pi\tau\cdot\Delta\nu \ll 1$, the RIN is given by

$$\mathrm{RIN}(t) = 16\pi R^2\Delta\upsilon\tau 2\mathrm{sinc}^2(f\tau). \tag{17.109}$$

This describes the limit of coherent small phase fluctuations, since the delay affects the phase very little, and the maximum RIN is proportional to the laser linewidth. The incoherent case is when the phase delay satisfies $2\pi\tau\cdot\Delta\nu \gg 1$. Under these conditions, Eq. (7.108) becomes

$$\mathrm{RIN}(t) = \frac{4R^2}{\pi}\frac{\Delta\nu}{\Delta\nu^2 + f^2}. \tag{17.110}$$

The conclusion is to select a narrow linewidth laser. Professional CATV transmitters utilize yttrium-aluminum-garnet (YAG) lasers with a narrow linewidth. Furthermore, this justifies the reason for EM transmitters, and thus the above-mentioned limits are negligible. DML, however, results in significantly increased RIN, and chromatic dispersion and chromatic RIN are increased.

### 17.4.3 Effects on QAM and AM-VSB CATV channels

The above-described effects decrease the quality of AM-VSB CATV in terms of CNR and QAM BER due to the decrease in CNR and increase of RIN in DML-type transmitters. Link tests are done by applying a transmitter and controlled impairments in order to evaluate performance.[25,36] The link is in accordance to CATV-channel planning of 61 AM-VSB and 16 QAM. OMI is adjusted as shown in Fig. 17.22. Reflections are controlled by variable back-reflectors with double polarization control. AWGN was added to control CNR. To achieve fiber effects, a 4-km SMF was used. QAM is generated by an externally triggered pattern generator that provides a 20 Mb/s, $2^{15} - 1$ pseudo-random beat sequence (PRBS) to the QAM that generates 16 QAM. The experiment was compared DML-DFB against EM-YAG laser (Fig. 17.23).

The experiment showed a significant drop of CNR versus reflections on a link based on 1310-nm DML DFB against EM-YAG laser. The channels monitored under this experiment were 55.25, 205.25, and 439.25 MHz. The OMI for this experiment was 3.3%.

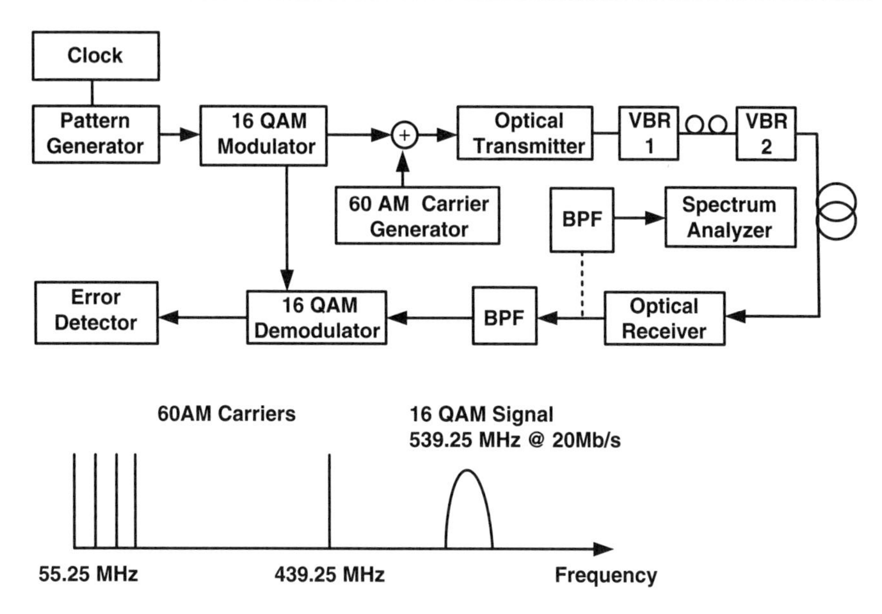


**Figure 17.22** A controlled experiment block diagram examining reflection effects on AM-VSB and QAM channels.[25] (Reprinted with permission from the *Journal of Photonics Technology Letters* © IEEE, 1994.)

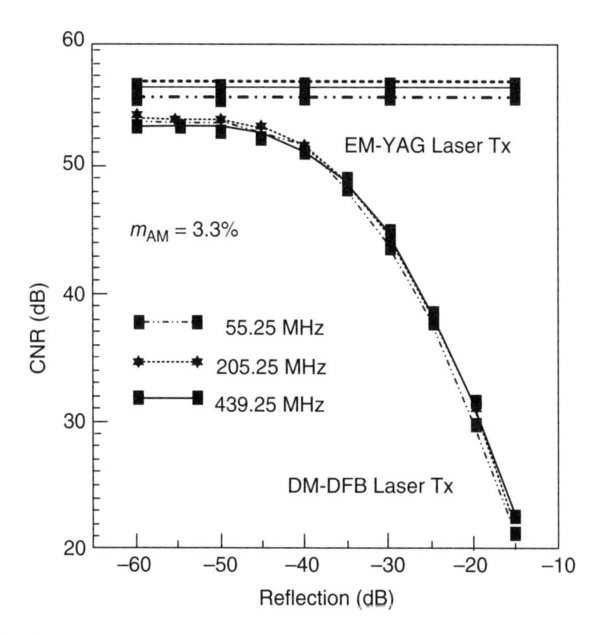


**Figure 17.23** CNR vs. reflection, comparing 1310 nm DML-DFB against 1319-nm EM-YAG laser. OMI equals 3.3%.[36] (Reprinted with permission from the *Journal of Lightwave Technology* © IEEE, 1998.)

The second experiment (Fig. 17.24) was made on a 1550-nm EM DFB with 10.8 km SMF with an OMI of 3.2%. Although this is an external modulated laser, the DFB CNR showed a strong frequency dependency versus reflections. This is because DFB linewidth is wider than the negligible linewidth of the YAG. Thus, the RIN of the DFB increases, as in Eq. (17.109). The next part of the link-susceptibility test examined the reflection effects on an uncoded 16-QAM transport versus the QAM OMI applied on a DML, as shown in Fig. 17.25.[36] This link was measured while having AM-VSB with 3.3% OMI/ch, as shown in the test set-up in Fig. 17.22. It can be seen that for −60-dB reflections, BER performance is close to the ideal case of no reflection. AM-VSB affects QAM BER in two ways: CNR degradation and clipping effects.

The next step in this CATV link test was to study coded 256-QAM transport sensitivity to SMF reflections, while the QAM OMI is 0.8%/ch. This experimental set-up is shown in Fig. 17.27 and compared with a DML-DFB, an EM-DFB, and a EM-YAG laser. Figure 17.26 shows the DM-DFB laser transmitter with an AM-modulation index of 3.3%, Fig. 17.26, a EM-YAG laser transmitter with an AM-modulation index of 3.6%, and Fig. 17.26, a 1550-nm DEM-DFB laser transmitter with an AM-modulation index of 3.2%. Thus, the EM-YAG laser transmitter–based system has the best immunity to multiple optical reflections for both the AM and the QAM channels.

Figure 17.28 depicts measured and calculated CSO at channel 2, 12 and 40 vs. the fiber span for 42 AM-VSB video channels. The system uses DML (direct modulated laser) of DFB type with frequency chirp of 720 MHz/mA for both experiments, over the SMF and DSF (dispersion shifted fiber). The system requirement is CSO< − 65 dBc. Hence for DML-DFB, SMF limits the distance compared to DSF. This is the main reason for not using DML-DFB of 1550 nm for

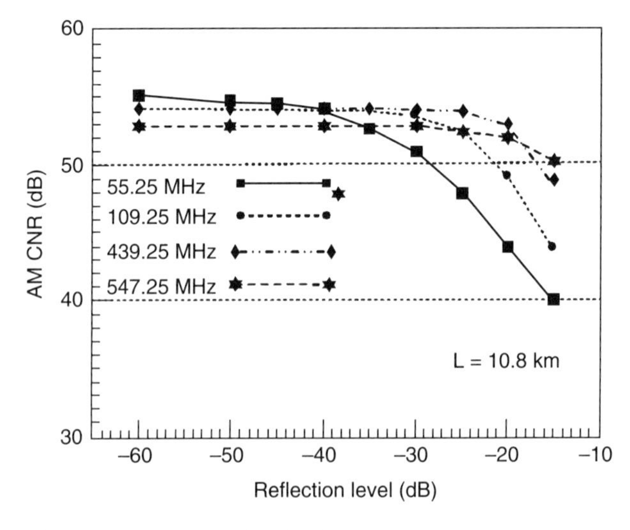


**Figure 17.24** CNR vs. reflection of 79-AM-VSB channels modulating 1550-nm EM-DFB on a link of a 10.8-km-SMF, channel frequency is a parameter. OMI equals 3.2%.[36] (Reprinted with permission from the *Journal of Lightwave Technology* © IEEE, 1998.)

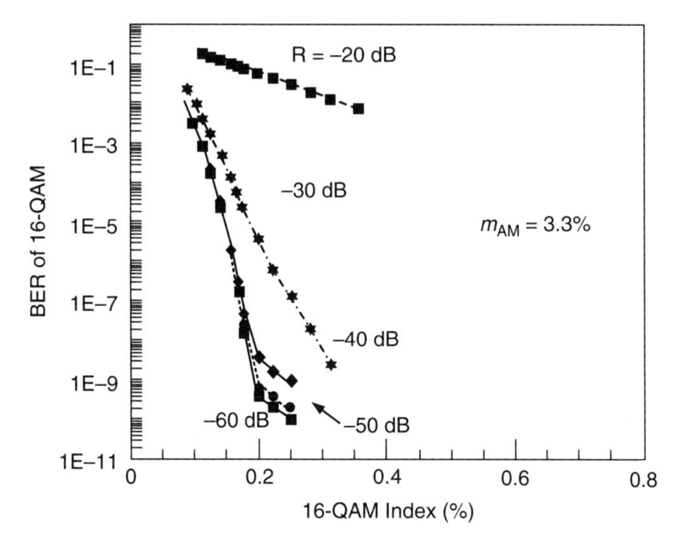


**Figure 17.25** Uncoded 16-QAM BER vs. the QAM OMI percentage for various reflection indices for a DML 1310-nm DFB-laser transmitter, the AM-VSB channels are at 3.3% OMI/ch.[36] (Reprinted with permission from the *Journal of Lightwave Technology* © IEEE, 1998.)

long distance transport over SMF. To overcome this problem of chirp when using a DFB, an external modulator (EM) is used rather than the DML method. An EM modulator solves issues of dispersion but not CSOs as explained in Sec. 17.2.1.[29] The other method to combat chromatic dispersion is using DSF.

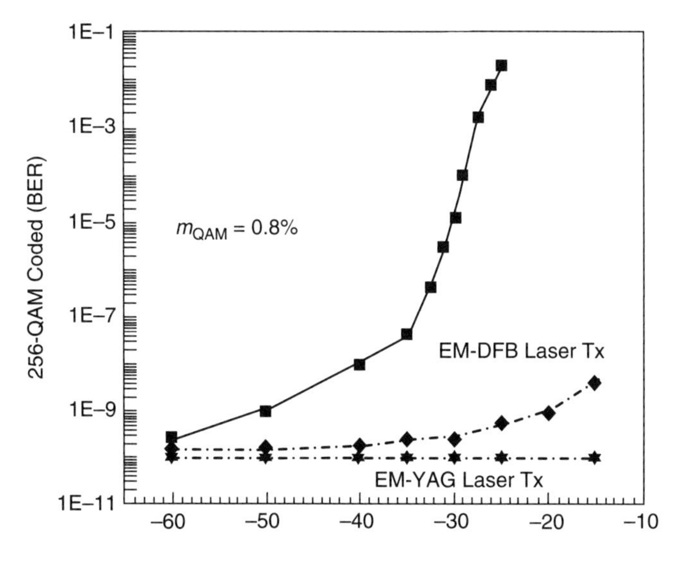


**Figure 17.26** The measured 256-QAM-coded BER vs. the reflection level at a QAM index of 0.8% for a DM-DFB laser transmitter, an EM-YAG laser transmitter, and a 1550-nm EM-DFB laser transmitter.[36] (Reprinted with permission from the *Journal of Lightwave Technology* © IEEE, 1998.)

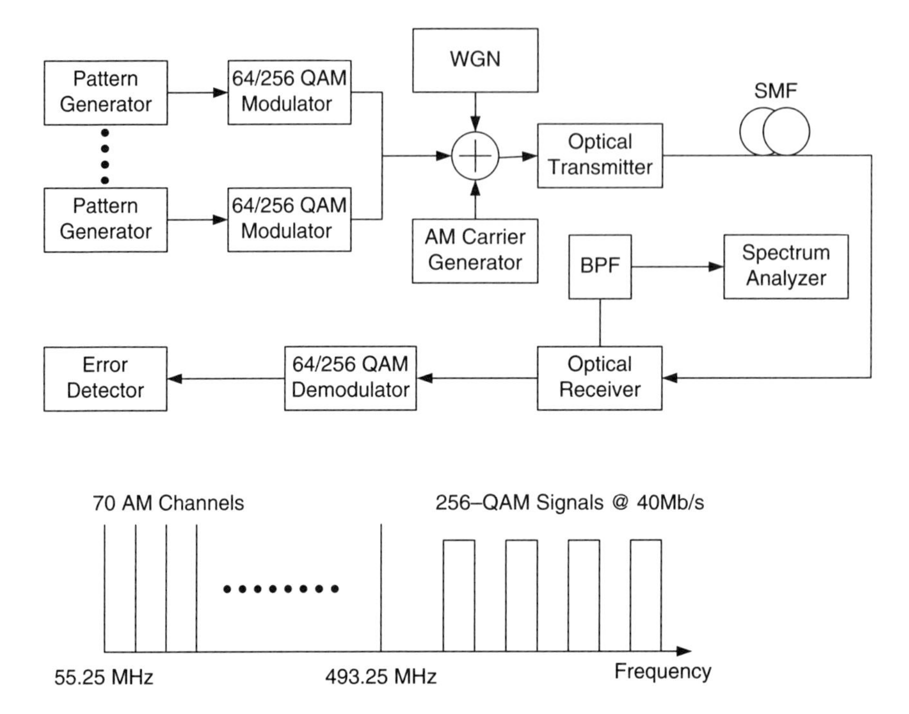


**Figure 17.27** The experimental set-up and frequency allocation plan for a hybrid multichannel AM-M-QAM video lightwave transmission system.[36] (Reprinted with permission from the *Journal of Lightwave Technology* © IEEE, 1998.)

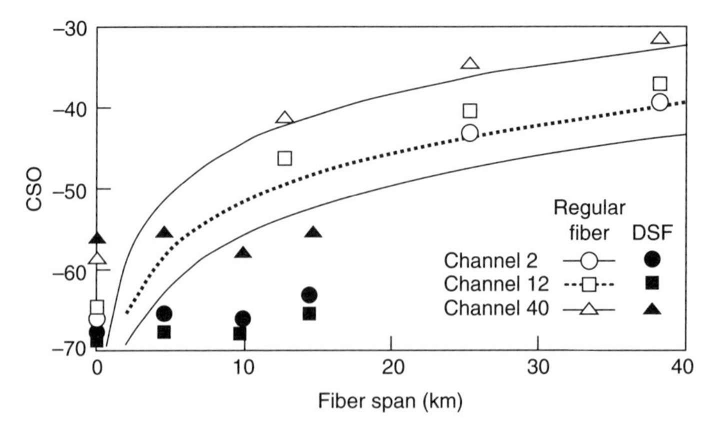


**Figure 17.28** CSO distortions vs. SMF and dispersion-shifted fiber (DSF) length. Measured (symbols) and calculated lines results are shown for channels 2, 12, and 40. For a DML transmitter with a chirp of 720 MHt/mA.[41] (Reprinted with permission from the *Journal of Photonics Technology Letters* © IEEE, 1991.)

## 17.5 Dispersion-Induced Nonlinear Distortions

This section explains the last passive characteristics of an SMF, which is degrading the NLD performance of an optical link due to its dispersion, as explained in Sec. 4.1.[44] The first mechanism was reviewed in Sec. 17.4 and is related to multiple reflections. Multiple reflections create beat-note distortions as well as eye degradation that appears in jitter. The jitter is due to the delay in the arrival of data and appears as shadow eye, thus eye-opening is affected.[43] The second mechanism is dispersion due to DML laser chirp and residual chirp in EM, which generates the chromatic dispersion. Moreover, it was explained that the laser suffers from mode hopping. Thus, at a given instance, the laser emits several wavelengths, with one dominant wavelength that hops to another wavelength, and so on. Thus, besides the chirping effect in DML, there is a mode-hopping effect that cannot be ignored in the case of EM. The EM laser combats modulation chirp, but it is not a mode-hopping cure. One of the methods to examine NLD due to dispersion is to use a two-tone test. By using these tones on a DFB and directly modulating it, the optical power at the SMF input can be described as an AM[39]

$$P_{\rm in} = P_0\left\{1 + {\rm OMI}\left[\sin\left(\omega_1 t + \phi_1\right) + \sin\left(\omega_2 t + \phi_2\right)\right]\right\}, \tag{17.111}$$

where $P_0$ is the averaged optical power, $\omega_1$ and $\omega_2$ are the angular frequencies of the subcarriers, and $\phi_1$ and $\phi_2$ are the phase differences between laser chirp and intensity modulation. The instantaneous electric field is given by

$$E_{\rm in} = \sqrt{2P_0}\sqrt{1 + {\rm OMI}\left[\sin\left(\omega_1 t + \phi_1\right) + \sin\left(\omega_2 t + \phi_2\right)\right]}. \tag{17.112}$$

Using the Taylor series, Eq. (17.112) can be expanded, and its spectral distribution given by the Fourier transform showing its FM components.[39,41,42] However, a straightforward analysis can be made by observing the wave propagation through the SMF and how it is affected by its losses and its group delay variation due to the fiber's dispersion. Hence, the optical power at the fiber output is given by[40]

$$P_{\rm T\times SMF} = \alpha\frac{{\rm d}(t - \tau)}{{\rm d}t}P_{\rm in}\left[t - \Delta\tau_{\rm g}(I)\right], \tag{17.113}$$

where $\Delta\tau_{\rm g}(I)$ is the group delay variation as a function of the modulating current of the DFB laser, which results due to the SMF dispersion, and $\alpha$ is the SMF attenuation factor, as explained in Sec. 17.4.1. The group delay variation as a function of the modulating current is given by

$$\Delta\tau[I(t)] = DL\Delta\lambda(t) \cong DL\frac{{\rm d}\lambda}{{\rm d}f}\frac{{\rm d}f_{\rm opt}}{{\rm d}I}I(t) - -DL\frac{\lambda^2}{c}\frac{{\rm d}f_{\rm opt}}{{\rm d}I}I(t), \tag{17.114}$$

where $f_{\rm opt}$ is the optical frequency, $\lambda$ is optical wavelength, $c$ is the speed of light, $D$ is the fiber dispersion, $L$ is the fiber length, ${\rm d}f/{\rm d}I$ is the frequency chirp due to direct modulation, and $I(t)$ is the harmonic modulation current of the two subcarriers, which is given by

$$I(t) = I_{\text{peak}}\left[\sin(\omega_1 t + \varphi_1) + \sin(\omega_2 t + \varphi_2)\right]. \tag{17.115}$$

The power term variation in Eq. (17.113) can be expanded by the first term of Taylor series:

$$P_{\text{in}}\left[t - \Delta\tau_g(I)\right] = P_{\text{in}}(t) - \Delta\tau_g(I)\frac{dP_{\text{in}}(t)}{dt}. \tag{17.116}$$

Substituting Eqs. (17.114) into (17.116) and then into Eq. (17.113), the second-order CSO with respect to the carrier are provided by

$$\frac{IM_2}{C} = \left[D \cdot L \cdot I_{\text{peak}}\left(\frac{\lambda^2}{c}\right)\frac{df_{\text{opt}}}{dI}\right]^2 (\omega_1 + \omega_2)^2, \tag{17.117}$$

where $C$ is the carrier power level and $IM_2$ is the second-order distortion level. Since CSO is composite second order rather than discrete second order (DSO), all products should be counted. The CSO frequency would be denoted by $\omega_d$. Hence Eq. (17.117) becomes

$$\text{CSO} = N_{\text{CSO}}\left[D \cdot L \cdot I_{\text{peak}}\left(\frac{\lambda^2}{c}\right)\frac{df_{\text{opt}}}{dI}\right]^2 (\omega_d)^2. \tag{17.118}$$

The CSO is proportional to the square of the distortion frequency, the frequency chirp, and the laser-driving current as well as the wavelength. Thus one way to combat these induced chromatic distortions is by EM. In this case, the CSO and CTB are given by[23]

$$\text{CSO} = 10\log\left[\frac{\text{OMI} \cdot D \cdot \lambda_c^2 L \cdot f}{4c}\sqrt{16(\Delta\tau)^2 + \frac{4\lambda_c^4 D \cdot L^2 \pi^2 f^6}{c^2}}\right] + 10\log N_{\text{CSO}} + 6, \tag{17.119}$$

$$\text{CTB} = 10\log\left\{\frac{9 \cdot \text{OMI}^2 \cdot D^2 \cdot \lambda_c^4 L^2 \cdot f^2}{4c}\left[4(\Delta\tau)^2 + 4\pi^2 f\right]\right\} + 10\log N_{\text{CTB}} + 6, \tag{17.120}$$

where $f$ is the RF frequency, $\lambda_c$ is the optical carrier wavelength, and $N_{\text{CSO}}$ and $N_{\text{CTB}}$ are the product counts of the CSO and CTB products. It can be seen the CSO/CTB is proportional to $10\log(f^4)/10\log(f^3)$.

## 17.6 Optical Fiber and Optics Nonlinear Effects

This section reviews nonlinear effects resulting from the SMF. Even though SMF seems to be a passive device, its characteristics at high optical power levels resemble those of an active device when producing NLD products. These phenomena are major considerations when designing a hybrid fiber coax (HFC) WDM or DWDM system for a long haul. Such long fiber deployments with multiusers, as in the FTTx, require amplification using EDFAs and splitting. High optical power might reach the SMF upper bound of optical power injection. These levels generate several fiber NLDs, which are SBS[7], SPM, and external phase modulation (EPM).[8] When having a multiwavelength system such as WDM or DWDM, additional nonlinearities appear, such as SRS and XPM. Additionally, the optical front-end contributes to NLD products owing to PDD effects. All these impairments affect the link budget and performance due to CNR degradation, carrier to X-talk ratio, and carrier-to-distortion ratio, which affect BER performance and eye opening.

### 17.6.1 Stimulated Brillouin scattering effect

Stimulated Brillouin scattering effect is an acoustic phenomenon in which the transmitted optical power in the SMF is converted into a backward scattered signal. Consequently, this limits the maximum optical power that can be launched into the fiber. This effect is the result of the high optical power momentum, $h\nu$, being partially absorbed by a phonon causing its vibration. This vibration affects the refractive index by generating an acoustical fiber Bragg grating (FBG). In order to create reflections, its period should satisfy $\Lambda = \lambda/2n$, where $n$ is the refraction index. For SMF, with $n = 1.5$ and wavelength of 1550 nm, the period $\Lambda \approx 0.516\ \mu\text{m}$. Assuming the acoustic ultrasonic wave, generated due to the optical excitation, and energy and momentum conservation, propagates at 6 km/s, then phonon vibration is at the frequency of 6 [km/s]/0.516 [μm] = 11.6 GHz. The vibration frequency is lower than the optical frequency. The overall momentum is preserved; thus, the result is a backscattered light beam at a lower frequency. This interaction between optical wavelength at a given power and the SMF is called the Brillouin–Stokes frequency offset. Therefore, there is a threshold power for SBS to occur.[47,48]

In optical fibers, SBS can be observed only in the backward direction, since the Brillouin shift in the forward direction becomes zero. The Brillouin shift is given by:

$$f = \frac{2n}{\lambda} \nu_B, \tag{17.121}$$

where $\nu_B$ is the velocity of sound waves, and $f$ is frequency shift. The SBS threshold for CW light, $P_{CW}$, may be written according to the Smith condition:[47,52]

$$P_{CW} = 21 \frac{A_e K}{g_B L_e}, \tag{17.122}$$

where $L_e$ is the effective interaction length given by

$$L_e = \frac{1 - \exp(-\alpha L)}{\alpha}, \tag{17.123}$$

and $A_e$ is the effective core area of the fiber, $\alpha$ is the fiber loss in $m^{-1}$, $L$ is the length of the fiber, $K$ is the polarization factor, $1 \le K \le 2$, and $g_B$ is the peak Brillouin gain coefficient. With these values, the SBS threshold is given by

$$P_{CW} = 21 \frac{A_e K}{g_B L_e} \frac{(\Delta\nu_p \otimes \Delta\nu_B)}{\Delta\nu_B}, \tag{17.124}$$

where $\otimes$ denotes convolution, $\nu_P$ is the pump linewidth, and $\nu_B$ is the spontaneous Brillouin BW. For Gaussian profiles, the convolution is given by

$$\nu_p \otimes \nu_B = \sqrt{\nu_p^2 + \nu_B^2}. \tag{17.125}$$

For Lorentzian profiles, the convolution is given by

$$\nu_p \otimes \nu_B = \nu_p + \nu_B, \tag{17.126}$$

and $g_B$ is given by Ref. [48]:

$$g_B = \frac{2\pi n^2 P_{12}^2 K}{C\lambda^2 \rho_0 V_a \Delta\nu_B}, \tag{17.127}$$

where $\rho_0$ is the material density, $P_{12}$ is the elasto-optic coefficient, and $V_a$ is the acoustic velocity.

SBS affects both analog and digital transports. AM-VSB and QAM signal are affected by the CSO, CTB, and CNR.[49] Equations (17.128) and (17.129) provide an approximation for the CSO and CTB products:[53]

$$\text{CSO[dB]} = 10 \log\left\{ N_{CSO} \left[ \frac{(1-\sigma)m}{4\sigma} \right]^2 \right\}, \tag{17.128}$$

$$\text{CTB[dB]} = 10 \log\left\{ N_{CTB} \left[ \frac{3(1-\sigma)m}{16\sigma} \right]^2 \right\}, \tag{17.129}$$

where $m$ is the OMI, $N$ refers to the CSO and CTB products' count, and the variance, $\sigma$, which is the fractional transmission coefficient for the SMF, is given by the empirically estimated formulation driven from the ratio of the backward scattered power to the injected power $R_{BS}$:

$$\sigma = \frac{1}{2}\left(1 - \frac{R_{BS}}{0.85}\right). \tag{17.130}$$

CSO distortions of SMF should be below −62 dBc. Hence, the backward scattered power for a 14-dBm drive should be lower than −20 dBc.[54,55]

Eye-opening relates to digital NRZ ASK[45–47] and affects BER[50] eventually; the higher the SBS threshold power, the shorter the fiber. Figure 17.29 demonstrates OC48 data rate BER versus the receiver's optical power, where the launched power is a parameter. Figure 17.30 demonstrates the SBS threshold dependence on fiber length. Figure 17.31 demonstrates the SBS occurrence point, where it can be seen that below the SBS threshold, $P_{out}/P_{in}$ is the fiber loss, then it saturates. Figure 17.32 shows that fiber efficiency depends on its length. The experiment to produce those results was as follows.[50] An original light signal of 1557 nm was amplitude modulated at 622 Mb/s and launched though a 6-km SMF together with a 1551-nm pumping light. The wavelength converted light that generated through the fiber was selected by optical filters. Then BER was tested, as shown in Fig. 17.33, where circles denote the original wavelength BER. Figure 17.34 shows the wavelength conversion.

### 17.6.2 Self- and external phase modulation effects

The self-phase modulation effect is a result of interaction between the fiber's chromatic dispersion and the spectrum of the modulated beam of light is optical signal propagating through the fiber. Multiple narrow, single side-band SCMS in an analog

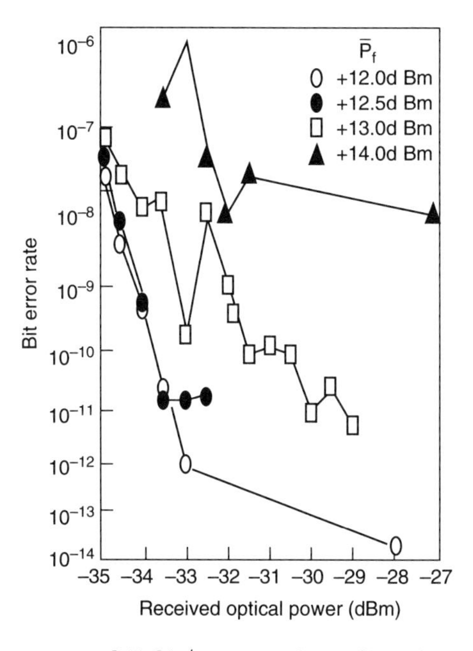


**Figure 17.29** BER curves at 2.5 Gb/s vs. various fiber launch powers. The BER penalties are due to the efficient conversion of low-frequency components of the digital signal, primarily strings of "1" in the forward-propagating signal.[45] (Reprinted with permission from the *Journal of Lightwave Technology* © IEEE, 1993.)

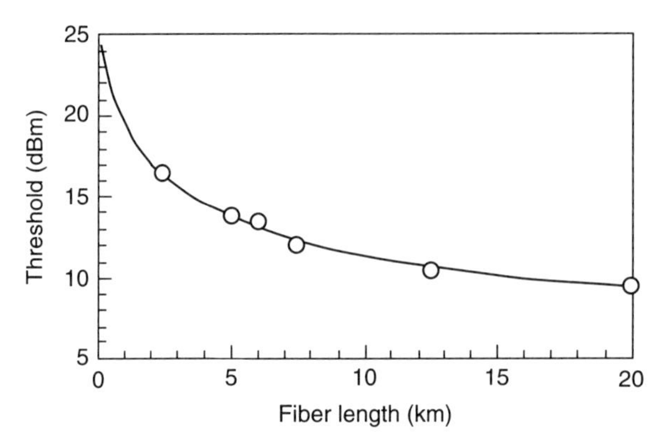


**Figure 17.30** SBS threshold vs. fiber length.[50] Reprinted with permission from *Photonics Technology Letters* (© IEEE, 1995.)

CATV transporter of 10 or 20 Gb/s with a DWDM-system transmission limitation through SMF is affected by referring to the CATV channel, resulting in SPM due to CSO and CTB, which can be analyzed and obtained by using the wave envelope equation in a lossy dispersive and nonlinear mediums, as will be demonstrated in this section. CSO and CTB result when DML-DFB or EM lasers with high power, which is below the Brillouin threshold, are launched through dispersive SMF.[60–67] The self-phase modulation phenomenon occurs when the traveling lightwave signal through the SMF modulates its own phase.

Assume an optical power, $P(z, t)$, propagating through the SMF that is proportional to the square of an electrical field envelope $E(z, t)$. Assume a backward-modulated propagating wave:

$$E(z,t) = \sqrt{x(z,t)}\exp[jy(z,t)], \tag{17.131}$$

where its intensity modulation envelope information is given by $x(z, t)$, and phase is given by $y(z, t)$.

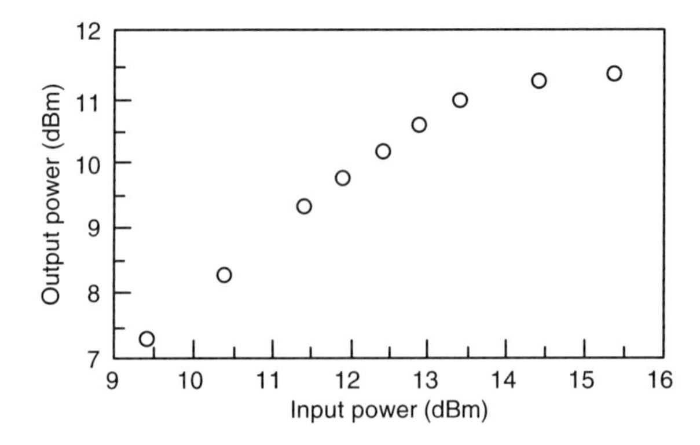


**Figure 17.31** The Input–output characteristics of the fiber used for BER test.[50] (Reprinted with permission from *Photonics Technology Letters* © IEEE, 1995.)

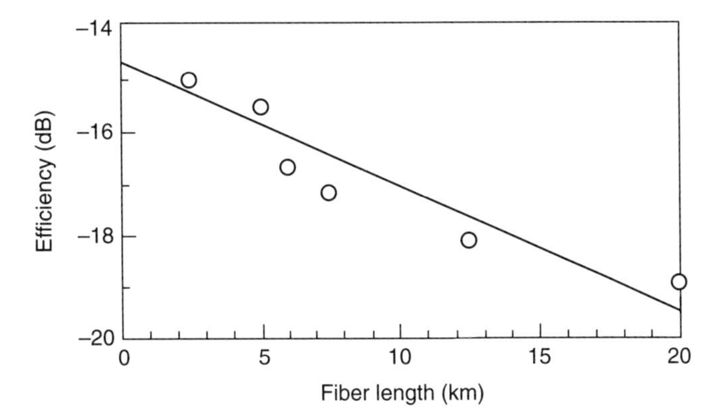


**Figure 17.32** Fiber efficiency vs. fiber length, pump power is set to SBS threshold.[50] (Reprinted with permission from *Photonics Technology Letters* (© IEEE, 1995.)

For dispersion to occur and AM by fiber to happen, the refraction index of the fiber should vary with power. Hence, $n(E) = n_0 + n_1|E|^2$. Since the SMF has loss, its loss coefficient is given by $\alpha$. Now there is a need to define the wave propagation index, $\beta$, which is affected by the refractive index. Dispersion results from a group delay, reflect the various propagation speeds of different wavelengths, but also vary due to the refractive index change versus power:

$$\frac{1}{V_g} = \beta_1 = \frac{d\beta}{d\omega}. \tag{17.132}$$

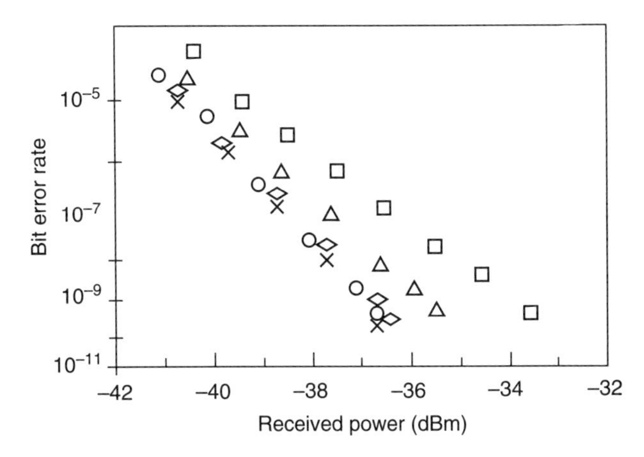


**Figure 17.33** BER of wavelength-converted light, pump power is a parameter: 14.75 dBm, □; 14.5 dBm, Δ; 14.0 dBm, ◇; 13 dBm, ×; BER of original signal, ○.[50] (Reprinted with kind permission from *Photonics Technology Letters* © IEEE, 1995.)

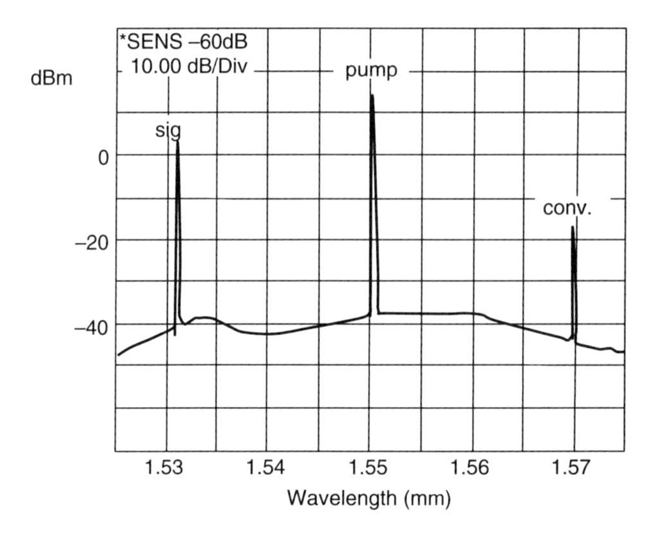


**Figure 17.34** Wavelength conversion using 2.5 km SMF. Pump power is set to SBS threshold.[50] (Reprinted with permission from *Photonics Technology Letters* © IEEE, 1995.)

The variance of a group delay versus frequency is given by the second derivative of the wave propagation index:

$$\beta_2 = -\left(\frac{1}{V_g}\right)^2 \frac{dV_g}{d\omega} = \frac{d^2\beta}{d\omega^2} = -\frac{D\lambda^2}{2\pi c}, \tag{17.133}$$

where $D$ is the fiber dispersion coefficient, $c$ is the speed of light, and $\lambda$ is the wavelength.

The envelope equation of a lossy nonlinear medium is given by[76]

$$\frac{\partial E}{\partial z} + \beta_1 \frac{\partial E}{\partial t} - \frac{j}{2}\beta_2 \frac{\partial^2 E}{\partial t^2} + \alpha E = -jkn|E|^2 E, \tag{17.134}$$

where $k = 2\pi/\lambda$.

Using a slow varying envelope approximation, and neglecting second-order derivatives, we get[66]

$$\frac{\partial E}{\partial z} + \beta_1 \frac{\partial E}{\partial t} + \frac{\alpha}{2} E = -j\frac{1}{A_{\text{eff}}} kn|E|^2 E, \tag{17.135}$$

where $A_{\text{eff}}$ is the effective core cross-section area of the SMF.

Assuming multicarrier frequencies $\Omega_i$ directly modulating a laser for single wavelength transport with OMI equal to $m$. Also assume that launched power at the SMF input located at $z = 0$ equals to $P_0$ and phase $\phi_i$, hence the light intensities and phases are given by:

$$x(0,t) = P_0\left[1 + m\sum_{i=1}^{N} \cos\left(\Omega_i t + \phi_i\right)\right], \tag{17.136}$$

$$y(0,t) = G(I_b - I_{th})m\int_0^t dt' \sum_{i=1}^{N} \cos(\Omega_i t' + \phi_i)$$
$$= \gamma m \int_0^t dt' \sum_{i=1}^{N} \cos(\Omega_i t' + \phi_i), \tag{17.137}$$

where $\gamma$ is the laser chirp factor, $I_b$ is the laser-bias current, and $I_{th}$ is the laser threshold current. For an EM laser, the phase is given by:

$$y(0,t) = \beta \sum_{i=1}^{N} \cos(\Omega_i t' + \phi_i). \tag{17.138}$$

Performing all substitutions in these relations gives CSO NLD that result from SPM:[24,62]

$$\mathrm{CSO_{SPM}}[\Omega] = N_{\mathrm{CSO}}\left\{\frac{1}{2} m \frac{\lambda^2}{2\pi c} Dkn \frac{P_0}{A_{\mathrm{eff}}} \Omega^2 \frac{[\alpha L + \exp(-\alpha L) - 1]}{\alpha^2}\right\}, \tag{17.139}$$

where $L$ is the fiber length, and $N_{\mathrm{CSO}}$ is the CSO product count.

Dispersion effects should not be neglected when transmission distance is very long, the phase index $\beta$ is large, and the phase-modulating tone frequency is very high. This was confirmed by an experiment demonstrated in Ref. [62]. Also, by avoiding the use of DML and minimizing chirp with an EM laser, CSO products that result from SPM can be optimized. Figure 17.35 depicts the worst case CSO in channel 78 as a function of launched optical power from an EDFA into a fiber

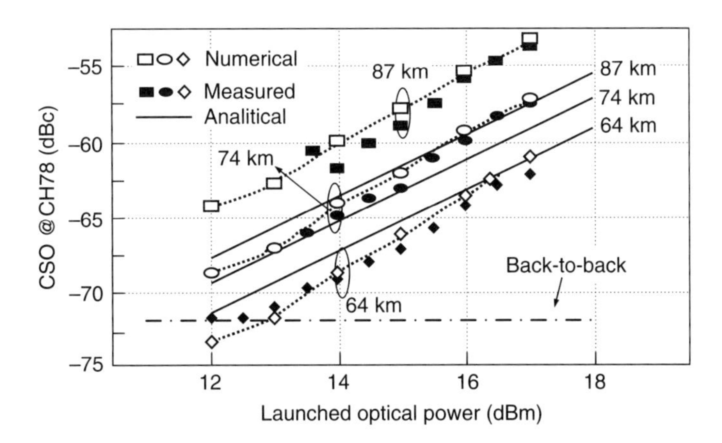


**Figure 17.35** CSO at 78 channels as a function of launched optical power into repeater-less AM-CATV system with three different transmission distances of 64, 74, and 87 km. Experiment parameters are: OMI/ch = 2.8%, $D = 17$ ps/nm/km, $\lambda = 1551.7$ nm, $\alpha = 0.2$ dB/km, $A_{\mathrm{eff}} = 90\ \mu\mathrm{m}^2$, and $n = 2 \times 10^{-20}\ \mathrm{m}^2/\mathrm{W}$ $\beta$, for the three tones at 1.9, 3.8, 5.7 GHz are 3.9, 3.9, and 1.3 respectively. CSOs resulted from both SPM and EPM. (Reprinted with permission of *IEEE* ©).[62]

spool. Three transmission distances were observed, 64 km, 74 km, and 87 km. Results show good match to analysis.

### 17.6.3 Stimulated Raman scattering effect

Chapter 1 described the structure of passive optical networks (PONs), and WDM and DWDM multiplexer devices and related technologies are described in Chapters 4 and 5. Chapter 2 provides an introduction to the optical network unit (ONU) and subscriber modules. This section explains an important system performance parameter resulting from optical X-talk of two or more high-power WDM channels due to an SRS effect.[47] In WDM-FTTx-HFC systems, each wavelength is individually modulated by a high-frequency data signal.[63,68] In the analog channel, the CATV modulates the data from the laser at a given wavelength with a defined OMI, while the digital-channel at different wavelengths and modulation depth is defined by ER. These parameters define the link's SNR. When multiple wavelength–carrying SCM signals propagate in a single fiber, fiber nonlinearities can lead to X-talk between subcarriers of different wavelengths. X-talk is measured by CNR and BER degradation. Stimulated Raman scattering causes power transfer from shorter-wavelength WDM channels to longer-wavelength WDM channels. The power transfer increases as the wavelength difference between any two channels until about 100 nm. Therefore, the SRS power transfer cannot become the maximum within EDFA BW.

The brief analysis presented here is based on Refs. [47], [63], and [68], while the performance analysis and statistics is from Refs. [69] and [70]. Assume a case of two WDM channels, where each optical carrier is modulated by $M$ subcarriers. The optical power at the fiber input, $z = 0$, is given by

$$P_k(t, z = 0) = P_0\left[1 + \sum_{i=1}^{M} m\cos(\Omega_i t + \theta_{k,i})\right], \tag{17.140}$$

where $k = p,s$ represents the pump with the shorter wavelength, and the signal with the longer wavelength, respectively. The index "$i$" accounts for all modulating tones up to a total of $M$, with the frequencies $\Omega_i$. It is further assumed that all subcarriers have the same OMI, $m$. The phase $\theta_{k,i}$ of any given subcarrier $i$ is totally random, i.e., $\theta_{kp}$, $\theta_{ks}$.

SRS is proportional to $g_{\mathrm{ps}}P_{\mathrm{p}}P_{\mathrm{s}}$,[75] where $g_{\mathrm{ps}}$ is the Raman efficiency between the pump wavelength with the power $P_{\mathrm{p}}$ and signal channels with the power $P_{\mathrm{s}}$.

From Eq. (17.140), it can be observed that the total X-talk for subcarrier $i$ in the pump channel can be viewed as being comprised of three terms:

- $mg_{\mathrm{ps}}P_0^2\cos(\Omega_i t + \theta_{\mathrm{p},i})$: The first term states that this expression is due to the SRS interaction between subcarrier $i$ and the optical carrier channel. This is just the optical power loss, $mg_{\mathrm{ps}}P_0^2$, of the subcarrier $i$ and is analogous to pump depletion in conventional on–off keying (OOK) WDM systems.

- $mg_{ps}P_0^2\cos(\Omega_i t + \theta_{s,i})$: The second term states that this expression is attributable to SRS interaction between the optical carrier in the pump channel and the subcarrier $i$ in the signal channel. Note that this term is not found in OOK WDM systems and is the most significant SRS X-talk term in WDM SCM systems.
- $\sqrt{M/8}(m^2 g_{ps}P_0^2)$: The third term is negligible compared to the previous two and is an order of magnitude lower. The reason is that in order to keep the clipping noise low, the OMI index, $m$, of each subcarrier is so set that $\mu = \sqrt{m^2M/2} \approx 0.48$ for a number of channels $M \geq 10$. This condition states that it is about one order of magnitude lower in terms of CNR.

Since the SRS interaction between optical carriers results in only a dc optical power loss or gain, they do not contribute to X-talk at the subcarrier frequencies.

For subcarrier $i$ in the signal channel, the X-talk is the same, except that now the first term contributes optical power gain $mg_{ps}P_0^2$ instead of loss. Thereby, the shorter-wavelength channel still suffers more from SRS X-talk than the longer-wavelength channel.

A formalistic approach to determine X-talk levels is by analytically solving the governing equations of an SRS interaction in an optical fiber[63,69] and assuming a linear regime case:

$$\frac{\partial P_s}{\partial z} + \frac{1}{V_{gs}}\frac{\partial P_s}{\partial t} = (g_{ps}P_p - \alpha)P_s, \tag{17.141}$$

$$\frac{\partial P_p}{\partial z} + \frac{1}{V_{gp}}\frac{\partial P_p}{\partial t} = (-g_{ps}P_s - \alpha)P_p, \tag{17.142}$$

where $V_{gs}$ is the group velocity of the transmitted light with wavelength $\lambda_s$, and $V_{gp}$ is the group velocity of pump signal with wavelength $\lambda_p$, $\alpha$ is the fiber loss, and $g$ is the standard Raman-gain coefficient divided by the effective area of fiber. The convention is to first solve Eq. (17.141), neglecting $g$ and then substituting $P_s$ into Eq. (17.142). After all substitutions are done,[63,69] SRS X-talk is given by

$$XT_i^{sp} = \frac{(2\,g_{ps}P_0)^2\left[\left(\frac{\alpha L_{eff}}{2}\right)^2 + \exp(-\alpha L)\sin^2\left(\frac{d_{ps}\Omega_i L}{2}\right)\right]}{\left[\alpha^2 + (d_{ps}\Omega_i)^2\right](1 \pm 2\,g_{ps}P_0L_{eff})^2}, \tag{17.143}$$

where $L_{eff}$ is the effective fiber length given in Eq. (17.123), and $d_{ps}$, which is the group velocity mismatch between the pump and the signal wavelength, is given by

$$d_{ps} = \frac{1}{V_{gp}} - \frac{1}{V_{gs}}. \tag{17.144}$$

The signs of the denominator term, $1 \pm 2 g_{ps} P_0 L_{eff}$, are as follows: "−" is for the pump channel and " + " is for the signal channel. By setting $d_{ps}$ to zero, it is clear that SRS X-talk is independent of the subcarrier frequency if the fiber dispersion is negligibly small. When operating at a 1.3-μm wavelength window, dispersion effects are ignored. Figure 17.36 explains the experimental test set-up for an SRS test. Figure 17.37 illustrates the SRS X-talk level for electrical versus optical power per WDM channel at 1.3 μm.

Figures 17.38 and 17.39 provide SRS X-talk plots for 1.3 and 1.5 μm as a function of number of channels when power level is a parameter. For WDM SCM systems operating in the 1.5 μm window, SRS X-talk will be lower for high subcarrier frequencies (≥400 MHz), since dispersion is significant and may result in walk-off between RF subcarriers at different optical wavelengths. (Waveform distortion due to SRS walk-off is elaborated in Ref. [59], [68]). SRS X-talk can be minimized. The method for minimizing these effects is done by X-talk cancellation technique using parallel transmission.[63]

The idea is straightforward. If one can transmit the same set of signals on two fibers and then combine them at the receiver, parameters can be arranged in such a manner that X-talk from the two fibers is cancelled when signals are added.

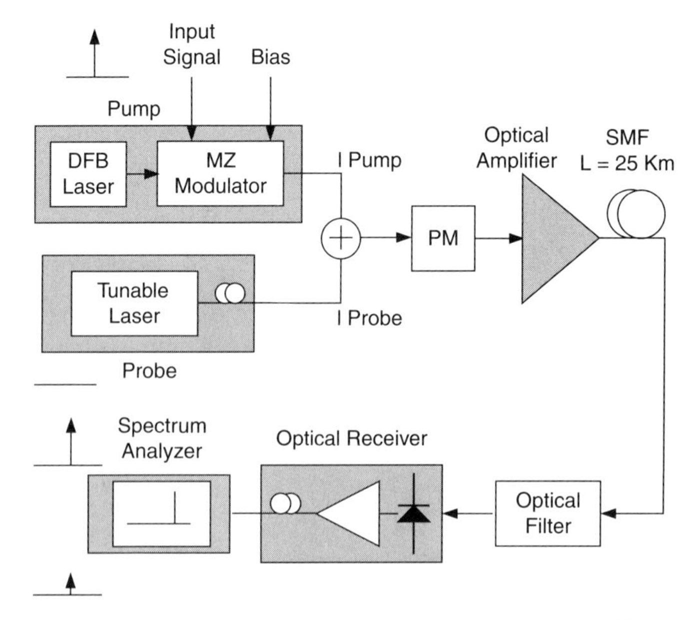


**Figure 17.36** Experimental set-up for SRS X-talk measurement.[63] (Reprinted with permission from *IEEE Journal of Lightwave Technology* © IEEE, 2000.)

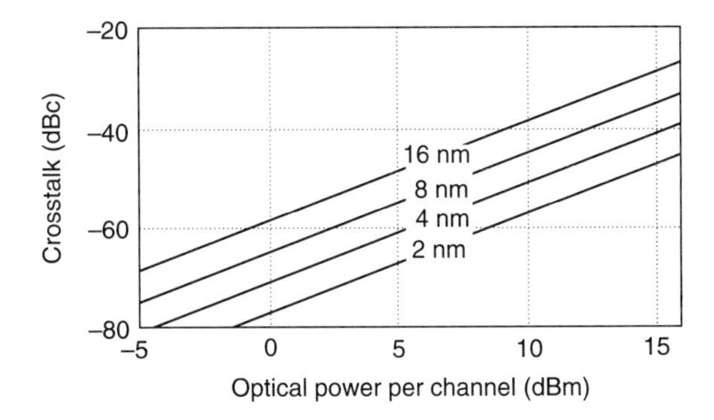


**Figure 17.37** The SRS X-talk level, electrical vs. input optical power per WDM channel for 1.3-mm WDM SCM systems, with two WDM channels and 20 km of standard SMF spools. The X-talk dependency for channel spacing of 2, 4, 8, and 16 nm, respectively, are provided.[69] (Reprinted with permission from *Photonics Technology Letters* © IEEE, 1995.)

The idea is to use a fiber with reverse dispersion characteristics or to operate the wavelengths of each fiber at opposite sides of the zero-dispersion wavelength of the same type of fiber.

### 17.6.4 Cross phase modulatione (XPM) effects

The previous section explained that an FTTx WDM SCM system carrying modulated subcarriers is susceptible to X-talk between wavelengths. Due to fiber dispersion, the phase of the transmitted signals is converted into intensity modulation; this phenomenon can be explained by investigating how laser phase noise is converted into intensity noise.[65] These interactions lead to so-called XPM, resulting in X-talk.[63,66,67]

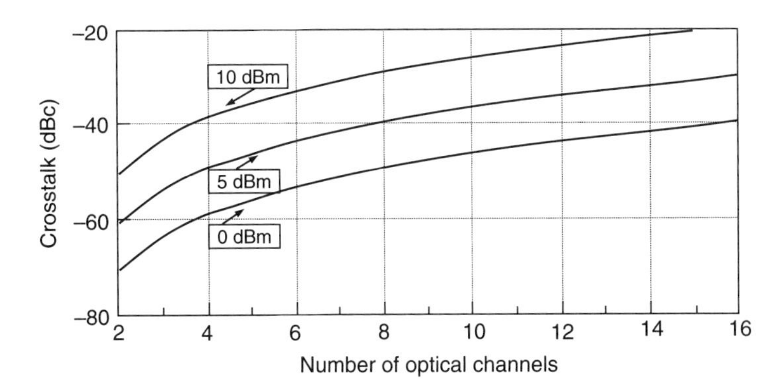


**Figure 17.38** The SRS X-talk level (electrical) vs. the number of WDM channels for 1.3-μm WDM SCM systems, with a channel spacing of 4 nm and 20 km of standard SMF. The X-talk levels for input optical power levels of 0.0, 5.0, and 10.0 dBm per WDM channel are shown.[69] (Reprinted with permission from *Photonics Technology Letters* © IEEE, 1995.)

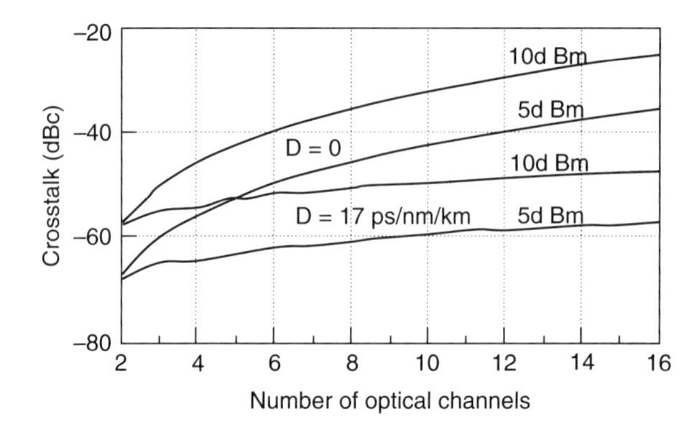


**Figure 17.39** The SRS X-talk level (electrical) vs. the number of WDM channels for 1.5 μm WDM SCM systems, with 2-nm channel spacing, 20 km of standard SMF, and a subcarrier frequency of 500 MHz. The input optical power per WDM channel is specified in the figure. SRS X-talk levels for $D = 0$ ps/nm/km are also shown for comparison.[69] (Reprinted with permission from *Photonics Technology Letters* © IEEE, 1995.)

Assume two optical waves with the same polarization copropagating in SMF, as in the previous section. Let the field $E_k(z, t)$ be a slow-varying complex-field envelope of each wave, with $k = 1$ for $\lambda_1$ and $k = 2$ for $\lambda_2$, where $\lambda_1 > \lambda_2$. When the slowly varying field over time is employed, second-order derivatives are negligible. Assuming normalization for power, $|E_k|^2 = P_k$.

The coupled equations that describe XPM under these assumptions are given by[63]

$$\frac{\partial E_1}{\partial z} + \frac{1}{V_{g1}}\frac{\partial E_1}{\partial t} = \left(-j2\gamma P_2 - \frac{\alpha}{2}\right)E_1, \quad (17.145)$$

$$\frac{\partial E_2}{\partial z} + \frac{1}{V_{g2}}\frac{\partial E_2}{\partial t} = \left(-j2\gamma P_1 - \frac{\alpha}{2}\right)E_2, \quad (17.146)$$

where $\gamma$ is the nonlinearity coefficient, $\alpha$ is the absorption coefficient of the fiber, i.e., loss, and $V_{g1}$ and $V_{g2}$ are the group velocities. Using the same technique for solving the SRS equations, the normalized XPM X-talk at $\lambda_1$ is given by

$$\mathrm{XT}_{\mathrm{XPM1}} = \frac{-2\ddot{\beta}_1\Omega^2\gamma P_c}{(\alpha - j\Omega d_{12})^2}\Big(\big\{\exp(-\alpha L)[\cos(\Omega d_{12}L) - 1 + \alpha L]\big\} + j\big\{\exp(-\alpha L)[\cos(\Omega d_{12}L) - \Omega d_{12}L]\big\}\Big) \quad (17.147)$$

For $\lambda_2$ XPM X-talk is given by

$$\mathrm{XT}_{\mathrm{XPM2}} = \frac{-2\ddot{\beta}_2\Omega^2\gamma P_c}{(\alpha - j\Omega d_{21})^2}\left(\left\{\exp(-\alpha L)(\cos(\Omega d_{21}L) - 1 + \alpha L)\right\}\right.$$
$$\left. + j\left\{\exp(-\alpha L)(\cos(\Omega d_{21}L) - \Omega d_{21}L)\right\}\right), \tag{17.148}$$

where $\ddot{\beta}_i = d^2\beta_i/d\omega^2$ and the index $i = 1$ or 2. Unlike SRS-induced X-talk analysis done in Sec. 17.6.3, Eqs. (17.147) and (17.148) have the same sign and the only change is indices of group velocity mismatch "$d$," also called the walk-off parameter between wavelength 1 to 2, and $\ddot{\beta}$.

The total X-talk is the sum of SRS and XPM. For a small $\Omega$ or large $\Delta\lambda = |\lambda_1 - \lambda_2|$, the X-talk is dominated by SRS. For high modulation frequencies or small $\Delta\lambda$, the X-talk is dominated by XPM effect. In between, phases of both mechanisms should be considered. It turns out that SRS and XPM are exactly in phase or out of phase, depending on $k$ and the sign of $\ddot{\beta}$. When optical phase modulation is converted to intensity modulation via group velocity dispersion, its phase will change somewhat but remain relatively close. As a result, assuming $\lambda_1 > \lambda_2$ stimulated Raman scattering cross talk (SRS-XT) and XPM-XT will add in-phase when $k = 1$ and out-of-phase for $k = 2$ per $XT_k = |\mathrm{XT}_{\mathrm{SRSk}} + \mathrm{XT}_{\mathrm{XPMk}}|^2$. Figures 17.40(a) and 17.40(b) demonstrate the theoretical X-talk level at $\lambda_1$ versus modulation frequency $f = \Omega/2\pi$, $\Delta\lambda = 4$ nm, $P_c = 17$ dBm, and $L = 25$ km. The solid line denotes the total X-talk, and the dashed lines denote the magnitudes of the individual contributions from SRS and XPM, respectively. Stimulated Raman scattering dominance is at low frequency and XPM is at high frequency. Where there is an in-phase addition, the overall result is 6 dB higher than each of the two contributors.

### 17.6.5 Polarization-dependent distortion (PDD) effects

Section 4.5 provided a glance at the isolator structure and light-beam propagation through it. It is a passive nonreciprocal device since it has a defined direction. Polarization distortion induces CSO and CTB products due to the multiple reflections created within the birefringent crystal.[71] A representative analysis is reported in Refs. [71]–[74]. Isolators are used in integrated triplexers as well as in CATV transmitters. There are other polarization mechanisms of these distortion types. The first is the combination of fiber-polarization mode dispersion (PMD) and laser chirp. The other mechanism is the combination of fiber PMD, laser chirp, and fiber polarization-dependent loss (PDL).[73] PMD and PDL effects on BER are analyzed in Ref. [73]. This section discusses PDDs created in the optical isolators, cascaded EDFAs, and fiber transmission lines.[71,74]

The generation mechanism of PDD in isolators is due to the inner structure of the isolator and the refraction index characteristics of the birefringent crystal. As can be seen from Fig. 17.41, the crystal is tilted at a slight angle to eliminate reflection from the crystal's front facet, which faces the output port. Although the

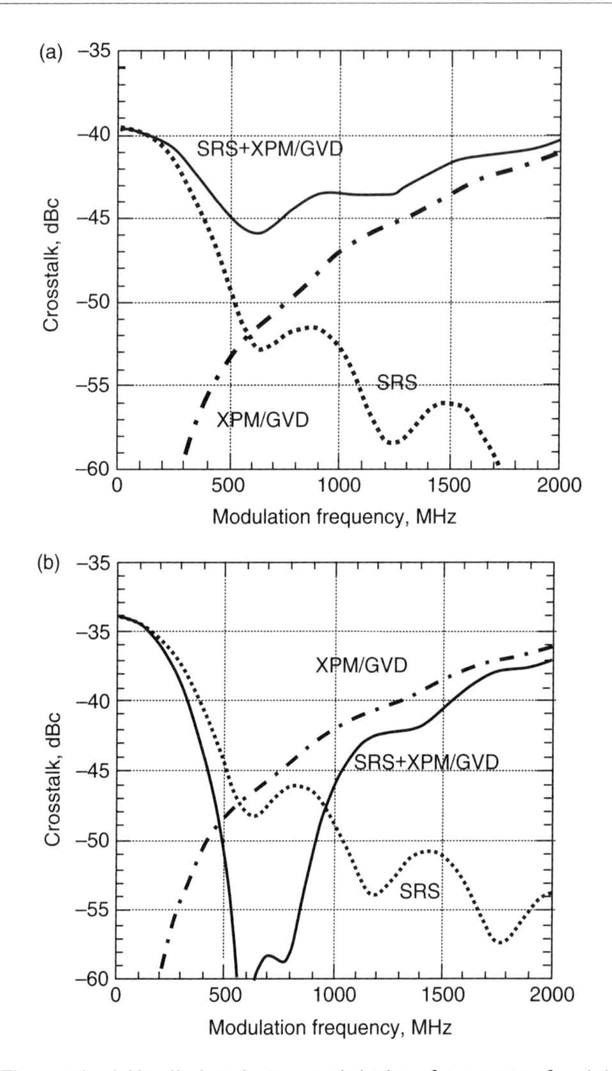


**Figure 17.40** Theoretical X-talk level vs. modulation frequency for (a) $\lambda_1$ (b) $\lambda_2$, for $\Delta\lambda = 4$ nm, $P_c = 17$ dBm, and $L = 25$ km. The solid line depicts the overall X-talk and the dashed lines are for XPM and SRS component contributions. Note the area of the in-phase (coherent) addition, where the total X-talk is higher by 6 dB than each contributor.[63] (Reprinted with permission from the *Journal of Lightwave Technology* © IEEE, 2000.)

birefringent crystal is set an angle, some multiple reflections are transferred to the input port from the plane of the back-facet: those reflections induce distortions. From Fig. 17.41, it is clear that the reflected light suffers a delay of $\tau$ due to its round trip between the birefringent-crystal facets' distance $L_d$, which is twice the thickness of the crystal and the tilt angle of the crystal. Thus $\tau = L_d/c$, where $c$ is the speed of light.

Figure 17.41 illustrates the mechanism of PDD as a function of reflection angles $\theta_A$ and $\theta_B$. These angles are related to the input light polarization states

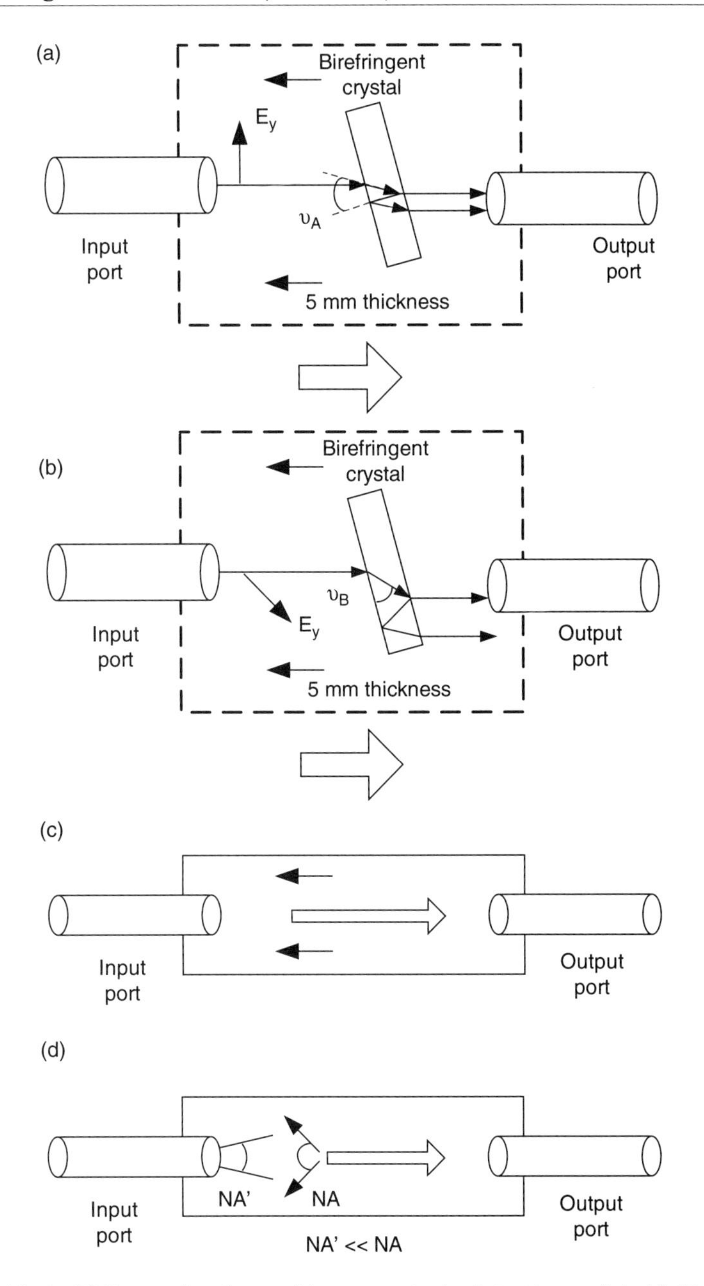


**Figure 17.41** A PDD mechanism with generated distortions. (a) Multiple optical reflections are sent into the SMF in a vertically polarized beam. (b) Multiple reflections are not sent into the SMF in a horizontally polarized beam. (c) Type A isolators (d) Type B isolator.[71] (Reprinted with permission from the *Journal of Lightwave Technology* © IEEE, 1994.)

and their interaction with the birefringent crystal. Such a crystal has two optical axes (see Sec. 4.5), known as slow and fast, with corresponding refractive indices that correspond to the reflection angles.

Such a reflective mechanism between the two facets creates multiple reflections; however, only the first-order reflection is referred to here. The round trip of the reflected beam creates a delay. Hence the receiver's detector experiences two tones with inteferometric distortions. When using DML–DFB, the modulation creates chirp. This results in CSO, due to the interaction polarization–induced multiple optical–reflection mechanism. One of the ways to overcome this is by using EM and minimizing the chirp effect. Figure 17.42 demonstrates the reduction of NLD when using an EM laser against DML for a type A isolator.

The experiment for examining this phenomenon is as follows. A screened 1.5-$\mu$m analog DFB LD integrated with a 30-dB isolator was used. Recall that SMF has a lower loss at 1.5 $\mu$m but higher dispersion compared to 1.3 $\mu$m. The laser was directly modulated by 40-CW channels, under the Japanese frequency plan, within the range of 91 to 421 MHz and an OMI of 4% per channel. The examined channel was at 421 MHz, where the CSO count, $N_{CSO} = 11$, is maximal, and the laser linewidth under the CW conditions was 78 MHz.[71,74] The laser output was connected to EM, so that two options were available to modulate the laser and examine chirp effects at the DML state. In order to validate the effects of the polarization states as the root cause for distortions created by the device under test (DUT) isolator, a polarizer consisting of $\lambda/2$ and $\lambda/4$ plates was connected to the laser output. The $\lambda/4$ produces linear polarized light, and the $\lambda/2$ rotates the polarization axis. In this manner, linearized light with a known rotation angle could be input to the DUT. Figure 17.45 illustrates the abovementioned set-up.

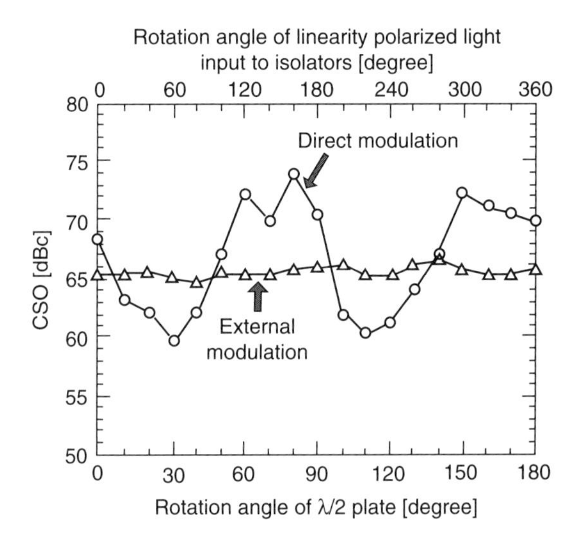


**Figure 17.42** CSO measured results vs. rotation angle of $\lambda/2$ plate for type A-1 isolator for DML and EM modulations.[71] (Reprinted with permission from the *Journal of Lightwave Technology* © IEEE, 1994.)

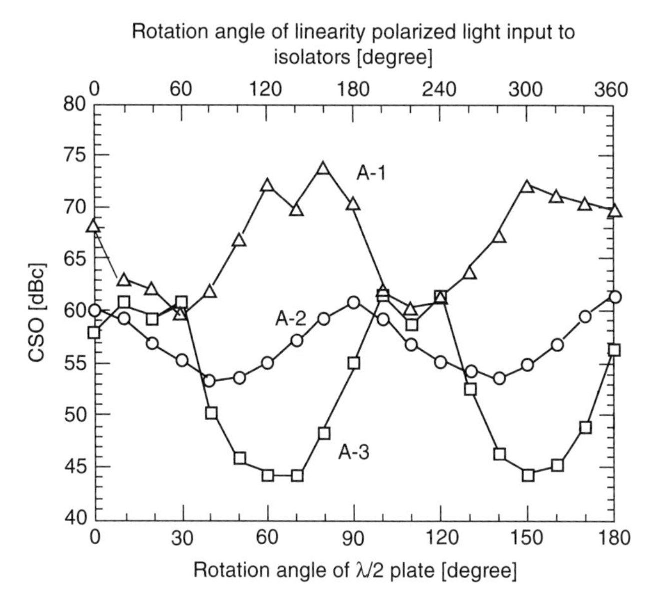


**Figure 17.43** CSO measured results vs. rotation angle of $\lambda/2$ plate for type A isolators. The polarization mode dispersion value $\tau$ equals 2.2 ps for A-1 and A-2 isolators, and it is 0.04 ps for an A-3 isolator; results are for DML.[71] (Reprinted with permission from the *Journal of Lightwave Technology* © IEEE, 1994.)

Two types of isolators A and B were evaluated against a reference PANDA (polarization maintaining) fiber. The PANDA fiber has a large and stable PMD value. In this case, a 2.8-m PANDA fiber with 2.5 ps was tested to determine whether the PMD effect is related to PDD. The type-A isolator optically separates the backward light and the input port, and type-B isolators, with a different numerical aperture (NA) compared to the fiber's NA, as in Figs. 4.33 and 17.41, were the DUTs. Under DML, the PANDA had shown no PDD. However, large PDD was observed with isolator type A3, as shown in Fig. 17.43, even though its PMD is only 0.04 ps. These results mean that PMD is not related to PDD in isolators and PANDA fibers. Further, type B had shown no sensitivity to the $\lambda/2$ plate rotator, as depicted in Fig. 17.44.

## 17.7 CATV/Data Transport Coexistence

Currently, FTTx transport systems are based on WDM and DWDM multiplexing methods. Data transferred over the fiber consists of CATV, AM-VSB, and QAM together with 155 Mb/s up to OC48.[48,51,57,58] Such a link requires stringent specifications for the CATV link, which is CTB $>63$ dB, CNR $>50$ dB, CSO $>65$ dB, and BER $<10^{-9}$ for a 80-km link. As was previously reviewed, this link exhibits X-talk due to SBS and due to the electrical X-talk explained in Chapter 20, as well as finite isolation of the optical demultiplexer (DEMUX) unit. To

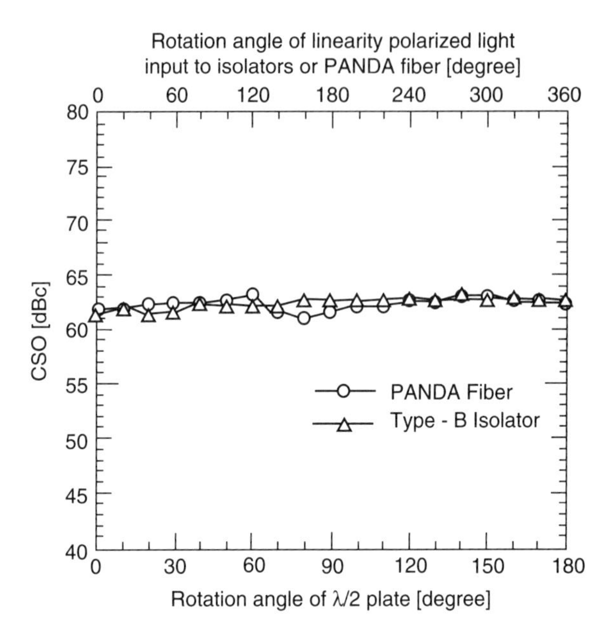


**Figure 17.44** CSO measured results vs. rotation angle of $\lambda/2$ plate for a type-B isolator with PMD $\tau = 0.82$ ps and 2.8 m PANDA fiber $\tau = 2.5$ ps.[17] (Reprinted with permission from the *Journal of Lightwave Technology* © IEEE, 1994.)

overcome SBS X-talk, a large-effective-area fiber (LEAF) is used. This type of fiber has higher SBS threshold power, low chromatic dispersion, and low attenuation. Those characteristics allow the launching of higher optical power, which enables a longer-distance digital lightwave system. Hai Han Lu and Wen Shing Tsai[57] demonstrated such a link performance experiment, which is described in Fig. 17.46. Kawata et al.[77] provided a report of video and IP coexistence for FTTH deployments. The wavelength plan was as follows: three wavelengths, $\lambda_1 = 1560.6$ nm, $\lambda_2 = 1559$ nm for AM-VSB, $\lambda_3 = 1557.4$ nm for QAM, were modulated by a highly linear push–pull driver, and QAM, with 40.2 Mb/s, which was converted from 750 to 860 MHz by a modulator. This way, each laser is loaded with a lower OMI, which reduces the probability of clipping, and thus the intrinsic BER and CNR are improved. The remaining wavelengths, $\lambda_4 = 1560.6$ nm, $\lambda_5 = 1559$ nm, $\lambda_6 = 1557.4$ nm, $\lambda_7 = 1557.4$ nm, were modulated by OC48. The video signal was amplified by an EDFA, summed with the digital data, and its power set by a variable optical attenuator. These wavelengths were launched into the 40-km LEAF, amplified by EDFA, and were again transmitted through a 40-km LEAF, and then demuxed by a DWDM DEMUX. The demuxed wavelengths were measured for CNR and BER performances. These results can be compared to the reference, where each wavelength operates exclusively.

The comparison of BER performance of the OC48 link versus optical input power is shown in Fig. 17.47. The received signal for BER of $10^{-9}$ is $-33.5$ dBm. Since the received signal for that link is $-20$ dBm, BER is $<10^{-9}$ for an 80-km link. The link RF spectrum is shown in Fig. 17.48: with a ripple of $\pm 1.5$ dB. This is due to the half-split band as well as LEAF use.

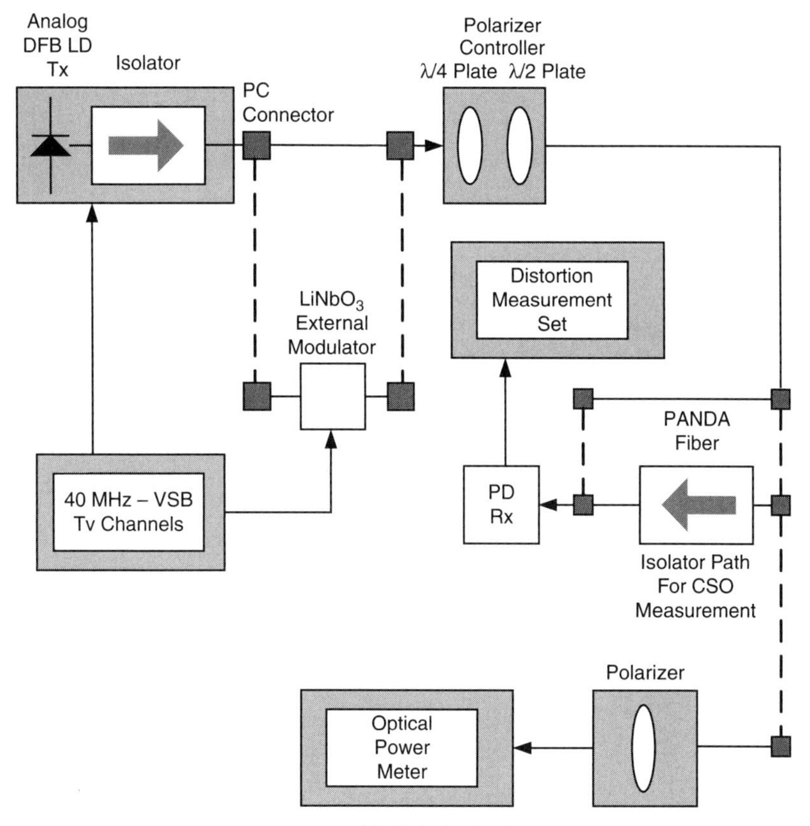


**Figure 17.45** The test set-up to measure PDD of PANDA fiber and isolators.[71] (Reprinted with permission from the *Journal of Lightwave Technology* © IEEE, 1994.)

Linear X-talk arises from the incomplete isolation of the channels at the DEMUX, which is expressed by

$$XT_{\text{linear}}[\text{dB}] = 20\log\left[\frac{K_{12}(P_{20}/P_{10})}{1+K_{12}(P_{20}/P_{10})}\right], \tag{17.149}$$

where $P_{10}$ is the average power of channel one, $P_{20}$ is the average power of channel two, and $K_{12}$ is the ratio of the power transmission of channel two to the power transmission of channel one at DEMUX power one.

Multichannel video and IP are the typical transport of FTTH defined by the International Telecommunications Union Telecommunications Standardization Sector (ITU-T) recommendation G.983.3. The broadband PON system, which is based on the wave allocation of ITU-T G983.3, can provide high-speed IP service as well as CATV service, and bidirectional service such as video on demand. A cost-effective system uses coarse WDM, where the wavelength spacing is 20 nm. This was proposed by the ITU-T recommendation G.694.2.

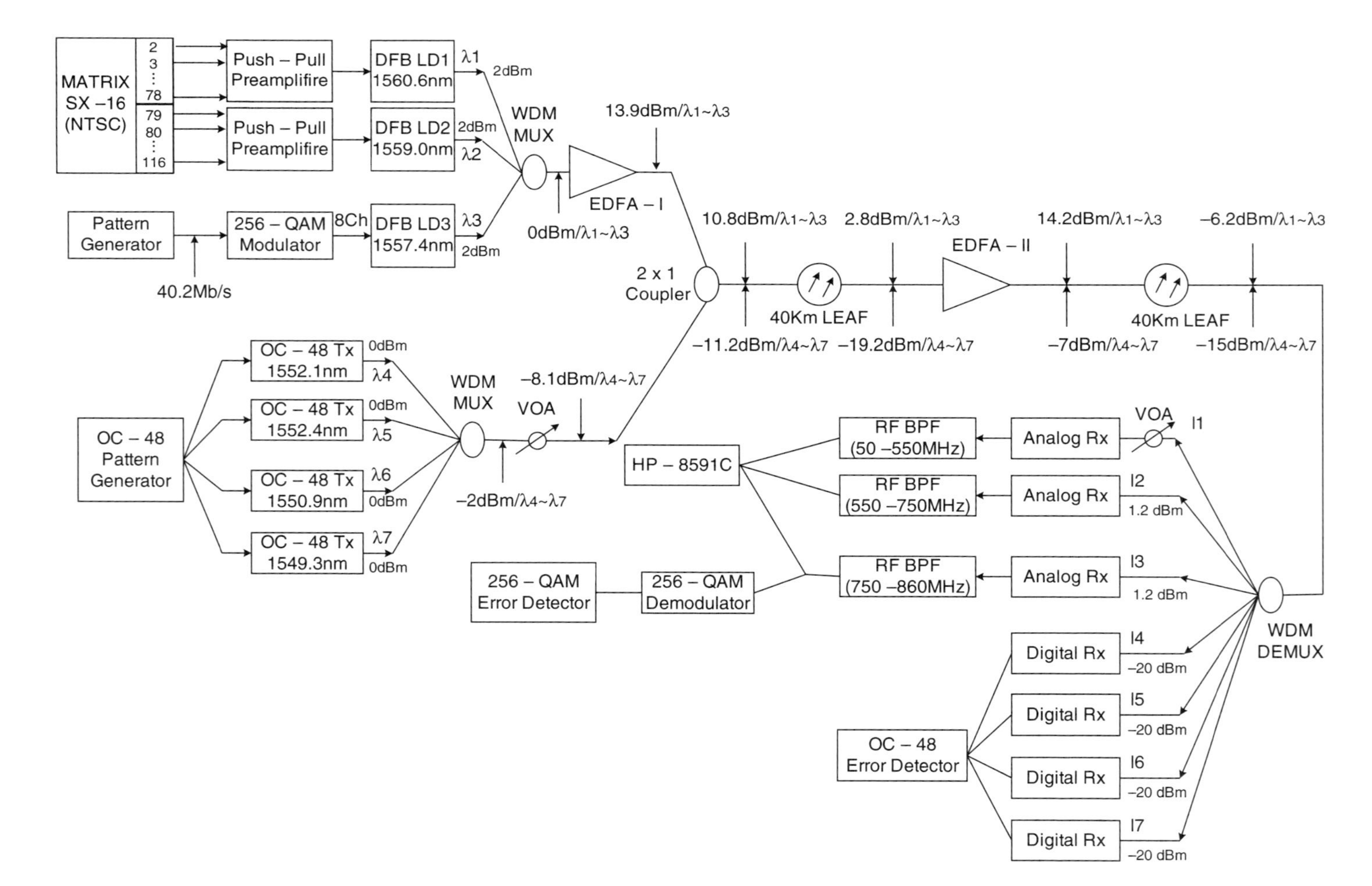


**Figure 17.46** The DWDM X-talk test using a 80-km LEAF fiber.[57] (Reprinted with permission from *Transactions on Broadcasting* © IEEE, 2003.)

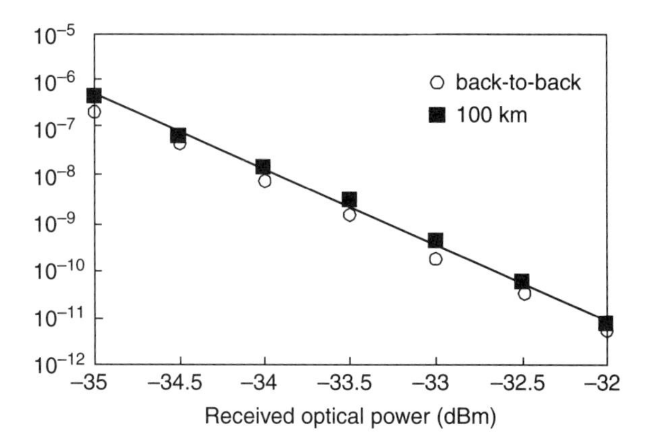


**Figure 17.47** OC48 BER after 80-km LEAF fiber is applied for DWDM link.[57] (Reprinted with permission from *Transactions on Broadcasting* © IEEE, 2003.)

Such a system is described in Fig. 17.49, and the wavelength allocation is shown by Fig. 17.50.[78]

The DEMUX, as was previously demonstrated, must sufficiently suppress the leakage signal, which becomes additional noise. At first approximation for the digital signal, it is assumed that the noise created by the amplitude distribution of the detected photocurrent of the analog signal, which is AM subcarrier multiplexing, has a Gaussian distribution. Since the number of video channels is large, the power penalty $PP$ [dB] is given by[78,79]

$$PP = -10\log(1 - \varepsilon), \tag{17.150}$$

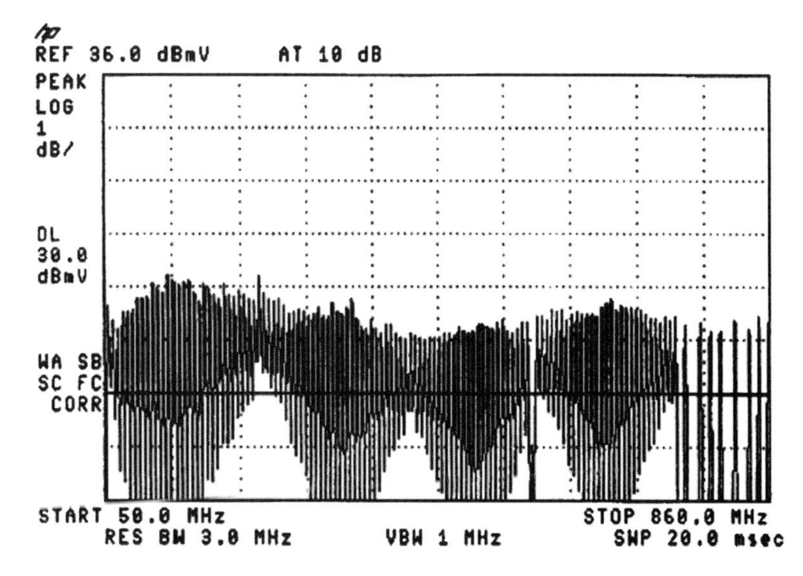


**Figure 17.48** The CATV spectrum after 80-km LEAF fiber is applied to a DWDM link.[57] (Reprinted with permission from *Transactions on Broadcasting* © IEEE, 2003.)

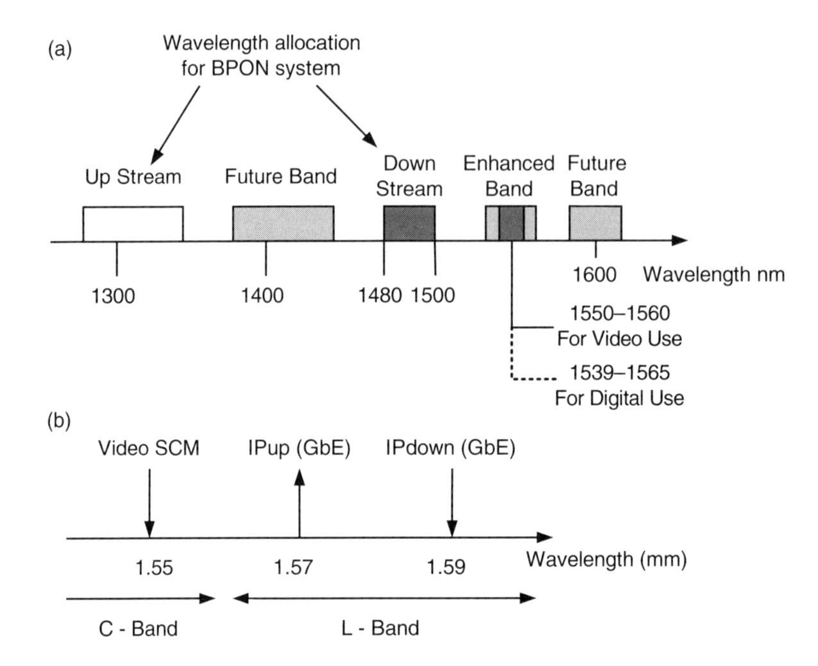


**Figure 17.49** (a) G983.3 wavelength allocation and application examples. (b) Proposed wavelength allocation, multichannel video 1550 nm, IP upstream 1570 nm, and downstream 1590 nm.[78] (Reprinted with permission from the *Journal of Lightwave Technology* © IEEE, 2004.)

where ε is the total X-talk power. For the analog signal, the total CNR is given by

$$\frac{1}{\mathrm{CNR_{tot}}} = \frac{1}{\mathrm{CNR_{sys}}} + \frac{1}{\mathrm{CNR}_i}, \tag{17.151}$$

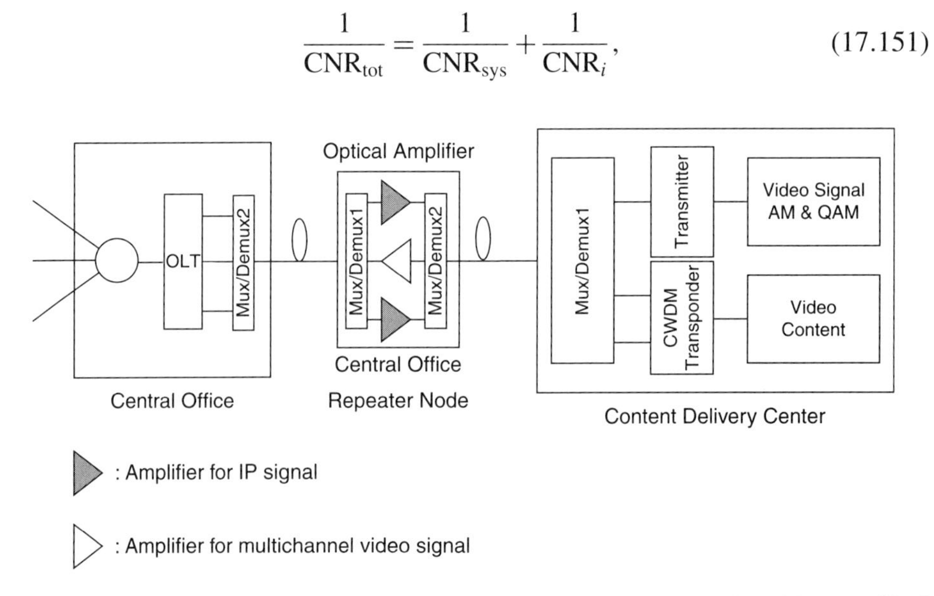


**Figure 17.50** Proposed multiplexing system configuration. Each signal is amplified separately.[78] (Reprinted with permission from the *Journal of Lightwave Technology* © IEEE, 2004.)

where $CNR_{sys}$ is the system's CNR only when an analog signal is transmitted, and $CNR_i$ is $C/N_i$ when those values are given by [78,79]

$$C = \frac{1}{2}(mr_d P_d)^2, \quad (17.152)$$

$$N_i(P_i) = 2R_b \left[ r_i P_i \left( \frac{ER - 1}{ER + 1} \right) \right]^2 \int_{f^-}^{f^+} \frac{\sin^2(\pi f / R_b)}{\pi^2 f^2} df, \quad (17.153)$$

where $m$ is the OMI, $r_d$ is the photodiode responsivity for the analog signal wavelength, $P_d$ is the analog power being received, $P_i$ is the received optical power of the leakage digital signal, $R_i$ is the responsivity of the analog receiver for the digital wavelength leakage, $R_b$ is the bit rate, ER is the extinction ratio of the digital leakage, and $f^-$ and $f^+$ are the lower and the upper limits of the VBW. For example, when $CNR_{tot}$ degradation for $CNR_{sys}$ is to be less than 0.1 dB, the total X-talk from the leakage digital signal must be less than −37.5 dB against the analog signal. The typical values are: $CNR_{sys} = 52$ dB, $P_d = 0$ dBm, $r_d = R_i = 0.95$, $R_b = 1.25$ Gb/s (OC12), OMI = 5%, $\varepsilon = 9$ dB, $f^- = 88.25$ MHz, and $f^+ = 94.25$ MHz.

## 17.8 Main Points of this Chapter

1. To calculate the link budget, there is a need to know the input modulation current, input impedance, output current, and load impedance for calculating the link gain.

2. The link gain is affected by laser-slope efficiency or transmitter conversion efficiency in milliwatts per milliampere, fiber loss, and receiver's conversion factor or photodiode responsivity in milliampere per milliwatt.

3. The laser modulation efficiency is affected by the matching network.

4. An additional parameter that affects the link performance is effective input noise current, which defines the overall link CNR.

5. Noise sources are laser RIN, photodiode shot noise, transmitter's amplifier, and receiver's electronics.

6. The analog link's is dynamic range is defined by the P1 decibel and the SFDR, which is defined by the IP3.

7. Clipping noise results from the limiting effect of the LD, which rectifies the input signal peaks.

8. Clipping noise statistics become Gaussian as the number of channels increase.

9. Laser clipping is a single-sided clipping; the EM clipping model is double-sided limiting.

10. AM-VSB and QAM data OMI is optimized to be 4% and 2% for CNR and BER due to the clipping effect.

11. The clipping effect results in CSO and CTB, which reduce the link carrier to distortion.

12. The clipping effect of AM-VSB results in NLD, which reduces the QAM BER performance.

13. An additional mechanism in lasers that induces NLD is turn-on delay, or memory effect. The memory effect results in a charge that distorts the modulating signal as it is used to discharge the previous charge state. Thus, the signal is self-modulated.

14. One of the methods to prevent clipping is by having an active bias point for the laser. When a large signal swing appears, the bias is increased before the signal reaches the laser. The peak is detected, the bias is increased, and the delayed modulating signal is applied on the laser with minimum clipping.

15. A different approach for predistortion is based on the same concept; the modulating signal's clipping peak is sensed by a threshold detector, inverted, and combined with the original delayed modulating signal. The preclipped signal is applied to the laser, resulting in modulation with minimal clipping. See Fig. 17.18.

16. Clipping NLD, such as CSO, appears in bursts, called "burst-on NLD." It is a statistical process.

17. Another impairment of an optical link is multiple optical reflections resulting at the fiber surface of the connection point. It has a Rayleigh statistic known as "double Rayleigh backscattering."

18. DRB affects the laser RIN, since the DML linewidth and reflection delay increase the laser jitter, which is converted to RIN. The laser's main propagation when it beats with the reflected light converts the laser phase noise into RIN.

19. DRB is directly related to the reflections level in decibels. The larger this reflection value is, the lower the CNR is, and the BER performance of a QAM link gets worse.

20. DRB is directly related to the laser's Lorentzian linewidth. Thus, a YAG laser has an advantage over a regular DFB laser, since it has a narrower linewidth.

21. Direct modulation of a laser affects the DFB laser linewidth. Therefore, one of the ways to overcome DRB is by using an EML.

22. Dispersion-induced NLD in SMF are a consequence of the SMF chromatic dispersion interaction with the DML's resultant wavelength chirp. This delay spreads the FM sidebands, creating CSO. The longer the fiber, the higher the NLD products.

23. Dispersion-induced CSO, distortions are proportional to $40 \log f$, and CTB distortions are proportional to $30 \log f$, where $f$ is the RF frequency of the modulating tone.

24. Stimulated Brillouin scattering (SBS) is a result of quantum momentum transfer from the photon, according to De Broglie $p = h\nu$ relation, the phonon lattice. The vibrated lattice varies the refraction index of the fiber. The lattice vibrates (phonon) at the microwave frequency.

25. The SBS refractive index varies periodically as an acoustical fiber Bragg grating (FBG).

26. Due to the momentum preservation, there is a back-reflected lightwave at a lower frequency than the transmitted wavelength. This frequency shift is called Brillouin–Stokes frequency shift.

27. The FBG is formed at the speed of the propagating sound wave. Thus, the Brillouin–Stokes frequency shift, $\Delta f = V/\Lambda$, where $V$ is the acoustical wave velocity and $\Lambda$ is the FBG period.

28. The SBS optical power threshold is directly related to its effective active area and losses coefficient, and inversely to its Brillouin gain.

29. The fiber Brillouin gain is related to the material density, refractive index, and elasto-optic coefficient, which are material properties.

30. SBS effects can be countered by phase modulation at a frequency twice the channel frequency, or by frequency dithering to increase the laser linewidth. This will affect other link parameters such as chromatic dispersion.

31. SBS results in CNR degradation and CSO and CTB distortions due to chromatic dispersion degradation effects that result from the variation in the refractive index.

32. Stimulated Raman scattering effects result in X-talk between short wavelengths in a WDM channel and a longer one. This reduces CNR and thus affects BER performance.

33. Self-phase modulation is a result of the interaction between the fiber's chromatic dispersion and the modulated beam's spectrum propagating through the fiber.

34. The self-phase modulation phenomenon occurs when the traveling light-wave signal through the single-mode fiber (SMF) modulates its own phase. This is also defined as eternal phase modulation (EPM).

35. Self-phase modulation results in CSO, which is directly related to the fiber dispersion OMI and optical power.

36. Self-phase modulation can be minimized by using an EML.

37. The cross-phase modulation effects result in X-talk between subcarriers modulating the beam of lights, which are on different wavelengths.

38. PDD results from the tilt angle of the birefringent crystal within an optical isolator, resulting in a reflected beam of light. This reflected light is delayed with respect to the main beam; delay is due to its round trip over the crystal thickness, thus resulting in CSO. The walk-off on crystal is due to polarization of light.

39. PDD can be minimized using an EML and type-B isolator, which does not reflect polarized light into the fiber.

40. X-talk between wavelengths in DWDM FTTx systems may come from finite isolation in the DWDM DEMUX.

41. Large effective area fiber (LEAF) can be used to overcome the SBS effect.

42. A LEAF has a higher SBS threshold power, low chromatic dispersion, and low attenuation, which can minimize all other fiber impairments reviewed here.

## References

1. Maeda, M., and M. Yamamoto, "FM-FDM optical CATV transmission experiment and system design for MUSE HDTV signals." *IEEE Journal on Selected Areas in Communications*. Vol. 8 No. 7, pp. 1257–1267 (1990).
2. Frigo, N.J., M.R. Phillips, and G.E. Bodeep, "Clipping distortion in light-wave CATV systems: Models, simulations, and measurements." *Journal of Lightwave Technology*. Vol. 11 No. 1, pp. 138–146 (1993).
3. Mazo, J.E., "Asymptotic distortion spectrum of clipped DC-biased, gaussian noise." *IEEE Transactions on Communications*. Vol. 40 No. 8, pp. 1339–1344 (1982).
4. Ovadia, S., *Broadband Cable TV Access Networks: From Technologies to Applications*. Prentice Hall PTR (2001).
5. Application Note. *RF and Microwave Fiber-Optic Design Guide*. ORTEL a Division of EMCORE (2003).

6. Katznelson, R., "Statistical properties of composite distortions in HFC systems and their effects on digital channels." NCTA Tech-nical Papers, Broadband Innovations Inc. San Diego, CA (2002).
7. Jaworski, M., and M. Marciniak, "Initial pulse modulation method for SBS counteracting in long distance optical fiber CATV link." *Proceedings of XXVIIth General Assembly of the International Union of Radio Science URSI* (2002).
8. Radmacher, R., and J. Seidenberg, "Long distance transmission of 1550 nm CATV signals on different optical fiber types." *Proceedings of the Microwave and Optoelectronics Conference*, pp. 359–362 (2001).
9. Garrett, L., and M. Engelson, "Use the spectrum analyzer's zero-span setting," *Microwaves & RF* (1996).
10. Muys, T., "Clipping induced distortion in lightwave CATV transmission systems: numerical simulation and experiments." *Proceedings of IEEE*, pp. 596–600 (1995).
11. Maeda, K., and S. Morikura, "Study of BER of 64 QAM signal and OMI window for feasible operation in analog/digital hybrid SCM transmission system." *IEEE Journal of Lightwave Technology*. Vol. 17 No. 6, pp. 1011–1017 (1999).
12. Lu, H.H., and W.S. Tsai, "A hybrid CATV/256—QAM/OC48 DWDM system over an 80 Km LEAF transport." *IEEE Transactions on Broadcasting*. Vol. 49 No. 1, pp. 97–102 (2003).
13. Braye, G., J.-C. Froidure, and M. Blondel, "Impact of clipping distortions on hybrid AM—VSB/16 QAMCATV systems." *Proceedings of IEEE*. pp. 1031–1033 (1996).
14. Green, R.J., and O. Pan, "Clipping noise effects on the performance of digital baseband signal in lightwave hybrid digital and analogue transmission system." *IEE Proceedings—Optoelectron*. Vol. 145 No. 6, pp. 335–338 (1998).
15. Shi, Q., and S. Ovadia, "Effects of clipping induced impulse noise in externally modulated multichannel AM/M—QAM video transmission systems." *IEEE Transactions on Communications*. Vol. 46 No. 11, pp. 1448–1450 (1998).
16. Betti, S., E. Bravi, and M. Giaconi, "Effect of turn on delay of a semiconductor laser on clipping impulsive noise." *IEEE Photonics Technology Letters*. Vol. 9 No. 1, pp. 103–105 (1997).
17. Pan, Q., and R. Green, "Performance analysis of preclipping AM/QAM hybrid lightwave systems." *IEEE Journal of Lightwave Technology*. Vol. 15 No. 1, pp. 1–5 (1997).
18. Lu, X., G.E. Bodeep, and T.E. Darcie, "Clipping induced impulse noise and its effect on bit error performance in AM—VSB/64 QAM hybrid lightwave systems." *IEEE Photonics Technology Letters*. Vol. 6 No. 7, pp. 866–868 (1994).
19. Germanov, V., "The impact of CSO/CTB distortions on BER characteristics by hybrid multichannel analog/QAM transmission systems." *IEEE Transactions on Broadcasting*, Vol. 45 No. 1, pp. 88–92 (1999).

20. Pham, K., J. Conradi, G. Cormack, B. Thomas, and C. Anderson, "Impact of noise and nonlinear distortion due to clipping on the BER performance of a 64–QAM signal in hybrid AM–VSB/QAM optical fiber transmission system." *IEEE Journal of Lightwave Technology*. Vol. 13 No. 11, pp. 2197–2201 (1995).
21. Lai, S., and J. Conradi, "Theoretical and experimental analysis of clipping induced impulsive noise in AM—VSB subcarrier multiplexed lightwave systems." *IEEE Journal of Lightwave Technology*. Vol. 15 No. 1, pp. 20–30 (1997).
22. Yang, P., and W.I. Way, "Ultimate capacity of laser diode in transporting multichannel M—QAM signals." *IEEE Journal of Lightwave Technology*. Vol. 15 No. 10, pp. 1914–1924 (1997).
23. Han, H., W.S. Tsai, C.Y. Chen, and H.C. Peng, "CATV / radio on fiber transport systems based on EAM and optical SSB modulation technique." *IEEE Photonics Technology Letters*, Vol. 16 No. 11, pp. 2565–2567 (2004).
24. Wu, M.C., C.H. Wang, and W.I. Way, "CSO distortions due to combined effect of self and external phase modulations in long distance 1550 nm AM CATV systems." *IEEE Photonics Technology Letters*. Vol. 11 No. 6, pp. 718–720 (1999).
25. Ovadia, S., L. Eskildsen, C. Lin, and W.T. Anderson, "Multiple optical reflections in hybrid AM/16 QAM multichannel video lightwave transmission systems." *IEEE Photonics Technology Letters*, Vol. 6 No. 10, pp. 1261–1264 (1994).
26. Ovadia, S., "The effect of interleaver depth and QAM channel frequency offset on the performance of multichannel AM—VSB/256 QAM video lightwave transmission systems." *Photonics Technology Letters*, Vol. 10 No. 8, pp. 1174–1176 August (1998).
27. Darcie, T.E., G.E. Bodeep, and A.A.M. Saleh, "Fiber reflection induced impairments in lightwave AM—VSB CATV systems." *IEEE Journal of Lightwave Technology*. Vol. 9 No. 8, pp. 991–995 (1991).
28. Ho, K.P., and J.M. Kahn, "Optimal predistortion of Gaussian inputs for clipping channels." *IEEE Transactions on Communications*. Vol. 44 No. 11, pp. 1505–1513 (1996).
29. Muys, W., and M.F.R. Mortier, "Numerical simulation of clipping induced distortion in externally modulated lightwave AM—SCM systems." *IEEE Photonics Technology Letters*, Vol. 6 No. 6, pp. 747–749 (1994).
30. Wagner, S.S., T.E. Chapuran, and R.C. Menedez, "The effect of analog video modulation on laser clipping noise in optical video distribution networks." *IEEE Photonics Technology Letters*, Vol. 8 No. 2, pp. 275–277 (1996).
31. Anderson, T., and D. Crosby, "The frequency dependence of clipping induced bit error rate in subcarrier multiplexed systems." *IEEE Photonics Technology Letters*, Vol. 8 No. 8, pp. 1076–1078 (1996).
32. Kanazawa, A., M. Shibutani, and K. Emura, "Preclipping method to reduce clipping induced degradation in hybrid analog/digital subcarrier multiplexed optical transmission systems." *IEEE Photonics Technology Letters*, No. 7, pp. 1069–1071 (1995).

33. Pan, Q., and R.J. Green, "Pre-clipping AM/QAM hybrid lightwave systems with band stop filtering." *IEEE Photonics Technology Letters*, No. 8, pp. 1079–1081 (1996).
34. Kuo, C.Y., and S. Mukherjee, "Clipping reduction for improvement of analog and digital performance beyond clipping limit in lightwave CATV systems." *Proceedings of OFC*, Post Deadline paper PD81–1 (1996).
35. Pophillat, L., "Optical modulation depth improvement in SCM lightwave systems using dissymmetrization scheme." *IEEE Photonics Technology Letters*, No. 6, pp. 750–753 (1994).
36. Ovadia, S., and C. Lin, "Performance characteristics and applications of hybrid multichannel AM-VSB/M-QAM video lightwave transmission systems." *Journal of Lightwave Technology*, Vol. 16 No. 7, pp. 1171–1186 (1998).
37. Wan, P., and J. Conradi, "Impact of double rayleigh backscatter noise on digital and analog fiber systems." *IEEE Journal of Lightwave Technology*, Vol. 14 No. 3, pp. 288–297 (1996).
38. Way, W.I., C.E. Zah, L. Curtis, R. Spicer, and W.C. Young, "Multiple reflection induced intensity noise studies in a lightwave system for multichannel AM—VSB television signal distribution." *IEEE Photonics Technology Letters*, Vol. 2 No. 5, pp. 360–362 (1990).
39. Kaneko, S., A. Adachi, J. Yamashita, and H. Watanabe, "A compensation method for dispersion induced third order intermodulation distortion using an etalon." *IEEE Journal of Lightwave Technology*, Vol. 14 No. 12, pp. 2786–2792 (1996).
40. Crosby, D.B., and G.J. Lampard, "Dispersion induced limit on range of octave confined optical SCM transmission systems." *IEEE Photonics Technology Letters*, Vol. 6 No. 8, pp. 1043–1045 (1994).
41. Bergmann, E.E., C.Y. Kuo, and S.Y. Haung, "Dispersion induced composite second order distortion at 1.5 mm." *IEEE Photonics Technology Letters*, Vol. 3 No. 1, pp. 59–61 (1991).
42. Kuo, C.Y., "Fundamentals second order nonlinear distortions in analog AM CATV transport systems based on single frequency semiconductor lasers." *IEEE Journal of Lightwave Technology*, Vol. 10, pp. 235–243 (1992).
43. Konrad, B., K. Petermann, J. Berger, R. Ludwig, C.M. Weinert, H.G. Weber, and B. Schmauss, "Impact of fiber chromatic dispersion in high-speed TDM transmission systems." *IEEE Journal of Lightwave Technology*, Vol. 20 No. 12, pp. 2129–2135 (2002).
44. Papnnareddy, R., and G. Bodeep, "HFC/CATV transmission systems." Arris International, Purdue University, Lafayette, IN (2004).
45. Fishman, D., and J.A. Nagel, "Degradations due to stimulated brillouin scattering in multigigabit intensity modulated fiber optic systems." *IEEE Journal of Lightwave Technology*, Vol. 11 No. 11, pp. 1721–1728 (1993).
46. Djupsjobacka, A., G. Jacobsen, and B. Tromborg, "Dynamic stimulated Brillouin scattering analysis." *IEEE Journal of Lightwave Technology*, Vol. 18 No. 3, pp. 416–424 (2000).
47. Aoki, Y., K. Tajima, and I. Mito, "Input power limits of single mode optical fibers due to stimulated brillouin scattering in optical communication

systems." *IEEE Journal of Lightwave Technology*, Vol. 6 No. 5, pp. 710–719 (1988).

48. Mao, X.P., R.W. Tkach, A.R. Chraplyvy, R.M. Jopson, and R.M. Derosier, "Stimulated brillouin threshold dependence on fiber type and uniformity." *IEEE Photonics Technology Letters*, Vol. 4 No. 1, pp. 66–69 (1992).
49. Mao, X.P., G.E. Bodeep, R.W. Tkach, A.R. Chraplyvy, T.E. Darcie, and R.M. Derosier, "Brillouin scattering in externally modulated lightwave AM—VSB CATV transmission systems." *IEEE Photonics Technology Letters*, Vol. 4 No. 3, pp. 287–289 (1992).
50. Inoue, K., T. Hasegawa, and H. Toba, "Influence of stimulated Brillouin scattering and optimum length in fiber four wave mixing wavelength conversion." *IEEE Photonics Technology Letters*, Vol. 7 No. 3, pp. 327–329 (1995).
51. Chand, N., P.D. Magill, S.V. Swaminathan, and T.H. Daugherty, "Delivery of digital video and other multimedia services (.1 Gb/s Bandwidth) in passband over 155 Mb/s baseband services on a FTTx full service access network." *IEEE Journal of Lightwave Technology*, Vol. 17 No. 12 (1999).
52. Smith, R.G., "Optical power handling capacity of low loss optical fibers as determined by stimulated Raman and Brillouin scattering." *Applied Optics*, Vol. 11, pp. 2489–2494 (1972).
53. Philips, M.R., and K.L. Sweeney, "Distortion by stimulated brillouin scattering effect in analog video lightwave systems." *OSA TOPS System Technologies*, Vol. 12, pp. 182–185 (1997).
54. Williams, F.W., W. Muys, and J.S. Leong, "Simultaneous suppression of stimulated brillouin scattering and interferometric noise in externally modulated lightwave AM—SCM systems." *IEEE Photonics Technology Letters*, No. 6, pp. 1476–1478 (1994).
55. Williams, F.W., J.C. Van Der Plaats, and M. Muys, "Harmonic distortion caused by stimulated brillouin scattering suppression in externally modulated lightwave AM—CATV systems." *IEE Electronics Letters*, No. 30, pp. 343–345 (1994).
56. Cox, C., E. Ackerman, R. Helkey, and G.E. Betts, "Techniques and performance of intensity modulation direct detection analog optical links." *IEEE Transactions on Microwave Theory and Techniques*,Vol. 45 No. 8, pp. 1375–1383 (1997).
57. Lu, H.H., and W.S. Tsai, "A hybrid CATV/256- QAM/OC-48 DWDM system over an 80—Km LEAF transport." *IEEE Transactions on Broadcasting*, Vol. 49 No. 1, pp. 97–102 (2003).
58. Attard, J.C., J.E. Mitchell, and C.J. Rasmussen, "Performance analysis of interferometric noise due to unequally powered interferers in optical networks." *IEEE Journal of Lightwave Technology*, Vol. 23 No. 4, pp. 1692–1703 (2005).
59. Wong, K.Y., M.E. Marhic, M.C. Ho, and L.G. Kazovsky, "Nonlinear crosstalk suppression in a WDM analog fiber system by complementary modulation of twin carriers." *Proceedings of OFC*, pp. WV5-1–WV5-3 (2001).

60. Chen, W.H., and W.I. Way, "Multichannel single sideband SCM/DWDM transmission system." *IEEE Journal of Lightwave Technology*, Vol. 22 No. 7, pp. 1679–1693 (2004).
61. Chen, J.C., "Enhanced analog transmission over fiber using sampled amplitude modulation." *IEEE Transactions on Microwave Theory and Techniques*, Vol. 49 No. 10, pp. 1940–1944 (2001).
62. Wu, M.C., C.H. Wang, and W.I. Way, "CSO distortions due to the combined effects of self and external phase modulation in long distance 1550 nm AM—CATV systems." *IEEE Photonics Technology Letters*, Vol. 11 No. 6, pp. 718–720 (1999).
63. Yang, F.S., E. Marhic, and L.G. Kazovsky, "Nonlinear crosstalk and two countermeasures in SCM WDM optical communication system." *IEEE Journal of Lightwave Technology*, Vol. 18 No. 4, pp. 512–520 (2000).
64. Betti, S., E. Bravi, and M. Giaconi, "Nonlinear distortions due to the dispersive transmission of SCM optical signals in the presence of chirping effect: an accurate analysis." *IEEE Photonics Technology Letters*, Vol. 9 No. 12, pp. 1640–1642 (1997).
65. Yamamoto, S., N. Edagawa, H. Taga, Y. Yoshida, and H. Wakabayashi, "Analysis of laser phase noise to intensity noise conversion by chromatic dispersion in intensity modulation and direct detection optical fiber transmission." *IEEE Journal of Lightwave Technology*, Vol. 8 No. 11, pp. 1716–1722 (1990).
66. Chiang, T.K., N. Kagi, M.E. Marhic, and L.G. Kazovsky, "Cross phase modulation in fiber links with multiple optical amplifiers and dispersion compensators." *IEEE Journal of Lightwave Technology*, Vol. 14 No. 3, pp. 249–260 (1996).
67. Bigo, S., G. Bellotti, and M.W. Chbat, "Investigation of cross-phase modulation limitation over various types of fiber infrastructures." *IEEE Photonics Technology Letters*, Vol. 11 No. 5, pp. 605–607 (1999).
68. Norimatsu, S., and T. Yamamoto, "Waveform distortion due to stimulated raman scattering in wide band WDM transmission systems." *IEEE Journal of Lightwave Technology*, Vol. 19 No. 2, pp. 156–171 (2001).
69. Wang, Z., A. Li, J. Mahon, G. Jacobsen, and E. Bodtker, "Performance limitations imposed by stimulated raman scattering in optical WDM SCM video distribution systems." *IEEE Photonics Technology Letters*, Vol. 7 No. 12, pp. 1492–1494 (1995).
70. Ho, K.P., "Statistical properties of stimulated Raman crosstalk in WDM systems." *IEEE Journal of Lightwave Technology*, Vol. 18 No. 7, pp. 915–921 (2000).
71. Kikushima, K., K. Suto, H. Yoshinaga, and E. Yoneda, "Polarized dependent distortion in AM—SCM video transmission systems." *IEEE Journal of Lightwave Technology*, Vol. 12 No. 4, pp. 650–657 (1994).
72. Yan, L.S., Q. Yu, Y. Xie, and A.E. Willner, "Experimental demonstration of the system performance degradation due to the combined effect of polarization dependent loss with polarization mode dispersion." *IEEE Photonics Technology Letters*, Vol. 14 No. 2, pp. 224–226 (2002).

73. Huttner, B., C. Geiser, and N. Gisin, "Polarization induced distortions in optical fiber networks with polarization mode dispersion and polarization dependent losses." *IEEE Journal of Selected Topics in Quantum Electronics*, Vol. 6 No. 2, pp. 317–329 (2000).
74. Kikushima, K., "Polarization dependent distortions caused by isolators in AM—SCM video transmission systems." *IEEE Photonics Technology Letters*, Vol. 5 No. 5, pp. 578–580 (1993).
75. Chraplyvy, A.R., "Optical power limits on multi–channel wavelength division multiplexed systems due to stimulated raman scattering." *Electronics Letters*, Vol. 20, pp. 58–59 (1984).
76. Phillips, M.R., T.E. Darcie, D. Marcuse, G.E. Bodeep, and N.J. Frigo, "Nonlinear distortion generated by dispersive transmission of chirped intensity–modulated signals." *IEEE Photonics Technology Letters*, Vol. 3 No. 5, pp. 481–483 (1991).
77. Kawata, H., T. Ogawa, N. Yoshimoto, and T. Sugie, "Multichannel video and IP signal multiplexing system using CWDM technology." *IEEE Journal of Lightwave Technology*, Vol. 22 No. 6, pp. 1454–1462 (2004).
78. Kawata, H., T. Ogawa, N. Yoshimoto, and T. Sugie, "Multichannel video and IP signal multiplexing system using CWDM technology." *IEEE Journal of Lightwave Technology*, Vol. 22 No. 6, pp. 1454–1462 (2004).
79. Muys, W., J.C. Van Der Plaats, F.W. Willems, and P.H. Van Heijningen, "Mutual deterioration of WDM–coupled AM CATV and digital B–ISDN services in single fiber access networks." *IEEE Photonics Technology Letters*, Vol. 5 No. 7, pp. 832–834 (1993).
80. Desurvire, E., *Erbium Doped Fiber Amplifiers Principles and Applications*. John Wiley & Sons (2002).
81. Desurvire, E., *Erbium Doped Fiber Amplifiers*. John Wiley & Sons (2002).
82. Shi, Q., "Asymptotic clipping noise distribution and its impact on M-ary QAM transmission over optical fiber." *IEEE Transactions on Communications*, Vol. 43 No. 6, pp. 2077–2083 (1995).
83. Gimlett, V.L., and N.K. Chaing, "Effects of phase on intensity noise conversion by multiple reflections on gigabit-per-second DFB laser transmission system." *IEEE Journal of Lightwave Technology*, Vol. 7 No. 6, pp. 888–895 (1989).

# Part 5

# Digital Transceiver Performance

# Chapter 18

# Introduction to Digital Data Signals and Design Constraints

The goal of this chapter is to provide digital designers of FTTx transceivers with basic system fundamentals. The design limitations and impairments determine the component specifications. Such parameters are generally bit error rate (BER) and its degradation due to extinction ratio (ER) trade-offs, mode-partition noise, jitter, and relative intensity noise (RIN). In this chapter, these parameters are presented first. The next section deals with digital transport through analog networks such as ladder filter and receiver control field (RFC) coils. The goal here is to explain how the pseudo-random bit sequence (PRBS) signal is affected by the dispersion of these elements in the aspect of pattern dependent jitter. Fiber impairments were already dealt with in Chapter 17; so both sides, electronics and optics, are covered. Digital formats such as nonreturn to zero (NRZ), Manchester, and return to zero (RZ) are briefly reviewed. In this background, the clock-and data-recovery (CDR) concept is explained. Additional information about amplitude control loops needed for digital transceivers, which are critical in burst mode, is provided in Chapters 12 and 13. In Chapter 12, the foundation in control theory for CDR is provided and phase locked loops (PLL) are discussed. Burst-mode automatic-gain control (AGC) techniques are explained in Chapter 19. A basic theory is provided here about CDR operation, which is a PLL. This information is important to understand the limitations of fast synchronization and acquisition in burst-mode operation. Trade-offs between CDR with a low-jitter transfer function and fast acquisition are also discussed. This review is a prerequisite one must have prior to starting basic design of any digital transceiver.

## 18.1 Eye Analysis and BER

There are several ways to measure receiver sensitivity versus quality of service, such as defining signal-to-noise and distortion (SINAD), signal to-noise ratio (SNR), or carrier-to-noise ratio (CNR) in an analog receiver for proper detection. Another way is to define the BER in a digital receiver.[6] An introduction to photodetection and noise limits, which affect the system performance envelope, is provided in Chapters 8 and 10. The sum of all of these noise-power limits defines the system SNR. These noise limits include shot noise, thermal noise, RIN, and mode-partition noise. In Chapter 7, a review of the mechanisms of RIN and mode-partition is provided. In addition, sampling jitter is a factor for performance degradation as well. An introduction about jitter and phase noise is provided in Chapters 14 and 15. All of these lead to the need for an additional SNR definition for digital transport of optical transceivers. These definitions are BER versus received optical level, BER versus jitter, BER versus extinction ratio (ER), and BER versus laser noise types.

The second definition for sensitivity degradation has the same meaning; it defines the minimum SNR for proper digital bit or symbol detection. Generally, specifications define BER in the following manner: $\text{BER} \leq 1 \times 10^{-9}$@ for received power of $P_{\text{RX}} = -25$ dBm. It means that the BER should be less than or equal to an average of one error per 1 Gb at a given received power state. In Fig. 18.1, the detected signal received by the decision circuit, which samples it at a decision instance of $t_{\text{D}}$, is shown. The sampled value of $I_1$ fluctuates from bit to bit around an average value $I_1$ or $I_0$ depending on the logic level value 1 or 0. The decision circuit compares the value with a threshold value $I_{\text{D}}$. The detection probability is also illustrated in Fig. 18.1.

It is clear from Fig. 18.1 that when the margin between the 1 logic level and the 0 logic level is increased, the probability of error is minimized. It means that the dark crossover area is minimized. The decision conditions for detection are 1 if $I_{\text{D}} \leq I_1$ and 0 if $I_0 \leq I_{\text{D}}$. Violation of these conditions results in an error. If the level of noise is high enough, then the signal fluctuations can reach the decision area and result in an error. As a consequence, a large margin between the logic levels can help in preventing errors. Thus, having a higher ER and eye opening provides higher immunity to errors.

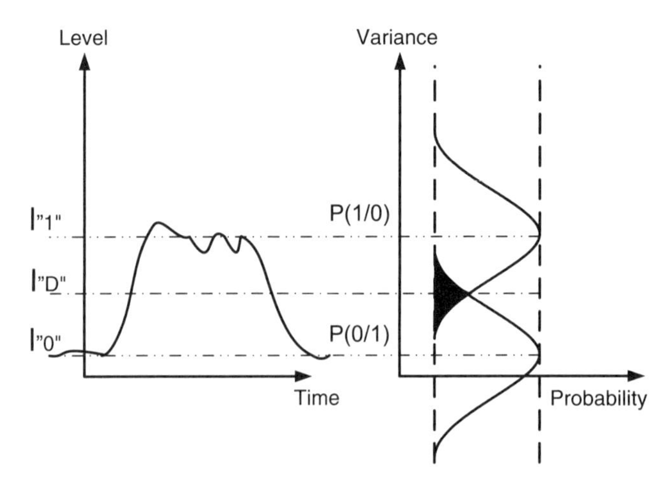


**Figure 18.1** Fluctuating signal (left), Gaussian probability densities of the levels 1 and 0 (right). The crossover is the area of error.

The probability of error, also called BER, is defined as

$$\mathrm{BER} = p(1)P\left(\frac{0}{1}\right) + p(0)P\left(\frac{1}{0}\right), \tag{18.1}$$

where $p(1)$ and $p(0)$ are the probabilities of receiving 1 and 0, respectively, $P(0/1)$ is the probability to decide 0 when 1 received, and $P(1/0)$ is the probability to decide 1 when 0 received. The probabilities of an event of 1 or 0 are identical, hence $p(1) = p(0) = 1/2$; as a result, the BER function is defined as

$$\mathrm{BER} = \frac{1}{2}\left[P\left(\frac{0}{1}\right) + P\left(\frac{1}{0}\right)\right]. \tag{18.2}$$

The probability density functions $P(0/1)$ and $P(1/0)$ depend on the probability function $p(I)$ of the sampled value $I$, and $p(I)$ depends on the noise statistics that cause the current fluctuations; namely, the thermal noise, its equivalent thermal current $i_{\mathrm{T}}$ and the shot noise with its equivalent shot current $i_{\mathrm{S}}$. Both noise mechanisms have an approximate Gaussian process with zero mean in PIN detectors. Hence, the noise variance is the square sum of both the processes. Moreover, the sum of two Gaussian processes results in a Gaussian process:

$$\sigma^2 = \sigma_{\mathrm{S}}^2 + \sigma_{\mathrm{T}}^2. \tag{18.3}$$

However, both variances are different for each state since the photocurrent depends on optical power; hence, shot noise has different averages at different levels [Eqs. (18.4)–(18.6)]. According to Bayes rule, the conditioned probability of error $P(0/1)$ is given by

$$P\left(\frac{0}{1}\right) = \frac{1}{\sigma_1\sqrt{2\pi}}\int_{-\infty}^{I_{\mathrm{D}}} \exp\left[-\frac{(I - I_1)^2}{2\sigma_1^2}\right] \mathrm{d}I = \frac{1}{2}\mathrm{erfc}\left(\frac{I_1 - I_{\mathrm{D}}}{\sigma_1\sqrt{2}}\right). \tag{18.4}$$

In this case, $\sigma_1$ is the 1 state noise, which is composed of shot noise RIN and additive white-Gaussian noise (AWGN), which refers to the illumination power at the high logic level Eq. (18.13). On the other hand, at the zero-state level the probability of error is given by

$$P\left(\frac{1}{0}\right) = \frac{1}{\sigma_0\sqrt{2\pi}}\int_{I_{\mathrm{D}}}^{\infty} \exp\left[-\frac{(I - I_0)^2}{2\sigma_0^2}\right] \mathrm{d}I = \frac{1}{2}\mathrm{erfc}\left(\frac{I_{\mathrm{D}} - I_0}{\sigma_0\sqrt{2}}\right). \tag{18.5}$$

Replacing $P(0/1)$ and $P(1/0)$ in Eq. (18.2) by those in Eqs. (18.4) and (18.5) results in the BER term

$$\mathrm{BER} = \frac{1}{4}\left\{\mathrm{erfc}\left[\frac{(I_1 - I_{\mathrm{D}})}{\sigma_1\sqrt{2}}\right] + \mathrm{erfc}\left(\frac{I_{\mathrm{D}} - I_0}{\sigma_0\sqrt{2}}\right)\right\}. \tag{18.6}$$

In this case, $\sigma_0$ is the 0 state noise variance that is composed of shot noise RIN and AWGN, which refers to the low optical power illumination power for low logic level state. Equation (18.13) provides the current levels for 0 and 1 states. For a low optical power state, shot noise and RIN are at a quantum-limit state.

In practice, the decision threshold level $I_D$ is optimized to the center as an average between the high-level-logic state and the low-level logic state. Hence, in order to have equal probability and centered $I_D$, the following condition of $P(1/0) = P(0/1)$ results in

$$\frac{I_1 - I_D}{\sigma_1} = \frac{I_D - I_0}{\sigma_0}. \tag{18.7}$$

Solving Eq. (18.7) results in the condition of $I_D$, which is the decision threshold limit:

$$I_D = \frac{\sigma_0 I_1 + \sigma_1 I_0}{\sigma_0 + \sigma_1}. \tag{18.8}$$

The above result means that in case of large amount of noise at logic state 1, the decision threshold is shifted down in order to have an equal swing between the decision threshold line states corresponding to 1 and the decision threshold line states corresponding to 0. In the case of $\sigma_0 = \sigma_1$, $I_D = (I_1 + I_0)/2$; however, $\sigma_0 < \sigma_1$ is always true. Moreover, it shows that when SNR gets worse, the eye is closed because $I_D$ shifts down and $\sigma_1$ gets larger. Thus, BER performance gets worse. Defining the ratio at Eq. (18.7) as $Q$ simplifies, the notation for BER becomes

$$\frac{I_1 - I_D}{\sigma_1} = \frac{I_D - I_0}{\sigma_0} = Q. \tag{18.9}$$

Hence, the probability of error can be written as

$$\mathrm{BER} = \frac{1}{2}\mathrm{erfc}\left(\frac{Q}{\sqrt{2}}\right) \approx \frac{\exp(-Q^2/2)}{Q\sqrt{2\pi}}. \tag{18.10}$$

Substituting Eq. (18.8) into Eq. (18.7) results in a simple expression for $Q$:

$$Q = \frac{I_1 - I_0}{\sigma_1 + \sigma_0}. \tag{18.11}$$

As a conclusion, $Q^2$ defines the system SNR, which is directly related to the eye diagram opening and modulation extinction as is explained in the next section.

## 18.2 Extinction Ratio

$Q$ factor is a ratio that represents eye opening to total noise, in fact it is root square of SNR. Generally speaking, ER is used as a measure for eye opening, and average optical power $P_{\mathrm{ave}}$ is used in fiber links calculations. Hence, it is much more convenient to describe BER as a function of $P_{ave}$ and ER. The average power of an eye pattern is given by

$$P_{ave} = \frac{P_1 + P_0}{2}. \tag{18.12}$$

Additionally, the currents at each logic level and optical state are given by

$$\begin{aligned} I_1 &= rP_1, \\ I_0 &= rP_0, \end{aligned} \tag{18.13}$$

where $r$ is the photodiode responsivity. ER is defined as

$$\mathrm{ER} = \frac{I_1}{I_0} = \frac{P_1}{P_0} = 1 + \frac{\Delta P}{P_0}; \tag{18.14}$$

hence, $1 \leq \mathrm{ER} \leq \infty$. Substituting Eq. (18.12) into Eq. (18.14) and then into Eq. (18.11) results a new expression for $Q$ as a function of ER and noise variance:

$$Q = r2P_{ave}\left(\frac{\mathrm{ER}-1}{\mathrm{ER}+1}\right)\left(\frac{1}{\sigma_1 + \sigma_0}\right). \tag{18.15}$$

This term is important and shows the $Q$ asymptotic limit when $\mathrm{ER} \to \infty$:

$$Q = r2P_{ave}\left(\frac{1}{\sigma_1 + \sigma_0}\right) = \frac{rP_{ave}}{(\sigma_1 + \sigma_0)/2}. \tag{18.16}$$

As was expected, it is the ratio between the average current to the average noise current. Hence, $Q^2$ refers to the ratio of powers and is the SNR limit. The average noise in the eye diagram is the average ISI + noise. More impairments such as mode partition and mode hopping should be included and are reviewed in the next section. Jitter effects are explained in Sec. 15.8.6. Several more relations and trade-offs can be concluded from the above relations. For instance, in order to maintain a constant level of BER, SNR should be preserved.

In case of a finite ER, $Q$ is defined by Eq. (18.15). However, if ER gets lower, $Q$, which is the SNR, gets lower and $P_{ave}$ should be increased. The reason behind that is to preserve SNR in order to have the same BER value.

A figure of merit can be determined by the ratio of $Q(\mathrm{ER})$ against $Q(\mathrm{ER} \to \infty)$, i.e., a ratio between the ideal SNR and the finite value SNR. The ratio between $Q(\mathrm{ER})$ and $Q(\mathrm{ER} \to \infty)$ is defined as the power penalty $\delta_{\mathrm{ER}}$, showing the additional power needed to preserve a constant BER level by preserving SNR:

$$\delta_{\mathrm{ER}} = 10\log\left(\frac{\mathrm{ER}+1}{\mathrm{ER}-1}\right). \tag{18.17}$$

In the above discussion, there are some approximations to which attention should be paid. The noise variances are affected by the changes in ER, since shot noise and RIN are directly dependent on optical power levels. However, both variances can be approximated to thermal noise in case the thermal noise of the receiver is dominant. Hence, the following approximation is applicable:

$$\sigma_0 = \sigma_1 = \sigma_T. \qquad (18.18)$$

In Fig. 18.2, BER is shown as a function of SNR in units of dB. The relation between SNR and $Q$ can be written as

$$\mathrm{SNR(dB)} = 20\log(Q). \qquad (18.19)$$

In order to meet the criterion BER $< 10^{-10}$, the SNR should be higher than 16 dB. The power penalty versus ER is illustrated in Fig. 18.3. The higher the ER, the lower the penalty because the SNR is improving. It is observed that the function has an asymptotic nature. As ER $\to \infty$, the power penalty goes to 0 dB. In real life, there is always a power penalty, since ER has a final value. Generally speaking, ER values are around 8 dB; hence, the power penalty is 1.5 dB, which means that an additional 1.5 dB is required to preserve the SNR value for a given BER performance.

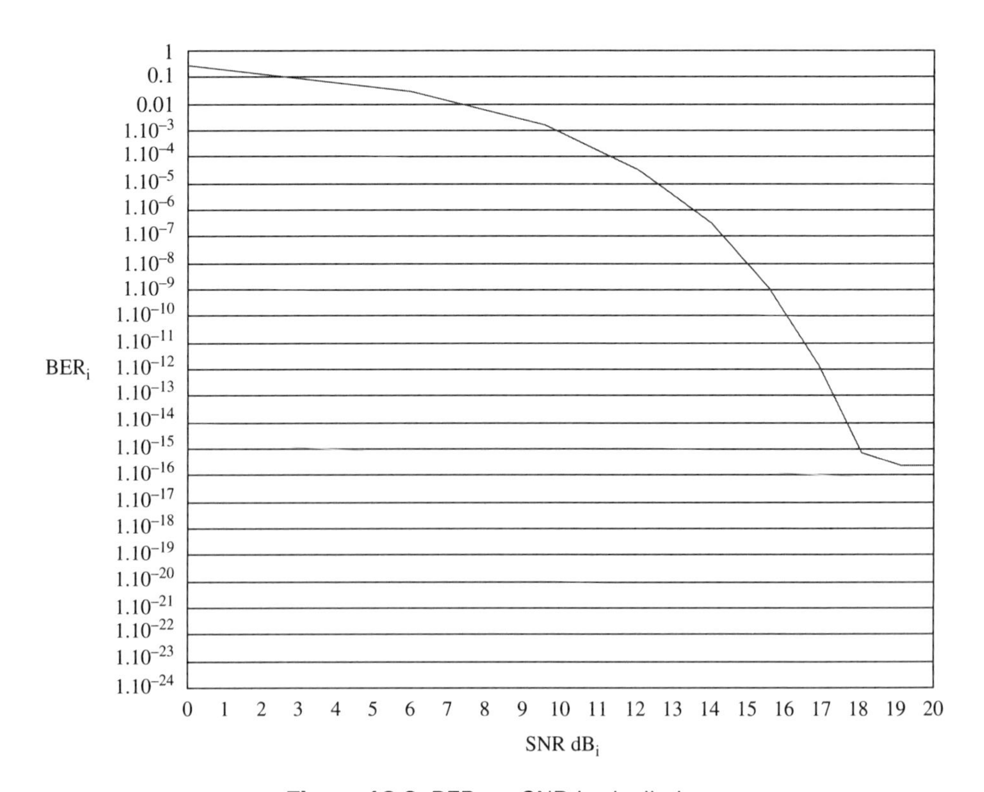


**Figure 18.2** BER vs. SNR in decibels.

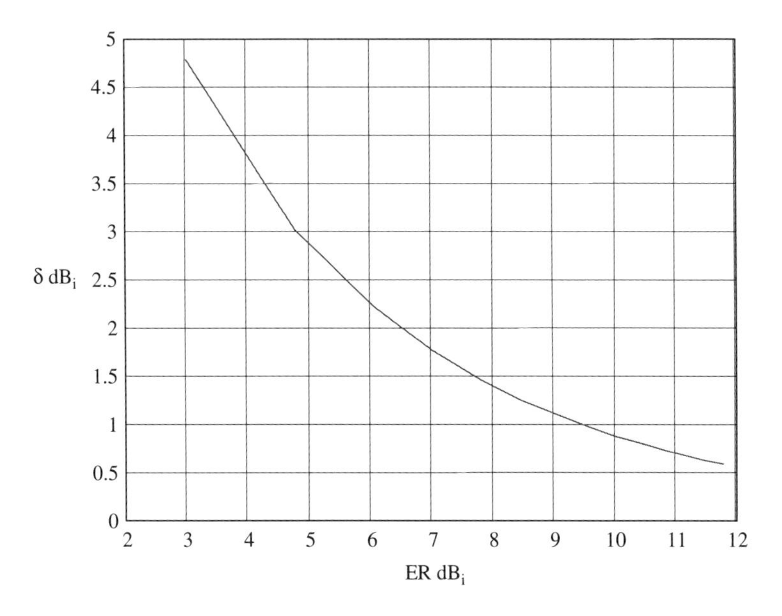


**Figure 18.3** Power penalty vs. extinction ratio, ER.

## 18.3 Mode Partition

An introduction to mode-partition noise, denoted here as $\sigma_{PC}$, is provided in Sec. 7.4.3. Since the normalized carrier or signal equals unity, Eqs. (7.85) and (7.86) can be modified into the following form:[1]

$$\frac{1}{\text{SNR}} = \frac{N}{S} = \sigma_{\text{PC}}^2 = \frac{1}{2}(\pi B)^4(A_1^4\sigma^4 + 48A_2^4\sigma^8 + 42A_1^2A_2^2\sigma^6), \tag{18.20}$$

where $B$ is the data rate, the parameters $A_1$ and $A_2$ can be found from Eq. (18.23), and $\sigma$ is the wavelength rms deviation variance. Equation (18.20) indicates that the ratio of signal-to-partition noise depends on half width of the laser diode spectrum and the chromatic dispersion of the fiber. Additionally, Eq. (18.20) shows that SNR is independent of signal power. As a consequence, by providing an unlimited large signal power, the overall system SNR cannot be improved beyond the limit imposed by the partition noise. This results in the error rate asymptote in the error rate versus input optical power, which is determined by the signal to the mode-partition noise. For a given error rate asymptote, the mode partition noise must satisfy Eq. (18.21), which defines the SNR for the desired BER as shown in Fig. 18.2:

$$\frac{1}{\sigma_{\text{PC}}^2} \geq Q^2. \tag{18.21}$$

The target value of $Q$ in dB (SNR) is given in Fig. 18.2. In the case of silica fiber, the wavelength range between 1000 and 1600 nm and the group delay change is given by

$$\frac{\mathrm{d}\tau}{\mathrm{d}\lambda} = \frac{s(\lambda_C - \lambda_0)}{c\lambda^2}, \tag{18.22}$$

where $s$ is a material constant, $C$ is the speed of light, $\lambda_0$ is minimum dispersion wavelength. The coefficients $A_1$ and $A_2$ are given by

$$\begin{cases} A_1 = \dfrac{s(\lambda_C - \lambda_0)z}{c\lambda_C^2}, \\ A_2 = \dfrac{sz}{2c\lambda_C^2}. \end{cases} \tag{18.23}$$

Observing Eq. (18.23) applied into Eq. (18.20) and the limit condition given by Eq. (18.21), the product $B \cdot z$ (GB $\times$ s$^{-1}$ $\times$ km) can be evaluated versus wavelength. This product is defined as the bit rate distance and can be written in terms of noise limits:

$$\frac{1}{Q^2} \geq \sigma_{PC}^2 = \frac{(\pi Bz)^4}{2}\left\{\left[\frac{s(\lambda_C - \lambda_0)}{c\lambda_C^2}\right]^4 \sigma^4 + 48\left(\frac{s}{2c\lambda_C^2}\right)^4 \sigma^8 + 42\left[\frac{s(\lambda_C - \lambda_0)}{c\lambda_C^2}\right]^2 \left(\frac{s}{2c\lambda_C^2}\right)^2 \sigma^6\right\}, \tag{18.24}$$

where $\sigma$ is half rms width of the spectrum. From Eq. (18.24) it can be seen that the mode partition noise is directly related to the wavelength rms deviation variance and increases with bit rate product. Moreover, it is not affected by the signal power. An ideal link with $1/\mathrm{SNR} = 1/Q^2$ degraded by mode partition noise will have SNR degradation as

$$\frac{1}{\mathrm{SNR_{eq}}} = \frac{1}{Q_{eq}^2} = \frac{1}{Q^2} + \mathrm{S}\frac{1}{\mathrm{CNR_{PC}}} = \frac{1}{Q^2} + \sigma_{PC}^2. \tag{18.25}$$

This equation can be presented as the ideal SNR, multiplied by the asymptotic error coefficient. After some algebraic manipulation the mode partition carrier to noise is given by

$$\sigma_{PC}^2 = \frac{1}{Q_{eq}^2} - \frac{1}{Q^2} = \left(\frac{1}{Q^2}\right)\frac{Q^2 - Q_{eq}^2}{Q_{eq}^2}. \tag{18.26}$$

Thus, to ensure the desired BER performance Eq. (18.26) is modified to

$$\frac{1}{\sigma_{PC}^2} \geq Q^2 \frac{Q_{eq}^2}{Q^2 - Q_{eq}^2} = \frac{Q_{eq}^2/Q^2}{1 - (Q_{eq}^2/Q^2)} = Q^2 \frac{K}{1-K} = Q^2\rho, \tag{18.27}$$

where $0 < K < 1$ is the SNR penalty because of the mode-partition noise. $K$ describes the amount of SNR correction, or signal increment needed to compensate for the $\mathrm{SNR_{eq}}$ performance degradation of $Q$ due to mode-partition noise. However, observing Eq. (18.25) carefully, it can be seen that mode-partition noise sets the overall system SNR limit. Hence, even though the optical power is increased and the intrinsic ideal SNR, denoted by $Q$, $1/Q^2 \to 0$, is improved, the overall SNR is limited by the mode-partition noise, $\sigma_{PC}^2$. Thus, this power penalty describes the SNR loss or the equivalent carrier power loss in the case of no mode partition noise. The statement in Eq. (18.27) is that SNR caused by mode-partition noise $\sigma_{PC}^2$ should be below the worst-case SNR degradation quantity so that the equivalent SNR will satisfy $Q_{eq}$ for the desired BER performance. For clarity see Eq. (18.26). It is clear that the condition $\sigma_{PC}^2 - 1/Q^2 \geq 1/Q_{eq}^2$, hence $Q_{eq}^2 > 1/[\sigma_{PC}^2 - (1/Q^2)]$. In other words, the higher the $1/\sigma_{PC}^2$, ($\sigma_{PC}^2 \to 0$), the higher the equivalent SNR, and the power penalty is minimized. The power penalty in decibels is given by

$$\alpha = 10 \log \sqrt{K}. \tag{18.28}$$

An alternate penalty expression is provided by Ref. [1]:

$$\frac{1}{\sigma_{PC}^2} \geq Q^2 \frac{10^{2\alpha/10}}{10^{2\alpha/10} - 1} = Q^2 \left( \frac{Q_{PC}^2}{Q^2} \right). \tag{18.29}$$

This equation states that for a given mode partition SNR, $Q_{PC}^2$, and ideal SNR for a given BER performance, $Q^2$, without partition noise, in order to ensure the BER performance, the mode-partition SNR, $1/\sigma_{PC}^2$, should be better than the ideal SNR after the power penalty correction; i.e., its contribution should be below the SNR correction amount. Moreover, the mode partition noise sets the BER limit, beyond which BER cannot be improved. This is equal to the phase-noise BER limit presented in Sec. 15.8.6.

This carrier-to-noise relation can be presented by useful design plots. For instance, Eq. (18.24) states that the mode-partition noise generates dispersion, which increases partition noise as the fiber length $z$ increases. Thus, the bit-rate distance product $B \cdot z$ is a way to measure SNR degradation caused by mode-partition noise dispersion. From Fig. 18.2, BER of $10^{-9}$ is guaranteed for SNR of 15.5 dB or $Q = 6$. For a laser with a peak wavelength deviation, $\sigma$, of 1 to 2 nm, the bit rate product would be varying and would set a link budget limit for the desired BER performance. This limit states the maximum distance between the transmitter and the receiver. This limit sets the data rate versus fiber length as a function of mode partition noise due to chromatic dispersion. The higher the data rate, the shorter the distance between the receiver and the transmitter. As mode partition or wavelength deviation becomes higher, the bit rate distance product becomes lower. As the data rate increases, the absolute fiber length becomes shorter. This performance envelope guarantees the desired BER of $10^{-9}$. It can be seen that at the center wavelength of $\lambda_0 = 1310$ nm, the $B \cdot z$ product is maximized since there is no chromatic dispersion (the material $s$ parameter value was assumed to be 0.05):

$$Bz = \sqrt[4]{\frac{2}{\pi Q^2\left\{\left[s(\lambda_C - \lambda_0)/c\lambda_C^2\right]^4\sigma^4 + 48(s/2c\lambda_C^2)^4\sigma^8 + 42\left[s(\lambda_C - \lambda_0)/c\lambda_C^2\right]^2(s/2c\lambda_C^2)^2\sigma^6\right\}}}. \tag{18.30}$$

The curve in Fig. 18.4 can be used as follows. Assume that the data rate is 1.25 GB/s. Then for 1050 nm $B_2 = 10$, the maximum distance will be 10 km/1.25 = 8 km for $\sigma = 1$ nm and for $\sigma = 2$ nm it will be $6/1.25 = 4.8$ km.

Using Eq. (18.30) under a specific wavelength variance, but variable $Q$, results in a different bit rate distance product with respect to a specific $Q_0$ at a given BER asymptote. The plot in Fig. 18.4 is for $Q_0 = 6$ to satisfy a BER of $10^{-9}$. In Fig. 18.5, the data-rate correction factors for a different SNR are provided. For that purpose, the coefficient $\eta$ is determined. This factor is the ratio between the

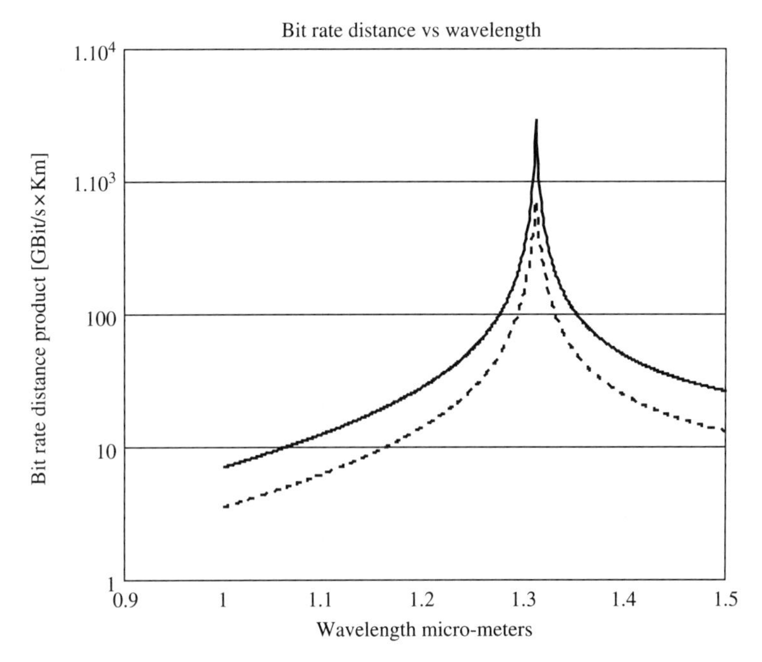


**Figure 18.4** Bit rate distance (GB/s × km) vs. wavelength $\lambda_c$ ($\mu$m) for BER = $10^{-9}$ and signal-to-noise asymptote $Q = 6$, the center wavelength $\lambda_0 = 1310$ nm, material constant is $s = 0.05$, and rms wavelength deviation variances are $\sigma = 1$ nm (solid line) and $\sigma = 2$ nm (dashed line).

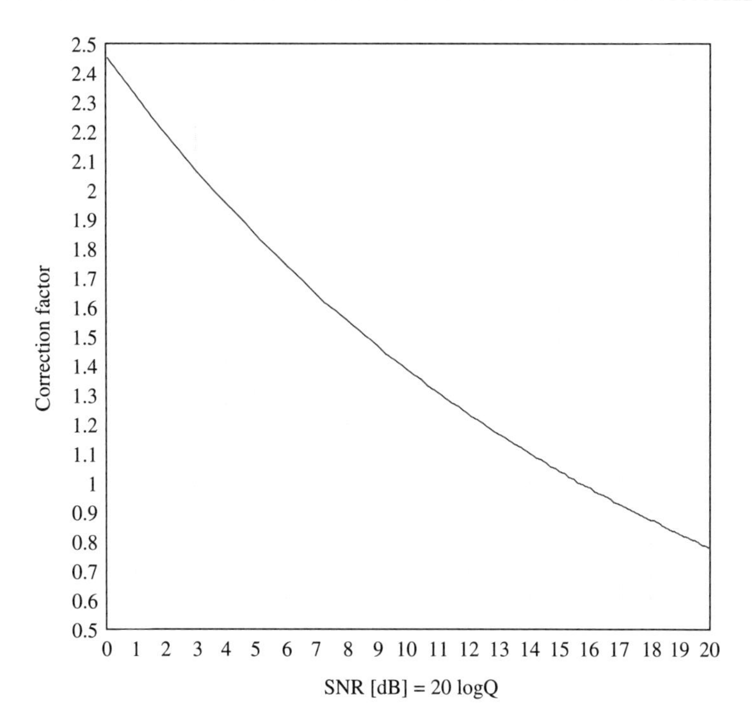


**Figure 18.5** Normalized bit rate distance $\eta$ calculated with respect to a BER of $10^{-9}$ at $Q = 6$ (SNR = 16 dB) for a wavelength of 1250 nm and wavelength rms deviation variance $\sigma = 2$ nm.

bit-rate-product function *Bz*, where $Q$ is varying to *Bz* value in which the $Q$ value is a given SNR for a specific BER. In this case the bit-rate product function *Bz* values are $Q = 6$ for BER of $10^{-9}$. It can be seen that as the data rate increases, SNR decreases, and vice versa. Hence, to preserve BER performance one should compromise the following: for high data rates reduce distance, or for low data rates increase distance. In Fig. 18.6, BER performance versus the correction factor $\eta$ is shown. As $\eta$ increases above unity, where $\eta = 1$ equals the state of $Q = 6$ and $20 \log 6 \approx 15.5$ dB SNR, the noise due to mode partitioning becomes higher, hence, the SNR becomes lower and BER performance is reduced, and vice versa.

In conclusion, mode-partition noise reduces SNR and closes the eye; in addition, it adds chromatic dispersion. Moreover, it sets the limit of SNR for a given BER asymptote and limits the data-rate transport due to chromatic dispersion.

This analysis shows that there is a trade-off among the desired data rate, fiber length, and wavelength deviation. The *Bz* factor demonstrates that due to the SMF chromatic dispersion effect, there is a limit of data rate versus distance due to a wavelength modal-noise effect. Higher data rates are more susceptible to wavelength deviation. Hence, long distance results in large dispersion for high rates. Therefore, a high rate has to have a short distance and vice versa. Ideally, a laser that has no wavelength deviation due to mode partitioning would provide the best *Bz* results as shown in Fig. 18.4.

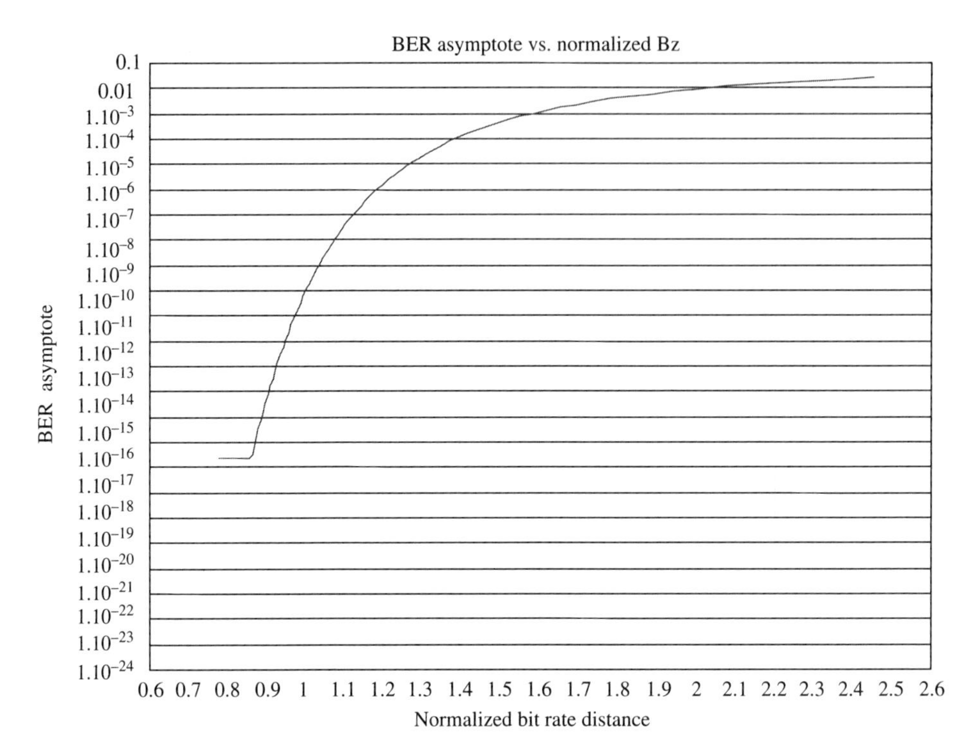


**Figure 18.6** BER vs. normalized bit rate distance $\eta$ (correction factor), which is calculated with respect to a BER of $10^{-9}$ for $Q = 6$ (SNR $\approx$ 15.5 dB) at a wavelength of 1250 nm, and wavelength RMS deviation variance $\sigma$ of 2 nm. Note that $\eta = 1$ for a BER of $10^{-9}$. When $\eta < 1$, it means that SNR is getting values higher than the values at $Q = 6$ and therefore BER is improving. That is due to the definition of $\eta$.

## 18.4 Timing Jitter

An elaborated review about phase noise, its relation to time domain jitter,[2] and the phase error it creates, denoted by $\Delta\theta_{\text{rms}}$, which reduces the SNR, is given in Chapters 14 and 15. As a consequence, the sampling error, marked as $\Delta t$, is random and BER performance is degraded according to the jitter statistics. This is mainly due to sampling time fluctuations from bit to bit. Again, as in the explanation for mode partition noise in the previous section, the phase noise sets the SNR limit above which BER cannot be improved [this is demonstrated by Eq. (15.85)]. Timing jitter in the digital transceiver corresponds to amplitude fluctuations due to sampling error statistics. Hence, under the assumption of biasing the laser below the threshold, the current amplitude should be $I = I_1 - \langle i \rangle$, where $\langle i \rangle$ is the average current-error value. For a PIN detector, which is dominated by thermal noise variance, $\sigma_T$, Eq. (18.11) can be written as

$$Q = \frac{I_1 - \langle \Delta i_j \rangle}{\sqrt{\sigma_T^2 + \sigma_j^2} + \sigma_T}. \qquad (18.31)$$

The noise variances under the root square represent the noise at 1 state when the laser is on. The variance $\sigma_j$ represents the jitter noise during the period of $\Delta t$. The second variance $\sigma_T$ represents only thermal noise without jitter noise at the off state of the laser. This is the intrinsic thermal noise at the PIN photodiode. Note that the dark shot current is negligible. Thermal noise remains the same at both 1 and 0 states. As was explained in Sec. 15.8.6 using analog design terminology of phase noise and FM noise, misalignment in the sampling clock creates timing jitter. Assuming a data rate of $B$, the sampling period can be calculated as $T = 1/B$. Therefore, the current fluctuations can be described in terms of $\Delta t$ misalignment. Thus, the current shape is governed by a pulse-shape function $h_{\mathrm{OUT}}(t)$. As a result, the current fluctuations can be written as

$$\Delta i_{\mathrm{j}} = I_1[h_{\mathrm{OUT}}(0) - h_{\mathrm{OUT}}(\Delta t)]. \tag{18.32}$$

The ideal sampling case occurs at $t = 0$. Using the same approach presented in Sec. 15.8.6, it can be found that $h_{\mathrm{OUT}}(t) = \cos^2(\pi Bt/2)$. Hence at $t = 0$ and $\Delta t = 0$, $\Delta i = 0$. For low values of $\Delta t$, $B\Delta t \ll 1$, $h_{\mathrm{OUT}}(t)$ can be approximated using the Taylor expansion as

$$\Delta i_{\mathrm{j}} = \left(\frac{2\pi^2}{3} - 4\right)(B\Delta t)^2 I_1. \tag{18.33}$$

The timing jitter distribution is assumed to be Gaussian with zero mean. For that reason, the timing-jitter-probability density is provided by

$$p(\Delta t) = \frac{1}{\tau_{\mathrm{j}}\sqrt{2\pi}} \exp\left(-\frac{\Delta t^2}{2\tau_{\mathrm{j}}^2}\right), \tag{18.34}$$

where $\tau_{\mathrm{j}}$ is the rms variance of $\Delta t$. Equation (18.34) can be modified to the current fluctuation probability density by applying Eq. (18.33). Note that $\Delta i_{\mathrm{j}}$ is proportional to $\Delta t^2$. Remembering that the normalized Gaussian distribution is given by

$$G(x) = \frac{1}{\sigma\sqrt{2\pi}} \exp\left(-\frac{x^2}{2\sigma^2}\right), \tag{18.35}$$

The current fluctuation probability density function can be found by

$$p(\Delta i_{\mathrm{j}}) = \frac{1}{\sqrt{\pi[(4\pi^2/3) - 8](B\tau_{\mathrm{j}})^2 I_1}} \exp\left\{-\frac{\Delta i_{\mathrm{j}}}{[(4\pi^2/3) - 8](B\tau_{\mathrm{j}})^2 I_1}\right\}. \tag{18.36}$$

The current jitter variance $\sigma_j$ is given by

$$\sigma_j = \sqrt{\langle(\Delta i)^2\rangle} = \frac{[(4\pi^2/3) - 8](B\tau_j)^2 I_1}{\sqrt{2}}. \quad (18.37)$$

The mean value is given by

$$\langle \Delta i_j \rangle = \frac{\sigma_j}{\sqrt{2}} = \frac{mI_1}{\sqrt{2}}. \quad (18.38)$$

To simplify the notation, $m$ is defined as

$$m = \left(\frac{4\pi^2}{3} - 8\right)(B\tau_j)^2, \quad (18.39)$$

where it should be noted that $m$ contains the timing variance. Using Eqs. (18.31) and (18.37) and with the aid of the $m$ parameter, the average optical power received by a photodetector with responsivity value $r$ can be calculated as

$$\overline{P}_{Rx}(m) = \left(\frac{\sigma_T Q}{r}\right)\frac{1 - (m/2)}{[1 - (m/2)]^2 - (m^2Q^2/2)}, \quad (18.40)$$

where the relation between the dc current at 1 to the average optical power is given by $I_1 = 2r\overline{P}_{Rx}$. When there is no jitter, $m = 0$ because the timing error variance $\tau_j$ is equal to 0. Therefore, it is possible to define jitter power penalty parameter, $\delta_j$, in dB, as the ratio between the averages of optical power at $m \neq 0$ and $m = 0$ states. The power penalty parameter describes the amount of power increment needed in the receiver's path to keep the same BER performance due to jitter degradation:

$$\delta_j = 10\log\left[\frac{\overline{P}_{Rx}(m)}{\overline{P}_{Rx}(0)}\right] = 10\log\left\{\frac{1 - (m/2)}{[1 - (m/2)]^2 - (m^2Q^2/2)}\right\}. \quad (18.41)$$

Assume an asymptotic SNR value with $Q = 6$, SNR $= 10 \log 36$, to guarantee a BER value of $10^{-9}$ as per Fig. 18.2. The jitter power penalty curve can be presented as a plot shown in Fig. 18.7. The drawbacks of this approximation are that it is for small fluctuations and that the jitter distribution is not Gaussian. Hence, Eq. (18.41) can provide a ballpark number that is an underestimate of the accurate power penalty.

The plot in Fig. 18.7 can be used as follows. Assume the data rate is 625 MB/s and the rms jitter variance is 80 ps. Therefore, the timing jitter parameter is given by $B\tau_j = 625$ MB/sec $\times$ 80 psec $= 0.05$. This results in a 0.3-dB optical power penalty. It is equal to a reduction in $Q$ because the noise is increased. If there is no jitter accompanying the signal at the input optical power of $P_o$ (dBm),

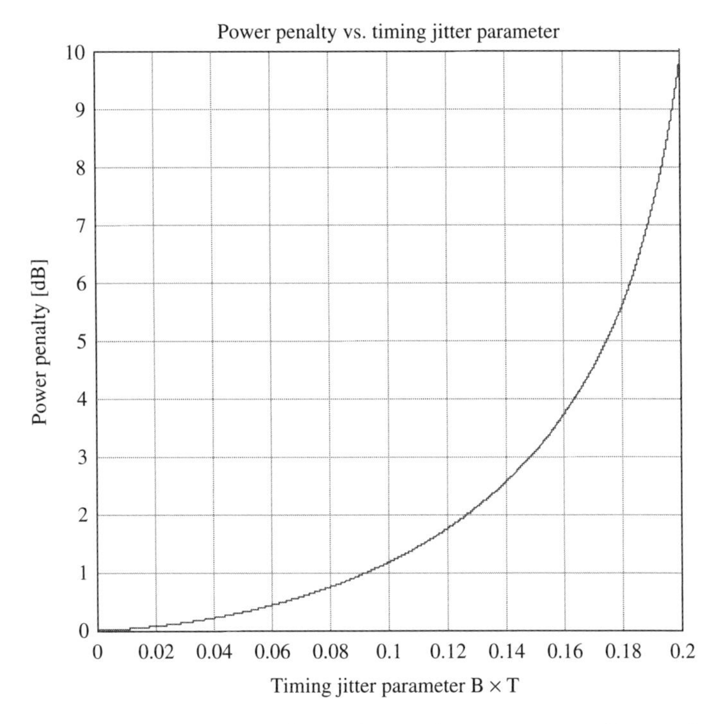


**Figure 18.7** Power penalty ($\delta$) vs. timing jitter parameter $B\tau_j$ for $Q = 6$.

the SNR parameter is $Q$. Jitter reduces $Q$ and a power penalty is the amount of power needed to preserve the same $Q$ value for getting the same BER performance as there is no jitter. For signal analysis, it is convenient to mark the current as

$$I_1 = r10^{0.1P_o - 3}. \tag{18.42}$$

Due to power penalty causing SNR degradation, it is equal to current reduction for a fixed noise level. Thus, the new degraded current is

$$I_2 = r10^{0.1P_o - \delta - 3}. \tag{18.43}$$

Knowing that $Q = I/\sigma$, where $\sigma$ is preserved and $I$ is reduced, the new $Q$ factor due to jitter is

$$Q_2 = \frac{I_2}{\sigma} = \frac{r10^{0.1P_o - 0.1\delta - 3}}{\sigma} = \frac{r10^{0.1P_o - 3 - 0.1\delta}}{\sigma} = \frac{I_1 10^{-0.1\delta}}{\sigma} = Q10^{-0.1\delta}. \tag{18.44}$$

Thus, for this example $Q$ is reduced by 0.933 from 6 to approximately 5.6. Using Fig. 18.2, the SNR was reduced from 10 log 36 = 15.5 dB to 10 log 31.36 = 14.9 dB and the BER becomes $10^{-8}$.

This example shows the BER degradation calculation method, which can be used with the same concept to evaluate the BER performance reduction for all link impairments explained here such as RIN, mode-partition noise, and ER. It can also provide the designer a measure of how much to correct the optical level in order to improve SNR until the limit bound is set by the noise impairment.

## 18.5 Relative Intensity Noise

The theory behind RIN is provided in Sec. 7.4.4, while CNR limits are dealt with in Sec. 10.7. This section provides a useful tool for estimating BER reduction as a function of the power penalty as a result of RIN effects. The overall noise variances for SNR calculations are given by the sum of thermal noise, shot noise, and RIN variances:

$$\sigma^2 = \sigma_{\mathrm{T}}^2 + \sigma_{\mathrm{S}}^2 + \sigma_{\mathrm{RIN}}^2, \tag{18.45}$$

where the RIN noise is given by

$$\sigma_{\mathrm{RIN}} = r\sqrt{\langle(\Delta P_{\mathrm{OPT}}^2)\rangle} = rP_{\mathrm{OPT}}R, \tag{18.46}$$

where $r$ is the responsivity of the photodiode, and $R$ is a constant defined by

$$R = \frac{\sqrt{\langle(\Delta P_{\mathrm{OPT}})^2\rangle}}{P_{\mathrm{OPT}}}. \tag{18.47}$$

This parameter depends on the transmitter's laser quality and is defined as

$$R^2 = \frac{1}{2\pi}\int_{-\infty}^{\infty} \mathrm{RIN}(\omega)\mathrm{d}\omega, \tag{18.48}$$

where $\mathrm{RIN}(\omega)$ is given by Eq. (7.83) in Sec. 7.4.4. When observing the performance of a transmitter at a given BW of interest, as indicated by Eq. (18.48), the SNR at the output can be a 20-dB signal to RIN. Hence, the coefficient $R$ represents the inverse SNR. Thus for a 20-dB SNR, $R = 0.01$ or $-20$ dB.

Using the same bias conditions for a laser as in Sec. 18.5, and referring to the calculations to the average received optical power by the relations of $I_1 = rP_1 = 2r\overline{P}_{\mathrm{Rx}}$, the signal's current-to-noise variance is given by

$$Q = \frac{2r\overline{P}_{\mathrm{Rx}}}{\sqrt{\sigma_{\mathrm{T}}^2 + \sigma_{\mathrm{S}}^2 + \sigma_{\mathrm{RIN}}^2} + \sigma_{\mathrm{T}}}, \tag{18.49}$$

where the noise variances are given by

$$\sigma_S = \sqrt{4qr\overline{P}_{Rx}\Delta f}, \tag{18.50}$$

$$\sigma_{RIN} = 2rR\overline{P}_{Rx}, \tag{18.51}$$

$$\sigma_T = \sqrt{\frac{4KT\Delta fF}{R_L}}, \tag{18.52}$$

where $F$ is the noise factor, $k$ is the Boltzmann constant, $\Delta f$ is the bandwidth and $R_L$ is the load resistance, which is the real part of the input impedance $Z_L$ of receivers. Changing the subject of Eq. (18.49) provides an expression for the input power as a function of RIN parameter denoted as $R$:

$$\overline{P}_{Rx}(R) = \frac{Q\sigma_T + Q^2 q\Delta f}{r(1 - R^2Q^2)}. \tag{18.53}$$

It can be observed that the power increases as $R$ increases in order to preserve the same BER. If there is no RIN, then the received optical power is at its minimum. This is because there is no power penalty and degradation in $Q$. Therefore, there is no requirement to compensate for RIN noise by increasing the power in order to preserve SNR or $E_b/N_0$ that guarantees the same BER performance. Consequently, it is convenient to define the RIN power penalty as the ratio between the ideal case of $R = 0$ and a nonideal case of $R \neq 0$. Based on the above definition, the power penalty in decibels is given by

$$\delta_{RIN} = 10\log\frac{\overline{P}_{Rx}(R)}{\overline{P}_{Rx}(0)} = -10\log\left(1 - R^2Q^2\right). \tag{18.54}$$

This parameter can be presented by a plot given in Fig. 18.8 for Eq. (18.54) for $Q = 6$, which guarantees a BER of $10^{-9}$. It shows the amount of optical power correction needed in order to preserve BER performance. From Eq. (18.46), it can be written as

$$Q^2 = \frac{\left(2r\overline{P}_{Rx}\right)^2}{\sigma_T^2 + \sigma_S^2 + \sigma_{RIN}^2 + \sigma_T^2}. \tag{18.55}$$

Thus

$$\begin{aligned}\frac{1}{Q^2} &= \frac{1}{\left[2\sigma_T^2/(2r\overline{P}_{Rx})^2\right] + \left[\sigma_S^2/(2r\overline{P}_{Rx})^2\right]\left[\sigma_{RIN}^2/(2r\overline{P}_{Rx})^2\right]} \\ &= \frac{1}{(2/\mathrm{CNR_T}) + (1/\mathrm{CNR_S}) + (1/\mathrm{CNR_{RIN}})}.\end{aligned} \tag{18.56}$$

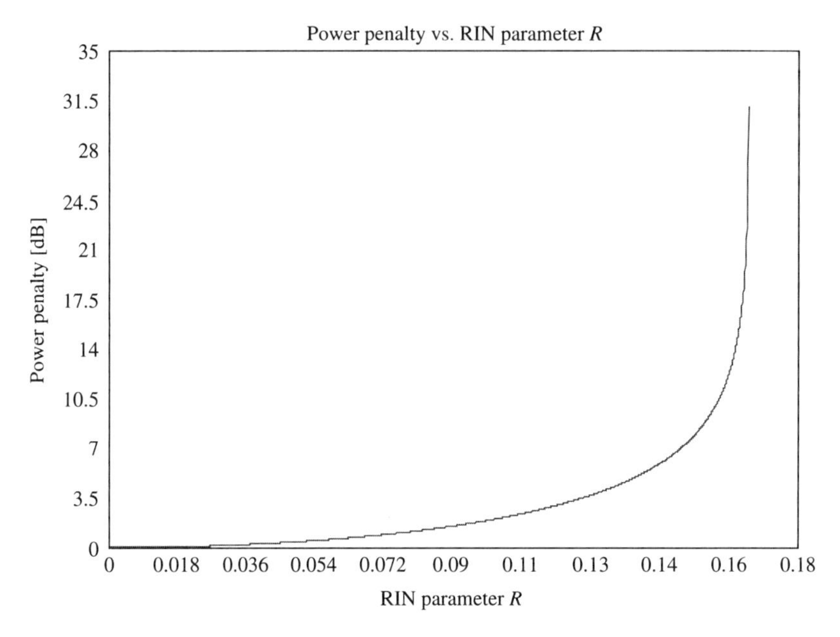


**Figure 18.8** Power penalty vs. RIN parameter $R$.

It can be observed that the lowest SNR performance would set the limit of the overall equivalent SNR. Hence, there is a limit beyond which the increase in optical power would not improve the BER. Moreover, it was explained by Eq. (7.83) that as the data rate exceeds the relaxation oscillation frequency of the laser, the RIN noise is severely increased. This is related as well to rise and fall time, eye overshoots, and ringing. The calculation is done assuming an ideal single mode laser is in use and there is no chromatic dispersion. However, real lasers suffer from additional modes and mode hopping, which creates chromatic dispersion as well as mode-partition noise. Thus, the SNR and BER performances are further degraded as was explained in Sec. 18.4.

## 18.6 Minimum Detectable Signal and Optical Power

In analog channel-detection thresholds defined as minimum detectable signal (MDS), a similar definition exists for digital transceivers as well. However, in digital transport this quantity is measured by BER versus $Q$. The goal is to determine the minimum optical power for a specified BER performance using Eq. (18.10). It is assumed that the laser bias conditions for 0 level are below the threshold and $P_0 = 0$, where the bias current $I_0 = 0$, and for 1 the optical power level equals to $P_1$ at bias current $I_1$. To make this case more general, it is assumed that the photo detector gain value is $M$. Hence at the 1 state the current is given by

$$I_1 = MrP_1 = 2Mr\overline{P}_{\mathrm{Rxave}}, \tag{18.57}$$

where

$$\overline{P}_{\mathrm{Rxave}} = \frac{P_1 + P_0}{2}. \tag{18.58}$$

Using the same assumptions as before, the photo diode is illuminated at 1, thus the input noise is AWGN and shot, while at 0 the noise is AWGN. It is assumed also that the dark-current shot noise is negligible. Thus the noise variances for shot and AWGN are given by

$$\sigma_{\mathrm{S}}^2 = 2qM^2F_{\mathrm{A}}r\left(2\overline{P}_{\mathrm{Rxave}}\right)\Delta f, \tag{18.59}$$

$$\sigma_{\mathrm{T}}^2 = \frac{4kT}{R_{\mathrm{L}}}F_n\Delta f, \tag{18.60}$$

where $F_{\mathrm{A}}$ is the APD excess noise discussed in Sec. 10.8. Using these terms, $Q$ is given by

$$Q = \frac{I_1}{\sigma_1 + \sigma_0} = \frac{2Mr\overline{P}_{\mathrm{Rxave}}}{\sqrt{\sigma_{\mathrm{T}}^2 + \sigma_{\mathrm{S}}^2} + \sigma_{\mathrm{T}}}. \tag{18.61}$$

For given BER specifications, $Q$ is determined. Hence, the average optical power can be extracted by

$$\overline{P}_{\mathrm{Rxave}} = \frac{Q}{r}\left(qF_{\mathrm{A}}Q\Delta f + \frac{\sigma_{\mathrm{T}}}{M}\right). \tag{18.62}$$

Equation (18.62) shows how the received optical power at a given BER depends on various receiver parameters. Moreover, APD gain compensates for the thermal noise and improves the SNR by setting lower optical power for a given BER. For a PIN photodetector, $M = 1$ and the thermal noise it governs. Therefore, the minimum input power for a given BER is given by

$$\overline{P}_{\mathrm{Rxave}} \approx \frac{Q}{r}\sigma_{\mathrm{T}}. \tag{18.63}$$

Hence, $\overline{P}_{\mathrm{Rxave}} \propto \sqrt{B}$, where $B = 2\Delta f$; this is because Eq. (18.63) depends on the variance $\sigma_{\mathrm{T}}$ given by Eq. (18.52). It is clear that for, PIN photodetector, the minimum optical power is set by the thermal noise and increases proportionally to $\sqrt{B}$, where $B$ is the bit rate and bit rate BW defines the noise BW.

Equation (18.62) can be modified by replacing $F_{\mathrm{A}}$ with Eq. (10.81) from Sec. 10.8. Taking the derivative of Eq. (18.62) versus $M$ and solving for $M$

$\partial \overline{P}_{\text{Rxave}}/\partial M = 0$ will provide an optimum minimum for $\overline{P}_{\text{Rxave}}$:[14]

$$M_{\text{OPT}} = \frac{1}{\sqrt{k_{\text{A}}}}\sqrt{\frac{\sigma_{\text{T}}}{Qq\Delta f} + k_{\text{A}} - 1} \approx \sqrt{\frac{\sigma_{\text{T}}}{k_{\text{A}}Qq\Delta f}}. \tag{18.64}$$

Using the above optimal result, and Eqs. (10.81) and (18.82) will provide the minimum optical value for an APD receiver:

$$\overline{P}_{\text{Rxave}} = Q^2\frac{2q\Delta f}{r}[k_{\text{A}}(M_{\text{OPT}} - 1) + 1]. \tag{18.65}$$

It is observed that the optical level depends on the APD ionization coefficient $k_{\text{A}}$. The sensitivity improves as $k_{\text{A}}$ gets lower. For InGaAs, sensitivity improves about 6 to 8 dB. In addition, it can be observed that $\overline{P}_{\text{Rxave}} \propto B$, which is linearly dependent compared to PIN photodiode. The linear dependency of the received optical power on BW is common to shot-noise-limited receivers. In special cases, the thermal variance $\sigma_{\text{T}} = 0$ and $M = 1$. It shows that the sensitivity degradation, which requires higher optical power for preserving a specified BER performance, is limited by the APD excess noise. In the ideal case of an APD with $M = 1$ and $\sigma_{\text{T}} = 0$, the minimum optical power needed is given by

$$\overline{P}_{\text{Rxave-ideal}} = Q^2\frac{2q\Delta f}{r}. \tag{18.66}$$

Using the definition of responsivity provided in Chapter 8, the shot noise limit for SNR is given by

$$Q = \sqrt{\text{SNR}} = \frac{rP_1}{2q\Delta f} = \frac{\eta P_1}{2\,h\nu\Delta f} = \frac{r\overline{P}_{\text{Rxave}}}{q\Delta f} = \frac{\eta\overline{P}_{\text{Rxave}}}{h\nu\Delta f}, \tag{18.67}$$

where $\eta$ is quantum efficiency of the photodetector. The photon energy is related to the light power pulse during a given time slot. It is known that $B = 2\Delta f$, where $B$ is the data rate. Thus, the duration per each bit pulse is $1/B$. Therefore, the energy pulse of the photons is given by

$$E_{\text{p}} = P_{\text{in}}\int_{-\infty}^{\infty} h_{\text{p}}(t)\,\text{d}t = \frac{P_{\text{in}}}{B}. \tag{18.68}$$

On the other hand, during a one-bit duration, the energy is defined by the number of photons $N_{\text{p}}$ impinging on the photodetector:

$$E_{\text{p}} = N_{\text{p}}h\nu. \tag{18.69}$$

Thus, the optical power results in

$$P_{\text{in}} = N_{\text{p}}h\nu B. \tag{18.70}$$

Applying this relation into Eq. (18.67) and using the BER equation from Eq. (18.10) provides the BER as a function of quantum efficiency and the number of photons incident on the photodiode:

$$\mathrm{BER} = \frac{1}{2}\mathrm{erfc}\sqrt{\frac{\eta N_{\mathrm{P}}}{2}}. \tag{18.71}$$

Using $\eta = 1$ in the ideal case would provide the number of photons to be 36 which are needed for BER of $10^{-9}$. However, the actual number is much higher, about 1000, due to the thermal-noise degradation effect.

## 18.7 Digital Through Analogue

Prior to dealing with low pass and filter limitations on digital transport, it is important to remember Nyquist's theorem. It states that for synchronous impulses that have a symbol rate of $f_s$, the response to these impulses can be observed independently, that is, without inter-symbol interference (ISI). This is interpreted to mean that a filter must have a bandwidth equal to the symbol rate or data rate. Nyquist further describes a filter meeting this criteria. However, other filters such as the common LC, RC filters, and Bessel or Gaussian filters are also used. These filters are often used with 3-dB bandwidth less than the Nyquist bandwidth. They remove all undesirable harmonics. However, the phase response of such filters may affect long PRBS data streams. These effects may create pattern-dependent jitter.

The output spectrum of NRZ created by maximal-length BRBS shift register generates a sinc($x$) function shape. In this function, the first null occurs at the clock frequency, which corresponds to the data rate. The longer the PRBS series, the more spectral lines it is going to have. The spectral lines are the Fourier transform of the PRBS NRZ. Detailed information is provided in Sec. 20.3. The unfiltered power envelope spectrum of such a PRBS series is provided in Refs. [3] and [20]:

$$P = \left\{\frac{\sin[\pi(f/f_{\mathrm{CLK}})]}{\pi(f/f_{\mathrm{CLK}})}\right\}^2 \left(\frac{A}{Z}\right)^2, \tag{18.72}$$

where $A$ is the signal amplitude and $Z$ is the impedance. The spectral line separation frequency, $\Delta f$, depends on the PRBS series length $n$ defined by the linear shift register (LSFR) and is given by

$$\Delta f = \frac{f_{\mathrm{CLK}}}{2^n - 1}. \tag{18.73}$$

The requirement of ISI filtering suggests Bessel or RC low-pass filters, which are realized by using ladder networks. Such simple RC LPF consists of a series resistor $R_1$ and a shunt capacitor $C_1$, where $R_s$ is the source output impedance and $R_L$ is the load as described by Fig. 18.9.

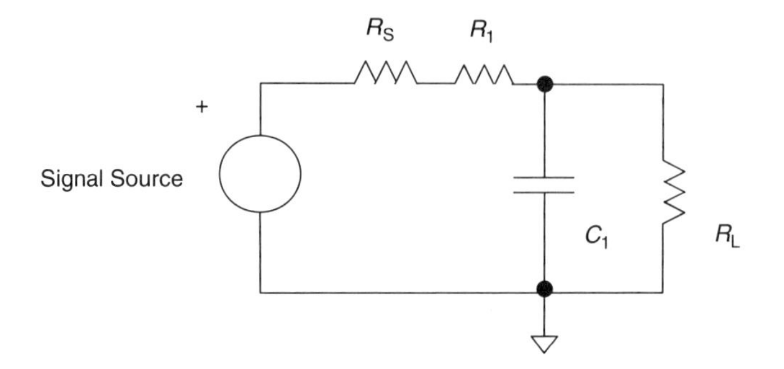


**Figure 18.9** A simple LPF ISI filter.

Simple analysis provides the output-voltage transfer function:

$$\begin{aligned} V_{\mathrm{OUT}} &= \left(\frac{R_{\mathrm{L}}}{R_{\mathrm{S}}+R_1+R_{\mathrm{L}}}\right)\frac{1}{1+sC_1R_{\mathrm{L}}[(R_{\mathrm{S}}+R_{\mathrm{L}})/(R_{\mathrm{S}}+R_1+R_{\mathrm{L}})]} \\ &= K\frac{1}{1+s\tau\beta}. \end{aligned} \tag{18.74}$$

The phase response of such an LPF filter is a function of frequency, where $s = j\omega$ and is given by

$$\varphi = -\mathrm{arctg}\,\omega\tau\beta. \tag{18.75}$$

The group delay of the LPF is equal to $d\varphi/d\omega$:[4]

$$\frac{d\varphi}{d\omega} = \frac{\tau\beta}{1-(\omega\tau\beta)^2}. \tag{18.76}$$

It can be seen that as the LPF reaches its cutoff and the group delay starts to increase rapidly. Due to finite $Q$, the phase response depends on the circuit damping factor and so does the group delay. The Taylor series expansion of Eq. (18.76) can provide a good estimate of the filter BW necessary to meet the desired group delay. This trade-off is a compromise between noises filtering to deterministic jitter that results because of group delay. The group delay effect of the LPF LSI filter near its 3-dB frequency creates dispersion on the high-spectral-line frequencies. Too narrow or sharp of an LPF shape would start to generate dispersion by affecting the PRBS series spectral components earlier. In Fig. 20.2, PRBS sinc($x$) spectral shape is displayed as a function of PRBS LSFR length.

A similar problem exists for isolation coils, which are used to isolate dc from the data signal. In this case, the RFC coils are connected on the ac side via a shunt; thus, the equivalent circuit is a high-pass filter (HPF), as shown in Fig. 18.10. The concern is to have high impedance at a frequency as low as possible. In other words, the HPF should be of a very low cutoff frequency to ensure that the low

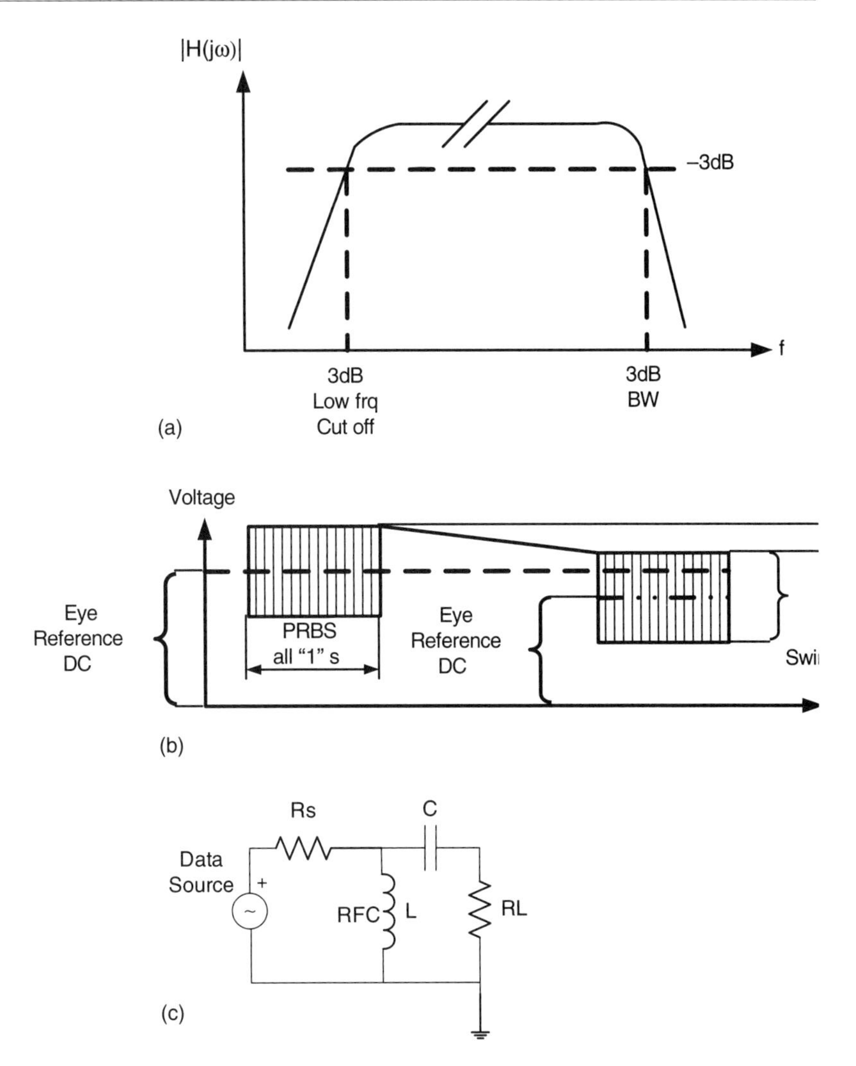


**Figure 18.10** Low-frequency cutoff effect: (a) low-frequency cutoff definition and maximum BW, (b) dc wander effect due to PRBS series charging the ac coupling capacitor, and (c) equivalent circuit HPF, showing the RFC coil and coupling capacitor *C*.

PRBS frequency spectral components are not dispersed or shorted to ground. Such PRBS sinc($x$) spectral lines are displayed in Fig. 20.2. This frequency boundary is defined as low-frequency cutoff.[16]

Another problem is the ac-coupling capacitor between the TIA and the main amplifier (post amplifier) as shown in Fig. 18.10. When receiving a long train of consecutive bits of data in a long PRBS, this may cause a base line wander of the eye. As a result, this increases BER and pulse-width distortions, which also vary with base-line wander.

The next step is to evaluate the power penalty that results from the low-frequency cutoff and determine how low the low-frequency cutoff $f_{LF}$ should be.

The question is what is the longest string of zeros or ones in a data stream. In SONET/SDH systems, which use scrambling as a line code, the run length potentially is unlimited. However, tests of SONET/SDH are checking the system limits under long consecutive-identical-digit (CID) immunity. This checks both the CDR and the pattern-dependent jitter.

Thus, long PRBS patterns are used with more than 2000 bits with a 50% mark density followed by 72 consecutive bits of zero, then another 2000 bits followed by 72 consecutive bits of one. Hence, it is reasonable that the run length is $n = 72$. In gigabit ethernet systems, where eight-bit 10-bit is used, the number of zeros is limited to $n = 5$.

For the case of a linear HPF with a single pole, the drift voltage represents the capacitor charge and is given by:[16]

$$V_{\text{drift}} = \frac{V_{\text{O}}^{\text{PP}}}{2}\left[1 - \exp\left(-2\pi f_{\text{LF}}\frac{n}{B}\right)\right] \approx \frac{V_{\text{O}}^{\text{PP}}}{2}\frac{n}{B}2\pi f_{\text{LF}}, \tag{18.77}$$

where $V_{\text{O}}^{\text{PP}}$ is the output signal swing and $B$ is the bit rate. The Taylor approximation is valid if $1/2\pi f_{\text{LF}} \gg n/B$. The drift voltage causes the power penalty (PP) as described by

$$PP = 1 + \frac{2V_{\text{drift}}}{V_{\text{O}}^{\text{PP}}}. \tag{18.78}$$

Substituting the approximation in Eq. (18.77) for the drift voltage results in

$$f_{\text{LF}} \leq (PP - 1)\frac{B}{2\pi n}. \tag{18.79}$$

For example, for a SONET system with a power penalty of 0.05 dB, PP = 1.0116, and a data rate of 2.5 GB/s (OC – 48), where $n = 72$, the lowest data rate should satisfy the following condition:

$$f_{\text{LF}} \leq (0.0116 \times 2.5\,\text{GB/s})/(6.28 \times 72) = 64\,\text{kHz}.$$

For 10 GB/s,

$$f_{\text{LF}} \leq (0.0116 \times 10\,\text{GB/s})/(6.28 \times 72) = 257\,\text{kHz}.$$

In conclusion, the low-frequency cutoff for a SONET system should be 40,000 times lower than the bit rate. For a gigabit ethernet, where $n = 5$, the specification is more forgiving.

## 18.8 Data Formats

Data transmission can be either synchronous as in SONET, in which symbols are transmitted at a regular periodic rate, or asynchronous, where spacing between word or message segments is no longer regular. Asynchronous transmission is often given in the descriptive name start/stop. Asynchronous communication requires that symbol synchronization be established at the start of each message segment or code word. This requires a great deal of overhead. In the synchronous mode, symbol timing can be established at the very beginning of the transmission, and only minor adjustments are needed.

One of the synchronous digital communication's methods to encode the data and clock together in a series-bit format. The main motivation behind this is to create enough data transitions to provide the data-clock-recovery PLL the ability to synchronize, lock, and restore the clock signal.[18] This problem becomes critical as a long series of constant-level consecutive bits are transmitted. Another problem resulting in large digital data strings is dc creep or wander,[16,17] as was explained in Sec. 18.8. The next section provides a brief review as a preparation for CDR in Sec. 18.10.

### 18.8.1 Nonreturn to zero

Nonreturn to zero (NRZ) data (see Fig. 18.11) is shown in the format that is used within a computer or other data source. A digital one logic level is shown as a high level and a digital zero is shown as a low level. Each bit in a random NRZ sequence has a 50% probability of being a 1 or a 0, regardless of the state of the preceding bits; therefore, large sequences of consecutive identical bils (CIDs) are possible. Designing high-speed systems that can work with random data can be difficult due to the low-frequency content from the long sequences of CIDs in the data signal.

Additionally, this method creates a "dc creep," or "wander,"[16,17] which has plagued designers since the start of digital transmission. As the number of high levels varies, the average dc voltage also varies. Thus, in order to have an average zero voltage, the number of one-level bits should be lower than the number of zero-level bits. A higher number of one logic-level bits will increase

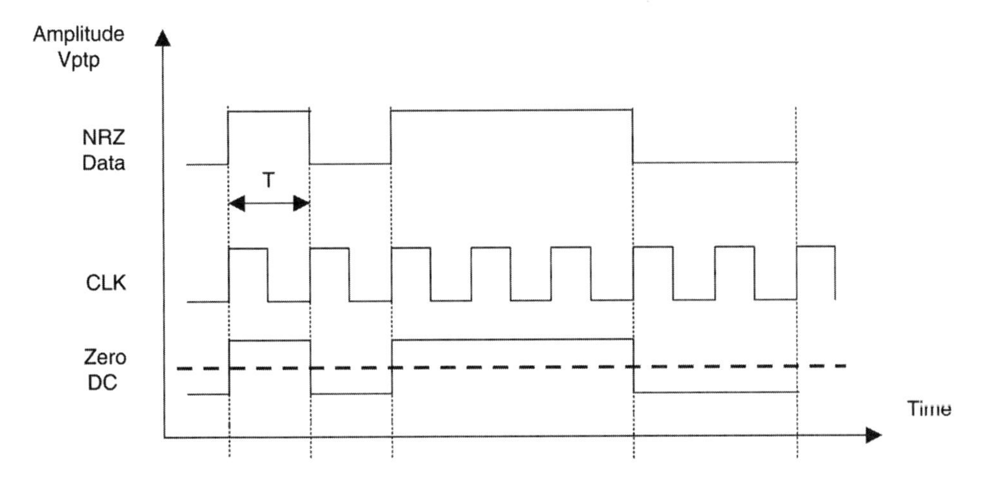


**Figure 18.11** A time domain representation of NRZ data showing the dc average at its ideal position of zero dc. The clock duration *T* represents a single bit cell.

the dc creep voltage. To remove the dc voltage and permit transmission through transformers, the original positive swing from zero volts to a higher voltage (0 to 1) is replaced by positive and negative voltage levels. The dc creep is still there. However, in the case of an equal number of 1 and 0, the average dc is zero volts. On the other hand, on a unipolar level the average dc is half of the swing. Various changes have been made in the data pulse shape to reduce dc creep. The dc and CID results in BER and affects CDR synchronization.

### 18.8.2 Return to zero

Return to zero (RZ) is a modification of the NRZ pattern. NRZ was modified as follows. The pulse level returns to a zero level after a half clock cycle for 1 level only. There is no pulse for a digital zero (Fig. 18.12). The dc creep still exists, but is reduced in level. Observing the time domain of both NRZ and RZ there is a problem of long periods without transitions. Note that for RZ, the problem is for a long string of 0s and for NRZ it is for both levels 1 or 0. This problem results in CDR loss of synchronization. It becomes severe at long strings. RZ coding is done by an AND gate between the data and clock signals. Thus the 1 becomes narrower and the dc creep is minimized compared with that in NRZ.

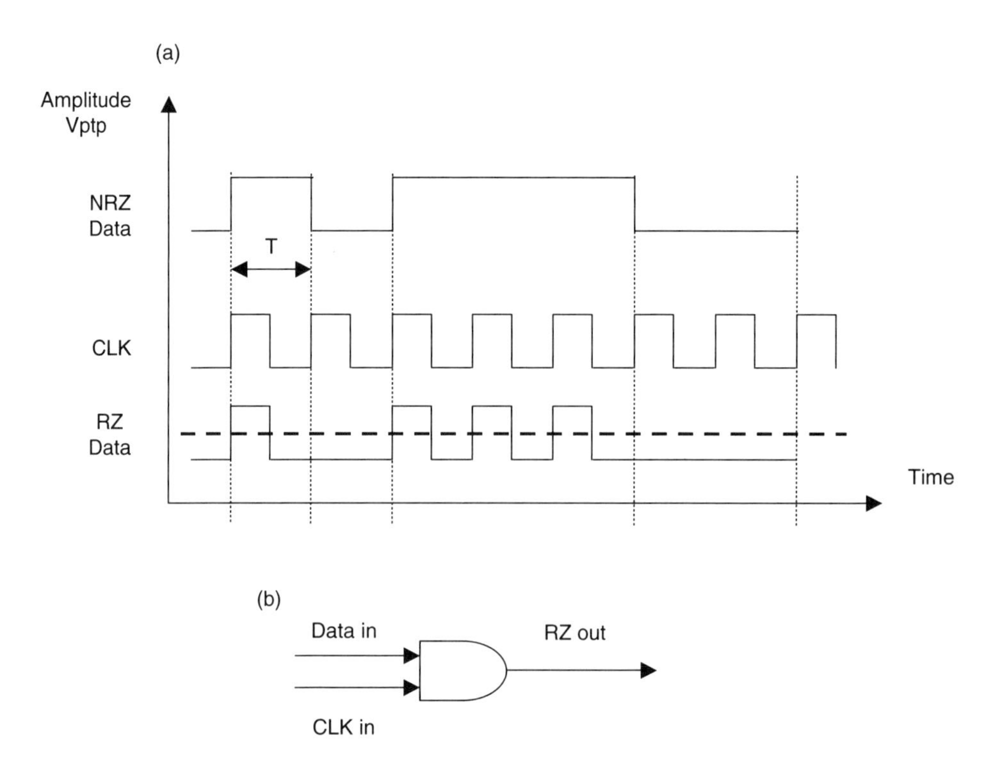


**Figure 18.12** (a) A regular NRZ data-level clock and RZ levels at the ideal dc level. (b) An RZ encoder.

The drawback of RZ compared with NRZ is its BW, which is twice as much compared with that of NRZ. An intuitive way to understand this is by comparing it to the mixing process, which is an AM process. The data is the information signal and the clock is the carrier. The AND gate is the mixer and the RZ data is the up-converted signal. Thus, the clock harmonics are observed and the NRZ spectrum appears on both sides of the clock harmonics. Hence, it is equal to the double-sided AM with the carrier. Another way to understand it is to use the Fourier transform from the time domain to frequency domain. In this case, time domain multiplication becomes frequency-domain convolution. Therefore, clock (CLK) convolution with NRZ produces wider BW, which is twice the NRZ BW. For these reasons, it makes the RZ sequence easier to recover the clock by a PLL than NRZ, which is similar to a single side-band-suppressed carrier.

### 18.8.3 Manchester coding

Both previous patterns suffer from two main problems: dc creep and few transitions when a long series of consecutive bits with the same logic level is transmitted. To overcome these two problems the Manchester code is used. The manchester code used with an IEEE 802.3 Ethernet is another type of code used to reduce the dc creep. It is a widely used method for wireless local area networks (LANs) and in some short-range wireless transmission methods. The difference between a 1 and 0 is in the polarity of the change with respect to the clock. The method utilizes the data clock and NRZ data together. The NRZ data reverses the clock polarity. The encoding device is a simple XOR (exclusive OR) gate (Fig. 18.13); thus, the dc creep becomes almost zero. Moreover, one of the main advantages of Manchester encoding is that it is centered around the clock as a carrier. Thus, it has no dc components and it solves the problem of CID, which is presented by the NRZ or RZ line coding. Manchester coding is also useful to overcome the low-frequency cutoff problem. This coding method has more

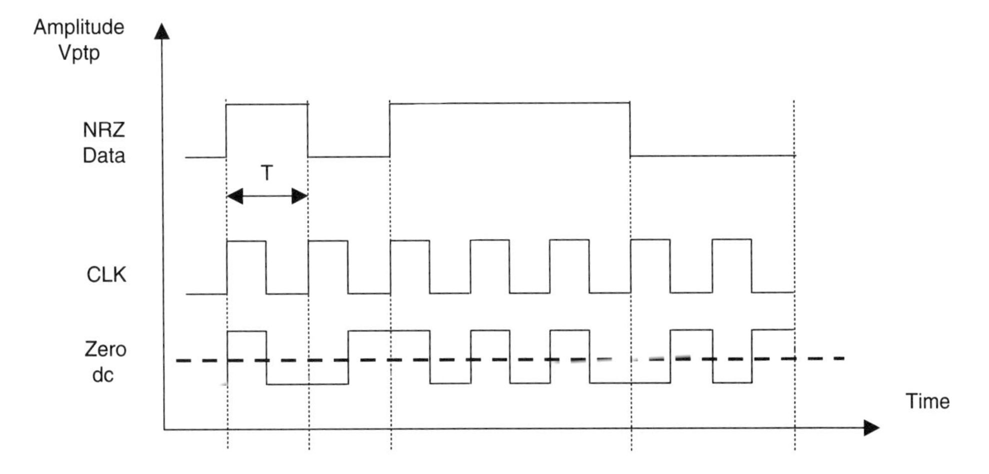


**Figure 18.13** The Manchester code used with an Ethernet made by inverted XOR.

transitions compared with RZ and NRZ. The Manchester coding spectrum has a lower dc energy and is centered around the clock frequency. Its drawback is its BW, which is wider than the regular NRZ for the same reasons that were presented for RZ coding.

### 18.8.4 8-Bit 10-bit

The 8-bit 10-bit family of line encoding was initiated by IBM.[19] The main goal here is to solve two problems: dc creep and synchronization. This code is particularly well suited for high-speed LANs, where the information format consists of variable length packets from about a dozen up to several hundreds of 8-bit bytes. This code translates each source byte into a constrained 10-bit binary sequence. The main rule is to have an equal number of zeros and ones. In this manner, the dc wander is avoided since the average of all 10 bits is zero.

## 18.9 Clock and Data Recovery

The clock and data recovery device is based on a linear PLL or Costas loop,[7] and is commonly used in SONET applications. There are several advantages in clock restoration. The entire system-jitter transfer function would be better referred to only as one-clock-source phase noise or jitter. Second, it can be used for coherent detection, buying better BER performance versus $E_b/N_o$. A typical PLL structure is displayed in Fig. 18.14. The main building blocks of a PLL are the phase detector with its gain $K_d$ measured in units of voltage per radian; the loop filter, which is the integrator that sets the loop dynamics; the BW and jitter transfer function; and a voltage controlled oscillator (VCO), which is equivalent to an ideal integrator with a single pole at the origin.[9,10] The VCO is characterized by the gain constant $K_{VCO}$ in units of frequency versus voltage.

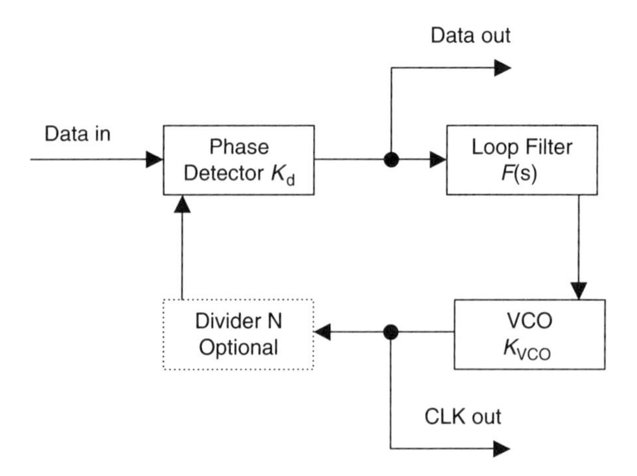


**Figure 18.14** A phase-locked loop (PLL) basic block diagram showing active loop filter $F(s)$.

Using the Laplace notation, the following second-order second-type CDR PLL transfer function can be written, with loop order and type as defined in Sec. 12.2.4. Assuming an input signal has a phase of $\theta_i(t)$ and the VCO has an output phase of $\theta_o(t)$, then the error voltage generated by the phase detector is given by

$$V_d(t) = K_d[\theta_i(t) - \theta_o(t)] \Leftrightarrow V_d(s) = K_d[\theta_i(s) - \theta_o(s)]. \tag{18.80}$$

This error voltage is filtered by the loop filter, which sets the proper dc control voltage to the VCO in order to minimize the phase error. Since, in theory, the loop amplifier has infinite gain, and the control voltage has a fixed value, and loop amplifier is an integrator, then, at a steady state, the error voltage input is zero. Practically, there is no zero-error voltage since the signal is always corrupted by noise, thus there is a finite value of error voltage.

The VCO frequency depends on its output-phase changes and is described by

$$\frac{d\theta_o(t)}{dt} = K_{VCO}V_C(t). \tag{18.81}$$

Using the Laplace transform it becomes, as expected, an ideal integrator:

$$\theta_o(s) = \frac{K_{VCO}V_C(s)}{s}. \tag{18.82}$$

The control voltage at the VCO input is given by

$$V_C(s) = V_d(s)F(s) = K_d[\theta_i(s) - \theta_o(s)]F(s). \tag{18.83}$$

Using the block diagram in Fig. 18.14, the following loop transfer functions are derived:

$$\begin{aligned}\frac{\theta_o(s)}{\theta_i(s)} = H(s) &= \frac{K_d K_{VCO} F(s)}{s + K_d K_{VCO} F(s)} \\ &= \frac{K_d K_{VCO} F(s)/s}{1 + [K_d K_{VCO} F(s)/s]} = \frac{\mathrm{LT}(s)}{1 + \mathrm{LT}(s)}.\end{aligned} \tag{18.84}$$

Equation (18.84) describes the closed-loop transfer function $H(s)$, where $\mathrm{LT}(s)$ is the open-loop transfer function. Note that the VCO is an ideal integrator with one pole at the origin. Hence, the minimum type of a closed loop can be one. The phase error describes the closed-loop phase-detector's transfer function and is given by

$$\begin{aligned}\frac{\theta_i(s) - \theta_o(s)}{\theta_i(s)} = \frac{\theta_e(s)}{\theta_i(s)} &= \frac{s}{s + K_d K_{VCO} F(s)} \\ &= \frac{1}{1 + [K_d K_{VCO} F(s)/s]} = \frac{1}{1 + \mathrm{LT}(s)} = 1 - H(s).\end{aligned} \tag{18.85}$$

Equation (18.85) shows that the PLL error transfer function is complementary to the closed-loop transfer function. Since the PLL closed-loop transfer function represents the low-pass filter response (LPF), the error represents the high-pass-filter response (HPF). The next transfer function is for the control voltage:

$$V_{\mathrm{C}}(s) = \frac{sK_{\mathrm{d}}F(s)\theta_{\mathrm{i}}(s)}{s + K_{\mathrm{d}}K_{\mathrm{VCO}}F(s)} = \frac{s\theta_{\mathrm{i}}(s)}{K_{\mathrm{VCO}}}H(s) = \frac{\theta_{\mathrm{i}}(s)}{K_{\mathrm{VCO}}/s}H(s). \tag{18.86}$$

Equation (18.86) demonstrates that the control voltage tracks the input phase changes and the fast transitions of the input phase are filtered out. On the other hand, Eq. (18.82) shows that fast enough transitions in the phase error output of the phase detector are not filtered by the PLL and are transparent. This is useful for data recovery or data tracking as depicted by Fig. 18.14.

The next step of this analysis deals with a second-order second-type PLL loop commonly used in CDR circuits. For an integrator, as provided in Fig. 18.15, where the operational amplifier gain is infinite, $F(s)$ is given by[10]

$$F(s) \approx -\frac{sCR_2 + 1}{sCR_1} = -\frac{s\tau_2 + 1}{s\tau_1}, \tag{18.87}$$

where $\tau_1 = R_1C$ and $\tau_2 = R_2C$. Applying Eq. (18.87) into Eq. (18.84) results in

$$H(s) = \frac{K_{\mathrm{d}}K_{\mathrm{VCO}}(s\tau_2 + 1)/\tau_1}{s^2 + sK_{\mathrm{d}}K_{\mathrm{VCO}}(\tau_2/\tau_1) + K_{\mathrm{d}}K_{\mathrm{VCO}}(1/\tau_1)}, \tag{18.88}$$

or in the frequency domain, where $s = j\omega$,

$$H(s) = \frac{2\zeta\omega_{\mathrm{n}}s + \omega_{\mathrm{n}}^2}{s^2 + 2\zeta\omega_{\mathrm{n}}s + \omega_{\mathrm{n}}^2}, \tag{18.89}$$

where $\omega_{\mathrm{n}}$ is the natural frequency of the PLL loop, and $\zeta$ is the damping factor of the loop, which indicates about its dynamics and stability. These values are

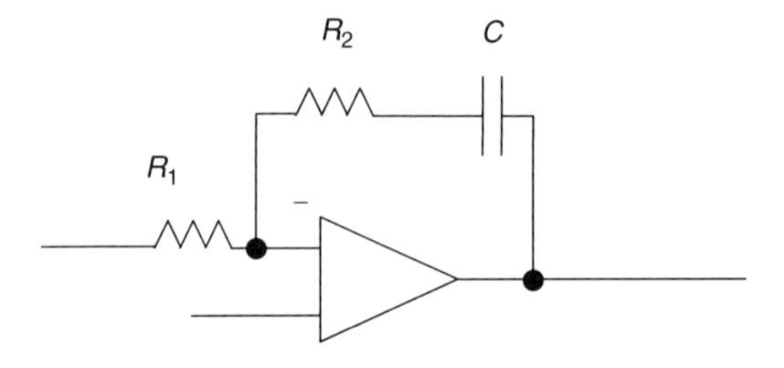


**Figure 18.15** A second-order integrator with inverting gain.

given by

$$\omega_n = \sqrt{\frac{K_{VCO}K_d}{\tau_1}}, \tag{18.90}$$

$$\zeta = \frac{\tau_2}{2}\sqrt{\frac{K_{VCO}K_d}{\tau_1}} = \frac{\tau_2\omega_n}{2}. \tag{18.91}$$

The loop BW is defined by a 3-dB power-drop frequency. By solving the condition of Eq. (18.89), $|H(s)|^2 = 1/2$, the 3-dB frequency is given by

$$\omega_{3\,dB} = \omega_n\sqrt{2\zeta^2 + 1 + \sqrt{(2\zeta^2+1)^2 + 1}}. \tag{18.92}$$

It can be seen that for fast acquisition and timing recovery, the loop BW should be increased;[9,10] thus, the phase-detector's gain can be increased during the training burst, or $\tau_1$ can be minimized by changing $R_1$ using a digital potentiometer, or by controlling the capacitance $C$ using a varactor. After acquisition is accomplished, the loop BW is decreased to provide a better jitter-transfer function.[5,11,12]

Using the expression of $F(s)$ from Eq. (18.87) and applying it into Eq. (18.85) result in the error transfer function $\theta(s)$, which is an HPF:

$$\frac{\theta_e(s)}{\theta_i(s)} = 1 - H(s) = \frac{s^2}{s^2 + 2\zeta\omega_n s + \omega_n^2}. \tag{18.93}$$

Jitter transfer of CDR is analyzed in the same manner as a regular synthesizer.[11,12] In this case, CDR is a phase-locked oscillator (PLO), where its division ratios are $N = 1$ and $M = 1$. The parameter $N$ is the feedback divider that sets the frequency and $M$ is the reference setting. All noise components in a typical CDR PLL are illustrated in Fig. 18.16. Using Fig. 18.16, the CDR jitter transfer function from $\theta_{in}$ to $\theta_{out}$ of each noise variance can be analyzed.[11]

The input-data jitter transfer function is given by the output-noise variance $\sigma_{n\_data}$:

$$\begin{aligned}\sigma_{n\_data} &= \varphi_{n\_data}\frac{K_{VCO}K_dH(s)/s}{1 + [K_{VCO}K_dH(s)/sN]} \\ &= \varphi_{n\_data}\frac{2\zeta\omega_n s + \omega_n^2}{s^2 + 2\zeta\omega_n s + \omega_n^2},\end{aligned} \tag{18.94}$$

where $\varphi_{n\_data}$ is the input jitter variance to the CDR.

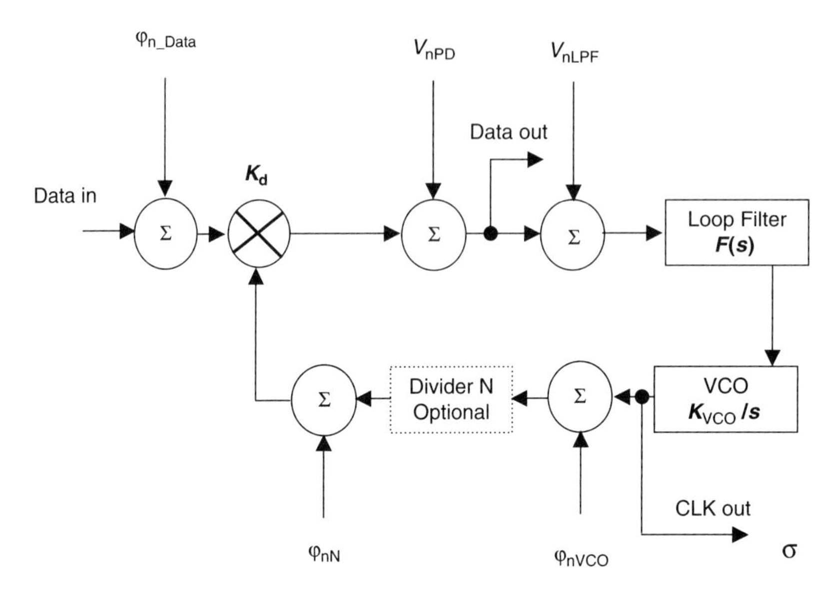


**Figure 18.16** Noise components in a typical CDR PLL based on Ref. [11]. (Reprinted with permission from *Microwave Journal* © 1999.)

The phase-detector noise variance contribution is given by

$$\sigma_{\mathrm{nPD}} = \frac{V_{\mathrm{nPD}}}{K_{\mathrm{d}}} \frac{K_{\mathrm{VCO}} K_{\mathrm{d}} H(s)/s}{1 + [K_{\mathrm{VCO}} K_{\mathrm{d}} H(s)/sN]}. \tag{18.95}$$

Two interesting conclusions can be derived from these equations. First, both the data-jitter and the phase-detector jitter transfer functions are the same having a low-pass frequency response. In addition, a higher phase-detector gain, $K_{\mathrm{d}}$, reduces the phase-detector noise variance $V_{\mathrm{nPD}}$ contribution to the overall CDR jitter. Another parameter, which has a significant effect on the CDR jitter, is the VCO. Its jitter transfer function is given by

$$\sigma_{\mathrm{nVCO}} = \varphi_{\mathrm{nVCO}} \frac{1}{1 + [K_{\mathrm{VCO}} K_{\mathrm{d}} H(s)/sN]} = \varphi_{\mathrm{nVCO}} \frac{s^2}{s^2 + 2\zeta\omega_{\mathrm{n}} s + \omega_{\mathrm{n}}^2}. \tag{18.96}$$

Equation (18.96) demonstrates how the CDR suppresses the VCO phase noise variance $\varphi_{\mathrm{nVCO}}$ close to the carrier due to its HPF response. The VCO phase-noise variance close to the carrier contributes a small amount to the overall $\Delta T$ jitter because the PLL suppresses it. It is clear from Eq. (18.85) that if the PLL loop filter is steep, then the complementary HPF filter will be as steep. Thus, a high-order PLL loop filter results in a better rejection of the VCO phase noise within the loop BW. CDR VCOs are realized by distributed RC-feedback tank circuits with a relatively low $Q$. These types of VCOs are called ring oscillators. Therefore, it is important to suppress their close-to-carrier phase noise, which can be relatively higher than the input data jitter. Note that the VCO phase-noise variance at the output of the CDR is calculated by multiplying the VCO phase-noise variance by the phase-error transfer function provided in Eq. (18.93). Since at large, frequency offset from carrier VCO phase noise is relatively low, lower than the CDR phase noise within the PLL BW, its contribution to the overall CDR jitter is also low.

The last contributor to the CDR overall jitter transfer function is the loop amplifier filter. However, its effects are relatively lower compared with that of the previous CDR building blocks. The CDR-loop-filter jitter contribution to the overall jitter variance is given by

$$\sigma_{\mathrm{nLPF}} = \frac{V_{\mathrm{nLPF}}}{K_{\mathrm{d}}} \frac{K_{\mathrm{VCO}} K_{\mathrm{d}} H(s)/s}{1 + [K_{\mathrm{VCO}} K_{\mathrm{d}} H(s)/sN]}. \tag{18.97}$$

It can be seen that the loop-filter jitter transfer function has a low-pass jitter response like the phase detector. It can also be observed that the loop-filter jitter energy is compensated by the phase-detector gain. Therefore, the higher the phase-detector gain, $K_{\mathrm{d}}$, the lower the loop-amplifier noise variance $V_{\mathrm{nLPF}}$ at the CDR output.

In the case of $N \neq 1$ and the VCO is compared to a lower reference clock in aided-acquisition fast-locking CDR,[13] the feedback divider noise variance, $\varphi_{\mathrm{nN}}$, transformed to the output is given in Ref. [11]. Such a CDR is described by Fig. 18.17:

$$\sigma_{\mathrm{nN}} = \varphi_{\mathrm{nN}} \frac{K_{\mathrm{VCO}} K_{\mathrm{d}} H(s)/s}{1 + [K_{\mathrm{VCO}} K_{\mathrm{d}} H(s)/sN]}. \tag{18.98}$$

Divider noise variance has a LPF response and defines the CDR PLL noise floor. At high gain, the feedback value is equal to $N$ and thus the output noise variance becomes $N\varphi_{\mathrm{nN}}$. In conclusion, it can be seen that when $N \neq 1$, all jitter variances degrade the CDR noise floor by $20 \log N$. So, it is essential to operate with $N = 1$. However, in aided locking during preamble, the loop BW is increased. In this manner, it becomes faster and acquisition is faster, which enables fast synchronization. Wider CDR has higher jitter and integrated phase noise. Hence, fast locking and jitter are trade-offs and a design compromise. Some receivers with CDR have tunable CDR BW for these reasons.

The overall CDR jitter transfer function over frequency is given by the rms sum of all noise variances:

$$\sigma = \sqrt{\sigma^2_{\mathrm{n\_data}} + \sigma^2_{\mathrm{nPD}} + \sigma^2_{\mathrm{nVCO}} + \sigma^2_{\mathrm{nLPF}} + \sigma^2_{\mathrm{nN}}}. \tag{18.99}$$

From this analysis, it is desirable to have a CDR with a relatively high phase-detector gain. This design approach minimizes the jitter transfer function of the CDR loop, since the loop filter and phase-detector noise variance improve per

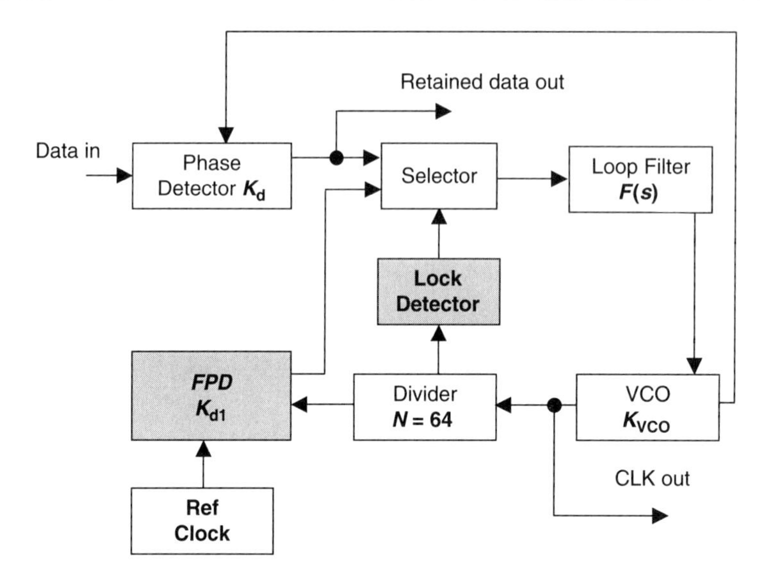


**Figure 18.17** An aided-acquisition receiver CDR, using CMOS and bipolar technologies.[13] (Reprinted with permission from the *IEEE Communications Symposium* ©IEEE, 2000.)

Eqs. (18.95) and (18.97). In addition, it can be seen that the CDR loop filter determines the jitter transfer function's shape and noise suppression of the CDR VCO. The data-jitter transfer function is directly related to the loop filter, which determines the CDR 3-dB cutoff frequency. Narrow CDR BW improves and purifies the data jitter, and suppresses the VCO jitter close to the carrier. However, a narrow-frequency-response loop slows the CDR during acquisition. Thus, a variable BW is sometimes used in order to accomplish fast clock and timing recovery during the preamble period.

Sometimes CDR jitter is characterized by phase noise in the frequency domain and by timing jitter in the time domain, or by integrated phase error denoted as $\Delta\theta_{rms}$. All of these values represent the same quantity of FM noise and its phase variance over time. The translation procedure from each one is provided in Secs. 14.5 and 15.8.6.

When transceivers operate in burst mode, the receiver time slot should be efficient. Hence, CDR locking and acquisition processing should be fast. Previously it was explained that the CDR is designed to operate in a narrow-band loop for accomplishing low jitter. Hence, the training series should be long enough to enable locking. Cycle slips due to large phase error increase locking time even further.[9] For that purpose, aided-acquisition circuit topology is required. Aided-acquisition circuit response should bring the CDR VCO close to the locking region of the loop pull in the frequency range. The search should be slower than the loop BW.[9,10,11] When long PRBS data streams with a large amount of consecutive bits are used, the CDR requires large acquisition time. Thus, if preamble is saved, the CDR can be locked initially to a reference clock that is similar to the data clock the CDR needs to recover. When the phase error is low enough, within the pull range of the CDR PLL, it can be switched into the data-recovery mode.[13] To accomplish the decision of data recovery or prelocking, a lock-detection circuit is used. When the CDR is out of lock or has drifted too far, the auxiliary loop indicates that it is out of lock. Thus, the selector switches the phase-detector, feeding the loop filter from the CDR phase detector marked as $K_d$, to the auxiliary phase and the frequency detector marked as FPD. The PLL locks the VCO to the reference clock. At the locking instance, the lock detection indicator switches back the loop-filter input to the CDR phase detector. The VCO is already centered to the system clock and is within the pull-in range of the CDR locking. In this manner, the CDR retains the data and recovers the system clock. This aided-acquisition method helps to accelerate the acquisition time and prevent CDR drifts. Two kinds of technologies are used: bipolar for the CDR loop and CMOS for the aided-acquisition lock detector and FPD. High-frequency VCOs are divided by $N$ to be locked to the auxiliary loop. In CDR mode, they feed directly the main loop's phase detector, while the auxiliary phase-lock monitors proper CDR process. In Fig. 18.17, such a receiver CDR PLL is illustrated. Hence, in burst mode, during the idle period of a receiver slot, the CDR is locked to the reference clock. As data appears, it is switched to receiver mode and the receiver guard time is reduced since the CDR is already locked.

VCO realization in high-data-rate CDR is based on a push–push differential VCO.[8,13] These VCOs operate as a frequency multiplier and oscillate at an output frequency of $2f$. The fundamental frequency $f$ is suppressed at 20 dB lower, but can be used to lock the auxiliary loop, as shown in Fig. 18.17. This may simplify the divider design, since the frequency to the phase detector is lower.

So far the discussion has mainly been on continuous-signal PLL. However, CDR used for clock recovery, called DPLL (digital PLL),[15] is synchronized on a random digital-data stream. These data streams are sent synchronously with a clock signal. The task of DPLL CDR is to extract the clock out of a random data bit stream. Assume a bit stream of an arbitrary stream of 0s and 1s, which is synchronized to a clock with frequency $f_{CLK}$. Assume at first case, that there are enough transitions between logic levels, as shown in Fig. 18.18. Now assume that at a given instance there are several 1 levels and 0. If this bit stream is compared to the original clock, it can be said that this data signal is a clock with a varied frequency. It can be interpreted as a kind of FM that has its highest frequency, is the case of transitions, alternate between 0 and 1. This stream has half of the clock frequency, since the duration of each bit is a whole clock cycle. Therefore, this fast FM change can be detected at the output of the phase detector as indicated by Eq. (18.85), since this is a high pass characteristic to the PLL output. However, because the CDR PLL loop BW is narrow, the fast changes of $\theta_i$ are filtered out as indicated in Eq. (18.83). At the instance both the VCO phase and the input-bit stream phase coincide, the CDR is locked and the PLL output frequency of the VCO becomes the clock frequency. Hence,

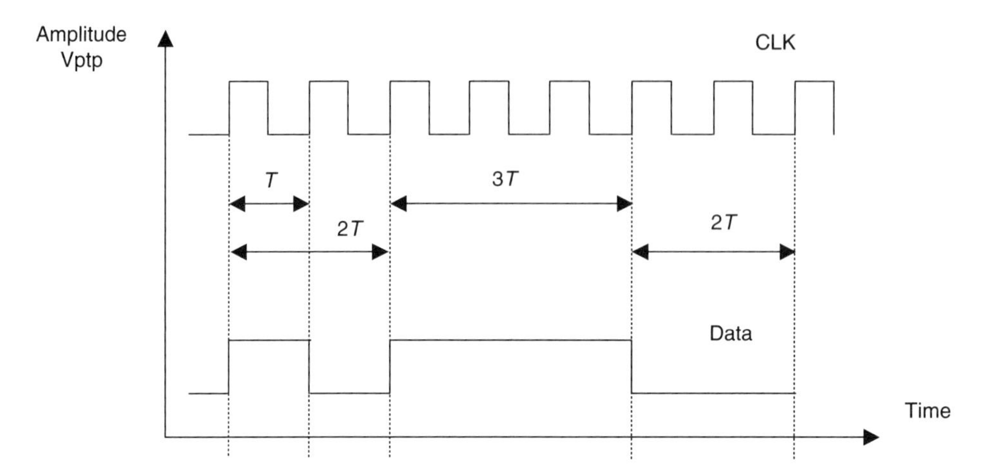


**Figure 18.18** Data and clock signals demonstrating that the data duration is a variation of frequency with respect to the system clock and hence can be interpreted as FM. Note that the time alignment between the data and clock is a result of rhe phase-locking process.

CDR PLL is considered to be FM and PM detectors. Generally, CDR VCO is centered closely to the baud rate of the data signal. The PLL is synchronized due to the transitions of the data signal.

The drawback of CDR PLL lies in the limitation of the amount of consecutive bits with the same logic level. Too long of a data series with a fixed logic level, which exceeds the PLL time constant, brings the loop to the out-of-lock condition. This is because the interpretation is at a dc level at the CDR input since there are no signal transitions. Thus, the CDR VCO, which is the recovered clock, starts to drift and the receiver loses synchronization. Hence, there is a need for data codes, which generate many transitions for a long series of consecutive bits with the same logic levels. The second goal is to have an efficient spectrum coding, and the last goal is to minimize dc drift or wander. These data patterns were briefly reviewed in Sec. 18.9.

## 18.10 Main Points of this Chapter

1. The BER evaluation function approximation of $Q$ assumes a Gaussian distribution for both AWGN and shot noise.

2. The SNR is equal to $Q^2$.

3. The other assumption made for a simplified BER calculation is that the laser is biased below threshold for the 0 logic level. Hence, the only noise at that state is AWGN.

4. The higher the ER, the better the SNR or the value of $Q^2$, thus the BER is improved.

5. For infinite ER, the SNR becomes the ideal limit of the average optical power multiplied by the responsivity $r$ and divided by the average noise. The ideal case is $r = 1$.

6. Mode partition noise that results from laser mode hopping sets the SNR limit for BER, which cannot be corrected even though the signal level is increased.

7. One of the results of mode-partition noise is dispersion. The larger the wavelength rms deviation variance $\sigma$, the larger the chromatic dispersion. Thus, the chromatic jitter increases.

8. Chromatic dispersion sets the limit for the data rate at a given fiber length. The higher the rate, the larger the chromatic dispersion and the fiber length is shorter. This parameter is called bit-rate distance product in $(\text{GB/s}) \times \text{km}$.

9. Mode partition acts on both dimensions of the eye: eye closing due to SNR reduction and jitter due to chromatic dispersion.

10. Timing jitter results from the transition statistics between logic levels of 0 and 1. This creates signal-sampling fluctuations and sampling error. Hence, the SNR is reduced by the amount of sampling phase misalignment. This is also referred to as the signal-phase noise or FM noise.

11. RIN noise (laser AM noise) sets the SNR limit in the same manner as was explained for mode-partition noise.

12. The parameter power penalty is used in order to describe BER degradation because of the above-mentioned impairments. The power penalty reduces the SNR and, as a result, BER performance is reduced. This problem can be solved by increasing optical power up to the SNR limits imposed by these impairments.

13. The minimum detectable signal and optical power set the system threshold for the desired BER specifications.

14. One of the main problems of digital-data transport is the digital-through-analog network such as filters or RFC inductors. These networks are equal to LPF and HPF filters applied on a wide band signal.

15. The LPF network affects the digital signal at its high-frequency components by dispersion. Dispersion is due to nonlinear phase response of the filter near to its 3-dB frequency, which results in group delay.

16. LPF networks result in pattern-dependent jitter because of the dispersion of the high-order harmonics of a digital signal.

17. HPF networks affect the low-frequency components of a wide-band digital signal.

18. The longer the PRBS of a digital signal is, the more low-frequency and high-frequency components it will have.

19. One more important parameter in HPF networks is low-frequency BW limit, which defines the figure of merit to transport the long PRBS series with a large number of consecutive bits that have the same logic level.

20. HPF networks may due to the capacitor charging generate dc wander, while long PRBS series are applied through.

21. Several codes are used to overcome dc wander, have a fixed average dc level for digital series, and have the ability to restore the system clock using a PLL. The most common codes are NRZ, RZ, Manchester, and 8B 10B.

22. The goal of line codes is to have no dc offset which depends on the bit pattern, i.e., the number of 1 and 0 and to be efficient in spectrum. PLL should have sufficient transitions to enable the CDR, to restore the clock.

23. NRZ format is the most commonly used pattern where 1 is high voltage and 0 is low level. Its disadvantages are dc creep, because of the dependence of average dc on the number of 1 and 0, the difficulty to synchronize for long PRBS with large CIDs. Additionally, it has large dc energy and it is spectrally efficient. Its spectral response is given by sinc($x$), where the null is at the clock frequency.

24. RZ is a modification of NRZ in order to create a large number of transitions for synchronization and to minimize dc wander. Its drawback is the BW that is twice as much compared with that of NRZ.

25. RZ is generated by multiplying the NRZ data with the CLK, using an AND gate. Hence, it is similar to AM and the clock signal appears in the spectrum.

26. The Manchester code is ideal for CDR and synchronization, since it has a large number of transitions even at a large amount of consecutive identical digits. It is realized by XOR of the NRZ data with the CLK signal.

27. The Manchester code's spectrum is centered at the CLK frequency and therefore has low dc energy. As a result, it has low dc wander. It has a wider spectrum than the NRZ.

28. 8B, 10B is an IBM code that answers to both the dc drift problem and transitions even at a large of consecutive identical digits. Thus, it is easier to synchronize than on NRZ.

29. The CDR circuit is a PLL, which consists of a phase detector, loop filter (integrator), VCO, and a feedback path from the VCO to the phase detector.

30. CDR VCO is generally centered at the system clock or data rate.

31. CDR input is the data signal and its outputs are clock restored from the VCO and data from the phase detector.

32. The CDR transfer function is an LPF. The phase-detector transfer function's frequency response is high pass with the following relation HPF = 1 − LPF.

33. CDR jitter and the jitter transfer function depend on the phase-detector noise-voltage variance, loop-filter noise-voltage variance, VCO phase noise, and divider noise variance, in the case used in the CDR.

34. The output jitter of a CDR composed from the VCO phase noise at a large offset from the carrier because it has an HPF response; phase-detector-close-to-carrier jitter because it has an LPF response; loop filter and divider with the same jitter response of an LPF; and data jitter with LPF jitter response.

35. High phase-detector gain reduces the jitter effect of the phase detector and loop filter.

36. At a high CDR loop gain, the feedback value of the PLL is $N$ if the divider with a division ratio of $N$ is used.

37. The divider reduces the CDR jitter response of the phase-detector's VCO data and loop filter by $+20 \log N$.

38. There are several methods to synchronize a CDR. First is by using a preamble training series with many transitions. Second is by having the CDR locked to a reference clock and transfer the CDR to data-mode synchronization. This method is called aided acquisition.

39. PLL is both a PM and FM detector.

40. CDR-design check-list considerations are:

    40.1 Eye margin

    - 40.1-1. How much noise can be added to input while maintaining target BER (voltage margin, gain margin)?
    - 40.1-2. How far can clock-phase alignment be varied while maintaining-target BER (phase margin)?
    - 40.1-3. How does the static phase error vary versus frequency, temperature, and process variation?
    - 40.1-4. Are input amplifier gain, noise, and offset sufficient?

40.2 Jitter characteristics

40.2-1. What is the jitter generation (VCO, phase noise, phase detector charge pump)?

40.2-2. What is the jitter transfer function (overshoot peaking and BW)?

40.2-3. What is the jitter tracking tolerance versus frequency?

40.3 Pattern dependency

40.3-1. How do long runlengths affect system performance?

40.3-2. Is BW sufficient for individual isolated bit pulses?

40.3-3. Are there other problematic data patterns (resonances)?

40.3-4. How does PLL, jitter and stability change versus data transition density?

40.4 Acquisition time

40.4-1. What is the initial power on lock time?

40.4-2. What is the phase lock acquisition time when input source is changed?

40.5 How is precision achieved?

40.5-1. Are there any needed external capacitors and inductors?

40.5-2. Does CDR require any external reference frequency?

40.5-3. Is there any required laser trimming or highly precise IC process?

40.6 Input/output impedance

40.6-1. Is S11/S22 maintained through the frequency band (input output impedance, reflections on eye, deterministic jitter)?

40.6-2. Are the reflections large enough to close the eye, creating a shadow eye?

40.7 Power supply

40.7-1. What is the power-supply-rejection ratio (PSRR)?

40.7-2. Does the CDR reflect noise to supplies?

40.7-3. What are the power supply effects on the jitter (is the PSRR sufficient enough)?

40.8 False-lock susceptibility

40.8-1. Is there any phase ambiguity?

40.8-2. Does the phase detector have dead zone?

40.8-3. Is there any condition or data pattern that can create false lock?

40.8-4. Can VCO leakage to the phase detector create injection locking?

## References


1. Ogawa, K., "Analysis of mode partition noise in laser transmission systems." *IEEE Journal of Quantum Electronics*, Vol. 18 No. 5 (1982).
2. Agarwal, G.P., *Fiber–Optic Communication Systems*. John Wiley & Sons (2002).
3. Horowitz, P., and W. Hill, *The Art of Electronics, Second Edition*. Cambridge University Press (1989).
4. Fogiel, M., *Handbook of Mathematical Scientific and Engineering Formulas, Tables Functions, Graphs, Transforms*. Research & Education Assoc. (1994).
5. Cvijetic, M., *Coherent and Nonlinear Lightwave Communications*. Artech House (1996).
6. Bures, K.J., "Understanding timing recovery and jitter in digital transmission systems, part I." *RF Design*, pp. 42–53 (1992).
7. Ransijn, H., and P. O'Conor, "A PLL based 2.5 Gb/S GaAs clock and data regenerator IC." *IEEE Journal of Solid State Circuits*, Vol. 26 No. 10, pp. 1345–1353 (1991).
8. Voinigescu, S., P. Popescu, P. Banens, M. Copeland, G. Fortier, K. Howlett, M. Herod, D. Marchensan, J. Showell, S. Szilagyi, H. Tran, and J. Weng, "Circuits and Technologies for Highly Integrated Optical Networking IC's at 10 Gb/s to 40 Gb/s." Quake Technologies Inc. Ottawa, Ontario, Canada (2001).
9. Meyr, H., and G. Ascheid, *Synchronization in Digital Communications: Phase Frequency Locked Loops and Amplitude Control, Vol. 1*. John Wiley & Sons (1990).
10. Gardner, F.M., *Phaselock Techniques*. John Wiley & Sons (2005).
11. Brillant, A., "Understanding phase-locked dro design aspects." *Microwave Journal*, pp. 22–42 (1999).
12. Kroupa, V.F., "Noise properties of PLL systems." *IEEE Transactions on Communications*, Vol. 30, pp. 2244–2252 (1982).

13. Friedman, D.J., M. Meghelli, B.D. Parker, J. Yang, H.A. Ainspan, A.V. Rylyakov, Y.H. Kwark, M.B. Ritter, L. Shan, S.J. Zier, M. Sorna, and M. Soyuer, "SiGe BiCMOS integrated circuits for high speed serial communications links." *IBM Journal of Research and Development*, Vol. 47 No. 2/3, pp. 259–282 (2003).
14. Smith, R.G., S.D. Personick, and H. Kressel, *Semiconductor Devices for Optical Communications*. Springer Verlag, New York, N.Y. (1982).
15. Best, R.E., *Phase Locked Loops, Design Simulation and Application*. McGraw–Hill (2003).
16. Sackinger, E., *Broadband Circuits for Optical Communications*. John Wiley & Sons (2005).
17. Bellamy, J.C., *Digital Telephony*. John Wiley & Sons (1982).
18. Roden, M.S., *Digital Communication System Design*. Prentice Hall (1988).
19. Widmer, A.X., and P.A. Franaszek, "A DC balanced partitioned block, 8B/10B transmission code." *IBM Journal of Research and Development*, Vol. 27 No. 5, pp. 440–451 (1983).
20. Feher, K., et al., Telecommunications Measurements, Analysis, and Instrumentation. Nobel Publishing Corp. (1996).

# Chapter 19

# Transceivers and Tunable Wavelength Transceiver Modules

In this chapter, digital transceiver structure is reviewed. The concept of digital transceivers has been briefly introduced in Sec. 2.4. However, in this section, detailed design considerations and design topologies of burst-mode control, modulation, heat dissipation, housing, transmit-optical subassembly (TOSA), and receiver-optical sub-assembly (ROSA) blocks' inside structure, and wavelength locker are explained. A preliminary review of digital transceiver design approaches is given in this chapter. Burst mode is reviewed in Sec. 19.1, a glance on tunable wavelength transmitters is provided in Sec. 19.2, and design approaches for transceiver integration into small form packages are dealt with in Sec. 19.3.

## 19.1 Burst Mode

In the continuing endeavor toward broadband networking and combined multimedia services such as interactive video, voice, and fast internet, the systems have become BW thirsty. Applications of voice-over-internet protocol (VOIP), high-definition-TV-over-internet protocol (HDTVOIP), and internet traffic have been increased. Hence, there is an increasing demand for broadband-access networks for FTTx and FTTC that can operate in time division multiple access (TDMA) bursts.[2,5] The ethernet passive-optical network (EPON) is a prospective solution for very high data rate transport to the end user in the FTTx, FTTP, and FTTH configurations.[2,3,42,54] A general architecture of an EPON is illustrated in Fig. 19.1. The main idea of this approach is to have multiple optical-network units (ONUs), which are connected to an optical-line terminal (OLT) through a single optical fiber and a tree network, which can be a star coupler of 1:$N$. The upstream is managed in time-division-multiplexing (TDM) technology. In this manner, each ONU has its own dedicated transmit time slot. ONUs are at the offices or homes on the user's end of the network. The OLT is the network central. Slots

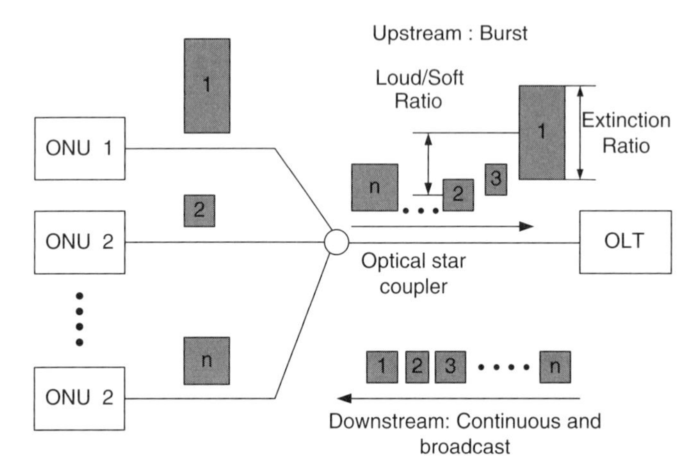


**Figure 19.1** An EPON architecture.[2] (Reprinted with permission from the *Journal of Solid-State Circuits* © IEEE, 2004.)

from ONU are synchronized, so the packets from one ONU do not coincide and collide with that of another, once data transport is coupled to the fiber. The OLT broadcasts downstream with a header to identify the relevant ONU that should receive a packet.

The ONU extracts the clock from downstream, using clock data recovery (CDR) that synchronizes the whole system. To accomplish this multiuser coexistence and synchronization, the packet stream always starts with a training series called the preamble as shown in Fig. 19.3. The preamble concept for synchronization was explained in Sec. 18.10. Additionally, to prevent data collisions, ONU transmissions are separated by a guard time. At this stage, no signal is transmitted from any ONU, so it is in a dark state. Since network deployment is such that each ONU distance from the OLT is different, the optical power level arriving to the OLT varies from burst to burst. Therefore, the second role of the preamble series is to set the correct level for the decision circuit and bring automatic gain control (AGC) to its proper gain position. Extreme power levels between packets are design challenges that the burst-mode receiver (BMR) should handle. Data burst comprises short guard and preamble times, as depicted in Fig. 19.3. Thus, the receiver should accomplish a high sensitivity for low signal reception as well as a good response for a strong signal. As a consequence, BMR must have a large dynamic range. Moreover, BMR starts to operate at high sensitivity maximum gain and converges to its gain because of the preamble training series. On the transmitter side, ONU transmitters should have a fast automatic-power-control (APC) settling time prior to the short preamble burst transfer. This requirement ensures that the OLT BMR receiver is ready for real data transport.[5–8,52,53] Fast APC and laser-bias-point stabilization guarantee that there are no residual amplitude distortions, while modulating the laser because of power fluctuations, until the APC loop had been stabilized. Burst-mode receivers are reviewed in Sec. 19.1.1, while burst-mode transmitters are explained in Sec. 19.1.2.

### 19.1.1 Burst-Mode Receivers

#### 19.1.1.1 Burst-mode receiver concepts

Burst-mode transceivers are commonly used in passive-optical network (PON), full-service-access network PON (FSAN PON), asynchronous-transfer-mode PON (ATM PON), EPON, broadband PON (BPON), and gigabit PON (GPON) applications.[1–3,5–8,40–42,55–57] Such networks typically use TDM protocol, in which information is transmitted from the remote nodes, such as multiple ONUs, that are connected to a single fiber, such as tree network of $N$:1, to the central office, such as OLT in a PON, or from one node to another on an optical bus sharing the same wavelength. The upstream traffic is managed by TDM technology in which time slots are dedicated to the ONUs. Time slots are synchronized so the upstream packets from the ONUs do not interfere with each other once the data is coupled into a single fiber. Between packets is the guard time in which no signal is transmitted from any ONU in the system. Those data bursts on an optical network that arrive at the OLT receiver exhibit large differences in optical power. Hence, it is necessary to have a receiver that can adapt to those optical power ranges. On the other hand, the transmit section in such a transceiver should have a clever APC loop that can control the laser power dynamically during a burst transmission. For a PON system, the receiver's wide dynamic range is accomplished by an AGC. In that manner, system deployment is much more flexible, since splitters can be inserted while the receiver still maintains its ability to detect signals with a decent BER. For power, the APC loop should recover fast at the moment of burst transmit and satisfy modulation parameters such as biasing and modulation depth, which define the overall extinction ratio. Furthermore, since such a transceiver contains CDR for data-clock recovery and reduces data dependent jitter,[39] which means that the system should have time to synchronize and lock the CDR.[2] The synchronization involves setting the AGC to the proper signal level at the decision circuit and preventing data distortions, which may affect the eye quality and BER, and synchronizing the CDR in order to restore the clock and recover the data for correct sampling timing. Sampling issues are explained in Sec. 15.8.6. So, a burst consists of guard time that allows termination of all setting transitions as well as having a preamble training series for the CDR loop and AGC threshold setting. After the guard time, the receiver is ready to accept valid data. The ratio between the burst duration and the preamble time may provide a measure about the burst efficiency. Consequently, the motivation is to reduce the training series length to a minimum.

Those above-mentioned system constraints influence the design of a transceiver and its performance trade-offs. In receiver section, there are two main approaches to the design of the PIN-detector-to-postamplifier coupling that follows a PIN TIA, which is the preamplifier. Those approaches are dc coupling, and sometimes pseudo-dc coupling that has a dc restoration circuit in conjugation to the ac circuitry, or ac coupling.[1] The motivation for dc coupling is its ability to restore the low-frequency components of the digitally received data. This is because it is a Fourier transform that shows low-frequency spectral lines of the burst-mode data as well as high-frequency energy. The capacitor might filter low-frequency energy since it is a high pass filter. This issue is explained in Secs. 18.8 and 20.1.

#### 19.1.1.2 Sensitivity penalty for burst mode

As explained in Chapter 18, BER statistics of continuous (nonburst) digital transport is affected by SNR, the statistics of which is Gaussian because of the noise statistic approximations are Gaussian. BMR analysis and BER degradation are calculated with respect to continuous-mode operation. That kind of analysis holds for any receiver for which the Gaussian-noise approximation holds. The performance-penalty calculation is derived by comparing quality of service, BER, of the burst mode with that of continuous receiver. The approach for analyzing the burst-operation penalty arises due to the fact that the first bit exhibits amplitude variations. Thus, having a threshold based on this exclusive bit results in a threshold voltage that has similar Gaussian statistics to those in Chapter 18.[1] The variations in threshold may result in BER degradation. This is because they affect the decision point value that becomes statistical and not deterministic, as was demonstrated in Fig. 18.1 for the continuous scenario. That degradation is defined as the BER penalty with respect to the continuous mode. Consequently, when using a longer preamble, the penalty becomes lower. In other words, in continuous mode, a large amount of consecutive bits of 1 are averaged to establish a threshold or adjust the gain. The averaging process is conducted by a low pass filter, which takes time.[1] The continuous mode establishes a solid threshold, while burst has a fluctuating threshold. A longer preamble resembles a pseudo-continuous-mode threshold, and a single-bit preamble results in a 3-dB penalty.

In digital systems, like the analog systems described in Chapter 12, there are two main concepts for realizing AGC[1,2,40] as illustrated in Figs. 19.2(a), (b), and (c). The first AGC concept is feedback AGC. In this configuration, the photodetector (PD) is dc coupled to a PIN-TIA preamplifier that is ac coupled to a gain-controlled postamplifier where the gain is controlled by an amplifier using feedback with a low-pass transmission. The GCA output is connected to a decision circuit, which is a comparator, with threshold reference voltage $V_t$. Gain control is accomplished by a signal that is proportional to the low-pass feedback output. This approach might be too slow for burst applications due to the low-pass integration time that is proportional to its BW; hence, a faster feedforward concept[1,40] is used as shown in Figs. 19.2(b) and (c). In this case, the PIN TIA is dc coupled to both the threshold sample and hold estimator and the decision circuit. Unlike the first approach, the threshold is dynamically affected by the preamble training series of 1. The peak detector is used for the sample and the hold, and the capacitor reset discharge is done by the FET transistor as shown in Fig. 19.2(c). The integration time and reference value are determined during burst guard and preamble. Guard time is used for reset and followed by preamble as the burst starts, causing the threshold to be dynamically updated by the training series. Proper choice of a constant RC time in the peak detector is possible to make the charging time correspond to $n$ bits of preamble. During charging, the circuit acts as an approximate integrator for the $n$ pulses of the preamble and establish

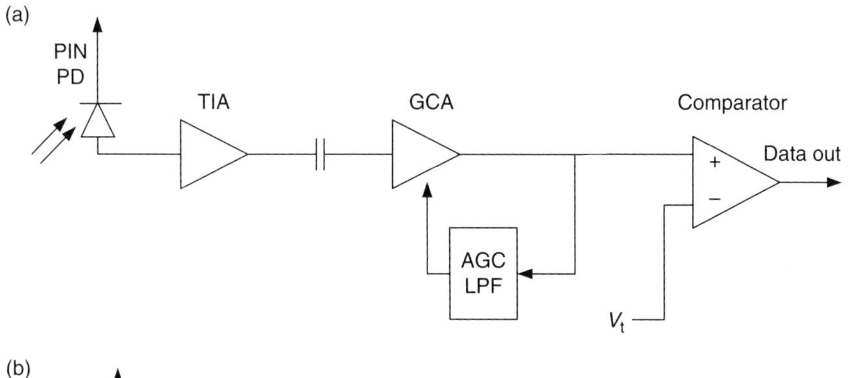


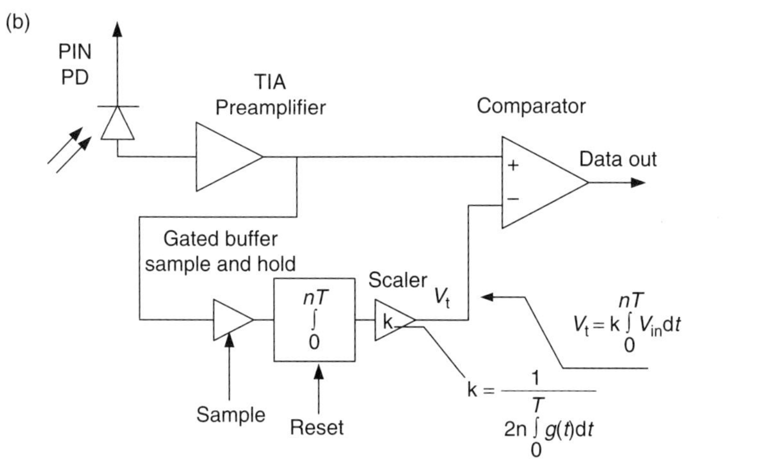


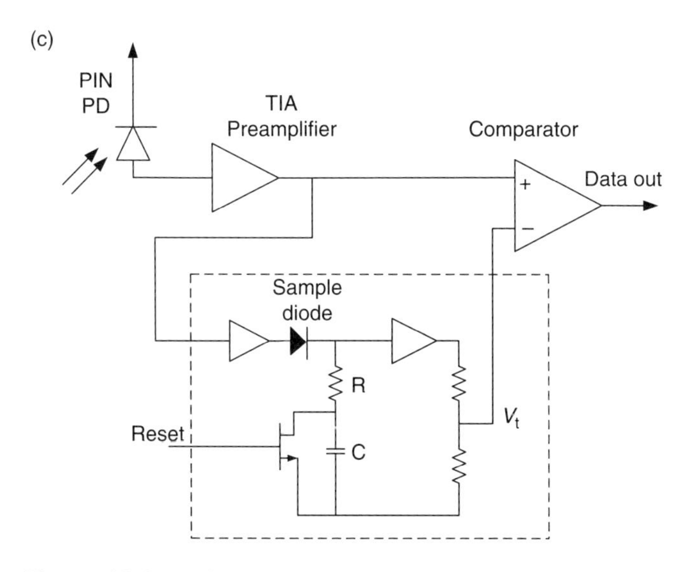


**Figure 19.2** (a) Continuous-data-stream feedback AGC; (b) sample and hold averaging reference gain control using feedforward, $g(t)$ is the normalized pulse shape and $T$ is the symbol duration; (c) simplified peak detector used as an approximation to the ideal burst-mode-feedforward circuit of (b).[1] (Reprinted with permission from the *Journal of Lightwave Technology* © IEEE, 1993.)

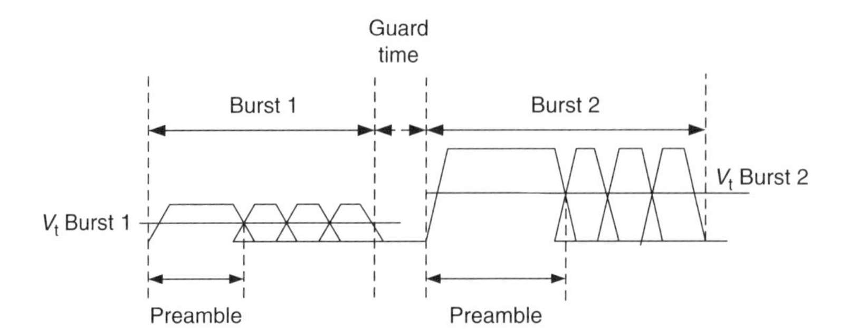


**Figure 19.3** A burst scenario from a far and near nodes amplitudes and powers are different. The reference is set by the preamble to half voltage between the two logic levels. Guard time is inserted between two bursts in which the reference $V_t$ is reset.[1] (Reprinted with permission from the *Journal of Lightwave Technology* © IEEE, 1993.)

the proper threshold by using a simple voltage divider at the output. The advantage of the circuit in Fig. 19.2(c) is that it has no sampling control due to the detector diode that operates as a catch diode, preventing the capacitor discharge for receiving 0 after it was charged with 1. During guard time, this capacitor is discharge by the parallel FET. As a consequence, $V_t$ would have the average value between the 0 and 1 states. For a weak signal burst, the absolute value is lower than that of the stronger signal $V_t$, and the threshold level is updated per burst preamble. In this way, the dynamic decision per burst improves BER performance, and the threshold level is updated per optical-power burst.

Prior to continuing with statistical analysis, several assumptions need to be made:

- Each burst contains a noise corrupted $n$-bit preamble consisting of a string of 1s, which is used to establish the threshold voltage $V_t$. $V_t$ sets a reference to determine the 1s and 0s in subsequent data streams. The $n$ bits in the preamble are not considered to be data and are not considered in BER calculations.

- Averaging $n$ bits in the preamble can be used to reduce the error in $V_t$ by using an ideal circuit to filter the noise in the preamble, resulting in an energy-to-noise power spectral-density ratio of $n \times E_b/N_o$, where $E_b/N_o$ is the energy-to-noise power spectral-density ratio for any single bit in the preamble or data stream.

- Receiver sensitivity is limited by additive white Gaussian noise (AWGN) in which the statistics for determination of 1 or 0 are equal. This will apply to all types of PIN receivers, TIAs with high or low impedance, and some types of APD receivers, which are far from the sensitivity limit and limited by the amplifier rather than the detector noise. That means the amplifier noise figure is a figure of merit.

The goal of the analysis is to find the statistics of the dynamic threshold voltage[1] and use it to calculate the equivalent probability of error based on modifications of Eqs. (18.2) and (18.6). The first modification refers to voltage rather than current since the TIA output is a voltage. This transformation is simply the transimpedance gain. Based on Chapter 18, the continuous probability of error equations with respect to decision threshold $V_t$ can be found Ref. [1].

The probability of error for a binary system is the probability of detection of a logical 1 when 0 was transmitted, plus the probability of the opposite scenario (Fig. 18.1). In the case of a fixed-threshold system, these quantities are determined by the probability density functions of the received signal $r$ when 0 or 1 are transmitted. Those are defined as $f(r|0)$ and $f(r|1)$, respectively. Hence, probability of error can be written as

$$P_e(1|0) = \int_{-\infty}^{V_t} f(r|0)\mathrm{d}r, \tag{19.1}$$

which is the probability of detecting 1 when 0 is transmitted.

$$P_e(0|1) = \int_{-\infty}^{V_t} f(r|1)\mathrm{d}r, \tag{19.2}$$

which is the probability of detecting 0 when 1 is transmitted, where $r$ is the received signal.

Using the Gaussian distribution that has the same probability considerations as in Chapter 18, the overall probability is given by

$$P_e = \frac{1}{2}\left\{\int_{V_t}^{\infty} \frac{1}{\sqrt{2\pi}\sigma_0} \exp\left[\frac{-(r - S_0)^2}{2\sigma_0^2}\right]\mathrm{d}r + \int_{-\infty}^{V_t} \frac{1}{\sqrt{2\pi}\sigma_0} \exp\left[\frac{-(r - S_1)^2}{2\sigma_0^2}\right]\mathrm{d}r\right\}. \tag{19.3}$$

Since fiber optics modulation is on–off keying (OOK), has a good extinction ratio of $>10$ dB with a symmetrical half-way value decision threshold, under same definitions as in Chapter 18, the probability of error becomes

$$P_{eOOK} = \frac{1}{2}\mathrm{erfc}\left(\frac{S_1}{2\sqrt{2}\sigma_0}\right) = \frac{1}{2}\mathrm{erfc}\left(\frac{E_b}{\sqrt{2}N_o}\right). \tag{19.4}$$

So far, the static probability is given by $E_b/N_o$, where $\sigma_0^2 = N_o$, $E_b$ is the energy per bit and $N_o$ is the noise-power spectral density. However, in burst mode, the reference voltage is corrupted by noise. Thus, the decision line between the two states in Fig. 18.1 becomes a Gaussian bell shape. Note that the threshold voltage is an average of the 1 and 0 train.

Where more than one bit is used to define the threshold, i.e., where preamble is used, the variance $\sigma_0^2$ decreases. Under ideal conditions, it is $\sigma_0^2/n$, where $n$ is the preamble length. Hence, the threshold voltage distribution is given by[1,40]

$$f(V_t) = \frac{1}{\sigma_0}\sqrt{\frac{n}{2\pi}}\exp\left\{\frac{-n[V_t - (S_1/2)]^2}{2\sigma_0^2}\right\}. \tag{19.5}$$

Substituting this value into Eq. (19.3) and performing a dual integral result in the burst-mode probability of error function. Note that $S_1/2$ is the half-way voltage between the two logic levels:

$$P_e = \frac{1}{2}\left\{\int_{-\infty}^{\infty} f(V_t)\int_{V_t}^{\infty}\frac{1}{\sqrt{2\pi}\sigma_0}\exp\left[\frac{-(r-S_0)^2}{2\sigma_0^2}\right]dV_t dr + \int_{-\infty}^{\infty} f(V_t)\int_{-\infty}^{V_t}\frac{1}{\sqrt{2\pi}\sigma_0}\exp\left[\frac{-(r-S_1)^2}{2\sigma_0^2}\right]dV_t dr\right\}. \tag{19.6}$$

This is a numerical integral. Assuming $f(V_t)$ is ergodic, the average burst-mode BER can be calculated from a sufficiently large number of bursts independently of the burst length. The conclusion is that the more accurate the threshold and its SNR, the sharper the decision boundary. As a result, the BER penalty delta between continuous and burst-mode operations gets smaller (Fig. 19.4). The solved curve for a 2-bit preamble in Fig. 19.5 shows a 3-dB penalty against a continuous-mode receiver with ideal threshold. This penalty decreases as the preamble length increases, as can be seen in Fig. 19.4.

The APD case is well analyzed in Ref. [41], which takes into account the APD gain factor, preamble length, extinction ratio, and reference voltage dc offset: the fluctuation in the decision boundary voltage value.

#### 19.1.1.3 Burst mode receiver dynamics

As was previously explained, ac-coupling topology involves a loss of low-frequency energy. Hence, codes such as 8B10 or 16B18B are used and preferred to NRZ.[64] The ac-coupling method is feasible in this manner because its frequency-response limitation can be overcome. Thus, the receiver is adaptive to various optical powers between packets. This was demonstrated in the previous section. Such receivers contain CDR and can produce clocking within 10 ns in the case of 10 GB/s.[4] The main problem in BMR is recovery time. Dynamic analysis of BMR is reported by Rotem and Sadot.[4] From Fig. 19.6, it is clear that the ac-coupled signal with a decision circuit has a time constant of $\tau = RC$.

Using previous definitions given by Eqs. (8.7) and (18.17), the output voltage from the TIA, $V(t)$, can be written as[4]

$$V(t) = i_p[1 + \delta_{ER}D(t)]G, \tag{19.7}$$

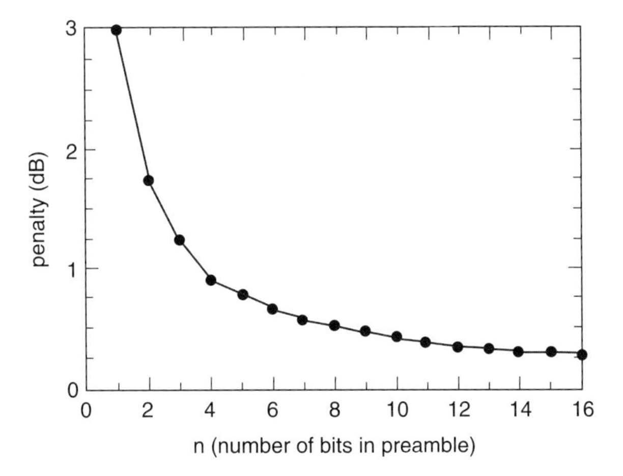


**Figure 19.4** Penalty for burst mode operation as a function of preamble length that is used to establish threshold.[1] (Reprinted with permission from the *Journal of Lightwave Technology* © IEEE, 1993.)

where $D(t)$ represents the digital bits train. $D(t) = 1$ for mark bit and $D(t) = -1$ for space bit, $i_p$ is the detected optical current, $G$ is the TIA transimpedance gain and $\delta_{ER} = (ER - 1)/(ER + 1)$, where ER is the extinction ratio, defined as the ratio between the optical powers of a mark bit and a space bit. Note that the ER power penalty is defined as $\eta = 1/\delta_{ER}$. From Fig. 19.6, it is clear that the comparator input voltage is lower due to the voltage drop on the capacitor $C$. Hence, it can be noted as

$$V^{+}(t) = V(t) - V_C(t). \tag{19.8}$$

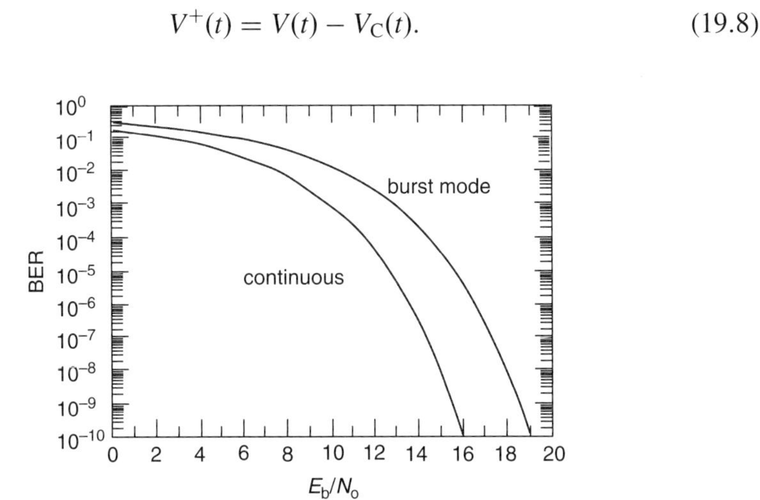


**Figure 19.5** BER vs. $E_b/N_o$ of continuous mode operation when the threshold is assumed to be determined without errors against burst mode. The threshold is determined from the first 1 received. The penalty of burst vs. continuous is 3 dB.[1] (Reprinted with permission from the *Journal of Lightwave Technology* © IEEE, 1993.)

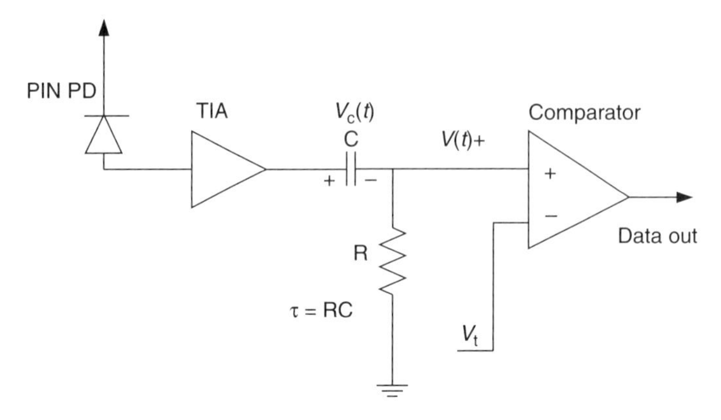


**Figure 19.6** A schematic diagram of an ac-coupled optical receiver. $V(t)$ is time varying voltage and $V_t$ is the threshold voltage, which can be zero.[4] (Reprinted with permission from *Transactions on Communications* © IEEE, 2005.)

Using Eqs. (19.8) and (19.7) with the states of $D(t)$ for 0 and 1 results in two conditions:

$$V_1^+(t) = i_p(1 + \delta_{ER})G - V_C(t) \tag{19.9}$$

and

$$V_0^+(t) = i_p(1 - \delta_{ER})G - V_C(t). \tag{19.10}$$

In case the optical power, $P$ is constant during a time period that is substantially longer than $\tau$, the coupling capacitor is charged with the dc level of the signal. Thus $V_C(t) = S$, where $S = i_P G$. This is a steady state denoted by the subscript. Hence, for the general case, steady-state voltage at the comparator is given by

$$V_{SS}^+(t) = S\delta_{ER}D(t). \tag{19.11}$$

Therefore, the mark and space voltages are given by

$$V_{1\_SS}^+(t) = S\delta_{ER} \tag{19.12}$$

and

$$V_{0\_SS}^+(t) = -S\delta_{ER}. \tag{19.13}$$

These equations indicate that the eye center is zero regardless of the optical power. So the eye center was shifted by the coupling capacitor from the average dc voltage $S$ at the TIA output to zero, while keeping its opening as shown in Fig. 19.7.

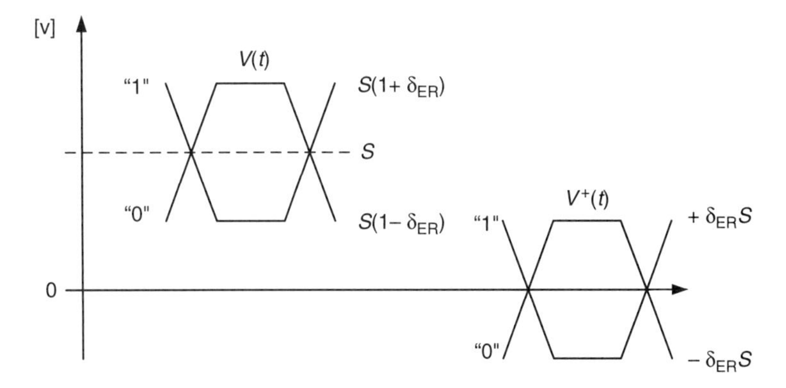


**Figure 19.7** Eye center shift between TIA output to comparator input.[4] (Reprinted with permission from *IEEE Transactions on Communications* © IEEE, 2005.)

In conclusion, this transition to zero may affect BER performance if the eye is not perfectly symmetrical around zero. This means that the likelihood of one state is preferred to the other. Therefore, there is a need for recovery time or settling time. One of the methods is preamble, as was explained previously, and AGC can be added to improve the dynamic range. This scenario is demonstrated by Fig. 19.8. The BMR sensitivity power $P_{\mathrm{SENS}}$ at the required BER ($\mathrm{BER_R}$) is defined by the voltage $V_{\mathrm{SENS}}$. The error probability, $p_e$, of a single bit is equal to $\mathrm{BER_R}$ when $V_0^+ = -V_{\mathrm{SENS}}$ is a space bit or $V_1^+ = -V_{\mathrm{SENS}}$ is a mark bit. It is assumed that the receiver is dominated by thermal noise, and other noise mechanisms, such as shot noise, resulting from the photodetector are neglected.

The recovery time definition is related to BER performance. Thus, $t_{\mathrm{Rx}}$ is the time required from the arrival of a new packet until all the bits satisfy the condition

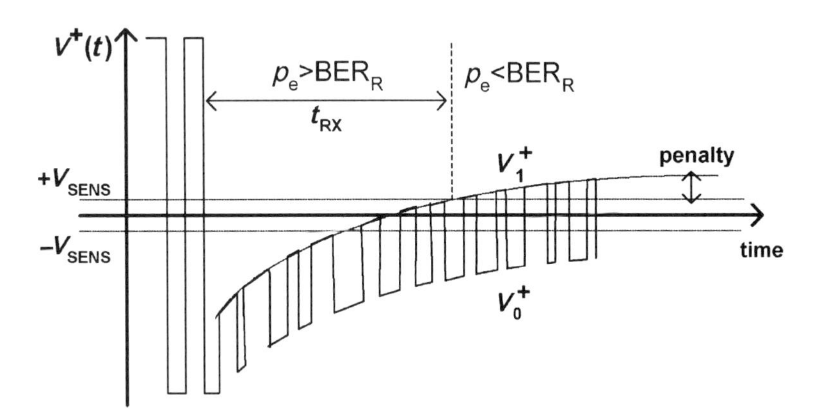


**Figure 19.8** Signal recovery after burst level change from maximum optical power to minimum optical power. The capacitor charge becomes negative biased and starts to decay.[4] (Reprinted with permission from *Transactions on Communications* © IEEE, 2005.)

$P_e \leq BER_R$; an additional 10 ns are required for CDR recovery, but these are omitted to simplifying the calculation. The maximal recovery time occurs when a packet with minimal optical power $P_{min}$ follows a packet with maximal power $P_{max}$, i.e., when optical power was changed from $P_{min}$ to $P_{max}$. The opposite case shows negligible recovery time. Immediately after that power transition, the capacitor is charged to the maximum average voltage $S_{max}$ of the previous packet, resulting in a negative voltage at the comparator input port. The average voltage starts to change during the exponential decay of charge from $S_{max}$ to $S_{min}$ and the eye returns to the zero-offset voltage as it was before. The capacitor voltage change versus time is described by

$$V_C(t) = S_{max} + (S_{min} - S_{max})\left[1 - \exp\left(-\frac{t}{\tau}\right)\right]. \tag{19.14}$$

Substituting Eq. (19.14) into Eq. (19.8) with all relations found results in

$$V_C(t) = (S_{min} - S_{max})\left[1 - \exp\left(-\frac{t}{\tau}\right)\right] + \delta_{ER} S_{min} D(t). \tag{19.15}$$

The solution of Eq. (19.15) occur when the sensing-voltage condition satisfies $V_1^+(t_{Rx}) = +V_{SENS}$. The optical dynamic range of a burst is proportional to the average dc-voltage range and is defined as $\beta$:

$$\beta = \frac{P_{max}}{P_{min}} = \frac{S_{max}}{S_{min}}. \tag{19.16}$$

The power penalty of a BMR is defined as the ratio between the weakest power level and the sensing level:

$$\alpha_B = \frac{P_{min}}{P_{SENS}} = \frac{\delta_{ER} S_{min}}{V_{SENS}} \tag{19.17}$$

Substituting the relations in Eqs. (19.16) and (19.17) into Eq. (19.15) results in

$$\frac{\delta_{ER} S_{min}}{\alpha_B} = (S_{min} - \beta S_{min}) \exp\left(-\frac{t}{\tau}\right) + \delta_{ER} S_{min}. \tag{19.18}$$

The solution provides the recovery time $t_{Rx}$:

$$t_{Rx} = \tau \, \ln\left[\eta(\beta - 1)\frac{\alpha_B}{\alpha_B - 1}\right], \tag{19.19}$$

where $\eta$ is the extinction ratio penalty and is equal to $1/\delta_{ER}$. It is clear that the equation is valid when the argument under the logarithm is larger than 1, otherwise it is an instantaneous recovery. The behavior of $t_{Rx}$ is shown in Fig. 19.9. As the

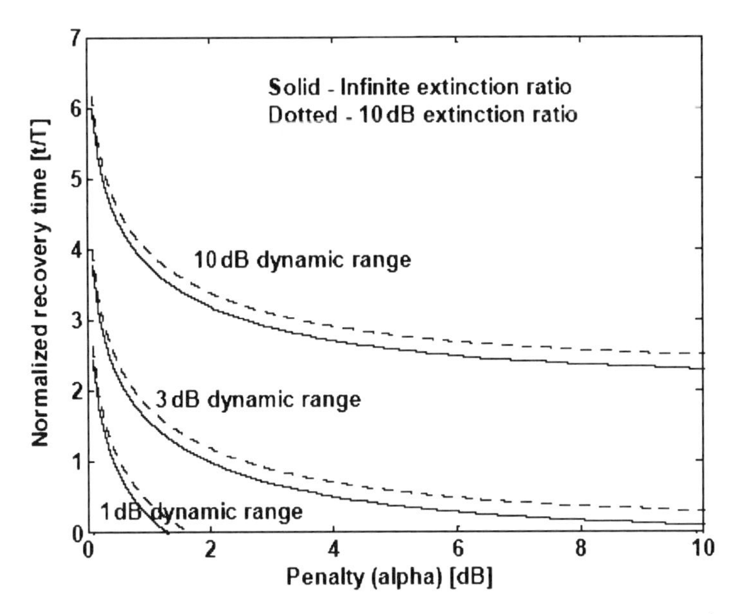


**Figure 19.9** Recovery time vs. power penalty for $P_{max}$ to $P_{min}$ transition.[4] (Reprinted with permission from *Transactions on Communications* © IEEE, 2005.)

penalty factor $\alpha_B$ goes to infinity, the normalized recovery time ($t_{Rx}/\tau$) reaches its asymptotic bound:

$$\frac{t_{RX}}{\tau} = \ln[\eta(\beta - 1)]. \tag{19.20}$$

The above postulation demonstrates that the recovery time is based on the stringent requirement in which the threshold $BER_R$ is maintained throughout the process. In fact, the average BER can be better than $BER_R$, but this BER analysis is for the worst case of overestimated power penalty. A different approach requires that the average BER should be equal to $BER_R$ involves packet structure and length considerations. Thus, at some instances, the probability of error might be higher than $BER_R$, i.e., $p_e > BER_R$ at the head of the payload, immediately after $t_{Rx}$ in Fig. 19.8, while the overall average can be lower.

The next step is to investigate the power penalty caused by ac-coupling cutoff.[4] This mainly affects the NRZ data with long PRBS as elaborated in Chapter 20 (Secs. 20.2 and 20.3) and in Chapter 18 (Sec. 18.8). Many components in an optical link, such as modulators, drivers, and receiver optical front end (OFE), which is a PD with TIA, are ac coupled. Ac coupling is an HPF filter and the main goal is to reduce its corner low-frequency cutoff as low as possible. In the case of periodic dc power, which coincides with a low-frequency-cutoff-time constant of the HPF, it would result in baseline wander. These drifts affect the horizontal baseline of the eye, resulting in BER performance degradation. Thus, balanced codes are used to null dc drift (Secs. 18.8 and 18.9). For NRZ data transport, the dc component can be defined by cumulative-bit difference (CBD), which describes the return difference between number of 1s and the number of 0s. The maximal baseline wander normalized by the eye opening is denoted by

$$\mathrm{BLW}_{\max} = 1 - \exp\left(-\mathrm{CBD}_{\max}\frac{T_{\mathrm{B}}}{\tau}\right), \tag{19.21}$$

where $T_{\mathrm{B}}$ is the bit period, and BLW is baseline-wandering factor. Hence, the stimulated power penalty can be written as

$$\alpha_{\mathrm{D}} = \exp\left(\mathrm{CBD}_{\max}\frac{T_{\mathrm{B}}}{\tau}\right). \tag{19.22}$$

These equations describe the average penalty and base line wander where an average CBD is substituted.

A regular SONET frame transport can consist of hundreds of consecutive 1s and 0s and has a CBD of that order. This may result in too high penalty in case a time constant of nanoseconds is required for BMR recovery. As a consequence, dc-balanced coded blocks are used such as 8B10B or 16B18B. This solution is at the expense of BW overhead, which is 25% and 12.5% for 8B10B and 16B18B, respectively. The maximal CBDs of 8B10 and 16B18B codes are 14 and 26, respectively.

Now both power penalties can be used to calculate the recovery time of a BMR. It can be observed from the above analysis that the nature of each penalty is contrasting with each other: $\alpha_{\mathrm{D}}$ decreases with $\tau$, while $\alpha_{\mathrm{B}}$ increases per modification of Eq. (19.18). This means that there is an optimum; an equivalent global constraint can be defined as $\alpha_{\mathrm{lim}} = \alpha_{\mathrm{D}} \times \alpha_{\mathrm{B}}$. The design goal is not to exceed the limit penalty per $\alpha_{\mathrm{D}} \times \alpha_{\mathrm{B}} \leq \alpha_{\mathrm{lim}}$. Using Eq. (19.22) for $\alpha_{\mathrm{D}}$, the recovery time of Eq. (19.19) can be modified with respect to $\alpha_{\mathrm{lim}}$ and CBD:

$$t_{\mathrm{Rx}} = \tau \ln\left\{\eta(\beta - 1)\frac{\alpha_{\mathrm{lim}}}{\alpha_{\mathrm{lim}} - \exp[\mathrm{CBD}_{\max}(T_{\mathrm{B}}/\tau)]}\right\}. \tag{19.23}$$

This expression describes the BMR recovery time as a function of CBD, $\tau$, and $\alpha_{\mathrm{lim}}$. Optimization of the recovery time is done by using the expression $\partial t_{\mathrm{RX}}/\partial\tau = 0$. However, it is useful to analyze it numerically and produce some plots as in Fig. 19.10 for 8B10B. In this case, the maximum CBD is 14, $\alpha_{\mathrm{lim}}$ is a parameter with values of 1, 2, and 3 dB. The bit period $T_{\mathrm{B}}$ is equal to 1/(data rate), the dynamic range $\beta$ is the optical range and $\eta$ is the extinction ratio penalty $1/\delta_{\mathrm{ER}}$.

So far, partial optimization was made only on the receiver; however, system-wise, the overall receiver and transmitter should be optimized. One of

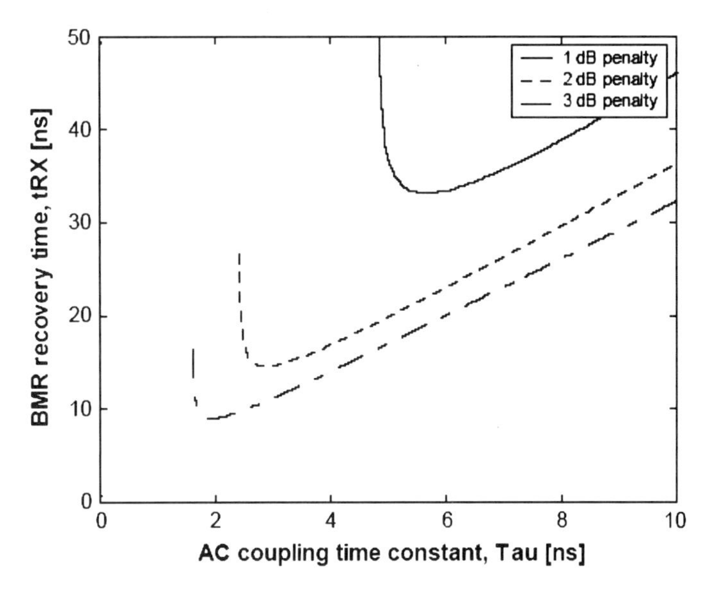


**Figure 19.10** Optimal recovery time in nanoseconds of a BMR vs. the ac-coupling time constant in nanoseconds. ER is 10 dB, penalty range is 1, 2, and 3 dB, data rates 12.5 GB/s, CBD is 14, and $CBD_{ave}$ is 1.23.[4] (Reprinted with permission from *Transactions on Communications* © IEEE, 2005.)

the interesting cases is when having a tunable laser transmitter or burst APC. Hence, a packet consists of three stages: laser tuning time $t_{Tx}$ or, in the case of single wavelength transmitter, APC setting time; BMR recovery time $t_{Rx}$, which is a preamble period; and the coded payload duration, which can be 8B10B or 16B18B. Thus, the link setup time $t_{LS}$ is the sum of $t_{Rx}$ and $t_{Tx}$.

Consider a situation where a tunable laser does not emit when it is set to its new wavelength. Then the ac capacitor in Fig. 19.6 discharges exponentially. Recalling Eq. (19.14), at the beginning of a new packet the voltage across the capacitor is $V_C = S_{max} \exp(-t_{Tx}/\tau)$. Hence, the dynamic range is reduced since the capacitor voltage is dropped. As a result, the recovery time is shortened in the rest of the calculation. Therefore, $\beta$ should be modified accordingly as an effective dynamic range parameter:

$$\beta_{eff} = \beta \exp\left(\frac{-t_{Tx}}{\tau}\right). \tag{19.24}$$

Substituting Eq. (19.24) into Eq. (19.23) and adding $t_{Tx}$ provide the link settling time:

$$t_{LS} = t_{Tx} + \tau \ \ln\left\{\eta\left[\beta \exp\left(\frac{-t_{Tx}}{\tau}\right) - 1\right]\frac{\alpha_{lim}}{\alpha_{lim} - \exp[CBD_{max}(T_B/\tau)]}\right\}. \tag{19.25}$$

Attention should be paid on two conditions regarding Eq. (19.25):

$$\alpha_{\lim} - \exp\left(\mathrm{CBD}_{\max}\frac{T_{\mathrm{B}}}{\tau}\right) > 0 \Rightarrow \tau > \frac{\mathrm{CBD}_{\mathrm{M}}T_{\mathrm{B}}}{\ln\ \alpha_{\lim}} \tag{19.26}$$

and

$$\left[\beta \exp\left(\frac{-t_{\mathrm{Tx}}}{\tau}\right) - 1\right] > 0 \Rightarrow t_{\mathrm{Tx}} < \tau\ \ln \beta. \tag{19.27}$$

If $t_{\mathrm{Tx}} > \tau \ln \beta$, the capacitor is already discharged, so that the analysis of the low-power packet arriving after a high-power packet is not relevant. Instead, what is relevant is where a high-power packet arrives after a low-power packet. So it becomes $t_{\mathrm{LS}} = t_{\mathrm{Tx}} + 0$ rather than $t_{\mathrm{LS}} = t_{\mathrm{Tx}} + t_{\mathrm{Rx}}$, since the BMR recovery time is negligible in this case. The other constraint is if $\tau < (\mathrm{CBD}_{\mathrm{M}}T_{\mathrm{B}}/\ln \alpha_{\lim})$ then, according to Eq. (19.22), $\alpha_{\mathrm{D}} > \alpha_{\lim}$, which is possible. The minimum tuning time is derived by calculating the extreme of Eq. (19.25) according to

$$\frac{\partial t_{\mathrm{LS}}}{\partial t_{\mathrm{Tx}}} = \frac{1}{1 - \beta \exp(-t_{\mathrm{Tx}}/\tau)}. \tag{19.28}$$

The derivative reaches a discontinuity of minus infinity at $t_{\mathrm{Tx}} = \tau \ln \beta$. Larger values of $t_{\mathrm{Tx}}$ are not allowed due to the condition in Eq. (19.27); it is concluded that this extreme is the optimal point. This results in a $\beta_{\mathrm{eff}}$ value of 1; hence, according to Eq. (19.24), no receiver adjustment is required, thus $t_{\mathrm{Rx}} = 0$. Eventually, the penalty is invested in the data-code-related penalty. Therefore, $\alpha_{\mathrm{D}} = \alpha_{\lim} = \exp[\mathrm{CBD}_{\max}(T_{\mathrm{B}}/\tau)]$, which results in $\tau = \mathrm{CBD}_{\max}T_{\mathrm{B}}/\ln \alpha_{\lim}$ and this provides the minimum link settling time, which is the optimal time:

$$t_{\mathrm{LS\,min}} = t_{\mathrm{Tx-opt}} = \mathrm{CBD}_{\max}T_{\mathrm{B}}\frac{\ln \beta}{\ln \alpha_{\lim}}. \tag{19.29}$$

In conclusion, at the optimal point, link settling is only affected by the transmitter, thus at long-wavelength tuning, the optimal time is equally long while the receiver does not add any more delay. A useful chart is provided in Fig. 19.11 for the tunable laser link, where 8B10B, 12.5-GB/s and 16B18B, 11.25-GB/s patterns are used. BMR with less than 10-ns recovery time can be realized with a small penalty. Moreover, in a synchronous burst switching system, based on tunable lasers or in any burst-switching system with a dark period between the bursts, the recovery time of a BMR is practically zero with a CDR recovery time of 10 only ns.[4,63]

#### 19.1.1.4 Burst-mode receiver topologies

Following aforementioned design considerations, there are several BMR chip topologies for METRO and FTTx.[3,5,24,42,43] The main challenges in BMR are the fast-adaptive AGC-hardware design[2] and automatic-threshold control

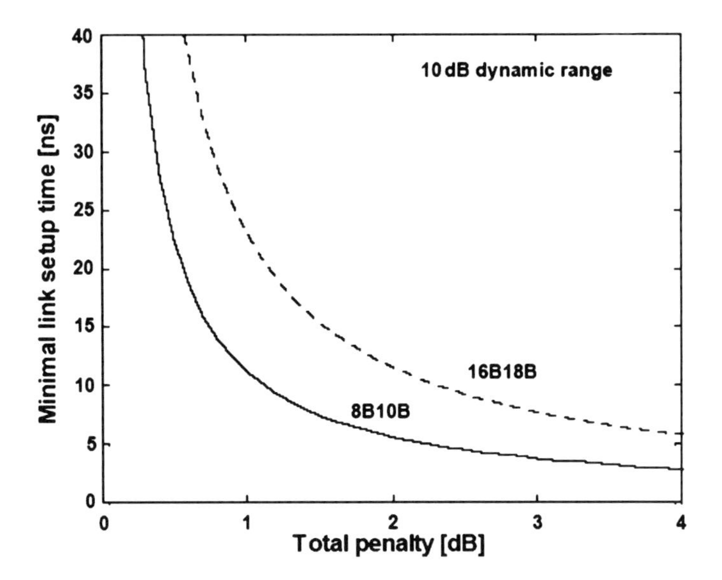


**Figure 19.11** Minimum link setup time vs. power penalty for 8B10B and 16B18B in a 10-dB dynamic range and with a 10-dB extinction ratio.[4] (Reprinted with permission from *Transactions on Communications* © IEEE, 2005.)

(ATC).[42] The technology used is SiGe BiCMOS and the standard calls for 5 V or 3.3 V for digital PECL output. Temperature range is −40 to 85°C and the power range is from −30 to −8 dBm. Another consideration when designing a receiver is the PD responsivity distribution, which affects the detected current and PD stray capacitance that dictate the bandwidth. Hence, the receiver's front end at high power should cover the PD range, typically between 0.65 and 1 A/W. This is critical when a weak burst is received. So the worst case for the dynamic range is between a weak burst with low responsivity and a high-power burst with high responsivity. For an extinction ratio of 10 dB, the total current dynamic range may reach 67 dB. In Fig. 19.12 a BMR receiver chip with an adjustable reference is illustrated, while the ATC concept is shown in Fig. 19.13.[3]

The external PIN PD is connected to a TIA, which converts the photocurrent into a voltage. This voltage is amplified further to have larger swing at lower optical inputs. The signal is limited (compressed) by postamplification that follows the PIN TIA in order to remove undesirable amplitude fluctuations. The second-stage postamplifier input is also fed by a dummy TIA to compensate for the dc variation of the main TIA, thereby reducing the dc offset overheating. A large capacitor, $C_F$, is connected to the dummy TIA to reduce noise. Those two stages are the receiver's front end. The output from the second amplifier, the postamplifier, feeds a high-speed comparator. The threshold level is set by an 11-bit digital-to-analog converter (DAC). As can be observed in Fig. 19.12, there are two paths feeding the selector. This is to accelerate the threshold value computation, thus relaxing the DAC settling time, which is about 100 ns with a voltage range between 3.8 and 0.8V. The main problem, as was presented, is how to handle bursts of different amplitudes. The ITU-T recommendation

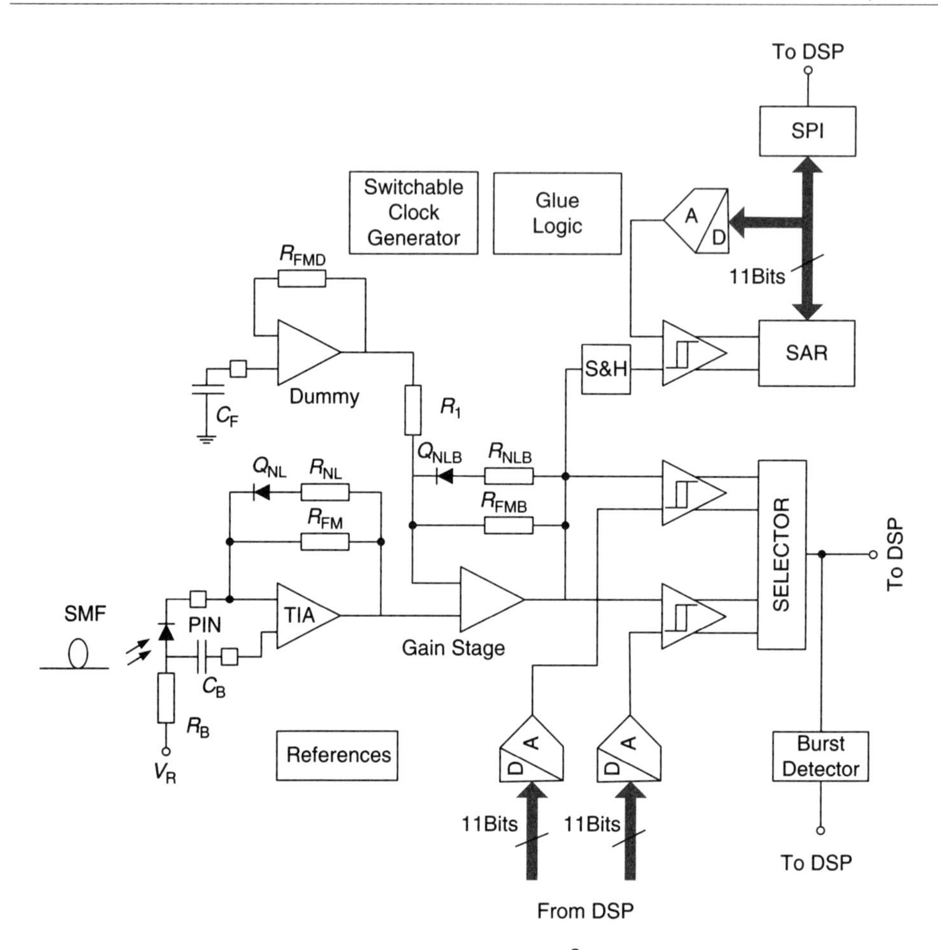


**Figure 19.12** A burst-mode receiver block diagram.[3] (Reprinted with permission from the *Journal of Solid State Circuits* © IEEE, 2002.)

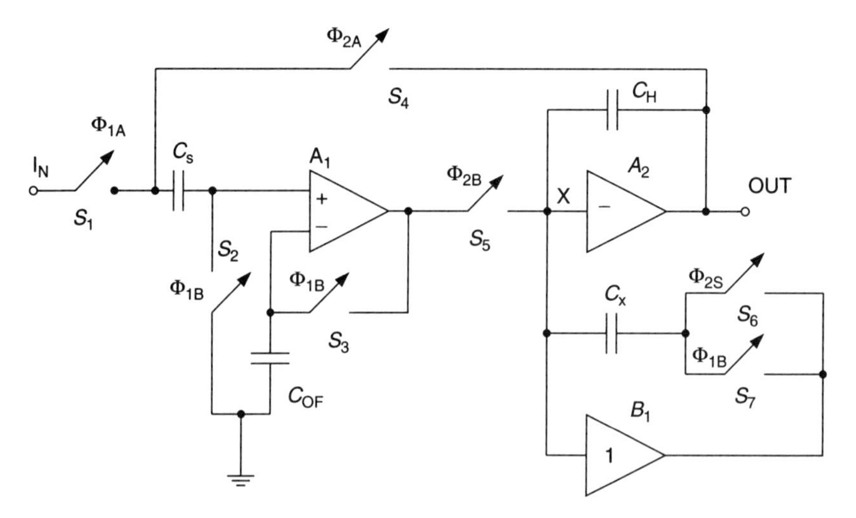


**Figure 19.13** A burst-mode sample and hold concept.[3] (Reprinted with permission from the *Journal of Solid State Circuits* © IEEE, 2002.)

G.983.1, which specifies the physical layer, allows splitting this problem into two tasks: the burst-amplitude measurement and threshold setting.

The measurement can be carried out in the so-called physical-layer operation and maintenance cells (PLOAM) available in the ATM protocol. This process includes a specific receiver-control field (RCF), which contains all 1 patterns. For this purpose, one of the DACs is used together with a comparator to implement a successive approximation ADC.

The reference level for the threshold based on measurements is estimated by using an algorithm applied on external digital signal processing (DSP). Measurements for that purpose are done using the on-chip 11 bit sample and hold (S & H) and ADC of the 1 levels of each ONU ($V_{li}$), during the ATM PLOAM cells and dark level, or voltage $V_{dark}$. The measurement of the dark level while remote transmitters are off is performed every second allowing compensation for the offset thermal drift.

Hence, one of the DACs is used to create a successive approximation ADC. That is, the corrective feedback between the S&H measurement and updating the threshold resets the threshold while converging iteratively. Note that the logic level of 1 refers to a certain optical level, while 0 refers to another. The difference between the two is the extinction ratio. That is why two nodes may have different values for high and low logic levels. Therefore, optical power levels for each result in different thresholds and the so-called loud–soft ratio that measures the difference between 1 levels for the extreme optical powers. Both optical levels related to logic levels differ from dark levels. This is because the laser is not biased below a threshold.

For these reasons, the concept of the architecture presented in Fig. 19.12 is called the alternate operation. When a path receives a signal from a particular ONU, the other path DAC is set for the other threshold voltage level for the next burst to come from another ONU; this way settling time is saved. In an ATM-PON system, each ONU can transmit only during its assigned time slot; hence, external digital chip knows which ONU is going to transmit the following burst. The resultant signal is routed out by the selector and feeds the chip output via a PECL output buffer.

Following this description, the threshold voltage tuning algorithm is given by Ref. [3]:

$$\begin{cases} V_{TH} = \dfrac{V_{li} + V_{dark}}{2} & \text{if} \quad V_{li} < V_{calib} \\ V_{TH} = V_{li} - \Delta V & \text{if} \quad V_{li} \geq V_{calib} \end{cases}, \tag{19.30}$$

where $V_{calib}$ and $\Delta V$ are determined during system calibration to minimize pulse width distortion (PWD):

$$\mathrm{PWD} = \frac{\mathrm{PW} - \mathrm{PW_N}}{\mathrm{PW_N}}, \tag{19.31}$$

where $\mathrm{PW_N}$ and PW are the nominal and actual pulse widths, respectively.

When the burst amplitude and arrival time are unknown, a burst detector is used. This is called ranging procedure. At that stage, the BMR operates with a minimum threshold voltage, just above dark level. The burst detector is triggered when a signal is higher than $V_{TH-\min}$ for a period of time longer than a given minimum value.

Sample-and hold-architecture, shown in Fig. 19.13, comprises seven switches, four storage capacitors, and three operational amplifiers using CMOS technology—the circuit implementation methods capable of mitigating the usual problems of clock feed-through, offset, and finite gain. The accuracy of such a circuit is better than 2 mV corresponding to 11 bits over 2.4-V dynamic range with a settling time of 250 ns.

When $S_1$ and $S_2$ are closed, the capacitor $C_s$ is charged to the input signal level. Sampling is accomplished when $S_1$ or $S_2$ goes to the off state. Next, after $S_4$ and $S_5$ are closed, the bottom plate of the sampling capacitor $C_s$ is connected to the output as a feedback capacitor and the output voltage is updated to give a new output sample. During this phase the capacitor $C_H$ performs the Miller pole-splitting compensation of the operational amplifier, which is made of a cascade of $A_1$ and $A_2$. At the end of $\Phi_2$, assuming the coincidence of slightly delayed clock phases $\Phi_{2A}$ and $\Phi_{2B}$, the amplifier output is settled and the switches $S_4$ and $S_5$ are opened. The output voltage is held by $C_H$, which operates as a memory element.

During the clock phase $\Phi_1$, switches $S_2$ and $S_3$ are on, thus $A_1$ is in unity-gain configuration and makes offset available across $C_{OF}$. Since the major contribution to the total input referred is due to the first stage $A_1$, a simple offset compensation is accomplished. Furthermore, if $S_2$ and $S_3$ are matched when they are turned off, they inject the matched charge into $C_s$ and $C_{OF}$. This allows compensation of the associated clock feedthrough if $C_s$ is equal to $C_{OF}$. To schedule the opening of $S_2$ prior to that of $S_1$ prevents charging of $C_s$ due to the input signal.

A significant aspect of S&H circuits is the charge injected by the holding switch, $S_5$ in this example, when it is turned off. Another approach to solve this case is implemented. A unity gain buffer $B_1$, a capacitor $C_x$, and two switches $S_6$ and $S_7$ are added to the circuit as shown in Fig. 19.13. Slightly prior to the end of phase $\Phi_{2B}$, $S_6$ is turned off in order to precharge $C_x$ with a fraction of its channel charge. Then, when $S_5$ is turned off, a fraction of its channel charge is injected to the node X, which is the input of $A_2$. If $S_5$ and $S_6$ are matched, a compensation of the injected charge is obtained by closing $S_7$ after opening $S_5$. The unity gain buffer sets the same dynamic biasing for $S_6$ as for $S_5$.

#### 19.1.1.5 Burst-mode AGC topologies

In previous sections, BMR dynamic analysis and receiver architecture are presented. The next challenge in realizing BMR is to have "fast-attack" AGC and reset the system[2] with a minimum signal distortion and a high dynamic range, and feedforward auto bias.[56] The distortions include duty cycle and require automatic-offset compensation (AOC) and auto-offset control (AOC).[42]

In 1998, ITU-T announced a global standard for the ATM-PON system as Recommendation G.983.1. The high-gain sensitivity specification of Class C was defined to construct further flexible and economical networks. In this system, the transmission loss varies for each path, thereby the burst cells in OLT have a large power difference. Additionally, the laser-diode bias current can be applied on burst cells to reduce cost of ONU since they are installed at the subscriber side. As a result, the receiver on the OLT side should exhibit high sensitivity, wide dynamic range between burst cells, and a high loud–soft ratio and receive signals with ER as low as 10 dB. Conventional preamplifiers that have bit-by-bit-controlled transimpedance suffer from the inability to receive large signals with low ER due to the log amplifier operation is mode nature. For that purpose AGC should handle a fast response time from the first bit of the cell, be able to receive signals with low ER, and high loud/soft ratio as shown in Fig. 19.14 without affecting and distorting the data swing. Such amplifiers realized in CMOS can handle 30-db loud–soft ratio, they have a sensitivity better than −39.3 dBm and the AGC has a fast response from the first bit. Such amplifiers satisfy the high sensitivity specification of Class C and are more flexible; thus, economical ATM PON can be constructed.[57]

In regular burst-mode configuration as in Fig. 19.14, the TIA converts the photocurrent into voltage to feed the input of the decision circuit. As previously explained, the decision circuit recovers the data per the generated threshold.

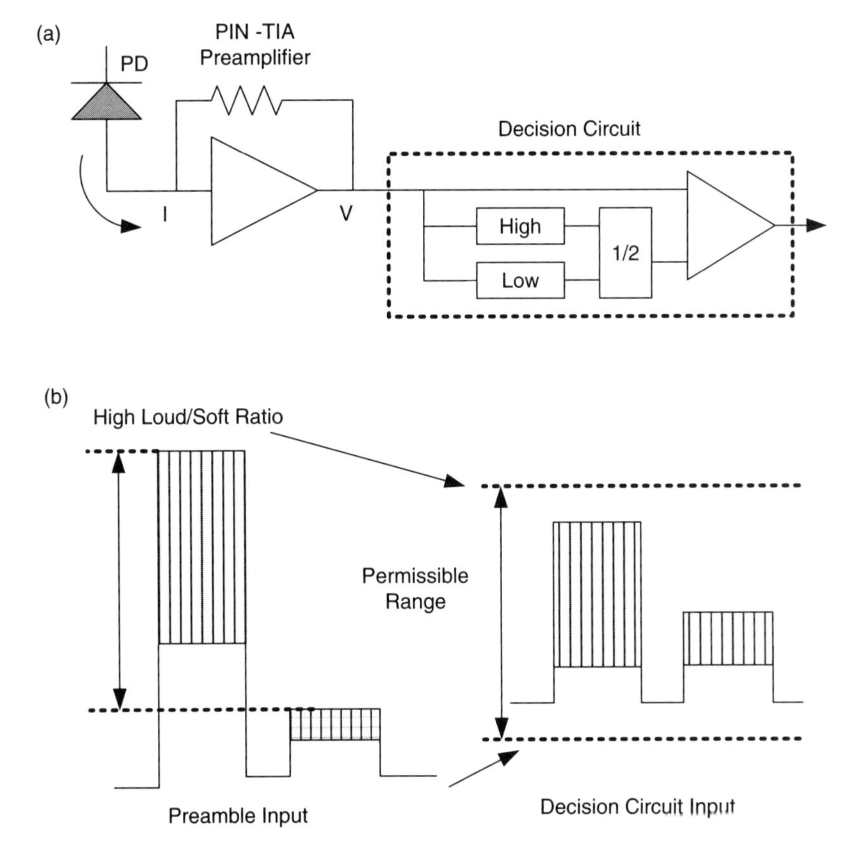


**Figure 19.14** (a) BMR configuration and (b) the required operation range of the preamplifier.[57] (Reprinted with permission from the *Journal of Solid State Circuits* © IEEE, 2002.)

The threshold setting is at half the magnitude of the signal swing. In order to realize a loud–soft ratio, the preamplifier TIA is needed to compress the dynamic range of the input signal into the allowable range of the decision circuit as, shown in Fig. 19.14. The strategy is to realize the TIA transimpedance gain to be controlled cell by cell according to the amplitude of the input signal. The gain strategy is to have high gain for small low-power signal, and low gain for a large signal. Additionally, the TIA should handle low ER signals and have high sensitivity.

Traditional TIA operates as shown in Fig. 19.15. In this case, a current bypass diode is connected in parallel with the resistive feedback of the TIA. As long as the signal is below the diode's on threshold, the TIA gain is defined by the resistive feedback. Thus, $V_{OPT} = R \times I_{OPT}$. When the current exceeds the diode's on voltage, the TIA output voltage drops. This type of TIA is called bit-AGC control since the TIA gain control is done by every bit. Moreover, if a low ER signal appears at large optical input power, it will be centered on the dc bias generated after detection by the PD and amplification of TIA. In this manner, the amplitude of the signal is clipped and ER is reduced. Consequently, it is harder to discriminate between 0 and 1. This is overcome with a cell-by-cell concept according to the input-signal amplitude. In that case, the AGC response is according to the input signal amplitude and the gain strategy is cell by cell. In the case of a low ER with a large signal, the transimpedance gain will be $G_1$. However, when the signal is weaker, the gain is increased to $G_2$ as shown in Fig. 19.16. This allows a better signal level discrimination and can be realized by having a FET in parallel to the feedback resistor. A FET is a channel device and operates as a gate-voltage-controlled variable resistor as shown in Fig. 19.17.

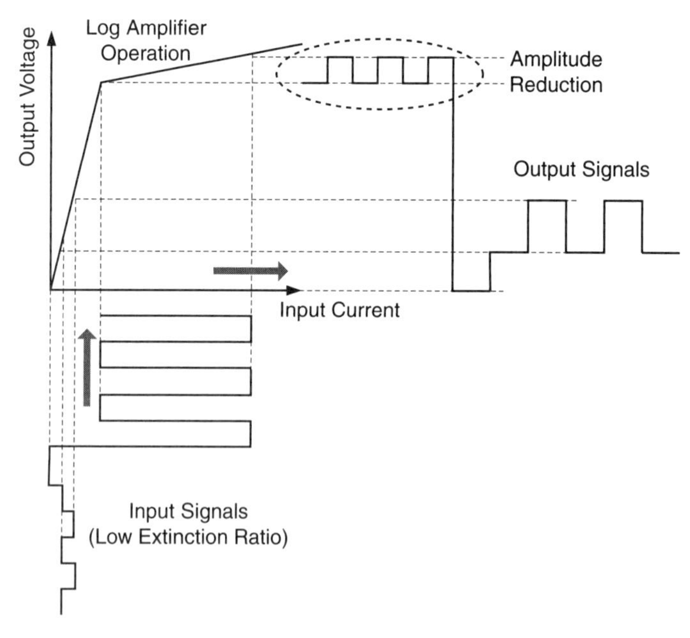


**Figure 19.15** Conventional bit AGC for burst-input signals with low ER.[57] (Reprinted with permission from the *Journal of Solid State Circuits* © IEEE, 2002.)

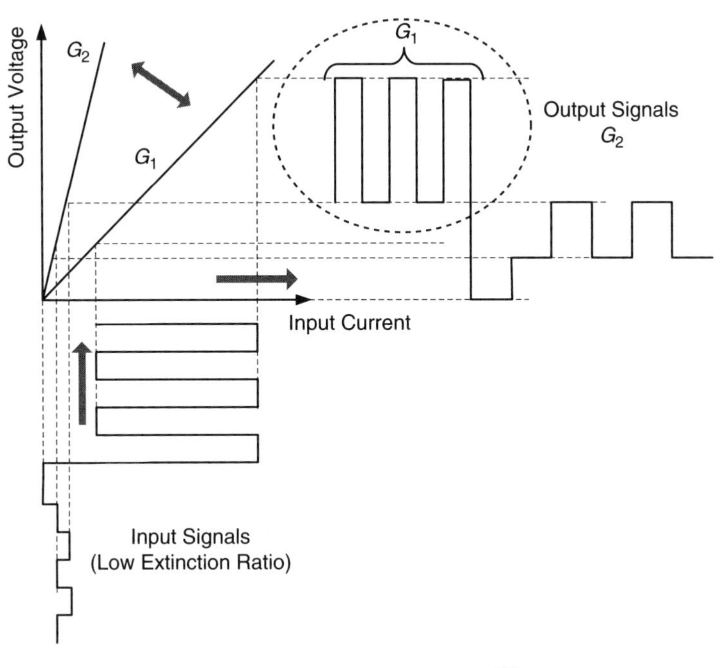


**Figure 19.16** Cell AGC for burst-input signals with low ER.[57] (Reprinted with permission from the *Journal of Solid State Circuits* © IEEE, 2002.)

At the instant a signal is detected, it goes through a high-gain state since the shunt FET $V_{gs}$ is low. The bottom-level detector (BLD) quickly detects the bottom level of Amp 3, shown in Fig. 19.17. A hold circuit in the BLD keeps this state. The gain-control circuit (GCC) receives this level and determines the new gate voltage for the shunt FET. The feedback resistance is maintained during the entire cell period. In case the input current at the PD output increases, i.e., a new cell arrives, $V_{gs}$ increases. Thus, the transconductance $g_m$ of the FET increases, thereby reducing the effective feedback resistance, as shown in Fig. 19.17. As a consequence, the transimpedance gain of the TIA goes from $G_1$ to $G_2$, as shown in Fig. 19.16 and better discrimination between 0 and 1 is accomplished. When the cell changes, the reset signal is triggered and launched into the BLD. Then, the GCC output and transimpedance are set to a normal state of maximum-gain high sensitivity, which is an off state for the shunt FET in Fig. 19.17. In this manner, the loud–soft ratio is accomplished where a low-extinction-ratio signal can be received without clipping. A similar TIA-AGC concept appears is in Fig. 19.18 and is described in Ref. [2]. Assuming a normal TIA with an equivalent input capacitance to the TIA, $C_{in}$, which is the PD capacitance plus the TIA input capacitance, then the gain of the TIA is given by

$$G = \frac{V_{out}}{I_{in}} = -\frac{A_V}{A_V + 1} \times \frac{R_F}{1 + s[R_F C_{in}/(A_V + 1)]} = -\frac{A_V R_F}{1 + A_V + R_F C_{in} s}, \quad (19.32)$$

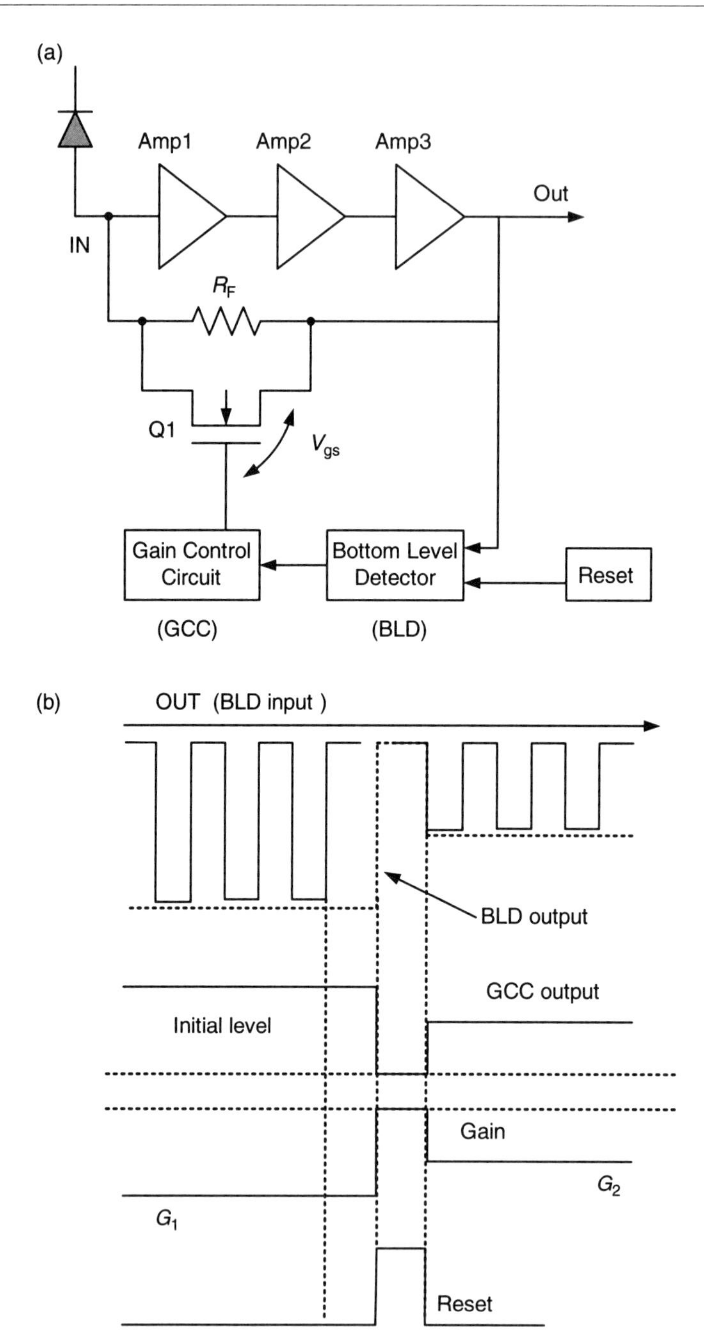


**Figure 19.17** (a) A gain-controlled circuit (GCC) with bottom-level detector (BLD), (b) operation phases.[57] (Reprinted with permission from the *Journal of Solid State Circuits* © IEEE, 2002.)

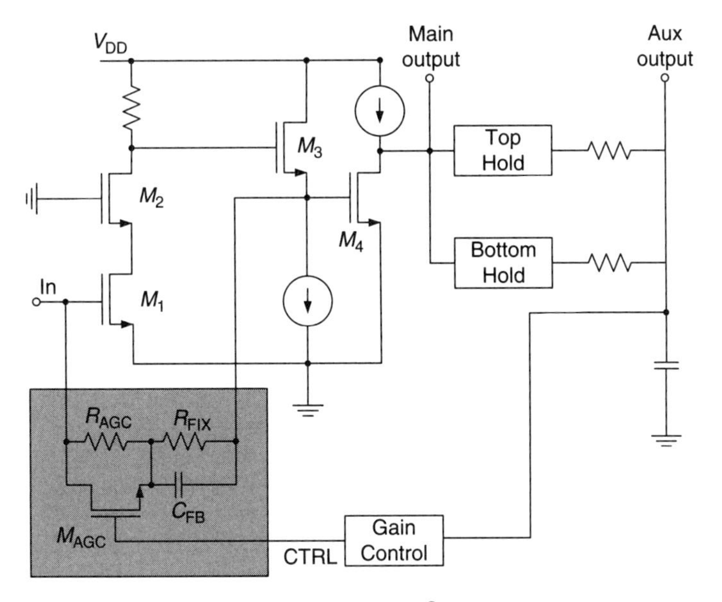


**Figure 19.18** TIA with AGC realized on feedback.[2] (Reprinted with permission from the *Journal of Solid State Circuits* © IEEE, 2002.)

where $s$ is the Laplace domain variable, $A_V$ is the voltage gain of the TIA, and $R_F$ is the feedback resistor. The 3-dB pole is given by

$$f_{3dB} = \frac{1}{2\pi} \times \frac{A_V + 1}{R_F C_{in}}. \tag{19.33}$$

$A_V$ has a limited BW represented by a pole, hence, the transfer function in Eq. (19.32) is a second-order polynomial denominator. This makes a peaking effect, which expands the whole TIA BW. Of course, this affects the eye by overshooting and undershooting transitions. Replacing the simple feedback resistor $R_F$ by a variable feedback resistor as in Fig. 19.18 provides more degrees of freedom for optimization. Thus, it can be optimized by proper selection of $R_{FIX}$, $C_{FB}$, $R_{AGC}$, and $M_{AGC}$. That is the cure for the overshooting problem, adding an additional time constant, thereby modifying the feedback Eq. (19.32). The equivalent feedback resistance, $R_F$, is given by Eq. (19.34). This feedback resistance drops at high frequency:

$$R_F = R_{AGC} + \frac{R_{FIX}}{1 + R_{FIX} C_{FB} s}. \tag{19.34}$$

Note that $M_{AGC}$ is absorbed within $R_{AGC}$. The 3 dB of the compensation pole of the peaking effect is given by Eq. (19.35). From Fig. 19.18, it can be seen that $R_{FIX}$ sets the minimum TIA gain for improving the TIA BW:

$$f_{3dB} = \frac{1}{2\pi} \times \frac{1}{R_{FIX} C_{FB}}. \tag{19.35}$$

In Fig. 19.19, a BLD configuration is shown. By detecting and holding the bottom level, using a CMOS diode and a hold capacitor, the circuit creates a control signal for the GCC. Since the circuit is CMOS, the diode uses the source contact as a junction. The CMOS pair resets the circuit by shorting the catch diode realized by a FET. In this manner, the storage hold capacitor is wholly charged by the current source to establish the hold level.

The GCC concept is provided in Fig. 19.20. As the BLD amplitude increases, the differential amplifier output also increases. This gradually turns on the FET switch denoted as SW. If the FET SW is off, $M_1$ operates exclusively as an active load. If the FET SW is on, then both $M_1$ and $M_2$ are operating together as an active load. Both FETs are designed to have the same operating point. By setting the impedance of $M_2$ lower than that of $M_1$, when they are operating as an active load, this circuit will have high gain for a small signal and low gain for a large signal.

Noise performance of AGC TIA is a major concern. Therefore, the short channel CMOS process reduces the channel resistance and, as a result, the thermal noise is lowered. As explained in Sec. 11.3, FET is a field device,

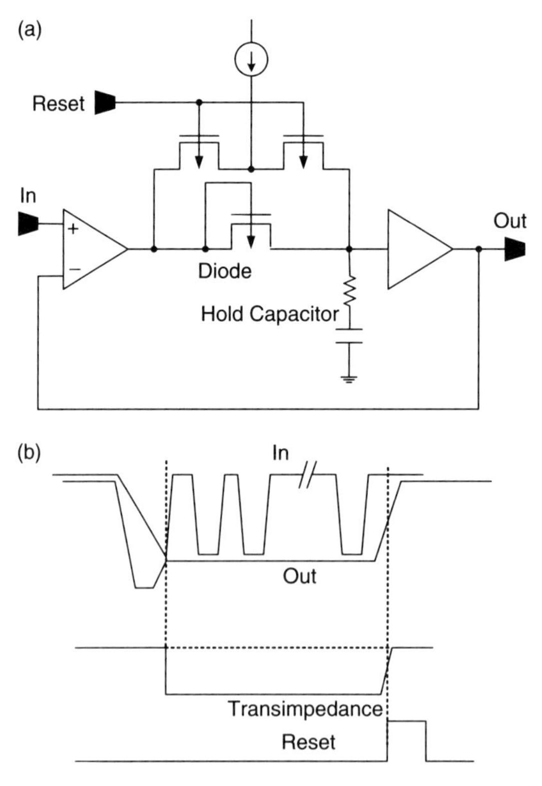


**Figure 19.19** BLD configuration: (a) circuit concept and (b) operation timing sequence.[57] (Reprinted with permission from the *Journal of Solid State Circuits* © IEEE, 2002.)

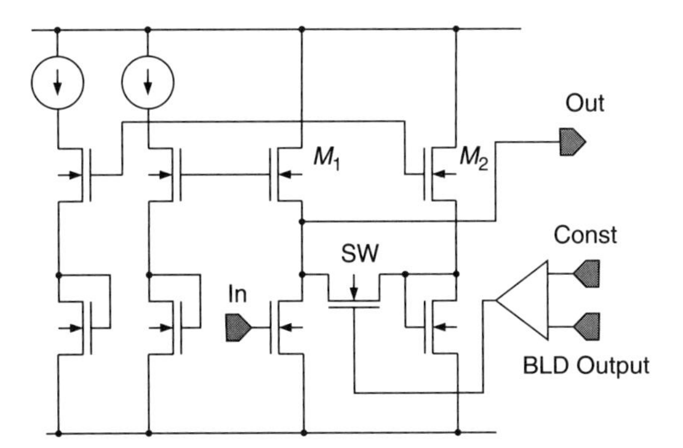


**Figure 19.20** The GCC configuration.[57] (Reprinted with permission from the *Journal of Solid State Circuits* © IEEE, 2002.)

which creates a charge drift current in the channel; thus its noise is characterized mostly by Johnson noise. The input noise current density of a TIA circuit, as shown in Fig. 19.17, can be expressed by[57]

$$\frac{i^2}{\Delta f} = 2qI_g + \frac{4kT}{R_F} + 4KT\left(\frac{2}{3}\right)g_m\left(\frac{1}{R_F^2} + \frac{C^2\omega^2}{g_m^2}\right), \tag{19.36}$$

where $I_g$ is the gate leakage current that provides the shot noise component, $g_m$ is the input FET transconductance, $R_F$ is feedback resistance, and $C$ is the input capacitance. The other two expressions in Eq. (10.36) are the thermal noise components that result from the feedback resistor and the FET. There are several design constraints for $R_F$. A high value of $R_F$ decreases the shot noise in the second term on the right-hand side (note that $C$ is the input capacitance and $g_m$ is the FET transconductance). However, the 3-dB BW imposes a limitation as per

$$f_{3dB} = \frac{1}{2\pi} \times \frac{G}{R_F C}, \tag{19.37}$$

where $G$ is the open-loop gain of the IC device. An additional variant in noise optimization of AGC TIA is the FET's $g_m$. As it goes to high values, the noise density decreases until it reaches a minimum point, above which the $C$ value starts to govern noise; thus, noise density starts to increase. The reason for that is that increasing $g_m$ increases FET size and thereby increases its input capacitance.

#### 19.1.1.6 Burst-mode receiver testing

In previous sections and in Sec. 18.10, it is explained that the lock acquisition of a BMR is an important matter since it defines the number of preamble bits required for CDR and AGC recovery. In order to increase transmission efficiency, it is essential to minimize the preamble, training series length. That is a consideration for designing a BMR test setup. Another matter is the estimation and the measurement of the CDR locking. CDR is a PLL and there are several methods available to test locking. One method is to monitor the voltage-controlled oscillator (VCO) and estimate the frequency accuracy; another method uses a mixer as a correlator and measures the beat note between the CDR and the targeted frequency that is locked to the same reference as the CDR input sequence. The drawback of these methods is the overestimation of acquisition; it is not necessary for the clock to be perfectly aligned with the data before the payload becomes valid. Increasing the acceptable percentage offset decreases the measured acquisition lock time. The question though is what offset percentage will decrease lock time. There is no direct relation between BER on the payload and the acceptable offset. The other practical problem is that the VCO tuning voltage is not accessible, and CDR is part of a chip or receiver IC.

The last challenge for a BMR is measuring its performance under burst transport. Burst synchronization consists of three main cycles: AGC preparation, decision-level preparation, and CDR synchronization. Such a setup consists of a bit-error-rate tester (BERT) and two packet generators with variable phase, frequency, and amplitude to emulate PON or bursty traffic. The goal of such a setup is to have BER measurements versus phase, frequency, or amplitude steps to emulate traffic from various ONUs. Lastly, the setup has to measure preamble length in order to guarantee a certain BER on the payload.[51]

In order to test the lock acquisition time of a BMR against packets that vary in phase, frequency, and amplitude, the packets must be created. There are several methods to realize such traffic. For that purpose some pulse-pattern generators (PPG) offer options for two alternating packets. However, the phase, frequency, and amplitude of each packet cannot be set independently. Some solutions offer the use of programmable delay lines (PDL), fiber delay lines (FDL), fast optical switches, fast tunable transmitters, and variable optical attenuators (VOA). The drawback of this strategy is the additional error it adds to the device under test's (DUT's) acquisition time result due to the reconfiguration time of the setup, which is a statistical variance by itself.

A BMR test setup is illustrated in Fig. 19.21(a). It contains two PPGs and a power combiner (PC) for the generation of a packet. There is no need for FDLs, PDLs, VOAs, fast optical switches, and fast tunable transmitters. The phase, amplitude, and frequency of each packet are set independently. The flexibility of this solution comes at the cost of synchronization complexity between the two PPGs. By sharing a common global clock, the two PPGs can achieve at best-bit synchronization. Thus, the two PPGs may end up transmitting overlapping bits. Packet synchronization had to be implemented in order to emulate PON and OPS network traffic. In order to overcome this problem and generating two nonoverlapping packets, a Matlab graphical user interface (GUI) monitors and controls the two PPGs. While one PPG is being transmitted, the other is set to send 0 so that it appears to be silent or NRZ. Assuming that the two PPGs generate the same frequency, the pattern loaded by the Matlab controller in each PPG has to be twice the length of each individual packet.

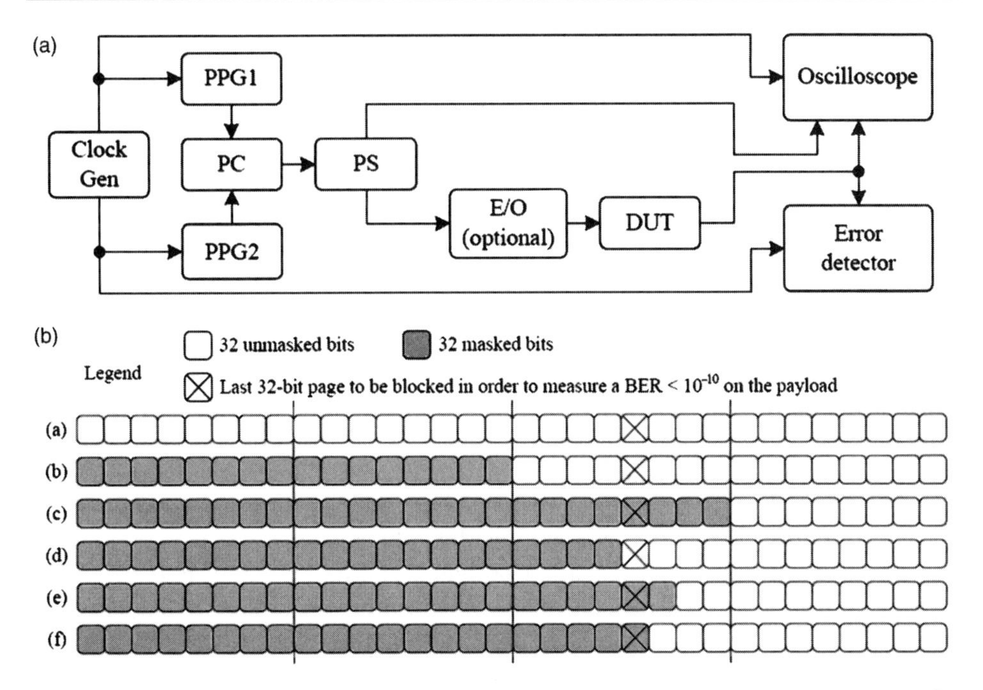


**Figure 19.21** (a) Burst-mode test setup containing a pulse-pattern generator (PPG), power combiner (PC), and power splitter (PS). (b) Preamble-length measurement using the nonlinear algorithm. The preamble length in this example is $21 \times 32 = 672$ bits.[51] (Reprinted with permission from *Lightwave Technologies in Instrumentations and Measurements* © IEEE, 2004.)

The PPGs used in this test setup are HP80000 and Anritsu MP1763B. The maximum delay range between the two packets is 5 ns, while the delay resolution is 1 ps. The delay range sets a lower limit for the bit rate of the generated packets. For testing convenience, the delay range is enough to cover a full bit period. A 5-ns-delay range covers the complete period of a 200 Mb/s. The delay resolution sets a higher limit on the bit rate of the generated packets. A 1-ps-delay resolution represents 1% of the bit period of a 10-Gb/s signal. The packet generator can generate up to 12.5 Gb/s. For testing the effect of amplitude steps on lock acquisition time, the amplitude of each packet can be set independently at the voltage range of 0.2 to 2 V.

The quality of service in digital communication is measured by BER. The design challenge is to modify a continuous-test setup into a burst-mode setup. There are two ways to establish such a tester: using the UUT CDR as a synchronization reference clock and using a global reference clock. This is the main issue in test setups as the clock is used as a sampling clock for the BERT error detector (ED). To establish a BER testing, the ED needs synchronization. The reference pattern can be either generated in hardware by the ED for PRBS test patterns, or stored in internal memory for user-defined patterns. Synchronization times can range from microseconds to minutes and depend on the pattern length, the bit rate, and synchronization method. Due to the fact that the synchronization is on the microsecond time scale, it is important for the ED not to lose synchronization from packet to packet. This is especially true for short packets. SONET data, since it is continuous, will not cause a synchronization loss. Synchronization is required when the link is first established and never lost during the BER test. In contrast, in burst-mode, synchronization may be lost if the ED is clocked by the CDR-recovered clock. The reasons for that are CDR lost synchronization, i.e., out-of-lock PLL, and the data signal is shifted with respect to the recovered clock, causing the incoming pattern not to correspond to the reference pattern programmed in the ED. Those scenarios occur when the CDR is subjected to a frequency or amplitude step. The other case of the above mentioned is synchronization loss due to a phase step. Phase step may cause the recovered clock to sample the incoming bits at the edge of the bit window. Under this condition, the CDR may output a bit either exactly when the ED expects it or one bit ahead of time or one bit behind, to advance or retard. This is a random process over the first few bits of a packet.

Thus, due to the synchronization problem, continuous setup based on CDR cannot be used for BMR evaluation. The main problem is that the signal under test and its recovered clock are also used as sampling clocks. The solution is to use a reference clock as a sampling clock. If the sampling clock is derived from the same clock used to generate the data, then the ED should never lose synchronization because of the aforementioned scenarios.

Prior to BER measurement, the ED has to be loaded with a properly shifted version of one of the two packets. Recall that the length of the reference pattern is twice the length of each individual packet, where a single common frequency is assumed. The first of the two packets, denoted as packet 1, is used as a dummy packet, which lets the CDR reach a steady state prior to a phase, frequency, or amplitude step is applied. Any error relating to packet 1 is omitted by masking the corresponding bit. BER measurements are performed on packet 2 only. The Matlab GUI applies circular shifts to each PPG in order to guarantee that packets 1 and 2 do not overlap. These circular shifts are used to derive the position of packet 2 in the reference pattern to be loaded in the ED. The Matlab GUI uploads the reference pattern to the ED and applies circular shifts.

The suggested concept of the BMR-BER tester enables the BER to be measured, while CDR loses locking due to the amplitude or phase step, or during the acquisition period. In fact, BER is a measure to evaluate the CDR synchronization and locking acquisitioning. This testing method enables the measurement of the BER in case of retarded or advanced bit.

The advantage of this test setup is in its ability to evaluate the preamble series and optimize its length. It was explained that the payload efficiency increases as the preamble-series bit count decreases. The idea is to reduce the length of the preamble by reducing the number of the masked bits in packet 1, while targeting a certain BER such as $10^{-10}$. In this example the Anritsu MP1764A ED, operating as PPG, allows for the masking of bits on a 32-bit resolution. There are several algorithms to find the preamble length, as follows.

- The linear algorithm shown in Fig. 19.21. The packet length is 1024 bits. Here only packet 2 is shown, this is the valid data packet. The gray and white squares represent 32 masked and unmasked bits, respectively. In this example, the 21st square, marked with an "X," represents the last 32 bit page that should be masked to guarantee the target BER on the payload. In order to find it automatically, the linear algorithm considers 22 block window patterns, which include the initial empty block window. Using the linear algorithm, 22 BER measurements are needed to determine the preamble length for a given phase, frequency, or amplitude step. The complete linear algorithm takes approximately 2 min to execute. Though 2 min is not a long time, bear in mind that the preamble length has to be measured many times to generate a plot of preamble length versus phase, frequency, or amplitude step. Generating such a plot with a 10-ps resolution and a 1000-ps delay range, i.e., 100 points, requires 3.3 hr. So it is a slow process and another method is needed.

- The nonlinear algorithm is the preferred way to overcome the drawbacks of the linear algorithm. It requires fewer trails, 6 rather than 22, and a preamble length can be found within 20 to 30 s. From Fig. 19.21, the search for preamble length stops when the incremental block window is 32 bits. More generally, the number of BER measurements required to determine the preamble length is predicted by $N_b$ BER measurements $= \log_2 N - 4$, where $N$ is the packet length and as shown in Fig. 19.21, $N$ is equal to 1024 and the number of BER tests is equal to six. Generating a plot of preamble length versus phase step, on a 10-ps resolution and 1000-ps delay range, i.e., 100 points requires 45 min compared with 3.3 hr of the linear algorithm. However, the linear algorithm performs better in situations where the preamble is short. This is because the nonlinear algorithm has six BER tests, while the linear algorithm has a variable number of tests.

### 19.1.2 Burst Mode Modulation Strategies

In Chapters 12 and 13, a deep theoretical review about power leveling loops is provided. However, it describes mostly a static condition and the continuous mode of operation in AGC and APC. The main challenge in dynamic APC-loop operation is somewhat similar to receiver-mode AGC preparations. During a continuous transport, APC is locked to a certain bias specification to guarantee the desired optical power launch from the laser. The main challenge in burst-mode upstream transmitter (US-TX) is fast turn off, the APC power-level storage state, and having a fast aided-acquisition of power APC.[8] Additional functionality required from any optical transmitter is end-of-life (EOL) detection. Typical burst-mode transmitter is shown in Fig. 19.22.

Generally, there are two methods of coupling data to laser ac and dc as well as two biasing methods: above threshold where the laser is already starting to emit

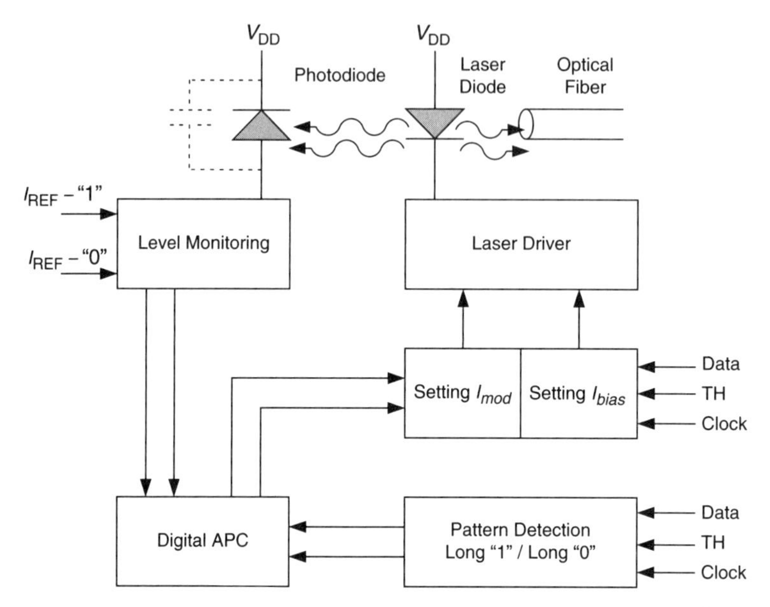


**Figure 19.22** Burst-mode optical-transmitter building blocks.[6] (Reprinted with permission from *IEE Electronics Letters* © IEE, 2004.)

and below threshold where the laser is not emitting or lasing.[6] In burst-mode operation, it is preferred that the laser be above threshold in order to accelerate turn-on time and limit the duty-cycle distortion caused by the turn-on delay explained in Sec. 17.2.6. On the other hand, burst-mode operation requires an instance where the whole system has no light emission from the ONU. Those requirements define the APC architecture and its operation sequence.

In GPON US-TX, it is required to have dc-coupled interface between the laser driver and the laser diode. The reason is that the guard time between bursts is limited to 32 bits or 25.6 ns at 1.25 Gb/s per the ITU G.948.2,[60] where the data launched after scrambling may reach 72 consecutive identical digits.

For proper operation of APC loops in burst mode, APC should have three phases for laser turn on. These turn-on phases are demonstrated in Fig. 19.23 and result due to APC dynamics (reviewed in chapter 13) and dictated by the GPON system structure[6,48]. APC is a power-control loop that has to be adjusted and prepared. The lasers' APC has to be trained for the data pattern and desired power. This is done as follows: first, prebias bits are provided as a preamble. The APC loop control logic settles and the laser is biased, which overlaps the training bits. Hence, the APC back-facet monitor delivers both the dc level for locking the laser to its desired power and the detected pattern adjusts modulation, which is done by controlling the extinction ratio. At the termination of APC settling and laser-bias delay to turn-on state, valid data is applied. At the turn-on cycle, APC should bring the laser to its proper bias under modulation, thus, preamble is used. The design challenge has a short preamble in order to reduce transmission overhead. The preamble bits overflow the laser turn-on-time to accomplish

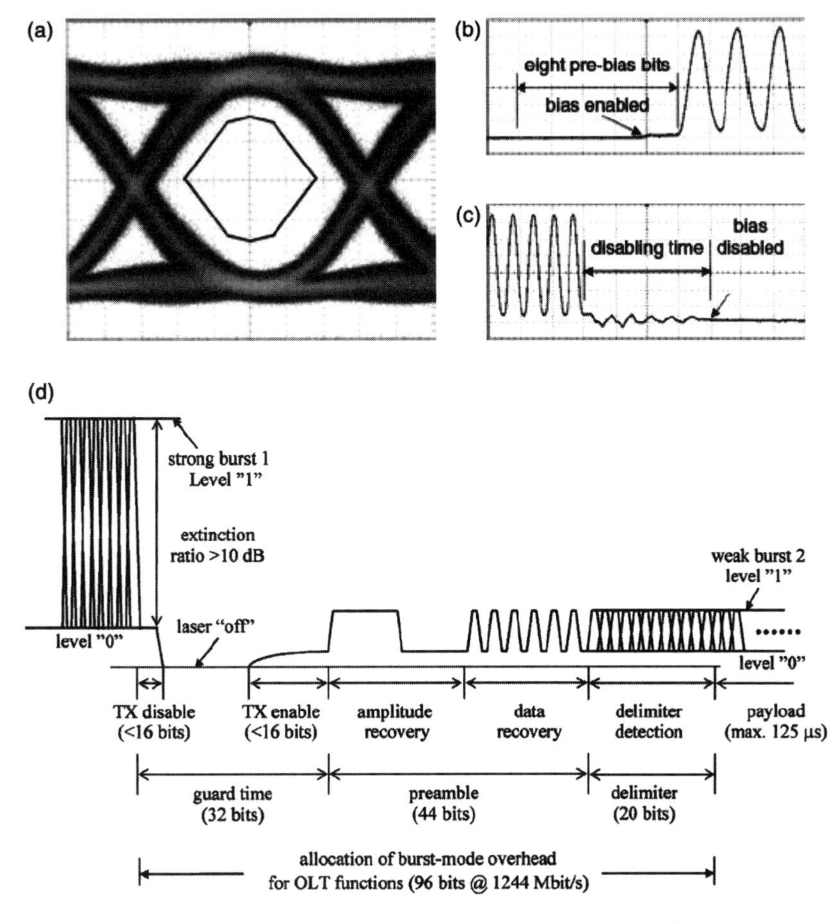


**Figure 19.23** An upstream transmitter (US-TX) waveform at 1.25 Gb/s. (a) An eye diagram at 1.25 Gb/s; (b) burst turned on 1.25 ns; (c) burst turned off 2.5 ns.[6] (Reprinted with permission from *IEE Electronics Letters* © IEE, 2004.) (d) specification of upstream burst mode over head showing 3 stages of OLT. (Reprinted with permission from *IEEE Journal of Lightwave Technology* © IEEE, 2004.)[48]

proper APC power locking and fast termination of all modulation-related effects of the laser biasing and locking to power. After the APC is set, valid data preamble is applied in order to order to train the burst-mode receiver's AGC and CDR. When the transmitting cycle ends, the laser is turned off. Turning it off consumes time until the laser is completely off. Figure 19.23 displays the transmitter's turn on phase in (b), turn on in (c), and an eye diagram is provided in (a). Those above mentioned processes define the burst-mode overhead for the OLT in Fig. 19.23(d). Assume a strong burst in the previous time slot; the laser turn off requires time until it is at a fully dark state. Thus, there is a guard time between slots lasting for a period of 32 bits. After that guard time, preamble time is inserted for the receiver to perform AGC and CDR setting and consuming the delimiter time. The ITU – T G484.2 calls for mandatory guard time of 32 bits, preamble of 44 bits and delimiter time of 20 bits. The total length of guard time is determined by the laser turn on and turn off.

An additional design constraint is a laser-performance variation and parameter distribution. Those parameters show temperature-dependent characteristics and a nonlinear relationship between the laser current and optical power. Thereby, both the laser-bias current and modulation current should be tuned according to the targeted optical power and extinction ratio.

In Fig. 19.22, a burst-mode laser-driver concept in a transceiver is depicted.[6] A similar design approach can be found in Ref. [7]. The operation of such a laser driver is as follows. In order to start a transmission cycle, first preamble data is transferred and synchronized with the clock. This is done during the guard-time stage, which is in the dark state. The laser driver is responsible for modulation and determines the ER, while the APC sets the power level. For that purpose, it should be "trained" by the preamble series, starting prior to the biasing state. This is the so-called aided acquisition. Hence, during the dark state, as well as during normal operation mode, power-detection block provides the digital-APC block to the pattern. At the dark state, it provides a starting point for bias current and modulation current based on the pattern detector. The role of the pattern detector, mainly for long consecutive bit streams, is to determine the average optical level at the APC and make the proper corrections in order to preserve the modulation-depth current and threshold bias, which determines the ER. The back-facet monitor photodiode senses the average power emitted from the laser. However, in order to determine and control the modulation depth, which is the ER, a peak detector is required to follow the peak optical power. Those, plus desired ER, can provide a good estimation for the APC setting. The other option is to have a level monitoring system that detects the data using current mirrors and determines the 0 and 1 states by comparing them with reference currents. These currents are fed to the APC block as well. Smart APC control is fed by the back-facet-monitor-detected data and pattern detection, and it sets the proper modulation current and bias point, thus controlling optical power and ER. Experimental data[6] shows the advantage of the prebias training series in Tables 19.1 and 19.2.

Earlier designs involved storing the APC state using analog memory as a means of operational amplifiers, capacitors, and analog switches.[8,52,53] In this case, since ER is known, the desired laser power bias is defined as well as the modulation depth. The laser acquires power locking, to a certain average point. The idea is to preserve this point for the next burst and aided acquisition (Sec. 13.1). This design concept is improved as follows.[8]

In Fig. 19.24, a block diagram of a burst-mode laser driver is shown.[8] This driver is designed to operate at 155 MB/s for ATM PON and FSAN applications.

**Table 19.1** Measured turn-on delay for different prebias bits length (G.984.2 length is 16 bits).

| Prebias bits | 0 | 2 | 4 | 6 |
|---|---|---|---|---|
| Turn-on delay (ns) | 1.21 | 0.52 | 0.14 | 0.00 |

**Table 19.2** Sum of launched optical power tolerance over −40 to 85°C, 0.5 dBm at −40 to 60°C.

| | | | |
|---|---|---|---|
| Pave calibrated at (dBm) | −5.5 | −2.5 | 0.5 |
| ER calibrated at (dB) | 11.0 | 14.0 | 17.0 |
| ΔPave (dB) | 0.8 | 0.6 | 0.5 |
| ΔER (dB) | 0.7 | 0.9 | 0.2 |

It can be used for narrow PON and π-PON (50 MB/s). The input signals can be applied in low voltage CMOS (LVCMOS) or low voltage PECL (LVPECL), depending on the desired bit rate and available format. The data is predistorted by the turn-on-delay-compensation (TODC) block prior to reaching the laser driver, denoted as the laser-driver stage (LDS). In this manner, pulse-width distortion of the optical signal is prevented. The emitted signal is detected by the back-facet monitor PD and is feeding a peak comparator (PC), which compares the signal with a reference value. The digital section (DIG) digitally controls the on current of the LDS as well as producing an EOL alarm. A clock signal for the peak comparator and DIG is derived from the framing and overhead-flag signals by the clock generator (CLK).

In Fig. 19.25, two laser-driving stages are shown. The first is a conventional differential pair. The modulation is done by the differential inputs to the FET stages, while the bias is set by the current source. It is commonly used, as it has low jitter, small turn-on delay, low power supply noise, and a good power-supply-rejection ratio (PSRR), which are all beneficial. However, it consumes a significant amount of power since $I_{MOD}$ as well as $I_{BIAS}$ is always on.

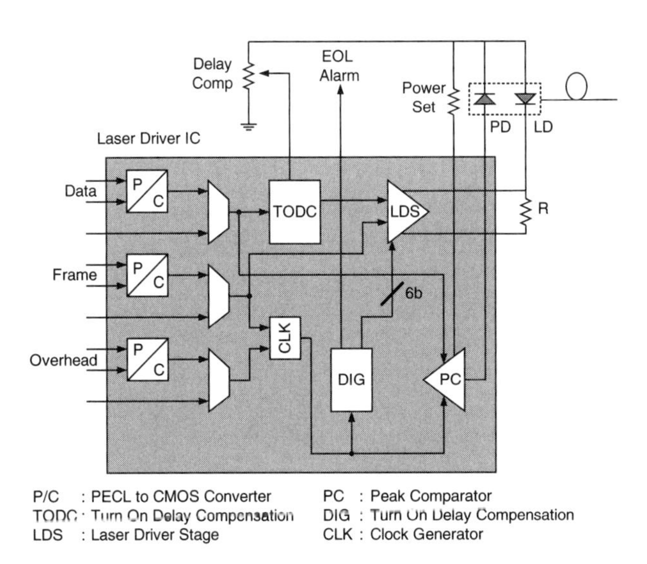


**Figure 19.24** A laser-driver-block diagram.[8] (Reprinted with permission from the *Journal of Solid State Circuits* © IEEE, 2000.)

Thus the power consumption is $P = V_{DD}(I_{MOD} + I_{BIAS})$. Hence, designers' goals are to reduce this idle current to a no-modulation-signal state. An optional solution is given in Fig. 19.25(b), where the current steering stage is replaced by a digitally controlled switch that draws current only during the transmission of active frames. Moreover, it draws current only at the transmission of 1. In this way, power saving is accomplished. The laser is operated without bias to save current and achieve larger ER. Therefore, the power consumption of that driver is given by

$$P = 0.5 \times V_{DD} \times I_{ON} \times S_{TDMA}, \tag{19.38}$$

where $I_{ON}$ is the current driven into the laser when 1 is transmitted, $S_{TDMA}$ is the TDMA slot usage, typically 1 out of 16, and 0.5 due to the average 50% mark density within a burst. Compared with that of a regular driver, the current is reduced by approximately 1/32.

Since the laser diode, generally 1310-nm Fabry Perot (FP) in a full-service-access network (FSAN) upstream, is unbiased, it exhibits a turn-on delay of

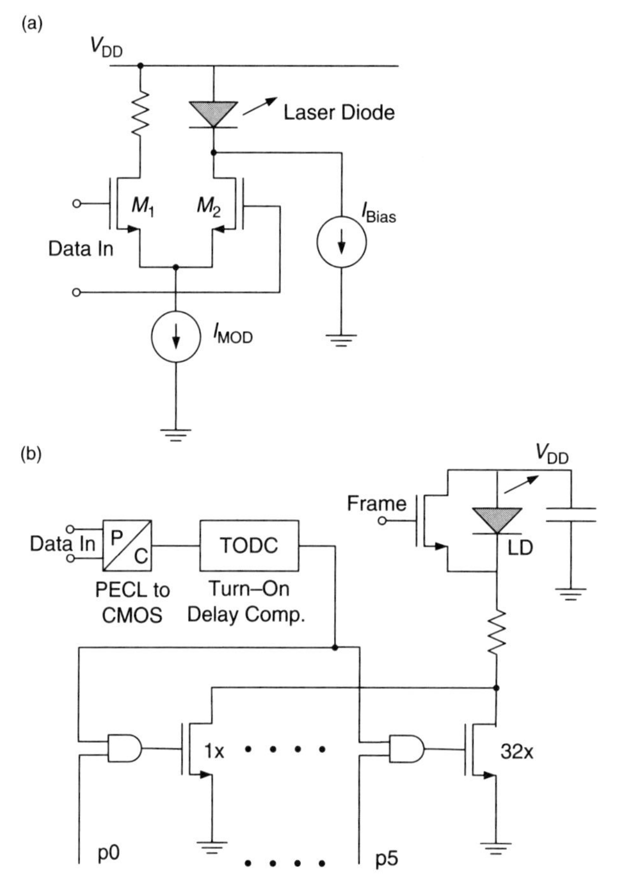


**Figure 19.25** A laser drive: (a) conventional and (b) low power with turn-on delay.[8] (Reprinted with permission from the *Journal of Solid State Circuits* © IEEE, 2000.)

about 1.3 ns, which results in a peak-to-peak jitter increase of 0.3 ns. Therefore, the turn-on delay is precompensated. This is solved as shown in Fig. 19.26(a). The solution is a low-pass filter that extends the data-pulse duration. Recall that the LPF is an integrator, thus, whatever slot of data enters to the laser driver, marked as data in Fig. 19.26(b), it ends by the amount of turn delay after the original data packet ended. The implementation is done by a conditionally loaded inverter. On falling data edges, node X is loaded by the capacitor $C_1$ until the voltage $V_{DEL} - V_{TH}$ is reached. At that point, $M_1$ is turned off, and $M_2$ prevents rising-edge-data transition since it isolates the capacitive load and slowly discharges the capacitor $C_1$. That way the data end time overflows and compensates the laser turn-on delay as shown in Fig. 19.27. The analog node denoted by X is restored to digital levels by the second-stage inverter. The output pulse width in this circuit is increased slightly with the temperature, but due to the small temperature coefficient of 3.2 ps/°C, it is of no practical concern. The $V_{DEL}$ voltage is provided by an external trim potentiometer as shown in Fig. 19.24.

The role of APC is to fix the laser's output power to a desired targeted power level. This is explained elaborately in Sec. 13.1. However, in burst mode, the average power differs from the peak value because the mark density varies from slot to slot depending on the burst activity. Conventional burst APC is shown in Fig. 19.28. The back-facet monitor converts the modulated light to the average power. The peak power is detected by the peak detector, which is fed by a TIA. The peak detector has to be fast and follow bit by bit; thus it consumes a high current. Moreover, the peak detector has a limited hold time; thus, it may result in an error for the next burst. A better burst-mode APC approach is shown in Fig. 19.29, where the APC operates burst by burst rather than bit by bit, thus has no need for fast power-consuming circuits. This design uses the PD parasitic capacitor $C_{PD}$ as a part of an integrator rather than a nuisance in a conventional

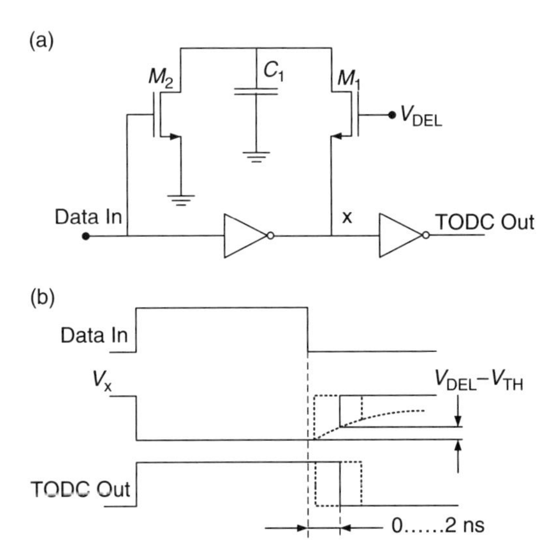


**Figure 19.26** (a) A turn-on delay compensation circuit, (b) corresponding timing.[8] (Reprinted with permission from the *Journal of Solid State Circuits* © IEEE, 2000.)

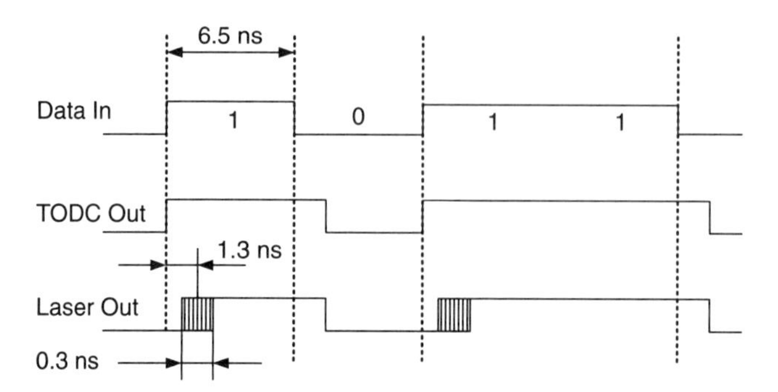


**Figure 19.27** Concept of turn-on delay timing compensation.[8] (Reprinted with permission from the *Journal of Solid State Circuits* © IEEE, 2000.)

approach. APC goal is to minimize the error voltage created between the peak value of the PD current $i_{PD}$ to the reference current $I_{REF}$. During the first data burst, the PD capacitor $C_{PD}$ is precharged to a known voltage $V_{PC}$. Then, during the following data burst, this capacitor is charged by the $i_{PD}$ current from the PD and simultaneously discharged by current pulses generated by the data switch and the reference current source $I_{REF}$. Finally, at the end of the burst, the voltage $v_X$ at the node marked as X is compared with the precharge voltage source $V_{PC}$ by a clocked comparator. Depending on the result, a counter that controls the laser output power is stepped up or down. For instance, when $v_X > V_{PC}$, which indicates that $I_{PD} > I_{REF}$, the counter is stepped down to reduce the laser power. Since the laser power is stored in a counter, it is preserved and not discharged as in a regular sample and hold peak detector. The counter is updated between bursts to prevent a disturbance in the transmitted data. This whole cycle can be described analytically.

The voltage at the node X is described by[8]

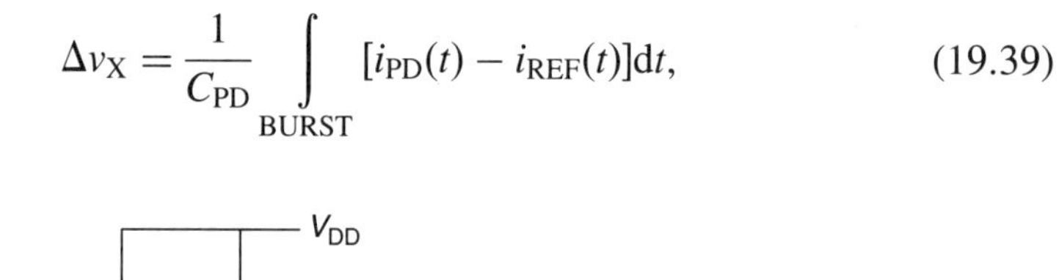

$$\Delta v_X = \frac{1}{C_{PD}} \int_{\text{BURST}} [i_{PD}(t) - i_{REF}(t)]\mathrm{d}t, \qquad (19.39)$$

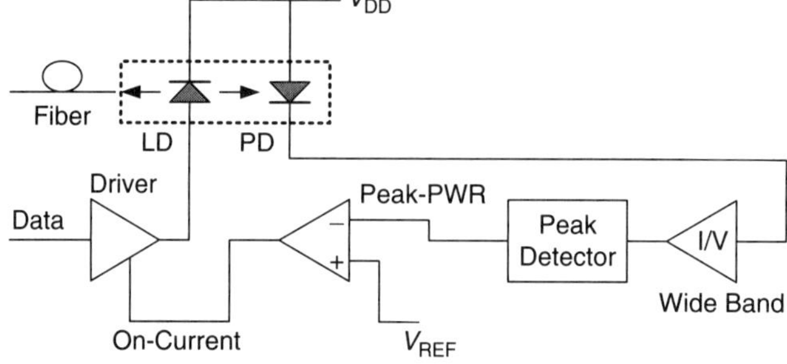


**Figure 19.28** Conventional burst mode power control.[8] (Reprinted with permission from the *Journal of Solid State Circuits* © IEEE, 2000.)

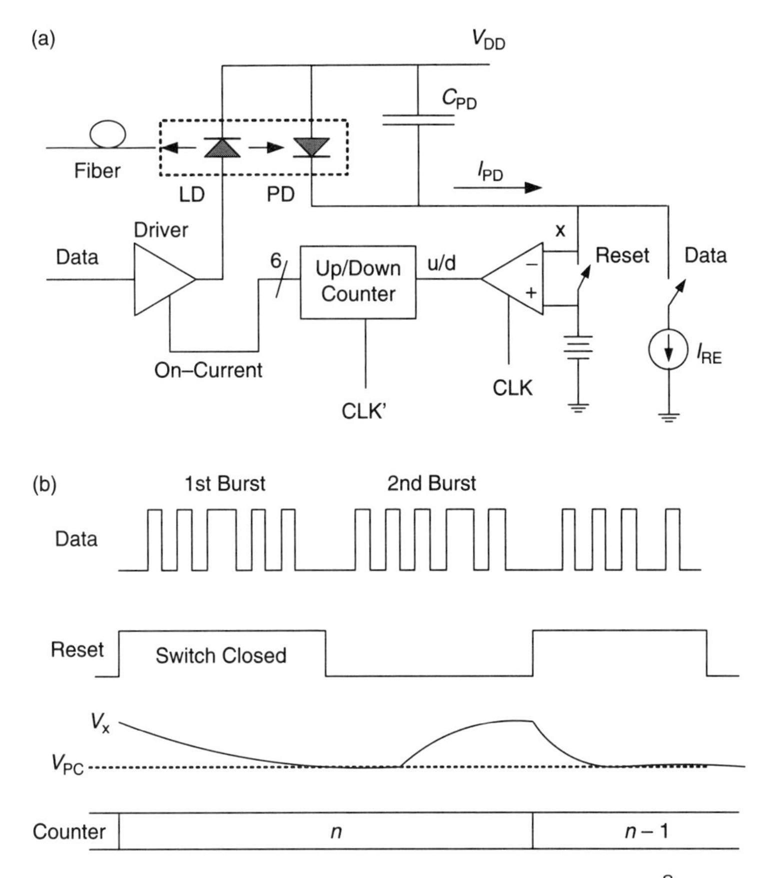


**Figure 19.29** (a) Low-power burst mode and (b) timing diagram.[8] (Reprinted with permission from the *Journal of Solid State Circuits* © IEEE, 2000.)

where $i_{PD}(t)$ is the PD current and $i_{REF}(t)$ is the modulated reference current. Substituting burst signals for current results in

$$\Delta v_X = \frac{1}{C_{PD}}(I_{PD} \times t_{PD} \times n_1 - I_{REF} \times t_{REF} \times n_1), \tag{19.40}$$

where $I_{PD}$ is the PD peak current, $t_{PD}$ and $t_{REF}$ are the pulse widths of the PD and reference currents, respectively, and $n_1$ is the number of 1 bits in the burst. Lastly $t_{PD} = t_{REF}$, which is satisfied due to the turn-on delay compensation results in

$$\Delta v_X = \begin{cases} \text{positive} & I_{PD} > I_{REF}, \\ \text{negative} & I_{PD} < I_{REF}. \end{cases} \tag{19.41}$$

In conclusion, from this analysis the decision regarding the PD peak current is made without ever measuring its value. Moreover, the PD parasitic capacitance value has no meaning or effect on the final outcome; hence it is not critical.

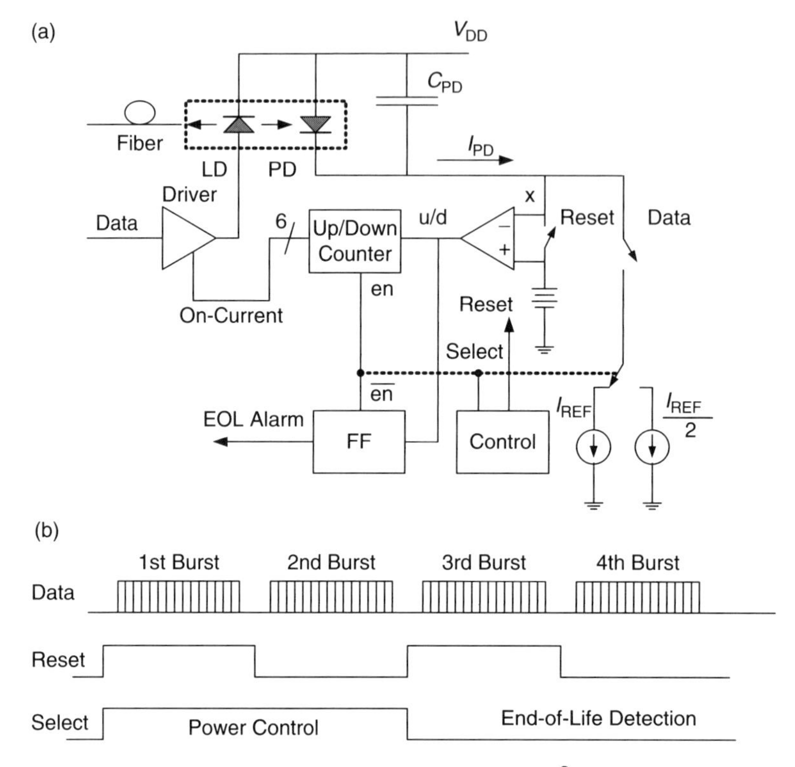


**Figure 19.30** An end-of-life detector and timing diagram.[8] (Reprinted with permission from the *Journal of Solid State Circuits* © IEEE, 2000.)

Delays between the PD and the reference current pulses, e.g., resulting from the turn-on delay compensation, have no impact on the decision either.

Additional circuitry is in the end-of-life (EOL) monitor. As a laser ages, its physical characteristics vary, resulting in a current increase for the same power level. As the laser's demand for current is above the maximum current available from the laser driver, the optical power starts to drop over time. This is observed by the back-facet monitor. If it drops below a certain fraction of the nominal value, which in this case is $1/2$ or $1/3$ of the nominal current, it determines the laser's EOL. In Fig. 19.30(a) such EOL-circuit topology is shown. Further, this circuit is an extension of the one shown in Fig. 19.28. For that purpose, a controller is added, which selects the first two successive bursts for APC and the following two for EOL by switching the current source to be the proper reference for EOL, i.e., $I_{REF}/2$ or $I_{REF}/3$ rather than $I_{REF}$.

## 19.2 Wavelength Lockers and Wavelength Control Loop

In Sec. 2.4, a top-level look into the structure of digital transceiver is provided, where some of its parts are common to tunable laser transponder in Sec. 2.6.

One of the key building blocks of such a transponder is the wavelength locker. The role of this device is to provide a measure of wavelength deviation from the desired wavelength for locking. An analogy to that device is the phase detector in PLL. However, unlike PLL, wavelength locking of the tunable laser is not done with respect to a reference crystal oscillator; thus the wavelength locker is not a phase detector but is more similar to a frequency modulation (FM) discriminator.

The idea behind this is as follows. The laser beam is split into two routes with a given splitting ratio $m{:}n$, for instance 33:67. The weak port of the splitter is directly connected to a reference photo detector (PD). This PD senses all time the laser power without any relation to its wavelength state. The stronger arm of the power splitter is connected to a band-pass filter (BPF) realized by a fiber Bragg grating (FBG). The FBG output is coupled to a second PD. This way, FBG losses are compensated by higher optical power from the splitter. When the laser is tuned to the desired wavelength, the voltage at the PD that is coupled to the FBG is optimal. This voltage is compared with that of the other PD port, which serves as a reference, using an operational differential amplifier. If the laser is at its desired wavelength, the error voltage is zero. When there is a wavelength deviation, then there is an error voltage, which is proportional to the wavelength offset. There are two laser states: the first is when the laser wavelength is above the reference wavelength and the other is when it is below the reference. The reference wavelength is defined by the FBG and set on its slope of FBG wavelength response as shown in Fig. 19.31. To create a voltage polarity detector, this reference wavelength is set at half of the filter slope as shown in Fig. 19.31. Therefore, as the laser wavelength deviates from the desired wavelength, say going into the pass band of the BPF, the error voltage goes to one direction, assume the positive direction since $V_{\text{laser}} > V_{\text{ref}}$. Now when the laser wavelength goes to the FBG stop band, the condition becomes $V_{\text{laser}} < V_{\text{ref}}$; thus, voltage polarity changes as a discriminator. This is because the current in the PD that is connected to the FBG tracks the FBG filter shape factor. Since the BPF filter is symmetrical, both of its roll-off sides serve as a discriminator, and there is another wavelength that can be locked on the same filter. Moreover, if this FBG is periodic, then there are many wavelengths that can be locked depending on filter wavelength repetition known as free spectral range (FSR).[44,58] Additionally, since the reference PD is not sensitive to a specific wavelength, it can be used for the APC power monitor as well.

With this device, one can build a locking circuit using corrective feedback for wavelength locking. The laser pigtail is connected to a directional coupler. The main arm transfers the signal, while the coupled arm delivers a portion of it to the wavelength locker. Its output is connected to an error amplifier and then to an integrator realized by a controller. In Fig. 19.32 wavelength-locking concept is illustrated.[13]

A large effort was invested to miniaturize and integrate wavelength lockers of small sizes with lasers.[11,16,17] For that purpose, an etalon is used rather than an FBG, while splitters made of mirrors replace the planar-lightwave-circuit (PLC) splitter as shown in Fig. 19.32.[13] Such an integrated wavelength locker is shown in Fig. 19.33.

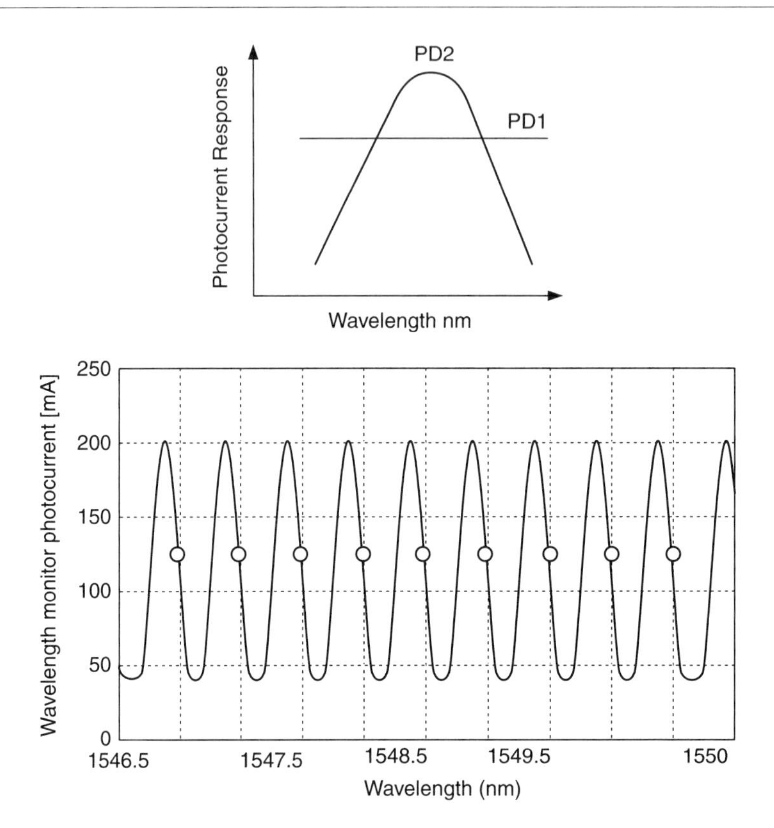


**Figure 19.31** An FBG or etalon is used as wavelength discriminator, resulting in a periodic transfer function that enables to lock various wavelengths per free spectral range (FSR) increments defined by the ITU grid. In this case a 50-GHz grid.[58] (Reprinted with permission from the *Journal of Lightwave Technology* © IEEE, 2004.)

Integrated wavelength lockers are divided into two types.[16] The first is an internal back locker, which is placed at the back of the laser, with the light beam emitted from the back facet that is used as an input for the wavelength locker of the laser. In this structure the incident beam is not collimated.

The second type is an internal front locker, which received the input light from the front facet of the laser. In both cases the reference channel depends only on the output power of the laser, while the etalon channel's signal depends on wavelength. Thus, both the power-monitoring and the wavelength-locking function can be accomplished by using signals that are provided by integrated flip chip photodetectors.

A third option of integrated wavelength monitor is reported in Refs. [44], [46], and [58]. In this case, the wavelength discriminator is based on a prism as a splitter and etalon as the wavelength discriminator. Beam collimation is accomplished by the lens. Thermistors are placed in order to keep a constant temperature, controlled by the thermo-electric cooler (TEC), preventing temperature effects on the wavelength discriminator that may result in locking drift (Fig. 19.34). With tunable laser transmitter building blocks, it is straightforward

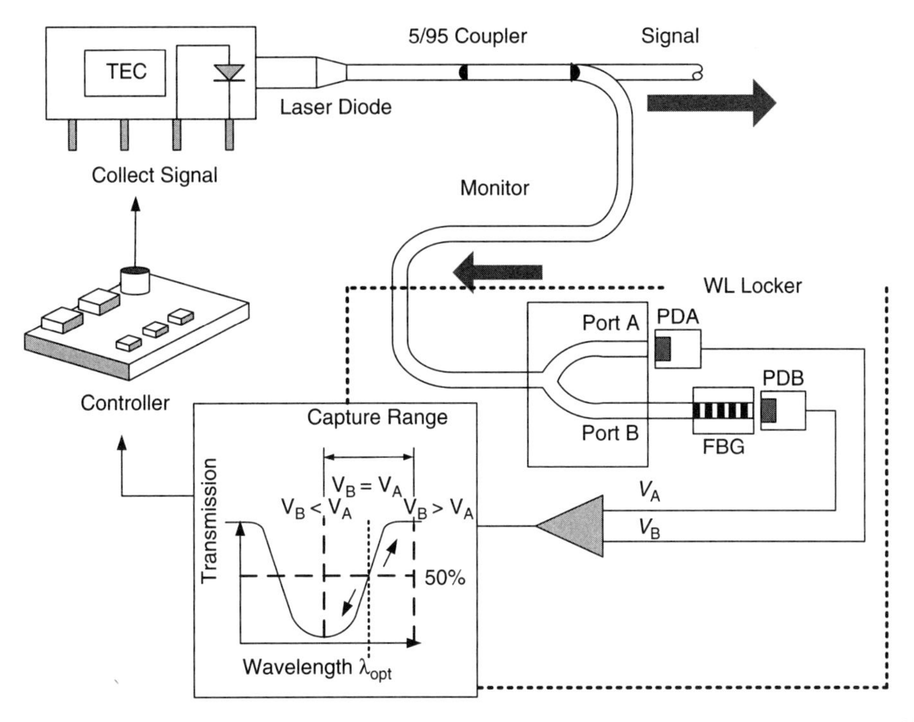


**Figure 19.32** A wavelength locker based on PLC and FBG used in the locking system.[13] (Reprinted with permission from the *Proceedings on Optical Passive Components* © IEEE, 2002.)

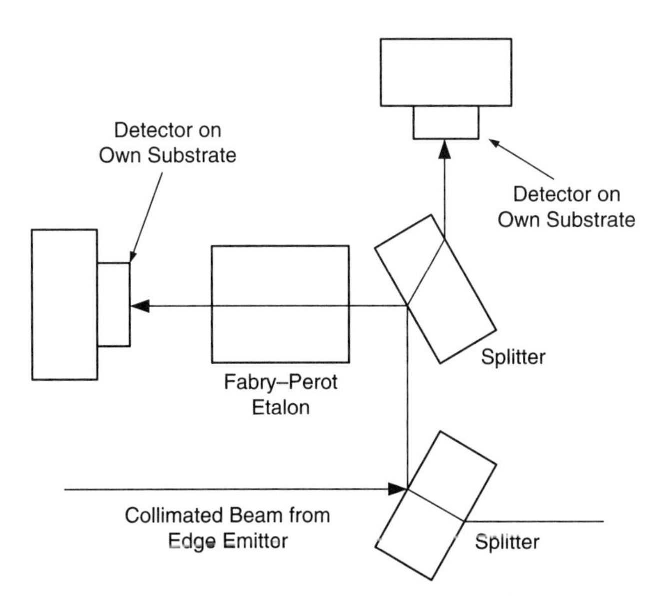


**Figure 19.33** An integrated locker realized by etalon and mirror beamsplitters.[17] (Reprinted with permission from *IEEE LEOS* © IEEE, 2002.)

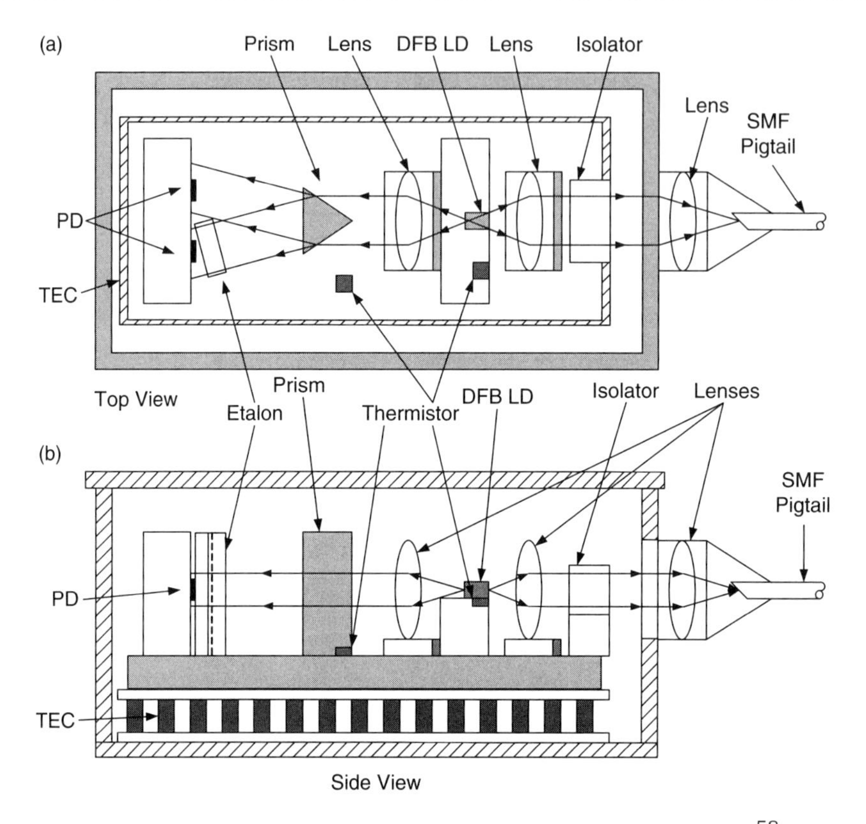


**Figure 19.34** A butterfly laser module with a wavelength locker and TEC.[58] (Reprinted with permission from the *Journal of Lightwave Technology* © IEEE, 2004.)

as to how the block diagram of such a transmitter will look. Further details are provided in Sec. 2.6.

Additional wavelength-locking techniques are not related to the tunable transponder realization that was reviewed here. Those are injection[10,23] and beat frequency locking.[21,22] Injection locking's advantage is the reduction of nonlinear distortion and chirp effects. Injection locking is done when a master laser is used to lock a slave laser. In the second method, lasers are locked to a reference laser. By using the square-law characteristic of the PD, locking is done. Since the PD detects the electrical field, having a PD that receives two fields, one of the reference and the other of the locked laser, would generate a beat current at a frequency equal to the wavelength difference. Thus, it operates as a mixer in an RF application. Therefore, in this case locking is similar to the PLL concept. In Sec. 8.5 a glance into the square-law concept of a PD is provided, where coherent detection is introduced. Further review can be found in the literature.

The next challenge in implementation of tunable wavelength transponders is to have a fast tunable transmitter[9,12,15,59] as well as a stabilizing wavelength.[14] Wavelength stabilization can be done with cooling provisions.[19] Tunable laser's control voltage or current affects cavity resonance. By applying control voltage

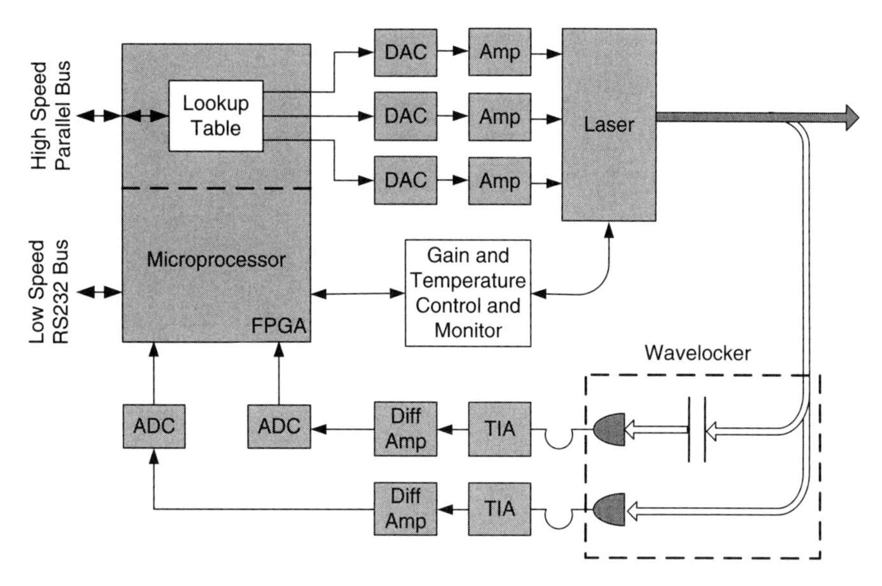


**Figure 19.35** Block diagram of fast tunable laser transponder. Used for 64 ITU channels, 50-GHz grid. Has FP etalon to stabilize the laser frequency. Locking accuracy is $\pm 3$ GHz within 50 ns.[15] (Reprinted with permission from *Photonics Technology Letters* © IEEE, 2004.)

on MEMS mirrors,[20] or more controls in case of the GCSR, which involves microcontroller's algorithms,[14,18] wavelength is changed. Tunable laser's switching between channel 20 to 31 other channels lasts 50 ns for 80% power and an error-free wavelength. Such a fast tunable transponder concept is shown in Fig. 19.35.[15]

Another method for wavelength locking is done using a wavelength selective photodetector (WSP).[50] In this concept, two photodetectors are connected as one chip back to back. One PD is a PIN and the other is a reversed PIN PD known as NIP. The photocurrent is related to the responsivity of each PD. Thus, there is a specific wavelength window, to which both PDs provide the same photocurrent. A TEC loop is locked on the WSP PD as a wavelength detector and thus wavelength stability is reached over temperature. Further elaboration on TEC loops and heat budgets is provided on Chapter 13.

## 19.3 Transceiver Housing TOSA ROSA Structure and Integration

Another design challenge for digital optical transceivers is packaging.[35] Packaging design considerations include housing, miniaturization of optics modules, high-density populated circuit design, thermal conduction, and EMI radiation.[25,32] Digital transceivers with a one-by-nine transmitter optical subassembly and a

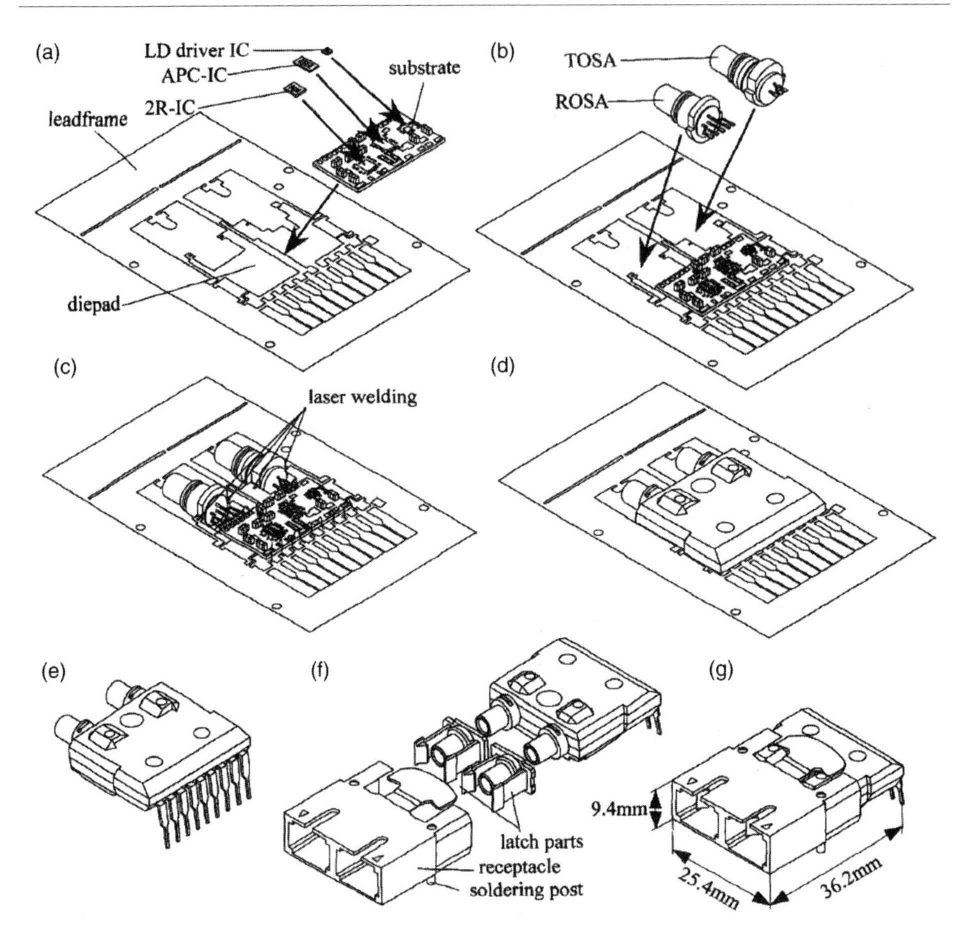


**Figure 19.36** An assembly process of old one-by-nine 1.25-Gb/s transceivers with TOSA ROSA.[30] (Reprinted with permission from Proceedings of *Electronic Components and Technology Conference* © IEEE, 1998.)

one-by-nine optical receiver subassembly (TOSA ROSA) as in Fig. 19.36 were reported in 1996 and 1998.[34,30] Those early transceivers with miniature bulk optics had to support a 9-dB extinction ratio, −19- to −3-dBm receiver power, −11.5- to −3-dBm launch power for a 1.25-GB/s baud rate with jitter more than 240 ns. The IEEE 802.3z published Spring 1998 set the requirement for size reduction and a low operating voltage of 3.3 V.

The evolution of digital transceiver packages started as old "one-by-nine" configuration as, shown in Fig. 19.36,[30,34] or GIBIC to small-size packages of SFF and XFP, while increasing the data rate and adding more diagnostics and controls SFP. The main challenge was in overcoming electrical X-talk between to receiving or transmitting, where receiving was (more analysis about X-talk and shielding theory is given in Chapter 20). This gave a push for the chip design of laser drivers and receivers with larger amount of functionalities. In this way smaller printed-circuit-board assemblies (PCBAs) with a denser population became feasible.

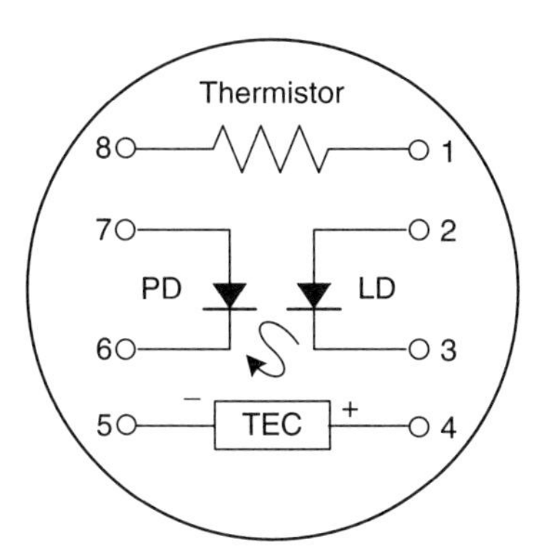


**Figure 19.37** A block diagram and pin assignment of cooled TOSA.[25] (Reprinted with permission from Proceedings of *Electronic Components and Technology Conference* © IEEE, 2005.)

TOSA ROSA modules comprise TO in which there is a laser diode and back facet monitor for APC.[30] Further advanced TOSA designs contain a thermistor and TEC in order to save space at the SFP PCBA and to simplify mechanical integration (Fig. 19.37 and Fig. 19.38).[25] TOSA ROSA replaced butterfly packages and have flexible PCBs rather than leads with data rates that could be increased up to 10 GB/s[28] (Figs. 19.45 and 19.46).

Modern SFP TOSA are hermetically sealed, laser welded, and contain TEC for wavelength stabilization to satisfy the 100-GHz ITU grid in DWDM.[25] Wavelength accuracies in such designs utilizing TEC reaches $\pm 0.01$ nm over a temperature range between $-5$ and 70°C. ROSA units contain PIN APD as well as PIN TIA in order to increase sensitivity. SFP modules reported in Ref. [31] accomplish a high optical signal-to-noise ratio (OSNR) with minimum receiver sensitivity of $-32$ dBm for back to back, $-32.6$ dBm after transmission over 80 km of standard SMF, and $-28$ dBm for OSNR level of 16 db at a BER of $1 \times 10^{-10}$.

Having such a high performance in small SFP package is a challenge. Modern SFP PCBAs support both the receiver and transmitter. Receiver IC supports the receiver portion by a limiting amplifier, postamplifier, following the PIN TIA in order to remove amplitude noise. An additional APD control bias is implemented by a dc/dc step-up power supply. The bias current of the APD is low, not more than 1 mA. Special optical-power overdrive-protection circuit is realized by a current mirror circuit that monitors the APD optical current. As the received power reaches too high, it turns off the APD.

On the transmit side there are two control loops at least: an APC loop to lock the laser power, and a TEC control with a pulse-width modulation (PWM) TEC driver, which is reviewed in Chapter 13. It is a feedback system, which is solved by linearization for optimization at the quiescent point.[49,50] The TEC loop is monitored by a CPU that updates the TEC driver's reference temperature for locking. This CPU interfaces the analog ports of the APD bias and the TEC reference temperature by DACs.

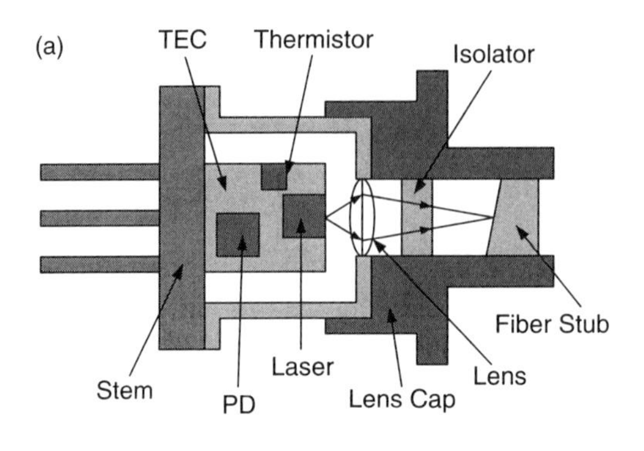


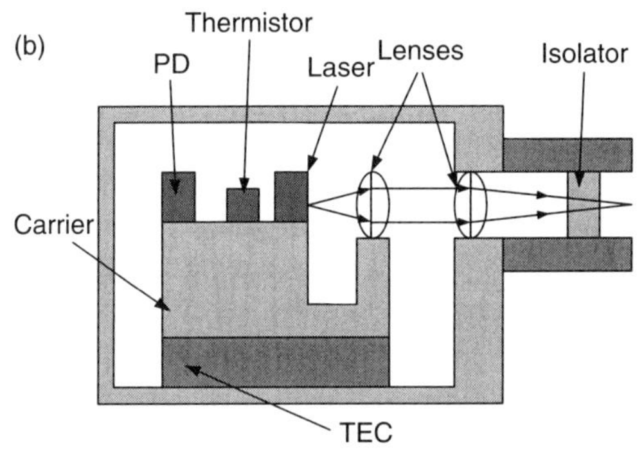


**Figure 19.38** Cross section of cooled TOSA against legacy Butterfly package, TOSA contains TEC thermistor and back facet monitor PD for APC loop. Such design operates at OC48.[25] (Reprinted with permission from Proceedings of *Electronic Components and Technology Conference* © IEEE, 2005.)

Early SFF transceiver designs used SMD potentiometers for calibrating the APC loop and ER.[33] Either they were replaced by digital potentiometer ICs or they became part of the transceiver IC. In this manner, SFP transceivers or other advanced transceivers are programmed using $I^2C$ BUS. Such a transceiver is illustrated in Fig. 19.39.

The above-mentioned small design raised two issues referring to packaging. The first concern is shielding (more about its design approach is given in Chapter 20) and the second concern is heat load and heat conduction.[25]

In order to prevent X-talk and desensitization of the receiver path by the transmitter, shielding and proper layout should be done. Receiver/transmitter separation should be made. Inner shielding is generally used in SFF modules. Older designs had two PCBAs: one for receiving and one for transmitting.[33] By having a dividing wall and flexible PCB for the TOSA ROSA, and improves sensitivity reduces X-talk. Such a design approach is shown in Fig. 19.40.

Additional challenges are to design light weight housing while reducing costs and light weight design effort. New epoxy composites using, a liquid crystal polymer (LCP) reinforced with long carbon fibers (LCF) are bringing shielding effectiveness (SE) to over 20 dB with a weight savings of 25%.

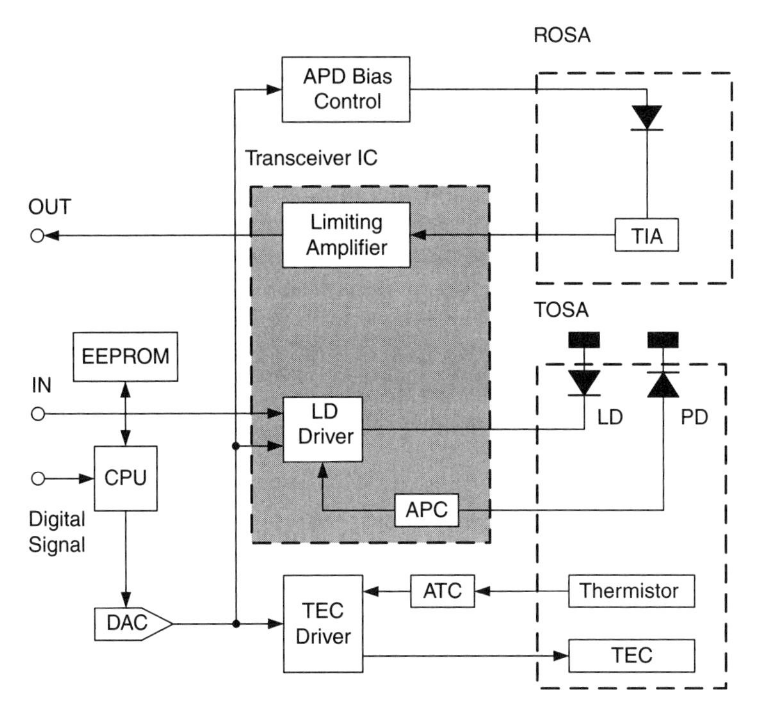


**Figure 19.39** A block diagram of a modern SFP transceiver with APC TEC and APD.[25] (Reprinted with permission from Proceedings of *IEEE Electronic Components and Technology Conference* © IEEE, 2005.)

Other techniques involve a woven continuous-carbon fiber (CCF) epoxy composite. By weaving the CCF into a balanced twill structure (BTS) with excellent conductive networks, SE is significantly increased while still having a low weight percentage of the carbon fiber. Thus, the SE of such a housing is higher than the CCF by 20 dB. Such packages are used for one-by-nine OC48 transceivers.[32]

Thermal load due to the power dissipation of laser driver, laser diode, and receiver section is a design challenge as well. The main heat sources in SFP designs are the TOSA (0.2 W), even cooled, the ROSA (0.26 W), and the transceiver IC (0.2 W). This brings the heat budget between 0.66 and 0.7 W. The analysis is based on dividing the SFP into three zones: the front receptacle, the center, and the rear areas.[25] The front contains the TOSA ROSA modules, the center is the PCBA with the transceiver IC, and the rear has no heat source. The heat inclines toward the front. Therefore, to overcome the heat, there is a need to efficiently conduct and transfer it to the outer package. It is better to have a smaller inclination of the temperature on the outer package.

Thermal-wise TOSA should be isolated from other building blocks that generate heat to prevent heat combination. TOSA with TEC is a temperature sensitive device. The SFP packaging concept is similar to SFF. It has an outer shielding metal and inner die-cast metal plate, generally made of aluminum or zinc alloys. These

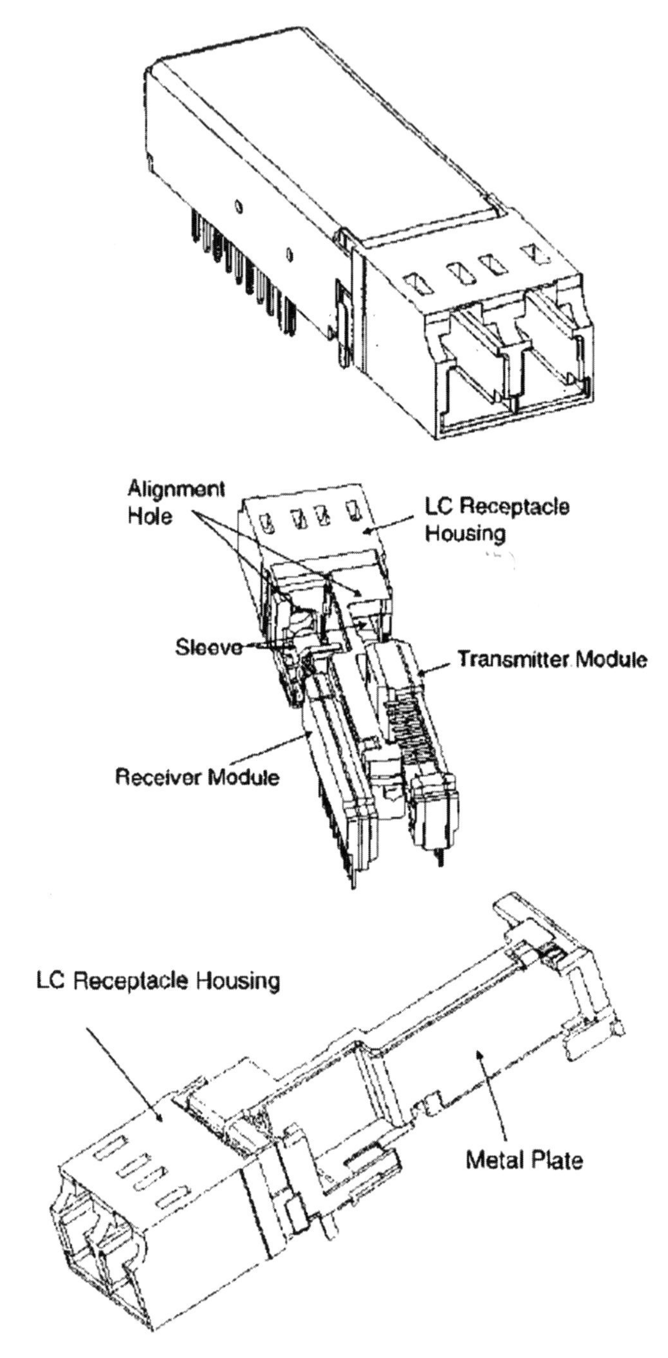


**Figure 19.40** An SFF two by nine assembly process showing shielding between the receiver and transmitter.[33] (Reprinted with permission from *Electronic Letters* © IEEE, 1999.)

alloys were chosen to radiate the thermal energy. However, the coefficient of thermal conductivity (CTC) of the material has been restricted to less than 200 mW/mK. Thus, the plate-working method is adopted where it serves as local heatsink at critical heat zones. The penalty is cost, but the gain is a higher CTC of about 350 mW/mK for a Cu alloy. The main idea is demonstrated in Fig. 19.41, where the radiating Cu plate is thermally connected to the transceiver IC via a thermal sheet and extends to the rear, making thermal contact to the push-out spring. Since the rear is at a lower temperature, a heat tunnel is generated, and heat is transferred to the back.

DWDM SFP is the most intelligent device among the SFP modules. Thermally, the TOSA is the most sensitive block in this design because the TEC is inside it. Thus, its mechanical design is most difficult. As mentioned, it radiates heat upward and downward, but the other significant problem is EMI RFI shielding.

As will be explained in Chapter 20, air slots in housings have the potential for emissions and EMI RFI susceptibility. There are two big openings at the TOSA ROSA receptacles, where optical fiber is inserted in a conventional SFP as

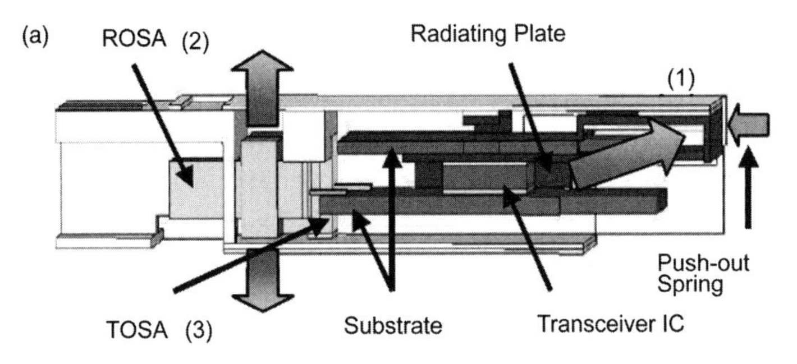


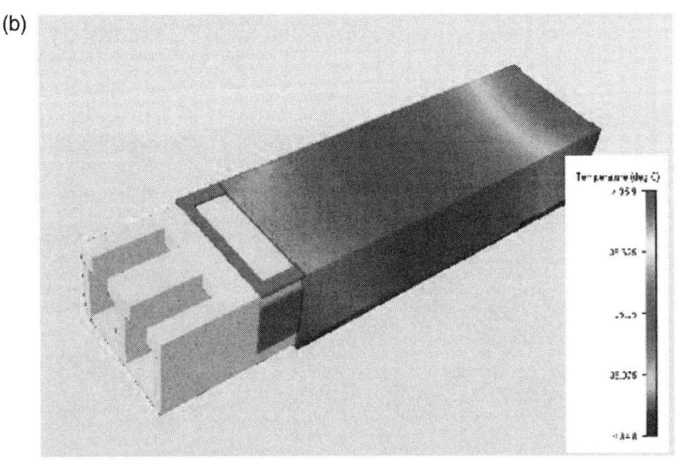


**Figure 19.41** (a) A sectional view of DWDM SFP showing: (1) a backward heat transfer, (2) an upward transfer, and (3) a downward transfer, and (b) thermal-heat simulation of the transceiver IC only.[25] (Reprinted with permission from Proceedings of *Electronic Components and Technology Conference* © IEEE, 2005.)

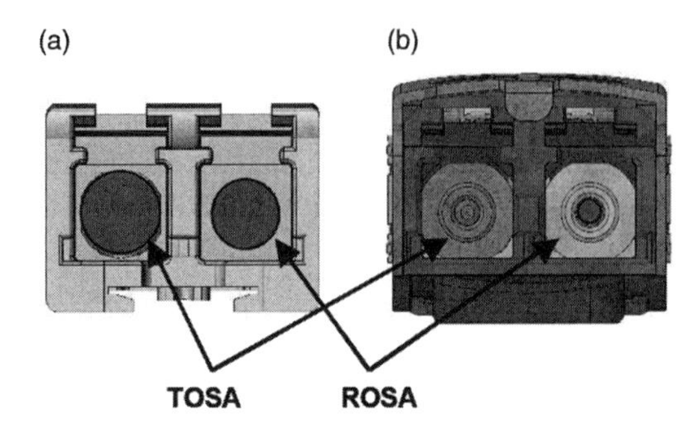


**Figure 19.42** The front view of an SFP: (a) standard and (b) DWDM SFP, which solves the EMI problem.[25] (Reprinted with permission from proceedings of *Electronic Components and Technology Conference* © IEEE, 2005.)

shown in Fig. 19.42. Moreover, the gap between the SFP housing and the inner cage is covered by EMI RFI spring fingers. This way the design combats emissions and X-talk between SFP modules in a router or a switch. It is known that the smaller these slots are, the better it is since slot dimensions define the cutoff frequency for emissions. To make these openings even smaller, TOSA ROSA are separated into some portion and the protruded metal part attached to the receptacle, which is called the sleeve and is isolated from the other metal parts of the TOSA package. The sleeve is tied to a stable potential like the chassis ground through the housing and the fingers of the DWDM SFP. The ROSA is designed with the same concept. As a result, only small holes for the optical beam exist as shown in Fig. 19.42(b).

All of those design concepts make the DWDM-SFP a state-of-the-art product. Such modules operate with a high wavelength accuracy according to the ITU grid, and have an APC loop, a burst-mode AGC, and an APC control as well as CDR, a

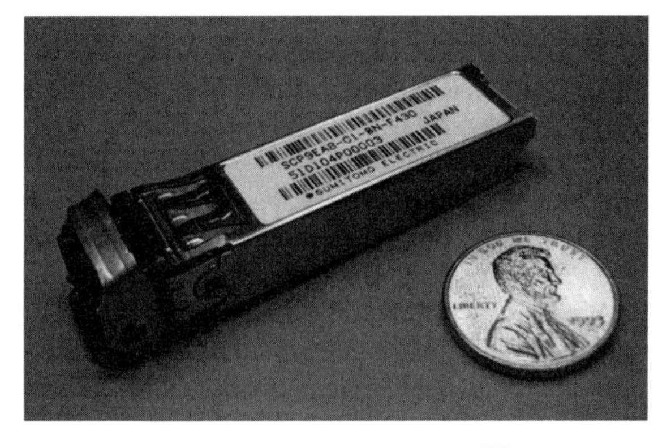

**Figure 19.43** A sate-of-the-art DWDM-SFP transceiver.[25] (Reprinted with permission from Proceedings of *Electronic Components and Technology Conference* © IEEE, 2005.)

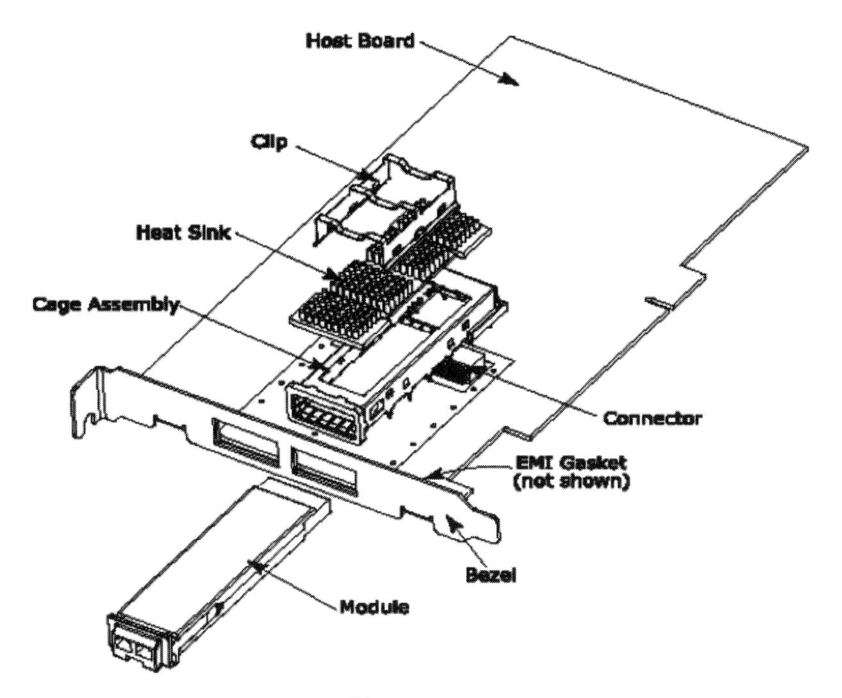


**Figure 19.44** An XFP motherboard.[62] (Reprinted with permission from *MSA* © 2002.)

full diagnostic of receiver optical power, ER calibration and status, temperature and current monitoring, EOL, and all other features mentioned in previous sections. Such a DWDM is illustrated in Fig. 19.43.

It is clear that in high-data-rate transport, the key is miniaturization of transceiver modules to reduce the parasitics and have proper shielding. The faster the data transport becomes, the more radiation it has in higher frequencies. Moreover it picks up noise when receiving and degrades performance. Thus, having sealed, tight packaging and narrow slots is crucial. In Fig. 19.44, a 10-GB/s XFP module

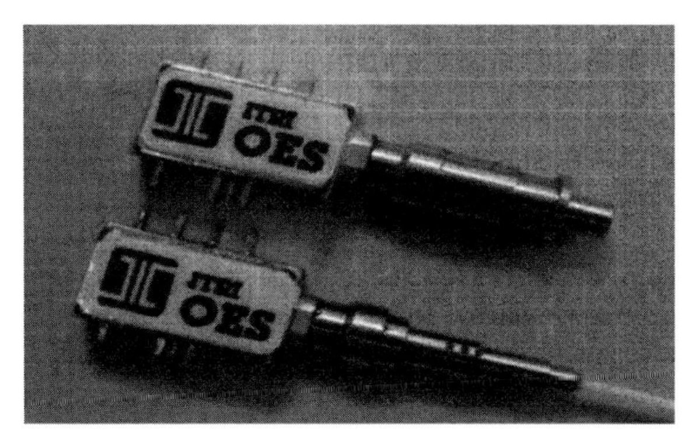


**Figure 19.45** A 10-GB/s 1310-nm mini-flat transmitter.[27] (Reprinted with permission from *IEEE LEOS* © 2004.)

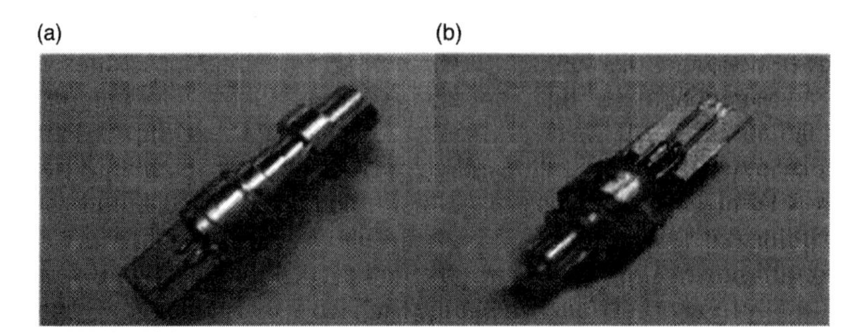


**Figure 19.46** A 10-GB/s (a) TOSA, (b) ROSA, both with interface PCB and transmission lines.[27] (Reprinted with permission from *IEEE LEOS* © 2004.)

insertion to the motherboard is shown. The heat sink and EMI cage are also shown.[62]

Other strategies of SFP modules and FTTx use PLC technology and have bidirectional (BiDi) transceivers.[24,47,52] For high speeds such as 10 GB/s, TOSA ROSA are integrated with a silicon optical bench, thus, it can solve

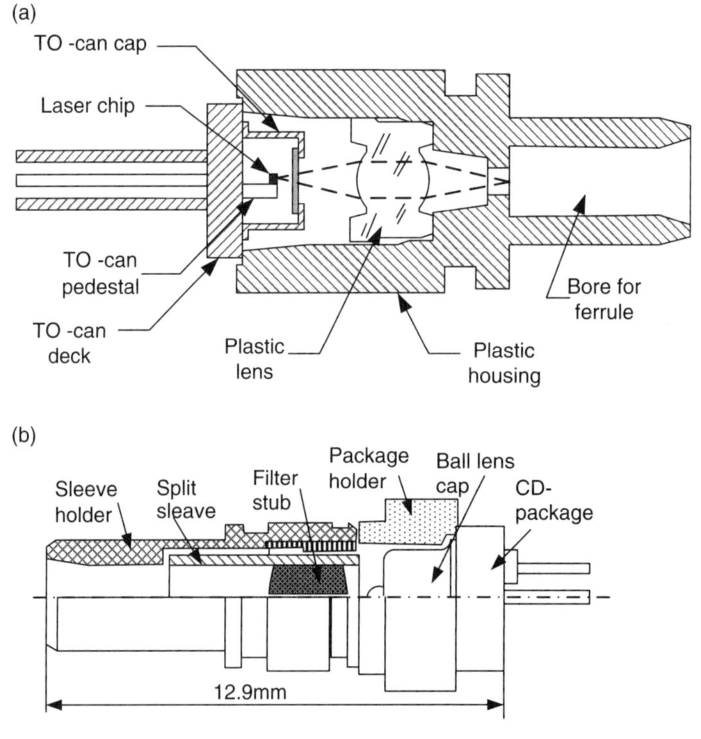


**Figure 19.47** A TOSA cross section showing the methods of TOSA designs.[30,61] (Reprinted with permission from proceedings of *Electronic Components and Technology Conference* © IEEE, 1998; and *Transactions on Components Packaging and Manufacturing* © IEEE, 1997.)

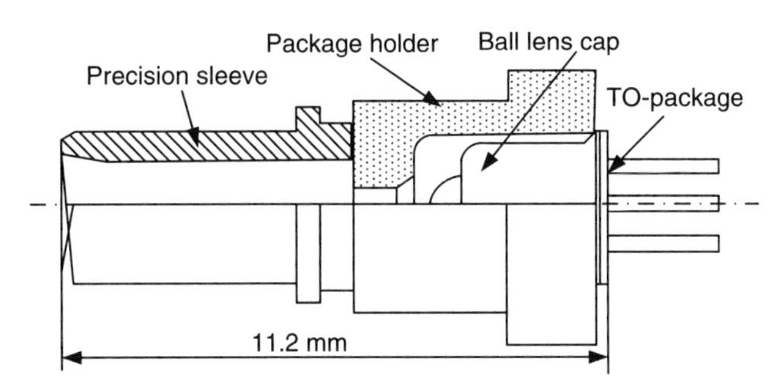


**Figure 19.48** A ROSA cross section showing inner design.[30] (Reprinted with permission from Proceedings of *Electronic Components and Technology Conference* © IEEE, 1998.)

pattern-dependent jitter with PRBS $2^{31-1}$ since parasitic inductance between optics and electronics are reduced.[27] This whole module becomes an SMT flat transmitter as can be seen in Figs. 19.45 and 19.46. The designs of TOSA ROSA are shown in Figs. 19.47 and 19.48.

Digital optical transceivers are becoming highly integrated. As was previously reviewed, there are two contradicting design constraints: functionality increase, such as wavelength stability in DWDM, power locking, AGC, and miniaturization according to MSA requirements for which a large effort is invested.

In FTTx design, mixed-signal systems on package (SOP) is designed.[45] These designs have integrated optical devices, such as a PLCs or Y-branch-polymer couplers rather than bulk optics; digital-transceiver ICs for OC48, including CDRs, laser drivers, TIAs: MUX DMUX that converts NRZ to parallel data and vice versa, and CMOS RF ICs. For example, a 10-GB/s transponder architecture is similar to the one that appears in Sec. 2.6, where in the receiver-path section data-dependent jitter equalizers are added.[39]

Design attention is now focused on the development of new packaging technologies, such as silicon bench.[29] In gigabit ethernet transceivers, CMOS integrated detectors are reported.[26] On the other hand, the number of functionalities and diagnostics is increased.

## 19.4 Main Points of this Chapter

1. BMRs require a training series known as preamble in order to prepare an AGC decision reference voltage and set CDR locking.

2. The preamble length defines the accuracy of the decision reference voltage.

3. The reference voltage setting is made by the feedforward method, where the received signal is split into two routes: one to the comparator and the other for reference.

4. Reference voltage varies between bursts because the link budget varies between various ONUs.

5. Bursts are isolated by guard time, during which period the previous reference voltage is reset.

6. Decision-reference-voltage statistics are Gaussian and affected by the preamble length. The longer the preamble, the more accurate is the reference, thus BER is improved.

7. Burst efficiency is determined by the ratio between the preamble length and the burst duration. The remaining time is dedicated to actual data, which is the payload of the burst. Thus, the aim is to have a relatively short preamble.

8. The sensitivity penalty is determined by the reduction in BER performance versus preamble length. The longer the preamble, the lower the sensitivity penalty, but the burst efficiency is reduced.

9. Generally, receivers are ac coupled. This affects the NRZ code because of the large dc energy it carries. Hence, the eye suffers from dc wander. As a result, the eye is shifted with respect to the decision reference voltage. Thus, the BER performance is affected. As a solution, balanced codes are used such as 8B10B or 16B18B with no dc.

10. Optimization of the burst-mode link consists of three stages: (1) laser-tuning time or APC-locking time; preamble-settling time, where the decision reference is set and the CDR is locked; and payload-time, where data is transferred.

11. The burst-mode receiver has an automatic gain control applied on the trans impedance amplifier. The method is to vary the transimpedance resistance value, thus controlling the current-to-voltage gain versus the optical-level input.

12. A modern BMR has two reference decision circuits that operate in parallel. When the preamble is received, a sample and hold circuit senses the bit train and converts it to digital by an ACD. Digital signal processing estimates the preamble average voltage and sets it to the comparator input as threshold until BER is optimal in a so-called successive approximation manner. There are two comparators in such a receiver. While one is readily waiting for the other burst, the first is adjusted and receives the current burst. Selecting between the two is done by a selector switch.

13. The AGC in the BMR has to handle the loud–soft ER ratio, which is the two extremes of the receive scenarios. The first is a low-fiber loss near ONU with a high optical power and the other is a low-optical power and a low SNR for the same ER. The delta between the powers of the

1 level is called the loud–soft ratio. If this delta (in decibels) is large, it is called the high ratio, indicating a large dynamic-range requirement.

14. The AGC system is a limiting process that prevents a large swing at the decision circuit.

15. The problem with a log amplifier AGC is the amplitude reduction of the strong swing for a high-power signal and the enhancement of the low-SNR low-power signal swing. The TIA gain is not a uniform slope throughout the input current range, thus the loud–soft ratio is affected.

16. The TIA feedback-impedance control saves the loud–soft ratio since the TIA gain slope is uniform and constant throughout the relevant optical power. It varies at the instant of the optical-power change.

17. The TIA-AGC concept is realized by a FET in shunt to the TIA feedback resistance. Since the FET is voltage controlled, resistor-transimpedance gain is continuously controlled.

18. The FET gain voltage is controlled by a GCC, which sets the gate voltage. As a result, since the FET is shunted to a resistor, when there is a high current the source potential is going down with respect to the gate, $V_{gs}$ increases and resistance decreases.

19. The GCC is controlled by a BLD. This circuit holds the bottom level and is reset at the guard time for the next burst.

20. The BMR tester generates the preamble series to check the AGC. It measures BER and the optimal preamble for minimum BER.

21. The burst-mode transmitter comprises the laser and laser driver that is optimized for a bias current and modulation to reach the desired ER. The APC is fed by the back-facet monitor to measure the average power, power level, and pattern sequence. The APC data feeds the laser driver. Thus, the APC can manage power in bursts. In this way, the transmitter preserves the optical level and the ER.

22. When a laser is biased below threshold, it consumes time at turn on. The preamble for bias is required in order to define the APC average power. This process starts during guard time to accelerate laser turn on at the correct bias, and, therefore, at the correct optical power.

23. The longer the transmission preamble series during guard time, the faster the turn-on convergence.

24. Since the laser has a turn-on delay, it might have a lag with respect to modulation data.

25. To solve the laser turn-on delay, data is delayed as well in order to align the data start with the laser turn on and the termination of data with laser turn off.

26. When laser is aging, its optical power at a given bias current declines.

27. The end-of-life circuit concept is to compare the sensed current from the back-facet monitor to a reference current, below which, an alarm flag indicates EOL.

28. The wavelength locker is not a phase detector but a frequency discriminator.

29. The wavelength locker has two routes. The first is a reference beam routed directly to a photo detector. The second route transfers the other portion of the split beam through an optical band-pass filter to a second detector. In case of wavelength-deviation, power is affected by the optical BPF. This results in an error voltage between the reference and the wavelength dependent voltage. This error voltage from the wavelength locker is used to tune the laser to the proper wavelength and lock it.

30. The tunable laser transmitter is called a transponder, since it has more functions than a regular transmitter.

31. A digital transmitter has optical modules called TOSA for transmitting and ROSA for receiving.

32. DWDM SFP contains a TOSA TEC in it for better temperature control.

33. Shielding and small slots in the housing are essential to prevent emissions, and a carbon epoxy housing is used to save weight.

## References

1. Eldering, C.A., "Theoretical determination of sensitivity penalty for burst mode fiber optic receivers." *IEEE Journal of Lightwave Technology*, Vol. 1 No. 12, pp. 2145–2149 (1993).
2. Le, Q., S.G. Lee, Y.H. Oh, H.Y. Kang, and T.H. Yoo, "A burst mode receiver for 1.25 Gb/S ethernet PON with AGC and internally created reset signal." *IEEE Journal of Solid State Circuits*. Vol. 39 No. 12, pp. 2379–2388 (2004).
3. Brigati, S., P. Colombara, L. D'Ascoli, U. Gatti, T. Kerekes, P. Malcovati, and A. Profumo, "A SiGe BiCMOS burst-mode 155 Mb/S receiver for PON." ACCO Microelectronica S.r.l., Pavia, Italy; *IEEE Journal of Solid-State Circuit*. Vol. 37 No. 7, pp. 887–894 (2002).

4. Rotem, E., and D. Sadot, "Performance analysis of AC-coupled burst-mode receiver for fiber-optic burst switching networks." *IEEE Transactions on Communications*, Vol. 53 No. 5, pp. 899–904 (2005).
5. Qiu, X.Z., F. Fredricx, and P. Vetter, "Burst mode transmission in PON access systems." *Proceedings of the European Conference on Networks and Optical Communication* (1995).
6. Bauwelinck, J., D. Verhulst, P. Ossieur, X.Z. Qiu, J. Vandewege and B. De Vos, "DC coupled burst mode transmitter for 1.25 GBit/S upstream PON." *IEE Electronics Letters*, Vol. 40 No. 8, pp. 501–502 (2004).
7. Bauwelinck, J., Y. Martens, P. Ossieur, K. Noldus, X.Z. Qiu, J. Vandewege, E. Gilon, and A. Ingrassia, "Generic and intelligent CMOS 155 MB/S burst mode laser driver chip design and performance." *Solid-State Circuits Conference ESSCIRC*, pp. 495–498 (2002).
8. Sackinger, E., Y. Ota, T.J. Gabara, and W.C. Fischer, "A 15-mW, 155-Mb/s CMOS burst mode laser driver with automatic power control and end of life detection." *IEEE Journal on Solid State Circuits*, Vol. 35 No. 2, pp. 269–275 (2000).
9. Bhardwaj, A., J. Gripp, J.E. Simsarian, and M. Zirngibl, "Demonstration of stable wavelength switching on a fast tunable laser transmitter." *IEEE Photonics Technology Letters*, Vol. 15 No. 7, pp. 1014–1016 (2003).
10. Renaud, C.C., M. Duser, C.F.C. Silva, B. Puttnam, T. Lovell, P. Bayvel, and A.J. Seeds, "Nanosecond channel-switching exact optical frequency synthesizer using an optical injection phase locked loop (OIPLL)." *IEEE Photonics Technology Letters*, Vol. 16 No. 3, pp. 903–905 (2004).
11. Ackerman, D.A., K.M. Paget, L.F. Schneemeyer, L.J.P. Ketelsen, F.W. Warning, O. Sjolund, J.E. Graebner, A. Kanan, V.R. Raju, L.E. Eng, E.D. Schaeffer, and P. Van Emmerik, "Low cost athermal wavelength locker integrated in a temperature tuned single frequency laser package." *IEEE Journal of Lightwave Technology*, Vol. 22 No. 1, pp. 166–171 (2004).
12. Simsarian, J.E., A. Bhardwaj, J. Gripp, K. Sherman, Y. Su, C. Webb, L. Zhang, and M. Zirngibl, "Fast switching characteristic of widely tunable laser transmitter." *IEEE Photonics Technology Letters*, Vol. 15 No. 8, pp. 1038–1040 (2003).
13. Ichioka, M., J. Ichikawa, Y. Kinpara, T. Sakai, H. Oguri, and K. Kubodera, "Athermal wavelength lockers using fiber bragg gratings." *IEEE, Lasers and Electro-Optics*, Vol. 1, pp. 208–212 (2001).
14. Sarlet, G., G. Morthier, and R. Baets, "Wavelength and mode stabilization of widely tunable SG-DBR and SSG-DBR lasers." *IEEE Photonics Technology Letters*, Vol. 11 No. 11, pp. 1351–1353 (1999).
15. Simsarian, J.E., and L. Zhang, "Wavelength locking a fast switching tunable laser." *IEEE Photonics Technology Letters*, Vol. 16 No. 7, pp. 1745–1747 (2004).
16. Han, H., B. Hammond, R. Boye, B. Su, J. Mathews, B. TeKolste, A. Cruz, D. Knight, B. Padget, and D. Aichele, "Development of internal wavelength lockers for tunable laser applications." *IEEE Electronic Components and Technology Conference*, pp. 805–808 (2003).

17. Hammond, B., B. Su, J. Mathews, E. Chen, and E. Schwartz, "Integrated wavelength locker for tunable laser applications." Lasers and Electro-Optics Society. *The 15th Annual Meeting of the IEEE*, Paper WO1. pp. 479–480 (2002).
18. Brodberg, B., P.J. Rigole, S. Nilsson, M. Renlund, L. Andersson, and M. Renlund, "Widely Tunable Semiconductor Lasers." *Optical Fiber Communication Conference, 1999, and the International Conference on Integrated Optics and Optical Fiber Communication. Proceedings of OFC/IOOC '99. Technical Digest*, Vol. 2, pp. 137–139 (1999).
19. Kawano, R., N. Yamanaka, S. Yasukawa, and K. Okazaki, "Compact 10 GB/S optical transceiver/receiver set with novel cooling structure for 640 GB/S WDM interconnection." *IEE Electronics Letters*, Vol. 34 No. 19, pp. 1876–1877 (1998).
20. Pezeshki, B., E. Vail, J. Kubicky, G. Yoffe, S. Zou, J. Heanue, P. Epp, S. Rishton, D. Ton, B. Faraji, M. Emanuel, X. Hong, M. Sherback, V. Agarwal, C. Chipman, and T. Razazan, "20 mW widely tunable laser mouble using DFB array and MEMs selection." Santur Corp Paper. *IEEE Photonics Technology Letters*, Vol. 14 No. 10, pp. 1457–1459 (2002).
21. Ahn, Y.S., S.Y. Kim, S.H. Han, J.S. Lee, S.S. Lee, and W.S. Seo, "Bidirectional DWDM transmission using a beat—frequency—locking method." *IEEE Photonics Technology Letters*, Vol. 13 No. 8, pp. 899–901 (2001).
22. Kim, S.Y., S.W. Huh, Y.S. Hurh, K.H. Seo, and J.S. Lee, "Accurate alignment of ultradense WDM channels using the beat frequency locking method." *IEEE Journal of Lightwave Technology*, Vol. 21 No. 11, pp. 2891–2894 (2003).
23. Chang, C.H., L. Chrostowski, and C.J.C. Hasnain, "Injection locking of VCSELs." *IEEE Journal of Selected Topics in Quantum Electronics*, Vol. 9 No. 5, pp. 1386–1393 (2003).
24. Kimura, H., T. Yoshida, and K. Kumozaki. "Compact PLC based optical transceiver module with automatic tunable filter for multi-rate applications." *IEE Electronics Letters*, Vol. 39 No. 18, pp. 1319–1321 (2003).
25. Ichino, M., S. Yoshikawa, H. Oomori, Y. Maeda, N. Nishiyama, T. Takayama, T. Mizue, I. Tounai, and M. Nishie, "Small form factor pluggable optical transceiver module with extremely low power consumption for dense wavelength division multiplexing applications." *IEEE Electronic Components and Technology Conference*, pp. 1044–1049 (2005).
26. Bockstaele, R., and P. De Pauw, "Fabrication of a low – cost module for gigabit Ethernet transceivers." Report by Phocon and Sipex, Belgium. *Proceedings of SPIE*, Vol. 4942, pp. 44–53 (2003).
27. Liu, Y.S., M.T. Chu, Y. Chu, and S.G. Lee, "High speed packaging technologies for 10 gigabit Ethernet applications." *Proceedings of the IEEE LEOS*, pp. 511–512 (2004).
28. Hogan, W.K., R.K. Wolf, A. Shukla, and P. Deane, "Low cost optical subassemblies for metro access applications." *IEEE Electronic Components and Technology Conference*, Vol. 1, pp. 203–207 (2004).

29. Iwase, M., T. Nomura, A. Izawa, H. Mori, S. Tamura, T. Shirai, and T. Kamiya, "Single mode fiber MT-RJ SFF transceiver module using optical subassembly with a new shielded silicon optical bench." *IEEE Transactions on Advanced Packaging*, Vol. 24 No. 4, pp. 419–428 (2001).
30. Go, H., N. Nishiyama, E. Tsumura, Y. Fujimura, H. Nakanishi, I. Tonai, and M. Nishie, "Low power consumption molded gigabit fiberoptic transceiver module." *IEEE Electronic Components and Technology Conference*, pp. 1192–1198 (1998).
31. Priyadarshi, S., S.Z. Zhang, M. Tokhmakhian, H. Yang, and N. Margalit, "The first hot pluggable 2.5 GB/S DWDM, transceiver in an SFP form factor." *IEEE Optical Communications*, pp. S29–S31 (2005).
32. Wu, T.L., W.S. Jou, W.C. Hung, C.H. Lee, C.W. Lin, and W.H. Cheng, "High shielding effectiveness plastic package for 2.5GB/S optical transceiver module." *IEEE Electronics Letters*, Vol. 40, No. 4, pp. 260–262 February (2004).
33. Kurashima, H., H. Go, E. Tsumura, T. Inujima, N. Nishiyama, Y. Mikamura, and M. Nishie, "Manufacturing technique of SFF transceiver." *IEEE Electronic Components and Technology Conference*, pp. 554–559 (1999).
34. Sumitomo Electric, "Technical specification for small form factor pluggable (SFP)." Specification: TS-S02D085B September (2002).
35. Robinson, S.D., F.C. Anigbo, and G.J. Shevchuk, "Advanced modeling technique for optical transceivers." *IEEE Electronic Components and Technology Conference*, pp. 1109–1115 (1996).
36. Small Form Factor Pluggable (SFP) Multi Source Agreement (MAS), inf-8074: http://www.schelto.com/SFP/.
37. Gigabit Interface Converter (GBIC) MSA, http://www.schelto.com/t11_2/GBIC/gbic.htm.
38. DWDM Pluggable Transceiver MSA http://www.hotplugdwdm.org/.
39. Buckwalter, J., and A. Hajimiri, "A 10 GB/S data dependent jitter equalizer." *IEEE Custom Integrated Circuit Conference*, pp. 39–42, 3-5-1 to 3-5-4 (2004).
40. Su, C., L.K. Chen, and K.W. Cheung, "Theory of burst-mode receiver and its applications in optical multiaccess networks." *IEEE Journal of Lightwave Technology*, Vol. 15 No. 4, pp. 590–606 (1997).
41. Ossieur, P., X.Z. Qiu, J. Bauwelinck, and J. Vandewege, "Sensitivity penalty calculation for burst-mode receivers using avalanche photodiodes." *IEEE Journal of Lightwave Technology*, Vol. 21 No. 11, pp. 2565–2575 (2003).
42. Lee, J., J. Park, J. Limy, K. Jung, S. Limy, Y. Choy, S.M. Park, and J. Choi, "A 1.25-Gb/s CMOS burst-mode optical receiver with automatic level restoration for PON applications." *The 47th IEEE International Midwest Symposium on Circuits and Systems*. pp. I-137–I-140 (2004).
43. Nakamura, M., N. Ishihara, and Y. Akazawa, "A 156 MB/s CMOS optical receiver for burst mode transmission." *IEEE Journal of Solid State Circuits*, Vol. 33 No. 8, pp. 1179–1187 (1998).

44. Nasu, H., T. Takagi, T. Shinagawa, M. Oike, T. Nomura, and A. Kasukawa, "Highly stable and reliable wavelength monitor integrated laser module." *IEEE Journal of Lightwave Technology*, Vol. 22 No. 5, pp. 1344–1350 (2004).
45. Iyer, M.K., P.V. Ramana, K. Sudharsanam, C.J. Leo, M. Sivakumar, B.L.S. Pong, and X. Ling, "Design and development of optoelectronic mixed signal system on package (SOP)." *IEEE Transactions on Advanced Packaging*, Vol. 27 No. 2, pp. 278–285 (2004).
46. Nasu, H., T. Mukaihara, T. Shinagawa, T. Takagi, M. Oike, T. Nomura, and A. Kasukawa, "Wavelength monitor integrated laser modules for 25 GHz spacing tunable applications." *IEEE Journal of Selected Topics in Quantum Electronics*, Vol. 11 No. 1, pp. 157–164 (2005).
47. Hashimoto, T., A. Kanda, R. Kasahara, I. Ogawa, Y. Shuto, M. Yanagisawa, A. Ohki, S. Mino, M. Ishii, Y. Suzuki, R. Nagasc, and T. Kitagawa, "A bidirectional single fiber 1.25 GB/S optical transceiver module with SFP package using PLC." *IEEE Electronic Components and Technology Conference*, pp. 279–283 (2003).
48. Qui, X.Z., P. Ossieur, J. Bauwelinck, Y. Yi, D. Verhulst, J. Vandewege, B.D. Vos, and P. Solina, "Development of GPON upstream physical media dependent prototypes." *IEEE Journal of Lightwave Technology*, Vol. 22 No. 11, pp. 2498–2508 (2004).
49. Ackerman, D.A., K.A. Paget, L.F. Schneemeyer, L.J.P. Ketelsen, O. Sjolund, J.E. Graebner, A. Kanan, F.W. Warning, V.R. Raju, L. Eng, E.D. Schaeffer, and P.V. Emmerik, "Low cost athermal wavelength locker integrated in a temperature tuned laser package." *Proceedings of OFC*, PD32-1–PD32-3 (2003).
50. Colace, L., G. Masini, and G. Assanto, "Wavelength stabilizer for telecommunication lasers: design and optimization." *IEEE Journal of Lightwave Technology*, Vol. 21 No. 8, pp. 1749–1757 (2003).
51. Faucher, J., M. Mony, and D.V. Plant, "Test setup for optical burst mode receivers." *IEEE Lightwave Technologies in Instrumentation and Measurement Conference*, Palisades, New York, pp. 123–128 (2004).
52. Okano, H., A. Hiruta, H. Komano, T. Nishio, R. Takahashi, T. Kanaya, and S. Tsuji, "Hybrid integrated optical WDM transceiver module for FTTH systems." Hitachi U.D.C. 621.372.88.029.72.049.779 (1998).
53. Sackinger, E., Y. Ota, T. J. Gabara, and W. C. Fischer, "Low power CMOS burst mode laser driver for full service access network application." *IEEE CLEO*, pp. 519–520 (1999).
54. Sackinger, E., Y. Ota, T.J. Gabara, and W.C. Fischer, "15 mW 155 Mb/s CMOS burst mode laser driver with automatic power control and end of life detection." *IEEE International Solid State Circuits Conference*, pp. 386–387 (1999).
55. Meerschman, B., Y.C. Yi, P. Ossieur, D. Verhulst, J. Bauwelinck, X.Z. Qiu, and J. Vandewege, "Burst bit error rate calculation for GPON system." *IEEE/LEOS Proceedings Symposium*, pp. 165–168 (2003).

56. Nakamura, M., N. Ishihara, Y. Akazawa, and H. Kimura, "An instantaneous response CMOS optical receiver IC with wide dynamic range and extremely high sensitivity using feed forward auto bias adjustment." *IEEE Journal of Solid State Circuits*, Vol. 30 No. 9, pp. 991–997 (1995).
57. Yamashita, S., S. Ide, K. Mori, A. Hayakawa, N. Ueno, and K. Tanaka, "Novel cell AGC technique for burst mode CMOS preamplifier with wide dynamic range and high sensitivity for ATM—PON system." *IEEE Journal of Solid State Circuits*, Vol. 37 No. 7, pp. 881–886 (2002).
58. Nasu, H., T. Takagi, T. Shinagawa, M. Oike, T. Nomura, and A. Kasukawa, "A highly stable and reliable wavelength monitor integrated laser module design." *IEEE Journal of Lightwave Technology*, Vol. 22 No. 5, pp. 1344–1351 (2004).
59. Brillant, I., "OPEN widely tunable laser technology." White Paper, GWS Photonics Publication (2004).
60. "Gigabit-capable passive optical networks (GPON): physical media dependent (PMD) layer specification." ITU-T Recommendation G.984.2.
61. Cohen, M.S., G.W. Johnson, J.M. Trewhella, D.L. Lacey, M.M. Oprysko, D.L. Karst, S.M. DeFost, W.K. Hogan, M.D. Peterson, and J.A. Weirick, "Low-cost fabrication of optical subassemblies." *IEEE Transactions on Components, Packaging, and Manufacturing Technology:Pt. B*, Vol. 20 No. 3, pp. 256–263 (1997).
62. XFP, "10 Gigabit Small Form Factor Pluggable Module," MSA Rev 092 (2002).
63. Sadot, D., and A. Mekkittikul, "4 Tb/s ATM interconnection through optical WDM networks." *Fiber and Integrated Optics*, Vol. 17 No. 2, pp. 95–112 (1998).
64. Han, S., and M.S. Lee, "Burst-mode penalty of AC-coupled optical receivers optimized for 8B/10B line code." *IEEE Photonics Technology Letters*, Vol. 16 No. 7, pp. 1724–1726 (2004).

Part 6

# Integration and Testing

# Chapter 20

# Cross-Talk Isolation

## 20.1 Introduction

One of the most difficult problems in a bidirectional communications system is coexistence, that is, desensitized by an a very sensitive receiver that is adjacent noisy transmitter. We may ask how much the receiver sensitivity is degraded due to the emitted noise that leaks from the transmitter into the receiver. From a different perspective, the problem is to define the reduction of a receiver's "noise-floor" performance at its threshold input as a consequence of the transmitter noise leakage.

In a narrow-band full duplex communications system, in which the transmitting and receiving channels do not overlap in band, the problem is solved by using a front-end receiving/transmitting duplexer filter, as demonstrated in Fig. 20.1. The requirements of the transmitting portion of the RF duplexer are to attenuate any kind of white noise, or other noise, emitted from the transmitter into the receiving front-end low-noise amplifier (LNA). The transmitter noise composed from a wideband with spurious and discrete intermodulation (IMD). Hence, the requirements from the transmitter-duplexer portion are to attenuate the transmitter out-of-band noise, spurious and harmonic IMD, which can leak into the receiver. The receiver portion of the duplexer filter attenuates the transmitter signal power, transmitter in-band noise, and prevents the transmitter power and noise from going into the receiver's front end, thus preventing the compression of the receiver. Leakage of any transmitter signal into the receiver section may prevent proper signal reception.

The similar concept of an RF front-end (RFFE) duplexer is used when community access television (CATV) channels and direct broadcast satellite (DBS) channels are fed from the same photodetector (PD). The PD is connected to the CATV RF chain via a low-pass filter (LPF) with a sharp elliptic response that rejects the DBS band, which occupies the 950–2100 MHz. The DBS section is isolated from the CATV band by an HPF with an ecliptic response that rejects the 50–870 MHz band filter. Design methods are given by Zverev.[19]

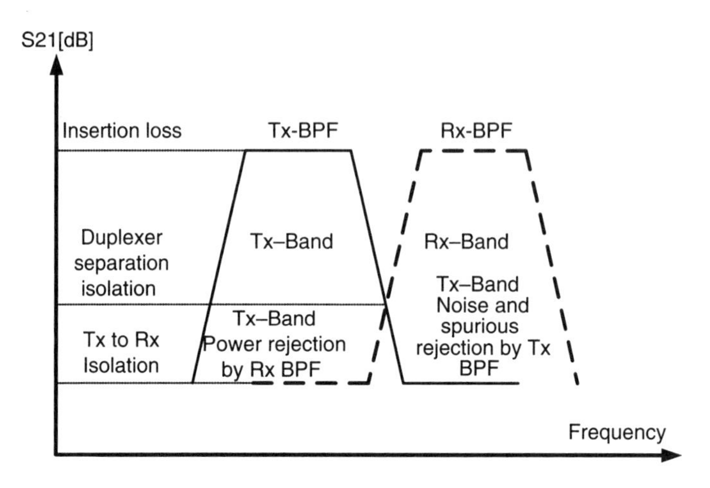


**Figure 20.1** An RFFE duplexer.

## 20.2 Desensitization in Wideband Systems with Overlapping Rx/Tx BW

In wideband systems such as digital- and analog-integrated modules, desensitization or X-talk is more severe and more difficult to isolate. Unlike other, narrowband systems, as was described previously, a CATV video receiver channel and the digital transmission bands overlap. The digital transmission carries nonreturn-to-zero (NRZ) pseudo-random bit sequence (PRBS), such as $2^7 - 1$ and a high-data rate such as OC12 or OC24. The video channel's bandwidth (BW) is 50–870 MHz. The spectral characteristic of an NRZ is sinc($x$) or $\sin(x)/x$, as depicted in Fig. 20.2, and its spectral power is given by $[\text{sinc}(x)]^2$, which is measured by a spectrum analyzer. The spectral picture is the Fourier transform of the NRZ PRBS train. The NRZ code represents a wideband data signal with BW occupation from the dc to its first null, which is the clock (CLK) frequency. For instance, the main lobe null of OC12 is approximately 617.76 MHz. The noise-energy-density distribution depends on the length of the PRBS series. For a short series such as $2^7 - 1$, there are 127 spectral lines in the main sinc($x$) lobe with a frequency spacing of 617.76 MHz/127 or 4.86425 MHz. For a longer series such as $2^{23} - 1$, there are 8,388,607 spectral lines in the main sinc($x$) lobe with a frequency spacing of 617.76 MHz/8388607 or 73.6 Hz.

However, the overall energy is the sum of all the spectral energy lines and is the same power for any rate. Therefore, PRBS $2^{23} - 1$, we may see less noise receiver on the noise floor due to X-talk, while PRBS $2^7 - 1$ will be harder to overcome.

A long PRBS interference has a greater spread over the spectrum with much less energy per spectral component compared to a shorter PRBS series. Thus, for a long PRBS series, spectral lines will hardly pop out of the noise floor of the CATV receiver if the isolation between receiver/transmitter is reasonable. In conclusion, in a longer PRBS series, there are many more dc components

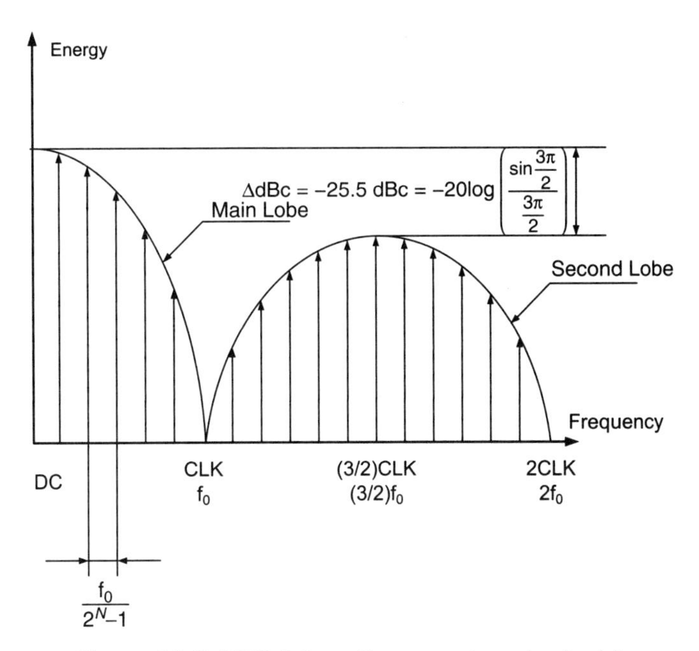


**Figure 20.2** NRZ data pattern spectrum is sinc(x).

compared to a shorter PRBS series. As a consequence, it is harder to defeat desensitization or X-talk in a shorter PRBS series.

The CATV BW is from 50 to 870 MHz. Hence, the frequency band above the CLK rate (OC12, 617.76 MHz) is occupied by the second lobe of the sinc($x$), which is also an interfering noise. The second lobe has a lower power but the coupling mechanism improves with frequency. Therefore, we may detect an interfering noise at a $3/2$ CLK frequency, which is the peak of the second lobe. In OC12, this peak would be at 926.64 MHz, but noise would start to show earlier due to an increase in coupling, at around 800 MHz for OC12. The reason for noise rising at 800 MHz is that it reaches the area of the second lobe energy peak. In such a case, with two overlapping channels, digital transport and CATV, and noise characteristics on the other hand (frequency-dependent noise or discrete interference), the isolation problem between the receiver and the transmitter becomes much harder to solve.

As a rule of thumb, the transmitter noise must be suppressed 10 dB below the noise floor of the CATV video-receiver section. In this way, it will keep a reasonable SNR in the CATV receiver section, without any significant effect on its spurious noise-floor performance. Otherwise, if the interference is too high, the input signal-to-noise and distortion (SINAD) ratios at the back-end receiver, a TV, becomes lower, affecting the picture quality. Thus, X-talk interference should be within the noise floor or lower than any composite second-order (CSO) or composite triple-beat (CTB) product defined by the SCTE standard.

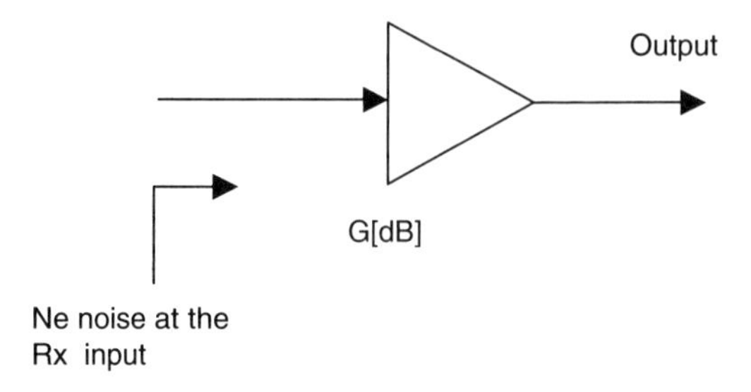


**Figure 20.3** Noise amplification.

Assume the following design problem and constraints analysis. The noise figure of a typical FTTx CATV receiver is generally at the level of 5 dB. The noise density at the receiver input is evaluated by using the relations:

$$Ne[\text{dBm/Hz}] = NF + 10\log KT + 30, \tag{20.1}$$

$$10\log KT + 30 = -174\,\text{dBm/Hz}, \tag{20.2}$$

$$Ne = -169\,\text{dBm/Hz}. \tag{20.3}$$

Assume that the receiver's net gain is $G \approx 17$ dB; then the noise density at the receiver output, as appears in Fig. 20.3, will be

$$Ne + G \approx -152\,\text{dBm/Hz}. \tag{20.4}$$

The result of $Ne + G$ is the noise floor of the receiver measured using the spectrum analyzer. In conclusion, the requirement is that the transmitter-noise leakage at the CATV receiver output will be below $-162$ dBm/Hz. This value satisfies the design goal of a standard system's CNR. The NTSC CATV channel has a 4-MHz BW. Hence, the total noise power at the FTTx CATV receiver output is

$$-152\ \text{dBm/Hz} + 10\log 4\,\text{MHz} = -86\,\text{dBm}. \tag{20.5}$$

If the requirement for the CATV signal tone is 14 dBmV or $-34.7$ dBm, then the CNR is 51.3 dB.

Hence, the X-talk interferences should be 60 dB below the carrier in order to maintain a decent CNR and SINAD, or comply with the minimum CSO and CTB spec of 60 dBc. In other words, the absolute level of the X-talk interferences should be below $-95$ dBm with respect to the signal. That means a high transmitter-receiver HPF-isolation response is needed, as will be explained later on. There is a need to remember that in this case, the transmitter noise is not a white Gaussian noise. It is a colored discrete noise with a frequency-dependency nature of sinc($x$), and the power leakage to the transmitter is proportional to $[\text{sinc}(f)]^2 \times \text{HPF}(f)$, where HPF($f$) is the X-talk frequency response of the isolation between the digital transmitter section or receiver section and the analog section, and $f$ represents frequency.

## 20.3 Wideband PRBS NRZ Interference Analysis

Wideband interference is created by a current or voltage waveform that has a sharp and fast rise and fall times, and a short level dwell time. Any sequential waveform can be converted by a Fourier transform into its spectral components, which are sine and cosine harmonics at different amplitudes. To be accurate, any sequence can be described by a Fourier series. To examine the coexistence of the digital channel with an analog CATV channel, the discussion is on the square wave.

Figure 20.4 shows a square wave NRZ data versus CLK, where PW is the pulse-width, $T_0$ is the pulse interval (CLK frequency), and $A$ is the amplitude. The envelope of the interference is given by

$$a = 2AT_0 \operatorname{sinc}(\pi f T_0), \tag{20.6}$$

where the frequency or the data rate is given by $f_0 = 1/T_0$. The area $AT_0$ represents the pulse energy.

The Fourier transform of the NRZ code is used to calculate each harmonic level and is given by:[1]

$$A_n = A\frac{1}{\text{PRBS}}\operatorname{sinc}\left(n\pi f\frac{T_0}{\text{PRBS}}\right). \tag{20.7}$$

For the CLK signal, the Fourier series is

$$C_n = A\frac{\text{PW}}{T_0}\operatorname{sinc}\left(n\pi f\frac{\text{PW}}{T_0}\right). \tag{20.8}$$

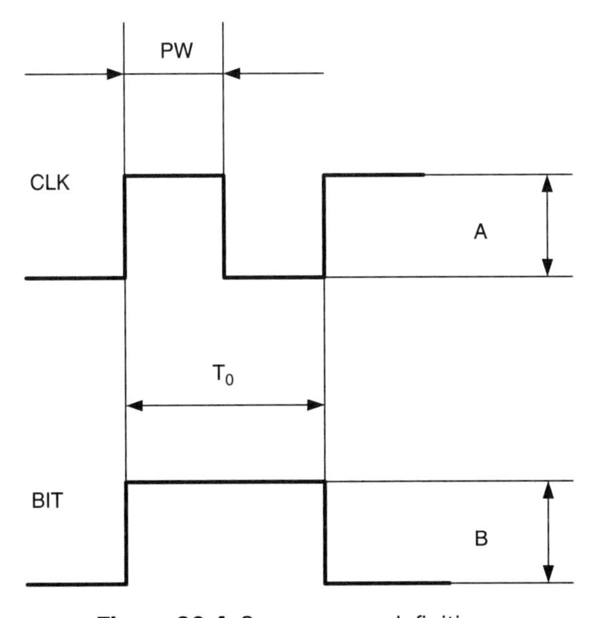


**Figure 20.4** Square wave definitions.

For a 50% duty cycle CLK, Eq. (20.8) becomes

$$C_n = A\,\mathrm{sinc}\left(n\pi f\,\tfrac{1}{2}\right). \tag{20.9}$$

The relation between the CLK and PRBS harmonic-frequency increments observed in the spectrum analyzer is given by

$$\Delta f = \frac{f_0}{\mathrm{PRBS}} = \frac{1}{T_0(\mathrm{PRBS})} = \frac{f_0}{2^N - 1}, \tag{20.10}$$

where $N$ is the number of the PRBS generator shift-register stages in its maximal length, for example PRBS of $2^7 - 1$, $2^{23} - 1$ etc.

This means that for series with $N$ states, the shift register has $2^N - 1$ states. This is equal to the division of the CLK frequency by $2^N - 1$ until the last bit is out. In fact, the shift register can be observed as a frequency divider. Thus, there are $2^N - 1$ spectral lines in the NRZ Fourier transform. Since the power is proportional to the square of the voltage amplitude $A$, the power of each spectral line is given by

$$P_n = \frac{A^2}{Z_0}\frac{\mathrm{PRBS}+1}{\mathrm{PRBS}^2}\,\mathrm{sinc}\left(n\pi f\,\frac{T_0}{\mathrm{PRBS}}\right)^2. \tag{20.11}$$

The power of each spectral line picture seen on the spectrum-analyzer screen is given by[1]

$$P_n = \frac{A^2}{Z_0}\frac{\mathrm{PRBS}+1}{\mathrm{PRBS}^2}\mathrm{sinc}\left(n\pi f\,\frac{T_0}{\mathrm{PRBS}}\right)^2 \sum_{n_{n\neq 0}=-\infty}^{\infty} \delta\left(f - \frac{n}{\mathrm{PRBS}}\cdot f_0\right) + \frac{A^2}{Z_0}\frac{1}{\mathrm{PRBS}^2}\,\delta(f). \tag{20.12}$$

However, since the system is causal, $n > 0$, there are not any frequencies below zero. The dc marker of the spectrum analyzer is $f = 0$. In conclusion, the strength of the interference is directly related to the length of the PRBS series. The number of harmonics is directly related to the PRBS series length. The discussion here refers to a theoretical case of a square wave; however, in real time, the wave has a trapezoidal shape with a finite rise and fall times. In that case, the Fourier transform envelope of the pulse is given by

$$b = 2AT_0\,\mathrm{sinc}(\pi f T_0)\cdot \mathrm{sinc}(\pi f t_r). \tag{20.13}$$

The signal envelope in this case behaves as described in Fig. 20.5. Several conclusions can be drawn from Fig. 20.5 and Eq. (20.13). The rise and fall time of the data pattern affects the roll-off of the interference. A fast rise time $t$ would have a

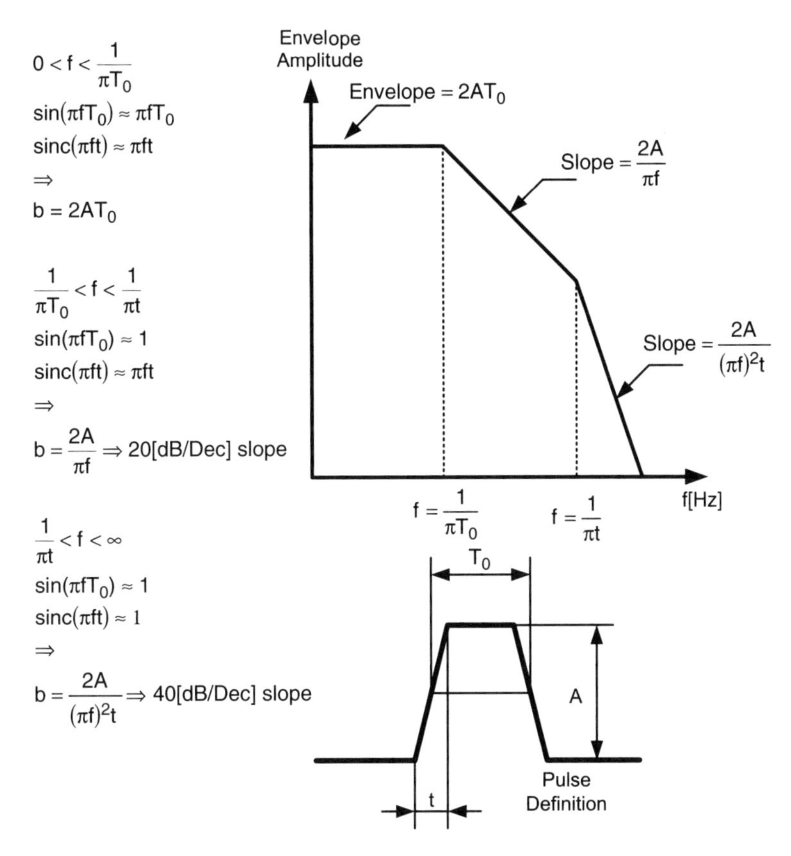


**Figure 20.5** Trapezoid pulse and its Fourier-transform envelope.

higher frequency 2nd pole $2A/t$ in the envelope of the interference. Moreover, it can be said that the main lobe's BW has a flat response until its first pole. This fact simplifies the calculation of the X-talk isolation requirements.

For a 600 mV OC12 (617.76 MHz) data stream, PRBS $2^7 - 1$ and the rise and fall times of 50 ps at the transmitter's input (BERT, bit error rate tester, output to the transmitter), the following signal parameters are obtained: the first pole frequency (Fig. 20.5) is 199 MHz. This pole frequency is of the OC12 CLK pulse width. It represents the main lobe BW peak energy. The second pole frequency is at 6.37 GHz, which is the rise and fall times of 50 ps of the BERT, as shown in Fig. 20.5.

Hence, the NRZ data's main lobe energy of a digital system with $Z_0 = 50\ \Omega$ and 600 mV amplitude, evaluated as per Eq. (20.12) and Fig. 20.5, is $V^2/Z_0$ and is given by

$$10\log\left[0.6^2 \cdot \frac{1}{(2^7 - 1)}\right] - 10\log 50 + 30 = -12.5\,\text{dBm} = 36\,\text{dBmV}. \quad (20.14)$$

As a result, the isolation requirement to meet the leakage of $-95$ dBm in an integrated triplexer (ITR) is $12.5\,\text{dBm}\ -(-95\,\text{dBm}) = 82.5$ to $90\,\text{dB}$

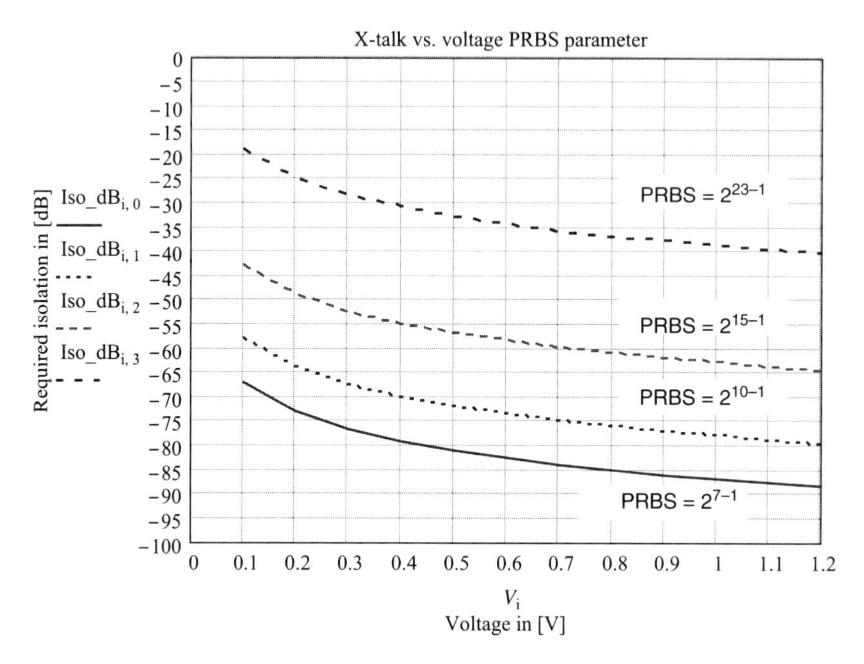


**Figure 20.6** The X-talk isolation requirement as a function of data voltage, while PRBS is the parameter. System noise floor: −85 dBm at 4 MHz and the discrete X-talk limit is 10-dB below the receiver's noise floor; i.e., maximum leakage level is −95 dBm.

approximately. Since X-talk coupling increases with frequency, the second lobe peak might leak. The second lobe peak of the PRBS NRZ is at $3/2 f_0$ (at OC12 $3/2 f_0 = 926.64$ MHz) and is 13.5 dB lower than the main lobe peak. The conclusion is that the X-talk coupling versus the frequency above 199 MHz should be below −70 dB, or the second lobe peak would leak.

Different PRBS series lengths result in a different power leakage and X-talk. The ratio between X-talk power levels is related to the NRZ PRBS length. For instance, the spectral power ratio between the two series $2^7 - 1$ and $2^{23} - 1$ is $10 \log(2^7 - 1/2^{23} - 1)$, and equals a 48-dB reduction in X-talk when a longer PRBS such as $2^{23} - 1$ is used instead of $2^7 - 1$. Figure 20.6 provides the isolation requirements for various PRBS series versus data voltage levels.

## 20.4 EMI RFI Sources Theory for Shielding

The source of emission is well described by two of Maxwell's equations:[5,15]

$$\nabla \times \mathbf{H} = \varepsilon \frac{\partial \mathbf{E}}{\partial t} + \mathbf{J} = j\omega\varepsilon \mathbf{E} + \mathbf{J}, \tag{20.15}$$

$$\nabla \times \mathbf{E} = -\mu_0 \frac{\partial \mathbf{H}}{\partial t} = -j\omega\mu_0 \mathbf{H}, \tag{20.16}$$

The first equation describes Ampere's law. The second equation describes Faraday's law. Faraday's law claims that the change in the magnetic-field flux creates voltage. Using Stock's theorem, Eq. (20.16) becomes

$$\oint \mathbf{E} d\mathbf{l} = -\mu_0 \int \frac{\partial \mathbf{H}}{\partial t} d\mathbf{s}, \tag{20.17}$$

Equation 20.17 claims that the magnetic flux versus time, for instance through a coil area or a closed loop, generates voltage. Equation (20.15) explains surface currents, denoted as **J**, created on conductive surfaces due to a time-varying current in the field-emitting wire. This means that the shield that protects the susceptible part and isolates the emitter part should have very high conductivity and very low impedance. The surface current generated on the shield is given by Ohm's law:

$$\mathbf{J} = \sigma \mathbf{E}, \tag{20.18}$$

where $\sigma$ is the material conductivity in units of $\Omega^{-1}/\text{m}$. At high frequencies, the induced current on a protective shield is a surface current with a penetration depth $\delta$ given by the skin effect:

$$\delta = \sqrt{\frac{2}{\omega\mu\sigma}}. \tag{20.19}$$

For an ideal shield where the conductivity $\sigma = \infty$, the skin depth $\delta$ goes to zero, and the fields inside the shield go to zero exponentially; $\mu$ is the magnetic permeability.

According to antenna theory, the surface current generated on the shield creates electromagnetic (EM) fields. Therefore, in order to minimize the voltage created on the shield or to minimize the current density, it is important to attach the shield to the lowest ground (GND) potential at several points. This is equal to several impedances connected in parallel. The result is that the shield potential is reduced. In case the shield is attached only to one GND point, it would create a high-current density at the current drain point, resulting in strong EM field radiation. These fields would create strong EMI interferences. If the shield is not connected to GND it floats, acting like an energy reflector due to the surface current generated on it. In conclusion, in order to make the shield effective and minimize the shield radiation, it should be connected to the lowest GND potential at several points. This way, induced currents, due to the emitted **E** field, are effectively drained to GND. This can be understood by the potential gradient equation where $\psi$ is the voltage potential:

$$\mathbf{E} = \vec{\nabla} \cdot \psi. \tag{20.20}$$

It also can be noted as a surface current equation, which describes Ohm's law:

$$\frac{\mathbf{J}}{\sigma} = -\vec{\nabla} \cdot \psi. \tag{20.21}$$

The current always flows to the lowest potential. The electric field **E** is always closed on the lowest potential.

## 20.5 GND Discipline Theory for Shielding and Minimum Emission

When designing mixed-signal communication systems with wideband noisy digital channels on the one hand, and low-noise sensitive wideband-analog receiver channels on the other hand, it is an ultimate requirement to separate the analog and digital GND. This way, noise leakage from the noisy system (emitter, source) to the quiet system (receiver, susceptible victim) is prevented.

An integrated optical module is generally composed of three GND planes: the receiver plane and transmitter plane for the digital section, and a third plane for the analog section. The receiver structure can be realized in several ways: a single printed-circuit-board (PCB) design containing all three signal disciplines in one-or two-PCB structure that has two floors: one for the analog and one for the digital. The bottom floor can be the digital card and a part of it can be used to mount the optics block. The top floor can be the video CATV receiver card section. The goal of GND design is to minimize the GND plane impedance and to prevent emissions. In case one of the GND planes has a higher impedance than the other, the EM fields emitted from the source GND plane would be closed on the lower potential GND plane: the target plane, which has the lowest GND impedance. In addition, it is common engineering practice to connect all GND planes at a single point called the "GND star point." This way, current loops between the different GNDs are prevented, therefore minimizing the X-talk between the analog module and digital sections. Figure 20.7 illustrates a

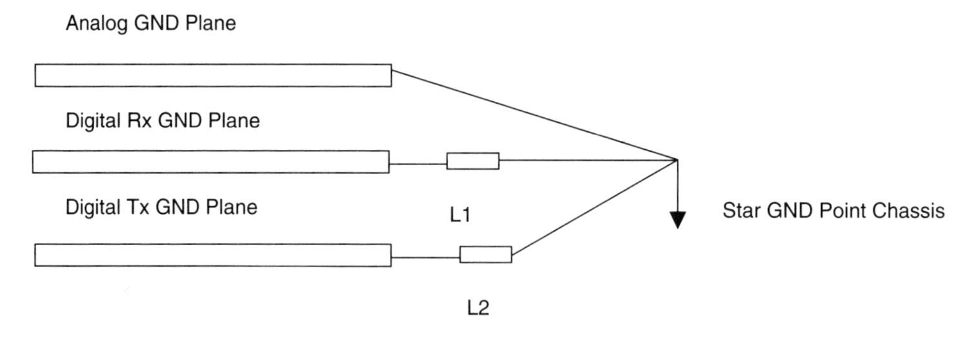


**Figure 20.7** GND path topology.

GND path using a star point. However, when connecting the GND planes at the star point, it is important to minimize the residual inductances L1 and L2, which are between the digital GNDs and the star point.

These residual inductances are as a result of the "header" GND pin, which is a parasitic inductance. Generally, the analog GND plane is the chassis GND and has many connections to the main GND, and so its residual inductance is very low compared to the digital GND. Moreover, it is a common error to use ferrite beads to isolate GND types at a test fixture or motherboard. Indeed, resistance wise, all GND planes are connected and have low dc resistance inbetween. But by analyzing the GND frequency response and emitted EM fields from digital GND due to digital bouncing interferences versus frequency, it might be observed that even though there is $\sim 0\ \Omega$ between all GNDs to the star point, the impedance between GND planes to the star point is frequency dependent and is given by $X_L = j\omega L$. Therefore, when the data rate is getting higher, or the PRBS series is shorter and is in a higher frequency, high-frequency energy components of the digital data would have a higher impedance of $X_L$ between the star point GND to the main GND plane of the PCB. Note that the analog PCB is the star point; the result would be emission from the digital GND into the analog GND, creating X-talk from the digital into the analog section. This is because the digital GND and analog GND do not have the same potential. In fact, the digital GND has a higher potential, as shown in Fig. 20.7. Additionally, there is a second coupling mechanism between the GNDs, which is capacitive coupling.

Both of the above described phenomena, radiation due to residual inductance and capacitive coupling, are characterized by a high-pass transfer function. This is due to the fact that in low frequency $X_L \approx 0\ \Omega$ and the capacitive coupling $X_C \approx \infty\ \Omega$; hence both digital GND and analog GND are in the same potential. As frequency increases, the impedance between the two GNDs also increases. Thus the digital GND is isolated from the chassis, it becomes in higher potential and starts to radiate. The analog GND is in lower potential, hence it becomes susceptible to the transmitter noise: the result is X-talk. Special attention should be paid to keeping the star point with low residual inductances and to minimizing its $X_L$. By doing so, GND current loops between the analog GND and the digital GND are avoided. This is due to the fact that the GND return path always gets lower in potential toward the power supply GND, which holds a neutral potential level of 0 V. Fig. 20.8 illustrates the isolation transfer function between the digital sections to an analog section in an integrated module.

## 20.6 Emission and Reception Mechanisms in Integrated Modules

The purpose of the antenna is to pick up an EM signal field. The higher the antenna impedance is, the higher the induced voltage at the antenna ports due to the EM fields. In the RFFE of an analog receiver, there is always a matching network composed of reactive components such as coil inductors and transformers. The purpose of this network is to match the first-stage LNA input impedance to the PD output

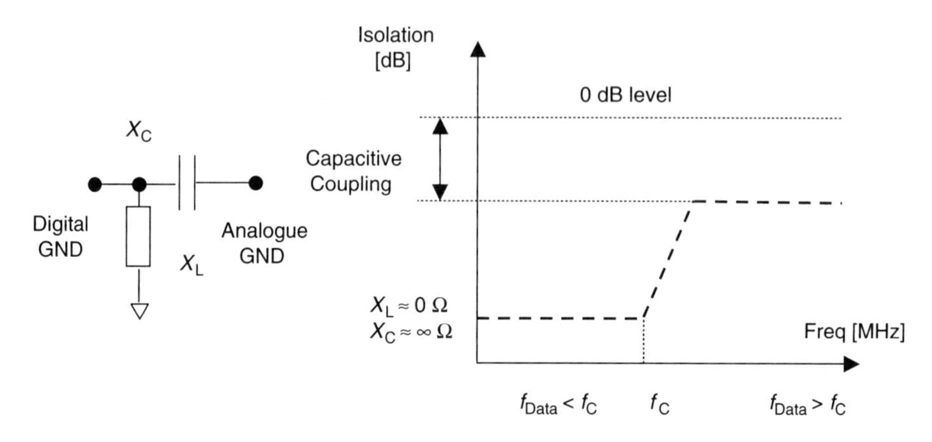


**Figure 20.8** Isolation between digital GND to analog GND. The goal is to minimize the residual inductance impedance $X_L$ and coupling capacitance impedance $X_C$ in order to render the X-talk cutoff frequency $f_c$ higher than the video analog frequency BW of operation.

impedance, which is high impedance. If lumped elements are used, such as an inductor in the RFFE matching, the higher its inductance $L$ the higher the induced voltage on the inductor, according to

$$\frac{\mathrm{d}\phi}{\mathrm{d}t} = -L\frac{\mathrm{d}i}{\mathrm{d}t}. \tag{20.22}$$

The inductor L1 in Fig. 20.9 is required to increase the BW and to compensate for the PD stray capacitance. The transformer has a 2:1 ratio and its primary has the higher number of turns. It is used to match the LNA input impedance to the PD output impedance. According to $Z_{in}/Z_o = (N_1/N_2)^2$, the impedance seen from the PD portion is 200 Ω, for 50 Ω input impedance to the LNA. The PD equivalent circuit is given in Fig. 20.10 and reflects high impedance. As a result, the matching inductor has high impedance at its two ports, which is equal to the photodiode

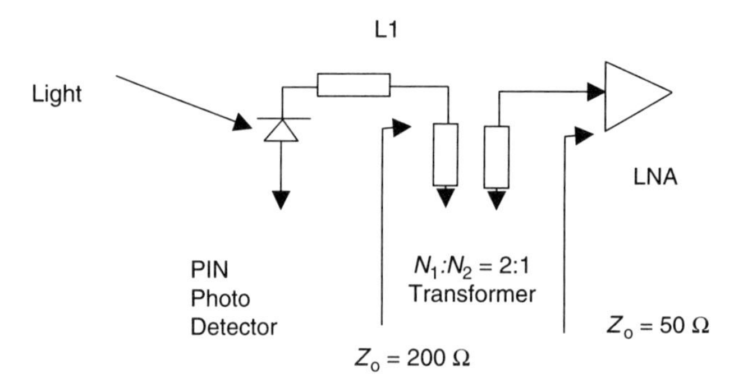


**Figure 20.9** RFFE input-matching network.

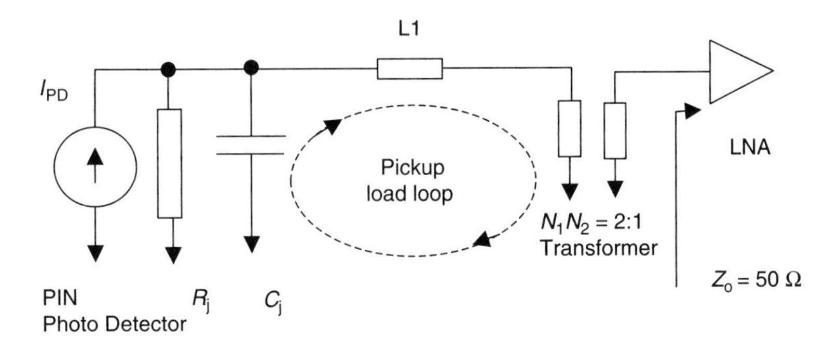


**Figure 20.10** A PD equivalent circuit showing the EM-pickup-load loop of the matching inductor L1. The current generated by the induced voltage over L1 is converted via the transformer as an interfering signal into the LNA input.

impedance in series with the load reflected to the matching transformer primary. This condition satisfies high-induced-voltage pickup by the matching inductor and a current loop as seen in Fig. 20.10. In the same manner, the transformer windings serve as a pickup of X-talk. The source for the X-talk is the laser modulation current of the transmit section, which is generally close to the PD due to the natural structure of an optical triplexer module, as was explained in Chapter 5.

Using traditional microwave design methodologies and creating isolation compartments for each side of the optical module may increase the isolation between the source of the interference and the receiver of the interference. Additionally, proper shielding of the high-impedance area in the analog RFFE would reduce even more the pickup of EM energy by the reactive matching components of the RFFE; hence, the induced voltage level would be reduced.[9]

A similar pickup mechanism and load loop occurs at any digital transceiver module such as a small-form pluggable (SFP) or any other kind. Generally, the observed phenomenon is sensitivity reduction during a BER test while the transmit section is on. BER reduction is pattern oriented and extinction ratio sensitive. The receiver suffers from desensitization due to the transmitting signal. In small-form-factor (SFF) designs, the distance between the emitter (source) and target (receiver) is on the order of near field. The same applies for the integrated triplexer (ITR). Generally, an SFF design is composed of receiver/transmitter PCB, TOSA (transmit optical subassembly), and ROSA (receive optical subassembly), and flexible transmission lines PCB to connect the receiver/transmitter PCB and the TOSA and ROSA optics. The flexible transmission line PCB carries the signal lines and biasing lines. Transmit-flexible PCB is connected to the TOSA, which is a laser diode with a varying impedance $Z[I(t)]$, where $I$ is the modulating current. During the ON state, when the transmission is at a high digital level of 1, the current is high, the laser is forward biased, and $Z(I(t))$ is at a low-impedance value. Hence, the transmitter's flexible PCB is a loop antenna excited by the modulation current. The receive section is connected via a flexible PCB to a ROSA module, which contains a PD sometimes with a PIN TIA. The output of the ROSA is differential and feeds the receive module. Observing the receiver's

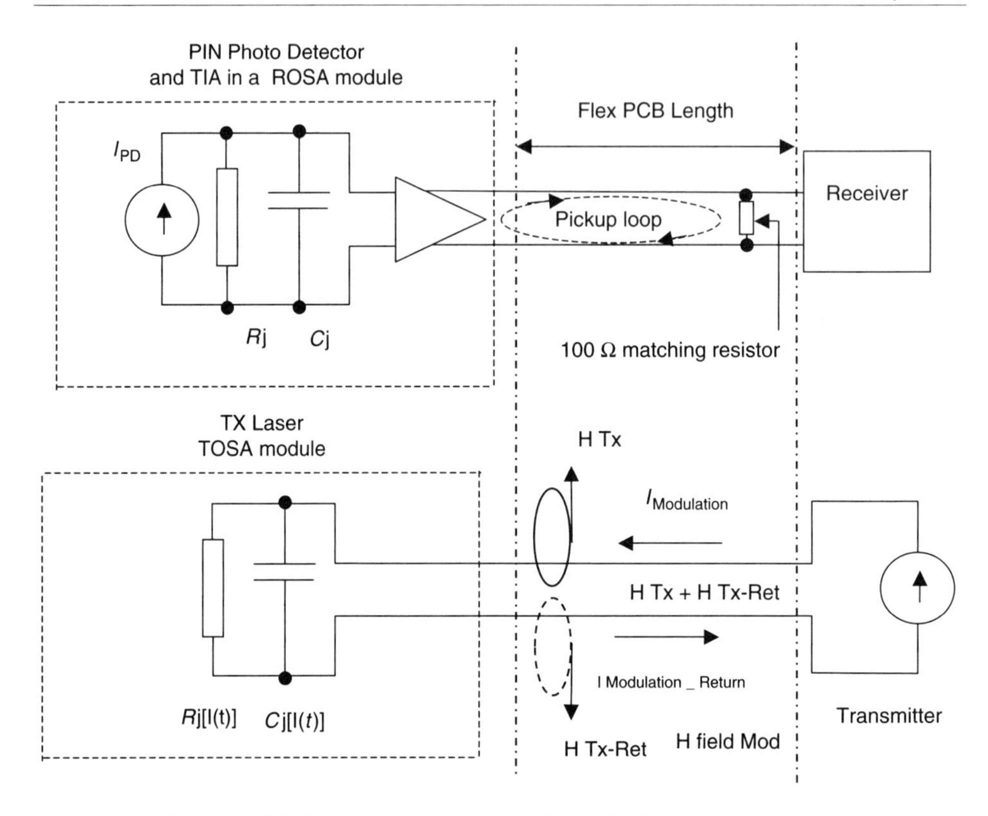


**Figure 20.11** A X-talk mechanism in a TOSA–ROSA SFP transceiver module.

flexible PCB transmission lines, one can see that they generate a loaded loop antenna that can pick up magnetic-field flux emitted from the laser's flexible PCB and create an induced voltage. That voltage is proportional to the modulation depth of the transmitter laser. Fig. 20.11 illustrates a TOSA–ROSA transceiver with a flexible PCB. The PIN diode is integrated with a differential PIN TIA inside the ROSA module and its output impedance is 50 Ω per port. The receiver section on the PCB is 100 Ω and appears in odd-mode propagation, which is the case of a differential signal; the 100 Ω are as 50 Ω in a series with virtual GND, as shown in Fig. 20.14. The laser modulation is single ended, with a signal data line and a signal return. Fig. 20.12 shows the flex PCB arrangement to minimize X-talk by using the GND plan of the flex as an additional shield and trying to cancel **H** fields by having the flexible PCB GND planes opposite to each other.

The laser modulation current generates a magnetic field perpendicular to the plane of the paper and travels toward it, according to the right-hand law. The laser return modulation current generates a magnetic field perpendicular to the plane of the paper and travels away from it, according to the right-hand law. Both fields are almost equal due to the proximity of both the modulation and the return current of the ROSA-flexible PCB. However, if the TOSA return is connected to the chassis or housing, X-talk increases, since the receiver's GND is bounced by the transmitter's return current. Moreover, the equivalent **H**

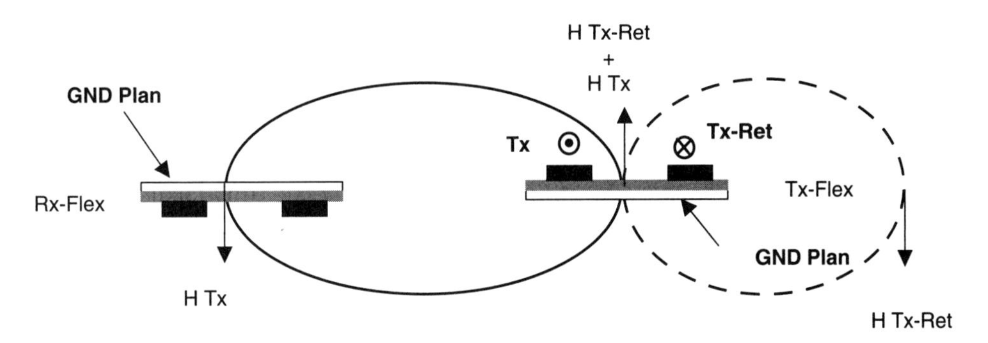


**Figure 20.12** GND usage as an additional X-talk reduction between the and transmitter's/receiver's flexible PCB.

field is stronger since the transmitter's return field does not cancel the **H** field induced by the transmitting line, and the receiver's flexible PCB pickup loop induces undesired interfering voltage. Since the flexible PCB is an RF microstrip transmission line, the bottom of it is a GND plane. Hence, the field cancellation is not enough to prevent the induced interference voltage in the pickup loop of the receive section, which is the receiver's flexible PCB. Fig. 20.12 shows the **H** fields.

The solid line shows the magnetic field **H** due to the transmitter's modulation current perpendicular to the plane of the paper and traveling away from it. The dashed line represents the magnetic field of the transmitter's modulating return current perpendicular to the plane of the paper and traveling toward it. For connecting the transmitter return to chassis GND at the TOSA, the transmitter return would bounce the receiver's GND and there is no canceling of the **H** field from the transmitter return. The strongest **H** field appears between the transmitter and the transmitter return lines flexible PCB. Due to the proximity of the transmitter to the transmitter return, the equivalent **H** field on the receiver's flexible PCB is minimal, since the transmitter and the transmitter return **H** fields are cancelled within the receiver flex pickup loop. This does not happen if the ROSA return is connected to the chassis.

## 20.7 Differential Signal

For differential signals, the inputs are excited in an odd mode. The signal swing amplitude doubles. Any random noise that hits the differential pair is correlated and is therefore at the input of the differential line receiver due to its common-mode rejection ratio (CMRR). This way, signal immunity to interfering noise as well as the signal-to-interfering-noise ratio are improved. In addition, since the "+" and "−" data currents are in opposite directions, the magnetic field $\mathbf{H}(t)$ that is varying in time is the highest between the lines and is cancelled in space, as was explained above. Therefore the differential pair is less noisy and less radiating (Fig. 20.13).

An additional advantage is a 100 Ω differential termination at the receiver input that is equal to two 50 Ω resistors and a virtual GND. This way, there are

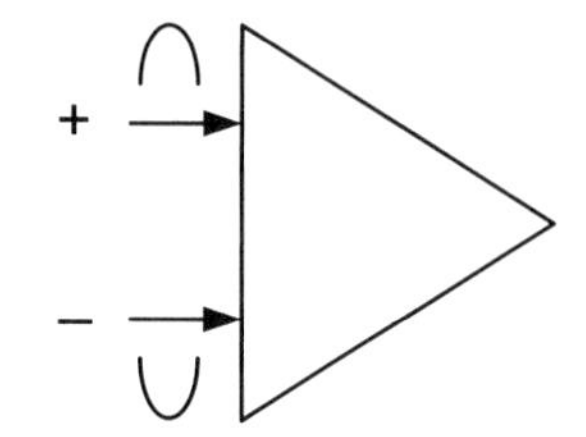


**Figure 20.13** Differential pair.

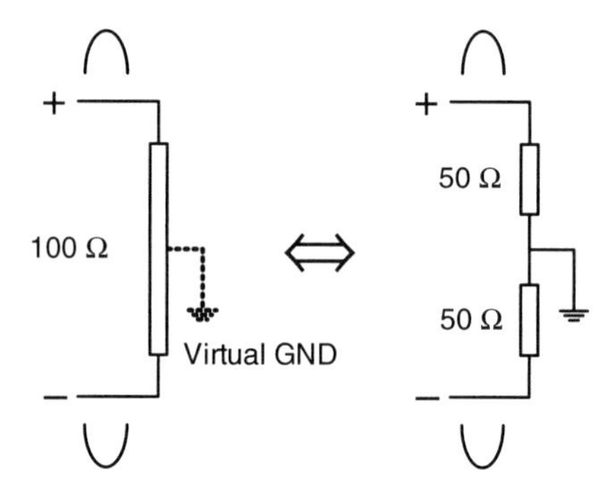


**Figure 20.14** Differential termination.

no current loops on the GND plane. Hence, a less noisy GND is achieved (Fig. 20.14).

The disadvantage of differential termination occurs when it is excited with a single mode (single-end). The termination is not 50 Ω but the load is 150 Ω. Thus the reflection factor is given by

$$\Gamma = \frac{Z_\mathrm{L} - Z_0}{Z_\mathrm{L} + Z_0} = \frac{150 - 50}{150 + 50} = 0.5, \quad S_{11} = 20 \log \Gamma = -6\ \mathrm{dB}. \tag{20.23}$$

This creates reflections and therefore much more radiation; thus the differential transmitter should not be excited as the single end, as shown in Fig. 20.15.

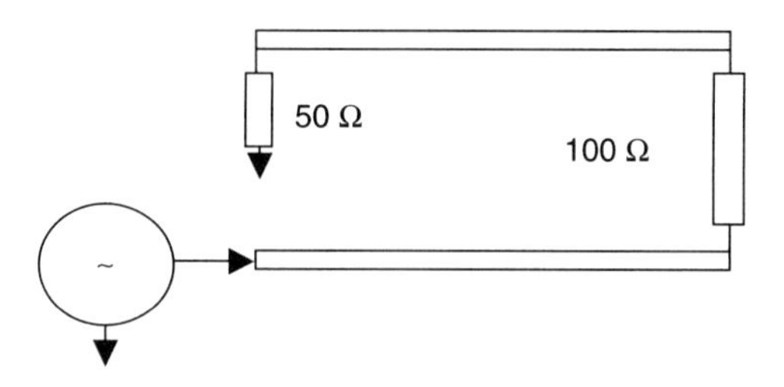


**Figure 20.15** Unbalanced excitation of a differential input.

## 20.8 Important Definitions and Guidelines

### 20.8.1 GND discipline

When designing mixed-signal communication systems with wideband noisy digital channels on the one hand and low-noise-sensitive wideband analog Rx channels on the other hand, it is an ultimate requirement to separate the analog GND and digital GND in order to prevent noise leakage from the noisy system (emitter) to the quiet system (receiver, susceptible target).

### 20.8.2 Distributed phenomenon

An element shall be called a distributed element when its physical length reaches $\lambda/4$, where $\lambda$ is the wavelength of the signal.

### 20.8.3 Near field

The definition of "near field" is complex and requires the understanding of Green's function; however, for the sake of simplicity, it can be defined by its radiation characteristics that are on the order of $1/R^n$, where $n > 2$. The near-field phenomenon occurs when proximity "$d$" to the emitter is within the condition of $d \ll \lambda/2\pi$. In this case, the electrical field impedance is higher than the free space impedance of $Z_0 = 120\pi$ and is given by $Z_0\lambda/2\pi d > Z_0$. This means a higher voltage is induced. Within this duality, the magnetic field impedance is lower at the near field, and its impedance is given by $Z_0 d/2\pi\lambda < Z_0$. The near field involves radial field components, while the far field is a planar field, where the **H** field is orthogonal to the **E** field, and both are orthogonal to the propagation vector **k**.

## 20.9 A Brief Discussion about Transmission Lines

### 20.9.1 Coplanar waveguide versus microstrip

A coplanar waveguide is a printed transmission line on a dielectric substrate $\varepsilon_r$ with GND planes on both sides. The wave propagation characteristic is quasi-TEM; hence, the wavefront propagates at different velocities. Part of the field propagates in free space and part of it within the dielectric material. The difference in propagating velocities causes dispersion at high frequencies. This is a radiating transmission line, but its advantage over a microstrip is that its field fringes are closed to the adjacent GND planes and there is less **E** field in free space compared to an equivalent microstrip line.

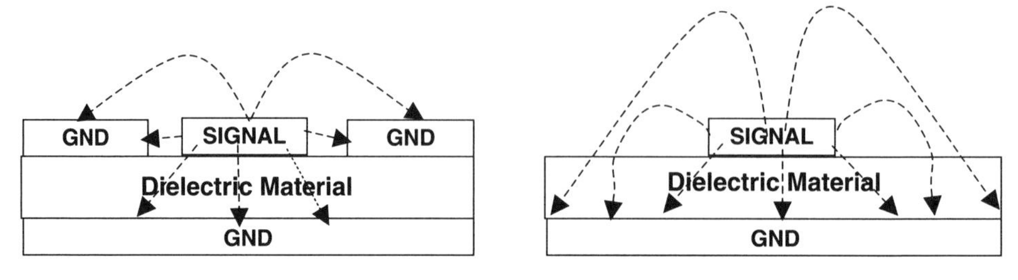


**Figure 20.16** Left: "coplanar waveguide;" right: "microstrip" cross sections and **E** fields.

The wave velocity in free space is:

$$C = \frac{1}{\sqrt{\varepsilon_0 \mu_0}}.$$

The wave velocity at the dielectric material is:

$$V = \frac{1}{\sqrt{\varepsilon_0 \varepsilon_R \mu_0}},$$

where $\varepsilon_0$ is the permittivity of free space, $\varepsilon_r$ is the relative permittivity of the dielectric, and $\mu_0$ is permeability of free space. Figure 20.16 depicts a microstrip versus coplanar transmission line.

The **H** field is a radial field around the signal strip and its direction is defined by the current direction. In case the potential at the signal strip is lower than the GND potential, then the **E**-field direction is from the GND to the signal strip.

## 20.9.2 Stripline

A stripline transmission structure, as shown in Fig. 20.17, is realized when the bearing signal transmission line is sandwiched between the two GND planes of the dielectric material $\varepsilon_r$. The electromagnetic (EM) energy propagates homogeneously since there is only a single velocity of the EM wave unlike the microstrip, where part travels in air and part in the dielectric. Hence, the stripline

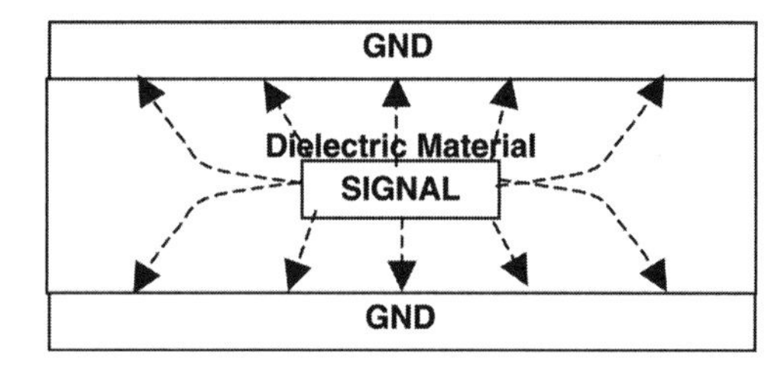


**Figure 20.17** Stripline and its **E** field.

is not depressive media. There is no radiation to space; however, access to the conductor is by PCB via holes. Hence it is harder to tune or add tuning elements.

The **H** field is a radial field around the signal strip and its direction is defined by the current direction. In case the potential at the signal strip is lower than the GND potential, then the **E**-field direction is from the GND to the signal strip.

## 20.10 Main Points of This Chapter

1. The nature of an NRZ digital signal is a wideband $\text{sinc}(x)/x$, which is its Fourier transform.

2. The main lobe of an NRZ signal is null at the CLK frequency.

3. The NRZ second lobe has a peak at $3/2$ CLK.

4. A PRBS signal is a pseudo–random pattern that repeats itself each $2^N - 1$, where $n$ is the depth of the generator's register.

5. The longer the PRBS, the more spectral lines it has; however, the energy per spectral line is lower compared to that of a shorter PRBS.

6. It is easier to isolate and combat X-talk of longer PRBS than shorter PRBS patterns, since the energy per spectral line for long PRBS is lower.

7. The frequency response of isolation between the source of radiation and a target is HPF.

8. It is desirable to have isolation HPF response with the highest cutoff frequency possible.

9. A high cutoff frequency of the isolation response is accomplished by minimizing the GND impedance between the emission source and the quiet GND.

10. GNDs are connected at a particular point, called the star point. The impedance to the star point should be the least possible.

11. At a high frequency, the connection impedance of the noisy GND increases, and the noisy GND starts to emit.

12. At a high frequency, the noisy GND is at a higher potential compared to the quiet GND; thus noisy GND fields are closed on the quiet GND, which is the target.

13. The source of radiation in ITR is from the digital transmitter modulation current.

14. The CATV receiver RFFE has a high impedance and closed loop inductive antenna; it is susceptible to induced voltage due to magnetic-flux change versus time.
15. One of the methods to prevent X-talk is by creating RF compartments that isolate the optical triplexer ports.
16. It is important to connect shielding to the lowest GND potential; that way, the induced surface currents on the shield are grounded.
17. Shielding that is not grounded develops surface currents that emit fields; thus it works against isolation rather than for it.
18. In SFF transceivers designs, the flexible PCB that feeds the TOSA laser from the laser driver, creating a current loop antenna that emits a magnetic field that is proportional to the extinction ratio.
19. The flexible PCB that connects the ROSA to the receiver creates a high-impedance pickup-loop antenna that senses the TOSA loop-antenna emission.
20. One of the methods to isolate the TOSA from the ROSA loops is to have opposing flexible PCBs, as shown in Fig. 20.12.
21. A differential pair is immune to amplitude interferences, since they are canceled by the CMRR of the receiver's postamplifier.
22. There are three types of commonly used transmission lines: microstrip, coplanar, and stripline.
23. A microstrip line is a dispersive quasi-TEM mode transmission line; thus, it emits interfering fields.
24. A coplanar transmission line is a less noisy transmission line and is thus commonly used in designs.
25. A stripline is the best PCB transmission line, but is hard to access and tune.
26. Differential transmission lines are coupled lines that are analyzed by the odd-and even-mode method.
27. The differential matching resistor in a differential pair would reflect $Z_0$ at odd mode and can be ignored at even mode.
28. The differential resistor value is $2Z_0$.
29. Operating a differential input transmitter as single ended would create X-talk, since the reflection factor $\Gamma$ is not optimal.

30. Modern SFP SFF transceivers use carbonated shielding rather than metal because it is cheaper and lighter and provides the same functionalities as a metal shield.

## References


1. Feher, K., et al., *Telecommunications Measurements, Analysis, and Instrumentation*. pp. 14–21 Nobel Publishing Corp. (1996).
2. Montrose, M.I., *EMC and the Printed Circuit Board Design, Theory, and Layout made Simple*. IEEE Wiley-Interscience (1998).
3. Mardiguian, M., *EMI Troubleshooting Techniques*. McGraw-Hill (2000).
4. Mardiguian, M., *EMI Troubleshooting Techniques*. McGraw-Hill (2000).
5. Ramo, S., J.R. Whinnery, and T. Van Duzer, *Fields and Waves in Communications Electronics*. John Wiley & Sons, USA Library of Congress (1994).
6. Hoffmann, R.K., *Handbook of Microwave Integrated Circuits*. pp. 267–376, Artech-House (1984).
7. Smith, C.E., M.D. Tew, and W.L. Wu, "Use SPICE to analyze crosstalk." *EMC Test & Design*, pp. 20–27 (1994).
8. Coombs, C.F., *Printed Circuits Handbook, 4th Edition*. McGraw-Hill (1995).
9. Brilliant, A., T. Ciplickas, and M. Heimbuch, "Methods and apparatus for high performance optical transceivers having reduced cross-talk," US Patent Pending. Luminent OIC.
10. Neu, T., "Designing Controlled Impedance vias." *EDN*, pp. 67–72 (2003).
11. Wang, T., R.F. Harrington, and J.R. Mautz, "Quasi-static analysis of microstrip via through a hole in ground plan." *IEEE Transactions on Microwaves Theory and Techniques*, Vol. 6 No. 6, pp. 1008–1013 (1988).
12. Wang, T., R.F. Harrington, and J.R. Mautz, "The excess capacitance of microstrip via in a dielectric substrate." *IEEE Transactions on Microwaves on Computer Aided Design*, Vol. 9 No. 1, pp. 48–56 (1990).
13. Oh, K.S., D. Kuznetsov, and J. Schutt-Aine, "Capacitance computations in a multilayered dielectric medium using closed-form special green's function." *IEEE Transactions on Microwaves Theory and Techniques*, Vol. 42 No. 8, pp. 1443–1453 (1994).
14. Oh, K.S., J. Schutt-Aine, R. Mittra, and B. Wang, "Computation of equivalent capacitance of a via in a multilayered board using the closed-form green's function." *IEEE Transactions on Microwaves Theory and Techniques*, Vol. 44 No. 2, pp. 347–349 (1996).
15. Gupta, K.C., R. Garg, and I.J. Bahl, *Microstrip lines and slotlines*. Artech-House (1996).
16. Johnson, H., and M. Grahm, *High-Speed Digital Design: A Handbook of Black Magic*, Prentice-Hall (1992).

17. Wadell, B.C., *Transmission Line Design Handbook*. Artech-House (1991).
18. Dutta, S.K., C.P. Vlahacos, D.E. Steinhauer, A.S., Thanawalla, B.J. Feenstra, F.C. Wellstood, and S.M. Anlage, "Imaging microwave electric fields using a near-field scanning microwave microscope." *Applied Physics Letters*, Vol. 74 No. 1, pp. 156–158 (1999).
19. Zverev, A., *Handbook of Filters Synthesis*. Chapter 5, "The catalog of normalized low pass filters." John Wiley & Sons (2005).

# Chapter 21

# Test Setups

## 21.1 CATV Power Measurement Units

In community access television (CATV), it is common to measure power by decibels relative to 1 mV in units of dBmV, rather than 1 mW in units of dBm. The ratios between decibels relative to 1 mV over 50 Ω and 75 Ω with respect to decibels relative to 1 mW systems are given by the following definitions:

$$0 \text{ dBmV} = 20 \log 1 \text{ mV} \tag{21.1}$$

$$\begin{aligned} P[\text{dBm}] \text{ at } 1 \text{ mV} &= 10 \log\{[(1 \text{ mV}/1000)^2/50\ \Omega] \times 1000\} \\ &\approx -47 \text{ dBm; thus } 0 \text{ dBmV} \Leftrightarrow -47 \text{ dBm} \end{aligned} \tag{21.2}$$

$$\begin{aligned} P[\text{dBm}] \text{ at } 1 \text{ mV} &= 10 \log\{[(1 \text{ mV}/1000)^2/75\ \Omega] \times 1000\} \\ &\approx -48.7 \text{ dBm; thus } 0 \text{ dBmV} \Leftrightarrow -48.7 \text{ dBm} \end{aligned} \tag{21.3}$$

- To convert dBmV to dBm: subtract 47 dB in a 50 Ω system, or 48.7 dB in a 75 Ω system.
- To convert dBm to dBmV: add 47 dB in a 50 Ω system, or 48.7 dB in a 75 Ω system.

## 21.2 OMI Calibration Using PD

When using a calibrated photodetector (PD) as an optical-modulation-index (OMI) calibration tool, the frequency dependency of the laser is taken into account. Furthermore, there is no need to measure the laser slope efficiency, and sometimes it is impossible. The PD is a light-controlled current source. However, it also provides two outputs, ac and dc, as shown in Fig. 21.1.

1. Photocurrent related to dc voltage measured on the internal PD (current source load) resistor $Z_1$ (Fig. 21.1). This current is designated as $I_{dc}$.

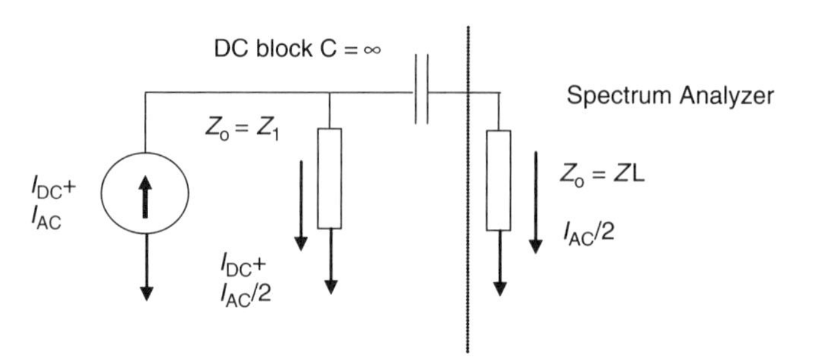


**Figure 21.1** Example of new-focus model 1414 PD equivalent circuit.

2. Alternating current is measured as power on $Z_0$ load [spectrum analyzer (SA)]. This current is designated as $I_{ac}$.

In a matched detector $Z_1 = Z_0 = Z_L$, where $Z_0$ can be 50 Ω or 75 Ω, depending on the application.

Using a reference PD, OMI is calculated and calibrated versus frequency, providing the laser frequency response. The OMI calibration procedure is listed below:

- Measure the RF power of a test tone using an SA.
- Calculate $I_{ac\text{-}rms}$ and evaluate the $I_{ac}$ peak on the load.
- With the $I_{ac}$ peak and measuring the $V_{dc}$ across the PD internal impedance, the OMI is evaluated by the relation between $I_{dc}$ and $I_{ac}$.

The relation between $I_{dc}$ and $V_{dc}$ is given by:

$$I_{dc} = V_{dc}/Z_1. \tag{21.4}$$

The general definition of an OMI is given by

$$\text{OMI} = \frac{V_{peak}}{V_{dc}} = \frac{I_{RF\text{-}peak} \cdot Z_0}{I_{dc} \cdot Z_0} = \frac{I_{RF\text{-}peak}}{I_{dc}}. \tag{21.5}$$

The peak current ratio to CW RF rms current test is given by

$$I_{RF\text{-}peak} = I_{RF} \cdot \sqrt{2}. \tag{21.6}$$

The RF power measured by the SA is the rms power; hence, $I_{RF}$ gets the rms value. Remember that $Z_L = Z_1 = Z_0$; only half of the rms RF current goes into the load (SA); hence $P_{RF}$ is given by

$$P[\text{dBm}] = 10\log[(I_{RF}/2)^2 Z_0 \cdot 1000]. \tag{21.7}$$

Changing the subject of the power formula gives the ratio between the rms RF current and $P_{RF}$[dBm]:

$$I_{RF} = \sqrt{\frac{10^{0.1P[dBm]} - 3 \cdot 4}{Z_0}}. \tag{21.8}$$

Hence, the peak current is given by

$$I_{RF\text{-peak}} = \sqrt{2}\sqrt{\frac{10^{0.1P[dBm]} - 3 \cdot 4}{Z_0}}. \tag{21.9}$$

Therefore the peak voltage $V_{peak}$ is given by

$$V_{RF\text{-peak}} = \sqrt{10^{0.1P[dBm]} - 3 \cdot 8 \cdot Z_0}. \tag{21.10}$$

OMI is the ratio between the peak RF voltage to the dc voltage measured at the PD output (Fig. 21.1). It is the same definition of the ratio between the peak RF current that modulates the laser and the laser bias current. OMI is amplitude modulated and is defined by

$$OMI\% = \frac{\sqrt{10^{0.1P[dBm]} - 3 \cdot 8 \cdot Z_0}}{V_{dc}} \cdot 100, \tag{21.11}$$

$$P[dBm] = 10 \log \frac{(OMI\% \cdot V_{dc})^2}{80 \cdot Z_0}. \tag{21.12}$$

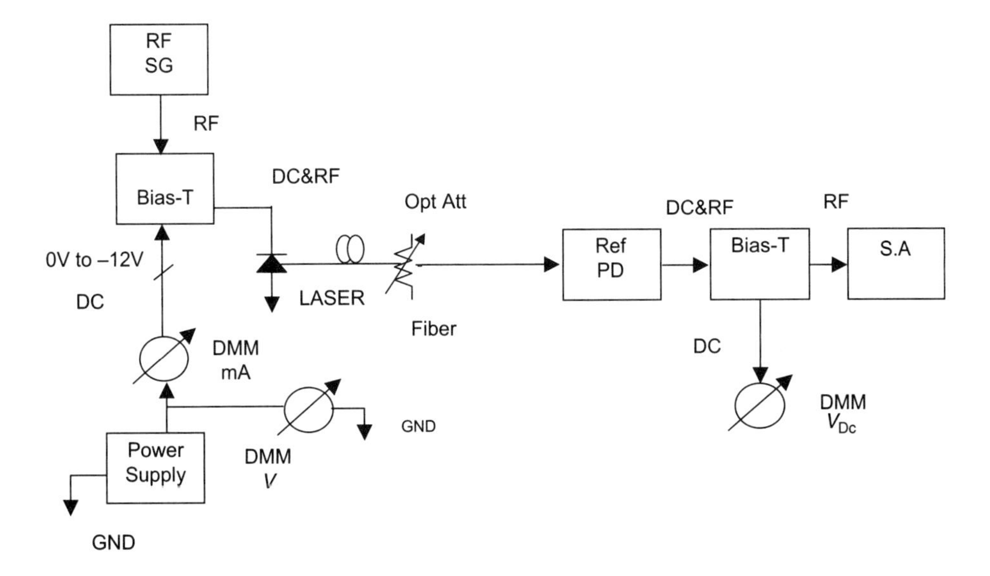


**Figure 21.2** OMI evaluation using a reference PD setup.

Hence, during calibration, the RF power that should be observed in an SA can be determined for a given OMI requirement and system impedance, as shown by Eqs. (21.11) and (21.12). This way, the RF power that drives the transmitter's laser can be adjusted until it meets the desired level on the SA screen. That point would be the desired OMI level. Figure 21.2 demonstrates such an OMI test setup.

## 21.3 Two-Tone Test

A two-tone test is commonly used in the evaluation of CATV receivers in order to estimate the performance of composite triple beat (CTB) and composite second order (CSO) distortions known as dual-tone second order (DSO) and dual-tone beat (DTB).[1,5–7] This method is much cheaper in production compared to a multitone test generator; however, it is less accurate since it is not as random as a multitone test. A two-tone test can provide a ballpark figure for performance evaluation during preliminary development in the assessment stages. Further information is provided by the Society of Cable Telecommunications Engineers (SCTE).[9]

Chapter 9 elaborates the theory and relationships between a dynamic range and the intercept point as well as relations to multitone loading. In order to avoid measurement degradation and error, a two-tone test is done by using two lasers with similar wavelengths and a separation of 5 nm between them; this way, an optical beat note between the lasers is avoided. By using two lasers, one per RF tone, the intermodulation that could be created by a single laser that is fed by two RF tones is prevented.

Special attention should be paid when calibrating the OMI of each laser. Since the PD has wide wavelength responsivity, it would generate an equivalent current that is the sum of two dc currents created by each laser or wavelength illumination. If we assume both lasers are biased to provide the same optical power at the output of the optical combiner, then both generated dc currents at the PD are equal. That assumption is correct if the two lasers are similar in wavelength and if we assume that within the range of a 5-nm separation, the PD has the same responsivity value $r$. In that case, the dc current is given by

$$I_{dc\_eq} = I_1 + I_2 = (P_1 + P_2) \times r = 2P \times r = P_{eq} \times r = 2 \times I_{dc}. \qquad (21.13)$$

Equation (21.14) describes the dc test condition at the PD of the receiver under test. The RF-peak-current level at the detector generated by each laser is given by

$$I_{\text{RF-peak}} = \text{OMI} \times I_{dc}. \qquad (21.14)$$

But at a test condition of optical level $P_{eq}$, the OMI for each tone is given by

$$\frac{I_{\text{RF-peak}}}{P_{eq}r} = \frac{I_{\text{RF-peak}}}{(P_1 + P_2)r} = \frac{I_{\text{RF-peak}}}{2 \times P \times r} = \frac{I_{\text{RF-peak}}}{I_1 + I_2}$$
$$= \frac{I_{\text{RF-peak}}}{2 \times I_{DC}} = \text{OMI} = 0.5 \times \text{OMI}_{\text{single-laser}}. \tag{21.15}$$

The explanation is as follows: since $I_{dc}$ represents a single laser result and the PD senses both wavelengths, it means the PD senses twice the optical level of a single laser. Another way to observe this is that if the test condition is at an optical level $P$, then each laser provides only half power, $P/2$. For this reason, OMI calibration of a laser in a two-tone setup should be done while the other laser is on with no RF signal. As a consequence, the RF power driving a laser in a two-tone test for a given OMI is higher by 6 dB, compared to a single laser with the same OMI and bias point. This is due to the additional dc level generated at the PD by the unmodulated laser in the two-tone setup. When each laser of a two-tone test represents half the channel load, then the OMI per laser can be derived by modifying Eq. (8.51) from Sec. 8.2.4 to Eq. (21.16), where $M$ is the total number of channels, $N$ is the number of NTSC-high-OMI channels, and $k$ is the ratio of the quadrature amplitude modulation (QAM) OMI to the analog OMI; generally $k = 0.5$:

$$\text{OMI}_{\text{eq per laser}} = \text{OMI}_{\text{ch-A}} \cdot \sqrt{\frac{1}{2}[N + k^2(M - N)]}. \tag{21.16}$$

Having the equivalent OMI value, calibration is done according to Sec. 21.2.

An additional important parameter that should be considered is the wavelength separation for a two-tone two-laser measurement of an optical triplexer or a duplexer, reviewed in Sec. 5.2. In that case, the wavelength separation between the two-test lasers should take into account the optical filter bandwidth (BW) at the analog-PD input. Both wavelengths should pass the optical filter and hit the analog PD with equal power. In practice, the optical combiner is not always balanced, and therefore, it would require driving the two lasers at two different bias currents in order to have the same optical power levels at the optical combiner output. Figure 21.3 illustrates such a two-tone test setup.

## 21.4 Multitone Test

When testing a CATV system and optical receivers for full performance, the test should be done under full-channel loading, where the first 79 channels are under an OMI of 3% to 4%, and the remaining 31 channels are with an OMI of 1.5% to 2%, respectively. Such a testing system should cover the optical dynamic range from

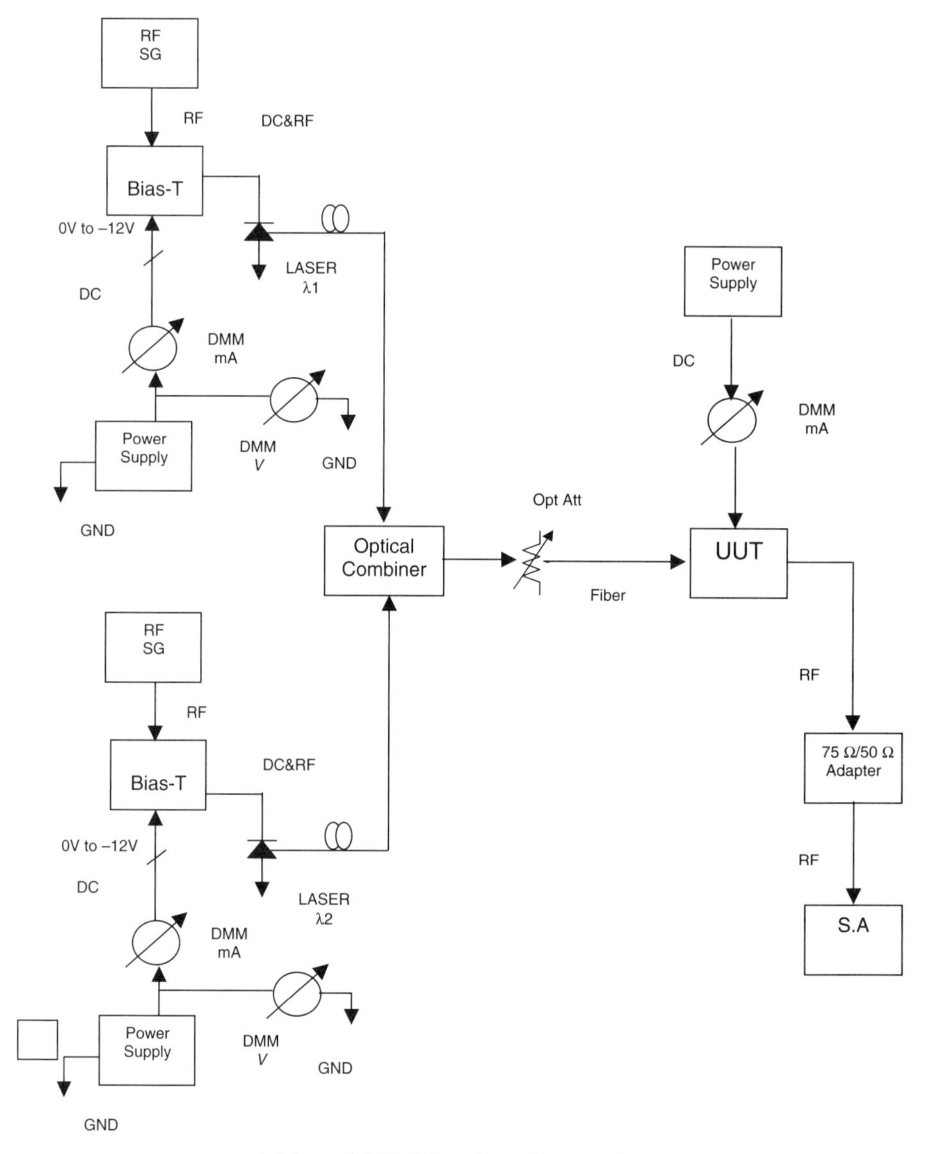


**Figure 21.3** A two-tone laser setup.

−6 dBm to +2 dBm. Within this optical range, the following parameters are evaluated:[1−4]

- *At high optical power*: CSO, CTB, and maximum RF power are observed in order to check that the gain is not too high and that automatic gain control (AGC) limits power, thus preventing output stage compression. Moreover, this test makes sure that the PD's output power does not compress the low-noise amplifier (LNA). Carrier-to-noise ratio (CNR) is observed as well in order to ensure that the AGC operates correctly without adding too high of attenuation. In the PIN type voltage variable attenuator (VVA), high attenuation means low current, thus a higher IMD potential. The CNR should be higher than at low optical power.

- *At low optical power*: CNR and minimum RF power are observed in order to check that there is enough gain and make sure the LNA compensates the RF-chain-noise figure, thus providing the minimum threshold CNR. The CNR should be lower than at high optical power. CSO and CTB are observed as well in order to ensure that the AGC operates well. In a low-attenuation state, AGC VVA should have lower IMD.
- The system also checks the PD-voltage monitor in correlation to the RF results. High optical power means a high-PD-monitor voltage and vice versa.

A multitone system is generally used for production and development. In production, such a system is fully automated and capable of producing statistics for test results and yield per pass-fail parameter. Such a system consists of a multitone source, where each tone is synthesized by a phase locked loop (PLL) with direct digital synthesis (DDS) as a reference. The DDS clock is generated from a crystal. There are two modes of operation for the frequency source: coherent, where all synthesizers are locked to the same crystal; or random, in which each DDS is locked to a deferent crystal. The output of each synthesizer is amplified and then all amplified test tones are combined using a multicoupler. This way, the linearity of the multitest-tone source is kept high, without degradation due to intermodulations (IMDs) if all sources were amplified by a single amplifier after the RF combiner. Moreover, each frequency source can be controlled exclusively; this way, by changing the amplitude of each RF test tone, OMI of each RF tone can be adjusted. Typical RF sources are matrix and Aeroflex–RDL.

The output of the multitone RF source is connected to a CATV predistorted and linearized laser transmitter. In order to prevent mismatch-related IMD products, an attenuation pad is inserted between the RF-multitone source's output and the laser transmitter's input. The laser transmitter structure and architecture is provided in Chapter 16. The output of the laser transmitter is connected to a directional coupler, where the coupled output is connected to an OMI setup, as explained in Sec. 21.2. This setup is HPIB controlled, and the OMI is adjusted for each tone. Advanced transmitters are HPIB controlled as well; thus laser APC and OMI are monitored. The OMI test setup is also useful to monitor the reference transmitter's relative intensity noise (RIN).

The direct arm of the optical coupler is connected to an erbium-doped-fiber amplifier (EDFA) to increase optical power. This amplification process does not affect OMI. The output of the EDFA is distributed by an optical star divider to the test stations.

A test station consists of an optical attenuator to cover the optical dynamic range input to the unit under test (UUT) and a power meter to verify the optical power value and to calibrate the setup.

The UUT pigtail is connected to the optical attenuator output and the UUT RF output is connected to several band-pass filters (BPFs) that are agreed upon for testing several CATV channels for pass-fail. The role of the BPF is to transfer only the channel under test to the LNA, which is connected between the BPF and the SA, as described in Fig. 21.5. This way, the LNA does not add any intermodulations due to the 110 channels which are representing the full RF loading. The LNA's purpose is to compensate for the SA noise floor, thereby reducing measurement error in low CNR products such as CSO and CTB, besides improving low CNR. The second role of the test setup LNA is to improve the accuracy of the noise-power measurement from the UUT as it does for CSO and CTB. Generally, the production test is done with the worst case CATV channels. These channels are generally those that have the highest CSO CTB beat-note counts. This way, a long test time is saved by not having to test all 110 channels. In the design validation test (DVT) or qualification test procedure (QTP) of a new product, sometimes all channels are tested. As was explained in Chapters 3, 9, and 10, those tests are CSO, CTB, and CNR. In order to observe CTB products, it is important to remove the channel tone in which the CTB is observed because CTBs fall at the same frequency of a channel. The CTB test is done by turning off the channel under test in the multitone tester. That is a limitation since such a described system supports several test racks. It could be that while one station asks for CTB tests, another tests CSO and is affected by CTB because the tone is missing. Further, during CTB test of a UUT, other units are not loaded with 110 channels. Thus, sometimes CTB is done at qualification only; otherwise, the production floor should be synchronized, which is not practical. Figure 21.4 describes a UUT spectrum per the SCTE. Figure 21.5 describes a multitone tester used for production as described above.

There are several nuances regarding CNR tests. Generally, when defining the CNR of a UUT, it refers to its dark CNR. However, during a dark CNR test, there is no RF signal. For UUTs with feedback AGC, turningoff the laser light affects the AGC attenuation because there is no signal, thus AGC jumps to minimum attenuation and maximum sensitivity. As a result, the state of the UUT is a minimum noise figure but it is not necessarily at the same AGC control voltage state. As a consequence, there is a mechanism called the "AGC defeat control" that maintains the same AGC conditions for a dark CNR test, which refers to a certain optical input power. The alternative is to examine the UUT at its AGC threshold state. But there are more practical options as described below. In feedforward AGC, as described in Chapter 12, this dark CNR test would change the PD-monitor voltage, bringing the UUT to its maximum gain. This is when a CNR test is measured at high optical levels. There are two options to overcome this problem as a conventional rule:

- Testing UUT CNR and dark CNR at the AGC threshold, where the AGC voltage for feedback AGC reaches its maximum value. Consequently, this does not affect the gain. In this manner, only CNR at the minimum optical gain is observed.

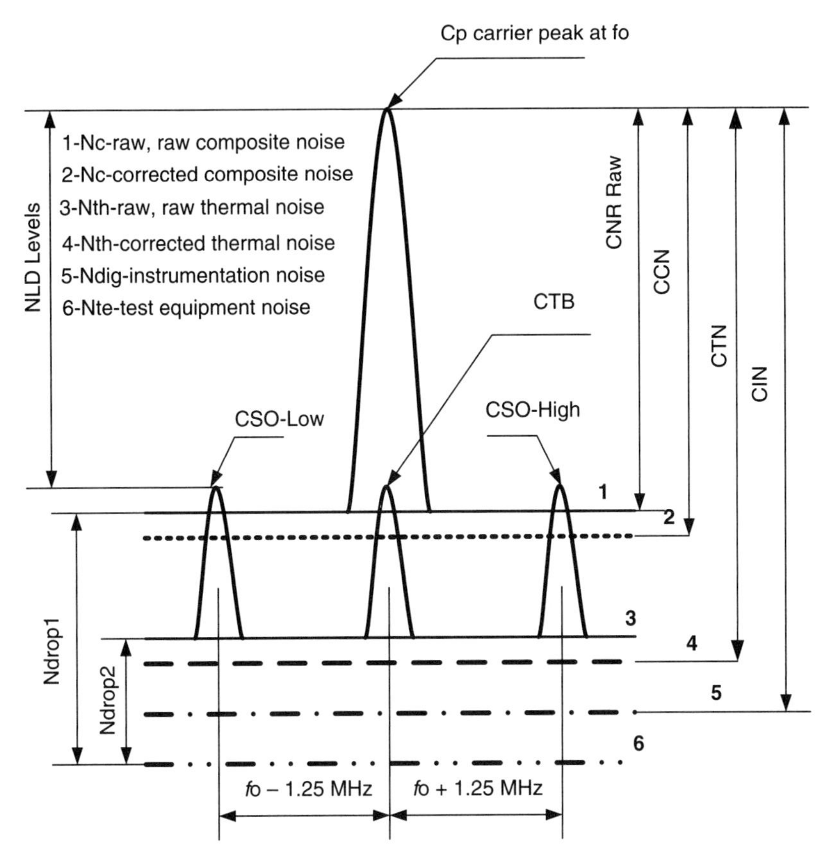


**Figure 21.4** Spectral definition of multitone CSO and CTB results according to the SCTE.[2] (Reprinted with permission from *SCTE* © IEEE, 2001.)

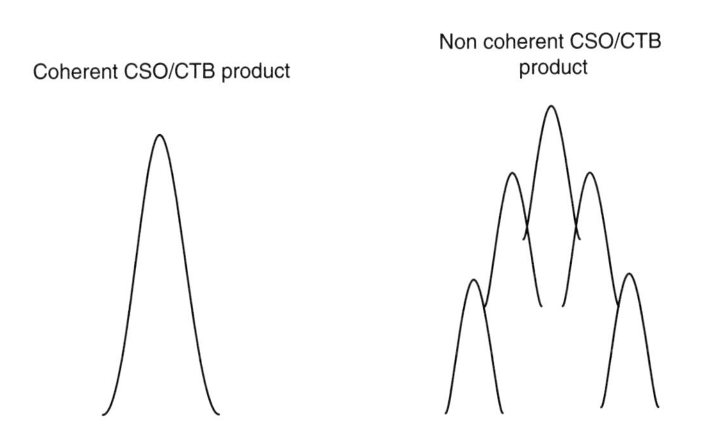


**Figure 21.5** Coherent CSO/CTB vs. noncoherent CSO/CTB.

- Testing UUT CNR at any optical level and extracting the shot noise and RIN contribution to the overall CNR. This can be done using the PD-monitor voltage observed on an accurate 1 KΩ resistor in series with the PD; such a front-end topology is shown in Figs. 2.1, 2.2, and 12.2. This way, PD responsivity is measured as well as the PD current at a given test point. By using Eqs. (10.64) and (10.75), dark CNR can be evaluated.

When performing an acceptance test procedure (ATP)/QTP the raw data should be processed. The test setup degrades the acceptance/qualification test results (ATR/QTR) data and performance may be slightly better. The SCTE[1,2] defines several CNR measurement values and correction factors (CORs), as shown in Fig. 21.4. These values are carrier to composite noise (CCN), carrier-to-thermal noise (CTN), and carrier-to-intermodulation noise (CIN). An additional correction refers to the SA. There are two parameters that should be considered. First, the filter-shape factor it is not an ideal brick-wall filter; because it has a Gaussian shape, thus, the noise BW should be corrected. Second is the detector amplifier. In HP SAs, this is the log-amplifier COR. In Rodeh-Schwarz spectrums, the COR is different, thus SCTE guidelines relate to HP SAs.

The SA's detector measures the signal and the noise accompanying the signal at a given BW. When examining low-level signals such as CSO and CTB, the test is done at a low CNR, and the CSO and CTB powers are corrected and noise effects calibrated. When measuring a large carrier, the CNR is larger than 10 dB and the COR is negligible. SCTE recommendations for SA settings appear in Table 21.1.

Prior to reforming any test-setup system, the noise floor should be measured; i.e., the SA noise floor should be recorded while the UUT is disconnected and the cable connected to the SA is terminated. This value is $N_{TE}$, marked as 6 in Fig. 21.4. The SA attenuator should be optimized to show a noise drop of more than 10 dB between UUT noise under the full-channel load and spectrum noise. If not, then the LNA should be used, where if noise drops below 10 dB, COR should be used.

**Table 21.1** SCTE recommendations for SA settings for NLD and CNR tests.

| Parameter | Test conditions |
|---|---|
| Center frequency | Carrier frequency under test |
| Span | 3 MHz (300 KHz/Div) |
| Detector | Peak |
| Resolution BW | 30 KHz |
| Video BW | 30 Hz |
| Input attenuation | ≥10 dB |
| Vertical scale | 10 dB/Div |
| Log detect Rayleigh COR | Subtract 2.5 dB |
| Video-filter shape-factor correction | Subtract 0.52 dB |

**Table 21.2** SCTE bandwidth correction factors for a home-passed spectrum analyzer.

| System BW | Normal Marker Correction | Noise Marker Correction |
|---|---|---|
| 4 MHz | 23.3 dB | 66.0 dB |
| 5 MHz | 24.3 dB | 67.0 dB |

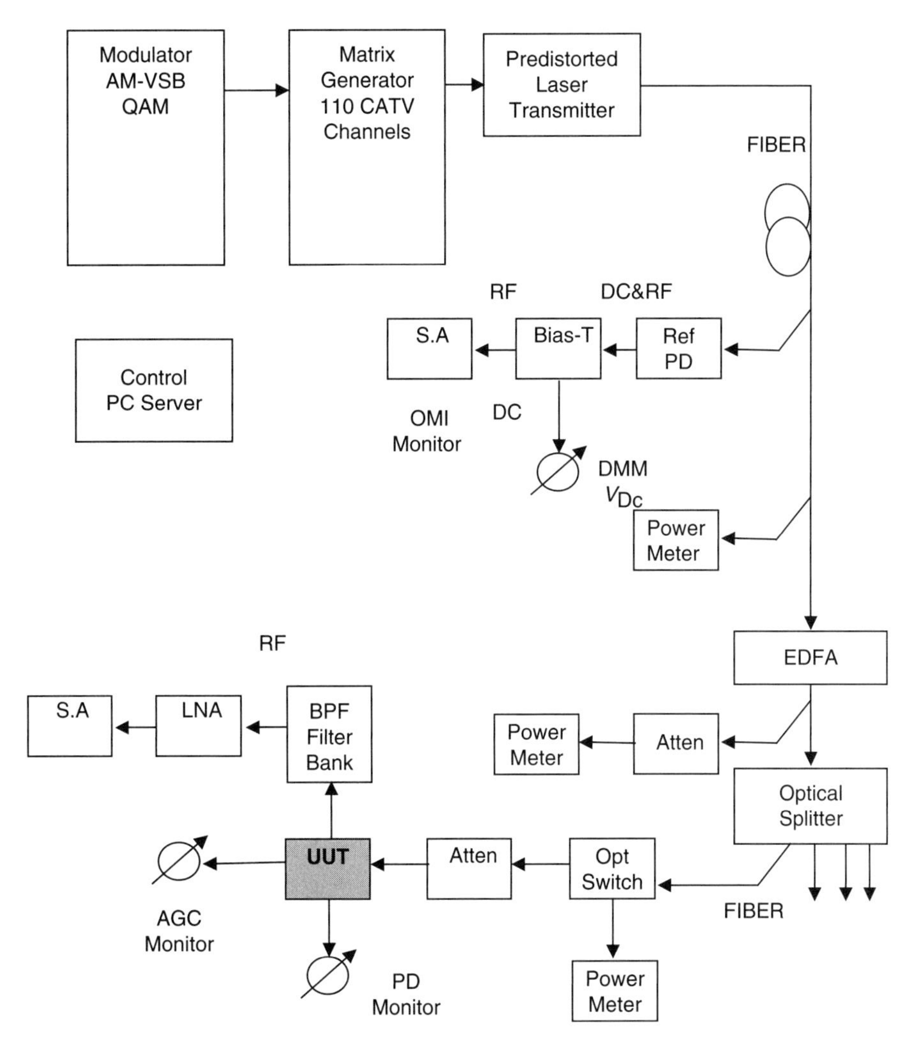


**Figure 21.6** A matrix-generating test setup for a large production with automatic OMI and power calibration, a BPF bank for CSO test to prevent LNA compression, an AGC monitor/defeat, control, and a PD monitor for responsivity reporting. All are HPIB controlled by the main server with a local-rack-control PC. Note that the LNA noise figure $\leq 10$ dB and gain is between 10 and 20 dB.

The SA has two methods to display the noise marker: normal marker and noise marker:

- *Normal marker*: This method uses the SA marker in normal mode. The marker will display the noise in the resolution BW being used.
- *Noise marker*: This method uses the SA marker set to noise mode. The marker will display the noise level normalized at a 1-Hz BW. That is the noise density.

Each noise marker method has a COR that appears in Table 21.2. Further elaboration is presented on the SCTE website.[2]

When testing full-channel loading using a multitone test environment, as shown in Fig. 21.6, it is important to emphasize that each tone is locked on a differently synthesized source. The synthesized source is a DDS that serves as a reference source to a PLL, while the DDS itself has its own reference, TCXO. Hence all sources are not coherent and each source has its own frequency accuracy and drifts defined by the reference crystal. Therefore the sources with respect to each other are at random phase. This is the real-case scenario. As a result, the CSO–CTB picture that appears on the SA is slightly different from the one depicted in Fig. 21.4, and looks like the one in Fig. 21.5. The variation is in the appearance of the CSOs and CTBs. Since all sources are not coherent and each PLL has its own accuracy and phase, the IMDs do not add coherently and resemble a wider IMD tone rather than being a CW look-alike. It appears as a random signal, like a modulated tone that fluctuates and beats according to the random errors developed in each PLL source. This is a much more forgiving CSO-CTB test case than the coherent scenario where all channels are locked to the same crystal. In that case, which is not realistic, all IMDs are added coherently, resulting in CW-look-alike CSO and CTB products. Those products are higher than the random scenario IMD by approximately 6 dB. Further elaboration about frequency synthesis is provided in Chapter 15 and Sec. 18.10.

## 21.5 AGC Calibration Using Pilot Tone

When dealing with feedback-AGC systems, while testing two-tones IP3 or IP2, there is a need to have a pilot tone that is equivalent to the sampled power by the AGC loop, as shown in Fig. 12.2, which shows that only a portion of the received CATV is sampled through an LPF. This, of course, is equivalent to a single pilot tone where the OMI is given by the number of sampled channels. In a coherent case where all channels are in phase, this pilot is expressed by Eq. (21.17), where $N_{\mathrm{AGC}}$ is the number of sampled tones for the AGC:

$$\mathrm{OMI}|_{\mathrm{p}} = \mathrm{OMI}_{\mathrm{ch}}\sqrt{N_{\mathrm{AGC}}}. \tag{21.17}$$

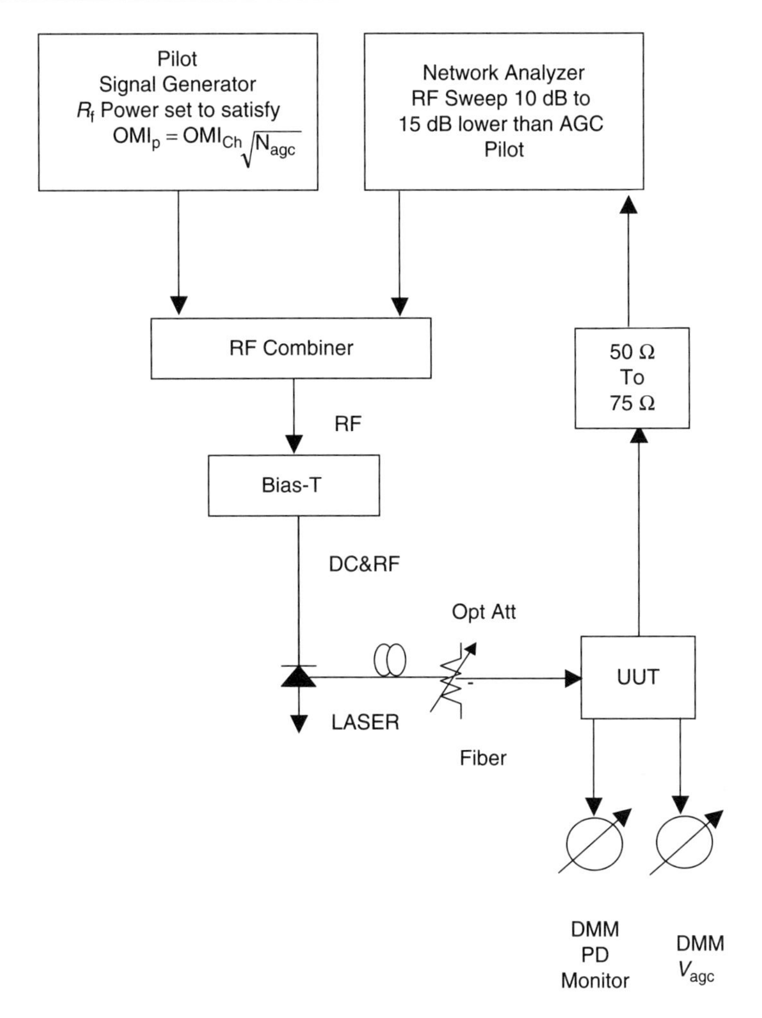


**Figure 21.7** An S21 test of a CATV optical receiver with an AGC pilot, showing AGC and PD monitoring vs. optical power. S21 plots are produced for various optical levels showing flatness vs. input power. Flatness is affected by AGC state.

Calibration of the OMI is done with an SA. The RF power equivalent, as observed in the SA, is calculated by Eq. (21.12). Figure 21.7 depicts an S21 test setup with pilot tone generator used for AGC.

## 21.6 Feedback AGC Performance Under a Video-Modulated Signal

One of the parameters an AGC system should not be sensitive to is X-modulation (cross modulation, X-mod). When a CATV video-signal carrier exhibits AM, there is a peak-to-rms ratio for which the AGC also has to be compensated. As was explained in Sec. 12.2.8, this value for a normal picture is about 4.6 dB. Assume that the video-signal carriers are modulated in deep AM indices, since there are sharp transitions from white level to black level. This, of course, would increase the value of peak-to-rms compensation. It is not a real life case, but it can be used as a test for evaluating the receiver's AGC response, as well as to provide a clue of the AGC performance under a X-mod test. One of the performances tested under X-mod[3] is the AGC time response due to the peak-to-rms changes. Moreover, assuming that the AGC-sampled channels are not modulated with a deep AM index but only with the channels above the AGC sampling BW, then it still may be affected due to X-mod products created by the deep modulating index applied on channels outside the AGC BW of sampling, which create X-mod products within the AGC sampling BW. (Chapter 9 provides a basic analysis of X-mod.) Therefore, the AGC loop BW should be slower than the AM rate. These are not true X-mod problems, as will be shown below, but are easily seen with this type of X-mod set-up. It is necessary to evaluate them with both visual and objective signals.

Figure 21.8 shows a standard NCTC test set-up for measuring X-mod. For visual tests, the 4-channel oscillator in the multifrequency generator is turned off, and the DVD player supplying the modulator is a substitute for the NTSC signal. In order to take measurements, the modulator is turned off and the multi-frequency generator supplies the 4-channel signal. For X-mod measurements, all channels are 100% modulated with a 15.750-kHz square wave, with the exception of the channel under test (channel 4 in this case), which is modulated with a CW signal at the same peak amplitude. The analysis is done on a feedback AGC receiver, with the following design discrepancies in order to emphasize the problems in a multicarrier environment and modulation effects as well as X-mod: the receiver's AGC loop has a wide BW, and there was no smoothing filter and RF-sampling prior to the detector, as recommended by Fig. 12.2.

For the first observation, channel 4 is fed by the CW signal and all other channels are fed by CW signals. The reference level is set on the SA and then modulation is turned on for all channels except channel 4. As a result, the second observation shows that all the signal powers increase by about 5 dB.

The reason is that the broadband AGC detector cannot, in practice, respond to the peak-carrier amplitude, as can a single-channel AGC detector. The AGC detector operates as a power meter detecting the average power. This is a fundamental

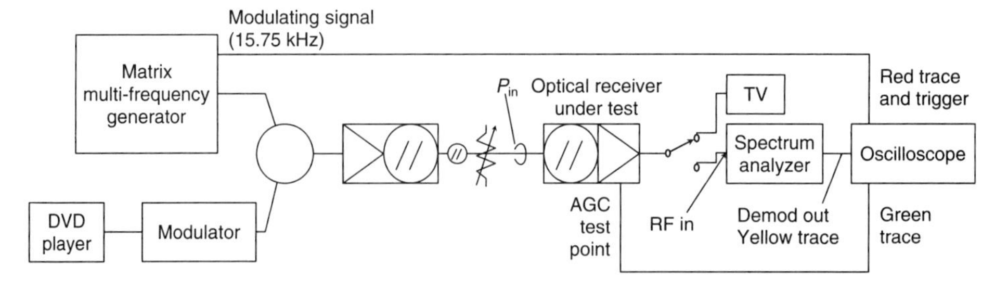


**Figure 21.8** The X-mod set-up.

limitation of the broadband AGC technique. It means that changes in amplitude are going to occur between the modulated and unmodulated carriers, as was analyzed in Sec. 12.2.8.

There is no easy fix for these amplitude changes and additional circuits are required. Fortunately the real world is not as bad as the test case. Therefore there is a need to establish an offset level to which the AGC is set in order to get the correct output amplitude with modulated carriers. This requires a peak-to-rms correction circuitry, and is a consequence of using a wideband AGC detector as well as detector log video detection. It also means that the detector must operate with about 3–5-dB lower signals compared to CW carriers due to the 50% duty cycle of the square wave. The AGC threshold has to be set a few lower compared to modulated signals, as was explained in Sec. 12.2.8.2. It also means that forcing the AGC voltage during a test is even more important. This results in an additional circuit for the AGC defeat control.

The third observation involves the pictures on the TV. The optical attenuator is adjusted for various levels of $P_{\mathrm{in}}$ and the image quality is observed.

Table 21.3 shows the result of the observations done on a too-wide BW AGC loop. At higher signal levels then, there is more unacceptable flashing in the picture. Acceptable picture quality would not be delivered until the signal drops below the level of $-1$ dBm. The explanation is that the AGC circuit in the receiver is too fast, resulting in apparent X-mod, and there are spikes getting through the AGC loop that exceed the BW of the loop's operational amplifier. They will have to be filtered prior to the operational amplifier. This is the reason for using a smoothing filter after the AGC RF-level detector, as shown in Fig. 12.2. In fact, the AGC tracks the resulting AM's residual AM over the video signal; hence, AGC should have narrow BW.

Figure 21.8 defines the traces shown in Fig. 21.9. In this case, the SA is tuned to channel 4 (the CW carrier under test), and set to 0 span and 300 kHz BW, as is done for X-mod measurements. The trace-2 in Fig. 21.9 is the 15.750 kHz modulation signal from the multifrequency generator. The trace-3 is the AGC voltage read at the AGC monitor test point, and the trace-1 is the demodulated output from the SA (video out).[8]

There are several phenomena happening here. Notice first the ringing in the trace-1 (1) following each transition of the trace-2. This is likely due to fast edges of the signal put out by the AGC detector in the receiver. These get through the operational amplifier in the AGC loop because the frequency

**Table 21.3** Picture quality vs. optical level into receiver, $P_{\mathrm{in}}$.

| $P_{\mathrm{in}}$ | Picture Quality |
|---|---|
| $+1$ dBm | Totally unacceptable |
| 0 dBm | Marginally usable but not competitive |
| $-1$ dBm | Slight noise in gray area of picture—operator would not accept |
| $-2$ dBm | OK |

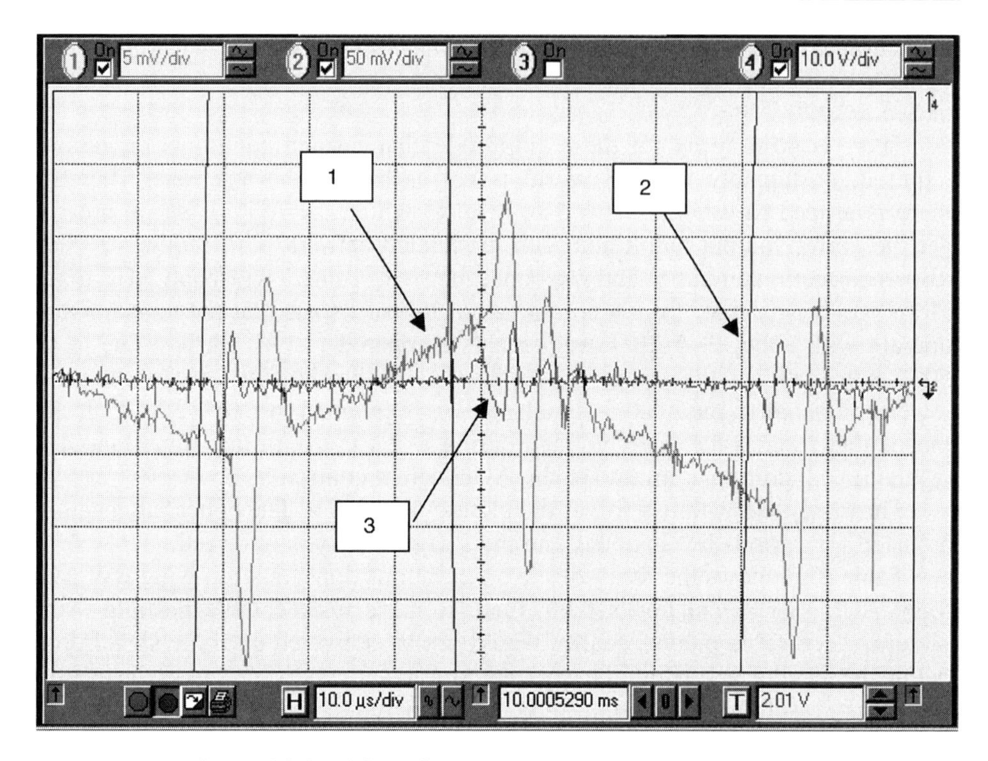


**Figure 21.9** AGC performance at +1 dBm optical input.

components exceed the amplifier's response, and so it cannot filter them. The only good solution is a suitable low-pass smoothing filter between the detector output and the operational amplifier, as was explained in Fig. 12.2.

Next, notice the trace-1 slope between transitions of trace-2. These are likely caused by the AGC trying to follow the modulation on the channels other than channel 4. The solution for this is to reduce the BW of the AGC loop. This way, the AGC would be open to fast amplitude transitions, as was explained at the beginning of Chapter 12. There is a need to reduce the loop-filter BW by increasing the integration time constant, as well as implementing a smoothing filter after the detector to remove the ringing. As was explained in Sec. 12.2.4, the poles of the smoothing filter should be one decade above the overall AGC BW to maintain the loop stability. The long time constant of the integrator reduces the AGC tracking of the modulated signal; therefore the loop does not track peak-to-rms changes. The implementation of the smoothing filter reduces the transition distortion described above. The real world situation will not be as severe as this test, but this is the way many operators evaluate the performance of a system.

Figure 21.10 shows the response with an optical input of −1 dBm. Here the ringing in the AGC voltage trace-3 is still observed, as well as the ringing and slope in the demodulated signal trace-1. This appears to be somewhere above the threshold of visibility, but is not as severe as in the +1 dBm case.

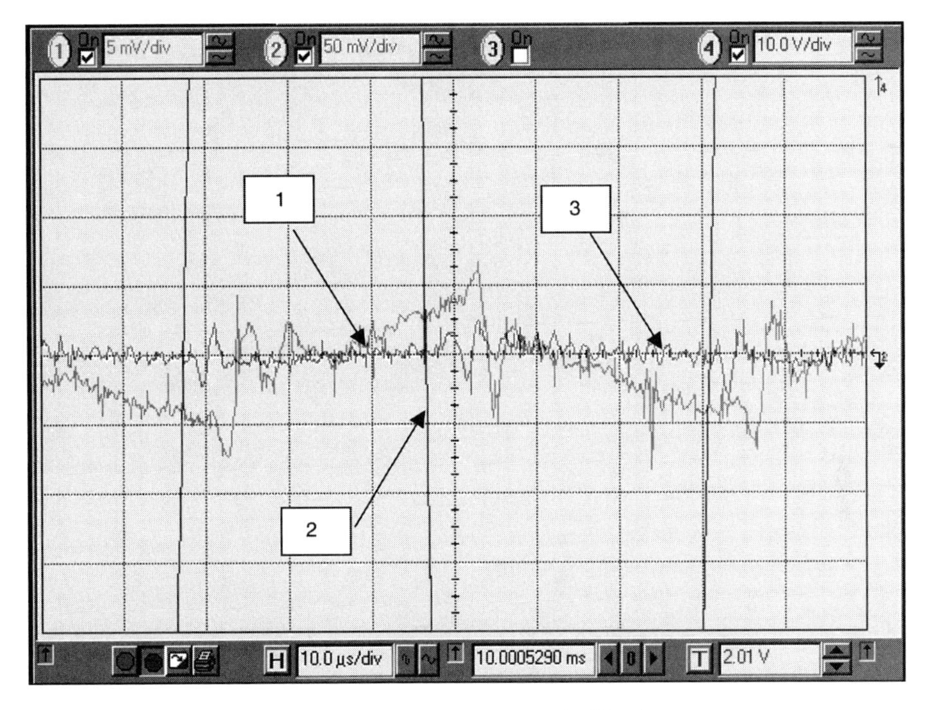


**Figure 21.10** Response at −1 dBm input.

Figure 21.11 illustrates the response with −3 dBm input, where the picture is judged acceptable. At this level, the X-mod can still be seen in the trace-1. This is what to expect X-mod will like a square wave. The delay between trace-2 and trace-1 is due to fiber optic delay. Notice that trace-3 shows some remnants of the ringing, but the amplitude is low. The is because of the load reduction in the RF channel. The net RF power at the input gain block is reduced (see Fig. 12.2); therefore, less third order and X-mod beat products are converted into the channel under test frequency, which is detected by the SA video detection. Additionally, less beat energy affects the RF detector. This is one more aspect that shows the need to have an RF filter before the AGC RF detector, in order to limit the detector BW and reduce the AGC sensitivity to X-mod That means that the RF signals are high enough than the X-mod IMD. As a result, the detector measures the envelope of the modulated signal only, and thus the residual AM is generated due to the control voltage fluctuations over the VVA since the AGC loop BW is too wide and operates as an AM detector. But additional X-mod errors are negligible at that point.

A way to measure and estimate AGC BW is to feed the receiver an AM modulated signal. At the beginning of the experiment, the modulation frequency should be high, higher than the AGC loop BW. Observation of the AGC-control-voltage should show as a dc line, at the proper level for the desired RF level under which the test is done. At the instant the control voltage starts to track the modulating-signal frequency, the AGC 3-dB corner frequency is above the modulating

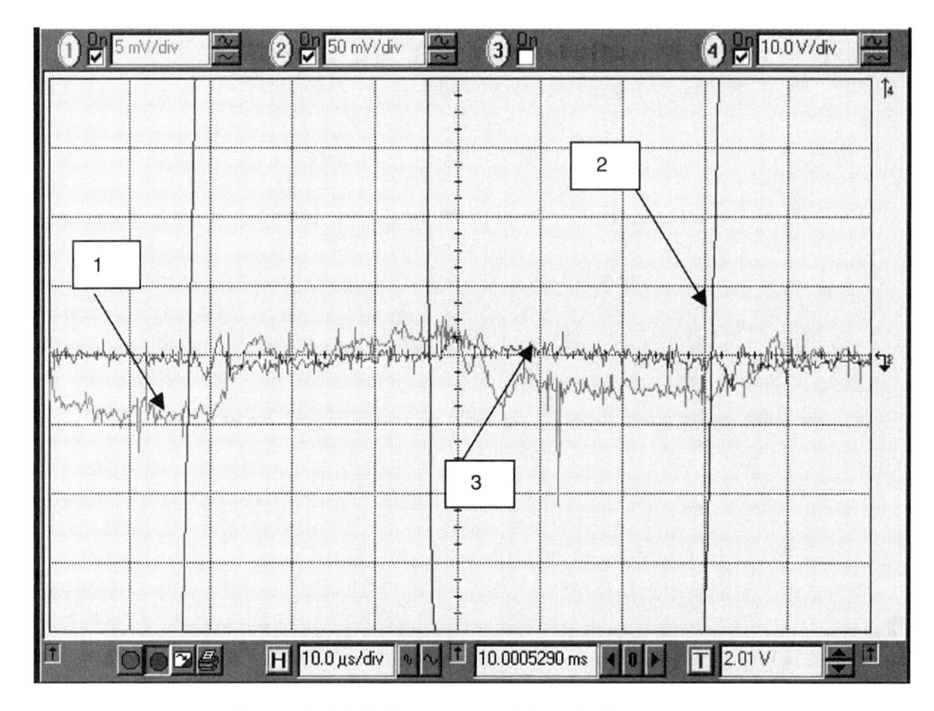


**Figure 21.11** Response with −3 dBm input.

signal frequency. Hence the AGC loop operates as an AM detector. This is similar to a phenomenon observed on a PLL when used as FM detector.

Figure 21.12 shows a different setup for the measurement of AGC under X-mod conditions by using two tones. Two optical transmitters are used: one of

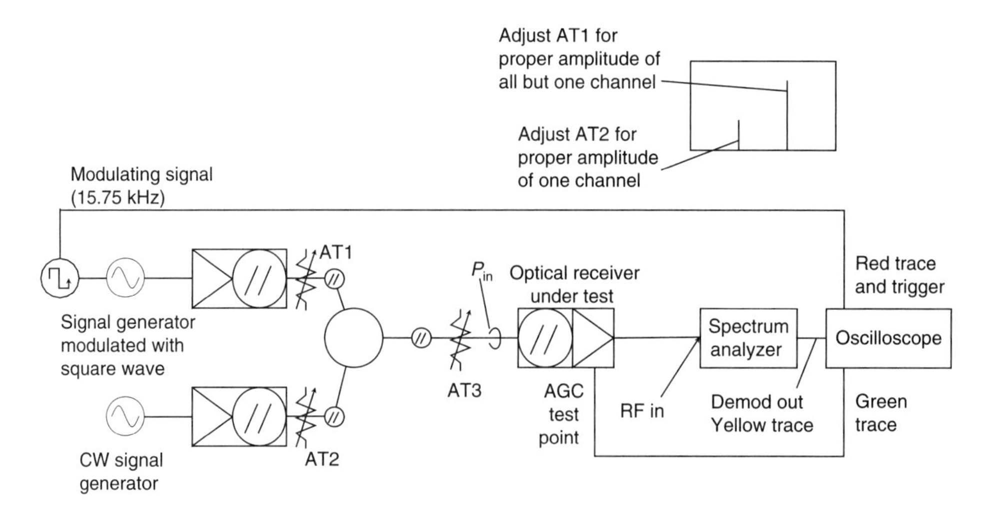


**Figure 21.12** Test circuit.

them is modulated by an RF generator at a convenient frequency such as 100 MHz. The generator is modulated with a 100% square wave that has a 50% duty cycle, and a frequency of about 15.75 kHz (the exact frequency is not essential, but this is what the standard NCTA X-mod test recommends). Note that most signal generators AC couple the square wave, meaning that as the modulation is adjusted, the peak envelope of the signal changes, as was explained in Sec. 12.2.8.1. The goal is to keep the peak envelope power at the desired level. This is usually easy to read on the SA, and so long as it is set to 300 kHz or a wider IF (resolution) and baseband (video) BW, the peak-to-rms measurement can be done. Having the power read, the peak-to-average (PAR) is compensated for the difference between a modulated and nonmodulated tone.

The second optical transmitter is supplied with a CW signal on a convenient frequency, for instance 67.25 MHz, the picture carrier for channel 4. Attenuators AT1 and AT2 are adjusted until the correct relative amplitudes of the two carriers are set; then AT3 is used to adjust the overall signal level into the optical receiver under test. For the test, the SA is used as a demodulator, under the following setup:[1–5,9]

- Zero span (span = 0)
- Resolution and video BW = 300 kHz (may be wider, but not narrower)
- Vertical axis: linear, not log mode.

Tune the analyzer to the CW-signal frequency, and place the carrier at the top of the screen.

The video output from the analyzer is used to feed the oscilloscope. The scope trigger should be on the square-wave modulation. With this two-tone set-up, as shown in Fig. 12.12, wave forms of Figs. 21.9–21.11 can be duplicated in a multitone environment.

## 21.7 Cross-Talk Test Methods

Chapter 20 provided a complete review about isolation and X-talk between digital and analog sections of an integrated triplexer (ITR). A good engineering practice is to measure S21 between the digital transmitter and the CATV receiver as shown in Fig. 21.13. For that purpose, a vector network analyzer (VNA) is used in order to increase isolation measurement sensitivity compared to the scalar network analyzer (SNA). The VNA S11 port sweep source output signal is connected to the digital-transmitter is inverting input while the noninverting input is terminated. The network analyzers output RF power level from S11 port is calibrated to meet the peak-to-peak voltage level as defined by the interface specifications of the digital transmitter input. The calculation is done per Eqs. (21.18) and (12.19). Note that the network analyzer produces a sine wave, and voltage units

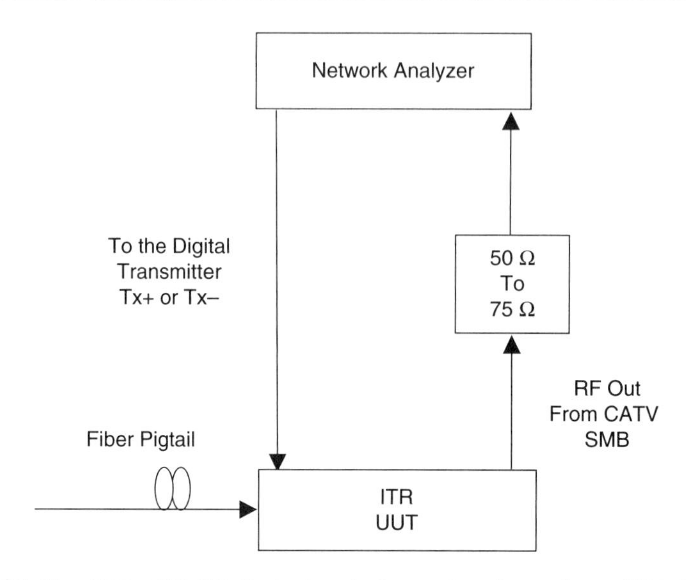


**Figure 21.13** The FTTx ITR X-talk test using a network analyzer S21 for isolation between digital Tx and CATV Rx. The network analyzer is sweep power is set to meet the digital data rms voltage. The pigtail is connected to a reflective connector, to increase reflections from Tx into Rx, in order to check isolation through the optical triplexer, while taking into account the electrical X-talk caused by amplification of the digital receiver.

are in millivolts. The output of CATV receiver's subminiature version B (SMB) connector that is connected to the VNA S22 port:

$$P[\text{dBm}] = 10\log\left(\frac{V_{\text{peak}}^2}{\sqrt{2}Z_0}\frac{1}{10^3}\right), \tag{21.18}$$

$$V_{\text{peak}}[\text{mV}] = 100\sqrt{\frac{Z_0 10^{0.1\times P_{\text{dBm}}}}{5}}. \tag{21.19}$$

After completion of S21 measurement of noninverting transmitter's port isolation, the inverting section is evaluated. The advantage of this measurement is that it is a relative measurement independent of power. Additionally, the frequency response of isolation between the digital transmitter and the CATV receiver is provided. This is helpful since it shows the high-pass filter's isolation response and its corner frequency. The corner frequency provides an indication regarding GND quality and its impedance to main chassis GND. Knowing the nonreturn-to-zero (NRZ) PRBS and transmitting power, the power for each spectral line, can be determined per Eq. (21.20), where HPF($f$) is the S21 isolation response, and $A$ is the PRBS voltage-amplitude power for the NRZ signal:

$$P_{\text{X-talk}} = \frac{A^2}{Z_0}\frac{\text{PRBS}+1}{\text{PRBS}^2}\,\text{sinc}\left(n\pi f\frac{T_0}{\text{PRBS}}\right)^2\text{HPF}(f). \tag{21.20}$$

This way the power leakage from the digital transmitter into the CATV receiver can be evaluated. Moreover, knowing NRZ and having a short PRBS, a specification for isolation, the frequency response is determined, as was explained in Sec. 20.3, and the isolation requirement is established per Fig. 20.6.

In this setup, the UUT's pigtail is connected to a reflective FC connector to increase reflections from the digital transmitter into the optical triplexer. This way, any digital transmission at 1310 nm is returned to the digital receiver's PD. If the triplexer is a 1310 nm, 1310 nm, and 1550 nm, as appears in Fig. 5.8, the digital receiver amplifies the received digital signal and reduces isolation by increasing X-talk to the CATV receiver. If the optical triplexer is FSAN, as appears in Fig. 5.8 in Chapter 5, the leakage of 1310 nm into 1490 nm will affect CATV receiver as well, depending, of course, on the optical filter within the triplexer.

A different approach is to observe the leakage of a short PRBS NRZ transmission of the digital transmitter section to the CATV receiver on an SA. Pass–fail is defined by CSO-CTB-CNR specifications of SCTE. This way, for 62-dBc-CNR specifications and for a 18-dBmV tone, the leakage should be below −54 dBmV. All those methods are commonly used for qualifications of FTTx ITR.

## 21.8 Understanding Analog Receiver Specification

The previous sections and chapters provided the overall design considerations for the analog FTTx CATV receiver. In order to comply to the FTTx system requirements, the ITR analog receiver section must satisfy performance parameters given in Table 21.4 as an example. Those parameters are reviewed in the next two sections.

### 21.8.1 Gain sensitivity and AGC

The analog receiver used for CATV reception is considered a state-of-the-art TIA operating from 47 to 870 MHz. In this specification, FTTH up-tilt of 3 to 5 dB is required since a coax cable is shorter and has fewer splitters and connections compared to FTTC. An FTTC receiver requires steeper tilt. The up-tilt compensates for the COAX down-tilt response.

Observing that RF output power is kept constant at the optical dynamic range of −5 to 1 dBm, one can understand that this receiver has an AGC with a 12-dB RF dynamic range as a minimum. The AGC threshold is at −5 dBm at minimum and its saturation is at 1 dBm at maximum.

From the optical responsivity data and optical power range at an RF output power of over 75 Ω, the TIA gain in ohms can be evaluated as well as the absolute power-gain ratio in the following manner. For transimpedance gain (TIG), RF voltage at the output is divided by the RF current developed at the input of the receiver. It is assumed that the PD impedance is high so that the current detection efficiency is 100%, since all RF current goes to the receiver's input. Hence, the

**Table 21.4** Examples of commercial FTTx ITR analog specifications.

| No | Parameter | Units | Min | Typ | Max | Conditions |
|---|---|---|---|---|---|---|
| 1 | Frequency range | MHz | 47 | | 870 | |
| 2 | RF output power | dBmV | | | | 3% OMI/channel |
| | 55 MHz | | 16 | | | 1.5% OMI/channel for frequency above 550 MHz |
| | 550 MHz | | 18 | | | |
| | 750 MHz | | 14 | | | |
| 3 | RF output tilt | dB | 3 | | 5 | Network test 47 to 870 MHz |
| 4 | S22 output return loss | dB | 14 | | 20 | 75 Ω |
| 5 | Distortions | dBc | | | | Matrix test at 1 dBm optical |
| | CSO | | −65 | | −55 | |
| | CTB | | −62 | | −59 | Power: above 550 MHz CSO/CTB 7 dB higher |
| 6 | Carrier to noise | dB | 48 | 52 | | At −5 dBm 3% OMI |
| 7 | Channel BW | MHz | | 4 | | |
| 8 | Channel spacing | MHz | | 6 | | |
| 9 | AGC time constant | s | 0.5 | | | |
| 10 | AGC RF sampling | | | | | 30 channels 3% OMI |
| 11 | PD monitor voltage | V | 0.85 | 0.9 | 1 | 1 KΩ resistor |
| 12 | PD monitor voltage accuracy | % | | | 1 | |
| 13 | Optical return loss | DB | 20 | | | |
| 14 | Receiver wavelength | Nm | 1550 | 1555 | 1560 | |
| 15 | Receiver's average optical power | dBm | −5 | | 1 | 3% OMI/Channel |
| | | | | | | 1.5% OMI/Channel for frequency above 550 MHz |
| 16 | Operating voltage | V | 11.7 | | 14 | |
| 17 | Current | MA | 150 | | 170 | |
| 18 | Video inhibition | V | | | | |
| | On | | 3 | | 5 | |
| | Off | | 0.2 | | 0.8 | |
| 19 | Video-inhibition isolation | dB | | | −50 | |
| 20 | X-talk | dBc | | | −65 | PRBS $2^7-1$ NRZ Digital receiving 1440–1500 nm High 2 V, low 0.6 V 622.08 MB/s Transmit 1260–1360 nm Input high 300–1200 mv Differential LVPPECL 155.52 MB/s |
| 21 | Optical transmitter to receiver X-talk | dB | | | −65 | 1310 nm transmitter to 1490 nm receiver |
| 22 | Optical receiver to receiver isolation | dB | | | −65 | 1490 nm receiver to 1550 nm receiver |
| 23 | Optical transmitter to receiver X-talk | dB | | | −65 | 1310 nm transmitter to 1550 nm receiver |
| 24 | Optical receiver isolation | dB | | | −65 | 1550 nm receiver to 1490 nm transmitter |

TIG can be written as follows. Equation (21.21) describes the input current at the matching transformer primary:

$$I_{\text{IN-rms}} = \frac{r \cdot P_{\text{opt}} \cdot \text{OMI}}{\sqrt{2} \cdot 100}. \tag{21.21}$$

Equation (21.22) describes the rms voltage developed at the ITR output OMI (in %):

$$\text{V}_{\text{out}} = \sqrt{Z_0 \frac{10^{0.1 \cdot P_{\text{dBm}}}}{1000}}. \tag{21.22}$$

$$\text{TIG}[\Omega] = \frac{V_{\text{out}}}{I_{\text{IN-rms}}} = \frac{\sqrt{20 \cdot 10^{0.1\text{dBm}} Z_0}}{r \cdot P_{\text{opt}} \text{OMI}}. \tag{21.23}$$

Equation (21.22) can be written in terms of decibels relative to one millivolt:

$$V = 10^{0.05\text{dBmV}^{-3}}. \tag{21.24}$$

Hence, TIG is given by Eq. (21.24) divided by Eq. (21.21):

$$\text{TIG}[\Omega] = \frac{V_{\text{out}}}{I_{\text{IN-rms}}} = \frac{1}{\sqrt{50}} \frac{10^{0.05\text{dBmV}}}{r \cdot P_{\text{opt}} \cdot \text{OMI}}. \tag{21.24}$$

Note that TIG is affected by the output impedance. In Eqs. (21.22), (21.23), and (21.24) it is clear from the notation, that the decibels relative-to-one-millivolt value is affected by impedance as per Eqs. (21.2) and (21.3).

TIG is a fixed value that is not affected by OMI even though it seems to be from the above expressions. That is because the specification defines that the output RF power is governed by the specific OMI. Thus an overall optical receiver is considered as a voltage-controlled source by an optically controlled current source. That is why when OMI is reduced by half for QAM, band power is reduced by 6 dB, owing to the relation of power to voltage and current. Moreover, the reason for less than a 6-dB drop from high to low channels during testing is due to the designed up-tilt of the receiver. Considering the RF power gain, the design goal is to maximize power efficiency by properly matching to the PD. With that in mind, the RF power gain is given by

$$G = \frac{20 \cdot 10^{0.1\text{dBm}}}{\left(r \cdot P_{\text{opt}} \text{OMI} \cdot N\right)^2 Z_{01}} = 10^{-2} \frac{\left(10^{0.05\text{dBmV}}\right)^2}{\left(r \cdot P_{\text{opt}} \text{OMI} \cdot N\right)^2 50 \cdot Z_{01} Z_0}, \tag{21.25}$$

where $Z_0$ is the output impedance at the SMB connector, $Z_{01}$ is the first RF-stage-input impedance, OMI is a percentage, and $N$ is the transformer's turn ratio. Note that the power gain has no units and is affected by impedance matching. Moreover, both TIG and power gain increase, as the input power lowers until it reaches the AGC-threshold level. Recall that $P_{\text{out}}$ is locked to a constant level by the AGC.

The second advantage of the transformer-matching method is the equivalent noise current density reduction as described in Chapter 10. Knowing responsivity and CNR is defined by the specification, the noise density can be found using the relevant plots from Chapter 10. The excessive shot noise and RIN should be removed by using Eqs. (10.43) and (10.51) in order to derive the dark CNR. From this specification, it is clear that the BW for the noise power calculation is 4 MHz or 66 dB.

From the AGC specifications, it is clear that this is a feedback AGC with a BW of approximately 2 Hz, in order to overcome blanking and X-mod effects described in Sec. 21.6. The AGC pilot OMI for testing is calibrated, as described in Secs. 21.2–21.5.

The receiver output has a decent return loss and thus it minimizes ripple resulting from mismatch between the SMB connector and the coax cable. This is well explained in Chapter 9.

The video-inhibit function describes the level of the CATV signal when the RF is off and PD bias is on. The signal level at that stage is 50 dB below normal reception power, which means the RF level in the range of −80 dBm at 4 MHz BW. This ensures that no signal is received in case of billing sanctions applied on the subscriber: CNR is almost 0 dB in that case.

### 21.8.2 ITR X-talk

X-talk in FTTx integrated platform is a painful problem, as was described in Chapter 20. However from the above specification, it can be seen what the test conditions are for electrical X-talk from the digital section into the CATV RF section. The test conditions call for an NRZ train of PRBS $2^7-1$ sequence. This pattern has high-energy spectral lines. It defines the data rates for both the receiver and transmitter to maximize the interference. From the CSO CTB performance, it is clear that X-talk leakage should be below 65 dB, or below the worst-case CSO CTB defined in specifications. From the optical specifications that call for optical isolation and X-talk, it is easy to see how to calculate the additional shot noise leaking into the analog PD by using the relations in Chapter 10. Note that RF–wise, 65 dB of optical isolation is equal to 130 dB RF and electrical isolation. The high optical isolation prevents any digital signal detection in the analog section. It is clear that high electrical isolation against conducted and radiated emissions from the transmitter on the digital side guarantees that the transmitter's NRZ pattern spectral lines are lower than the CSO CTB levels. Otherwise, it is conclusive that X-talk leakage into the CATV receiver is due to electrical conduction and emission rather than an optical path.

## 21.9 Main Points of this Chapter

1. Decibels relative to 1 mV is a power measurement unit that is affected by the load impedance.

2. Decibels relative to 1 mW is a power measurement unit that not affected by the load impedance.

3. OMI calibration requires a dc-coupled calibrated PD with standard $Z_0$ impedance of 50 or 75 $\Omega$.

4. The SA impedance for OMI calibration should match the PD impedance.

5. Spectrum impedance should be ac coupled to the reference PD.

6. In case spectrum analyzer (SA) impedance and PD impedance are not the same, ac analysis is not of an equal current division between current source and SA as in Fig. 21.1.

7. Two-tone test should be made with two lasers to prevent degradation due to laser-limited IP3 and IP2.

8. The wavelength separation between the two lasers is critical when testing an optical triplexer, since too large a wavelength separation might be filtered by the triplexer filter.

9. Too narrow a wavelength separation between the lasers might create an RF beat note within the RF BW of the receiver. Thus wavelength separation should be such that its beat-note frequency would be out of the RF BW of the UUT receiver.

10. When calibrating OMI of a laser in a two-lasers setup, both lasers should be at “on” state and only the laser under OMI calibration should have an RF signal.

11. A multitone testing laser is a predistorted laser transmitter with high linearity performances.

12. A multitone tester frequency source has two modes of operation, random and coherent; the coherent mode degrades CSO CTB results compared to the random mode.

13. Dark CNR test of an AGC system can be done in the threshold state, since AGC gain is not affected by lack of locking signal.

14. At high optical power, dark CNR is evaluated by removing RIN and shot noise contributions.

15. CTB measurement in the multitone tester is done by turning off the test tone at the frequency of interest.

16. The reason for LNA at the SA input is to compensate for the SA noise floor and to increase measurement accuracy.

17. The role of the sharp BPF between the UUT output and the LNA input is to prevent CSO CTB performance degradation due to the 110 channel load of the LNA.

18. SCTE defines the noise test results CCN, CTN, and CIN, and proper CORs for SA detector error, video filter, and all test-related factors. These are freely available on the SCTE website.

19. A pilot OMI is calculated per the number of RF tones sampled by AGC sampling LPF. It is assumed all tones are coherent.

20. Wideband RF sampling AGC and too wide an AGC loop is susceptible to X-mod since the RF detector is a true rms detector, which averages RF power. This results in peak-to-average error.

21. Peak-to-average and X-mod errors are worse when the AGC detector is a DLVA, since it is an envelope detector rather than a true rms detector.

22. X-talk testing can be done by a network analyzer measuring S21 between the digital transmitter's input and the CATV's RF output.

23. X-talk S21 response is HPF with a high insertion loss.

24. Good isolation is an S21 HPF response with high corner frequency. This means that low-frequency rejection is a wideband that nullifies the main and second lobes of the NRZ sinc.

25. FTTx ITR specifications should define output power related to OMI.

26. PD responsivity and output power define RF gain requirement from the ITR.

27. The electrical X-talk test should specifically define under what PRBS pattern it should be performed.

28. FTTx RF section can be describes as a TIA gain or as power gain.

## References

1. Engineering Committee Interface Practices Subcommittee, Society of Cable Telecommunications Engineers. *Composite Distortion Measurements (CSO & CTB)*, American National Standard, ANSI/SCTE (1999).
2. Engineering Committee Interface Practices Subcommittee, Society of Cable Telecommunications Engineers. *Test Procedure for Carrier to Noise (C/N, CCN, CIN, CTN)*, American National Standard, ANSI/SCTE (2001).
3. Engineering Committee Interface Practices Subcommittee, Society of Cable Telecommunications Engineers. *AM Cross Modulation Measurements*, American National Standard, ANSI/SCTE (2003).
4. Engineering Committee Interface Practices Subcommittee, Society of Cable Telecommunications Engineers. Measurement Procedure for Noise-Figure, American National Standard, ANSI/SCTE (2002).
5. Laverghetta, T.S., *Handbook of Microwave Testing*. Artech-House (1981).

6. Thomas, J.L., *Cable Television, Proof of Performance*, Prentice-Hall (1995).
7. Thomas, J.L., and M. Francis, *Digital Basics for Cable Television Systems*, Edgington, Prentice-Hall (1998).
8. Len-Garrett and M. Engelson, "Use the spectrum analyzer's zero-span setting." *Microwaves & RF*, March (1996).
9. Engineering Committee Interface Practices Subcommittee, Society of Cable Telecommunications Engineers. *Test Method for Reverse Path (Upstream) Intermodulation Using Two Carriers*, SCTE (2006).

# Index

## C

**D**

## G



## H

## W



## X



## Y



## Z

**Avigdor Brillant** is a senior staff engineer at Qualcomm in Israel. He is involved with wireless video IC architecture and system design for DVB-H / DVB-T T / T-DMB ISDB-T / FLO standards. Prior joining Qualcomm, he served as a Senior Technical Contributor in Marvell – DSPC, previously Intel–DSPC in Petach Tikwa, Israel, where he was involved with advanced architectures of CMOS RF ICs solutions for cellular applications for 2.5G and 3G. During his role in DSPC, Mr. Brillant was involved with PMIC (power management integrated circuit) architectures as well. From 1999–2005 he was with LuminentOIC (known today as Source–Photonics) a subsidiary company of MRV Communications in Chatsworth, California, USA. During this time he was the senior FTTx RF architect. He established a strong RF and FTTx team and led the analog RF group. Mr. Brillant was the designer and architect of the LuminentOIC integrated triplexer (ITR) and the ITC solution for higher RF power to the curb, bringing it from concept to mass production. This product was employed by Verizon in their largest FTTx deployment with Tell–Labs, AFC, and Motorola. He also helped to drive LuminentOIC to ISO process. From 1995–1999, Mr. Brillant was with Optomic Microwaves (known as Belcom today) in Migdal Ha'Emek, Israel, in a position of a senior engineer in the fields of VSAT and cellular design, and was in the management staff. He is currently on the advisory board of Belcom. From 1993–1995 Mr. Brillant was with MTI Technology and Engineering at Tel-Aviv in a position of system engineer where he was involved with WLL (wireless local loop) designs as well as COMINT and ELINT systems. Between 1986 and 1993, he was with MicroKim Ltd. (previously a subsidiary of M/A-COM) of Haifa, Israel, where he was in charge of the MIC amplification group, a member of the frequency sources group, and was the head of the Short Range Hunter RPV (remote platform vehicle) C-Band up-down-link program for the U.S. Naval Air branch. Mr. Brillant received his Electrical Engineering degree from the Technion Israel Institute of Technology in 1986. He is a Senior Member of the IEEE, and published several papers on related subjects. He is also an adjunct instructor in the Electrical Engineering Department Communication Laboratory of the Technion.